HANDBOOK OF
DRIVING SIMULATION FOR ENGINEERING, MEDICINE, AND PSYCHOLOGY

T0239958

HANDBOOK OF
DRIVING SIMULATION FOR ENGINEERING, MEDICINE, AND PSYCHOLOGY

EDITED BY
DONALD L. FISHER • MATTHEW RIZZO
JEFF K. CAIRD • JOHN D. LEE

CRC Press
Taylor & Francis Group
Boca Raton London New York

CRC Press is an imprint of the
Taylor & Francis Group, an **informa** business

MATLAB® and Simulink® are trademarks of The MathWorks, Inc. and are used with permission. The MathWorks does not warrant the accuracy of the text or exercises in this book. This book's use or discussion of MATLAB® and Simulink® software or related products does not constitute endorsement or sponsorship by The MathWorks of a particular pedagogical approach or particular use of the MATLAB® and Simulink® software.

Cover art: © Figge Art Museum, successors to the Estate of Nan Wood Graham/Licensed by VAGA, New York, NY.
Grant Wood (American, 1892–1942)
Death on the Ridge Road, 1935
oil on masonite
frame: 39 x 46 1/16 in. (99 x 117 cm)
Williams College Museum of Art, Williamstown, MA, Gift of Cole Porter, (47.1.3)

CRC Press
Taylor & Francis Group
6000 Broken Sound Parkway NW, Suite 300
Boca Raton, FL 33487-2742

First issued in paperback 2017

© 2011 by Taylor and Francis Group, LLC
CRC Press is an imprint of Taylor & Francis Group, an Informa business

No claim to original U.S. Government works

ISBN 13: 978-1-4200-6100-0 (hbk)
ISBN 13: 978-1-138-07458-3 (pbk)

Library of Congress Cataloging-in-Publication Data

Handbook of driving simulation for engineering, medicine, and psychology / edited by Donald L. Fisher ... [et al.].
    p. cm.
  Includes bibliographical references and index.
    ISBN 978-1-4200-6100-0
    1. Automobile driving simulators--Handbooks, manuals, etc. 2. Automobile driving--Physiological aspects--Handbooks, manuals, etc. 3. Automobile driving--Psychological aspects--Handbooks, manuals, etc. I. Fisher, Donald L.

TL152.7.D7H36 2011
629.28'3011--dc22
                        2010032406

Visit the Taylor & Francis Web site at
http://www.taylorandfrancis.com

and the CRC Press Web site at
http://www.crcpress.com

Contents

SECTION I Introduction

SECTION II Selecting a Driving Simulator

SECTION III Conduct of Simulator Experiments, Selection of Scenarios, Dependent Variables, and Evaluation of Results

Simulator Sickness

Independent and Dependent Variables

Analyses of the Data

SECTION IV Applications in Psychology

Experience and Maturity

SECTION V Applications in Engineering

Acknowledgments

The editors want to begin by acknowledging the very generous efforts of all of the many authors who contributed to the *Handbook*, their patience with our several changes in format along the way, and their unfailing good will when answering our questions and responding to our suggested changes. We also want to acknowledge up front someone without whose help this *Handbook* may never have made it to the publisher. Tracy Zafian gave hundreds of hours to the production of the *Handbook*, the detailed formatting of the document, the search for long-lost references, and the suggestion for more substantive changes. She did this all while raising two young children. We simply cannot thank her enough. Pamela Rivest has also devoted long hours to the *Handbook*'s production, literally coming in on the weekends when the deadline looked like it was going to prove the death of the *Handbook*. We also want to acknowledge the help of Bobbie Seppelt with the *Handbook* wiki, answering the same questions for the tenth time with the same level of aplomb she did the first time. Finally, we want to thank Cindy Carelli, Jill Jurgensen, and Richard Tressider at CRC Press and Taylor & Francis and Rajesh Gopalan at Amnet International who have helped us throughout the production process.

Jeff Caird is grateful to Peter Hancock for tasking him to build a driving simulator in 1989 as his first unsuspecting graduate student at the University of Minnesota. His fourth driving simulator was funded by the Canadian Foundation for Innovation (CFI) and operational support provided by the AUTO21 Network of Centres of Excellence (NCE), among others. He is indebted to all of his students who have taught him about torque motors, A/D, C code, ethics, protocol, glance durations and p values from innumerable driving simulation studies. John Lee would like to thank his colleagues at the University of Iowa and the National Advanced Driving Simulator for their constant flow of interesting ideas and dedication to interesting research. He is particularly grateful to all the students and staff he has worked with who have done the hard work of operating and maintaining the driving simulators. Matthew Rizzo is deeply grateful to Annie, Ellie, and Frannie for their enduring support. Don Fisher wants to thank Joe Goldstein, former Dean of the College of Engineering, and Fred Byron, former Vice Chancellor for Research at the University of Massachusetts Amherst, for their support when he first purchased a driving simulator back in the early 1990s, a half-million dollar leap of faith that seemed impossible at the time. He also cannot thank enough the many graduate students, postdocs, and visiting scientists who have again and again worked long hours into the night in order to keep the experiments moving forward, especially when his late wife was so sick. Their teamwork, good will and extraordinary dedication to their work is evidenced in the many sponsors that have funded research in this lab. Finally, he wants to thank Susan Alice Duffy, Annie James Duffy Fisher and Jennifer Duffy Fisher for their unfailing support and understanding of his peculiar work habits, and Susan Taylor Haas for the joy which she has brought into his life over these last five years, making it possible to move forward on the *Handbook* when it would have so easy to let it slip.

We also need to acknowledge the many agencies without whose support this *Handbook* would not have been possible. Grants to Jeff Caird from the Canadian Foundation for Innovation and AUTO21 Network of Centres of Excellence supported infrastructure and research, respectively, which made the University of Calgary Driving Simulator a Swiss Army knife for several generations of students. Support from the Federal Highway Administration, National Highway Traffic Safety Administration, and the National Institutes of Health (NIA R01 AG026027 and NIA RO1 AG 15071) have provided partial support to John Lee and the graduate students who make so many of the simulator-related developments possible. Matthew Rizzo's research has been supported by the National Institutes of Health as well (RO1 AG 17707, NIA R01 AG026027 and NIA RO1 AG 15071 and NHLBI RO1 HL091917). Grants from the National Institutes of Health (1R01HD057153-01) and National Highway Traffic Safety Administration provided partial support to Donald Fisher.

MATLAB® and Simulink® are registered trademarks of The MathWorks, Inc. For product information, please contact:

The MathWorks, Inc.
3 Apple Hill Drive
Natick, MA, 01760-2098 USA
Tel: 508-647-7000
Fax: 508-647-7001
E-mail: info@mathworks.com
Web: www.mathworks.com

Editors

Dr. Donald L. Fisher, is the head of the Department of Mechanical and Industrial Engineering at the University of Massachusetts Amherst, and the director of the Arbella Insurance Human Performance Laboratory in the College of Engineering. He has published over 150 technical papers, including recent ones in the major journals in transportation, human factors, and psychology. He is currently a principal or co-principal investigator on over 10 million dollars of research and training grants, including awards from the National Science Foundation, the National Institutes of Health, the National Highway Traffic Safety Administration, MassDOT, the Arbella Insurance Group Charitable Foundation, the State Farm Mutual Automobile Insurance Company, and the New England University Transportation Center. He has served on the editorial boards of the leading journals in human factors, has been a member of the National Academy of Sciences Human Factors Committee, chaired or co-chaired a number of TRB workshops, and served as a member of both the joint National Research Council and Institute of Medicine Committee on the Contributions from the Behavioral and Social Sciences in Reducing and Preventing Teen Motor Crashes and the State Farm® Mutual Automobile Insurance Company and Children's Hospital of Philadelphia Youthful Driver Initiative. Over the past 15 years, Dr. Fisher has made fundamental contributions to the understanding of driving, including the identification of those factors that: increase the crash risk of novice and older drivers; impact the effectiveness of signs, signals, and pavement markings; improve the interface to in-vehicle equipment, such as forward collision warning systems, back over collision warning systems, and music retrieval systems; and influence drivers' understanding of advanced parking management systems, advanced traveler information systems, and dynamic message signs. In addition, he has pioneered the development of both PC-based hazard anticipation training (RAPT) and PC-based attention maintenance training (FOCAL) programs, showing that novice drivers so trained actually anticipate hazards more often and maintain attention better on the open road and in a driving simulator. This program of research has been made possible by the acquisition in 1994 of more than half a million dollars of equipment, supported in part by a grant from the National Science Foundation. He has often spoken about his results, including participating in a congressional science briefing on the novice driver research sponsored several years

previous. Recently, the Human Performance Laboratory was recognized by the Ergonomics Society, receiving the best paper award for articles that appeared in the journal *Ergonomics* throughout 2009. The paper described the work in the Human Performance Laboratory on hazard anticipation. Dr. Fisher received an AB from Bowdoin College in 1971 (philosophy), an EdM from Harvard University in 1973 (human development), and a PhD from the University of Michigan in 1982 (mathematical psychology).

Dr. Matthew Rizzo is professor of neurology, engineering, and public policy at the University of Iowa. In neurology, he is vice-chair for translational and clinical research, director of the Division of Neuroergonomics (http://www.uiowa. edu/~neuroerg/) and its laboratories, which include an interactive state of the art driving simulator (the SIREN) and two instrumented vehicles (ARGOS and NIRVANA), a senior member of the Division of Behavioral Neurology and Cognitive Neuroscience, and a senior attending physician in the Memory Disorders Clinic. Dr. Rizzo also directs the University of Iowa Aging Mind and Brain initiative. He has contributed to many professional organizations and committees, including the U.S. Federal Drug Administration's Panel for Peripheral and Central Nervous System Drugs, the National Academy of Sciences Committee on Human-Systems Integration, and was appointed by the U.S. Secretary of Transportation to the U.S. Federal Motor Carriers Safety Administration's Medical Advisory Committee. He has advised the American Association of Automobile Administrators, the American Academy of Neurology, the states of California and South Carolina, and the governments of Australia, Canada, and Sweden, on developing licensing guidelines for impaired drivers, is a founder of the biannual Driving Assessment Symposium (http://drivingassessment.uiowa.edu/), and has advised the U.S. Army Research Laboratories on its neurosciences research program. Dr. Rizzo has long-standing research grant support from the U.S. National Institutes of Health and the Centers for Disease Control and Prevention. Dr. Rizzo's domains of research interest are outlined in his recent books, *Principles and Practice of Behavioral Neurology and Neuropsychology* (Rizzo and Eslinger 2004) and *Neuroergonomics: The Brain at Work* (Parasuraman and Rizzo 2007). He has also worked as a development executive and science advisor at a New York City-based media company, producing documentaries for major television venues. Dr. Rizzo is a graduate of Columbia University in New York City and the Johns Hopkins University School of Medicine.

Dr. Jeff K. Caird is a professor in the Department of Psychology and an adjunct professor in the Faculty of Kinesiology, the Departments of Anesthesia and Community Health Sciences at the University of Calgary. In 1994, he received his PhD in human factors from the University of Minnesota, where he is still an affiliated faculty member of the Center for Cognitive Science. He is director of the Cognitive Ergonomics Research Laboratory and the University of Calgary Driving Simulator (UCDS) funded by the Canadian Foundation for Innovation. Most recently, he opened and currently directs the Healthcare Human Factors and Simulation Laboratory in the Ward of the 21st Century Research and Innovation Centre. He is currently co-leader of the Teen and Novice Driver Network that is part of the AUTO21 Network of Centres of Excellence (NCE). He has co-edited a number of books on human-machine systems in addition to the *Handbook of Driving Simulation*. He was awarded a Killam Fellowship to study traffic safety and the Faculty of Social Sciences Distinguished Researcher Award. His undergraduate and graduate students have won eight national and international research awards. He is a member of a number of national and international transportation and healthcare committees in Canada, the United States, and the Netherlands, including the National Academy of Sciences and the Transportation Research Board. His broad areas of research are in transportation and healthcare human factors with research projects focusing on: older, teen, and novice drivers; vulnerable road users; in-vehicle intelligent transportation system evaluation and design; nomadic, integrated, and external sources of driver distraction; patient and driving simulation; perception, attention, and motor processes; ICU, surgery, and emergency medicine system design; medical device evaluation; pharmacy, chemotherapy, and anesthesia drug organization, design and use; and interruptions and fatigue in healthcare.

Dr. John D. Lee is the Emerson Electric Professor in the Department of Industrial and Systems Engineering at the University of Wisconsin, Madison, and director of the Cognitive Systems Laboratory. Previously, he was a professor at the University of Iowa and director of human factors research at the National Advanced Driving Simulator. Before moving to the University of Iowa, he was a research scientist at the Battelle Human Factors Transportation Center for six years. He is a co-author of the textbook, *An Introduction to Human Factors Engineering*, and is the author or co-author of over 170 papers. He recently co-edited the book, *Driver Distraction: Theory, Effects, and Mitigation*. Support for this research includes grants and contracts for basic and applied research from both government and industry, including: the National Science Foundation, the Office of Naval Research, the National Institutes of Health, the National Highway Traffic Safety Administration, the Federal Highway Administration, Intel, Nissan, GM, and Honda. His research focuses on the safety and acceptance of complex human-machine systems by considering how technology mediates attention. Specific research interests include trust in technology, advanced driver assistance systems, and driver distraction. He served on the National Academy of Sciences Committee on human system integration, the committee on electronic vehicle controls and unintended acceleration, and several other committees. He now serves on the editorial board of *Cognitive Engineering and Decision Making*; *Cognition, Technology and Work*; *International Journal of Human Factors Modeling and Simulation*; and is the associate editor for the journals, *Human Factors* and *IEEE-Systems, Man, and Cybernetics*. He received the Ely Award for best paper in the journal *Human Factors* (2002), and the best paper award for the journal *Ergonomics* (2005). Both these papers addressed simulator-based evaluation of collision warning systems. Dr. Lee received a BA in 1987 (psychology) and a BS in 1988 (mechanical engineering) from Lehigh University, an MS in 1989 (industrial engineering), and a PhD in 1992 (mechanical engineering) from the University of Illinois at Urbana–Champaign.

Contributors

Michelle L. Ackerman
Center for Research on Applied
    Gerontology
University of Alabama at Birmingham
Birmingham, Alabama

R. Wade Allen
Systems Technology, Inc.
Hawthorne, California

George J. Andersen
Department of Psychology
University of California, Riverside
Riverside, California

Linda S. Angell
Touchstone Evaluations, Inc.
Grosse Pointe Farms, Michigan

Karlene K. Ball
Center for Research on Applied
    Gerontology
University of Alabama at Birmingham
Birmingham, Alabama

Brent M. Barbour
Department of Psychiatry &
    Neurobehavioral Sciences
University of Virginia Health System
Charlottesville, Virginia

Michel Bédard
Centre for Research on Safe Driving
Lakehead University
Thunder Bay, Ontario, Canada

Myra Blanco
Virginia Tech Transportation Institute
Virginia Polytechnic Institute and State
    University
Blacksburg, Virginia

Mike Blommer
Research and Advanced Engineering
Ford Motor Company
Dearborn, Michigan

Linda Ng Boyle
Department of Industrial and Systems
    Engineering, and Department of Civil
    and Environmental Engineering
University of Washington
Seattle, Washington

Karel A. Brookhuis
Department of Psychology
University of Groningen
Groningen, the Netherlands

Wiebo Brouwer
Department of Neurology
University Medical Center Groningen
Groningen, the Netherlands

Thomas G. Brown
Douglas Mental Health Research
    Institute
Verdun, Quebec, Canada

Rens B. Busscher
Department of Neurology
University Medical Center Groningen
Groningen, the Netherlands

Jeff K. Caird
Cognitive Ergonomics Research
    Laboratory
University of Calgary
Calgary, Alberta, Canada

Alex Chaparro
Department of Psychology
Wichita State University
Wichita, Kansas

Jeremy R. Chapman
Department of Civil and Environmental
    Engineering
University of Wisconsin–Madison
Madison, Wisconsin

Judith Charlton
Accident Research Centre
Monash University, Clayton Campus
Victoria, Australia

Susan T. Chrysler
Human Factors Program
Texax Transportation Institute
College Station, Texas

Marcia L. Cook
Systems Technology, Inc.
Hawthorne, California

Joel Cooper
Texas Transportation Institute
Texas A & M University System
College Station, Texas

Daniel J. Cox
Department of Psychiatry &
    Neurobehavioral Sciences
University of Virginia Health System
Charlottesville, Virginia

Janet I. Creaser
ITS Institute
University of Minnesota
Minneapolis, Minnesota

Ragnhild J. Davidse
SWOV Institute for Road Safety Research
Leidschendam, the Netherlands

Gregory W. Davis
Office of Safety Research and Development
Federal Highway Administration
McLean, Virginia

Jeffrey D. Dawson
Department of Biostatistics
The University of Iowa
Iowa City, Iowa

Dick de Waard
Department of Psychology
University of Groningen
Groningen, the Netherlands

Paula A. Desmond
Department of Psychology
Southwestern University
Georgetown, Texas

Anna Devlin
Accident Research Centre
Monash University, Clayton Campus
Victoria, Australia

Frank A. Drews
Department of Psychology
University of Utah
Salt Lake City, Utah

Caitlin W. Duffy
Northwestern University
Evanston, Illinois
and
National Institute of Child Health and
    Human Development
National Institutes of Health
Bethesda, Maryland

Amanda K. Emo
Office of Safety Research and
    Development
Federal Highway Administration
McLean, Virginia

Johan Engström
Volvo Technology Corporation/SAFER
    Vehicle and Traffic Safety Centre
Chalmers University of Technology
Göteborg, Sweden

Douglas F. Evans
DriveSafety, Inc.
Salt Lake City, Utah

Lisa Fern
San Jose State University Research
    Foundation
U.S. Army Aeroflightdynamics
    Directorate
Moffett Field, California

Donald L. Fisher
Department of Mechanical and
    Industrial Engineering
University of Massachusetts, Amherst
Amherst, Massachusetts

John M. Flach
Department of Psychology
Wright State University
Dayton, Ohio

Gregory J. Funke
Department of Psychology
University of Cincinnati
Cinncinnati, Ohio

Thomas M. Granda
Federal Highway Administration (Retired)
McLean, Virginia

Timofey F. Grechkin
Department of Computer Science
The University of Iowa
Iowa City, Iowa

Jeffry Greenberg
Research and Advanced Engineering
Ford Motor Company
Dearborn, Michigan

Leo Gugerty
Department of Psychology
Clemson University
Clemson, South Carolina

Peter A. Hancock
Institute for Simulation and Training
University of Central Florida
Orlando, Florida

Richard J. Hanowski
Virginia Tech Transportation Institute
Virginia Polytechnic Institute and State
    University
Blacksburg, Virginia

Jeffrey S. Hickman
Virginia Tech Transportation Institute
Virginia Polytechnic Institute and State
    University
Blacksburg, Virginia

William J. Horrey
Liberty Mutual Research Institute for
    Safety
Hopkinton, Massachusetts

Vaughan W. Inman
Science Applications International
    Corporation
McLean, Virginia

Richard J. Jagacinski
Department of Psychology
The Ohio State University
Columbus, Ohio

Hamish Jamson
Institute for Transport Studies
University of Leeds
Leeds, United Kingdom

Barry H. Kantowitz
Department of Industrial and Operations
    Research
University of Michigan
Ann Arbor, Michigan

Bart Kappé
Netherlands Organization for Applied
    Scientific Research, TNO
Soesterberg, the Netherlands

Konstantinos V. Katsikopoulos
Max Planck Institute for Human
    Development
Berlin, Germany

Gerold F. Kauert
Department of Forensic Toxicology
Goethe University of Frankfurt
Frankfurt, Germany

Joseph K. Kearney
Department of Computer Science
The University of Iowa
Iowa City, Iowa

Michael A. Knodler, Jr.
Department of Civil and Environmental
    Engineering
University of Massachusetts, Amherst
Amherst, Massachusetts

John D. Lee
Department of Industrial and Systems
    Engineering
University of Wisconsin–Madison
Madison, Wisconsin

Mikael Ljung Aust
Volvo Car Corporation/SAFER Vehicle
    and Traffic Safety Centre
Chalmers University of Technology
Göteborg, Sweden

Michael A. Manore
Vispective Management Consulting, LLC
Hudson, Wisconsin

Michael P. Manser
ITS Institute
University of Minnesota
Minneapolis, Minnesota

Gerald Matthews
Department of Psychology
University of Cincinnati
Cinncinnati, Ohio

Gerald McGwin, Jr.
Department of Epidemiology, School of
 Public Health
University of Alabama
Birmingham, Alabama

Brian P. McKenna
Resilient Cognitive Solutions
Pittsburgh, Pennsylvania

Shaunna L. Milloy
Cognitive Ergonomics Research
 Laboratory
University of Calgary
Calgary, Alberta, Canada

Jane Moeckli
National Advanced Driving Simulator
The University of Iowa
Iowa City, Iowa

Manfred R. Moeller
Institute of Legal Medicine
Saarland University
Homburg, Germany

John A. Molino
Science Applications International
 Corporation
McLean, Virginia

Michael Mollenhauer, Jr.
Virginia Tech Transportation Institute
Blacksburg, Virginia

Henry Moller
University Health Network, Department
 of Psychiatry
University of Toronto
Toronto, Ontario, Canada

Justin F. Morgan
Virginia Tech Transportation Institute
Virginia Polytechnic Institute and State
 University
Blacksburg, Virginia

Nadia Mullen
Centre for Research on Safe Driving
Lakehead University
Thunder Bay, Ontario, Canada

Alicia A. Nelson
Texas Transportation Institute
Houston, Texas

David A. Noyce
Department of Civil and Environmental
 Engineering
University of Wisconsin–Madison
Madison, Wisconsin

Marie Claude Ouimet
University of Sherbrooke
Sherbrooke, Quebec, Canada

Yiannis Papelis
Virginia Modeling Analysis and
 Simulation Center
Old Dominion University
Suffolk, Virginia

Alexander Pollatsek
Department of Psychology
University of Massachusetts, Amherst
Amherst, Massachusetts

Harie Pot
Department of Neurology
University Medical Center Groningen
Groningen, the Netherlands

Anuj K. Pradhan
National Institute of Child Health and
 Human Development
National Institutes of Health
Bethesda, Maryland

Michael E. Rakauskas
ITS Institute
University of Minnesota
Minneapolis, Minnesota

Jan G. Ramaekers
Department of Neuropsychology &
 Psychopharmacology
Maastricht University
Maastricht, the Netherlands

Thomas A. Ranney
Transportation Research Center, Inc.
East Liberty, Ohio

Michelle L. Reyes
Public Policy Center
The University of Iowa
Iowa City, Iowa

Matthew Rizzo
Department of Neurology
Carver College of Medicine
The University of Iowa
Iowa City, Iowa

John Robinson
Delphi-MRC
Halifax, Nova Scotia, Canada

Theodore J. Rosenthal
Systems Technology, Inc.
Hawthorne, California

Dario D. Salvucci
Department of Computer Science
Drexel University
Philadelphia, Pennsylvania

Dyani J. Saxby
Department of Psychology
University of Cincinnati
Cincinnati, Ohio

Chris W. Schwarz
National Advanced Driving Simulator
The University of Iowa
Iowa City, Iowa

Thomas B. Sheridan
Department of Mechanical Engineering
Massachusetts Institute of Technology
Cambridge, Massachusetts

Bruce G. Simons-Morton
National Institute of Child Health and
 Human Development
National Institutes of Health
Bethesda, Maryland

Harsimran Singh
Department of Psychiatry &
 Neurobehavioral Sciences
University of Virginia Health System
Charlottesville, Virginia

Alison Smiley
Human Factors North
Toronto, Ontario, Canada

Matthew R. H. Smith
Advanced Driver Support Systems
Delphi Electronics & Safety
Kokomo, Indiana

Heather A. Stoner
Realtime Technologies, Inc.
Royal Oak, Michigan

David L. Strayer
Department of Psychology
University of Utah
Salt Lake City, Utah

Eef L. Theunissen
Department of Neuropsychology &
    Psychopharmacology

Maastricht University
Maastricht, the Netherlands

Jon M. Tippin
Department of Neurology
The University of Iowa Hospitals and
    Clinics
Iowa City, Iowa

Lana M. Trick
University of Guelph
Guelph, Ontario, Canada

Ergun Y. Uc
Carver College of Medicine
The University of Iowa
Iowa City, Iowa
and
Veterans Affairs Medical Center
Iowa City, Iowa

Peter C. van Wolffelaar
Department of Psychology
University of Groningen
Groningen, the Netherlands

Willem Vlakveld
SWOV Institute for Road Safety Research
Leidschendam, the Netherlands

Nicholas J. Ward
Western Transportation Institute
University of Montana
Bozeman, Montana

Joanne Wood
School of Optometry and Institute of
    Health and Biomedical Innovation
Queensland University of Technology
Brisbane, Queensland, Australia

I

Introduction

1

Handbook of Driving Simulation for Engineering, Medicine, and Psychology: An Overview

Donald L. Fisher
*University of Massachusetts,
Amherst*

Jeff K. Caird
University of Calgary

Matthew Rizzo
*Carver College of Medicine
The University of Iowa*

John D. Lee
University of Wisconsin–Madison

Abstract

The Problem. To date, there has been no single convenient and comprehensive source of information on driving simulation research being conducted around the world. Nor has there been a single repository for information regarding the numerous challenges that confront new simulator users or the broader challenges that confront the whole community. The *Handbook of Driving Simulation for Engineering, Medicine and Psychology* strives to put much of this critical information in one easily accessible place. *Role of Driving Simulators*. Driving simulation is now a key element in regional, national, and international efforts to probe the: (1) efficacy of novice, commercial, and older driver training programs; (2) fitness to drive in patients with performance declines due to aging and mild cognitive impairment, traumatic brain injury, and neurodegenerative disorders such as Alzheimer's and Parkinson's diseases, and sleep disturbances; (3) acute and chronic effects of many medications such as analgesics, antidepressants, psychostimulants, antidepressants, and cancer chemotherapy agents; (4) impact of alternative signs, signals, and pavement markings on drivers' behavior; and (5) relative advantages and/or disadvantages of technologies being used or introduced into the vehicle such as cell phones, left-turn and rear-end collision warning systems, advanced cruise control, navigation aids, and internet access. *Key Results of Driving Simulator Studies*. The chapters in the *Handbook* detail the many key results. In the broadest of terms, there has been a literal explosion in our understanding of the differences in vehicle and driver behaviors among different groups of people; of the impact of different traffic control devices, traffic scenarios, road geometries, and general lighting and weather conditions on vehicle and driver performance; of the effect of different types of distraction inside and outside the vehicle on various metrics of performance; and on the use of driving simulators as assessment and training tools. *Scenarios and Dependent Variables*. The chapters are replete with discussions of the scenarios and dependent variables that are critical for analysis purposes. Additionally, there are key sections specifically targeting scenario authoring and dependent variables. *Limitations*. There are still a number of "evergreen" issues. They include simulator adaptation syndrome, fidelity of the virtual environment, cross-platform comparisons among different simulators, development of standard scenarios for testing and training, and transfer of effects to the real world.

This is an exciting time for researchers using driving simulators. In the 1970s there were in the neighborhood of 20 research driving simulators throughout the US and Europe including many small, part-task devices used for training and licensing. (See chap. 2 by Allen, Rosenthal, & Cook in this book for a discussion of the history of driving simulator research.) Today there are hundreds if not thousands of driving simulators spread around the globe. International conferences have become well established, including, most notably, the Driving Assessment (held in the United States) and the Driving Simulation Conference (held in Asia, Europe, North America), with many others drawing large audiences as well. Governments and industry around the world are much more aggressively funding simulator research.

Now is a time to take stock, gather in one place the many different advances, and cast a critical eye on ourselves. Have we made any progress on mitigating simulator adaption syndrome? (See chap. 14 by Stoner, Fisher, & Mollenhauer in this book.) Are we confined to learning how drivers perform in the laboratory, rather than how they behave in the real world? (See chap. 9 by Ranney and chap. 13 by Mullen, Charlton, Devlin, & Bédard.) Now is also the time to provide what lessons we can for those wanting to get into the field. (See chap. 5 by Caird & Horrey.) The barriers to entry are no longer so formidable. Costs are coming down. However, there are still numerous challenges that confront new simulator users, ranging from selecting (see chap. 12 by Jamson; chap. 7 by Greenberg & Blommer; and chap. 8 by Andersen) and validating the system (see chap. 11 by Schwarz) to programming scenarios (see chap. 6 by Kearney & Grechkin) and selecting dependent variables (see chap. 10 by Angell; chap. 15 by McGwin; chap. 16 by Milloy & Caird; chap. 17 by Brookhuis & de Waard; chap. 18 by Fisher, Pollatsek, & Horrey and chap. 19 by Gugerty), and to completing data reduction and interpretation. (See chap. 20 by Reyes & Lee; chap. 21 by Boyle; chap. 22 by Dawson; and chap. 23 by Moeckli.)

The broad sweep of contributions in this *Handbook* is impressive. Driving simulation is now a key element in private and governmental efforts to probe the: (1) efficacy of behavioral interventions such as novice, commercial and older driver training programs; (2) fitness to drive in patients with visual, cognitive and motor impairments due to aging and neurological impairments (such as traumatic brain injury, sleep disturbances, and neurodegenerative disorders, including Alzheimer's and Parkinson's diseases); (3) acute and chronic effects of medications (such as analgesics, psychostimulants, antidepressants, and cancer chemotherapy agents); (4) impact of alternative traffic control devices—signs, signals, pavement markings—and geometric designs on drivers' behavior; and (5) risks and benefits of in-vehicle technologies and devices such as cell phones, adaptive cruise control, navigation aids, internet access, and left-turn, back-up and rear-end collision warning systems. As is discussed throughout the *Handbook*, simulation is an essential tool for these efforts because it offers a high degree of experimental control and allows investigators to understand the mechanisms of driver error that create the potential for a serious crash and would be too unsafe to investigate in other ways. The demand for driving simulation research

and the pool of dedicated researchers will continue to grow as the cost of driving simulators decreases and the applications proliferate.

To date, there has been no single convenient and comprehensive source of information on the driving simulation research being conducted around the world. The *Handbook of Driving Simulation for Engineering, Medicine, and Psychology* strives to put much of this critical information in one easily accessible place. The audience for this information consists of a broad collection of academics, professionals and students, including more specifically: (1) researchers in engineering, medicine and psychology with a wide range of simulator experience, from novice to expert; (2) private and governmental (e.g., including state, regional, provincial, territorial, and federal) program managers wanting to fund research in the areas listed above; (3) transportation engineers who must determine which signs, signals, pavement markings and geometric designs should be used in a variety of traffic environments; (4) automotive, computer and information systems engineers developing new in-vehicle technologies; (5) health professionals trying to advise patients, families and the state on whether the patients can drive safely or not; (6) pharmaceutical companies needing to evaluate the effects of various drugs on driving performance; and (7) faculties teaching human factors and applied cognitive psychology in graduate and undergraduate courses who need to expose their students to one or more of the topics covered in this book.

This *Handbook* provides a comprehensive resource to guide researchers, designers, program managers, and practitioners in: The history and future of driving simulation along with a discussion of its most pressing issues (Section I); the selection and validation of driving simulator technology (Section II); the conduct of simulator studies, including the selection of scenarios and dependent variables as well as the evaluation of results (Section III); and the use of simulators in studies relevant to psychology (Section IV), engineering (Section V) and medicine (Section VI). Finally, we have initiated a resource for the open sharing of scenarios, CARSS (Coordinated Assessment of Roadway Simulator Scenarios). This resource is necessary because, in the absence of a universal scene and scenario development tool, we must fall back on the actual programs used to generate the scenarios if we are ever going to replicate one another's research on different simulator platforms. While it is true that a uniform way of capturing the information would allow researchers around the world to more easily replicate one another's findings on the same or different platforms, such a common description is probably not forthcoming in the immediate future. The URL is http://www.drivingwiki.org or http://www.drivingwiki.com.

1.1 *Handbook* Web Site

Because of the restrictions on color reproductions, only grayscale figures were included within chapters, with a representative sample of the color figures shown in the *Handbook* insert. Copies of all the color equivalents can be found in the *Handbook* section on the University of Wisconsin driving wiki (http://www.drivingwiki.org or http://www.drivingwiki.com). The color equivalents are referenced in the individual chapters. Finally, we wanted to be able to include actual videos of drives through various simulated worlds. These are also included on the driving wiki and references are made to the videos in the following chapters.

1.2 Guide for Readers

We organized the sections of this *Handbook* so that readers with a variety of interests can readily identify relevant chapters and topics. However, as an aid to readers, we will indicate below what we believe are the most relevant chapters and sections for the various different groups of readers that will access the *Handbook*, including students and new researchers to driving simulation, researchers in psychology, engineering and medicine, research program managers, practicing transportation engineers, automotive engineers, health professionals, teachers and last, but not least, skeptics.

1.2.1 Students and Researchers New to Driving Simulation

Our students, who are more commonly in engineering or psychology, are easily attracted to research with driving simulators. However, the ascent of the learning curve in becoming competent in using a given driving simulator requires prolonged trial and error and/or training from students and faculty members who have not undertaken previous projects. With research on driving simulation, there is so much to be learned that starting a research project on a driving simulator can seem an almost insurmountable task. We hope to reduce the learning time, to help undergraduate and graduate students (and other researchers new to driving simulator research) take their first steps in becoming familiar with: the software that lets one construct the virtual worlds displayed to a participant and the scenarios within those virtual worlds, where cars, pedestrians and traffic devices are all interacting with one another; the hardware that is needed to support the entire enterprise; the independent and dependent variables one might consider; and the myriad other details that will be needed to complete successfully one's first experiment. In many driving simulation research laboratories, graduate students get their degrees and venture on to take excellent jobs with corporations and government agencies. However, when they leave, the knowledge that they have acquired through hard work with software programming, experimental protocol, and statistical analysis goes with them unless it is transferred to students who are just starting their research on driving simulators in a laboratory. We endeavor to capture and systematize a body of knowledge about the overall research enterprise so that the proverbial method wheel does not have to be completely reinvented with the entry of each new student.

We suggest that students and researchers who are new to driving simulator research start off by reading chap. 12 by Jamson ("Cross Platform Validation Issues") and chap. 5 by Caird and Horrey ("Twelve Practical and Useful Questions About Driving Simulation"). Even though you may already be in a situation where you have a simulator, knowing the full range of possibilities can help you understand both the strengths and limitations of your own equipment. If you are thinking of actually purchasing a driving simulator, you should also read chap. 6 by Kearney and Grechkin ("Scenario Authoring"). Conceptually, it is critical from the start to understand both how the visual world is constructed and how other vehicles, pedestrians and signals interact with each other and with the participant driver. There are large differences between how one does this on different simulators and understanding these differences helps make it much easier to communicate with other researches involved in a similar enterprise. We believe it is important to follow that by a close reading of the chapters dealing with physical (chap. 7 by Greenberg & Blommer, "Physical Fidelity of Driving Simulators"), perceptual (chap. 8 by Andersen, "Sensory and Perceptual Factors in the Design of Driving Simulation Displays") and psychological (chap. 9 by Ranney, "Psychological Fidelity: Perception of Risk") fidelity. How much physical and psychological fidelity one wants and needs is critical to consider from the outset. Finally, we believe that readers will find it most useful to read Section III (Conduct of Simulator Experiments), which deals with how to select scenarios and choose which independent variables to manipulate and dependent variables to measure, and how to analyze the data gathered.

Once researchers have acquired a new simulator and trained themselves and their students to use the new system, they must still develop an effective research agenda. Researchers often realize that justifying, acquiring and installing their simulator was relatively easy compared to what they must do to get their research going. There are common issues to all simulator studies (such as simulator sickness) that can create unforeseen problems (chap. 14 by Stoner, Fisher, & Mollenhauer, on Simulator Sickness). Then there are the almost limitless problems that less than complete fidelity creates, making it challenging to study driving at night (chap. 28 by Wood & Chaparro, "Night Driving"), reflectance off of traffic signs (chap. 36 by Chrysler & Nelson, "Design and Evaluation of Signs and Pavement Markings Using Driving Simulators"), and even responses to a lead vehicle braking. Additionally, there are the many problems that need to be addressed when answering particular research questions. For instance, does one need just braking data (chap. 17 by Brookhuis & de Waard, "Measuring Physiology in Simulators") or both eye movement and braking data? (chap. 18 by Fisher, Pollatsek, & Horrey, "Eye Behaviors"). And finally, someone starting out needs to be aware of what research is most pressing. Many kinds of studies have been done before and innumerable gaps in the research need to be addressed, provided that researchers know where these gaps exist. To date, most new driving simulation laboratories require several years of sustained projects and theses to adequately refine the questions that are asked and to acquire sufficient methodological sophistication. It is hoped that the accumulated wisdom, contained herein, of many of the world's leading driving simulator researchers will help shorten this time considerably (Section IV, "Applications in Psychology"; Section V, "Applications in Engineering"; Section VI, "Applications in Medicine").

1.2.2 Researchers in Psychology

Experimental psychologists have been trained to favor theory development and questions that are considered basic science in orientation (Nickerson, 1999). For these psychologists, pursuing driving as a "common everyday task" makes sense in so far as it can provide insight into theoretical issues in human perception and cognition (Groeger, 2000). Others in psychology have more applied interests. For these psychologists, focusing on driving is an opportunity to reduce the frequency of injuries and fatalities, say in left turns or rear-end crashes. However, it is also possible to pursue questions which depend on a symbiotic relationship between applied and basic research, a symbiotic relationship that occurs naturally within the context of driving (Leibowitz, 1996). For example, in order to address the underlying human limitations in certain circumstances one requires a deeper understanding of basic research, whereas to evaluate countermeasures to mitigate a fundamental limitation one requires a deeper understanding of applied research (Dewar & Olson, 2007). Driving research is inherently multidisciplinary as the chapters in the *Handbook* illustrate. Understanding unfamiliar areas of research may provide readers with important insights.

Bearing the above in mind, there is much of interest for psychologists that is spread throughout the *Handbook*, including not only cognitive and engineering psychologists, but social (e.g., chap. 24 by Ouimet, Duffy, Simons-Morton, Brown, & Fisher, "Understanding and Changing the Young Driver Problem") and clinical (e.g., chap. 47 by Moller, "Psychiatric Disorders and Driving Performance") psychologists as well. In Section II ("Selecting a Driving Simulator") fundamental, theoretical questions about perception are addressed. (Chap. 8 by Andersen, "Sensory and Perceptual Factors in the Design of Driving Simulation Displays") In Section III ("Conduct of Simulator Experiments"), the analysis not only of eye glances, but also of eye glance patterns, is described in detail. (Chap. 18 by Fisher, Pollatsek, & Horrey, "Eye Behaviors") In Section IV ("Applications in Engineering"), behavioral decision theories are used to describe drivers' route choice behavior. (Chap. 37 by Katsikopoulos, "Advanced Guide Signs and Behavioral Decision Theory") In this same section, cognitive architectures are laid out which can be used to predict lane-changing and car-following behavior (chap. 42 by Salvucci, "Cognitive Architectures for Modeling Driver Behavior") and more general control systems are described which can be used to predict a broad range of different driver behaviors. (Chap. 43 by Flach, Jagacinski, Smith, & McKenna, "Combining Perception, Action, Intention and Value: A Control Theoretic Approach to Driving Performance") In short, we believe that there is much of interest to psychologists throughout the *Handbook*, not just in Section IV ("Applications in Psychology").

1.2.3 Researchers in Engineering

Researchers in engineering that are interested in driving simulation typically include individuals with backgrounds in industrial, civil or electrical engineering. The goal of a particular research program might be to understand something about novice or older drivers' behavior that can guide interventions (often true of research initiated in industrial engineering departments), to investigate alternative signs, signals, pavement markings and roadway geometries (perhaps civil engineering), or to evaluate advanced in-vehicle devices (perhaps electrical engineering or industrial engineering). Because the research is often application-driven, three questions are likely to be common ones to any of the above efforts.

First, can a virtual world be built that incorporates the central elements of the application? A number of different applications are discussed. The applications include geometric design (see Granda, Davis, Inman, & Molino, chap. 34), signaling (see Noyce, Knodler, Chapman, Fisher, & Pollatsek, chap. 35), signs and pavement markings (see Chrysler & Nelson, chap. 36), advance guide signs (see Katsikopoulos, chap. 37), and highway-railway grade and transit crossings (see Caird, Smiley, Fern, & Robinson, chap. 38).

Second, just how close can the physical characteristics of the overall virtual world be made to resemble the real world? We believe that some chapters in Section II should go a long way towards answering this question. These chapters discuss in detail how closely visual displays (chap. 8 by Andersen, "Sensory and Perceptual Factors in the Design of Driving Simulation Displays"), scenarios (chap. 6 by Kearney & Grechkin, "Scenario Authoring"), vehicle dynamics (chap. 11 by Schwarz, "Validating Vehicle Models"), and motion, sound and lighting (chap. 7 by Greenberg & Blommer, "Physical Fidelity of Driving Simulators") can be made to maintain fidelity.

And third, just how much work needs to be put into the development of the virtual world and the driving tasks in that world in order to make the results obtained from a driving simulator apply to the real world? Chapter 39 by Manore and Papelis ("Roadway Visualization"), which discusses the use of driving simulation to visualize transportation improvements is an important one to read in this regard. Chapter 9 by Ranney ("Psychological Fidelity: Perception of Risk"), which deals with the psychological fidelity of driving simulators is also essential. And chap. 19 by Gugerty ("Situation Awareness in Driving"), which discusses situation awareness, provides an important perspective on what level of fidelity is required and what measures are needed to address this important issue. Finally, one can compare the studies that have been done on the driving simulator with studies in the real world to shed some light on the current question. The answers are discussed in chap. 13 written by Mullen, Charlton, Devlin and Bédard ("Simulator Validity: Behaviors Observed on the Simulator and on the Road"). For engineering applications, driving simulators must provide drivers with an experience representative of that of actual driving so that the resulting analysis provides a valid indicator of a driver's performance. These chapters address this complex issue from a variety of perspectives.

1.2.4 Researchers in Medicine

Driving simulators have been successfully applied to quantify driving performance and to study basic aspects of cognition in drivers with neurological, medical and psychiatric disorders. Driving simulation offers advantages over the use of road tests or driving records in assessments of driver fitness in those who may be at increased risk of a vehicle crash. It provides the only means to replicate exactly the experimental road conditions under which driving comparisons are made but with none of the risk of the road or test track. As we shall see, simulation has been successfully applied to assess performance profiles in drivers who are at risk of a crash due to a variety of conditions, such as: medical (chap. 46 by Rizzo, "Medical Disorders"), psychiatric (chap. 47 by Moller, "Psychiatric Disorders and Driving Performance") and sleep disorders (chap. 49 by Tippin, "Driving Simulation in Epilepsy and Sleep Disorders); fatigue and stress (chap. 29 by Matthews, Saxby, Funke, Emo, & Desmond, "Driving in States of Fatigue or Stress"); alcohol (chap. 44 by Creaser, Ward, & Rakauskas, "Acute Alcohol Impairment Research in Driving Simulators"), cannabis (chap. 45 by Ramaekers, Moeller, Theunissen, & Kauert, "Validity of Three Experimental Performance Tests for Predicting Risk of Cannabis–Induced Road Crashes"); Alzheimer's disease, Parkinson's disease and stroke (chap. 48 by Uc & Rizzo, "Driving in Alzheimer's Disease, Parkinson's Disease and Stroke"), or traumatic brain injury (chap. 50 by Brouwer, Busscher, Davidse, Pot, & van Wolffelaar, "Traumatic Brain Injury: Tests in a Driving Simulator as Part of the Neuropsychological Assessment of Fitness to Drive"). There are several different types of simulators (film, non-interactive, interactive, fixed versus motion-based, desktop, full cab) used in this medical research, all of which are discussed in the *Handbook* (e.g., chap. 2 by Allen, Rosenthal & Cook, "A Short History of Driving Simulation").

1.2.5 Research Program Managers

Many private, foundation and governmental programs fund research in transportation. Research program managers at these various private companies, foundations and government agencies will inevitably wonder whether a particular project that they have in mind could benefit from being undertaken in whole or in part on driving simulators. We hope that research program managers find relevant material in the sections describing applications in psychology (Section IV), engineering (Section V), and medicine (Section VI). More generally, program managers might well wonder whether the results from driving simulators apply in the real world. These concerns are addressed in part in chap. 13 by Mullen, Charlton, Devlin and Bédard ("Simulator Validity: Behaviors Observed on the Simulator and on the Road"); chap. 9 by Ranney ("Psychological Fidelity: Perception of Risk"); and chap. 19 by Gugerty ("Situation Awareness in Driving"). Program managers might wonder whether it makes much difference which type of driving simulator they select to fund a particular project.

Chapter 12 by Jamson addresses this concern ("Cross Platform Validation Issues"). Finally, program managers might wonder whether they could use less expensive alternatives to driving simulators. Chapter 10 by Linda Angell considers such alternatives ("Surrogate Methods and Measures").

1.2.6 Practicing Transportation Engineers

The section on applications in engineering speaks directly to transportation engineers who may need or want to consider alternatives to current signs, signals, pavement markings and roadway geometries. Some transportation engineering professionals may not think of the *Handbook* as having much to offer since simulation has played a relatively small role to date in the transportation engineering community. However, there has been a recent shift in the role of simulation in this community. Typically, all of the research that has led to changes in the *Manual on Uniform Traffic Control Devices* and other highway design guidelines was done in the field. However, this is now changing. For example, the standards for the phasing of protected/permissive left turn indications will now include a flashing yellow arrow, the result of research that was undertaken primarily on driving simulators (Noyce & Kacir, 2001; see also chap. 35 by Noyce et al., "Traffic Signals"). It was almost never the case that the signage for a major construction project would or could be evaluated on a driving simulator before the advent of powerful visual database tools, yet now major construction projects are being built in virtual worlds and the signs, signals and pavement markings tested in those worlds (e.g., the 163 lane miles of the Central Artery/Tunnel project in Boston, Massachusetts were all modeled on the simulator, as were the signs, signals and pavement markings, see Upchurch, Fisher, Carpenter, & Dutta, 2002; for a more general discussion of signage see chap. 36 by Chrysler & Nelson, "Design and Evaluation of Signs and Pavement Markings Using Driving Simulators," and chap. 38 by Caird, Smiley, Fern, & Robinson, "Driving Simulation Design and Evaluation of Highway-Railway Grade and Transit Crossings"). Simulators are also being used in geometric design (chap. 34 by Granda, Davis, Molino, & Inman, "The Use of High Fidelity Real-Time Driving Simulators for Geometric Design") and visualization (chap. 39 by Manore & Papelis, "Roadway Visualization"). Finally, simulators are increasingly being used to evaluate alternative intelligent transportation system sign designs (chap. 37 by Katsikopoulos, "Advanced Guide Signs and Behavioral Decision Theory").

1.2.7 Practicing Automotive and Software Engineers

Emerging technology will change the nature of cars and trucks. Already Adaptive Cruise Control (ACC), collision warning systems, and electronic stability control are making their way into many vehicles now in production. (See chap. 40 by Manser, "Advanced Warning Technologies: Collision, Intersection Incursion," and chap. 41 by Engström, & Ljung Aust, "Adaptive Behavior in the Simulator: Implications for Active Safety System

Evaluation"). Such technology promises to enhance driving safety, but only to the extent that its design is consistent with the very real constraints on drivers' ability to multitask. For example, some of these systems are designed to extend drivers' ability to detect hazards and augment their ability to control the vehicle. This they may do, but the ultimate benefit of this technology depends on how drivers will respond—an issue that simulators can help address. Technology is also changing what drivers can do while driving. Until recently, cars contained relatively few distractions, but cell phones, music players, internet access, and DVD players have since entered the car. (See chap. 27 by Strayer, Cooper, & Drews, "Profiles in Cell-Phone Induced Driver Distraction".) Some of these devices can pose a substantial distraction if drivers choose to use them while driving. A considerable body of simulator-based research has already demonstrated that simulators can provide useful insights into whether, and to what degree, these emerging technologies can distract drivers (Regan, Lee, & Young, 2008). The coming years will see major changes to vehicle technology and driving simulators can play a central role in ensuring that this technology is crafted in a manner consistent with driver limits and capabilities (see chap. 4 by Hancock & Sheridan, "The Future of Driving Simulation"). In addition to the ones highlighted above, this *Handbook* contains many chapters aimed at helping automotive software engineers address the challenges associated with new automotive technology, particularly the chapters in Section III ("Conduct of Simulator Experiments") and Section V ("Applications in Engineering"). These sections may be particularly useful for those who are designing technology that is used by drivers, but who have not been part of the automotive design community, such as those in the telecommunications, Internet, or computer industries.

1.2.8 Practicing Health Professionals

A host of medical, neurological and psychiatric disorders (for example, Alzheimer's and Parkinson's diseases, sleep disorders, hepatic disease, renal disease, diabetes, personality disorders, and effects of licit and illicit drugs) can impair the ability to drive. The *Handbook of Driving Simulation* outlines principles and a conceptual framework for approaching these problems, which are relevant to a wide range of practicing health professionals, including general physicians, specialists (e.g., neurologists, internists, and psychiatrists), physician's assistants, nurse practitioners, clinical psychologists, neuropsychologists, occupational therapists, and students of health care. Relevant chapters are not just restricted to those in Section VI ("Applications in Medicine"), and can be found in other sections. For example, in chap. 32, Singh, Barbour and Cox argue that the results of a number of studies offer compelling evidence for the effectiveness of driving simulators in providing driving re-training to people who have suffered a neurological and/or physical compromise ("Driving Rehabilitation as Delivered by Driving Simulation"). This could be of real interest to health care professionals involved in the development of rehabilitation medicine.

As another example, Ball and Ackerman in chap. 25 ("The Older Driver") discuss techniques one can use to assess and train older drivers (see also chap. 26 by Trick & Caird, "Methodological Issues When Conducting Research on Older Drivers"). The information in this chapter is of special interest to geriatricians. Finally, Pollatsek, Vlakveld, Kappé, Pradhan, and Fisher in chap. 30 discuss techniques one can use to reduce the likelihood that teen drivers will crash ("Driving Simulators as Training and Evaluation Tools: Novice Drivers"). This material is becoming increasingly relevant to pediatricians.

1.2.9 Teachers

Obviously for teachers looking for material in driving simulation, the *Handbook* is of direct relevance. However, we believe that the book is of a potentially much broader significance and could well be used for teaching many of the basic concepts that are often part of graduate and undergraduate courses in engineering psychology and cognitive ergonomics. Questions about experimental design, dependent and independent variables, and statistical analyses are treated at length (Section III). While it is true that the discussion of such questions is in the context of driving simulation, we believe that the treatment is general enough that it could prove useful in graduate courses with a broader focus than just driving simulation. Additionally, the issue of the generalizability of one's results from the laboratory to the field and, more generally, of the need for fidelity, is a continuing one and is covered in detail in some Section II chapters, specifically chap. 11 by Schwarz ("Validating Vehicle Models"); chap. 12 by Jamson ("Cross Platform Validation Issues"); and chap. 13 by Mullen, Charlton, Devlin and Bédard ("Behaviors Observed on the Simulator and on the Road"). The discussion of these issues applies to the increasingly common use of simulators and other naturalistic approaches to data collection in helping to understand human behavior in a wide range of application domains.

1.2.10 Skeptics

Finally, we realize that there are skeptics, who for one reason or another believe that driving simulation has weaknesses that cannot be overcome. Skepticism is important because commitment to a program of driving simulation is likely to command substantial time, effort and funds, and key relationships between simulated driving and real driving remain to be demonstrated. Simulation is not a panacea and is only one of several tools utilized to enhance transportation safety. Consequently, we have given what we hope is a fair and balanced treatment of the limitations of driving simulation, recognizing the pervasive question of whether results in a simulator reflect what happens on the open road, under naturalistic driving conditions, and even on other simulators. We have also asked the authors of each chapter, where appropriate, to address specific limitations of simulation and chap. 13 by Mullen, Charlton, Devlin and Bédard ("Behaviors Observed on the Simulator and on the

Road") is entirely devoted to issues of the validity, reliability and generalizability of driving simulation. Finally, in our discussion of evergreen issues in the very next section, we address questions of generalizability. Our conclusion—which we believe to be the conclusion of most of the authors included—is that there are some areas of investigation where generalizability, both absolute and relative, is very good; some where only relative generalizability is good; and some where it appears that the results on a driving simulator do not generalize to the open road. However, we leave this for you to decide for yourselves once you have read the chapters in the *Handbook*.

1.3 Evergreen Issues

1.3.1 Selecting and Using a Driving Simulator

The doubling of computing power every 18 months as specified by Moore's Law has lead to a dramatic increase in computing power and the associated capabilities of driving simulators over the last 20 years. In some ways this has made the task of selecting a driving simulator easier. Very capable driving simulators are now available for a fraction of the cost of their previous, less capable incarnations. Surprisingly sophisticated driving "simulators" are even available for smartphones, such as the iPhone. The increasing capacity and lower cost of driving simulators also makes the process of selecting and using a simulator more difficult than ever. With increasing capacity comes increasing complexity. Sophisticated simulators can impose a substantial demand on their users to learn scripting languages and understand options and settings that they might never need (for example, see chap. 6 by Kearney & Grechkin, "Scenario Authoring"). In the past, simulator selection primarily revolved around using the available budget to purchase a simulator with the highest fidelity. Although this is often still the case, those selecting a simulator might consider searching for the *lowest* fidelity simulator capable of meeting their needs rather than striving for the highest fidelity system.

The failure of Moore's Law to affect many elements of a driving simulator is one reason to seek out the lowest possible fidelity simulator that can meet the demands of the particular research and design program. A simulator that includes a full vehicle cab, a motion base, or multiple projectors is not appreciably less expensive than a similar simulator a decade ago (exclusive of the computing costs, which have come down substantially). Each of these three elements can significantly enhance the range of cues available to guide more realistic behavior, but the benefits may not be as great as the cost (e.g., see chap. 2 by Allen, Rosenthal, & Cook, "A Short History of Driving Simulation"; chap. 7 by Greenberg & Blommer, "Physical Fidelity of Driving Simulators").

Another important reason to search for the lowest fidelity simulator that can meet a user's needs is the somewhat hidden costs of ownership of a higher fidelity simulator. The operating costs of a more sophisticated simulator are often unexpectedly high and range from the mundane details of learning more functions of a more complex programming environment, to the process of "turning on" the simulator for data collection, and even the higher likelihood of simulator sickness caused in the driver because of the broader field of view and higher resolution graphics. Many of these costs are hidden by the extraordinary efforts of graduate students and technicians responsible for simulator operations. Higher levels of fidelity also demand more attention to scenario and visual database design than might be expected—higher resolution screens reveal imperfections that low-resolution screens mask. Reworking databases to remove these imperfections is expensive and often a surprising additional cost. Beyond the additional demands of operating a higher fidelity simulator, the ongoing maintenance costs can undermine a research program. These costs include additional software maintenance fees, which some vendors charge for each visual channel/projector. Similarly, the periodic replacement of five projectors rather than a single projector is costly. More subtly, maintaining the color balance and alignment of multiple projectors can represent a substantial burden.

The potential for a more sophisticated simulator to undermine the quality of the research is also possible. Efforts made by the graduate students and technicians to manage the additional complexity of the simulator operations may detract from the efforts that could have been directed towards the fundamental research question or to validating the data from the simulator. More directly, even subtle levels of simulator sickness induced by a wider field of view, or by poorly calibrated projectors, can change driver strategies and affect the measured responses to the events of interest. Even without these overt problems, the greater complexity of the auditory and visual environment makes it more likely that confounding factors will make their way into the experiment. For example, pedestrians in the simulated environment that "automatically" respond to a driver drifting towards the sidewalk might provide a critical cue about lane position that might confound the assessment of a lane departure warning system. One reason for using simulators is that they provide a degree of control that is not possible on the road. Paradoxically, as simulators become increasingly realistic and other traffic and pedestrians become automated, this control diminishes and researchers must make an additional effort to maintain control.

Knowing the full cost of a higher fidelity simulator is unlikely to convince many to do anything but buy the highest fidelity simulator within their budget. Despite the above comments, such a temptation can be well-justified, though not necessarily because of the increase in fidelity *per se*. Rather, it is because the increase in control of the virtual worlds that can be displayed is almost always coincident with the increase in fidelity. This is especially critical when one is interested in applications: Designing evaluations of novel in-vehicle collision warning systems, developing training programs, or evaluating novel signs, signals, and pavement markings. Low fidelity simulators often do not allow the user the full control of roadway geometry and scenario that is so critical to creating virtual worlds that mimic real-world scenarios. Control over the road geometry gives the user the opportunity to model transportation infrastructure projects that other low fidelity simulator users would find all but impossible (e.g., Kichhanagari, DeRamus Motley, Duffy, & Fisher, 2002; Upchurch et al., 2002). Control of the actual design of signs, signals and pavement markings can give the user the opportunity to participate in state-of-the-art research with federal, state and local transportation agencies (e.g., Dutta, Carpenter, Noyce, Duffy, & Fisher, 2002; Katsikopoulos, Duse-Anthony, Fisher, & Duffy, 2002; Laurie, Zhang, Mundoli, Duffy, Collura, & Fisher, 2004). Control over the details of the interactions between traffic, pedestrians and signals can provide the user with the power needed to evaluate the increasing range of in-vehicle devices and their effects on driver performance, both those devices which are designed to assist the driver (collision warning systems) and those whose function is primarily for communication (e.g., the cell phone).

One final note on simulator fidelity is in order here. Simulator fidelity is often defined by how closely the physical characteristics of the simulator match the system, with the degree of similarity being referred to as the physical fidelity. Because physical fidelity relies on engineering measurements, it provides a convenient method to make direct comparisons between simulators and the actual on-road driving environment. Unfortunately, physical fidelity may not be a very useful concept in helping to address the issues of efficiency and effectiveness. Functional fidelity may be a more useful way of describing simulator capabilities. This describes the degree to which simulator features induce a psychological experience similar to that induced by the actual domain of application. The meaning of "domain of application," "simulator features," and "psychological experience" must be carefully defined if this definition is to be useful. And, as noted above, fidelity itself may not be the only issue that should direct a user's selection of which simulators are the most effective. Rather, the control one has over the geometry and scenario is of real relevance in many situations, and has a direct impact on effectiveness. Chapters in this *Handbook* provide critical contributions along these lines: physical (chap. 7 by Greenberg & Blommer, "Physical Fidelity of Driving Simulators"), perceptual (chap. 8 by Andersen, "Sensory and Perceptual Factors in the Design of Driving Simulation Displays") and psychological (chap. 9 by Ranney, "Psychological Fidelity: Perception of Risk") fidelity along with geometry and scenario control (chap. 2 by Allen, Rosenthal, & Cook, "A Short History of Driving Simulation"; chap. 6 by Kearney & Grechkin, "Scenario Authoring"; chap. 33 by Evans, "The Importance of Proper Roadway Design in Virtual Environments"; and chap. 34 by Granda, Davis, Inman, & Molino, "The Use of High-Fidelity Real-Time Driving Simulators for Geometric Design").

Given the above considerations, in order to adequately justify the investment in a higher fidelity simulator, the users of the simulator should carefully consider the particular aspects of driving that are important to capture in the simulator and design it so that it can provide a set of cues and action opportunities that are representative of those aspects. There are many types of "high fidelity" simulators and it is critical to match the features of the simulator to the aspects of driving critical to the research program the simulator is meant to support.

More generally, the selection of a simulator ultimately depends on a balance between *efficiency*, which refers to the monetary, time, and personnel resources associated with a particular evaluation approach, and *effectiveness*, which refers to the ability of a particular evaluation approach to provide valid data that addresses the particular evaluation issue (Rouse, 1991). Linda Angell provides a clear example of how extremely low fidelity simulator alternatives can provide a very efficient means of addressing targeted elements of driver behavior (chap. 10, "Surrogate Methods and Measures"). The great range of simulators, each with substantially different capabilities and costs, makes it critical to select the appropriate simulator for each evaluation issue. The range of simulator cost and fidelity governs the efficiency and effectiveness of any pairing of evaluation issue and simulator.

In summary, defining the domain of application and the simulator features can be problematic for several reasons. To some extent, simulator validity is purpose-specific, making it difficult to formulate any general statements concerning the validity of any particular simulator. Validity depends on the ability of the simulator to replicate the purpose, functions, and physical characteristics of the actual driving environment for a particular evaluation issue. Therefore, the pairing of an evaluation issue with a particular simulator should explicitly consider the simulator capabilities from the perspective of the subset of the overall driving environment associated with the evaluation issue. In addition, this process should identify how simulator functions and features combine to realize this purpose—the whole is often more than the sum of the individual simulator components. These issues make selecting a simulator challenging, in part because they reflect broader considerations of validity and representativeness that are fundamental to science and engineering.

1.3.2 Replicating Virtual Worlds Across Simulators

For many experiments in the scientific world, the method section of an article contains all the information that is critical to replicating an experiment. However, for experiments on a driving simulator, this is all but impossible to realize, except when a participant drives through a virtual world consisting of only the most simple of scenes and scenarios, which can be rendered using the same software and hardware. The recognition of the problems that researchers face when trying to replicate one another's experiments on different driving simulators is not new (Fisher, Mourant, & Rizzo, 2005). In fact, it was what eventually led to the production of this *Handbook*: the desire to push the frontiers of science and engineering in the field of driving simulation, and in simulation more generally. There are five related issues that need discussing. At the end we propose a solution, far from perfect, but perhaps more workable in the short term than the alternatives.

First, one needs to ask whether the hardware can be described in ways that lead to repeatability across different platforms (e.g., projectors, screens, graphical processing units, and vehicle controls). We believe that the answer is now almost a certain yes for some hardware, but only a very qualified yes for other hardware. To start with, consider projectors: The behavior of projectors manufactured by different vendors is well characterized by the factors known to influence how individuals perceive the projected image, including the resolution, brightness, contrast ratio, and video update rate, as well as the type of projection used (CRT, LCD or DLP). Many of the same remarks could be made about displays and screens. There seems little reason to be concerned about replicability across these different types of simple hardware. Next, consider the hardware used to generate force feedback and motion cueing. Here, the situation is considerably different, perhaps because the function of such hardware varies considerably with the software used to drive it. For example, different ways of giving force feedback on the steering wheel/pedals exist, and for a given implementation, say with a particular type of torque engine, the control hardware/software may have a large effect on the "feel" of the steering wheel and, consequently, on the steering behavior of the simulated vehicle. The same holds for motion cueing. The different platforms (three degrees of freedom, six degrees of freedom, x/y tables) with different motion drive algorithms will have very different consequences for vehicle control and handling.

Second, one can ask whether the software that renders the scenes and scenarios can be described in ways that capture the differences between the rendering of virtual worlds in different simulator languages. At first, it might be thought that this is relatively easy. After all, there are only a few programming languages (e.g., C and Java), but the problem does not come at the level of the programming language. Rather, it comes in the almost infinite flexibility given to programmers in how they code both scenes (the static world) and scenarios (the dynamic world) and then in the rendering of those worlds.

Three examples will be given of the difficulties one faces when describing the scenes that are actually rendered. 1) The letters that we see on a sign, for instance, can either be rendered in vector graphics or as a texture (digital image). If rendered as vector graphics, the image stays crisp as one moves towards the sign until one eventually passes it. If the image is a texture (bitmap), then that image will be clear at whatever distance in the simulated world from the driver the original image is neither expanded nor compressed when projected on the screen. If the picture of the sign was taken at fifty feet, then farther than fifty feet, the image would need to be compressed. No image compression program is perfect. Clarity is lost. So, if one experimenter used vector graphics on one sign and another used textures, the differences in the legibility distance as measured on the simulator could be large. But no one ever talks about how the particular letters on each and every sign were created. Even if one did, there are different ways of implementing vector graphics and texture compression programs. So, it seems almost impossible at one level to compare the behaviors of drivers viewing the same signs rendered by the same software, but using different authors who might have different pictures of one and the same sign taken at different distances and perhaps different resolutions, let alone the same signs and

different software and authors. 2) A second example is antialiasing. When an image is rendered digitally, the boundaries will appear jagged if they are anything other than horizontal or vertical (this was of much more concern when the resolutions were lower than they typically are today). When the scene is viewed dynamically, edges that are not antialiased can appear to creep up the screen and become noticeable (thus creating a distraction). Even when antialiased, there are numerous different antialiasing algorithms. 3) Finally, there is the problem often referred to as z-buffering. Objects in the distance, say a crosswalk (one object) placed on a roadway surface (a second object), will compete for the same space as the driver approaches. This can cause a flickering (first one object, the crosswalk, is displayed and then the other object, the roadway surface, is displayed) which, again, is both a distraction and a potential source of confusion. Researchers never discuss this potential confusion, yet it is present in almost all scenes that are rendered unless one is very careful.

Similar problems in replicating scenarios occur when those scenarios are created in different scenario-generation tools. Just two examples of the difficulties that occur when replicating scenarios will be given here, although many others are possible. a) Autonomous traffic has been available in many scenario-development languages for some time. By definition the behavior of this traffic is not controlled by the user. Autonomous traffic is useful because it relieves the developer of the burden that comes with programming the behavior of each and every vehicle. However, there is no generally accepted definition of autonomous traffic across scenario-generation languages. Thus, what one participant sees in one virtual world may be very different from what the same participant sees in another virtual world. (This is especially relevant when distraction could be a critical factor in a driver's behavior and that distraction is not controlled). b) In many instances the paths of one or more vehicles involved in the construction of scenarios must be specified beforehand (the non-autonomous traffic). Exactly when these vehicles will become visible to the driver depends precisely on when each vehicle's motion is triggered by the participant's vehicle, what obstructions exist between the participant's vehicle and the other vehicles in the scenario, and where in space each of the various vehicles is located. All of this will influence the time it takes a driver to anticipate a hazard. Yet discussion of this is simply too detailed to include in a method section of a journal article.

The third consideration that researchers face is whether the network of roads and signalization can be described in ways that are standard across different implementations. The answer would appear to be yes. The network of roads and signals can be thought of as the "logical database" that many driving simulators use. They too come in different types and in different flavors. But there is now an open file format (OpenDRIVE®, http://www.opendrive.org/) for the logical description of road networks. It was developed and is being maintained by a team of simulation professionals with large support from the simulation industry (http://www.opendrive.org/). Its first public appearance was on January 31, 2006. At least in Europe it is being increasingly used by investigators in the simulation community. The OpenDRIVE file format provides the following features: XML format; hierarchical structure; analytical definition of road geometry; plane elements, elevation, crossfall, lane width, and so forth; various types of lanes; junctions including priorities, logical interconnection of lanes; signs and signals including dependencies; signal controllers (e.g., for junctions); road surface properties; road and roadside objects; and user-definable data beads. This is just a partial list, but it gives the reader some sense of the detail that is involved.

Fourth, one can ask whether the different measures of performance can be implemented in ways across platforms that are repeatable. It would seem that such could easily be done, but it does require some standard development like that which has begun for describing the logistics of road networks. As an example, investigators often report the mean and variance of the lateral position of the vehicle on the roadway, but in the absence of other information, this turns out to be surprisingly vague. Neither the reference position on the road nor the reference position on the vehicle is well defined. For example, it is not clear whether the lateral position is being measured with respect to the shoulder delineator, the central delineator, or the middle of the lanes, which are so many meters wide. Nor is it clear whether the vehicle position is being measured as the center of gravity of the vehicle, the position of the head, or the left or right front edge of the vehicle. Second, if that is all determined, then what about the measures used to summarize a lane change? It seems that during a lane change, the mean lateral position is not defined, so that part needs to be removed. Furthermore, the heuristics that investigators use to segment the lane change from a time series of lateral positions needs to be well defined, but it almost never is. Also, what about the measures used to summarize performance on curves? Drivers tend to cut a curve, changing mean lateral position, even when they do what is perfectly normal. Or, to make matters worse, some logical networks tend to begin a curve by connecting an arc directly to a straight line. That is interesting, as even perfect drivers will now automatically deviate from their normal mean lateral position, as they can never follow the instantaneous start of a curve with their vehicles. If the driver happens to be navigating a virtual world that uses clothoids (a clothoid is a spiral whose curvature is a linear function of its arc length; clothoids allow a smooth transition between straight roads and curves), then these curve entrances are smooth. In summary, standards are clearly needed for the definition of performance measures that take into account not only the recording of vehicle and driver position, but also of the geometry of the roadway.

The fifth consideration researchers face in replicating one another's work is whether an abstract description of the scenes, scenarios, road networks and performance measures can be developed. The answer appears to be that this can be done, if not right now then sometime in the not too distant future. Researchers at TNO in the Netherlands are leading the way here (Kappé, de Penning, Marsman, & Roelofs, 2009). They use e-learning and e-testing standards to describe simulator

scenarios at a meta-level. Similarly, they allow results to be labeled and stored in a standardized way.

The Sharable Content Object Reference Model (SCORM) is a mature standard for doing this (Advanced Distributed Learning, 2008). It features a set of XML based "tags" for describing content and results, allowing a scenario to be wrapped in a SCORM "package". SCORM is produced by ADL, a research group sponsored by the United States Department of Defense (DoD). The Question and Test Interoperability standard (QTI) describes a similar set of tags specifically designed for digital tests. The QTI was produced by the IMS Global Learning Consortium (IMS, 2009).

SCORM packages can be shared and managed by commercial and open-source Learning and Content Management Systems (LCMSs). They feature a rich set of capabilities for managing SCORM packages (storing, retrieving, sharing), for e-learning (presenting content to students, storing results), and for supporting the learning process (wikis, forums, Skype, etc.). The Modular Object-Oriented Dynamic Learning Environment (Moodle) is an example of open source software that was developed for this purpose (Nag, 2005) and supports SCORM content. Its development is undertaken by a globally diffused network of commercial and non-commercial users, streamlined by the Moodle company, which is based in Perth, Western Australia (http://www.moodle.com.au/).

The SCORM information needs to be accessible across different simulator platforms. The High Level Architecture (HLA) is the standard for interoperability across simulator platforms. Unfortunately, it does not include standards for learning content. So, the SCORM-Sim study group of the Simulation Interoperability Standards Organization (SISO, see http://www.sisostds.org) has worked on an initiative to relate SCORM and HLA. This initiative is now under development at TNO and is known as SimSCORM (de Penning, Boot, & Kappé, 2008). Briefly, SimSCORM allows simulator-based learning content to be developed and stored as SCORM content in any SCORM compliant LMS like Moodle. When this content is started, it can use SimSCORM to communicate with an HLA (or DIS) compliant simulator to initialize and start scenarios, monitor student activities, determine performance assessments and record the results in Moodle according to the SCORM data model.

It would appear that it will take some time before the above interoperability is fully in place. Moreover, this interoperability still does not do away entirely with the problems that stem from different renderings of scenes and scenarios as described above. Therefore, as an alternative, we are proposing the following as an intermediate step: that investigators place their scenes and scenarios on a common Web site. This would have the added advantage of speeding the development of scenes and scenarios in simulation labs around the world. For a simulator like STI (from Systems Technology, Inc.), this is as simple as uploading a text file and a video of the drive. For more complex simulators, such as the RTI (from Realtime Technologies, Inc.), this means uploading the scenes as one database, the scenarios as a second database, and then videos of drives through the virtual world. We have now done such on the University of Wisconsin driving wiki, including scenes and scenarios for both the RTI and STI simulators (http://www.drivingwiki.org).

1.3.3 Naturalistic Driving and Generalizability to the Open Road

Simulation is one of a range of tools that provides evidence on driver behavior and performance. Rizzo's chapter on medical disorders reviews the tradeoffs which are encountered in using these tools between the level of experimental control and the nature of the context in which the experiments occur (chap. 46). The tools range from simple tests of reaction time and sensory perception to standardized neuropsychological tests which are capable of measuring levels of cognitive performance in health or disease, but which are not as good at predicting real-world driving performance, to computer-based tests that may use real-world images (e.g., change detection and hazard perception tasks), to various types of simulators (task-focused, non-motion-based, multicapable motion-based), to driving in the real world (on a test track, on a state road test, in an instrumented vehicle), and naturalistic driving (in the driver's own car, over extended time frames). Each tool has its advantages and drawbacks and the approach depends on the individuals or populations being tested and the questions that are being asked.

In this section we want to focus on the most recent advances and outline both the advantages and limitations of even these most advanced tools. To begin with, multiple studies have used instrumented vehicles (IVs) in traffic safety research, starting as early as 1995 (e.g., Dingus et al., 1995; Dingus et al., 2002; Hanowski, Wierwille, Garness, & Dingus, 2000; Rizzo, 2004; Dawson, Anderson, Uc, Dastrup, & Rizzo, 2009). In most cases an experimenter is present, and drivers who are aware of being observed are liable to drive in an overly cautious and unnatural manner. Because total data collection times are often less than an hour and crashes and serious safety errors are relatively uncommon, these studies have been of limited value. In fact, no study, until recently, had captured pre-crash or crash data for a police-reported crash.

More recently, the experimenter has been removed from the vehicle. In some studies where the experimenter is removed from the vehicle, crashes and crash-related behavior are not the central focus. For example, internal networks of modern vehicles allow continuous detailed information from the driver's own car over extended time frames (Rizzo, Jermeland, & Severson, 2002). Modern vehicles report variables relevant to speed, emissions controls, and vehicle performance, and some vehicles allow more detailed reporting options via the vehicle's on-board diagnostics (OBD) port. Lane-tracking video can be processed to assess lane-keeping behavior. Radar systems in the vehicle can gather information on the proximity, following the distance and lane-merging behavior of the driver and other neighboring vehicles on the road (Pietras, Shi, Lee, & Rizzo, 2006). Global positioning systems (GPS) can show where and when a driver drives, takes risks, and commits errors. Cell phone use can be tracked without recording conversations to assess potential driver distractions and risk acceptance (i.e., choosing to be distracted). Wireless systems can check the instrumentation and send performance data to remote locations. These developments can provide direct, real-time information on driver strategy, vehicle use, upkeep, drive lengths, route choices, and decisions to drive during inclement weather and high traffic. There is no obvious limitation to such studies.

In other studies where the experimenter is removed from the vehicle, the focus of the study is on crash-related behaviors. In such studies, a person driving his or her own IV is exposed to the usual risk of the real-world road environment without the psychological pressure that may be present when a driving evaluator is in the car. Road test conditions can vary depending on the weather, daylight, traffic, and driving course. This is an advantage in naturalistic testing, because repeated observations in varying real-life settings can provide rich information regarding driver risk acceptance, safety countermeasures, and adaptive behaviors, as well as unique insights on the changing relationships between low-frequency/high-severity driving errors and high-frequency/low-severity driver errors.

Such "brain-in-the-wild" relationships (Rizzo, Robinson, & Neale, 2007) were explored in detail in a study of naturalistic driving performance and safety errors in 100 normal individuals, driving 100 total driver years (Dingus et al., 2006; Neale, Dingus, Klauer, Sudweeks, & Goodman, 2005). All enrolled drivers allowed installation of an instrumentation package into their own vehicles (78 cars) or drove a new model-year IV provided for their own use. Data collection provided almost 43,000 hours of actual driving data, with over 2,000,000 vehicle miles. There were 69 crashes, 761 near crashes, and 8,295 other incidents (including 5,568 driver errors) for which data could be completely reduced (Klauer, Dingus, Neale, Sudweeks, & Ramsey, 2006). Crash severity varied, with 75% being mild impacts, such as when tires strike curbs or other obstacles. Using taxonomy tools to classify all relevant incidents, the majority could be described as "lead vehicle" incidents; however, several other conflict types (adjacent vehicle, following vehicle, single vehicle, object/obstacle) occurred at least 100 times each. Driver inattention was deemed to be a factor in most of these incidents, as in Figure 1.1 (also see Web Figure 1.1). Perhaps the most controversial aspect of this study was its clear difference in the estimate that talking on the cell phone plays in crashes, the estimate here being considerably smaller than it is in other studies. It is difficult to argue with the naturalistic data.

A recent pilot study conducted at the University of Iowa (UI) used an event-triggered, in-vehicle electronic and video data recorder (known as the DriveCam system) to examine rural teen driver performance (McGehee, Carney, Raby, Lee, & Reyes, 2007a; McGehee, Raby, Carney, Lee, & Reyes, 2007b). Data collection began in March 2006 with 25 rural teen drivers and lasted 13 months. A DriveCam system was installed in each driver's car. The drivers covered more than 300,000 miles. Data were first collected for a two month period to test remote video downloading and coding and to identify expected baseline numbers of

(a)   (b)   (c)

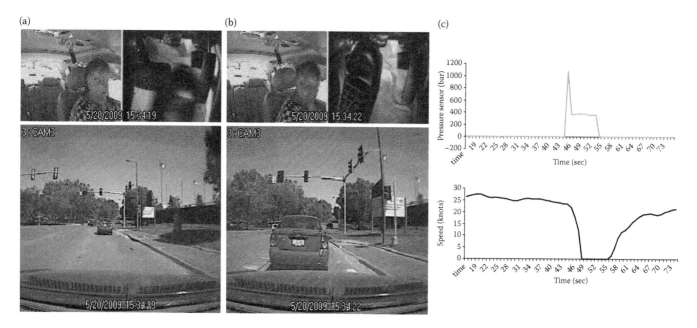

FIGURE 1.1   An inattentive older driver of the University of Iowa instrumented vehicle (NIRVANA) approaches a lead vehicle (panels a and b) and brakes hard (panel c upper plot) to avoid a rear-end collision. There is a corresponding abrupt dip in speed from over 30 mph (25 knots) to 0 mph, as shown by the electronic data from the instrumented vehicle (panel c, lower plot). The severity of the braking simultaneously triggered a DriveCam system to record video of the event.

recorded video segments. The goals of the baseline data collection period were to gather information on participants' everyday driving behavior, in order to identify the number of driving events that triggered video recording, and to identify the optimal period for baseline data collection.

All participants followed the data downloading procedures, which involved driving the car to the school parking lot and parking for no more than three minutes. Encrypted data were transferred to the UI where two researchers reviewed each recorded clip of a system-triggering event. Any video data captured while a non-consented driver was using a participant's vehicle were deleted. False triggers (e.g., hitting a pothole) were identified and reviewed to determine the likely cause. In some cases, the analysis of a false trigger revealed safety-related behavior (e.g., a driver talking on a cell phone when the vehicle hit a pothole). These clips were not included in the data set. Finally all "true" triggers were analyzed and scored using a standardized coding scheme. Each safety-related event was analyzed and coded based on a variety of factors, including demographic information, event variables, contributing factors, driver state variables, environmental factors, and infrastructure/environment. The results are being used to develop a clinical trial aimed at improving teen driver safety through an educational intervention that includes providing feedback to the teen drivers and family members on driver errors (Carney, McGehee, Lee, Reyes, & Raby, in press; McGehee et al., 2007a). Perhaps the most significant finding here was that the great majority of the triggered events were caused by a small minority of the drivers. Again, it is difficult to see how one can argue with these results, which raise serious issues about the utility of current driver education efforts.

Event-triggered data acquired from systems such as the DriveCam can capture the context associated with a safety error. This evidence allows clear contextual feedback in the form of video and audio of each episode captured, to raise the awareness of risky driving as part of an effective driving safety intervention. While drivers may rationalize their behavior and generate a host of reasons to explain away a vehicle's black box event data as recorder anomalies, this is much harder to do with video evidence (Carney et al., in press; Lee, 2007; McGehee et al., 2007a, 2007b). Two recent studies examined the effectiveness of an event-triggered video-based intervention in younger drivers (Carney et al., in press; McGehee et al., 2007a, 2007b; also see Lotan & Toledo, 2006). The findings suggest that pairing this technology with feedback to drivers and their families in the form of a weekly video review and graphical report card can improve driving safety among at-risk drivers of all ages. The naturalistic observations in older drivers fit with the U.S. Department of Transportation (DOT) roadmap on the need for real-world (ground "truth") measurements of driver behavior to augment laboratory studies, and the use of naturalistic data as feedback in behavioral interventions to reduce impaired driver errors (Hickman, Hanowski, & Ajayi, 2009). The effort complements simulator-based interventions aimed at improving driver safety (e.g., Roenker, Cissell, Ball, Wadley, & Edwards, 2003).

In summary, the use of instrumented vehicles without an experimenter present in the car is already providing critical information that clearly complements what has been available to date in simulator studies. At times this is contrary to what has been claimed from simulator studies. The long number of contextually relevant hours of data collection that are available with studies using instrumented vehicles makes it possible to target more precisely the population attributable risk of any given behavior. One simply cannot gather such information on simulators. At the same time, it is often difficult to isolate the various explanations for a given finding in a study using instrumented vehicles. For this, one needs the experimental control offered by simulators and the other tools discussed above.

1.3.4 Simulator Sickness

Simulator sickness is still a real issue, especially with older adults (chap. 14 by Stoner, Fisher, & Mollenhauer, "Simulator and Scenario Factors Influencing Simulator Sickness"). The problem has not been solved by any stretch of the imagination. Here, we want only to mention one relatively inexpensive alternative which looks more promising than most. This is the turning cabin simulator first described by Asano and Uchida (2005) and since implemented by Mourant at Northeastern University (Yin & Mourant, 2009; http://www.coe.neu.edu/Research/velab/velab/Vehicle_Turning_Driving_Simulator.html). Briefly, the cab of the vehicle is situated inside a cylindrical screen. The amount of vehicle rotation corresponds to the steering inputs from the driver. Yin and Mourant have shown that the optical flow when making left and right turns is some 50% to 100% greater in a fixed-base simulator as opposed to a turning cabin simulator and, when operating on curves, up to 200% greater in a fixed-base simulator than it is in a turning cab simulator. Not surprisingly, given the relation between optical flow and simulator sickness (Kennedy, Hettinger, & Lilienthal, 1988), Asano and Uchida reported a smaller incidence of simulator sickness on the turning cab simulator. Research is still underway at Northeastern.

1.3.5 Support for Simulator Research

Finally, it would be remiss not to mention the support that exists for simulator research around the world in private companies, public foundations and government and regional agencies. By no means can this list be exhaustive, but it can help the researcher get started.

Private companies financing initiatives in simulation have, or currently include, the major automotive companies from around the world (e.g., BMW, Daimler-Benz, Ford, General Motors, Honda, Renault). In the United States, they also include a number of companies that make various of the devices that are now being used inside the cabin of the automobile, including Mitsubishi Electric Research Labs (Cambridge, Massachusetts), Bose Corporation (Framingham, Massachusetts), and Calspan (Buffalo, New York) as well as companies that do actual research on simulators in their own laboratories, for example, the Liberty Mutual Research Institute for Safety (Hopkinton, Massachusetts). In Europe, there is SINTEF in Norway (http://www.sintef.no/Home/), TNO in the Netherlands (http://www.tno.nl/downloads/N_Driving%20Simulator1.pdf), and VTT in the Technical Research Centre of Finland (http://www.vtt.fi), among others. In Japan, Toyota has

a major driving simulator in its Higashifuji Technical Centre Japan similar to the National Advanced Driving Simulator (http://vodpod.com/watch/2607288-toyotas-driving-simulator-video). In China, they manufacture a number of simulators and use them in their driving schools, but the manufacturers do not appear to currently support a strong research program there.

Public foundations and government and regional agencies also support research on driving simulators. In the United States, foundations include the AAA Foundation for Traffic Safety and the Link Foundation for Simulation and Training. Governmental programs include, but are certainly not limited to, the Department of Transportation (Federal Motor Carrier Safety Administration, National Highway Traffic Safety Administration, University Transportation Centers), Transportation Research Board (National Highway Cooperative Research Program, Transit Cooperative Research Program, IDEA Program), National Science Foundation, National Institutes of Health, and Centers for Disease Control. At the lower levels of government, these include the Governor's Highway Safety Bureau, the various state departments of transportation, and the several regional state transportation consortia.

Similar organizations support simulator research in Canada, Australia, Asia and Europe. In Canada, the major government support comes from Transport Canada, AUTO21 Network of Centers of Excellence, Natural Sciences and Engineering Research Council of Canada and the Canadian Foundation for Innovation. In Australia, there is the Monash University Accident Research Center (also instrumented vehicle research, see http://www.monash.edu.au/muarc/projects/simulator/index.html). Also, Queensland University of Technology has recently announced an advanced simulator and program to be established there (http://www.news.qut.edu.au/cgi-bin/WebObjects/News.woa/wa/goNewsPage?newsEventID=16299). In Asia, the major government support would appear to come from the National Natural Science Foundation of China.

In Europe, the support is broader still. A European project on driver training and simulators is "Train All" (http://www.trainall-eu.org/). In the Netherlands, there is SWOV (the Dutch National Road Safety Research Institute; http://www.swov.nl/uk), in Sweden VTI (Swedish National Road and the Transport Research Institute, http://www.trainall-eu.org/), and in Norway TØI, the Institute for Transport Economics (http://www.managenergy.net/actors/A2246.htm). In England there is TRL (the Transport Research Laboratory, http://www.trl.co.uk/), in France INRETS (the National Institute on Transportation Research and Security, http://www.inrets.fr/), in the Czech Republic CDV (the Transport Research Institute, http://www.cdv.cz/english), and in Germany BASt (the Federal Highway Research Institute, http://www.bast.de/EN/e-Home/e-homepage__node.html?__nnn=true), among others.

1.4 Our Thanks and Hopes

We cannot thank enough all of the various authors of the chapters who have helped us produce an edited volume that we hope, first and foremost, will be of interest to our readers. Of course with any undertaking like this one—especially one this large—one also hopes that the effort made will set the stage for further advances in theory and/or practice. Towards this end we have tried to include material relevant to the broad range of audiences that will be needed to make these advances, including the basic researchers, the program managers, and the practitioners.

1.4.1 Applications

With respect to practice, we hope that this book both catalyzes various applied research efforts which either have long been languishing or are currently of pressing need. Among efforts where little progress has been made, yet the need has long been obvious, we think first and foremost of research on novice drivers. Novice drivers during the first six months of solo driving are much more likely to be in a crash than older teens (McCartt, Shabanova, & Leaf, 2003). Lack of experience, not immaturity, seems to be the basis for the much higher fatality rate (Vlakveld, 2005). Yet little has been done to develop and to evaluate alternative driver education programs over the last forty years which would give drivers the experience they need to anticipate hazards, maintain their attention, and respond appropriately when a crash is imminent (Nichols, 2003). Perhaps this was not possible until recently with the advent of relatively inexpensive graphics engines and displays. But now it is possible, in theory, to use simulation in the training of novice drivers with hopefully the same success that simulation has had in other similar training endeavors.

Among applied research efforts that are currently of pressing need, we think immediately of the importance of engaging in a national effort to understand the growing contribution to crashes of in-vehicle devices. Distractions have become one of the major contributing factors to the rise in the number of vehicle crashes (Klauer et al., 2006) In-vehicle distractions include the use of cellular phones, iPods and other MP3 players, eating, tuning the radio or CD player, monitoring a navigation aid, and adjusting the climate. The existing evidence clearly points to cell phones as a major contributor to crashes. Evidence is mounting that other in-vehicle activities, even some of the most common, are also just as hazardous (Klauer et al., 2006). Yet, we know very little about how to mitigate the problems presented by the proliferation of in-vehicle crashes and their role in distracting drivers.

1.4.2 Theory

Perhaps a little less obvious, but also potentially of more long term significance for our theoretical understanding of human behavior in all of its many varieties, we believe that research in the area of driving simulation can afford us a platform for studying many of the same issues that have been studied so successfully in relatively simple laboratory settings, but now in a much more dynamic and demanding setting, one that will give us a better opportunity to describe and understand the very behaviors that we use on a daily basis. Cognition, perceptual and motor skills are

almost infinitely flexible. What we have learned about such skills in the more simple settings may or may not transfer to the more complex settings. Until driving simulators became as widespread as they are today, it was all but impossible to undertake studies in more dynamic and meaningfully demanding environments. However, we now have at our fingertips the equipment needed to expand our existing theories of visual search, attention, learning and memory, and perception to such environments. We believe that in the long run our theories will be all the richer and more relevant, explaining not just behavior that bears directly on driving but also on a much broader range of behaviors that occur in similar environments as well.

1.4.3 Dedication

For the many of us who have experienced the loss of someone close to us on the roadways, there is a special meaning to this volume, but so too for everyone, for everyone is affected in some way by the tragic loss of lives on our highways. We dedicate this volume to those whose memories we keep alive in our efforts, large and small, to make the roadways a safer place for all.

Acknowledgments

The editors are grateful to Willem Vlakveld and Theresa Senserrick for their help in identifying agencies around the world that fund simulator research and to Bart Kappé for greatly adding to the discussion of ways to increase interoperability. Portions of this research were funded by grants from the National Institutes of Health (1R01HD057153-01) and the National Science Foundation (Equipment Grant SBR 9413733 for the partial acquisition of the driving simulator) to Donald Fisher.

Web Resources

The *Handbook* web site (http://www.drivingwiki.org) contains supplemental material for the chapter, including a color version of Figure 1.1.

References

Advanced Distributed Learning. (2008). ADL Guidelines for creating reusable content with SCORM 2004 (July 1, 2008, version 1.0). ADL: Alexandria, VA.

Asano, Y., & Uchida, N. (2005). *Improvement of driver's feeling by turning cabin driving simulator*. Proceedings of the Driving Simulation Conference North America. Iowa City, IA: Public Policy Center, University of Iowa.

Carney, C., McGehee, D. V., Lee, J. D., Reyes, M. L., & Raby, M. (2010). Using an event-triggered video intervention system to expand the supervised learning of newly licensed teen drivers. *American Journal of Public Health, 100*(6), 1101–1106.

Dawson, J. D., Anderson, S. W., Uc, E. Y., Dastrup, E., & Rizzo, M. (2009). Predictors of driving safety in early Alzheimer's disease. *Neurology 72*, 521–527.

de Penning, L., Boot, E., & Kappé, B. (2008). *Integrating training simulations and e-learning systems: The SimSCORM platform.* Proceedings of the Interservice/Industry Training, Simulation and Education Conference (IITSEC) Conference 2008, Orlando, Florida.

Dewar, R. E., & Olson, P. (2007). *Human factors in traffic safety* (2nd ed.). Tuscon, AZ: Lawyers and Judges Publishing.

Dingus, T. A., Klauer, S. G., Neale, V. L., Petersen, A., Lee, S. E., Sudweeks, J., et al. (2006). *The 100-car naturalistic driving study: Phase II-Results of the 100-car field experiment.* (Rep. No. DOT HS 810 593). Washington, DC: U.S. Department of Transportation, National Highway and Traffic Safety Administration.

Dingus, T. A., McGehee, D., Hulse, M., Jahns, S., Manakkal, N., Mollenhauer, M., et al. (1995). *TravTek evaluation task C3 camera car study* (Pub. No. FHWA-RD-94-076). Washington, DC: US Department of Transportation, Federal Highway Administration.

Dingus, T. A., Neale, V. L., Garness, S. A., Hanowski, R., Keisler, A., Lee, S., et al. (2002). *Impact of sleeper berth usage on driver fatigue* (Rep. No. 61-96-00068). Washington, DC: US Department of Transportation, Federal Motor Carriers Safety Administration.

Dutta, A., Carpenter, R., Noyce, D. A., Duffy, S. A., & Fisher, D. L. (2002). Drivers' understanding of overhead freeway exit guide signs: Evaluation of alternatives using an advanced fixed base driving simulator. *Transportation research record: Human performance models, intelligent vehicle initiative, traveler advisory and information systems (Safety and Human Performance), 1803,* 102–109.

Fisher, D. L., Mourant, R., & Rizzo, M. (2005, January). *The development of standardized descriptions of driving simulator scenarios: Human factors considerations* (TRB Workshop). Transportation Research Board Annual Meetings, Washington, DC.

Groeger, J. A. (2000). *Understanding driving: Applying cognitive psychology to a complex everyday task.* Hove, UK: Psychology Press.

Hanowski, R. J., Wierwille, W. W., Garness, S. A., & Dingus, T. A. (2000). *Impact of local/short haul operations on driver fatigue: Final report.* (Rep. No. DOT-MC-00-203). Washington, DC: US Department of Transportation, Federal Motor Carriers Safety Administration.

Hickman, J. S., Hanowski, R. J., & Ajayi, O. (2009). *Evaluation of an onboard safety monitoring device in commercial vehicle operations.* Proceedings of the Fifth International Driving Symposium on Human Factors in Driver Assessment, Training and Vehicle Design, Big Sky, MT. Iowa City, IA: Public Policy Center, University of Iowa.

IMS Global Learning Consortium. (2009). IMS question & test interoperability specification. Retrieved from December 20, 2009, from http://www.imsglobal.org/question/

Kappé, B., de Penning, L., Marsman, M., & Roelofs, E. (2009). *Assessment in driving simulators: Where we are and where we go.* Proceedings of the Fifth International Driving

Symposium on Human Factors in Driver Assessment, Training and Vehicle Design, Big Sky, MT. Iowa City, IA: Public Policy Center, University of Iowa.

Katsikopoulos, K., Duse-Anthony, Y., Fisher, D. L., & Duffy, S. A. (2002). Risk attitude reversals in drivers' route choice when range of travel time information is provided. *Human Factors, 44,* 466–473.

Kennedy, R., Hettinger, L., & Lilienthal, M. (1988). Simulator sickness. In *Motion and space sickness* (chap. 15, pp. 317–341). Boca Raton, FL: CRC Press.

Kichhanagari, R., DeRamus Motley, R., Duffy, S. A., & Fisher, D. L. (2002). Airport terminal signs: Use of advance guide signs to speed search time. *Transportation Research Record: Aviation, Airport and Air Traffic Economic and Operational Issues (Aviation), 1788,* 26–32.

Klauer, S. G., Dingus, T. A., Neale, V. L., Sudweeks, J. D., & Ramsey, D. J. (2006). *The impact of driver inattention on near-crash/crash risk: An analysis using the 100-car naturalistic driving study data* (Report No. DOT HS 810 594). Washington, DC: US Department of Transportation, National Highway Traffic Safety Administration.

Laurie, N. L., Zhang, S., Mundoli, R., Duffy, S. A., Collura, J., & Fisher, D. L. (2004). Evaluation of alternative DO NOT ENTER signs: Failures of attention. *Transportation Research Part F: Psychology and Behaviour, 7,* 151–166

Lee, J. D. (2007). Technology and teen drivers. *Journal of Safety Research, 38*(2), 203–213.

Leibowitz, H. W. (1996). The symbiosis between basic and applied research. *American Psychologist, 51*(4), 366–370.

Lotan, T., & Toledo, T. (2006). An in-vehicle data recorder for evaluation of driving behavior and safety. *Transportation Research Record, 1953,* 112–119.

McCartt, A. T., Shabanova, V. I., & Leaf, W. A. (2003). Driving experience, crashes and traffic citations of teenage beginning drivers. *Accident Analysis and Prevention, 35*(3), 311–320.

McGehee, D. V., Carney, C., Raby, M. Lee, J. D., & Reyes, M. L. (2007a). *The impact of an event triggered video intervention on rural teenage driving.* Proceedings of the Fourth International Symposium on Driver Assessment in Vehicle Design. Stevenson, WA.

McGehee, D. V., Raby, M., Carney, C., Lee, J. D., & Reyes, M. L. (2007b). Extending parental mentoring using an event-triggered video intervention in rural teen drivers. *Journal of Safety Research, 38*(2), 215–227.

Nag, A. (2005). Moodle: An open source learning management system. Retrieved December 20, 2009, from http://www.linux.com/archive/articles/44834

Neale, V. L., Dingus, T. A., Klauer, S. G., Sudweeks, J., & Goodman, M. J. (2005). *An overview of the 100-car naturalistic study and findings* (Paper No. 05-0400). Washington, DC: National Highway Traffic Safety Administration.

Nickerson, R. S. (1999). Basic and applied research. In R. J. Sternberg (Ed.). *The nature of cognition* (pp. 409–444). Cambridge, MA: MIT Press.

Nichols, J. L. (2003). *A review of the history and effectiveness of driver education and training as a traffic safety program.* Washington, DC: National Transportation Safety Board.

Noyce, D. A. & Kacir, K. C. (2000). Drivers' understanding of simultaneous traffic signal indications in protected left-turns. *Transportation Research Record, 1801,* 18–26.

Pietras, T. A., Shi, Q., Lee, J. D., & Rizzo, M. (2006). Traffic-entry behavior and crash risk for older drivers with impairment of selective attention. *Perceptual and Motor Skills, 102*(3), 632–644.

Regan, M., Lee, J., & Young, K. (2008). *Driver distraction: Theory, effects, and mitigation.* Boca Raton: CRC Press.

Rizzo, M. (2004). Safe and unsafe driving. In M. Rizzo, & P. J. Eslinger (Eds.), *Principles and practice of behavioral neurology and neuropsychology* (pp. 197–220). Philadelphia: W. B. Saunders.

Rizzo, M., Jermeland, J., & Severson, J. (2002). Instrumented vehicles and driving simulators. *Gerontechnology, 1,* 291–296.

Rizzo, M., Robinson, S., & Neale, V. (2007). The brain in the wild. In R. Parasuraman, & M. Rizzo (Eds.), *Neuroergonomics: The brain at work* (pp. 113–129). Oxford University Press.

Roenker, D. L., Cissell, G. M., Ball, K. K., Wadley, V. G., & Edwards, J. D. (2003). Speed-of-processing and driving simulator training result in improved driving performance. *Human Factors, 45,* 218–233.

Rouse, W. B. (1991). *Design for success: A human-centered approach to designing successful products and systems.* New York: John Wiley & Sons.

Upchurch, J., Fisher, D. L., Carpenter, C., & Dutta, A. (2002). Freeway guide sign design with driving simulator for Central Artery-Tunnel: Boston, Massachusetts. *Transportation Research Record: Traffic Control Devices, Visibility, and Rail-Highway Grade Crossings (Highway Operations, Capacity, and Traffic Control), 1801,* 9–17.

Vlakveld, W. (2005). *Young, novice motorists, their crash rates, and measures to reduce them: A literature study* (Technical Report R-2005-3). Leidschendam, the Netherlands: SWOV.

Yin, Z. & Mourant, R. R. (2009). *The perception of optical flow in driving simulators.* Proceedings of the Fifth International Driving Symposium on Human Factors in Driver Assessment, Training and Vehicle Design. Iowa City: Public Policy Center, University of Iowa.

2

A Short History of Driving Simulation

R. Wade Allen
Systems Technology, Inc.

Theodore J. Rosenthal
Systems Technology, Inc.

Marcia L. Cook
Systems Technology, Inc.

Abstract

The Problem. Driving is the most universal and ordinary task people perform every day as well as the most complex and dangerous. It requires a full range of sensory, perceptual, cognitive, and motor functions, all of which can be affected by a wide range of stressors and experience levels. The historical context here will provide appropriate perspective for simulating the driving experience. ***Role of Driving Simulators.*** Driving has measurable, real-world impacts and consequences for everyone, therefore methods are needed in order to safely quantify the driving experience. Experimental studies can always be conducted with on-road tests, however using a simulator is safer and more cost effective; provides for objective and repeatable measures of driver performance; allows for complete control of the driving environment (traffic, weather, etc.); and can be easily administrated in a laboratory setting. ***Key Results of Driving Simulator Studies.*** Since the earliest days of driving simulation, simulators have been used in a wide range of clinical studies in order to understand the driver, the vehicle, and the complex driving environment. From early studies that investigated traffic control devices and highway signage, to modern studies dealing with driver texting, cell phone use, and sedative hypnotic pharmaceutical compounds, driving simulation has been a leading research tool. The areas of human factors, medical research, vehicle dynamics, highway design, and more have all benefited from results obtained through driving simulation. ***Scenarios and Dependent Variables.*** One of the huge advantages for using driving simulation is the ability to create and repeat most conceivable driving situations, leading to documentation of how the driver performed. Within a simulator, all aspects of the driving environment can be controlled and specific tests or events that the driver navigates will lead to the collection of desired performance outcomes. Roadway environmental conditions such as weather, traffic patterns, and signal light timing can all be controlled and repeated over many trials. Anything within the simulated environment can be measured providing objective and repeatable measures that cannot be obtained during on-road testing. ***Platform Specificity and Equipment Limitations.*** Over the years driving simulators have come in all shapes and sizes with a variety of approaches used based on the technology available at the time. Depending on the needs of the researcher, simulators have ranged from a simple set of pedals that a driver reacted with when a light turned on, to entire facilities dedicated to creating the most realistic simulator by using actual car cabs strapped to moving platforms. Although driving simulation does not yet match the fidelity of real-world driving, depending on the questions being asked, there are, and have been for some time, numerous driving simulators that can answer each question in its own unique way.

2.1 Introduction

The history of driving simulation has been motivated by general advancements in technology, related to the various cueing systems including visual, auditory, and proprioceptive feedback, and the equations of motion or vehicle dynamics that translate driver control actions into vehicle motions. The history has also evolved from our knowledge of driver behavior and issues that we desire to address with simulation, including research, assessment, training, and demonstration. Early simulation involved analog electronics and driving scenarios based on physical models, simplified calligraphic (i.e., line-drawn) displays, or film and video. With the development of digital computers and computer graphics, simulators became more sophisticated. Now in the PC (personal computer) era, with the advent of extremely capable and affordable CPUs (central processing units) and GPUs (graphical processing units) the development of driving simulation has become primarily a matter of software improvements plus driver interface hardware advances for establishing face validity.

As a practical matter, driving simulation development started in the 1960s using analog computers, electronic circuits and various display technologies (Hutchinson, 1958; Sheridan, 1967; Rice, 1967; Sheridan, 1970; Weir & Wojcik, 1971; Kemmerer & Hulbert, 1975; Allen, Hogge, & Schwartz, 1977). The block diagram in Figure 2.1 (Web Figure 2.1 for color version) shows the functional elements of driving simulators that have steadily expanded and improved to this day. The simulation computer processing (SCP) block includes all computations required to indicate vehicle motion relative to the environment, including driver control actions, and aerodynamic and road surface inputs. The first driving simulators used electronic circuits and/or analog computers

for this function. Over the past two decades these functions have been mainly mechanized using digital computers. The SCP block then provides inputs to the sensory feedback generation (SFG) block which produces sensory cueing commands or inputs to the sensory display device (SDD) block. SFG was originally provided by very custom devices needed to provide visual, auditory and proprioceptive cueing. In the past two decades these functions have been mechanized by PC level devices including graphical processing units (GPUs), audio processing units (APUs) and digital algorithms for proprioceptive cueing, including steering feel and motion base commands. The SDD functions have mainly been implemented with commercial devices such as video monitors and projectors, sound equipment, torque loaders and various types of electric and pneumatic motion platforms. Given displayed sensory cues, the human operator (driver) then senses this information and, based on training and experience, produces control inputs that are fed back to the SCP. In virtual reality (VR) applications using head-mounted displays (HMDs), which have proliferated in the past decade, head orientation must also be provided to the SCP.

Driving simulator development has been motivated and supported by advances in electronics, computers and various display technologies. Understanding of driver and vehicle behavior has also motivated simulation developments along with the derivation of measurement algorithms for quantifying driver behavior and system performance. These developments have been advancing for over four decades, and with current technology advancing as it is, driving simulation development and refinement should continue for the foreseeable future. Researchers concerned about the evolution of driving simulation have, in general, been motivated to achieve a valid representation of the driving environment. This

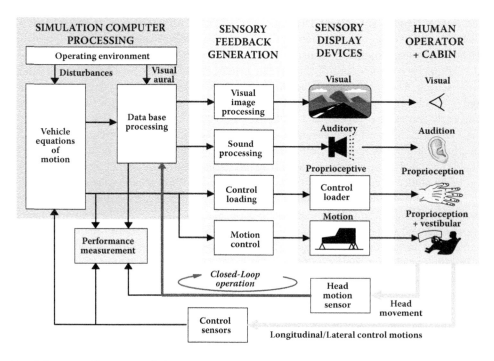

FIGURE 2.1   Functional elements of driving simulation.

involves fidelity in sensory cueing, vehicle handling (response to control inputs) and in the task environment which can be considered as cognitive fidelity (for more detailed discussions of fidelity see in this book, chap. 7 by Greenberg & Blommer; chap. 8 by Andersen, and chap. 9 by Ranney). The task environment involves all aspects of the driving environment, including traffic control devices (signs, signals, roadway markings and other delineators), traffic, pedestrians and various roadside elements including cultural features and flora. Early simulators implemented steering and speed control tasks and sign recognition that could be accomplished with slides or film projection. The ability to create a complex task environment requiring cognitive skills such as situation awareness, hazard perception and decision-making has been more fully realized with digital graphics, which permit the creation of complex 3D (three dimensional) scenes with control of the spatial and temporal properties of scene elements.

The major elements of a typical driving simulator as summarized in Figure 2.1 include: cueing systems (visual, auditory, proprioceptive, and motion), vehicle dynamics, computers and electronics, cabs and controls, measurement algorithms and data processing and storage. Cueing systems involve stimulation of all driver sensory and perceptual systems. In each of the cueing systems (visual, auditory, proprioceptive and motion) the appropriate stimulus resulting from the driver's control inputs must be computed and then accurately displayed to the driver. Cues such as steering feel are a direct consequence of the driver's control response and resulting vehicle reaction. Motion cues are a function of the vehicle's dynamic response to driver control inputs, with additional independent inputs due to roadway (e.g., road crown) and aerodynamic (e.g., wind gust) disturbances. Visual and auditory cues can result from driver/vehicle responses, but also have significant independent inputs due to dynamic roadway elements (e.g., traffic, pedestrians, and traffic control devices) in the driving scenarios. Vehicle dynamics have developed somewhat independent of the real-time simulation community but have been significantly employed in driving simulation through the years. The remaining elements have developed primarily outside of the real-time simulation community and have been adapted for use in quantifying driver behavior and system performance. These developments will be discussed subsequently.

Several functional elements in Figure 2.1 have been important and even critical to driving simulation's historical development but may not be obvious to the casual observer. These include vehicle dynamics, control and presentation of driving scenarios (road profiles, traffic control devices, traffic and pedestrians and roadside objects), and sensors and measurement algorithms. Ground vehicle dynamics started with fairly good models in the 1950s, followed by significant developments in the 1980s and 90s, particularly of tire models that produce the maneuvering forces that are essential to determine vehicle stability.

The control and presentation of driving scenarios has developed steadily over the years and is still an active area of development (Green, 2007). Measurement algorithm development has been spurred on by a combination of the need for quantifying driver behavior and system performance, and the development of fundamental mathematical and computational algorithms that allow desired and appropriate processing of time histories and transient events. Because of the visual nature of driving, one key area of sensor and measurement development has been eye movements.

To a large degree driving is primarily a visual task, so developments of visual cueing have been of vital importance. Figure 2.2 (see insert or Web Figure 2.2 for color version) shows some visual display effects as driving simulator display systems have improved over the years. Early simulators had electronically generated calligraphic displays (Figure 2.2a), model boards with video presentations, or film, but the introduction of digital graphics provided a rather dramatic step forward that has carried us to the sophisticated graphics systems that we are familiar with today. Figures 2.2b–d illustrate the evolution of digital graphics from primitive low count flat-colored polygons, through shaded polygons, and into today's technology, which allows texturing of high count polygonal models. Along the way display technology has also improved with monitors and projectors displaying increasing color and pixel resolution, contrast ratio, and brightness range and levels. Auditory system displays have always been fairly sophisticated because of high fidelity sound equipment. Developments in sound synthesis and sound generation and control cards have improved greatly along with software for generating realistic sound effects. Video game technology has led the way for many visual and sound effects through a combination of both software and hardware. Proprioceptive feedback such as steering feel has been available with properly controlled torque

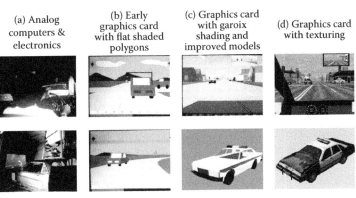

(a) Analog computers & electronics   (b) Early graphics card with flat shaded polygons   (c) Graphics card with garoix shading and improved models   (d) Graphics card with texturing

FIGURE 2.2 **(See color insert)** Evolution of driving simulators and PC graphics.

motors for the last couple of decades. Full motion systems have seen the most development, evolving to the current six degrees of freedom hexapod (Stewart Platform) systems with hydraulic and electrical actuators ("History of Video Games", n.d.) and for high-end facilities, the addition of horizontal motion tracks, (Garrott, Mazzae, & Goodman, 2006).

Historically, over the last four decades, developments in all of these areas have continually improved driving simulator capabilities. These developments have been applied to a full range of simulator configurations and applications, from large facilities to smaller setups; from research to testing and training of drivers. In the past, developments such as simulation computers and visual cueing systems were quite expensive and required significant equipment investment and thus tended to be implemented only in significant facilities. However, developments in the personal computer arena over the last decade have reduced the requisite computer hardware investment. The most significant outlay now resides in software development plus cabs and motion systems which still require significant hardware investment.

2.2 Early History

Driving simulation was originally developed to avoid the cost of field studies, achieve more control over circumstances and measurements, and safely present hazardous conditions. In the second half of the twentieth century, simulation was being successfully applied to aeronautical, rail and maritime operations. Passenger car and truck simulators were being used in studies of the driver (e.g., impairment, visual search patterns, training, etc.), vehicle characteristics (handling qualities, accident avoidance, design, etc.), and the environment (e.g., visibility, roadway characteristics, and design, etc.). In the late 1970s simulator design had evolved into five key areas: Visual and auditory display generation, kinesthetic cues (motion and control feel), driving scenario programming, computational vehicle dynamics, and the vehicle cab including controls and instruments. Several summaries of early driving simulation have been prepared over the years (e.g., Hutchinson, 1958; Sheridan, 1967; Rice, 1967; Sheridan, 1970; Weir & Wojcik, 1971; Kemmerer & Hulbert, 1975; Allen et al., 1977; O'Hanlon, 1977; Tu, Wu, & Lee, 2004; Garrott et al., 2006; Green, 2007; "History of Video Games", n.d.). Figure 2.2a shows an early simulator that included a calligraphic display generator, video projector and analog computation (Allen et al., 1977).

In the 1970s there were at least 20 research driving simulators throughout the US and Europe including many small, part task devices used for training and licensing. In these early simulators display generation and computations were done with parallel electronic circuits and analog computers so that high image frame rates (basically video frame rates) could be maintained to produce displays with good dynamic characteristics (Donges, 1975), and video or graphics projection systems were used to present large size (45°–60° FOV) visual displays (Donges, 1975; Gilliland, 1973). Digital computer-generated imagery was being developed to provide complex visual fields for car driving (Gilliland, 1973; Michelson, Niemann, Olch, Smiley, & Ziedman, 1978), however,

these efforts were limited by the relatively slow serial processing characteristics of digital machines, and the delay tended to be proportional to image complexity (update rates of 20 Hz or less and time delays of greater than 100 msec). Digital computational delays were understood to be a serious artifact in the dynamics of the operator's control task (Leslie, 1966), and later on compensation techniques were developed to offset this artifact (Ricard, Cyrus, Cox, Templeton, & Thompson, 1978).

The recognition of the consequences of computational delays was critical to simulation development, particularly as simulation components were evolving from analog electronics to digital computation. The general problem with time delays in human/machine systems was realized after its identification as a significant simulation artifact (Ricard et al., 1978; Hess, 1982; McFarland & Bunnell, 1990), and methods for measuring and compensating for visual delays were developed for driving simulation (Hogema, 1997). Computational delays are still an issue that must be considered with current technology for very complex 3D visual scenes, and must be dealt with carefully in order to avoid significant artifacts, such as Simulation Adaptation Syndrome (Rizzo, Sheffield, & Stierman, 2003). Compensation techniques involve various computational methods to provide anticipation or prediction into the display variables (e.g., Hogema, 1997).

Point light source or shadowgraph techniques provided an early alternate approach to display generation, but tended to be limited in their capability to reproduce photometric conditions (Shuttel, Schmacher, & Gatewood, 1971). This type of display implemented the roadway scene as a tinted Plexiglas model illuminated by the point light source with the image presented on a screen in front of the driver (see Green & Olson, 1989, for a description of a shadowgraph display). Model motion was controlled to represent the speed and heading of the simulated car. This approach was obviously limited by the difficulty of constructing and controlling large complex models.

Film-based motion picture simulators provided excellent detail, but were not truly interactive. Typically the driver's steering actions controlled the pan angle of one or more projectors which gave an impression of heading control, and speed was represented by varying the speed of the projector (Hutchinson, 1958). The detailed resolution of 35 mm film allowed the motion picture technique to present elements such as signs in great detail approaching real-world viewing conditions as shown in Table 2.1 (Templeton, n.d.; Computer Display Standard, n.d.). However, the arduous film production efforts were a serious drawback to this approach.

Scale models can represent complex geometric conditions including traffic and roadside objects. This approach was taken in simulation with a moving belt model with a closed circuit TV display (Weir & Wojcik, 1971). As shown in Figure 2.3 the model belt moved towards the video camera with belt speed representing the velocity of the vehicle and camera azimuth angle and lateral position representing vehicle heading and lateral position respectively. Vehicle dynamics were mechanized on an analog computer and could be set up to represent a range of vehicle characteristics, including articulated vehicles (i.e., trailer towing). However, this approach suffered from limited resolution, lighting, and depth of field. Large

TABLE 2.1 Resolution Capability of Various Media

Medium	Resolution
35 mm film	5300 × 4000
Human Eye (1 minute of visual arc)	30° HFOV = 1800 × 1350 arc minutes
	45° HFOV = 2700 × 2025 arc minutes
Typical Digital Display Resolutions	1024 × 768 (XGA)
	1280 × 768 (WXGA)
	1280 × 1024 (SXGA)
	1600 × 1200 (UXGA)
	1920 × 1080 (1080p HDTV)
	1920 × 1200 (WUXGA)
	2048 × 1536 (QXGA)
	2560 × 1600 (WQXGA)
	1920 × 1200 (WUXGA)

Source: Templeton, n.d.; Computer Display Standard, n.d.

terrain boards have been used in aeronautical simulators to provide scene complexity (Carlson, 2003), but these approaches are generally limited by the difficulty of constructing large, complex models, and the representation of environmental conditions (lighting and atmospheric effects). The transition to digital graphics generation ultimately supplanted the need for physical models.

The transition from analog to digital processing was accomplished in stages as more simulation elements were converted. For example, studies of driver decision-making have been accomplished with analog computer vehicle dynamics, analog electronics for generating roadway elements, and a paper tape programmer run at vehicle speed for controlling tasks and events in the driving scenario (Allen, Hogge, & Schwartz, 1975). With the availability of capable PC graphics cards, all digital versions of driving simulation were produced (Allen, Stein, Aponso, Rosenthal, & Hogue, 1990). The development and application of driving simulation has proliferated since with increased capability of PC-based CPUs and GPUs.

2.3 Cueing Systems

Cueing system developments for simulation have definitely benefited from technical progress in the graphics and audio industries. Visual feedback in driving simulation is the most compelling cueing system, and developments of display devices and graphics rendering have historically been the prime drivers of development (Allen et al., 2000). Visually rich cueing displays were originally presented by film projection (Kemmerer & Hulbert, 1975), with a transition to CRT (cathode ray tube) and video displays as these technologies developed and matured.

Digital 3D graphics development has been ongoing for the last four decades (e.g., Carlson, 2003). Much of the early work was too computationally intensive for real-time simulator applications. Real-time graphics were first produced on specialized and expensive display generators and workstations. More recently graphics cards and graphics processing units (GPUs) for desktop computer systems have made extremely fast, real-time, photorealistic graphics rendering available at reasonable costs (Crow, 2004; NVIDIA Corporation, n.d.). Display resolution has been an issue historically because of the rendering of traffic control devices, particularly roadway signs (see Chrysler & Nelson, this book, chap. 36 for a more detailed discussion). Table 2.1 summarizes display resolutions, based on the capabilities of the eye and various display media. Digital displays are just now getting up to the basic capability of the human eye for reasonable fields of view and are still far below the resolution of 35 mm film. Furthermore, high resolution display devices are still quite expensive.

Driving simulation started off in the 1970s with analog electronics, and various display concepts. An early concept with calligraphic displays and projectors is shown in Figure 2.2a (Allen et al., 1977). Processing delays were not an issue with the early analog electronics approach, but subsequently became a problem with early digital computers and computer graphics systems. Processing delays of more than 100 msec were found to be a problem with the human operator (Allen & DiMarco, 1984), and digital processors and visual image generators were developed that minimized this problem.

Moving model belt

Driver's video view

FIGURE 2.3 Video display system with model belt.

It wasn't long after the original IBM PCs were introduced that more capable graphics cards were developed for the PC bus. Some interesting historical background can be found on the Web regarding the development of computer graphics, including PC bus display adapters (e.g., the history by Carlson, 2003). Single chip GPUs started with the Texas Instruments Graphics Architecture (TIGA) ("TMS34010," n.d.), then readily began to evolve with the introduction of the 3DFX standard and have continued with current NVIDIA, ATI and Intel chips ("Graphics Processing Unit," n.d.). Current GPUs offer an impressive array of photorealistic effects that are made available through the user-programmable shading capabilities of the graphics chip ("Unified Shader Model," n.d.).

Audio cueing has been historically advanced by the audio entertainment industry and home audio equipment has long given more than adequate resolution for simulator aural cueing displays (Audio Engineering Society, 1999). More recently the digital recording and reproduction of music has provided a convenient standard for simulation sound systems. Computer standards for sound such as .wav files have resulted in adequate aural resolution for driving simulation. Editing software for digital sound files has emerged more recently, and the availability of digital sound libraries has proliferated. Although audiophiles complain about digital resolution and "warmth", there does not seem to be any reason to go beyond current digital music standards and 3D sound cueing (Audio Engineering Society, a list of standards, n.d.). Three dimensional sound is quite appropriate for the driving simulation environment, providing necessary spatial cueing (for example, a car horn during an ill-advised lane change) as well as to simulate effects of relative motion (Doppler effect). Digital sound production has recently made the simulation of 3D auditory effects practical, and surround sound electronics and displays have made this approach cost-effective.

Proprioceptive cueing includes the feel in the steering system, pedals and gear shifting. Active steering feel has traditionally been provided by torque motors and simulation of the elements such as tires, suspension and steering system, which contribute to steering torque as a function of the important maneuvering conditions (Norman, 1984). Steering feel is fairly critical to the driver's sense of vehicle handling (Adams & Topping, 2001), and this research has made a significant contribution to simulation steering feel fidelity (Brocker, 2006).

A wide variety of whole body motion concepts have been tried with driving simulation, with Stewart (1965/1966) as an early example. The motivation for motion cueing has been both to improve the realism and validity of the simulator experience, and to minimize simulator sickness effects. The development of motion systems for driving simulators has generally been based on motion cueing ideas developed for aircraft simulation. Aircrafts make coordinated turns, so that motion cues are primarily angular rates. Four-wheeled ground vehicles impart significant lateral and longitudinal accelerations, and for reasonable speeds yaw rate is a near threshold cue compared with lateral acceleration. This is because lateral acceleration is a function of longitudinal velocity squared, while yaw rate is only a linear function of velocity.

Lackner and DiZio (2005), in a review of contributions to spatial orientation, note that while the vestibular contributions to body orientation have long been recognized, the more recent contributions of proprioceptive and somatosensory signals have also proved to be significant in relation to specific force cueing. For lateral maneuvers such as lane changes, Grant, Artz, Blommer, Cathey and Greenberg (2002) found that classical motion cueing parameter sets which reduced roll errors at the expense of lateral acceleration errors resulted in higher perceived fidelity. Higher gains, which lead to large errors in the shape of both the roll and lateral motions, also resulted in the lowest subjective fidelity.

Haycock and Grant (2007) have shown that both acceleration and jerk (i.e., the derivative of acceleration) contribute significantly to the perceived strength of motion. In a number of experimental cases, the subjective measure of motion strength was larger for a lower level of acceleration when the jerk was larger by a sufficient amount. This suggests that increased levels of jerk in a simulator could lead to an impression of excessive simulator motion, or that scaled down accelerations could be augmented with additional motion jerk. More recent examples of motion systems are given in Nordmark, Jansson, Lidstrom and Palmkvist (1986), Drosdol and Panik (1985), and Greenberg, Artz and Cathey (2003) with the ultimate example being the National Advance Driving Simulator described in Allen et al. (2000). The most universal approach to motion cueing has been through use of the Stewart platform or so-called hexapod. A good summary of the history of the Stewart platform and the development of appropriate cueing algorithms is described in some detail in Tu et al. (2004). In addition to hardware, good cueing algorithms are critical to obtaining the appropriate motion feel, and considerable discussion has been devoted to this issue in the literature (Grant et al., 2002; Haycock & Grant, 2007; Nordmark et al., 1986; Drosdol & Panik, 1985; Greenberg et al., 2003; Romano, 2003a; Zywiol & Romano, 2003). Key issues in motion cueing algorithms have to do with artifacts that the driver perceives as not corresponding to visually perceived or control-induced motions, including washouts and proclivities of motion actuators such as turnaround bumps (i.e., when an actuator changes motion direction it momentarily sticks at the zero velocity position).

2.4 Vehicle Dynamics

Vehicle dynamics define the response of the vehicle to driver control inputs and external disturbances (road, aerodynamic). The perceived handling of the simulated vehicle and even steering feel depend on the modeling of the vehicle dynamics. There are various vehicle dynamics effects that are important from the driver's perception including speed sensitivity, understeer, maneuvering limits (lateral and longitudinal acceleration) and torque feedback to the steering system. Modeling and analysis of vehicle dynamics has been developed extensively over the years based on an increasingly good understanding of ground vehicle handling and stability. Developments have also been motivated by the needs of driving simulation for handling fidelity and limit performance maneuvering (tire saturation effects).

One of the first comprehensive vehicle dynamics models was developed by Segel (1956/1957) at Calspan. Subsequent modeling was carried forward by Weir, Shortwell and Johnson (1967), and Ellis (1969). A significant amount of analysis of early linear models was carried out in the frequency domain (i.e., Laplace and Fourier transforms: Spiegel, 1965). This is particularly insightful for understanding dynamic modes and stability properties (Ellis, 1969). The understanding of limit performance handling, and stability in particular, requires a nonlinear tire model (Allen, Rosenthal, & Chrstos, 1997; Pacejka & Bakker, 1993) which generally must be analyzed in the time domain. The first significant analysis effort of nonlinear computer simulation modeling was carried out at Bendix for NHTSA (Hartz, 1972). More recently, computer simulation modeling has been advanced significantly by the multibody modeling approach (Sayers, 1999; Romano, 2003b; Heydinger, Salaani, Garrott, & Grygier, 2002). However, multibody modeling tends to require a significant number of parameters to define detailed vehicle characteristics, and solution procedures require significant computing resources and lengthy computational times which can limit simulation fidelity. The validation of complex vehicle dynamics models has been addressed in order to ensure model fidelity (Heydinger et al., 2002; Allen, Chrstos, Howe, Klyde, & Rosenthal, 2002). Simplified nonlinear vehicle dynamics models (VDMs) have been developed to minimize the parameter specification and computational load (Allen et al., 1990).

As driving simulation has become more sophisticated, particularly with the addition of motion cueing, detailed vehicle dynamics models have become more important in providing the driver with appropriate cueing. Motion cueing will impart acceleration cues to the driver, so acceleration response to maneuvering is a key aspect of the VDM. The driver also feels a torque response in the steering wheel that is a function of vehicle maneuvering, tire characteristics, caster and steering system compliance, and the VDM must properly model these characteristics. Auditory cueing can give tire screeching sounds due to the amount of force saturation that is involved, which is a function of lateral and longitudinal acceleration. Auditory cues such as tire screeching have typically derived from recordings to minimize the computational load, even in critical applications such as auto industry NVH (noise, vibration, harshness) (Blommer & Greenberg, 2003). VDM characteristics affect visual cueing due to angular rates and velocities, but details of the VDM will not typically be displayed very sensitively in the visual display. (Additional discussion of vehicle dynamics models and their validation can be found in this book, chap. 11 by Schwarz.)

2.5 Driving Tasks and Scenarios

The basis for driving simulator research, training and clinical applications resides in the development of scenarios that produce the desired independent and dependent variables of interest. The independent variables include the tasks and events relevant to the driver behavior of interest, and the dependent variables consist of measures of this behavior and the related system performance. Various processes have evolved for specifying complex tasks and driving scenarios as the computational capability of simulation has evolved.

Task and scenario development is essential to real-time driving simulator applications. Scenarios create the visual, auditory and proprioceptive environments that are important for the face validity of the driving simulator, and provide for relevant situations which are critical to various driving applications. In research on driver behavior, the scenario must contain elements that stimulate relevant behavior. For training, the scenarios must promote the repeated use of critical skills in a variety of situations that will advance the learning process and make the driver proficient and safe. For driver assessment and prototype testing, relevant performance measures must be available and keyed to the scenario situations so that the driver's behavior can be properly quantified.

Task and scenario development are important for defining a range of elements found in the driving environment, including roadway geometry (e.g., horizontal and vertical curvature, intersections and traffic circles), TCDs (traffic control devices, including signals, signs and markings), interactive traffic and pedestrians, and roadside objects such as buildings and flora). Control of the timing of traffic, pedestrian movements, and signals is also important in order to present critical hazards to drivers. These capabilities have generally expanded as simulator capability has advanced, and these developments have, in fact, significantly improved the utility of driving simulation over the years.

Driving simulator tasks and scenarios have been developed to measure, train and assess driver competence in relation to tasks that are critical to performance and safety. The driver must exert behavioral skills in dealing with the complexity of the roadway environment. These skills, which include the perceptual, psychomotor and cognitive functions required in vehicle navigation, guidance and control, must be applied competently to maintain system safety and performance. Task and scenario design and programming have developed historically along with technological advancements in simulation. The earliest film and physical model-based simulations were limited in their ability to represent tasks and scenarios other than vehicle control and sign recognition (Kemmerer & Hulbert, 1975). These approaches basically gave a fixed scenario that could only be changed by re-filming or re-doing the model database.

The earliest efforts with programmable events and signs allowed for more flexibility in task and scenario specification (Lum & Roberts, 1983; Alicandri, 1994). With the addition of significant digital computational power, procedural methods for designing scenarios and visual databases were developed for making driving simulators more easily programmable and adaptable (Allen, Rosenthal, Aponso, & Park, 2003). Significant attention has been devoted to the development and application of procedural methods in recent years (Kaussner, Mark, Krueger, & Noltemeier, 2002; van Wolffelaar, Bayarri, & Coma, 1999). Procedural methods allow scenarios to be defined with script-based languages rather than in 3D database modeling programs, with the simulator assembling and drawing the 3D database. Databases for real-time simulation with 3D digital graphics systems were traditionally developed in graphics programs as composite 3D models. This approach requires extensive effort and experience with graphics modeling programs. Procedural methods have allowed scenarios to be developed more easily and have permitted scripting control of the spatial and temporal variables of task elements. (Park, Rosenthal, & Aponso, 2004. For a more detailed discussion see Kearney & Grechkin, this book, chap. 6.)

2.6 Performance Measurement

Performance measurement is a critical aspect of driving simulation, allowing for objective quantification of both driver and system behaviors. Performance measurement has evolved along with simulator development, motivated by the desire to better quantify specific driving behaviors and the driver/vehicle/environment system. Early driving simulators collected data with pen recorders and magnetic tape recorders. Most data processing was done offline. The sophistication of performance measurement has progressed quite dramatically with the advent of digital computation which has allowed for a range of statistical metrics and time series analysis of input/output algorithms.

Simple global measures were first implemented, including accidents, tickets, speed limit exceedances, lane and speed deviations, turn indicator usage, and so forth. This category has also included various measures of driver steering, throttle and brake control actions and associated vehicle responses including body axis accelerations and velocities (Sheridan, 1970; Weir & Wojcik, 1971; Kemmerer & Hulbert, 1975; Allen et al., 1977). These measures have been collected during entire simulator runs, and were subdivided into sections of driving scenarios where road geometry, vehicle and pedestrian interactions, traffic control devices and other task demands made them particularly relevant as the ability to program driving scenarios became more mature. Various algorithms were applied to these measures including distributions and moments (e.g., mean and standard deviation), power spectra, and more modern procedures such as wavelet analysis (Thompson, Klyde, & Brenner, 2001), which can quantify time variations in driver behavior. (For an additional discussion of independent variables in driving simulators see in this book, chap. 15 by McGwin, and chap. 17 by Brookhuis & de Waard; for a discussion of recent surrogate methods, see Angell, this book, chap. 10; and for a discussion of qualitative measures, see Moeckli, this book, chap. 23.)

With increasing computational capability in driving simulators more powerful measurement paradigms were employed where independent variables were more closely controlled and measurement algorithms quantified the relationship between dependent variables (i.e., driver response) and independent variables. For example, time series analysis methods have been used to quantify the relationship between driver response and road curvature, aerodynamic disturbances, and lead vehicle velocity changes (Allen & McRuer, 1977; Marcotte et al., 2005). These methods allowed for the analysis of driver time delay in responding to stimulus inputs, and the correlation of driver response to the stimulus input (additional discussion can be found in this book, chap. 21 by Boyle). With advancing capability in programming driving scenarios, driver response has also been quantified for more discrete and transient stimuli such as a traffic signals or conflicts with vehicles and pedestrians. These situations covered steering and/or speed control responses, and were analyzed in terms of driver decision-making and response time (Stein & Allen, 1986).

The measurement of human operator behavior, including driving, has been pursued for more that half a century, and is rooted in the general problem of modeling the human operator (Young, 1973; Sheridan & Ferrell, 1974). The early work dealt with the stable feedback control of vehicle dynamics in general, and a special conference, the Annual Conference on Manual Control, was held for over two decades and was devoted to the behavior and modeling of the human operator (Bekey & Biddle, 1967; Miller, 1970). Figure 2.4 generally illustrates the driving task and its three key components: driver, vehicle and environment. This conceptual model portrays several issues associated with modeling and measuring the performance of the driving task. The driver controls a vehicle, and this feedback process must be stable in a closed loop sense. Theories of linear feedback control have been applied to this problem, and a range of models have been proposed for its quantification (McRuer, 1980) that deal with stability either structurally, such as classical stability analysis (Weir & McRuer, 1973; Allen, 1982), or algorithmically, with procedures such as optimal control (MacAdam, 1981; Kleinman, Baron, & Levison, 1971; Thompson & McRuer, 1988). These two approaches raise the general issue of the computational procedures that are used in driver measurement, which have involved classical time series analysis and modern techniques such as wavelets (Thomspon et al., 2001). Higher level characteristics have been ascribed to the human operator (Goodstein, Andersen, & Olsen, 1988; Pew & Mavor, 1998), and cognitive functions such as risk perception, decision-making, and situation awareness are strongly factored into the driver's reaction to environmental inputs such as traffic, traffic control devices, and hazards in general.

Generally, driver models and measurements have been broadly categorized according to their control, guidance and navigation functions. Control concerns psychomotor functions that stabilize the vehicle path and speed against various aerodynamic and road disturbances. Guidance involves perceptual and psychomotor functions coordinated to follow delineated pathways, adhere to implied speed profiles, interact with traffic and avoid hazards. Navigation involves higher level cognitive functions applied to path and route selection and decisions regarding higher level

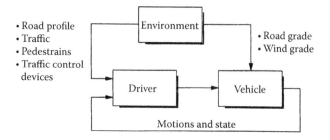

FIGURE 2.4 Performance measurement: driver, vehicle and system.

traffic interactions (e.g., avoiding congestion). Models have been developed that allow for the driver to respond to disturbances and commands in the driving environment (Allen & McRuer, 1979). These models have characterized the driver operating in so-called compensatory and pursuit modes. The compensatory mode relates to nulling out errors such as lane or speed deviations. Pursuit behavior arises when the human can perceive commands independently of errors, for example road curvature (McRuer, Allen, Weir, & Klein, 1977). Drivers will also respond to perceived vehicle motions and steering torque, which speaks to the importance of these cueing variables and accounting for them in the measurement process. Through vestibular and proprioceptive feedbacks the human operator can also respond to vehicle motions and control system forces which may be important in limit performance maneuvering (Young, 1973). (For more detailed discussions of driver models see in this book, chap. 42 by Salvucci, and chap. 43 by Flach, Jagacinski, Smith, & McKenna.)

Through the years there has been a trend in performance measures to switch from focusing on more driver-centered behavior (e.g., control activity, transfer functions, psychophysiological responses) to a more system-related performance involving safety and reaction to the driving environment. This trend has been influenced to a certain degree by the advancement of scenario design and control which have allowed more hazard-related events to be presented to the driver, and the desire to quantify safety in this context. Measurement is also somewhat specific to the application being addressed. For example, drowsiness and fatigue studies will typically focus on uneventful driving (Thiffault & Bergeron, 2003, see also Matthews, Saxby, Funke, Emo, & Desmond, this book, chap. 29), while training and assessment applications will focus more on situational awareness (e.g., see also Gugerty, this book, chap. 19) and how the driver responds to hazardous driving conditions (De Winter, Wieringa, Kuipers, Mulder, & Mulder, 2007; Johnson, Van der Loos, Burgar, Shor, & Leifer, 2001). When a range of measures are obtained, there have also been efforts to develop composite measures of performance using multivariate analysis approaches (Allen, Park, Cook, & Fiorentino, 2007; De Winter, 2009).

2.7 Simulator Sickness

Simulator sickness (SS) and its etiology have been a source of concern from the earliest days of simulator development and application (Reason, 1978; Casali & Frank, 1988). Much of the early work and concern revolved around aeronautical simulators, but driving simulators have also demonstrated similar issues (Rizzo et al., 2003). One issue that distinguishes driving simulation from aeronautical simulation is the same issue that impacts on motion cueing algorithms, that is, that aircraft mainly make coordinated turns while ground vehicle turns can induce large specific forces. In fact, as discussed under motion cueing above, specific forces dominate at high speeds, so the stimulation of drivers during maneuvering is much different than the stimulation of pilots. One historical rationale for improving simulator fidelity has been to minimize SS. This has led to efforts to improve cueing fidelity

in visual and motion systems on many occasions, but with minimal success. Questionnaires have been developed to quantify the effects of SS (Kennedy, Lane, Berbaum, & Lilienthal, 1993). SS has been related to various simulator design configurations (Draper, Viirre, Furness, & Gawron, 2001; Roe, Brown, & Watson, 2007), and to environmental conditions (Rizzo et al., 2003). SS rates under various conditions have been quantified in some simulators (Park et al., 2004). It would appear that with proper care in simulator design and with attention to the environment SS rates can be minimized, although the details of these specifications are still not well understood. (For a more detailed treatment see in this book, chap. 8 by Andersen, and chap. 14 by Stoner, Fisher, & Mollenhauer.)

2.8 Fidelity and Validity

Fidelity and validity are continuing concerns for all driving simulation applications. Fidelity relates to the sensory experience in driving simulators. Face validity is concerned with the subjective impression of the physical layout of the simulator (controls, displays and cabin surround). More general validity relates to the suitability of the simulation for its intended applications. In research applications the simulator should provide measures of driver behavior (e.g., psychomotor, cognitive) and system performance (e.g., speed and lane deviations) that are consistent with real-world behavior (see also Ranney, this book, chap. 9). For assessment applications (e.g., driver capability, licensing) the question is how well does a simulator relate to driver performance in the real world (see also Mullen, Charlton, Devlin, & Bédard, this book, chap. 13)? For training applications, the question is how well the trained behavior transfers to the real world (e.g., fewer crashes; see also Pollatsek, Vlakveld, Kappé, Pradhan, & Fisher, this book, chap. 30)?

Simulation validity is multidimensional, and can relate to behavioral and physical dimensions (Jamson, 1999) as well as to the perceived sensation of the subjective experience and objective performance (Fildes, Godley, Triggs, & Jarvis, 1997; Wade & Hammond, 1998). Leonard and Wierwille (1975) have set down a validation methodology that generally follows good experimental design (see also Ouimet, Duffy, Simons-Morton, Brown, & Fisher, this book, chap. 24). Simulator validity must also be considered task-dependent (Kaptein, Theeuwes, & Van Der Horst, 1996). An early approach for validating simulation followed the typical psychological measurement assessment validity theory (American Psychological Association, 1954; Tiffin & McCormick, 1965). However, this approach has been criticized as a validity assessment (Ebel, 1961), as, more specifically, it relates to simulation (McCoy, 1963).

In general, validation procedures relate to simulation features or applications of interest which should be clearly stated in any such effort. For example, simulator speed profiles have been validated against data from real-world road tests for highway engineering studies (Bella, 2005). Furthermore, in decision-making research, driver stopping decisions at signal lights show reasonable similarities to real world observational data (Allen, Rosenthal, & Aponso, 2005). There are also issues associated with the driving task and driver motivation in the simulator versus the real world that can significantly affect driver behavior and system performance (Allen et al., 2005).

The Figure 2.1 block diagram illustrates the components of a simulation that factor into fidelity and validity. As discussed previously, the major elements of a typical driving simulator include: Cueing systems (visual, auditory, proprioceptive, and motion), vehicle dynamics, computers and electronics, cabs and controls, measurement algorithms and data processing and storage. Cueing systems involve stimulation of all driver sensory and perceptual systems. In each of the cueing systems (visual, auditory, proprioceptive and motion) the appropriate stimulus resulting from the driver's control inputs must be computed and then accurately displayed to the driver. Cues such as steering feel are a direct consequence of the driver's control response and the resulting vehicle reaction. Motion cues are a function of the vehicle's dynamic response to the driver control inputs, with additional independent inputs due to roadway (e.g., road crown) and aerodynamic (e.g., wind gust) disturbances.

There are three methods in general for validating simulator characteristics. First, if there is some absolute criterion for validating simulator components such as display resolution, then the characteristics can be measured. A second validation method involves comparing simulation measurements with results obtained in real vehicles under controlled experimental conditions, e.g., validating a component such as vehicle dynamics (Allen et al., 2002) or validating driver/vehicle performance (Jamson, 1999; Bella, 2005). A third method, which might be considered the highest form of validation, involves the comparison of simulator behavior with real-world results obtained under uncontrolled observational conditions. In this third case, if combined operator/vehicle behavior is being validated, operators are presumably performing under real-world conditions with appropriate motivation, for example, with regards to driver performance (Bella, 2009; Allen et al., 2005), or transfer of training for training simulators (Blaiwes, Puig, & Regan, 1973; Rose, Evans, & Wheaton, 1987).

As has been pointed out in the past (Allen, Mitchell, Stein, & Hogue, 1991), it is not possible to completely validate a simulator with one set of experiments or measures. For example, a simulation that is validated for speed production (Bella, 2005) may not have adequate resolution for sign reading and therefore would not be useful in evaluating the recognition distance of signs. Another example might be a fixed-base simulation that is useful for presenting complex scenarios requiring situation awareness and decision-making, but is not suitable for evaluating control actions because of the lack of motion cues (Siegler, Reymond, Kemeny, & Berthoz, 2001). Clearly, establishing the fidelity and validity of driving simulation is a multidimensional problem, and much work is required to establish the fidelity and validity of each of the simulator components summarized in Figure 2.1.

There are some efforts at more formal procedures for the verification and validation of modeling and simulation. Sargent (1998) describes different approaches for various validation techniques including data validity, and also notes that there is no specific set of tests to determine which techniques or procedures should be used. Balci (1997) describes a more prescribed approach, and addresses the life cycle of applying verification and validation to modeling and simulation projects, but notes the difficulty of formally applying these procedures. Balci deals primarily with accuracy and certification that a model of simulation is acceptable for use for a specific purpose. As simulations are applied to training and assessment, the question of accuracy will become more and more important. In this regard some efforts are being concentrated on the verification and validation of data bases submitted to regulatory authorities (Clinical Data Interchange Standards Consortium, Data Standards Team, 2005).

2.9 The Future

The future of driving simulation will generally be influenced by a combination of improvements in cueing systems and computational capacity along with a better understanding of how the driver reacts to the driving environment. Visual cueing will benefit from ongoing advancements in GPU development (Wilson, 2007; Gschwind, n.d.) software for creating a wide range of visual effects (Lankes, Strobl, & Huesmann, 2008; Williams, Chou, & Wallick, 2005; Ellis & Chalmers, 2005) and display devices that will improve resolution, contrast and brightness. These improvements should lead to better night and inclement weather scenes, and generally cover a wider range of the important visual variables such as brightness, contrast, resolution and field of view. Improvements in GPU and CPU capability will allow for increased visual complexity in scenes that will make them more photorealistic and also provide for more complex road and traffic environments which will be important for training and assessment. There is also the possibility of binocular displays (Law, 2009) that might enhance the rendering of close traffic conditions.

Sound cueing will most likely continue to take advantage of commercial PC computer-based solutions (Heitbrink & Cable, 2007). Motion cueing will advance with ongoing developments in motion base hardware and cueing algorithms (Colombet et al., 2008; Brünger-Koch, Briest, & Vollrath, 2006). A key issue here will be to minimize artifacts that are not consistent with visual inputs and control actions. Because of the participant safety problem, effective motion systems will typically impose facility requirements that increase initial purchase, maintenance and logistics costs. Hardware cost will also be a consideration here as larger, more costly motion platforms also provide more capability.

In the future, driving simulator applications will expand beyond research and development to more applied uses. Software will extend the capabilities of driving scenarios and performance measurement, and will permit a wide variety of simulator training and assessment applications. For new simulator applications the key will be user interfaces that make the simulator convenient to use for clinicians. Software development will allow training and assessment applications to be more directly suited to operators and participants (Parkes, 2003; Akinwuntan et al., 2005). New applications will develop, such as simulator assessment of highway designs and traffic engineering problems (Bella, 2009, Qiao, Lin, & Yu, 2007). Regarding software development for highway design simulation, an open standard has been proposed for roadway descriptions (Dupuis & Grezlikowski, 2006). Given low cost desktop simulators, it is possible that highway and traffic engineers will be able to routinely visualize and assess their designs before investing significantly in real-world facilities. A key issue here will be the capability of importing road design CAD models into simulator rendering systems, and a convenient means for adding relevant traffic control devices (signals, signs, markings and delineators). These efforts are under active consideration by the US Transportation Research Board Visualization in Transportation committee (Transportation Research Board, n.d.). (For related discussions of the future of driving simulation, see in this book, chap. 4 by Hancock & Sheridan; chap. 34 by Granda, Davis, Inman, & Molino, and chap. 39 by Manore & Papelis.)

Key Points

- Development has taken place in a significant number of simulator components including the rendering and display of sensory cues, the development and production of driving scenarios, and the measurement of driver behavior and driver/vehicle system performance.
- Hardware development outside of the simulation field has had a significant impact on simulator development and advancements, including motion cueing systems, computer processing units (CPUs), and graphical processing units (GPUs).
- The ultimate capability of simulators, given adequate hardware, depends on the software that controls the creation of the driving environment and the rendering of sensory cueing.
- Through software advancements, the ability to specify and control elements of the driving environment has increased dramatically over the years. These elements include roadway profiles, roadway traffic, roadside structures, flora and fauna, and traffic control devices including signs, signals, markings, and other delineation elements.
- Display resolution is still a limiting issue for tasks such as reading signs. It remains a practical challenge to achieve the limits of visual resolution (i.e., one minute of visual arc).

Keywords: Driving Scenarios, Fidelity and Validity, Performance Measurement, Sensory Cueing, Simulator Sickness

Web Resources

The *Handbook* web site contains supplemental material for the chapter, including color versions of Figure 2.1 and Figure 2.2.

Web Figure 2.1: Functional elements of driving simulation (color version of print Figure 2.1).

Web Figure 2.2: Evolution of driving simulators and PC Graphics (color version of Figure 2.2).

Key Readings

Akinwuntan, A. E., De Weerdt, W., Feys, H., De Vooght, F., Devos, H., Baten, G., et al. (2007). Training of driving-related attentional performance after stroke using a driving simulator. *Proceedings of the fourth international driving symposium on human factors in driver assessment, training and vehicle design* (pp. 112–118). University of Iowa, Iowa City.

Drosdol, J., & Panik, F. (1985). *The Daimler-Benz driving simulator a tool for vehicle development* (SAE Paper 850334). Society of Automotive Engineers International Congress and Exposition, Detroit, MI.

Kaptein, N. A., Theeuwes, J., & Van Der Horst, R. (1996). Driving simulator validity: Some considerations. *Transportation Research Record, 1550*, 30–36.

Park, G., Rosenthal, T. J., & Aponso, B. L. (2004). Developing driving scenarios for research, training and clinical applications. *International Journal of Advances in Transportation Studies*, Special Issue, 19–28.

Sheridan, T. B. (Ed.) (1967, September 28–29). *Mathematical models and simulation of automobile driving.* Conference Proceedings. Cambridge, MA: Massachusetts Institute of Technology.

References

Adams, W. K., & Topping, R. W. (2001). The steering characterizing functions (Scfs) and their use in steering system specification, simulation, and synthesis. (SAE Tech. Paper 2001-01-1353), In J. P. Chrstos, W. R. Garrott, & G. J. Heydinger (Eds.), *Vehicle dynamics and simulation 2001.* Warrendale, PA: Society of Automotive Engineers.

Akinwuntan, A. E., De Weerdt, W., Feys, H., Pauwels, J., Baten, G., Arno, P., et al. (2005). Effect of simulator training on driving after stroke: A randomized controlled trial. *Neurology, 65*(6), 843–850.

Alicandri, E. (1994). HYSIM: The next best thing to being on the road. *Public Roads, 57*(3).

Allen, R. W. (1982). *Stability and performance analysis of automobile driver steering control* (SAE Paper 820303). Proceedings of SAE International Annual Congress & Exposition. Warrendale, PA: Society of Automotive Engineers.

Allen, R. W., Chrstos, J. P., PhD, Howe, J. G., Klyde, D. H., & Rosenthal, T. J. (2002). Validation of a non-linear vehicle dynamics simulation for limit handling. *Journal of Automobile Engineering, Special Issue on Vehicle Dynamics, Proceedings of the Institution of Mechanical Engineers, 216*(D4), 319–327.

Allen, R. W., & DiMarco, R. J. (1984). Effects of transport delays on manual control system performance. *Proceedings of the 20th annual conference on manual control* (Vol. 1, pp. 185–201; NASA CP-2341). Moffett Field, CA: NASA Ames Research Center.

Allen, R. W., Fancher, P. S., Levison, W. H., Machey, J., Mourant, R., Schnell, T., et al. (2000). Simulation and measurement of driver and vehicle performance. *Transportation in the New Millennium Perspectives from Transportation Research Board Standing Committees*. Washington, DC: Transportation Research Board.

Allen, R. W., Hogge, J. R., & Schwartz, S. H. (1975). An interactive driving simulation for driver control and decision-making research (NASA TM X-62,46). *Proceedings from the 11th annual conference on manual control* (pp. 396–407). Moffet Field, CA: NASA Ames Research Center.

Allen, R. W., Hogge, J. R., & Schwartz, S. H. (1977, January). *A simulator for research in driver, vehicle, and environment interaction*. Paper presented at the 56th Annual Meeting of the Transportation Research Board, Washington, DC: National Research Council.

Allen, R. W., & McRuer, D. T. (1977). The effect of adverse visibility on driver steering performance in an automobile simulator. *1977 SAE Transactions* (Vol. 86, pp. 1081–1092; SAE Paper 770239). Warrendale, PA: Society of Automotive Engineers.

Allen, R. W., & McRuer, D. T. (1979). The man/machine control interface—pursuit control. *Automatica, 15*(6), 683–686.

Allen, R. W., Mitchell, D. G., Stein, A. C., & Hogue, J. R. (1991). Validation of real-time man-in-the-loop simulation. *VTI Rapport, 372A*, Part 4, 18–31.

Allen, R. W., Park, G. D., Cook, M. L., & Fiorentino, D. (2007). A simulator for assessing older driver skills. *Advances in Transportation Studies An International Journal*. Special Issue, 23–32.

Allen, R. W., Rosenthal, T. J., & Chrstos, J. P., PhD. (1997). *A vehicle dynamics tire model for both pavement and off-road conditions* (SAE Paper 970559). Presented at Society of Automotive Engineers International Congress & Exposition, Detroit, MI.

Allen, R. W., Rosenthal, T. J., Aponso, B. L., & Park, G. D. (2003). *Scenarios produced by procedural methods for driving research, assessment and training applications*. Proceedings of the Driving Simulator Conference 2003 North America Conference, Dearborn, MI.

Allen, R. W., Rosenthal, T. J., & Aponso, B. L. (2005). Performance measurement paradigms for driving simulator research. *Advances in Transportation Studies An International Journal*, Special Issue, 45–58.

Allen, R. W., Stein, A. C., Aponso, B. L., Rosenthal, T. J., & Hogue, J. R. (1990). A low cost, part task driving simulator based on microcomputer technology. *Transportation Research Record, 1270*, 107–113.

American Psychological Association. American Educational Research Association, & National Council on Measurements Used in Education (U.S.). (1954). *Technical recommendations for psychological tests and diagnostic techniques*. Washington: American Psychological Association.

Audio Engineering Society. (1999). *An audio timeline*. Retrieved October 2, 2008 from http://www.aes.org/aeshc/docs/audio.history.timeline.html

Audio Engineering Society. (2009). *A list of standards in print*. Retrieved March 27, 2009 from http://www.aes.org/publications/standards/list

Balci, O. (1997). *Verification, validation and accreditation of simulation models*, Proceedings of the 1997 Winter Simulation Conference. New York: Association of Computing Machinery, Special Interest Group on Simulation.

Bekey, G. A., & Biddle, J. M. (1967, March). *The effect of a random sampling interval on a sampled-data model of the human operator* (NASA SP-144). Third Annual NASA–University Conference on Manual Control, Los Angeles, CA.

Bella, F. (2005). Validation of a driving simulator for work zone design, *Transportation Research Record*, 1937.

Bella, F. (2009). *Can driving simulators contribute to solving the critical issues in geometric design?* (TRB Paper No. 09-0152). Proceedings of the Transportation Research Board 88th Annual Meeting. Washington, DC: National Research Council.

Blaiwes, A. S., Puig, J. A., & Regan, J. J. (1973). Transfer of training and the measurement of training effectiveness, *Human Factors, 15*(6), 523–533.

Blommer, M., & Greenberg, J. (2003). *Realistic 3D sound simulation in the virttex driving simulator*. Proceedings of DSC North America 2003. Dearborn, MI.

Brocker, M. (2006). New control algorithms for steering feel improvements of an electric powered steering system with belt drive. *Vehicle System Dynamics, 44*(Supplement), 759–769.

Brünger-Koch, M., Briest, S., & Vollrath, M. (2006). Virtual driving with different motion characteristics—braking maneuver analysis and validation. *Proceedings of driving simulation conference DSC Europe* (pp. 69–78). Paris.

Carlson, W. (2003). A critical history of computer graphics and animations. Retrieved March 3, 2009 from http://design.osu.edu/carlson/history/lessons.html

Casali, J. G., & Frank, L. H. (1988). Manifestation of visual/vestibular disruption in simulators: Severity and empirical measurement of symptomatology. *Motion cues in flight simulation and simulator induced sickness, AGARD conference proceedings*, 433, pp. 11.11–11.18.

Clinical Data Interchange Standards Consortium, Data Standards Team. (2005). *Study data tabulation model implementation guide: Human clinical trials*. Retrieved July 9, 2009, from http://www.cdisc.org

Colombet, F., Dagdelen, M., Reymond, G., Pere, C., Merienne, F., & Kemeny, A. (2008). Motion cueing: What's the impact on the driver's behavior? *Proceeding of the Driving Simulator Conference Europe 2008* (pp. 171–181). Monaco.

Computer display standards. (n.d.). Retrieved from Wikipedia: http://en.wikipedia.org/wiki/Computer_display_standard

Crow, T. S. (2004). *Evolution of the graphical processing unit*. Unpublished Master's thesis, University of Nevada. Retrieved March 27, 2009 from http://www.cse.unr.edu/~fredh/papers/thesis/023-crow/GPUFinal.pdf

De Winter, J. (2009). *Advancing simulation-based driver training*. Doctoral dissertation, Technical University of Delft, the Netherlands.

De Winter, J. C. F., Wieringa, P. A., Kuipers, J., Mulder, J. A., & Mulder, M. (2007). Violations and errors during simulation-based driver training. *Ergonomics, 50*(1), 138–158.

Donges, E. (1975). Experimental investigation of human steering behavior in a simulated road driving task. *ATZ Automobiltechnische Zeitschrift, 77*(5/6), 141–146.

Draper, M. H., Viirre, E. S., Furness, T. A., & Gawron, V. J. (2001). Effects of image scale and system time delay on simulator sickness within head-coupled virtual environments. *Human Factors, 43*(1), 129–146.

Drosdol, J., & Panik, F. (1985). *The Daimler-Benz driving simulator a tool for vehicle development* (SAE Paper 850334). Society of Automotive Engineers International Congress and Exposition, Detroit, MI.

Dupuis, M., & Grezlikowski, H. (2006). *OpenDRIVE: An open standard for the description of roads in driving simulations.* Proceedings of Driving Simulation Conference Europe 2006. Paris.

Ebel, R. L. (1961). Must all tests be valid? *American Psychologist, 16*(10), 640–647.

Ellis, J. R. (1969). *Vehicle dynamics.* London: Business Books.

Ellis, G., & Chalmers, A. (2005). *Generation of high fidelity graphics for simulator environments.* Proceedings of the Driving Simulator Conference North America 2005, Orlando, FL.

Fildes, B. N., Godley, S. T., Triggs, T., & Jarvis, J. (1997). *Perceptual countermeasures: Simulator validation study.* (Rep CR 169 (FORS)). Australian Federal Office of Road Safety: Canberra ACT 2601.

Garrott, W. R., Mazzae, E. N., & Goodman, M. J. (2006). *NHTSA's National Advanced Driving Simulator Research Program.* Proceedings of the 19th International Technical Conference on the Enhanced Safety of Vehicles, Washington, DC.

Gilliland, M. G. (1973, Spring). *Applications of computer-generated imagery to driver training, highway research, and design.* Paper presented at the North Carolina Symposium on Highway Safety, Simulation: Its Role in Driver Research and Highway Design, Chapel Hill, NC.

Goodstein, L. P., Andersen, H. B., & Olsen, S. E. (Eds.). (1988). *Tasks, errors and mental models.* London: Taylor and Francis.

Grant, P. R., Artz, B., Blommer, M., Cathey, L., & Greenberg, J. (2002, September). *A paired comparison study of simulator motion drive algorithms.* Driving Simulation Conference Europe 2002, Paris.

Graphics Processing Unit. Retrieved May 1, 2008 from http://en.wikipedia.org/wiki/Graphics_processing_unit

Green, P. (2007). Why driving performance measures are sometimes not accurate (and methods to check accuracy). *Proceedings of the fourth international driving symposium on human factors in driver assessment, training, and vehicle design* (pp. 394–400). Iowa City, IA: University of Iowa.

Green, P., & Olson, A. (1989). *The development and use of the UMTRI driving simulator* (Rep. UMTRI-89-25). Ann Arbor, MI: University of Michigan, Transportation Research Institute.

Greenberg, J., Artz, B., & Cathey, L. (2003). *The effect of lateral motion cues during simulated driving.* Proceedings of Driving Simulation Conference North America 2003, Dearborn, MI.

Gschwind, M. (n.d.). *The cell architecture.* Retrieved 2009 from IBM Research Website: http://domino.research.ibm.com/comm/research.nsf/pages/r.arch.innovation.html

Hartz, J. R. (1972). *Computer simulation of vehicle handling Vol. I—Summary Report & Vol. II—Technical Report* (No. 6358, Vol. I & Vol. II): Benedix Corporation/Research Laboratories Division (RLD).

Haycock, B., & Grant, P. R. (2007, September). *The influence of jerk on perceived simulator motion strength.* Proceedings of the Driving Simulation Conference North America 2007, University of Iowa, Iowa City, IA.

Heitbrink, D. A., & Cable, S. (2007). *Design of a driving simulation sound engine.* Proceedings of the Driving Simulation Conference North America 2007, University of Iowa, Iowa City, IA.

Hess, R. A. (1982, August 9–11). *The effects of time delays on systems subject to manual control.* (AIAA 82-1523). Proceedings of the AIAA Guidance and Control Conference, San Diego, CA.

Heydinger, G. J., Salaani, M. K., Garrott, W. R., & Grygier, P. A. (2002). Vehicle dynamics modeling for the National Advanced Driving Simulator. *Journal of Automobile Engineering, Proceedings of the Institution of Mechanical Engineers, 216* (D4), 307–318.

History of video games. (n.d.). Retrieved March 13, 2009 from Wikipedia: http://en.wikipedia.org/wiki/History_of_video_games

Hogema, J. H. (1997). Compensation for delay in the visual display of a driving simulator. *Simulation, 69*, 27–34.

Hutchinson, C. H. (1958). *Automobile driving simulator feasibility study* (Rep. YM-1244-V-6). Buffalo, NY: Cornell Aeronautical Laboratory, Inc. (CAL).

Jamson, A. H. (1999). Curve negotiation in the Leeds Driving Simulator: The role of driver experience. In D. Harris (Ed.), *Engineering psychology and cognitive ergonomics* (Vol. 3, pp. 351–358). Ashgate.

Johnson, M. J., Van der Loos, H. F. M., Burgar, C. G., Shor, P. C., & Leifer, L. J. (2001). Design and evaluation of driver's SEAT: A car steering simulation environment for upper limb stroke therapy. *Robotica, 21*(1), 13–23.

Kaptein, N. A., Theeuwes, J., & Van Der Horst, R. (1996). Driving simulator validity: Some considerations. *Transportation Research Record, 1550*, 30–36.

Kaussner, A., Mark, C., Krueger, H. -P., & Noltemeier, H. (2002). *Generic creation of landscapes and modeling of complex parts of road networks.* Proceedings of Driving Simulation Conference DSC Europe. Paris.

Kemmerer, R. H., & Hulbert, S. F. (1975). UCLA Driving Simulation Laboratory: With a 360 degree scene around a full size car. Simulators and simulation—design, applications, and techniques, *Proceedings of the Society of Photo-Optical Instrumentation Engineers, 59*, 158–170.

Kennedy, R. S., Lane, N. E., Berbaum, K. S., & Lilienthal, M. G. (1993). A simulator sickness questionnaire (SSQ): A new method for quantifying simulator sickness. *International Journal of Aviation Psychology, 3*, 203–220.

Kleinman, D. L., Baron, S., & Levison, W. H. (1971). A control theoretic approach to manned-vehicle systems analysis. *IEEE Transactions on Automatic Control, AC-16*(6), 824–832.

Lackner, J. R., & DiZio, P. (2005). Vestibular, proprioceptive, and haptic contributions to spatial orientation. *Annual Review of Psychology, 56*, 115–147.

Lankes, R., Strobl, M., & Huesmann, A. (2008). *Advanced real-time rendering techniques in driving simulation.* Proceeding of the Driving Simulator Conference Europe 2008. Monaco.

Law, L. (2009). Game-changing technology. *Computer Graphics World, 32*(3, March 24).

Leonard, J. J., & Wierwille, W. W. (1975). *Human performance validation of simulators; theory and experimental verification.* 19th Annual Meeting of the Human Factors Society, Dallas, TX.

Leslie, J. M. (1966). *Effects of time delay in the visual feedback loop of a man-machine system* (Rep. NASA CR-560). Stanford, CA: Stanford University.

Lum, H. S., & Roberts, K. M. (1983). A highway simulation analysis of background colors for advance warning signs. *Public Roads, 47*(3).

MacAdam, C. C. (1981). Application of an optimal preview control for simulation of closed-loop automobile driving. *IEEE Transactions on Systems, Man, and Cybernetics, SMC-11*(6), 393–399.

McCoy, W. K. (1963). Problems of validity of measures used in investigating man-machine system. *Human Factors, 5*, 373–377.

McFarland, R. E., & Bunnell, J. W. (1990). *Analyzing time delays in a flight simulation environment* (AIAA 90-3174). AIAA Flight Simulation Technologies Conference and Exhibit, Dayton, OH, September 17–19.

Marcotte, T. D., Rosenthal, T. J., Corey-Bloom, J., Roberts, E, Lampinen, S., & Allen, R. W. (2005). *The impact of cognitive deficits and spasticity on driving simulator performance in multiple sclerosis.* Proceedings of the Driving Assessment 2005: 3rd International Driving Symposium on Human Factors in Driver Assessment, Training, and Vehicle Design. University of Iowa, Iowa City, IA.

McRuer, D. T. (1980). Human dynamics in man-machine systems. *Automatica, 16*(3), 237–253.

McRuer, D. T., Allen, R. W., Weir, D. H., & Klein, R. H. (1977). New results in driver steering control models. *Human Factors, 19*(4), 381–397.

Michelson, S., Niemann, R., Olch, R., Smiley, A., & Ziedman, K. (1978). *A driving simulator for human performance studies using all digital techniques.* Southern California Research Institute (SCRI).

Miller, D. C. (1970). *Human performance in time-optimal state regulation tasks* (NASA SP-215). Fifth Annual NASA-University Conference on Manual Control, Cambridge, MA, March 27–29.

Norman, K. D. (1984). *Objective evaluation of on-center handling performance* (SAE 840069). Society of Automotive Engineers International Congress and Exposition, Detroit, MI, February 27–March 2.

Nordmark, S., Jansson, H., Lidstrom, M., & Palmkvist, G. (1986). *A moving base driving simulator with wide angle visual system* (VTI Reprint 106). Linkoping, Sweden: Swedish Road and Transport Research Institute.

NVIDIA Corporation. (n.d.). *Graphics processing unit (GPU).* Retrieved March 27, 2009 from http://www.nvidia.com/object/gpu.html

O'Hanlon, J. F. (1977). *Directory of research laboratories operating driving simulators and research vehicles in North America.* Human Factors Research, Inc.

Parkes, A. M. (2003). Truck driver training using simulation in England. In J. D. Lee, M. Rizzo, and D. V. McGehee (Eds.), *Proceedings of the second international driving symposium on human factors in driver assessment, training and vehicle design.* (pp. 59–63). Iowa City: University of Iowa.

Pacejka, H. B., & Bakker, E. (1993). The magic formula tire model. *Vehicle System Dynamics, 21*(Supplement).

Park, G., Rosenthal, T. J., & Aponso, B. L. (2004). Developing driving scenarios for research, training and clinical applications. *International Journal of Advances in Transportation Studies,* Special Issue, 19–28.

Park, G., Rosenthal, T. J., Allen, R. W., Cook, M. L., Fiorentino, D. D., & Viirre, E. (2004). *Simulator sickness results obtained during a novice driver training study* (STI-P-637). Proceedings of the Human Factors and Ergonomics Society 48th Annual Meeting. New Orleans, LA.

Pew, R. W., & Mavor, A. S. (Eds.). (1998). *Modeling human and organizational behavior.* Washington, D.C.: National Academy Press.

Qiao, F., Lin, X., & Yu, L. (2007). *Using driving simulator for advance placement of guide sign design for exits along highways.* Proceedings of the Driving Simulation Conference North America 2007. University of Iowa, Iowa City, IA.

Ricard, G. L., Cyrus, M. L., Cox, D. C., Templeton, T. K., & Thompson, L. C. (1978). *Compensation for transport delays produced by computer image generation systems* (IH-297/AFHRL-TR-78-46). Orlando, FL: Human Factors Laboratory, Naval Training Equipment Center.

Rice, R. S. (1967). *Methods of improved point-light source techniques for application to automobile driving simulators* (Rep. YM-1783-V-1). Buffalo, NY: Cornell Aeronautical Laboratory, Inc. (CAL).

Reason, J. T. (1978). Motion sickness adaptation: A neural mismatch model. *Journal of the Royal Society of Medicine, 71*, 819–829.

Rizzo, M., Sheffield, R., & Stierman, L. (2003). Demographic and driving performance factors in simulator adaptation syndrome. In M. Rizzo, J. D. Lee, & D. McGehee, (Eds.).

Proceedings of driving assessment 2003: The second international driving symposium on human factors in driver assessment, training and vehicle design (pp. 201–208), Park City, UT. Iowa City, IA: The University of Iowa.

Roe, C. R., Brown, T., & Watson, G. (2007). *Factors associated with simulator sickness in a high-fidelity simulator.* Proceedings of the Driving Simulation Conference North America 2007. University of Iowa, Iowa City, IA.

Romano, R. (2003a). *Non-linear optimal tilt coordination for washout algorithms* (AIAA 2003-5681). AIAA Simulation and Modeling Conference, Austin, TX, August 11–14.

Romano, R. (2003b). *Real-time multi-body vehicle dynamics using a modular modeling methodology* (SAE 2003-01-1286). Society of Automotive Engineers International Congress and Exposition, Detroit, MI, March.

Rose, A., Evans, R., & Wheaton, G. (1987). Methodological approaches for simulator evaluations. In S. M. Cormier, & J. D. Hagman (Eds.), *Transfer of learning.* San Diego, CA: Harcourt.

Sargent, R. G. (1998). *Verification and validation of simulation models.* In D. J. Medeiros, E. F. Watson, J. S. Carson, & M. S. Manivannan (Eds.), Proceedings of the 30th Winter Simulation Conference. Washington, DC.

Sayers, M. W. (1999). Vehicle models for RTS applications. *Vehicle System Dynamics: International Journal of Vehicle Mechanics and Mobility, 32*(4), 421–438.

Segel, L. (1956–1957). Theoretical prediction and response of the automobile to steering control. *Research in Automobile Stability and Control in Tire Performance, Proceedings of the Automobile Division, Institution of Mechanical Engineers, 7,* 26–46.

Sheridan, T. B. (Ed.) (1967 Sept 28–29). *Mathematical models and simulation of automobile driving.* Conference Proceedings. Cambridge, MA: Massachusetts Institute of Technology.

Sheridan, T. B. (1970). Big brother as driver: New demands and problems for the man at the wheel. *Human Factors, 12*(1), 95–101.

Sheridan, T. B., & Ferrell, W. R. (1974). *Man-machine systems: Iinformation, control, and decision models of human performance.* Cambridge, MA: MIT Press.

Shuttel, H. H., Schumacher, S. P., & Gatewood, R. D. (1971). *Driver training simulators, ranges and modified cars: A review and critique of the experimental literature* (AIR-86400-7/71-IR). American Institutes for Research (AIR).

Siegler, I., Reymond, G., Kemeny, A., & Berthoz, A. (2001). Sensorimotor integration in a driving simulator: Contributions of motion cueing. *Elementary driving tasks. Proceedings of driving simulation conference* (pp. 21–32). Sophia-Antipolis, France.

Spiegel, M. R. (1965). *Schaum's outline of theory and problems of Laplace transforms.* New York: McGraw-Hill.

Stein, A. C., & Allen, R. W. (1986). *The effects of alcohol on driver decision making and risk taking.* Proceedings of the 30th Conference of the American Association for Automotive Medicine (AAAM). Montreal, Quebec, Canada: American Association for Automotive Medicine.

Stewart, D. (1965-1966). A platform with six degrees of freedom *Institution of Mechanical Engineers Proceedings, 180* (Part 1, No.15), 371–386.

Templeton, B. (n.d.). *How many pixels are there in a frame of 35 mm film?* Retrieved from http://pic.templetons.com/brad/photo/pixels.html

Thiffault, P., & Bergeron, J. (2003). Monotony of road environment and driver fatigue: a simulator study. *Accident Analysis & Prevention, 35*(3), 381–391.

Thompson, P. M., & McRuer, D. T. (1988). Comparison of the human optimal control and crossover models. *Proceedings of AIAA Guidance, Navigation and Control Conference* (Vol. 2, pp. 1083–1090; AIAA 88-4183). Minneapolis, Minnesota.

Thompson, P. M., Klyde, D. H., & Brenner, M. J. (2001, August 6–9). *Wavelet-based time-varying human operator models* (AIAA 2001-4009). AIAA Atmospheric Flight Mechanics Conference and Exhibit, Montreal, Canada.

Tiffin, J., & McCormick, E. J. (1965). *Industrial psychology.* Englewood Cliffs, NJ: Prentice Hall.

TMS34010. (n.d.). Retrieved May 1, 2008 from Wikipedia: http://en.wikipedia.org/wiki/TMS34010

Transportation Research Board, Visualization in transportation committee. (n.d.). Retrieved from http://www.trbvis.org/MAIN/TRBVIS_HOME.html

Tu, K. -Y., Wu, T. -C., & Lee, T. -T. (2004, 10-13 Oct). *A study of Stewart platform specifications for motion cueing systems.* Paper presented at the 2004 IEEE International Conference on Systems, Man and Cybernetics.

Unified shader model. (n.d.). Retrieved May 1, 2008 from: http://en.wikipedia.org/wiki/Unified_shader_model

van Wolffelaar, P., Bayarri, S., & Coma, I. (1999) Script-based definition of complex scenarios. *Driving Simulation Conference Europe 1999,* Paris.

Wade, M. G., & Hammond, C. (1998). *Simulation validation: Evaluating driver performance in simulation and the real world.* Final Report. (MN/RC-1998-28). Minneapolis, MN: University of Minnesota.

Weir, D. H., & McRuer, D. T. (1973). Measurement and interpretation of driver/vehicle system dynamic response. *Human Factors, 15*(4), 367–378.

Weir, D. H., Shortwell, C. P., & Johnson, W. A. (1967). *Dynamics of the automobile related to driver control.* (SAE 680194). Aerospace Vehicle Flight Control System A-18 Committee Meeting. Warrendale, PA: Society of Automotive Engineers.

Weir, D. H., & Wojcik, C. (1971). Simulator studies of the driver's dynamic response in steering control tasks. *Highway Research Record, 364,* 1–15.

Wilson, R. (2007, Oct 11). *NVIDIA, ATI/AMD look beyond GPUs towards unified gaming engines.* Retrieved March 25, 2009 from Electronics Design Strategy News, http://www.edn.com/blog/1690000169/post/1110015711.html

Williams, J. R., Chou, T., & Wallick, B. L. (2005). *Advanced rendering cluster for highway experimental research* (ARCHER). Proceedings of the Driving Simulator Conference North America 2005. Orlando, FL.

Young, L. R. (1973). Human control capabilities. In J. F. Parker, V. R. West, Jr. & P. Webb (Eds.), *Bioastronautics data book* (2nd ed., pp. 751–806). Washington, DC: National Aeronautics and Space Administration (NASA).

Zywiol, H. J., & Romano, R. A. (2003). *Motion drive algorithms and simulator design to study motion effects on infantry soldiers.* Paper presented at the Driving Simulation Conference, North America 2003, Dearborn, MI.

3

Using Driving Simulators Outside of North America

Barry H. Kantowitz
University of Michigan

Abstract

The Problem. While driving-simulator hardware is similar the world over, simulators are used differently in different regions of the world. I argue that the safety culture of a region determines what problems should be addressed. The Vision Zero safety culture of Europe promotes different simulator research than the military safety culture of the United States. *Role of Driving Simulators.* While regional safety culture determines what problems have priority, local opinion about features of driving microworlds such as tractability, realism, engagement, and validity determines how simulator research is accomplished. *Key Results of Driving Simulator Studies.* European studies have demonstrated that driving simulators are a useful tool for improving road safety. *Scenarios and Dependent Variables.* Specific examples of European simulator research in the areas of highway design, driver constraints, and driver distraction/workload are reviewed. Researchers from three institutions outside North America explain their philosophy of simulator use in their own words. *Platform Specificity and Equipment Limitations.* All simulator users must solve administrative issues allocating resources to a simulator facility. Simulators with different fidelity levels and different costs all can be successfully applied to improving road safety.

3.1 Introduction

At first glance, there is little difference between driving simulators of the same fidelity. All have a vehicle cab connected via computer to a screen that displays roadway scenes. Driver control inputs in the cab alter the displayed microworld. Thus, a comparison of driving simulators inside and outside of North America would initially seem to be limited to a discussion of power supplies, since the 110 volt 60 Hz AC standard used in the United States is generally not used outside of North America.

Therefore, a discussion of simulator hardware around the world might not be very productive. Indeed, the hardware and software of the National Advanced Driving Simulator in the United States is duplicated, with some additional improvements, in the VTI Driving Simulator in Sweden. However, a discussion of how driving simulators are utilized in different countries might be quite illuminating. Each country has its own culture of driving safety. Since this is a treatise on simulators, not sociology, I will not formally define a safety culture, other than to state that it is a framework that sets boundaries for the goals of safety research. Within those boundaries, certain endeavors are more likely to prosper and other undertakings may not even be considered. The AASHTO-FHWA scan tour report (Kantowitz, 2005) discussed later in this chapter emphasized the benefits of the European safety culture. I was able to report these findings at a subsequent AAA Foundation workshop in Washington, DC during October 2005, which led AAA to include safety culture as a topic in their research portfolio (AAA Foundation for Traffic Safety, 2007).

I have selected three topics within the boundaries of driving safety research that illuminate cross-cultural differences in selecting appropriate topics for simulator research. Having the simulator world of research in three parts divided, like Caesar's Gaul, allows me to select two areas with minimal consideration in North American research and one area that is tilled in common. The common research area is driver distraction and cognitive load. The disparate topics are using simulators both to design highways and to place constraints on drivers by systems such as intelligent speed adaptation. I will use examples from these three research areas to demonstrate how driving simulators are utilized outside of North America.

I have limited my selection of driving simulators to institutions I have been able to visit personally. This has allowed me the pleasure of having interesting discussions with the researchers and administrators responsible for defining and conducting programs of simulator research. Institutions differ in the amount of resources (e.g., hardware costs, space, support personnel, etc.) devoted to driving simulator research; since the driving simulator is only one instrument in the toolbox of driving safety research, it is reasonable that reliance upon that tool will differ across various institutions. Indeed, some (otherwise) excellent research institutions lack a modern driving simulator. Obviously, these institutions are not considered herein. I believe that the institutions discussed in this chapter represent the best simulator practices and research philosophies outside of North America and hope that my North American colleagues will benefit from increased awareness of such research efforts. Review of North American research is specifically eschewed since this chapter is devoted to efforts outside North America. So, for example, while European research on driver behavior in tunnels is discussed, complimentary American efforts with the Boston tunnel (e.g., Upchurch, Fisher, Carpenter, & Dutta, 2002) are omitted as they are beyond the scope of this chapter.

The body of this chapter starts with a discussion of safety culture in Europe. Since simulator hardware and software are much the same the world over, it is more enlightening to focus upon how simulators are used rather than upon how they are assembled.

Safety culture determines the types of research questions that are addressed. It is easy to accept one's own safety culture as a given, without pondering the constraints and benefits it supplies. I hope that my discussion of European safety culture encourages North American researchers to examine their own safety culture more fully. Since simulator research varies over many dimensions, I will then review the concept of a microworld as a framework for comparing and organizing characteristics of how simulator research is accomplished. This overview yields the tools that allow me to contrast different research studies on a small number of salient dimensions. For example, the microworld construct of realism has strong implications for generalization of simulator results to actual roadway design and operation. Having laid the groundwork for framing simulator research as the joint product of safety culture and microworld features, I will proceed to examine specific simulator efforts in three research arenas: highway design, placing constraints on driver behavior, and driver distraction/cognitive workload. Finally, to illustrate the effects of safety culture and microworld features on research, in a manner unfiltered by my own safety culture and microworld preferences, I have asked foreign researchers to summarize, in their own words, the goals and techniques of simulator research in their respective institutions.

3.2 European Safety Culture

This section contrasts the safety culture in the United States with that in Europe. This comparison may be unfair because I will assess a middle-view of United States practice against a cutting-edge view from Scandinavia. My goal is not only to set the stage for a discussion of simulator use, but also to advocate for the European culture, which I believe is more consistent with and appreciative of the principles of ergonomics than is the American culture. My opinions were shaped by the opportunity to act as Facilitator for an AASHTO-FHWA scanning tour (Kantowitz, 2005) of roadway human factors and behavioral safety in Europe where I first encountered these new concepts. This was a whirlwind tour of eight research institutions in six countries that was both fascinating and exciting.

Safety culture influences what kinds of simulator research is funded and, as suggested by a reviewer, this may indirectly influence what simulator capabilities are developed. For example, the U.S. has many more road signs than Europe which might encourage further development of hardware and software aimed at displaying highly legible signage. The European emphasis on social responsibility of drivers might produce more capability to combine more than a single driver within a simulator network. I have not evaluated this hypothesis since this chapter focuses upon empirical simulator research and not simulator development.

Roadway design and operation in the U.S. is governed by a military safety culture that accepts a certain number of fatalities and injuries as normal, albeit unfortunate. Forty thousand or so roadway deaths per year are a high price for the mobility demanded by society, but perhaps research and deployment of improved systems could lower this to only 20,000 or 30,000 deaths. Cost-benefit analysis is used to determine which improvements are paramount. Any new federal transportation safety initiatives in the U.S. are likely to focus upon how to decrease fatalities by X% by improvements in technology, such as intelligent transportation systems. Even on a municipal level, it may require a well-publicized traffic fatality before a new stop light or warning sign is added to the street system. Highway improvement funds are limited and so tend to be used to fix safety defects, discovered retroactively by crash rates, rather than to practively prevent incidents. As is the case with military success, surface transportation success must accept some collateral damage and loss of friendly troops as the everyday price of doing business. American transportation safety policies are designed to keep the number of fatalities and congestion from increasing as traffic vehicle-hours increase, or at best, when the sun is shining and the stars are in proper alignment, to yield modest decrements in fatality rates.

A decade ago the Swedish parliament passed a road safety bill, familiarly known as Vision Zero, which expressed the goal that no person would die or suffer grievous injury on a Swedish roadway. The bill was based upon the ethical principle that human life must be preserved, with safety more important than cost. In the Netherlands, SWOV has espoused a similar goal. Vision Zero implies that the best technical solution to preserve life must be implemented rather than the most cost-effective solution—and certainly not the solution of ignoring a safety problem until some threshold has been crossed.

In the United States, responsibility for safe operation of motor vehicles rests with the driver. If a driver on a two-lane rural road crosses the center line and hits an oncoming vehicle, the fault lies with the driver who made the error. The blameless victim who was operating the vehicle correctly in the opposite lane, or their survivors if the crash was intense, can try to remedy the fault by suing the errant driver in a civil court and the local police may issue a traffic citation. It is unlikely that legal action against the designer, the builder, or the operator of the roadway would be successful since responsibility for safe vehicle operation falls upon the driver.

Good ergonomic practice does not automatically blame the operator for errors. Instead, it assumes that errors will inevitably occur and tries to provide system designs that eliminate or mitigate the costs of driver error. For example, no driver can go through a stop light at a roundabout because it has no stop light. Furthermore, drivers who err and enter the roundabout inappropriately will crash at a small angle, which is far less damaging than the perpendicular collision that occurs when a stop light is ignored at an intersection. Vision Zero places a demand upon roadway designers and operators to create fault-tolerant infrastructure. A head-on collision between vehicles moving in opposite directions becomes impossible when a cable guardrail has been installed on the roadway. For example, the cable barrier in a European 2+1 roadway (see Federal Highway Administration [FHWA], 2005) not only prevents incursions into oncoming traffic but also protects the errant driver by absorbing the energy of a collision and trapping the vehicle so it does not bounce back into traffic.

Vision Zero is an active on-going commitment in Europe. A conference "Safe Highways of the Future" held in Brussels in February 2008 was devoted to the goal of eliminating, and not merely reducing, death on the roads of Europe. It included such topics as "safer vehicles by design" and "safer roads by design." The goal of the conference was: "Safe Highways of the Future will explore the technologies and challenges required to enable zero deaths on Europe's roads to become a reality."

Although there are nascent attempts to copy Vision Zero in North America, it seems clear to me that there is a substantial gap between the European safety culture and the American safety culture. Because the European culture is grounded in an ethical set of values that places human life above all other factors, substantial economic costs become acceptable to reach the goals of Vision Zero. By contrast, the American culture demands that some threshold value of utility be exceeded before highway improvement funds can be allocated. Vision Zero also implies a greater investment in simulator research on transport ergonomics. Since every avenue that has the potential to enable zero deaths needs be explored, the European safety culture pursues behavioral means to alter driver behavior through appropriate design and modification of both vehicle and highway technologies. For example, research discussed later in the chapter on in-vehicle speed governors is accepted and utilized in Europe whereas similar research in the U.S. is either not performed at all or would not be utilized.

This section has discussed how the safety culture of a region can affect simulator and other research by altering the probability that a given research topic is deemed worthy of investigation. But research is also heavily influenced by the local research culture and its view of simulator utility. The next section concerns virtual microworlds used to study driving. It explains how characteristics of microworlds relate to simulator utility. The section gives the background needed to support my hypothesis that decisions on how to employ driving simulators are formed jointly by the safety culture and the world-view of driving microworlds that are advocated in each research institution.

3.3 Driving Microworlds

Driving simulators have some characteristics of highly-controlled laboratory settings and some characteristics of less-controlled observational field studies. The driver can access any system state space provided by the simulated environment, including spaces off the roadway and spaces that intersect other vehicles. This freedom increases the generalizability of results, but at the cost of decreased experimental control relative to a typical experimental psychology laboratory setting where the operator is limited to pressing buttons when stimuli are illuminated. Anyone programming a driving simulator will have experienced the frustration of the driver not behaving exactly as anticipated, so that programmed triggers are not executed. For example, it can be difficult to establish a window such that a following vehicle under human control is preceded by a controlled simulated vehicle that suddenly brakes when a set of parameters (speed, headway, etc.) is satisfied; on some trials the lead vehicle may never brake. This outcome is similar to observational studies where thousands of trials may be required to observe some desired critical event. So using a driving simulator may be viewed as a kind of compromise whereby researchers navigate a narrow course between the Scylla of lack of experimental control found in field studies and the Charybdis of sterile well-controlled laboratory studies that are difficult to generalize.

Some years ago I was privileged to present a keynote address at the first international driving symposium in Aspen Colorado, which discussed how microworlds could be used in driving research (Kantowitz, 2001). Driving simulators are the prime example—but not the only example—of microworlds used in transportation research. Thus, it is worthwhile and instructive to restate some previously discussed features of microworlds that will introduce terminology to help compare different simulator applications.

Microworlds (Brehmer & Dörner, 1993) are computer-generated artificial environments that are complex (have a goal structure), dynamic (operate in real-time) and opaque (the operator must make inferences about the system). Thus, they can avoid results of limited generality—for example, some laboratory research on stimulus-response mappings is not helpful to human factors practitioners (Kantowitz, Triggs, & Barnes, 1990)—while maintaining a satisfactory degree of experimental control. Microworlds have been used to study topics such as process control (Howie & Vicente, 1998), extended spaceflight (Sauer, 2000), fighting forest fires (Omodei & Wearing, 1995), air traffic control (Bramwell, Bettin, & Kantowitz, 1989), stock market trading and internet shopping (DiFonzo, Hantula, & Bordia, 1998), and submarine warfare (Ehret, Gray, & Kirschenbaum, 2000).

Ehret et al. (2000) have identified three useful dimensions that allow researchers to compare microworlds and other simulated task environments: tractability, realism, and engagement. *Tractability* relates to how the researcher can effectively use the simulated environment. For example, Bramwell et al. (1989) used Seattle firefighters as subjects in their air traffic control microworld because the major aim of the study was to see how experienced teams with leaders of a known management style interacted. The goal was not to study air traffic control per se and the microworld created was an amalgam of enroute and terminal control. Two controllers (North and South) sat at different screens, which contained the same airport. In order for them to successfully route traffic, they had to coordinate their efforts because they could not see the other controller's screen. A team leader, who had no screen and thus was dependent upon the two controllers for information about traffic flow, was responsible for this coordination. This microworld exhibited tractability because it was simple enough for the controllers to learn both quickly and easily how to command their aircraft, but was still sufficiently complex to require teamwork and coordination. *Realism* refers to matching experience in the real and simulated worlds. For example, process-control microworlds should obey the same laws of thermodynamics as physical plants. *Engagement* refers to the willing suspension of disbelief on the part of experimental participants. Researchers want their subjects to act "naturally" and to produce the same behavior as in the real world (see, for example, chap. 9, "Psychological Fidelity" and chap. 41, "Adaptive

Behavior in the Simulator: Implications for Active Safety System Evaluation"). Engagement is a joint property of people and the simulated environment, and the same microworld could be engaging for one person but not for another. Professionals who are highly motivated and knowledgeable might accept an unrealistic simulation because they can fill in missing details. For example, professional airline pilots will cheerfully fly a lower-fidelity Link trainer. Professionals might also reject a realistic simulation if it conflicts with their philosophical world-view. Military pilots might not wish to participate in studies of side-stick controllers because they believe that such devices do not belong in airplanes. I vividly recall one commercial truck driver who stormed out of our simulator after a crash because he had never experienced an accident in his entire career. Did this mean he was too engaged?

These three concepts, tractability, realism, and engagement give us the vocabulary to compare microworlds with each other and with the real world. Driving simulators are used in research with varying levels of tractability, realism and engagement. The three concepts characterize the use of a particular simulator, not the simulator itself. The same simulator could be utilized for different experiments that varied across these three concepts.

I do not believe that concerns about tractability, realism, and engagement vary systematically according to safety culture. Instead, such variations occur on a more local level with each research institution selecting its own parameters for these dimensions. Thus, safety culture influences *what* is researched on a regional level, while *how* research is conducted is decided on a local level according to institutional judgments about driving microworlds.

Finally, to complete the picture I must embrace a fourth concept: validity. Validity is a familiar concept in research (see chapter 7 in Elmes, Kantowitz, & Roediger, 2006 for a discussion of different types of validity). In general, validity is related to the truth of observations and conclusions. Since no driving simulator can duplicate all the perceptual cues of the real world, it is important that the relevant cues be present with sufficient quality and quantity to allow generalization of results beyond the microworld. Successful microworlds need not require complete realism. Indeed, simulator researchers have known for some time that psychological fidelity is as important as physical fidelity (Kantowitz, 1988; Rankin, Bolton, Shikiar, & Sadri, 1984). The same driving simulator may be useful and valid for some applications while invalid for others. Thus, with each new application of a driving simulator, a new validation study should be conducted.*

A review of driving simulator validity (Kaptein, Theeuwes, & Van der Horst, 1996) identified two important kinds of validity:

---

* Since those who fund research are usually more interested in obtaining results, rather than validating them, I commend any researcher who has been able to conduct validation studies of their simulator. It would be interesting to perform a literature review that compared the number of articles about how to build simulators and what a specific simulator will be able to do once completed to the number of articles that validated existing simulators by comparing them to real vehicles and to other simulators: In a perfect world validation studies would outnumber simulator construction articles by an order of magnitude.

absolute and relative. A simulator with absolute validity produces results and effect sizes that are identical to the real world. (Actually, this is my definition; Kaptein et al. (1996), used the safer word "comparable" rather than identical, but I don't like hedging.) Relative validity means that treatment in the simulator produced the same rank order as in reality. They concluded that most simulator results produced relative validity, which is still worthwhile and useful.

Although this is a valuable distinction, I prefer to think of simulator validity more in terms of regression. How well does the simulator predict an outcome on the road? This allows for outcomes that are not absolute but are still better than relative validity. There is no need to settle for an ordinal scale when an interval scale might be achieved. Regression analysis also offers a metric that explains how well simulators predict reality so that different users can make their own judgments about the sufficiency of the fit for their own design purposes. Further elucidation of the regression approach to validation can be found in Kantowitz (2001).

This section has discussed features of driving microworlds: tractability, realism, engagement, and validity. The remainder of this chapter is devoted to specific examples of driving microworlds created outside of North America within a regional safety culture. The safety culture influences what is studied while the microworlds establish how it is studied. The first concrete example is highway design (pardon the pun).

3.4 Using Simulators for Highway Design

It seems obvious to researchers that simulators are ideal test beds for new highway design. Various roadway features such as signage, striping, and geometrics can easily be programmed in a simulator. Fixing a design error using a simulator microworld is relatively fast and quite inexpensive compared to repairing an error encapsulated in concrete. However, most highway designers in the U.S. are not familiar with driving simulation as a design tool. Indeed, with the exception of a very limited set of pioneering studies, such as those conducted jointly between state roadway designers and the simulator facility at the Federal Highway Administration Turner-Fairbanks Highway Research Center, the driving simulator has not been a viable tool for highway engineers in the U.S. Even those civil engineers who have been exposed to simulators often regard them as lacking in face validity when compared to observational field studies. The undergraduate training of civil engineers provides no mention of simulators and most graduates have had minimal exposure to ergonomics, although the advent of new technologies, such as intelligent transportation systems, makes it more likely that current undergraduates will be aware of ergonomics.

3.4.1 Designing Curved Roadways

In my opinion, driving simulators exhibit high tractability in this application. Indeed, it is possible to conduct studies that could never be realized due to time and cost limitations on concrete roadways. For example, one simulator study compared orthogonal combinations of curve radius and deflection angle (see Kantowitz, 2001) and obtained validated effects of both on curve entry speed. While many observational studies were familiar with the effects of radius, none had been sufficiently sensitive to find the effects of the deflection angle. Even if one could locate a sufficient set of radii and angles on concrete highways—and this would be unlikely—a meta-study combining observational data would probably lack the statistical power and sensitivity to detect differences due to the unavoidable presence of many confounding variables. Simulators also exhibit good to excellent realism and the accurate depiction of roadway features is well within the technical capabilities of driving simulators. While it can be financially expensive to provide high levels of realism for visual features of the roadway such as pavement markings and signage, the safety culture of Vision Zero permits such expenditures. Thus, the ability to provide high levels of realism in driving microworlds is far more a financial issue than a technical problem. Finally, simulators can provide high realism for the driving public (with the possible exception of some senior civil engineers in the United States); indeed, local authorities are starting to use 3D and 4D computer simulations to present new projects to the public. Thus, there are no technical reasons for not using simulators to design highways in the U.S. The remainder of this section reviews Scandinavian simulator studies of highway design that have been quite successful. I believe this success is related to the European safety culture and conversely, that the lack of general success in this application in the U.S., is due to its safety culture.

3.4.2 Designing Long Highway Tunnels

SINTEF (Foundation for Scientific and Industrial Research) in Trondheim, Norway, is the largest independent private research organization in Scandinavia. It has been a pioneer in using simulators to investigate drivers' perception of long road tunnels. Its first efforts in 2001 were in support of the Norwegian Laerdal tunnel, which is the world's longest road tunnel at 24.5 km (Figure 3.1).

The simulator was used to study four possible lighting models (Figure 3.2). The rock crystals lighting design was most successful in reducing driver anxiety and increasing driver comfort. The Laerdal project has won two European lighting awards.

More recently, SINTEF has been doing design research for the Quinling Shongan Tunnel in China and the world's longest subsea tunnel (Rogfast) in Norway (Flø & Jenssen, 2007). For the China tunnel, three design alternatives were compared in the simulator. Objective measures of driving performance showed no differences. However, subjective ratings showed a clear preference for one of the lighting designs. The China tunnel opened in January 2007. The E39 Rogfast tunnel, not yet built, is planned to be 25 km long and nearly 400 m deep. SINTEF research

FIGURE 3.1 The Laerdal tunnel project. (Courtesy of SINTEF [Foundation for Scientific and Industrial Research].)

showed that drivers have greater anxiety in sub-sea tunnels relative to land tunnels, so that artistic tunnel designs that remind drivers of the ocean should be avoided. SINTEF has also used their simulator to investigate driver response to a fire in the tunnel—clearly an activity more safely performed in the simulator than in the actual tunnel. Furthermore, comparisons of lateral position and speed revealed correlations between the simulator and road values (Giaever, 2006).

A more detailed validation of driving behavior in a tunnel and a simulator has been published in English using the VTI (Linkoping, Sweden) simulator (Tornros, 1998). Speed in the simulated tunnel was 8.6 km/h faster, a statistically significant result which was interpreted as a lack of absolute validity.

Lateral position in the real tunnel was 13 cm further from the side line than in the simulator on straight sections; interactions were found on curved sections. Test-re-test reliability was excellent in the real tunnel but only good (0.87 for speed and 0.78 for lateral position) in the simulator. The author concluded that the simulator provided good relative validity, but unsatisfactory absolute validity. My own interpretation of these data is even more positive. I suspect that if a regression analysis had been done to predict road speed and position from simulator speed and position, the results would be favorable. Because simulators lack the optical flow of real roads, it not surprising that drivers tend to drive faster in simulators. Furthermore, so long as the correction is robust with a high correlation, the lack of

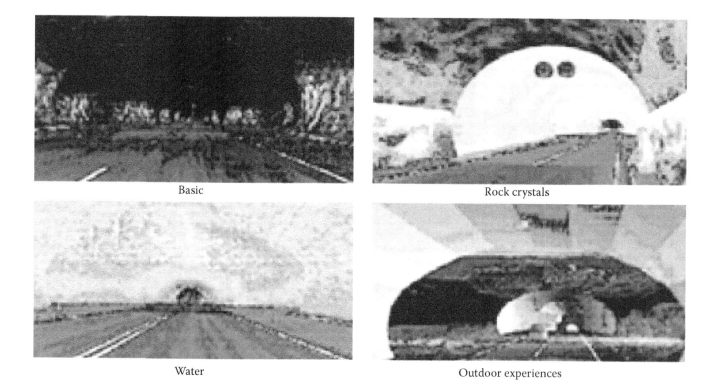

Basic

Rock crystals

Water

Outdoor experiences

FIGURE 3.2 Lighting models for the Laerdal tunnel. (Courtesy of SINTEF [Foundation for Scientific and Industrial Research].)

absolute validity is unimportant: *The critical goal is to predict road speed from simulator speed*, not to duplicate road speed in the simulator.

I believe these European studies clearly demonstrate that driving simulators are useful tools for designing long road tunnels. I think these types of studies would be desirable in the United States and, in fact, some are now being undertaken (Upchurch et al., 2002).

3.5 Using Simulators to Place Constraints on Driver Behavior

Drivers in the United States generally demand great freedom in exercising their indigenous rights to mobility. In some Western states boredom is relieved by using roadway signs for target practice. In congested areas, discourtesy and even road rage are not rare. Although it has been many years since the federal government unsuccessfully mandated the use of seatbelt buzzers to discourage drivers from not buckling their seat belts, the roaring tide of driver protest is remembered still by transportation professionals. The buzzers were disconnected by many drivers and eventually this legislation was repealed. American drivers take their independence seriously and strenuously resist government attempts to place constraints on their driving behavior.

Thus, because of this aspect of American safety culture, studies on how to constrain driver behaviors are uncommon. However, the safety culture in Europe is more consistent with the goal of constraining driver behavior to achieve a safety benefit for all drivers. This section discusses an example of such research on driver constraints: intelligent speed adaptation (ISA). ISA is a set of technologies that limits driver speed. In my opinion, driving simulators exhibit high tractability for testing different forms of ISA. It is easy to change how the technology is applied in a simulator, for example, advisory information versus compulsory speed limits. Realism is also high since the ISA technologies work the same way in the simulator and on the road; nevertheless, few would advocate applying simulator ISA results without validation studies on the roadway. Finally, engagement will also be high, so the driving simulator provides a useful microworld for testing ISA technology.

TNO (Netherlands Organization for Applied Scientific Research) has conducted a study of ISA using their driving simulator at Soesterberg (Rook & Hogema, 2005). Speed information was presented to the driver both visually by the speedometer and by tactile vibration from the gas pedal. The gas pedal in the dead throttle condition did not function above the set speed limit, that is, depressing the pedal would not increase vehicle speed above the limit. Three ISA conditions were investigated:

- ISA with a tactile gas pedal that warned of excessive speed;
- ISA with a dead throttle (including vibration) that prevented excessive speed; and
- No ISA control condition.

A set of driving scenarios that increased the realism of the experiment was used:

- Free driving—no sharp curve or lower limit segments;
- Sharp curves on a rural road;
- Car following, no overtaking possible;
- Car following, overtaking possible;
- Built-up area, same speed limit as for segments above;
- Car following, lead car half speed (35 km/h), overtaking possible; and
- Lower speed limit.

Results showed that the mean free-driving speed was reduced with ISA, with the dead throttle producing greater speed reduction than the tactile pedal. Reductions in the standard deviation of free-driving speed also followed this pattern. However, while this same pattern was observed for mean free-driving speed in straight segments in the sharp-curve scenario, there was no effect of ISA on speed for curved segments; drivers selected curve speeds below the ISA limit in all three ISA conditions. The tactile pedal was preferred to the dead throttle. Almost half (44%) of the 32 participants wanted this ISA in their own vehicle versus 25% who wanted the dead throttle.

These results will be used by the European Commission to help determine policy for ISA and implementation of road speed management methods. It is an interesting thought experiment to ponder how such results might be used by American policy makers.

3.6 Using Simulators to Study Driver Distraction and Cognitive Workload

The topics of cognitive workload and driver distraction have been studied extensively inside and outside North America. Kantowitz and Simsek (2000) have provided a review and methodological criticism of some of this work. In terms of tractability, in my opinion this topic rates high because it is easy to program all manner of workload and distraction in the driving simulator. I would rate realism very good, but not necessarily high, because of tendencies to program secondary tasks that are either very simple or offered as surrogates for actual tasks without performing any validation tests for the secondary tasks. Note that this difficulty can also accompany field operational tests because it is a property of the set of secondary/distraction tasks used, rather than a property of the driving simulator. Furthermore, the relative allocation of attention between the primary task of driving and both in-vehicle and out-of-vehicle secondary tasks can be suspect relative to a roadway environment because no driver has ever died in a simulator crash. Careful use of single-task control conditions is required in order to detect possible allocation difficulties (Kantowitz & Simsek, 2000). Thus engagement can range from very good to high, depending on the contingencies associated with errors on the primary driving task. The driving simulator is a reasonable microworld for studying distraction and cognitive workload, provided validation studies on the road are also executed. The most valuable use of the driving simulator in this domain is to efficiently decrease the number of possibilities that need be examined on the road where research takes more time than simulator studies and is also far more expensive. Simulator microworlds excel at efficiently evaluating and comparing sets of in-vehicle tasks and formats that may provide different levels of cognitive workload.

3.6.1 Secondary Task Selection

To illustrate this topic I have selected one of the best recent simulator experiments I could find (Ostlund, Nilsson, Tornros, & Forsman, 2006). The goal of this experiment was to evaluate and compare effects of cognitive and visual load in simulated and real driving. The first issue these researchers faced was selecting appropriate in-vehicle secondary tasks: Should a real in-vehicle information system (IVIS) or a surrogate system (S-IVIS) be used?

The researchers carefully listed the advantages and disadvantages of the surrogate task. Using S-IVIS, one finds:

1. The surrogate task allows pure insertion of a specific mental activity;
2. The level of task demand can be manipulated systematically;
3. The task is new so that subjects have no prior learning; and
4. The task is readily analyzed having been selected to produce quantitative data.

Using IVIS, one finds:

1. The ecological validity is high by definition;
2. An embedded secondary task makes strategic sense to the driver; and
3. Allocation of attention is more realistic.

The VTI researchers and their HASTE (Human machine interface And the Safety of Traffic in Europe) colleagues chose S-IVIS. I agree with this research strategy for several reasons. First, I think that ecological validity can be overrated. It is far more important that the underlying psychological processes be valid than that the situation appear to be valid. Let me illustrate this with an ancient example from animal learning (Sidman, 1960). Imagine that we wanted to understand why people hoard resources. We could train monkeys to store gold coins in their bedding. A naïve observer would claim high ecological validity: The monkeys are behaving just like people who hide money in mattresses. But the underlying reinforcement contingencies (e.g., psychological processes) would be vastly different between monkey and human hoarders. The apparent high ecological validity is misleading in this situation.

Second, an embedded task is defined as a task already present within a larger system. But embedded tasks are only embedded when operators have prior experience with them. Any completely new IVIS—think of the first time you used an in-vehicle GPS navigation system—is as arbitrary as an S-IVIS to the subject.

Third, allocation of attention in a microworld is complex and not fully controlled by the selection of a real IVIS. Since an S-IVIS is almost always less complicated to operate than an IVIS, it may be easier for a driver to control allocation of attention with the S-IVIS.

Fourth, ergonomics as a discipline makes better progress when it can relate data to models (e.g., Kantowitz, 1998). An S-IVIS, especially when selected to relate to some specific model as was done by the VTI researchers, offers a much greater opportunity to improve our models. By contrast, the IVIS impacts the driver (and the model) in several confounded ways, making it difficult to explain why one IVIS yields better results so that an even better next generation IVIS can be designed and implemented.

3.6.2 Task Description

The visual task was a search task requiring the detection of an arrow pointing up in an array of arrows. Subjects responded by pressing "Yes" or "No" on a touch screen. Both accuracy and latency of response was reported. If I had been analyzing these data, I probably would not have resisted the temptation to calculate signal detection parameters such as d'; indeed d'^2/RT can be a useful index similar to transmitted information rate (bits/sec). However, the more important outcome was that the researchers correctly presented results for the visual task performed in isolation and also in concert with the driving task.

The cognitive task demanded short-term memory and required subjects to count aloud how many times specific target sounds were played in an array of 15 sounds. Task difficulty was manipulated by varying the number of target sounds to count: two, three or four. Correct and false responses were measured.

A variety of routes, events and road complexities were programmed in the driving simulator. I will not bother to summarize these other than to note that appropriate combinations of single- and dual-task conditions were examined. Many measures of driving performance were studied: one subjective report, seven measures of lateral performance, three physiological measures of workload, and twelve measures of longitudinal performance. While the ability to record multiple measures of driver performance easily is an advantage of simulators, technically doing many comparisons using traditional analysis of variance (ANOVA) may inflate the possibility of Type I statistical errors. If I had been analyzing these data I probably would also have included a multifactor analysis of variance (MANOVA), as recommended by textbooks when multiple dependent variables are analyzed (e.g., Kantowitz, Roediger, & Elmes, 2008).

3.6.3 Results

Results, of course, are the most important outcome of experiments. This experiment reports all manner of outcomes for driving performance with and without secondary tasks and secondary-task performance with and without driving are presented in substantial detail. Again, I will not attempt to summarize all these results. The visual task had a greater influence on driving performance than the cognitive task and effects were greater on simulated rural roads than on simulated motorways. Despite the artificiality of the S-IVIS tasks, there was no evidence that the secondary task was abandoned so that greater attention could be devoted to the primary driving task.

3.6.4 Field Validation

Finally, an instrumented vehicle on-road study was conducted using the S-IVIS tasks, except that the visual task was simplified to minimize crash risk; the cognitive task remained identical to that used in the simulator study. Since there were substantial differences between the simulator and road studies (e.g., different subjects, different scenarios, and changes to the visual task) the on-road study is at best a partial validation of the simulator study. However, given that most simulator studies offer no confirming on-road follow-up studies, I am favorably impressed by the partial validation. Simple effect estimates were used to compare results relative to the baselines. Results were complex and I found it particularly disappointing that the visual task was much better performed on the road because the task had been simplified for safety reasons; thus, the partial validation failed due to the different forms of the visual task. However, the cognitive task yielded reduced speed control, similar to results obtained in the simulator. In general, speed was reduced on the road versus in the simulator, a common outcome discussed earlier in this chapter. The authors concluded that simulator validity was high for tactical and operational levels but not for strategic levels. My own speculation is that a regression analysis of validity (discussed earlier in this chapter) might have revealed favorable validities for all three levels.

In summary, this research is an excellent example of how driving simulator studies should be conducted. It related the research plan and results to a model of driving behavior. It carefully included necessary single-task baseline conditions for both primary and secondary tasks. Finally, it attempted a partial validation using an instrumented vehicle.

The preceding sections have discussed the results of simulator experiments conducted in three research areas: highway design, driver constraints, and driver workload. While it is possible to induce some general properties of simulator use outside of North America from these examples, it seems more effective to ask foreign researchers to explain, in their own words, how simulator research is best accomplished. I refer to the goals, constraints, and strategies controlling simulator utilization as simulator philosophy.

3.7 Simulator Philosophy in Australia and Europe

Simulator philosophy refers to the frame of mind of the researchers and administrators responsible for providing resources and executing research using driving simulators. It is, of course, strongly shaped by the safety culture of a region and the microworld view of each research institute. Had I been writing this chapter 15 years ago I would have ignored the role of administrators, since it is well-known, especially among younger researchers, that administration is a function that appeals primarily to aging researchers with diminishing technical skills. However, after serving in two major administrative roles as Director of the University of Michigan Transportation Research Institute and Director of the Battelle Memorial Institute's Human Factors Transportation Research Center, I have come to appreciate how administrative decisions can facilitate or retard simulator research. Accordingly, I asked colleagues using simulators outside of North America to provide me with written statements about their institutional philosophies guiding simulator use. Since few people have such written statements at hand my request was a major inconvenience and I am extremely grateful to the three institutions that most graciously responded: Monash University Accident Research Centre (MUARC), VTI and TNO.

3.7.1 The TNO Driving Simulator as One of the Tools for Conducting Human Factors Research in Road Traffic

(This section was written by Richard Van der Horst of TNO Defense, Security and Safety in the Netherlands.)

3.7.1.1 Introduction

The successful introduction of advanced driver-support systems, dynamic traffic management systems, or complex road designs depends strongly on how capable are the road users or how willing the road users are to deal with these changes. This requires knowledge of people's performance and behavior in complex and dynamic task environments. Especially, in the design and development of Intelligent Transportation Systems (ITS), the end user of such systems should be taken into account. For many ITS systems, this is the driver of a car or truck. In all stages of ITS development, (from conceptual design all the way through large-scale implementation), empirical research is needed to investigate issues such as driving behavior, human-machine interaction, workload and acceptance. Human Factors research offers a wide range of methods, each having their own strengths and weaknesses (Alferdinck, Hogema, Theeuwes, & Van der Horst, 2002). These methods range from mathematical models and lab experiments on the one hand to observations in real situations, such as accident analysis, video observations, research with instrumented vehicles, and Field Operational Tests on the other. The choice for the most suitable method involves a trade-off between similarity to the real world on the one hand, and flexibility and interpretability of experiments on the other. Typically, interactive driving simulators are positioned about "in the middle" on both aspects (Alferdinck et al., 2002).

Driving simulators have the advantage of having control over experimental conditions. In a field study, rainy weather and traffic jams can arbitrarily occur and consequently interfere with an experiment. In a driving simulator, the presence and behavior of surrounding traffic can be controlled. Vehicles can be programmed to brake at a specific time or traffic situation, and traffic can cut-in. Furthermore, a simulator allows easy recording of many behavioral measures. These include performance indicators (speed, time-to-collision, etc.), visual attention (number of glances, etc.), workload (peripheral-detection-task, subjective measures, etc.), driving comfort (e.g., lateral acceleration, etc.) and driver acceptance (questionnaires).

Simulators allow prototype systems (conceptual ITS designs) to be evaluated with human drivers, in an early stage. Similarly, driving simulators enable a dynamical assessment of new infrastructural designs such as tunnels, analyzing road-user behavior in relation to their traffic-management measures, and in interaction with other traffic (Van der Horst, Hoekstra, & Theeuwes, 1995). Sometimes, safety is an issue when choosing whether to undertake a simulator study. Consider, for example, the state of a driver due to a behavioral change caused by drowsiness, alcohol or diabetes (Stork, 2006).

3.7.1.2 Examples of Applications

We conduct a wide variety of driving simulator studies ranging from a focus on fundamental perceptual and cognitive research questions (e.g., Alferdinck, 2006; Martens, 2005, 2007; Hoedemaeker, Hogema, & Pauwelussen, 2006) and the assessment and evaluation of complex road designs and traffic safety measures (e.g., Van der Horst & De Ridder, 2007) to the design, development and testing of various kinds of in-car driver-support systems and related HMI issues (e.g., Hogema & Rook, 2004; Rook & Hogema, 2005; Van Driel, Hoedemaeker & Van Arem, 2007).

3.7.1.3 Trends in Driving Simulation Research

A recent trend is to combine several driving simulators, allowing drivers to interact in a single, shared, virtual environment. In such a configuration, not only are driving simulators coupled, but other traffic and transport-related research tools are coupled as well. The TNO research program SUMMITS combined many of these tools in the so-called SUMMITS Tool Suite (Driessen et al., 2007). In a recent demonstration of Tool Suite elements, a coupled simulation of the VW fixed-base and the BMW simulator has been demonstrated using HLA (High-Level Architecture) techniques. HLA is the de facto standard for distributed simulation. This IEEE standard defines both the services provided by the HLA interoperability layer and the structured development process that needs to be followed for each simulator or simulation joining a shared simulation.

Another important trend is the shared use of vehicle (sub-) models and Advanced Driver Assistance Systems (ADAS) models via Matlab-Simulink. This environment is suitable for developing, testing and tuning models. Subsequently, these models can be applied in various ways in driver-in-the-loop simulators. For example, a detailed steering system sub-model can easily be incorporated in the BMW driving simulator (Hogema, Hoekstra, & Verschuren, 2003). An ADAS-like Intelligent Speed Adaptation (ISA) can be developed, tuned and tested offline, and then the same ADAS model can be used directly in the driving simulator, as has been done in the EU project PROSPER. This project studied HMI effects of various types of ISA on driving behavior and acceptance (Hogema & Rook, 2004, 2005). In a similar way,

vehicle (sub-) models are being shared by the BMW moving base simulator and the driving simulator on Desdemona.

3.7.1.4 Conclusions

The driving simulation facilities of TNO cover a wide range, from low-cost, fixed-base, (via much more immersive moving base) to high-end motion systems. As with any research method, they all have specific advantages for a certain research question, and it is up to the researcher to select or develop a method that is most suitable for the research question at hand. Each experiment will have its own requirements, with specific needs for different levels of the driving task. Studies that focus on effects at the control level generally need motion cueing. For relatively straight highway driving, a hexapod platform will do (e.g., for studying subtle effects of driving-support systems such as an ACC). For control tasks that rely on linear acceleration cues, such as curve driving, and hard accelerations or decelerations, a more sophisticated motion platform seems to be required (to answer questions about comfort, appreciation of acceleration profiles, and driver behavior in more extreme motions). Research questions at the behavioral level—for instance, to find out whether or not a system is used—could be performed in a fixed-base driving simulator. A specific type of research question that requires, for example, the interaction between drivers in a common virtual world, could be answered by coupled driving simulators. Furthermore, sharing sub-models between traffic tools may increase the efficiency in research tasks that cover multiple domains (e.g., human factors and road capacity).

Systematic control over the experimental conditions with respect to road design elements, traffic management, other traffic, environmental conditions, and so forth,. makes human factors research in a driving simulator attractive, efficient, and effective. However, empirical research in real-world driving remains necessary, be it for validating, calibrating, or verification of driving simulator research results. The issue of driving simulator validity continues to be as important today as it was ten years ago (Kaptein et al. 1996).

3.7.2 MUARC Philosophy for Simulator Use

(*This section was written by Brian Fildes and Michael Regan who have been intimately involved with the driving simulator research at the Monash University Accident Research Center in Australia.*)

MUARC's mid-range driving simulator was custom-built in the mid-1990s, by a local Melbourne-based company, to specifications developed by MUARC. It is the most advanced driving simulator in the Southern Hemisphere and was designed to support and stimulate academic and commercial research on road safety topics.

The simulator was initially built to support commercial research on young driver performance and safety (see below), although it is also available for use by students as part of their graduate studies programs. As MUARC is predominantly a self-funded research center, the amount of resources available for ongoing use and maintenance of the simulator is minimal. Thus, the Center has to find the funds to sustain the day-to-day operations of the facility. There is always a degree of competitiveness about who gets access to the simulator, and when.

The mid-range simulator has been used for a variety of purposes—for prototyping, designing and evaluating in-vehicle driver assistance systems and road infrastructure treatments, for designing and evaluating training programs, and for basic research on driving performance and fitness to drive (e.g., when distracted or intoxicated). It is used almost exclusively for applied research, for clients who require a high degree of confidence in the reliability and validity of research outcomes. A lower-fidelity simulator is also owned and operated by MUARC. It is used by students who require a less expensive and more accessible research platform for undertaking both basic and applied research.

The mid-range simulator, which has a partial motion platform, induces simulator discomfort, and the design of scenarios is always a trade-off between good experimental design and minimization of motion discomfort. All users are advised of the dangers of motion discomfort, and are screened for their susceptibility to its effects and must indemnify the university in writing prior to use of the facility in case of any adverse side-effects.

To date, the mid-range driving simulator has been managed by MUARC researchers. However, the Center employs two technicians to maintain the facility and program experimental scenarios. A managed booking system is used to regulate competition for its use. As the simulator is in great demand, there needs to be a rational system adopted to ensure its ongoing availability, which means striking a balance between student and commercial research. Moreover, with increased demand for its use, the need to adopt a more commercial approach to its use is apparent to ensure its survival. Hence, there have been calls recently for it to be managed more as a commercial operation, but with some guaranteed access for students.

3.7.3 VTI Guiding Philosophy of Simulation

(*This section was written by Lena Nilsson, Staffan Nordmark and Jonas Jansson, who describe the simulator at VTI in Sweden.*)

The factor of success for VTI within the simulator area is the close cooperation between researchers from technical, behavioral, and human factors sciences. The technical development of the simulators has always been driven by the requirements, issues and problems raised in the applied research projects. The people around the simulators and their competence are key factors, and much more important than the tool itself.

The simulated virtual environment is an illusion. The real traffic system exists out there, and should not be fully recreated in simulators for its own sake. In our view, the reasons (and benefits) for using simulators are mainly:

* *Human in the loop: Real drivers* are needed to cover the complete dynamic task of driving. Human (driver) behavior cannot be modeled well/accurately enough to be simulated.

* *Controllability*: A controlled experimental environment and controlled conditions and prerequisites are critical. Variable(s) of interest can be varied systematically while keeping all other factors constant.

* *Comparative studies*: Identical scenarios can be used for all participants (drivers). Fewer participants are required for statistical power in results.

* *Conceptual solutions*: Quick comparisons can be made between different alternatives and solutions, including conceptual approaches, i.e., before expensive physical in-vehicle prototypes or road constructions have been implemented. The relative impact and effectiveness of various system solutions can be determined. Road layouts like lining, infrastructure, tunnels, curvature, and black spot measures can be studied. Simulation is useful as a planning (projecting) and development tool.

* *Dangerous situations and conditions*: Experiments which are not possible to conduct in real life/traffic, or even on closed test tracks (risks, critical situations, driver state, environmental/weather conditions, ethics), can be undertaken. Driver behavior under demanding conditions can be investigated to the point of criticality and even crashing or driving off the road can be studied. Actions or lack of actions preceding critical situations and crashes (In what situations do drivers need support? What type of support? How should it be designed?) can be analyzed. The effects of fatigue, alcohol, and drugs can be identified.

* *Accelerated testing*: In some cases, where track tests or driving in traffic with a real vehicle is an option, a simulator study may provide a more cost-effective testing method and/or shorter lead time. However, it should be stressed that not all types of testing are suitable for simulator studies.

Validation is important, including measuring road characteristics (Road Surface Tester—RST-vehicle), lateral position, occurrence of other traffic, noise and vibrations, and implementing these in the simulator. All parts in the simulator must be realistic and perform as intended. This means that the vehicle model used must be validated as an open loop digital program against standard field test maneuvers. This validation is normally performed in close cooperation with automotive companies that alternatively can supply validated models directly like CarSim or similar. All input data to the simulator must be correct as far as is possible and beside the pure vehicle data this also encompasses road unevenness and friction properties between the tires and road. VTI has highly-specialized equipment for that data collection (tire testing equipment, road surface testing).

Finally, all these data are gathered together in the simulator with test subjects and a validation process starts comparing the performance of the test subjects in the simulator with the performance in real traffic. This validation process is not carried out for every simulator experiment but the validation experiments actually performed are extensive enough that further validation is not felt to be necessary and similar scenarios may be performed and evaluated.

Using the VTI driving simulator has taught us some valuable lessons, namely:

- The critical selection of applications, and the courage to renounce unsuitable ones (e.g., dynamic route guidance studies requiring many complex junctions and other choice points, and studies in high contrast) are the basis for useful results.
- Critical selection and evaluation help find measures (indicators) that really reflect aspects with implications for traffic safety, driving performance, driver behavior and acceptance, usability and so forth.
- Confounding effects that result from participants' unfamiliarity with simulator driving, vehicle characteristics, and the specific experimental tasks, can be avoided (or at least minimized) by letting the subjects practice in the simulator.
- Unrealistic driver behavior (playing with the simulator), because the subjects are not exposed to any real danger, can be prevented by carefully instructing and retraining them to drive as they usually would do under comparable conditions in real traffic.
- Interpretation can be facilitated by not covering "all" aspects in an extensive study but by performing many "small" studies and focusing on one specific aspect in each study, thus allowing researchers to benefit from keeping the conditions constant between studies and, in the end, to connect together the pieces of what is a large and complex puzzle.

3.7.4 Summary of Simulator Philosophies

It is not surprising that researchers from the three institutions discussed above have independently touched upon some of the same topics. This section reviews a few key topics.

First is the issue of research administration. Unless institutional administrators provide simulator resources, there will be no simulator research staff. Although equipment costs for simulators have decreased greatly as prices for computing power—and especially display hardware—have diminished, the commitment to build and maintain a mid-range (or better) driving simulator is substantial with recurring maintenance and staffing costs. For example, at VTI there was much discussion before deciding that a major simulator facility would be a vital long-term asset. As the comments from MUARC reveal, simulators housed in universities need to balance the educational demands of serving students and faculty against the needs of a client base that pays for simulator support. Since universities recover much lower overhead costs than do commercial or non-profit independent research institutes, there is little financial slack to cover shared simulator costs.

Administrators must also consider whether their institution is better served by a single multi-purpose simulator or a range of simulators with different capabilities and costs. While some research, such as vehicle dynamics, demands an expensive full-mission moving-base simulator, other research problems are amenable to lesser simulators. For example, SINTEF has been very successful with a single fixed-base simulator applied to a narrow range of problems. On the other hand, TNO offers clients several platforms that can be matched to project requirements.

All institutions acknowledge the importance of validation. Each new application demands a new validation. While validation is not required for every single simulator experiment, each class of experiment calls for concurrent on-road data collected under conditions that match the simulator microworld as closely as can be managed. While the constructs of absolute and relative validity recommended over a decade ago by TNO are a good starting point, more research is needed to examine other quantitative metrics of validity. How should we measure the ability of the simulator microworld to predict on-road behavior?

Finally, it is my opinion that the kinds of simulator research that have been performed outside North America reflect the safety cultures of their parent countries. I am not surprised that this was not explicitly mentioned above by my colleagues. Just as a fish tends not to notice the water that surrounds it, the quotidian research efforts that occur tend to focus on lower-level technical issues. Laboratories in North America operate the same way. Most working researchers do not reflect on cultural limits of research on a daily basis. Yet I hope this chapter will increase discourse on the importance of safety culture in defining critical research problems.

3.8 Conclusions

While simulators, as machines, are much the same the world over, the ways in which they are utilized vary. The safety culture of a country determines what problems are examined using driving simulators. The local research institution decides what problems to investigate based on the explicit or implicit safety culture. Roadway and tunnel design is seldom attempted in North America using simulators, while European researchers are more likely to explore this topic. Similarly, using simulators to investigate how to constrain drivers is more likely outside of North America and rare within North America. However, driver distraction issues fall within a world-wide shared definition of important safety problems.

There are many ways to design and conduct research using driving simulators. Such technical decisions are made at a local institutional level based upon the training and experience of the research and support staff. Thus, while the regional safety culture determines what issues are worthwhile, the local researchers find out how their simulator can be most effective. This judgment depends directly upon the simulator resources available to the research staff, and all simulator users, both within and without North America, must solve the administrative issues involved in allocating resources to a simulator facility.

Key Points

- Simulator research is the joint product of safety culture and driving microworld features.
    - Regional safety culture determines what issues are researched and what issues are ignored.
    - Local institutional preferences for driving microworlds determine how simulator research is conducted.
    - Driving microworlds vary in tractability, realism, engagement, and validity.
- Driving simulators are excellent test beds for highway design.
- Driving simulators are useful for examining constraints on driver behavior.
    - Driving simulators are useful for studying driver distraction and workload.
    - Simulator philosophies vary across countries and institutions.

Keywords: Engagement, Realism, Safety Culture, Tractability, Validity, Vision Zero

Glossary

Driver workload: An intervening variable, similar to attention, that modulates the tuning between the capabilities of the driver and the demands of the vehicle-roadway environment.

Driving microworld: A virtual computer-generated environment that is complex, dynamic and opaque, and is used to study driving.

Safety culture: A framework that sets boundaries for the goals of safety research.

Simulator philosophy: The framework of values used by researchers and administrators responsible for providing resources and executing research using driving simulators.

Vision zero: A road safety bill passed by the Swedish parliament which expressed the goal that no person would die or suffer grievous injury on a Swedish roadway.

Key Readings

Brehmer, B., & Dörner, D. (1993). Experiments with computer-simulated microworlds: Escaping both the narrow straits of the laboratory and the deep blue sea of field study. *Computers in Human Behavior, 9,* 171–184.

Federal Highway Administration. (2005). *Roadway human factors and behavioral safety in Europe.* (Report FHWA-PL-05-005). Retrieved from http://www.international.fhwa.dot.gov

Kantowitz, B. H. (2001). Using microworlds to design intelligent interfaces that minimize driver distraction. *Proceedings of the first international driving symposium on human factors in driver assessment, training and vehicle design* (pp. 42–57). Aspen, CO.

Kaptein, N. A., Theeuwes, J., & Van der Horst, A. R. A. (1996). Driving simulator validity: Some considerations. *Transportation Research Record, 1550,* 30–36.

References

AAA Foundation for Traffic Safety. (2007). *Traffic safety culture in the United States: The journey forward.* Washington, DC: AAA Foundation for Traffic Safety.

Alferdinck, J. W. A. M. (2006). Target detection and driving behavior measurements in a driving simulator at mesopic light levels. *Ophthalmic and Physiological Optics, 26,* 264–280.

Alferdinck J. W. A. M., Hogema J. H., Theeuwes J., & Van der Horst A. R. A. (2002). *Methods for measuring perception and driving behavior in darkness and fog.* Proceedings of the First International Congress of Vehicle & Infrastructure Safety Improvement in Adverse Conditions and Night Driving (VISION). Suresnes, France: Société des Ingénieurs de l'Automobile SIA.

Bramwell, A., Bettin, P., & Kantowitz, B. H. (1989). *The effect of leadership style on team performance of a simulated air traffic control task.* Paper presented at the annual meeting of the Western Psychological Association. San Francisco, CA.

Brehmer, B., & Dörner, D. (1993). Experiments with computer-simulated microworlds: Escaping both the narrow straits of the laboratory and the deep blue sea of field study. *Computers in Human Behavior, 9,* 171–184.

DiFonzo, N., Hantula, D. A., & Bordia, P. (1998). Microworlds for experimental research: Having your (control and collection) cake and realism too. *Behavior Research Methods, Instruments, & Computers, 30*(2), 278–286.

Driessen, B., Hogema, J., Wilmink, I., Ploeg, J., Papp, Z., & Feenstra, P. (2007). *The SUMMITS tool suite: Supporting the development and evaluation of cooperative vehicle–infrastructure systems in a multi–aspect assessment approach.* (TNO memo 073401–N17). Delft: TNO Traffic and Transport.

Ehret, B. D., Gray, W. D., & Kirschbaum, S. S. (2000). Contending with complexity: Developing and using a scaled world in applied cognitive research. *Human Factors, 42*(1), 8–23.

Elmes, D. G., Kantowitz, B. H., & Roediger, H. L. (2006). *Research methods in psychology* (8th ed.). Belmont, CA: Thomson Learning.

Federal Highway Administration. (2005). *Roadway human factors and behavioral safety in Europe.* (Report FHWA-PL-05-005). Available from http://www.international.fhwa.dot.gov

Flø, M., & Jenssen, G. (2007). *Drivers' perception of long tunnels.* Fourth International Conference on Traffic and Road Safety in Tunnels. Hamburg, Germany.

Hoedemaeker, D. M., & De Ridder, S. N. (2003). *The Dutch experience with LDWA systems.* (TNO Report TM-03-C020). Soesterberg, the Netherlands: TNO Human Factors.

Hoedemaeker, D. M., Hogema, J. H., & Pauwelussen, J. (2006). *Rijsimulatorstudie naar het effect van omgevingscomplexiteit op de werkbelasting* [A driving simulator study on the effect of motorway environment characteristics on work-load]. (Report TNO-DV 2006-C 244). Soesterberg, the Netherlands: TNO Human Factors (in Dutch).

Hogema, J. H., Hoekstra, W., & Verschuren, R. M. A. F. (2003). *A driving simulator vehicle model with an extended steering system sub-model.* (TNO Report TM-03-D004). Soesterberg, the Netherlands: TNO Human Factors.

Hogema, J. H., & Rook A. M. (2004). *Intelligent Speed Adaptation: The effects of an active gas pedal on driver behavior and acceptance.* (TNO Report TM-04-D011). Soesterberg, the Netherlands: TNO Human Factors.

Howie, D. E., & Vincente, K. J. (1998). Measures of operator performance in complex, dynamic microworlds: Advancing the state of the art. *Ergonomics, 41*(4), 485–500.

Kantowitz, B. H. (1988). Laboratory simulation of maintenance activity. *Proceedings of the 1988 IEEE 4th conference on human factors and nuclear power plants* (pp. 403–409). New York.

Kantowitz, B. H. (1998). Computational models for transportation human factors. *Proceedings of the human factors and ergonomics society 42nd annual meeting* (pp. 1220–1221).

Kantowitz, B. H. (2001). Using microworlds to design intelligent interfaces that minimize driver distraction. *Proceedings of the first international driving symposium on human factors in driver assessment, training and vehicle design* (pp. 42–57). Aspen, CO.

Kantowitz, B. H. (2005). Best European highway safety practices: The AASHTO-FHWA human factors scan tour. *Proceedings of the human factors and ergonomics society, 49,* 1970–1974.

Kantowitz, B. H., Roediger, H. L., & Elmes, D. (2008). *Experimental psychology* (9th ed.). Belmont, CA: Cengage.

Kantowitz, B. H., & Simsek, O. (2000). Secondary-task measures of driver workload. In P. Hancock & P. Desmond (Eds.), *Stress, workload, and fatigue* (pp. 395–408). Mahwah, NJ: Erlbaum.

Kantowitz, B. H., Triggs, T. J., & Barnes, V. (1990). Stimulus-response compatibility and human factors. In R. W. Proctor & T. Reeves (Eds.), *Stimulus-response compatibility* (pp. 365–388). Amsterdam: North-Holland.

Kaptein, N. A., Theeuwes, J., & Van der Horst, A. R. A. (1996). Driving simulator validity: Some considerations. *Transportation Research Record 1550,* 30–36.

Martens, M. H. (2005). *Responding to changes in the traffic situation: Does experience make a difference?* (TNO Report DV3 2005 D018). Soesterberg, the Netherlands: TNO Human Factors.

Martens, M. H. (2007). *The failure to act upon important information: Where do things go wrong?* (Doctoral thesis). Amsterdam: Vrije Universiteit.

Omodei, M. M., & Wearing, A. J. (1995). The Fire Chief micro-world generating program: An illustration of computer-simulated microworlds as an experimental paradigm for studying complex decision-making behavior. *Behavior Research Methods, Instruments, & Computers, 27,* 303–316.

Ostlund, J., Nilsson, L., Tornros, J., & Forsman, A. (2006). *Effects of cognitive and visual load in real and simulated driving.* (VTI Report 533A). Lindkoping, Sweden: Swedish Road and Transport Research Institute (VTI).

Rankin, W. L, Bolton, P. A., Shikiar, R., & Sadri, L. M. (1984). *Nuclear power plant simulators for operator licensing and training.* (Report NUREG/CR- 3725, PNL-5049). Washington, DC: US Nuclear Regulatory Commission.

Rook, A. M., & Hogema, J. H. (2005). *Intelligent speed adaptation: The effects of a tactile gas pedal and a dead throttle on driver behavior and acceptance.* (TNO Report DV3 2005 D010). Soesterberg, the Netherlands: TNO Human Factors.

Sauer, J. (2000). The use of micro-worlds for human factors research in extended spaceflight. *Acta Astronautica, 46*(1), 37–45.

Sidman, M. (1960). *Tactics of scientific research.* New York: Basic Books.

Stork, A. (2006). *Diabetes and driving; performance decision making and legal aspects,* Doctoral dissertation. Utrecht, the Netherlands: Utrecht University.

Tornros, J. (1998). Driving behavior in a real and a simulated road tunnel—A validation study. *Accident Analysis & Prevention, 30,* 497–503.

Upchurch, J., Fisher, D. L., Carpenter, C., & Dutta, A. (2002). Freeway guide sign design with driving simulator for central artery-tunnel: Boston. *Transportation Research Record: Traffic Control Devices, Visibility, and Rail-Highway Grade Crossings, 1801,* 9–17.

Van der Horst, A. R. A., & De Ridder, S. (2007). *The influence of roadside infrastructure on driving behavior: A driving simulator study.* Transportation Research Board 86th Annual Meeting (CD-ROM) (Paper 07-1434). Washington, DC: Transportation Research Board.

Van der Horst A. R. A., Hoekstra, W., & Theeuwes, J. (1995). Visualization of the Ekeberg Tunnel for assessing human factors aspects in the design phase. In *Pioneers in 3D visualization and simulation for transportation. 1993 and 1994 Transportation Research Board meetings.* Washington, DC: Transportation Research Board.

Van Driel, C. J. G., Hoedemaeker, M., & Van Arem, B. (2007). Impacts of a congestion assistant on driving behavior and acceptance using a driving simulator. *Transportation Research Part F: Traffic Psychology and Behavior, 10*(2), 139–152.

4

The Future of Driving Simulation

Peter A. Hancock
University of Central Florida

Thomas B. Sheridan
*Massachusetts Institute
of Technology*

Abstract

The Past to Present. The simulation of the driving experience has been used for both research and most especially for driver training. In large part, the state of the art in driving simulation has been contingent upon wider developments in simulation science. The challenge of ground-vehicle simulation provides stiffer challenges than simulation of airborne vehicles. Current advances have seen high-fidelity, multi-million-dollar facilities. The advantage is that they provide capacities now coming very close to the Turing test for simulated reality. The disadvantage is that they are so expensive as to be almost unique and so no replicable science is conducted on them. *The Present to the Near Future.* Simulation not only improves with the technical capacities of the age, it also diversifies. Thus, in modern simulations there are options associated with game-playing, full virtual environments, and augmented forms of reality as well as improvements on the traditional fixed and motion-based facilities. We anticipate that such branches of development will further diversify as new and innovative methods of rendering surrogate surroundings continue to proliferate. *Worlds to Come.* The fundamental function of simulation is to augment current reality with programmable objects or entities or to replace the whole environments with a surrogate experience. However, our whole world of experience is represented in the brain. Thus, all external technologies only serve to generate a pattern of brain stimulation. Our further future is thus headed toward direct brain stimulation. External facilities of the sort we see today at the most advanced facilities will be replaced by direct brain stimulation portable packages. Dangerously, one will be able to chose one's own reality and may therefore become confused about just what reality is. At such a juncture simulation and reality may no longer be distinguishable. Thus, the end point of all forms of simulation will be a philosophical paradox.

4.1 Philosophical Perspectives

"I never make predictions, especially about the future." (Yogi Berra).

4.1.1 The Purposes of Driving Simulation

In trying to distill what the future of driving simulation will be, the central concern must be a direct examination of what we predict such simulations will be used for. This requires that we first step back and look at the larger picture of what motivates transportation in the first place, what motivates it now, and our expected future motivations for transportation in general, but road transportation in particular. In many European and old world countries, personal vehicles are a pleasant luxury but not a prime necessity of life in the same way that they are in the United States. Indeed, it is evident from even a brief visit to any of Europe's main conurbations that they were not conceived, designed, or constructed to deal with the mass of personal and commercial road transport which they are now required to accommodate. In this sense, contemporary roadway transportation systems in these older centers of habitation are modern creations overlaid on a palimpsest of previous forms of infrastructure. Now, major European cities such as London are being forced to introduce congestion charging to try to reduce the level of traffic density on roads that simply cannot handle the volumes created by contemporary demand. Pressures on space and the inability to re-engineer the local infrastructure to any significant degree directs political emphasis to other forms of transportation, and many modern countries are now seeking to respond strategically to these evolving demands. Thus, in many locations in Japan, Europe, and the near Middle East, a personal highway vehicle is something that one might like to have, whereas in countries like the United States and Canada such a vehicle is something that one almost cannot do without. The respective differences between desirable versus almost obligatory ownership affect each respective nation's transportation policies and therefore their need, desire, and use for driving simulation, both now and in the future. Nor should we ignore the vast increase in demand for personal vehicles in China where much of the expansion of the coming years will inevitably be focused (see Evans, 2004). Although our discussion focuses largely on personal transportation, the argument based on the supply and evolution of demand can easily be broadened to the consideration of the commercial sector of road transport and beyond to other intrinsically limited facilities such as ship, rail, and air transport, which themselves have their own respective unique simulation needs. We do not intend the present work to be a discourse on these respective societal requirements; we just wish to establish that there is a spectrum of different constituencies for transportation and, thus, varying sources of motivation for driving simulation.

In ground transportation, these various constituencies can, to a degree, be divided according to their emphasis on how fast any particular journey needs to be accomplished. Speed of transit is largely motivated by the economic imperative to improve movement efficiency. In contrast, the safety of the driver, passengers, and whatever other goods a vehicle is carrying, is largely motivated by public health needs to avoid collisions and resultant injury and fatality rather than any direct economic needs associated with speed of transit. In some countries, where there is a strong emphasis on social development and well-being, driving simulation is primarily motivated by the desire to facilitate this overall safety and thus to enhance collision prevention. This motivation, for example, is very evident in Sweden (e.g., Tingvall, 2009; see also, this book, chap. 3 by Kantowitz, "Using Driving Simulators Outside of North America"). However, for almost all nations, economic concerns are the primary motivators for both users and vehicle manufacturers and thus largely dictate the nature of simulation work undertaken therein. Like other practical sciences, simulation tends to respond to immediate needs and reacts to evaluate transportation innovations as they arise.

Within the past decades the major change in ground transport systems has been the introduction of any number of new advanced technologies into highway and vehicle operations. This has served to partially change their character by making them more into places of work and entertainment (or more widely information assimilation) as opposed to pure sources of transport. Such additions have posed any number of new problems and demands associated with multi-tasking and obligatory vehicle control performance (see Hancock, Lesch, & Simmons, 2003; Hancock, Simmons, Hashemi, Howarth, & Ranney, 1999; Strayer & Drews, 2004; Strayer & Drews, 2007; Trivedi & Cheng, 2007). This has led to a significant growth in simulation-based research (e.g., Horrey & Wickens, 2004) and thus reflects the response of the simulation community to the ebb and flow of social demands for necessary services. Therefore, to understand what current driving simulation is predominantly used for and, more pertinently, what it is likely to be used for in the near future, we have to ask who is paying for the simulation facility, and why. As one wise Washington commentator is reputed to have said "follow the money." Although money implies a financial motive, contemporary simulation tends more to lend itself to safety research by its very nature (and to some extent by the persuasion of the majority of scientists involved with it). Thus safety and its linkage to vehicle design and sales is also a very potent stream of funding for simulators. And, although we have tended to contrast the two major sources of motivation—safety and performance—it is certainly true that there is much overlap and that much research is undertaken which can well fit under both umbrella motivations. Indeed, advances in either area may well facilitate the goals of the other. After all, crashed vehicles tend to not be very efficient in terms of transit time. And exceptionally high-speed freeway vehicles (while evidently the aim of certain technological "autopians") are still more conceptual than practical in these first decades of the twenty-first century. Before we attempt to predict the near and longer-term future of driving simulation, it is important to comment on the issue of predicting the future in general.

4.1.2 On the Nature of Future Prediction

"The main value of prediction is the amusement it will give to those who live in the future."

We all want to know about the future. From the research physicist examining the tracks of elementary particles to the everyday consumer scanning the astrological columns of their local newspaper, we would each like to be more certain about what is to come. We can all make some accurate but fundamentally puerile assertions about the near future. A prediction that the sun will rise tomorrow is very likely to be true, but at the same time it is rather uninformative. In fact, Shannon's established concept of information is that the informational value of almost any prediction tends to co-vary with its improbability, such that for any longer range prediction, the more precise or detailed and therefore (potentially) informative that a particular prediction becomes the less likely it is to actually occur. Although the specificity-improbability relationship is a continuum, even scientists tend to make a distinction between their quantitative models and theories generated in the realm of science and their more generalized predictions, even of the most informed of individuals (see Bartlett, 1962; Bush, 1945; Teilhard de Chardin, 1964). Our notions for the present chapter are very much set in this latter, more general realm of discourse. We look to elucidate largely the qualitative trends in the future of driving simulation. Although we do provide some examination of evolving physical configurations for simulation, we do not engage in specific quantitative predictions in terms of the precise nature of supportive computational capacity (e.g., Moore, 1965), or the evolving cost comparisons of these respective systems (e.g., see Hancock, Caird, & White, 1990).

The cynic can point out the advantages of making only these more generalized predictions. If the prediction proves to be simply wrong then one advantage is that most people rapidly forget it and one's reputation as a prognosticator is hardly affected. Indeed, if one is careful not to make any prediction too precise, one can then interpret appropriately selected portions of almost any range of actual events as a match to an underspecified prediction. These so-called "hits" are then trumpeted to the world as evidence of prescience. This strategy has been ruthlessly exploited across many centuries by those who purport to be psychically endowed, from Nostradamus through Edgar Cayce to the present-day exponents of this doubtful "art" (Randi, 1980). Understanding and exploiting the flaws in common human reasoning has always been a profitable industry across the millennia.

For the purpose of discussion, however, there are two general characteristics of the future upon which we can rely in respect to our assertions about what is to come. The first characteristic is the degree to which the future is like the past. The second is the degree to which the future is not like the past. In what immediately follows we use this differentiation first: To identify currently evident trends and extrapolate them into the near and more distant future. Trends of this type are often used by futurists to derive their particular vision (see Kurzweil, 2005). However, it is the elements of the unknown or non-linear quirks in development that make the future really interesting. In the following section we seek to examine and explicate one of these interesting, non-linear possibilities. We focus first on the effect on driving simulation of simulating the driver, as compared to the more prevalent approach of simulating the vehicle and the environment.

4.1.3 Modeling the Driver Versus Simulating the World

As we conceive of the future of driving simulation, it is almost inevitable that our minds first spring to the physical entities which compose the assemblages with which we work. This list includes scene generation projectors and associated screens combined with computational software; motion bases or ride motion effectors; and in-vehicle displays with the necessary analog-to-digital conversion technologies. All of these are the items that we think of when we look to understand how improvements are to be made in future driving simulations. Yet, important as these technologies are, they will not be the most important things in the future, or even, in fact, in current driving simulations. The most important component is the driver. (See also, this book, chap. 9 by Ranney, "Psychological Fidelity: Perception of Risk".) Not only do we fabricate the artificial circumstances of our surrogate driving worlds to convince human drivers that they are undergoing a veridical experience, it is the very responses of those drivers which represents the subject matter of why we are engaged in driving simulation in the first place. The logical question now becomes: Can we do without all the paraphernalia of representative world generation and focus directly on the simulation of the drivers themselves? The question devolves to the following: What is the present state-of-the-art in human performance modeling?

The answer to the above question is limited at the present time but is cautiously hopeful for the future (see Ness, Tepe, & Ritzer, 2004). It is certainly the case that a variety of constituencies are strongly in pursuit of more veridical and reliable human performance models. Among these groups, the military is probably the highest in its profile and its financial support. There are any number of military modeling efforts, such as IMPRINT, which are trying to use ever-more sophisticated modeling advances to help predict the capabilities, capacities and training potential of their individual and collective human resources (e.g., Allender et al., 1995). There are, of course, many other interested agencies beyond the military pursuing the same basic goal. Added to these fundamentally pragmatic efforts there are the ongoing projects of many groups of research scientists who are using human performance models for their own purposes in order to understand more about the basic science of human behavioral response. Outstanding in this realm are computational models such as ACT-R (see Anderson, 1996; Salvucci, 2006) which seek to wed knowledge in psychology and neuroscience and advanced computational techniques in order to generate structured hypotheses as to the organization of cognition. As either separated projects or welded together as hybrid structures, these steps towards progress now begin to render it more likely that context-contingent models of human response capacities in specific conditions such as driving become ever more feasible.

Within the community of transportation research itself, we have any number of important theoretical advances such as Boer's "driving entropy" conception whose insights can serve to inform and constrain the form and operation of applicable models which would make outcomes from a successful "model driver" most useful to driving researchers. (See also, in this book, chap. 42 by Salvucci, "Cognitive Architectures for Modeling Driver Behavior", and chap. 43 by Flach, Jagacinski, Smith, and McKenna, "Coupling Perception, Action, Intention and Value: A Control Theoretic Approach to Driving Performance".) This and similar valuable advances have been the topic of purpose-directed research meetings over the last few years (see McGehee, Lee, Boyle, & Rizzo, 2007). In reality, many of the questions for which we construct driving simulators are motivated by the particular practical need of some company or organization. And even if such pragmatic concerns do not drive the actual fabrication of the simulator itself, they most often support its daily operations. Given that this is so, we must then assess when and where any improving "model driver" can replace the requirement for human

FIGURE 4.1 The National Advanced Driving Simulator (NADS) of the National Highway Traffic Safety Administration, housed in Iowa City, Iowa. At present, this is probably the most advanced facility in the world for driving simulation. (Reprinted with permission of NADS.)

subjects to be brought into facilities which currently represent the highest achievement in the field of driving simulation (see, for example, Figures 4.1 and 4.2; color versions available on the *Handbook* web site as Web Figure 4.1 and 4.2).

4.1.4 For Future Consideration

If advances in driver modeling represent one potential source of non-linear development, what other advances are possible which could radically alter driving simulation? To provide at least a shadow of an answer to this largely impenetrable question, we would have to go back to the very nature of simulation itself (see Hancock, 2009). Simulation is a surrogate phenomenon that seeks to generate and control both proximal and distal perception-action experiences to create an alternative reality for

FIGURE 4.2 The VTI driving simulator in Linkoping, Sweden. This also is one of the most advanced, state-of-the-art motion-based driving simulators in the world. (Photograph: P.A. Hancock, reproduced by permission of VTI.)

the exposed individual(s). Of course, we are focusing here almost exclusively on real-human-in-the-loop simulation, since a wider definition of simulation in general, while valid, is beyond our present purview. Given our present functional definition then, the vital issue is the immediate experience of the human participant. If this is correct, then one of the possible non-linearities of the future would seem to arise from innovative solutions to the fundamental questions of consciousness itself. Two questions now thrust themselves to the fore: First, is the reliance on proximal stimulation of the various senses an appropriate line of progress? For example, in the same way that the philosopher George Berkeley argued for the superfluity of matter (Berkeley, 1710), can we find better ways to directly stimulate the appropriate brain structures so as to create an experience that is indistinguishable from reality but still a simulated one? The answer, from recent research in brain-machine interfaces, neuro-ergonomics, and the virtual relocation of self (Ehrsson, 2007; Hancock & Szalma, 2003; Parasuraman, 2003), seems to be that yes we can. The question which follows concerns the individual phenomenological experience under these conditions of direct stimulation. Will we be in danger of creating a "schizophrenic" episode for such an individual? And how would they empirically re-establish contact with what we like to call reality after such exposure? If direct brain stimulation experiences can fully satisfy the "Turing" test for reality, we will have created potentially dangerous conditions, and not just for driving simulation. These exciting developments are likely to be the topic of discussion and exploitation in the coming decades. Even partial perfection of these technical issues may render the present, multi-million dollar facilities, such as those shown in Figure 4.1 and 4.2, completely obsolete.

Among our predictions for the future of driver-centered driving simulation, we must also consider the reasonable possibility of its complete demise. This depends upon just how far are we from the goal of a totally automated transportation system. Such an achievement might still seem more in the realm of science fiction than fact. However, as the mixture of inventory between partially-manual and fully-automated vehicles begins to favor the latter, can human-in-the-loop control be sustained as a practical feasibility? Soon, the operational time-horizons for the automated vehicles will have grown so short that human intervention may be not only practicably impossible but perhaps even disastrous if attempted. The vision that most probably represents an unfortunate fiction is the generation of lightning-quick human reactions as epitomized by the modern movie hero. In reality, it cannot be much longer before human response capacities are exceeded in practice, as well as in theory. If this vector of progress continues, it may well be that all human-centered ground simulation is rendered null and void because all ground transportation will exclude direct human control. These respective scenarios as to momentary versus supervisory control and adaptive human-machine systems have been the topic of extensive discussion elsewhere and is thus not pursued in detail here, although the interested reader is directed to our own work and that of others (see Hancock, 1997; Parasuraman & Mouloua, 1996; Sheridan, 1991, 1992, & 2002, among others).

Of course, as we are discussing non-linear potentials there is no barrier to speculation that alternative forms of transportation (of the Star Trek kind) will be discovered. Although this seems unlikely, the development of remote presence technologies may well obviate the need for many human journeys that we now consider obligatory. This "being there" (Clark, 1997) question is one that may well dominate our thinking over the next few decades (see Ehrsson, 2007), especially as we now seem to have recently passed the "peak oil" threshold (Campbell, 1997) and the pressure to search for transportation alternatives grows heavier. If these issues occupy our long-term concerns, what then of the developments of the immediate decades to come? It is to these issues that we now turn.

4.2 Specific Needs and Possibilities

Having proffered some skeptical, and sometimes pessimistic, perspectives about simulating the driving environment, we now suggest why certain aspects of the environment that we deem critical to safety research and driver training will still need to be simulated in the immediate future. In respect of these immediate needs, we describe a novel and emerging way in which to do this. This novel technological innovation can best be approached by first considering the role of motion systems in advanced driving simulation.

4.2.1 The Need for Motion Simulation

In understanding the emergence of innovative driving simulation systems, it is instructive to compare and contrast flight simulation and driving simulation with respect to both the current state of the art and what is likely to be needed in the immediate future

for training and research. Human-in-the-loop simulation began in the aviation domain with the Link trainer and some French counterparts relatively early in the last century. These early simulators had no motion base per se (see Figure 4.3; Web Figure 4.3). However, with the development of hydraulic and electric motor control systems for aiming large guns on World War II ships and the invention of the Stewart platform or hexapod (six independent legs whose lengths are actuated by hydraulic pistons or ballscrews) it became possible to generate small motions in six degrees of freedom to support cabs that included both the pilot trainee(s) and the cockpit instrument mockup. Sustained accelerations lasting more than a second were possible only by tilting the platform so that the human experiences the sensation of longitudinal or lateral acceleration; in these situations sustained vertical acceleration was not possible. Such motion simulators were quickly adopted by the military for training fighter pilots where the rapid onset of G-forces combined with higher frequency vibrations and "seat-of-the pants" cues were significant to aircraft control and training. The latter cues included vestibular otolith linear acceleration, and semicircular canal rotational acceleration, as well as muscle and tendon joint sensing. Soon vendors were able to sell such hexapod motion-base simulator technology to airlines for training and currently all of the large airlines use these for "full mission" simulation training. Interestingly, only recently has evidence revealed that pilot training for commercial carrier aircraft is not significantly improved by training in motion-base simulators over training in corresponding fixed-base (non-motion) simulators (Bürki-Cohen & Go, 1995). Apparently, while turbulence can be reproduced nicely by hexapods, such disturbances are not useful in training, and the very low frequency small magnitude acceleration cues, which are barely sensed by commercial pilots, seem to make little difference as action feedback for controlling large aircraft in mild maneuvers (as compared to fighter aircraft).

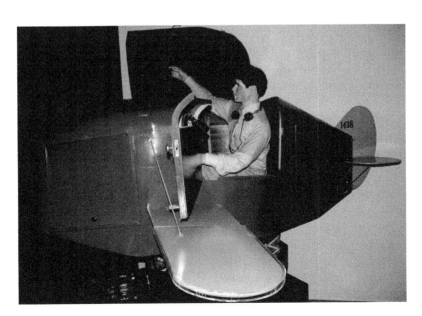

FIGURE 4.3 The original Link Trainer or "blue box" as it was affectionately known. Although there is no true motion base per se, the facility did sit on a platform which provided some movement.

The situation with driving simulators is a stark contrast. The hexapod technology, and computer-graphic visual displays that accompany it, are commercially available; however the driving safety research and driver training communities are not as wealthy as the airlines and military, the principal purchasers of motion-base simulators for aviation. In one evident exception, the U.S. Department of Transportation National Highway Traffic Safety Administration did finance the most sophisticated motion-base simulator (at roughly 80 million dollars), the National Advanced Driving Simulator (NADS) at the University of Iowa in Iowa City (see Figure 4.1). This simulator incorporates a feature not found in aircraft simulators: A roughly 25×25 meter X-Y translational platform which carries the hexapod. The X-Y platform allows for large amplitude low frequency longitudinal and lateral translational movements so critical to simulating braking and steering maneuvers. There are several other hexapod motion simulators built by automobile manufacturers for their own use in both the U.S. and Europe. Unfortunately these simulators are, by and large, not available to driving safety researchers from universities and/or small firms (but see Figure 4.2) since the usage fees are often simply prohibitive.

Perhaps the most important contrast between aviation and driving with regard to simulators is the need in relation to the use. While airlines have regularly used hexapod motion simulators for many years, their usefulness for training is questionable—as we have noted—because there is relatively insignificant motion feedback for the slow maneuvers in commercial aircraft, and the high frequency turbulence accelerations do not provide feedback to any intended pilot actions. On the other hand, in driving, the higher amplitude, higher frequency motion cues from rapid control maneuvers are absolutely critical driver feedback for avoiding obstacles and preventing crashes. Complex visual feedback without acceleration cues is well known to lead to over-steering as well as motion sickness (and for work on simulation sickness see Kennedy, Lane, Berbaum, & Lilienthal, 1993; see also, this book, chap. 14 by Stoner, Fisher, & Mollenhauer). Any control engineer can explain why higher time-derivative sensing is essential to good transient control, and the human organism is no exception.

4.2.2 A New Augmented Reality Paradigm for Driving Simulation

Fixed-base driving simulators have been used effectively for studying driver distraction, fatigue, car-following behavior, signage placement and other issues which do not involve sudden onset acceleration cues. Driving safety researchers certainly want to attack critical crash-avoidance maneuvering, which, after all, is the most life-threatening aspect of driving. However, all current high-end motion-base simulators are largely too expensive or cannot even faithfully recreate the acceleration patterns that occur in crash-avoidance maneuvering in all six degrees of freedom in actual highway vehicles (except at an almost prohibitive price-tag). In concluding a report on research needs for the future, Smith, Witt and Bakowski (2007, p.8) stated: "We need to find repeatable methodologies that allow us to replicate single-exposure imminent-collision warning trials on test tracks, where subjects (falsely) perceive that they are at risk of an imminent collision. To assess safety benefit directly the worst cases must have virtual collisions..." This necessity obviously holds true whether the goal is driver training or driver maneuvering in crash-avoidance.

There is, however, a way to achieve this desired end that can provide the needed fidelity in both vision and motion *and* is within reach economically. As we noted earlier, new technologies of sensors, computer graphic software, and displays have each made very rapid advances in recent years. These collective steps now enable advances in driving simulation that permit experimental subjects to safely experience dynamic hazards (scenarios with other vehicles, pedestrians, highway geometry, etc.) while still experiencing full motion cues—in fact, perfect motion cues—all at a cost and convenience much improved over high-end motion-base driving simulators. This is achieved by having the driver drive an actual vehicle (e.g., a car or truck) on a test track while viewing an out-the-windshield scene which is the actual environment *except for* the object that poses the pending collision hazard. The collision hazard is a virtual one. It is generated by a computer and is continuously sized and oriented on the roadway to correspond to the moment-to-moment position and orientation of the vehicle as if it were real. It is thus displayed continuously at the proper location relative to the driver's viewpoint. This technological assemblage is a prime example of what is termed an applied Augmented Reality (AR) system (see Goldiez, Ahmad, & Hancock, 2007; Sheridan, 2007).

All the software to model and generate convincing, rapidly-changing images of vehicles, people or other collision hazards is already available. The technological challenge now is how best to measure the position and orientation of the vehicle relative to the environment, and how to superimpose the virtual image on the real roadway image so that it is properly sized, oriented, aligned with the environment, and moving in a way that is perceived to be natural. In the present context both the measurement and superposition can be done in several ways. One means of achieving continuous measurement of vehicle position relative to the environment (e.g., test track) is to add real objects to the real environment that can be seen and recognized by a video camera mounted on the vehicle (and appearing in the camera field at locations corresponding to their position relative to the vehicle). Another way is to use a gyroscope or accelerometer whose signal can be integrated to determine position. A third method is to set initial conditions into a mathematical model of vehicle response to steering and braking a few seconds prior to the (virtual) appearance of the impending collision, where measurement of steering wheel angle and brake actuation continuously update the model throughout the duration of the virtual image generation and display.

Superposition of collision objects onto the view of the actual forward environment may also be accomplished in several ways. One is to use a half-silvered mirror mounted in front of the

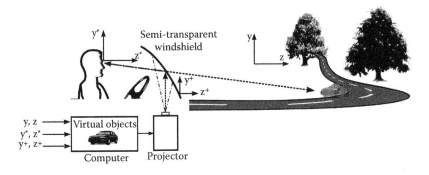

FIGURE 4.4   Road scene viewed directly through semi-transparent mirror. (Head-Up Display.)

driver, where the virtual image is projected from below (Figure 4.4; Web Figure 4.4). An alternative is to use a see-through head-mounted display, the equivalent arrangement except that the half-silvered mirror, or prism, is reduced in size (Figure 4.5; Web Figure 4.5). Sufficiently high resolution and sufficiently wide field of view head-mounted displays have not been available until recently, and they are still tens of thousands of dollars in cost, but the technology and price are improving rapidly. In either of these methods the virtual image may appear somewhat transparent because light from objects "behind" the virtual image will still be detectable (though for suddenly appearing hazards it probably will make little difference in the startle reaction of the driver). Also, with either of the above methods the range of alternative levels of brightness of the virtual image may be insufficient to match the wide-ranging brightness levels of the actual environment.

One means of making the virtual image seem to more naturally "belong" to the actual environment view is to observe the environment view with a video camera and superpose the virtual image on the video electronically (Figure 4.6; Web Figure 4.6), which is now standard technology. In this case the brightness and resolution (which are improving all the time) between virtual and actual environment images can be the same. Test drivers have had little difficulty driving through virtual environments in conventional driving simulators with screens approximately ten feet away, so in principle drivers should have no difficulty with a display a smaller number of feet away, provided suitable optics are used to make the screen appear farther away.

Driving an actual vehicle when observing the actual environment through video is quite acceptable to most drivers, provided that the video is of sufficient resolution and the field of view is scaled 1:1 with the field of view that the driver would have were the video intermediary not used. The second author has tried this, and discovered that a distortion in the field size was much more troublesome than driving through video per se. Of course any test track should have no actual collision hazards on or near the roadway, and the video system must be reliable. Concerns for the safety of driving in this mode can be mitigated by having a second driver and second set of controls in the passenger seat (and thus the situation should be no less safe that if one is riding as a passenger with that second driver). The video superposition method can be used with a head-mounted display as well, in this case not requiring the see-through feature (Figure 4.7; Web Figure 4.7). Finally, using the vehicle-mounted display or the head-mounted display the environment, as well as the hazard, can be entirely virtual (computer-generated) (Figure 4.8, Web Figure 4.8). The advantage of having the real environment's image projected by means of video, is that it is by far the greatest component of the displayed image, the hazard object being only a small part (far fewer pixels). Therefore all the issues of graphic modeling and refresh rate are mitigated in the video image case.

Tests with the video image and virtual hazard superposition have been conducted with a standard vehicle in a school parking lot (Sheridan & Askey, 2005). Measurement of the vehicle position relative to the environment was accomplished by means of several computer-recognizable red traffic cones placed alongside the test track. Figure 4.9 (Web Figure 4.9) shows a (virtual) truck approaching from the front, slightly out of its lane. Figure 4.10 (Web Figure 4.10) shows another such test condition with the truck converging from the side.

It is evident that this general approach to human-in-the-loop driving simulation, namely driving an actual vehicle on a test track, modified to the extent of the computation and display technology to superpose computer-generated virtual images onto the normal driver forward view, and programming those images to move in ways that appear normal and in alignment

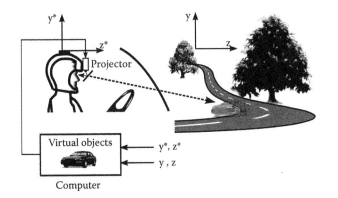

FIGURE 4.5   Optical-see-through approach with a head-mounted display. (HMD.)

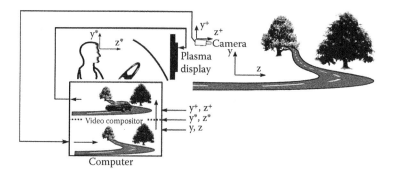

FIGURE 4.6 Video mixed-image approach with vehicle-mounted display.

with the background scenery, promises a new simulation paradigm. Most of the situation involves the normal actual reality: The driven vehicle along with its controls; and the roadway and environment (except that it is likely to be a test track for safety purposes). Only the hazard objects (one or more vehicles, people or other objects) need be artificial. This can allow for enormous saving in both time and money in building and managing a special simulation test facility; savings in motion-base technology (the greatest share of the cost of a motion-base simulator); savings in vehicle modeling and simulator programming for generating faithful motion cues; and savings in modeling and displaying the moving environment (except for the hazard objects which occupy only a small fraction of the display). However, since the foundation of this technology requires the use of real-world conditions, it is limited by the surroundings which the experimenters can conveniently access.

4.3 Summary and Conclusions

In many ways we expect that the near future will probably look very much like the present. Although there will be predictable advances in technology, the basic version 1.0 human being is unlikely to experience any evident, natural change in the immediate future.

Of all the elements of modern driving simulation, what we have seen change most rapidly in our own era is the computational capacity associated with visual scene representation. In keeping with Moore's Law (Moore, 1965), we have seen enormous gains in graphical capacities and scene-generation systems, which less than two decades ago cost millions of dollars but can today be bought for a fraction of the cost at the local computer store. This progress is evident in the fact that the vast majority of the most sophisticated driving simulation facilities use PCs and their associated graphics cards to generate pictorial scenes. In one sense there becomes a functional limit as to just how good visual scene representation can become and this limit appears to be set by the resolution of the human eye (and see Hancock, 2009; see this book, chap. 2 by Allen, Rosenthal, & Cook, "The History of Driving Simulation", Table 2.1). However, this apparent barrier need not be an immutable one and a further generation of "super-simulation" is both conceivable and feasible as we understand more about the interactive role of attention and visual physiology. Super-simulation implies sensory and perceptual capacities that exceed any unaided individual but look strongly like those abilities possessed by our fictional super-heroes (see Hancock, 2009; Hancock & Hancock, 2008).

If highly sophisticated visual graphics are now within the range of every scientist's pocket-book, conventional motion-bases are certainly not, except by the means we have suggested in Section 2. Unlike the economic impetus for better computational facilities, conventional motion systems have neither the constituency nor the market to drive down investment costs. And indeed, a poor motion system might actually hinder the transfer

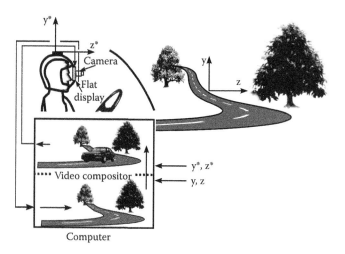

FIGURE 4.7 Video mixed-image approach with head-mounted display.

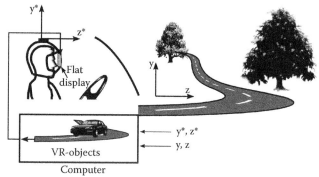

FIGURE 4.8 Full VR approach with head-mounted display.

FIGURE 4.9 Two images from a video clip of the driver's display in a test maneuver in a school parking lot. A (virtual) truck approaches in a near head-on collision. The driver swerves his (actual) vehicle. The background is the real environment. The traffic cones were used as fiduciary landmarks.

of outcome results to real-world circumstances. Surely one also has to ask of what real use is accurate reproduction of motion cues. The answer, we believe, is for simulating critical braking and steering maneuvers to avoid near collisions (Hancock & de Ridder, 2003). We have suggested what we believe to be a viable approach, though much further development needs to be achieved to realize and exploit its full potential.

We expect to see advances in general systems modeling, including significant steps in human-operator modeling. A much more cohesive program would then see rapid, synthetic simulation studies in which all elements are encoded in software and periodic checks with live drivers and actual vehicles will be used as discrete points of confirmation from the output of large collective models. In this manner, all the present material components of driving simulation will metamorphose into software agents in which fast-time processing will allow for "virtual" years of testing to be conducted in moments. How such outcome information is to be validated and verified will represent one of the next great challenges to the driving simulation community.

Key Points

- Driving simulation has been used for research and training.
- The development of driving simulation has been contingent on the development of flight simulation.

- Modern technologies have largely dissociated this dependence.
- Such technical capacities not only enhance simulation they diversify it.
- Crucial forms of driver simulation can now be achieved by augmented forms of reality.
- Gradually, all simulations evolve toward the threshold of the Turing Test for reality.
- Passing such a threshold poses fundamental questions about the nature of reality itself.
- Creating surrogate realities may therefore not always be beneficial or even moral.

Keywords: Augmented Reality, Driver Modeling, Driving Simulation, Future Trends, Virtual Reality

Acknowledgments

The authors gratefully acknowledge Prof. Matthias Roetting of Technical University Berlin, the Liberty Mutual Insurance Company, Dr. Stephen Ellis, NASA Ames Research Center, David Askey, and the Volpe National Transportation System Center for discussions, preliminary tests and some of the graphics used to describe the augmented reality simulation. Completion of this chapter was facilitated by grants from the Army Research Laboratory and the Florida Department of Transportation. The views expressed are those of the authors and do not necessarily represent those of any named agency. Since this chapter is concerned with the future it is important to note it was written in 2008.

FIGURE 4.10 Two images from a video clip of the driver's display in test maneuver on a country road. A (virtual) truck approaches from the right in a near collision and the driver brakes. The background is the real environment. The white signs on the trees were used as fiduciary landmarks.

Web Resources

The *Handbook*'s web site contains supplemental materials for the chapter including all of the chapter's figures in color versions.

Web Figure 4.1: The National Advanced Driving Simulator (NADS) of the National Highway Traffic Safety Administration, housed in Iowa City, Iowa. At present, this is probably the most advanced facility in the world for driving simulation. (Reprinted with permission of NADS). (Cover version of Figure 4.1).

Web Figure 4.2: The VTI driving simulator in Linkoping, Sweden. This also is one of the most advanced, state-of-the-art motion-based driving simulators in the world. (Color version of Figure 4.2). (Photograph: P.A. Hancock, reproduced by permission of VTI).

Web Figure 4.3: The original Link Trainer or "blue box" as it was affectionately known. Although there is no true motion base per se, the facility did sit on a platform which provided some movement (Figure 4.3).

Web Figure 4.4: Road scene viewed directly through semi-transparent mirror (Head-Up Display) (Figure 4.4).

Web Figure 4.5: Optical-see-through approach with a head-mounted display (HMD) (Figure 4.5).

Web Figure 4.6: Video mixed-image approach with vehicle-mounted display (Figure 4.6).

Web Figure 4.7: Video mixed-image approach with head-mounted display (Figure 4.7).

Web Figure 4.8: Full VR approach with Head-Mounted Display (Figure 4.8).

Web Figure 4.9: Two images from a video clip of the driver's display in a test maneuver in a school parking lot. A (virtual) truck approaches in a near head-on collision. The driver swerves his (actual) vehicle. The background is the real environment. The traffic cones were used as fiduciary landmarks (Figure 4.9).

Web Figure 4.10: Two images from a video clip of the driver's display in test maneuver on a country road. A (virtual) truck approaches from the right in a near collision and the driver brakes. The background is the real environment. The white signs on the trees were used as fiduciary landmarks (Figure 4.10).

Key Readings

Clark, A. (1997). *Being there: Putting brain, body and world together again.* Boston, MA: MIT Press.

Hancock, P. A. (2008). The future of simulation. In D. Vicenzi, J. Wise, M. Mouloua, & P. A. Hancock (Eds.), *Human factors in simulation and training* (pp. 169–188). Boca Raton, FL: CRC Press.

Hancock, P. A., & de Ridder, S. (2003). Behavioral response in accident-likely situations. *Ergonomics, 46*(12), 1111–1135.

Ness, J. W., Tepe, V., & Ritzer, D. R. (Eds.). (2004). *The science and simulation of human performance.* Amsterdam: Elsevier.

Sheridan, T. B. (2002). *Humans and automation: System design and research issues.* John Wiley & Sons.

References

Allender, L., Kelley, T. D., Salvi, L., Lockett, J., Headley, D. B., Promisel, D., . . . Feng. T. (1995). Verification, validation, and accreditation of a soldier-system modeling tool. *Proceedings of the annual meeting of the human factors and ergonomics society, 39,* 1219–1223.

Anderson, J. R. (1996). ACT: A simple theory of complex cognition. *American Psychologist, 51,* 355–365.

Bartlett, F. C. (1962). The future of ergonomics. *Ergonomics, 5,* 505–511.

Berkeley, G. (1710). *A treatise concerning the principles of human knowledge. (And) Three dialogues between Hylas and Philonous [1713],* R. Woolhouse (Ed.). New York, NY: Penguin (1988).

Bush, V. (1945). As we may think. *Atlantic Monthly,* July.

Bürki-Cohen, J., & Go, T. H. (2005). *The effect of simulator motion cues on initial training of airline pilots.* AIAA-2005-6109. Proceedings of the American Institute of Aeronautics and Astronautics Modeling and Simulation Technologies Conference. San Francisco, CA.

Campbell, C. J. (1997). *The coming oil crisis.* Brentwood Essex, England: Multi-Science Publishing Co.

Clark, A. (1997). *Being there: Putting brain, body and world together again.* Boston, MA: MIT Press.

Ehrsson, H. H. (2007). The experimental induction of out-of-body experiences. *Science, 317,* 1048.

Evans, L. (2004). *Traffic safety.* Bloomfield Hills, MI: Science Serving Society.

Goldiez, B., Ahmad, A. M., & Hancock, P. A. (2007). Effects of augmented reality display settings on human way-finding performance. *IEEE Transactions on Systems, Man, and Cybernetics, Part C: Applications and Reviews, 37*(5), 839–845.

Hancock, P. A. (1997). *Essays on the future of human-machine systems.* Eden Prairie, MN: Banta Information Services Group.

Hancock, P. A. (2009). The future of simulation. In D. Vicenzi, J. Wise, M. Mouloua, & P. A. Hancock (Eds.), *Human factors in simulation and training* (pp. 169–188). Boca Raton, FL: CRC Press.

Hancock, P. A., Caird, J. K., & White, H. G. (1990). *The use of simulation for the assessment, training, and testing of older drivers* (Technical Report NIA 90-01). Washington, DC: National Institute on Aging, National Institutes of Health.

Hancock, P. A., & de Ridder, S. (2003). Behavioral response in accident-likely situations. *Ergonomics, 46*(12), 1111–1135.

Hancock, P. A., & Hancock, G. M. (2008). Is there a super-hero in all of us? In R. S. Rosenberg & J. Canzoneri (Eds.), *The psychology of super-heroes* (pp. 105–117). Dallas, TX: Benbella Books.

Hancock, P. A., Lesch, M., & Simmons, L. (2003). The distraction effects of phone use during a crucial driving maneuver. *Accident Analysis & Prevention, 35*(4), 501–514.

Hancock, P. A., Simmons, L., Hashemi, L., Howarth, H., & Ranney, T. (1999). The effects of in-vehicle distraction upon driver response during a crucial driving maneuver. *Transportation Human Factors, 1*(4), 295–309.

Hancock, P. A., & Szalma, J. L. (2003). The future of neuroergonomics. *Theoretical Issues in Ergonomic Science, 4*(1), 238–249.

Horrey, W., & Wickens, C. D. (2004). Driving and side task performance: The effects of display clutter, separation, and modality. *Human Factors, 46*(4), 611–624.

Kennedy, R. S., Lane, N. E., Berbaum, K. S., & Lilienthal, M. G. (1993). Simulator sickness questionnaire: An enhanced method for quantifying simulator sickness. *International Journal Aviation Psychology, 3*(3), 203–220.

Kurzweil, R. (2005). *The singularity is near: When humans transcend biology.* New York, NY: Penguin.

McGehee, D., Lee, J., Boyle, L., & Rizzo, M. (Eds.). (2007). *Proceedings of the fourth international driving symposium on human factors in driving assessment, training and vehicle design.* Iowa City, IA: University of Iowa, Public Policy.

Moore, G. E. (1965). Cramming more components onto integrated circuits. *Electronics, 38*(8), 19.

Ness, J. W., Tepe, V., & Ritzer, D. R. (Eds.). (2004). *The science and simulation of human performance.* Amsterdam: Elsevier.

Parasuraman, R. (2003). Neuroergonomics: Research and practice. *Theoretical Issues in Ergonomics Science, 4*(1), 5–20.

Parasuraman, R., & Mouloua. M. (Eds.). (1996). *Automation and human performance: Theory and applications.* Hillsdale, NJ: Erlbaum.

Randi, J. (1980). *Flim-flam.* Amherst, NY: Prometheus Books.

Salvucci, D. D. (2006). Modeling driver behavior in a cognitive architecture. *Human Factors, 48*(2), 532–542.

Sheridan, T. B. (1991). *Human factors of driver vehicle interaction in the IVHS environment* (DOT HS 807–837). Washington, DC: National Highway Traffic Safety Administration, U.S. Department of Transportation.

Sheridan, T. B. (1992). *Telerobotics, automation, and human supervisory control.* Cambridge, MA: MIT Press.

Sheridan, T. B. (2002). *Humans and automation: System design and research issues.* John Wiley & Sons.

Sheridan, T. B. (2007). *Vehicle operations simulator with augmented reality,* U.S. Patent 7,246,050, dated July 17, 2007.

Sheridan, T. B., & Askey, D. (2005). *Augmented reality driving system.* US DOT Volpe National Transportation Systems Center (Project NA-23), Boston, MA: US DOT.

Smith, M. R., Witt, G. L., & Bakowski, D. L. (2007, January 25). *The application of real-time distraction monitoring to driver safety systems.* Paper presented at the NHTSA Human Factors Forum, Washington, DC.

Strayer, D. L., & Drews, F. A. (2004). Profiles in driver distraction: Effects of cell phone conversations on younger and older drivers. *Human Factors, 46*(4), 640–649.

Strayer, D. L., & Drews, F. A. (2007). Cell phone-induced driver distraction. *Current Directions in Psychological Science, 16*(3), 128–131.

Teilhard de Chardin, P. (1964). *The future of man.* New York: Harper & Row.

Tingvall, K. (2009, September). *Vision zero and distraction.* Plenary paper presented at the First International Conference on Driver Distraction and Inattention, Gothenburg, Sweden.

Trivedi, M. M., & Cheng, S. Y. (2007). Holistic sensing and active displays for intelligent driver support systems. *Computer, 4*(5), 60–68.

5

Twelve Practical and Useful Questions About Driving Simulation

Jeff K. Caird
University of Calgary

William J. Horrey
Liberty Mutual Research Institute for Safety

Abstract

The Problem. A number of practical questions or concerns are frequently expressed about driving simulators. For example, what can you use a driving simulator for? What are the advantages and disadvantages of using a simulator? Probably the most important question though is whether driver behavior in a simulator mimics that which is exhibited while driving in the real world. These and other epistemological questions are discussed. *Role of Driving Simulators.* The first driving simulator research paper appeared in 1934. Since then, driving simulators have evolved into a flexible means to measure a variety of variables while drivers drive in a wide range of traffic environments. Empirically, driving simulation allows researchers to exert control over confounding variables that are common in actual driving. Testing of drivers in crash-likely conditions is also possible, which for ethical reasons cannot be done in the real world. Uses of driving simulators for research are discussed. *Key Results of Driving Simulator Studies.* Some have claimed that driving simulator research has not contributed to the progression of knowledge about driving performance, behavior, or safety. An inspection of the chapters in this *handbook* attests to the scope and scale of contributions. Further, the corpus of research using simulation has been constant and is now increasing. Areas of contributions are analyzed. *Scenarios and Dependent Variables.* The selection of scenarios that address certain research questions is dependent on the joint capability of a simulator and research team, which often evolves with experience. A list of common dependent variables is provided. *Platform Specificity and Equipment Limitations.* A realistic appraisal of a simulator's capabilities in light of the results produced is required, including limitations therein.

5.1 Introduction

A number of commonly and not so commonly asked questions about driving simulators are here addressed. Some of the questions are relatively easy to answer, whereas others do not necessarily have an answer. Other questions are long-standing criticisms of driving simulation (e.g., Dingus et al., 2006; Evans, 2004) while others have yet been overlooked. The answers, in some cases, are a list of pointers to others who have addressed the criticism or have a solution. Specifically, the following questions are addressed:

1. What are the origins of driving simulation?
2. What can a driving simulator be used for?
3. What advantages and disadvantages are there in using a driving simulator?
4. Where are driving simulators that are used for research located throughout the world?
5. When using driving simulators, what special considerations should be discussed when applying to an ethics committee or institutional review board (IRB)?
6. What dependent variables can be collected using a driving simulator?

7. How do you map the necessary realism or fidelity of a driving simulator to the specific research questions?

8. To what degree are the results from a simulator similar to those measured in the real world?

9. How well do the results from simulator, on-road and more basic laboratory studies correspond with one another?

10. What are common threats to internal and external validity?

11. Is there evidence that driving simulators have contributed to the advancement of knowledge since the 1970s?

12. Based on the progress of software, hardware and projection capabilities, where will the future of driving simulation lead?

The first questions deal with historical information, the practical uses of driving simulators and the prevalence of simulators today. The remaining questions, which parallel the progression of research activities in a study, focus on critical issues, concerns and threats in using simulators in scientific endeavors. While simulators carry tremendous flexibility and potential utility, we believe that it is important that all users of simulators be cognizant of the inherent issues and limitations.

5.2 The Twelve Questions

1) What are the origins of driving simulation?

Driving simulators have been used to explore aspects of driving since the 1960s (see this book, chap. 2 by Allen, Rosenthal, & Cook, "A Short History of Driving Simulation"). While this epoch represents the period in which they became more prevalent, the origin of the concept of driving simulation dates much further back. For many years, the static simulator, shown in Figure 5.1, developed by De Silva (1936) was considered one of the potential originators

of driving simulation. For instance, Gibson and Crooks (1938) cite De Silva's 1935 research at the State College of Massachusetts in their seminal paper on a theory of driving.

The first author (JKC), while engaged in a game of scholarship, tracked an even earlier reference to driving simulation. This was the result of a challenge issued by Hancock & Sheridan (authors of "The Future of Driving Simulation" this book, chap. 4) some 20 years ago to find the original or first citation pertaining to driving simulation. (It is worth mentioning to the current generation of would-be scholars that library stacks are not searchable like e-journals, at least not until all known publications are scanned and accessible.) The reference for perhaps the first driving simulator is shown in Figure 5.2. This 75 year-old citation has not been shown for a number of generations until now.

Miles and Vincent built an apparatus to create the illusion of driving circa 1934. The purpose of the driving apparatus was to develop compulsory tests for drivers to combat the public indignation about the high rates of road traffic deaths and injuries. The specifics of the device are as follows:

> An apparatus has been constructed at the Institute [U.K.] which gives a very strong illusion of actual driving, in which the conditions can be kept absolutely constant, and which keeps a printed record of the track and speed of the driver. The subject sits in a dummy car, with the usual controls and 'drives' (see [the figure; here Figure 5.2]). The illusion of movement is given by a moving picture on the projector mounted on the chassis of the dummy car. The motor driving the projector is mounted on the chassis of the dummy car, and the vibration it causes varies with the apparent speed, adding greatly to the illusion. The driver can go where he [or she]

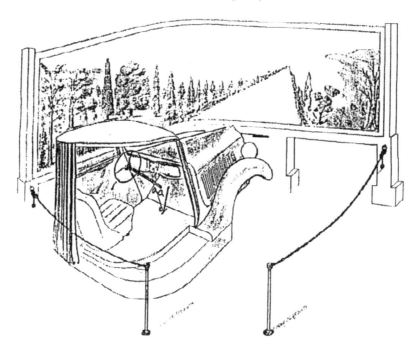

FIGURE 5.1 An early device that resembled a driving simulator appeared in the journal *The Human Factor* in 1936 (De Silva). The mural was a static scene.

FIGURE 5.2 The earliest citation of a system that resembled simulated driving appeared in the journal *The Human Factor* in 1934 (Miles & Vincent). This journal was published in the U.K. for about 10 years during the 1920s and 1930s. The bottom arrow points to a set of miniature models that were projected by a device (top arrow) onto the screen before a driver.

likes—on the wrong side of the road, on the grass, into the hedge—he can take any turning he likes or he can swing his car around, if the road is wide enough, and go back the way he came. Behind the screen is a miniature landscape with roads an inch or so wide and trees and hedges on the same scale. A projector passes over the roads of this model, throwing an enlarged moving picture upon the screen. The observer sees just the view he would get if he were actually driving his car along the roadway. At the same time a pen is tracing out the exact track on a plan. The track of the pen is a dotted line, the dots being made at the rate of ten per second. It is thus possible to calculate the speed of any part of the track by measuring the distance between the dots. The subject is first allowed ten minutes to accustom himself to the 'feel' of the apparatus, and then a number of track records are taken (Miles & Vincent, 1934, pp. 253–255, brackets [inserted information]).

The presentation of miniature models continued to be used for many years (cf., see Allen, Rosenthal, & Cook, this book, chap. 2; Blaauw, 1982; Sheridan, 1970; Wierwille & Fung, 1975). In addition to predating De Silva's device (Figure 5.1), Miles and Vincent built a driving apparatus that was interactive. With De Silva's device in Figure 5.1, the imagination of the driver would have been required to produce a sense of driving while viewing the static mural. Today we still tacitly ask our participants to make believe that they are driving even though they are not. As an operational assumption, researchers generally accept that drivers believe they are driving when using a simulator, although this is open to considerable debate.

To place the "driving apparatuses" of Miles and Vincent and De Silva into historical context, the Link flight simulator was introduced in 1929 and put into extensive use during WWII as a means to train pilots to fly by instrumentation (in Wikipedia, search on *<Flight Simulation>*, and Web Figure 5.1). Interestingly,

simulator sickness was prevalent when using the Link trainer, which is illustrated in a cartoon from 1942 (Web Figure 5.2). Flying and driving are very different constellations of tasks. Pilots are a highly specialized, homogenous group (e.g., Lee, 2005; Tufano, 1997). There is no analog in driving for instrument flight rules (IFR), where an aircraft is piloted using the instrument panel alone (i.e., no visual information from outside the cockpit). A large, heterogeneous population of drivers use complex visual information to control their vehicles, which was first theoretically described by Gibson and Crooks (1938) and extended by Schiff and Arnone (1995). As such, a mass-produced, easily accessible vehicle made the need to train in a surrogate environment largely irrelevant for economic, task and access reasons.

2) What can a driving simulator be used for?

Many researchers and practitioners have a general idea about what a driving simulator can be used for. We compiled a comprehensive list of actual and possible uses of driving simulators:

- Perform research on traffic safety.
- Conduct a structured training curriculum.
- Assess training and education effectiveness.
- Assess those who are about to be licensed or require re-licensing.
- Evaluate vehicle designs.
- Test traffic control devices.
- Understand the effects of basic human limitations on driving.
- Examine the sources and consequences of driver impairments.
- Develop and evaluate new in-vehicle and co-operative infrastructure technologies.
- Assess roadway geometries before they are built.
- Play a game.
- Develop the software and hardware skills of students and technicians.

- Determine the capability of the simulator technology.
- Visualize crashes and crash contributors.
- Understand the effects of disease conditions and various medications on driving.
- Demonstrate the advanced technology of a country, institution or corporation.

The details of each of these uses are more complicated than what a simple list can convey and go beyond the desired scope of this chapter. Fortunately, many of these topics are eloquently grouped and described in other chapters in this *Handbook* (see Table of Contents). However, one topic in particular is often overlooked; namely, the last one on the list. Driving simulators are used for public relations purposes. For instance, doing demonstrations, or demos, is an activity that both of this chapter's authors, along with their colleagues and their students have done for many years for the University of Calgary (and the University of Minnesota before that) and the University of Illinois. I (JKC) have felt an obligation to talk about the research that we do and why it is important to society because we receive public funding. In fact, the mission of those associated with the University of Calgary Driving Simulator (UCDS) is to reduce traffic fatalities and injuries through the creation and dissemination of basic and applied knowledge. Dissemination of knowledge through peer-reviewed publications is only one of many ways to teach researchers, allied professions, the community, students, and drivers. In addition, many researchers who get into simulation are unprepared for the demands placed on them to perform activities that are seemingly unrelated to research.

Driving simulators naturally attract interest from people of all ages. Collectively, we have given well over 100 formal demos over the past eight years to: government officials, media, corporate heads, engineers of all stripes, student organizations, high school classes, class laboratories (e.g., Introduction to Human Factors), professional organizations, conferences, donors, and other researchers and scientists (including Nobel Laureates). Group size matters. For everyone to be able to try the simulator and ask dozens of questions in about an hour, a group of 8 to 12 people is ideal. Restricted laboratory space made one visit from a group of 30 important persons into a circus.

Once a group comes, we try to teach our visitors something about research, traffic safety and driving simulation before we let them go for a drive. The message is often tuned to the level of the audience. After a 5 to 10 minute slide show about the research activity done with the UCDS, visitors take turns, if they would like to, traveling through a demonstration drive that we developed for this purpose. Most of the demonstration time is unstructured while each visitor gets an opportunity to try out the simulator. While this is going on, we try to answer all the questions that we get, which come quickly once our visitors see and experience the possibilities. Our enthusiasm to do this public service, which is desired by institutions but not rewarded or recognized, has waned somewhat over the years. Saying no to requests, especially while studies are underway, is more common. Contrary to conventional wisdom, demonstrations have a low relationship to funding opportunities.

3) What advantages and disadvantages are there in using a driving simulator?

A number of known advantages and disadvantages about driving simulators are shown in Table 5.1. Arguments marked with an asterisk * are expanded or debated in greater depth in Questions 5 and 11, respectively. Disadvantages marked with a ‡

TABLE 5.1 Advantages and Disadvantages of Driving Simulators

Advantage	Disadvantage
• *Has the capability to place drivers into crash likely situations without harming them, such as when they are using drugs, fatigued, engaging in police pursuits, during extreme weather, using new technologies, among other dangerous activities.	• *Simulated crashes do not have the same consequences as a real crash and may affect subsequent behavior. Crashes in a simulator may have an unknown psychological impact on participants.
• Many confounding variables that occur in on-road driving can be controlled when driving simulation is used (e.g., weather, traffic, lighting, frequency of vulnerable road users, wind, potholes, proportion of vehicle types, irrational or unexpected behavior of other drivers, and so forth).	• These confounding or interacting variables that occur in the real world also need to be understood and, since they cannot be fully recreated in simulators, are not necessarily amenable to testing (as yet). In other words, understanding driver behavior is in the interacting details.
• All of the sensory details of the real world are not used by drivers anyway. Perceptual information (Gibson, 1986) for driving is knowable and can be faithfully reproduced using simulators.	• The real world can never be perfectly reproduced (for now). The important combinations of real-world information and feedback that are important to driving are not completely known.
• Events or scenarios can be identically repeated for each participant.	• ‡Each exposure or trial affects responses to subsequent exposures.
• Simulators offer cost savings through flexible configurability so that a wide range of research questions can be addressed (see Jamson, this book, chap. 12).	• High-end simulators, such as NADs, require considerable hardware and software development to address a limited number of research questions.
• Even low-cost, low-fidelity simulators in the right hands can address a wide variety of interesting research questions.	• Low-cost simulators can be imprecise and inflexible and therefore do not address all needs.
• Driving simulation is compelling and elicits emotional reactions from drivers that are similar to those of actual driving.	• Drivers do not believe in the authenticity of the simulation at a fundamental level and responses are based on this perception.
• *Simulators are good at assessing *driver performance* or what a driver *can do* (Evans, 2004).	• *Simulators are not able to address questions of *driver behavior*, which is what a driver does do in their own vehicle (Evans, 2004).
• A structured driver training curricula can be set up and run for new drivers and for some skills, transfers to the open road (see Pollatsek et al., this book, chap. 30)	• The extent that the driver training transfers to on-road skills is not known nor is the relative cost-effectiveness of such programs (see Jamson, this book, chap. 12).

are addressed in Question 10, which deals with common threats to internal and external validity.

To justify using a simulator, researchers frequently mention the advantages of doing so over other methods such as naturalistic observation, instrumented vehicles or laboratory studies. (And the same is true of using other methods over driving simulation.) Often the counterargument is overlooked, but in Table 5.1 the advantages and contrasting views are paired together. The use of "advantage" and "disadvantage" only loosely applies to the conceptual comparisons. The intent of providing contrasting views is to encourage critical thought about driving simulators and alternate methods. Even the most capable and proficient researchers cannot offset all of the potential disadvantages in conducting simulator research; however, we believe that all researchers and users of simulators should *at least* be cognizant of these issues.

One obvious way to use Table 5.1 is to emphasize the disadvantages of driving simulation to justify pursing an alternative research agenda such as naturalistic observation. Emphasizing the disadvantages of a methodological approach and ignoring the advantages (or vice versa) is a form of self-deception that may be practiced because of the immense efforts required to build, operate and effectively use expensive and complex methods. However, the abstract relationships among the advantages and disadvantages and the subtleties of each statement require a much deeper consideration that transcends simplistic scorecard thinking.

4) Where are driving simulators that are used for research located throughout the world?

A partial list of research simulators throughout the world is kept by INRETS (French National Institute for Transport and Safety Research, http://www.inrets.fr/ur/sara/Pg_simus_e.html). The list of countries with research simulators includes Australia, Canada, France, Germany, Japan, China, The Netherlands, New Zealand, Poland, Spain, Sweden, the U.K., and the U.S.

All things considered, checking out who has what can lead to simulator envy. That being said, the hardware and software used to create driving simulation is less than one-half of the capital necessary to create research "excellence and innovation". Researchers, students and technicians comprise the most important ingredients of a successful simulation facility. Inspection of representative publications from various facilities should moderate initial impressions of various kinds of interesting driving simulators. The authors in this *Handbook* represent an obvious starting place to establish who is known for what.

5) When using driving simulators, what special considerations should be discussed when applying to an ethics committee or institutional review board (IRB)?

To perform studies with human participants requires an ethics certification. Each study, and substantive modification, requires interaction with an IRB or Ethics Committee. A typical list of issues that need to be addressed when an ethics application is submitted includes: Informed consent, consent to use participant image, negative transfer, simulator sickness, parental consent (if needed), screening criteria, recruiting procedures, questionnaires, experimental script or verbal protocol, payment, and debriefing. Many of these considerations, including simulation sickness, are discussed by Trick and Caird, this book, chap. 26. However, negative transfer and use of participant images are not.

Negative transfer may occur after a driver has been exposed to driving in a consequence-free environment (also one that may not faithfully replicate real-world driving conditions). Following the experiment, he or she may not be adequately (or fully) recalibrated to real-world driving and may operate his or her vehicle with the same disregard for consequences. At best, the importance and impact of negative transfer from simulation is an empirical question that remains unresolved. In aviation, pilots are required, in some companies but not others, to wait 24 hours after using a flight simulator before flying a plane. However, the empirical basis for the rule, despite years of research, has not been found.

This topic is not often discussed in the context of simulation research or IRB protocol. In general, it is advisable to err on the side of caution and keep participants in the laboratory for a period of time following their involvement in the simulator, filling out questionnaires or the like, in order to afford them a longer time to readjust to the real world. In some situations, it may be advisable to provide transportation home for participants.

The use of participant images for presentation purposes, once a study is over, is an extension of informed consent procedures. Images can include still pictures and videos of participants which have been captured during the course of experimentation. Uses of images at conferences include illustrating certain patterns of performance or behavior such as crashes, facial reactions, and interaction with nomadic devices. The principles of full disclosure, choice and confidentiality apply. Practically, we have used a second informed consent form that outlines and illustrates how images will be used and treated confidentially. A participant may choose to be in a study and not have his or her image used. Accidental dissemination of images by sharing presentation slides or videos should be avoided because control over the images is no longer possible. Careful consideration of the illustrative value of using a video clip or picture is also recommended.

6) What dependent variables can be collected in a driving simulator?

A number of researchers convinced us of the importance of a handy dependent variable list for driving simulation. Investigators should have a good idea about the scope of variables that can be collected and interpreted. A number of common questions arise about which dependent variable can be measured and how to interpret each relative to past use. The selection of dependent variables occurs based on prior use, simulation capability, researcher expertise, practical and applied generalities, and theoretical considerations. More succinctly, the choice of dependent variables is made based on the questions being asked. Table 5.2 lists common, but not necessarily agreed upon, groupings of dependent variables. Other measures have been used that are not listed here, so this is not an exhaustive list. In addition, new and interesting measures should be developed, so the purpose of this list is to provide initial guidance and easy reference.

TABLE 5.2 Driving Simulation Dependent Variables, Descriptions and References

Variable Classification	Variable	Description	Sample Reference
Longitudinal Control	Speed	Travel speed in km/h or mph.	HASTE (n.d.); Tijerina, Barickman, & Mazzae (2004)
	Speed Variability	Standard deviation of speed.	
	Time or Distance Headway	Time or distance to the rear bumper of the lead vehicle.	
Reaction Time	Perception Response Time (PRT)	PRT is the time to ease the foot off accelerator and initially press brake.	Green (2000); Lamble, D., Kauranen, T., Laakso, M., & Summala, H. (1999, p. 620); Olson & Farber (2003); Summala (2000)
	Brake Response Time (BRT)	BRT is the time from a hazard appearance to the brake onset.	
	Time to contact (TTC)	$TTC = ((v2 + 2Ad)^{0.5} \wedge - v_r)/A$ where d is distance, vr is relative velocity of the vehicle ahead, and A, is the deceleration of the vehicle ahead.	
Crash	Crash	The boundary of a driver's vehicle overlaps with other vehicles or objects, or the current vehicle control dynamics exceed those allowed by the vehicle equations of motion (e.g., rollover).	Caird et al. (2008); Horrey & Wickens (2007); Strayer & Drews (2004)
Lateral Control	Lateral Position	Continuous location of a vehicle with respect to a lane reference.	Green et al. (2004); HASTE (n.d.); Godthelp, H., Milgram, P., & Blaauw, G. L. (1984); van Winsum, W., Brookhuis, K. A., & de Waard, D. (2000)
	SDLP	Standard deviation of lateral position.	
	Lane exceedances (LANEX)	LANEX is the proportion of time outside of a lane or a frequency count of the number of times the lane threshold is exceeded.	
	Time to lane crossing (TLC)	TLC is the time to cross a lane boundary at current steering angle and speed.	
	Reversal rate (RR)	RR is the number of steering wheel direction changes per time or distance value.	
Eye Movements	Glance	All consecutive fixations on a target plus preceding transitions (i.e., saccades).	Green (2007); ISO (2002); SAE (2000); Fisher et al., (this book, chap. 18); Horrey et al. (2006)
	Eyes-off-road-time	The sum or proportion of all of the time associated with all glances not directed to the road.	
	Fixation	Momentary direction of the pupil's gaze (separated by saccades).	
	Percent Dwell Time (PDT)	Percent dwell time to a given area of interest (AOI; e.g., instrument panel).	
Workload, Subjective	NASA-Task Load Index (TLX)	A multi-dimensional subjective workload measure composed of six sub-scales and overall workload.	Hart & Staveland (1988)
	Rating Scale Mental Effort (RSME)	Paper-and-pencil instrument that measures workload on a continuous unidimensional scale.	Young & Stanton (2005); Zijlstra (1993)
	Situation Awareness Global Assessment Technique (SAGAT)	Stopping operator (or driver) activity to ask questions about dynamic information needs.	Endsley (2000); Jones & Kaber (2005)
	Driving Activity Load Index (DALI)	A modified workload assessment tool specific to driving activities.	Pauzié (2008)
Workload, Physiological	Heart Rate (HR)	Number of beats for a time period (usually one minute).	Brookhuis & de Waard (this book, chap. 17); Mulder, L. J. M., de Waard, D., & Brookhuis, K. A. (2005, p. 20–1)
	HR Variability	Standard deviation of HR.	
	Respiration	Breaths per minute.	
	Electroencephalography (EEG)	Evoked potential amplitude measured from central nervous system.	
	Skin Conductance	Electrical resistance of the skin.	
Other Measures	Entropy	Prediction error of vehicle signals.	Boer (2000); HASTE (n.d.); Horrey & Simons (2007)
	Safety Margins	Amount of space that drivers maintain around their vehicle. Includes but is not limited to headway.	
	Navigation	Assessment of driver's wayfinding ability or memory for trip directions.	
	Other higher-order or aggregate measures…		

Simulation studies do not necessarily focus on crashes although traffic safety epidemiology does. Crashes in simulation are of sufficient importance to elaborate. A common aphorism is that no one has died in a simulator crash and a fundamental question is to understand what crashes mean as an ordinal measurement variable. Evans considered simulator crashes as a thought experiment some time ago:

> Consider a make-believe simulator consisting of an actual car, but with the remarkable property that after it crashes a reset button instantly cancels all damage to people and equipment. What experiments could be performed on such make-believe equipment which would increase our basic knowledge about driving? (Evans, 1991, p. 127)

Two related questions can be circumscribed that do not dismiss driving simulation as a method outright, which appear to be the intent of the rhetorical arguments. First, because crashes in driving simulation have few consequences, does driving performance or behavior change towards more risk taking in the simulator? Second, does the absence of consequences necessarily negate results obtained using driving simulators? The first question is an empirically testable question that surprisingly, to our knowledge, has not been addressed. One group is put into a crash and another is not. Measures of performance are compared. Short of having results from this important experiment, in some situations, a simulated crash can be quite compelling—even frightening. For instance, older drivers, when involved in a simulated crash, are often quite concerned about the well-being of the other driver or pedestrian, which are virtual entities, and express concern over being reported to the licensing authorities. Novice drivers, similarly, are concerned that their parents might be informed when a crash occurs. The loss of privileges with a family vehicle seems to underlie the concern.

Crash experience coupled with information about the outcomes of a crash may be an effective way to offset the absence of any physical consequences. For example, once a crash occurs, the simulation can be stopped and drivers can be given information that estimates the extent of their injuries (including fatalities in more severe circumstances) along with estimated repair costs, based on the collision angles and impact velocities. Crashes may also become adverse to drivers in simulators by imposing some form of financial penalty for crash involvement, a strategy that might be particularly effective in university-age samples (see Ranney, this book, chap. 9). However, by replacing the utilitarian goals for driving with alternative incentives and/or punishments, participants may modify their performance so that it is dissimilar to normal or everyday driving. We note that a number of social motivations such as avoiding an embarrassing crash or showing off for others may also be in play in monitored environments, which include simulation and naturalistic studies. Predictable elicitation of the behavior and describing the impact of it when it occurs on crash risk represent important research directions. The impact of video gaming, and driving video games in particular, on driving simulation performance

and actual driving is another area that requires systematic research (Fisher, Kubitzki, Guter, & Frey, 2007).

Creating circumstances in which it is exceptionally difficult for drivers to avoid a crash in a simulator is rather trivial. Placing hazards into a driver's path with little space and time to respond is easy to achieve, though it is important that these hazards or critical events are not unavoidable, such that attentive and responsive drivers can safely avoid them. Tuning the constraints of a particular event so that a range of abilities is accommodated requires some pilot testing. Creating an array of difficult events to contend with is the basis of determining the limitations of drivers to respond while engaged in distraction tasks (e.g., Chisholm, Caird, & Lockhart, 2008). If crashes are unavoidable, there may be greater repercussions for the driver in subsequent experimental blocks or trials such as modifying glances and speed to anticipate hazards.

Another important point of discussion is the issue of repeated crash experience in a simulator. While simulators are effective at generating many crash-likely situations (and therefore many crashes), real-world crash exposure or experience is quite low. The extent to which driving simulation crash-likely protocols compress long-term exposure into crashes requires firmer footing in injury epidemiology (Green, 2008). At the individual or practical level, does repeated exposure to crashes in a simulator create unrealistic expectations on the part of drivers? Should a simulator session with a participant be terminated if more than one crash has occurred (see Ranney, this book, chap. 9)? Wickens (2001) discusses the psychology of surprise and the implications for research that tries to employ more than one surprising event for a single participant. Unfortunately, the reality is that you can only truly surprise a research participant a single time (and the experimental setting itself may dampen this prospect at the outset!). Subsequent responses after experiencing a surprising event hasten response times, depending on participant age, by about a half-second (Olson & Sivak, 1986).

Evans (2004, p. 188, *italics* in original) has also thought about the effect of expectancy on reaction times:

> Enthusiasm for driving simulators ignores some of the most basic understanding about the nature of traffic crashes. The discussion above on reaction time showed the primacy of expectancy. Even in experiments using actual instrumented vehicles, reaction times are substantially shorter than in normal driving. Any reliance by traffic engineers on reaction times determined on a simulator no matter how realistic; could produce unfortunate results. However, the reason that simulators are unlikely to produce knowledge relevant to traffic safety is more fundamental than this.
>
> Simulators measure *driving performance*, what the driver *can do*. However, safety is determined primarily by *driver behavior* or what a driver *chooses to do*. It is exceedingly unlikely that a driving simulator can provide useful information on a driver's tendency to speed, drive while intoxicated, run red lights, pay attention to non-driving distractions, or not fasten a safety belt. Twenty-year-olds perform nearly all tasks on simulators better than the 50-year-olds, but it is the 50-year-old who has sharply lower crash risks.

Several important questions are raised in this passage. First, do driving simulators provide reliable reaction time results? Second, can any knowledge about driver behavior be obtained by using a driving simulator? One way to determine if reaction times differ across methods is to compare the results from available research to determine if there are systematic differences. For instance, Caird, Chisholm, Edwards, and Creaser (2007) examined reaction times to yellow lights over the past 50 years. Across studies and methods, there is a close correspondence in reaction time results. Observational, test track, experimental, field and simulator are represented in the collection of studies. This issue is discussed further in Questions 8 and 9.

With respect to drivers' expectations, still others have examined patterns of systematic variance such as when a driver expects or does not expect an event (Olson & Farber, 2003). Green (2000) and Summala (2000) debate the implications on response time for different types of events: unexpected and expected events as well as for surprising intrusions. While one cannot truly surprise a participant more than once in a session, the use of different types of scenarios and event configurations can at least reduce the drivers' ability to anticipate these particular events. A continuum of surprise can be conceptualized depending on the frequency with which events are ordinarily experienced on a day-to-day basis. So, while the participant may quickly come to expect certain types of events, at least they can be made temporally and geometrically uncertain.

Regarding the second issue raised by Evans (2004), is there evidence or methods from driving simulation that shows aspects of driver behavior based on choice or self-paced interaction? Does having a prescribed task to perform necessarily negate results if the rate of a self-paced behavior is not necessarily known? For example, previous research on driver multitasking and distraction often will employ tasks that are outside of the driver's control. That is, the investigators usually prescribe the conditions under which the tasks are performed. While this technique may be useful in examining the interference from concurrent activities, it does not capture the adaptive potential of drivers (see Lee & Strayer, 2004; Horrey & Lesch, 2009 for further discussion). Thus, it is very possible that certain methodologies and—by extension—driving simulation may create artificial situations that are not completely reminiscent of real-world situations. Alternative methods that allow a repertoire of behaviors to occur, such as engaging in distracting activities when stopped at stop lights, are often difficult to categorize and analyze (Caird & Dewar, 2007; Stutts et al., 2005). For example, Yan, Adbel-Aty, Radwan, Wang, and Chilakapati

(2008) observed drivers who did not come to a complete stop at an intersection in the simulator and related it to the history of crashes at the actual intersection.

Consideration of behaviors that occur in naturalistic driving studies as a precursor to experimentation would surely reduce the occurrence of simulation studies that are not interesting; when this is done, research should have the potential to change our understanding of driving phenomena. Researchers should also carefully consider, in their discussion of study limitations, how their results may or may not generalize to real-world behavior.

7) How do you map the necessary realism or fidelity of a driving simulator to the specific research questions?

Naïve realism, or the misplaced desire for the precise replication of perceived reality, tends to be the default assumption of lay researchers about many forms of simulation such as medical, flight and driving (Gaba, 2004; Lee, 2005; see Jamson, this book, chap. 12, respectively). Fidelity or realism is the degree to which a driving simulation matches aspects of driving on-road (Meister, 1995). By making a simulation look and feel exactly as real tasks and environments do, one forces a set of requirements onto hardware, software and scenarios. By asking what needs require sufficient fidelity to achieve a certain level of verisimilitude, certain requirements can be relaxed. Within cost constraints, the operational assumption of less experienced researchers is to err on the side of higher levels of fidelity. For example, a signals engineer may be concerned about making sure that the traffic lights in a simulation function according to the correct timing cycles and algorithms. However, the traffic lights may be only relevant when a driver passes through several intersections on a drive. In general, the engineering disciplines tend towards requiring an exactness of the simulation even if these physical and virtual aspects of the simulation are not relevant to the tasks of the driver. Knowing which traffic environment features and tasks are relevant to the driver and to the research question being posed is central to developing an effective experimental design.

The typical approach to solving the mapping between the level of fidelity required in each of the various dimensions and the research agenda is to propose a taxonomy or matrix that lists simulator fidelity in columns and research questions or simulator components as rows. The start of one of these lists is shown in Table 5.3 and is left for the reader to complete (i.e., …). A complete table would describe a general classification scheme for simulators that often is a variant of

TABLE 5.3 Simulation Fidelity by Component or Research Question

	Low	Moderate	High
Moving Base	Fixed	Fixed or limited motion cuing	Moving base
Screen Width	20 degrees	150 degrees	360 degrees
Screen Resolution	.	.	.
.	.	.	.
Sign Legibility	Poor	Fair	Good
Night-time Visibility	No capability	Poor	Fair
.	.	.	.

low, medium and high fidelity. For example, relatively recent pictures of low, medium and high fidelity simulators are provided in Jamson, this book, chap. 12 and Shinar (2007). An excellent discussion of physical fidelity, or the matching of physical components in a simulator with vehicle capabilities, is provided by Greenberg & Blommer, this book, chap. 7. One of the limitations with this matrix approach is that simulator categories migrate from right to left as technology progresses (see Allen, Rosenthal, & Cook, this book, chap. 2). For instance, the graphics of the low-fidelity simulators of today are well beyond the capabilities of driving simulators a decade ago. Knowing where the state of technology is (or was) is important to interpreting the inherent limitations of various studies across time.

The mapping of simulation fidelity to a research question, in part, assumes that a simulator can be prescribed at the time that a particular question is addressed. In reality, the justification for a range of research problems is offered at the time that simulator funding is sought and when the simulator is purchased (see Question 2). Researchers hope that they have anticipated or planned adequately for a research agenda that corresponds to the first few years of activity; that is, if operational costs are also adequately anticipated. If a simulator is funded and the researchers have limited experience using one, several years of lag typically occur before papers begin to appear at conferences. During this time, students and technicians figure out, often by trial and error, how to translate research questions into programming details, to develop complex experimental designs and to analyze vast quantities of data. As technical and experimental design experience is acquired, the scope of questions that can be addressed tends to be resolved. Thus, the actual fit of a simulator to research questions will be somewhat dependent on the understanding the researcher has when applying for funding and the eventual accumulation of research expertise using a simulator.

A mid-fidelity simulator is not needed for every research application. For instance, a very high-definition visual system could be connected to a gaming steering wheel, brake and accelerator. Such a high-low fidelity simulator might be used for visibility or sign comprehension research where vehicle response characteristics are not necessarily important measures. If a set of research questions can be framed with precise task requirements or fidelity, those physical components can be selected at higher levels of fidelity, whereas other simulator equipment not central to important tasks and measures is made to be nominally functional.

The fidelity of specific tasks that are the focus of a study often requires greater approximations of the real thing than a simulator can achieve. Simulators with fixed physical properties that cannot be appreciably modified have limited fidelity. In some cases, research can or cannot address a question based on inherent limitations. More likely, the researcher can address a question, but the results represent an approximation and may be open to criticism if peer reviewed. Questions that require specific lighting conditions (e.g., driving at dusk, dawn or night), high resolution of detail (e.g., complex signs), important response properties (e.g., electronic stability control) or tracking in the periphery (e.g., the approach of a high-speed train or merging in a work zone) may simply not be able to be addressed by a simulator with limited projection luminance, graphics engines, vehicle control models coupled to motion bases, and field of view.

How aspects of driving simulation with insufficient fidelity affect measures requires either comparison to on-road results or qualification with respect to the limitation. For instance, insufficient visibility of a sign due to resolution limitations will require that drivers be closer to the sign in order to read it. The overall time to view a sign would be expected to be less, which would also affect comprehension and legibility distance too. Comparison of simulator results to on-road testing is likely to reveal measures that are truncated in predictable ways as a function of the simulator limitation.

8) To what degree are the results from a simulator similar to those measured in the real world?

Probably the most often repeated yet also most difficult question about driving simulators is whether measures in a simulator mimic those measured when driving in the real world. This multi-level correspondence problem has broadly been called simulation validation, which has been defined as the replication of simulator and on-road tests to determine the extent to which measures correspond across contexts. Simulation validation has been a concern for at least 25 years (Blaauw, 1982). A number of important aspects of this topic are addressed by several chapters in this *Handbook*; namely, performance validity (see Mullen, Charlton, Devlin, & Bédard, this book, chap. 13) and cross-platform validation issues (see Jamson, this book, chap. 12). A number of authors have also reviewed simulation validation at different times (e.g., Blaauw, 1982; Godley, Triggs, & Fildes, 2002; Kaptein, Theeuwes, & Van der Horst, 1996).

A number of types of simulation validity are important. *Relative validity* indicates that the direction of change of a variable is in the same direction as a corresponding manipulation and measure in the real world (Kaptein et al., 1996). For example, if speed is measured at 5 km/h greater in a simulator than on-road across a range of speeds, the pattern of data would represent relative validity because it is in the same direction. *Absolute validity* is the extent that a manipulation in the real world when manipulated in the simulator produces the same or equivalent numerical change in the same measure (Blaauw, 1982, p. 474). For instance, for a specific condition, 50 km/h in the simulator for a given stretch of roadway is also measured at 50 km/h on-road.

Given the variability of driver performance, the exact correspondence of on-road and simulation measures is unlikely. Some authors suggest that finding measures that reflect absolute validity are not as important as finding measures that produce consistent changes in the dependent variable or relative validity (e.g., Kaptein et al., 1996; see in this book, chap. 3 by Kantowitz and chap. 13 by Mullen et al.; Törnos, 1998; Yan et al., 2008). Others have further argued that simulator companies should produce studies that establish validity for their products. For various reasons, the responsibility of validating simulators has fallen to researchers. Published studies on simulation validation are obviously a small fraction of operating simulators, especially if each simulator type represents a unique need to validate. A sizable body of gray literature on simulation validation appears in conference proceedings, technical reports and internal reports.

Sampling from published studies, simulation validation studies have generally been focused on findings of relative validity. For example, simulators have been found to have relative validity for speed (e.g., Blaauw, 1982; Törnos, 1998; Yan et al., 2008). Higher speeds were found in the tested simulators than in on-road segments. Lane-keeping measures were slightly more variable in the simulator compared to on-road measures in several studies (e.g., Blaauw, 1982; Reed & Green, 1999). However, the effects tend to vary depending on the simulator being evaluated and additional independent variables (e.g., experience, age, etc.). For example, higher speeds on-road were found in Godley et al. (2002) and lateral position in a tunnel shifted 13 cm more to the left in the actual tunnel (Törnos, 1998). Still others have used unique validation measures such as safety surrogates (Yan et al., 2008), which attempted to relate risk-taking maneuvers in a simulated intersection with the crash history of the actual intersection. Reed and Green (1999) measured many variables while drivers dialed a phone. The correlation between on-road and simulation variables ranged from 0.18 to 0.76. Thus, the use of some variables is consistent across simulation validation studies, whereas other variables are unique and cannot be compared across studies. The uniformity of relative validity is somewhat variable depending on the measure. Finally, compared to the number of variables represented in the dependent variable list of Table 5.2, relatively few have been used (or reported) in simulation validation studies, although this appears to be changing (see Mullen et al., this book, chap. 13).

The general comparison of variables in simulation validation can be abstractly illustrated. For a given measure (or aggregation of measures), a plot of obtained values from a simulator can be plotted against the values obtained for on-road performance under similar driving conditions. Thus, Figure 5.3 represents the correspondence of simulation and on-road measures. An exact correspondence of measures from road and simulator indicates that each is measuring exactly the same thing (i.e., absolute validity). Each line on the graph can be also thought of as the best fitting line of a scatterplot of data. The degree or extent that a given measure departs from an on-road variable is also plotted (relative validity). Several lines of relative validity are plotted (i.e., A & B). One illustrates a situation where on-road measures are consistently higher than those from simulation A. (The reverse is also plotted, B, where the measures are consistently lower.) Attribution as to why relative validity is higher or lower in either setting would then require systematic elimination of potential causes. For example, the finding that a simulator may yield higher lane variability may indicate that steering control is insufficiently tightly coupled. The situation becomes more complicated in circumstances depicted by line C of Figure 5.3, where some parameter yields higher values in on-road (versus simulator) measurement for a certain range of values (left-hand side of figure), yet lower values in another range (right-hand side). Whether line A is more valid than B and the extent to which line C is invalid are unresolved questions that require further empirical support and discussion.

Statistical tests are one way to determine if measures taken from on-road and simulation differ with respect to absolute validity or from other tests where relative validity is established, but

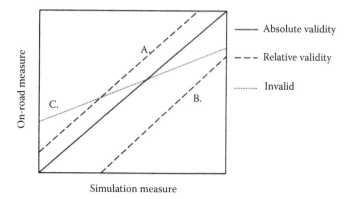

FIGURE 5.3 Absolute and relative validity expressed as a function of the correspondence of measures from on-road and simulator for identical contexts. See text for details.

what is the practical significance of a measure that differs more or less from absolute validity? Statistical testing of significant differences between simulation and on-road measures can be affected by unbalanced within-subjects treatment orders and strategies (e.g., boredom may lead to speed increases) (see Godley et al., 2002 for discussion). Measures are often collected over a limited range of values and are not necessarily treated as continuous variables. The relative validity of a simulator should be qualified relative to the range over which measures are collected. In addition, the relative validity of a collection of measures requires consideration.

9) How well do the results from simulator, on-road and more basic laboratory studies correspond with one another?

Researcher beliefs about the validity of driving simulation results range from advocacy to dismissive. Some scientists do not seem to be swayed by consistent evidence obtained from studies that compare on-road and simulator results that establish relative validity. In addition to direct comparisons within a study that determine simulation relative validity, measures can be compared across studies of on-road, simulator and laboratory. The correspondence of results across different experimental approaches is related to the previous question although comparisons are not restricted to simulation and on-road.

A number of recent studies have sought to statistically compare results arising from different methods. For instance, Caird, Willness, Steel, and Scialfa (2008) meta-analyzed studies from simulator, laboratory and on-road settings where the impact of cell phone conversations on driving performance were measured. Meta-analysis is a means to statistically compare the results of studies using effect sizes (Elvik, 2004; Rosenthal & DiMatteo, 2001). The categories of laboratory, simulator and on-road were defined. A laboratory study had at least an approximation of a tracking task ($N = 14$ studies). Driving simulator studies had a physical steering wheel coupled to an approximation of a visual traffic environment ($N = 18$). On-road studies were conducted on a test track or on ordinary roadways ($N = 5$). (Horrey & Wickens (2006) performed a similar analysis and found similar results but collapsed simulator and laboratory categories together.) For conversations while driving with hand-held or hands-free phones, on-road, simulator and laboratory settings all produced similar effects on reaction time and speed (i.e., slower reaction time and slightly slower speeds when conversing). As such, the belief that a single method—whether on-road, simulation, or laboratory—is best for approximating performance decrements does not seem to hold, at least for these tasks and measures. Approximating task demands (i.e., conversation) across settings produced similar effect sizes. The use of surrogate tasks to approximate actual performance is also addressed by Angell, this book, chap. 10.

Studies of in-vehicle information system interactions while driving have also compared results across laboratory, simulation and real-world settings (Santos, Meret, Mouta, Brookhuis, & de Waard, 2005). The abstract task that was created, and installed on a central screen, was a visual search task (called S-IVIS). The three research settings were able to detect, using lateral control measures, when drivers were looking down at the task, but simulation and field studies were also able to discriminate levels of search difficulty too. Comparisons across settings for other variables such as speed, speed variation, and distance headway were more complicated to interpret based on methodological differences across settings. The authors discuss a number of cautions about blanket generalizations.

As part of the same European project (called HASTE), Engström, Johansson and Östlund (2005) reported results of cognitive and visual tasks across real and simulated driving settings. The same visual search task was used (i.e., S-IVIS) and also an auditory memory task (i.e., ACMT), which was responded to by interacting with a touch screen at the top of the central column. Of interest, fixed (Volvo) and moving-based (VTI) simulators results for longitudinal, lateral, physiological, and eye movement measures were compared. The visual search task affected lateral control and reduced speed, whereas the cognitive memory task reduced lateral variability but not speed. Eye movements were concentrated on the center of the road when engaged in the cognitive task and to the search screen when engaged in the visual task. The pattern of measurement differences across settings provided similar results with some differences. Lateral variation was greater in the fixed simulator than the moving base simulator. In spite of this difference, replication of results across methods corresponds well when tasks are similar and method variance is minimal.

Validity and generalizability are both important experimental properties. The selection of a method that can provide an approximation of a generalizable effect of interest is fundamental to experimentation. Many years ago, Chapanis (1967) made the observation that laboratory tasks are rough approximations of real-life situations. Since then, human factors beliefs have wandered from accepting laboratory tasks as having external validity, which is the extent that results obtained from participants, environments and tasks necessarily relate to other participants, environments or tasks in real-life (Abelson, 1995). If task, environmental and population samples that are integrated into a study are similar to the same in real-life, generalizability is more likely to be achieved. Cost and time may dictate the methods or analytical approaches available to the researcher (Carsten & Brookhuis, 2005) and these decisions may limit the generalizability of a study. The danger in assuming that an effect in the laboratory or

simulator holds to the same extent in the real world is that many other variables, not tested, may dilute or reduce the relative importance of a found effect (Chapanis, 1967). For example, many have wondered why more crashes do not occur when using cell phones when driving. One disturbing possibility, besides drivers not confessing to using a cell phone when a crash does occur, is that other conditions not tested may interact and mitigate the decrements of conversation on reaction time and speed in the real world.

Another related question concerns situations where data is available or are combined from different methodological approaches. What is the relative weight that should be given to driving simulation versus real-world results? The bias of some individuals is to place "real world" results on the highest platform of validity with other results relegated to lesser platforms. This naïve view is based on the intuitive belief that "real world" results, if obtained, have the highest validity. Consideration of the strengths and weaknesses of a given method—in essence, overcoming these potential biases—requires greater thought, scientific training and experience.

If a simulator study found similar results as a naturalistic study, which one is more valid? One finds a causal relationship between or among manipulation and measure(s), whereas the other describes an event or pattern of observations. In situations where the two approaches show convergent findings, the distinction between approaches may be less critical. However, if the results from a simulator conflicted with those found in a naturalistic study, which one purports the greater truth? Naturalistic studies are compelling because real behavior and crashes are captured and the videos can be seen, although it takes a while to capture these events. Driving simulation is relatively efficient at describing performance in a specific context and establishing causality among manipulations and measures. Observations made using the latter methodology represent a potential reality which may generalize to a limited set of conditions in the real world. That said, either method can yield low or high quality results (Nickerson, 1999). Judgment of research quality is an important expertise of researchers that can sort out how much weight or credibility to assign to a particular set of results (Wortman, 1994).

10) What are common threats to internal and external validity?

After reviewing hundreds of manuscripts, we have identified a number of recurrent threats to internal and external validity when conducting driving simulation studies (see Table 5.4). The purpose of listing the threats, solutions and references is to encourage researchers to take a careful pause before starting a study. Recognizing a design weakness or flaw before collecting data is preferable to realizing this prior to submitting a manuscript. Many threats to internal and external validity can also be learned from the school of "ego-crushing" manuscript reviews. Obviously reviewers could adopt a guiding or mentoring style when reviewing manuscripts that have been submitted.

These are high-level problem and solution descriptions and the details of implementation for a given study require further investigation and consideration by researchers. Issues of statistical analysis, reporting and interpretation are comprehensively addressed by others (see Boyle, this book, chap. 21; Cohen, 1990; see Dawson, this

TABLE 5.4   Common Threats to Internal and External Validity When Using Driving Simulation

Threat	Description	Solution	Reference
Failure to adequately screen participants.	Vision or health problems (among others) of certain individuals may contribute to experimental error that masks or distorts effects.	Test for outliers. Use appropriate tests to screen drivers. Use covariate analyses to remove offending variance.	Trick & Caird (this book, chap. 26); Rizzo (this book, chap. 46); Tabachnick & Fidell (2006)
Generalization issues.	Tasks, population samples and environments are not similar to whom or what you wish to generalize.	Qualify results according to generalizability limitations. Include similar tasks, samples, and environments to desired generalizations.	Abelson (1995); Kaptein et al. (1996)
Drop out due to simulator sickness.	Properties of the simulator or activities in the simulator cause participants to become sick.	Carefully screen at-risk participants. Reduce maneuvers that require a sweeping motion such as left or right turns (among many other remediation strategies).	Stoner et al. (this book, chap. 14); Trick & Caird (this book, chap. 26)
Non-randomization of participants, treatments or events.	Treatments, participants or events are not randomly assigned to levels of the independent variable. Events are predictably located within drives.	Randomize. Check for order effects.	Abelson (1995); Wilkinson et al. (1999)
Range or carry-over effects.	Multiple treatments are experienced by the same participant. The order of treatment and experience causes asymmetric effects.	Use between-subjects designs for different treatment levels. Limit the number of treatments and counterbalance. Check for order effects and qualify results accordingly.	Poulton (1982)
Low number of participants or observations per cell.	Few participants or observations per cell reduce the stability of results.	Run more participants or collect more observations. Conduct power analysis. When possible, consider more efficient experimental designs (e.g., within-subjects).	Cohen (1992); Wilkinson et al. (1999)
Visual fidelity distortions.	Incorrect luminance levels, missing or distorted monocular and binocular cues.	Qualify results according to visual limitations.	Andersen (this book, chap. 8); Wood & Chaparro (this book, chap. 28)
Control fidelity distortions.	The quality of steering, brake and acceleration is unlike that of actual vehicle performance.	Qualify results according to control limitations. Provide sufficient practice such that drivers meet some criterion level of performance or competence.	Flach et al., (this book, chap. 43); Greenberg & Blommer (this book, chap. 7)
Greater than 5% loss of data.	Equipment or subject problems cause a loss of data.	Use appropriate data substitution methods and report rationale. Drop participants, fix equipment and re-run. Use appropriate statistical tests if assumptions are violated.	Tabachnick & Fidell (2006); Siegal & Castellan (1988); Wilkinson et al. (1999)

book, chap. 22; Tabachnick & Fidell, 2006; Wilkinson et al., 1999). The final caveat of this list is that we have probably omitted a number of threats to internal and external validity so researchers should be vigilant about improving the quality of their experimentation.

11) Is there evidence that driving simulators have contributed to the advancement of knowledge since the 1970s?

Evans (2004, p. 190) makes the following argument:

Driving simulators are far from new. A 1972 article refers to an earlier 1970 article listing 28 devices then in use, 17 of them in the U.S. Since the 1960s, driving simulators have incorporated moving bases and multiple movie projectors to provide visual information, including to the rear view

mirror. The list was published in 1972. The research literature provides scant evidence that [the] research agenda was advanced by simulators, neither by those in existence in 1970, nor by the much larger number of far more expensive and sophisticated simulators that have been built.

To clarify, the list (or research agenda) that appears in Evans' book on page 191 includes traffic control devices, drug effects, the driver as a control element, vehicle characteristics, highway design, and driving conditions. To distill out a question, has driving simulation contributed to the development of these or other areas of knowledge since the 1970s?

In Figure 5.3, we plot the progression of studies that have used driving simulation as a method since 1970 in two relevant journals: *Accident, Analysis and Prevention (AAP) and Human*

Factors (HF). These journals were selected to get a general indication of international driving simulation research activity over a span of four decades. HF has been published by the Human Factors and Ergonomics Society since 1959 and is representative of activities in the U.S. and other countries. AAP was originally a European journal that has more recently (circa 2002) systematically expanded the number of studies published to take advantage of online dissemination. Both journals publish studies that focus on the driver and, to a lesser degree, on the underlying simulation technology.

Both journals were searched for studies that used driving simulation. Titles, abstracts and full papers were searched for the occurrence of <driving simulation> and other variants. Study inclusion criteria are modified from Caird et al. (2008) and Horrey and Wickens (2006) who, in their meta-analyses of cell phone studies, established categories for different methodological approaches; namely, laboratory, simulation and on-road. The simulation category included part-task, low-fidelity and high-fidelity simulators. In the present analysis, however, studies that fractured the coupling between perception and action are not included because a participant does not actively control anything. For instance, ratings or responses to picture or video presentations are not included.

How much research has been produced? As shown in Figure 5.4, the number of published studies using driving simulation in *AAP* has increased dramatically in the past four years. (The year 2008, which had 22 simulator studies, includes studies in press that may appear in 2009.) This rise seems to parallel the increase in the number of manuscripts accepted by *AAP*. Attribution of the increase to a growth in driving simulation research activity may be only partly correct. The increase probably reflects a modest rise in research activity, but also possible shifts in acceptance criteria at *AAP*. The increase in driving simulator studies in *HF* has been less dramatic over the entire 40-year span, though there is a recent upwards trend since the mid-90s.

For many years, research centers within various countries (e.g., TNO, NHTSA) developed driving simulators and productivity (i.e., peer-reviewed papers) was a reflection of the activity of these centers. Many technical difficulties in hardware and software had to be overcome to execute each research study. The current generation of driving simulators is significantly more flexible and allows researchers to develop and execute research on a much shorter time scale compared to previous generations of simulators. The cost of mid-fidelity simulation has declined as graphics and computation capabilities increased many fold in commonly available systems. The enhanced sophistication of simulators in this respect may also be an important contributor in the increase in published research using simulators.

What kinds of research have been produced? The purpose of Table 5.5 is to aggregate the general pattern of research activity in *AAP*. The subcategories roughly capture measures and/ or manipulations of particular studies. A study was counted in several categories if multiple measures or manipulations captured the purpose of each study. For example, speed perception was counted under Speed and Perception/Attention. Clearly driving simulators have been used to evaluate devices for distraction, new road geometries, and traffic control as well as to understand lifespan individual differences among many other uses.

The list in Table 5.5 only partially resembles the list discussed by Evans (2004, p. 189), which indicates how research simulators were being used in 1972. To be fair, many of the studies that appear in Table 5.5 have been published since the original criticism by Evans. Finding research gaps is as simple as combining categories or factors until that research has not been performed previously. Some questions or combinations are obvious, whereas others are deeper and transcend independent variable combinations.

There are some limitations to simply counting and classifying studies from journals. Many driving simulation studies are conducted for a sponsoring agency (e.g., NHTSA). The results

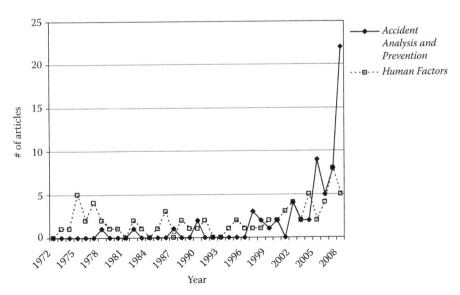

FIGURE 5.4 The number of simulator studies by year published in *Accident, Analysis and Prevention* and *Human Factors* from 1970 to 2008. (Not plotted *HF*: 1964 = 1, 1968 = 1.)

TABLE 5.5 Types and Subcategories of Driving Simulation Studies Published in *Accident, Analysis and Prevention* From 1969 to 2008

Study Type	Subcategories
Alcohol (4) or Drugs (1)	On-board detection (1), Age (1), Skilled drivers (1), Other (2)
Distraction (11)	Cell phone (5), Conversation (1), Feedback (1), Eating (1), Multiple sources (2), MP3 (1)
Devices (2)	Warnings (1), Audio Speakers (1)
Fatigue or Sleepiness (10)	Napping (1), Other intervention (1), Simple versus complex measures (1), Sleep apnea (2), Physiological measures (1), Other (4)
General Individual Differences (8)	Expert/Novice (1), ADHD (1), Gender (2), Police (1), Truck Drivers (1), Teens (1), Training (1)
Older Drivers (8)	Self ratings (1), Individual differences testing (4), Pedal errors (1), Other (2)
Pedestrians (4)	Unfamiliar traffic direction (1), Age/development (2), Timing (1)
Perception/Attention (7)	Visibility (3), Fog (2), Speed selection (1), Visual fields (1)
Roadway (12)	Curves (2), Rural two-lane (1), Intersections (3), Horizontal curves (1), Tunnels (2), Width (2) Traffic lights (1), Lane treatments (1)
Simulation Validation (4)	Intersections (1), Tunnel (1), Speed (1), Older drivers (1)
Speed (5)	Two-lane rural road (1), Perception (1), Gender (1), Fog (1), Curves (1)

are put into a technical report and may or may not be published in a journal. For instance, Meister (1995) provides pointers to many of these flight and driving technical reports that appeared from the 1950s to the 1980s and reference backtracking reveals many others. Technical reports often have considerably more details and analyses than condensed peer-reviewed publications, including simulator and methods descriptions. Thus, reliance on journals such as *AAP* and *HF* probably underestimates the true scope and overall activity that driving simulation has contributed to knowledge. Moreover, some editors may choose to exclude certain methods if they feel the results are not valid.

The contribution of *AAP* and *HF* to knowledge advancement based on published driving simulation studies is open to criticism. At the time that a study is published, reviewers and editors may not be aware whether or not a particular study advances the state of knowledge over that which existed prior to publication. Awareness of research worldwide is daunting but facilitated by search engines, citation indexes and custom publication alerts. Determination of the relative contribution of a given study to the collective body of knowledge is fallible and the relative influence or impact of a given study varies. The collective contribution of the body of research across time does not simply reduce to a journal's impact factor multiplied times the number of studies on driving simulation that appear in it (among other metrics) (e.g., Addair & Vohra, 2003). The qualitative dimension of influence on subsequent thought, method, policy and countermeasure is not captured by quantitative reductions. Suffice it to say that the collective impact of these papers has been constant and is now accelerating.

12) Based on the progress of software, hardware and projection capabilities, where will the future of driving simulation lead?

While this chapter was being written, U.S. and Canadian government policy has decided to bail out the Detroit 3 automakers as a long-term recession progressed worldwide. A quarter previously, worldwide vehicle manufacturing was looking towards India's Nano by Tata, a low-cost vehicle, and China's increasing demand for automobiles for the middle class. Globally this outlook has completely changed. Without question the current financial oscillations will affect budgets for research and development within corporations, local and national governments. In turn, the use of driving simulation to address a range of research and development agendas will likely be curtailed especially as large-scale commitments to naturalistic studies have been made in the U.S. and Europe.

Beyond the uncertainty of the current global recession (or depression), future directions will resemble past and current technological vectors. First, computer processor speed, in both graphics and operations, will increase in accord with Moore's Law, which predicts the cost of processing power halving every 18 months. However, the translation of this speed into graphics engines and usable software for driving simulators is imperfect compared to the same translation into computer games. An off-the-shelf gaming system running a driving game looks significantly better than the graphics of most moderate-fidelity driving simulators. Although sitting in an actual car surrounded by screens is compelling, simulation graphics capabilities are impoverished compared to what can be played at home. The reason for this obvious discrepancy is the close relationship that has evolved between game developers and graphics chip producers. Driving simulator companies are not part of this relationship.

Second, the usability of most driving simulator software frequently requires a computer programmer to overcome numerous idiosyncratic interactions, scripting and programming requirements to execute important functions. This becomes a perennial cost center in addition to software licensing and hardware upgrades. In contrast, easy-to-use experimenter interfaces provide a means of addressing more research questions by a wider range of students and researchers. A large visual database and an easy-to-use interface probably facilitate the production of more research over the half-life of a simulator. The limited marketplace for driving simulators prevents small, undercapitalized companies from adequately designing software, visual databases, graphics engines and user interfaces among other aspects of driving simulators. Market forces such as the adoption of training curricula or driver testing using driving simulation may create a larger market for driving simulators, but these simulators may not necessarily be of a flexible variety because curriculum modules will become part of the sale.

Third, increases in storage media size will permit researchers to collect more variables at higher sampling rates, including multiple digital video streams, all at lower price points. However, what to do with all the data will remain an issue. Reduction of data and the conversion of it into useful and insightful results will require that researchers know what to look for, where to find it and how to convey it coherently. Some argue that only theory can guide researchers on where to look. If only it was so simple (Meehl, 1993). Only a bricolage of observations and theories will sufficiently describe and explain driver performance and behavior across contexts (e.g., Dewar & Olson, 2007; Groeger, 2000; Shinar, 2007). Current and future generations of simulation may allow for greater insight into more and more subtle combinations of relationships. For instance, dynamic visualizations of complex variable relationships and re-creations of actual crashes (e.g., National Transportation Safety Board, 2003) are important directions that need to be pursued.

Fourth, research with driving simulators should make a difference and address important gaps in knowledge (e.g., Pedean et al., 2004; Sivak et al., 2007). In particular, understanding why and how drivers are killed and injured is an important endeavor (Evans, 2004) as is determining if countermeasures are effective (Elvik, 2004). Identifying where and when the greatest number of drivers are killed or injured is a form of research triage that can prioritize a research agenda. Current hot topics in research, for instance driver distraction, can influence researchers away from considering important crash contributors that are not constantly in the media. In addition, the business of attracting contract or grant funding to feed simulator operational and soft-funding budgets can obscure the importance of pursing traffic safety questions which do not necessarily fit into funding research agendas of the day.

Finally, will researchers in the future look back at the technology of driving simulators today with the same detached curiosity we have when examining Figures 5.1 and 5.2? Will they exclaim, "How did they do what they did with such primitive technology?" Hopefully, "Why did they do that (kind of research)?" will be less of a concern. Will driving simulation be indistinguishable from real driving (see Hancock & Sheridan, this book, chap. 4)? Will our children's children still drive at all (or swim)?

Key Points

- Studies that have used driving simulation since the 1970s have contributed to a number of important research areas including human factors and traffic safety.
- Driving simulators are inherently neither bad nor good methodologically. How simulators are used by researchers can result in high-quality research or results that are open to criticism.

- Threats to internal and external validity in driving simulation can, in part, be addressed by attention to a number of methodological details.
- Driving performance in simulators may provide an optimistic view of drivers' capabilities relative to actual behavior.
- In the future, the cost of driving simulators will decrease, realism will increase, and applications are likely to expand.

Keywords: Epistemology, History, Philosophy of Science, Simulation Validation

Acknowledgments

Don Fisher, Mary Lesch, Marvin Dainhoff, Andrew Mayer, Shaunna Milloy, Pierro Hirsh, Lana Trick, & Matt Rizzo provided helpful guidance, comments and edits to early versions of this chapter. The Canadian Foundation for Innovation (CFI) funded the University of Calgary Driving Simulator (UCDS) and the AUTO21 Networks of Centers of Excellence (NCE) funded numerous research studies using it.

Glossary

Absolute validity: The extent to which a manipulation of a variable in the real world produces the same or equivalent change in the same measure when manipulated in a driving simulator (Blaauw, 1982, p. 474).

Comprehensiveness: The extent to which operational functions and environmental characteristics, etc., are reproduced in a simulator (Meister, 1995, p. 204).

External validity: The extent to which results obtained from participants, environments and tasks obtained in one setting necessarily relate to other participants, environments or tasks in other, broader settings (Abelson, 1995).

Fidelity or realism: The degree to which reality is matched in a simulation (Kantowitz, this book, chap. 3).

Interestingness: To change what people believe about an important topic (Abelson, 1995, p. 13).

Internal validity: The conclusion that a manipulation of an independent variable results in a corresponding effect on a dependent variable and does not result from some other known or unknown influence (Abelson, 1995).

Physical fidelity: The extent to which a physical variable in a simulator (e.g., roll) corresponds to its operationally equivalent component in the real world (Lee, 2005, p. 88; see also, Greenberg & Blommer, this book, chap. 7).

Psychological fidelity: The degree to which the simulation task is perceived by participants as being a duplicate to the operational task (Meister, 1995, p. 206).

Relative validity: The direction of change of a variable is in the same direction as a corresponding manipulation and measure in the real world (Kaptein et al., 1996).

Simulation validation: The replication of simulator and on-road tests to determine the extent to which measures correspond across contexts.

Web Resources

The *Handbook*'s web site contains supplemental materials for this chapter including two color figures:

Web Figure 5.1: Link flight simulator (left) and terrain map (right) displayed at the Smithsonian Air and Space Museum in Washington, D.C.

Web Figure 5.2: Link flight simulator comic showing simulator sickness or vertigo circa 1942 displayed at the Canadian Air Force Hall of Fame Museum in Wetaskiwin, Alberta.

Key Readings

Abelson, R. P. (1995). *Statistics as principled argument*. Hillsdale, NJ: Lawrence Erlbaum.

Evans, L. (2004). *Traffic safety*. Bloomfield Hills, MI: Science Serving Society.

Gibson, J. J., & Crooks, L. E. (1938). A theoretical field-analysis of automobile driving. *American Journal of Psychology, 51*(3), 453–471.

Shinar, D. (2007). *Traffic safety and human behavior*. Amsterdam: Elsevier.

References

Abelson, R. P. (1995). *Statistics as principled argument*. Hillsdale, NJ: Lawrence Erlbaum.

Addair, J. G., & Vohra, N. (2003). The explosion of knowledge, references and citations: Psychology's unique response to a crisis. *American Psychologist, 58*, 15–23.

Blaauw, G. J. (1982). Driving experiments and task demands in simulator and instrumented car. *Human Factors, 24*, 473–486.

Boer, E. R. (2001). Behavioral entropy as a measure of driving performance. *Proceedings of the first international driving symposium on human factors in driving assessment, training and vehicle design* (pp. 225–229). Aspen, CO.

Caird, J. K., Chisholm, S., Edwards, C., & Creaser, J. (2007). The effect of yellow light onset time on older and younger drivers' perception response time (PRT) and intersection behavior. *Transportation Research, Part F: Traffic Psychology and Behavior, 10*(5), 383–396.

Caird, J. K., Chisholm, S., & Lochhart, J. (2008). The effect of in-vehicle advanced signs on older and younger drivers' intersection performance. *International Journal of Human Computer Studies, 66*(3), 132–144.

Caird, J. K., & Dewar, R. E. (2007). Driver distraction. In R. E. Dewar & P. L. Olson (Eds.), *Human factors in traffic safety* (2nd ed., pp. 195–229). Tucson, AZ: Lawyers & Judges Publishing.

Caird, J. K., Willness, C., Steel, P., & Scialfa, C. (2008). A meta-analysis of cell phone use on driver performance. *Accident Analysis & Prevention, 40*, 1282–1293.

Carsten, O., & Brookhuis, K. (2005). The relationship between distraction and driving performance: Towards a test regime for in-vehicle information systems. *Transportation Research, Part F: Traffic Psychology and Behavior, 8*(2), 75–77.

Chapanis, A. (1967). The relevance of laboratory studies to practical situations. *Ergonomics, 10*(5), 557–577.

Chisholm, S. L., Caird, J. K., & Lockhart, J. (2008). The effects of practice with MP3 players on driving performance. *Accident Analysis & Prevention, 40*, 704–713.

Cohen, J. (1990). What I have learned (so far). *American Psychologist, 45*(12), 1304–1312.

Cohen, J. (1992). A power primer. *Psychological Bulletin, 112*(1), 155–159.

De Silva, H. R. (1936). On an investigation of driving skill. *The Human Factor, 10*, 1–13.

Dewar, R., & Olson, P. (2007). *Human factors and traffic safety* (2nd ed.). Tucson, AZ: Lawyers and Judges.

Dingus, T., Klauer, S. G., Neale, V. L., Petersen, A., Lee, S. E., Sudweeks, J., et al. (2006). *The 100-car naturalistic driving study, Phase II—Results of the 100-car field experiment* (Rep. No. DOT HS 810 593). Washington, D.C.: National Highway Traffic Safety Administration.

Elvik, R., & Vaa, T. (2004). *The handbook of road safety measures*. Amsterdam, the Netherlands: Elsevier Science.

Endsley, M. R. (2000). Direct measurement of situation awareness: Validity and use of SAGAT. In M. R. Endsley & D. J. Garland (Eds.), *Situation awareness and measurement*. Mahwah, NJ: Lawrence Erlbaum.

Engström, J., Johansson, E., & Östlund, J. (2005). Effects of visual and cognitive load in real and simulated motorway driving. *Transportation Research, Part F: Traffic Psychology and Behavior, 8*(2), 97–120.

Evans, L. (1991). *Traffic safety and the driver*. New York: Van Nostrand Reinhold.

Evans, L. (2004). *Traffic safety*. Bloomfield Hills, MI: Science Serving Society.

Fisher, P., Kubitzki, J., Guter, S., & Frey, D. (2007). Virtual driving and risk-taking: Do racing games increase risk-taking cognitions, affect, and behaviors? *Journal of Experimental Psychology: Applied, 13*(1), 22–31.

Gaba, D. (2004). The future vision of simulation in health care. *Quality Safety in Health Care, 13*, 2–10.

Gibson, J. J. (1986). *The ecological approach to visual perception*. Hillsdale, NJ: Lawrence Erlbaum.

Gibson, J. J., & Crooks, L. E. (1938). A theoretical field-analysis of automobile driving. *American Journal of Psychology, 51*(3), 453–471.

Godley, S. T., Triggs, T. J., & Fildes, B. N. (2002). Driving simulator validation for speed research. *Accident Analysis & Prevention, 34*, 589–600.

Godthelp, H., Milgram, P., & Blaauw, G. L. (1984). The development of a time-related measure to describe driving strategy. *Human Factors, 26*(3), 257–268.

Green, M. (2000). "How long does it take to stop?" Methodological analysis of driver perception-brake times. *Transportation Human Factors, 2*(3), 195–216.

Green, P. (2007). Where do drivers look while driving (and for how long)? In: R. E. Dewar & R. Olsen (Eds.), *Human Factors in Traffic Safety* (2nd ed., pp. 57–82). Tucson, AZ: Lawyers & Judges Publishing.

Green, P. (2008). Developing complex crash warning simulations for human factors evaluations. *Proceedings of the human factors and ergonomics society 52nd annual meeting* (pp. 1865–1869). Santa Monica, CA: Human Factors and Ergonomics Society.

Green, P., Cullinane, B., Zylstra, B., & Smith, D. (2004). *Typical values for driving performance with emphasis on the standard deviation of lane position: A summary of literature* (Tech. Rep. SAVE-IT, Task 3a). Ann Arbor, MI: University of Michigan Transportation Research Institute (UMTRI).

Groeger, J. A. (2000). *Understanding driving: Applying cognitive psychology to a complex everyday task.* Philadelphia, PA: Taylor & Francis.

Hart, S. G., & Staveland, L. E. (1988). Development of the NASA-TLX (Task Load Index): Results of empirical and theoretical research. In P. A. Hancock & N. Meshkati (Eds.), *Human mental workload* (pp. 139–183). Amsterdam, the Netherlands: North Holland.

Human Machine Interface and the Safety of Traffic in Europe (HASTE). (n.d.). *Internal deliverable: WP2 Pilot Dependent Variables.*

Horrey, W. J., & Lesch, M. F. (2009). Driver-initiated distractions: Examining strategic adaptation for in-vehicle task initiation. *Accident Analysis & Prevention, 41*, 115–122.

Horrey, W. J., & Simons, D. J. (2007). Examining cognitive interference and adaptive safety behaviors in tactical vehicle control. *Ergonomics, 50*(8), 1340–1350.

Horrey, W. J., & Wickens, C. D. (2006). Examining the impact of cell phone conversations on driving using meta-analytic techniques. *Human Factors, 48*(1), 196–205.

Horrey, W. J., & Wickens, C. D. (2007). In-vehicle glance duration: Distributions, tails, and a model of crash risk. *Transportation Research Record, 2018*, 22–28.

Horrey, W. J., Wickens, C. D., & Consalus, K. P. (2006). Modeling drivers' visual attention allocation while interacting with in-vehicle technologies. *Journal of Experimental Psychology: Applied, 12*(2), 67–78.

International Standards Organization. (2002). *Road vehicles—Measurement of driver visual behavior with respect to transport information and control systems—Part 1: Definitions and parameters,* ISO committee Standard 15007-1, Geneva, Switzerland: International Standard Organization.

Jones, D. G., & Kaber, D. B. (2005). Situation awareness measurement and the situation awareness global assessment technique. In N. Stanton, A. Hedge, K. Brookhuis, E. Salas, & H. Hendrick (Eds.), *Handbook of human factors and ergonomics methods* (pp. 42-1–42-8). New York: CRC Press.

Kaptein, N. A., Theeuwes, J., & Van der Horst, R. (1996). Driving simulator validity: Some considerations. *Transportation Research Record, 1550*, 30–36.

Lamble, D., Kauranen, T., Laakso, M., & Summala, H. (1999). Cognitive load and detection thresholds in car following situations: Safety implications for using mobile (cellular) telephones while driving. *Accident Analysis & Prevention, 31*, 617–623.

Lee, A. T. (2005). *Flight simulation.* Aldershot, United Kingdom: Ashgate.

Lee, J. D., & Strayer, D. L. (2004). Preface to the special section on driver distraction. *Human Factors, 46*(4), 583–586.

Meehl, P. (1993). Theoretical risks and tabular risks: Sir Karl, Sir Ronald and the slow progress of soft psychology. In C. A. Anderson & K. Gunderson (Eds.), *Selected philosophical and methodological papers of Paul E. Meehl* (pp. 1–42). Minneapolis, MN: University of Minnesota Press.

Meister, D. (1995). Simulation and modeling. In J. R. Wilson & E. N. Corlett (Eds.), *Evaluation of human work* (2nd ed., pp. 202–228). London: Taylor and Francis.

Miles, G. H., & Vincent, D. F. (1934). The Institute's tests for motor drivers. *The Human Factor, VIII* (7–8), 245–257.

Mulder, L. J. M., de Waard, D., & Brookhuis, K. A. (2005). Estimating mental effort using heart rate and heart rate variability. In N. Stanton, A. Hedge, K. Brookhuis, E. Salas, & H. Hendrick (Eds.), *Handbook of human factors and ergonomics methods* (pp. 20-1–20-8). Boca Raton: CRC Press.

National Transportation Safety Board. (2003). *Ford Explorer Sport collision with Ford Windstar Minivan and Jeep Grand Cherokee on Interstate 95/495 near Largo, Maryland, February 1, 2002* (Rep. No. NTSB/HAR-03/02). Washington, DC: National Transportation Safety Board.

Nickerson, R. S. (1999). Basic and applied research. In R. J. Sternberg (Ed.), *The nature of cognition* (pp. 409–444). Cambridge, MA: MIT Press.

Olson, P. L., & Farber, E. (2003). *Forensic aspects of driver perception and response* (2nd ed.). Tucson, AZ: Lawyers and Judges, Inc.

Olson, P. L., & Sivak, M. (1986). Perception-response time to unexpected roadway hazards. *Human Factors, 28*, 91–96.

Pauzié, A. (2008). A method to assess the driver mental workload: The driving activity load index (DALI), *IET Intelligent Transport Systems, 2*(4), 315–322.

Peden, M., Scurfield, R., Sleet, D., Mohan, D., Hyder, A. A., Jarawan, E., & Methers, C. (2004). *World report on road traffic injury prevention.* Geneva: World Health Organization.

Poulton, E. C. (1982). Influential companions: Effects of one strategy on another in the within-subjects designs of cognitive psychology. *Psychological Bulletin, 91*(3), 673–690.

Reed, M. P., & Green, P. A. (1999). Comparison of driving performance on-road and in a low-cost simulator using a concurrent telephone dialing task. *Ergonomics, 42*(8), 1015–1037.

Rosenthal, R., & DiMatteo, M. R. (2001). Meta-analysis: Recent developments in quantitative methods for literature reviews. *Annual Review of Psychology, 52*, 59–82.

Santos, J., Meret, N., Mouta, S., Brookhuis, K., & de Waard, D. (2005). The interaction between driving and in-vehicle information systems: Comparison of results from laboratory, simulator and real world studies. *Transportation Research, Part F: Traffic Psychology and Behavior, 8*, 135–146.

Schiff, W., & Arnone, W. (1995). Perceiving and driving: Where parallel roads meet. In Hancock, P. A., Flach, J., Caird, J. K., & Vicente, K. (Eds.), *Local applications of the ecological approach to human machine systems* (pp. 1–35). Hillsdale, NJ: Lawrence Erlbaum.

Sheridan, T. B. (1970). Big brother as driver: New demands and problems for the man at the wheel. *Human Factors, 12*(1), 95–101.

Shinar, D. (2007). *Traffic safety and human behavior.* Amsterdam: Elsevier.

Siegal, S., & Castellan, N. J. (1988). *Nonparametric statistics for the behavioral sciences* (2nd ed.). New York: McGraw-Hill.

Sivak, M., Luoma, J., Flannagan, M. J., Bingham, C. R., Eby, D. W., & Shope, J. T. (2007). Traffic safety in the U.S.: Re-examining major opportunities. *Journal of Safety Research, 38*, 337–355.

Society of Automotive Engineering. (2000). *Definition and measures related to the measurement of driver behavior using video based techniques* (SAE Recommended Practice J2396). Warrendale, PA: Society of Automotive Engineering.

Strayer, D. L., & Drews, F. A. (2004). Profiles in driver distraction: Effects of cell phone conversations on younger and older drivers. *Human Factors 46*(4), 640–649.

Stutts, J. C., Faeganes, J., Reinfurt, D., Rodgman, E., Hamlett, C., Gish, K., et al. (2005). Drivers' exposures to distractions in their natural environment. *Accident Analysis & Prevention, 37*(3), 1093–1101.

Summala, H. (2000). Brake reaction times and driver behavior analysis. *Transportation Human Factors, 2*(3), 217-226.

Tabachnick, B. G., & Fidell, L. S. (2006). *Using multivariate statistics* (5th ed.). Boston, MA: Pearson, Allyn, & Bacon.

Taylor, J. L., Kennedy, Q., Noda, A., & Yesavage, J. A. (2007). Pilot age and expertise predict simulator performance: A 3-year longitudinal study. *Neurology, 68*, 648–654.

Tijerina, L., Barikman, F. S., & Mazzae, E. N. (2004). *Driver eye glance behavior during car following* (Rep. DOT HS 809 723). Washington, D.C.: National Highway Traffic Safety Administration.

Tijerina, L., Johnson, S., Parmer, E., & Winterbottom, M. D., & Goodman, M. (2000). *Driver distraction with wireless telecommunications and route guidance systems* (Rep. No. DOT HS 809-069). Washington, D.C.: National Highway Traffic Safety Administration.

Törnos, J. (1998). Driving behavior in a real and a simulated road tunnel—A validation study. *Accident, Analysis and Prevention, 30*, 497–503.

Tufano, D. R. (1997). Automotive HUDs: The overlooked safety issues. *Human Factors, 39*(2), 303–311.

van Winsum, W., Brookhuis, K. A., & de Waard, D. (2000). A comparison of different ways to approximate time-to-lane crossing (TLC) during car driving. *Accident Analysis & Prevention, 32*(1), 47–56.

Wickens, C. D. (2001). *Attention to safety and the psychology of surprise.* Proceedings of the 2001 Symposium on Aviation Psychology. Columbus, OH: Ohio State University.

Wickens, C. D. (2002). Multiple resources and performance prediction. *Theoretical Issues in Ergonomic Science, 3*(2), 159–177.

Wickens, C. D., Todd, S., & Seidler, K. (1989). *Three-dimensional displays: Perception, implementation and applications* (Tech. Rep. CSERIAC SOAR 89–001). Wright-Patterson Air Force Base, OH: US Air Force, Armstrong Aerospace Medical Research Laboratory (AFAAMRL).

Wierwille, W. W., & Fung, P. P. (1975). Comparison of computer generated and simulated motion picture displays in a driving simulation. *Human Factors, 17*(6), 577–590.

Wilkinson, L., & Task Force on Statistical Inference (1999). Statistical methods in psychology journals: Guidelines and explanations. *American Psychologist, 54*(8), 594–604.

Wortman, P. M. (1994). Judging research quality. In: H. Cooper & L. V. Hedges (Eds.), *Handbook of research synthesis* (pp. 97–109). New York: Russell Sage Foundation.

Yan, X., Adbel-Aty, M., Radwan, E., Wang, X., & Chilakapati, P. (2008). Validating a driving simulator using surrogate safety measures. *Accident Analysis & Prevention, 40*, 274–288.

Young, M. S., & Stanton, N. A. (2005). Mental workload. In N. Stanton, A. Hedge, K. Brookhuis, E. Salas, & H. Hendrick (Eds.), *Handbook of human factors and ergonomics methods* (pp. 39-1–39-9). Boca Raton: CRC Press.

Zijlstra, F. R. H. (1993). *Efficiency in work behavior. A design approach for modern tools.* Doctoral dissertation, Delft University of Technology. Delft, the Netherlands: Delft University Press.

II

Selecting a Driving Simulator

6

Scenario Authoring

Joseph K. Kearney
The University of Iowa

Timofey F. Grechkin
The University of Iowa

Abstract

The Problem. Simulators offer the opportunity to study driving behavior in a safe, controlled environment that puts the driver in realistic situations comparable, in many ways, to what they experience on natural roadways. To effectively use a driving simulator, researchers must be able to create a scenario that tests the hypotheses of interest. *Role of Driving Simulators.* Researchers use simulators to study how a wide range of factors influence both normative characteristics of driving (such as the size of gaps accepted) and how drivers respond to exceptional circumstances (such as abrupt braking by a lead vehicle or unexpected incursions of pedestrians and vehicles into the path of the driver). A key step in designing a simulation experiment is the specification of the scenario that determines what happens, when it happens, and where it happens during a trial. This chapter examines what a scenario is, why it is important, and what makes it difficult to create realistic, replicable scenarios in a driving simulator. *Key Aspects of Driving Simulator Systems.* The scenario authoring components of four driving simulation systems are presented. Methods to place objects in the simulated environment and to coordinate the behaviors of objects to create desired conditions are highlighted. An online companion section presents case studies describing how three prototypical scenarios can be implemented in the four systems. *Platform Specificity.* Scenario systems are designed with specific purposes in mind that are reflected in the tools provided for authoring scenarios. When selecting a scenario system, it is important to consider how well the tool set provided by the system matches the goals of the project.

6.1 Introduction

The scenario control system of a driving simulator is responsible for choreographing the action of a simulation. It determines what happens in a simulation by specifying where dynamic objects are to be placed, what they are to do, and when they are to do it. The aim of scenario control is to create a predictable experience for the human driver operating the simulator. This is particularly important for experimental investigations of driving behavior. Simulators offer the promise of conducting experiments with the realism of the natural world, but with the control of a laboratory study. In order to compare performance among groups of subjects,

it is critical that the essential aspects of conditions be replicated from trial to trial. The key challenge for scenario control is to make the right things happen at the right time and place, while giving the subjects the impression that the events occur spontaneously.

Two factors make scenario control a very difficult problem. First, driving behavior is complicated and not well understood; this makes it difficult to simulate realistic traffic. Simulators typically include animated vehicles that are programmed to drive on the virtual roadways. These intelligent vehicles serve multiple purposes. One purpose is as role players to create the conditions under which experimenters want to study driver behavior. Another purpose for intelligent vehicles is to provide a backdrop of ambient traffic. The level of sophistication and competence of the autonomous driving behaviors varies from simulator to simulator. This influences both the realism of the simulation and the range of conditions that can be created.

The second factor that makes scenario control difficult is that human driving behavior is highly variable. Individual drivers continuously vary their speed and lane position as they adjust to road conditions and surrounding traffic. They vary how they drive based on mood, sense of urgency, and in attempts to avoid boredom. In addition to this within-driver variability, there is substantial between-driver variability in preferred speed, following distances, and risk taking. The ever-changing nature of drivers and driving means that every run of a simulation is unique. Over the course of a simulation trial, the accumulation of small differences can lead to large variations in the time it takes to arrive at a particular point even when drivers follow the same route. This makes it very difficult to design rigidly scripted scenarios that provide a similar experience for all drivers. As a consequence, most simulators provide some means of conditional, event-driven activation of scenario events that adapts to the driving behavior of the human driver.

In this chapter we discuss what a scenario is, why it is important, and what makes it difficult to create realistic, replicable scenarios in a driving simulator. We describe the approaches to authoring scenarios in four different simulation systems. In an online companion section we present case studies in which we describe our experience in implementing three prototypical scenarios in the four systems. We conclude the chapter by considering some of the recurring challenges in designing effective scenarios.

6.2 What Is a Scenario?

Broadly speaking, a scenario determines the setting and circumstances of a simulation. There is some variance in the usage of the word "scenario" in the driving simulation literature. It is sometimes used to encompass both the layout of the environment and the activities during a simulation trial or training session. Others use "scene" to describe the static structure of the virtual environment (e.g., the terrain, road network, vegetation, and buildings) and reserve "scenario" to cover the dynamic characteristics of a simulation (e.g., ambient traffic and critical events). With one exception, the scenario control systems we examine in this chapter explicitly separate the specification of the layout of the environment from the specification of what happens in the environment. This chapter will primarily focus on the ways scenario systems allow authors to specify what happens in a simulation. We will use "scenario" loosely to refer to the specification of dynamic characteristics of a simulation and the binding of activities to places in the scene.

Scenarios are commonly organized as a series of episodes, during which conditions are tightly controlled, that are interspersed among periods of free driving. Often these episodes mimic situations that are thought to be contributing factors in crashes as surmised from epidemiological studies. For example, abrupt braking by a lead vehicle and unexpected incursions into the driver's line of travel (by vehicles or pedestrians) are common precipitants of crashes which have frequently been created in simulation-based experiments to investigate how various factors, such as driver experience or attentiveness, influence the likelihood of a crash in these circumstances (Donmez, Boyle, & Lee, 2007; Strayer, Drews, & Crouch, 2006; Rizzo, Reinbach, McGehee, & Dawson, 1997; Jamson, Westerman, Hockey, & Carsten, 2004; Lee, McGehee, Brown, & Reyes, 2002; Horrey & Wickens, 2004).

The exceptional events that contribute to crashes occur with relatively low frequency in the real world. Simulators allow us to make these rare events commonplace. However, they must do so without heightening the awareness of the driver or prematurely alerting the driver to their occurrence. The event should naturally arise out of the normal course of driving. This is particularly difficult in a simulator because the novelty of being in a simulator may raise the drivers' attentiveness and curiosity (see also Ranney, this book, chap. 9). Often drivers are given secondary tasks to divert their attention. For example, they may be asked to adjust the radio, push a button on a control panel when it becomes lit, or do mental arithmetic.

Scenarios often populate the roads with ambient traffic during the interstitial periods between scenario episodes to provide a sense of realism and to conceal critical events. It is important to consider the relationship between the ambient traffic and actions of entities that play roles in a scenario episode. The surrounding traffic can have an important influence on how the driver behaves before and during a critical situation. For example, one of the reactions that drivers may have to abrupt braking by a lead vehicle is to steer into an adjacent lane to avoid a collision. Oncoming traffic on a two-lane road can present the threat of a head-on collision that prevents a driver from swerving into to the left lane, thus reducing the options available to the driver. To ensure consistency from trial to trial, scenario designers must identify what factors influence driving behavior and control surrounding traffic during critical events to replicate the essential circumstances on each trial.

In addition to providing a platform for examining how drivers respond in critical situations, simulators offer a controlled means to study how drivers perform routine driving tasks such as road tracking, car following, or gap acceptance. By precisely controlling road properties and the patterns of traffic, it is possible to estimate normative properties of driving behaviors in ways that are difficult to do in observational studies of real driving. For example, to examine the size of gaps that child and adult bicyclists would accept in crossing a stream of traffic, we designed a scenario in which a continuous stream of cross traffic was structured in logical blocks (Kearney, Grechkin, Cremer, & Plumert, 2006). Each block contained a random permutation of five different gap sizes. This ensured that subjects had a fair chance to see all gaps and allowed us to determine what size gaps subjects were willing to cross and to examine how they timed their motion to cross through the gap (Plumert, Kearney, & Cremer, 2004).

There has been substantial discussion in the simulation community about establishing standards for scenarios to facilitate cross-platform validation (Allen et al., 2007; Caird, Rizzo, & Hancock, 2004). One of the difficulties in establishing standards is determining what matters and what does not in a scenario episode. How similar do the conditions have to be from trial to trial and from experiment to experiment in order to compare results? For example, McGehee et al., 1996 found that the presence of a lead vehicle influenced driver behavior as they approached an intersection on a throughway with stopped traffic on the crossroad (McGehee et al., 1996). Seeing the leader drive through the intersection gave the drivers confidence that the cars waiting to enter the throughway from the crossroad were paying attention and respecting the right of way of vehicles on the through street. When no lead vehicle was present, drivers approached the intersection more cautiously. Scenario designers must be attentive to all aspects of a scenario and how they might influence driver behavior.

6.3 The Means and Mechanisms to Author Scenarios

In this section, we examine the means that simulator systems provide to author scenarios. Our focus will be on the author's perspective, examining the constructs for placing objects in the simulation environment and for controlling the behaviors of dynamic objects contingent on subject behavior. To explain how these constructs work and the impact they have on system performance, we also examine the underlying computational

mechanisms that implement author specifications. In the next section, we describe how the scenario authoring techniques we present in this section are incorporated into four simulator systems.

6.3.1 Object Creation and Placement

The first step in defining a scenario is usually to determine where the scenario is to take place. In most simulation systems the author either selects a scene from a library of preconstructed scenes or assembles one from prefabricated tiles through a graphical user interface (Figure 6.1, Web Figure 6.1 for color version). Tiles are modular scene segments (e.g., a freeway interchange, a stretch of rural highway, or a city block) that can be cloned and pieced together to create an integrated scene as a mosaic of connected tiles.

In most systems the scene can be customized to some extent, for example, to change the location or appearance of buildings, to modify signage, or to alter road markings. In addition, systems provide some way to place dynamic objects in the scene and select important locations.

Scenario authoring interfaces support object placement in variety of ways:

- *Placement on a Map* – A graphical interface allows authors to place objects on a map of the scene.

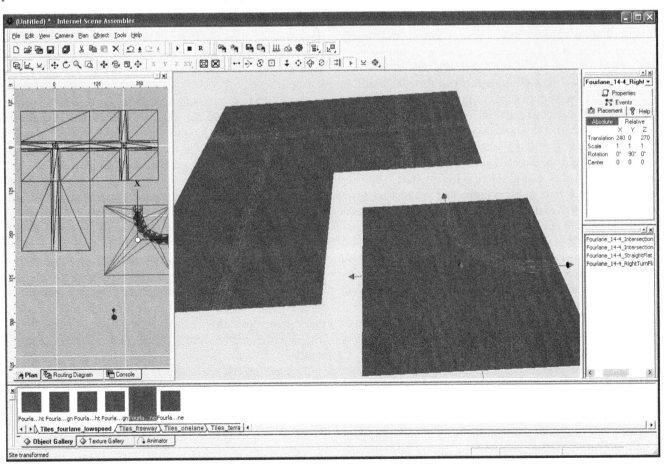

FIGURE 6.1 **(See color insert)** Assembling scene from tiles in SimVista.

- *Placement by Address* – Objects are placed through textual statements that specify a location on a specific road at which to place an object.
- *Placement by Route* – Objects are placed with reference to the route of the human driver. Locations are specified in terms of the distance along the route driven by the subject.

While the three methods of placing objects are compatible with one another, systems typically offer only one of the three methods.

6.3.2 Orchestrating Activities and Critical Events

Simulation systems provide a variety of ways to create the circumstances for scenario episodes. We first consider approaches to generate and control the characteristics of ambient traffic. Typically, this involves a process of injection of new vehicles (or recirculation of a fixed pool of vehicles) to maintain the appearance of natural, steady streams of traffic. We next consider tools to choreograph object behaviors in order to create situations and events of interest to experimenters. The timing of these episodes is usually tied to the movement of the human driver and may involve fine-grained synchronization and coupling between the actions of pedestrians, vehicles, and traffic lights and the human driver's vehicle.

Most simulators provide some means to generate and manage the flow of vehicles on roadways. To conserve on computation, scenarios usually attempt to simulate traffic only in the vicinity of the human driver. By strategically shifting the locations where vehicles are produced and consumed, scenario designers can create the illusion that traffic is everywhere, at all times.

In some systems, the scenario author must take explicit responsibility for specifying where and when vehicles are to be injected into the simulation and where and when they are to be removed or placed into a dormant state. This may involve writing scripts to place individual vehicles on roadways. Alternatively, simulators may provide scenario objects called *sources* and *sinks*, which serve as producers and consumers of vehicles and pedestrians. Using the tools for placing objects described in the previous section, the scenario author determines where traffic is injected and removed from the simulation by placing sources and sinks. The timing of production may be linked to the driver's position or to other scenario activities. Care must be taken to place sources out of the driver's sight so that vehicles do not suddenly appear as though from thin air.

Sources may be tuned to produce a traffic stream with certain statistical properties. Multiple streams, generating traffic at different locations, may be coordinated to produce synchronized streams of traffic.

Some simulators offer mechanisms that automatically generate traffic in the neighborhood of the driver, thus relieving the author from the responsibility of managing traffic generation (Olstam, Lundgren, Adlers, & Matstoms, 2008). These systems create a moving bubble centered on the driver's vehicle within which roadways are populated with traffic. Outside the bubble, there is no traffic or traffic is coarsely simulated. The bubble should be sufficiently large that the edges are out of sight. Traffic is usually produced at the boundaries of the bubble where roads intersect the edges. As vehicles leave the area of the bubble they are removed, deactivated, or simulated at a much lower level of fidelity. The bubble may move continuously with the driver or may move periodically in discrete jumps.

Event-driven controllers are commonly used to craft specific situations required for scenarios. The standard approach is to use conditionally-activated elements called *triggers* or *sensors* that direct the simulation for a period of time after they are activated. Typically, these triggers are situated in the environment and are activated when the driver or a simulated object enters their vicinity. Triggers are usually placed in the environment in the same way that other objects are placed (Figure 6.2 or Web Figure 6.2 for color version). Most systems allow some flexibility in the triggering condition. The simplest approach is to define circular or rectangular regions with adjustable dimensions. The trigger can be set to respond when objects enter this region, leave this region, or whenever an object is inside the region. Normally, the trigger can be tailored to respond to all or a subset of moving objects (e.g., the driver, all passenger vehicles, all trucks, etc.). More complicated triggers may depend on conditions such as the estimated time of arrival of the subject at a location. For example, a trigger may fire five seconds before the subject is expected to arrive at an intersection based on the vehicle's position and speed.

In addition to specifying where the trigger is and what conditions activate it, the author must specify what the trigger does when it fires. Triggers interact with the simulation by creating new objects or by modifying the behavior of objects that already exist in the simulation. For example, a trigger may cause the lead vehicle to suddenly brake or a vehicle waiting at a stop light to ignore the light and enter the intersection. In addition to directing objects to initiate some action, triggers may instruct objects to link their behavior to another object for a period of time. For example, a trigger may direct the lead vehicle to couple its motion with the motion of the driver in order to maintain a time headway of two seconds. These compliant behaviors are important for creating consistent preconditions for critical events. In some systems, objects have published interfaces through which they receive directions to modify their behavior. Essentially, the trigger pushes buttons and sets dials on the dashboard of the vehicle.

The experimenter/operator may also play a role in determining how a scenario unfolds. A simulator console may provide an interface which allows the experimenter to trigger events that influence the simulation. As with programmatically-controlled directors, this may involve directing vehicles, pedestrians, or traffic lights to modify their behavior. It may also involve directly manipulating object variables to, for example, steer a simulated vehicle.

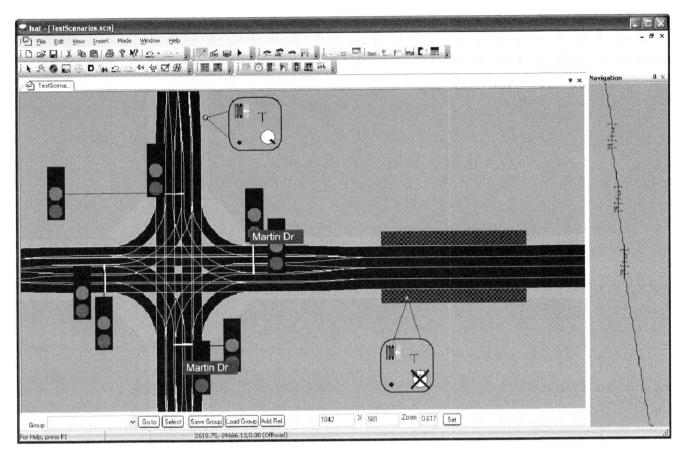

FIGURE 6.2   Example of graphical user interface for authoring scenarios as implemented in ISAT.

Through judicious distribution of ambient traffic and careful placement of triggers along the intended route of the subject, an author can create a (reasonably) predictable series of scenario episodes for each driver. Even with careful planning, scenarios take substantial tuning to create the intended experience. Unanticipated interactions between ambient traffic, objects with scripted behaviors, and the human driver can disrupt the timing of the actions and can create conflicts that interfere with the performance of their roles within the scenario.

Whenever a scenario is tied to the behavior of the human driver, there is the possibility that the scenario will fail if the subject behaves in an unexpected way. It is important for experimenters to be aware that scenarios have limits and to understand what range of subject behaviors can be tolerated. For example, one of the prototypical scenarios we examine in the case studies in our online companion section involves abrupt braking by a lead vehicle. To compare performance from trial to trial, it may be important to ensure that the gap between the lead vehicle and the subject vehicle at the time of braking events is the same in all trials. One way to accomplish this is to couple the behavior of the lead vehicle to the subject's vehicle so that it adjusts its speed to maintain a fixed size (either temporal or spatial) gap. The size of the gap to be achieved can have an important influence on subject behavior. In one experiment with elderly drivers (who tend to prefer larger gaps), the drivers

continuously reduced their speed in an effort to increase gap size just as the lead vehicle was reducing its speed in an effort to reduce the gap size. Eventually some subjects came to a stop on a freeway. In this extreme case, it was impossible to proceed with the scenario.

6.4 Case Studies

Probably the best way to understand scenario design is to see how it is practiced. In this section, we present scenario authoring tools from four different simulation systems. We envisioned this study as neither comprehensive nor as a competitive evaluation based on "performance" scores, but rather as an opportunity to learn from the diverse approaches employed by the developers of these systems to create scenarios. With that in mind, we have selected systems that represent different flavors of driving simulators (see Table 6.1).

In the following sections we introduce the four scenario authoring systems selectively emphasizing distinctive features to illustrate the general concepts described above; the full description of the systems can be found in the related publications and technical documentation. An online companion section on the *Handbook* web site introduces three sample scenarios which we used to help us explore the functionality of these systems and describes how we implemented these scenarios in each of the systems.

TABLE 6.1 Scenario Authoring System

	STISIM Drive	SimVista	Hank	ISAT
Developer	Systems Technology, Inc.	Realtime Technologies, Inc,	Hank Project, Dept. of Computer Science, The University of Iowa	The University of Iowa, NADS
Type	Commercial, low-cost, small-scale	Multi-purpose, commercial simulator	In-house, small-scale research simulator	Large-scale research simulator
Web site	www.stisimdrive.com	www.simcreator.com	www.cs.uiowa.edu /~hank	www.nads-sc.uiowa.edu
Interface	Text-based	GUI	Text-based	GUI
Object Placement	By route	On map	By address	On map
Scene	Integrated development; text-based	Integrated development; tile-based	Developed separately; text-based "logical" scene + visual model	Integrated development; tile-based
Ambient Traffic	Created manually by individual vehicle events	Automatically generated using "bubble-based" algorithm	Generated by sources	Generated by sources
Critical Events	Controlled by triggers built into individual events	Controlled by special scenario objects - sensors	Controlled by triggers built into other scenario objects	Controlled by special scenario objects - triggers

6.4.1 STISIM Drive

STISIM Drive is a low-cost simulator produced by Systems Technology Inc. (STI) (Allen, Rosenthal, & Aponso, 1998). It is aimed at driver training, assessment, and research and development applications. STISIM Drive is sold in a wide variety of configurations including turnkey systems with one or more screens and custom designed systems.

STISIM Drive scenarios (Allen, Rosenthal, Aponso, & Park, 2003; Park, Rosenthal, & Aponso, 2004) are defined in scripts that specify the layout of the environment, the locations of vehicles, and the occurrence of scenario events. All of these aspects of the scenario are considered events and are defined in a uniform framework tied to the human driver's progress through the simulation. Statements in the script have a general form:

```
Distance, event, parameter 1, parameter 2,
..., parameter n
```

Where distance is the distance traveled by the human driver relative to the starting position in the simulation; event is a general construct that covers the layout and properties of the road network and the locations of vehicles and pedestrians. Parameters of the events are used to define such characteristics as the curvature and cross section of a road or the speed at which a vehicle is to drive.

The tight connection of scene and scenario to the human driver's path is distinctive and has several important advantages. By organizing the road layout and scenario activities around the driver's path, STISIM Drive ensures consistency from run to run, independent of the human driver's speed and even independent of the choices the operator makes for turns at intersections. The roadway is constructed on the fly to match the specification of the scenario script.

For example, consider a scenario that specifies that the operator should encounter a four-way intersection at mile four of a drive, followed by another four-way intersection with cross traffic at mile five of the drive. On each drive through the environment, the driver will encounter two four-way intersections with the same cross traffic, regardless of the direction (straight, right turn or left turn) that the driver chooses at the first intersection. As a result, the driver is given the illusion of freedom to navigate, but a consistent experience is ensured by presenting the same order of events on every trial. In contrast, many simulations presume that the driver follows a prescribed path. Deviations from this path would cause events to be missed and could expose backstage elements not meant to be seen such as road terminations and incomplete structures.

A potential limitation of this approach is that the scene lacks topographical consistency. For example, if the driver makes three 90 degree right turns in quick succession, they would expect to cross the road on which they started. Since a globally consistent road network is not constructed, the driver will never cross roads previously traversed. This makes it impossible for a human driver to explore the scenario and draw a map of the environment, and means that this approach may not be appropriate for experiments where spatial orientation or map learning are important.

The event mechanism in STISIM Drive covers scene layout, road network configuration, placement of objects, and coordination of scenarios. Road events allow the author to place a segment of the road and define its curvature, elevation profile, markings, visual appearance, and so forth. Separate events are defined for static objects such as trees, buildings, and traffic signs. Finally, each occurrence of the surrounding traffic in STISIM Drive is also defined as an event.

Traffic generation events fall into one of the three categories: vehicles traveling in the same direction with the human driver (Vehicle), vehicles traveling in the opposite direction of the human driver (Approaching Vehicle), and cross traffic on an intersection (Cross Traffic). Vehicle behavior models can function either in the autonomous "smart" mode or in a tightly-controlled, scripted mode. In smart mode they recognize other vehicles and adjust their speeds to avoid collisions, obey the traffic signs and lights, and generally follow the rules of the road. In the scripted mode, vehicle behavior is specified by the scenario author. Scripted vehicles are unresponsive to road signs and road lights as well as other vehicles. This mode is designed for scripting events that require tight control over traffic and can be activated using a built-in trigger. Vehicles switch between the two modes automatically: Once the scripted routine is over, the vehicle goes back to the default smart behavior.

The scripted behaviors include actions such as overtaking the human driver's vehicle and changing speed. Each event definition can include up to five actions. An action is defined by four parameters: an action trigger, the change in lateral position, the change in speed, and the time interval for the desired transition. Actions can be triggered by conditions such as the position of a scenario vehicle or conditions tied to the behavior of the human driver, such as time to collision.

Cross traffic events provide a way to create smart vehicles traveling on a crossroad. These events also provide an option for coupling the motion of a scenario vehicle to the motion of the human-controlled vehicle. When such a compliant relationship is activated, the cross traffic vehicle will automatically adjust its speed to arrive at the intersection in front of the human driver, simultaneously with the driver, or right behind him or her. Additionally, it can also maintain a speed proportional to the speed of the human driver.

Finally, STISIM Drive provides a mechanism to encapsulate a set of events into a reusable component called a Previously Defined Event (PDE). A PDE specifies a group of events to be executed, each at a specified distance relative to the location of the PDE. As with other events, a PDE is assigned a position on the driver's path some distance from the driver's starting point. The PDE's distance is added to the distance of each event contained in it. The PDE also supports parameter passing from the call statement to the events in PDE. A scenario script can include multiple instances of the same PDE at different distances with different parameters.

For example, if a scenario requires a sequence of five intersections with the same critical events occurring on each of them, a PDE can define a single intersection and related events. Then, the scenario can be built as a composition of five instantiations of the PDE.

6.4.2 SimVista

SimVista is a scenario design and authoring tool developed by Realtime Technologies Inc. as a companion for their real-time simulation and modeling system—SimCreator (Romano, 2000). Constructed as an extension for an interactive 3D editor of VRML models called the Internet Scene Assembler (ISA), SimVista provides an integrated visual environment for authoring both the scene and the scenario.

SimVista uses a tile-based approach to design a scene. Figure 6.1 illustrates the graphical user interface (GUI) for assembling tiles. The composite scene can be enriched by adding a wide selection of static objects including buildings, traffic signs, parked vehicles, trees and bushes, road construction objects, billboards, etc. In cases where the selection of standard tiles is insufficient, SimVista provides a mechanism for importing externally designed tiles created in a third party 3D modeling package (for example, Presagis' Multigen Creator).

Scenario control objects are added using the same GUI-based interface (Figure 6.1 in Section 6.3.1 and Web Figure 6.1). These objects are visible in the authoring interface, but are not rendered on the simulator. Scenario control objects can be broadly classified into groups by their function; the most important of these groups are *sensors* and *markers*.

Sensors act to trigger actions of other objects (Mollenhauer, Morrison, & Romano, 2006). The actions triggered by a sensor are written in JavaScript code which can be shared between sensors and re-used in other projects. SimVista provides three types of sensors distinguished by the type of trigger-condition: proximity, time, and time-to-collision (TTC) sensors.

Proximity sensors are activated when a dynamic object enters or leaves the sensor's volume. Proximity sensors come in several different shapes including spheres, cubes and cylinders. They can be used, for example, to change a vehicle's speed once it reaches a certain position on the road. Time sensors are triggered by the simulation clock. They can be set to fire at a particular time or to fire repeatedly at regular intervals. The length of these intervals is controlled by the sensor script and, therefore, can be adjusted according to simulation requirements at run time. Time sensors can be useful to trigger vehicle movements at a certain point of time during the simulation or to periodically modify parameters of the vehicle behavior. Finally, TTC sensors define a collision plane and trigger action when the projected trajectory of a dynamic object intersects this plane and the estimated time for it to reach the plane crosses a given threshold. This is handy for synchronizing the timing of critical events with the approach of the driver (e.g., the triggering of lane intrusions as the human driver approaches a certain point on the road). See the online companion to this chapter for practical examples of how sensors are used to implement our sample scenarios.

Markers are primarily used as reference points for scripting the behavior of dynamic objects. To serve their purpose, markers provide an interface that allows them to be accessed or referenced from scenario scripts and the simulator itself. For example, a single instance of a special marker object must be specified in every scene to determine the starting position of the human driver when a simulation is initialized.

Another important use of markers is to define paths to be tracked by computer-controlled dynamic objects. A path is defined by selecting a single starting marker and a sequence of path markers. The minimum number of markers required to form a path is two. Some of the markers, such as the driver's starting point, have an associated direction vector.

Vehicles and pedestrians are represented as dynamic objects. SimVista creates two pools of dynamic objects when a simulation is initialized: one for vehicles and one for pedestrians. The current version of SimVista does not allow dynamic objects to be created at run time via a scenario control script nor does it allow them to be placed directly into the scene using the GUI. This means that the maximum number of dynamic objects is fixed when the simulation starts and requires that the author be careful when budgeting the allocation of dynamic objects to ensure that there are not conflicting demands. The scenario author, however, can determine the size of the object pool and pick a visual model to be used to represent each of the vehicles and pedestrians in it.

Instead of creating a new dynamic object at run time, the author moves an existing object to a marker. All objects in the pre-defined pools have unique names which can be used in scenario scripts. The dynamic object can be directed to traverse a path associated with a selected marker. Roads contain implicitly defined paths for vehicles, so a vehicle placed on the road network will travel according to the appropriate rules of the road.

SimVista includes a generator for ambient traffic, which uses a bubble-like generation model. It defines two concentric "bubbles" around the human driver such that the border of the smaller bubble is just out of the driver's sight. New vehicles are placed randomly on the road network on the border of the larger bubble traveling toward the human driver. This is done in order to ensure that there is enough time for the dynamic model to be initialized properly before the human driver sees the vehicle. Vehicles inside the smaller bubble cannot be moved because they are within driver's sight. Once they exit the smaller bubble, however, they can be re-used by the ambient traffic generator and placed in a new location.

Scenario scripts associated with sensors can modify the properties of dynamic objects. Each dynamic object has a JavaScript interface through which it can be directed to change its behavioral state, speed, acceleration rate, and other parameters. The objects in both the vehicle and pedestrian pools have unique names, which consist of a generic type name and index. This makes it easy to reference any of the objects. A drawback to this

approach is that the pool of vehicles is shared between the ambient traffic generator (if it is active) and scenario scripts; thus careful planning is required to make sure that two mechanisms do not interfere with each other.

6.4.3 Hank

Hank is a research simulator developed in the Computer Science Department at the University of Iowa. The primary goal of Hank is to provide a flexible platform to study scenario control methodologies. Hank forms the software foundation of an immersive, interactive bicycling simulator used to study how child and adult cyclists make decisions about crossing traffic-filled roads (Plumert, Kearney, & Cremer, 2004). Hank is meant to be open and malleable in order to allow researchers to explore new techniques for modeling the behaviors of autonomous vehicles and to coordinate their actions to create conditions of interest to experimenters.

Scenarios in Hank are expressed in text-based commands. Scenario commands are linked to a pre-defined road network through references to specific locations that identify a road by name (e.g., "MainStreet_200Block") and a place on the road through road-based coordinates. Each road is assigned a curvilinear axis (usually, the centerline of the road) around which the road surface can twist to bank the road on curves. A coordinate system is attached to each road that begins at one end of the road axis. Points on the road are expressed in distance along the road axis and offset on the road surface from the road axis. This representation, based on mathematical ribbons, is convenient for modeling scenarios on roads with complex shapes and facilitates the programming of driving behaviors (Willemsen, Kearney, & Wang, 2006). Lanes are indexed relative to the axis of the road with positive indices to the left of the axis and negative indices to the right of the axis when facing in the positive direction of the road coordinate system. For example, the following statement places an autonomous vehicle on the road named "MainStreet_200Block", 100 feet from the beginning of the road segment, in the lane with index −2 (i.e., the second lane to the right-side of the road axis), facing in the positive direction of the road axis, and sets its initial speed to 25 miles per hour:

```
create Vehicle(speed = 25.0,location = sdl.
locator("MainStreet_200Block", 100.0, -2,
"pos"));
```

Hank includes an infrastructure for creating modular behaviors based on Hierarchical Concurrent State Machines (HCSM) and support for dynamic creation and deletion of objects. This core software for behavior programming takes advantage of the ribbon-based road database by providing fast, robust queries to determine the layout of the road network, the shape of road surfaces, and the locations of moving objects in road-based coordinates. The current set of autonomous behaviors includes lane tracking, following, stopping, lane-changing, and intersection negotiation (Wang, Kearney, Cremer, & Willemsen, 2005). The modular nature of behaviors makes it simple to turn components on and off individually, creating a wide range of possible behavior models. This approach facilitates tailoring of autonomous vehicle behaviors to meet the demands of a particular scenario task.

Sources and sinks are used to produce flows of traffic as required for experimental studies. Sources create a sequence of vehicles with gaps generated according to a specific rule (Kearney et al., 2006); sinks destroy dynamic objects in their vicinity. Sources and sinks are often paired to produce a stream of vehicles that originate at the source and disappear at the sink.

Hank includes a variety of sources that produce traffic in one or two lanes, driving in the same or opposite directions, with traffic streams moving at the same or different speeds. Sources are triggered to start producing traffic based on the proximity of the human driver and shut off when the driver recedes from the source. By adjusting the placement of sources and the placement and activation radius of sinks, scenario designers can control traffic flows to meet the needs of experiments without overtaxing the system with unnecessary traffic.

While the library of scenario objects in Hank is relatively small compared to the other systems we examined, its internal flexibility allows us to provide a similar range of functional capabilities. In fact, due to the small size of our team we often prefer to create new objects and customize existing ones for the needs of a particular scenario rather then maintain a comprehensive library of objects. This approach can also be seen as taking advantage of the in-house nature of the system, where source code is readily available for customization.

6.4.4 ISAT

The Interactive Scenario Authoring Tool (ISAT) is the scenario authoring tool created for the National Advanced Driving Simulator (NADS) developed by the National Highway Traffic Safety Administration and located at the University of Iowa.

ISAT runs as a standalone application on a conventional PC. Scenarios constructed with ISAT can be executed on a variety of simulators at NADS or any other simulator that supports the NADS scenario specification format.

ISAT scenes are constructed using a tile-based modeling system similar to SimVista's. Dynamic, static, and scenario control objects are specified through a map-based GUI. As in SimVista and Hank, scenario control objects share many of the properties of material objects; however, they are not visible during simulation (Papelis, Ahmad, & Watson, 2003; Papelis, Ahmad, & Watson, 2005).

Material objects in ISAT have a "chameleon" character in that the visual representation and behavior model are fully decoupled. The system provides a set of behavior models and a set of visual models (vehicle, truck, pedestrian, etc.), which can be matched together to create dynamic objects. The combinations of behavior and visual models can be restricted to pairings that make sense. We will now briefly describe the set of behavior models.

The simplest dynamic behavior model is Deterministic Dynamic Object (or DDO). As suggested by its name, the behavior of this type of object is completely scripted by the scenario author. It follows a path determined by the sequence of control points, which can be located on-road or off-road. Each control point also specifies orientation and desired velocity. The motion of the DDO is computed by interpolating the control points.

Dependent Deterministic Dynamic Objects (or DDDOs) also follow predefined paths. However, their speed is based on a compliant relationship with another dynamic object or human-controlled vehicle. This compliant relationship is based on three parameters: the controlling object, its target location, and the DDDO's own target control point. On each step of the simulation the DDDO computes the projected arrival time of the controlling object to its target location. The DDDO's behavior model adjusts its own speed to ensure that DDDO will arrive at its target control point when the controlling object is expected to arrive at its target location.

The most complex type of behavior model in ISAT is Autonomous Dynamic Object (or ADO). It is designed primarily to control autonomous vehicles and is capable of complex behaviors including following, lane tracking, lane changing, and negotiating intersections. ADOs are designed to interact with one another as well as with the human driver in the simulation.

In addition to having a visual representation and a behavioral model, all objects in ISAT have a set of common properties which can be used for creating scenarios more effectively as well as for managing system resources. These properties include creation radius, activation delay, and lifetime. The creation radius specifies a condition on when the object will become active. If the radius is set, then the object will be activated only when the human driver enters the radius. The activation delay sets a timer to activate the behavior of an object after a fixed period of time has elapsed since the object was created. Finally, the lifetime parameter sets a self-destruction timer which destroys an object after a fixed period of time has elapsed since the object was created.

The ISAT toolset includes a number of scenario control objects including traffic sources, triggers, and a traffic light manager. The latter is a global mechanism for coordinating light changes of multiple traffic light objects on an intersection.

The concept of traffic sources in ISAT is very similar to the one used by Hank; new vehicles are generated at a specified location (or set of locations). The time interval between creating two vehicles is determined by a source's parameter settings. Sources can be created, activated, deactivated, and deleted using common object parameters and/or through the use of triggers.

Triggers in ISAT are implemented as a classical condition-action pair. Activation conditions can be based on global time, entrance or exit of objects from a predefined area on a road called a road pad, time-to-arrival to some control point, or the evaluation of an arbitrary logical expression. In general, whenever the condition becomes true, the trigger performs an associated action; in ISAT this behavior can be further modified by one of the following options:

- The trigger can be declared as one shot trigger.
- The action can be delayed for a specified period of time, known as fire delay.

- Finally, the trigger might have a down time after each execution, creating a minimum delay between two consecutive executions of the trigger's action.

The actions are also very diverse. Triggers can create and delete other objects and modify behavior parameters of other objects. In addition, the actions can simulate failures of the human driver's vehicle, place phone calls to a human driver, and so on.

6.5 Discussion

The four different simulation systems illustrate a variety of approaches for conceptualizing scenarios. Each was designed with specific purposes in mind which are reflected in the tools provided for authoring scenarios. When selecting a scenario system, it is important to consider how well the toolset provided by the system matches the goals of the project. A number of common threads run through all of the systems. Every system provides some means to place objects in the environment (by map, by address, or by the human driver's route); every method provides some means to dynamically manipulate the configuration of objects around the human driver (either through creation of new objects or through transporting objects from a pre-made pool); and every method includes some facility for conditional activation of events contingent on the behavior of the human driver.

In an online companion section we describe how three common scenario episodes can be implemented in each of the four simulation systems using the mechanisms described above. We were able to implement the three episodes in every scenario system. However, in some cases we had to relax or alter some of our requirements to find a workable solution. For example, the gap acceptance scenario calls for a stream of traffic that continues until the driver crosses through the gap. Some systems provide mechanisms to create an indefinite number of vehicles at run time as needed contingent on the driver's behavior. For other systems, we had to specify the number of vehicles in the stream prior to run time in the scenario script. This example illustrates a common point of consideration in comparing scenario systems. It is often the case that a capability can be provided in different ways that offer different levels of adaptability. In general, methods that allow decisions to be made at run time allow greater flexibility to adapt the variations in subject behavior as compared to methods that must be completely specified before the simulation is run. However, with this flexibility there is often a price in terms of complexity and unpredictability. The scenario programmer must consider the range of possible subject behaviors in order to specify the relationship between the subject behavior and actions in the scenario. In addition, the experimenter must accept that the scenario will vary from subject to subject

The scenario authoring systems in the four systems provide a rich set of tools for adaptively orchestrating scenarios. However, scenario authoring still requires meticulous crafting to coordinate the interactions among scripted objects, ambient traffic, and the human driver. Extensive testing must be done to assure that scenarios unfold as expected with tolerance for a range of subject behaviors. Each of the systems that we examined contains some form of the scenario review mode to facilitate such testing. Because of the effort required to create effective scenarios it is also important that authoring systems provide ways to re-use previously developed scenarios in making new ones. Each of the four systems described includes some mechanism to package elements of a scenario into re-usable components.

Scenario control methodologies will continue to develop and evolve in order to meet the needs of experimenters. As the collective base of knowledge and experience with driving simulators grows, it is important to determine the critical factors in scenario design and establish effective principles for orchestrating complex driving situations.

Key Points

- Simulation systems provide a variety of ways to specify scenarios.
- The key operations of the scenario system include object placement, traffic generation, and conditional activation of events.
- When selecting a scenario system, it is important to consider how well the tool set provided by the system matches the goals of the project.

Keywords: Ambient Traffic, Critical Event, Scenario, Sensor, Source, Trigger

Acknowledgments

This material was based on research supported by the National Center for Injury Prevention and Control (R49.CCR721682) and the National Institute of Child Health and Human Development (R01-HD052875).

Glossary

Ambient traffic: Ambient traffic comprises the virtual vehicles in the driving simulation which do not directly participate in critical events. The ambient traffic primarily serves to enhance the realism of a simulation. While the presence of surrounding vehicles might be an important factor in the behavior of the human driver during the critical event, ambient traffic is typically much less tightly controlled than vehicles participating in the critical event itself. Often the ambient traffic is generated only in the proximity of the human driver in order to conserve computational resources.

Critical event: A critical event is an exceptional event that forms the core of an episode in a scenario. Critical events often model situations that are commonly associated with (and considered as possible precipitants of) crashes such as abrupt lead vehicle braking or the unexpected intrusion of a vehicle into the human driver's lane of travel.

Scenario: A scenario specifies the dynamic characteristics of a simulation. This includes what is to happen and where it is to happen. Thus, it binds together activities and places. A scenario is typically defined as a series of episodes with tightly controlled critical events interspersed with periods of free driving.

Scenario object: A scenario object is a special virtual object designed to coordinate events and activities within a scenario. Scenario objects are typically not visible to human participants. Sensors and Traffic Sources are good examples of scenario objects.

Scene: The scene specifies the static characteristics of the simulation including the terrain, roadways, signage, buildings, vegetation, and cultural features.

Traffic source: A traffic source is a scenario object, designed to create new vehicle objects at simulation run time in order to generate a stream of virtual traffic which follows specific patterns or has certain statistical properties.

Trigger or Sensor: A trigger or sensor is a mechanism to coordinate critical events with the human driver's progress through the simulation scenario. Typically triggers contain conditional statements based on spatial position, proximity (approach time), or the simulation clock. The triggering conditions can be based on the motion of the human driver or on the motion of simulated objects such as vehicles. When a trigger fires it can create or delete simulated objects, initiate actions, change the behaviors of objects, or otherwise alter the virtual environment.

Web Resources

The *Handbook* web site contains supplemental materials for this chapter, including a discussion of the implementation of three different sample scenarios using three different driving simulator platforms STSIM Drive, SimVista, and Hank.

The three sample scenarios:

- Lead vehicle braking scenario
- Intersection incursion scenario
- Gap acceptance scenario

Related figures:

Web Figure 6.1: The SimVista tile-based scene authoring graphic user interface.

Web Figure 6.2: The ISAT scenario authoring graphic user interface. The boxes labeled with a "T" are triggers.

Web Figure 6.3: Implementing sample scenarios in SimVista: (top) lead vehicle braking, (center) intersection intrusion, and (bottom) gap acceptance.

Web Figure 6.4: Implementing sample scenarios in ISAT: (top) lead vehicle braking, (center) intersection intrusion, (bottom) gap acceptance.

Key Readings

Kearney, J. K., Grechkin, T., Cremer, J. F., & Plumert, J. M. (2006). Traffic generation for studies of gap acceptance. *Proceedings of driving simulation conference (DSC) 2006 Europe* (pp. 177–186).

Mollenhauer, M. A., Morrison, M. M., & Romano, R. A. (2006). *Sensor functionality for ground vehicle simulation authoring.* Proceedings of IMAGE Conference, Scottsdale, AZ.

Olstam, J. J., Lundgren, J., Adlers, M., & Matstoms, P. (2008). A framework for simulation of surrounding vehicles in driving simulators. *ACM Transactions on Modeling and Computer Simulation, 18*(3), 1–24.

Papelis, Y., Ahmad, O., & Watson, G. (2003). *Developing scenarios to determine effects of driver performance: Techniques for authoring and lessons learned.* Proceedings of Driving Simulation Conference (DSC) 2003 North America.

Park, G., Rosenthal, T. J., & Aponso, B. L. (2004). Developing driving scenarios for research, training and clinical applications. *Advances in Transportation Studies,* Special Issue, 19–28.

References

Allen, R. W., Boer, E., Evans, D., Rizzo, M., & Jamson, H., Kearney, J. K., & Lee, J. (moderators). (2007). *Science or art: Replicating, validating, and standardizing simulator research.* Panel session at the Driving Simulation Conference (DSC) North America 2007.

Allen, R. W., Rosenthal, T. J., & Aponso, B. L. (1998). *Low-cost simulation for safety research, prototyping and training.* International Technical Conference on the Enhanced Safety of Vehicles (ESV). National Highway Transportation Safety Administration: Washington, DC.

Allen, R. W., Rosenthal, T. J., Aponso, B. L., & Park, G. (2003). *Scenarios produced by procedural methods for driving research, assessment and training applications.* Proceedings of Driving Simulation Conference (DSC) North America 2003.

Caird, J., Rizzo, M., & Hancock, P. (2004). Critical issues in driving simulation methods and measures. *Human factors and ergonomics society annual meeting proceedings* (pp. 2252–2255). Surface Transportation.

Donmez, B., Boyle, L. N., & Lee, J. D. (2007). Accounting for time-dependent covariates in driving simulator studies. *Theoretical Issues in Ergonomics Sciences,* 1–11.

Horrey, W. J., & Wickens, C. D. (2004). Driving and side task performance: The effects of display clutter, separation, and modality. *Human Factors, 46*(4), 611–624.

Jamson, A. H., Westerman, S., Hockey, G. R. J., & Carsten, O. M. J. (2004). Speech-based e-mail and driver behavior: Effects of an in-vehicle message system interface. *Human Factors, 46*(4), 625–639.

Kearney, J. K., Grechkin, T., Cremer, J. F., & Plumert, J. M. (2006). Traffic generation for studies of gap acceptance. *Proceedings of driving simulation conference (DSC) 2006 Europe* (pp. 177–186).

Lee, J. D., McGehee, D. V., Brown, T. L., & Reyes, M. L. (2002). Collision warning timing, driver distraction, and driver response to imminent rear-end collisions in a high-fidelity driving simulator. *Human Factors, 44*(2), 314–334.

McGehee, D. V., Hankey, J. M., Dingus, T. A., Mazzae, E., Garrott, W., Grant, A., & Reinach, S. (1996). Examination of drivers' collision avoidance behavior using conventional non-antilock brakes. In *NADS Cooperative Project Order No.5. Final Report* (Project NRD-20-95-0806). National Highway Traffic Safety Administration.

Mollenhauer, M. A., Morrison, M. M., & Romano, R. A. (2006). *Sensor functionality for ground vehicle simulation authoring.* Proceedings of IMAGE Conference.

Olstam, J. J., Lundgren, J., Adlers, M., & Matstoms, P. (2008). A framework for simulation of surrounding vehicles in driving simulators. *ACM Transactions on Modeling and Computer Simulation, 18*(3), 1–24.

Papelis, Y., Ahmad, O., & Watson, G. (2003). *Developing scenarios to determine effects of driver performance: Techniques for authoring and lessons learned.* Proceedings of Driving Simulation Conference (DSC) North America 2003.

Papelis, Y., Ahmad, O., & Watson, G. (2005). Driving simulation scenario definition based on performance measures. *Proceedings of driving simulation conference (DSC) North America 2005* (pp. 128–138).

Park, G., Rosenthal, T. J., & Aponso, B. L. (2004). Developing driving scenarios for research, training and clinical applications. *Advances in Transportation Studies,* Special Issue, 19–28.

Plumert, J. M., Kearney, J. K., & Cremer, J. F. (2004). Children's perception of gap affordances: Bicycling across traffic-filled intersections in an immersive virtual environment. *Child Development, 75*(4), 1243–1253.

Rizzo, M., Reinach, S., McGehee, D., & Dawson, J. (1997). Simulated car crashes and crash predictors in drivers with Alzheimer's disease. *Archives of Neurology, 54*(5), 545–551.

Romano, R. A. (2000). *Real-time driving simulation using a modular modeling methodology* (SAE Tech. Paper 2000-01-1297). Warrendale, PA: Society of Automotive Engineers.

Strayer, D. L., Drews, F. A., & Crouch, D. J. (2006). A comparison of the cell phone driver and the drunk driver. *Human Factors, 48*(2), 381–391.

Wang, H., Kearney, J. K., Cremer, J. F., & Willemsen, P. (2005). Steering behaviors for autonomous vehicles in virtual environments. *Proceedings of IEEE virtual reality conference* (pp. 155–162).

Willemsen, P., Kearney, J. K., & Wang, H. (2006). Ribbon networks for modeling navigable paths of autonomous agents in virtual environments. *IEEE Transactions on Visualization and Computer Graphics, 12*(3), 331–342.

7
Physical Fidelity of Driving Simulators

Jeffry Greenberg
Ford Motor Company

Mike Blommer
Ford Motor Company

Abstract

The Problem. Driving simulators are often used to explore important facets of the transportation landscape. They have been used to measure driver response to new technologies like crash avoidance systems or to assess the effect of distracting tasks on situation awareness. Both the authors and the readers of these studies hope that the results obtained in the simulator can be generalized to the real world. But in deciding how or even if such a generalization is warranted we are confronted with questions about the fidelity of the simulation. Did the simulator accurately represent the critical aspects of the driving experience? In addition to the visual scene, did the driver receive other cues for audio, motion, or tactile perception? How accurately were these cues delivered? And most critically, how might the known errors in the simulation affect the results? *A Framework for Understanding Fidelity.* An important way to begin answering these questions is to gain an understanding of what contributes to the physical fidelity of a driving simulator. We introduce a four-step framework that describes how any physical aspect of the driving experience (e.g., the acceleration that occurs when a vehicle changes lanes) is transformed from an abstract representation into a sensation. The framework allows us to examine the types of errors and distortions that occur at each stage of the cue-rendering process. *Deciding When Fidelity Matters.* Assessing the importance of these errors for any given experimental (or training) goal requires an estimation of the experimental artifacts that may result. If the artifacts are controllable, the fidelity of the simulator is sufficient. This viewpoint turns the unanswerable question "Is my simulator valid?" into a more practical one: "Do I understand the errors in my simulator well enough to explain their effect on my experimental data?"

The treatment of cueing systems in this chapter is not meant to be comprehensive. Specific aspects of the motion, steering, and audio systems are explored in some depth in order that simulator users can carry out the four-step process for their particular applications. In spite of its obvious importance, the visual system is not considered in detail here but has been left to more knowledgeable authors.

7.1 Introduction—What Do We Mean by Physical Fidelity?

A driving simulator attempts to reproduce some portion of the experience of driving an automobile. To the extent that the experience in the simulator seems to correspond to our sense memory of the actual event we often say that the simulation was "high-fidelity". When our expectations are not met we sometimes describe the simulation as "low-fidelity". These are convenient descriptors but the lack of precision underlying their use has made the concept of "fidelity" in driving simulators a problematic one. In practice, virtually every simulator is described as "high fidelity", especially if the developers have spent sufficient funds to make the label of "low-fidelity" unpalatable.

Physical fidelity—as opposed to psychological or perceptual fidelity—is a simpler concept which is perhaps more amenable to proper definition. A driving simulator is designed to deliver perceptual cues to the driver in order to further some experimental end (see Figure 7.1; Web Figure 7.1 on the *Handbook* web site for color version). But the source of these cues is generally grounded in a representation of a physical system. This representation can be wholly mathematical or it can be via physical analog or even a combination of the two. For example, the source of cues about the vehicle's motions comes from a model which accounts for the performance and handling properties of a vehicle. Vehicle handling models are generally pure mathematical constructs that calculate a vehicle's accelerations in response to a given set of inputs.

Control systems in a simulator often have multiple sources for their representation. The steering wheel is usually physically represented—often by an actual automotive steering wheel. But the response of the steering wheel to a driver input may be governed by a mathematical model that accounts for the action of the vehicle's power-assisted steering system. A brake system can be represented in a wide variety of ways. A physical brake pedal normally provides the driver interface. The feel of the pedal can be provided by a mechanical system like a spring; by a hydraulic "hardware-in-the-loop" system where the actual brake hydraulics form part of the simulator; or by a programmable actuator which actively generates pedal forces based on a mathematical model.

Cues for the roadway and environment in a simulator are almost universally based on a three-dimensional graphic model of the environment, perhaps augmented by a specialized method of describing the interaction of the vehicle's tires with the road. The representation of traffic is generally a combination of three-dimensional graphical models with a mathematical description of the way that surrounding vehicles move along the roadway and interact with the simulator driver and with each other.

7.1.1 The Rendering of Cues

In each of these cases, the output of an underlying representation or model must be *rendered* by the simulator systems into a cue for the driver. The rendering process involves both a mathematical transformation of the output of the model into a form

suitable for the simulation environment and a physical transformation of that output into a perceptible form. The following four-step framework can be used to describe how any physical aspect of the driving experience is transformed from an abstract representation into a sensation:

Step 1. Model the physical system.
Step 2. Compute the outputs of model in Step 1.
Step 3. Transform the model outputs into signals or commands that the simulator can use.
Step 4. Transform the commands into physical sensations using the simulator hardware.

The steps involved in rendering vehicle motion are shown in Figure 7.2; Web Figure 7.2 for color version. Step 1 occurs "off-line", that is, it happens before the actual simulation begins. In Step 1, the performance of the vehicle chassis and powertrain are abstracted into a mathematical form. Steps 2 through 4 occur in "real-time". They happen while the simulation is in progress. In Step 2, the model is given an input and produces an output. In this case the output is the vehicle's expected lateral acceleration in response to a rapid movement of the steering wheel. In Step 3 the lateral acceleration is scaled and filtered to make it compatible with the simulator hardware. These steps ensure that the commanded accelerations do not attempt to move the simulator motion hardware beyond its physical limitations. In Step 4, these transformed motion commands are sent to the simulator, which responds with motions that are directly felt by the driver.

Each step in the rendering process introduces some measurable distortion or error into the overall result. The magnitude of the error introduced can vary greatly depending on the type of maneuver and the nature of the cueing systems that are available. For the example in Figure 7.2 we can examine each of the errors in more detail.

Step 1. ***Modeling Errors***. The powertrain and handling response of the vehicle must be abstracted into a mathematical model. Many models have been developed that are suitable for use in a driving simulator (e.g., Bae & Haug, 1987; Sayers & Mousseau, 1990; Allen et. al., 1997). For many types of vehicle maneuvers the error introduced by the model is near the experimental error with which the vehicle response can be measured. However, there are certain types of responses such as high frequency noise or vibration that are not predicted accurately by the typical low-order simulator model.

Step 2. ***Real-Time or Sampling Errors.*** While the simulation is in progress, a continuous process of measurement and computation must take place. Errors are introduced to this process in three general ways. Quantization errors arise because inputs like steering wheel angles are measured with finite precision. Estimation errors occur because some inputs, perhaps steering wheel velocity, are not directly sensed but must be approximated from other available data. The

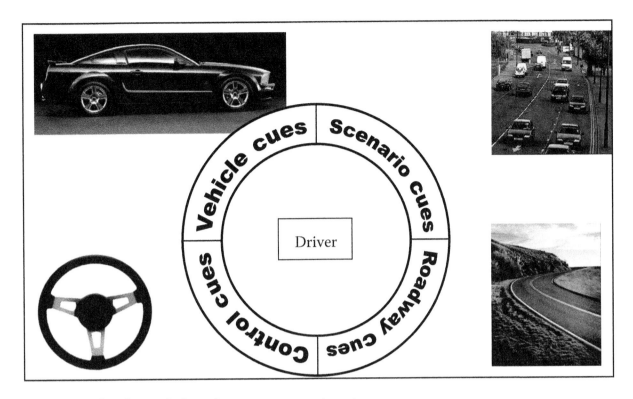

FIGURE 7.1 The simulator driver is the focus of numerous cue-generation systems.

estimation process can also be affected by sensor calibration errors (e.g., a poorly calibrated steering wheel sensor may indicate the wheel is turned to the right when in fact it is not). Temporal errors occur because the simulator systems require a finite period of time to complete the calculation cycle. For a vehicle dynamics model like the one shown in Figure 7.2, the computation time is only a few milliseconds and the resultant time delay is small compared to the response time of the human operator. But for other simulator systems—especially the graphics computers that provide the visual simulation—the computation delay can be larger. This effect has been reduced in modern simulators as computing power has increased but it should not be ignored. Time delays greater than 50 milliseconds have been shown to adversely affect vehicle control and operator comfort (Allen & DiMarco, 1984; Frank, Casali, & Wierwille, 1988).

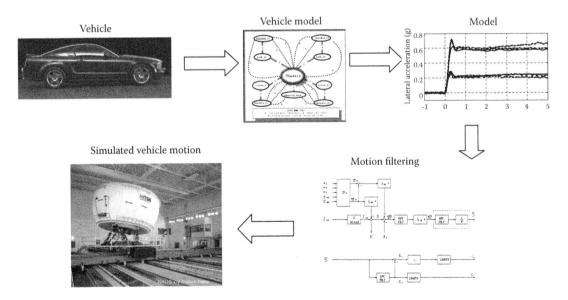

FIGURE 7.2 Motion rendering in a driving simulator.

Step 3. **Cue-Generation Errors.** The output from a model is generally not directly usable by the simulator cuing systems. A vehicle acceleration produced by the dynamics model must be transformed into commands to the simulator actuators. An audio system output (e.g., the sound of a passing car) must be transformed into discrete signals for individual speakers. The torque felt on the steering wheel in response to a wind gust must be transformed into a steering actuator command. In each case, a model output must be translated into a form that is specific to the simulator hardware. This transformation is often a radical one that gives rise to large errors.

For example, in a fixed-base simulator the transformation of lateral acceleration by the cue-generation system is multiplication by zero! Even in a large motion-based simulator as shown in Figure 7.2, the transformation is considerable (Reid & Grant, 1993). Accelerations must be scaled or reduced in value by a process that may be non-linear. A filtering process is also implied which can be non-linear in nature. These scaled and filtered accelerations need to be such that they describe motions that are compatible with the simulator constraints. Given that automobile maneuvers can result in hundreds or thousands of meters of travel, and that the largest driving simulators in existence have motion ranges of no more than a few tens of meters it is clear that the scaling and filtering process will dramatically alter the model outputs (see Section 4.3 for a further discussion of the effect of motion control software on fidelity).

Step 4. **Presentation Errors.** The final state of the rendering process is to present these mathematically-generated cues to the driver by means of the simulator hardware systems. These systems reproduce the cues with varying degrees of faithfulness. In the case of motion simulation, it is clear that the actuators used have a finite bandwidth and accuracy. But there are other effects that can be equally important. Motion systems act via an often complex mechanical path that connects the actuators to the driver. Compliances in this path can introduce distortion that changes the character of the presented motion. Resonances can be severe enough to make the visual scene appear to shake or in an extreme case, actually impact the integrity of the simulation structure (Pagnottta, Callahan, & DiEdwardo, 2000).

7.1.1.1 Vehicle Model Fidelity

Physical fidelity can be quantified by measuring the error introduced in each of the four steps in the rendering process. Steps 2, 3 and 4 can all be characterized by comparing the input and output of each stage at any point in time. The Modeling Errors in Step 1, however, must be assessed by examining the global characteristics of the simulation model in relation to the target vehicle. Bernard and Clover (1998) have proposed that the validation of a vehicle dynamics model requires answering three questions:

1. Is the model appropriate for the vehicle and maneuver of interest?
2. Is the simulation based on equations that faithfully replicate the model?
3. Are the input parameters reasonable?

Question 1 is particularly important in a driving simulator because the type of maneuvers being tested may be quite different from those that were considered when the simulator was initially constructed. Because simulators offer the illusion that any type of "driving" is permitted it is possible to construct simulator scenarios for which the underlying vehicle models are not appropriate. Speeds which are very high or very low, limit-handling maneuvers, tire blowouts, driving on ice and snow are all situations where specialized models may be needed to ensure a valid simulation.

Question 2 is best handled by examining the model output in various test conditions. Table 7.1 from Bernard and Clover (1998) suggests tests which can help validate the model implementation. Each row in Table 7.1 lists a type of test and the corresponding model errors that can be found. Many of these tests can be performed with simple techniques. For example, the first entry in the table requires that the experimenter use standard handling equations (e.g., Gillespie, 1992) to predict the response of the simulated vehicle during normal maneuvering conditions (< 0.3 g). Calculations like the lateral acceleration gain (the

TABLE 7.1 Error-Checking Procedures for Vehicle Models

Methodology	Remarks
Linear range (up to 0.3 g on a dry surface): Compare with closed form solution.	Finds errors in mass, simple geometry, cornering and aligning stiffness, roll steer, steering compliance, and lateral load transfer.
Mid-range nonlinear calculations (up to 0.5 g on a dry surface), no brakes: Compare with closed form estimates of lateral load transfer.	Finds errors in front to rear lateral load transfer distribution, roll stabilization.
Limit maneuvers (tire saturation occurring), no braking: Check individual tire forces as a function of kinematics and normal load.	Finds errors in the implementation of the tire models, including shear force limits, overturning moments; severe test for suspension kinematics.
Straight line braking: Compare with estimates of load transfer and wheel angular acceleration.	Finds errors in fore-aft load transfer, brake torque, wheel spin algorithms, tire shear force limits.
Braking in a turn: Check individual longitudinal and lateral tire forces.	Attempts to find errors in the tire model.

Source: Bernard, J. E., & Clover, C. L. (1994). *Validation of computer simulations of vehicle dynamics* (SAE Paper 940231). Warrendale, PA: Society of Automotive Engineers. With permission.

amount of lateral acceleration that results from a given steering angle) are simple algebraic expressions involving key vehicle parameters such as vehicle mass and length. These parameters should be known from the simulator documentation or they can be obtained from the simulator manufacturer (or from the simulation engineer in the case of an internally developed simulator code.) The value of acceleration gain calculated using these parameters should be very close to the result obtained by running the simulator at a fixed speed and recording the simulator's output for lateral acceleration under constant steering input. If the calculated value does not agree with the output of the simulator test, this is a sign that one or more of the elements listed in Table 7.1 (those listed in the second column, i.e., mass, geometry, etc.) contains an error and should be revised.

How important such errors are depends on the purpose of the testing. For vehicle design applications it is clearly critical that the model implementation be consistent and that it correctly represent the targets vehicle. It is less important for testing where the simulator is intended to represent a "generic" vehicle. But even in the generic case, it is helpful to know that the simulator's behavior is really appropriate to the class of vehicle being represented. If a test subject is told that he or she will be driving a simulated passenger car during an experiment they should not experience handling appropriate to a heavy truck!

The third question, parameter validation, is a difficult problem even for vehicle modeling specialists. Gruening and Bernard (1996) suggest confirmation tests which can help find erroneous parameter values. The procedures used are similar to the one outlined above. Simple measurements are made on the existing simulator software and these are compared directly to known data about the target vehicle. Often gross errors can be uncovered by referring to simple performance tests such as stopping distances or acceleration times. These tests are easy to run in the simulator and the results should compare favorably to data in publicly available reviews of new cars and trucks.

7.1.2 The Problem of Extrapolation

Even when a simulator has been properly validated for a specific use, there is always a danger of operating the simulator outside of its region of validity. Simulators are designed to give the illusion that they are vehicles and it is tempting to think that they can be driven in all of the varied ways that a real vehicle would allow. This is not so—it is quite possible for a simulator to be "high fidelity" in one operating regime and for the same device to be "low fidelity" in another. A vehicle is a physical system. So is a driving simulator but the underlying mechanisms in the two devices are radically different. The design and operation of a simulator motion system does not resemble the design of a vehicle chassis. A simulator visual display presents images that find their way to the driver's retina but those images are created by a display device rather than by reflections from natural objects. These differences mean that simulators can offer no assurance of correct operation beyond an explicitly validated operating regime.

An experiment using vehicles on a roadway may call for maneuvers that never exceed a threshold—for example, 0.3 g lateral acceleration. Through driver or experimental error the actual accelerations could conceivably be larger. Perhaps a large fraction of the tests occurred at 0.5 g peak acceleration. Such an error using real vehicles may invalidate the test—this depends on the objectives of the research—but it does not call into question the validity of the vehicle response. The vehicle simply responds according to the laws of physics.

This type of error is more serious in an artificial environment like a simulator. An experimenter cannot assume that the simulator will respond "correctly" in a new performance regime unless its operation in that regime has been independently validated. In a motion simulator, for example, there may be hardware or software that limit the maximum acceleration so that a "small change" from 0.3 g to 0.5 g results in a drastically different performance. In some cases the simulation may halt; in others it may simply be grossly inaccurate. In still other cases, the performance may be unaffected. But there is, in general, no way to be certain of the behavior without explicit verification. Whenever the operation of a simulator is extended to a new driving regime the model verification steps outlined in Section 3.1 must be carried out anew.

7.1.3 Experimental Artifacts

We have seen that errors can accumulate at each stage of the cue rendering process. We also need some way to determine if these errors are important to the experimental purpose of the simulator. This, of course, depends on the simulator application. Using a driving simulator to assess chassis design requires establishing the physical fidelity of the simulation through all stages of the rendering process. This is a form of "virtual prototyping" where the ability of the simulator to accurately represent a particular vehicle design takes the highest precedence.

On the other hand, using a simulator to assess proposed highway signage has a different set of requirements. Chassis or steering dynamics are only important in so far as the vehicle needs to be drivable without undue stress. Visual rendering of the roadway design and signs now takes precedence. But what if the chassis dynamics in the simulator are such that drivers must pay unrealistically close attention to lane-keeping just to maintain vehicle control? Wicken's Multiple Resource Theory (Wickens, 2002) suggests that a very high level of visual or cognitive demand from the primary driving task may well influence performance on the sign-recognition task that is the object of the experiment. In an extreme case the simulator drivers may no longer have the capacity to read and comprehend signage which would not be problematic if encountered in actual driving. In this case, a lack of fidelity in the simulator has created an *experimental artifact*.

The *American Heritage Dictionary* defines an artifact as "An inaccurate observation, effect, or result, especially one resulting from the technology used in scientific investigation or from experimental error." Controlling artifacts is one of the major tasks in proper experimental design. For an experimentalist or a training specialist an examination of expected simulator-induced artifacts is essential to determine the suitability of the simulator for a particular task.

For the vast majority of research the simulator is (more or less) a given. The relevant question is then one of correspondence. The simulator is simulacrum—it was designed to approximate some portion of the driving experience. It does so with varying degrees of success which can be characterized by examining the cue-rendering errors for each major cueing system. The research problem probes some aspect of driving: distraction by hand-held devices in urban settings, car following behavior, a driver's ability to avoid a crash by last-second action, and so forth. If we can identify a meaningful area where these two domains overlap—where the simulator and reality appropriately correspond—then we have some hope of designing a good experiment that is free of experimental artifacts.

But the approximate nature of simulation complicates this problem. Simulator designers are creating a virtual world but they are working in a very physical reality. Espié, Gauriat and Duraz (2005) summarized the situation in a paper on driving simulator validation. They point out that motion systems cannot reproduce accelerations at unity scale factors; visual systems fall short in resolution, luminosity and colorimetry. These shortfalls give rise to artifacts of the simulation. The following table lists a few common measures of physical fidelity and the type of artifacts that may arise as a result of imperfect simulation.

An example may help illustrate the issues involved. Imagine an experiment designed to measure the effect of various in-vehicle tasks on lane-keeping performance. The tasks vary in their duration and in the amount of visual attention they require (e.g., cellular phone dialing, reading the speedometer, changing the cabin temperature, tuning the radio, etc.) The simulator to be used is a fixed-base device with a 180° horizontal field of view. Because of maintenance issues, the steering wheel control loader will not be operational and will be turned off for the duration of the experiment. What artifacts might be expected to arise?

Several entries in Table 7.2 seem applicable to this scenario. In the Motion Characteristics section we note that a restricted motion range (zero in this case) can be expected to affect vehicle

TABLE 7.2 Common Simulation Artifacts

	Physical Characteristic	Key Measure	Most Likely Artifacts
Visual System Characteristics	Horizontal Field of View	Angular extent	FOV too small: Not enough visual information to see or drive scenario, especially at intersections. Improper mirror scanning behavior and visual allocation if the FOV does not include rearward areas that require the use of mirrors. FOV too large: Increased potential for simulator sickness has been observed as horizontal FOV is increased, especially in fixed-base simulators.
	Perceived resolution	Angular extent of the smallest resolvable optical line pair	Too Large: Human "eye-limiting" resolution in the fovea is between 1.5 and 2.0 arcmins/optical line pair. Typical simulator resolutions range from 3 to 6 arcmin/optical line pair or greater. Large angular extents result in a host of artifacts related to a lack of acuity. Threats are identified at closer distances. Drivers have difficulty seeing roadway markings and signage. Eyestrain & fatigue result from looking at blurry images.
	Transport delay	Time delay between operator action and visual system response	Too large: Degraded vehicle control resulting in steering oscillations and path-following errors. Increased simulator sickness.
	Accommodation	Distance at which eye accommodates	Too Small: Eyestrain, poor depth perception and distance judgments, simulator sickness.
Motion Characteristics	Useful motion range (<3 Hz)	Maximum useful acceleration scale factor per degree of freedom	Too Small: Poor vehicle control, simulator sickness, visual demand of the primary driving task unrealistically high.
	Small Signal bandwidth	3dB frequency 90 degree phase	Too Small: Incorrect ride dynamics, poor road feel, "magic carpet" ride.
	Roughness	Acceleration noise (g's)	Too Large: False cues, poor on-center control, discomfort, anxiety.
Control Characteristics	Steering loader torque	Torque as a function of vehicle speed and steering wheel angle	Too Small: Poor vehicle control. Too Large: Difficulty steering, potential safety concern.
	Steering loader damping	Free response to small step input	Too Small: Instability, potential safety issue. Too Large: Steering feels "dead", poor steering control during lane changes.
	Brake pedal stiffness	Force per unit distance	Too Small: Low initial braking forces. Too Large: High initial braking forces, cannot modulate brake pedal, increased simulator sickness.

control. It is also noted that a lack of motion cues may result in "unrealistically high" visual demand for the primary driving task. In the Control Characteristics section for low steering wheel torque it is also noted that poor vehicle control may be expected. Taken together, these artifacts suggest that our imaginary experimenter may be expected to find poor lane-keeping throughout the test. It is reasonable to ask if the variability in lane-keeping that is expected to result from the lack of steering and motion cues may actually obscure any effect of secondary task performance. More serious is the possibility that the high visual demand in the primary driving task caused by the lack of simulator cues may saturate the driver's visual channel and provoke unrealistically high degradation of lane-keeping performance for some secondary tasks (those with high visual demand) but not for others (those with low visual demand.) Just such an interaction between motion cues and secondary task type has been documented in simulator studies conducted under both motion and no-motion conditions (Greenberg, Artz, & Cathey, 2003). This type of interaction is particularly troublesome because it is not a simple level shift and cannot be accounted for by reporting relative results.

The estimation of possible artifacts is only a rough, albeit useful, guide when evaluating the suitability of a particular simulator to a given experimental purpose. The extent to which the artifacts in Table 7.2 are actually present in a given simulator needs to be established empirically by the research team operating the facility. But when the artifacts are known to be large, the experimental design should be altered to reduce their impact and thereby improve the generality of the results.

7.2 Simulation of Vehicle Motions

Eyesight provides drivers with the ability to detect a path and guide a vehicle along an acceptable trajectory. This provides the most basic form of driving simulation and for many simulators visual simulation is the primary cue provided. But the experience of driving is also an experience dominated by the sensation of forces. As human beings we feel a vehicle accelerate, brake and turn. That feeling of motion, delivered through our vestibular and kinesthetic senses can be a pleasurable one. Indeed much of the advertising of passenger cars is dedicated to extolling the "fun to drive" character of automobiles.

For human operators, the feeling of motion provides more than just a pleasurable sensation. It also provides leading information that drivers use to improve control. For this reason some simulators go to great lengths to physically move the driver during the simulation. The amount and character of that movement, however, is quite different from what happens in an actual vehicle. The most obvious difference is one of displacement. A short drive in an automobile might still cover several kilometers. Even a relatively large simulation laboratory measures only some tens of meters in any dimension. This difference has large implications for the fidelity of the motion cue.

To limit the required displacement, simulators rely on a combination of kinematic and perceptual attributes. From kinematics we know that only accelerations can produce forces which the driver will directly experience. Travel at constant velocity requires no special action on the part of the simulator motion system (apart from road vibration) because the driver experiences the same forces at constant speed as they would at rest. This is promising but not as helpful as it might appear because it is rarely acceptable to begin a simulation at constant speed or to end it abruptly without bringing the simulation back to zero speed. The act of simulating the initial acceleration not only creates a displacement but also a velocity in the simulator that must be eliminated before the limits of the simulator hardware are reached. A simple example shows how formidable these kinematic considerations are for even a mild acceleration and a very large simulator.

A typical vehicle acceleration of 0.2 g to a speed of 50 kph requires more than seven seconds and a displacement of nearly 50 m. But although the forces felt by the driver stop once the acceleration is finished, the simulator cab is now traveling at 50 kph, or nearly 14 m/s. To reduce this velocity without alerting the driver requires a constant deceleration that must be below the 0.005 g threshold of human detection. Bringing the velocity back to zero at this rate requires an additional 1,968 m or nearly 2 km of travel! Clearly such a brute-force approach is impractical for even the largest simulator motion systems.

For this reason simulators make use of some attributes of human perception, most notably our inability to accurately sense absolute levels of acceleration. This fact offers two opportunities. First, the accelerations that would be experienced in the actual vehicle can be reduced by a scale factor without changing the driver percept. The minimum size of the scale factor depends on the stimulus and on the purpose of the study but the practical lower limit on scale factor is generally taken to be 0.5 in flight simulations (Reid & Nahon, 1986). Second, if the driver has no visual access to the laboratory frame of reference, tilt-coordination can be used. In tilt-coordination, the driver, vehicle, and visual scene are rotated together so that gravity can substitute for the feeling of acceleration. For example, tilting the cabin backward by 11.5 degrees provides the illusion of constant 0.2 g acceleration. Of course the driver also feels that "gravity" has been reduced by a small amount, but if the amount of tilt is not too great the discrepancy is not noticeable. These effects and others are used by a simulator's motion control algorithm to distort the vehicle accelerations and limit the motion base displacement.

7.2.1 Types of Motion Systems Found in Driving Simulators

The world of flight simulation long ago settled on a preferred solution for motion-based facilities. Almost all flight simulators intended for training use a variant of the Stewart platform. Not so for driving simulation, where a large variety of motion system designs, including Stewart platforms (Figure 7.3); Web Figure 7.3, have been employed. The reason for this difference can be found in the different types of motions that are important for aircraft and ground vehicles.

FIGURE 7.3 A Stewart platform.

Commercial aircraft commonly execute turns in a coordinated manner which involves simultaneously yawing and banking the airframe. In this type of turn passengers feel no apparent lateral acceleration because the centripetal acceleration arising from the turn is exactly cancelled by the banking. This is good for airframe control and good for keeping the passengers happy because a cup of coffee inside an airplane making a coordinated turn will not slosh or spill. Passengers are often unaware that such a turn is even taking place. For this reason, among others, emulating flight dynamics in commercial pilot training does not require the ability to generate large, sustained lateral accelerations, at least during normal maneuvers. Vertical motion, however, is very important to emulate turbulence and landing. Large roll and pitch motions are also useful. The Stewart platform meets these requirements.

In a ground vehicle things are different. Ground vehicles do not bank appreciably so turns are not coordinated (hence there is an entire industry devoted to "spill-proof" beverage containers for use in cars). Lateral accelerations can and do affect the driver and passenger during turns and lane changes. Longitudinal accelerations are important for braking and acceleration. This gives rise to a need to sustain large lateral and longitudinal accelerations for several seconds.

A wide variety of motion base designs have been used on driving simulators to meet these needs. Some of these are shown in Table 7.3 (Web Table 7.1). A version of this table also appears in the color insert. The Stewart platform designs have the advantage of being mature technology. They are also efficient in the sense that virtually all of the moving mass of a Stewart platform (aside from payload) is contained in the actuators. This contrasts favorably with the large displacement systems where gimbals and support structures add significantly to the moving mass. The biggest disadvantages of the Stewart platform are the limited displacement and awkward workspace. This results in a Stewart platform design occupying a relatively large volume for a given amount of displacement. An operational concern with the Stewart design is that the degrees of freedom are coupled so that the amount of displacement available in any one degree of freedom depends on the current state of all six actuators. This can complicate the control of the platform and makes it difficult to execute motions like braking in a turn where several degrees of freedom have simultaneous large displacements.

The large displacement designs attempt to overcome these difficulties by decoupling the lateral and/or longitudinal degrees of freedom from vertical displacement and rotational motions. The low-payload designs have attractive price/performance characteristics but the payload limitation places serious constraints on the type of cabin or dome that can be used in the simulator. The large payload, large-displacement designs have the fewest compromises but also the highest cost.

7.2.2 Characterizing Simulator Motion Performance

The scale factor for any degree of freedom in a simulator is simply the fraction of the calculated vehicle acceleration that the motion system attempts to reproduce. For example, a scale factor of 0.5 in the longitudinal direction means that the simulator motion system will only try to deliver 50% of the actual vehicle acceleration to the driver. The most important characteristic of a simulator motion system is the *maximum useful scale factor* (MUSF) in each degree of freedom. The MUSF is the largest scale factor value that a simulator can use in a given degree of freedom, for a given maneuver. The MUSF is limited by the simulator, the specific maneuver, the degree of freedom under consideration and the maximum acceptable distortion of the motion cue. Values of MUSF can be between 0.5 and 1.0 for moderate (<0.6 g) lateral maneuvers on a large motion system like the NADS and are often less than 0.05 for longitudinal maneuvers on a small Stewart platform. MUSF is, by definition, identically zero for fixed-base simulators.

For simulated on-road driving there is often an asymmetry in MUSF for the lateral and longitudinal degrees of freedom. When driving on-road, the road width provides a limit on the maximum expected lateral displacements needed to reproduce vehicle motion. Simulators can take advantage of this fact to use larger MUSFs for lateral motion. Longitudinal motion, in contrast, is usually unconstrained by the driving scenario and MUSFs in this direction are therefore lower.

An example of scaled motion is shown in Figure 7.4 Web Figure 7.4) from Nordmark, Jansson, Palmkvist and Sehammar (2004).

In this case, a double lane change is being executed in the simulator. The lateral accelerations from the vehicle model have been scaled by 0.7 and are plotted as the "Input Acceleration" on the figure. The ability of the simulator hardware to reproduce this motion is shown by the "Measured Acceleration" plot. Because the motion system apparently has enough lateral displacement in this case to reproduce the motion with only a simple scaling transformation, the motion system has been commanded to follow the Input Acceleration directly. The Measured Acceleration shows that, except for some high-frequency noise, the motion system hardware is quite capable of executing the maneuver. For this simulator and this maneuver, the MUSF for the lateral degree of freedom is at least 0.7.

TABLE 7.3 Driving Simulator Motion Systems **(See color insert)**

Type	Typical Example(s)	Notes
Small Stewart Platform	Moog 12 Inch Series 	Electric actuator design. Displacements: X: +/− 0.25 m Y: +/− 0.25 m Z: +/− 0.18 m Roll: +/− 21 degree Pitch: +/−22 degree Yaw: +/−22 degree Payload: 1000 Kg
Large Stewart Platform	Ford VIRTTEX Simulator	Hydraulic actuator design. Displacements: X: +/−1.6 m Y: +/−1.6 m Z: +/− 1.0 m Roll: +/−20 degree Pitch: +/−20 degree Yaw: +/−40 degree Payload: ~6000 Kg
Asymmetric Large Displacement Systems	VTI Driving Simulator III	Electric actuator design: X or Y: +/−3.75 m Roll: −9 degree to +14 degree Pitch: +/−24 degree Vibration: Z: +/−0.06 m X: +/−0.06 m Roll: +/−6 degree Pitch: +/−3 degree
Symmetric Large Displacement Systems (Payload < 1500 Kg)	Renault Ultimate Driving Simulator 	Electric actuator design: X: 6.0 m Y: 6.0 m Z: +/−0.2 m Roll: +/−15 degree Pitch: +/−15 degree Yaw: +/−15 degree Payload: ~1,000 Kg

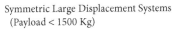

TABLE 7.3 (Continued)   Driving Simulator Motion Systems (**See color insert**)

Type	Typical Example(s)	Notes
Symmetric Large Displacement Systems (Payload > 1500 Kg)	National Advanced Driving Simulator (NADS)	Hydraulic/Electric actuator design: X: +/−9.75 m Y: +/−9.75 m Z: +/−0.6 m Roll: +/−25 degree Pitch: +/−25 degree Yaw: +/−330 degree There are also 3 vibration degrees of freedom. Payload: ~10,000 Kg
Asymmetric Large Dispalcement Systems (Payload > 1500 Kg)	Toyota Driving Simulator	Hydraulic/Electric actuator design: X: +/− 17.5 m Y: +/− 10 m Z: +/− 0.6 m Roll: +/− 25 degree Pitch: +/− 25 degree Yaw: +/− 330 degree There are also 3 vibration degrees of freedom. Payload: ~10,000 Kg

A similar plot for longitudinal acceleration in Figure 7.5 (Web Figure 7.5) shows the result for braking with a scale factor of 0.5. Despite the smaller scale factor, longitudinal accelerations are reproduced less accurately than in the lateral case. In fact, the motion system does not have sufficient displacement to reproduce the longitudinal acceleration with only a simple scaling transformation. The motion control algorithm first computes a "Commanded Acceleration" that represents the result of both scaling and filtering operations. This Commanded Acceleration is then used to control the motion system. The initial or onset accelerations that occur immediately upon braking are reproduced

well but the maximum level of the acceleration cue falls short of the target. The acceleration targets peak around −4 m/s² but because these are scaled by 0.5, the actual vehicle accelerations for this maneuver were approximately −8 m/s². The peak acceleration provided by the motion system is between −2 m/s² and −3 m/s². Such performance is typical of longitudinal maneuvers in a very large simulator and may provide an acceptable motion simulation.

MUSF provides a simple way to specify how a motion system is being used in a particular application. It has the advantage of a clear physical interpretation and an easy method of assessment. To use a simulator, scale factors must be chosen and presumably

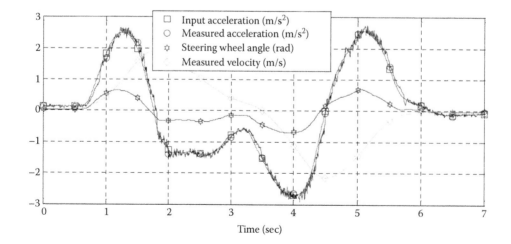

FIGURE 7.4   Lateral acceleration during a lane change maneuver (scale factor 0.7). (From Nordmark, S., Jansson, H., Palmkvist, G., & Sehammar, H. (2004). *The new VTI driving simulator. Multi purpose moving base with high performance linear motion.* Proceedings of Driving Simulation Conference Europe 2004. Paris. With permission.)

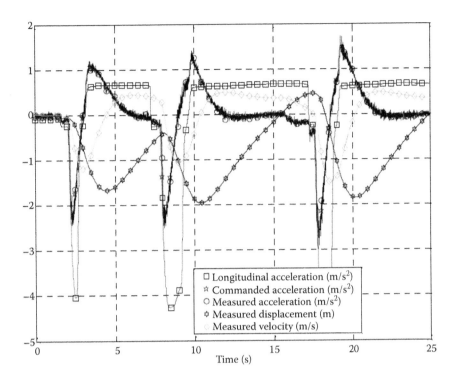

FIGURE 7.5 Longitudinal acceleration simulated by the linear motion (scale factor 0.5). (From Nordmark, S., Jansson, H., Palmkvist, G., & Sehammar, H. (2004). *The new VTI driving simulator. Multi purpose moving base with high performance linear motion.* Proceedings of Driving Simulation Conference Europe 2004. Paris. With permission.)

these will be set to maximize the use of the motion system. Therefore the MUSF (or something very close to it) is readily available and known for every simulator and application.

The disadvantage of MUSF is that it depends on the simulation scenario and on the precise definition of "useful." As we have seen the MUSF for a steering maneuver is not necessarily the same as the one for braking. Similarly, an experiment that requires only very mild braking scenarios may be able to use a larger MUSF than one that requires heavier braking. It is also the case that the amount of distortion that can be tolerated in a given maneuver may vary from experiment to experiment depending on factors such as the frequency of the maneuver, the sensitivity of the test subjects to simulator sickness and the particular goals of the experiment. Nonetheless, when the MUSF is reported as part of the experimental documentation, the effect of the simulator motion base on driver cueing becomes much clearer.

It is possible to assess simulator motion fidelity in a more objective way. Grant, Artz, Greenberg and Cathey (2001) and Schwarz, Gates and Papelis (2003) have characterized driving simulator motion systems using procedures based on AGARD Report AR-144 (Lean & Gerlach, 1977). Various simulator attributes such as transfer functions, noise levels, signal-to-noise contours and dynamic thresholds can be measured in a way that is completely independent of any simulation scenario. These performance measurements provide a method for making detailed comparison between motion systems. They also allow an individual motion system's characteristics to be documented in a way that is independent of any particular scenario or usage. However,

examples of such measurements in the literature are disappointingly rare.

Figure 7.6 shows an X-axis transfer function measurement from the VIRTTEX simulator (Grant et al., 2001). The amplitude ratio defines the ratio of the input signal to the output response as a function of input frequency, plotted in dB. The phase response characterizes the linear delays inherent in the simulator response.

Figure 7.7 shows X-axis crossbeam noise measurements for the NADS simulator (Schwarz et al., 2003). Unwanted accelerations can be characterized as "noise". By driving the motion system at a known frequency with various amplitudes, the noise level can be quantified at the various amplitudes.

Noise can be further characterized by calculating the signal-to-noise ratio (SNR) over a variety of frequencies and velocities and plotting the results. The SNR can be defined as:

$$SNR = \frac{\sigma_d}{\sigma_{tot}}$$

Where σ_d is the variance of the driving signal and σ_{tot} is the variance of the noise components. Both Grant and Schwarz give calculation procedures for determining the SNR. The results can be summarized in a table or contour plot.

An example SNR contour is shown in Figure 7.8 Web Figure 7.6. This plot is the result of 31 SNR measurements that have been fit to contour lines of constant SNR. It shows, for example, that the high velocity regions in this simulator have the lowest SNR

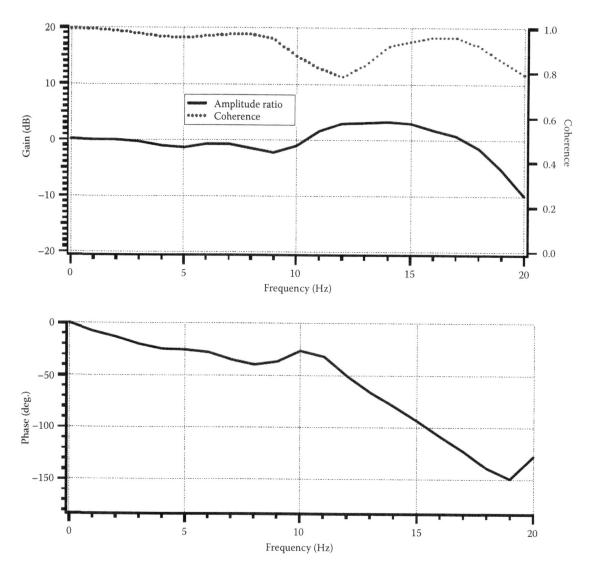

FIGURE 7.6 X-axis transfer functions in VIRTTEX. (From Grant, P., Artz, B., Greenberg, J., & Cathey, L. (2001). *Motion characteristics of the VIRTTEX motion system.* Proceedings of the 1st Human-Centered Transportation Simulation Conference. Iowa City, IA. With permission.)

and that motions near the center of the motion regime have the highest SNR.

Dynamic threshold measures the latency in the motion system. A square wave input is applied in a given degree of freedom and the output response is measured. The time between the presentation of the input signal and the output signal reaching 63% of the commanded value is defined as the dynamic threshold. Figure 7.9 shows the X-crossbeam dynamic threshold for the NADS simulator. The threshold value is 30 ms for this degree of freedom.

7.2.3 The Effect of Motion Drive Algorithms

A wide variety of Motion Drive Algorithms (MDAs) have been proposed and implemented for use in driving simulators. An MDA takes a desired input motion (usually an acceleration generated from a vehicle dynamics model) and generates commands which can be sent to the motion system. A detailed discussion of these

algorithms is beyond the scope of this chapter but there are certain characteristics that all of the MDAs have in common. MDAs generally contain multiple parameters which can be adjusted to change the way that they distort the output of the vehicle dynamics model. For a given scenario the MDA and its parameters are chosen in such a way as to balance three competing objectives:

1. Maximize the useful scale factor in key degrees of freedom.
2. Minimize or eliminate "false cues". These are cues that are actually the opposite of those expected to occur in a real vehicle. For example, pressing the accelerator pedal and feeling the vehicle actually slow down is a type of false cue.
3. Ensure that the motion system remains within its limits of operation. Safety systems will generally abort a simulation if commanded motions drive the simulator out of the normal operating regime. These aborts are disruptive to the

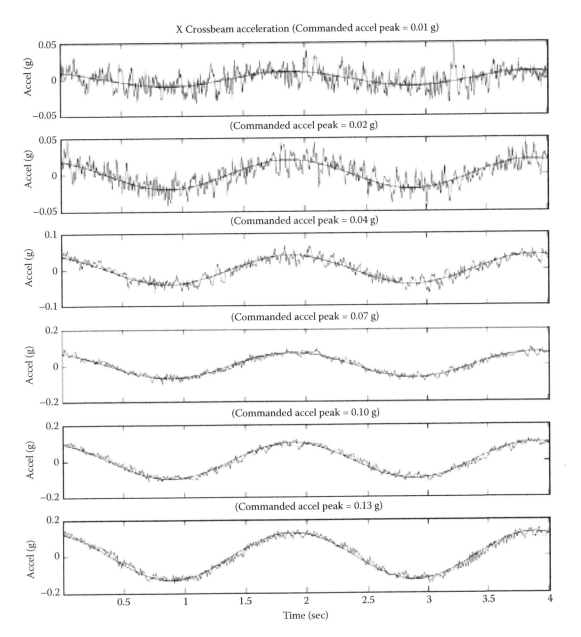

FIGURE 7.7   X-axis crossbeam 1/2Hz noise in the NADS. (From Schwarz, C., Gates, T., & Papelis, Y. (2003). *Motion characteristics of the National Advanced Driving Simulator.* Proceedings Driving Simulation Conference North America 2003. Dearborn, MI. With permission.)

purpose of the simulation and they are often unpleasant for the drivers.

All of the algorithms share the characteristic that they severely limit the maximum displacements associated with the motion cues. For this reason any assessment of motion fidelity must include the effect of both the MDA and the specific parameters used in the simulator (e.g., Curry, Greenberg, & Kiefer, 2005).

7.3 Steering Simulation

A steering simulation typically consists of a physical steering wheel mounted in front of the driver's seat. The wheel is fitted with a position transducer that senses the wheel rotation and is connected to some means of generating a feedback torque. More sophisticated systems can add additional sensors for torque and/ or rotational velocity. In most modern simulators the feedback torque is generated by an electric motor which is controlled by a mathematical model.

A typical rack and pinion steering system of a real vehicle is shown schematically in Figure 7.10. The steering wheel is connected via a shaft and Cardan joints to the pinion gear. The pinion gear drives the rack which, in turn, is connected at each end to the wheels by means of a tie-rod linkage. Forces and moments generated by the tires are thus transmitted through the rack and eventually back to the driver's hands by way of the steering wheel. In systems that have hydraulic power assistance, the pinion gear

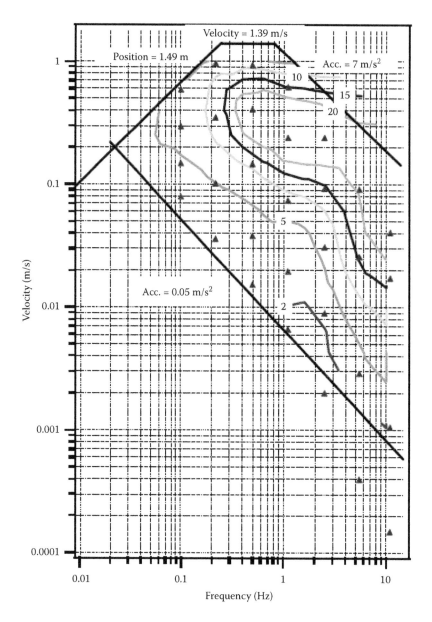

FIGURE 7.8   X-axis SNR contour for the VIRTTEX simulator. (From Grant, P., Artz, B., Greenberg, J., & Cathey, L. (2001). *Motion characteristics of the VIRTTEX motion system.* Proceedings of the 1st Human-Centered Transportation Simulation Conference. Iowa City, IA. With permission.)

contains a compliant element known as a torsion bar. Torque which is transmitted through the pinion gear deflects the torsion bar which opens a hydraulic valve. High-pressure hydraulic fluid flows through the valve, completing a feedback loop that generates a boost-force and reducing the effort required to move the rack.

For small steering wheel movements the resulting torque felt by the driver is simply proportional to the steering wheel angle. At larger angles non-linear effects become important. Sliding and sticking friction also play major roles in determining steering feel because the perceived wheel torque is never exactly zero as long as the steering wheel is in motion. Typical results for a passenger vehicle are shown in Figure 7.11 (Web Figure 7.7).

Despite the linear relationship between steering angle and steering wheel torque at small angles, it is not appropriate to

model steering feel as a simple spring with constant stiffness. The effective stiffness in the linear region increases with vehicle speed due to the presence of higher tire forces at higher speeds (see Salaani, Heydinger, & Grygier, 2004). Toffin, Reymond, Kemeny and Droulez (2003) showed that drivers are able to adapt to a variety of steering feedback strategies without measurably changing their ability to control a vehicle through curves. However, they found that "unusual" steering feedback configurations (e.g., no feedback force, or feedback in the "wrong" direction) resulted in poor performance or even an inability to maintain vehicle control. These results suggest that all driving simulators should strive to achieve typical steering feedback forces to ensure that drivers can adequately adapt their behavior. However, simulators that are not intended for the detailed study

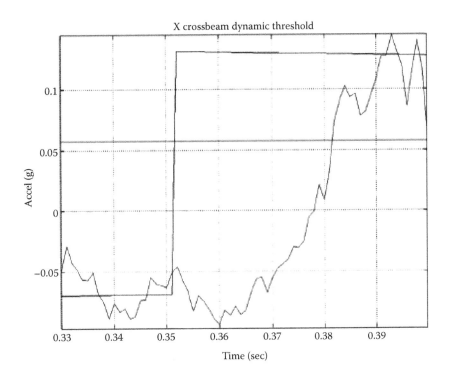

FIGURE 7.9 X-crossbeam dynamic threshold measured in the NADS. (From Schwarz, C., Gates, T., & Papelis, Y. (2003). *Motion characteristics of the National Advanced Driving Simulator.* Proceedings Driving Simulation Conference North America 2003. Dearborn, MI. With permission.)

of steering or vehicle dynamics may not require an extremely accurate steering feel system.

Models that account for the kinematics of the steering system and the dynamic forces that appear on the steering wheel have been developed for a variety of applications (see Post & Law, 1996; Salaani et al., 2004). The accuracy of such models is highly dependent on the vehicle dynamics and tire models because of the crucial role that tire-generated forces play in determining what the driver actually feels. Artz, Cathey, Curry, Grant, and Greenberg (1999) showed that a real-time steering model could generate steering torque versus lateral acceleration curves that closely match experimental measurements.

In Figure 7.12, data for a 1997 Ford Contour sedan (labeled "cdw27 data") is plotted against the output of a model at various vehicle speeds (note that Figure 7.12 uses a different sign convention for steering wheel torque than Figure 7.11). Plotting the steering torque

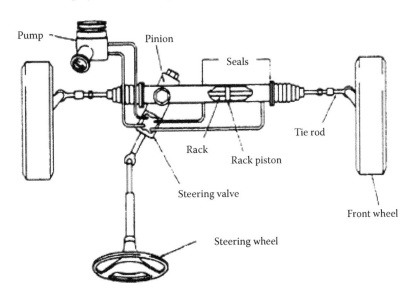

FIGURE 7.10 Rack and pinion steering components. (From Howe, J. G., Rupp, M. Y., Jang, B., Woodburn, C. M., Guenther, D. A., & Heydinger, G. J. (1997). *Improving steering feel for the National Advanced Driving Simulator* (SAE Paper 970567). Warrendale, PA: Society of Automotive Engineers. With permission.)

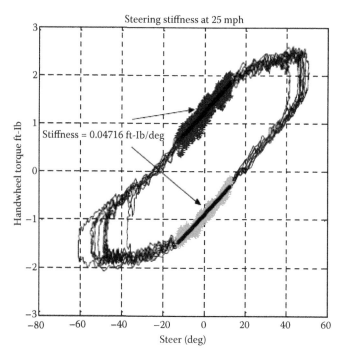

FIGURE 7.11 Handwheel torque—Steer angle curve for the 1997 Jeep Cherokee at 25 mph. (From Salaani, M. K., Heydinger, G. J., & Grygier, P. A. (2004). *Experimental steering feel performance measures* (SAE Paper 2004-01-1074). Warrendale, PA: Society of Automotive Engineers. With permission.)

against lateral acceleration instead of steering wheel angle is sometimes called a "feel plot". Although the simulated and tested vehicles had tires from different manufacturers the agreement between the model and test data is very good at speeds below 45 mph. At higher speeds and higher lateral accelerations, discrepancies between the model and vehicle results become more pronounced.

The torque calculated by a steering model must be rendered during the simulation. This is usually done by means of a servo motor attached directly to the steering wheel. A wide variety of motors have been used depending on cost, performance and packaging requirements. In general, such a motor should have low inertia, low cogging torque and high bandwidth (see Salaani et al., 2004).

7.4 Audio Simulation

While some driving simulators for the automotive industry focus primarily on audio simulation to produce extremely accurate representations of the vehicle's acoustic environment in order to study noise, vibration, and harshness (NVH) issues, audio simulation is usually one of the lower fidelity components in most other driving simulators. Many audio simulation systems simply consist of one engine sound and one road/wind sound that are typically looped for playback, with their level and "pitch" modified as a function of vehicle and engine speed. It is often the belief that realistic audio simulation requires significant programming effort, or is cost-prohibitive—especially after paying for the visual and motion systems.

This section focuses primarily on non-NVH driving simulators, providing an overview of audio simulation characteristics and describing how relatively inexpensive interactive three-dimensional (3D) sound simulation systems can now be realized in driving simulators because of recent advances in binaural (two-eared) recording, PC-based sound simulation, and professional audio hardware components. Binaural recordings of sounds at the driver's position are easily acquired, largely due to increased focus by vehicle manufacturers on customer satisfaction with vehicle sound quality. The recent explosion in popularity of PC-based multi-media games has led to cost-effective hardware and software methods to simulate effects such as Doppler shift, reverberation, and spatialization. The ever-expanding types of professional audio hardware components now make it easy to filter, mix, and route multi-channel audio data.

7.4.1 Purpose of Audio Simulation in Driving Simulators

At a minimum, audio simulation provides additional context to fully immerse drivers into a virtual world. Wind and road noise increase with increasing vehicle speed, and engine noise indicates how hard the engine is working, and when the transmission is shifting. With some effort to understand the acoustic environment inside the driving simulator, along with timing and update rates for the sound playback system, the range of applications for a driving simulator can increase dramatically. Presentation of sounds with accurate timing, spectral content, and sound level makes it possible to carry out meaningful studies for active safety systems, in-vehicle information systems (e.g., cellular telephones, navigation systems), and vehicle calibration (e.g., engineers use sound of the engine rpm to make decisions on powertrain parameters).

7.4.2 Aspects of the Soundscape

The soundscape can be broken down into three primary groups: (i) "ownship", or "host" vehicle sounds, (ii) external sounds, and (iii) experiment sounds.

"Ownship" or "host" vehicle sounds are generated by the vehicle interacting with the environment, or sounds generated by vehicle systems. Vehicle interactions with the environment are generally associated with a vehicle's NVH characteristics. Sounds include both steady-state and transient powertrain, road, and wind noises. Examples of transient sounds include engine starter, tire squeal, and road potholes or bumps.

Examples of sounds generated by vehicle systems include warnings for the vehicle state (e.g., chimes associated with key in ignition, door ajar, and lights on), warnings for the vehicle interacting with the environment (e.g., active safety warnings associated with lane departure, or forward collision), or sounds from vehicle information systems (e.g., navigation system).

External sounds are generated by external sources such as other vehicles or people. Examples include the vehicle pass-by (i.e., "whoosh") sounds, another vehicle's horn, and road construction sounds (e.g., jackhammer).

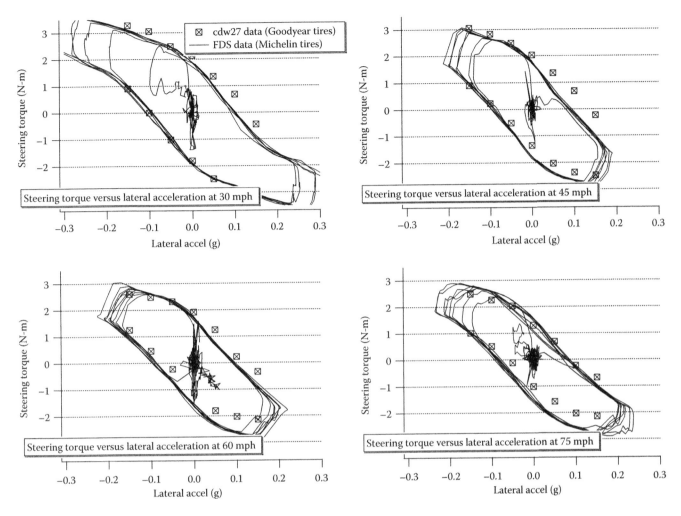

FIGURE 7.12 Steering wheel torque versus vehicle lateral acceleration. (From Artz, B. E., Cathey, L. W., Curry, R. E., Grant, P. R., Greenberg, J. (1999). Using the Ford driving simulator to develop metrics for on-center steering feel. *Proceedings of driving simulation conference 1999* (pp. 443–458). Paris. With permission.)

Experiment sounds include experimenter/subject communication, and abstract sounds. The abstract sounds are typically tones or beeps to indicate the start or end of an experiment task. Pre-recorded experimenter/subject communication, or voice-overs, are often overlooked in driving simulation, but play an important role in maintaining precise and consistent timing of experiment tasks across drivers (e.g., a simulation scenario program initiates a lead car deceleration exactly X seconds after the voice-over completes a request for the driver to carry out a secondary task). They also prevent any change in voice tone or inflection to indicate something "different" might happen next (e.g., a surprise lead-car deceleration).

7.4.3 Some Aspects of 3D Sound Reproduction

The methods currently used to render audio in driving simulators vary greatly (e.g., Colinot & Belay, 2000; Goetchius, Ketelhut, Smallwood, & Eaton, 2001; Heidet, Warusfel, Vandernoot, Saint-Loubry, & Kemeny, 2001; Allman-Ward, et al., 2003; Blommer & Greenberg, 2003; Heitbrink & Cable, 2007; Gauduin & Boussard, 2009). Some simulators play sounds over speakers without regard to the acoustics of the listening environment, whereas other simulators use headphone playback in order to avoid the complex acoustic field inside the vehicle cabin. For discussion purposes, the following subsections focus on audio simulation using a sound database. As such, the audio simulation process in a driving simulator is fairly common and is implemented by a sound module, or program, which carries out the following steps:

1. Receives vehicle and scenario state information from the driving simulation;
2. Selects appropriate pre-recorded sounds from a database for powertrain, wind, road, and background noise;
3. Appropriately pitch-bends (including Doppler effect) and blends selected sounds to generate sounds corresponding to the vehicle and scenario states: Transient sounds are generated by receiving timing information from the vehicle and scenario states; continuous sounds are generated by appropriately looping sounds from the database; and
4. Plays sounds over the simulator buck speaker system or headphones.

The key aspects of the audio simulation process are recording/synthesizing sounds, generating a sound database, possibly modifying the sounds based on the current state of the driving simulator (e.g., vehicle speed, engine rpm, Doppler effects for exterior objects, etc.), and then playback of the sounds to the listener. These aspects are detailed in the following subsections.

7.4.3.1 Recording/Synthesizing Sounds

Sounds can be recorded or synthesized using a variety of techniques. Binaural recordings at the driver's position in a vehicle cabin will better represent what a driver actually hears. The binaural head recording contains most characteristics of the acoustic filtering by a listener's head, torso, and pinnae (Blauert, 1997). These characteristics include diffractions, reflections, and linear distortions of the incoming acoustic waves, which result in amplitude and phase modifications of the waves as a function of both frequency and spatial location. The human auditory system "decodes" these amplitude and phase modifications into spatial locations of the sound sources.

By measuring these head-related transfer functions (HRTFs) from a spatial location to the ear canal of the right and left ears, one can also incorporate binaural effects when synthesizing sounds. Simply filter the synthesized sound by a set of HRTFs that correspond to desired spatial locations (Begault, 1994).

7.4.3.2 Sound Database

Wind and road sounds are recorded as a function of vehicle speed, with the sounds recorded separately or simultaneously. Separate measurements can be made in a wind tunnel and on chassis rolls, respectively. Simultaneous measurements can be made by turning the engine off at various speeds on different test track surfaces. Sounds measured in 16 kph (10 mph) increments and of 10–20 seconds in duration are usually sufficient for realistic road and wind noise simulation. Sounds are prepared to prevent clicks/pops during looping playback.

Powertrain sounds can be recorded on chassis rolls at a number of rpm/load conditions. The rolls and tires are selected to minimize their contribution to the powertrain recording. Genuit and Bray (2001) recommend engine rpm sounds measured in 500 rpm increments in order to simulate the powertrain without perceptible artifacts. Similar to the wind and road sounds, the powertrain sounds are 10–20 seconds in duration and are prepared to prevent clicks/pops during looping playback.

Most other sounds (e.g., rumble strips, car horns, voice-overs, etc.) are recorded either with a binaural head or monophonically. Appropriate HRTFs or panning algorithms are used if the sounds are recorded monophonically.

7.4.3.3 Realtime Modification of Sounds

Generating interactive effects such as echo, reverberation, and Doppler shifts are fairly standard in software today. For example, many Windows®-based software applications generate these effects using Microsoft® DirectSound®. These effects can greatly enhance the perception of 3D sound, with an example being the Doppler shift due to a passing vehicle.

Simulation of wind and road noise at speeds between those in the database can be done through appropriate sound-level mixing of the database sounds. Pitch-bending and sound-level mixing are used to simulate powertrain sound at rpm/load conditions which are not in the database.

7.4.3.4 Rendering Sound Field

It is generally preferred to render 3D sound using loudspeakers instead of headphones because drivers do not wear headphones in real vehicles. For loudspeaker playback, there are various techniques to "expand" the acoustic image in an attempt to render 3D audio. The most straightforward technique is to use many loudspeakers and play sounds from the appropriate speaker in order to create the 3D perception. A reduced version of this technique is surround sound, which typically involves four speakers placed about the listener and one "center-channel" speaker.

For two-speaker playback, crosstalk cancellation or stereo-dipole techniques are used to expand the acoustic image. A pictorial representation of crosstalk cancellation is shown in Figure 7.13. The loudspeakers are usually placed 30–60 degrees to each side of the listener. The goal is to eliminate the crosstalk from the right speaker to the left ear (and the left speaker to the right ear) so that the loudspeakers essentially act as a set of headphones. That is, the right channel of the stereo sound goes only to the right ear, and the left channel goes only to the left ear. Simple models are typically used to represent the transfer functions from the speakers to the ears. Drawbacks for crosstalk cancellation include a relatively small "sweet spot" (i.e., where the expansion effect works well), and also insensitivity to front/back source location.

The stereo-dipole technique is somewhat similar to crosstalk cancellation, except the speakers are placed 5–10 degrees apart and in front of the listener. Stereo-dipole playback usually has a larger sweet spot than crosstalk cancellation, but is also insensitive to front/back source location.

The crosstalk cancellation and stereo-dipole playback methods can be enhanced with the use of HRTF equalization. That is, the standard implementation of these playback methods assume a point-receiver model for each ear of the listener and do not take into account the acoustic filtering of the head, torso, and pinnae. By including HRTF equalization in lieu of the point-receiver models, the expansion effect is improved. However, the computation requirements can be greatly increased with HRTF equalization, which may not make it appropriate in all interactive audio environments.

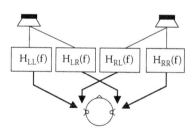

FIGURE 7.13 Crosstalk cancellation.

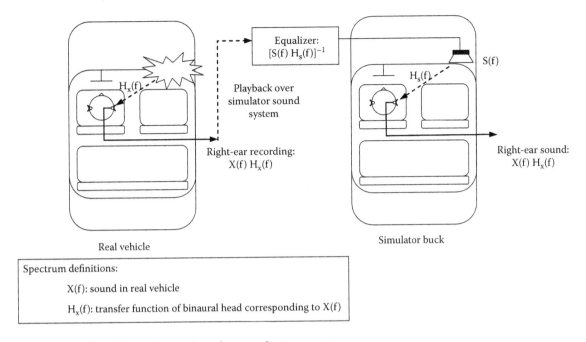

FIGURE 7.14 Overview of vehicle recording and simulator equalization process.

The binaural recording and equalization technique (Xiang, Genuit, & Gierlich, 1993; Krebber, Gierlich, & Genuit, 2000; Genuit & Bray, 2001) uses standard loudspeaker playback in combination with HRTF equalization to enhance the perception of a 3D acoustic environment. The basic idea is to place a binaural head at the driver's location in the simulator buck, and measure the transfer function consisting of the right simulator speakers, simulator buck cabin, and right ear of the binaural head. The inverse of this transfer function is calculated and programmed into equalizers corresponding to the right simulator speakers. A similar process is carried out for the left simulator speakers. The driver will then experience realistic vehicle sounds when binaural recordings are played over the simulator sound system. Figure 7.14 shows an overview of the

vehicle recording and simulator equalization process. For the sake of simplicity, only the process for the front right speaker is shown.

Figure 7.15 (Web Figure 7.8) shows the 1/3-octave responses of the driver-side speakers to the driver outboard ear (DOE). Notice that these responses vary by more than 20 dB, which must be flattened by the equalizers. Figure 7.16 (Web Figure 7.9) shows the *equalized* 1/3-octave responses of the driver-side speakers to DOE. That is, these responses include an equalizer in the measurement chain with equalizer gains set to flatten out the response of Figure 7.15. The equalized frequency responses are within +/−3 dB from 40 Hz to 10 kHz.

Crosstalk for the binaural equalization technique is defined as the acoustic energy that reaches the DOE from the passenger-side

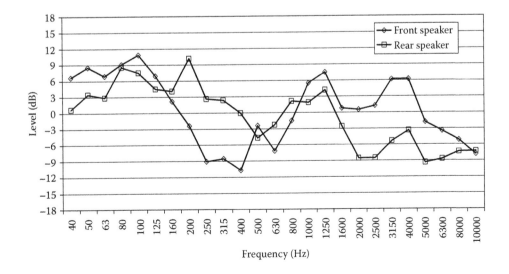

FIGURE 7.15 1/3-octave responses corresponding to driver-side speakers and DOE.

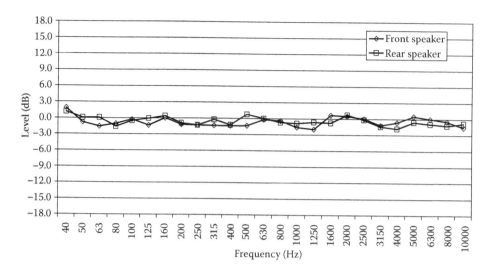

FIGURE 7.16   Equalized 1/3-octave responses corresponding to driver-side speakers and DOE.

speakers, and the acoustic energy that reaches the driver inboard ear (DIE) from the driver-side speakers. Crosstalk is reduced with the binaural equalization technique by relying on acoustic shadowing effects of the driver's head (Blauert, 1997). That is, the head provides acoustic attenuation, or "shadowing" from the passenger-side speakers to the DOE, and from the driver-side speakers to the DIE.

7.4.4 Hardware and Software

Many sound-rendering APIs currently exist for Windows-based PCs that include support for 3D sound and DSP effects (e.g., FMOD, BASS, OpenAL). Additionally, there are complete products available for 3D sound rendering (e.g., GENEIS, V+ with SMx Audio System, & H3S). The hardware system can be based on relatively inexpensive "off-the-shelf" pro-audio components that interface with a Windows-based PC. As an example, Figure 7.17 shows the layout of the hardware components for the VIRTTEX driving simulator. There are speakers located at the four "corners" of the simulator buck, along with a subwoofer and a seat shaker. The seat shaker is typically excited with lowpass-filtered road noise and its level is subjectively tuned to provide contextual high-frequency (>20 Hz) vibration to the driver.

7.5 Concluding Remarks

Characterizing the physical fidelity of a given simulator is a complex task that requires an assessment of both the simulator and the purpose for which the simulation is being conducted. Simple labels such as "high fidelity" or "low fidelity" are devoid of clear meaning precisely because they do not take into account the complex interplay between a simulator's physical capability and its suitability for a particular experimental purpose. As we have seen, a simulator that is highly suited for one type of experimental application may be a poor choice in another.

Everything that a driver experiences in a simulator begins as a representation (often abstract) of a physical phenomenon that must be transformed into a sensation by a rendering process consisting of four stages: modeling the phenomena, output generation, transformation of outputs to desired sensations, delivery of the outputs as a physical stimulus. Each of these cue rendering stages introduces errors or distortions that can be quantified.

The importance of the rendering errors, however, can only be assessed in relation to a specific experimental or training purpose. When the inaccuracies in a simulation give rise to experimental artifacts, the experimental design or the simulator (or both) must be altered. In practice it is not uncommon for an experimental plan to be abandoned if an analysis of the expected artifacts casts sufficient doubt on the utility of the results. However, even when the expected artifacts do not pose such a large threat to validity, it is critical that they be clearly understood by the experimental team and properly documented as an aid in interpreting the findings.

No cueing system in a simulator is error-free. All systems introduce some level of distortion to the cues that the driver experiences. We have examined several systems in detail and in each case the nature of the cueing errors is not apparent from a simple physical description of the system. For a motion system, for example, a description of the hardware and displacement limits does not shed light on the quality of the motion cues presented to the driver. Documenting the scale factors actually used in an experiment, however, provides useful information about the magnitude of the motion cues present during the simulation. A similar characterization of the steering and audio cues is also required to assess the real impact of these cueing systems on driver perception. Physical system description is not enough.

The proper approach to fidelity in driving simulators requires a process of constant reassessment. Whenever the simulator or the experimental purpose changes, the same questions must

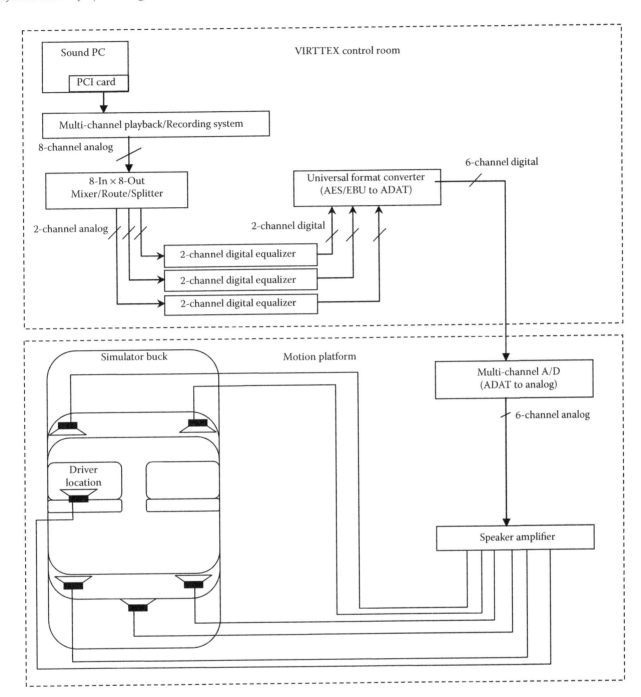

FIGURE 7.17   Layout of VIRTTEX sound system hardware.

be asked anew: What are the rendering errors? What artifacts will they engender? How should these effects be controlled and documented? The answers to these questions define the physical fidelity of a driving simulator.

Key Points

- Everything that a driver experiences in a simulator begins as a representation (often abstract) of a physical phenomenon that must be transformed into a sensation by a rendering process. By examining each stage in the process we can quantify the errors or distortions present in the simulator.

- The importance of the rendering errors should be assessed by estimating the experimental artifacts that are likely to result from them in the context of the current experiment (or training scenario).

- To assess a motion system it is important to describe the size and accuracy of the motion cues that are actually being delivered in a particular experiment. A mechanical description of the motion system properties is rarely sufficient. A detailed motion system performance characterization is very helpful but when this is not possible a simpler approach is to document the motion scale factors that were used.

- To assess the audio system it is important to understand the acoustic environment inside the driving simulator, along with timing and update rates for the sound playback system. Presentation of sounds with accurate timing, spectral content, and sound level makes it possible to carry out meaningful studies for a variety of applications. Relatively inexpensive interactive three-dimensional (3D) sound simulation systems can now be realized in driving simulators because of recent advances in binaural (two-eared) recording, PC-based sound simulation, and professional audio hardware components.

Keywords: Artifacts, Audio, Fidelity, Motion, Rendering

Glossary

Artifact: An inaccurate observation or result that arises because of flaws in the simulation.

Cue rendering: The process of transforming a representation of a physical system into a sensation that can be directly experienced in the simulator.

Head related transfer function (HRTF): A mathematical relationship that defines most characteristics of the acoustic filtering by a listener's head, torso, and pinnae.

Maximum useful scale factor (MUSF): The maximum scale factor for each degree of freedom that was used in a simulator during a given experiment.

Motion drive algorithm (MDA): Software that transforms computed vehicle accelerations into commands that appropriate for the simulation hardware.

Web Resources

FMOD. Melbourne, Victoria (Australia): Firelight Technologies Pty, Ltd. Available from http://www.fmod.org.

BASS. un4seen. Available from http://www.un4seen.com.

Open AL. San Jose, CA: Creative Labs. Available from http://www.openal.org.

GENESIS. Aix En Provence, France: GENESIS. Available from http://www.genesis-acoustics.com.

V+ with SMx Audio System. Tampa, FL: SimPhonics Incorporated. Available from http://www.simphonics.com.

H3S. Herzogenrath, Germany: HEAD acoustics. Available from http://www.headacoustics.com.

Lean, D., & Gerlach, O. H. (1977). *Dynamic characteristics of flight simulator motion systems* (AGARD Advisory Rep. 144). Advisory Group for Aerospace Research and Development. North Atlantic Treaty Organization (NATO).

Web Table 7.1: Driving Simulator Motion Systems (color version of Table 7.3).

Web Figure 7.1: The simulator driver is the focus of numerous cue-generation systems (color version of Figure 7.1).

Web Figure 7.2: Motion rendering in a driving simulator (color version of Figure 7.2).

Web Figure 7.3: A Stewart platform (color version of Figure 7.3).

Web Figure 7.4: Lateral motion (color version of Figure 7.4).

Web Figure 7.5: Longitudinal motion (color version of Figure 7.5).

Web Figure 7.6: X-axis SNR contour for the VIRTTEX simulator (color version of Figure 7.8).

Web Figure 7.7: Handwheel torque—Steer angle curve for the 1997 Jeep Cherokee at 25 mph. (color version of Figure 7.11).

Web Figure 7.8: 1/3-octave responses corresponding to driver-side speakers and DOE (color version of Figure 7.8).

Web Figure 7.9: Equalized 1/3-octave responses corresponding to driver-side speakers and DOE (color version of Figure 7.9).

Key Readings

Bernard, J. E., & Clover, C. L. (1994). *Validation of computer simulations of vehicle dynamics* (SAE Paper 940231). Warrendale, PA: Society of Automotive Engineers.

Blommer, M., & Greenberg, J. (2003). *Realistic 3D sound simulation in the VIRTEX driving simulator.* Proceedings of Driving Simulation Conference North America 2003. Dearborn, MI.

Frank, L. H., Casali, J. G., & Wierwille, W. W. (1988). Effects of visual display and motion system delays on operator performance and uneasiness in a driving simulator. *Human. Factors, 30*(2), 201–217.

Greenberg, J., Artz, B., & Cathey, L. (2003). *The effect of lateral motion cues during simulated driving.* Proceedings of Driving Simulation Conference North America 2003. Dearborn, MI.

References

Allen, R. W., & DiMarco, R. J. (1984). Effects of transport delays on manual control systems. *Proceedings of the 20th annual conference on manual control* (pp. 185–210). Washington, DC: National Aeronautics and Space Administration.

Allen, R. W., Rosenthal, T., Klyde, D., Anderson, F., Hogue, J., & Christos, J. (1997). *A vehicle dynamics model for low-cost, PC-based driving simulations.* Proceedings of the Third Driving Simulation Congress. Paris: ETNA.

Allman-Ward, M., Williams, R., Cockrill, M., Distler, H., Crewe, A., & Heinz, T. (2003). *The interactive NVH simulator as a practical engineering tool.* Proceedings of SAE Noise & Vibration Conference (SAE Paper 2003-01-1505). Traverse City, MI: Society of Automotive Engineers.

Artz, B. E., Cathey, L. W., Curry, R. E., Grant, P. R., & Greenberg, J. (1999). Using the Ford driving simulator to develop metrics for on-center steering feel. *Proceedings of driving simulation conference 1999* (pp. 443–458). Paris.

Bae, D., & Haug, E. (1987). A recursive formulation for constrained mechanical system dynamics: Part I. Open loop systems. *Mechanics Based Design of Structures and Machines, 15*(3), 359–382.

Bae, D., & Haug, E. (1987). A recursive formulation for constrained mechanical system dynamics: Part II. Closed loop systems. *Mechanics Based Design of Structures and Machines, 15*(4), 481–506.

Begault, D. (1994). *3D sound for virtual reality and multimedia.* Cambridge, MA: AP Professional.

Bernard, J. E., & Clover, C. L. (1994). *Validation of computer simulations of vehicle dynamics* (SAE Paper 940231). Warrendale, PA: Society of Automotive Engineers.

Bernard, J. E., & Clover, C. L. (1998). *Tire modeling for low-speed and high-speed calculations* (SAE Paper 950311). Warrendale, PA: Society of Automotive Engineers.

Blauert, J. (1997). *Spatial hearing. The psychophysics of human sound localization.* Cambridge, MA: MIT Press.

Blommer, M., & Greenberg, J. (2003). *Realistic 3D sound simulation in the VIRTEX driving simulator.* Proceedings of Driving Simulation Conference North America 2003. Dearborn, MI.

Colinot, J., & Belay, G. (2000). A multifunctional sound generator for the PSA's sherpa driving simulator. *Proceedings of driving simulation conference (DSC) 2000* (pp. 165–179).

Curry, R. C., Greenberg, J. A., & Kiefer, R. J. (2005, August). *NADS versus CAMP closed-course comparison examining "last second" braking and steering maneuvers under various kinematic conditions* [pdf] (DOT HS 809 925). Performed by Crash Avoidance Metrics Partnership (CAMP), Contract DTFH61-01-X-00014. Washington, DC: National Highway Traffic Safety Administration. Retrieved from http://www.nrd.nhtsa.dot.gov/pdf/nrd-12/acas/FCWTask4.pdf

Espié, S., Gauriat, P., & Duraz, M. (2005). *Driving simulators validation: The issue of transferability of results acquired on simulator.* Proceedings of Driving Simulation Conference North America 2005. Orlando, FL.

Frank, L. H., Casali, J. G., & Wierwille, W. W. (1988). Effects of visual display and motion system delays on operator performance and uneasiness in a driving simulator. *Human Factors, 30*(2), 201–217.

Gauduin, B., & Boussard, P. (2009). *High fidelity sound rendering for car driving simulators.* Proceedings of Driving Simulation Conference 2009 Europe. Monaco.

Genuit, K. & Bray, W. (2001). *A virtual car: Prediction of sound and vibration in an interactive simulation environment.* Proceedings of SAE Noise & Vibration Conference (SAE Paper 2001-01-1474). Traverse City, MI: Society of Automotive Engineers.

Gillespie, T. D. (1992). *Fundamentals of vehicle dynamics.* Warrendale, PA: Society of Automotive Engineers.

Goetchius, G., Ketelhut, C., Smallwood, B., & Eaton, C. (2001). *Subjective evaluation of NVH CAE model predictions using an operator-in-the-loop driving simulator.* Proceedings of SAE Noise & Vibration Conference (SAE Paper 2001-01-1590). Traverse City, MI: Society of Automotive Engineers.

Grant, P., Artz, B., Greenberg, J., & Cathey, L. (2001). *Motion characteristics of the VIRTTEX motion system.* Proceedings of the 1st Human-Centered Transportation Simulation Conference. Iowa City, IA.

Greenberg, J., Artz, B., & Cathey, L. (2003). *The effect of lateral motion cues during simulated driving.* Proceedings Driving Simulation Conference North America 2003. Dearborn, MI.

Gruening, J., & Bernard, J. E. (1996). *Verification of vehicle parameters for use in computer simulation* (SAE Paper 960176). Warrendale, PA: Society of Automotive Engineers.

Heidet, A., Warusfel, O., Vandernoot, G., Saint-Loubry, B., & Kemeny, A. (2001). *A cost-effective architecture for realistic sound rendering in the Scanner II driving simulator.* Proceedings of the 1st Human-Centered Transportation Simulation Conference. Iowa City, IA.

Heitbrink, D., & Cable, S. (2007). *Design of a driving simulation sound engine.* Proceedings of Driving Simulation Conference North America 2007. Iowa City, IA.

Howe, J. G., Rupp, M. Y., Jang, B., Woodburn, C. M., Guenther, D. A., & Heydinger, G. J. (1997). *Improving steering feel for the National Advanced Driving Simulator* (SAE Paper 970567). Warrendale, PA: Society of Automotive Engineers.

Krebber, W., Gierlich, H., & Genuit, K. (2000). Auditory virtual environments: Basics and applications for interactive simulations. *Signal Processing, 80*(11), 2307–2322.

Lean, D., & Gerlach, O. H. (1977). *Dynamic characteristics of flight simulator motion systems* (AGARD Advisory Rep. 144). Advisory Group for Aerospace Research and Development. North Atlantic Treaty Organization (NATO).

Nordmark S., Jansson H., Palmkvist G., & Sehammar H. (2004). *The new VTI driving simulator. Multi-purpose moving base with high performance linear motion.* Proceedings of Driving Simulation Conference Europe 2004. Paris.

Pagnotta, M., Callahan, J., & DiEdwardo, A. (2000). Structural design issues of motion-based simulators. *AIAA Modeling and Simulation Technologies Conference and Exhibit.* Denver, CO.

Post, J. W., & Law, E. H. (1996). *Modeling, characterization and simulation of automobile power steering systems for the prediction of on-center handling* (SAE Paper 960178). Warrendale, PA: Society of Automotive Engineers.

Reid, L. D., & Grant, P. R. (1993). Motion algorithm for large-displacement driving simulator. *Transportation Research Record, 1403*, 98–106.

Reid, L. D., & Nahon, M. A. (1986). *Flight simulation motion-base drive algorithms: Part 2—Selecting the system parameters* (UTIAS Rep. 307). Toronto, Canada: University of Toronto Institute for Aerospace Studies.

Salaani, M. K., Heydinger, G. J., & Grygier, P. A. (2004). *Experimental steering feel performance measures* (SAE Paper 2004-01-1074). Warrendale, PA: Society of Automotive Engineers.

Sayers, M. W., & Mousseau, C. W. (1990). Real-time vehicle dynamic simulation obtained with a symbolic multibody program. *Transportation Systems 1990* (pp. 51–58). American Society of Mechanical Engineers.

Schwarz, C., Gates, T., & Papelis, Y. (2003). *Motion characteristics of the National Advanced Driving Simulator.* Proceedings Driving Simulation Conference North America 2003. Dearborn, MI.

Toffin, D., Reymond, G., Kemeny, A., & Droulez, J. (2003). *Influence of steering wheel torque feedback in a dynamic driving simulator.* Proceedings Driving Simulation Conference North America 2003. Dearborn, MI.

Wickens, C. D. (2002). Multiple resources and performance prediction. *Theoretical Issues in Ergonomics Science, 3*(2), 159–177.

Xiang, N., Genuit, K., & Gierlich, H. (1993, Oct 7–10). *Investigations on a new reproduction procedure for binaural recordings.* Paper presented at the 95th Convention of the Audio Engineering Society. New York.

8

Sensory and Perceptual Factors in the Design of Driving Simulation Displays

George J. Andersen
University of California, Riverside

Abstract

The Problem. Driving simulators involve the presentation of visual information (along with other information) to examine driving skill and performance. In order to optimize driving skill and performance, simulator displays should be designed in accordance with the sensory and perceptual capabilities of the driver. *Role of Driving Simulators.* The effect of different ways of displaying visual information on drivers' performance in a driving simulator is itself the object of study. *Key Results of Driving Simulator Studies.* Driving simulator studies have indicated the types of visual information important for performing a variety of tasks. These types include sensory factors (acuity, contrast, luminance, and motion) and perceptual factors (texture, optical flow, depth and distance information, and cue conflicts). Relevant results from psychophysics studies are discussed that can be used to optimize the presentation of visual information for driving tasks. In addition, minimum standards are proposed to optimize research and training using simulators. Finally, both the role of visual factors that can contribute to simulator sickness and the role of procedures that can be used to minimize risk of simulator sickness are also discussed. *Scenarios and Dependent Variables.* Simulator scenarios that are highly dependent on visual factors include steering, collision detection and avoidance, car following, braking, speed regulation, and the identification/recognition of targets (i.e., roadway signs, pedestrians, traffic lights). *Platform Specificity and Equipment Limitations.* The present chapter will discuss how differences in the platforms and equipment in driving simulators determine what visual information is displayed, which in turn has an impact on driving performance.

8.1 Introduction

A primary goal of driving simulation research is to understand performance under conditions where visual information is manipulated or controlled to perform a specific task. The presentation of visual stimuli in a simulator is an approximation of the information available in real-world driving. For example, the resolution available in a simulator when viewing roadway signs is considerably lower than the resolution available under similar conditions in real-world driving. Luminance levels of daytime visual displays are below the luminance available under real-world conditions. A critical assumption of driving simulation research is that the presentation of visual information in the simulator is sufficiently rendered to minimize any differences between performance on the driving task under investigation in the simulator and performance on the same task in real-world driving. Any violation of this assumption can lead to incorrect generalizations from performance on the simulator task to performance in real-world driving, and may have a negative impact on the design of new technologies to improve driving safety or the use of simulators for training.

It is unlikely that simulators can have fidelity equivalent to real-world vision. For example, perceiving detail in real-world vision is limited by the resolution of the photoreceptor lattice in the retina. In contrast, perceiving detail in driving simulation displays is limited by the resolution of the monitor or projection system, which is considerably lower than resolution in real-world vision. This limitation as well as others (i.e., contrast, motion, texture, etc.) raises several pertinent questions regarding simulator design and the external validity of simulators for understanding driving skill and performance. Do these limitations represent a problem for driving simulation research? If optimal fidelity cannot be obtained then what minimum standards should be used to optimize external validity? Is it necessary to use simulators with the highest fidelity? Finally, what is the potential impact on performance when minimum conditions are not used?

The purpose of this chapter is to address these questions. First, I will present an overview of what is known about the visual information used to perform different driving tasks, including discussions of both the sensory information (e.g., spatial resolution, temporal resolution, luminance, and contrast) and perceptual information (e.g., perception of motion, perception of a 3D scene, egocentric distances and depth). Second I will discuss why limitations in the capabilities of driving simulators make it likely that the *sensory information* presented in simulated driving tasks is oftentimes inadequate and can lead drivers to perform a task in ways than are different from what they might do in the real world. Third, I will take up a similar discussion of the how limitations in the capabilities of driving simulators can make it likely that the *perceptual information* a driver is using differs from what might be used in the real world. Fourth, I will discuss perceptual conflicts that can exist in driving simulators and the potential impact such conflicts can have on driving performance. Finally, I will discuss the relationship between fidelity issues in simulators and the incidence of simulator sickness.

8.2 Visual Factors and Driving Tasks

Driving is a skill that is critically dependent on visual information. As we drive our visual system extracts information from the surrounding scene that we use to perform different tasks (steering control, braking, etc.). Understanding what visual information is used to perform these tasks is critical to determining what information should be present in a driving simulator. These tasks can be categorized, for the sake of convenience, into the following performance issues: identification/recognition of targets, steering control, collision detection and avoidance, braking, car following and speed regulation. This list is based on the types of tasks that have been studied in driving research and is not intended to be an exhaustive list. But it does provide a basis from which we can evaluate what is known from research regarding the information used to perform the tasks and the sensory and perceptual information used to perform the tasks. A thorough review of the visual information used to perform these tasks is beyond the scope of the present chapter. However, the review in the present chapter will focus on those issues likely to be of relevance for optimal design of driving simulation displays.

8.2.1 Detection/Recognition of Targets

In this section, the detection and recognition of targets such as roadway signs will be discussed. An extensive body of literature exists concerning the design of signs for optimal legibility on roadways (e.g., Kline & Fuchs, 1993; Mori & Abdel Halim, 1981; Shinar & Dori, 1983). A detailed review of this literature is beyond the scope of this chapter. However, the research literature on signage and driving has primarily focused on issues of font type, size, and contrast. In simulation displays, the visibility of signage is limited by the display resolution, luminance and contrast of the monitor/projection system.

Most simulation displays operate with a resolution of 1024 by 768 and include anti-aliasing (algorithms to blend features across pixels to avoid jagged edges). Lower resolution settings result in greater pixilation of the image, resulting in a decreased ability to see signage detail at increased simulated distances. The degree of pixilation will depend on the size of the display, the viewing distance from the driver to the screen, and the simulated distance of the character in the simulation.

8.2.2 Steering Control

Research on steering control suggests that drivers use two different types of visual information. One source of information is optical flow—the perspective transformation of the visual world during motion of the observer. An extensive body of literature has examined the usefulness of optical flow for determining the instantaneous direction of observer motion (e.g., Gibson, 1966; Warren, Morris, & Kalish, 1988; Dyre & Andersen, 1997). Optical flow is characterized by the collection of velocity vectors projected to the retina. In terms of driving simulation displays, accurate depiction of optical flow will be determined by the spatial resolution of the display (a lower resolution results in less precision in specific local velocity information) and the temporal resolution of the display (lower temporal resolution results in less accuracy in the magnitude of specific local velocity information). The second source of information used by drivers is a representation of the 3D driving scene. Recent research (Hahn, Andersen, & Saidpour, 2003) showed that observers use static scene information in determining the instantaneous direction of heading. More recently, Andersen and Enriquez (2006b) showed that static scene information is based on an allocentric representation of landmarks in the scene (egocentric distances between the driver and landmarks or features in the scene). Optimal information for driving simulation displays would involve the presentation of a sufficient number of landmarks in the driving scene and the availability of rich visual information for determining egocentric distance. Indeed, studies have found asymptotic performance in car following when nine landmarks were present (Andersen & Saidpour, 1999). Other studies (Land & Lee, 1994) have shown that the visibility of the curvature of the 3D roadway is used for steering along a curved path. Their results suggest that observers use the tangent of the curve to estimate the future path of the driver's vehicle. This research suggests that visibility of the roadway is important in steering along a curved path.

8.2.3 Collision Detection and Avoidance

Detection and avoidance of collision is a complex and varied task. The range of conditions is quite varied when one considers all possible combinations (constant and changing velocity; linear and curvilinear paths) of moving objects and a moving vehicle. Furthermore, a related topic that has received considerable focus in the driving literature is determining the time-to-contact (TTC) of an impending collision. Studies examining time-to-contact (Lee, 1976; Tresilian, 1991; Wann, 1996) as well as collision detection (Andersen & Kim, 2001; Andersen & Enriquez, 2006a) have shown that the ability to detect collisions is based on optical expansion (i.e., approaching objects increase in size over time).

Research has also examined the optical information for detecting collisions for different trajectories of motion. Consider the case of an approaching object on a linear trajectory. If the object is on a collision path with the driver the bearing will be constant, whereas if the object is not on a collision path (i.e., will pass by the driver) the bearing will vary over time. If the approaching object is on a circular trajectory then an object on a collision path will have a constant rate of bearing change whereas if the object is not on a collision path the rate of bearing change will vary. Previous studies have demonstrated sensitivity of drivers to use constant bearing information for detecting linear path collision events (Andersen & Kim, 2001; Andersen & Enriquez, 2006a) and constant rate of bearing change for detecting circular path collision events (Ni & Andersen, 2008).

Optical expansion is one type of optical flow information (Koenderink, 1986) and involves the perception of motion.

Optimal information for the perception of motion would be determined by the spatial and temporal information in the display. Bearing information (the projected position of an object in the visual field) would be determined by the spatial resolution of the display. Change in bearing (change in the projected position of an object in the visual field) would be determined by the projected velocity of the object in the display. Thus, optimal presentation of these information sources in driving simulation displays would be based on the spatial and temporal resolution of the displays.

8.2.4 Braking

Previous research on optical information for braking suggests that drivers use two different information sources. As noted earlier, previous studies have examined time-to-contact judgments in driving (Lee, 1976, 1980; Schiff, Oldak, & Shaw, 1992; DeLucia, Bleckley, Meyers, & Bush, 2003). Time-to-contact information is specified by the inverse rate of expansion of the approaching object. Lee (1976) also showed that the time derivative of TTC information can be used for regulating braking behavior. Several studies have shown that drivers use this information for braking (see Andersen & Sauer, 2004, for a review). Other studies have shown that drivers use egocentric distance information (Andersen, Cisneros, Saidpour, & Atchley, 1999, 2000). Because braking involves the use of expansion information (which is dependent on the perception of motion) it also is dependent on spatial and temporal information.

8.2.5 Speed Regulation

To date, only one study has examined the visual information that is used for the perception of vehicle speed. Larish and Flach (1990) examined the ability of human subjects to perceive the speed of self motion. They used a magnitude estimation methodology in which subjects rated the perceived speed of forward motion. Their results indicate that observers use global optical flow rate (the rate at which optical texture passes a particular location in the visual field). These results indicate that scene texture is a necessary source of information for determining vehicle speed, as well as spatial and temporal information for the perception of motion.

Other studies have also suggested the important role of texture in the perception of speed. Nguyen-Tri and Faubert (2007) found evidence that texture and motion interact in order to determine speed. Manser and Hancock (2007) found that variations in the texture pattern of a simulated tunnel wall during driving could result in variations in driver speed in a simulator. They found that drivers increased speed when the width of the texture pattern was decreased, and decreased speed when the width was increased. These results are consistent with earlier studies by Denton, which found that decreasing the spacing of horizontal lines across a roadway resulted in decreased speed of approach to a traffic circle (Denton, 1980).

In most real-world settings, texture is present at several different levels of scale. For example, consider a roadway consisting of concrete sections. At one level of scale, the expansion joints between the concrete sections represent a pattern of texture. At a smaller level of scale, the spacing of gravel within the concrete section represents a different level of scale (see Gibson, 1979 for a discussion). An important limitation in driving simulation displays is that texture is presented at one level of scale, or a subset of the possible levels of texture scale present in real-world scenes. This limitation may have a profound impact on performance and may represent a problem in relating driving simulation to real-world driving. For example, the limited texture in simulated 3D scenes has been shown to result in a compression of the perceived space along the depth axis (Andersen & Braunstein, 1998a). This finding suggests that limited texture in driving simulation scenes will lead to a misperception of distances in the scene, which will likely impact driving tasks such as collision detection and avoidance.

8.2.6 Car Following

Research on the visual information used to car follow (successfully maintain a fixed distance behind a lead vehicle) has examined visual angle and change in visual angle (associated with expansion or contraction of the rear-end of the lead vehicle). In a series of experiments Andersen and Sauer (2007) examined car following performance to variations in lead vehicle speed defined by sinusoidal and ramp function profiles. Based on this research they proposed a model based on visual angle (used to assess a desired following distance) and change in visual angle (used to estimate instantaneous change in relative speed between the driver and lead vehicle). Data from real-world car following performance was tested using model parameters determined from driving simulation studies. The results indicated that the model was highly predictive of car following performance under real-world driving conditions. Indeed, for real-world driving data the DVA model accounted for 10% more variability in driving speed and 35% more variability in distance headway.

The review of previous research on driving suggests that several different sources of information are important for performing the variety of tasks involved in driving. These information sources can be categorized as sensory information (spatial resolution, temporal resolution, luminance, and contrast) and perceptual information (perception of motion, perception of a 3D scene, egocentric distances and depth). In the next sections we will discuss sensory and perceptual issues in the design of simulation displays. We will discuss those factors important for the driving tasks discussed above, as well as perceptual factors normally not considered in driving simulation displays which can have important consequences for performance.

8.3 Sensory Issues in Simulator Research

Sensory information is important for several aspects of driving performance. For example, reduced visibility due to decreased luminance (present during night driving) or decreased contrast (present during driving in fog) has been shown to be associated with increased risk of a crash (Evans, 2004). Given the role of sensory information in driving tasks an important issue is whether the reduced fidelity of driving simulation scenes, as compared to real-world driving scenes, might have an impact on performance in driving tasks in a simulator. In the next sections we will consider this issue for spatial resolution, temporal resolution, luminance, and contrast.

8.3.1 Spatial Resolution

As introduced earlier the display resolution of a computer monitor or projection system falls well below the actual spatial resolution of the visual system. Under optimal luminance and contrast conditions, the spatial resolution of the visual system, assessed using sine wave grating patterns, is approximately 60 cycles per degree, or 0.016 degrees visual angle (Campbell & Robson, 1968). In contrast, consider a driving simulation display projected to a 60 degree by 40 degree visual angle. If the display resolution is 1024 by 780 (a resolution common for most driving simulators) then the dimensions of a single pixel will be 0.058 by 0.051 degrees visual angle. Thus, the driving simulation display has a resolution that is nearly four times lower the optimal resolution of the visual system.

There are three possible effects of decreased resolution of simulation displays on driving performance. One effect is that the lower resolution of simulation displays will result in decreased visibility of targets that are located at greater distances in the simulation. To illustrate this effect, consider the visibility of a roadway stop sign with 33 cm high lettering. In addition, assume that a minimum of 4 pixels is needed to depict the vertical dimension of the letter S in the word "stop". Under these conditions the sign would be detectable at a distance of 93.7 m. If we apply the same conditions to real-world vision (assuming 4 times the minimum resolution of 0.016 degrees) the sign would be visible at a distance of 295 m. Under most driving conditions the effects of the difference in maximum distance visibility would be negligible. However, this difference does become relevant when considering driving conditions at higher speeds. For example, consider a driving speed of 40 km/h. We can calculate the amount of time a driver has to read the sign by determining how many seconds of driving time it takes to pass the sign given the maximum distance that it can be seen. We will refer to this value as the *distance reading time*. At a vehicle speed of 40 km/h the distance reading time for the driving simulation display is 7.7 seconds whereas in real-world driving the reading time is 24.4 seconds. One would expect that both of these reading times are more than adequate to read the sign regardless of the tasks being performed by the driver. Now consider the reading time when the driving speed is 105 km/h. The distance reading time for the driving simulation display is 2.9 seconds whereas the distance reading time for real-world driving is 9.3 seconds. These reading times suggest that at highway speeds the reduced distance reading time may be problematic for drivers in the simulator, particularly if the driver is engaged in performing multiple tasks or if the driving scene is highly cluttered with other stimuli which might distract the driver. The point of these calculations is not to argue specific limitations in driving simulation displays. The actual distance reading times can vary across simulator displays (e.g., variations in visual angle, size of signage, etc.). The point of these calculations is to demonstrate, with a specific set of parameters (display resolution, visual angles, at signage size), the difference between driving simulation displays, real-world vision conditions, and the potential effects on performance.

A second effect concerns the precision of information for the perception of motion. The perception of motion involves the spatial displacement of information over time (Braddick, 1974). A reduction in spatial resolution will result in a reduction in the accuracy with which motion is specified. This reduction can alter the perceived motion of objects in the driving scene as well as the perception of the direction and speed of the driver's vehicle. Although this issue should typically not be a problem in most simulation displays with resolutions of at least 1024 by 780, it does become an issue when the viewing distance to the screen or monitor is reduced. For example, consider a 1024 by 780 projected image viewed at a distance of 2 meters and subtending a horizontal visual angle of 60 degrees. Under these conditions the horizontal extent of a pixel is 0.058 degrees visual angle. Reducing the distance to 1 meter to produce a larger visual angle of the display will result in a projected horizontal extent of the display of 73.8 degrees. Under these conditions the horizontal extent of a pixel is 0.072 degrees. This represents a 23% reduction in the display resolution (in projected visual angle) which will result in less accurate information for the spatial displacement of motion.

A third effect concerns the visibility of simulation display pixels. As noted above, under optimal viewing conditions there is an approximately fourfold decrease in resolution for simulation displays as compared to real-world viewing conditions. Since the minimum visual angle in a simulation display is well above the threshold of human vision, then the individual pixels of a display may be visible. Anti-aliasing algorithms (image processing procedures which average luminance across pixels) are used in simulation displays to reduce the visibility of individual pixels and minimize the appearance of jagged edges for diagonal lines. Despite the use of anti-aliasing algorithms some degree of pixilation is visible in simulation displays. The problem with pixel visibility is that it provides a texture cue that is inconsistent with the 3D simulation of the driving scene. More specifically, the size of pixels is approximately uniform in the vertical and horizontal dimensions. As a result, the pixels represent a texture pattern that the image is frontal parallel and the driving scene has no depth. However, this information contradicts the motion and pictorial cues in the display which specify that the driving scene extends into 3D space. Information of this type has been referred to in the literature as a flatness cue (e.g., Eby & Braunstein, 1995). One effect of flatness cues is to compress the perceived depth of the scene. Gogel (1954) found that the perceived depth of adjacent objects was compressed to the average depth across a scene. He referred to this effect as the adjacency principle and found that the use of this principle by human observers resulted in misperceived

distance (Gogel, 1972), size (Gogel, 1998), and speed (Gogel, 1982) of objects in a scene. As a result the egocentric distances of objects in the scene (i.e., other vehicles, signs, pedestrians, crosswalks, etc.) will appear closer than the distances that are simulated. The compression of visual space may have a profound effect on driving performance.

8.3.2 Temporal Resolution

The primary issue of temporal resolution and driving simulation concerns the effects of the refresh rate of the display and the perception of motion. Previous research has shown that the perception of motion declines when the presentation rate of individual frames is slower than one every 40 msec (Baker & Braddick, 1985). In terms of the refresh rate of the display this rate translates to 25 frames per second. Most research on depth perception and motion perception uses a minimum refresh rate of 30 frames per second. These studies suggest that the refresh rate should not fall below 30 frames per second.

More problematic, however, is the relationship between refresh rate in the simulation and simulation display detail. Simulation research necessarily involves real-time computations for producing closed loop control. As a result, the computer is performing calculations in real-time based on control inputs and the rendering of the 3D graphics presented in the display. The amount of computation (number of calculations per unit time) can vary according to several factors including the complexity of the driving scene and complexity of the simulation of vehicle dynamics. More specifically, the greater the amount of detail to be presented in the display (e.g., number of buildings, texturing, objects, moving vehicles, etc.) the greater the number of calculations that must be performed. The variation in the number of calculations can have a profound effect on the refresh rate of the display, resulting in a display refresh that varies over time. The change in refresh rate over time can dramatically alter the perception of motion that is simulated. For example, if the refresh rate changes from 30 frames per second to 10 frames per second, the change in temporal information can dramatically alter the perception of motion. Given the 25 frames per second limit (Baker & Braddick, 1985) this type of variation will lead to a perception of smooth motion at the high frame rate to a perception of jittery motion or no motion at the low frame rate. Thus, the smoothness of the perception of motion will vary as the frame rate changes over time. The variation in smoothness of motion is often perceived as a variation in velocity that is not simulated in the driving scenario but is an artifact of the variation in frame rate (Palmisano, Burke, & Allison, 2003). This change in velocity can have a profound effect on driving performance in the simulator. Indeed, variations in perceived velocity that are not consistent with the simulation (e.g., accelerations or decelerations of driver responses) may be a leading cause of simulator sickness (Palmisano, Bonato, Bubka, & Folder, 2007; Bonato, Bubka, & Palmisano, 2006).

A simple solution to this problem is to set minimum temporal rates for the simulation. Variations in the refresh rate above

30 frames per second are less likely to be detected by the human visual system because all motion above 30 frames per second is perceived as smooth motion (Baker & Braddick, 1985). However, frame rates that fall below 25 are likely to be perceived as jittery motion. These findings suggest that all simulation studies should have the refresh rate set to a value no lower than 30 frames per second (a value well above the 25 frames per second limit).

8.3.3 Luminance and Contrast

Visual displays in simulators have luminance and contrast levels well below luminance and contrast in real-world vision. For example, consider the levels of luminance that are optimal for normal human vision. Campbell and Robson (1968) found the greatest sensitivity to variations in contrast occurred for stimuli with average luminance of 500 cd/m². Luminance levels under the best conditions in simulators are approximately 50 cd/m². In general, these differences should not result in performance differences between real-world and simulator driving. However, the lower luminance levels can limit the ability of simulators to create problematic conditions for drivers such as glare (e.g., McGregor & Chaparro, 2005; Wood, Tyrell, & Carberry, 2005) or the range of luminance present under fog conditions (e.g., Broughton, Switzer, & Scott, 2007; Buchner, Brandt, Bell, & Weise, 2006).

8.4 Perceptual Issues in Simulator Displays

8.4.1 Motion and Optic Flow

Of all the sources of visual information used in driving probably no single source is as important as the perception of motion and optic flow. As reviewed above, most driving tasks require the perception of motion and recovery of information from optical flow (i.e., optical expansion). Problems can occur in simulation displays for the perception of motion and optical flow when the spatial and temporal resolution is low. A reduction in the spatial resolution of a display will result in less precision of the information present to a driver in specifying the direction of local display velocity. Similarly, a reduction of temporal resolution (display refresh) can result in jittery motion than can lead to misperceived magnitude of local velocity information. These reductions can have a significant effect on perceiving optical flow information important in driving. For example, a severe reduction in spatial and temporal rates may lead to decreased ability to determine the rate of expansion—information important for determining the time to an impending collision.

As noted earlier, temporal limitations in simulators can be addressed if the display refresh rate is maintained at 30 frames per second. Spatial limitations only become an issue for performance if the display resolution is low or if the viewing distance from the driver to the monitor or screen is small. Previous research on motion perception have typically used viewing distances of one meter or more with display resolutions of 1024 by 780 or higher. This suggests that problematic display resolution

may only occur with very low-fidelity simulators or with simulator displays viewed at close distances (less than one meter).

8.4.2 3D Texture

The primary limitation of texture in simulation displays, compared to real-world vision, is the relationship between texture and scale. In real-world scenes, texture occurs at several different levels of scale. For example, consider a roadway scene of an urban area containing city blocks with buildings. In this scene there are several different scales of texture. One level of texture is available from the repeated pattern of the buildings located along the roadway. A second level of texture is available from the pattern of windows on a building. A third level of texture is available from the materials used for the exterior of the building (e.g., granite or concrete). Thus, in real-world driving scenes information is available from several types of texture at different levels of scale.

Often in computer simulations the levels of texturing are limited. For example, texture information might be available from the first or second level listed above (e.g., pattern of buildings or windows on a building) but may not include the more detailed texture from the third level. The decision to include different levels of texture is often determined in part by the computational requirements to incorporate multiple levels of texture in the driving scene and its impact on the refresh rate of the display.

The point of raising this issue is to argue that: (1) current driving simulation scenes do not include the full range of texture information available from a real-world scene; and (2) the potential impact of including a limited range of texture on driving performance in a simulator is unclear. The limited range of texture scale in simulators may have an impact on driving performance. For example, previous research has shown the edge rate (the number of texture units that pass a location in the visual field per unit time) is used to determine speed of forward motion (Larish & Flach, 1990) and has suggested that edge rate information is important in driving tasks such as car following (Andersen, Sauer, & Saidpour, 2004). What is not known is the effect of multiple levels of texture scale on edge rate information. An important issue for future research will be to determine the role of different levels of texturing scale on driving performance.

8.4.3 Stereopsis/Binocular Disparity

Several studies have suggested the importance of egocentric distance information for driving performance. For example, Andersen, Cisneros, Saidpour and Atchley (1999, 2000) suggested that drivers use distance information to regulate braking behavior. One source of information for distance perception is stereopsis or binocular disparity. Stereopsis provides information about distance as a result of each eye receiving a unique view of the world. The difference in relative positions in the projections to each eye is used to determine distance. Previous research has found that disparity information is effective at specifying depth for distances of up to 100 meters (see Cutting & Vishton, 1995, for a review).

The importance of binocular disparity information on driving simulation displays is twofold. First, most driving simulation research does not include binocular information. An extensive body of literature has shown the role of binocular information in perceiving depth information when other sources of depth information (e.g., texture, motion, pictorials cues) are present (see Howard & Rogers, 2002, for a detailed review). The problem is that when disparity information is not present it may affect the use of these information sources for distance perception. The lack of disparity information may have a profound effect for some driving tasks. For example, the perception of motion in depth is important for recovering time-to-contact information for an impending collision. Previous research has shown that disparity information interacts with motion information (expansion) to produce the most accurate estimates of time-to-contact (see Gray & Regan, 2004, for a review). If disparity information is not present then time-to-contact estimates are less accurate (Gray & Regan, 1998). Indeed, studies have found a decrease in accuracy (when comparing monocular with monocular and binocular information) in estimating time-to-contact of between 30 to 50% (Cavallo & Laurent, 1988; Heuer, 1993). Thus driving performance that includes time-to-contact will not match the performance in real-world driving when disparity information is present. A similar problem exists when estimating the speed of approaching objects. Harris and Watananiuk (1995) found greater accuracy in estimating speed of approach when disparity information was present. Thus, the lack of disparity information in a driving simulator may have a profound effect on drivers' estimates of the speed of approaching objects.

A more serious problem exists in driving simulators when disparity information is absent. The driver views the simulator display (either monitor or projection screen) with both eyes. Under these conditions the disparity information to the eyes is consistent with a flat surface (i.e., the screen or monitor). But the simulation is specifying that the driving scene is three-dimensional. Thus, disparity information when viewing with both eyes provides conflicting information for depth and distance that is depicted from motion and pictorial cues. Since disparity is useful for distances of up to 100 meters this problem exists for any simulation display where the distance between the driver and display (either projection screen or monitor) is less than 100 meters—a limit violated in every driving simulator.

There are no straightforward solutions to this issue. Although stereo systems exist that can be utilized in driving simulation displays, the most common apparatus involves shutter glasses that must be worn by the driver. This can be problematic if the driver wears prescription glasses. Furthermore, shutter glasses can diminish the apparent brightness of the display (at any given moment one lens is filtered and thus dark; when the visual system fuses information from both eyes it averages the dark image with the visible light image of the screen, thus reducing overall luminance of the fused image). Driving simulation research should be conducted with an understanding that conflicting disparity information might impact driving performance.

8.4.4 Eye Convergence and Accommodation

Convergence refers to the change in the optical angle between the two eyes to bring an object into focus. The feedback from the muscles in the eyes provides feedback regarding egocentric distance. Accommodation refers to the change in the optical focus of the lens to bring an object into focus. The muscle feedback to change the shape of the lens to alter focus also provides information for egocentric distance. Convergence can provide depth information for distances of up to 10 meters (Cutting & Vishton, 1995). Accommodation can provide depth information for distances of up to 2 meters (Cutting & Vishton, 1995). Both convergence and accommodation can provide information to a driver that they are viewing a flat screen rather than a 3D display. The presence of this conflicting information can compress the perceived distance when viewing a display. For example, Andersen and Braunstein (1998b) found that accommodation reduced the perception of the intervals in depth in a computer-generated scene. Since convergence can provide information of a flat screen for distances up to 10 meters it is a potential problem for viewing most driving simulation displays (the notable exception is large dome simulators). This problem is also present in head-mounted display systems, which include binocular disparity and defocus the image to limit accommodation information, but cannot alter the angular eye position from convergence. Since accommodation provides distance information for displays up to 2 meters, viewing any projection system or monitor at distances closer than 2 meters can result in flatness information from accommodation.

False flatness information from accommodation can be reduced or removed using optical techniques. For example, a Fresnel lens system can be used to reduce accommodative focus. A related technique often utilized in flight simulation displays is to use parabolic mirrors for projecting the display to the pilot. A final method for removing accommodative focus is to use a large plano-convex lens (Andersen & Braunstein, 1998b).

8.4.5 Flatness Cues

The review of the issues discussed above indicates that there are a number of visual information sources that can provide conflicting information for the perception of a 3D scene. These include disparity, convergence, accommodation and visibility of display/ image pixels. In addition to these perceptual conflicts the visibility of the frame of the display or projection screen can serve as a cue to flatness. Previous research (Eby & Braunstein, 1995) found that the presence of a surrounding rectangular frame resulted in a compression of depth of a 3D scene. The compression of depth occurred regardless of whether the 3D scene was viewed in a darkened room or in a fully illuminated room. Many driving simulation displays are presented with the frame of the monitor or projection screen visible. As a result, in driving, if a task involves the use of depth information, it will be altered because of the compression of perceived space along the depth axis. The presence of this information may limit the conclusions that can be reached from simulations studies. To avoid this problem simulation displays should be presented such that the surrounding frame of the display is occluded from the driver's view.

8.5 Visual Displays and Simulator Sickness

A serious problem with driving simulators, as well as other types of simulator, is that exposure to visual stimuli can result in simulator sickness. The incidence of simulator sickness is quite common and occurs regardless of whether the simulator has a fixed base (the simulator platform remains stationary) or variable base (the platform is allowed to move to simulate g-forces and roll, pitch and yaw). Simulator sickness has occurred with displays that are presented on a monitor, projection system or head-mounted system. Reason and Brand (1975) proposed that motion sickness was the result of perceptual conflicts between different sensory systems. This theory, often referred to as the cue conflict theory, is relevant to simulator sickness because of fidelity differences between different types of sensory information in a simulator. The type of motion and accelerations/decelerations that can be presented visually is virtually unlimited because motion can be adequately presented in visual displays. However, vestibular and kinesthetic information is severely limited in a simulator. Fixed-based simulators provide little or no information regarding motion and accelerations and decelerations. Variable-based simulators cannot precisely reproduce all the information available during motion under real-world conditions. Thus, a conflict exists between the visual information and kinesthetic and vestibular information. To better predict conditions that result in simulator sickness one needs to consider the different types of information available from different sensory systems for the perception of observer motion. The visual system can provide information for constant speed, changing speed, constant direction, and changing direction (see Watanabe, 1998, for a detailed review). In contrast, the vestibular system can only provide information for changing speed and changing direction (see Mergner, Rumberger, & Becker, 1996). This difference suggests that any simulated conditions in which a change in direction or a change in speed can occur are conditions that might result in simulator sickness.

One might assume that the conflict would be greater in a fixed-based simulator, but several studies suggest that the incidence of simulator sickness is much higher in variable-based simulators (see Hettinger, Berbaum, Kennedy, & Dunlap, 1990). This finding suggests that no input from vestibular/kinesthetic information in a fixed-based simulator is preferable to information that is a low-fidelity replication of physical movement of a driver in a variable base simulator.

8.5.1 Sensory Conflicts and Simulator Sickness

The Reason and Brand theory of motion sickness suggests that conflicts across different sensory modalities (e.g., vision and the vestibular system) produce sickness. However, the specific

conditions that result in simulator sickness are not known, nor is there a well accepted theory that predicts the conditions under which simulator sickness occurs. Despite the lack of a theory of simulator sickness, there are several well-known conditions that, if present, increase the likelihood of simulator sickness. In the next sections we will review the conditions known to increase the likelihood of simulator sickness.

8.5.2 Display Field of View

One of the most common conditions that can result in simulator sickness is a visual display with a large field of view. The increased incidence of simulator sickness with a large field of view suggests that stimulation of the peripheral visual field is critical. Previous research has shown that motion in the peripheral visual field is important for producing vection—the perception of observer motion through space (see Andersen, 1986). Although stimulation of a small area of the central visual field (15 degrees) is sufficient to produce vection (as well as motion sickness; Andersen & Braunstein, 1985) it is likely that motion in the periphery leads to a more compelling impression of vection (Andersen, 1986). The compelling impression of vection—that the observer/driver is in motion—contradicts the vestibular and kinesthetic information in a simulator. This conflict would thus increase the likelihood of simulator sickness.

Any conditions which increase the magnitude of the conflict will increase the likelihood of simulator sickness. For example, any driver motion involving a curved path will not be accompanied by appropriate stimulation of the vestibular system that the driver is physically moving along a curved path. Indeed, previous research (Mourant, Rengarajan, Cox, Lin, & Jaeger, 2007) has found greater simulator sickness when driving curved as compared to straight roadways. The increased incidence of simulator sickness on curved roadways suggests that minimizing the frequency of large changes in vehicle path (e.g., right or left hand turns) should be considered in designing scenarios for driving research in order to reduce the likelihood of simulator sickness.

8.5.3 Simulator Design Eye

A common cause of simulator sickness is an inconsistency between the simulated projection point in the computer model of the 3D driving scene (referred to in computer graphics as the simulator design eye) and the viewpoint of the driver relative to the monitor/projection screen (see Hettinger et al., 1987). This inconsistency is due to differences in the perspective of the computer simulation and the perspective of the driver's viewpoint to the display, which result in severe distortions of the driving scene when viewed. In the computer simulation, perspective is determined by the geometry of a simulated eye point relative to an image plane (i.e., the ratio of the distance between the eye point and the image plane to the distance between the eye point and the ground). The inconsistency occurs when the geometry of the driver to the display does not match the geometry in the simulation of the eye point to the image plane. When the geometry is inconsistent the driver's view is incorrect, resulting in considerable distortion of the 3D world.

8.5.4 Display Alignment

Finally, an additional cause of simulator sickness is misalignment of multiple displays (see Hettinger et al., 1987, for a discussion). Simulators that present a wide field of view will often use multiple screens or monitors to produce a large visual angle. Each screen/monitor is presenting a different view of the simulated scene. If these displays are not correctly aligned the visual system will infer that the misalignment is a change in viewpoint. To illustrate this issue, consider a horizon line presented across three projection screens. If one of the screens is misaligned relative to the other screens the horizon line will not appear as a single line across the screens. Instead, the horizon line in the misaligned display will appear above or below the horizon line in an adjacent display. When a driver is scanning across the three displays the misalignment across the displays will be perceived as a sudden change in viewpoint. The change in viewpoint will not be consistent with the simulation, and thus will result in simulator sickness. To avoid this problem screens and monitors in simulators must be properly aligned.

8.6 Summary

Driving is a skill that is heavily dependent on visual information that is processed and responded to by the driver. The goal of this chapter was to discuss the limitations of visual displays used in simulators and how these limitations might impact driving research. To understand these limitations it is important to note that the types of visual information used in driving will vary according to the type of driving task. For example, the information used in car following will be different to the information used in detecting and avoiding a collision. In this chapter I have reviewed the visual information (both sensory and perceptual) used to perform different driving tasks including detection/recognition of targets, steering control, collision detection and performance, braking, speed regulation, and car following.

Driving simulators present visual information that is an approximation of the information present under real-world driving conditions. For example, in a simulator visual motion is presented using a series of discrete presentations of static images whereas in real-world driving visual motion is a continuous projection of spatial displacement over time. The impact of simulator limitations in presenting visual information can have a serious effect on driving performance. For example, if the presentation rate or display refresh falls below 30 Hz the perceived motion of the driver and of objects in the scene will be altered. In this chapter I have reviewed the potential impact of spatial resolution, temporal resolution, luminance, contrast, texture, stereopsis, convergence and accommodation on driving performance in a simulator. An additional concern often not considered in simulator research is the presence of flatness cues that can alter the perceived depth of the driving scene. Finally, visual factors that can increase the incidence of simulator sickness and how these factors can be minimized or eliminated are discussed.

Driving simulators are an important tool for understanding driving skill and performance. Simulators allow researchers to examine performance issues that otherwise would not be possible using real-world driving (e.g., collision detection and avoidance). Understanding the limitations discussed in this chapter and the methods available to minimize or remove these limitations should lead to better use of simulators in driving research.

Key Points

- Driving performance is highly dependent on visual information.
- Different driving tasks (e.g., steering control, car following, collision detection and avoidance) involve different sources of visual information. Driving simulation displays should be designed to optimize the presentation of these information sources.
- Sensory issues that can impact performance in driving simulators include spatial resolution, temporal resolution, luminance, and contrast.
- Perceptual issues that can impact performance in driving simulators include motion, optic flow, texture, and stereopsis.
- A serious issue often overlooked in driving simulation is the presence of flatness cues from conflicting perceptual information (e.g., stereopsis, accommodation, convergence) or the visibility of the frame of the display.
- Factors that result in simulator sickness are quite varied and include sensory conflicts, a wide field of view, improper position of the driver given the simulation, and display misalignment.

Keywords: Driving Tasks, Simulator Display Design, Simulator Sickness, Visual Perception

Acknowledgments

This research was supported by NIH AG13419-06 and NEI EY18334-01.

Key Readings

Andersen, G. J., & Sauer, C. W. (2004). Optical information for collision detection during deceleration. In H. Hecht & G. J. P. Savelsbergh (Eds.), *Time-to-Contact* (pp. 93–108). Amsterdam, the Netherlands: Elsevier Science.

Andersen, G. J., & Sauer, C. W. (2007). Optical information for car following: The DVA model. *Human Factors, 49,* 878–896.

Cutting, J. E., & Vishton, P. M. (1995). Perceiving layout and knowing distances: The integration, relative potency, and contextual use of different information about depth. In W. Epstein & S. Rogers (Eds.), *Perception of space and motion. Handbook of perception and cognition* (2nd ed., pp. 69–117). San Diego, CA: Academic Press.

Eby, D. W., & Braunstein, M. L. (1995). The perceptual flattening of three-dimensional scenes enclosed by a frame. *Perception, 24*(9), 981–993.

Manser, M. P., & Hancock, P. A. (2007). The influence of perceptual speed regulation on speed perception, choice, and control: Tunnel wall characteristics and influences. *Accident Analysis & Prevention, 39*(1), 69–78.

References

Andersen, G. J. (1986). Perception of self-motion: Psychophysical and computational approaches. *Psychological Bulletin, 99*(1), 52–65.

Andersen, G. J., & Braunstein, M. L. (1985). Induced self-motion in central vision. *Journal of Experimental Psychology: Human Perception and Performance, 11*(2), 122–132.

Andersen, G. J., & Braunstein, M. L. (1998a). The perception of depth and slant from texture in 3D scenes, *Perception, 27*, 1087–1106.

Andersen, G. J., & Braunstein, M. L. (1998b). Effect of collimation on perceived layout in 3D scenes. *Perception, 27*, 1305–1315.

Andersen, G. J., Cisneros, J., Saidpour, A., & Atchley, P. (1999). Speed, size and edge rate information for the detection of collision events. *Journal of Experimental Psychology: Human Perception and Performance, 25*, 256–279.

Andersen, G. J., Cisneros, J., Saidpour, A., & Atchley, P. (2000). Age-related differences in collision detection during deceleration. *Psychology & Aging, 15*(2), 241–252.

Andersen, G. J., & Enriquez, A. (2006a). Aging and the detection of observer and moving object collisions. *Psychology and Aging, 21*(1), 74–85.

Andersen, G. J., & Enriquez, A. (2006b). Use of landmarks and allocentric reference frames for the control of locomotion. *Visual Cognition, 13*(1), 119–128.

Andersen, G. J., & Kim, R. D. (2001). Perceptual Information and attentional constraints in visual search of collision events. *Journal of Experimental Psychology: Human Perception and Performance, 27*, 1039–1056.

Andersen, G. J., & Saidpour, A. (1999). Optical information for the control of steering: A control theory analysis. In D. H. Harris (Ed.), *Cognitive Ergonomics and Engineering Psychology* (Vol. 3, pp. 359–367). Aldershot, England: Ashgate.

Andersen, G. J., & Sauer, C. W. (2004). Optical information for collision detection during deceleration. In H. Hecht & G. J. P. Savelsbergh (Eds.), *Time-to-contact* (pp. 93–108). Amsterdam, the Netherlands: Elsevier Science.

Andersen, G. J., & Sauer, C. W. (2007). Optical information for car following: The DVA model. *Human Factors, 49*, 878-896.

Andersen, G. J., Sauer, C. W., & Saidpour, A. (2004). Visual information for car following by drivers: The role of scene information. *Transportation Research Record, 1899*, 104–108.

Baker, C. L., & Braddick, O. J. (1985). Temporal properties of the short-range process in apparent motion. *Perception, 14*(2), 181–192.

Bonato, F., Bubka, A., & Palmisano, S. (2006). Changing and steady vection effects on simulator sickness. *Journal of Vision, 6*, 383.

Braddick, O. (1974). A short-range process in apparent motion. *Vision Research, 47*, 519–527.

Broughton, K. L. M., Switzer, F., & Scott, D. (2007). Car following decisions under three visibility conditions and two speeds tested with a driving simulator. *Accident Analysis & Prevention, 39*(1), 106–116.

Buchner, A., Brandt, M., Bell, R., & Weise, J. (2006). Car backlight position and fog density bias observer-car distance estimates and time-to-collision judgments. *Human Factors, 48*(2), 300–317.

Campbell, F. W., Robson, J. G. (1968). Application of Fourier analysis to the visibility of gratings. *Journal of Physiology, 197*, 551–566.

Cavallo, V., & Laurent, M. (1988). Visual information and skill level in time-to-collision estimation. *Perception, 17*, 623–632.

Cutting, J. E., & Vishton, P. M. (1995). Perceiving layout and knowing distances: The integration, relative potency, and contextual use of different information about depth. In W. Epstein & S. Rogers (Eds.), *Perception of space and motion. Handbook of perception and cognition* (2nd ed., pp. 69–117). San Diego, CA: Academic Press.

DeLucia, P. R., Bleckley, M. K., Meyer, L. E., & Bush, J. M. (2003). Judgments about collisions in younger and older drivers. *Transportation Research Part F: Traffic Psychology & Behaviour, 6*(1), 63–80.

Denton, G. G. (1980). The influence of visual pattern on perceived speed. *Perception, 9*, 393–402.

Dyre, B. P., & Andersen, G. J. (1997). Perception of heading: Effects of conflicting velocity magnitude and trajectory information, *Journal of Experimental Psychology: Human Perception and Performance, 23*, 546–565.

Eby, D. W., & Braunstein, M. L. (1995). The perceptual flattening of three-dimensional scenes enclosed by a frame. *Perception, 24*(9), 981–993.

Evans, L. (2004). *Traffic safety.* Bloomfield Hills, MI: Science Serving Society.

Gibson, J. J. (1966). *The senses considered as perceptual systems.* Boston, MA: Houghton Mifflin.

Gibson, J. J. (1979). *The ecological approach to visual perception.* Boston, MA: Houghton Mifflin.

Gogel, W. C. (1954). Perception of the relative distance position of objects as a function of other objects in the field. *Journal of Experimental Psychology, 47*, 335–342.

Gogel, W. C. (1972). Depth adjacency and cue effectiveness. *Journal of Experimental Psychology, 92*, 176–181.

Gogel, W. C. (1982). Analysis of the perception of motion concomitant with a lateral motion of the head. *Perception & Psychophysics, 32*, 241–250.

Gogel, W. C. (1998). An analysis of perceptions from changes in optical size. *Perception & Psychophysics, 60*, 805–820.

Gray, R., & Regan, D. (1998). Accuracy of estimating time-to-collision using binocular and monocular information. *Vision Research, 38*, 499–512.

Gray, R., & Regan, D. (2004). The use of binocular time-to-contact information. In H. Hecht & G. J. P. Savelsbergh (Eds.), *Time-to-contact* (pp. 303–325). Amsterdam, the Netherlands: Elsevier B. V.

Harris, J. M., & Watamaniuk, N. J. (1995). Speed discrimination of motion-in-depth using binocular cues. *Vision Research, 35*, 885–896.

Hahn, S., Andersen, G. J., & Saidpour, A. (2003). Static scene analysis for the perception of heading. *Psychological Science, 16*, 543–548.

Hettinger, L. J., Berbaum, K. S., Kennedy, R. S., & Dunlap, W. P. (1990). Vection and simulator sickness. *Military Psychology, 2*(3), 171–181.

Hettinger, L. J., Nolan, M. D., Kennedy, R. S., Berbaum, K. S., Schnitzius, K. P., & Edinger, K. M. (1987). Visual display factors contributing to simulator sickness. *Proceedings of the 31st annual meeting of the human factors society* (pp. 497–501).

Heuer, H. (1993). Estimates of time to contact based on changing size and changing target vergence. *Perception, 22*, 549–563.

Howard, I. P., & Rogers, B. J. (2002). *Seeing in depth.* Toronto, Canada: University of Toronto Press.

Kline, D. W., & Fuchs, P. (1993). The visibility of symbolic highway signs can be increased among drivers of all ages. *Human Factors, 35*(1), 25–34.

Koenderink, J. J. (1986). Optic flow. *Vision Research, 26*(1), 161–179.

Land, M. F., & Lee, D. N. (1994). Where do we look when we steer? *Nature, 369*(6483), 742–744.

Larish, J. F., & Flach, J. M. (1990). Sources of information useful for perception of speed of rectilinear motion. *Journal of Experimental Psychology: Human Perception and Performance, 16*, 295–302.

Lee, D. N. (1976). A theory of visual control of braking based on information about time to collision. *Perception, 5*(4), 437–459.

Lee, D. N. (1980). The optic flow field: The foundation of vision. *Philosophical Transactions of the Royal Society of London Series B, 290*, 169–179.

Manser, M. P., & Hancock, P. A. (2007). The influence of perceptual speed regulation on speed perception, choice, and control: Tunnel wall characteristics and influences. *Accident Analysis & Prevention, 39*(1), 69–78.

McGregor, L. N., & Chaparro, A. (2005). Visual difficulties reported by low-vision and non-impaired older adult drivers. *Human Factors, 47*(3), 469–478.

Mergner, T., Rumberger, A., & Becker, W. (1996). Is perceived angular displacement the time integral of perceived angular velocity? *Brain Research Bulletin, 40*(5–6), 467–471.

Mori, M., Abdel Halim, H. A.-H. (1981). Road sign recognition and non-recognition. *Accident Analysis & Prevention, 13*, 101–115.

Mourant, R. R., Rengarajan, P., Cox, D., Lin, Y., & Jaeger, B. K. (2007). The effect of driving environments on simulator sickness. *Proceedings of the 51st annual meeting of the human factors and ergonomics society* (pp. 1232–1236).

Nguyen-Tri, D., Faubert, J. (2007). Luminance texture increases perceived speed. *Vision Research, 47*(5), 723–734.

Ni, R., & Andersen, G. J. (2008). Detection of collision events on curved trajectories: Optical information from invariant rate of bearing change. *Perception & Psychophysics, 70*, 1314–1324.

Palmisano, S., Burke, D., & Allison, R. S. (2003). Coherent perspective jitter induces visual illusions of self-motion. *Perception, 32*, 97–110.

Palmisano, S., Bonato, F., Bubka, A., & Folder, J. (2007). Vertical display oscillation effects on forward vection and simulator sickness. *Aviation, Space, and Environmental Medicine, 78*, 951–956.

Reason, J. T., & Brand, J. J. (1975). *Motion sickness.* Oxford, England: Academic Press.

Schiff, W., Oldak, R., & Shah, V. (1992). Aging persons' estimates of vehicular motion. *Psychology & Aging, 7*(4), 518–525.

Shinar, D., & Drory, A. (1983). Sign registration in daytime and night-time driving. *Human Factors, 25*, 117–122.

Tresilian J. R. (1991). Empirical and theoretical issues in the perception of time to contact. *Journal of Experimental Psychology: Human Perception and Performance, 17*, 865–876.

Wann, J. P. (1996). Anticipating arrival: Is the tau margin a specious theory? *Journal of Experimental Psychology: Human Perception and Performance, 22*(4), 1031–1048.

Warren, W. H., Morris, M. W., & Kalish, M. (1988). Perception of translational heading from optical flow. *Journal of Experimental Psychology: Human Perception and Performance, 14*, 646–660.

Watanabe, T. (Ed.) (1998). *High-level motion processing: Computational, neurobiological and psychophysical perspectives.* Cambridge, MA: MIT Press.

Wood, J. M., Tyrrell, R. A., & Carberry, T. P. (2005). Limitations in drivers' ability to recognize pedestrians at night. *Human Factors, 47*(3), 644–653.

9

Psychological Fidelity: Perception of Risk

Thomas A. Ranney
*Transportation Research
Center, Inc.*

Abstract

The Problem. High-fidelity driving simulators provide a realistic and compelling experience for research participants. However, the credibility of research results from simulator studies continues to be challenged. The fidelity of the driving experience appears insufficient to overcome criticisms concerning the lack of psychological fidelity, defined as the extent to which the risks and rewards of participation in the experiment correspond to real-world risks and rewards. *Key Results of Driving Simulator Studies.* Experimental studies eliminate the injury risk associated with driving. They also typically eliminate the trip purpose, which influences all components of real-world driving. Unfortunately, researchers typically give little consideration to this problem, often instructing participants to drive as they normally would. In the absence of a well-defined driving context, such instructions can be confusing to some participants. *Tools Available to Researchers.* Well-designed driving simulator experiments eliminate confusion about driving motives by creating constrained situations to elicit specific behaviors. Researchers must identify the driving components that have been eliminated by the simulation and attempt to replace them through the use of instructions and performance incentives. Instructions define the performance space and driving task components; incentives define the relative priorities associated with the task components. The effects of incentives on performance are determined by some combination of (1) the nature of the incentive; (2) task characteristics; (3) aspects of performance selected for measurement; and (4) individual differences. Incentives are likely to improve certain aspects of performance, while degrading others at the same time, implying that care must be taken in matching incentives to performance measures. *Scenarios and Dependent Variables.* Reward/penalty schemes are used to incorporate performance incentives into driving simulator studies. Practical issues associated with their use include: simulating the effects of significant negative outcomes (i.e., crashes); multiple crashes; effects of incentives over time; assessment of reward/penalty systems; and non-independence of performance measures. Detailed examples of the use of reward/penalty systems are presented. *Platform Specificity and Equipment Limitations.* Problems of psychological fidelity apply to all platforms. Improving psychological fidelity eliminates unwanted variability due to individual differences in driving, which result from uncertainty about the experimenter's priorities.

9.1 Psychological Fidelity: Perception of Risk

The availability of sophisticated driving simulators provides an increasingly realistic driving experience for research participants. The current generation of driving simulators allows researchers to create myriad situations with complex roadway geometry, realistic surrounding traffic, pedestrians and traffic control devices. Among the most advanced simulators, projections of the virtual world completely surround the driver, images from the mirrors are realistic, and drivers feel the effects of their steering or braking inputs. As simulator capabilities have become more affordable, the number of experimental studies using driving simulators has increased. For example, numerous studies have been conducted addressing the question of whether cell phones are sufficiently disruptive to driving to be considered a safety hazard. Despite the fact that these studies provide relatively consistent results (Horrey & Wickens, 2006; Caird, Willness, Steel, & Scialfa, 2008), their findings are often challenged or discounted, relative to those of epidemiological or observational studies. For example, McCartt, Hellinga and Braitman (2006) reviewed 54 studies that utilized driving simulators or instrumented vehicles to assess the impact of cell phone use on driving behavior or performance. Their overall assessment was that the observed "changes in performance of experimental tasks have uncertain implications for real-world driving" (p. 92). While these authors raised a number of methodological issues, they identified one concern that is most relevant to the present discussion—namely, the authors question the lack of realism depicted by experimental studies.

Goodman et al. (1997) discussed the limitations of driving simulators, including lack of realism in the visual display, absence of motion among fixed-base simulators, and most importantly (in the context of cell phone research) the simulator's effects on drivers' priorities. They suggested that drivers may be more inclined to devote an unrealistic amount of attention to the secondary phone task because "there are no serious consequences with driving errors in the simulator" (p. 86). Interestingly, the authors continued by suggesting: "the use of high fidelity simulators such as the National Advanced Driving Simulator (NADS) will greatly enhance our ability to address such concerns." These authors apparently felt that the increased fidelity of the driving experience would lead to a more realistic allocation of attention between driving and secondary (cellular phone) tasks. Implicit in this conclusion is the idea that participants will be drawn into a more compelling experience, which in turn will encourage them, presumably without much thought, to revert to their natural driving behavior.

More than 10 years have passed since that optimistic projection was made and the NADS has been used for a variety of experimental studies, including several studies of drivers' responses to cell phones (Ranney et al., 2005). Unfortunately, there have been no comparative studies addressing the effects of increased fidelity on participants' behavior. Thus, there is no direct way to test whether the assertion made by Goodman et al. (1997) is true; however,

according to Caird et al. (2008), the findings of the NADS cell phone research are relatively consistent with those of other studies that have used lower-fidelity simulators. Effect sizes appeared to be in the same range as those derived from studies conducted using simulators with less fidelity. Thus, advanced simulators have not been associated with changes in the patterns of results that would indicate that the compelling driving experience is sufficient to overcome the problems identified by Goodman et al. (1997). Increased fidelity of the driving experience does not address the problem of poor psychological fidelity.

9.2 What Is Psychological Fidelity?

Researchers are typically proud of the fidelity of their driving simulators. However, because they are often intimately involved in their development, they may overlook the peculiarities of their experimental setups and assume that research participants can readily make the leap of imagination necessary to behave in the simulator as they would while driving in the real world. It may be difficult for researchers to consider how the simulator is viewed by members of the community, particularly those removed from the university or research environment. For example, when confronted with an unemployed truck driver, whose participation in a simulator experiment is intended simply to make a few dollars between jobs, or a busy mom whose main concern is that she completes the experiment to be home in time to greet her kids when they get off the school bus, the researcher may conclude success if the participants appear to engage in the simulated driving task without complaint. While the credibility of the driving experience is an important hurdle, it unfortunately ignores the more fundamental problem that is at the heart of psychological fidelity, namely that when we drive in the real world, we do so for a purpose (Duncan, 1990). Even the most sophisticated simulator cannot overcome the fact that driving simulator studies alter the driving task fundamentally, eliminating this most basic component of driving, namely, the trip purpose. Thus, while experimenters may specify a trip purpose, the fundamental artificiality of the experimental setting, including the destination, remains.

Drivers' motives are significant determinants of on-road driving behavior (Näätanen & Summala, 1976; Duncan, 1990; Ranney, 1994). Although empirical studies typically do not address this issue directly, drivers may change their on-road behavior depending on the purpose of their trip. Getting to work on time may evoke different driving behavior than embarking on a recreational trip with no time constraints. Drivers' motives may also change within a given trip, leading to changes in driving behavior. For example, some drivers may increase their speed and alter their decision-making when they realize that they will be late for an appointment or planned event. Emergent driving situations may also influence drivers' momentary motives and their driving behavior. For example, drivers may alter their attention entirely when they find themselves trapped behind a slow-moving vehicle in dense traffic. Together, the global trip purpose and the driver's momentary motives play a significant part in determining much of their on-road behavior, including speed, following distance, and decision-making in accepting gaps in passing and entering traffic. In the context of Michon's (1985) hierarchical model, motives at the strategic level (e.g., trip purpose) combine with those at the tactical (e.g., gap acceptance) and vehicle control (e.g., speed selection) levels to influence driving behavior. As a particular example, we can look to the effect that teen passengers have on the behavior of teen drivers. It has been observed that teen drivers in the presence of male passengers drive faster than the general traffic and have shorter headways (Simons-Morton, Lerner, & Singer, 2005). Here, both the tactical and vehicle control levels are being influenced by the driver's motives.

Well-designed driving simulator experiments eliminate confusion about driving motives by creating constrained situations to elicit specific behaviors. An example is the slow-moving lead vehicle scenario described above; however, without a defined trip purpose, there is no reason to expect that drivers will consistently experience the momentary frustration assumed to motivate them to give high priority to extricating themselves from these situations in real-world driving. Removing the trip purpose leaves a void in the participant's motivation that must either be defined by the experimenter or left to the participants to fill. Depending on the experimental objectives, ignoring this potential problem can serve to introduce a significant amount of unwanted variability due to individual differences in priorities (Edwards, 1961). Unfortunately, according to Zeitlin (1996), who examined 106 published research simulation studies, researchers have given little consideration to this potential problem.

Current trends in research funding provide another way of considering the importance of psychological fidelity to driving behavioral research. Recently, the "100-Car Naturalistic Driving Study" was conducted by Virginia Tech Transportation Institute (VTTI) (Klauer, Dingus, Neale, Sudweeks, & Ramsey, 2006). This study represents an emerging research methodology in which drivers are observed in their own vehicles during everyday driving. Based on the success of this and other naturalistic studies, the second Strategic Highway Research Program (SHRP 2) has committed a significant portion of their Safety Program resources to the implementation of a large-scale naturalistic study. The stated rationale for this undertaking derives from the desire to better understand pre-crash behavior (Transportation Research Board, 2008). But instead of focusing on crashes per se, the study will follow drivers for several years on the assumption that some of them will become involved in crashes during this period. The inefficiency of this approach for studying pre-crash behavior, relative for example to one that utilizes a high-fidelity driving simulator (such as the NADS), becomes evident when one considers the fact that on average an injury crash occurs less than once per million vehicle miles of vehicle travel (National Highway Traffic Safety Administration, 2008). In a simulator, over the course of a year researchers can expose thousands of drivers to carefully calibrated crash-imminent situations. Situational dynamics can be systematically varied to evaluate their effects on the likelihood of crashes. In contrast, a naturalistic observational study will discard thousands of hours of driving data for each crash and the circumstances of each crash will likely be so different as to challenge conclusions about the generality of pre-crash behaviors. If one accepts the validity of the NADS's realism, it makes little sense to devote relatively huge amounts of research resources to an undertaking as relatively inefficient as a naturalistic study. Why then is this being done? Sadly, we must conclude that validity of the NADS realism is not accepted among policymakers and that naturalistic observation is preferred, even given the significant associated inefficiencies, because it involves real rather than artificial driving. The fact that the funding pendulum is swinging strongly in this direction can be viewed as an endorsement of the importance of (naturalistic) realism in the study of driving behavior. Clearly, to compete in this environment, researchers using simulated driving environments must address this credibility issue by means other than improving the fidelity of their simulators.

9.3 How to Improve Psychological Fidelity

The fact that researchers trade realism for experimental control is a cornerstone of the scientific method. Making this trade is necessary to test specific hypotheses, but the loss of realism implies a loss of psychological fidelity. There are tools at the researcher's disposal that can help improve the psychological fidelity of the simulator experiment. Most generally, this requires identifying the components of driving that have been eliminated by the simulation and attempting to replace them through the combination of instructions and performance incentives. The focus of the remainder of this chapter is on the use of these tools.

9.3.1 Simulation of Performance Versus Behavior

Researchers are sometimes imprecise in describing their experimental objectives. For example, they may use the terms driving *behavior* and *performance* indiscriminately or interchangeably. However, these terms have different meanings in the context of driving behavioral research (Evans, 1991; Näätänen & Summala, 1976; Ranney, 1994) and understanding this distinction will help determine how best to improve the psychological fidelity of the simulation. Performance refers to drivers' responses at the limits of their ability, or what the driver can do; behavior refers to the typical unconstrained on-road driving, most of which involves a level of effort significantly below the driver's limiting ability. Driving simulators have been used for both purposes, yet researchers typically do not refer to this distinction, nor discuss its implication for the generalizability of their results. The results of driving performance studies generalize most directly to critical on-road situations, which typically occur when task demands increase unexpectedly and drivers are required to respond at or near the limits of their abilities to avoid a crash. Examples include vehicles unexpectedly encroaching into the travel lane or unexpected patches of slippery roads. Early theories of crash causation assumed that crashes were primarily caused by such failures and this led to an emphasis on identifying the limits of drivers' skills (Ranney, 1994). However, the convergence of theoretical models that emphasize motivational factors (e.g., Näätanen & Summala, 1976) with research results demonstrating an inverse relation between driving skill and crash involvement (Williams & O'Neill, 1974) led to increased interest in the study of errors in non-critical situations (e.g., inadequate safety margins) as a means to better understand and develop theories of driving behavior (e.g., Brown, 1990). To improve psychological fidelity, researchers must first understand whether they want to study crash-avoidance performance in limiting situations or more typical behavior in routine non-limiting situations. This distinction has significant implications for the design of experimental protocols and for developing participant instructions.

9.3.2 Instructions

Participation in experimental research studies differs from everyday experience; participants typically enter the experimental situation with very few expectations. For this reason, participants are usually very attentive and actively attempt to construct expectations by observing experimenters' behaviors. Researchers who study driving performance typically want participants to perform as well as possible. To accomplish this, they may rely on the participant's inherent desire to do well, or the assumption that participants bring an attitude of deference into what they perceive to be a "testing" environment, as sufficient motivation to ensure peak performance. Alternately, they may instruct the participants to perform "as well as possible without making errors." While this instruction is widely used in laboratory tests, particularly those involving a speed-accuracy tradeoff, it is fundamentally a contradictory combination of instructions, as pointed out by Pachella (1974). Specifically, it is not possible to know precisely how well one can perform "without making errors," unless some are actually made. Edwards (1961), although reviewing research from a different era, noted that ambiguous or internally contradictory instructions are not uncommon in psychological experimentation.

In contrast, researchers who study driving behavior may instruct participants to "drive as you normally would." Excluding naturalistic situations, the use of this instruction is generally an indication that little consideration has been given to the psychological fidelity of the experimental protocol. As discussed above, drivers may alter their "normal" driving behavior depending upon the trip purpose and their momentary motives. If the experimental protocol fails to address these critical determinants of driving behavior, it is not surprising that some participants may be confused by such an instruction. Researchers would do well to consider the generic instruction to "drive as you normally would" as being the first part of an instruction that requires further specification. Examples of more complete instructions would include:

"Drive as you normally would when you are late for a job interview" or

"Drive as you normally would when you just finished your last exam."

Finally, an implicit part of the instruction to "drive as you normally would" contains a corollary problem, namely the questionable assumption that drivers normally drive in a consistent manner across days or across situations in a given drive. The anticipated emergence of accessible naturalistic data bases may provide an opportunity to examine this assumption, particularly given designs that provide data over an extended time period for each participant.

Many studies combine these two approaches, asking participants to drive normally and subsequently surprising them with unexpected situations requiring avoidance maneuvers (e.g., lead-vehicle braking). Several assumptions are implicit in this scenario. First, researchers assume that participants' inherent desire to avoid crashes will transfer intact from the roadway to the experimental situation, resulting in realistic crash-avoidance behavior. To the extent that scenarios elicit drivers' immediate and automatic crash-avoidance responses, this may be true. However, it is also possible that some participants may want to take advantage of the fact that driving in a simulator will ensure that there are no serious consequences for their driving errors. These adventuresome drivers, although representing a small minority of typical research participants, may adopt unusually risky behaviors to increase the likelihood of experiencing a crash. Second, if the researchers are primarily interested in the drivers' responses to the unexpected events, they may fail to consider whether the strategies adopted by drivers during the (normal) driving influenced their subsequent responses to the unexpected events (e.g., Brown, Lee, & McGehee, 2001). For example, if the researcher fails to define the overall context of the simulated trip, it is reasonable to expect the drivers to adopt different speeds and headways for a variety of unknown reasons. These differences can be expected to influence drivers' responses to the unexpected event, adding variance to the experimental design that may reduce statistical power for addressing the questions of primary concern. Although researchers may argue that leaving the context undefined preserves realism by allowing for individual differences in driving styles, this argument is based on the untested and probably erroneous assumption that drivers' real-world driving practices transfer intact to the simulated driving environment. At a minimum, drivers in experimental studies need enough information to guide their speed selection. However, providing the speed limit may not be enough information, drivers must also understand how it will be enforced and the consequences of exceeding the speed limit.

Many contemporary driving simulator experiments utilize a dual-task paradigm, in which participants perform secondary tasks while driving. A significant issue for this type of study with respect to psychological fidelity is the drivers' allocation of attention between primary and secondary tasks, which requires some judgment as to the relative importance of the two tasks. Edwards (1961) pointed out that when the implicit instruction for an experiment is to "do the best you can," this essentially is an instruction to maximize or minimize some mathematical function defined by the components of the experiment. However, in a dual-task situation, such as driving while performing a secondary task, this instruction becomes troublesome because it implies that the participant is intended to maximize or minimize two functions simultaneously. It is very unlikely, Edwards continues, that two functions will have maxima or minima that can be jointly achieved, which makes it impossible for the subject to do their best possible on both components simultaneously. In this situation, it is incumbent on the researcher to provide specific information concerning the relative importance of the concurrent tasks in this paradigm. Edwards (1961) advocated providing a complete payoff matrix to participants to avoid misinterpretation of instructions. Thus, instead of simply instructing the driver concerning the relative priorities of the primary and secondary tasks, the researcher would present a matrix that defines all possible outcomes based on the combinations of task (primary or secondary) and level of task performance (e.g., good, acceptable, poor) and defines the specific rewards associated with each task combination.

9.3.3 Incentives and Driving Simulation

Incentives are fundamental to learning theory and there is much, albeit mostly older, research on how they affect performance or behavior. Zeitlin (1996) distinguished among three categories of incentives that can influence performance or behavior in simulators: (1) *Consequential incentives* are real-world consequences associated with simulator performance, such as a certification or job licensing. These incentives relate primarily to the use of simulators for testing and it is generally assumed that because the test outcome has real-life implications the participants will strive to perform to the best of their ability. (2) *Intrinsic incentives* are features of the simulation that inherently motivate participants to perform at a desired level. These incentives include the entertainment value or challenge associated with the simulated events. According to Eysenck (1983), intrinsic motivation to perform a task will be present to the extent that performing the task increases an individual's feeling of competence and self-determination. Increasing intrinsic incentives is typically not compatible with research or training objectives because the specific elements that must typically be incorporated into driving simulations to increase the intrinsic motivation are likely to be features of games that are not typically found in real-world driving (e.g., real-time scoring system, exciting scenario events). (3) *Extrinsic incentives* are rewards and penalties associated with different aspects of performance or behavior, including monetary rewards and penalties, praise, or food. As will be discussed below, this category offers an opportunity to motivate behavior that approaches on-road driving in research studies using driving simulators. However, the use of extrinsic incentives will serve to restrict the participant's performance or behavior to the activities associated with the delivery of contingent rewards (Eysenck, 1983).

9.3.3.1 Incentives and Performance

Generally, incentives are considered to have the same effects as arousing stimuli (Eysenck, 1983) and the Yerkes-Dodson model (1908) is used to describe the effects of incentives on performance. According to this model, there is an inverted U-shaped relationship between arousal or motivation and performance, with intermediate levels of motivation being optimal for performance. Within this general framework, Eysenck (1982, 1983) has suggested that the measured effects of incentives on performance are determined by some combination of the following factors: (1) nature of the incentive; (2) task characteristics; (3) aspects of performance selected for measurement; and (4) individual differences. The nature of the incentive includes the amount of the incentive and the probability of attaining the incentive. The interaction of these two factors may be most important in determining the level of motivation. For example, a very large incentive combined with a near-zero probability of success is likely to provide a relatively low level of motivation. Task characteristics include task complexity and duration. Most generally, according to Eysenck (1982), performance on complex tasks, including those involving problem-solving or requiring creativity, is more likely to be adversely affected by incentives than performance on simple, particularly speed-based tasks, such as response time. This generalization reflects the conclusion that incentives encourage relatively narrow, focused thinking, which may not be optimal for tasks that require cognitive flexibility. Using multiple performance measures is preferable when feasible for assessing effects of incentives. According to Eysenck (1982), when the effects of incentives on two or more aspects of performance are considered, it is common to find improvement on one measure and impairment on another. Finally, individuals differ in the extent to which they are motivated by incentives or frustrated by non-reward. Relevant individual characteristics include both personality traits (e.g., introversion/extraversion) and transient states or moods. Participants may also differ in their response to a particular incentive. For example, the perceived value of monetary rewards may vary according to socio-economic status.

Among the paradigms used to study incentive effects, the dual-task studies most closely resemble the task demands of driving. Most generally, incentives lead to reallocation of attentional resources in dual-task studies, with greater resources invested in the task designated as the main task, and a corresponding decrease in resources to the task designated as subsidiary or secondary (Eysenck, 1983). Incentives are thus likely to improve certain aspects of performance, while degrading others at the same time (Eysenck, 1982). Among studies that consider the interaction between incentive and stressor effects, incentives are found to enhance performance more for sleep-deprived than for non-sleep-deprived individuals (Broadbent, 1971). Finally, for tasks in which participants can trade speed for accuracy, incentives are likely to increase speed and reduce accuracy.

Several experimental studies have shown that transient changes in drivers' motivation can produce changes in driving behavior. In a series of experiments in which participants provided speed choice decisions while viewing static pictures of driving situations, Delhomme and Meyer (1997) examined the effect of transient motivational state on speed selection. They manipulated motivation by deceiving participants about their performance on tasks unrelated to driving. They hypothesized that earlier task failure would lead to perceptions of transient control loss, which in turn would increase participants' motivation to regain control. In turn, the increased control motivation would limit risk-taking, which would be reflected by limited speed choice. Their data supported this hypothesis: Participants in the "failure" group (high-control motivation) chose slower speeds than those in the "success" group (low-control motivation). Delhomme and Meyer (1997) also explored the interaction between transient motivational factors and driving experience. They hypothesized that motivational effects related to loss of control should affect inexperienced drivers in situations with a heavier cognitive load. They found the novices' performances were more dependent on their motivational state than the more experienced drivers, who based their decisions on a more detailed analysis of the available visual information. However, their data were insufficient to determine whether this effect reflected differences between experienced and inexperienced drivers either in the degree to which driving task components are automatized or in the generally greater difficulties novice drivers have regulating their behavior.

Desmond and Matthews (1997) examined the interaction between drivers' motivational state and fatigue-related performance decrements. They manipulated motivational state by providing instructions that the participants' driving was being evaluated during certain portions of a simulator drive and measured the effects of these instructions on driving performance. They found changes in driving performance (heading error) associated with the motivating instruction, but only in the latter part of the driving task and only on straight road (i.e., low-demand) segments. Interestingly, the motivating instruction influenced performance more for subjects in their task-induced fatigue condition.

9.3.3.2 Reward/Penalty Systems in Driving Simulation

Stein, Allen and Schwartz (1978) provided a rationale for the use of monetary reward/penalty systems in driving simulation. Based on their conceptualization, reward/penalty parameters and computational algorithms were incorporated into the STISIM driving simulator software (Rosenthal, Parseghian, Allen, & Stein, 1994). Central to their model is a reward for completion time, which either adds or subtracts a prorated amount of money depending upon the amount of time required to complete a driving scenario, relative to a pre-established criterion. Faster completion times yield rewards, while slower times result in penalties. At the same time, speeding is discouraged by monetary penalties associated with speeding tickets, which are issued based on a probabilistic scheme that can be varied by the experimenter. Crashes are also associated with monetary penalties, which are larger than those associated with tickets. Crashes also influence completion time, because there is a delay during which the simulator is reset and the vehicle must start from a stop.

The conceptual model underlying this reward/penalty system represents speed selection as a tradeoff between two motives: (1) the desire for timely arrival, and (2) the desire to avoid speeding tickets and crashes. Specifically, providing a monetary reward for timely arrival can create a sense of urgency associated with the simulated drive, which can motivate the driver to increase speed. In contrast, monetary penalties for speeding, crashes, or other violations can simulate the costs that deter speeding. Together, these monetary rewards and penalties have the potential to effectively simulate the tension that motivates speed selection in real-world driving. The credibility and effectiveness of such reward/penalty systems would be increased by empirical studies that examined the effects of different parameter values on driving behavior.

Because extrinsically-motivated participants can be expected to restrict their performance to the activities associated with rewards and penalties (Eysenck, 1983), reward/penalty systems will determine the driver's allocation of attentional resources among driving task components. Reward/penalty systems must therefore be sufficiently comprehensive to represent all aspects of the driving task, which the experimenter expects participants to monitor. Michon's (1985) three-level hierarchy, together with the assumption that drivers actively decide how to allocate resources among strategic, tactical, and operational levels of control (Ranney, 1994), can serve as a useful starting point. To maximize ecological validity, this implies that researchers should assign extrinsic incentives to each of the hierarchical levels in proportions that reflect experimental objectives. The speed-selection model described above is consistent with this recommendation. Adjusting the reward for timely arrival addresses the strategic motivation (i.e., trip purpose), while the issuance of speeding tickets motivates tactical (e.g., adequate gap acceptance as in passing) and operational (speed selection) decision-making. Embedding visible enforcement targets into the scenario could be a way to elicit the automatic responses made by speeding drivers when they see a patrol car alongside the road.

9.3.3.3 Issues Relating to the Use of Reward/Penalty Systems

Incentives and automatic behaviors. Aspects of driving behavior or performance that are highly automatic may not be amenable to the effects of performance incentives. The amount of time available between stimulus and response is a key determinant; the more time that drivers are given to decide how to respond to a situation, the more malleable their responses are likely to be. For example, one would not expect a driver surprised by the sudden intrusion of a vehicle to respond any differently in a simulator than in real driving. In contrast, passing decisions, which drivers may have considerable time to plan and execute, would likely be more amenable to such manipulation. Thus, to the extent that operational-level behaviors operate in consistently shorter time frames than either tactical or strategic behaviors, they may generally be less susceptible to the effects of incentives or instructions.

Simulating significant negative outcomes. Human subjects' committees (Institutional Review Boards), responsible for ensuring the safe and ethical treatment of human participants in research studies, will typically not approve of studies that allow participants to sustain significant negative outcomes. This creates a problem for researchers studying driving behavior, who would like participants to attempt to avoid simulator crashes as they would avoid real-world crashes. Attempting to simulate this real-world expectation with a monetary reward/penalty system would require a penalty of significantly greater magnitude than penalties associated with all other errors, such that the unfortunate crash victims would be required to pay significant sums of money to the experimenter. This creates a significant obstacle for the simulation of realistic consequences, because it is highly unlikely that experimenters would be permitted by Institutional Review Boards to collect significant sums of money from participants in research studies. A specific example of an attempt to simulate and enforce realistic consequences associated with crash involvement is presented in the final section.

Multiple crashes. In the real world, a single collision will either terminate a trip or create a delay, the duration of which will depend upon the level of injury and/or property damage. In contrast, many experimental simulation protocols allow multiple collisions, with minimal delay. The rationale for continuing experiments following collisions typically is based on the experimenter's desire to complete all planned data collection, thus avoiding the need to compensate for unbalanced data sets. However, when multiple collisions are allowed in an experimental protocol, it is certainly possible that the subject's behavior following a crash will be different from the behavior observed before the crash. This can create difficulties for data analyses, if the statistical procedure is based on the assumption that behavior during one part of a driving run is independent from behavior during another part of the run. Clearly, the most realistic way of addressing this problem is to terminate the experimental session following a crash. Short of that, researchers can explicitly test for differences before and after crashes; however, if differences are found, alternative analytical strategies may be required. For example, we have used the "time into the run before the first collision" as a summary measure of overall driving alertness. This measure eliminates the effects of the potentially unrealistic behavior resulting from portions of runs in which drivers accumulated a large number of crashes. We found this measure to be more sensitive than crash frequency as a measure of impairment due to fatigue (Ranney, Simmons, Boulos, & Macchi, 2000).

We also found that when participants became significantly impaired, they would sustain multiple crashes within a relatively short time interval. We defined a criterion, based on a certain number of crashes within a specified time interval, and interpreted clusters of crashes satisfying this criterion as reflecting the point in time at which the driver would have stopped driving, either voluntarily or involuntarily, in a real-world setting.

Coercive completion bonus. Human subjects' considerations require that participants be able to terminate participation at any point in an experiment and that no coercion be used to encourage participation when a participant wants to stop. Excessive completion bonuses may be coercive to a participant who faces the conflict between wanting to terminate participation and realizing that perseverance for another hour will yield a significant monetary payoff. Institutional Review Boards must determine what amount of completion bonus is potentially coercive for a given experimental protocol and participant population.

Effects of incentives over time. In an experimental protocol in which participants were required to participate for several days, we observed fairly consistent changes in drivers' attentiveness over time (Ranney & Pulling, 1989). Upon first arrival, the participants were highly attentive and generally appeared to feel that they were in a "testing" environment. This may have been because many of the subjects were elderly and may have been sensitive to the possibility that their skills were being evaluated. However, on the second or third day of participation, participants had clearly determined that there was no significant threat and that the task requirements were more tedious than challenging. At this point, we began to observe increases in apparent lapses of attention. As the participants became more comfortable with the task requirements, they abandoned the hyper-vigilance they initially brought to the experiment. We concluded that the changes in attentiveness were due at least in part to the lack of specificity of the reward/penalty system. Specifically, participants were paid an hourly rate plus a relatively small increment for "acceptable performance." In this study, the incentive increment was most likely too small and ill-defined, since it was not tied to specific outcomes. One possible solution to avoid changes in performance over time is to design experiments that require only a single session.

Assessment of reward/penalty systems. Monetary rewards influence performance, but how does an experimenter know when the rewards and penalties are having the desired effect on driver behavior? The most direct approach is to include reward/penalty parameter values as independent variables in the experimental design and compare performance at different levels. While this will increase the required data collection, the resulting accumulation of information concerning the effects of incentives on driving behavior will help experimenters better understand the ecological validity of their research paradigms. A body of research supporting their use will also improve the credibility of reward/penalty systems.

Non-independence of performance measures. Components of reward/penalty systems may not be independent, which can create problems for statistical analysis. For example, if a delay is associated with penalized outcomes such as crashes, completion time may be correlated with crash frequency. Similarly, analyses using the total amount of money received as a dependent measure may preclude additional analyses of component measures that comprise the overall measure. In this situation, researchers must choose which measure is more consistent with study objectives.

9.3.3.4 Representing Rewards and Penalties With Decision-Making Models

Although classical decision theories may not be appropriate for describing the processes involved in real-world decision-making (e.g., Beach & Lipshitz, 1993), they are useful for structuring simple driving decisions required of participants in experimental contexts. Decision representation requires specification of four basic elements (Lehto, 1997), including: (1) the potential actions (A_i) to choose among, (2) the events (E_j) that occur as a result of the decision, (3) the consequences (C_{ij}) associated with each combination of action and event, and (4), the probability of occurrence (P_{ij}) associated with each combination of action and event. For example, consider the decision whether to stop or continue through an intersection when the traffic signal changes from green to yellow. The potential actions include braking to a stop or continuing through the intersection. There are essentially four categories of events that may occur as a result of these actions, based on a matrix that crosses the decision (stop or go) with the outcome (success or failure). A successful "stop" occurs when there is insufficient time to clear the intersection before the traffic signal changes to red. When there is sufficient time to clear the intersection, the "stop" is unsuccessful, because the driver's progress is delayed unnecessarily. A successful "go" decision occurs when there is sufficient time to clear the intersection. When there is insufficient time to clear the intersection the "go" decision is unsuccessful, because of the possibility of receiving a ticket on being involved in a crash. The consequences associated with the successful decisions include continued progress toward the destination without adverse consequences.

Following Lehto (1997), the expected value (*EV*) of each action A_i can be calculated by weighting the various consequences C_{ij} over all events j, by the probability associated with each event which follows action A_i. A value function $V(C_{ij})$ is used to transform the consequences C_{ij} into values, which in our example are monetary values. The expected value for a given action then becomes:

$$EV[A_i] = \sum_j P_{ij} V(C_{ij})$$

Subjective expected utility (SEU) theory provides a normative model to represent decision-making under uncertainty (Lehto, 1997). SEU theory emphasizes the distinction between the (objective) value of an outcome, typically expressed in currency, and the (subjective) utility, which reflects the usefulness of the outcome to the individual. While the relationship between value and utility is generally monotonic, it may not be linear (e.g., Tversky & Kahneman, 1981). However, in an experimental situation in which subjects are assumed to be motivated to maximize the amount of monetary reward, their behavior will likely be neither risk-aversive nor risk-seeking with respect to the types of decisions required. In this situation, the value function is linear. Moreover, when the utilities are defined in terms of monetary values, utilities can be assumed to be equivalent to values and therefore SEU is equivalent to expected value (*EV*) theory. Allen, Stein and Schwartz (1981) represented decision-making at a yellow traffic signal using *EV* theory, as shown below:

$$\text{SEV(Go)} = V(F|Go) * SP(F|Go) + V(S|Go) * SP(S|Go)$$

$$\text{SEV(Stop)} = V(F|Stop) * SP(F|Stop) + V(S|Stop) * SP(S|Stop)$$

where

V is the value of the outcome
SP is the subjective conditional probability of the outcome
F is the outcome fail
S is the outcome success

Allen et al. (1981) used this model to develop risk acceptance functions, which they found to predict driver decision-making at a signalized intersection in an experimental study. The model structure demonstrates the importance of incorporating rewards and penalties for predicting decision-making behavior in driving experimentation. Clearly, if there are no consequences (values) associated with the various decision outcomes, there is no basis for predicting decision-making behavior and thus no reason to expect subjects to prefer one choice over another.

In a separate study, Stein et al. (1978) varied the monetary reward/penalty values associated with tickets for speeding or red-light violations and found that drivers responded by adopting slower speeds and modifying their decision-making at signalized intersections. These results show how variations in reward/penalty structures can influence driver decision-making in experimental settings.

9.4 Examples of Effects of Reward/ Penalty Systems in Simulator Studies of Driving Behavior

This section includes two examples of driving simulation experiments that incorporated reward/penalty systems. These studies had slightly different objectives, which necessitated slightly different reward/penalty parameters. Both studies were conducted on a fixed-base driving simulator, based on STISIM simulation software (STISIM, v. 7.03), which was developed by Systems Technology Inc. (STI). Drivers manipulated standard vehicle controls while sitting in a mock-up truck cab. The roadway scene was projected onto a wall-mounted screen. Scenario events and performance measures were different for each experiment. What is instructive about these two examples is just how difficult it can be to identify a reward/penalty system in the driving simulator that corresponds to the system which is functioning in the real world.

Example 9.1: Adaptive Warnings for Collision-Avoidance Systems

The first experiment (Lehto, Papastavrou, Ranney, & Simmons, 2000) examined the effects of different warning system thresholds and visibility levels on drivers' decisions whether or not to pass slow-moving vehicles ahead. Fifteen subjects completed three sessions, consisting of a control run (no warning) and two one-hour driving runs (different warning thresholds). During each run, participants encountered a number of passing opportunities, some of which, if attempted, were very likely to result in a crash, due to the presence of an oncoming vehicle. At the point in time at which the driver was required

TABLE 9.1 Components of Passing Decision-Making Task

		STATE OF THE WORLD	
		Oncoming Vehicle	No Oncoming Vehicle
DRIVER RESPONSE	Attempt to Pass	Miss (*M*)	Correct Rejection (*CR*)
	No Attempt	Detection (*D*)	False Alarms (*FA*)

Source: Reproduced with permission from Lehto, M. R., Papastavrou, J. D., Ranney, T. A., & Simmons, L. A. (2000). An experimental comparison of conservative versus optimal collision avoidance warning system thresholds. *Safety Science, 36,* 185–209.

to make the passing decision, the information available to the driver concerning the presence of the oncoming vehicle was incomplete, as it might be in a real-world situation with restricted visibility, such as in heavy fog. In particular, the driver could see an object of varying brightness in the distance, but was uncertain about whether or not the object was an oncoming vehicle (three levels of visibility: dim, medium, and bright). A warning system, when present, provided information to the driver concerning the probability that the object was actually a vehicle. The warning display was either a red bar plus an auditory signal, indicating that the warning system had concluded that it was not safe to pass, or a green bar, indicating that it was safe to pass. The driver was required to combine the information available visually with the information provided by the warning system to decide whether or not to pass the slower vehicle ahead.

Participants were instructed to make their passing decisions as if they were driving a real vehicle and to avoid risky passing attempts. There were 51 passing events (the lead vehicle slowed from 60+ mph to 30 mph in a straight section). The probability of an oncoming vehicle was set at 19/51; that of no oncoming vehicle at 32/51. A system of performance incentives and penalties was implemented to elicit behavior that approached realistic driving. Specifically, each participant was given an hourly base pay ($6.00 per hour) for each hour of participation in the study. This amount was not influenced by performance on the task. In addition, each participant was given a daily allotment of $20.00 at the beginning of the session. To this allotment, $0.20 was added for each successful pass and $0.10 was subtracted for each missed passing opportunity. We subtracted $10.00 for each unsafe passing attempt, which typically resulted in a crash. A siren was sounded to indicate issuance of a ticket for unsafe passing and crashing noises (breaking glass, screeching tires) were sounded if a crash occurred, either with an oncoming vehicle or if the vehicle ran off the roadway. The participants were instructed

that they would be allowed to keep the total of the daily allotment and the driving performance rewards and penalties, and that this amount would vary between $0.00 and $30.00. Therefore, in effect, their hourly pay could increase from $6.00 per hour to approximately $16.00 per hour. Moreover, because repeated tickets or crashes could result in a negative balance, it was stipulated that, if the subject had a negative balance at the end of any session, the subject would not be allowed to participate further in the experiment. This stipulation was added to the protocol in an attempt to simulate the significant negative consequences associated with a collision in the real world, including the disruption or termination of travel plans.

The components of the rewards and penalties associated with the various decisions and outcomes are represented in a decision matrix in Table 9.1. The expected monetary value (*EV*) associated with different decision-making strategies under the control condition (no warning) and two warning system conditions (with different thresholds) was calculated using the following equation:

$$EV = (N_M \times R_M) + (N_{CR} \times R_{CR}) + (N_D \times R_D)$$
$$+ (N_{FA} \times R_{FA})$$

where *N* is the frequency associated with each respective outcome [Miss (***M***), Correct Rejection (***CR***), Detection (***D***), and False Alarm (***FA***)] and *R* is the monetary reward/penalty associated with the particular outcome (Table 9.2). For example, with the control condition the optimal decision is to pass only when there is a dim stimulus (no oncoming vehicle) since when there was a medium stimulus a car was present on 2 of 17 trials. This means that there were no misses and 19 detections (the 19 trials in which there was an oncoming car all had a medium or bright stimulus). Of the remaining 32 trials in which there was no oncoming car, there was a dim stimulus

TABLE 9.2 Reward/Penalty Scheme

Driver Behavior	Outcome Category	Amount of Reward/Penalty ($)
Safe no-pass decision	Detection (*D*)	0
Fail to pass	False Alarm (*FA*)	−.10
Unsafe passing attempt	Miss (*M*)	−10.00
Safe Pass	Correct Rejection (*CR*)	+0.20

Source: Reproduced with permission from Lehto, M. R., Papastavrou, J. D., Ranney, T. A., & Simmons, L. A. (2000). An experimental comparison of conservative versus optimal collision avoidance warning system thresholds. *Safety Science, 36,* 185–209.

in 17 cases (correct rejection; driver passes) and a medium stimulus in 15 cases (false alarm; driver does not pass). Thus, the expected value is:

$$EV = (0 \times -10.00) + (17 \times 0.20) + (19 \times -0.10) + (15 \times -0.10)$$

$$= \$1.90$$

The expected value under the optimal policy gains with the other two warning systems are, respectively, $4.30 and $6.40. The larger point here is that individuals could come away with either $27.90 ($6.00 + $20.00 + $1.90), $30.30 or $32.40 if they followed the optimal policy.

Despite these monetary incentives, 5 (33%) out of 15 participants were not permitted to participate beyond the first session, due to their inability to meet the performance criterion of maintaining a positive monetary balance at the end of each session. Discussions with these participants revealed that they typically had adopted decision strategies which would have resulted in disastrous consequences in real-world driving. Specifically, even though they knew there was a chance they could crash in a specific condition and that this might lead to the loss of all incentive pay, they risked this loss in an attempt to earn greater monetary rewards. This finding led us to consider the expectations that participants brought to the experiment. In particular, the population from which participants for this study were drawn tended to include a significant proportion of unemployed people, who were motivated to earn money. One might have expected them to be particularly cautious, given that individuals are usually risk averse in the domain of gains. Yet, they were more willing to take risks. Moreover, we have found that individual participants' situations may change day-to-day, leading some participants to drop out in the middle of a multiple-day study because they have found a longer-term job. Some participants were thus clearly motivated to maximize their single-day pay. This is quite different from the traditional use of young undergraduate students, who are motivated to participate either because of their enthusiasm for psychology or as a course requirement. Nevertheless, one would again expect participants in the current experiment to be risk averse.

This example raises important questions concerning the effects of reward/penalty systems on drivers' simulator behavior or performance. In particular, for those drivers who adopted excessively risky passing strategies we do not know to what extent their behavior reflects an attempt to take advantage of the inherent artificiality of the simulator setting versus the effects of the reward/penalty system alone. One might argue that the reward/penalty system elicited unrealistic behavior; however, it could also be argued that the lack of significantly negative consequences associated with simulator driving motivated the unexpected decision-making. There is no way to know without additional experimentation and unfortunately the funding to address such methodological issues is virtually non-existent. What can be concluded is that anomalous behavior does occur, at least among some participants in simulator experiments. Moreover, the use of the reward/

penalty system provided not only a mechanism to identify such behavior, but also a model within which to shape participants' behavior to conform more closely with expectations based on real-world decision-making models. Finally, the elimination of drivers from the protocol provides a means of simulating the significantly negative consequences of a crash with some amount of face validity.

Example 9.2: Effect of an Afternoon Nap on Driving Performance

The second study (Ranney et al., 2000) evaluated the effects of an afternoon nap on overnight driving performance. Eight professional drivers completed two replications of a two-day (43–47 hour) protocol, each including eight hours of overnight driving following a truncated (five-hour) sleep period on the previous night. One replication included a three-hour nap on the afternoon before the overnight driving; the other replication involved overnight driving with no preceding nap. The overnight driving consisted of four two-hour runs, separated by half-hour breaks. The driving task included vehicle control on straight and curved roads, detection of pedestrians appearing alongside the roadway and targets in the mirrors, and avoidance of obstacles and oncoming vehicles.

The monetary reward/penalty system used in this experiment was a two-tiered hierarchical system, with macro and micro components. The macro reward/penalty system had three components. First, the participants were paid a daily rate for their participation in the experiment, including their sleeping time and time when not driving. Second, because the experimental design required participants to complete the (40+ hour) protocol twice, they were given a significant bonus for completing both parts of the experiment. Third, a micro reward/penalty system was in effect during each driving simulator run. Specifically, drivers were rewarded for timely arrival at the destination, which was defined as a specific number of miles. Based on pilot testing, we developed a reference time for the pre-established distance. Drivers were then given $1.00 per minute (pro-rated) for each minute or portion thereof faster than the reference time in which they completed the drive. They were penalized the same amount for each minute, or portion thereof, they arrived after the deadline. In addition, drivers were penalized for crashes and speeding tickets and were rewarded for each target detected.

The experimental protocol was intended to stress the participants to the point of psychological fatigue and to elicit micro-sleep episodes while driving. The micro reward/penalty structure was intended to motivate truck drivers to complete each two-hour drive under nighttime conditions. The effects of the (micro) reward/penalty structure alone were assessed in a series of pilot studies conducted in preparation for this work. Specifically, we began pilot studies without a reward/penalty structure, using alert drivers not subjected to sleep deprivation. We found that most drivers became drowsy after approximately 30 minutes of simulated nighttime driving; however, we were surprised by how quickly they recovered their alertness when the two-hour drive ended. We then implemented the micro reward/penalty structure, which included the elements described above and found that the pilot drivers maintained their alertness during the entire drive. We concluded that the reward/penalty structure was effective in motivating realistic nighttime driving. We also concluded that the drowsiness observed among the initial pilot drivers had nothing to do with fatigue; rather, it reflected the combination of the uneventful nighttime scenario and the lack of a trip purpose.

When the micro reward penalty structure was used with the sleep deprivation protocol, we found that the macro reward/penalty structure, which included the completion bonus, was successful in simulating the conflicting incentives that motivate truck drivers to continue driving after the point at which they should have stopped driving. As participants became significantly drowsy, they sustained an increasing number of crashes, sufficient to ensure that no monetary reward would be received for a particular run. However, most participants decided to persevere, despite sustaining penalties for crashes and tickets, because they were motivated by their desire to complete the experiment and receive the larger hourly rate and completion bonus.

This example allows several conclusions about the use of reward/penalty systems. First, while the simple (micro) reward/penalty structure was sufficient to motivate attentive driving among alert drivers in a simple protocol, it was insufficient for this purpose among sleep-deprived participants when a longer more complex protocol was used. The macro reward/penalty structure, which combined the micro structure with the significant completion bonus, was necessary to motivate the perseverance that pushed drivers beyond the limits at which normally they might stop driving. Researchers must therefore consider how a reward/penalty system interacts with the other manipulations in the experiment. Second, the behaviors elicited from drivers that were beyond their apparent limits led us to conclude that some portions of the driving trials were not representative of real-world driving. Specifically, we searched for a way to define the point in time at which a driver would have been unable to continue safely, so that we could eliminate behaviors that were not representative of actual on-road driving. For example, we determined the point in time at which the first crash occurred in each two-hour drive and used it as a performance measure. Thus, when reward/penalty systems are in use, researchers need to be vigilant to the possibility that some portion of the recorded behavior will be confounded by multiple crashes or other events that are not sufficiently independent for analytical purposes. Some portion of a balanced design may need to be sacrificed to increase the psychological fidelity of the overall protocol.

studies typically eliminate the trip purpose, which influences all components of real-world driving. It is therefore incumbent on researchers to provide direct guidance to participants concerning their expectations about the simulated trip that they are engaged in. Without such guidance, research participants may have considerable difficulty interpreting the often-used instruction to "drive as you normally would," lacking an appropriate frame of reference. However, providing an artificial trip purpose is likely to be insufficient unless accompanied by specific details and consequences.

In dual-task situations, in which participants are asked to engage in secondary tasks such as a phone conversation or destination entry, researchers must provide clear guidance to help participants determine the relative importance of the component tasks. Failure to provide such guidance can be expected to introduce significant unwanted variability into performance data due to individual differences in the assignment of priorities.

The main tools available for improving psychological fidelity are task instructions and reward/penalty systems. Specification of rewards and penalties in relation to a target completion time can be used to manipulate the urgency of trip completion. Specification of penalties for speeding and other violations using a probabilistic delivery schedule can simulate the potential costs of violating traffic laws. Together, these two mechanisms can create the tension that exists in real-world driving between timely arrival and avoidance of tickets. Institutional Review Boards make it difficult to realistically simulate the consequences of the ultimate negative outcome, namely a crash. The increase in psychological fidelity gained through use of specific instructions and reward/penalty systems may create problems that are inconsistent with balanced designs. Examples include termination of participants who perform below a specified criterion or elimination of portions of a protocol due to multiple crashes.

Data pertaining to the effects of instructions and reward/penalty systems are sorely needed. The increasing preference for naturalistic methods makes it essential that laboratory researchers acknowledge the importance of psychological fidelity and address it directly. The incorporation of psychological fidelity into experimental designs would facilitate the demonstration that it can be manipulated and thus better understood. The resulting accumulation of a body of research results on this topic would better define the tools needed to reduce unwanted variability from research designs while improving the credibility of simulator studies. Approaching psychological fidelity directly would allow researchers to confidently assert the role of driving simulator studies in the quest to understand driving behavior and performance.

9.5 Conclusions

Some amount of artificiality is inherent in all experiments that use driving simulators. An often overlooked component of this artificiality is psychological fidelity, defined as the extent to which the risks and rewards of participation in the experiment correspond to real-world risks and rewards. Experimental

Key Points

- Driving simulator studies continue to be criticized for lack of realism, despite significant advancements in the fidelity of the driving experience.
- Some amount of artificiality is inherent in all experiments that use driving simulators. An often-overlooked

component of this artificiality is psychological fidelity, defined here as the extent to which the risks and rewards of participation in the experiment correspond to real-world risks and rewards.

- Experimental studies typically eliminate the trip purpose, which influences all components of real-world driving. Researchers must therefore provide direct guidance to participants concerning their expectations about the simulated trip.
- Failure to define the driving context is likely to result in a significant amount of unwanted variability in performance measures, reflecting a combination of confusion and different assumptions made by participants about the priorities in the experiment.
- Instructions and performance incentives are the main tools available to researchers for improving psychological fidelity of their experiments. Instructions define the performance space; incentives provide guidance about priorities.
- Reward/penalty systems can be used to represent the real-world tradeoff between the desire for safe and timely arrival and the desire to avoid speeding tickets and crashes. Their credibility would be increased if researchers would incorporate reward/penalty system components as independent variables in experimental designs and examine their effects on driving behavior.
- Institutional Review Boards, responsible for ensuring the safe and ethical treatment of human participants in research studies, will typically impose limits on researchers' use of performance incentives and penalties. Excessive rewards (e.g., completion bonuses) may be considered coercive while limits on penalties make it virtually impossible to accurately simulate the significant negative consequences associated with crash involvement.

Keywords: Driver's Motives, Instructions, Psychological Fidelity, Performance Incentives, Risk Perception, Rewards and Penalties

Glossary

Extrinsic incentives: Rewards and penalties associated with different aspects of performance, including money, praise, or food. Extrinsic incentives are typically used in driving simulator studies to motivate the behavior desired by experimenters.

Hierarchical model of driving behavior: Driving behavior consists of concurrent activity at strategic (e.g., trip purpose), tactical (e.g., gap acceptance) and vehicle control (e.g., speed selection) levels. Instructions provided to participants of simulator studies should provide guidance for decision-making at all levels in this model.

Psychological fidelity: Realism of the simulator experience includes the extent to which the risks and rewards of

participation in the experiment correspond to real-world risks and rewards.

Reward/Penalty systems: In a driving simulator study, drivers are typically rewarded for timely arrival at a destination and penalized for speeding tickets and crashes. The combination of these rewards and penalties comprises the reward/penalty system.

Key Readings

Duncan, J. (1990). Goal weighting and the choice of behavior in a complex world. *Ergonomics, 33,* 1265–1279.

Edwards, W. (1961). Costs and payoffs are instructions. *Psychological Review, 68,* 275–284.

Eysenck, M. W. (1982). *Attention and arousal: Cognition and performance.* Berlin: Springer-Verlag.

Lehto, M. R., Papastavrou, J. D., Ranney, T. A., & Simmons, L. A. (2000). An experimental comparison of conservative versus optimal collision avoidance warning system thresholds. *Safety Science, 36,* 185–209.

Näätanen, R., & Summala, H. (1976). *Road-user behavior and traffic accidents.* Amsterdam, the Netherlands: North-Holland Publishing Company.

References

Allen, R. W., Stein, A. C., & Schwartz, S. H. (1981). *Driver decision-making analysis* (Rep. No. 301). Hawthorne, CA: Systems Technology Inc.

Beach, L. R., & Lipshitz, R. (1993). Why classical decision theory is an inappropriate standard for evaluating and aiding most human decision-making. In G. A. Klein, J. Orasano, R. Calderwood & C. E. Zsambok (Eds.), *Decision-making in action: Models and methods.* Norwood, NJ: Ablitz Publishing Corporation.

Broadbent, D. E. (1971). *Decision and stress.* London: Academic Press.

Brown, I. D. (1990). Drivers' margins of safety considered as a focus for research on error. *Ergonomics, 33,* 1307–1314.

Brown, T. L., Lee, J. D., & McGehee, D. V. (2001). Human performance models and rear-end collision avoidance algorithms. *Human Factors, 43*(3), 462–482.

Caird, J. K., Willness, C. R., Steel, P., & Scialfa, C. (2008). A meta-analysis of the effects of cell phones on driver performance. *Accident Analysis & Prevention, 40,* 1282–1293.

Delhomme, P., & Meyer, T. (1997). Control motivation and driving experience among young drivers. In T. Rothengatter & E. C. Vaya (Eds.), *Traffic & transport psychology: Theory and application* (pp. 305–316). Amsterdam: Pergamon.

Desmond, P. A., & Matthews, G. (1997). The role of motivation in fatigue-related decrements in simulated driving performance. In T. Rothengatter & E. C. Vaya (Eds.), *Traffic & transport psychology: Theory and application* (pp. 325–334). Amsterdam: Pergamon.

Duncan, J. (1990). Goal weighting and the choice of behavior in a complex world. *Ergonomics, 33,* 1265–1279.

Edwards, W. (1961). Costs and payoffs are instructions. *Psychological Review, 68,* 275–284.

Evans, L. E. (1991). *Traffic safety and the driver.* New York: Van Nostrand Reinhold.

Eysenck, M. W. (1982). *Attention and arousal: Cognition and performance.* Berlin: Springer-Verlag.

Eysenck, M. W. (1983). Incentives. In G. R. J. Hockey (Ed.), *Stress and fatigue in human performance.* Chichester, England: John Wiley & Sons.

Goodman, M., Bents, F. D., Tijerina, L., Wierwille, W. W., Lerner, N., & Benel, D. (1997). *An investigation of the safety implications of wireless communications in vehicles* (DOT HS 808 635). Washington, DC: National Highway Traffic Safety Administration, US Department of Transportation.

Horrey, W. J., & Wickens, C. D. (2006). Examining the impact of cell phone conversations on driving using meta-analytic techniques. *Human Factors, 48,* 196–205.

Klauer, S. G., Dingus, T. A., Neale, V. L., Sudweeks, J. D., & Ramsey, D. J. (2006). *The impact of driver inattention on near-crash/crash risk: An analysis using the 100-car naturalistic driving study data* (DOT HS 810 594). Blacksburg, VA: Virginia Tech Transportation Institute.

Lehto, M. (1997). Decision-making. In G. Salvendy (Ed.), *Handbook of human factors and ergonomics* (2nd ed.), New York: John Wiley & Sons.

Lehto, M. R., Papastavrou, J. D., Ranney, T. A., & Simmons, L. A. (2000). An experimental comparison of conservative versus optimal collision avoidance warning system thresholds. *Safety Science, 36,* 185–209.

McCartt, A. T., Hellinga, L. A., & Braitman, K. A. (2006). Cell phones and driving: Review of research. *Traffic Injury Prevention, 7,* 89–106.

Michon, J. A. (1985). A critical view of driver behavior models. What do we know, what should we do? In L. E. Evans & R. Schwing (Eds.), *Human behavior and traffic safety.* New York: Plenum Press.

Näätanen, R., & Summala, H. (1976). *Road-user behavior and traffic accidents.* Amsterdam, the Netherlands: North-Holland Publishing Company.

National Highway Traffic Safety Administration. (2008). *2007 traffic safety annual assessment—Highlights* (DOT HS 811 017). Washington, DC: National Center for Statistics and Analysis, National Highway Traffic Safety Administration.

Pachella, R. G. (1974). The interpretation of reaction time in information-processing research. In B. H. Kantowitz (Ed.), *Human information processing.* Hillsdale, NJ: Earlbaum Associates.

Ranney, T. A. (1994). Models of driving behavior: A review of their evolution. *Accident Analysis & Prevention, 26,* 733–750.

Ranney, T. A., & Pulling, N. H. (1989). Performance differences on driving and laboratory tasks between drivers of different ages. *Transportation Research Record, 1281,* 3–9.

Ranney, T. A., Simmons, L. A., Boulos, Z., & Macchi, M. M. (2000). Effect of an afternoon nap on nighttime performance in a driving simulator. *Transportation Research Record, 1686,* 49–56.

Ranney, T. A., Watson, G. S., Mazzae, E. N., Papelis, Y. E., Ahmad, O., & Wightman, J. R. (2005). *Examination of the distraction effects of wireless phone interfaces using the National Advanced Driving Simulator—Final report on a freeway study.* (DOT HS 809 787). Washington, DC: National Highway Traffic Safety Administration, US Department of Transportation.

Rosenthal, T. J., Parseghian, Z., Allen, R. W., & Stein, A. C. (1994). Appendix E: Reward/penalty computations, *STISIM user's guide.* Hawthorne, CA: Systems Technology, Inc.

Simons-Morton, B., Lerner, N., & Singer, J. (2005). The observed effects of teenage passengers on the risky driving behavior of teenage drivers. *Accident Analysis & Prevention, 37*(6), 973–982.

Stein, A. C., Allen, R. W., & Schwartz, S. H. (1978, April 25–27). *Use of reward-penalty structures in human experimentation.* Paper presented at the 14th Annual Conference on Manual Control, University of Southern California.

Transportation Research Board. (2008). *Asking why to learn how: 2007 annual report of the second strategic highway research program.* Washington, DC: Transportation Research Board.

Tversky, A., & Kahneman, D. (1981). The framing of decisions and the psychology of choice. *Science, 211,* 453–458.

Williams, A. F., & O'Neill, B. (1974). On-the-road driving records of licensed race drivers. *Accident Analysis & Prevention, 6,* 263–270.

Yerkes, R. M., & Dodson, J. D. (1908). The relation of strength of stimulus to rapidity of habit formation. *Journal of Comparative Neurological Psychology, 18,* 459–482.

Zeitlin, L. R. (1996, January). *Vehicle simulation research: What is and is not being simulated.* Paper presented at the Human Factors Workshop at the Annual Meeting of the Transportation Research Board: Washington DC.

10

Surrogate Methods and Measures

Linda S. Angell
Touchstone Evaluations, Inc.

Abstract

The Problem. Beginning in about 2000, a need emerged for methods with which to assess the effects of advanced information systems on driver workload early in product development—surrogate methods that were valid, practical, and applicable before on-road driving studies could be done with instrumented vehicles. Surrogate methods that emerged included techniques such as visual occlusion, peripheral detection tasks, the lane-change test, and other techniques. As these methods have been studied, a variety of questions emerged about their possible use and role in the context of driving simulation. ***Role of Driving Simulators.*** Driving simulators are an important tool for examining responses of drivers to the driving scenarios they present. Simulators are themselves a kind of surrogate—a surrogate for the overall driving experience on the road, in scenarios—of interest. In that context, this chapter explores what role *other* surrogate methods might play—how they are different, as well as how they can be used together with simulation—in the study of driving behavior. ***Key Results of Studies for Appropriately Applying Surrogates and Simulations.*** Studies in different venues (lab, simulator, track, and road) have been done which provide key findings that are salient for how best to use surrogates and simulations together in research. These findings are reviewed, with an emphasis on building robustness into research designs. ***Scenarios and Dependent Variables.*** Examples of the sorts of surrogates that can be incorporated in simulation are discussed, along with surrogates that are implemented as simulations. Issues are identified, along with opportunities for innovation through simulation. ***Generality.*** Both surrogates and simulations can be powerful tools in their own right, but the key to unlocking their power lies in the clarity of the conceptualization that underlies the research question driving their application. This is a general and fundamental point, but a very critical one to advancing the start of the art.

10.1 Introduction

Definition of "Surrogate":

"One that takes the place of another; a substitute."
"To put in the place of another;. . . replace."

(From Answers.com)

Definition of "Simulate":

1. a. To have or take on the appearance, form, or sound of; imitate.

b. To make in imitation of or as a substitute for. See . . . imitate.

2. To make a pretense of; feign: simulate interest.

3. To create a representation or model of (a physical system or particular situation, for example).

(From Answers.com)

Since 2000, one focus of research on driving behavior has been on understanding driver distraction, its causes, and the sources from which it arises. Early in this research on driving distraction, it was recognized that on-road evaluations of the effects

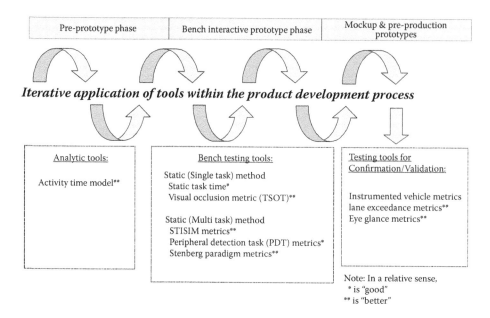

FIGURE 10.1 An illustrative toolkit of surrogate methods, used up-front in product development. (Developed by the CAMP Driver Workload. Metrics Project, Angell, L., et al., DOT HS 810 635, Washington, DC: National Highway Traffic Safety Administration, US DOT, 2006.)

of secondary tasks during driving were time-, labor-, and cost-intensive. However, because it was desirable to perform evaluations of many tasks done by drivers (on conventional secondary systems, newly emerging information systems, and devices carried into vehicles by drivers for use during driving), the need was substantial for methods or measures that were faster to apply, required less staffing, and less money to conduct than on-road studies. Therefore, a search for "surrogate methods" or "surrogate metrics" was initiated in that part of the scientific community interested in these topics (e.g., Angell, Young, Hankey, & Dingus, 2002; Young & Angell, 2003; Alliance of Automotive Manufacturers, 2003; Mattes 2003; Noy & Lemione, 2000; among many others).*

This search had, as its objective, the goal of finding or developing methods or metrics which could be used early in product development to perform assessments of the extent to which secondary in-vehicle tasks interfere with driving (at a time when a new device or subsystem for the vehicle may not be developed to a point where it exists in hardware or is integrated into a drivable vehicle for full-scale on-road driving assessment). Furthermore, since these tests are done early in a product's development, and may be repeated at several times during the course of its evolution, before it becomes integrated with a vehicle, it was necessary that any "surrogate" test or battery

of surrogate tests, yielded answers about the driver workload associated with each task that *were the same as* answers obtained from full-scale on-road testing. To the extent that answers from surrogate tests were different from those in on-road testing, it would imply that the surrogates lacked validity, reliability, or the ability to discriminate between types or magnitudes of attention or resources required by tasks—leading perhaps to errors in design and engineering decisions during development.

Thus, it was not so much that surrogate methods and metrics were envisioned as a complete replacement for on-road experiments and evaluations, as it was that they were envisioned as a "front-end" complement for the product development and validation process (to which on-road testing could be a final conclusion, if needed, and which should confirm the answers from earlier surrogate testing). The role that surrogate methods could play in a typical development process is illustrated above in Figure 10.1, from the CAMP Driver Workload Metrics Final Report (Angell et al., 2006).

At first blush, the effort to develop surrogate methods or metrics for identifying distracting tasks in new in-vehicle systems seemed straightforward and feasible. In reality, it has proven much more difficult and complex than perhaps anyone initially expected it to be. Many questions have arisen in this effort, but those of interest in the context of simulation include the following:

- Does the term "surrogate method" necessarily refer to a simulation?
- Can a simulation serve as a surrogate method? (In other words, is driving simulation, itself, a surrogate method?)
- Are the terms "surrogate method" and "surrogate metric" different ways of referring to the same thing?

---

* The surrogate methods and metrics discussed in this chapter are all related to driver workload and driver distraction. However, surrogates have been developed for evaluating effects in other arenas of driving performance (such as effects of drowsiness and fatigue, alcohol, and so on), but are not discussed here. Readers interested in development of surrogates in areas of impairment such as these are referred to the work of Kennedy and colleagues (e.g., Kennedy, Turnage, & Lane, 1997) as a starting point.

- Can a simulation employ an embedded surrogate method?
- Are there any key issues that must be considered in the application of surrogate methods during simulated driving?

This chapter explores answers to these and other questions, and provides a discussion of conceptual issues that should be considered when developing a surrogate method or metric in the context of a driving simulation environment.

10.2 Surrogate Methods in the Context of Driving Simulation

10.2.1 "Surrogate Methods" Are Not Always Simulations

Surrogate methods are intended to be methods that can be used "in place of another" (usually more resource-intensive) method—typically, "in place of" measuring driving behavior effects on the road through the use of a fully-instrumented vehicle. Although driving simulation is a type of surrogate method (one which emulates key cues of the driving environment and is used to create "a driving-like" experience for a driver), there are surrogate methods which are not simulations—and which are not creating a representation of the driving experience either in part or whole. These other surrogate methods seek instead to evoke a set of processes in the human operator that are similar to those elicited during the actual behavior/s of interest—as a means of making these processes observable and measurable.

An example of a surrogate method that is not intended as a simulation is the surrogate method utilizing "visual occlusion" (e.g., International Organization for Standardization, 2007; also, Gelau & Krems, 2004). In this method, a human operator performs a task while wearing special goggles which are equipped with a computer-controlled shutter. The shutter may be pre-programmed to open and close repeatedly, according to a fixed cycle (e.g., 1.5 second open period, followed by a 2.0 second closed period). During the shutter-open periods, the participant can obtain a brief glimpse of the task/s he or she is performing. The shutter makes observable how many glimpses, or how much total shutter-open time, is needed for a participant to complete a task using these brief viewing intervals (conceptualized loosely as resembling glances to the task). The method provides several metrics, depending on its use.

When the shutter-closed periods in this process are conceptualized to represent glances away from a secondary task (to view the road, for example), one relevant metric is Total Shutter Open Time (TSOT), because it represents the time the eyes are away from the road and viewing the task In such a conceptualization, Total Shutter Open Time is used as a surrogate metric for the visual demand that a secondary task imposes during driving—correlating with, for example, the number of task-related glances a secondary task requires to complete, and the total glance time required to complete it, as well as some correlation with task duration, and percent of trials with a lane exceedance.

Interestingly, in the early work of Senders, where the use of visual occlusion was pioneered, its application was conceptualized differently (cf. Appendix B in Senders, 1983, on the attentional demand of automobile driving). In this seminal work, the shutter-closed period was conceived to represent glances away from the road—with the intent to measure consequences to lane-keeping and other driving performance measurements that are dependent upon visually sampling the road scene at frequent enough intervals of sufficient length to extract information needed for controlling path and avoiding obstacles. In either case, visual occlusion is not meant to emulate or simulate a set of driving conditions—but rather, to evoke the processes associated with glances to the roadway (or to other regions) that demand visual attention. The occlusion goggles are used to "interrupt" vision as a means of making the consequences of a glance away, a blink, or an eye movement, observable.

Another example warrants discussion here, to emphasize the importance of clearly conceptualizing the questions to be examined when a method is selected. Peripheral Detection Task (PDT) methods comprise a class of techniques in which a small stimulus (usually a light) is briefly illuminated in the visual periphery—and the driver (or research participant) is to respond when it is detected (often with a button press). Such techniques are more generally called Detection Response Tasks (DRT), to include applications in which stimuli are presented at locations that are not peripheral, or stimuli which are presented through another modality (such as tactile) (Engström, Åberg, Johansson, & Hammarblick, 2005; Engström, 2008). Metrics include time to respond in reporting detections, and percent missed detections.

This class of surrogate has been used in two different ways; the differences are substantive enough that they may be considered distinct methods. In one application, the DRT is used as a surrogate for detecting objects or events on the roadway and, in this implementation, the method of presenting stimulus lights at unpredictable moments or locations is used as an emulation, or representation, of events to be detected on the road (cf. CAMP Driver Workload Metrics Project, Angell et al., 2006). In the other application, PDT is applied in a different way. This application is illustrated well in studies that have built upon work initiated by Van Winsum, Martens and Herland (1999), Martens and Van Winsum (2000), and others.[*] In this research, PDT (and its more recent variants, Visual Detection Tasks [VDT] and Tactile Detection Tasks [TDT]) is used to assess cognitive load—and is

---

[*] The class of techniques used in this research actually had its earliest origins in work on attention and dual task paradigms (for example, as discussed by Kahneman, 1973; and Allport, 1980). The use of a detection-task as a secondary task in a dual task paradigm for assessing the attentional demands of (primary) tasks was subsequently explored in *many* types of research over the ensuing years, including research done in the transportation arena to examine the demands of tasks done while driving.

not intended to be a simulation or emulation of events to-be-detected on the road. (It is, in fact, seen as a *third* task [a sustained vigilance task] done *in addition to* the primary task/s of driving [controlling vehicle position, path, speed, and assuring obstacle avoidance] and any secondary tasks done concurrently by the driver [e.g., using in-vehicle devices] but is nonetheless one that is ideal because it produces little interference with driving.) Engström (2008) and Victor, Engström and Harbluk (2008) have clarified that the use of DRT to examine cognitive load is quite different from other applications. The research which applies the method in this way uses a specific hardware configuration intended to invoke processes that make cognitive inattention effects observable (and to exclude any effects that would be attributed to visual scanning or to head turning). This difference is reflected in the way in which the DRT method is actually instrumented. In this application, when cognitive load alone is being measured, the stimulus light is mounted on the head so that when it is illuminated it appears in the visual field in such a way that it can *always* be seen whenever it is illuminated, regardless of head turns, and regardless of the direction of the driver's gaze in the field of view. Hence, the need to move the eyes to find and fixate on an event *cannot* influence whether the stimulus is detected or not in this application of DRT. Only attentional processes, as affected by cognitive loading, may influence responses. (Note how this method contrasts with the use of a detection task to measure response to events in the driving environment, an application in which the stimulus light or event is *not* mounted on the participant's head—but appears [or is illuminated] "in the world"—and *does* require drivers to both (1) move their eyes to the light and (2) attend to it, in order to detect it.)

Though the distinction here is somewhat fine, it is, nonetheless, important to make. In the one case, only cognitive attentional processes are intended to be revealed by the method (when it is used as a surrogate method for assessing cognitive load) while in the other application *both* visual attention (and visual sampling patterns, including eye movements and head turns) *as well as* cognitive attentional processes can play a role in affecting a driver's response. In the first instance, the purpose is to understand cognitive load, whereas in the second instance the purpose is to understand processes associated with responding to events on the roadway.

Not surprisingly, different empirical findings are associated with the two implementations of this method—in one case a surrogate (for cognitive load) and in the other case a simulation (of event detection). With the cognitive loading implementation, empirical findings reveal general attentional interference across the field of view as a function of load, and no interaction with task type (effects are not larger for visually demanding tasks). This pattern of results has been described by Engström (2008) and by Victor et al. (2008). However, with the "event detection" implementation of DRT, which often employs both central and peripheral light positions, empirical results show that: The breadth of visual scanning across the field of view narrows under higher load (e.g., mirrors are sampled less); event

detection is degraded, especially in the periphery, as task-workload increases; and these effects are much larger for visual–manual secondary tasks than for auditory–vocal secondary tasks.* Further, the empirical findings for this application show differential effects on both glance patterns and event detection for the two task types, visual–manual and auditory–vocal, as shown in Figure 10.2 (Angell et al., 2006; see also Young & Angell, 2003; and Angell, 2007).

For each metric in the figure, the "just driving" task is plotted along with the average auditory–vocal task and the average visual–manual task. For the percent of task time spent looking at the road (upper left graph), the auditory–vocal tasks led to a slight increase (approximately seven percentage points) compared with "just driving," while the visual–manual tasks were associated with a decrease of about forty percentage points in the percent of task time spent looking at the road. In this case, the visual–manual task effect is different from auditory–vocal tasks in both direction and magnitude: The forty percentage point drop means that during these visual–manual tasks drivers spent over five times less time-per-task looking at the road (on a percentage basis) than was typical during the auditory–vocal tasks (which showed the seven percentage point increase in looking at the road relative to "just driving"). More subtle differences are depicted for the percent of task spent looking at mirrors metric (upper-right graph). This graph shows that about 14% of the "just drive" task duration was spent looking at the mirrors versus 11% for the auditory–vocal tasks, a difference of only three percentage points. By comparison, only about 8% of the visual–manual task duration was spent looking at the mirrors. In this latter case, the change in mirror-scanning for visual–manual tasks compared to "just drive" is a larger drop of seven percentage points (due primarily to time spent looking at the visual–manual secondary tasks instead of the road and mirrors). The percent missed CHMSLs (a centrally presented light-to-be-detected), shown in the lower left quadrant, and the

---

* The tasks examined in this research are fully described in Angell et al. (2006), particularly in Appendix B. Due to constraints on length here, suffice it to say that that the task set used in the on-road condition included seven visual–manual tasks, seven auditory–vocal tasks, one mixed-mode task, and a "just drive" task (consisting of just driving on the highway). The tasks within each type varied in workload (from low to high), and in length, though most of the auditory–vocal tasks were constructed to be 2 minutes in length. The visual–manual tasks included a: CD task (find specified CD out of a set, insert, find track seven, initiate play), Cassette task (insert cassette for play on specified side and initiate play), Coins task (select specified amount of money from coin cup, e.g., for a toll), HVAC (adjust mode, temperature, and fan level to specified settings), Manual Dial of handheld cell phone, Radio Easy (tune to specified station not far from current setting, ~15–20 increments), Radio Hard (turn radio, change band, and tune to specified station located further away on band, ~35–37 increments). The mixed-mode task was Voice Dial of a handheld phone. The auditory–vocal tasks were: Route Instructions (listen to, remember, and paraphrase instructions for route to be taken on a set of errands), Route Orientation (listen to a series of turns and report direction of travel after last turn), Sports Broadcast (listen to 2 minute sports broadcast for a specific team and identify who they played and who won), Travel Computations (series of mental math problems involving distances on route, fuel needed, etc.), Biographical Question-and-Answer, Book-on-Tape Listen, and Book-on-Tape Summary. With the exception of tasks requiring listening (the auditory–vocal tasks), tasks were self-paced.

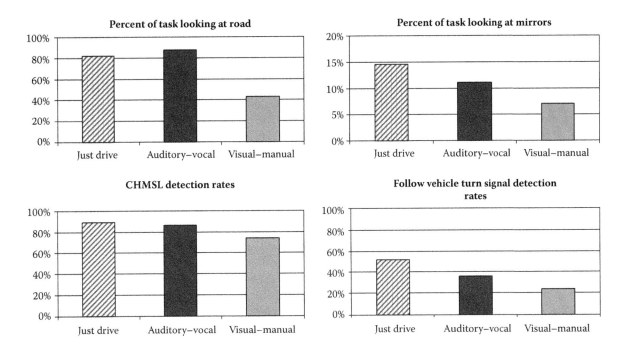

FIGURE 10.2 Visual–manual tasks produced larger decrements in breadth of visual scanning to the mirrors, detection of central events and especially detection of peripheral events when compared with "just driving" and auditory–vocal tasks. CHMSL=centrally presented light (center-high-mounted stop lamp); follow vehicle turn signal = detectable in outside rear-view mirror, a peripheral location. (From CAMP Driver Workload Metrics Project, Angell, L., et al., DOT HS 810 635, Washington, DC: National Highway Traffic Safety Administration, US DOT, 2006.)

follow-vehicle-turn-signal (FVTS) events (a peripherally-located light to-be-detected in the outside rear-view mirror), shown in the lower right quadrant, show similar results, in that the decrement on driving performance due to visual–manual tasks was much larger than for the auditory–vocal task effects on driving performance.*

The point of this discussion is that regardless of whether a surrogate method or a "simulation" is used, it is vital that the research question be carefully defined so that a method can be properly selected and properly applied in the research. Otherwise, mistaking surrogates for simulations can lead to misinterpretations of findings (in this illustration, measures of

cognitive load can be "misunderstood" as measures of event detection when, in fact, different underlying processes have been tapped). The analysis by Engström (2008) has been particularly lucid in bringing clarity to methods used for assessing cognitive load, and distinguishing them from other techniques (often called by the same name, but used for different purpose).

10.2.2 "Driving Simulation" Can Be a Surrogate Method

Driving simulation itself can be considered a surrogate method, one that attempts "to emulate" the conditions of driving to a sufficient extent that it evokes processes in the driver which are used to perform the driving task under "real" conditions. Driving simulations vary in the extent to which they provide a representation of the driving task, from part-task simulations to full-task simulations.

Throughout this continuum, however, there are two issues to bear in mind. First, there is the issue of the extent to which the driving simulator captures elements of driving. It is possible to form the impression that a driving simulator captures more elements of driving situations than it does. Yet simulators often represent a fairly small set of situations or scenarios, each constructed for a special purpose. A scenario is, in some respects, like a surrogate method. It is specially developed for a purpose and intended to elicit specific processes of interest in the driver so that their effects on driving can be observed in that setting. It is necessary that findings be understood and interpreted within this context. Second, there is the issue that driving simulators

---

* It should be pointed out that the metrics used for all quadrants of Figure 10.2 were selected as a means of fairly comparing tasks which differed greatly in length. It is for this reason that gaze is computed as a "percent of task time spent looking at the road," (or "at mirrors"), rather than as a raw measure of time spent looking (which would depend on a task's length, and would unfairly penalize short tasks). For measures of event detection (CHMSL and follow vehicle turn signal) there was one event opportunity that occurred per task (regardless of its length), so opportunities to detect were comparable across tasks of all lengths. Also, it may be important to know that drivers were instructed to drive on the freeway (at about 55 mph) with top priority on safety at all times, which was their responsibility (for themselves, their passengers, and the vehicle). They were told to perform their secondary task only when it was safe to do so, or continue doing so. Although task performance was requested at specific times during driving, the conditions of these task requests met pre-specified criteria to keep the conditions of task performance similar across tasks. Further, the criteria for these conditions were identified from analyses of conditions under which drivers have used devices during driving that gave rise to distraction-related crashes. See Angell et al. (2006) for more complete details on the method.

engage more than a single perceptual, cognitive, or response process (since they emulate driving and driving itself consists of multiple, coordinated tasks). This can make interpretation of the data more complex, and high-quality experimental designs become all the more important. If the simulator's engagement of multiple processes in the driver is not recognized and anticipated, interpretation of the data can be done inappropriately. This contrasts with the simpler surrogate methods, where data on a single metric may be much more straightforward to interpret (e.g., "task completion time").

Just as other surrogate metrics are tested for their validity against on-road driving, data obtained from driving simulations should also be compared with on-road driving data to gauge the extent to which the simulations succeed in eliciting driving behaviors that they were intended to elicit (and are similar to those observed on the road under similar conditions). Typically, this needs to be done on a scenario-by-scenario basis. Generally speaking, surrogate methods should not be used to "validate" each other—with the term "validation" perhaps best reserved for comparisons between a simulation (or a surrogate) and the "ground-truth" it is striving to "predict" or "emulate"; namely, the on-road experience and behaviors. While on-road research methods themselves are not perfect, it is on-road behavior that is the ground-truth to which surrogates ideally would be anchored (including simulator scenarios) and which can serve to help establish "criterion validity" for measures taken under laboratory or simulator conditions. It may, however, still be of interest to examine not only the relationships between surrogate measures and on-road data but also to compare surrogates and simulations to each other. Patterns of correlation among surrogates and simulations can offer converging evidence regarding what is being measured by each (as a means of establishing "construct validity"), and can lead to new research hypotheses and insights.

Unlike simpler surrogate methods, driving simulation is characterized by a higher degree of complexity and may be associated with a higher degree of resource demands (e.g., monetary expense, staffing, knowledge, time) to successfully use than most simple surrogate methods. As a result, during the design process (such as that illustrated in Figure 10.1), simpler tools may be applied early in design (during the pre-prototype and bench interactive prototype phases, when iterative changes are many and frequent and the design may not yet be fully embodied). After a system has undergone some level of refinement, and some of the straightforward issues in device design have been resolved, it moves into the mockup and pre-production prototype phase, where it can be represented in testable software (and hardware). At that point, it is often possible for it to be integrated with a driving simulator compartment (or a part-task buck) and, at that point, it then becomes possible to perform evaluations in a driving simulation. By this stage of development, many of the larger issues in a device or its interface may have been addressed, leaving a smaller set of issues to be examined in a driving simulation. Often these remaining issues are more complex and require more sophisticated treatment than can be given by the simpler surrogates administered earlier in development. Some of these issues may relate to safety, and addressing them in the safe environment of the simulator (prior to any final on-road product validation) can be especially beneficial. In fact, because driving simulation provides a safe environment in which to do research on questions which may potentially pose risks if done on the road, it can be a powerful tool for exploring these questions and providing an opportunity to resolve them in a device—as a precursor to any final product validation testing that may be done on the road.

Further, by using tools (from the simple surrogates to the more complex part-task and full-scale simulators, and even on-road instrumented vehicles) within a systematic development process, tools can be applied according to their strength (simple issues addressed early by simple surrogate methods; more complex issues addressed later by more sophisticated tools when the design has evolved at least into a functioning software prototype, if not hardware prototype). A systematic process allows those issues needing focused attention to emerge through rounds of iterative testing, and the process as a whole can still be cost-efficient (since the most cost-intensive tools are not applied to issues that do not require their power, but only to those issues for which they are needed).

It should also be noted that there are many applications for simulation besides the evaluation of different in-vehicle information systems for distraction. For example, during the innovation stages of product development (which occur in advance of product engineering) there may be a large set of new and novel user interfaces or systems that have been generated. It may be desirable in such a case to obtain a holistic assessment of driver-in-the-loop responses to them (obtaining not just driving performance metrics in salient scenarios, but also emotional, esthetic, and experiential responses to the integration of a new system). Such a holistic assessment might be done to advantage in a simulator that provides a virtual driving experience (and can often accept the integration of software prototypes of new systems, since these novel systems would usually not be available in hardware, nor integrated with drivable vehicles at this stage). Data from a driving simulation study of this sort could assist in narrowing a field of many ideas down to a few worth developing more completely, or as a means of determining which areas of research on the systems most deserve allocation of limited resources. Apart from product development and innovation, there are many opportunities for simulations in the research domain to push forward the state of knowledge about driving and drivers, and the importance of simulation to this area of application should not be underestimated. However, much of this *Handbook* elucidates these applications, and as this chapter's focus is on surrogate measurement and its relationship to simulation, it is only mentioned here, but it is mentioned with emphasis.

Sometimes surrogate techniques are themselves implemented as part-task driving simulations. An example of this is the Lane Change Test (LCT), developed by Mattes and colleagues as part of the ADAM project (see Mattes, 2003; Mattes & Hallen, 2008).

This surrogate utilizes a simulated driving task in which a scenario of approximately three minutes in length is driven at fixed speed, and signs appear at a controlled distance from the driver to indicate that a lane change should be made, whether it is a single or double lane change maneuver, and the direction of the maneuver (right or left). The simulation acquires data that allows the driver's steered path to be examined under baseline versus conditions of task load. The road signs—which indicate the lane change maneuver to be done—were intended to incorporate an element of event detection into the experience (see Benedict, Angell, & Diptiman, 2006, for exploratory findings on correlations between the LCT surrogate metric MDEV, and event detection measures from the CAMP Driver Workload Metrics Project), which, though very preliminary, were consistent with the notion that this surrogate does capture some of the variance associated with event response during driving and multitasking. While much research on this surrogate method is still underway, it shows promise as a tool for examining effects of secondary tasks on driver performance. Interested readers are referred to the extensive work of Mattes, and many other researchers, contributing to the development of an ISO Draft International Standard (DIS) for this method (ISO, 2008), balloted and approved in 2009 (to be published in 2010).

10.2.3 Differences Between the Terms "Surrogate Method" and "Surrogate Metric"

The terms "method" and "metric" are different, and it is useful to distinguish between them. Surrogate "methods" refer to the procedures, equipment, and protocols through which data are acquired. Surrogate "metrics" refer to specific measures of performance or behavior that are generated by a method. In addition, a metric may be associated with decision criteria, or "redlines" that represent values on that metric above or below which performance is deemed "unacceptable." Such decision criteria are applied in order to discover when a device intrudes upon driving to such an extent that design changes are needed. An example of this would be the "15 second rule" of SAE J2364 (Society of Automotive Engineers, 2004). The "15-second rule" is a decision criterion based on the metric of "task completion time" as measured in the laboratory (not on the road). It suggests that visual–manual navigation-related tasks longer than 15-seconds are "too long" and should be changed in design to reduce unwanted intrusion on driving performance. However, it is useful and important to separate decision criteria, or redlines, from the metrics themselves. A metric can be useful, but there can be a lack of data, or a lack of consensus, within a user community regarding where an appropriate redline should be drawn. Different issues surround the definition and use of redlines than surround the use of metrics and for that reason alone it is important to distinguish between them.

Often a surrogate method will generate several surrogate metrics. However, it is also possible for a metric to be a composite of several measures. In such a case, the metric would typically allow categorization, ranking, or even scaling of the measured attribute, dimension, or construct. For example, a metric may allow only sorting of tasks into nominal categories of "higher" or "lower" workload demand. Or it may allow ordinal ranking of tasks in terms of the magnitude of their visual demands on drivers. Or it may allow scaling and quantification of the magnitude of visual demand placed on drivers. As an example, the occlusion surrogate method generates several surrogate metrics: Total Shutter Open Time to complete a task (TSOT), and number of shutter openings to complete a task. TSOT is the most frequently used metric, because it correlates significantly for visual–manual tasks with median task duration, median standard deviation of lane position, percent of trials with a lane line crossing, median speed difference within a task's duration (the difference between maximum and minimum speed within the task-performance period), the number of glances to complete a task, and total glance time to complete a task. (TSOT does not correlate significantly with average glance length, nor with any measure of event detection on-road).

10.2.4 A Surrogate Can Be Embedded Within a Simulation

Within the context of a well-developed simulation, it is very possible to embed techniques or measures which, when used by themselves, would be considered "surrogates." It is useful, within this context, to use the distinction between "surrogate methods" and "surrogate metrics".

Often a surrogate metric can be obtained from a simulation, without any perceptible change to the driver's experience. For example, a Task Completion Time metric (for secondary tasks) can often be acquired through instrumentation or through the use of video recorded from the user's interactions with an in-vehicle device, in a way that is completely transparent to study participants. And, as a metric, it is a reasonably good surrogate (when applied to visual–manual tasks) for lane-keeping, speed-keeping, and glance behavior measures during the task (predicting over 50% of the variance associated with direct measures taken on the road of lane, speed, and glance during visual–manual tasks). Such metrics can play a useful role in analyzing data. For example, a metric of this type can be used as a converging measure to confirm or add confidence to a finding. In addition, an easy-to-reduce measure of this type can be useful in pilot phases of a study, when it is desirable to shape a scenario (but undesirable to spend large amounts of resources analyzing more time-consuming measures). Early iterative testing and refinement of simulator scenarios might be done using carefully selected surrogate metrics which are embedded in the data acquisition system.

When surrogate methods are embedded, or integrated with a simulation, they are likely to be directly experienced by the driver as part-and-parcel of the virtual experience given by the simulation or as naturally-occurring within the simulated scenario. For example, signal detection tasks have been combined with simulator scenarios, though often they have been used to obtain measures of a driver's responsiveness to visual events during driving, rather than as a pure measure of cognitive loading. As already noted above, if used as a pure measure of cognitive loading, the driver wears a head-mounted device that presents the peripheral stimuli while driving in the simulator. If the stimuli are instead intended to represent visual events which occur at unpredictable times/locations on the roadway (to emulate, for example, brake lights illuminating suddenly on a lead vehicle), then the lights would actually be embedded in the simulation to emulate or simulate the events to be detected. Note that the particular way in which a surrogate method is embedded in a simulation task depends fundamentally upon the conceptualization of the surrogate task, its implementation as a non-embedded surrogate, and its interpretation. In a simulation (but not always a surrogate task), the event to be detected can be made into an object (or set of objects) which occur naturally in driving scenes and which are controlled for salience and visual contrast with the background in order to create a more natural driving experience, and to reduce the extent to which the event detection task is experienced as an artificial task done in addition to driving and using secondary devices. In fact, software to create simulator scenarios is sufficiently flexible today that it often offers opportunities for *innovation* in creating new variants of surrogate techniques.

For example, one might imagine embedding the Modified Sternberg surrogate task (reported by the CAMP Driver Workload Metrics Project) within a simulation. In the CAMP version of this task, road signs are shown (either highway markers, bearing numerals of highways, or junction signs, showing the configuration of an intersection). A driver is shown a memory set to remember (e.g., those highways or intersections which define the route they will drive). Then, during the course of a secondary task that they perform (for example, searching for a point-of-interest on a navigation display), visual probes (signs bearing a highway numeral or junction) appear in front of them at random intervals and they must indicate whether it was one of the to-be-remembered highways/junctions. While this surrogate Sternberg task was studied as a stand-alone task, it might be a surrogate that could be embedded within a simulation where the signs appeared along the roadside in locations where signs might normally appear. Of course, the validity of this methodological improvement would need to be verified, but the point is that driving simulation offers many opportunities for innovation when it comes to embedding surrogate methods within a scenario.

10.2.5 Key Issues to Consider in Applying Surrogate Methods as Part of Simulated Driving

When a surrogate method is employed within a driving simulation for specific research purposes, there are key findings that have emerged from recent research that may bear upon their use. Some of these (many drawn directly from the CAMP

Driver Workload Metrics Project, Angell et al., 2006) include the following.

10.2.5.1 Workload-Induced Distraction Is Multidimensional

States of driver workload which produced overload or interference with driving performance were manifest not on just one underlying dimension of performance, but on several simultaneously affected ones, confirming that workload-induced distraction is multidimensional in nature (see Figure 10.3, for an illustration). If a surrogate technique is used to isolate or identify effects of a specific type, it may nonetheless be important to still acquire measurements of other dimensions of driving performance so that effects can be seen in full context, and properly interpreted. The point here is that it is important to collect several different independent measures to obtain a complete picture of a task's effect on driving performance. For example, if one looked just at Standard Deviation of Lane Position (SDLP) in Figure 10.3, only a very small difference between task types could be identified (visual–manual versus auditory–vocal tasks). Yet by looking at variables related to the duration of glances on the forward roadway, number of glances to mirrors, miss rates for centrally-placed CHMSL events and peripherally-placed FVTS events, large differences between the task types emerge and a more complete picture emerges of both task types. The picture that materializes is one in which the auditory–vocal tasks, as a whole (most of which lasted two minutes in length and included tasks of great difficulty), allowed drivers to keep their eyes and attention on the road more, as evidenced by the fact that the star charts in Figure 10.3 show that the duration of glances to the road were longer, glances to the mirrors more numerous, and miss rates for events lower (that is, *fewer* events were missed at both central and peripheral locations during these tasks). In contrast, for visual–manual tasks, as a whole, eyes and attention were off the road more frequently, as evidenced in the star charts by the fact that the duration of glances at the road were shorter, glances at the mirrors were fewer, number of glances at the secondary task were more numerous, total number of glances was higher (indicating more up-and-down looking between the secondary task and road), and miss rates for visual events was higher (that is, *more* events were missed at both central and peripheral locations during these tasks).

10.2.5.2 Multiple Task Types Should Be Included to Benchmark Effects

Specific patterns of effects have been found that are related to task type (e.g., visual–manual versus auditory–vocal), as just discussed in the prior paragraph. (Again, these effects are depicted in Figure 10.2 and Figure 10.3). Visual–manual tasks led to more pronounced intrusion on driving than did auditory–vocal tasks, as evidenced by the values shown in the star charts of Figure 10.3 on the radials for duration of glances to the road (shorter), number of glances to mirrors (fewer), number of glances to the (secondary) task (increased), percentage of miss rates for CHMSL

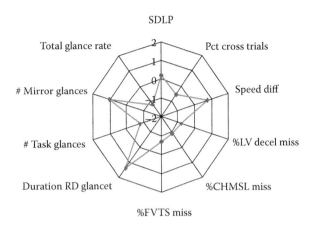

FIGURE 10.3 Effects of multitasking on driving are multidimensional and differ for each task type. (Reprinted from Angell, L., Auflick, J., Austria, P. A., Kocchar, D., Tijerina, L.,Biever, W., ... Kiger, S. (2006). *Driver workload metrics task 2 final report & appendices*. (Report: DOT HS 810 635). Washington, DC: National Highway Traffic Safety Administration, US Department of Transportation.) (Note: Points displayed on radials represent standardized scores where the mean is the average of all tasks equals zero.)

and FVTS events (elevated). The intrusion on driving from multitasking secondary tasks was discriminable from "just driving" for both types of tasks, but was much more pronounced for visual–manual tasks and much more subtle for auditory–vocal tasks. In any experimental design examining multitasking, it may therefore be very important to include as a benchmark at least one task of each major type so that any effects of the task-of-interest can be properly interpreted relative to some key anchor points along each dimension of driving performance. (It is not unusual in the literature to see claims about effect sizes being made on the basis of studying a single task compared to "just driving." Yet, if that task's effects could be placed in the context of other commonly performed tasks done while driving, the interpretation would sometimes be very different.) For example, suppose a new and novel in-vehicle task is introduced to the vehicle. Suppose further that it is studied in an on-road

experiment using an event detection technique to assess attention to the roadway situation and is found to differ reliably from the task of "just driving" by five percentage points on a measure of "percent CHMSLs not detected." How serious an issue would such a finding be for this novel task? Should the task be redesigned, or its use curtailed when the vehicle is in motion? The answers to these questions are not as straightforward as is often assumed and relate to the reasons that redline decision criteria are so difficult to develop.

One part of the answer lies in developing and using an understanding of the *range* of differences that are characteristic of in-vehicle device use during driving. This entails placing a newly-found difference within the context of a range of at least some other tasks or task types done while driving. Certainly a difference of five percentage points in event detection performance (relative to "just driving") can be understood in another way if it is placed within a context like that shown in Figure 10.4, where three *types* of tasks are shown and 15 individual tasks (in addition to "just drive") – than if it is interpreted only in the context of the "just drive" point and itself.

In addition, another part of the answer lies in understanding how differences in performance relate to crash risk. It is not the case that all measurable differences in performance from "just driving" are of sufficient magnitude or are of a type that elevate crash risk. Further, there are many intervening variables that modulate the effects of performance changes on crash risk, such as the frequency of task engagement, the conditions under which tasks of a given type are initiated, and the probability with which external events may co-occur during task performance. This issue, however, is one that can only be addressed in naturalistic studies of driving and/or in studies of crashes, near-crashes, and incidents. The main point of this section, though, is that all of these issues emphasize the importance of including at least one exemplar of each major task type in a study whenever possible, in

addition to the task/s of particular interest, as a means of providing reasonable context for interpretation of performance effects.

10.2.5.3 More Than One Task of Each Type Should Be Included When Possible

Different patterns of effects have emerged not just for *types* of tasks, but also for *individual tasks* within type, which were unique to the demands that each imposed on drivers. This suggests that specific structural interference between concurrently performed tasks continues to be an important theoretical construct in understanding what gives rise to performance degradation (Wickens, 1980; Wickens & Hollands, 2000). Yet, within a type of task, the level of resource demanded by a particular task, and the distribution of those demands over the timeline of the task's performance, tend to be unique to each task, producing a unique "thumbprint" for it within type (that is still generally consistent with the pattern that is prototypical for type of task, but is not identical for every task of that type). This finding, too, emphasizes the importance of including more than one task per type to be studied in an experimental design. The point here is that tasks should be sampled for the same reasons that subjects are sampled—there is individual variation between them, even within task type (cf., Hammond, Hamm, & Grassia, 1986).

10.2.5.4 Measures of Eye Glance and Event Detection Are Key

Interference with driving manifests itself differently (i.e., on different measures of performance) with tasks of different types (auditory–vocal versus visual–manual). However, for both types of tasks, eye glance measures and event detection measures are both key in evaluating the extent of intrusion on driving performance. This is useful to consider when embedding a surrogate within a simulation if eye glance and/or event detection can also be obtained.

10.2.5.5 Event Detection Methods Can Affect Eye Glance Patterns

Event detection methods that are used experimentally affect eye glance patterns in important ways, a finding that has both theoretical and methodological significance. Specifically, when events are presented within a paradigm, they alter where the driver looks and for how long. Angell et al. (2006) reported changes in visual scanning that depended upon the type of event detected. When a central event was detected, drivers engaged in increased scanning to the periphery, and this occurred even though no more events were to be presented during the task. When a peripheral event was detected, the duration of glances to the forward roadway lengthened (glances to task were shortened). These types of findings mean that an assessment which is used to quantify how much visual demand is associated with secondary tasks (using a metric such as "eyes off road time," for example) will yield *different answers* if its paradigm includes events to be detected during device use than if it does not. Thus, from a methodological point of view, it is important to take steps to clearly document what approach is taken and to assure clean

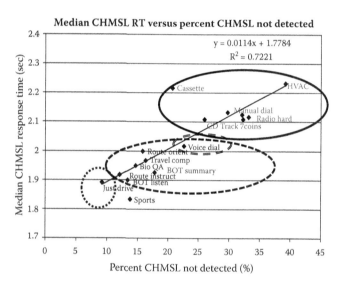

FIGURE 10.4 Depiction of the *range* of tasks along the regression-line relating percent of CHMSLs not detected by drivers during task performance under on-road driving conditions to median response times to CHMSL events that were detected. Overlapping task names illustrate that effects of some tasks are very similar to one another, especially *within* task type, and this underscores the importance of placing any single task within the larger context of a range before drawing conclusions. In considering this figure, it may be most useful to examine how task types (identified by larger circled areas) fall across the range. Dotted line circle in lower left identifies "just drive" task, dashed line oval identifies cluster of "auditory–vocal tasks", solid line oval in upper right identifies region of "visual–manual" tasks, with small double-dashed line circle in middle of range identifying the one mixed-mode task tested on-road, a voice-dial task (which had both visual–manual and auditory–vocal components). (Reprinted from Angell, L., Auflick, J., Austria, P. A., Kocchar, D., Tijerina, L., Biever, W., . . . Kiger, S. (2006). *Driver workload metrics task 2 final report & appendices*. (Report: DOT HS 810 635). Washington, DC: National Highway Traffic Safety Administration, US Department of Transportation).

measurement (of glance behavior separate from event detection, as well as glance behavior in the context of event detection). From a theoretical point of view, a much deeper understanding, if not model, is needed of how eye movements are controlled (through both top-down expectation-driven inputs about where it is important to look next, and bottom-up exogenous inputs), so that data can be properly interpreted, and methodologies can evolve.

The implications of these findings, taken together, include the facts that: (1) methods used to study event detection may affect the behavior of interest; and (2) it is important when evaluating the visual demand of tasks in an advanced information system or in-vehicle device that multiple test trials be conducted—some *with* and some *without* event detection. Any trials which are used to quantify the visual demand of a task (in order to make a decision regarding whether a redline on a metric is met by a task, such as "total glance time") should not include events to be detected in order to obtain clean measurements of glance behavior, free from the influence of co-occurring events. Today's industry redlines for glance behavior have been established through methods that have not included events to be detected. Therefore, if a study makes its measurements with a method which does include events, then its quantification of glance metrics will be altered as a result, and the results would be judged improperly relative to redlines that were developed for conditions free from event detection.

10.2.5.6 Sources of Information on Surrogate Methods Are Available to Guide Method Selection

A set of surrogate measures has been identified for major task types which can be used at each phase of common product development processes (e.g., the CAMP Driver Workload Metrics Project, Angell et al., 2006). These surrogate measures were found to be repeatable, meaningful (i.e., were valid for predicting driving performance effects), and enabled discrimination of high- from low-workload tasks (within the framework of the study). A toolkit of methods was described that would equip organizations to apply these measures in the evaluation of driver workload during development of advanced in-vehicle systems. Two summary tables of surrogate metrics are provided, for interested readers, in the CAMP Driver Workload Metrics final report, one for surrogates that are relevant for use with visual–manual tasks, and one for surrogates that apply to auditory–vocal tasks (Angell et al., 2006, Chapter 8, Table 8-1 and 8-2). It should also be noted that several other large-scale research efforts also provide excellent sources of information on surrogate methods and their use—such as the ADAM project (Breur, 2003), HASTE (see, for example, Carsten et al., 2005; Victor, 2004; Harbluk, Noy, Trbovich, & Eizenman, 2007), AIDE (e.g., Mattes, Föhl, & Schindlhelm, 2007), and SafeTE (Engström & Märdh, 2007).

10.2.5.7 Be Aware: Eye Glance Patterns in Lab/ Simulator Can Differ From Those On-Road

Eye glance patterns in the lab and simulator differ somewhat from those on the road. Generally, Total Glance Time (or Total

Eyes Off Road Time) to a secondary task will remain fairly stable across venue (lab, simulator, and road); however, in the lab or simulator (where there is a perception of safety from actual crash), drivers tend to make longer glances (but slightly fewer of them) than they do on the road. Again, Total Glance Time (to task) remains about the same but the bulk of glances to devices during on-road driving tend to be less than about 1.5 seconds (as first noted by Walter Wierwille and colleagues, e.g., Dingus, Antin, Hulse, & Wierwille, 1989; Wierwille, 1993). Thus, during actual driving, drivers use shorter, more frequent glances to complete a visually-intense task. In contrast, in the lab, they tend to use longer less frequent glances (though this depends on the specific demands of a simulator setup and the scenarios used). Glances in a higher-fidelity, fuller task simulators may begin to look more like those used during driving. For example, the distribution of glance durations in the field (Wikman, Nieminen, & Summala, 1998) and on a higher-fidelity driving simulator (Chan, Pradhan, Pollatsek, Knodler, & Fisher, 2009) on related in-vehicle tasks were very similar to one another. This is a phenomenon to be attentive to in interpreting and applying data from surrogates and simulators.

10.3 Concluding Comments

Surrogate methods and simulation methods can be powerful research techniques, with the key to their power being the clarity of the underlying conceptualizations which are used to guide the research of which they are a part. An understanding of driver perceptual, cognitive, and motoric processes, together with key research questions, will benefit the choice and implementation of method (surrogate or simulation) and metric.

Key Points

- There are different types of surrogate methods, and careful distinctions between types must be made when designing experiments.
  - Some seek to emulate key cues of the driving environment, and are used to create a "driving-like" experience for a driver. (These would usually be considered simulations.)
  - Others seek instead to evoke a set of processes in the human operator that are similar to those elicited during the actual driving behavior/s of interest—as a means of making these processes observable and measurable. (These are surrogate methods which do not qualify as simulations.)
- Surrogate methods may be embedded within simulations, and used to advantage.
- Simulations may themselves be used as surrogate methods.
- The terms "method" and "metric" are different, and it is useful to distinguish between them. Surrogate "methods" refer to the procedures, equipment, and protocols through which data are acquired. Surrogate "metrics" refer to specific measures of performance or behavior that are

generated by a method (e.g., time to complete a task, total glance time to perform a task). In addition, a "metric" may be associated with decision criteria, or "redlines," that represent values on that metric above or below which performance is deemed "unacceptable." Such "redlines" are often used in setting requirements for product design.

- Having a clear conceptualization with which to guide research using simulations or surrogates is the key to harnessing the power of these techniques. A thorough understanding of driver perceptual, cognitive, and motoric processes—together with key research questions—will benefit the choice and implementation of method (surrogate or simulation) and metric.

Keywords: Metrics, Product Development, Product Testing, Redline, Surrogate Methods

Acknowledgments

The material in this chapter draws heavily on the work done by the technical team and consortium of the Driver Workload Metrics Project, which was completed under a cooperative agreement between the US DOT and the Crash Avoidance Metrics Partnership in 2006. The efforts of the entire team and management group are recognized with gratitude, most especially Wayne Biever, Steven Kiger, Louis Tijerina, Al Austria, Jack Auflick, Tuhin Diptiman and Jim Hogsett.

Key Readings

Angell, L., Auflick, J., Austria, P. A., Kocchar, D., Tijerina, L., Biever, W., . . . Kiger, S. (2006). *Driver workload metrics task 2 final report & appendices* (Report: DOT HS 810 635). Washington, DC: National Highway Traffic Safety Administration, US Department of Transportation.

Engström, J. (2008, October). *The peripheral detection task & related methods: Theoretical & methodological issues.* Paper presented at the Driver Metrics Workshop. San Antonio, TX.

Victor, T. W., Engström, J., & Harbluk, J. L. (2008). Distraction assessment methods based on visual behavior & event detection. In M. A. Regan, J. D. Lee, & K. Young (Eds.), *Driver distraction: Theory, effects, and mitigation* (pp. 135–168). Boca Raton: CRC Press.

References

Alliance of Automotive Manufacturers. (2003). *Statement of principles, criteria, and verification procedures on driver interactions with advanced in-vehicle information and communications systems* (Version 2.1). Southfield, MI: Author.

Allport, A. (1980). Attention and performance. In G. Claxton (Ed.), *Cognitive psychology: New directions.* London: Routledge.

Angell, L. S. (2007). *Effects of secondary task demands on drivers' responses to events during driving: Surrogate methods & issues.* Proceedings of the Fourth International Driving Symposium on Human Factors in Driver Assessment, Training and Vehicle Design: Driving Assessment 2007 Conference. Stephenson, WA.

Angell, L., Auflick, J., Austria, P. A., Kocchar, D., Tijerina, L., Biever, W., . . . Kiger, S. (2006). *Driver workload metrics task 2 final report & appendices* (Report: DOT HS 810 635). Washington, DC: National Highway Traffic Safety Administration, US Department of Transportation.

Angell, L. S., Young, R. A., Hankey, J. M., & Dingus, T. A. (2002). *An evaluation of alternative methods for assessing driver workload in the early development of in-vehicle information systems.* SAE Proceedings, 2002-01-1981.

Benedict, D., Angell, L. S., & Diptiman, T. (2006, October 2–3). *Exploration of the lane change test.* Paper presented at the Driver Metrics Workshop sponsored by the Alliance of Automotive Manufacturers, Society of Automotive Engineers, and International Organization of Standardization. Ottawa, Canada.

Breur, J. (2003, July 18). *Effects of secondary tasks on driving performance (The ADAM project).* Paper presented to the Alliance of Automotive Manufacturers.

Carsten, O., Merat, N., Janssen, W., Johansson, E., Fowkes, M., & Brookhuis, K. (2005). *HASTE final report* (European Commission, Project No. GRD1/2000/25361). Leeds, UK: University of Leeds. Available from http://www.its.leeds.ac.uk/projects/haste/deliverable.htm

Chan, E., Pradhan, A. K., Pollatsek, A., Knodler, M. A., & Fisher, D. L. (2009, in press). Evaluation on a driving simulator of the effect of distractions inside and outside the vehicle on drivers' eye behaviors. *Human Factors.*

Dingus, T. A., Antin, J. F., Hulse, M. C., & Wierwille, W. W. (1989). Attentional demand requirements of an automobile moving-map navigation system. *Transportation Research Record, 23A*(4), 301–315.

Engström, J. (2008, October). *The peripheral detection task & related methods: Theoretical & methodological issues.* Paper presented at the Driver Metrics Workshop, San Antonio, TX.

Engström, J., Åberg, N., Johansson, E., & Hammarblick, J. (2005). *Comparison between visual and tactile signal detection tasks appled to the safety assessment and in-vehicle information systems (IVIS).* Proceedings of the Third International Driving Symposium on Human Factors in Driver Assessment, Training, and Vehicle Design. Rockport, Maine.

Engström, J., & Märdh, S. (2007). *SafeTE final report* (Tech. Rep. 2007:36). Swedish National Road Administration.

Gelau, C., & Krems, J. F. (2004). The occlusion technique: A procedure to assess the HMI of in-vehicle information and communication systems. *Applied Ergonomics, 35*(3), 185–187.

Hammond, K. R., Hamm, R., & Grassia, J. (1986). Generalizing over conditions by combining the multitrait-multimethod matrix and representative design of experiments. *Psychological Bulletin, 100*, 257–269.

Harbluk, J. L., Noy, Y. I., Trbovich, P. L., & Eizenman, M. (2007). An on-road assessment of cognitive distraction: Impacts on drivers' visual behavior and braking performance. *Accident Analysis & Prevention, 39*, 372–379.

International Organization for Standardization. (2007). *Road vehicles: Ergonomic aspects of transport information and control systems – Occlusion method to assess visual demand due to the use of in-vehicle systems* (ISO 16673). Author.

International Organization for Standardization. (2008). *Road vehicles: Ergonomic aspects of transport information and control systems – Occlusion method to assess visual demand due to the use of in-vehicle systems*. Draft (TC22/SC13 WG8).

Kahneman, D. (1973). *Attention and effort*. Englewood Cliffs, NJ: Prentice Hall.

Kennedy, R. S., Turnage, J. J., & Lane, N. E. (1997). Development of surrogate methodologies for operational performance measurement: Empirical studies. *Human Performance, 10*(3), 251–282.

Martens, M. H., & Van Winsum, W. (2000). Measuring distraction: The peripheral detection task. *Driver Distraction Internet Forum*. Available at: http://www-nrd.nhtsa.dot.gov/departments/nrd-13/driver-distraction/Topics 013040229.htm

Mattes, S. (2003). The lane-change-task as a tool for driver distraction evaluation. In H. Strasser, K. Kluth, H. Rausch, & H. Bubb (Eds.), *Quality of work and products in enterprises of the future* (pp. 57–60). Stuttgart, Germany: Ergonomia Verlag.

Mattes, S. Föhl, U., & Schindhelm, R. (2007). *Empirical comparison of methods for off-line workload measurement* (AIDE Deliverable 2.2.7, EU Project 1ST-1-507674-IP).

Mattes, S., & Hallen, A. (2008). Surrogate distraction measurement techniques: The lane change test. In M. A. Regan, J. D. Lee, & K. Young (Eds.), *Driver distraction: Theory, effects, and mitigation* (pp. 107–121). Boca Raton: CRC Press.

Noy, Y. I., & Lemione, T. (2000). *Prospects for using occlusion paradigm to set ITS accessibility limits*. Presentation given at the 2000 Occlusion Workshop. Turin, Italy.

Society of Automotive Engineers. (2004). *Navigation and route guidance function accessibility while driving*. (SAE Recommended Practice J2364). Warrendale, PA: Author.

Senders, J. W. (1983). *Visual scanning processes*. Tilburg, Netherlands: University of Tilburg Press.

Van Winsum, W., Martens, M., & Herland, L. (1999). *The effect of speech versus tactile driver support messages on workload, driver behavior and user acceptance* (Report TNO TM-99-C043. Soesterberg, the Netherlands: Netherlands Organization for Applied Scientific Research (TNO).

Victor, T. (2004, May 13–14). *Driving support from visual behavior recognition (VISREC)—Evaluations of real-time attention support functionality*. Presentation given at the International Workshop on Progress and Future Directions of Adaptive Driver Assistance Research. National Highway Traffic Safety Administration, US Department of Transportation.

Victor, T. W., Engström, J., & Harbluk, J. L. (2008). Distraction assessment methods based on visual behavior and event detection. In M. A. Regan, J. D. Lee, & K. Young (Eds.), *Driver distraction: Theory, effects, and mitigation* (pp. 135–168). Boca Raton: CRC Press.

Wickens, C. J. (1980). The structure of attentional resources. In R. S. Nickerson (Ed.), *Attention and performance VIII* (pp. 239–257). Hillsdale, NJ: Lawrence Erlbaum.

Wickens, C. D., & Hollands, J. G. (2000). *Engineering psychology and human performance* (3rd ed.). Upper Saddle River, NJ: Prentice-Hall.

Wierwille, W. W. (1993). An initial model of visual sampling of in-car displays and controls. In A. G. Gale, I. D. Brown, C. M. Haslegrave, H. W. Kruysse, & S. P. Taylor (Eds.), *Vision in vehicles IV* (pp. 271–279). Amsterdam: Elsevier Science.

Wikman, A. S., Nieminen, T., & Summala, H. (1998). Driving experience and time-sharing during in-car tasks on roads of different width. *Ergonomics, 41*(3), 358–372.

Young, R. A., & Angell, L. S. (2003). *The dimensions of driver performance during secondary manual tasks*. Proceedings of The Second International Symposium on Human Factors in Driving Assessment, Training, and Vehicle Design, Driving Assessment 2003. Park City, UT.

11

Validating Vehicle Models

Chris W. Schwarz
The University of Iowa

Abstract

The Problem. The models used in driving simulators must be verified to be free from defect and validated against the system being modeled to give confidence in the simulation. Additionally, engineering judgment must be used to create or select models of the appropriate fidelity for the simulation application. *Role of Driving Simulators.* The driving simulator application restricts the vehicle models that can be employed. The models for the vehicle and all its subsystems must run faster than real time, yet faithfully represent the fundamental behaviors of the vehicle. *Key Results of Driving Simulator Studies.* A large body of literature is available on the verification and validation of vehicle models for simulators. Most of the general validation literature describes models that are stochastic in nature; however most vehicle models are deterministic. Modeling data ought to always be independent from validation data, otherwise nothing of value is learned through the validation process. Many standards, tests, principles, and procedures have been developed to take a car from the drawing board to the road. Many of the same techniques can be profitably utilized to take it from the road to the simulator. *Scenarios and Dependent Variables.* Several scenarios for validating vehicle models are discussed, along with important variables for measuring performance and validity. The breadth of validation scenarios that is required is a function of the types of study scenarios in which the model will be used.

11.1 Introduction

One of the top reasons for acquiring a driving simulation capability is to do training, evaluation, or research with a human driver in the simulation loop. Because people in modern society have such a high degree of familiarity with the driving task, there is a minimum level of fidelity and accuracy required so as to not destroy the usefulness of the simulation. Getting a simple parameter such as the steering ratio incorrect can result in a lateral handling response that is much too sensitive or not sensitive enough, and which an average driver would find difficult to keep stable on the road. A more subtle example is the steering wheel deadband (or play), which, if not included in the model, could artificially add to the workload of the primary driving task to the detriment of accurately assessing the workload of a secondary task.

The entire process of creating a model should include verification and validation in some form or fashion. The aim of this chapter is to expose the reader to the tip of the iceberg that is validation and to apply it to the specific domain of vehicle dynamics. There are a few key points to keep in mind. Much of the validation literature deals with stochastic models. Although vehicle models are often deterministic, closing the loop with a human driver adds the random element back in. Vehicle models are made up of several independent systems; therefore each must be validated for the whole to be judged valid. Finally, the field of vehicle dynamics is rich and many standardized tests have been developed to fully characterize a vehicle. The adoption of these tests makes the validation process more accessible.

11.2 Definitions

Several terms are encountered in the literature on the topic of validation. We provide definitions of some of these terms, including the closely-related concepts of verification and accreditation. A model is defined as a representation or abstraction of an entity, system, or idea; and a simulation as the act of experimenting with or exercising a model (Balci, 2003). Sargent (2000) summarizes the modeling process and differentiates between the problem entity, the conceptual model, and the computerized model. Figure 11.1 concisely describes the fundamental relationship between the three.

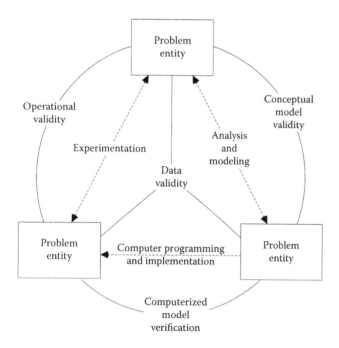

FIGURE 11.1 Simplified version of the modeling process. (From Sargent, R. G. (2000). Verification, validation and accreditation of simulation models. *Proceedings of 32nd conference on winter simulation* (Vol. 1, pp. 50–59). Orlando, FL. With permission.)

The procedure of determining the correctness and accuracy of a model or a simulation is the process of verification and validation. The verification process has been well documented (Balci, 2003; Kleijnen, 1995; Sargent, 2000). Sargent defines verification as "ensuring that the computer program of the computerized model and its implementation are correct" (Sargent, 2000, p. 50). It involves the elimination of programming bugs in the translation of the conceptual model to the computerized model.

Validation, on the other hand, has more to do with testing the conceptual model and the computerized model against the problem entity. "Model Validation deals with the assessment of behavioral or representational accuracy of the model and addresses the question of 'Are we creating the right model?'" (Balci, 2003, p. 150).

Depending on the type of organization that is using the model or simulation, some official judgment on the validity of the model may be required. This is called either certification or accreditation. It includes the concept of defining the specific purpose for which a model is created, and the appropriateness of the model for that purpose (Balci, 2003; Gass, 1993; Kleijnen, 1995).

Various combinations of the above terms have been combined to define comprehensive strategies to ensure that a model is appropriate, accurate, and robust, including Verification and Validation (V&V) (Balci, 2003), Verification, Validation and Accreditation (VV&A) (Balci, 1998), Verification, Validation and Testing (VV&T) (Balci, 1998), and Independent Verification and Validation (IV&V) (Sargent, 2000).

11.3 Assumptions

The validation methods presented in this chapter are largely independent of the model formulation as long as there is sufficient ability to probe the model for variables of interest. The fidelity of the vehicle model should meet certain minimum requirements for some of the vehicle testing procedures to be applicable. For example, tires do not actually generate force unless they slip against the terrain; and there will always be a slip ratio and slip angle between the tires and the ground. If a purely kinematic tire model is used, then slip is neglected and tests of lateral directional control, such as those published in SAE J266, cannot be correctly applied to the model.

A second assumption relates to the independence of vehicle subsystems. The longitudinal vehicle tests in particular can be divided along subsystems. Acceleration tests use the powertrain, braking tests use the brakes, and coastdown tests consider the tires and aerodynamics. The powertrain and brakes are independent systems with separate parameters. The vehicle model should reflect this reality by providing separate sets of equations and parameters for the different vehicle subsystems.

Finally, the model must meet certain commonsense requirements so that it does not violate the laws of physics and behave in unexpected ways. The tires provide an example here too. The friction circle is a way to express the fact that tire traction force is applied in some direction to resist its tendency to slip, up to the coefficient of friction. Thus the presence of longitudinal tire slip reduces its lateral traction capability as well. This interaction between lateral and longitudinal traction must be modeled if correct behavior is desired in extreme conditions. Likewise, the model must interact with environmental inputs, such as hills and wind gusts, in physically appropriate ways.

11.4 Standards

If one is comparing simulation results to vehicle dynamics test data that has been collected and published by a testing agency, it is critical to understand the collection and calculation methods used to arrive at the reported numbers. Fortunately these methods are readily available in published standards. If in doubt of which standard to consult, ask the testing agency that produced the data what test procedures they follow.

Later on, in Section 11.7.2.2, the directional control of a vehicle is discussed. Consulting the appropriate standard on directional control testing (SAE J266) will give the reader the theory behind the tests, typical vehicle instrumentation needed to measure the data, protocols and data sheets for running the tests and recording the data, and equations for analyzing the data. Equipped in this manner, the reader could then either commission tests at a test track facility, or conduct the tests for themselves. Moreover, one merely has to use the standard as a blueprint for setting up virtual tests on the model. The virtual tests are easily compared with the test track data for the purpose of validation.

11.4.1 Society of Automotive Engineers (SAE)

SAE standards are created by one of more than 700 technical committees in the society, and document broadly-accepted engineering practices and specifications for materials, products, processes, procedures and test methods. SAE standards are probably the best first resource for information on how to test, analyze, and interpret experimental vehicle data.

Basic vehicle dynamics terminology is defined in SAE J670e. This standard may serve as a good first resource to familiarize oneself with the field of vehicle dynamics from which many of the methods described in the chapter are borrowed. Directional control tests are outlined in SAE J266, which develops the concepts of understeer gradient and characteristic speed among others. More general test guidelines for vehicle and suspension parameters are given in SAE J1574.

11.4.2 International Organization for Standardization (ISO)

The ISO is a network of the national standards institutes from 157 countries, coordinated from Geneva, Switzerland. ISO standards facilitate the international exchange of goods and services and bridge the gap between the public and private sectors. Hundreds of ISO standards have been written specifically for road vehicle testing; though not all are useful for model validation. A few salient examples are listed in the Table 11.1

TABLE 11.1   ISO Road Vehicle Standards

ISO 13674-1:2003	Road Vehicles – Test method for the quantification of
ISO 13674-2:2006	on-centre handling – Parts 1 and 2
ISO 21994:2007	Passenger cars – Stopping distance at straight-line braking with ABS – Open loop test method
ISO 10521-1:2006	Road vehicles – Road load – Part 1: Determination under reference atmospheric conditions
ISO 3888-2:2002	Passenger cars – Test track for a severe lane-change maneuver – Part 2: Obstacle avoidance
ISO 15037-1:2006	Road vehicles – Vehicle dynamics test methods – Part 1: General conditions for passenger cars

11.4.3 Military Standards

The U.S. Department of Defense is the largest consumer of models in the world. The U.S. Army Training and Doctrine Command (TRADOC) Models and Simulation (M&S) and Data Management office has published regulation 5-11 to document policies, procedures, and responsibilities for the development and management of M&S tools, as well as data.

The Developmental Test Command (DTC) in the U.S. Army Test and Evaluation Command (ATEC) manages Test Operation Procedures (TOPs). There are several TOPs for vehicle testing; for example TOP 02-2-610 describes Gradeability and side-slope performance, and TOP 02-2-610 concerns Standard Obstacles testing.

11.5 Model Parameters and Test Data

A minimal set of parameters and data is required to create and validate a vehicle model. Fundamental vehicle parameters include the mass, inertia, wheelbase, track width, spring rates, and so forth. When these parameters have been used to create a computerized model of the vehicle, additional data is needed to validate the model. This data may be in the form of time histories of acceleration, orientation, velocity, and other variables; or it may be in the form of vehicle dynamics metrics such as understeer gradient, steering ratio, steering sensitivity and the like. The possession of raw data from vehicle testing makes available to the modeler many more options for validation analyses.

The boundary between modeling data and validation data is often blurry; but one must be sure to keep them separated. If the same data is used to validate a model as was used to create it, then nothing has been learned. It may be that the modeler has at their disposal a set of test track data from which model parameters will be identified. For example, if brake torque and vehicle deceleration time histories are available, then the mass of the car can be estimated. In this situation, some portion of the data should be set aside for the model validation task, so that independence is preserved. The reader will find that this can be quite challenging given limited data availability.

11.5.1 Data Validity

Data validity (Sargent, 2000) is a pervasive concern throughout the process of modeling, verification and validation. If bad data are used in any of these tasks, the validity of the model may be compromised. Data might have been collected from diverse sources or from several model years of a vehicle, in which case the question of data validity must also be considered. Data validity is closely tied to model fidelity. The more complex the model, the more detailed and extensive the data must be to implement the model; and the harder it is to acquire valid data. When working with a high-fidelity model, it is even more important to employ the full spectrum of validation methods described in this chapter to effectively converge towards a valid model.

11.5.2 Published Data

There are several sources of published vehicle dynamics data that can profitably be used for modeling and validation. So ubiquitous are these data that most of today's auto consumers are savvy enough to want to know detailed specifications about the vehicles they buy. Extensive information can often be obtained from manufacturer's brochures and web sites. Vehicle reviews in car magazines and web sites also give valuable information including vehicle test results. Finally, journal and conference articles about vehicle models may be available for the vehicle you are interested in validating.

11.5.3 Experimental Data

Experimental data collected from an instrumented vehicle doing test track maneuvers is not commonly available; and it is expensive to commission such a study. Such data is the ultimate resource for the simulation practitioner trying to validate a model.

The deliverables from a vehicle test facility typically include the raw data collected from vehicle instrumentation, as well as reduced data included in a final report. It can be difficult to make use of the raw data because several layers of processing are required. Sensor readings must be converted to engineering units. Sensor locations must be known to transform the measurements to points of interest in the model, such as center of gravity or head point. Vehicle orientation may have to be calculated from optical sensor data that measures distance to the ground; and so the sensor locations must be known precisely. Hopefully, the testing facility will also supply the processed data; nevertheless, working with experimental data represents the most expensive and time-consuming approach to model validation.

If there is not enough time and money in the budget to work with experimental data directly, one can restrict one's attention to the calculated vehicle dynamics performance measures. Many of these tests and performance measures are described in the following section.

11.6 Analysis Techniques

The process of determining a model's validity depends greatly on the type of validation data available and on the manner of testing. The computerized model of a vehicle for a driving simulator is assumed to be deterministic, consisting of physical equations of motion, look-up tables, and integration solvers. While the model may be quite complex, the outputs are predictable and repeatable. As noted previously, it is the human driver that adds randomness to the simulation.

Test drivers were likely used to generate experimental data for the validation; and human drivers may also be used in the simulator to generate model data. In that case, an experimental design may be considered, and test runs replicated, so that statistical tools can be brought to bear on the validation problem. However, if the simulation data can be created with computer-generated inputs, or playback files that exactly reproduce the test driver's inputs, then the source of randomness has been eliminated and the statistical analysis can be avoided.

We can offer a real-life example of using playback files to validate a model against test data. One way to test the steady-state directional response of a car is to drive in a constant radius circle and slowly increase the speed. Figure 11.2 shows the input (road wheel steer angles) as well as two outputs (roll and lateral acceleration) for both test track data as well as model output data. No model is perfect, so it is not expected that the model outputs match the test data exactly. The traces do overlay to overlay quite nicely, giving confidence that this real-world test is faithfully reproduced via simulation.

Several specific remarks can be made that pertain to issues raised in this chapter. Signals recorded from road or track data will never be smooth, although there are robotic steering mechanisms available that can be programmed to create smooth inputs. The difference between the left and right road wheel angles is one aspect of the model that can be validated from this figure. The lateral acceleration was measured using a three-axis accelerometer (called the cube) affixed to the floor of the car cab. Simulated lateral acceleration is shown at the center of gravity (CG), the head point, and at the estimated position of the cube. Such reference point considerations are common when dealing with instrumented signals. Finally, observe the linear region of the lateral acceleration response between about −0.1 g and −0.4 g. Then the lateral acceleration saturates a little below −0.4 g. This common directional response test is discussed in more detail in Section 11.7.2.2.

11.6.1 Validation Techniques

Several validation methods have been compiled by Sargent (2000), and are considered here in the context of vehicle dynamics validation.

11.6.1.1 Animation

Animation can be a valuable tool for validating vehicle models. Everyone knows basically how a vehicle should behave and many errors can be spotted right away by viewing an animation of a vehicle conducting a maneuver. Two examples of driving simulation animations are available for viewing on the *Handbook* web site (see Web Videos 11.1 and 11.2). The wheel loader example shows a high-fidelity model undergoing multiple types of inputs. The stability control example shows loss-of-control scenarios and the influence of electronic stability control (ESC).

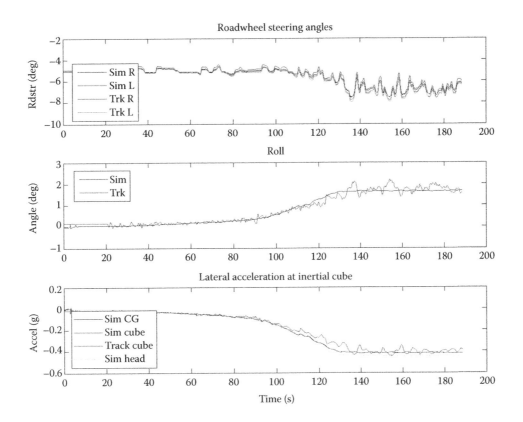

FIGURE 11.2 Constant radius increasing speed test.

On the other hand, the lack of obvious faults in an animation is not proof that none exist. It would be folly to declare a model valid purely on the basis of watching a vehicle animation. Experience with driving simulators has shown, for example, that lack of horizontal field-of-view limits the optical flow perceived by the driver, which distorts the perception of vehicle speed, among other things.

11.6.1.2 Operational Graphics

Some performance measures can be graphically displayed in an animation to add information or bring certain aspects of vehicle performance to the forefront. An example from CarSim™ is that arrows may be displayed at the tire contact patches that show the direction and magnitude of the tire forces. These arrows offer an effective way to observe weight shifts in the vehicle as it rounds a curve. Prudent use of operational graphics can decrease the time and cost of validation. The ESC animation linked to above also includes operational graphics to indicate the activation of the stability control system.

11.6.1.3 Face Validity

Face validity is an oft-maligned method of validation that simply involves asking people knowledgeable with the system whether the model seems valid. Nevertheless, it can be a useful test of validity, especially with vehicle models, with which most people have some level of expertise. The caution regarding face validity is the same as for animation. It may be a necessary condition, but it is by no means sufficient to show the validity of a model.

11.6.1.4 Comparison to Other Models

The abundance of vehicle dynamics models and third-party simulation packages means that there may be a validated model one can use as a basis for comparison. When one buys a third-party application like MSC CarSim™ or LMS Virtual Lab™, one is paying for peace of mind that the verification problem has been solved. There is also the added benefit of a potentially large user base and the opportunity to acquire validated models from other users.

11.6.1.5 Fixed Values

Some of the simplest validation tests involve using fixed values as inputs. In fact, they may border on being considered more as modeling tasks than validation but they are especially useful if the person doing the validation did not create the model. As an example, one may simulate the vehicle at rest (zero inputs), and measure the weights on each wheel to validate the proportioning of vehicle mass between the front and rear axles. Such basic vehicle characteristics must be correct for the more complicated dynamics measures to be valid.

11.6.1.6 Degenerate Tests

The degeneracy of a model is tested by exploring its failure modes with the proper selection of input values and parameters. The appropriate selection of degenerate tests requires a thorough understanding of the purpose of a model. An example from the steering system is the lock-to-lock steering range. What happens if one tries to provide a steering input outside this range? Does the software automatically limit the steering angle? Does the vehicle dynamics model allow the front wheels to rotate all the way around, or provide hard limiting from the steering rack and tie-rod assembly?

11.6.1.7 Extreme Condition Tests

Much of the discussion of degenerate tests applies equally well to extreme condition tests. If the model is to be used in extreme conditions, then it should be tested in a broad range of unlikely scenarios to ensure the model is robust and gives plausible results. An example is the directional stability test in SAE J266 that determines either the characteristic or critical speed of a vehicle. Such testing is important if the model will be used in situations that could result in loss of control.

11.6.1.8 Parameter Variability-Sensitivity Analysis

One of the major pitfalls of validation is optimizing a model's parameters against a limited quantity of validation data. The danger is that accuracy of the model can fall off in conditions that differ from the validation case. This can happen when the model's behavior is sensitive to a parameter that was not validated.

The potential for too narrow validation in vehicle models is mitigated by the wealth of tests that have been devised to fully characterize a vehicle's performance. The recommendation to the reader is to use as many of those tests as possible to validate the vehicle model, thus reducing the chances of over-optimizing to a narrow range of conditions.

11.6.1.9 Historical Data Validation

The availability of historical data can drastically change the way a vehicle model is validated. Historical data may include multiple drives of a test making statistical methods applicable. As was discussed earlier, historical data may be used for model parameter estimation as well as validation. In such cases, the modeler should be sure to separate the modeling data from the validation data, rather than using the same dataset for both tasks.

11.6.1.10 Traces

Tracing a signal through the model from input to output is especially useful when one is looking for the cause of model inconsistencies. Consider the lateral dynamics of the vehicle as controlled by the steering wheel. The model can be probed at the steering input, rack position, road wheel angles, camber and caster angles, slip angle, and lateral tire force to determine if there is a problem in the steering system, the tire formulation, or somewhere else in the model.

11.6.1.11 Internal Validity

Our definition of internal validity in the vehicle validation context differs from Sargent (Sargent, 2000), as his explanation focused on stochastic models and vehicle dynamics models are usually deterministic in nature. However, they can be quite complicated, being divided into different subsystems. Vehicle subsystem models can include the chassis, tires, steering, brakes, engine, transmission, drivetrain, and aerodynamics. We propose a deterministic definition of internal validity as the validation of individual vehicle subsystems or components. For example, a brake system modeler could use measured brake line pressure and wheel torque to validate the simulation of the brake calipers and pads.

11.6.2 Graphical Methods

Graphical methods offer subjective measures of validity that are widely applicable, even with data that does not lend itself to more rigorous statistical analysis (Sargent, 1996; Sargent, 2000). There are several reasons why data may not be amenable to full-fledged statistical analysis. The necessary assumptions of independence may not be met, there may be too few samples, or the model may be changing over time.

The graphical methods described here are tools that can be applied to most of the validation techniques described in the previous section. As long as there are multiple data sets (e.g., several runs of a specific maneuver), then graphical methods can potentially help with the analysis of historical data, traces, parameter variations, and other numerical data. One would not normally be examining many data sets in techniques such as animation and face validity.

11.6.2.1 Histograms and Box Plots

The key to using graphical methods for validation is to generate the reference distribution from the model, rather than from the theoretical t or F distributions favored in statistical analyses (Box, Hunter, & Hunter, 1978; Sargent, 1996). The data may be statistically dependent (i.e., correlated). The data from the model merely has to be identically distributed, as does the data measured from the system.

Two ways to generate reference distributions from model data are to use histograms and box plots. Sargent recommends generating at least 50 data points from the model to create the reference histogram or box plot. The number of system data points is typically not under the control of the modeler, and may be as few as one in a histogram. At least 10 system data points should be available for the box plot approach.

The test is then to see if the system data falls within the model reference distribution. For the box plot, look to see if the twenty-fifth percentile line of one sample exceeds the median line of the other (Walpole, Myers, & Myers, 1998). It should be noted that comparisons of this nature can show differences between the model and system, but cannot prove they are the same. See Figure 11.3 for examples of a histogram and box plot. In the histogram example, the system data does fall within the model reference distribution, indicating that the model seems to generate accurate results. Alternatively, one can see in the box plot example that the median line of one dataset falls outside the twenty-fifth percentile line of the other. Thus, the box plot comparison indicates a problem with the model.

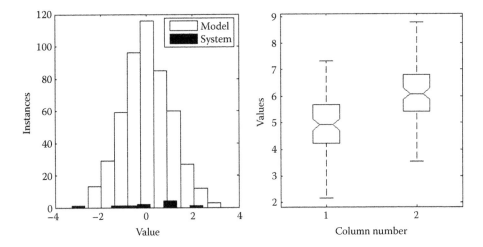

FIGURE 11.3 Histogram and box plot comparisons.

11.6.2.2 Behavior Graphs

Behavior graphs show the relationship between two variables in a model or system. A behavior graph is simply a scatterplot on which is shown both model and system data. The plot can then be visually analyzed to see if the datasets are clustered together or separated into different regions. They can be used with deterministic models as well as stochastic ones, even if the model and system data is nonstationary or correlated. While relatively few data points are required from a deterministic model and system, many points are desired in the case of a stochastic model and system (Sargent, 1996). An example behavior graph generated using random data points is shown in Figure 11.4. The obvious correlation between the two data sets indicates a positive validation result.

11.6.3 Hypothesis Tests

Hypothesis testing is a statistical method that can be used to accept or reject a model (Balci & Sargent, 1982; Kleijnen, 1995; Sargent, 2000). Various forms of the test are available with differing assumptions. Usually, independent samples from the same normal distribution are required. The standard deviation of the samples may be known or not, influencing which type of the test is used. If the statistical requirements are met for the system data one has available for model validation, then the test can be used to determine if a model's output behavior has a specified range of accuracy.

There are two types of risk in hypothesis testing. The first is called "model builder's risk" and is the risk of incorrectly rejecting a valid model (typically called a Type I error). The second type of risk is the" model user's risk", which is the risk of accepting an invalid model (typically called a Type II error). The second type of risk should be minimized as much as possible. The model user's risk goes down as the number of observations increases. One can estimate the required number of samples by calculating the power level of the test.

11.6.4 Time Domain

It is sometimes difficult to move beyond a simple eyeball comparison of two time series, but there are simple techniques that can be used to process this kind of data. One method to judge the

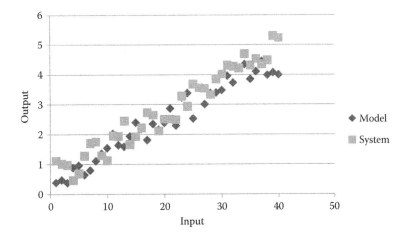

FIGURE 11.4 Sample behavior graph.

validity of a model signal's behavior relative to some system data is to use confidence bounds. Suppose that a time history is available of deceleration during a braking test. Obviously, there will be several causes of noise and variation in the data. One may define a confidence interval at each sample that spans from 95% of the sample value to 105% its value. A confidence region has therefore been defined for the entire time history; and if a simulation produces data that falls within the region, one may say the model's deceleration is accurate to within 5% of the system.

Time series data can be treated as random output variables and subjected to statistical comparisons. One would expect the two system and model signals to be highly correlated for a valid model. Graphically, this would present in a scatterplot as a narrow band of points along a line of unity slope. Numerically, one would also expect the correlation coefficient between the two signals to be positive and close to one (Lomax, 2007).

11.6.5 Frequency Domain

Those vehicle tests that elicit responses rich in frequency content are amenable to frequency domain analysis. An example of such a test would be a step input in the steering. Step inputs contain many frequencies, while slowly changing inputs contain far fewer frequencies. A spectral analysis can yield estimates of the model's natural frequencies; that is, those frequencies a model vibrates at best under no load conditions. Moreover, the bandwidth of the response is a simple measure of the model's fidelity. Bandwidth is defined as the difference between the highest and lowest frequencies present in a model.

The concept of correlation can be extended into the frequency domain by estimating the power spectrum (Proakis & Manolakis, 1988) of the time series, as well as their cross-spectral density. Then the coherence between the two time series measures the correlation between them at each frequency. The equation for coherence, $C_{xy}(\omega)$, is given by (Lomax, 2007) as:

$$C_{xy}(\omega) \omega \frac{\left|S_{xy}(\omega)\right|^2}{S_{xx}(\omega) S_{yy}(\omega)}$$

where the cross-spectral density of the two signals, $x(t)$ and $y(t)$, is written as $S_{xy}(\omega)$; and the power spectral densities of $x(t)$ and $y(t)$ are written as $S_{xx}(\omega)$ and $S_{yy}(\omega)$ respectively. The frequency, ω, is expressed in radians per second. For example, a system's natural frequency may be missing in the model, or shifted up or down on the frequency scale. In that case, one would expect to see the coherence value drop at the natural frequency.

11.7 Vehicle Dynamics Tests

The literature is replete with books and papers describing various validation methods and techniques and the reader is encouraged to explore these resources. The challenge is always to try and apply the methods to one's specific domain. The happy circumstance of the vehicle modeler is easy access to an equally rich body of work on vehicle dynamics. The wealth of techniques and methods in vehicle dynamics modeling and testing can and should be used to guide the validation process.

11.7.1 Longitudinal Tests

Longitudinal maneuvers provide relatively simple and stand-alone validation tests. Acceleration tests exercise the powertrain and drivetrain; braking tests the brakes; gradeability the tire forces and mass distribution; and coastdown primarily tests the rolling resistance and aerodynamics.

11.7.1.1 Acceleration

Basic validation numbers for many makes and models of vehicle can be found in car magazines or online. Car reviews often report the time it takes to accelerate from 0 to 60 mph; and vehicles in a common class will have similar acceleration times.

If more test data is available, then the acceleration curve's shape can be compared to that of the model. The peak jerk and peak acceleration can be computed and used as comparison points.

Armed only with velocity and acceleration data versus time, along with the model parameters of the powertrain and drivetrain, the shift points can be estimated. Computing vehicle jerk, the derivative of acceleration, aids in identifying the shift points and their harshness. Better estimates can be achieved with the additional knowledge of engine speed. If engine torque is also available, the efficiency of the drivetrain can also be estimated.

11.7.1.2 Braking

Braking test results, similar to acceleration times, can be found in many car reviews. They are usually reported as stopping distance from some speed with a maximum application of the brake pedal. The availability of a deceleration against time plot provides minimal additional information since brake torque is fairly constant through the maneuver. More useful data would be the individual brake torque or pressure, since that would illuminate the proportioning of brake pressure between the front and rear axles.

Experimental data that includes a measurement of tire slip can aid in validating the wheel lock-up behavior, as well as the performance of the ABS model, if it exists.

11.7.1.3 Coastdown

Coastdown tests are performed in neutral gear; thus, they remove the powertrain and braking systems from consideration, leaving only the sources of road load in the vehicle system. The major sources of loss that contribute to coastdown are aerodynamic drag and tire rolling resistance. The total road load force is expressed as (Gillespie, 1992):

$$R_{RL} = f_r W + 1/2 \rho V^2 C_D A + W \sin(\theta),$$

where f_r is the coefficient of rolling resistance, W the vehicle weight, ρ the air density, V the air velocity hitting the vehicle, C_D the aerodynamic drag coefficient, A the frontal area of the vehicle, and θ the road grade angle. The air velocity is obtained by adding the vehicle speed to the atmospheric wind speed. The rolling resistance coefficient can be reasonably estimated by a linear function of the velocity (Gillespie, 1992).

If the entire velocity profile from a coastdown test is available, then coefficients for the constant, linear, and squared velocity terms can be estimated. If only coastdown times with initial and final velocities are given, then reasonable coefficients can be guessed and tweaked to satisfy the test results. Performing this type of test with the vehicle in gear can provide information about the additional losses present in the engine and transmission.

11.7.1.4 Gradeability

Gradeability is typically measured only for military vehicles, although some commercial vehicles designed for off-road use may also include this test. Gradeability is sensitive to the tire friction properties, but also to a great extent on the distribution of mass on the axles. The more even the distribution of mass as a grade is mounted, the better able the tires are to share the load and keep the vehicle on the slope. Expressions for the axle loads are given in Gillespie (1992). The reader is cautioned that Gillespie's consideration of grades assumes small angles, while military vehicles are often capable of climbing 60% grades.

11.7.2 Lateral Tests

The maneuvers that are used to test the lateral behavior of the vehicle are more numerous and varied, for several reasons. Lateral dynamics are nonlinear with respect to vehicle speed, so lateral tests need to be replicated at different speeds. Moreover, the gradients of variables such as roll and steering angle with respect to lateral acceleration are well approximated by linear curves only up to lateral acceleration values of 0.3–0.4 g. Lateral tests can be divided into three distinct classifications: On-center handling tests, directional control tests, and frequency response tests.

11.7.2.1 On-Center Handling

On-center handling refers to the response of the car to small changes in steering input, typical of normal highway driving. On-center handling characteristics can be captured by the on-center weave and steering transition tests.

The on-center weave test is conducted by maintaining a constant speed and driving the steering wheel angle with a pure sine wave of about 2 Hz. There will be some degree of steering dead band in the system, and it can be measured by comparing the steering wheel angle with the road wheel angles. The lateral acceleration, roll angle, and yaw rate can also be measured for their amplitude and phase lag from the input (Salaani, Heydinger, & Grygier, 2004).

The steering transition test is started by driving straight and then slowly ramping up the steering wheel angle to about 50 degrees. It provides a very precise picture of the steering dead band sensitivity (Salaani et al., 2004).

11.7.2.2 Directional Control

Directional control tests measure the ability of the vehicle to respond to steering inputs as the lateral acceleration levels increase. A simple test is the slowly increasing steer in which the steering wheel angle is linearly increased. A rate of about 10 deg/sec (Garrott et al., 1997) may be used and the test is normally done in both directions. Eventually, the lateral acceleration level will saturate and after that point is reached, the test may be ended.

The measured dynamics variables can be used to measure several useful characteristics, such as steering ratio, steering sensitivity (Salaani et al., 2004), and roll gradient. Steering ratio—the ratio between the steering wheel angle and the average road wheel angles—can also be measured with the vehicle at rest. It should be noted that steering ratio at high velocities can differ drastically from the statically measured steering ratio (Gillespie, 1992).

11.7.2.2.1 Steady-State Directional Control

The slowly increasing steer test fits into the SAE J266 standard, which describes the various tests for steady-state directional control in passenger cars and light trucks. The standard presents several tests that can be used to measure directional control including: constant radius, constant steering wheel angle, constant speed/variable radius, and constant speed/variable steer. These four tests are equivalent and will give the same results provided they span the same variation of speed-steer-radius steady-state conditions. A response gain/speed test is also described for the measurement of yaw rate gain, lateral acceleration gain, and sideslip gain.

It is relatively difficult to follow a constant radius curve. The two easiest test procedures are to set the cruise control while slowly increasing the steering wheel angle or input a constant steering wheel angle while slowly increasing the throttle. If the former test is run using discrete levels of steering wheel angles, it may also satisfy the lateral acceleration requirement of the response gain test.

Figure 11.5 shows the yaw rate gain versus speed. Lines of constant understeer or oversteer are plotted on this figure. For oversteer vehicles the critical speed marks the velocity at which the oversteer curve goes to infinity. For understeer vehicles the characteristic speed marks the maximum of the understeer curve. Both critical and characteristic speed are defined using the same formula as a function of the understeer/oversteer gradient magnitude, K, in deg/g, as

$$V = \sqrt{\frac{57.3 \, gL}{|K|}}.$$

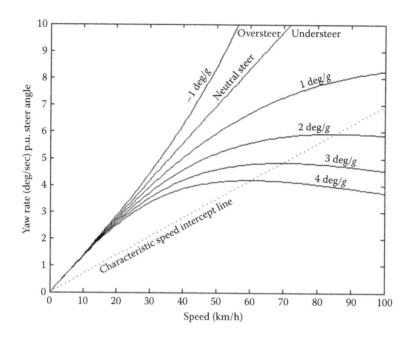

FIGURE 11.5   Response gain/speed test. (From SAE Standard J2666. With permission.)

The wheelbase, L, in meters, and the gravity constant, g, in meters per second squared, also appear in the equation. A published value of understeer gradient gives the modeler key information about whether the car will spin (oversteer) out or saturate its lateral acceleration (understeer) in a stable manner, and at what velocity this will occur. The modeler can then tune the properties of the chassis and tires to achieve this behavior.

11.7.2.3 Frequency Response

Linear system inputs that can yield frequency response information include the impulse, step, and chirp signals. The corresponding vehicle tests are the pulse steer, J-turn, and sine-swept steer tests respectively. The linear region extends until the lateral acceleration exceeds 0.3–0.4 g. Because of the nonlinearities in vehicle dynamics, the results of these tests are also analyzed in the time domain. Handling dynamics are frequency dependent (Garrott et al., 1997), making a frequency domain analysis an informative tool.

The pulse steer and J-Turn tests are a good way to get a complete picture with one maneuver. The steering input rate can reach 500 deg/sec (Garrott et al., 1997). In the linear region, in which the lateral acceleration scales linearly with steering input, the frequency response from these tests can be studied for the natural frequencies of the car. In the nonlinear region, the outputs can be measured for peak acceleration, yaw rate, and roll angle and can be compared to their corresponding system performance measures. Additionally, the timing of the peaks in the system and model response can be compared for phase delay errors.

A sine swept steering input is one that applies a sinusoid of some amplitude and slowly increases the frequency to some maximum value. Once the highest desired frequency of steering

is reached, the driver gradually reduces the frequency back down to its minimum. In this way, a swept sine excites the full range of frequencies in the bandwidth of the vehicle (Salaani et al., 2004), but is more spread out in time than the pulse or J-turn inputs. It is an ideal test for calculating the frequency response, and perhaps estimating the transfer function of the lateral dynamics. The tester will find that there exists some frequency past which the sinusoidal path of the car will converge and roll is the only discernable effect of the fast steering. This marks the upper bound on the bandwidth of the vehicle.

11.7.3 Vertical Tests

Vertical tests are harder to perform in principle then longitudinal or lateral tests. This is simply because the car is designed to move forward and turn, but not move up and down, except in response to disturbances from the terrain. The two approaches for measuring vertical characteristics on a test track are to glean the vertical response from roll and pitch excitations, and to drive the car over uneven terrain inputs.

The natural frequencies of a ground vehicle will center on the modes of the suspension, usually around 2–3 Hz, and the tires, around 10 Hz. All roads have a vibration profile that excites the tires with a speed-dependent frequency response. A benefit of simulation is that a road input which estimates white noise can be constructed for a more accurate identification of a vehicle's modes.

Military tests include more severe inputs such as the Belgian Blocks course. Military and off-road vehicles also introduce the concept of mobility which is a measure of the energy that goes into longitudinal motion in proportion to the energy that is lost in suspension jounce and rebound, loss of tire contact, and other factors.

11.8 Conclusion

It is hoped that the importance of verification and validation in the model building process has been adequately conveyed to the reader. Many general principles of validation were summarized from the literature; however, they were complemented with specific techniques that have developed over time in the automotive industry. The information presented ought to be useful to the engineer endeavoring to build a model that faithfully reproduces the behavior of a specific vehicle as well as the researcher who is given a black box model and wishes to put it through its paces and determine its usefulness for their project.

The simulation practitioner must be aware that models do not exist in a vacuum; rather, they are built for specific applications. This fact, more than any other, determines the appropriate level of effort to put into building, verifying, validating, and perhaps certifying a model. On the other hand, some minimum standard of fidelity and accuracy should be met for driving simulator models since the average person has a great deal of familiarity with everyday vehicle performance and handling.

Let the practices and standards of the automotive industry be your guide in devising validation tests for vehicle models. This approach will facilitate the direct use of vehicle performance test data in the development of a validation testing regimen.

Key Points

- Verification and validation are important steps in the modeling process that should not be neglected.
- There is a wealth of information on vehicle dynamics testing and characterization to guide the modeler in the validation process.
- Adding complexity to a vehicle model also makes the validation process correspondingly more complex.
- Experimental data is the best, but also the most time-consuming, way to validate a vehicle model.

Keywords: Modeling, Simulation, Validation, Vehicle Dynamics, Verification

Acknowledgments

The author would like to acknowledge the work of Robert Sargent in validation, Thomas Gillespie in vehicle dynamics, and the staff of the Vehicle Research and Test Center in East Liberty, Ohio for their work in modeling and testing. Their expertise was mined heavily for this chapter and their publications are commended to the reader for further investigation.

Glossary

Coastdown: A test that measures the time it takes for a car to coast in neutral gear from a set starting speed to a specified finish speed.

Gradeability: The steepest slope, or grade, that a vehicle can successfully traverse driving forward, typically in units of percent. (A 100% grade equals a 45 degree slope.)

On-center handling: The response of a car to small changes in steering input, typical of normal highway driving.

Understeer/Oversteer gradient: A measure of directional control that indicates the propensity of a vehicle to veer off a curved trajectory, either to the inside or outside.

Validation: The process of comparing a model against a problem entity to rate its accuracy and appropriateness.

Verification: The process of ensuring that the model implementation is correct and works as intended.

Web Resources

Two examples of driving simulation animations are available for viewing on the *Handbook* web site.

Web Video 11.1: The wheel loader example. This video shows a high-fidelity model undergoing multiple types of inputs.

Web Video 11.2: The stability control example. This video shows loss-of-control scenarios and the influence of electronic stability control (ESC).

Key Readings

Balci, O. (2003). Verification, validation, and certification of modeling and simulation applications. *Proceedings of the 35th conference on winter simulation* (Vol. 1, pp. 150–158). New Orleans, LA.

Garrott, W. R., Grygier, P. A., Chrstos, J. P., Heydinger, G. J., Salaani, K., Howe, J. G., & Guenther, D. A. (1997). Methodology for validating the national advanced driving simulator's vehicle dynamics (NADSdyna). (SAE Paper 970562). *Proceedings of the 1997 Society of Automotive Engineers (SAE) International Congress & Exposition*, 1228, 71–83.

Gillespie, T. D. (1992). *Fundamentals of vehicle dynamics.* Warrendale, PA: Society of Automotive Engineers.

Salaani, M. K., Heydinger, G. J., & Grygier, P. A. (2004). Experimental steering feel performance measures. *SAE Transactions, 113*(6), 665–679.

Sargent, R. G. (1996). Some subjective validation methods using graphical displays of data. *Proceedings of the 28th conference on winter simulation* (pp. 345–351). Coronado, CA.

References

Balci, O. (1998). Verification, validation, and accreditation. *Proceedings of the 30th conference on winter simulation* (Vol. 1, pp. 41–44). Washington, DC.

Balci, O. (2003). Verification, validation, and certification of modeling and simulation applications. *Proceedings of the 35th conference on winter simulation* (Vol. 1, pp. 150–158). New Orleans, LA.

Balci, O., & Sargent, R. G. (1982). Some examples of simulation model validation using hypothesis testing. *Proceedings of the 14th conference on winter simulation* (Vol. 2, pp. 621–629). San Diego, California.

Box, G. E. P., Hunter, W. G., & Hunter, J. S. (1978). *Statistics for experimenters: An introduction to design, data analysis, and model building*. New York: John Wiley & Sons.

Garrott, W. R., Grygier, P. A., Chrstos, J. P., Heydinger, G. J., Salaani, K., Howe, J. G., Guenther, D. A. (1997). Methodology for validating the national advanced driving simulator's vehicle dynamics (NADSdyna). (SAE Paper 970562). *Proceedings of the 1997 Society of Automotive Engineers (SAE) International Congress & Exposition*, 1228, 71–83.

Gass, S. I. (1993). Model accreditation: A rationale and process for determining a numerical rating. *European Journal of Operational Research, 66*(2), 250–258.

Gillespie, T. D. (1992). *Fundamentals of vehicle dynamics*. Warrendale, PA: Society of Automotive Engineers.

Kleijnen, J. P. C. (1995). Verification and validation of simulation models. *European Journal of Operational Research, 82*(1), 145–162.

Lomax, R. G. (2007). *An introduction to statistical concepts* (2nd ed.). Mahwah, NJ: Lawrence Erlbaum Associates.

Proakis, J. G., & Manolakis, D. G. (1988). *Introduction to digital signal processing*. New York; London: Macmillan; Collier Macmillan.

Salaani, M. K., Heydinger, G. J., & Grygier, P. A. (2004). Experimental steering feel performance measures. *SAE Transactions, 113*(6), 665–679.

Sargent, R. G. (1996). Some subjective validation methods using graphical displays of data. *Proceedings of the 28th conference on winter simulation* (pp. 345–351). Coronado, CA.

Sargent, R. G. (2000). Verification, validation and accreditation of simulation models. *Proceedings of 32nd conference on winter simulation* (Vol. 1, pp. 50–59). Orlando, FL.

Walpole, R. E., Myers, R. H., & Myers, S. L. (1998). *Probability and statistics for engineers and scientists* (6th ed.). Upper Saddle River, NJ: Prentice Hall.

12

Cross-Platform Validation Issues

Hamish Jamson
University of Leeds

Abstract

The Problem. A driving simulator is simply an apparatus to aid in the investigation of driver behavior, to consider engineering design solutions, or to facilitate driver training. Its attraction is its capacity to allow such studies to take place under safe and controllable conditions. However, driving simulators vary wildly in terms of their sophistication and therefore their individual ability to faithfully recreate the complexities of the driving environment, making it challenging for the simulator user to select the appropriate system for the task in hand. ***Role of Driving Simulators.*** The choice of a particular technical specification and configuration of a driving simulator should be dependent on the nature of the driving task required, be that shaped by research objectives or training needs. As a rule of thumb, the greater the investment in a facility, the greater its ability to realistically mimic a wide range of driving tasks. As such, driving simulators are often classified according to investment cost. However, very limited research has been performed on what the cost-benefit structure of simulator investment really is. This chapter attempts to address some of these issues. ***Key Results of Driving Simulator Studies.*** First, the theory of perception in a virtual environment is reviewed: the closed-loop feedback through the driver's visual, vestibular, and auditory sensory systems. ***Scenarios and Dependent Variables.*** Next, the practical implications of the design choices forced upon driving simulator developers is considered and how those choices may manifest themselves in terms of the quality of the overall simulation and validity of the driving data extracted from it. ***Platform Specificity and Equipment Limitations.*** Finally, the role of varying classifications of driving simulators is discussed in terms of scenario realism and simulator validity. Conclusions are drawn as to what level of driving simulator is most suitable in a range of typical applications.

12.1 Introduction

Driving simulators are designed and constructed to suit a number of aims and objectives. Their use spans a broad spectrum: From scientific evaluations of driver behavior, road safety and traffic engineering to the enhancement and development of driver training. Such evaluations may demonstrate both the negative (e.g., the distraction of in-vehicle devices such as telecommunications, navigation systems or even talkative passengers) and positive (e.g., the benefit of safety systems such as collision warning or lane keeping).

Indeed, the use of a driving simulator has several benefits over corresponding real-world evaluations using instrumented vehicles on the open-road or on closed-road test tracks. One of the main advantages of simulation is its versatility. Driving simulators can be easily and economically configured in that virtual scenarios can be created to exactly match the requirement of a particular investigation. Environmental conditions can be manipulated such as day/night operation, weather conditions and state of the road surface. The parameters of the driven vehicle can be altered: For example, suspension design, tire construction and steering characteristics can be matched to an existing or prototype vehicle. New and novel road schemes, methods of signage and highway infrastructure can be modeled virtually and evaluated without the logistical problems of actually modifying large areas of roadway. Furthermore, simulators have ethical advantages over real-world investigations in that they provide an inherently safe environment for drivers; neither they nor other road users are actually put in danger. This makes simulators particularly useful for studies into fatigue and impairment along with hazardous training applications, such as police driver pursuits.

However, there are also a number of disadvantages. Whatever the development budget for a driving simulator, its validity is always arguable since the complexity of the real world can never be replicated in its entirety. Simulation will always be an approximation of reality and even with recent technological advancements in image generation, projector design and motion cueing, the debate is likely to rage for some time as to whether the thorny issue of its validity will prevent simulators from becoming the panacea to all issues whether driver behavioral or training-related. Driver motivation may be lower in a simulator: Since a driver knows that their behavior does not directly influence their safety; this may manifest itself in more risky driving strategies than may be apparent in reality. Furthermore, some people, particularly older drivers, can develop simulator sickness when driving, with symptoms ranging from mild disorientation to severe ataxia and vomiting (see also, this book, chap. 14 on simulator sickness by Stoner, Fisher, & Mollenhauer). This can limit the number of available users of simulation facilities, a severe handicap in a typical training application. Finally, and most importantly with reference to this chapter, some simulators, especially those with large motion systems or high-end display systems, can be prohibitively expensive to set up.

A cost-benefit analysis of a driving simulator with regard to its validity is notoriously difficult to undertake in the majority of applications. A simulator designer has an enormous number of decisions to make on the degree of fidelity of each of the subsystems that make up the overall facility. A higher level of subsystem fidelity is likely to bring with it a higher price tag. For example, a perennial debate with the flight simulation community concerns the cost-benefits of expensive motion systems in the role of pilot training. Szczepanski and Leland (2000) undertook a meta-analysis from a number of aviation studies, concluding that motion is necessary, particularly when the real-world task includes motion stimuli that must be interpreted accurately for the pilot to make appropriate control inputs. However, a similar analysis carried out by Bürki-Cohen, Soja and Longbridge (1998) concluded that simulator motion did not have a significant effect on training efficacy. These contradictory findings are not uncommon in this area and the reasons behind this are equally applicable within the driving domain. The sheer scale of positively validating every potential task and maneuver is demanding, access to test equipment is limited and expensive, the research is disruptive to the normal operation of the facility and, as argued by Boldovici (1992), transfer of training studies can be flawed since the training applications most likely to be

influenced by motion (e.g., aircraft system failures) cannot all be safely tested in a real aircraft in order to make a comparison between virtual and real-world training environments.

However, in the majority of driving simulator applications, it is unlikely that such extreme maneuvers form the basic driving task in hand. In such a case, Wickens and Hollands (2000) suggest a generic cost-benefit analysis based on transfer of training by examining the cost-effectiveness of a training program. First the Transfer Effectiveness Ratio (TER) is calculated as the amount of savings divided by the transfer group time in the training program. For example, a control group learns a particular target task, achieving a satisfactory performance in ten hours. A new training program is developed that requires an additional four hours to undertake, but reduces the time for the group to reach satisfactory performance from ten to eight hours. The TER is $(10-8)/4 = 0.5$. A TER of less than one suggests that the new training program is inefficient (in this case since the additional training is longer than the saving in time to reach satisfactory performance). Conversely, a TER exceeding one suggests the new training program is efficient.

However, TER considers only time benefits. The new training program may be attractive since it is cheap. In this case, the Training Cost Ratio (TCR) needs to be evaluated, defined as the training cost in the target environment divided by the training cost in the new training program. If the new training program is cheap, the higher the TCR can become. The overall cost-effectiveness is defined as the product of the TER and the TCR. If this exceeds one, then the new training program is cost-effective. However, it is possible to still employ a training program that is not cost-effective, should safety considerations and ethics dictate.

While Wickens and Hollands' cost-benefit analysis technique is clearly applicable to driving simulators utilized in a training role, cost-effectiveness is much more difficult to define for those employed in a research environment. In such a situation, a simulator designer must address the demands of the actual driving tasks required in order to ensure that the facility possesses the necessary capabilities to fully simulate those tasks. The next section looks at the role of task analysis in the selection of a driving simulator system.

12.1.1 The Problem: Comparison of Differing Driving Simulator Platforms

The terminology "driving simulator" covers a broad range of possible solutions. In November 2007, Toyota unveiled what is, at present, the world's most advanced (and expensive) driving simulator. Heavily based on the National Advanced Driving Simulator at the University of Iowa, the simulator consists of a Lexus car cab located inside a 7.1-meter-diameter dome, housing a 360 degree projection system, all mounted on a 35 m × 20 m 9DoF motion platform. While its development costs have not been made public, most estimates exceed $100 million. At the other end of the scale, a mere $10 k will also pay for a driving simulator, for example the STISIM Drive system. In its simplest form, STISIM Drive is a PC-based system in which the driver is seated with pedal and steering controls in front of a single channel monitor display. To varying degrees, both systems replicate a range of visual, audio and tactile feedback required to create a simulated driving experience. In both it is possible to simulate typical scenarios such as car following or hazard avoidance, but of great debate within the driving simulator community is whether hugely expensive facilities, such as the Toyota Driving Simulator are truly worth their outlay. Maybe society is better served by 10,000 STISIMs rather than one Toyota simulator? This is *the* question when it comes to a debate on cross-platform validity.

The most important question that a simulator user can ask him/herself is "What task do I need my simulator to achieve?" Simulator designers will always strive to reproduce high quality visual, auditory and kinesthetic cues within their facilities in order to artificially recreate a realistic driving environment. However, financial or logistical constraints may limit a simulator's capabilities and in turn moderate its ability to fully stimulate the full range of drivers' sensory modalities. Clearly, these limitations have the potential to influence the efficacy of a particular investigation in terms of the reliability of driving data extracted from the simulator. But, if the task demands of the simulator are kept to the fore, it will be possible to develop a facility that meets the requirements of that particular task. Important questions to address during task analysis are:

- What is the role of human perception in performing these tasks in the real situation?
- How can the available resources and the simulator's characteristics (hardware and software) be best optimized to re-create these perceptions?

A poorly-designed simulator may well invoke unrealistic driver behavior that, in turn, is likely to lead to poor quality driver behavioral research or inadequate transfer of training. But whatever the results of developing a driving simulator using a task-based approach, the true efficacy of the simulator will always be debatable, since neither a standardized technique nor an international legislative body exists to fully assess or classify driving simulator validity. While the flight simulation market is tightly regulated by both the Joint Aviation Authority and the Federal Aviation Authority, the same is not true in the driving domain. Driving simulator validation is currently sporadic, under-funded and under-researched and is severely lacking common consistent and robust validation procedures. As such, making a true assessment of cross-platform validity that is accepted by the driving simulator community at large is, at best, challenging. However, there is a large body of literature available to help us make some judgments in considering the issues of cross-platform validity. To be able to make reasonable comparisons, we need first to define a driving simulator classification along with the key components that form their make-up. How the characteristics (and price) of these components affect a driver's perception in a simulated driving environment is instrumental in the cost-benefit and task analyses that should

precede and justify the system make-up in the design of any driving simulator facility.

12.2 Cross-Platform Validation: The Theory

In order to compare across driving simulator platforms, it is important to review what effect varying the characteristics of the key components has on the perceptions of the driver through the main sensory channels used in driving. By considering the task that a given simulator is required to perform, addressing the perceptual issues will allow the best optimization of the simulator's characteristics within the available development budget. There are three main sensory modalities with which drivers perceive their performance within a virtual environment and the key components are designed in order to excite these sensory channels. According to Kemeny and Panerai (2003), in order of "importance", these are:

- The visual system
- The motion system and kinesthetic feedback through control loading
- The sound system

12.2.1 Typical Driving Simulator Platform Classifications

It is important here to distinguish between a simulator's *utility* and its *usability*. Utility (or fidelity) describes the degree to which the simulator's characteristics replicate the driving task faithfully. Its usability, on the other hand, describes how versatile the simulator is in terms of ease of reconfiguration from study to study. Research simulators would ideally have good usability, whereas training simulators may focus on strong utility.

The financial outlay in the development of a driving simulator is often intrinsically linked to its utility and the ability of the simulator to excite the three main sensory modalities. This often leads to a classification based on cost. A similar classification is *low-level, mid-level* and *high-level* driving simulators (Weir & Clark, 1995).

Typically, low-level driving simulators are cheap since they are of limited utility, providing a fairly basic level of fidelity in the provision of visual, kinesthetic and audio cues to the driver. The image quality of the virtual environment need not be compromised in terms of frame rate, scene content (texture use, lighting models and depth complexity) and display resolution. The high cost associated with visual simulation throughout the 1980s and 1990s was typically due the high-cost graphical workstations, such as those manufactured by Silicon Graphics Inc., that were required to achieve the challenging graphical demands of both pixel fill rate and polygonal content. Similar scene complexity is now possible using the standard off-the-shelf graphics cards of most higher-end desktop PCs. However, the display of the visual scene in a low-level simulator will commonly consist of a relatively narrow horizontal field of view, often displayed on a single monitor in front of the driver. Kinesthetic feedback will be limited to the provision of some degree of torque feedback through a games-style steering wheel, providing the main interface between the driver and the simulator. Low-level simulators are particularly cost-effective, but are limited in their utility. A typical low-level simulator in shown in Figure 12.1.

Mid-level driving simulators, while enjoying the same usability of their low-cost cousins, typically have greater utility. They employ more advanced display techniques that typically form a wide horizontal field of view. While this can be achieved using a Head-Mounted Display, most applications will use a bank of projectors displaying images onto a cylindrical or spherical display surface. The individual images that are cast are distorted electronically to create a single, composite image. Such a display is not truly dynamic since commonly the head position of the observer is assumed to be fixed, thereby creating a consistent viewpoint (eye point) onto the virtual scene. The vast majority of such mid-level facilities do not employ the use of stereo displays, creating what is in effect a 2D representation of the 3D world and thereby losing any perceptual benefits gained from the disparity of the projected image. This disparity can help distance estimation, especially for objects close to the observer, aiding accurate stopping behaviors.

Drivers will usually be seated within a facsimile of the vehicle cab that the simulator is mimicking, with instrument (dashboard) displays and driver controls operating as in the real vehicle. While not always the full vehicle bodyshell, an actual vehicle cab creates a highly tangible and physical immersion within the virtual environment with high bandwidth torque motors

providing suitable kinesthetic feedback though the steering wheel and foot pedals. However, motion cues will not normally be represented to any great degree. If present at all, they will most likely be limited to high-frequency simulations of vehicle vibration and road roughness, often with actuators located at the suspension points or under the driver's seat. In some cases, minimal tilt cues can also be provided. A typical mid-level simulator is shown in Figure 12.2.

High-level driving simulators provide an almost 360 degree horizontal field of view, a logical extension of a front-projection system typical to its mid-level cousin. The vehicle cab will be complete with, at minimum, the dashboard operation and kinesthetic functionality provided by a mid-level simulator. The vehicle cab will be mounted on an extensive, high bandwidth motion system including at least six degrees of freedom with capabilities in the simulation of high-frequency onset (initiation of a maneuver) and low-frequency sustained cues (long, sweeping curve). It is the motion system that normally accounts for a large proportion of the financial outlay on a high-level system (Figure 12.3).

12.2.2 Key Components in the Perception of Driving in a Simulated Environment

12.2.2.1 Effects of the Visual System

The visual system is well covered in this book, chapter 8 by Andersen on sensory and perceptual factors in the design of driving simulation displays. However, it is briefly revisited here for the sake of comprehensiveness.

Under natural conditions, visual cues provide all the necessary inputs to allow an observer to form his/her perception of 3D space. However, under simulated conditions, the inherently inferior display characteristics (image resolution, update frequency, field of view, etc.) bring about a reduction in the quality of these cues. A driver's use of these cues is important for the estimation of vehicle speed, distance to objects, vehicle heading and lateral control.

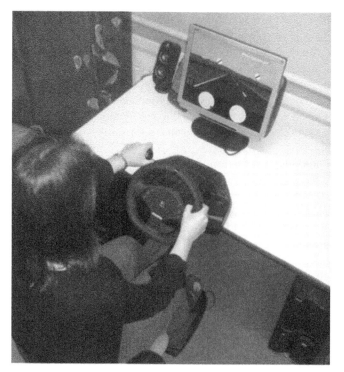

FIGURE 12.1 Low-level simulator: Typified by unsophisticated visual display, rudimentary driver control with little driver feedback.

FIGURE 12.2 Mid-level simulator: Typified by high-end visual display, facsimile of vehicle cab but lacking dynamic cues.

FIGURE 12.3 High-level simulator: As mid-level but including large amplitude motion system to maximize auditory, visual and vestibular driver feedback.

Early work of Gibson (1950) suggests that drivers' visual perception of their environment is primarily based on disparity and optic flow. Disparity refers to the relative lateral displacement of the retinal images in the left and right eyes of the same object in space. It is an effective binocular cue to depth at short distances. Optic flow, on the other hand, describes the dynamic pattern of retinal motion that the central nervous system uses to decipher an apparent visual motion. In driving, both are thought to play dominant roles, both in the control of heading (Lappe, Bremmer, & van den Berg, 1999) and in collision detection (Lee, 1976; see also Andersen, this book, chap. 8 for a discussion of optic flow and heading control).

Binocular cues are not normally present in driving simulators, since to achieve this acceptably requires either stereo projection or the use of Head-Mounted Displays. While it is generally accepted that the effectiveness of binocular convergence as a cue to absolute distance is limited to a few meters (van Hofsten, 1976), the effectiveness of binocular disparity may be up to 30 m (Loomis & Knapp, 1999). Given that the majority of objects within a driving environment are positioned beyond this range, disparity is normally less important to the qualities of a driving simulator display system than optic flow.

The evaluation of vehicle speed and the estimation of inter-vehicle distance is an essential skill in real-life driving. Maneuvers such as overtaking and collision avoidance require

such abilities. In a driving simulator, these skills require the accurate representation of self-motion from both optic flow and egocentric direction (Gogel & Tietz, 1979)—the direction of an object in space relative to the observer. Optic flow cannot give information about either absolute speed or distance but can be used to compare relative changes of the observer's perceived position in space, central to the accurate estimation of time-to-contact (Lee, 1976). Studies into speed perception have shown that observers tend to underestimate their velocity in simulated environments (Howarth, 1998; Groeger, Carsten, Blana, & Jamson, 1997). This effect is also sensitive to image contrast (Blakemore & Snowdon, 1999), the amount of texture (Blakemore & Snowdon, 2000), projector brightness (Takeuchi & De, 2000) and the overall field of view (Jamson, 2001). In each, the greater the extent, the weaker the underestimation becomes.

Distance estimation is also based on a number of reliable cues, such as optic flow (Bremmer & Lappe, 1999), disparity (Howard & Rogers, 1995) and motion parallax (Rogers & Graham, 1979). Motion parallax describes the differential motion of pairs of points as a result of their different depths relative to the fixation point and to the motion of the observer. It provides robust estimates of absolute egocentric distance when combined with additional visual information describing an observer's self-motion. In a recent study using a driving simulator, it was shown that the central nervous system is able to combine these two cues to calibrate the retinal image motion and infer absolute distance just as efficiently in a virtual environment as it does under natural conditions (Panerai, Cornilleau-Peres, & Droulez, 2002).

It has also been demonstrated that both optic flow and motion parallax are crucial for correct interpretation of heading and its control (Crowell, Banks, Shenoy, & Andersen, 1998). However, it is important to note that during curve negotiation, drivers tend to fixate points along their path (Land & Lee, 1994). These active gaze strategies play an important role in heading control (Land & Horwood, 1995). However, more recent studies have proposed that accurate heading control can be achieved through a combination of optic flow and visual egocentric cues (Wann & Land, 2000). This was demonstrated experimentally by Harris and Bonas (2002) in a study of human walking. When road markings were apparent, visual egocentric cues alone provided enough information to allow walkers to maintain accurate heading control. However, performance did not degrade when the markings were missing. It was concluded that in this case an optic flow strategy was dominant.

12.2.2.2 Platform Classification and the Visual System

Due to financial and computational limitations, the image quality presented in driving simulators is a trade-off between resolution, pixel density and field of view. Typically, low-level simulators will contain a narrow field of view displayed on a monitor. This presents problems to the viewer in terms of a reduction in the optic flow presented within the virtual environment compared to that typically experienced with a more expansive display system. As such speed can be significantly underestimated, steering performance can be more erratic and following distance can be overestimated (Jamson & Jamson, 2010). Monitor displays will also produce images of the virtual scene that are scaled down from life-size. In a car-following situation, the change in viewing angle is a significant cue towards looming and a driver making an accurate estimate of the time-to-contact to the lead vehicle. Smaller than life-size images will hamper the observer's ability to estimate looming and hence an overestimation of time-to-contact will be made.

In mid- and high-level simulators, the display system is commonly made up of a projected image over a number of channels to create wide field of views. While widening the field of view has benefits in terms of improving the validity of speed choice and lane position between simulated and real-world driving conditions (Jamson, 2001), the resolution of a projector's display will define the clarity of the visual image. Both technological constraints in the manufacture of affordable projectors and the requirement to maintain consistently high visual refresh rates makes achieving eye-limiting resolution currently impossible. Most display systems in mid- and high-level simulators do not exceed 1600×1200 per channel, which when viewed at a distance of around 2 m, gives a pixel density of around 3–4 arcmin/pixel. Given that the generally accepted threshold of human visual perception is some 1 arcmin (Fredericksen, Verstraten, & van de Grind, 1993), driving simulators struggle

to allow distant objects to be detected as in reality. Projectors with native resolutions that approach eye-limiting resolution are often beyond the reach of even high-end simulators. Their sheer size makes them impossible to use in a front projection system as they would occlude light from each other, thus casting shadows onto the projection surface. This may change in the future as laser projectors become more affordable. Laser projectors can also create much brighter virtual scenes, going some way towards achieving luminance levels in virtual environments that approach those in reality.

12.2.2.3 Effects of the Motion System

In a stationary observer, visually-induced self motion (vection), usually takes several seconds to establish itself (Melcher & Henn, 1981). The latency of this vection can be reduced by the addition of inertial motion cues (Groen, Howard, & Cheung, 1999). Motion cueing is widely used in flight and driving simulators to enhance the simulator's fidelity. However, due to its limited travel, a motion-base can only approximate the motion of the real vehicle.

Developments using motion cueing in vehicle simulation applications started relatively recently (Nordmark, Lidström, & Palmkvist, 1984; Drosdol & Panik, 1985) compared to the initial hexapod design of flight simulators (Stewart, 1966). The main reason for this is the much more demanding nature of ground vehicle dynamics compared to those of commercial airliners. In commercial flights, turns are often coordinated and g loads do not normally exceed 0.1–0.2 g for relatively short periods of time. In driving, where turns are uncoordinated, lateral accelerations can reach up to 0.4–0.5 g for much longer periods (e.g., a long, sweeping curve). Longitudinal accelerations during emergency braking can reach up to 1 g.

The familiar hexapod system is designed to allow motion in all six degrees of freedom (pitch, roll, yaw, heave, surge and sway). Motion cueing with a hexapod comes from two sources. First, displacement of the actuators establishes a translation of the motion system and gives the occupant an accurate, high-frequency onset cue of movement. However, when the actuators of the motion system approach their full extension, no more linear acceleration from translation is physically possible. At this point, the motion system takes advantage of the perceptual ambiguity and exchanges the sustained linear acceleration for a rotation in order to "fool" the vestibular system. This simulation technique is often referred to as "tilt coordination" (Nahon & Reid, 1990). When a visual scene is simultaneously presented to the driver depicting an accelerated transition, the relative variation of the gravity vector (or tilt) is interpreted by the central nervous system as a linear acceleration. The transition from high-frequency onset to low-frequency sustained acceleration needs to be performed at a rotation rate unperceivable to the occupant, approximately 3°/s (Groen & Bles, 2004). During tilt coordination below perceptual threshold, the optic flow that is presented simultaneously to display the same linear acceleration makes it impossible for the observers to determine whether the perceived motion arises from tilt or translation (Angelaki,

McHenry, Dickmann, Newlands, & Hess, 1999). However, in more expansive maneuvers, vehicle accelerations tend to build up much quicker than the establishment of imperceptible tilt-coordination permits, hence motion-cueing becomes a compromise between giving accurate rapid response and sustained acceleration cues without false cues (accelerations perceived in the wrong direction) or sag (a reduction in the perceived acceleration during the transient region of a maneuver). In essence, for small motion systems, the tuning engineer is faced with a choice between a simulator that feels responsive but has excessive body roll, or one that feels sluggish but with a more realistic tilt response. To extend this physical movement, thereby reducing sag, high-end motion systems can feature large amplitude (10 m+) linear motion to more faithfully reproduce the onset (immediate) and transient (longer term) cues from a driver control input.

While the theory behind large amplitude motion in driving simulation is well established, in practice it is only recently that such systems have been commercially available. To date, no research has been published comparing the effects of large amplitude translational motion on driver behavior. However, there has been some published work on more familiar hexapod motion systems, suggesting that the introduction of motion cues has a positive effect on simulator validity (Wierwille, Casali, & Repa, 1983; Reymond, Kemeny, Droulez, & Berthoz, 2001). Álm (1995) also compared real and simulated driving both with and without motion cues, reporting that significant differences in speed variance were observed between the natural and simulated conditions without motion, as well as between simulated conditions with and without the motion system functioning. Drivers also reported increased mental workload while driving the simulator compared to handling the corresponding instrumented vehicle. Further evidence to suggest that the vestibular cues excited by a motion system are used by drivers to better control steering and regulate speed is provided by van Winsum and Godthelp (1996). Additional validation studies performed using moving-base simulators show that drivers perform wider turns when lateral cues are present compared to those when only visual information is available (Siegler, Reymond, Kemeny, & Berthoz, 2001).

12.2.2.4 Platform Classification and Kinesthetic Feedback

Motion systems that achieve more than simple high-frequency buffet are the domain of high-level simulators, due to their significant cost. Many validation studies have reported higher observed speeds and speed variation in simulated compared to natural conditions (Alicandri, Roberts, & Walker, 1986; Riesmersma, van der Horst, & Hoekstra, 1990; Harms, 1996; Duncan, 1995; Blana, 1999). However, the addition of motion cues tends to reduce this speed differential (Soma, Hiramatsu, Satoh, & Uno, 1996; Reymond et al., 2001; Greenberg, Artz, & Cathey, 2002). Similarly, differences in braking performance have been observed in simulators both with and without motion (Boer, Kuge, & Yamamura, 2001). With motion, braking is

smooth while a multi-modal profile of brake applications is observed in the absence of dynamic cues. Blana (2001) suggested that motion cues improves the accuracy of drivers' perception of speed, the motion system creating an overall greater sense of immersion and hence realism within the simulated environment.

Greater variation in lane position has been observed for simulator drivers compared to those experiencing similar but real conditions (Harms, 1996; Duncan, 1995; Blana, 2001). The addition of motion cues reduces this variation (McLane & Wierwille, 1975; Reymond et al., 2001; Greenberg et al., 2002). This effect is not surprising as the main intention of the rendering of motion cues is to improve the dynamic "feel" of the vehicle. In the flight simulation domain, several authors have developed models describing tracking behavior based on visual and vestibular feedback (Hess, 1990; Hosman & Stassen, 1999). Within such models, the vestibular system typically shows a much shorter latency to perceived motion than the visual system. Given that steering control is a closed-loop process where drivers are constantly assessing the impacts of their control actions (van Winsum & Godthelp, 1996), drivers in a moving-based simulator will receive more timely feedback than those in a fixed-base simulator. Consequently, steering is easier to control and lane variation decreases. Interestingly, in real driving, a similar effect of misperceived motion cues occurs. It has been reported that inappropriate signals delivered to the central nervous system by diseased or disordered vestibular organs, an effect similar to the lack of motion cues in a simulator, cause inappropriate steering adjustments (Page & Gresty, 1985).

The actual vehicle cab in which the driver sits is important to the face validity of the simulation. For low-level simulators this may simply be a small games-style steering wheel with foot pedals and gear shift, but typically in mid- and high-level facilities will consists of either a full or cut-down real vehicle. If the cab is mounted on a motion system, a lightweight vehicle mock-up will often be used to minimize payload. Either way, mid- and high-level simulators will boast fully operational dashboard instrumentation, usually to give the driver indications of driving speed, engine speed, fuel level and other information concerning the vehicle's operation. In a high-end system with full motion, the whole cab may reside within a simulation dome. This is normally a stiff but lightweight construction, often glass-fiber layered into a strong honeycomb structure. The dome prevents unwanted extraneous sounds and light from polluting the simulation environment. The inside surface of the dome also doubles as a screen onto which the projection system displays the visual image. It is essential that the natural frequency of the dome is sufficiently high to avoid structural resonance at the typical frequencies of operation of the motion system.

The "feel" of a simulated drive is important in order to create a sense of realism in a driving simulator; hence, it is important that the vehicle controls have the same characteristics as the actual vehicle that the simulator is mimicking. The main feedback to a simulator driver is through the steering system and ultimately, the steering wheel. The forces and movements imposed on a steering system emanate from those generated at the tire-road interface. The resultant steering wheel torque is often produced by a motor located at the end of the steering column, controlled via the steering model within the vehicle dynamics module. In a low-cost simulator using a games-style force feedback wheel, it is unlikely that the bandwidth of the torque motor will be high enough to achieve a sufficiently high-frequency response for a reasonable feel.

Control feedback through the foot pedals—for example, maintaining the correct resistance to movement on both the clutch and accelerator—is also important. However, it is the feel of the foot brake that is arguably the most critical. Without the assistance of a brake booster or servo, the pedal effort required to operate the brakes fitted to the majority of existing vehicles would be impossibly high. These boosters normally use a pressure differential created by an engine-supplied vacuum to assist the driver in generating pressure within the system. The presence of a brake booster reduces the required brake pedal effort by up to ten times of that required without any assistance. Correctly simulating the brake booster ensures that the driver has the expected brake pedal feel.

There appears to be no published work scientifically assessing the effect of vehicle cab on simulator validation, but, anecdotally, the more realistic the cab, the more immersive the simulator has the capability of becoming.

12.2.2.5 Effects of the Sound System

The quality of the reproduction of auditory cues within a simulated environment is assumed to play an important role in its perceived realism; however, research is lacking in this area. It is typical in current driving simulators to render high quality, spatialized audio by "expanding" the acoustic image in an attempt to render 3D audio.

Current methods used to render audio in driving simulators vary greatly. The majority of simulators play sounds over speakers without regard to the acoustics of the listening environment (e.g., Colinot & Belay, 2000), whereas some use utilize transfer functions of both the vehicle interior and the driver's head in order to represent a more complex acoustic field inside the cabin (e.g., Blommer & Greenberg, 2003).

McLane and Wierwille (1975) investigated the effects of presence or absence of speed-related sounds in a driving simulator. The authors reported that the existence of audio cues had no significant effect on either driving speed or lane control, but they acknowledged that the audio rendering had the advantage that irrelevant sounds emanating from the various simulator subsystems were effectively masked, improving the simulator's "face" validity. Similarly, some 20 years later, Davis and Green (1995) debated the effect of sound, demonstrating unchanged drivers' rating of realism with and without audio cues. Given that most modern cars are well sound-proofed and the driver exists within a relatively quiet environment, perhaps the role of sound is more one of filling the simulation experience rather than as a perceptual input used to assess the current driving conditions?

12.2.2.6 Platform Classification and Its Role on the Sound System

Reproduction of good quality, spatialized sound is now relatively cheap and so simulator classification tends to play less of a role here. So as long as a simulator goes some way towards achieving the full acoustic range of audible frequencies and dynamic sound spectrum, along with mimicking a range of engine, transmission and environmental noises, there is little evidence to suggest that the varying sounds associated with varying simulator classification plays any role in the overall validity of the driving simulator.

12.3 Cross-Platform Validity: The Evidence

The essence of this chapter is to review the differences between various classifications of driving simulator. However, considering that so many driving simulators now exist worldwide, only a handful of studies have been published acting as evidence of the relative validity of individual systems (e.g., Bella, 2008; Yan, Abdel-Aty, Radwan, Wang, & Chilakapati, 2008; Godley, Triggs, & Fildes, 2002). In general, the current evidence suggests that, in the simulator, speeds tend to be greater, lane keeping behavior tends to be more varied, and the mental effort required to perform the driving task tends to be higher, than within the corresponding real-world equivalent.

Studies involving direct comparisons of driving simulators classifications are even more uncommon. This may have something to do with the fact that so many subsystems make up an individual simulator. This gives researchers many independent variables to control and makes true cross-platform validation studies extremely challenging. For example, if a comparison of behavior shows differences between simulator A and simulator B, this could be down to a number of reasons. Varying vehicle dynamics models, complexities of the visual scene, simulator latencies, image qualities and control feel can all potentially influence perception of the virtual environment and hence the behavior within it. In addition, varying tasks demanded of the driver are likely to be influenced by the vagaries of different subsystems. For example, a simulator with a long transport delay or latency is likely to be difficult to steer; however, this delay is unlikely to influence speed perception.

Onto the published evidence: Between 2003 and 2006, the European Union funded a project to investigate the main effects and interaction of varying complexities of both primary (driving) and secondary (in-vehicle) tasks. The primary task scenarios involved lane keeping and car following. Two secondary tasks were developed which produced either a visual or cognitive load on the driver (Merat, 2003). The study was designed in such a way that drivers were required to perform the same primary and secondary tasks in low-level, mid-level and high-level simulators.

While not comparing the baseline (no secondary task) conditions between the various levels of simulator classification,

Santos, Merat, Mouta, Brookhuis and de Waard (2005) reported that the same relative effects of the dual-task conditions were displayed. This involved more erratic steering behavior when performing the visual task and a reduction in lane position variation when performing the cognitive task. Both tasks involved a reduction of speed (increase in following headway) although the reduction was greater with the addition of the visual task. The authors reported that while these effects were observable in all three simulator classifications, the higher quality the simulator, the more sensitive the findings in terms of effect size. This is important as it would suggest that the higher the quality of the simulator, the more consistent the driver control of vehicle was in the single task condition (i.e., just driving). This would have provided a more stable baseline dataset, and so the effect of the secondary tasks thus became more powerful.

Jamson and Mouta (2004) compared the same primary and secondary tasks in a separate study involving two different simulators, one low- and one mid-level. Although participants drove significantly quicker in the low-level simulator, simulator development cost was not a significant factor in demonstrating the well established effect of a reduction in mean speed while drivers performed dual tasks. The mid-level simulator, however, tended to show more sensitivity to this effect in that greater speed reduction was observed with increasing complexity of a secondary task. This effect was not apparent in the low-level simulator where dual-task speed reduction was consistent regardless of the secondary task difficulty. In terms of lateral performance, steering reversal rate increased and lane tracking performance was worse in the low-level simulator regardless of secondary task modality. The authors concluded that the mid-level driving simulator was worth the extra investment over its low-cost cousin due to its increased sensitivity to subtle changes in driver performance, particularly those related to lateral performance and workload.

These two studies both involved abstract secondary tasks, not normally associated with real in-vehicle systems. In a similar study, but this time using identical simulator software running in two separate mid- and low-level simulators, Jamson and Jamson (in press) compared similar varying levels of driving demand in conjunction with interaction with a real navigation system. The secondary tasks were ranked according to the levels of visual/manual effort and included items such as: reading a list of directions (low demand), changing the display preference settings (medium demand), and changing the destination address using hand-typed data entry with a stylus (high demand). The speed reduction associated with dual-tasking was again observed in both environments. Once more, the mid-range simulator showed a more sensitive effect of linear speed reduction with the increase in secondary task demand. By their self-reports of driving performance, drivers rated themselves as being worse when they drove the low-level simulator, which was reflected in two separate components of their behavior that were found to deteriorate with distraction—headway keeping and lateral control. Drivers of the low-level simulator were more inclined to spend longer periods of time at shorter headways and exhibited poorer lateral control.

These studies have begun to build up an evidence base to demonstrate an effect of simulator classification on the quality of behavioral data that it brings about, but this is very much limited to a specific distraction study. Neither do any of the investigations touch on a cost-benefit evaluation, an area in which is impossible to currently find any publicly available material. It is this cost-benefit evaluation that is vital to the justification of the use of higher level systems.

12.4 Typical Applications According to Driving Simulator Classification

In this chapter, driving simulators have been already classified as low-, mid- and high-level facilities and some theoretical evidence has been presented to indicate how the absolute validity of the driving task differs across this classification. However, is an absolutely valid simulator really necessary in all applications? There is a need to look more deeply into their actual use and how the three arbitrary classifications can match the requirements of a particular investigation. In other words, what simulator should be used in what environment?

First, it must be once again acknowledged that achieving perfect simulation is impossible. High-level facilities with expansive motion capabilities can come close, but vehicles, unlike airplanes, turn in an un-coordinated manner achieving relatively high, low-frequency accelerations which would take an enormous motion space to simulate translational accelerations fully with a one-to-one representation. More important is to employ a bottom-up approach, where the development of the simulator is tailored towards its specific requirements. This ensures that development of simulator goals, and hence its classification, are approached in the most cost-effective manner. In order to tailor the classification of a simulator towards its goal, let us now consider the three typical spheres of operation for driving simulators: driver behavioral research, vehicle design and engineering, and driver training.

12.4.1 Driver Behavioral Research

This area is possibly the most common application of driving simulators and probably offers the most flexibility in terms of simulator classification. Researchers have long argued (e.g., Tornros, 1998) that it is the relative validity of a system that is critical, as opposed to a simulator's ability to extract absolute correspondence of behavior between the virtual and real environments. A typical behavioral study will compare baseline driving against that with some experimental factor present. Since the use of the simulator is common to both conditions, the effect of that experimental factor can be tested as a relative difference in a particular experimental metric under observation with and without the factor present. This design has commonly been used, for example, in the effect of cell phone use on driving. The inference is that this relative difference can be extrapolated to an expected relative difference in the real world. Depending on the investigation and driving task in question, applying this argument justifies the utilization of low-level simulators in small-scale research studies, where the perception of the visual scene is the key aspect. Studies might include:

- Product design. Allowing a user to experience a vehicle-related system (e.g., a new interface to an in-vehicle entertainment system) and draw conclusions about its usability in a driving environment.
- System demonstrator. A low-level simulator provides a platform allowing new users to become familiar with new concepts; for example, modern driver-assistance systems such as lane keeping assistance or advanced cruise control.
- Procedural training. Low-level simulators create an environment that allows simple procedural training to take place; for example, to aid a learner driver in understanding basic vehicle controls.

Depending on the application, more advanced research that require accurate perception of the road environment (in order for a driver to make appropriate control actions) are likely to require more elaborate simulations. Mid-level simulators are used in such studies, with examples such as:

- Impairment. A typical scenario may be a close-distance vehicle incursion where a braking response is needed. Using a metric such as brake reaction time only requires the initial perception of the unfolding traffic conditions as opposed to the control of the vehicle beyond this point.
- Distraction. By careful selection of scenarios and metrics that provide an onset cue to an unfolding traffic situation, for example, response to a traffic signal changing to stop, estimates of distraction from an in-vehicle system can be inferred.

Finally, those studies in which it is essential for a driver to actually feel the response of the vehicle require good quality motion cueing, associated with high-level simulators. These could include:

- Tactile road feedback. For example, behavioral studies involving speed humps, road surface roughness and dynamic comfort. Clearly, the bandwidth of the motion system must be capable of performing the frequency of motion required.
- Advanced driver assistance systems. In the study of systems that particularly take active control of the vehicle from the driver, it is vital for the driver to receive the dynamic cues that provide him/her with feedback as to how that system is actually controlling the vehicle.

12.4.2 Vehicle Design and Engineering

Perhaps the most challenging area for the use of driving simulators (and hence the most uncommon) is in the area of vehicle design and engineering. The quality of the visual system is relatively minor and the requirements become focused with the addition of high quality motion cues. Steering feedback is also extremely important, making heavy demands on an active control loading system. As such, this area is the domain of high-end simulators only.

As with any application, the success of the simulator depends on the driving task that is required. To use a simulator as a virtual test track in order to evaluate prototype vehicle ride and handling characteristics involves testing with a range of maneuvers. Some high-frequency, short-duration maneuvers, such as a single lane change, will require extensive translational motion and a low-latency system in order to simulate well, but a high-end simulator can provide a very useful test-bed in such a situation. Others, particularly low-frequency, large amplitude maneuvers, such as braking in a curve, require such huge motion envelopes that they become unrealistic to perform appropriately. The use of tilt coordination for high g maneuvers is flawed due to the low levels of perceivable tilt, making such evaluations even beyond the scope of the largest simulators worldwide.

12.4.3 Driver Training

This is also a highly demanding application for driving simulators. Low- and mid-level simulators can be employed acceptably if the requirement is relatively unchallenging and is limited to the ascertaining of appropriate procedures. However, to achieve high quality simulator training of specific driving tasks, as might be expected in specialist driver training, realistically only high-end facilities can be employed.

It is worth remembering here that early simulators were developed during World War I in order to familiarize new pilots with the controls of the airplane they were about to take to sky for the first time. And it is from within the field of flight simulation that the majority of driving simulator development has grown. Traditionally, training pilots is expensive and hazardous, and the requirement for realistic simulators can often be argued on the grounds of cost-effectiveness. However, driving is relatively inexpensive and the development of an equivalent simulation facility for use in training can be far more expensive than its real-world alternative. The only argument for the use of high-end simulators in this situation is when the training is impossible to carry out in reality due to safety reasons. The use of Wickens and Hollands' (2000) cost-benefit evaluation becomes extremely useful in such a case, where the training task requirement is fundamental to the justification of the use of a high-level simulator.

12.5 Concluding Remarks

This chapter has attempted to scrutinize the current aspects of driving simulator platform classification, assess the reliability of the varying qualities of the subcomponents that make up a platform, consider the potential validity of the driving data obtained and give some indication of the role that this validity has in the numerous roles that driving simulators currently undertake. This has been done, in the main, from a theoretical viewpoint, reviewing the available literature and current applications of simulators worldwide. But when revisiting the question posed in Section 12.2.1 (what is the relationship between acquisition cost and validity?), there really is no easy answer. While a reasonable body of work has been carried out in the areas of visual and motion perception, few studies have addressed the absolute validity of a simulator, correlating driver performance in real and artificial conditions. Fewer still have specifically assessed how the data from one classification of simulator platform compares with that of another. To some extent, this is because the number of applications is so extensive when one considers the vast breadth of potential driver behavioral research, engineering application and training opportunities.

In essence, what this chapter has tried to demonstrate is that, while the full complexities of the real world can never be replicated in their entirety, it is only through a full appreciation of the task requirements demanded from the simulator platform that will lead to an appropriate selection. In many evaluations, a driving simulator may provide an invaluable tool; in others it may not be the ideal apparatus. Task demands will drive this decision process. However, even if a driving simulator is used, the market is currently unregulated and no body of authority is in the position to insist on specific classifications being used in specific environments. It is for each individual user of a driving simulator to make the platform selection based on their own assessment of the task in hand, the validity of the platform available, and the available budget. To coin a well-known phrase, "You pays your money, you takes your choice". Hopefully, this chapter has gone some way towards making that choice somewhat easier to make.

Key Points

- An attempt is made to classify and compare typical driving simulator platforms and their underlying characteristics, with respect as to how perception of driving in a simulator is affected by this classification.
- Consideration is given to both driving simulator utility (faithfulness to the driving task) and usability (ease of configuration).
- The role of simulator validity in the selection of the appropriate system is discussed based on the individual research/training requirements of the task in hand.

Keywords: Cost-Benefit Analysis, Design and Classification, Driving Simulator Characteristics, System Selection

Key Readings

Blakemore, M. R., & Snowdon, R. J. (1999). The effect of contrast upon perceived speed: A general phenomenon, *Perception, 28*(1), 33–48.

Jamson, A. H., & Mouta, S. (2004). More bang for your buck? A cross-cost simulator evaluation study. *Proceedings of the driving simulation conference* 2004 (pp. 321–332). Paris, France.

Kemeny, A., & Panerai, F. (2003). Evaluating perception in driving simulation. *Trends in Cognitive Sciences, 7*(1), 31–37.

Santos, J., Merat, N., Mouta, S., Brookhuis, K., & de Waard, D. (2005). The interaction between driving and in-vehicle information systems: Comparison of results from laboratory, simulator and real-world studies. *Transportation Research Part F: Traffic Psychology and Behavior, 8*(2), 135–146.

Wickens, C. D., & Hollands, J. G. (2000). *Engineering psychology and human performance* (3rd ed.). Upper Saddle River, NJ: Prentice Hall.

References

Alicandri, E., Roberts, K., & Walker, J. (1986). *A validation study of the DoT/FHWA simulator (HYSIM)* (FHWA/RD-86/067). Washington, DC: US Department of Transportation, Federal Highway Administration.

Álm, H. (1995). *Driving simulators as research tools—a validation study based on the VTI driving simulator.* Unpublished internal VTI report. Linköping, Sweden: VTI, Swedish National Road and Transport Research Institute.

Angelaki, D. E., McHenry, M. Q., Dickmann, J. D., Newlands, S. D., & Hess, B. J. M. (1999). Computation of inertial motion: Neural strategies to resolve ambiguous otolith information. *Journal of Neuroscience, 19*(1), 316–327.

Bella, F. (2008). Driving simulator for speed research on two-lane rural roads. *Accident Analysis & Prevention, 40,* 1078–1087.

Blakemore, M. R., & Snowdon, R. J. (1999). The effect of contrast upon perceived speed: A general phenomenon? *Perception, 28*(1), 33–48.

Blakemore, M. R., & Snowdon, R. J. (2000). Textured backgrounds alter perceived speed. *Vision Research, 40*(6), 629–638.

Blana, E. (1999). Behavioral validation of a fixed-base driving simulator. *Proceedings of the driving simulation conference 1999* (pp. 227–241). Paris, France.

Blana, E. (2001). *The behavioral validation of driving simulators as research tools: A case study based on the leeds driving simulator.* Doctoral dissertation, Institute for Transport Studies, University of Leeds, U.K.

Blommer, M., & Greenberg, J. (2003). *Realistic 3D sound simulation in the VIRTTEX driving simulator.* Proceedings of the Driving Simulation Conference North America 2003. Dearborn, MI.

Boer, E. R., Kuge, N., & Yamamura. T. (2001). *Affording realistic stopping distance behavior: A cardinal challenge for driving simulators.* Proceedings of the First Human-Centered Transportation Simulation Conference. Iowa City, Iowa: University of Iowa.

Boldovici, J. A. (1992). *Simulator motion* (ARI Tech. Rep. No. 961). Arlington, VA: US Army Research Institute for the Behavioral and Social Sciences.

Bremmer, F., & Lappe, M. (1999). The use of optical velocities for distance discrimination and reproduction during visually simulated self-motion. *Experimental Brain Research, 127*(1), 33–42.

Bürki-Cohen, J., Soja, N. N., & Longbridge, T. (1998). Simulator platform motion—The need revisited. *The International Journal of Aviation Psychology, 8*, 293–317.

Colinot, J., & Belay, G. (2000). A multifunctional sound generator for PSA's Sherpa driving simulator. *Proceedings of the driving simulation conference 2000* (pp. 165–179). Paris, France.

Crowell, J. A., Banks, M. S., Shenoy, K. V., & Andersen R. A. (1998). Visual self-motion perception during head turns. *Nature Neuroscience, 1*(8), 732–737.

Davis, B. T., & Green, P. (1995). *Benefits of sound for driving simulation: An experimental evaluation* (Tech. Rep. UMTRI-95-16). Ann Arbor, MI: University of Michigan.

Drosdol, J., & Panik, F. (1985). *The Daimler-Benz driving simulator: A tool for vehicle development* (SAE Tech. Paper 850334). Warrendale, PA: Society of Automobile Engineers.

Duncan, B. (1995). *Calibration trials of the TRL driving simulator* (TRL report PA/3079/95). Crowthorne, England: Transport Research Laboratory.

Fredericksen, R. E., Verstraten, F. A. J., & van de Grind, W. A. (1993). Spatio-temporal characteristics of human motion perception. *Vision Research 33*(9), 1193–1205.

Gibson, J. J. (1950). *The perception of the visual world.* Westport, CT: Greenwood Press.

Godley, S. T., Triggs, T. J., & Fildes, B. N. (2002). Driving simulator validation for speed research. *Accident Analysis & Prevention, 34*, 589–600.

Gogel, W. C., & Tietz, J. D. (1979). A comparison of oculomotor and motion parallax cues of egocentric distance. *Vision Research, 19*(10), 1161–1170.

Greenberg, J., Artz, B., & Cathey, L. (2002). *The effect of lateral motion cues during simulated driving.* Proceedings of the Driving Simulation Conference North America 2003. Dearborn, MI.

Groeger, J. A., Carsten, O. M. J., Blana, E., & Jamson, A. H. (1997). Speed and distance estimation under simulated conditions. In A. G. Gale, I. D. Brown, C. M. Haslegrave, & S. P. Taylor (Eds.), *Vision in vehicles VII.* Amsterdam: Elsevier Sciences.

Groen, E. L., Howard, I. P., & Cheung, S. K. (1999). Influence of body roll on visually induced sensations of self-tilt and rotation. *Perception, 28*(3), 287–297.

Groen, E. L., & Bles, W. (2004). How to use body tilt for the simulation of linear self-motion. *Journal of Vestibular Research, 14*(4), 375–385.

Harms, L. (1996). Driving performance on a real road and in a driving simulator. In A. G. Gale, I. D. Brown, C. M. Haslegrave, I. Moorhead, & S. P. Taylor (Eds.), *Vision in vehicles V* (pp. 19–26). Amsterdam: Elsevier Sciences.

Harris, J. M., & Bonas, W. (2002). Optic flow and scene structure do not always contribute to the control of human walking. *Vision Research, 42*(13), 1619–1626.

Hess, R. A. (1990). A model for the human's use of motion cues in vehicular control. *Journal of Guidance, Control and Dynamics, 13*(2), 476–482.

Hosman, R. J. A. W., & Stassen, H. (1999). Pilot's perception in the control of aircraft motions. *Control Engineering Practices, 7*(11), 1421–1428.

Howard, I. P., & Rogers, B. J. (1995). *Binocular vision and stereopsis.* Oxford University Press.

Howarth, P. A. (1999). Oculomotor changes with virtual environments. *Applied Ergonomics, 30*(1), 59–67.

Jamson, A. H. (2001). Image characteristics and their effect on driving simulator validity. *Proceedings of the first international driving symposium on human factors in driver assessment, training and vehicle design* (pp. 190–195). Aspen, CO.

Jamson, A. H., & Mouta, S. (2004). More bang for your buck? A cross-cost simulator evaluation study. *Proceedings of the driving simulation conference 2004* (pp. 321–332). Paris, France.

Jamson, S. L., & Jamson, A. H. (2010). The validity of a low-cost simulator for the assessment of the effects of in-vehicle information systems. *Safety Science, 48*, 1477–1483.

Kemeny, A., & Panerai, F. (2003). Evaluating perception in driving simulation. *Trends in Cognitive Sciences, 7*(1), 31–37.

Land, M. F., & Lee, D. N. (1994). Where we look when we steer. *Nature, 369*(6483), 742–744.

Land, M. F., & Horwood, J. (1995). Which parts of the road guide steering? *Nature, 377*(6547), 339–340.

Lappe, M., Bremmer, F., & van den Berg, A. V. (1999). Perception of self-motion from visual flow. *Trends in Cognitive Sciences, 3*(9), 329–336.

Lee, D. N. (1976). A theory of visual control of braking based on information about time-to-collision. *Perception, 5*(4), 437–459.

Loomis, J. M., & Knapp, J. M. (1999). Visual perception of egocentric distance in real and virtual environments. In L. J. Hettinger & M. W. Haas (Eds.), *Virtual and Adaptive Environments.* Mahwah, NJ: Lawrence Erlbaum Associates.

McLane, R. C., & Wierwille, W. W. (1975). The influence of motion and audio cues on driver performance in an automobile simulator. *Human Factors, 17*(5), 488–501.

Melcher, G. A., & Henn, V. (1981). The latency of circular vection during different accelerations of the optokinetic stimulus. *Perception and Psychophysics, 30*(6), 552–556.

Merat, N. (2003). *Loading driver to their limit: The effect of increasing secondary task on driving.* Proceedings of the Second International Driving Symposium on Human Factors in Driver Assessment, Training and Vehicle Design. Park City, Utah.

Nahon, M. A., & Reid, L. D. (1990). Simulator motion-drive algorithms: A designer's perspective. *AIAA Journal of Guidance, Control and Dynamics, 13*(2), 356–362.

Nordmark, S., Lidström, M., & Palmkvist, G. (1984). *Moving base driving simulator with wide angle visual system* (SAE Tech. Paper 845100). Warrendale, PA: Society of Automobile Engineers.

Page, N. G., & Gresty, M. A. (1985). Motorists' vestibular disorientation syndrome. *Journal of Neurology, Neurosurgery and Psychiatry, 48*(8), 729–735.

Panerai, F., Cornilleau-Peres, V., & Droulez, J. (2002). Contribution of extra-retinal signals to the scaling of object distance during self-motion. *Perception and Psychophysics, 64*(5), 717–731.

Reymond, G., Kemeny, A., Droulez, J., & Berthoz, A. (2001). Role of lateral acceleration in curve driving: Driver model and experiments on a real vehicle and a driving simulator. *Human Factors, 43*(3), 483–495.

Riesmersma, J. B. J., van der Horst, A. R. A., & Hoekstra, W. (1990). The validity of a driving simulator in evaluating speed-reducing measures. *Traffic Engineering and Control, 35*, 416–420.

Rogers, B., & Graham, M. (1979). Motion parallax and an independent cue for depth perception. *Perception, 8*(1), 125–134.

Santos, J., Merat, N., Mouta, S., Brookhuis, K., & de Waard, D. (2005). The interaction between driving and in-vehicle information systems: Comparison of results from laboratory, simulator and real-world studies. *Transportation Research Part F: Traffic Psychology and Behavior, 8*(2), 135–146.

Siegler, I., Reymond, G., Kemeny, A., & Berthoz, A. (2001). Sensorimotor integration in a driving simulator: Contribution of motion cueing in elementary driving tasks. *Proceedings of the driving simulation conference 2001* (pp. 45–55). Nice, France.

Soma, H., Hiramatsu, K., Satoh, K., & Uno, H. (1996). *System architecture of the JARI driving simulator and its validation.* Proceedings of the Symposium on the Design and Validation of Driving Simulators. Valencia, Spain.

Stewart, D. (1966). A platform with six segrees of freedom. *The Institution of Mechanical Engineers Proceedings 1965-66, 180 Part 1*(15), 371–386.

Szczepanski, C., & Leland, D. (2000, August). *Move or not to move? Continuous question* (AIAA No. 2000-4297). Paper presented at the AIAA Modeling and Simulation Technologies Conference, Denver, CO.

Takeuchi, T., & De, V. (2000). Velocity discrimination in scotopic vision. *Vision Research, 40*, 2011–2024.

Tornros, J. (1998). Driving behavior in a real and a simulated road tunnel: A validation study. *Accident Analysis & Prevention, 30*(4), 497–503.

van Hofsten, C. (1976). The role of convergence in visual space perception. *Vision Research, 16*(2), 193–198.

van Winsum, W., & Godthelp, H. (1996). Speed choice and steering behavior in curve driving. *Human Factors, 38*(3), 257–268.

Wann, J., & Land, M. J. (2000). Steering with or without the flow: Is the retrieval of heading necessary? *Trends in Cognitive Sciences, 4*(8), 319–324.

Weir, D. H., & Clark, A. J. (1995). *A survey of mid-level driving simulators* (SAE Tech. Paper 950172). Warrendale, PA: Society of Automobile Engineers.

Wickens, C. D., & Hollands, J. G. (2000). *Engineering psychology and human performance.* (3rd ed.). Upper Saddle River, NJ: Prentice Hall.

Wierwille, W. W., Casali, J. G., & Repa, B. S. 1983. Driver steering reaction time to abrupt-onset crosswinds as measured in a moving-base driving simulator. *Human Factors, 25*(1), 103–116.

Yan, X., Abdel-Aty, M., Radwan, E., Wang, X., & Chilakapati, P. (2008). Validating a driving simulator using surrogate safety measures. *Accident Analysis & Prevention, 40*, 274–288.

13

Simulator Validity: Behaviors Observed on the Simulator and on the Road

Nadia Mullen
Lakehead University

Judith Charlton
Monash University

Anna Devlin
Monash University

Michel Bédard
Lakehead University

Abstract

The Problem. Driving simulators offer a safe, convenient alternative to measuring driving performance on-road. However, the results of simulator studies may not generalize to driving in the real world if the simulator lacks behavioral validity. Behavioral validity refers to the extent to which the simulator elicits the same driving behaviors that occur when driving in the real world. *Role of Driving Simulators.* Validation is important to generate and maintain simulator use, acceptance, and credibility, and is vital when simulator performance influences real-world outcomes, such as road or vehicle design, or whether drivers retain their license. *Key Results of Driving Simulator Studies.* A review of studies evaluating the behavioral validity of simulators showed that simulators provide a valid tool for assessing a variety of driving performance measures such as speed, lateral position, brake onset, divided attention, and risky traffic behaviors. Simulators also appear sensitive to age-related changes in driving performance and cognition. Measures for which simulators do not appear valid are discussed, in addition to factors influencing validity, such as driving ability. Overall, the evidence reviewed in this chapter indicates that simulator driving behavior approximates (relative validity), but does not exactly replicate (absolute validity), on-road driving behavior. This is sufficient for the majority of research, training, and assessment purposes for which simulators are used. However, where absolute values are required, on-road measures will generally be necessary. *Scenarios and Dependent Variables.* Validation studies involve consideration of factors such as the research question, task conditions, and dependent measures, each of which can affect validity. We discuss these methodological considerations, as well as statistical techniques used to establish validity. *Platform Specificity and Equipment Limitations.* Assumptions about driving simulator validity are critically dependent on the specific experimental conditions under which the driving behaviors are compared. Variations across simulator equipment, software, and environment may affect the generalizability of validation results. Therefore, each simulator should be validated for its ability to measure the driving behavior of the cohort for which it is to be used.

13.1 Introduction

This chapter addresses the topic of the behavioral validity of simulators, that is, the extent to which the simulator induces the same driving behaviors that occur when driving in the real world. The issue of validation is important in terms of generating and maintaining simulator use, acceptance, and credibility. It is especially vital when simulator performance outcomes are used to influence real-world outcomes, such as road or vehicle design, or whether drivers retain their license.

Driving simulator validation studies have generally assessed validity in terms of physical and behavioral validity (Blaauw, 1982; Blana, 1996). *Physical validity* refers to the extent to which the physical components of the simulator vehicle correspond to the on-road vehicle, including the simulator layout, visual displays, and dynamics such as the feel of braking and steering controls (see Table 13.1, and see in this book, chap. 8 by Andersen and chap. 7 by Greenberg & Blommer). *Behavioral validity* refers to the level of correspondence between the driving behaviors elicited in the simulator and on real roads (see also Ranney, this book, chap. 9). In some contexts, both physical and behavioral validity have been used to describe the simulator's *external validity* (Blana & Golias, 2002; Klee, Bauer, Radwan, & Al-Deek, 1999; Reed & Green, 1999; Törnros, 1998) or the extent to which driving behavior in the simulator can be generalized to real driving behavior (see Table 1; Hoskins & El-Gindy, 2006).

Behavioral validity has been further defined in terms of *absolute* and *relative validity* (Blaauw, 1982). While absolute validity requires that the two driving environments produce the same numerical values, relative validity is established when the differences between the two environments are in the same direction, and of the same or similar magnitude (see Table 1; Godley, Triggs, & Fildes, 2002). Absolute validity is rarely established in driving simulator studies due to a number of driving simulator characteristics that are not found in real on-road settings. Indeed, it is argued that absolute validity is not necessary for simulators to be useful research tools; rather, establishing relative validity is necessary and sufficient (Törnros, 1998). This is particularly so when the researcher is interested in comparing changes in driving patterns under different treatment conditions across simulator and real-world settings. In these circumstances, it is more important to establish that the treatment resulted in the same kind of behavior change (e.g., speed reduction) compared with a control condition, in both settings, and less critical to establish that the magnitude of change (e.g., amount of speed reduction) is precisely matched across settings.

13.2 Methodological Considerations

13.2.1 The Research Question, Task Conditions, and Dependent Measures

The concept of simulator validity has relevance only in the context of a specific research question and it is likely that the level of similarity of driving behaviors across simulator and real driving settings may be highly task-dependent. Participant characteristics may also influence validity, such as participants' experience of simulator discomfort and driver motivation (Blana, 1996), driver age and experience, and the presence of medical conditions. That is to say, while a simulator may be described as a valid instrument for training novice drivers to undertake hazard-avoidance maneuvers in complex city traffic, the same device may not be valid for evaluating the performance of experienced drivers or drivers with impairments on a different task, such as highway driving with little or no traffic and no hazardous events. Indeed, Kaptein and colleagues caution that "any use of driving simulators should be preceded by questioning whether the simulator is sufficiently valid for the task or ability to be investigated" (Kaptein, Theeuwes, & van der Horst, 1996, p. 31).

Assumptions about driving simulator validity are also critically dependent on the specific experimental conditions under which the driving behaviors are compared. Variations across simulator equipment (e.g., number and size of display monitors), software (e.g., simulator program), and environment (e.g., temperature, noisiness) may affect the generalizability of validation results. Several simulator-based factors have been implicated in influencing simulator validity, including the fidelity of proprioceptive information, the presence of a motion platform, and the quality of image resolution of the display of road and traffic environment (Kaptein et al., 1996).

Similarly, the operational definition of real-world driving (i.e., the precise conditions under which real-world driving performance is measured) can influence the authenticity of validation data. Validation studies (i.e., studies that compare real-world with simulated performance) reviewed in this chapter differed widely in their operational definitions of real-world driving. Some studies examined on-road driving in naturalistic or "uncontrolled" conditions (e.g., driving in real traffic with a standard vehicle); in other studies, on-road driving was examined in more controlled conditions (e.g., driving an instrumented vehicle on a test track with an experimenter as a passenger; Blana, 1996).

Assessments of simulator validity may produce widely varying results, depending on which driving measures are compared. For example, high levels of similarity for one measure of driver behavior indicating validity across simulator and real-world settings, such as for speed, may not necessarily be observed for other measures, such as braking, steering, or lane position. Kaptein et al. (1996) emphasized the importance of selecting measures to adequately assess simulator validity; not all variables measured during a simulator drive accurately reflect the same measure in an on-road drive. In a study involving older drivers, Lee, Cameron and Lee (2003) correlated age with individual behavior measures obtained from the simulator to establish criterion-related validity (i.e., to check that each simulator measure showed expected age-related declines in driving performance) prior to analyzing the variables of interest. A key point here in relation to study design is that not all measures may be necessary or relevant for the task and research question under investigation.

Variations in data collection methods may play a role in influencing data accuracy and authenticity (Blana, 1996). For example,

TABLE 13.1   Examples of Evaluating Validity

Type of Validity	Source of Definition	Possible Methods of Testing for this Validity	Examples and Expected Results when Validity is Established[a]
Behavioral Validity			
Absolute validity (established when the simulated and on-road drives produce the same numerical values)	Blaauw, 1982		*Example:* Speed measured during a simulated and an on-road drive is compared.
		ANOVA	*Expect:* Means will not differ significantly.
			e.g.,
			mean simulator speed = 50 km/h, mean on-road speed = 53 km/h, $p > 0.05$
		Visual inspection + correlation	*Expect:* Graphed data will have similar shape and may overlap, and correlation will be significant.
Relative validity (established when the simulated and on-road drives produce numerical values similar in magnitude and in the same direction)	Törnros, 1998		*Example:* Speed measured during a simulated and an on-road drive, on both urban and rural roads, is compared.
		ANOVA	*Expect:* Means may differ significantly between the simulated and on-road drive, but they will be in the same direction on each level of the independent variable, and of similar magnitude.
			e.g.,
			Urban roads: Mean simulator speed = 50 km/h, mean on-road speed = 58 km/h, $p < 0.05$
			Rural roads: Mean simulator speed = 80 km/h, mean on-road speed = 89 km/h, $p < 0.05$
		Visual inspection + correlation	*Expect:* Graphed data will have similar shape and be non-overlapping, a similar distance will separate the lines at each level of the independent variable, and correlation will be significant.
Interactive absolute validity (established when the simulated and on-road drives produce the same numerical values over time)	Godley et al., 2002		*Example:* Speed measured at 5 points along the same 50 m section of a simulated and an on-road drive is compared.
		ANOVA	*Expect:* Means at each of the 5 points will not differ significantly.
			e.g.,
			Point 1: Mean simulator speed = 50 km/h, mean on-road speed = 53 km/h, $p > 0.05$
			Point 2: Mean simulator speed = 55 km/h, mean on-road speed = 59 km/h, $p > 0.05$
			Point 3: Etc.
		Visual inspection + correlation	*Expect:* Graphed data will have similar shape and may overlap at all time points, and correlation will be significant.
Interactive relative validity (established when the simulated and on-road drives produce numerical values similar in magnitude and in the same direction over time)	Godley et al., 2002		*Example:* Speed measured at 5 points along the same 50 m section of a simulated and an on-road drive is compared.
		ANOVA	*Expect:* Means at each of the 5 points may differ significantly between the simulated and on-road drive, but they will be in the same direction at each time point, and of similar magnitude.
			e.g.,
			Point 1: Mean simulator speed = 50 km/h, mean on-road speed = 58 km/h, $p < 0.05$
			Point 2: Mean simulator speed = 55 km/h, mean on-road speed = 64 km/h, $p < 0.05$
			Point 3: Etc.
		Visual inspection + correlation	*Expect:* Graphed data will have similar shape and be non-overlapping at all time points, a similar distance will separate the lines at each time point, and correlation will be significant.

(continued)

TABLE 13.1 (Continued)   Examples of Evaluating Validity

Type of Validity	Source of Definition	Possible Methods of Testing for this Validity	Examples and Expected Results when Validity is Established[a]
Physical Validity			
Physical validity (the extent to which the physical components of the simulator correspond to on-road vehicles)	Blaauw, 1982	Visual inspection of the simulator components and layout, and its dynamic characteristics such as the response of braking and steering controls	*Example:* The physical features of a simulator are compared with a real vehicle. *Expect:* A high degree of similarity between the simulator and on-road vehicle. *e.g.,* A simulator with a full cab, motion-base, and projected 360 degree view has higher physical validity than a simulator with a car seat only, fixed-base, and view presented on a 17" computer monitor.
External Validity			
External validity (the extent to which the results obtained in the simulator can be generalized to driving on the road; can include physical and behavioral validity)	Kaptein et al., 1996	ANOVA; Visual inspection + correlation	Combination of behavioral and physical validity examples.
Internal (Criterion) Validity			
Internal validity (the extent to which causal inferences about the impact of an experimental treatment [e.g., the introduction of speed limits resulting in a speed reduction] can be made with confidence; can include physical and behavioral validity)	Reimer et al., 2006	ANOVA; Visual inspection + correlation	Combination of behavioral and physical validity examples.

[a] Significant differences mentioned in this column refer to statistical significance and do not imply practical significance.

some simulator measures are generated by the computing facility of the simulator, while other measures rely on observational techniques. In some research contexts, it may be pertinent to establish consistency across these different methods of data collection. For example, using a STISIM 400 simulator, Bédard and colleagues established a significant correlation ($r = 0.83$) between simulator-recorded driving errors and real-time researcher-recorded demerit points in the same simulator setting (Bédard, Parkkari, Weaver, Riendeau, & Dahlquist, 2010). They also showed a high level of consistency across driver behavior measures conducted in real-time with scores determined from later video analyses ($r = 0.83$). These results suggest that for the driving errors considered in Bédard et al. (2010) study, the various methods of collecting data did not affect the simulator's validity. However, as noted above, these findings may not generalize to other measures in this simulator (e.g., speed, braking responses) or to measures of driving errors in other simulators because any difference (e.g., in equipment, surroundings, or measurement technique) could affect validity. Furthermore, the relationships between the measures reported here for simulator driving behaviors will not necessarily hold for real-world driving; direct comparison of measures across the two settings would be required to determine this.

13.2.2 Analysis

Typically, simulator validation studies have used conventional statistical approaches for group comparisons, such as analysis of variance (ANOVA), to examine differences between simulator and on-road driving performance measures (Klee et al., 1999; McGehee, Mazzae, & Baldwin, 2000; Reed & Green, 1999). For example, absolute validity is claimed if group means for a driving behavior (e.g., speed) measured during the simulated and on-road drives do not significantly differ (Table 13.1). If mean performance measures differ significantly between the simulated and on-road drives, but are in the same direction across levels of another independent variable (e.g., speeds on urban roads being slower than speeds on rural roads during both simulated and on-road drives), relative validity will have been established (Table 13.1). These statistical procedures have been used to establish both absolute and relative validity for key driving variables such as mean speed, braking response time, variability in lane position, or steering wheel angle, measured at specific points of interest in the drive or expressed as an average across a designated section of the drive. Table 13.2 provides a summary of the studies reviewed in the following section of this chapter, and presents information concerning the statistical approach taken to assess simulator validity for each driving measure.

Another approach for evaluating validity relies on a comparison of specific aspects of the driving profile (such as speed or error measures) for a key region of a drive or across an entire drive, determining a simulator's *interactive* (or *dynamic*) *relative validity* (see Table 13.1). This approach captures the collective dynamic nature of driving and allows the comparison of patterns of performance *throughout* a driving task, in drivers responding

TABLE 13.2 Summary of Driving Simulator Validation Studies Cited in the Chapter

Validation Studies	Driving Simulator	Data Collection Technique	n/Age	Statistical Approach	Comments/Key Findings — Absolute Validity Established	Comments/Key Findings — Relative Validity Established
Bédard, 2008	STISIM	Sim. and on-road; physiological responses	N = 24, 21–57 yrs	ANOVA		Perceived threat from unexpected events
Bédard et al., 2010	STISIM	Sim. and on-road; cognitive performance	N = 8–38, 67–81 yrs	Correlation		Sim. driving and on-road demerit points
Bella, 2005	Inter-Uni. Research Center for Road Safety (CRISS)	Sim. and on-road	N = 35, 24–45 yrs	Comparison of means—bilateral Z test for non-matched samples		Speed
Bella, 2008	CRISS	Sim. and on-road	N = 40, 23–60 yrs	Comparison of means—bilateral Z test for non-matched samples	Speed for non-demanding road configurations	Speed and complexity of maneuver
Bittner et al., 2002	Uni. of Washington Real Drive	Sim. and on-road	N = 12, 21–34 yrs	Generalized Youden plots		Curve entry speeds
Blaauw, 1982	Institute for Perception TNO	Sim. and on-road	N = 48, 18–36 yrs	ANOVA, correlation	Speed	Lateral displacement
Blana & Golias, 2002	Leeds Advanced Driving Simulator (LADS)	Sim. and on-road	N = 100, M age = 36 yrs	Independent samples *t* test		No – lateral displacement
Charlton et al., 2008	Monash Uni. Accident Research Centre (MUARC)	Sim. and on-road	N = 30, 29–89 yrs Vision impaired and controls	ANOVA, correlation		Speed and gaze direction
Fisher et al., 2007	Uni. of Massachusetts-Amherst advanced fixed-base driving simulator	Sim. and on-road	N = 12–24, 18–21 yrs	Independent samples *t* test		Training effects and gaze direction
Godley et al., 2002	MUARC	Sim. and on-road, speed countermeasures	N = 24, 22–52 yrs	ANOVA; correlations and canonical correlations		Speed countermeasures
Hakamies-Blomqvist et al., 2001	VTI	Sim. and on-road	N = 35, 66–80 yrs	ANOVA		Steering wheel angle and lane position
Hoffman et al., 2002	Iowa Driving Simulator (IDS)	Sim. and on-road	N = 16, 25–55 yrs	Descriptive statistics: Comparison of means		Mean braking onset
Klee et al., 1999	Uni. of Central Florida (UCF)	Sim. and on-road	N = 30, 17–65 yrs	A bilateral Z test for non-matched samples		Forward speed
Lee, Cameron, et al., 2003	STISIM	Sim. and on-road	N = 129, 60–88 yrs	Correlations and principal component analysis		Sim. driving performance and on-road driving performance
Lee et al., 2007	STISIM	Sim. and on-road	n = 50 PD patients, n = 150 controls, 60–80 yrs	Pearson correlation, stepwise linear regression		Control participants' overall driving performance in sim. and on-road
Lee et al., 2005	STISIM	Sim., cognitive performance, driver violation points 2 year interval	N = 129, 65–88 yrs	Hierarchical Poisson regression analysis		The prediction of older driver crash history from driving sim. performance
Lee, Lee, & Cameron, 2003	STISIM	Sim. and measure of visual attention	N = 129, 60+ yrs	Repeated measures ANOVA		Visual attention skill and age

(continued)

TABLE 13.2 (Continued) Summary of Driving Simulator Validation Studies Cited in the Chapter

Validation Studies	Driving Simulator	Data Collection Technique	n/Age	Statistical Approach	Comments/Key Findings	
					Absolute Validity Established	Relative Validity Established
Lee, Lee, Cameron, & Li-Tsang, 2003	STISIM	Sim. and crash history	N = 129, 60–88 yrs	Logistic regression		Driving performance and previous crash history
Lew et al., 2005	STISIM	Sim. and on-road, observational data	n = 11 TBI, n = 16 controls, 18–58 yrs	Correlations		No – TBI individuals' driving performance
McAvoy et al., 2007	Doron Precision Systems Inc. AMOSII	Sim. and on-road	N = 127, 18–70 yrs	ANOVA		No – speed through a work zone
McGehee et al., 2000	IDS	Sim. and on-road	N = 120, 25–55 yrs	ANOVA and 95% confidence interval	Total brake reaction time	
Philip et al., 2005	Divided Attention Steering Simulator	Sim. and on-road; self-rated fatigue; reaction time	N = 12, 19–24 yrs	ANOVA		Inappropriate line crossings due to fatigue
Reed & Green, 1999	Uni. of Michigan Transportation Research Institute (UMTRI)	Sim. and on-road and a concurrent manual dialling phone task	N = 12, 20–30 yrs and 60+ yrs	ANOVA	Speed control	Effects of a phone task and age
Reimer et al., 2006	Instrumented full cab 2001 Volkswagen Beetle	Validated sim. with respect to self-report questionnaires	N = 48, 16–55 yrs	A multi-trait-multi method correlation matrix of driving behavior		Self-reported survey items and sim. performance
Riemersma et al., 1990	Daimler-Benz	Sim. and on-road	N = 24, 20–35 yrs	Descriptive statistics: Comparison of a reduction in mean speed		Speed reduction measures
Shinar & Ronen, 2007	STISIM	Sim. and on-road	N = 16, 28–30 yrs	Correlation and regression analysis		Changes in speed
Slick et al., 2006	DriveSafety DS-600c motion-based simulator	Sim. and on-road; physiological responses	N = 22, M age = 16 yrs	ANOVA	Physiological responses during turns at intersections	Physiological responses during entire drive
Törnros, 1998	VTI driving sim.	Sim. and on-road through a tunnel	N = 20, 23–52 yrs	ANOVA		Speed and lateral position
Toxopeus, 2007	STISIM	Sim.	N = 79, 18–30 yrs and 60–80 yrs	ANOVA		Reaction time (to brake and press horn) and age; speed and age
Volkerts et al., 1992	TS2 driving simulator	Sim. and on-road	N = 18, 25–31 yrs	MANOVA, correlation and multiple regression		No – sedative drug induced impairment
Wade & Hammond, 1998	Wrap around sim. (WAS) Uni. of Minnesota	Sim. and on-road	N = 26, 18–34 yrs	Descriptive statistics: Raw data charts of lane position from centerline		Lane position
Watts & Quimby, 1979	Transport and Road Research Laboratory	Sim. and on-road; physiological responses and perceived hazard risk ratings	N = 60, Age range unknown	Correlation coefficient	Perceived risk of hazards on the road and in the sim.	
Yan et al., 2008	UCF	Sim. and on-road; crash history	N = 241, 15–45+ yrs	ANOVA		Risky behaviors at signalized intersections

to different events or treatments. Godley et al. (2002) investigated interactive validity by comparing patterns of driver speed control in a simulator and on a real road, in response to rumble strips. In addition to their examination of absolute and relative validity, using conventional measures, the authors examined speed profiles over the entire data collection area (derived from one meter averages) and applied a correlation analysis, based on canonical correlation, to evaluate patterns of speed management in response to treatment and control (no rumble strip) sites on the road and in the simulator (shown in Figure 13.1, respectively). Participants drove faster at both treatment and control sites in the simulator compared with the on-road setting (arguing against absolute validity). However, when averaged over the total data collection area and across both experiments, speed at the treatment site stop sign approach was significantly slower than at the control site. Treatment site speeds were slower in both the simulator and on-road settings. Thus, relative validity

was established for the stop sign approach speed. Moreover, the authors noted that the *pattern of speeds for the treatment site relative to the control site* was also similar in both simulator and on-road settings throughout most of the stop sign approach, as demonstrated by a significant correlation ($r = 0.40$), suggesting support for interactive relative validity.

Finally, as is evident from Table 13.2, the association between the on-road and simulator data when establishing interactive relative validity need not be confined to a linear one. Validity is established when any mathematical function significantly relates the on-road and simulator data. This type of validity, which may be thought of as *functional validity*, allows for consideration of a range of mathematical relationships between the variables, including both linear and non-linear associations. Functional validity is a type of behavioral validity, and linear associations are synonymous with absolute or relative validity. However, there are many types of functions that may explain

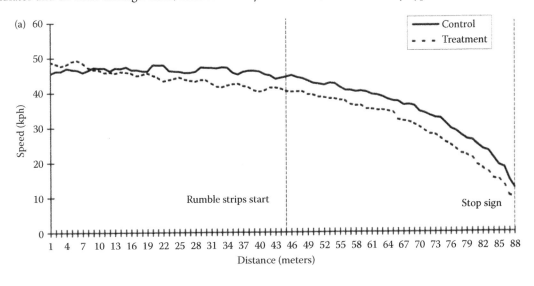

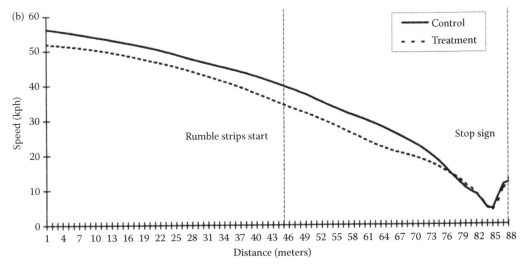

FIGURE 13.1   Mean approach speed to a stop sign in the (a) on-road drive, and (b) simulator drive. (Godley, S. T., Triggs, T. J., & Fildes, B. N. (2002). Driving simulator validation for speed research. *Accident Analysis & Prevention, 34*, 589–600. Copyright 2002 by Elsevier Science. Reprinted with permission.)

the relationship between on-road and simulator data (e.g., linear, quadratic, exponential, hyperbolic), not all of which may indicate an adequate level of validity.

13.3 Review of Studies Investigating Behavioral Validity

Despite the importance that is placed on the simulator as a tool for understanding driving performance and driver competence, an extensive literature search revealed a limited number of studies specifically evaluating the behavioral validity of simulators (Table 13.2). Most of the available literature on simulator validation has focused on measures of speed, lateral position, and braking responses, with a few studies examining more complex driving behaviors (e.g., performance while dialing a car phone) or behaviors of specific driver groups (e.g., older drivers). The following section explores the literature on this topic with a particular emphasis on the selection of performance measures, the tasks/conditions under investigation, and the evidence provided for behavioral validity of specific simulators and simulated driving behaviors.

13.3.1 Speed

One of the most commonly studied measures of behavioral validity of driving simulators is driver speed (Bella, 2005; Bella, 2008; Bittner, Simsek, Levison, & Campbell, 2002; Blaauw, 1982; Blana, & Golias, 2002; Godley et al., 2002; Klee et al., 1999; Törnros, 1998). Overall, studies have consistently demonstrated relative (but not absolute) validity for speed data. Research by Godley et al. (2002), presented above, shows strong evidence for both relative and interactive relative validity for speed on the Monash University Accident Research Centre (MUARC) driving simulator, but absolute values of speed differed across the simulated and real-world driving settings. Similarly, recent work by two of the authors using the same simulator has shown preliminary evidence for relative coupling of speed profiles in a simulator and controlled on-road driving event in drivers with both visual field loss and age-matched healthy controls (Charlton, 2008). The driving task involved several hazardous events including a car (target) exiting a driveway, stopping just short of the driver's lane. The timing and location of the hazardous events in the on-road drive was controlled as tightly as possible to match the simulated drive conditions using a system of road markers and use of interactive signals between the test vehicle and target vehicle. Inspection of the data in Figure 13.2 shows individual Case and Control participant profiles with a relatively tight coupling of the start point and rate of the deceleration in response to the unexpected incursion of a car at the side of the road next to the driver's lane. Case and Control participants reached their slowest approach speeds slightly earlier (within 25 m) in the simulator than the on-road task.

Klee et al. (1999) examined the validity of the fixed-base simulator at the University of Central Florida with respect to forward speed. Thirty drivers (aged 17–65 years) drove the same section of campus road on the simulator and in an instrumented vehicle. Longitudinal speed was recorded at 16 locations along the track.

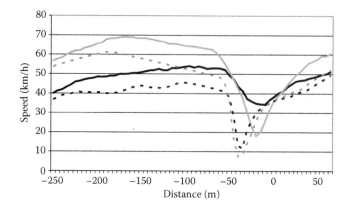

FIGURE 13.2 Individual Control (black) and vision-impaired Case (grey) participants' speed profiles for simulator (broken line) and controlled on-road (solid line) driving event involving a car exiting from a driveway next to the driver's lane at location 0 m.

Speeds in the two driving environments were similar at 10 locations, while at the remaining six locations participants drove at a slower speed in the simulator, leading the authors to conclude that absolute validity was not demonstrated. While the authors make no specific claims about relative validity, the results show that at the majority of data sites (10 out of 16) participants drove approximately 5–10 km/h slower in the simulator than on-road, and the speed differential was the same, implying relative validity. Bella (2005) provided further evidence for relative coupling of speed in a simulator and an on-road environment in a highway driving context. Thirty-five drivers (aged 24–45 years) completed a 12 km highway drive through a work zone from which speed recordings were obtained every five meters. A bilateral Z-test for non-matched samples was performed in order to estimate whether the driving speeds in the field were significantly different from those in the simulator. Consistent with Klee et al. (1999) findings, Bella (2005) reported that the simulator speeds tended to be lower than the speeds recorded in the field drive. However, the difference between the mean speeds in the simulator compared to the mean speeds in the field were not significantly different at each measurement site, demonstrating, at a minimum, interactive relative validity of the interactive static base simulator for assessing speed.

Yan, Abdel-Aty, Radwan, Wang and Chilakapati (2008) investigated the validity of the University of Central Florida's driving simulator for evaluating aspects of traffic safety at signalized intersections. A real-world intersection was replicated in a simulator, with eight intersection scenarios. Participants' speed (in four scenarios) and risky traffic behaviors (in four scenarios) were recorded. Absolute validity was demonstrated for the majority of intersection approach speed measures. Comparisons between participants' approach speed in the four real and simulated scenarios showed that all speed distributions followed normal distributions, there were no statistically significant differences in three of the four scenarios for mean approach speed (the fourth scenario approached significance at $p = 0.051$), and no differences in two of the four scenarios for speed variance. Additionally, Yan et al. (2008) found evidence for relative validity for selected measures of risky driving behavior. Comparisons of simulator driving behavior at two right-turn lanes—one with a real-world low rear-end crash record (two crashes) and one with a real-world high rear-end crash record (16 crashes)—revealed a higher number of risky behaviors (e.g., higher deceleration rate, faster speed at the stop line, closer following distance) in the lane with the higher real-world crash rate. That is, the same performance patterns observed in the real-world low and high risk intersections were upheld in the simulator.

Rather than using more conventional ways of measuring speed as an indicator of behavioral validity, Shinar and Ronen (2007) examined two speed-related measures: speed estimation (i.e., driver estimations of the vehicle's current speed) and speed production (i.e., adjustment of the vehicle's speed to achieve a predetermined speed). Sixteen participants (aged 24–30 years) completed an on-road and a simulated drive (in a full-cab simulator with STISIM software) with the speedometer shielded from view. During each drive, participants were required to estimate and produce seven different speeds (40, 50, 60, 70, 80, 90, and 100 km/h). The authors found that speed estimation performance in the simulator only differed significantly from on the real road for the speeds of 40 km/h (where higher speeds were estimated in the simulator) and 100 km/h (where lower speeds were estimated in the simulator). For speed production, produced speeds were consistently approximately 25 km/h faster in the simulator than on-road. This study shows absolute validity for most speeds for speed estimation, and relative validity for speed production.

13.3.2 Lateral Position

Another common measure for validation is the extent to which the lateral position of the vehicle is matched across simulator and real driving contexts. Blana and Golias (2002) investigated the validity of the fixed-base Leeds Advanced Driving Simulator for assessing vehicle lateral displacement (measured as the distance from the front tire on the passenger side of the vehicle to the white line at the edge of the road) on straight and curved sections of the road. One hundred on-road drivers were observed (with video cameras positioned on the ground and mounted on lamp posts) driving on one straight and two curved sections of a road. A further 100 participants (mean age of 36 years) completed a simulated version of the on-road drive. The results showed that for both straight and curved sections, the mean lateral displacement was significantly larger on the real road than in the simulator (suggesting that on-road drivers positioned their vehicles further from the road edge, closer to the centerline), while the standard deviation (SD) of lateral displacement was significantly smaller on the real road than in the simulator (suggesting that on-road drivers maintained a more consistent lane position). However, speed influenced the results. For straight sections, the mean and SD of lateral displacement only differed between the real and simulated roads when speeds were greater than 70 km/h. For curved sections, the mean lateral displacement differed significantly only when speeds were

less than 60 km/h (speed did not affect results for the SD of lateral displacement on curved sections). Although the direction of the difference in lateral displacement between the on-road and simulated drives was consistent across all speeds (i.e., the on-road mean lateral displacement was always larger than the simulated drive, and the on-road SD of lateral displacement was always smaller), the amount of difference in lateral displacement varied depending on speed, and thus neither absolute nor relative validity was demonstrated.

Wade and Hammond (1998) also examined simulator validity for measuring lane deviation (measured as the vehicle's distance from the centerline). They assessed the driving performance of 26 participants (aged 18–34 years) on a real-world driving route and a comparable virtual driving route programmed on the University of Minnesota's Human Factors Research Laboratory wrap-around simulator. The mean deviation from the centerline was statistically significantly larger in the virtual than the on-road environment (suggesting that on-road drivers positioned their vehicles closer to the centerline, further from the road edge), thus absolute validity was not demonstrated. However, mean deviations on four different types of roads showed a similar pattern in the two environments (e.g., the road with the highest mean deviation in the real world was also the road with the highest mean deviation in the virtual environment), thus demonstrating relative validity. It is important to note, however, that data were collected only on straight road segments with maximum speed limits of 35 mph (approximately 56 km/h) and where drivers would have a constant speed. Whether speed would influence validity for measures of lateral position (as found in Blana & Golias, 2002) remains unknown. Wade and Hammond (1998) also observed that drivers who displayed an aggressive driving style on-road (e.g., hard acceleration, hard braking, and speeding) showed similar aggressive behaviors in the simulator. Furthermore, similar habitual behaviors were observed in the on-road and simulated setting; participants who steered away from the centerline when approaching a real stop sign also displayed this behavior in the simulator. Participants also displayed reflexive responses to deceleration (i.e., they flexed their arm and shoulder muscles) when braking in the simulator. Reflexive responses in real-world driving prevent drivers moving forward due to inertia when they brake. There was no inertia in the simulator, and hence reflexive responses were not required, yet participants performed this habitual behavior. It is clear that the simulator evoked many behaviors of drivers in the real world, suggesting the simulator was behaviorally valid. Overall, these studies suggest that simulators can show relative validity for measures of lateral position, but they are unlikely to show absolute validity; variation in lane position is likely to be greater in simulated than real-world environments, and drivers are likely to travel further from the centerline in the simulated world than in the real world.

13.3.3 Braking Responses

In some studies, braking responses offer an obvious measure for simulator validation, particularly for driving tasks where rapid and timely braking is critical for safety to avoid a crash. Hoffman, Lee,

Brown and McGehee (2002) investigated the braking responses of drivers in an on-road and a simulator scenario to validate the Iowa Driving Simulator for braking. Sixteen participants (aged 25–55 years) were instructed to brake normally or hard to avoid a collision with a leading vehicle. The time headway (the time between the two vehicles when the lead vehicle began to break) was fixed at 1.7 seconds in the simulator. The time headway on the on-road course could not be fixed, but drivers had a mean time headway of 1.6 seconds, suggesting that the fixed simulator headway of 1.7 seconds was representative of real driving conditions. Performance was visually compared on several measures of brake onset, brake profile, and brake completion. The general pattern of results in the simulated and on-road courses was similar (most measures showed a difference between simulator and on-road performance of similar magnitude), and the driver's initial speed (30 or 60 mph) and the deceleration rate of the leading vehicle (0.15 or 0.40 g) affected braking responses similarly in the two driving environments (i.e., most means for simulator and on-road performance were in the same direction for each level of the independent variables), suggesting evidence of behavioral validity. However, it is difficult to confidently evaluate validity using only visual inspection to describe observed differences. In this study, simulator data were compared with data previously collected on-road; hence detailed data were available only for the simulated drives. Analyses were limited to graphed means for on-road data, and graphed means with error bars for simulator data. Although this enables some conclusions to be drawn regarding the simulator's validity, it is preferable to statistically analyze data, and the absence of statistical analyses should be considered a limitation of the study.

Regarding the effect of the braking instructions (normal or hard), drivers in Hoffman et al. (2002) study responded differently in the two environments. For the hard braking instruction, data were similar in the simulator and on the road. However, for the normal braking instruction, participants in the simulator responded in the same way as they did to the hard braking condition. Given that the simulator did not provide the same cues associated with normal braking, including haptic and vestibular feedback, participants may have found it difficult to implement the different degrees of braking required.

Braking and accelerator usage were also studied by McGehee et al. (2000) to evaluate the validity of the Iowa Driving Simulator for a crash-avoidance task. Participants completed a simulator drive that involved an intersection crash scenario, and an on-road drive that included a similar crash scenario. The on-road drive required participants to maintain a headway of two seconds behind a lead vehicle on approach to an intersection. Participants drove through the intersection a number of times while navigating through real-world traffic. The crash scenario in the on-road drive was constructed using a foam vehicle that replaced one of the real cars on the final lap and was propelled in front of the driver's vehicle. The time period from when the driver released the accelerator and put maximum pressure on the brake did not significantly differ between the simulator and on-road drives, nor did the point at which drivers first used the steering wheel to avoid crashing. However, the time to accelerator

release was significantly greater in the on-road drive compared to the simulator drive. The findings suggest that absolute validity was upheld for the timing of the drivers' transition from accelerator to maximum braking but not for the initial response time for foot off accelerator. The authors suggested that this might be due to methodological differences between the conditions, with participants completing three laps of a course in the on-road drive and one lap of a track in the simulator drive. Drivers in the on-road course might therefore have been more familiar with the intersection event, as they would have already driven through the intersection two times before the crash scenario occurred.

13.3.4 Validity of Using Simulators for Assessing Road Safety Countermeasures in the Real World

The validity of simulators for assessing the effects of road design and traffic control devices has also been examined. Riemersma, van der Horst, Hoekstra, Alink and Otten (1990) examined the validity of the Daimler-Benz advanced driving simulator for assessing speed reduction methods. A real-world study investigated the effect of infrastructure changes on drivers' speed when entering a Dutch village. The infrastructure changes to reduce speed included a median strip, colored asphalt, and a portal gate. Comparisons of vehicle speeds prior to and following installation of these infrastructure changes showed that the average speed of drivers upon entering the village decreased following installation, and faster drivers reduced their speed more than average-speed drivers. Riemersma et al. (1990) replicated this scenario in their simulator. Twenty-four male participants (aged 20–35 years) drove for approximately one hour around a course that required them to pass through the entrance to the village 12 times. Each approach consisted of a different combination of infrastructure changes. Comparing results from the real-world and simulated studies showed that on approach to the entrance of the village (i.e., 400 m from the village entrance), participants drove at a faster speed in the simulator than drivers in the real-world environment. However, simulator participants then reduced their speed to a greater extent than drivers in the real world, so that their speed at the entrance point to the village was slightly slower than drivers in the real world. The infrastructure changes produced larger speed reductions in the simulated than the real-world environment, with the mean entrance speed decreasing 25.7% in the simulator (from 78.3 km/h to 58.2 km/h) and 8.6% in the real world (from 72.4 km/h to 66.2 km/h), thus demonstrating relative but not absolute validity. In addition, the difference in entrance speed between on-road fast and average-speed drivers (approximately 10 km/h) was similar to the difference in entrance speed of simulator drivers when instructed to drive "as quickly as the conditions would allow" versus "in a relaxed and unhurried manner" (p. 418). The authors concluded that the simulator was an effective tool for evaluating speed reduction methods. Similarly, as described above, Godley et al. (2002) demonstrated the relative validity of the MUARC simulator for examining the effect of rumble strips as a speed-reduction method. In contrast,

McAvoy, Schattler and Datta (2007) found that their Doron Precision Systems, Inc. AMOS II simulator was not valid for evaluating night traffic speed control devices at construction zones (see also Wood & Chaparro, this book, chap. 28). Six real-world work zones were incorporated into a driving simulator scenario. Radar guns measured the speed of motorists travelling through the real-world work zone sites at night. Mean speeds in these field sites differed according to the presence or absence of steady-burn warning lights mounted on drums; however, the warning lights did not significantly affect mean speeds in the simulator. The simulator failed to meet standards for absolute or relative validity, perhaps because it failed to induce the perception of risk associated with driving through real-world work zones at night.

13.3.5 Validity of Using Simulators for Assessing Complex Driving Behaviors in the Real World

The validity of simulators for assessing more complex behavior, such as performance during divided attention tasks, has also been examined. Reed and Green (1999) investigated the validity of the University of Michigan Transportation Research Institute driving simulator for assessing decrements in driving performance during a manual car phone task. Younger (20–30 years) and older (>60 years) participants were instructed to dial a manual car phone while driving in an on-road setting and a simulator setting. Results showed that speed control (i.e., SD of speed and SD of throttle position) was similar in the two driving environments but lane keeping (i.e., mean lateral speed and SD of steering wheel position) varied more in the simulator. For seven of ten variables measuring lane position, speed, steering wheel angle, and throttle position, medium to high correlations were found between on-road and simulated performance ($r = 0.43$ to $r = 0.76$). For both age groups in both driving environments, the phone task resulted in decrements in speed control and lane keeping performance; however decrements were larger in the simulator. In addition, older drivers displayed greater decrements in driving performance than younger participants. Admittedly this is a simplified summary of their results (simulator fidelity was a further independent variable), but overall, absolute validity was demonstrated for speed control, while relative validity was shown for assessing the effects of dialing a phone and age on driving performance. Unfortunately, results for only four of the ten measured variables were presented in Reed and Green (1999) report (following analysis of all ten variables, one variable was selected to illustrate the full results for lane position, speed, steering wheel angle, and throttle position). There are few other validation studies investigating complex behavior, suggesting a need for further research in this area.

13.3.6 Validity of Using Simulators for Assessing Driving Behavior of Specific Driver Groups

Driving simulator validation studies typically involve young to middle-aged drivers. Increasingly, attention has been focused on studying driving patterns of specific sub-groups of drivers at high risk, based on variables such as age, medical conditions, and impairments. For vulnerable and potentially high risk drivers, the safety afforded by the simulator as a research tool is highly valuable. However, the usefulness of this research for understanding the impact of aging and impairments on driving behaviors hinges on the assumption that performance in the simulator reflects real-world driving. The accuracy of this assumption is especially critical if simulator performance outcomes are to be used for decisions about fitness to drive (e.g., see Ball & Ackerman, this book, chap. 25). It is possible that simulated driving conditions may differentially affect drivers with certain kinds of conditions—such as visual field loss—in a way that is not evident in a naturalistic driving setting; hence the importance of demonstrating the validity of the simulator, not just for specific driving tasks and outcome measures, but potentially for specific groups of drivers.

Hakamies-Blomqvist, Östlund, Henriksson and Fildes (2001) investigated the validity of the VTI simulator as a tool to study driving behavior of older drivers, taking into account participants' histories of accidents and incidents, and the difficulty experienced driving in stressful situations. One issue with assessing simulator validity for older drivers is variation in driver performance. Older drivers can be expected to exhibit greater variation in driving performance than younger drivers due to age-related declines in cognition, psychomotor function, vision, and health. Thirty-five drivers (aged 66–80 years) completed an on-road drive and a corresponding simulator drive. Participants drove in a similar way in the two settings, although in the simulator the mean speed was lower and there were greater variations in speed, lateral distance, and steering wheel movements. Participants also used the brake more frequently and with greater force in the simulator. Relative validity was obtained for operative measures such as the steering wheel angle and lane position. Results also showed that validity was strongest for drivers who reported the fewest number of driving crashes and incidents, and who reported no difficulty driving in stressful situations. These results suggest that the less competent the driver, the less valid the simulator is as a screening tool (see also, in this book, chap. 50 by Brouwer et al., and chap. 26 by Trick & Caird). This has implications for testing cohorts with poor driving ability (e.g., some drivers recovering from a stroke).

Recent research with the STISIM 400 simulator found that the on-road driving behavior of eight older adults (aged 67–81 years) was highly correlated ($r = 0.74$) with driving behavior in the simulator (Bédard et al., 2010). Although this analysis lacks power with a small sample size, there was a significant association ($p = 0.035$) between demerit points recorded during the on-road drive and a simulated replica. Lee and colleagues also conducted a number of studies involving older adults using the STISIM driving simulator. Lee, Cameron, et al. (2003) found that behavioral measures recorded from a driving simulator were highly correlated with measures recorded during an on-road drive, indicating that the STISIM simulator was a valid tool for assessing older driver behavior.

In subsequent research, Lee and colleagues examined the correspondence between simulator driving tasks and self-reported traffic violations and crash history (Lee & Lee, 2005; Lee, Lee, Cameron, & Li-Tsang, 2003). In a retrospective study, Lee, Lee, Cameron and Li-Tsang (2003) assessed older adults' driving simulator performance and crash history. One hundred and twenty-nine participants (aged 60–88 years) participated in a 45-minute driving simulator session, which was monitored by an assessor. Driving measures recorded during the simulated drive included compliance to traffic signs, driving speed, correct use of the indicator, ability to drive through a T intersection, working memory, speed limit compliance, rapid decision-making ability, functional reaction time, ability to perform simultaneous tasks, and knowledge of—and compliance with—traffic regulations. Prior to the drive, participants were interviewed about their crash history, with over 60% reporting involvement in at least one crash in the previous year. Logistic regression analysis showed that working memory, speed limit compliance, and rapid decision-making ability were significant predictors of reporting a prior crash. All associations were negative, such that participants were more likely to report prior crash involvement if they had poorer working memory (assessed by reporting back five street signs after ten minutes), poorer speed limit compliance, and poorer ability to make rapid decisions and judgments to avoid dangerous traffic situations in the simulated drive.

Following the retrospective study, Lee and Lee (2005) conducted a prospective study to determine whether simulator driving behavior could identify potentially at-risk older drivers. The same 129 drivers from Lee, Lee, Cameron and Li-Tsang (2003) completed a 45-minute simulated driving task, an interview, and a feedback session. Driving measures were recorded as per Lee, Lee, Cameron and Li-Tsang (2003). Participants were asked about their prior vehicle crashes and traffic violations, and gave permission for the researchers to access their driving records for the following three years. All participants received at least one traffic violation during the three-year period. As expected, driving task performance was negatively associated with age. The results also revealed that working memory ability and appropriate use of the indicator were negatively associated with the number of traffic violations over the three years. The authors attributed these findings to an age-related decline in psychomotor coordination and reaction time, and concluded that the simulator could identify older drivers at risk of future traffic violations.

Regarding the validity of simulators for evaluating the effects of training-specific cohorts, research with novice drivers shows that training effects measured on simulators correspond with effects measured on-road. Fisher, Pradhan, Pollatsek and Knodler (2007) report two studies (an on-road and a simulated study) where the same PC-based training program was used to train novice drivers (aged 18–21 years) to identify potential hazards. A mobile eye tracker collected eye movement data, so that participants' scanning behavior could be examined. In the on-road study, during a post-training on-road drive, 12 participants who completed the training program scanned critical areas significantly more often than 12 control (untrained) participants

(mean scanning time of 64.4% and 37.3% respectively). For the simulated study, post-training performance was tested on the University of Massachusetts at Amherst advanced fixed-base driving simulator. During the post-training simulated drive, six trained participants scanned critical areas of the road significantly more often than six untrained participants (mean scanning time of 77.4% and 40.0% respectively). Although only a simplified version of the results are reported here (scanning in different types of scenarios was a further variable examined), the results demonstrate a degree of correspondence between training effects measured on the simulator and those measured on-road. The magnitude of training effects measured in each drive did not statistically significantly differ, although training effects seemed larger when measured on the simulator (training improved scanning by a mean of 37.4% on the simulator versus a mean of 27.1% on-road). This may be due to the limitations of the study such as small sample sizes, and having different participants complete the on-road and simulated drives.

13.3.7 Crash and Infringement History

Validation methods based on comparisons of direct measures of driving behaviors in simulated driving against on-road driving behavior are arguably more robust than methods based on comparisons of driving behaviors in simulated driving with driving behavior questionnaires and crash history data (Anstey, Wood, Lord, & Walker, 2005). Crash history data and participant reports of traffic violations are not always adequate representations of individual driving skills. External factors can contribute to driving accidents, and people may not report all previous traffic violations (Lew et al., 2005). Despite their less robust design, the results of studies assessing simulator validity using comparisons of simulator behavior and behavior reported in questionnaires are encouraging (Lew et al., 2005; Reimer, D'Ambrosio, Coughlin, Kafrissen, & Biederman, 2006). Reimer et al. (2006) found high correlations between behaviors observed in a driving simulator and participant self-reports of driving behavior and accident history (valid measures included speeding, velocity, passing, traffic weaving, pause time at stop signs, and accidents). Furthermore, participants diagnosed with Attention-Deficit Hyperactivity Disorder (ADHD; who are more likely to report involvement in previous crashes than individuals without ADHD) were more likely than control participants to be involved in a crash on the simulated drive. However, the authors did acknowledge one issue with using questionnaires—there was some discrepancy between some questionnaire items and the corresponding simulator measure.

To ensure simulators are useful tools for assessment, researchers have examined the ability of simulators to predict future driving behavior and to distinguish between safe and unsafe drivers. Lew et al. (2005) conducted a prospective study comparing participants' simulator performance (recorded by the simulator and an observer) and on-road performance (recorded by an observer) with their on-road driving performance 10 months later. Both simulator- and observer-recorded measures of simulator driving performance predicted on-road driving performance 10 months following the simulated drive. However, simulator-recorded measures were more sensitive and accurate than observational measures. The predictive efficiency of the simulator-recorded measures was 82%, with sensitivity of 100% (i.e., all participants who failed the simulator drive were classified as failing the on-road drive 10 months later) and specificity of 71% (i.e., 71% of participants who passed the simulator drive were classified as passing the on-road drive 10 months later). On-road performance was not correlated with on-road performance 10 months later. These results suggest that driving simulator measures can be better predictors of future driving performance than data obtained from on-road drives. Other studies have similarly found that simulator driving measures can identify drivers at risk of future traffic violations (e.g., Lee & Lee, 2005) and distinguish between drivers of differing abilities (e.g., Lee, Lee, Cameron, and Li-Tsang 2003; Patomella, Tham, & Kottorp, 2006.)

13.3.8 Physiological Measures

Physiological responses have also been used to examine behavioral simulator validity. As noted above, while this is a less direct way of measuring driving behavior, physiological responses can be used to determine the participant's awareness level in the simulated environment, and thus provide some useful insights into behavior across simulated and real-world driving contexts. Valid simulations should induce similar physiological responses as in the real world. Watts and Quimby (1979) validated the Transport and Road Research Laboratory simulator using a measure of skin conductance. Participants assessed risks during an on-road drive and while viewing a 30-minute film of the same drive on the simulator. The film showed 45 hazardous scenarios from the driver's perspective and participants indicated the amount of perceived danger on a 10-point scale. During the film, participants' levels of skin conductance were measured. Risk assessment ratings were similar in the two environments and perceived risk rankings were highly associated with levels of skin conductance ($r = 0.78$). In another validation study with physiological measures, Bédard (2008) reported on the measurement of heart and respiration rates of participants exposed to threatening situations during a drive on the STISIM 400 simulator. The threatening situations included two green traffic lights that changed to amber as the driver approached the intersection, and a car that pulled out from the shoulder of the road in front of the driver. Each threatening situation increased participants' mean heart rate and two of the three situations increased participants' mean respiration rate, indicating that the simulator induced a high degree of presence. A third study included measures of both skin conductance and heart rate during an on-road drive and a simulated drive on the DriveSafety DS-600c motion-based simulator (Slick, Evans, Kim, & Steele, 2006). During four specific driving tasks (right and left turns at stop signs and traffic lights), Slick et al. (2006) found that skin conductance and heart rate did not significantly differ between the simulated and on-road drives. Similar results were found for mean changes in skin conductance and heart

rate during the entire drive for both males and females, with the exception that female participants' mean change in heart rate was significantly smaller in the simulator drive than the on-road drive. Overall, these studies suggest that physiological measures generally demonstrate the validity of simulators.

13.3.9 Ecological Validity

As an alternative to directly measuring and statistically comparing simulator and on-road behavior, some studies have examined simulator validity by measuring only simulator performance (e.g., the speed at which male and female participants drive in a simulator), and comparing this behavior with pre-established on-road behavior (e.g., the general trend for men to drive faster than women on-road). This examines a form of *ecological validity*, defined here as the degree to which simulator behavior reflects real-life on-road behavior patterns displayed over extended periods of time (as opposed to examining simulator and on-road performance at a single point on a drive; Lew et al., 2005). In this way, simulators can be validated for general patterns of behavior, without directly measuring on-road performance as part of the study. This method is a less direct and robust means of establishing validity than those previously discussed. Nonetheless, it can be a useful first step in assessing a simulator's validity, enabling conclusions regarding the overall pattern of behavior in the simulator compared with behavior in the real world. There is a vast amount of research on ecological validity and a full account of this literature is beyond the scope of this chapter. The following studies illustrate the general concept.

The effect of age on driving is well established in the real world. For example, younger drivers generally react more quickly than older drivers, and they tend to drive at faster speeds. Studies conducted on the STISIM 400 driving simulator have found similar age effects. Toxopeus (2007) examined braking responses to a stop sign in younger ($M = 22.1$ years) and older ($M = 69.1$ years) participants. A stop sign was displayed at the end of a 45-minute simulator highway drive and participants were required to stop as quickly as possible. Younger participants ($M = 0.92$ seconds) pressed the brake pedal significantly more quickly than older participants ($M = 1.01$ seconds), demonstrating behavioral validity for age effects on braking responses. Younger drivers ($M = 1.27$ seconds) also took less time to press the horn in response to a visual cue than older drivers ($M = 1.66$ seconds). Furthermore, when asked to maintain a speed of 55 mph, younger adults ($M = 56.4$ mph) drove faster than older adults ($M = 53.9$ mph), and there was a trend towards males ($M = 56.1$ mph) driving faster than females ($M = 54.7$ mph; $p = 0.055$).

Also using the STISIM driving simulator, Lee, Lee and Cameron (2003) examined visual attention ability in older adults. They had 129 adults aged over 60 complete a 45-minute simulated drive. A visual stimulus was presented 14 times during the drive, and participants were instructed to engage the turn indicator as soon as they saw the stimulus appear. Results showed a positive association between age and reaction time, with older seniors taking more time to respond than younger seniors. As a decline in visual attention was consistent with the aging literature, Lee, Lee and Cameron (2003) concluded that the STISIM simulator could be used to investigate visual attention ability in older drivers.

Demographic effects were also found by Yan and colleagues (2008) using the University of Central Florida driving simulator. In the simulated intersection scenarios, males drove faster than females, and participants in the 20–24 years age group drove faster than participants in older age groups. The consistency of these findings across simulators and studies suggests simulators do reflect similar behavior patterns as those seen in real-world driving.

Another means of evaluating simulator validity without on-road performance measurement is to relate simulator behavior with cognitive measures known to predict on-road driving performance (see also Ball & Ackerman, this book, chap. 25). For example, performance on the Useful Field of View® test (UFOV) is a good predictor of performance (pass/fail) in on-road driving evaluations (Myers, Ball, Kalina, Roth, & Goode, 2000), and is also associated with the frequency of on-road at-fault crashes (Clay, 2005). Bédard et al. (2010) reported two studies demonstrating strong correlations between simulator and cognitive measures such as UFOV. Participants (aged 18–83 years) drove a simulated road test route (i.e., a simulated version of the Manitoba road test, which is used for driver licensing) through their local city. The simulator recorded their driving errors and an observer recorded their demerit points. Performance on two cognitive tests—UFOV and Trail Making Test Part A (Trails A)—was significantly correlated with simulator-recorded driving errors and observer-recorded demerit points (*r*'s ranged from .57 to .74). Wald and Liu (2001) also found that simulator driving performance was correlated with the Trails A scores. Performance on the Attention Network Test (a cognitive test that has good concurrent validity with UFOV; Weaver, Bédard, McAuliffe, & Parkkari, 2009) has also shown significant associations with simulator driving performance (Mullen, Chattha, Weaver, & Bédard, 2008; Weaver et al., 2009). However, validity assessments should not assume that all relationships are linear (the reader will recall our earlier discussion of functional validity). Weaver et al. (2009) considered both linear and quadratic associations between simulator driving performance and UFOV scores, and found the latter relationship was stronger (linear *r* = 0.73, quadratic *R* = 0.79). Figure 13.3 similarly demonstrates the importance of considering alternative associations.

13.3.10 Summary

In summary, there is sufficient evidence to suggest that simulators provide a valid tool for assessing a variety of driving performance measures such as speed, lateral position, and risky traffic behaviors. They also appear valid for assessing the effects of divided attention on driving performance. Furthermore, measures from simulator drives can identify older drivers at risk of future traffic violations, and simulators are sensitive to age-related changes in driving performance and cognition. Simulators have been validated for assessing age effects on speed,

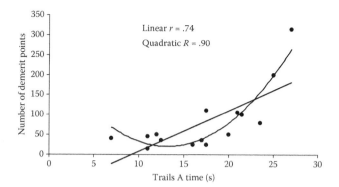

FIGURE 13.3 Relationships between simulator-recorded demerit points and Trails A.

braking responses, reaction time, and some operative measures (e.g., vehicle position on the road), although the evidence is less convincing for older drivers who experience difficulties with driving. Regarding braking responses, validity has been demonstrated for brake onset but not for braking force (i.e., braking normally versus hard). Although simulators appear valid for assessing the effects of some traffic control methods (e.g., rumble strips), they are not valid for all devices (e.g., lights at construction zones). The majority of validated measures show relative validity but fail to meet requirements for absolute validity.

The driving ability of participants can compromise simulator validity, such that simulators may be less valid for poorer drivers. In addition, variability in driving performance measures has a tendency to be greater in simulator than on-road drives. Hence researchers, driving evaluators, and other simulator users should remain aware that simulators do not always provide an accurate picture of on-road driving behavior.

13.4 Future Research Directions

Although Yan et al. (2008) established absolute validity for mean intersection approach speed, speed variability was comparable at just two intersection approaches with lower speed limits (45 mph). At two intersection approaches with higher speed limits (50 mph), simulator speeds had larger variability than on-road speeds. These results suggest that future research should investigate the variability of data in addition to measures of central tendency because these factors can affect validity conclusions.

The usefulness of simulators as a tool for assessment and training depends on simulator validation for the cohort being assessed or trained. Simulators offer a safe alternative to on-road training for cohorts who could be hazardous to themselves or other road users in real-world environments. For example, fatigue effects and the effects of medications on driving performance can be safely monitored in a simulator environment. Philip and colleagues (2005) were interested in the effects of sleep deprivation on driving behavior in an on-road and a simulated driving environment (Philip et al., 2005). Twelve men aged 19–24 years were divided into a regular sleep condition of eight hours per night and a restricted sleep condition consisting of two hours of sleep. Measurements included self-rated fatigue, simple reaction time, and the number of inappropriate line crossings. Participants' self-rated sleep scores, reaction times, and inappropriate line crossings were significantly higher in the simulator than on the road, but the effect of sleep deprivation was comparable in the simulated and on-road drives. In contrast, Volkerts and colleagues (1992) found the TS2 driving simulator was a less sensitive measure of sedative drug-induced impairment than an on-road drive (Volkerts, van Laar, van Willigenburg, Plomp, & Maes, 1992). Oxazepam (50 mg) and lormetazepam (1 mg) impaired on-road driving performance, but did not impair performance in the simulator. These findings highlight the need for further research to establish the validity of driving simulators for measuring the effects of medications.

While there are some validation studies examining older drivers, there is a need for future research to examine cohorts with specific medical conditions (e.g., stroke, Alzheimer's disease, acquired brain injury). Assessment of driving skills on a simulator may influence decisions concerning whether a driver's license may be retained or revoked, hence the importance of establishing validity with these cohorts (see also Brouwer et al., this book, chap. 50). Lee et al. (2007) examined the simulator and on-road driving performance of participants with and without Parkinson's disease (PD). Participants with PD had significantly poorer driving performance than healthy participants in both the on-road and simulator environments. However, simulator performance explained 68% of the variability in on-road performance for healthy participants but only explained 39% of the variability for participants with PD. These results suggest that the simulator may be less valid for participants with PD, and highlight the need for future research to examine the effect of other medical conditions on validity.

Simulator discomfort (SD) remains an unresolved issue for all cohorts, even those for whom validity has been established (see also Stoner, Fisher, & Mollenhauer, this book, chap. 14). SD refers to a range of symptoms involving stomach, oculomotor, and orientation disturbances (e.g., nausea, headache, dizziness; Kennedy, Lane, Berbaum, & Lilienthal, 1993); one suggested explanation, but not the only one, is that it results from the conflict between current sensory information (e.g., visual information suggesting the individual is moving and vestibular information suggesting the individual is stationary) and stored sensory expectations based on past experience (Reason & Brand, 1975). Symptoms can be sufficiently severe that participants are unable to complete the simulator drive. Alternatively, affected participants may be able to complete the drive but the effect of SD on their driving performance is unknown. Although SD can potentially affect any simulator user, studies have found that variations among individuals, simulator set-ups, and simulator tasks influence susceptibility. For example, Park, Cook, Rosenthal, Fiorentino and Allen (2006) found a higher incidence of SD in a scenario involving high speed driving in a complex environment (city streets with high traffic density and multiple turning maneuvers) than in slower, less complex scenarios with

fewer maneuvers. In a sample of older drivers (284 drivers aged 60–99) referred for driving evaluation, Freund and Green (2006) found almost 11% reported experiencing SD, with reports significantly greater among females than males. More than half (57%) of the participants who reported experiencing SD were unable to complete the drive. SD prevents otherwise valid simulators from being valid for all individuals and scenarios. Therefore, to improve simulator validity, future research should investigate methods to reduce or eliminate the occurrence of SD.

Although absolute and relative validity are the most common types of behavioral validity examined, and hence have been discussed throughout this chapter, Kantowitz (2001) offers another view of behavioral validity (see also, Kantowitz, this book, chap. 3). While acknowledging that the distinction between absolute and relative validity is valuable, he states:

> . . . I prefer to think of simulator validity more in terms of regression. How well does the simulator predict an outcome on the road? This allows for outcomes that are not absolute but still are better than relative validity . . . . Regression analysis also offers a metric that explains how well simulators predict reality so that different users can make their own judgments about the sufficiency of the fit for their own design purposes (Kantowitz, 2001, p. 51).*

We encourage researchers to consider this approach when examining simulator validity. Establishing whether a simulator exactly replicates on-road performance (absolute validity) is less important than determining *how well* a simulator predicts on-road performance; and this approach would provide a numerical measure. Regression analysis would also allow for examination of a range of relationships between simulator and on-road performance (e.g., linear, quadratic; as previously mentioned in our discussion of functional validity).

13.5 Conclusion

Many studies have investigated the validity of simulators for examining driving behavior. Most studies support the use of simulators, finding that driving behavior in simulators approximates (relative validity), but does not exactly replicate (absolute validity), on-road driving behavior. Simulator driving performance shows medium to strong correlations with a range of on-road driving performance measures, in addition to cognitive and physiological measures. This is sufficient for the majority of research, training, and assessment purposes for which simulators are used. However, a simulator shown to be valid in most settings is not guaranteed to be valid in the next setting. There can be critical differences (e.g., the environment, equipment, or protocol used), and when safety recommendations are based on simulator research, simulators almost necessarily require validation because the settings and scenarios are always changing.

Following simulator validation, adherence to the validated protocol is necessary to ensure measurements remain valid. Deviations from the protocol (e.g., altering the simulator set-up or changing performance measurements) will require validity reassessment. Validity has also generally been established with healthy cohorts. When using simulators with cohorts that deviate from this standard, validity will need to be established prior to evaluating their driving performance. In addition, where absolute values are required, on-road measures will generally be necessary. As simulators continue to advance technologically, their validity is only likely to broaden in terms of performance measures and cohorts, ensuring that simulators increase their role as a tool for driving performance measurements.

Key Points

- Simulators need to be validated to ensure the results of simulator studies generalize to driving in the real world. Behavioral validity refers to the extent to which the simulator induces the same driving behaviors that occur when driving in the real world. Behavioral validity has been further defined in terms of absolute and relative validity. While absolute validity requires that the two driving environments produce the same numerical values, relative validity is established when the differences between the two environments are in the same direction, and of the same or similar magnitude. Absolute validity is rarely established in driving simulator studies, but establishing relative validity is both necessary and sufficient for simulators to be useful research tools in most, though not all, cases.

- Methodological considerations for validation studies include factors such as the nature of the research question, task conditions, and dependent measures. Participant characteristics, simulator equipment, and how driving behaviors are measured in the simulator and real world can also affect validity. Statistical techniques used to establish validity include analysis of variance, correlation, and descriptive analyses of observed measures (e.g., speed profiles) of behavior measured on-road and during simulated drives.

- Simulators appear to provide a valid tool for assessing a variety of driving performance measures including speed, lateral position, brake onset, divided attention, and risky traffic behaviors. The majority of validated measures show relative validity but fail to meet requirements for absolute validity. Simulators do not appear valid for measures of braking force or for assessing the effects of some traffic control methods. Simulators can also induce similar physiological responses to those expected in the real world. Simulators appear sensitive to age-related changes in real-world driving performance and cognition, and can identify older drivers at risk of future traffic violations.

- Future research should investigate the variability of data in addition to measures of central tendency and methods

to reduce the occurrence of simulator discomfort, because these factors can affect validity conclusions. There is also a need to establish the reliability of using driving simulators with specific cohorts, such as drivers with specific medical conditions; research to date has generally established validity with healthy cohorts. Regression modeling offers a useful approach to exploring the question of validation and the extent to which variables measured in the simulator predict real-world driving.

- Most studies support the use of simulators, finding that simulator driving behavior approximates (relative validity), but does not exactly replicate (absolute validity), on-road driving behavior. This is sufficient for the majority of research, training, and assessment purposes for which simulators are used. However, simulator users should remain aware that simulators do not always provide an accurate picture of on-road driving behavior, and validity should not be assumed; each simulator set-up should be validated for its ability to measure the driving behavior of the cohort for which it is to be used. Furthermore, where absolute values are required, on-road measures will generally be necessary.

Keywords: Behavioral Validity, Dependent Measures, Real-world Driving Outcomes, Simulator Validity

Acknowledgments

Nadia Mullen acknowledges the support of a grant from the Natural Sciences and Engineering Research Council of Canada. Judith Charlton and Anna Devlin acknowledge the support of a grant from the Australian Research Council (Project ID LP0560272) and industry partners: GM Holden, VicRoads, Road Safety Trust (New Zealand), the Swedish Road Administration, and Zeiss. Michel Bédard is a Canada Research Chair in Aging and Health (http://www.chairs.gc.ca); he acknowledges the support of the Canada Research Chair Program. This research was also supported in part by grants from the Canadian Institutes of Health Research, the Canada Foundation for Innovation, AUTO21: Network of Centres of Excellence, and the Ontario Neurotrauma Foundation.

Glossary

Absolute validity: A type of behavioral validity; established when the simulated and on-road driving environments produce the same numerical values.

Behavioral validity: The extent to which a simulator elicits the same driving behavior that occurs when driving on real roads.

Ecological validity: The degree to which simulator behavior reflects real-life on-road behavior patterns displayed over extended periods of time.

External validity: The extent to which driving behavior in a simulator generalizes to real driving behavior.

Functional validity: A type of behavioral validity; established when the simulated and on-road driving environments produce numerical values that are associated with a mathematical function (linear or non-linear).

Interactive validity: A type of behavioral validity; established when the simulated and on-road driving environments produce the same numerical values (interactive absolute validity), or values similar in magnitude and in the same direction (interactive relative validity), over time.

Physical validity: The extent to which the physical components of a simulator vehicle correspond to on-road vehicles (includes consideration of the simulator layout, visual displays, and dynamic characteristics).

Relative validity: A type of behavioral validity; established when the simulated and on-road driving environments produce numerical values that are not identical, but are of similar magnitude and in the same direction.

Key Readings

Blaauw, G. J. (1982). Driving experience and task demands in simulator and instrumented car: A validation study. *Human Factors, 24,* 473–486.

Blana, E. (1996). *Driving simulator validation studies: A literature review.* Institute of Transport Studies, University of Leeds, UK.

Godley, S. T., Triggs, T. J., & Fildes, B. N. (2002). Driving simulator validation for speed research. *Accident Analysis & Prevention, 34,* 589–600.

Kaptein, N. A., Theeuwes, J., & van der Horst, R. (1996). Driving simulator validity: Some considerations. *Transportation Research Record, 1550,* 30–36.

Törnros, J. (1998). Driving behavior in a real and a simulated road tunnel—A validation study. *Accident Analysis & Prevention, 30,* 497–503.

References

Anstey, K. J., Wood, J., Lord, S., & Walker, J. G. (2005). Cognitive, sensory and physical factors enabling driving safety in older adults. *Clinical Psychology Reviews, 25,* 45–56.

Bédard, M. (2008, September). *Driving simulators: Are they useful tools?* Paper presented at the STISIM users group meeting, Québec City, Canada.

Bédard, M., Parkkari, M., Weaver, B., Riendeau, J., & Dahlquist, M. (2010). Assessment of driving performance using a simulator protocol: Validity and reproducibility. *American Journal of Occupational Therapy, 64,* 336–340.

Bella, F. (2005). Validation of a driving simulator for work zone design. *Transportation Research Record, 1937,* 136–144.

Bella, F. (2008). Driving simulator for speed research on two-lane rural roads. *Accident Analysis & Prevention, 40,* 1078–1087.

Bittner, A. C., Simsek, O., Levison, W. H., & Campbell, J. L. (2002). On-road versus simulator data in driver model development: Driver performance model experience. *Transportation Research Record, 1803,* 38–44.

Blaauw, G. J. (1982). Driving experience and task demands in simulator and instrumented car: A validation study. *Human Factors, 24,* 473–486.

Blana, E. (1996). *Driving simulator validation studies: A literature review.* Institute of Transport Studies, University of Leeds, UK.

Blana, E., & Golias, J. (2002). Differences between vehicle lateral displacement on the road and in a fixed-base simulator. *Human Factors, 44,* 303–313.

Charlton, J. L., Fildes, B., Oxley, J., Keeffe, J., Odell, M., Verdoon, A., et al. (2008, September). *Driving performance and visual search associated with visual field loss.* Paper presented at the International Conference on Traffic and Transport Psychology, Washington, DC.

Clay, O. J., Wadley, V. G., Edwards, J. D., Roth, D. L., Roenker, D. L., & Ball, K. K. (2005). Cumulative meta-analysis of the relationship between useful field of view and driving performance in older adults: Current and future implications. *Optometry & Vision Science, 82,* 724–731.

Fisher, D. L., Pradhan, A. K., Pollatsek, A., & Knodler, M. A., Jr. (2007). Empirical evaluation of hazard anticipation behaviors in the field and on driving simulator using eye tracker. *Transportation Research Record, 2018,* 80–86.

Freund, B., & Green, T. R. (2006). Simulator sickness amongst older drivers with and without dementia. *Advances in Transportation Studies,* Special Issue, 71–74.

Godley, S. T., Triggs, T. J., & Fildes, B. N. (2002). Driving simulator validation for speed research. *Accident Analysis & Prevention, 34,* 589–600.

Hakamies-Blomqvist, L., Östlund, J., Henriksson, P., & Heikkinen, S. (2001). *Elderly car drivers in a simulator—A validation study.* Linköping, Sweden: Swedish National Road and Transport Research Institute.

Hoffman, J. D., Lee, J. D., Brown, T. L., & McGehee, D. V. (2002). Comparison of driver braking responses in a high-fidelity simulator and on a test track. *Transportation Research Record, 2,* 59–65.

Hoskins, A. H., & El-Gindy, M. (2006). Technical report: Literature survey on driving simulator validation studies. *International Journal of Heavy Vehicle Systems, 13,* 241–252.

Kantowitz, B. H. (2001). Using microworlds to design intelligent interfaces that minimize driver distraction. *Proceedings of the international driving symposium on human factors in driver assessment, training and vehicle design, 2001* (pp. 42–57).

Kaptein, N. A., Theeuwes, J., & van der Horst, R. (1996). Driving simulator validity: Some considerations. *Transportation Research Record, 1550,* 30–36.

Kennedy, R. S., Lane, N. E., Berbaum, K. S., & Lilienthal, M. G. (1993). Simulator sickness questionnaire: An enhanced method for quantifying simulator sickness. *International Journal of Aviation Psychology, 3,* 203–220.

Klee, H., Bauer, C., Radwan, E., & Al-Deek, H. (1999). Preliminary validation of driving simulator based on forward speed. *Transportation Research Record, 1689,* 33–39.

Lee, H. C., Cameron, D., & Lee, A. H. (2003). Assessing the driving performance of older adult drivers: On-road versus simulated driving. *Accident Analysis & Prevention, 35,* 797–803.

Lee, H., Falkmer, T., Rosenwax, L., Cordell, R., Granger, A., Vieira, B., et al. (2007). Validity of driving simulator in assessing drivers with Parkinson's disease. *Advances in Transportation Studies,* Special Issue, 81–90.

Lee, H. C., & Lee, A. H. (2005). Identifying older drivers at risk of traffic violations by using a driving simulator: A 3-year longitudinal study. *American Journal of Occupational Therapy, 59,* 97–100.

Lee, H. C., Lee, A. H., & Cameron, D. (2003). Validation of a driving simulator by measuring the visual attention skill of older adult drivers. *American Journal of Occupational Therapy, 57,* 324–328.

Lee, H. C., Lee, A. H., Cameron, D., & Li-Tsang, C. (2003). Using a driving simulator to identify older drivers at inflated risk of motor vehicle crashes. *Journal of Safety Research Senior Transportation Safety and Mobility, 34,* 453–459.

Lew, H. L., Poole, J. H., Lee, E. H., Jaffe, D. L., Huang, H. C., & Brodd, E. (2005). Predictive validity of driving-simulator assessments following traumatic brain injury: A preliminary study. *Brain Injury, 19,* 177–188.

McAvoy, D. S., Schattler, K. L., & Datta, T. K. (2007). Driving simulator validation for nighttime construction work zone devices. *Transportation Research Record, 2015,* 55–63.

McGehee, D. V., Mazzae, E. N., & Baldwin, G. H. S. (2000). *Driver reaction time in crash avoidance research: Validation of a driving simulator study on a test track.* Paper presented at the meeting of the Human Factors and Ergonomics Association/International Ergonomics Association, Santa Monica, CA.

Mullen, N. W., Chattha, H. K., Weaver, B., & Bédard, M. (2008). Older driver performance on a simulator: Associations between simulated tasks and cognition. *Advances in Transportation Studies,* Special Issue, 31–42.

Myers, R. S., Ball, K. K., Kalina, T. D., Roth, D. L., & Goode, K. T. (2000). Relation of useful field of view and other screening tests to on-road driving performance. *Perceptual and Motor Skills, 91,* 279–290.

Park, G. D., Cook, M. L., Rosenthal, T. J., Fiorentino, D., & Allen, R. W. (2006). Simulator assessment of older driver proficiency. *Advances in Transportation Studies,* Special Issue, 75–86.

Patomella, A. H., Tham, K., & Kottorp, A. (2006). P-drive: Assessment of driving performance after stroke. *Journal of Rehabilitation Medicine, 38,* 273–279.

Philip, P., Sagaspe, P., Taillard, J., Valtat, C., Moore, N., Akerstedt, T., et al. (2005). Fatigue, sleepiness, and performance in simulated versus real driving conditions. *Sleep, 28,* 1511–1516.

Reason, J. T., & Brand, J. J. (1975). *Motion sickness.* London: Academic Press.

Reed, M. P., & Green, P. A. (1999). Comparison of driving performance on-road and in a low-cost simulator using a concurrent telephone dialling task. *Ergonomics, 42,* 1015–1037.

Reimer, B., D'Ambrosio, L. A., Coughlin, J. E., Kafrissen, M. E., & Biederman, J. (2006). Using self-reported data to assess the validity of driving simulation data. *Behavior Research Methods, 38,* 314–324.

Riemersma, J. B. J., van der Horst, A. R. A., Hoekstra, W., Alink, G. M. M., & Otten, N. (1990). The validity of a driving simulator in evaluating speed-reducing measures. *Traffic Engineering and Control, 31,* 416–420.

Shinar, D., & Ronen, A. (2007). Validation of speed perception and production in a single screen simulator. *Advances in Transportation Studies,* Special Issue, 51–56.

Slick, R. F., Evans, D. F., Kim, E., & Steele, J. P. (2006, May). *Using simulators to train novice teen drivers: Assessing psychological fidelity as a precursor to transfer of training.* Paper presented at the Driving Simulation Conference—Asia/Pacific, Tsukuba, Japan.

Törnros, J. (1998). Driving behaviour in a real and a simulated road tunnel—A validation study. *Accident Analysis & Prevention, 30,* 497–503.

Toxopeus, R. (2007). *Differences between elderly and young drivers: Driving vigilance in two tasks.* Unpublished master's thesis, Lakehead University, Thunder Bay, Ontario, Canada.

Volkerts, E. R., van Laar, M. W., van Willigenburg, A. P. P., Plomp, T. A., & Maes, R. A. A. (1992). A comparative study of on-the-road and simulated driving performance after nocturnal treatment with lormetazepam 1 mg and oxazepam 50 mg. *Human Psychopharmacology: Clinical & Experimental, 7,* 297–309.

Wade, M. G., & Hammond, C. (1998). *Simulator validation: Evaluating driver performance in simulation and the real world* (Report No. 28). Minneapolis, MN: Minnesota Department of Transportation.

Wald, J., & Liu, L. (2001). Psychometric properties of the driVR: A virtual reality driving assessment. *Studies in Health Technology and Informatics, 81,* 564–566.

Watts, G. R., & Quimby, A. R. (1979). *Design and validation of a driving simulator for use in perceptual studies* (Report No. 907). Berkshire, UK: Transport and Road Research Laboratory.

Weaver, B., Bédard, M., McAuliffe, J., & Parkkari, M. (2009). Using the Attention Network Test to predict driving test scores. *Accident Analysis & Prevention, 41,* 76–83.

Yan, X., Abdel-Aty, M., Radwan, E., Wang, X., & Chilakapati, P. (2008). Validating a driving simulator using surrogate safety measures. *Accident Analysis & Prevention, 40,* 272–288.

III

Conduct of Simulator Experiments, Selection of Scenarios, Dependent Variables, and Evaluation of Results

14

Simulator and Scenario Factors Influencing Simulator Sickness

Heather A. Stoner
Realtime Technologies, Inc.

Donald L. Fisher
University of Massachusetts, Amherst

Michael Mollenhauer, Jr.
Virginia Tech Transportation Institute

Abstract

The Problem. The consequences and implications of simulator sickness for the validity of simulation can be severe if not controlled and taken into account (Casali, 1986). Many of today's driving simulators are used to perform research, training, or proof of design activities. A prerequisite to generalizing the results found in research conducted in a simulator is an understanding of the validity of the resulting experience. Without question, simulator sickness is a factor that can affect the validity of research simulators. Given the potential consequences of simulator sickness, it is difficult to assess the value of the results obtained from a simulator study known to have significant sickness problems. *Role of Driving Simulators.* There are alternatives to driving simulators for studying most, if not all, issues. However, these alternatives are often unsafe, do not provide a well-controlled environment, and require large sums of money to implement. Thus, driving simulators are necessary and the associated issues of simulator sickness need to be addressed. *Key Results of Driving Simulator Studies.* Simulator sickness can affect a driver's performance in a variety of negative ways due to inappropriate behaviors, loss of motivation, avoidance of tasks that are found disturbing, distraction from normal attention allocation processes, and a preoccupation with the fact that something is not quite right. On the positive side, simulator selection, participant screening, scenario design, and control of the environment can all reduce the incidence of simulator sickness. *Scenarios and Dependent Variables.* Examples of the sorts of scenarios that lead to extremes of simulator sickness are discussed. Additionally, the various measures that have been used against simulator sickness are highlighted, including some with predictive validity. *Platform Specificity and Equipment Limitations.* Simulator sickness appears to be most extreme in fully immersive environments and when head-mounted displays are used. A motion base does not necessarily reduce simulator sickness symptoms.

14.1 Introduction

Simulator sickness or the report of ill feelings associated with the use of simulation devices has been a persistent challenge for simulator-based research. Simulator sickness is not limited to any one type of driving simulator. It is has been documented on both floor and head-mounted simulators (Draper, Viirre, Furness, & Gawron, 2001; Draper, Viirre, Furness, & Parker, 1997; Ehrlich, 1997); and among floor-mounted simulators it has been observed in both motion-base and fixed-base simulators. Casali (1986) noted that documentation of simulator sickness can be found in reports by Havron and Butler as early as 1957 in a helicopter flight training simulator. In these early reports, the phenomenon was reported as motion sickness or the result of exposure to low frequency, whole body motion. Both motion sickness and simulator sickness can result in an array of symptoms including eye strain, headache, postural instability, sweating, disorientation, vertigo, pallor, nausea, and vomiting (Ebenholtz, 1992; Pausch, Crea, & Conway, 1992). Kennedy et al. (1987) provide a full categorization of the symptoms associated with simulator sickness.

Although the symptoms are common between motion and simulator sickness, they are not identical. Casali (1986) makes the distinction based on research conducted by Money (1970), suggesting that stimulation of the vestibular system is required to induce motion sickness. Consistent with this, there are many reports of simulator sickness and related symptoms in fixed-based simulators that include no physical motion cues. Therefore, a distinction between motion and simulator sickness is useful because it is not only the actual physical motion that can cause sickness. It appears that some visual stimuli, likely perceived motion or vection, can also contribute to simulator sickness (Kennedy, Hettinger, & Lillenthal, 1988). Indeed, there are a number of factors that contribute to simulator sickness, a fact that led Kennedy and Fowlkes (1992) to describe simulator sickness as a syndrome because it has many complex contributing causes and manifests itself with many potential symptoms. A good discussion of contributing factors can be found in Kolasinski (1995).

The consequences and implications of simulator sickness on the validity of simulation can be severe if not controlled and accounted for (Bittner, Gore, & Hooey, 1997; Casali, 1986; Frank, Casali, & Wierwille, 1988). Many of today's driving simulators are used to perform research, training, or design evaluation. A prerequisite to generalizing the results found in research conducted in a simulator is an understanding of the validity of the resulting experience (see Ranney, this book, chap. 9). Without question, simulator sickness can undermine the validity of simulator data. Simulator sickness can affect an operator's performance (Uliano, Lambert, Kennedy, & Sheppard, 1986), although it need not always do so (Warner, Serfoss, Baruch, & Hubbard, 1993). Simulator sickness affects performance in a variety of different ways including the execution of inappropriate behaviors, loss of motivation (often including cessation), inability to concentrate (Kennedy et al., 1987), avoidance of tasks that are found disturbing, modification of behaviors to reduce symptoms (Silverman & Slaughter, 1995), distraction from normal attention allocation processes (e.g., closing eyes during turns, Silverman & Slaughter), and a pre-occupation with the fact that something is not quite right. Given the potential consequences of simulator sickness, it is difficult to assess the generalizability of the results obtained from a simulator if sickness is not carefully monitored and managed.

In addition to problems with validity created by simulator sickness, there is potential danger to the participants both during and long after an experiment has been completed. As one might expect, the effects of simulator sickness on participants' performances are most likely to occur in initial exposures to a simulator, particularly when there are high rates of optic flow and frequent changes in acceleration (Hettinger & Riccio, 1992). This creates a huge challenge for creators of driving simulation systems where such factors are necessary component of the simulated task.

Additionally, due to lingering reactions, there can be effects on participants using a driving simulator long after the simulation experience. Blurred vision, postural instability, nausea, and general discomfort are the types of lingering symptoms that can be experienced (Johnson, 2005). Flight simulators, particularly those used for training high-g maneuvers, are more likely to produce long-term after-effects. Even for these simulators, only 4.6% of Navy pilots suffered from symptoms 24 hours or more after simulator-based training (Ungs, 1989). Kennedy, Fowlkes and Lilienthal (1993) conclude that the most dangerous potential after-effects are disturbances in locomotor and postural control. These effects can last for hours, or potentially much longer. This can occur in high-g simulators but extended effects have been reported on other simulators as well. For example, when exposed to long periods of rotation, usually not a problem in most simulators, effects have been measurable three or four days after exposure (Fregly & Kennedy, 1965). Even longer-lasting effects have been reported (Berbaum, Kennedy, Welch, & Brannan, 1985; Guedry, 1965; Goodenough & Tinker, 1931). Care must be taken by simulation users to understand the impact of simulation exposure on the participant and protect him or her from potential danger.

Given its impact on a researcher's ability to undertake and complete an experiment, and on a practitioner's ability to undertake and complete training programs, simulator sickness has been widely researched and is the subject of a number of comprehensive reviews, most often for military applications (e.g., Crowley & Gower, 1988; Goldberg & Hiller, 1995; Johnson, 2005). We cannot hope to cover the breadth and depth of the work that has been done. However, we will do our best to introduce the reader to the extensive literature on the topic. To begin, it is important to understand the underlying mechanisms and processes that cause simulator sickness. A thorough understanding should allow for better design of scenarios, techniques for tuning the simulator, and novel experimental techniques to help reduce simulator sickness. The following section provides a brief explanation of the theoretical basis that explains the role of the visual and vestibular systems in simulator sickness and the physiological mechanisms associated with simulator sickness. A discussion follows of different possible measures of simulator sickness and then the methods one can use to prevent such sickness.

14.2 Theories of Simulator Sickness

There are several theories behind the concept of simulator sickness. The three most prominent theories are: cue conflict theory, poison theory, and postural instability. More recently, it has been hypothesized that it is not a conflict in the cues per se that is responsible for simulator sickness, but rather a conflict in the rest-frames that correspond to those cues. This theory will also be discussed because it suggests a range of actions that might be taken to mitigate simulator sickness.

14.2.1 Cue Conflict Theory

Cue conflict theory is the primary theory used to describe the etiological processes that occur with simulator sickness (Reason & Brand, 1975). (The reader is also referred to chap. 8, "Sensory and Perceptual Factors in the Design of Driving Simulation Displays," by Andersen in this book.) The main premise of the theory is that sickness occurs due to mismatches between what the sensory systems expect based on previous experience and what actually occurs in the simulator. The mismatch causes internal conflict that cannot be resolved and eventually results in the symptoms associated with simulator sickness. An example of this conflict can be found in a fixed-base simulator where visual cues are presented to indicate linear acceleration but because the driver is not actually moving, no corroborating or correlated vestibular cues are detected. Drivers of real vehicles have learned to expect that with visual cues of acceleration there will also be a corresponding vestibular cue of acceleration. Therefore, a conflict will be detected and simulator sickness could result. Examples also occur in motion-base and head-mounted simulators as well. For example, in a motion-base simulator where both visual and vestibular cues are delivered, there may be too long between an onset in the motion or visual cues. And in a head-mounted display (fixed-base), the vestibular cues delivered by head movements may not correspond to the actual changes in the visual world displayed to the participant which are naturally produced by such head movements.

There are a number of types of cue mismatch that can lead to cue conflict in driving simulators. To begin with, there are two broad categories of conflict: intermodal (e.g., conflict between the cues given by the vestibular and visual systems) and intramodal (e.g., conflicts between the cues given by the semicircular canals and the cues given by the otolith organs within the vestibular apparatus; Griffin, 1990). Additionally, within each category, there are three types of conflict that could occur: 1) signals exist from two separate cueing systems, say A and B, and they provide contradictory information; 2) signals are present from A, but not from B; and 3) signals are available from B but not from A.

Perhaps the most salient conflict is the conflict between the intermodal cues generated by the visual system and the vestibular systems. The coupling between these is quite close given their importance to spatial orientation and the rapid exchange of information that is required to support balance and locomotion. Examples can be generated for each of the three types of conflict that could occur between the visual and vestibular systems. 1) Conflict would exist between

the visual and vestibular system cues if a display was head-mounted and the signal from the head-mounted display were either noisy or inaccurate. 2) Signals from the visual system could be present, but those from the vestibular system absent, in a fixed-base simulator. In particular, these two systems provide potentially different information when the speed is changing or the vehicle is turning. 3) Finally, in a head-mounted system, signals from the vestibular system would be present, but those from the visual system absent, if a low display update rate were used. Other conflicts will be discussed below.

There does appear to be a relationship between the level of experience an individual has performing a real-world task outside of a simulator and the incidence of sickness seen while performing the task in a simulator (Pausch et al., 1992). The more experience an operator has, the more likely that he or she is to experience symptoms. This supports cue conflict theory in that the more intimate the operator is with the types of sensory responses he or she should be receiving, the more likely he or she will be to either consciously or unconsciously recognize when something does not match.

An important finding in motion sickness research is that a necessary requirement for experiencing sickness is a working vestibular system (McCauley & Sharkey, 1992, reviewing Howard, 1986). Such a finding is certainly consistent with cue conflict theory since there can be no conflicts between the visual and vestibular systems. Even though cue conflict theory is the most widely accepted theory of simulator sickness, there are several problems with it that have led some to question its utility as an explanation for simulator sickness (Stoffregren & Riccio, 1991). The first issue is that the theory does not allow for effective prediction of simulator sickness. There is no reliable formula based on sensory inputs and conflicts that can be used to determine which situations will produce sickness and which will not (Draper et al., 2001; Stoffregren & Riccio, 1991). This has led some investigators to conclude that in its present form the theory may be untestable (Ebenholtz, Cohen, & Linder, 1994, p. 1034). Second, according to the theory, lack of cue redundancy—such as either no motion or inadequate motion as reported by the vestibular system and motion reported by the visual system—is a major determinant of when sickness will occur. However, there are many environments where sensory cueing is not redundant which do not produce sickness. For example, no redundancies are present when an individual is sitting in a room watching a chase video. Yet, individuals do not experience sickness in such environments. Therefore, having no-motion or inadequate motion is not a clear predictor of simulator sickness rates. Third, there is no explanation for why simulator sickness is prevalent at first exposures and then will tend to disappear after repeated exposure. Last, there has been no physiological explanation of why cue conflict will result in a nauseogenic response. There are no known neural processing centers that would account for such a response and it is unlikely that there is an undiscovered neural processing center that is dedicated to this particular response.

Even with its potential drawbacks, available experimental data do tend to support the cue conflict theory fairly well and it remains the most widely accepted view. In terms of experimental design, scenarios which reduce cue conflict between the vestibular and visual systems, either by attenuating the expected inputs from the vestibular system (i.e., scenarios with few sharp turning movements or low calculated acceleration forces) or attenuating the optic flow in the visual system (i.e., scenarios which maintain an adequate distance from landscape features such as trees and signs and scenes which contain relatively few elements), will result in a lower sickness rate.

14.2.2 Poison Theory

The poison theory attempts to explain simulator sickness from an evolutionary point of view (Treisman, 1977). With this theory, it is believed that the types of sensory stimulation artifacts found in virtual environments such as blurred vision, temporal instability, and lack of sensory coordination caused by low visual resolution and improper or scaled motion cueing, are similar to the symptoms one experiences as a result of poison or intoxication. One of the body's most automatic responses to poison is vomiting in order to empty the contents of the stomach. There the premise of this theory is that the effects of virtual environments lead the body to believe that it has ingested poison and the body reacts to rid itself of the problem. As with the cue conflict theory, there are also problems with the poison theory. There is no way to predict when or how fast individuals will elicit an emetic response and thus there are no obvious recommendations for mitigating simulator sickness that derive from the theory. There is also no explanation as to why some individuals, such as pregnant women, are affected more than others, especially in the case of experience with the real-world task. Due to these limitations, it is hard to evaluate this theory. However, this theory could easily be layered on top of cue conflict theory to explain the nauseogenic response.

14.2.3 Postural Instability

The postural instability theory of simulator sickness was developed as an ecological alternative to cue conflict theory. The theory is centered on a premise that the sensory systems are constantly attempting to maintain postural stability in our environment. Postural stability is a state where uncontrolled movements attempting to correct perceived variance from normal postural states are minimized (Riccio & Stoffgren, 1991). So our perceptual and action systems are continually attempting to maintain our postural stability in our environment. Sickness occurs when an individual is attempting to maintain stability under a set of new environmental conditions when they have not yet learned strategies for accomplishing the task. The key to this statement is the new environment. Experienced drivers may become sick in a simulator because it is a new environment where they are trying to apply the skills acquired on the road. This may be best typified by passengers more likely to become sick than drivers in both simulators and actual driving situations (Rolnick & Lubow, 1991). These same phenomena have been observed for small aircraft.

In support of this theory, Stoffregen and Riccio argue that postural instability not only precedes sickness (Stoffregen & Riccio, 1991) but is also necessary to produce symptoms. They also note that in both vehicles and laboratory whole body motion platforms,

motion sickness is most likely to occur when periodic motion is imposed at frequencies between 0.08 Hz and 0.4 Hz, which is similar to the range of frequencies that characterize walking. The interaction of the imposed oscillations with the body's natural oscillations could lead to wave interference and the resulting severe disturbances. Although there is no explanation for how the lack of postural stability ultimately results in an emetic response, the theory does provide some basis for the diminishing effects of sickness as the individual learns the environment. And it would be consistent with efforts to expose individuals more gradually over time to a driving simulator in order to reduce symptoms to a minimum. Finally, because postural instability precedes simulator sickness, it could be used as a way to predict and potentially reduce such sickness (also see later discussion).

14.2.4 Rest-Frame Hypothesis

Recently, Prothero, Draper, Furness, Parker and Wells (1999) presented evidence that suggested that it was the conflict between the rest-frames implied by the constellation of visual cues available to an individual—and not the cues themselves—that was creating much of the observed simulator sickness. So, for example, if the participant is sitting in a fixed-base, visually-immersive simulator, then the rest-frame defined by the elements is the room itself (if the chair in the room upon which the driver is sitting is at rest, the driver is at rest); whereas the rest-frame defined by the displays on the screen is the virtual world (if the participant is moving through the virtual world, the participant is in motion). The rest-frame is the particular reference frame which an observer takes to be stationary, largely to reduce calculations of relative motion (Prothero, 1998). It seems as if these two are inevitably in conflict with a fixed-base simulator. However, interestingly, Prothero et al. (1999) argued that the virtual environment could be parsed into two elements—the content and the independent visual background (IVB) upon which that content is displayed. If the rest-frames implied by the IVB and the inertial frame of the participant (the room) could be linked, then this should reduce the conflict between the visual and vestibular cues since the rest-frames are aligned.

To test their hypothesis, Prothero et al. (1999) asked participants to wear a head-mounted display while standing. Measures were taken of postural disturbance and simulator sickness. In the semi-transparent mode, the participants could see the laboratory wall behind them through the lenses of the display. In the opaque mode, they had no independent visual background. Prothero et al. reported less postural disturbance and simulator sickness when the IVB was visible. Duh, Parker and Vanesse (2001) extended the results of Prothero et al. Again participants were standing; this time, however, the display was presented on a screen and the independent visual background was a simple grid superimposed over the display. There were three grid conditions: bright, dim and none. Again, the IVB reduced postural disturbance which is known to be associated with simulator sickness. Finally, an interesting extension of this into the arena of driving simulation was recently made by Lin, Abi-Rached, Kim, Parker, & Furness, (2002). Briefly, they asked participants sitting in the passenger seat in a driving simulator to report levels of sickness and enjoyment in one of four conditions: (1) no avatar, (2) an earth-fixed visually-guided avatar (a plane hovering in the sky), (3) an earth-fixed visually-guided avatar which indicated to the passenger in which direction the car would turn, and (4) a non-earth-fixed visually-guided avatar which again indicated turning directions. In the second and third conditions, the participant had independent information about the visual background. There were decreases in sickness in both conditions, but they were not statistically significant in the second condition (which provided an IVB, but no prediction about where the car would be turning). It is not clear, however, whether an IVB such as a visually-guided avatar, would actually help the driver (as opposed to a passenger) in a fixed-base simulator.

Although none of the competing theories fully explain the simulator sickness phenomenon, we can take a conservative approach to simulation design by working to create an environment, simulator, scenes and scenarios which will, within the context of each theory, reduce simulator sickness. Below, we look specifically to the visual and vestibular systems as areas that are both the primary causes of simulator sickness and the source of possible remedies.

14.3 The Visual Systems

In this section, the focus is on the visual systems, the potential such systems have for producing cue conflict, and the steps one can take to reduce such conflict. As noted above, other theories have been used to explain the existence of simulator sickness. Where obviously relevant, these theories will be introduced as well.

The visual system is a very complex and heavily researched sensory system. It is not within the scope of this chapter to provide a full description of the anatomy and processing that make up visual perception. Other reference materials such as Goldstein (1989) provide good explanations of the visual system. Within the scope of this chapter, it is important to understand key characteristics of the visual system that have some influence on the development of simulator sickness.

14.3.1 Central Versus Peripheral Vision

As our eyes move to acquire a target and then process it, they are working to focus the target image on the retina within the area of the fovea. The resulting area of perceived vision has been referred to as central vision. The receptors responsible for central vision are good at maintaining a sustained response which means they will continue to fire as long as the stimulus is present. Because the image must often be stabilized for some period of time while the perceptual processing occurs (e.g., when then head is moving, or the entire individual is moving), movements of the eye exist to support this stabilization including saccades, smooth pursuit, the vestibule-ocular reflex (VOR), and the optokinetic reflex (OKR). Of these eye movements, VOR and OKR are of particular interest when considering the effects of virtual environments and resulting simulator sickness; and they will be discussed later in further detail.

The area surrounding the fovea is not well adapted for seeing specific targets but is good at detecting moving objects and plays a central role in the perception of self-motion. The perceived vision from this area is called peripheral vision (Leibowitz, 1986). The receptors outside the fovea are much better suited for detecting transient stimuli and will fire as they detect a stimulus but will not continue to fire. Therefore, the peripheral sensors are sensitive to moving objects and also changes in orientation of the individual. Information about changes in orientation is believed to feed back into the part of the brain that determines posture, balance, and self-motion, acting almost as a proprioceptive sense. Information about changes in the location of objects in peripheral vision over time provides information about how an observer is moving through his or her environment. Movements of an observer through the real world (or of a virtual world around an observer) are coded by peripheral vision as optic flow.

14.3.2 Optic Flow

Here we talk about the contribution of optic flow to simulator sickness. To repeat, optic flow is created by the movement of elements in the optic array that occur as an observer moves relative to his or her environment (Goldstein, 1989). A simple example of this can be found as you ride in a vehicle and you fix your gaze in the direction you are traveling. All objects within the field of view will appear to move away from the center of your destination or point of expansion (POE). Figure 14.1 illustrates the directions that objects will appear to move as you move through the environment.

Optic flow provides information relevant to steering and turning. Optic flow also provides information about our speed relative to the environment. The more rapidly objects move along the flow lines, the faster the observer perceives their motion. Thus, immersive display devices and screens that display an image to the side as well as to the front and both well above and below the POE will emphasize the optic flow and, consequently, will contribute to the conflict between the visual and vestibular systems and simulation sickness. Human perception of changes in optic flow appears to be quite sensitive and often occurs without conscious thought or effort. Optic flow that specifies movements that do not coincide with vestibular cues can produce sensory conflict which can induce simulator sickness. Consequently, one way to reduce simulator sickness is to reduce optic flow, which can be achieved by decreasing the field of view and by removing elements in a scene that contribute to optic flow.

14.3.3. Perceived Self-Motion

Optic flow also contributes to perceptions of self-motion (much of the following description of perception of self-motion is taken from LaViola, 2000). Under various circumstances, individuals that are static with respect to their environment may experience a compelling illusion of self-motion. This effect is known as vection. Vection is typically measured by asking subjects to rate its magnitude (Prothero, 1998). Vection can occur in naturalistic environments such as looking out the window of a vehicle and feeling motion due to the movement of an adjacent vehicle even though no self-motion is present. These effects have often been seen in virtual environments as well. "Immersive" virtual

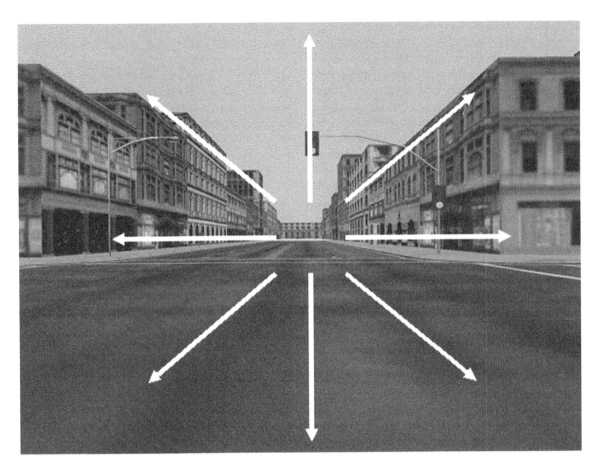

FIGURE 14.1 Optic flow example.

environments with wide field of view displays or helmet-mounted displays where fewer references to a static world exist are prone to causing this effect. In the case of a fixed-base simulator, these effects are being generated by changes in the optic flow.

The strength of vection can be influenced by several factors. Larger fields of view (greater than 30 degrees) have been shown to produce greater perception of motion. This is likely due to the increased information provided in the peripheral field of view which has been shown to have a greater influence on perception of self-motion (Johnson, 2005). Wider fields of view are often found in modern driving simulators because there are many instances in driving where full left and right scanning is required to negotiate the environment. Intersections are a good example where you must be able to look 90 degrees left and 90 degrees right in order to check for traffic, pedestrians, etc., before proceeding. Thus, it can be difficult to reduce the field of view and the resulting experience of vection while still maintaining scenarios that can test one's hypotheses.

Also important to the experience of vection is the rate of optic flow where increased flow rates equate to greater perception of speed of vection. Unlike an aircraft simulator, a typical driving scene has a high rate of optic flow because the observer's eye point is close to the road surface (low altitude). In addition, there are typically many more features in the scene that are close to the driver such as other vehicles, buildings, signs, roadside vegetation, etc. Measures can be taken in the design of a scene to reduce the elements that contribute to vection without necessarily compromising the validity of the study. For example, one could easily replace a picket fence along the side of the road with a rail fence.

As noted above, the changes in optic flow provided by the visual system provide both translational and rotational information. In a standard environment, these changes in optic flow would be accompanied by corresponding vestibular information. However, in a virtual environment, the vestibular information is not available for inter-sensory corroboration. As noted above, it is the result of this effect that forms the basis for the sensory conflict theory of motion sickness. However, it should be made clear the sensory conflict exists even without the perception of self-motion. The conflict may be enhanced if the participant also feels like he or she was in motion, but it is not known whether this is the case. In either case, factors that reduce vection reduce optic flow and therefore, presumably, cue conflict.

14.3.4 Perception of Depth

Our perception of depth comes from a number of sources including oculomotor cues, pictorial cues, motion-produced cues, and binocular disparity. The oculomotor cues are those given by the position of our eye and tension on the muscles within the eye. Pictorial cues are those that could be extracted from a still picture. Motion-produced cues depend on the motion of the observer or the objects in the environment. Binocular cues come from the fact that slightly different scenes are formed on the retina of our eyes. Oculomotor and motion produced cues are the only two sources of information likely to produce cue conflict.

The oculomotor cues include convergence and accommodation. Both operate by giving proprioceptive feedback to the brain about where an object is located upon which an observer is focusing. Convergence is the inward angular positioning of the eyes to keep an object focused on the fovea as the object is moved closer to the observer. The closer the object is to the observer, the greater is the muscular input required to keep it positioned on the fovea. Accommodation is the process of flexing muscles in the eye to change the shape of the lens as an image is brought into clear focus on the retina. The closer the object, the more muscle tension is required to bulge the lens of the eye. These effects typically only occur when the target object is within a distance of 5–10 feet. Objects further away are normally focused without any adjustments to the orientation of the eyes. Thus, there is the potential for cue conflict between the kinesthetic and visual systems when the display is located more than 5–10 feet from the driver since the oculomotor cues (convergence and accommodation) are not present, but the visual cues are present, as the driver focuses on an approaching vehicle.

Motion-produced depth cues include motion parallax and accretion and deletion. As an observer moves through an environment, objects that are further away appear to move slowly in the direction of the observer's movement. Closer objects appear to move more rapidly in the direction opposite the observer's movement. The apparent angular velocities of the objects will be inversely proportional to their distance from the observer. Accretion and deletion are related to motion parallax and interposition. If two surfaces are at different distances from the observer, any movement in the observer that causes one surface to cover another will give cues to depth. The covering surface is seen to be closer then the covered surface. In fixed- and motion-base simulators, motion-produced cues such as parallax are well reproduced when the vehicle's direction changes heading, but not when the driver himself or herself moves around in the cabin, leading to potential for cue conflict. In head-mounted simulators, these intramodal cue conflicts are not present.

14.3.5 Optokinetic Reflex

The optokinetic reflex (OKR) is one of several eye movements that function to identify a target in a visual scene, to position the target on the fovea, and to keep it positioned there. The OKR works by evaluating information from the entire retina to determine if image slip is occurring (e.g., the target is moving with respect to a fixed observer). If there is an image slip, a corresponding movement is made in the eye position to eliminate it, thus stabilizing the image. An example of this process at work is when we look out the window of a vehicle (assuming the head is still; otherwise with the head in motion the vestibulo-ocular reflex is also at work; see discussion below). As the optokinetic reflex detects slippage in the image, it applies a compensating movement to the eye with a gain equal to the motion and direction of the optic flow. The small differences between the eye and the image generator of the simulator can be at odds with each other. The human eye does not process information in a discrete number of frames per second, but the image generator in a simulator does. Whether the OKR adjusts perfectly for the stepwise slip of a digitally projected moving image is not clear; but if it does not, there will be intramodal conflicts between what the eye perceives and what it expects to perceive based on the faulty adjustment applied by the optokinetic reflex. Arguably, the validity of the OKR adjustment will improve as the frame rate increases.

14.4 Vestibular Systems

There are both central and peripheral vestibular systems. It is the peripheral vestibular system which interests us the most. Specifically, we focus on the potential such systems have for producing cue conflict, and the steps one can take to reduce such conflict. The peripheral vestibular system rests in an area of the inner ear called the labyrinth. It is made of up a series of tubes (semicircular canals) and sacs (utricle and saccule). The semicircular canals are primarily responsible for detecting angular acceleration in each of the three planes in which motion can occur. They are quite sensitive and can measure angular accelerations as low as 0.1 deg/s^2 (Gianna, Heimbrand, & Gresty, 1996). The utricle and saccule are responsible for detecting linear acceleration. The utricle is oriented to be able to detect motion in the horizontal plane; the saccule is oriented to detect motion in the vertical plane and fore-aft plane. Once the brain receives the impulses from the entire vestibular system, it uses the information for perception of motion and also transmits information to the visual system. More discussion of this process will be included in the following sections.

14.4.1 Vestibulo-Ocular Reflex

There is a clear relationship between the vestibular and visual systems where angular acceleration information about head movement is supplied to the visual system. The visual system interprets this information and makes a corresponding eye movement to stabilize the visual image on the retina. The process is called vestibulo-ocular reflex (VOR). A simple example of this effect can be shown by holding a piece of paper with some printed text in front of your eyes. If you move your head while holding the paper stationary, you will be able to read the text with some level of effort. If you hold your head stationary and move the paper, it is much more difficult or perhaps even impossible to read the text. When the head moves, the vestibular system is providing information to the eyes allowing them to stabilize the image of the shaking paper on the retina. In the case where the paper was moving, there was no information about how the paper was moving that could be sent to the eye and so the image could not be stabilized.

14.4.2 Vestibulo-Ocular and Optokinetic Relfexes

Often, both the head and the image are moving. In this case, the vestibular-ocular reflex (VOR) and the optokinetic reflex (OKR, as discussed earlier) work together synergistically to maintain a stable retinal image regardless of the type of motion being experienced (Zacharias & Young, 1981). The VOR is a very fast-acting reflex which serves to compensate for head movements in the 1–7 Hz range. However, the VOR is much less accurate at lower frequencies and has less than perfect gain. The OKR has the opposite performance characteristics. It has a longer latency due to the required evaluation of visual information to determine a response and has near unity gain at low (<0.1 Hz) frequencies. Between 0.1 and 1.0 Hz frequencies, the OKR begins to lose gain and also develops a phase lag due to inherent response latency. The two reflexes working in unison are able to provide stable retinal images through a wide range of frequencies. Cue conflict occurs primarily with head-mounted displays (Draper, 1996). Although not so true with today's head-mounted displays, slow sensors and/or slow scene rendering will create a mismatch between what it is expected to be seen (based on the VOR and OKR synergies) and what is actually seen (based on what is displayed).

14.4.3 VOR Adaptation

It has been demonstrated that the VOR response is adaptable, in that gain values will be adjusted to accommodate different sensory arrangements. An example is provided by the case of looking through magnified optics such as scuba goggles. VOR will adapt its gain to match the amount of eye movement required to stabilize the image even under the modified conditions. In a study to evaluate the effects of visual scale factor on VOR, Draper (1998) found that visual magnifications of 2× and 0.5× did result in corresponding VOR adaptations and that the visual adaptation was correlated with simulator sickness. One explanation is that OKR provides a tight feedback loop of information to the VOR adaptation process, allowing it to tune itself to the given conditions. VOR adaptation, or speed of VOR adaptation, has been hypothesized to be a predictor of simulator sickness potential where individuals who adapt faster are less likely to experience sickness symptoms (Draper et al., 1997). This link might explain how subtle artifacts of poor simulator engineering might delay the VOR adaptation process either through inconsistent feedback or by altering the performance of the OKR through visual anomalies.

14.5 Measures of Simulator Sickness

There are a number of different measures of simulator sickness that have been used. Below, we discuss several, including the most common subjective rating scale, as well as several alternatives, including postural stability and physiological measures. Additionally, we discuss the issue of when to measure just how sick a simulator makes an individual during a particular session and whether to use indices of simulator sickness as a covariate in measures of performance.

14.5.1 Simulator Sickness Questionnaire

The Simulator Sickness Questionnaire (SSQ) developed by Kennedy, Lane, Berbaum, and Lilienthal (1993) is perhaps the most widely used instrument, cited in over 300 publications since

it first appeared. It is derived from the Pensacola Motion Sickness Questionnaire (MSQ) developed some 30 years previously (Kennedy & Graybiel, 1965). The MSQ had a total of 28 different symptoms that were evaluated by participants. The SSQ was developed from the pre- and post-test assessments of motion sickness using the MSQ. A total of 1,119 pre- and post-test pairs from the MSQ were evaluated, the pairs collected on 10 different simulators. Because the goal was to determine which symptoms showed changes from pre- to post-test, those symptoms were excluded that showed little change. For example, vomiting was experienced in less than 1% of participants and was therefore excluded. Additionally, signs that were observed on only some simulators and appeared irrelevant to simulator sickness were excluded, e.g., boredom. A final set of 16 symptoms was included. They are listed in Table 14.1.

In order to determine whether there were sets of symptoms in the SSQ which were correlated with each other, a principal factors analysis was performed. Three subscales were identified, one related to nausea (N in Table 14.2), one related to oculomotor problems (O), and one related to disorientation (D). If, in the factor analysis, the loading on a factor of a given symptom was greater than 0.30, the symptom was included in the *total factor score* (marked as a 1 in the appropriate column in Table 14.2); otherwise it was not included. The total score on each factor could then be computed. It was equal to the weighted sum of the symptom scores for a factor. The symptom score was 0, 1, 2 or 3 as rated by an individual participant (Table 14.1). So, for

TABLE 14.1 Simulator Sickness Questionnaire

Directions: Circle one option for each symptom to indicate whether that symptom applies to you *right now*.				
1.	General Discomfort	None	Slight	Moderate	Severe
2.	Fatigue	None	Slight	Moderate	Severe
3.	Headache	None	Slight	Moderate	Severe
4.	Eye Strain	None	Slight	Moderate	Severe
5.	Difficulty Focusing	None	Slight	Moderate	Severe
6.	Salivation Increased	None	Slight	Moderate	Severe
7.	Sweating	None	Slight	Moderate	Severe
8.	Nausea	None	Slight	Moderate	Severe
9.	Difficulty Concentrating	None	Slight	Moderate	Severe
10.	"Fullness of the Head"	None	Slight	Moderate	Severe
11.	Blurred Vision	None	Slight	Moderate	Severe
12.	Dizziness with Eyes Open	None	Slight	Moderate	Severe
13.	Dizziness with Eyes Closed	None	Slight	Moderate	Severe
14.	Vertigo[a]	None	Slight	Moderate	Severe
15.	Stomach Awareness[b]	None	Slight	Moderate	Severe
16.	Burping	None	Slight	Moderate	Severe

Source: R. S. Kennedy, N. E. Lane, K. S. Berbaum, & M. G. Lilienthal., Simulator sickness questionnaire: An enhanced method for quantifying simulator sickness. *International Journal of Aviation Psychology, 3*(3):203–220, 1993. With permission.

[a] Vertigo is experienced as loss of orientation with respect to vertical upright;
[b] Stomach awareness is usually used to indicate a feeling of discomfort which is just short of nausea.

TABLE 14.2 Simulator Sickness Questionnaire

SSQ Symptoms		Weight	
General Discomfort	1	1	
Fatigue		1	
Headache		1	
Eye Strain		1	
Difficulty Focusing		1	1
Increased Salivation	1		
Sweating	1		
Nausea	1		
Difficulty Concentrating	1	1	
Fullness of Head			1
Blurred Vision		1	1
Dizzy (eyes open)			1
Dizzy (eyes closed)			1
Vertigo			1
Stomach Awareness	1		
Burping	1		

Source: R. S. Kennedy, N. E. Lane, K. S. Berbaum, & M. G. Lilienthal., Simulator sickness questionnaire: an enhanced method for quantifying simulator sickness. *International Journal of Aviation Psychology*, 3(3):203–220, 1993. With permission.

example, if a participant rated the seven symptoms under nausea as, respectively, 3, 2, 1, 2, 3, 2, 1 the unweighted nausea factor score would be 14. The weighted nausea score (N_s) would be 14 × 9.54. The weights for O and D were, respectively, 7.58 and 13.92. The *total score* was equal to the sum $N_s + O_s + D_s$ × 3.74.

14.5.2 Times of Administration

A potential problem with self-rating measures of simulator sickness has recently been addressed. In particular, participants, alerted to the possibility of simulator sickness in a pre-questionnaire (or other instructions) may thereby experience more simulator sickness than would someone not informed. In order to test this hypothesis, 30 participants between the ages of 20 and 45 years old were given the SSQ, either before and after exposure to a virtual environment, or just after exposure to that environment (Young, Adelstein, & Ellis, 2007). A head-mounted display was used. The average post-test SSQ (11.0) score of participants administered both the pre-test and post-test was roughly 80% higher than the post-test score (6.0) of participants administered just the post test, a difference which was statistically significant. The authors note that it is not clear whether the participants actually experienced more simulator sickness in the group that was administered both a pre-test and post-test or simply reported more simulator sickness at the end. This is certainly something that should be addressed in future research. The use of postural and physiological measures such as those discussed below could answer this question. Nevertheless, it is clear that comparisons of SSQ scores across experiments must take into account what was told to the participants prior to exposure.

14.5.3 Postural Stability

As noted previously, postural instability theory was proposed as an ecological alternative to cue conflict theory (Riccio & Stoffregen, 1991). It follows that measures of postural instability might be used to predict motion and simulator sickness. Towards this end, Stoffregen and Smart (1998) exposed standing participants to the very small oscillatory motions that are typical of walking, using a moving room (Stoffregen, 1985). Symptoms of simulator sickness were preceded by changes in postural sway, measured here as changes in the variability, velocity and range of head movements.

The above results are not strictly applicable to a virtual environment. Thus, Stoffregen, Hettinger, Haas, Roe and Smart (2000) used a fixed-base flight simulator to determine whether indices of postural sway preceded symptoms of simulator sickness in a virtual environment. The outside world was projected on a dome. The star field and a spacecraft projected on the star field oscillated in the roll axis on the experimental trials and remained constant on the control trials. Each participant was in the simulator for approximately two hours. The SSQ (long and short forms) was administered before testing, immediately after testing on the nauseogenic stimuli (sum-of-sines scenarios), and then one hour post-test. Head movement was measured throughout the testing using a magnetic tracking device. A total of 14 participants were run in the experiment. Six became sick; eight did not, as measured by the SSQ and the experimenters.

Prior to the testing, there were a number of differences in postural motion in the sick and well groups. For example, there were significant differences in the velocity of head movements in the yaw and roll axes. In each case velocity was greater in the sick group. Interestingly, it is estimated that these two measures accounted for roughly one-third of the variance, more than can currently be accounted for by physiological variables (Kennedy, Dunlap, & Fowlkes, 1990). Perhaps more importantly, postural motion predicted sickness not only for the strongly nauseogenic stimuli (the sum-of-sines scenarios), as indicated by significant increases in head velocity in those who later became sick, but also for the less nauseogenic stimuli (the 0.2 Hz oscillation). This prediction was made during the first 10 minutes of the experiment (Trial 5, a sum-of-sines trial). Note that the predictors of simulator sickness in this experiment (velocity of postural motion) were not the same as the predictors of motion sickness in an earlier experiment (variability of postural motion; Stoffregen & Smart, 1998). Kennedy et al. (1993) argue that this is not problematic for the postural instability theory (Riccio & Stoffregen, 1991) because no one parameter enjoys dominance in the theory. Practically, this means that investigators would need to measure variability, velocity and range of postural motion in all six degrees of freedom.

14.5.4 Physiological Measures

Physiological measures have not received the attention one might have expected given their clear relation to the symptoms of simulator sickness. Perhaps this is because very few clear relationships have emerged in most prior studies (Biaggioni, Costa,

& Kaufmann,1998; Bolton, Wardman, & Macefield, 2004; Collet, Vernet-Maury, Miniconi, Chanel, & Dittmar, 2000; Espié, 1999; Gianaros et al., 2003; Mullen, Berger, Oman, & Cohen, 1998). However, two recent studies seem to suggest that there may indeed be a relation (Bertin et al., 2004; Min, Chung, Min, & Sakamoto, 2004). We will focus here on one such study (Bertin et al., 2004).

In that study, drivers sat in an actual car. The virtual world was presented on three large screens subtending 150 degrees of visual angle. The continuous physiological data recorded included skin potential (SP), skin resistance (SR), skin temperature (ST) and heart frequency (HF). A continuous psychophysical measure of simulator sickness was also recorded by asking participants to indicate their condition by moving a cursor along a visual analog scale containing 10 stops, anchored at one end by "all is fine" and at the other end by "I'm about to vomit". The participant could control the movement of the cursor using levers on the steering column. The visual analog scale was projected low in the visual field. Strong reliable correlations were reported between simulator sickness scores and three of the four physiological variables (SR, ST and HF). It is not clear why some studies, such as this one, find such strong correlations. One possibility is that simulator sickness is a constellation of factors, as reflected in the three dimensions that underlie the SSQ. Clearly, it would be useful to better understand what controls the strength of the correlations, because physiological indicators could provide a very valuable tool in managing simulator sickness.

14.5.5 Use as a Covariate

Simulator sickness presents several challenges in the process of data collection and interpretation. Most obvious among the challenges is the discomfort and distress it can cause participants. It can also lead to lost data when a participant withdraws from a study before completing the experiment. Such lost data can jeopardize the integrity of the analysis because it might affect some experimental conditions more than others and some populations of participants more than others.

A more subtle effect concerns the situation in which a participant does not withdraw from the study, but drives differently as a result of feeling ill. Drivers might behave differently to minimize the cues that are causing the ill feelings, such as closing one's eyes when negotiating a turn. Feeling ill can also pose a distraction that might have more diffuse effects. Obviously the ideal approach is to design the simulators, scenarios, and protocols to avoid simulator sickness. Another is to statistically assess the influence of simulator sickness by using the SSQ or other measures as a covariate in an ANCOVA. Such an analysis identifies the potential confounding effect of simulator sickness and offers the potential of adjusting for its effect (Bittner et al., 1997).

14.6 Factors Associated With Simulator Sickness: Preventive Measures

The factors associated with simulator sickness can neatly be divided into simulator, task, and individual characteristics. There is much that can be done to alter the simulator and task characteristics. These measures include the simulator hardware itself, the scenes and scenarios that the participant needs to negotiate, the environment within which the simulator is located, the monitoring of ongoing levels of simulator sickness, and the careful adaptation of a driver to the simulator environment. A number of authors have published guidelines to reduce simulator sickness (Braithwaite & Braithwaite, 1990; Crowley & Gower, 1988; Kennedy et al., 1987; Kolasinski, 1995; Lilienthal, Kennedy, Berbaum, Dunlap, & Mulligan, 1987; McCauley, 1984; Naval Training Systems Center, 1988; Wright, 1995). Arguably the most thorough set of guidelines are those by Kennedy et al. and Wright. The reader may find additional suggestions in these guidelines specific to his or her particular situation. Although there is little if anything that can be done to alter the individual differences prior to an experiment (beyond screening high risk participants), exposing especially high risk individuals to a virtual environment in a slow, stepwise fashion is always possible as a way of potentially decreasing such individuals' risk (Johnson, 2005).

14.6.1 Simulator Design Factors

We have talked about some of the factors due to the design and construction of a simulator that impact simulator sickness. Below, we give a more complete list of these factors, including the type of simulator, field of view, display alignment, image resolution, graphics update and refresh rates, motion system, calibration of eye height, and transport delay. It goes without saying that a simulator whose fidelity matched that of the real world is the gold standard. Even with such a simulator, motion sickness could still be a problem, just as it is for some drivers on some roads; however, no existing simulator can do this. Moreover, what at first seems like a higher fidelity simulator can sometimes perversely increase simulator sickness. The best example is the addition of a motion base. Clearly, a simulator with motion has the potential to be of higher fidelity than a simulator without motion. But, if the visual and motion cues are still in conflict, then nothing is gained and poorly correlated motion could confront drivers with a greater perceptual mismatch than no motion. Regardless, there are still things that the researcher can do.

14.6.1.1 Types of Simulators

Prior to purchase, a researcher needs to decide whether to get a fixed-base, motion-base, or head-mounted simulator. The choice is an extremely difficult one when it comes to evaluating which simulator will significantly reduce sickness. We fully realize that this is not the only issue when deciding among simulators or even the primary issue; but it is the only one which we will discuss here. Ideally, there would be a common set of scenarios and one could simply compare the sickness rates for the simulators of interest. But, such a database does not exist.

If simulator sickness is likely to be of real concern to a researcher, either because of the participants being selected (e.g., older drivers) or the scenarios being used, then one may need to exercise special care when adding motion or using a head-mounted display. As a general rule, motion cues appear to decrease simulator sickness or leave it unchanged. Various studies on the effectiveness of motion cuing are discussed in the section on motion cueing. As for head-mounted displays, typically they are fully immersive, making it difficult for participants to maintain the correct rest-frame. Moreover, additional demands are placed on the hardware because head movements now need to be tracked precisely. These characteristics make simulator sickness more likely with head-mounted displays. The issues for head-mounted displays are different than those for fixed- or motion-based displays and are discussed below in their own section.

14.6.1.2 Field of View

As noted above, field of view has long been implicated as a contributing factor to simulator sickness (Kennedy, Lilienthal, Berbaum, Baltzley, & McCauley, 1989; Casali, 1986; Kolasinski, 1995; Pausch et al., 1992). While the effects of vection and resulting simulator sickness have been reported in fields of view as narrow as 15 degrees (Andersen & Braunstein, 1985), a greater field of view tends to be an elevating factor (Ijsselsteijn, Ridder, Freeman, Avons, & Bouwhuis, 2001). This is because wide fields of view have the potential to stimulate more of the peripheral visual system, which in turn results in greater optic flow and vection (Kennedy et al., 1988; for a review, see Andersen, 1986). However, there is no study of which we are aware that specifically quantifies the relationship between increases in the field of view and increases in sickness. There is one interesting finding discussed below which may deserve more investigation. In particular, at least in some circumstances, it would appear that providing special glasses which restrict the field of view in a wide field of view simulator leaves performance unchanged (Van Erp & Kappé, 1977; Pepper, 1986; Spain, 1988). If this is the case, then one can reduce optic flow and potentially the associated simulator sickness without affecting performance by simply designing a simulator with a relatively narrow field of view.

The simulation designer is faced with a trade-off where the visual system should have just enough field of view to support the requirements of the task being performed but not so much that optic flow becomes a problem (Stanney, Mourant, & Kennedy, 1998). The field of view required for driving ground vehicles varies depending on the experimental tasks being performed. In normal highway driving, the driver needs to be able to scan the environment ahead to determine the physical shape and orientation of the roadway in addition to acquiring self-motion information from the optic flow. At intersections, the driver needs less information about self-motion but has an additional requirement to scan left and right looking for potential hazards and checking for traffic. In this case, a 180-degree forward field of view would be ideal to safety negotiate an intersection. The requirements of off-road driving are also likely to include an ability to scan a wide field of view as the driver searches for hazards and looks into and around tight, winding roads.

14.6.1.3 Calibration of Eye Height: Simulator Design Eye

Care should be taken to position the viewer in the appropriate location such that the image presented from the eye point in the simulator (the simulator design eye) matches the eye point of the actual operator. Typically the height of drivers will not have a significant affect on where the displays need to be placed; however, a slight change in the driver location in the vehicle can have a significant effect on the optic flow and visual information. The slight change can come from something as simple as a head movement, a real problem in flight simulators (Kennedy et al., 1987). This is not as much of a problem for drivers in a ground vehicle simulator. However, the passenger seated in the simulator will not receive the same cues as the driver given the position in the vehicle (the simulator design eye and the passenger's point of view are different in this case) and will be much more susceptible to simulator sickness (Riccio & Stoffregen, 1991).

14.6.1.4 Display Alignment

Many simulators use multiple screens to project a virtual world that surrounds the driver. If misaligned, the misalignment can cause simulator sickness (see Hettinger et al., 1987; and Kennedy et al., 1987, for a more detailed discussion). Briefly, there are three reasons that simulator sickness could increase. First, unless viewed from the geometric center of projection, displays of the 3D world introduce distortions (Rosinski, 1982). Second, if the displays are not aligned, then the participant would experience the same scene but as if from different viewpoints (see also Andersen, this book, chap. 8). This would lead to intrasensory cue conflict. Finally, if the scenes have different virtual distances then they would be at different accommodative focuses. This could result in fatigue and headaches from the constant switching.

14.6.1.5 Scale

Images in the virtual world often appear as scaled (minified or magnified). The geometric field of view is defined as the angle subtended by the near or far plane in the viewing frustum. If the observer were sitting at the computed center of projection and the far plane was displayed on a screen just equal to the vertical and horizontal boundaries of the far plane, the image would be perfectly scaled. However, if the screen is closer it will be magnified; and if it is further it will be minimized. In order to test whether image scale had an affect on simulator sickness, Draper, et al. (2001) asked participants to navigate a virtual world which was minimized (0.5), to scale (1.0), or magnified (2.0). A head-mounted display was used. Thus, the feedback from the vestibulo-ocular reflex in the minification and magnification conditions was in conflict with the change in the visual information in the scene. Participants were asked to complete the SSQ both before and after being exposed to the virtual environment. The absolute values on the pre-test and post-test SSQ were much smaller in the neutral condition. Moreover, both minification and magnification led to larger changes in both the pre-test and post-test scores.

It seems clear that one should do as much as possible to ensure that there is no magnification or minification.

14.6.1.6 Image Resolution

Image resolution can have a marked effect on task performance and may also contribute to simulator sickness. A healthy human eye can perceive an image that subtends an angle of about 1 arc minute (arcmin) onto the foveal part of the retina. One arcmin/pixel resolution roughly equates to about 20/20 vision. Many driving simulators today have effective resolutions of about 3–5 arcmin/pixel (Kemeny, 2000; Jamson, 2001) which equate to 20/60 – 20/100 vision. The Federal Aviation Administration (FAA) requires their aviation training simulators to have effective resolutions of 3 arcmin/pixel or less. A simulator with a 180-degree field of view and a 3 arcmin/pixel geometric resolution would require three projectors that each have a horizontal pixel count of at least 1,200. Geometric and effective resolutions are two different measurements of image resolution. Effective resolution is determined by the geometric resolution along with what is known as a Kell factor. The Kell factor accounts for losses in the projector itself, cabling, lens quality, brightness, contrast ratio, etc. (Robin, 2003). A geometric resolution to reach vision limits (1 arcmin/pixel) would require a minimum of 27 such projectors to display the same visual field.

The resulting effects of limited resolution can include drivers missing key features that they should be able to perceive in the environment and potentially causing some amount of eye strain as the eyes attempt to resolve images that cannot be brought into focus (Govil et al., 2004). Moreover, with poor resolution there are potential trade-offs between required contrast, luminance, and resolution (Pausch et al., 1992), leading to potential complications with flicker fusion. Finally, it is possible that the higher resolution displays actually increase optic flow. Very high resolution displays now exist (e.g., 4096 × 2160, Sony Electronics, 2008) and could be used to evaluate the effect of the greatly improved resolution on the development of simulator sickness. However, direct links of the resolution level to simulator sickness have not been made.

Image resolution would appear to have a more direct impact on task performance than on simulation sickness. However, attempts should be made to increase resolution to a level that is sufficient for reducing eye strain while making it possible to extract task-relevant information from the scene.

14.6.1.7 Graphics Update Rate and Refresh Rate

The graphics update rate is the rate at which the display is updated based on the most recent interpretation of information concerning the vehicle state within the virtual environment (it influences the transport delay, discussed separately below). The graphics update rate is typically a function of the capability of the graphics generation hardware/software and the complexity of the visual scene and moving models. The relationship is one of inverse proportion whereby higher levels of complexity typically result in lower sustainable update rates and dropped frames. Decreased update rates can result in increased lag between a given control input and the presentation of the corresponding update of the state of the simulation system. A system running at 30 Hz without prediction

algorithms that interrogates the dynamics and extrapolates the eye point to a future location, will add a minimum of 32 ms to the total lag of the system without even accounting for the time it takes to process the information to provide the viewpoint for the graphics subsystem. Many simulators provide information on graphics refresh rate as a measure of performance. When possible this should be checked to ensure that complex scenes or that addition of many vehicles have not caused the frame rate to drop below 30 Hz. Ideally, the scene should have a graphics refresh rate of 60 Hz.

Frank et al. (1988) found that delays in the update of visual information were more disconcerting to simulator drivers than were delays in the update of the motion system. A graphic update rate of 10 Hz presenting an out-the-window view of a vehicle driving at 55 mph will only be updated every 8 feet. The resulting presentation appears "jerky" and has the potential to be perceived as flicker (Casali, 1986). The update rate would need to be at 30–60 Hz for typical vehicle operating speeds.

Refresh rate is the rate at which the display system re-draws the graphic view generated by the image generation system. Refresh rate is independent of the vehicle simulation and the rate at which it processes. Each refresh of the visual scene will present the current state of the graphical output from the image generator. So if the image generator was running at 30 Hz and the display system was capable of running at 60 Hz, each of the 30 Hz graphics frames would be drawn twice by the display system. Refresh rate has the potential to impact sickness if the rate is not constant or if it is slow enough that flicker can be detected (e.g., Harwood & Foley, 1987; Pausch et al., 1992; Rinalducci & MacArthur, 1990). Today's hardware and software are typically able to maintain consistent refresh rates of 60 Hz mono or 48 Hz stereo. At normal illumination levels, the refresh rate should not have an impact on simulator sickness in a modern driving simulator.

14.6.1.8 Transport Delay and Vehicle Control

Transport delay refers to the amount of time it takes to detect an operator input, process the new state of the simulator based on the input, and return to the operator the resulting changes in the state of the simulation above and beyond the lag in the vehicle being simulated (delays in the graphics update rate are just one example of transport delay). In the world of flight simulation it is given a precise definition (Federal Aviation Administration, 1994): "It is the overall time delay incurred from signal input until output response. It does not include the characteristic delay of the airplane simulated (p. 3)". The effect of transport delays (either in motion or visual information) is believed to cause additional sensory conflict between the visual and vestibular systems that might lead to simulator sickness and performance decrement (Draper, 1996; Frank et al., 1988; Pausch et al. 1992). In addition, in driving simulators, visual delays combined with missing vestibular cues can also cause self-induced steering oscillations which can exacerbate the problem through increases in visual artifacts caused by yaw rotation in the display. When simulators lack haptic cues they eliminate "lead" from the input to the driver, undermining stability. An important consequence of the reduced control is increased steering input and consequently more vection and greater mismatch between visual and vestibular cues as the car swerves back and forth. This instability might be an important cause of simulator sickness and could serve as an early warning of potential simulator sickness.

Cunningham, Chatziastros, von der Heyde and Bultoff (2001) manipulated visual transport delays to steering inputs on a high-fidelity, floor-mounted driving simulator. They evaluated steering performance as a function of these delays as drivers negotiated a curved route at fairly high speeds. Their goal was to determine how drivers adapt to the delays and if the adaptation transfers to other driving conditions. The delay values they used were 130, 230, and 430 ms. In their first experiment, they found that drivers did learn to adapt to the delays, but the longer the delay, the longer the adaptation period. In addition, they found that a subsequent removal of the delay resulted in a renewed decrement in performance. In their second experiment, they determined that the adaptation or learning accomplished in the first experiment generalized to a variety of different road types. So, while subjects can adapt and learn to drive with significant transport delay, their speed of learning and subsequent unlearning will depend on the magnitude of the delay. The longer the delay, the longer the time period required to adapt. A threshold of how small transport delay must be to maintain real-world (non-adapted) driving performance is not yet known. At least one author (Kemeny, 2000, citing Bloche, Kemeny, & Reymond, 1997) indicates that the value must be less than 50 ms.

Frank et al. (1988) performed a driving simulator study to determine the impact of both motion and visual delay and found that visual delay was more disconcerting than motion delay. They concluded that both visual and motion delay should be minimized but it was more important to minimize visual delays if trade-offs needed to be made between the two.

The effect of transport delay on driving simulator drivers is not well understood, or even calculated in most cases (Kemeny, 2001). Yet, it is perhaps the most important measure of simulator performance.

14.6.1.9 Motion Cueing

Motion systems have been added to many modern driving simulators in hopes of increasing realism and the validity of operator responses while also reducing simulator sickness. There are several different types of motion that are used. These types are vibration, small amplitude deceleration cues (1–4 inch movement), large amplitude tilting (hexapod), and large amplitude track. It is difficult to determine what type of motion is the best in terms of reducing simulator sickness because so few simulators are equipped to produce the requisite experimental conditions.

Even with the most capable motion-base available, it is impossible to duplicate the large accelerations felt in an aircraft or in ground vehicles. Other strategies must be used, such as scaled cueing and washout algorithms. Scaled cueing is a technique where a scaling value is applied to the forces being applied to the driver in the simulator. At a scaling factor of 0.25 and a real-world deceleration of 0.4 g, the driver of the simulator would experience a 0.1 g deceleration. Scaling allows for proportional acceleration inputs without extending the simulator beyond the limits of the motion hardware.

When motion is used in driving simulation, the impact on performance depends somewhat on the maneuvers being performed. Advani and Hosman (2001) state that driving skill-based behaviors are affected much more by motion cues than knowledge-based behaviors. Therefore, motion cues will have greater impact on vehicle disturbance and recovery maneuvers than on lane tracking tasks. The driver relies on the quality of the motion cues and close-coordination of corroborating visual information to make appropriate responses.

There are a number of studies where positive results on simulator sickness have been found from adding motion cueing (Casali, 1986; Curry, Artz, Cathey, Grant, & Greenburg, 2002). For example, Curry et al. conducted a study to compare their fixed-based simulator to their 6 DOF motion-base simulation. Their fixed-base simulator has a 140-degree horizontal field of view system and the motion-base system has a 180-degree horizontal front field of view plus 125 degrees to the rear dome on a 6 DOF motion-base. After conducting similar driving tasks for an equal amount of time, they reported lower simulator sickness questionnaire (SSQ) scores for those subjects that drove the 6 DOF motion-base simulator. Several authors (Sharkey & McCauley, 1992; Barrett & Thornton 1968) indicate that perhaps less expensive, higher frequency vibration transducers mounted on the occupant seat might help mask some of the proprioceptive and vestibular cues that might conflict with visually implied motion. In addition, real-world driving applications typically include some amount of higher frequency vibration, which may be an important cue to the perception of vehicle velocity. The mismatch between the true motion of the vehicle and the motion produced by the simulator actuators can induce simulator sickness, with a mismatch in the frequency range of 0.06 to 0.07 Hz being most critical.

However, it is important to understand that motion cues may lead to no discernable differences in simulator sickness (Sharkey & McCauley, 1992; Barnes, 1987; Hettinger, Berbaum, Kennedy, & Dunlap, 1990; Kennedy et al., 1993), or may actually make things worse. The exact factors that contribute to the success or failure of using physical motion cues to reduce cue conflict do not appear to have been determined, and there are few motion systems able to produce non-scaled cues.

14.6.1.10 Head-Mounted Displays

Head-mounted displays (HMD) offer a number of potential advantages to driving simulation applications. First, there is a freedom from visual field of view restrictions experienced when implementing traditional fixed-display technologies. Wider fields of regard are required to perform appropriate visual search and monitoring tasks while driving. With appropriate head tracking technologies, the effective field of regard could be as much as 360 degrees. Second, there is much less infrastructure required to support HMD-based systems due to the elimination of the physical display medium. In addition, the reduced overall footprint of HMD-based simulation systems make them more portable, increasing their applicability to a wide variety of driving applications. Lastly, the cost and complexity of HMD-based systems might also be lower due to elimination of physical display infrastructure and also a reduction in the required graphics generation requirements. Where in some traditional simulator implementations several graphics generators or channels are required to create a wide field of view visual scene, the HMD-based system would only require a single graphics generator. Even though there are a number of compelling potential benefits to applying HMD technologies in driving simulation, there are also some potential drawbacks.

While current HMD technologies provide unlimited field of regard, there are serious restrictions on instantaneous field of view (the field of view visible when the head is still). Most systems offer fields of view that are 50 to 25 degrees horizontally, a width which can be expanded by modifying the amount of ocular overlap. (For a more complete review of HMDs, see a recent article by Patterson, Winterbottom, & Pierce, 2006). Human eyes have an approximate 120-degree horizontal overlap between their fields of view. An ideal head-mounted display system will allow both eyes to clearly see what the other can see within this overlapping region. Most HMDs have two independent channels, one for each eye, and some designs do not fully support this overlapping region in a way that makes sense to the brain. Therefore, it is important to understand the viewing requirements of the simulation and the locations of objects that will need to be observed. Without 100% overlap, objects close to the viewer may cause disorientation as the eyes cannot see the images as the brain expects them to be seen. For instance, partial overlap can lead to visual illusions such as the appearance of a curved moon at the monocular border where binocular rivalry is greatest. The most appropriate modification of ocular overlap for a general driving simulation application has yet to be determined, but will likely be something less than 100%. Regardless, the failure of ocular overlap could lead to symptoms of simulator sickness.

To understand the impacts of reducing the field of view from our unmodified capability on performance and simulator sickness, we must refer back to the basic functions of the anatomy of the eye. Recall from the earlier section on central versus peripheral vision that central vision is good for static viewing and identifying what something is; peripheral vision is good for motion sensation, spatial orientation, and supporting gaze stability (Leibowitz, 1986). With respect to driving, Leibowitz notes that experienced drivers tend to use peripheral vision for steering the vehicle while using central vision for identifying potential hazards in the world. If instantaneous field of view is limited with the HMD, there may be an effect on the driver's ability to effectively steer the vehicle. For instance, Wood and Troutbeck (1994) found that with narrow fields of view it is more difficult to drive a vehicle in a straight line down a straight road. This can easily contribute to simulator sickness since there are no corresponding vestibular cues.

In an evaluation of several display types with pilots performing a flying task, de Vries and Padmos (1998) found that operator performance was worse with the HMD than with head-slaved or full screen displays. However, they attributed the performance reduction to the considerable image delay (190 ms) and heavy weight of their system as opposed to field of view reduction. They came to this conclusion because a limited field of view head-slaved option did not result in a corresponding reduction in performance. They go on to recommend adding vehicle references when using HMDs to help provide a stable reference from which adjustments of orientation can be made. This could also potentially reduce symptoms of simulator sickness if the rest-frame theory of simulator sickness explains why drivers develop such symptoms (Prothero, 1998). Kappe and Padmos (2001) performed a study similar to the one above in order to evaluate the effects of HMD, widescreen, and head-slaved displays on ground vehicle driving performance. They found similar results where the HMD resulted in a negative effect on driving performance.

In an assessment of a fixed-base driving simulator that makes use of an HMD, Mourant and Thattacherry (2000) found that subjects reported more oculomotor discomfort symptoms on an SSQ than what has typically been found in driving simulation studies using the SSQ. They attribute the shift from more nauseogenic symptoms to oculomotor symptoms to advances in virtual environment technology.

Burns and Saluaar (1999) conducted an evaluation of driver behavior using an HMD in a driving simulator as they negotiated their way through intersections and ensuing turns. They found that drivers with the HMD made longer glances but also made the same number of glances as did drivers in a real vehicle. They also found differences in driver's speed after turns, lane keeping ability, and subjective workload where use of the HMD decreased performance and increased workload. In a more theoretical study evaluating perception of self-rotation with an HMD, a widescreen, and a widescreen with field of view limiting blinders, Schulte-Pelkum, Riecke, and von der Heyde (2003) found that, in general, subjects tended to underestimate the amount of rotation they had experienced, but underestimated to a greater extent with the head-mounted display. They concluded that the effect had to do with something other than field of view given the significant difference between the performances in the HMD versus the widescreen with limited field of view blinders. In a second study, Schulte-Pelkum, Riecke, von der Heyde and Bülthoff (in press) evaluated the effects of curved versus flat screens in perception of eco-rotation through visual stimuli. They found that subjects underestimated rotation with curved screens but overestimated rotation with flat screens presenting the same field of view. They attribute the differences to subjects perceiving rotation as translational movement with the flat screen displays. It is not clear how these differences affect simulator sickness, if at all. The cue conflict associated with expected and actual rotation may contribute to simulator sickness, particularly if it also contributes to steering over corrections and high levels of vection.

Ruspa, Scheuchenpflug and Quattrocolo (2002) evaluated two simulator designs that were to be used for ergonomic vehicle evaluation. The first configuration was a 100-degree horizontal field of view fixed display system and the second used an HMD with 40-degree horizontal field of view. Data collected with these systems was compared with some data collected in actual vehicles. The key finding was that the subjects did not necessarily make use of the additional field of regard that was afforded by the HMD. In a backing task in the real vehicle, 28 of 36 subjects turned around to look while backing. In the HMD condition, only 1 of the 36 subjects turned around to look while backing. Others have reported a reduction in head movements while using HMDs (de Vries & Padmos, 1998; Burns & Saluaar, 1999). Wells and Venturino (1990) conducted a study of subject's performance on a target detection task with wide and narrow field of view HMDs. With the wider field of view, subjects moved their heads less but at faster rates when they did. The reduction in normal head movement might be caused by several factors. The weight and inertia of the hardware itself might be enough to cause some not to move their heads often. If display lags or tracking errors exist, some may not move their heads to avoid the "penalty" of experiencing the feelings of discomfort that these effects can bring. If HMDs do result in a reduction in voluntary head movement, it would likely result in reduced performance on driving tasks, especially in environments where a lot of lateral scanning is required. Note that while such a reduction affects the generalizability of the results obtained on an HMD, it would actually act to reduce simulator sickness since there are fewer chances for conflicts cues produced by the vestibulo-ocular reflex and the visual display.

There appears to be a trend in the literature to date that would indicate that driving performance will be worse with HMDs. Several studies have evaluated theoretical HMDs where a widescreen simulator system is used but a field of view restriction is placed on the driver through special glasses or masks (Van Erp & Kappé, 1977; Pepper, 1986; Spain, 1988). These represent "perfect" HMDs in that there is no latency or head tracking error and the weight of the head-mounted hardware is minimal. These studies have failed to find any differences between their "perfect" HMDs and widescreen simulation display. Therefore, this has caused some to hypothesize that it is not the field of view restriction that negatively impacts performance but rather it is the visuomotor interference which is caused by tracking latency and error that is the culprit. The real question is whether technical advances such as faster processors, more accurate tracking, and better prediction algorithms can solve or partially eliminate performance disparity. Given the potential benefits of being able to use HMD technologies including reduced overall costs, smaller footprint, etc., the issue certainly deserves more investigation and research. From the standpoint of simulator sickness, this leads to the interesting hypothesis that special glasses could reduce simulator sickness by augmenting floor-mounted wide field of view simulators. However, this needs further research and empirical evidence.

The relationship between HMDs and simulator sickness was referred to at several different points in the discussion. We now want to address this relationship more directly. The total impact of HMDs is as yet unknown. As noted previously, it has been shown that an increase in simulator field of view and the resulting increases in peripheral stimulation cause increases in simulator sickness (Kolasinski, 1995). Therefore, it is possible that the field of view limitations caused by HMDs might actually reduce simulator sickness (Pausch et al., 1992). Of course there are a number of other less optimistic factors that need to be considered as well. Most lighter weight HMDs make use of LCD technologies. Image smear caused by phosphor decay in rapidly moving images from LCD displays has been theorized to be a contributing factor to simulator sickness.

Additionally, HMDs require head tracking in order to present the appropriate orientation of view. Thus the transport delays described above can be severe where there is latency and error in the data being fed to the visual system; and the more severe the delays the more likely is simulator sickness to occur. Specifically, latency affects the visuomotor system in that it triggers a change in the vestibular ocular reflex response in order to accurately stabilize the image on the retina. The adaptation does occur naturally but will take some period of time to accomplish—anywhere from five minutes to several hours, depending on how much adaptation is required and how consistent the change. Variance and error in latency response can cause a prolonged adaptation period (Draper, 1996). This finding indicates that if you are going to be "off" with the tracking values, it is better to be consistently off so the visuomotor system can adapt to the error. The longer the subject experiences the sensory stimulus without adaptation, the greater the potential for sickness. To forgo the adaptation process, it would be necessary to reduce latencies in the head tracking processes down to around 50 ms (Kemeny, 2000).

Several HMD hardware design factors can have an impact on potential simulator sickness. The weight and inertia of HMDs has also been implicated as a potential cause of simulator sickness. HMD weight can affect the body's interpretation of the mass of the head and subsequent movements will distort the signaling produced by the otoliths responsible for perceiving tilt (DiZio & Lackner, 1992). This will in turn create a conflict between the proprioceptive and vestibular systems. Controlling for all other factors, Dizio and Lackner (1992) found that the weight of the head-mounted gear alone is enough to trigger sickness symptoms without consideration of any visual stimuli. HMDs with a weight as light as 600 g have been shown to cause sickness.

Inter-pupillary distance (IPD) is a design parameter or an adjustment setting associated with HMDs. The idea is that you adjust the width of the lensing or displays in the HMD to more closely match the individual's natural IPD. With respect to the effects of IPD supported by the HMD, Kolasinski (1995) summarized a study by Regan and Price. They hypothesized that individuals with departures from the design IPD would suffer eye strain, headaches, and visual system problems. They found instead that only those with IPDs greater than the design IPD suffered ocular problems. The majority of persons in their study had IPDs smaller than the design IPD. In those cases, it appears that the eyes are able to converge using normal binocular visual response without discomfort. However, those required to diverge their eyes would experience greater discomfort because this is not typically the way eyes move to resolve an image. Therefore, on systems where the IPD is not adjustable to the individual, it is necessary to make sure the design IPD is greater than the subject population's IPDs. The best approach might be to adjust for each individual and slightly bias towards setting it too narrow.

HMDs offer exciting advantages as display solutions for driving simulators. However, as shown in the discussion above, there are a number of areas where they appear to exacerbate simulator sickness and these areas need further research before the full potential of HMDs can be exploited.

14.6.2 Scene and Scenario Design Factors

Scene and scenario design may offer researchers who have already purchased a driving simulator the largest area for improvement. There are a number of things one can do and these are discussed below.

14.6.2.1 Scene Design

The basic rule of thumb when designing a scene is to reduce the cues in the scene that enhance the perception of optic flow (Figure 14.1) and vection. Scene enhancements such as the use of trees, buildings, or other static objects help give cues of motion to the driver and thus are part of the necessary furniture within the environment. However, the addition that they provide in realism needs to be counterbalanced by the knowledge that these cues are the very ones that create cue conflict for the driver. Researchers who can populate the environment with objects that they can position and then texture have the advantage of being able to adjust the optic flow. So, for example, the trees along the side of road could be placed further back from the road and made sparser. Or, the buildings on an urban street could be covered with as much unbroken wall surface as possible instead of covered with windows that themselves had mullions and other textured elements. The bottom line is that a totally featureless environment has no optic flow and therefore will not produce simulator sickness. But such an environment gives drivers no cues as to location, roadway and speed and therefore is not useful. To our knowledge, no one has explored just how featureless an environment could be and still provide the necessary visual cues to make it possible to generalize the results of the experiment from the laboratory to the real world. One reason for this is that the need for scene complexity depends on the specific driving tasks and research questions: Studies of speed perception will require more sickness-inducing detail than studies of driver distraction.

14.6.2.2 Scenario Design

Movement within the scene is another factor to consider when designing a scenario. In general it is recommended that one should minimize the rapid changes in direction and the number of sharp decelerations. Consistent with this, curves with larger radii and fewer roadside objects produce less simulator sickness than do tighter curves with densely packed objects (Chrysler & William, 2005). In a similar vein, 90-degree left and right turns are definitely known to increase the likelihood that the driver experiences simulator sickness (Edwards, Creaser, Caird, Lamsdale, & Chisholm, 2003; Mourant, Prasanna, Cox, Lin, & Jaeger, 2007; Park, Allen, & Fiorentino, 2006; Watson, 1995; Watson, 1998). Arguably, one can reduce the nausea produced at right angle turns by making such turns into Y intersections as the turn is less sharp. And for most simulators this has the added advantage of making it possible to see traffic approaching from the left or the right.

Interestingly, we also know anecdotally that simulator sickness can be reduced in an HMD by up to 30% if a vehicle does not pitch down as a driver brakes.* Note that such a downward movement of the front end of the vehicle does noticeably occur in the real world, but in the simulated world it does not need to occur. If it does not occur, then the horizon stays fixed as the driver is braking and so the cue conflict is reduced. However, in a fixed or motion-based simulator the opposite has been found.† When the amount of a vehicle pitch corresponds to the appropriate decelerations simulator sickness has been shown to be reduced.

14.6.2.3 Independent Visual Background

The role of an independent visual background was discussed above. Several ways of introducing such a background were mentioned: making the room behind the display visible through the display (Prothero et al., 1999); placing a fixed grid over the display (Duh et al., 2001); or adding an earth-fixed avatar to the display (Lin et al., 2002), one which indicated directions as well as the true rest-frame. Only the latter manipulation was run on a driving simulator. It seems impractical in most studies because the avatar was a plane that was in the upper part of the display, thereby necessarily creating a distraction. However, there seem to be many other possibilities that still need to be explored. For example, would an earth-fixed cloudy sky not only reduce the visual flow, but also help the driver keep in the forefront the correct rest-frame? Would a narrowed field of view in the vertical axis allow for the placement of a grid-like border around the virtual world which kept the driver from being seduced into the rest-frame implied by the cues in the virtual world? Even designing the virtual world so that much of the horizon is visible much of the time might be effective. It seems that there are many possibilities, none of which would be a cure-all, but each may be beneficial.

14.6.3 Environmental Conditions

Temperature has long been thought to contribute to simulator sickness and it is recommended to keep the cab temperature cool

---

* Personal communication with Konstantin Sizov, President and CEO of DriveSquare.
† Personal communication with James W. Stoner, Professor, University of Iowa.

(e.g., Kennedy et al., 1987). Several physiological changes occur as a result of simulator sickness. Heart rate, blood pressure, respirations, and skin temperature all increase as a result of experiencing a virtual environment (Jang et al., 2002). The relationship that ambient room temperature has with these physiological changes is not yet known but is thought to elevate the magnitude and rate of awareness of simulator sickness symptoms. As a precaution, it is recommended that adequate ventilation and temperature control be built into any virtual environment laboratory (e.g., preferred operating range being less than 70 degrees Fahrenheit). The ventilation and temperature control should take into account the heat produced by the hardware (computers, projectors, etc.) and the number of people generally in the room at the time of operation.

14.6.4 Screening

There are a number of questions one might use to identify participants who are at a greatly increased risk of developing simulator sickness. (Below, in a separate section, we discuss the broader range of individual differences and their often somewhat weaker relationship to simulator sickness.) In general, individuals who have fatigue or sleep loss, a hangover, an upset stomach, head colds, ear infections, ear blockages, pregnancy, or an upper respiratory illness, or who have recently taken medications or alcohol should postpone a session on the simulator (Johnson, 1995; Kennedy et al., 1987). Similarly, individuals who have been sick recently and are not fully recovered should be screened from participating in simulator studies or training.

Individuals should consider not participating if they have ever experienced motion sickness. Such individuals are more likely to experience simulator sickness as well (Allen & Reimer, 2006; Reason & Graybiel, 1972). For example, in one study, 25 healthy participants (21–59 years old, 41.36 years old on average) drove a fixed-based driving simulator (Fagbemi & Peffer, 2006). Nine participants experienced severe symptoms of simulator sickness, 16 did not. Six of these 9 (67%) had reported prior motion sickness whereas only 2 of the 16 (12.5%) in the well group reported previous signs of motion sickness. Exposure is most frequently measured by the Motion Sickness Questionnaire (Kellogg, Kennedy, & Graybiel, 1965).

Finally, as noted above one might want to measure postural stability. Specifically, it will be recalled that Stoffregen et al., (2000) found that prior to the testing, there were a number of differences in postural motion in the sick and well groups. However, these differences did not explain as much variance as do the differences in prior motion sickness, so one would want to use them with some caution.

14.6.5 Online Monitoring

Ideally, screening would be supplemented with online monitoring of an individual to determine whether simulator sickness was developing in those individuals who had passed the screening criterion. Given that no screening criterion is perfect, it makes sense to employ online monitoring as well as screening where possible.

14.6.5.1 Physiological Monitoring

In the study described above by Bertin et al. the continuous monitoring of symptoms of simulator sickness allowed the investigators to determine whether the changes in the physiological variables reliably preceded changes in the level of simulator sickness. Analysis of the results indicated that there were reliable decreases in skin temperature and skin resistance before the quick rise in self-reported simulator sickness. If one is going to use such information in an ongoing research project, one would normally need more information than has been provided to date in the extant studies. At the very least one would need to know something about the number of misses (e.g., individuals with decreasing skin temperature who did not get sick) and false alarms (e.g., individuals with constant or increasing skin temperature who did get sick) in order to identify the exact level of a physiological variable which maximizes the expected gain, however defined. Such studies have not been performed to date but clearly are of merit. Even if physiological instrumentation is not available, some physiological responses are visible to the experimenter, such as pallor, sighs and pronounced swallowing. These cues can be a useful indicator of discomfort and can be used to query the participant about feelings of discomfort.

14.6.5.2 Postural Stability

Postural stability, if monitored intermittently, can also help an investigator predict who is and who is not likely to develop symptoms of simulator sickness. Smart, Stoffregen and Bardy (2002) show that there is a strong correlation between changes in various indices of postural stability early in a simulator experiment and later sickness. Specifically, standing participants were exposed to an optical simulation of body sway. The symptoms of motion sickness were explained to participants prior to the experiment and they were told to end their participation as soon as any of these symptoms appeared. Postural stability was measured throughout. Changes were noted in measures of the variability, range and velocity of postural motion that preceded changes in the simulate sickness scores. The particular subset varies from one experiment to another (Stoffregen & Smart, 1998; Stoffregen, Hettinger, Haas, Roe, & Smart, 2000; Smart et al., 2002), but this is to be expected.

14.6.6 Breaks and Task Time

One general rule to follow is that the total simulation exposure should not last more than two hours (Johnson, 2005). The longer the period of performance in the simulator the more likely the discomfort level experienced will increase. Frequent breaks between drives are also recommended with a single drive lasting no more than an hour. Also, the more aggressive the scene and scenario is the shorter should be the duration of the driving session. Typically researchers use the guidelines where drives should last between 5 and 25 minutes with 10 minute breaks. To our knowledge there is no set rule or study that quantifies these exposures, but it is the general practice in the industry.

14.6.7 Simulator Practice and Adaptation

The human nervous system is a very complex set of mechanisms and processes but is also highly adaptable. This is evident from examples of micro processes discussed earlier, such as the adaptation of the vestibulo-ocular reflex and optokinetic responses with variations of input stimuli. At the same time, it is also generally accepted (Kennedy, Stanney, & Dunlap, 2000) that simulator sickness increases with time within a session and decreases over successive sessions. These effects have been confirmed and may vary as a function of scenario intensity as measured by scene complexity and number of moving models, and consistency of the cue presentation factors (Watson, 1997; Watson, 1995).

In a study to quantify adaptation as a function of scenario intensity and motion cueing, Watson found that SSQ total sickness, disorientation and ocular discomfort scores dropped by as much as 2/3 from the first to the third exposure. However, nausea subscale scores only showed a decline after the sixth exposure resulting in a recommendation of five or more sessions to allow subjects to become adapted. Watson also recommends limiting scenario intensity during the first few exposures to help facilitate adaptation (see also Kennedy et al., 1987). McCauley and Sharkey (1992) make similar recommendations including keeping exposure durations short and limiting aggressive maneuvers.

The issue of adaptation raises some interesting questions with regards to exposure and validity of application results. Applications of driving simulation such as research, training, design validation, etc., are typically challenged when it comes to available simulation resources. Cost and logistical constraints often result in users trying to get the most from the simulation in the shortest period of time. This conflicts to some degree with the recommended practices of allowing simulator drivers multiple (relatively benign) sessions to adapt before getting to the experimental phase of the simulation application. Without understanding the effects of simulation exposure on driver performance and motivation, it is difficult to generalize research results in the simulator to real driving. Early driver training scenarios have the potential to result in less transfer of training simply because drivers are learning to drive the simulator as opposed to focusing on the lessons that the scenarios hold. Regardless of the application, the users of simulation should strive to understand the effects of exposure and adaptation on their expected results.

14.6.8 Individual Differences

A number of individual differences are known to influence simulator sickness. These include susceptibility to motion sickness, current health status, age, concentration level, ethnicity, experience with the real-world task, experience with a simulator (adaptation), flicker fusion frequency threshold, gender, illness, personal characteristics, mental rotation ability, perceptual style, and postural stability. A review of the relationship of each of these individual differences to simulator sickness can be found in Goldberg and Hiller (1995). We focused above on the individual differences which serve as the standard screening questions one would use to identify those with a greatly increased risk of developing simulator sickness. Here we want to focus on one additional individual difference, age, because of the critical importance of understanding the behavior of older drivers in a society with that has an increasing amount of older drivers, many with a greater risk of crashing.

Although age is perhaps the largest individual difference of relevance, there does not appear to be a recent, comprehensive review of how the sickness rates vary with age (Johnson, 2005). Perhaps the largest database that was reviewed was back in the early 1990s. Hein (1993) analyzed the results from 22 separate studies, all undertaken on the fixed-base driving simulator owned by the Hughes Aircraft Company. A total of 469 participants were involved in the studies, the age range varied considerably. Hein reports that "older drivers … are severely susceptible to simulator sickness (p. 611)". Having said that, the sickness rates across studies vary widely even when controlling as much as possible for field of view, stops and starts, and frequency of turns.

For example, in one recent study, 57 men and 127 women between the ages of 60 and 99 (average of 77) were enrolled (Freund & Green, 2006). Participants in this study sat in an adjustable car seat, used standard accelerator and brake pedals, and had a standard size steering wheel mounted on a dash. The virtual world was projected on three 4 by 3 foot screens in front of the cab subtending 130 degrees of visual angle side to side. The participants had to drive for 30 minutes through urban scenarios which required left and right turns at four way intersections and changes in speed, including coming to a complete stop. In short, the scenes and scenarios were ones which should lead to a relatively high rate of simulator sickness; yet, only 10.6% of the participants became sick, as indicated by reports of light-headedness, dizziness, nausea or vomiting.

Contrast this with a study run by Edwards et al. (2003). Twelve older drivers between the ages of 65 and 83 (average of 71.4) and twelve younger drivers between the ages of 19 and 25 (average of 20.7) were enrolled. The simulator here was similar to the one in the study above, except that participants now sat in an actual vehicle. The screens were slightly larger, subtending 150 degrees of visual angle side to side. However, the drives (as best we could tell) were almost identical and included intersections, signals, pedestrians and traffic. Yet, even with such similar scenarios and simulator design, fully 40% of the older adults became sick as opposed to only 10% in the above study. It is not clear what differences between these two studies explain the dramatic difference in simulator sickness rates.

As predicted by the evidence provided above, one would expect to find—and one does find—that decreases in the number of turns decreases the level of sickness. So, for example, in a study involving older adults on straight roads only 12.5%

became sick (Edwards et al., 2003). This suggests that one can obtain acceptable sickness rates with older adults, but with some real constraints on the types of scenarios one can use to evaluate driver performance.

14.7 Conclusions

Simulator sickness has been an important concern from the first application of simulators over 50 years ago. Although many strategies can help reduce simulator sickness, even the most carefully tuned simulator can make participants feel ill. The motion sickness that some people feel on some types of roads demonstrates this challenge. Four theories reflect the most common explanations of simulator sickness: sensory cue conflicts, the body's response to position, postural instabilities, and rest-frame inconsistencies. These theories offer suggestions for minimizing simulator sickness which include tuning the simulator design, adjusting the scenarios and protocol, and monitoring and screening participants:

- Operate with a narrow field of view if possible; the wider field of view, motion-base, and higher resolution screens of higher fidelity simulators have the potential to increase simulator sickness if visual and vestibular cue conflicts are not resolved and vection and optic flow are not managed.
- Calibrate the eye height, align screens, maintain an adequate frame rate (>30 Hz) and ensure minimal transport delay (<50 ms).
- Pay special attention to head-mounted displays because of the potential lags in head tracking, the absence of any obvious earth centered rest-frame, and the special tuning they require.
- Design scenarios that minimize 90-degree turns, tight curves, abrupt braking, and unnecessary optic flow (e.g., picket fences and many roadside objects).
- Keep the simulator cab cool and well-ventilated.
- Use short drives and allow people to adapt to the simulator with an uneventful drive in which they follow a lead vehicle for several minutes.
- Acclimatize people over as many as six sessions to help minimize simulator sickness.
- Monitor simulator sickness during an experiment if at all possible by observing the participants carefully as they drive, and after the experiment using the SSQ.
- Screen participants to avoid those who are particularly susceptible to simulator sickness, such as those with the following conditions: fatigue or sleep loss, a hangover, an upset stomach, head colds, ear infections, ear blockages, pregnancy, an upper respiratory illness, or those or who have recently taken medications or alcohol.

Key Points

- There is a difference between motion sickness and simulator sickness. While the symptoms of motion and simulator

sickness overlap, there are clear differences in the causes of these two different types of sickness.
- There are several different theories of simulator sickness. These include theories that refer to inter- and intramodal sensory cue conflicts, the body's response to position, postural instabilities, and rest-frame inconsistencies. No theory has yet explained or predicted simulator sickness completely.
- Arguably, conflicting cues from the vestibular and visual systems influence simulator sickness the most. The features of each system that are most often in conflict in a simulator are discussed.
- There are several well-validated measures of simulator sickness that could be used in almost any study where simulator sickness is expected as a problem. Because simulator sickness can affect all aspects of driving, without such measures one cannot safely generalize results from a simulator to real driving.
- There are various preventive measures for simulator sickness. These methods such as screening participants, controlling environmental conditions, and scene and scenario design should be used when possible to help reduce sickness.

Keywords: Simulator Sickness, Vestibular System, Vection, Visual System

Acknowledgments

The initial review of literature was performed in support of a Small Business Innovative Research (SBIR) contract with the United States Army Tank-Automotive and Armaments Command (TACOM). The project is titled "Integrating a Motion Base into TARDEC'S CAVE Automatic Virtual Environment." Contract number DAAE07-02-C-L002. A portion of the support for this review was provided by a grant from the National Institutes of Health (1R01HD057153-01) to the second Author.

Key Readings

Casali, J. (1986). *Vehicular simulation-induced sickness, Volume I: An overview* (IEOR Tech. Rep. 8501). Virginia Polytechnic Institute and State University, VA.

Kennedy, R., & Fowlkes, J. (1992). Use of a motion sickness history questionnaire for prediction of simulator sickness. *Aviation, Space and Environmental Medicine, 63*, 588–593.

Kennedy, R., Stanney, K., & Dunlap, W. (2000). Duration and exposure to virtual environments: Sickness curves during and across sessions. *Presence: Teleoperators & Virtual Environments, 9*(5), 463–472

McCauley, M., & Sharkey, T. (1992). Cybersickness: Perception of self-motion in virtual environments. *Presence: Teleoperators & Virtual Environments, 1*, 311–318.

Pausch, R., Crea, T., & Conway, M. (1992). A literature survey for virtual environments: Military flight simulator visual systems and simulator sickness. *Presence: Teleoperators & Virtual Environments, 1*(3), 344–363.

References

Advani, S., & Hosman, R. (2001). Integrated motion cueing algorithm and motion-base design for effective road vehicle simulation. *Proceedings of the driving simulation conference 2001* (pp. 263–271). Sophia Antipolis, France.

Allen, R. W., & Reimer, B. (2006). New approaches to simulation and the older operator. *Advances in Transportation Studies: An International Journal,* Special Issue, 3–8.

Andersen, G. J. (1986). Perception of self-motion: Psychophysical and computational approaches. *Psychological Bulletin, 99*(1), 52–65.

Andersen, G. J., & Braunstein, M. L. (1985). Induced self-motion in central vision. *Journal of Experimental Psychology: Human Perception and Performance, 11*(2), 122–132.

Barnes, A. (1987). *Operating experience of a small six axis motion system inside a dome with a wide angle visual system* (AIAA Tech. Paper 87-2437). American Institute of Aeronautics and Astronautics.

Berbaum, K. S., Kennedy, R. S., Welch, R. S., & Brannan, J. K. (1985). *Space adaptation syndrome: Reducing symptomatology through perceptual training* (Final Rep. NAS 9-17278). Houston, TX: Johnson Space Center, National Aeronautics and Space Administration.

Bertin, R. J. V., Guillot, A., Collet, C., Vienne, F., Espié, S., & Graf, W. (2004). *Objective measurement of simulator sickness and the role of visual-vestibular conflict situations: a study with vestibular-loss (a-reflexive) subjects* (poster). Proceedings of the Society for Neurosciences 34th Annual Meeting. San Diego, CA. Retrieved February 20, 2009 from http://cogprints.org/3928/1/sfn2004.pdf

Biaggioni, I., Costa, F., & Kaufmann, H. (1998). Vestibular influences on autonomic cardiovascular control in humans. *Journal of Vestibular Research, 8,* 35–41.

Bittner, A. C. J., Gore, B., & Hooey, B. (1997). Meaningful assessments of simulator performance and sickness: Can't have one without the other. *Proceedings of the 41st human factors and ergonomics society annual meeting* (Vol. 2, pp. 1089–1093). Santa Monica, CA: Human Factors and Ergonomics Society.

Bolton, P. S., Wardman, D. L., & Macefield, V. G. (2004). Absence of short-term vestibular modulation of muscle sympathetic outflow, assessed by brief galvanic vestibular stimulation in awake human subjects. *Experimental Brain Research, 154,* 39–43.

Braithwaite, M. G., & Braithwaite, B. D. (1990). Simulator sickness in an Army simulator. *Journal of the Society of Occupational Medicine, 40,* 105–110.

Burns, P., & Saluaar, D. (1999). *Intersections between driving in reality and virtual reality (VR).* Proceedings of the Driving Simulation Conference 1999. Paris, France.

Casali, J. (1986). *Vehicular simulation-induced sickness, Volume I: An overview* (IEOR Tech. Rep. 8501). Virginia Polytechnic Institute and State University, VA.

Chrysler, S. T., & Williams, A. A. (2005). Driving performance in a simulator as a function of pavement and shoulder width, edge line presence, and oncoming traffic. *Proceedings of the third international driving symposium on human factors in driver assessment, training and vehicle design* (pp. 370–375). Rockport, ME.

Collet, C., Vernet-Maury, E., Miniconi, P., Chanel, J., & Dittmar, A. (2000). Autonomic nervous system activity associated with postural disturbances in patients with perilymphatic fistula: Sympathetic or vagal origin. *Brain Research Bulletin, 53,* 33–43.

Crowley, J. S., & Gower, D. W. (1988). Simulator sickness. *United States Army Aviation Digest, 1-88-11,* 9–11.

Cunningham, D., Chatziastros, A., von der Heyde, M., & Bülthoff, H. (2001). Driving in the future: Temporal visuomotor adaptation and generalization. *Journal of Vision, 1,* 88–98.

Curry, R., Artz, B., Cathey, L., Grant, P., & Greenburg, J. (2002). Kennedy SSQ results: Fixed- vs. motion-base Ford simulators. *Proceedings of the driving simulation conference 2002* (pp. 289–299). Paris, France.

de Vries, S., & Padmos, P. (1998). Steering a simulated unmanned aerial vehicle using a head-slaved camera and HMD: Effect of HMD quality, visible vehicle references and extended stereo cueing. *SPIE conference on helmet and head mounted displays III* (SPIE Vol. 3362). Orlando, Florida.

DiZio, P., & Lackner, J. R. (1992). Influence of gravitoinertial force level on vestibular and visual velocity storage in yaw and pitch. *Vision Research, 32,* 111–120.

Draper, M. (1998). The effects of image scale factor on vestibulo-ocular reflex adaptation and simulator sickness in head-coupled virtual environments. *Proceedings of the human factors and ergonomics society 42nd annual meeting* (pp. 1481–1485).

Draper, M. (1996). *Can your eyes make you sick?: Investigating the relationship between the vestibule-ocular reflex and virtual reality.* Unpublished thesis, Human Interface Technology Lab, University of Washington.

Draper, M. H., Viirre, E. S., Furness, T. A., & Parker, D. E. (1997, May 26–28). *Theorized relationship between vestibulo-ocular adaptation and simulator sickness in virtual environments.* Paper presented at International Workshop on Motion Sickness. Marbella, Spain.

Draper, M., Viirre, E. S., Furness, T. A., & Gawron, V. (2001). Effects of image scale and system time delay on simulator sickness within head-coupled virtual environments. *Human Factors, 43,* 129–146.

Duh, B., Parker, D., & Furness, T. (2001). An "independent visual background" reduced balance disturbance evoked by visual scene motion: implication for alleviating simulator sickness. *Proceedings of CHI 2001 conference on human factors in computing systems* (pp. 85–89).

Ebenholtz, S. M. (1992). Motion sickness and oculomotor systems in virtual environments. *Presence: Teleoperators & Virtual Environments, 1*(3), 302–305.

Ebenholtz, S. M., Cohen, M. M., & Linder, B. J. (1994). The possible role of nystagmus in motion sickness: A hypothesis. *Aviation, Space and Environmental Medicine, 65*, 1032–1035.

Edwards, C. J., Creaser, J. I., Caird, J. K., Lamsdale, A. M., & Chisholm, S. L. (2003). Older and younger driver performance at complex intersections: Implications for using perception-response time and driving simulation. *Proceedings of the second international driving symposium on human factors in driver assessment, training and vehicle design* (pp. 33–38). Iowa City: University of Iowa.

Ehrlich, J. (1997). Simulator sickness and HMD configurations. In Stein, M. R. (Ed.), *Proceedings of SPIE: telemanipulator and telepresence technologies IV* (Vol. 3206, pp. 170–178). Bellingham, WA: SPIE.

Espié, S. (1999). *Vehicle-driven simulator versus traffic-driven simulator: the INRETS approach.* Proceedings of the Driving Simulation Conference 1999. Paris, France.

Fagbemi, O. S., & Pfeffer, K. (2006). The relationship between sleep patterns and the experience of simulator sickness and motion sickness. *Advances in Transportation Studies: An International Journal,* Special Issue, 63–70.

Federal Aviation Administration. (1994). *Advisory circular, helicopter simulator qualification* (AC120-63).

Frank, L., Casali, J., & Wierwille, W. (1988). Effects of visual display and motion system delays on operator performance and uneasiness in a driving simulator. *Human Factors, 30*(2), 201–217.

Fregly, A. R., & Kennedy, R. S. (1965). Comparative effects of prolonged rotation at A0 RPM on postural equilibrium in vestibular normal and vestibular defective human subjects. *Aerospace Medicine, 1965, 36*(12), 1160–1167.

Freund, B., & Green, T. R. (2006). Simulator sickness amongst older drivers with and without dementia. *Advances in Transportation Studies: An International Journal,* Special Issue, 71–74.

Gianaros, P. J., Quigley, K. S., Muth, E. R., Levine, M. E., Vasko, R. C. J., & Stern, R. M. (2003). Relationship between temporal changes in cardiac parasympathetic activity and motion sickness severity. *Psychophysiology, 40*, 39–44.

Gianna, C., Heimbrand, S., & Gresty, M. (1996). Thresholds for detection of motion direction during passive lateral whole-body acceleration in normal subjects and patients with bilateral loss of labyrinthine function. *Brain Research Bulletin, 40*(5–6), 443–447.

Govil, V., Lovell, S., Piriyakala, S., Wu, O., Yan. G., & Mourant, R. (2004). A high-resolution wide-screen display for simulators and virtual reality. *Proceedings of the human factors and ergonomics society 48th annual meeting* (pp. 2131–2133). Santa Monica, CA: Human Factors and Ergonomics Society.

Goldberg, S. L., & Hiller, J. H. (1995). *Simulator sickness in virtual environments* (Tech. Rep. 1027; Army Project Number 2O262785A791, Education and Training Technology). Alexandria, VA: US Army Research Institute

for the Behavioral and Social Sciences. Retrieved on February 21, 2009 from http://www.cyberedge.com/info_r_a%2Bp05_ss.html

Goldstein, E. (1989). *Sensation and perception* (3rd ed.). Belmont, CA: Wadsworth Publishing.

Goodenough, F. L., & Tinker, M. A. (1931). The retention of mirror-reading ability after two years. *Journal of Educational Psychology, 22*, 503–505.

Griffin, M. J. (1990*). Handbook of human vibration.* London: Academic Press Limited.

Guedry F. E., Jr. (1965). Habituation to complex vestibular stimulation in man: Transfer and retention of effects from twelve days of rotation at 10 RPM. *Perceptual and Motor Skills, 21*, 459–481.

Harwood, K., & Foley, P. (1987). Temporal resolution: An insight into the video display terminal (VDT) "Problem". *Human Factors, 29*(4), 447–452.

Hein, C. M. (1993). Driving simulators: Six years of hands-on experience at Hughes Aircraft Company. *Proceedings of the human factors and ergonomics society 37th annual meeting* (pp. 607–611).

Hettinger, L. J., Berbaum, K. S., Kennedy, R. S., Dunlap, W. P., & Nolan, M. D. (1990) Vection and simulator sickness. *Military Psychology, 2*(3), 171–181.

Hettinger, L., & Riccio, G. (1992). Visually-induced motion sickness in virtual environments. *Presence: Teleoperators & Virtual Environments, 1*(3), 306–310.

Hettinger, L. J., Nolan, M. D., Kennedy, R. S., Berbaum, K. S., Schnitzius, K. P., & Edinger, K. M. (1987). Visual display factors contributing to simulator sickness. *Proceedings of the human factors society 31st annual meeting* (pp. 497–501).

Ijsselsteijn, W., Ridder, H., Freeman, J., Avons, S., & Bouwhuis, D. (2001). Effects of stereoscopic presentation, image motion, and screen size on subjective and objective corroborative measures of presence. *Presence: Teleoperators & Virtual Environments, 10*(3), 298–311.

Jamson, H. (2001). *Image characteristics and their effect on driving simulator validity.* Proceedings of Driving Assessment 2001. Snowmass, CO.

Jang, D. P., Kim, Y., Nam, S. W., Wiederhold, B. K., Wiederhold, M. D., & Kim, S. I. (2002). Analysis of physiological response to two virtual environments: Driving and flying simulation. *Cyberpsychology & Behavior, 5*(1), 11–18.

Johnson, D. M. (2005). *Introduction to and review of simulator sickness research* (Rep. 1832). Arlington, VA: US Army Research Institute for the Behavioral and Social Sciences.

Kappe, B., & Padmos, P. (2001). *Headslaved displays in a driving simulator.* Proceedings of the First Human-Centered Transportation Simulation Conference. The University of Iowa, Iowa City, Iowa (ISSN 1538–3288).

Kellogg, R. S., Kennedy, R. S., & Graybiel, A. (1965). Motion sickness symptomatology of labyrinthine defective and normal subjects during zero gravity maneuvers. *Aerospace Medicine, 36*, 315–318.

Kemeny, A. (2000). Simulation and perception of movement. *Proceedings of driving simulation conference 2000* (pp. 13–22). Paris, France.

Kemeny, A. (2001). Recent developments in visuo-vestibular restitution of self-motion in driving simulation. *Proceedings of the driving simulation conference 2001* (pp. 15–18). Sophia Antipolis, France.

Kennedy, R., & Fowlkes, J. (1992). Use of a motion sickness history questionnaire for prediction of simulator sickness. *Aviation, Space and Environmental Medicine, 63,* 588–593.

Kennedy, R., Fowlkes, J., & Lilienthal, M. (1993). Postural and performance changes following exposures to flight simulators. *Aviation, Space, and Environmental Medicine, 64*(10), 912–920.

Kennedy, R., Hettinger, L., & Lilienthal, M. (1988). Simulator sickness. In *Motion and space sickness* (pp. 317–341). Boca Raton, FL: CRC Press.

Kennedy, R., Lilienthal, M., Berbaum, K., Baltzley, D., & McCauley, M. (1989). Simulator sickness in US Navy flight simulators. *Aviation, Space, and Environmental Medicine, 60*(1), 10–16.

Kennedy, R., Stanney, K., & Dunlap, W. (2000). Duration and exposure to virtual environments: Sickness curves during and across sessions. *Presence: Teleoperators & Virtual Environments, 9*(5), 463–472.

Kennedy, R. S., Berbaum, K. S., Lilienthal, M. G., Dunlap, W. P., Mulligan, B. E., & Funaro, J. F. (1987). *Guidelines for alleviation of simulator sickness symptomatology* (NAVTRASYSCEN TR-87-007). Orlando, FL: Naval Training Systems Center.

Kennedy, R. S., Dunlap, W. P., & Fowlkes, J. E. (1990). Prediction of motion sickness susceptibility. In Crampton, G. H. (Ed.), *Motion and Space Sickness* (pp. 179–216). Boca Raton, FL: CRC Press.

Kennedy, R. S., Lane, N. E., Berbaum, K. S., & Lilienthal, M. G. (1993). Simulator sickness questionnaire: An enhanced method for quantifying simulator sickness. *The International Journal of Aviation Psychology, 3*(3), 203–220.

Kolasinski, E. (1995). *Simulator sickness in virtual environments.* (Tech. Rep. 1027). Army Research Institute.

LaViola, J. (2000). A discussion of cybersickness in virtual environments. *SICCHI Bulletin, 32*(1), 47–56.

Leibowitz, H. (1986). Recent advances in our understanding of peripheral vision and some implications. *Proceedings of the 30th annual human factors and ergonomics society meeting* (pp. 605–607). Santa Monica, CA: Human Factors and Ergonomics Society.

Lilienthal, M. G., Kennedy, R. S., Berbaum, K. S., Dunlap, W. P., & Mulligan, B. E. (1987). *Vision/motion-induced sickness in Navy flight simulators: Guidelines for its prevention.* Proceedings of the 1987 Image Conference IV (pp. 275–285).

Lin, J., Abi-Rached, H., Kim, D., Parker, D., & Furness, T. (2002). A "natural" independent visual background reduced simulator sickness. *Proceedings of the human factors and ergonomics society 46th annual meeting* (pp. 2124–2128). Santa Monica, CA: Human Factors and Ergonomics Society.

McCauley, M. E. (Ed.). (1984). *Research issues in simulator sickness.* Proceedings of a workshop. Washington, DC: National Academy Press.

McCauley, M., & Sharkey, T. (1992). Cybersickness: Perception of self-motion in virtual environments. *Presence: Teleoperators & Virtual Environments, 1,* 311–318.

Min, B. C., Chung, S. C., Min, Y. K., & Sakamoto, K. (2004). Psychophysiological evaluation of simulator sickness evoked by a graphic simulator. *Applied Ergonomics, 35,* 549–556.

Money, K. (1983). *Theory underlying the peripheral vision horizon device.* Proceedings of the NASA Peripheral Vision Display Conference. Dryden Research Facility, Edwards Air Force Base.

Money, K. E. (1970). Motion sickness. *Physiological Review, 50,* 1–39.

Mourant, R., & Thattacherry, T. (2000). Simulator sickness in a virtual environments driving simulator. *Proceedings of the IEA 2000/HFES 2000 congress* (pp. 534–537).

Mourant, R., Prasanna, R., Cox, D., Lin, Y., & Jaeger, B. K. (2007). *The effect of driving environments on simulator sickness.* Proceedings of the 51st Annual Meeting of the Human Factors and Ergonomics Society. Santa Monica, CA: Human Factors and Ergonomics Society.

Mullen, T. J., Berger, R. D., Oman, C. M., & Cohen, R. J. (1998). Human heart rate variability relation is unchanged during motion sickness. *Journal of Vestibular Research, 8,* 95–105.

Naval Training Systems Center. (1988). *Simulator sickness field manual mod 3.* Orlando, FL: US Navy.

Park, G. R., Allen, W., & Fiorentino, D. (2006). Simulator sickness scores according to symptom susceptibility, age, and gender for an older driver assessment study. *Proceedings of the 50th annual meeting of the human factors and ergonomics society* (pp. 2702–2706). San Francisco, CA. Santa Monica, CA: Human Factors and Ergonomics Society.

Patterson, R., Winterbottom, M. D., & Pierce, B. J. (2006). Perceptual issues in the use of head-mounted visual displays. *Human Factors, 48*(3), 555–573.

Pausch, R., Crea, T., & Conway, M. (1992). A literature survey for virtual environments: Military flight simulator visual systems and simulator sickness. *Presence: Teleoperators & Virtual Environments, 1*(3), 344–363.

Pepper, R. (1986). Human factors in remote vehicle control. *Proceedings of the human factors society's 30th annual meeting* (pp. 417–421). Santa Monica, CA: Human Factors Society.

Prothero, J. (1998). *The role of rest frames in vection, presence, and motion sickness.* Doctoral dissertation. University of Washington, Human Interface Technology Lab.

Prothero, J. D., Draper, M. H., Furness, T. A., Parker, D. A., & Wells, M. J. (1999). The use of an independent visual background to reduce simulator side-effects. *Journal of Aviation, Space, and Environmental Medicine, 70*(3), 277–283.

Reason, J. T., & Brand J. J. (1975). *Motion sickness* (pp. 135–171). London: Academic Press.

Reason, J. T., & Graybiel, A. (1972). Factors contributing to motion sickness susceptibility: Adaptability and receptivity. In M. P. Lansberg (Ed.), *Proceedings of the AGARD aerospace medical panel symposium on predictability of motion sickness in the selection of pilots (AGARD DP-109)*. Glasgow, Scotland.

Riccio, G., & Stoffgregen, T. (1991). An ecological theory of motion sickness and postural instability. *Ecological Psychology, 3*(3), 195–240.

Rinalducci, E. J., & MacArthur, M. (1990). *Annotated bibliography on the effects of flicker on simulator sickness*. Unpublished manuscript. Orlando, FL: Institute for Simulation and Training, University of Central Florida, Division of Sponsored Research, Orlando, FL.

Robin, M. (2003). *Revisiting Kell*, Broadcast Engineering, May 2003.

Rolnick, A., & Lubow, R. E. (1991). Why is the driver rarely motion sick? The role of controllability in motion sickness. *Ergonomics, 34*, 867–879.

Rosinski, R. R. (1982). *Effect of projective distortions on perception of graphic displays* (Paper TR 82-1, Contract N00014-77-C-0679). Pittsburgh, PA: University of Pittsburgh, Office of Sponsored Research.

Ruspa, C., Scheuchenpflug, R., & Quattrocolo, S. (2002). Validity of virtual reality driving simulators for ergonomic assessment. In D. de Waard, K. Brookhuis, S. Sommer, & W. Verwey (Eds.), *Human Factors in the Age of Virtual Reality*. Maastricht, the Netherlands: Shaker Publishing.

Schulte-Pelkum, J., Riecke, B. E., & von der Heyde, M. (2003). *Influence of display parameters on perceiving visually simulated ego-rotations—A systematic investigation*. Tübingen Perception Conference (TWK). Tübingen, Germany.

Schulte-Pelkum, J., Riecke, B. E., von der Heyde, M., & Bülthoff, H. H. (2004). *Influence of display device and screen curvature on perceiving and controlling simulated ego-rotations from optic flow* (Tech. Rep. 122). Tübingen, Germany: Max Planck Institute for Biological Cybernetics.

Sharkey, T., & McCauley, M. (1992). *Does a motion base prevent simulator sickness?* (AIAA Tech. Rep. 92-4133-CP). American Institute of Astronautics.

Silverman, D. R., & Slaughter, R. A. (1995). *An exploration of simulator sickness in the MH-60G operational flight trainer, an advanced wide field-of-view helicopter trainer* (Rep. AL/HR-TR-1994-0173). Mesa, AZ: Aircrew Training Research Division, Human Resources Directorate.

Smart, L. J., Stoffregen, T. A., & Bardy, B. G. (2002). Visually induced motion sickness predicted by postural instability. *Human Factors, 44*, 451–465.

Sony Electronics Inc. (2008, Oct 28). *Sony Unveils Latest Generation Of 4K SXRD Projectors*. Press Release. Retrieved on April 10, 2009 from http://pro.sony.com/bbsc/ssr/micro-sxrdsite/

Spain, E. (1998). *Assessments of maneuverability with the teleoperated vehicle (TOV)* (AD-A191 584). Springfield, VA: National Technical Information Service.

Stanney, K., Mourant, R., & Kennedy, R. (1998). Human factors issues in virtual environments: A review of literature. *Presence: Teleoperators & Virtual Environments, 7*(4), 327–351.

Stoffregen, T., & Riccio, G. (1991). An ecological critique of the sensory conflict theory of motion sickness. *Ecological Psychology, 3*(3), 15–194.

Stoffregen, T. A., & Smart, L. J. (1998). Postural instability precedes motion sickness. *Brain Research Bulletin, 47*, 437–448.

Stoffregen, T. A., Hettinger, L. J., Haas, M. W., Roe, M. M., & Smart, L. J. (2000). Postural instability and motion sickness in a fixed-base flight simulator. *Human Factors, 42*, 458–469.

Treisman, M. (1977). Motion sickness: An evolutionary hypothesis. *Science, 197*, 493–495.

Uliano, K. C., Lambert, E. Y., Kennedy, R. S., & Sheppard, D. J. (1986). *The effects of asynchronous visual delays on simulator flight performance and the development of simulator sickness symptomatology* (NAVTRASYSCEN 85-D-0026-1). Orlando, FL: Naval Training Systems Center, US Navy.

Ungs, T. J. (1989). Simulator induced syndrome: evidence for long-term after-effects. *Aviation, Space, and Environmental Medicine, 60*(3) 252–255.

Van Erp, J., & Kappé, B. (1977). *Head-slaved images for low-cost driving simulators and unmanned ground vehicles* (Report TM-97-A041), Soesterber, The Netherlands: TNO Human Factors.

Warner, H. D., Serfoss, G. L., Baruch, T. M., & Hubbard, D. C. (1993). *Flight simulator-induced sickness and visual displays evaluation* (AL/HR-TR-1993-0056). Williams Air Force Base, AZ: Aircrew Training Research Division.

Watson, G. (1995). *Simulator effects in a high-fidelity driving* simulator *as a function of visuals and motion*. Proceedings of Driving Simulation Conference 1995, Sophia, France.

Watson, G. (1997). *Simulator adaptation in a high-fidelity driving simulator as a function of scenario intensity and motion cueing*. Proceedings of Driving Simulation Conference 1997. Paris, France.

Watson, G. (1998). *The effectiveness of a simulator screening session to facilitate simulator sickness adaptation for high-intensity driving scenarios*. IMAGE Conference, Scottsdale, AZ.

Wells, M., & Venturino, M. (1990). Performance and head movements using a helmet mounted display with different sized fields-of-view. *Optical Engineering, 29*(8), 870–877.

Wood, J. M., & Troutbeck, R. J. (1994). Investigation of the effect of age and visual impairment on driving and vision performance. *Transportation Research Record, 1438*, 84–90.

Wright, R. H. (1995). *Helicopter simulator sickness: A state-of-the-art review of its incidence, causes, and treatment* (ARI Rep. 1680). Alexandria, VA: U.S. Army Research Institute for the Behavioral and Social Sciences.

Young, S. D., Adelstein, B. D., & Ellis, S. R. (2007). Demand characteristics in assessing motion sickness in a virtual environment: or does taking a motion sickness questionnaire make you sick? *IEEE Transactions on Visualization And Computer Graphics, 13*(3), 422–428.

Zacharias, G. L., & Young, L. R. (1981) Influence of combined visual and vestibular cues on human perception and control of horizontal rotation. *Experimental Brain Research, 41*, 159–171.

15

Independent Variables: The Role of Confounding and Effect Modification

Gerald McGwin, Jr.
University of Alabama

Abstract

The Problem. The valid interpretation of observational and experimental research necessitates that the role of confounding and effect modification be taken into account. Confounding occurs when an association (or lack thereof) between a dependent and independent variable is due to a third variable that itself is associated with both the dependent and independent variables. Effect modification, a related phenomenon, is present when an association between a dependent and independent variable varies according to a third variable. *Key Implications of Study of Confounding and Effect Modification for Driving Simulator Studies*. This chapter will provide a general overview of the principles of the concepts of confounding and effect modification in observational and experimental research including techniques for detecting and accounting for these issues. The role of confounding is an important consideration in observational research; this is also true, though to a lesser extent, of experimental studies, particularly those with small numbers of participants, such as is the case in driving simulator studies. *Scenarios and Dependent Variables*. Confounding and effect modification can exert their effects independently from the scenario being displayed or dependent variable being measured. *Generality*. Confounding and effect modification are important concepts that can undermine the validity of an otherwise well-designed and conducted study. As such they should be prominent in the minds of investigators during the conception and design of a study. Studies that do incorporate these concepts into their design should ensure that they are properly accounted for throughout, particularly during the analytical phase.

15.1 Introduction

Experimental and observational research in the social sciences, medicine, and other related fields is frequently hypothesis-driven; though exploratory, or hypothesis-generating, research is also common. Research seeking to address a given hypothesis is generally focused on a specific relationship between a dependent (or outcome) variable and an independent (or explanatory) variable. In experimental study designs this independent variable is assigned, often in a randomized manner, by the investigator or in the case of the crossover study design the study participant is subjected to each experimental condition

at separate points in time. In observational designs, the independent variable is a characteristic inherent to the study participant; for example, age, gender, or a medical condition (such as Parkinson's disease). Regardless of the study design, the valid interpretation of a study's results is dependent upon a number of factors related to proper design, execution, and analysis. Two such factors are confounding and effect modification. These concepts involve the role that secondary or extraneous variables have on the interpretation of the main dependent and independent variables of interest and require an appreciation for complex interrelationships as they relate to the hypothesis under investigation. Failure to do so may result in conclusions that lack validity. The objective of this chapter is to present the concepts of confounding and interaction and describe their role in driving simulator research.

15.2 The Role of "Secondary" Variables: Confounding and Effect Modification

15.2.1 Confounding

15.2.1.1 Definition

Confounding occurs when an observed association between a dependent and independent variable can be wholly or partly attributed to a third variable, which itself is associated with both the dependent and independent variables. This third or extraneous variable is referred to as a confounder. As an example from the field of epidemiology, MacMahon, Yen, Trichopoulos, Warren and Nardi (1981) reported a strong positive association between coffee consumption and cancer of the pancreas, a highly lethal disease (MacMahon et al., 1981). However, the results of this study were criticized for not adequately accounting for cigarette smoking (Tavani et al., 2000), which is a widely recognized risk factor for pancreatic cancer (Ghadirian, Lynch, & Krewski, 2003) and a behavior more common among coffee drinkers (Swanson, Lee, & Hopp, 1994). This criticism was ultimately borne out when subsequent research failed to replicate the original study's findings. In this example, cigarette smoking served as a confounder for the relationship between coffee consumption and pancreatic cancer, and failing to adequately account for it resulted in an inappropriate conclusion being drawn. It should be noted that confounding is not omnipresent to the same degree in all studies evaluating the same relationship (Miettinen, 1974). It should also be noted that confounding is of lesser concern for experimental studies wherein the independent variable of interest is often randomly assigned. Such randomization will create an even distribution of all characteristics, measured and unmeasured, thereby theoretically removing the opportunity for confounding. However, practically speaking, differences between the experimental groups can still occur, particularly with small numbers of study subjects. Thus, regardless of the study design, it is vital during the study design phase that investigators are aware of those characteristics which are likely to represent confounders and collect pertinent information on them.

15.2.1.2 Criteria for Judging Confounding

The definition of confounding provided above is a general one and while a confounder must be related to both the dependent and independent variables, this alone does not define a confounder. There are three criteria that should be used to determine whether a given characteristic should be considered a confounder: (1) A confounding factor must be associated with the dependent variable. (2) The association between the dependent variable and the confounding factor must be the same

across all levels of the independent variable. (3) A confounding factor cannot be affected by the independent or dependent variable. These criteria remove from consideration those factors that fall in the causal pathway between the dependent and independent variables, often termed intermediate factors. As an example, consider a study on the effect of varying levels of alcohol intoxication on simulated driving performance wherein the investigators also collect information on several judgment-based tasks. Performance on these tasks will likely be associated with the level of intoxication and, as a result, also be associated with driving performance. However, task performance is not a confounder, as it mediates the relationship between intoxication and driving performance. While the judgment task will be associated with the driving task, this association will depend on the level of alcohol intoxication—low with low levels of intoxication and high with high levels of intoxication—and judgment will be affected by the level of intoxication (though not by performance on the driving task). While documenting the presence of a confounder is made simpler with actual data, such documentation cannot occur if the proper data has not been collected. Thus, identifying potential confounders is as much a theoretical process as it is analytical.

15.2.1.3 Techniques to Correct for Confounding

15.2.1.3.1 Study Design Techniques

There are two general approaches to correct for confounding: in the study design and in the analysis. With respect to the former, perhaps the most effective approach is randomization. By randomly distributing study participants into experimental groups the potential for confounding factors to be related to the independent variable (i.e., experimental group) is largely removed. This is particularly attractive because it is not only true for measured but also unmeasured characteristics; that is, both known and unknown potential confounders. Riphaus, Gstettenbauer, Frenz and Wehrmann (2006) were interested in whether propofol sedation for routine endoscopy, an outpatient procedure, was associated with driving skills (Riphaus et al., 2006). For patients undergoing this procedure current recommendations are that they avoid driving for 24 hours, which may be unnecessary given the short-acting nature of propofol. A total of 100 patients undergoing routine gastroscopy and colonoscopy were randomly assigned to receive either propofol or midazolam/pethidine for sedation. A comparison of driving simulator performance before sedation and two hours following sedation revealed no differences for the propofol group but significant impairment for the midazolam/pethidine. These findings could not be attributed to differences in age, gender, benzodiazepine use, or alcohol consumption as these characteristics were shown to be equally distributed between the two groups. In theory, randomization seeks to ensure an equal distribution of all characteristics—both measured and unmeasured—between the treatment groups, thereby diminishing the potential for confounding. In practice, the likelihood that this is achieved is, in large part, a function of sample size. As the sample size increases, the likelihood that the distribution of all characteristics in any one of the treatment groups is the same as the distribution in any of the other treatment groups goes to one.

Despite the attractiveness of randomization there are certain caveats to consider. Imbalances between experimental groups can still occur either by chance or due to small sample sizes. Thus, information regarding potential confounders still must be collected to document that a lack of group differences cannot be attributed to confounding using analytical techniques.

Matching is perhaps one of the most frequently-used study design techniques to address the issue of confounding. Matching refers to the selection of study participants in such a way that the experimental groups are equally distributed with respect to one or more characteristics. Unlike randomization, where the distributions of measured and unmeasured characteristics in a small sample can be very different from one another due to chance factors, matching ensures that the one or more matched characteristics are equally distributed. For example, in a study evaluating driving performance in people with obstructive sleep apnoea, Juniper, Hack, George, Davies and Stradling recruited 12 obstructive apnoea patients and matched them to 12 controls on the basis of age, gender and driving experience (Juniper et al., 2000). Each of these characteristics is, or at least theoretically should be, associated with both obstructive sleep apnoea and driving performance. By creating two groups wherein these characteristics are equally distributed they cannot act as confounders. These authors mention several other characteristics including alcohol consumption, proficiency with video games and visual problems as being potential, though unmatched, confounders. This brings up an important consideration with respect to matching and that is overmatching. It is possible to match on too many factors so that both statistical and resource efficiency are harmed. With respect to the former, matching on a factor strongly associated with the primary independent variable but not associated with the dependent variable will provide little additional statistical information with respect to the effect of the primary independent variable. And, with respect to resource efficiency, this dearth of statistical information will come at the cost of identifying and enrolling subjects that meet a more refined set of criteria.

Finally, when analyzing data from a study design that includes matching, it is important that the matching is incorporated into the statistical analysis. For example, in a study to determine if patients undergoing unilateral first metatarsal osteotomy had impaired brake response times, 28 such patients were matched to 28 healthy volunteers with respect to age, sex, and driving frequency (Holt, Kay, McGrory, & Kumar, 2008). The investigators describe using the chi-square test and analysis of variance for between group comparisons. Given the pair matching used in this study these were not the appropriate statistical tests; rather McNemar's test and repeated measures analysis of variance (or a paired t-test) should have been used. Failure to consider the matched nature of the study design in the analysis can result in biased results and therefore inappropriate conclusions (Thompson et al., 1982).

15.2.1.3.2 Analytical Techniques

If randomization, matching or other study design approaches (e.g., restriction) are not feasible then the confounding influence of variables can be accounted for analytically. This process is often referred to as statistical adjustment or control; though the latter term connotes a level of completeness that is rarely achieved. The specific statistical tools necessary to perform statistical adjustment vary, depending upon several factors including the nature of the dependent and independent variables. That is, it depends on whether the dependent variable, independent variable and confounder(s) are categorical (ordinal versus otherwise) or continuous, normally distributed or not. Consider a simple situation wherein the dependent and independent variables, as well as the confounder, are all binary. In this case the initial approach would be to estimate the association between the dependent and independent variables according to the levels or strata of the confounder. The precise nature of the measure of association (e.g., odds ratio, risk ratio) depends upon a number of factors and is beyond the scope of this chapter. This initial step of examining the pattern of results across the various strata is important because it is the point of departure between confounding and effect modification (which will be discussed in the next section). Assuming that the observed association is indeed uniform or constant across strata then a weighted, summary, or pooled estimate can be calculated using a number of techniques including the maximum likelihood method and the Mantel-Haenszel method. The resulting pooled estimate will reflect the association between the dependent and independent variables independent of the influence of the confounding variable. The extent of the confounding influence can be evaluated by comparing the crude or un-stratified association to the adjusted or pooled association. While this approach is generalizable, the statistical tools will vary depending upon the nature of the variables. For example, if the dependent and independent variables are continuous and the confounder is categorical, stratum-specific correlation coefficients can be calculated and, if determined to be homogeneous, partial correlation coefficients can be calculated.

Consider the following example: Investigators are interested in whether adults with Attention-Deficit Hyperactivity Disorder (ADHD) have impaired driving performance as measured on a driving simulator. They recruit a total of 200 adults to participate in the study, 120 with ADHD and 80 controls. Among the dependent variables of interest is the occurrence of a motor vehicle collision. The data from the study appears in Table 15.1. Based upon this data, an odds ratio can be calculated revealing that the odds of a collision are over nine times higher in the ADHD group compared to the control group. Upon consideration of their results, the investigators are concerned that this association might partly reflect the differences in driving experience between the groups. Having collected information on this potential confounder the investigators separate the study

TABLE 15.1 Data From a Hypothetical Study on the Association Between ADHD and Driving Performance

	Group		
	ADHD	Control	Total
Collision			
No	36	64	100
Yes	84	16	100
Total	120	80	200
Odds Ratio = 9.3			

participants into two groups: those that drive regularly and those that do not.

Table 15.2 presents the data in Table 15.1 stratified according to driving experience. Note that despite the strong overall relationship between ADHD and collision involvement, there is no association in either strata. Using the Mantel-Haenszel method to adjust the association in Table 15.1 for driving experience, which assigns weights to each stratum-specific estimate, produces an adjusted odds ratio of 1.01. When compared to the unadjusted or crude odds ratio of 9.3 it is apparent that not only was confounding present but that failing to consider its effect would have resulted in a dramatic misinterpretation of the results. Stratification according to a confounding variable provides valuable insight regarding the nature of the relationship between a dependent and independent variable. However, this insight is limited to confounders that are categorical in nature. In some instances, continuous variables (for example age) can be categorized such that stratification is possible. Even so, stratification can be cumbersome with large numbers of strata for a given variable and/or large numbers of potential confounders, particularly in situations where sample sizes are small. Consider a study with three potential confounders: age (<30, 31–50, >50), gender (male, female) and race (black, white, other). This would require 18 (i.e., $3 \times 2 \times 3$) strata to represent the possible combinations of these variables. Even with a relatively large number of study participants, it is possible that many of the strata would contain few participants.

The most common technique used to adjust for confounding is multivariable regression, wherein a dependent variable is expressed as a mathematical function of a set of independent variables. A large number of mathematical models have been developed, each with unique characteristics related to its application. Linear and logistic regression models are the most familiar tools though many others also exist (e.g., Poisson regression, Cox proportional hazards regression). For linear regression models the dependent variable is continuous; whereas for logistic regression models the dependent variable is categorical, frequently binary. While the nature of the dependent variable is a consideration when deciding which regression model is best suited to a given situation, other factors must also be considered, specifically the assumptions associated with each model. Once an appropriate model is selected, techniques for selecting and modeling variables require careful consideration; however, these issues are beyond the scope of this chapter.

Uc, Rizzo, Anderson, Shi and Dawson investigated whether drivers with mild Alzheimer's disease are at increased risk of rear-end collisions or demonstrated risky avoidance behaviors (e.g., abrupt slowing) (Uc et al., 2006). Sixty-one participants with mild Alzheimer's disease and 115 neurologically normal participants underwent driving simulator assessment. The participants with Alzheimer's disease were approximately four years older and drove half as much as the normal participants. As these characteristics are associated with driving performance, their confounding influence should be accounted for in the analysis. The odds of abrupt slowing was 4.0-times higher among the participants with Alzheimer's disease compared to the normal participants; however, once adjusted for age and weekly mileage the increased odds were 2.3-times higher. Interestingly, the unadjusted and adjusted odds ratios for premature stopping were similar (2.3 versus 2.3, respectively) whereas the odds ratio for swerving increased from 1.9 to 3.9 following adjustment.

Some of the fluctuation between the unadjusted and adjusted estimates described above can be attributed to small sample size, which can produce unstable estimates. Thus, while the use of regression models to adjust for confounding has advantages over stratification; the success of this approach is highly dependent upon the available sample size, particularly in relation to the number of independent variables in the model. It has been generally accepted that for logistic regression models there should be at least 10 events per independent variable (Concato, Peduzzi, Holford, & Feinstein, 1995; Peduzzi, Concato, Feinstein, & Holford, 1995; Peduzzi, Concato, Kemper, Holford, & Feinstein, 1996). However, recent work suggests that this number might be

TABLE 15.2 Data From a Hypothetical Study on the Association Between ADHD and Driving Performance

	Regular Drivers				Irregular Drivers		
	Group				Group		
	ADHD	Control	Total		ADHD	Control	Total
Collision				Collision			
No	27	3	30	No	9	61	70
Yes	83	9	92	Yes	1	7	8
Total	110	12	122	Total	10	68	78
Odds Ratio = 1.02				Odds Ratio = 0.97			

an overestimate and that as few as five events per independent variable may be justified (Vittinghoff & McCulloch 2007).

15.2.2 Effect Modification/Interaction

15.2.2.1 Definition

The term "effect modification" will be familiar to those acquainted with the field of epidemiology. Effect modification is also referred to as heterogeneity of effect, non-uniformity of effect and effect variation (Rothman & Greenland, 1998). A similar, and perhaps more familiar, concept is that of interaction or statistical interaction. Though there is debate as to whether these concepts are interchangeable (Rothman & Greenland, 1998; Kleinbaum, Kupper, & Morgenstern, 1982), the nuances are beyond the scope of this chapter. In general, both concepts reflect the idea that the relationship between a given dependent and independent variable is *differential* according to a third variable or an effect modifier. That is, it is not possible to understand the relationship between a dependent and independent variable without considering the role of a third variable.

As an example from the field of epidemiology, Høidrup, Grønbaek, Pedersen, Lauritzen, Gottschau and Schroll, investigated the association between hormone replacement therapy and the risk of hip fracture (Høidrup et al., 1999). They reported that women who reported current use of hormone replacement therapy had a lower risk of hip fracture. However, they observed that this association was not consistent across all study participants. For example, while there was a lower risk of hip fracture associated with hormone replacement therapy use among sedentary women, there was no association among physically active women. The authors explained this finding by suggesting that the protective effect of hormone replacement therapy use was offset by the higher bone mass and muscle strength associated with exercise.

In the above example the primary relationship of interest is between hormone use and hip fracture while physical activity plays a largely ancillary role. If there is any distinction between effect modification and interaction, it is perhaps in this regard. Again using an example from the field of epidemiology, the association between smoking and asbestos and lung cancer is often used to illustrate the principle of interaction. Since the 1950s and 1960s it has been observed that smoking and asbestos fibers are

associated with lung cancer. However, the combined exposure to both agents appeared to impart a greater risk than would be expected by the sum of their individual effects (Erren, Jacobsen, & Piekarski, 1999; Lee, 2001). This notion of characteristics operating synergistically (in this case, in other cases antagonism is also possible) reflects another perspective on effect modification and interaction. Despite the aforementioned debate surrounding the exact meaning of these terms, the general concept of differential effects according to two independent variables remains the central issue.

15.2.2.2 Assessing the Presence of Effect Modification

In contrast to confounding, wherein the goal is to remove its effect, effect modification (or interaction) is frequently viewed as something of interest which should be reported and explained. Also, effect modification is an equally important consideration for observational and experimental studies whereas in experimental studies involving participant randomization, the role of confounding is minimized. Yet like confounding, detecting effect modification relies upon having the appropriate information available to document its existence. This requires that those characteristics likely to act as effect modifiers be identified during the study design process. This is not to suggest that all effect modification is hypothesized a priori; in some instances (perhaps many), it may be detected as a result of exploratory statistical analysis. Also like confounding, the steps to detect the presence of differential effects are similar to the initial steps to detect confounding and are perhaps best illustrated with an example. Consider a study evaluating the combined effect of sleep restriction and low-dose alcohol on driving simulator performance. In this study each participant completed a 60-minute simulated driving session under four experimental conditions: normal sleep with and without alcohol, and abnormal sleep (sleep restriction) with and without alcohol. Among the many driving performance measures were crash events. The data from this hypothetical study is shown in Table 15.3. There are a number of different comparisons that can be made when evaluating the data in Table 15.3. Taking a similar approach to that done in Table 15.2, one can evaluate the role of alcohol under normal and restricted sleep conditions. The data suggests that the effect of alcohol appears to be similar under both sleep conditions; that is, the proportion of participants that crashed is approximately

TABLE 15.3 Data From a Hypothetical Study on the Effect of Sleep Restriction and Alcohol on Driving Performance

	Experimental Conditions			
	Normal Sleep		Sleep Restriction	
	No Alcohol	Alcohol	No Alcohol	Alcohol
% Crashed	2.5	10.0	7.5	40.0
Risk ratio 1	Reference	4.0	Reference	5.3
Risk ratio 2	Reference	4.0	3.0	16.0

four to five times higher. Thus, alcohol appears to increase the likelihood of crash involvement to a similar degree whether one is sleep-deprived or not.

However, consider another comparison based upon this same data wherein three of the experimental conditions (i.e., normal sleep, alcohol; restricted sleep, no alcohol; restricted sleep, alcohol) are compared to the normal sleep and no alcohol condition. Such a comparison allows one to address the question: Do sleep restriction and alcohol act synergistically on the likelihood of crash involvement? Here we see that both alcohol and sleep restriction alone increase the likelihood of crash involvement (4-fold increase versus 3-fold increase, respectively) though the latter to a lesser degree, while exposure to both conditions results in a 16-fold increase. Regardless of whether one views these risks as either additive or multiplicative (for more information regarding these concepts see Rothman & Greenland, 1998), the combined effect is much greater than what one would expect based upon the sum of the individual effects. Thus, sleep restriction and alcohol involvement appear to act synergistically with respect to their impact on driving performance. While this may at first appear in conflict with the conclusion reached based upon the first comparison, it is important to note that each comparison is focused on a specific aspect of *differential* effects. The first comparison is primarily interested in isolating the effect of alcohol whereas the second comparison is interested in the joint effects of both alcohol and sleep restriction. In the majority of published driving simulator research, it is the latter comparison that appears to be the favored approach.

The above example illustrates the situation for binary dependent and independent variables. For continuous dependent variable and categorical independent variables a similar approach can be taken yet comparing means instead of proportions. When the dependent variable is continuous and one of the independent variables is continuous and the other is categorical, an analysis of covariance model can be applied to the data. A test of "parallelism" in this model will provide evidence as to the presence of effect modification/interaction.

15.3 Sample Size and Power Considerations

Reducing the impact of confounding can be accomplished via study design and/or analytical techniques. While a randomized study will generally remove the need for analytical solutions, residual confounding may exist in matched studies necessitating stratification or regression adjustment. Unlike confounding, effect modification—whether pre-specified in the primary hypothesis or planned as part of a secondary, exploratory analysis—is largely an analytical matter. This is not to suggest that addressing these concepts should be relegated to the analysis phase of a study, even if the decision to neither randomize nor match has been made at the outset. As has been mentioned above, it is important to identify the role of potentially confounding or modifying factors early in the study design phase such that information can be obtained on them. It is also important to consider the role of confounding and effect

modification during the design stage so that they can be taken into account when performing sample size and power calculations.

The proper selection, use, and interpretation of sample size and power calculations is discussed elsewhere in this book. The majority of textbooks and software focus on situations wherein there is a single dependent and independent variable; the impact of confounding and interaction on these calculations is rarely discussed. Yet, devoid of these considerations, it is likely that the estimates derived from standard power and sample size calculations are reasonable. This is particularly true in light of the fact that in most instances power and sample size estimates are based upon a number of uncertainties. For example, the anticipated effect size may be difficult to surmise. As such, instead of providing a single sample size estimate, investigators will often produce a range of sample sizes under a variety of assumptions. It is prudent to be conservative in these situations and choose sample sizes that incorporate a buffer against the level of uncertainty. Similarly, it is possible to indirectly address the role of confounding in these situations by selecting sample sizes based upon post-adjustment effect sizes. For example, in the study described previously by Uc et al., the odds ratio for the association between Alzheimer's disease and abrupt slowing was 4.0 but once adjusted for age and weekly mileage it was reduced to 2.3 (Uc et al., 2006). Conversely, for another outcome the odds ratio increased following adjustment from 1.9 to 3.9. Had the investigators anticipated these results they likely would have opted for a sample size based upon an odds ratio between approximately 2.0 and 2.3.

For those seeking more quantitative approaches there are formulae and software that incorporate the role of confounding and interaction into power and sample size calculations. Demindenko (2007, 2008) has derived formulas for logistic regression with binary covariates and their interaction and has developed online tools for their use (http://www.dartmouth.edu/~eugened/power-samplesize.php). Similar tools for linear regression and analysis of variance have been described by Lenth (2007) and are available online (http://www.cs.uiowa.edu/~rlenth/Power/). Many commercially available software packages will perform power and sample size calculations for 2- and 3-way analyses of variance models (http://www.biostat.ucsf.edu/sampsize.html). A general rule is that the stronger the relationship between the primary independent variable and the confounder, the larger the required sample size (Lui, 1990; Wilson & Gordon, 1986; Smith & Day, 1984). One reason for this is that, upon adjustment for the confounder, the association between the dependent variable and the primary independent variable may decrease. It is also possible that, upon adjustment for the confounder, the main association will improve (i.e., get stronger) thereby decreasing the required sample size. However, even in this case, the precision of the estimate may decrease. It is perhaps important to note that in the context of logistic regression, even when the confounder and the primary independent variable are independent, the power will improve and sample size decrease (Demidenko, 2007). This paradoxical situation does not occur in the contact of linear regression. Thus, it is imperative to evaluate the potential impact of confounding and effect modification during the study design phase. It is prudent to perform a number of simulations that consider not only the presence of a confounder but also vary its association with the dependent and primary independent variable.

Consider a study evaluating brake response time among patients that have undergone total knee arthroplasty compared to a control group. Based upon prior research it might be estimated that the total knee arthroplasty group has three times the odds (odds ratio = 3.0) of failure on a brake response time test compared to the control group. In this case it can be estimated that approximately 214 subjects (107 in each group) would be needed to detect this association as significant with an $\alpha = 0.5$ and $\beta = 0.2$. Now consider gender as a potential confounder. If gender is equally distributed between the groups (i.e., odds ratio = 1.0) and is independently associated with failure on a brake response time test (e.g., odds ratio = 2.0) the required sample size decreases to 170 (85 in each group) to detect that same odds ratio of 3.0. In this situation the sample size requirements will decrease as the strength of the association between the confounder and the dependent variable increases but only up to a point, at which point the sample size will rise. However, if the converse situation occurs (i.e., gender is associated with group status but not test failure), the sample size increases to 220 (110 in each group). And, in contrast to the prior situation, as the association between gender and group status increases, the sample size will continue to increase. Finally, in the situation where gender is associated with both group status and test failure (odds ratios = 2.0), the total sample size required to detect an odds ratio of 3.0 between the total knee arthroplasty group and brake response time test failure is 175 (~88 in each group). This example illustrates the point that power and sample size calculations should be performed under a number of realistic scenarios so that a range of estimates is produced. In situations where the determination of such realistic scenarios is not possible, techniques to conduct sample size re-assessments once a study has begun have been developed; however, these techniques are not without controversy (Friede & Kieser, 2006; Schafer, Timmesfeld, & Muller, 2006; Coffey et al., 2008).

15.4 Summary and Conclusions

The role of confounding is an important consideration in observational research; this is also true, though to a lesser extent, of experimental studies, particularly those with small numbers of participants. The validity of a study's results may be weakened if the role of confounding is not taken into account. This chapter has presented several study design and analytical tools that can be used to document the presence of and correct for confounding. For these techniques to be successful one must have an appropriate understanding of the etiologic mechanisms underlying the hypothesized relationships being investigated. Inappropriate application of matched study designs to accommodate confounding will decrease the efficiency of a study. For analytical solutions to be successful, information on potential

confounders must be appropriately collected during the study. In contrast to population-based research, sample sizes in driving simulator-based research are smaller. Thus particular attention must be paid to the impact of confounding on sample size and power.

The concept of differential effects, expressed using terms such as effect modification and interaction, represents an opportunity to gain valuable insight regarding etiologic relationships. The synergy (or antagonism) between two (or more) independent variables may reflect an a priori hypothesized relationship or a relationship that is discovered incidentally. In the former case, like confounding, proper planning with respect to study design and analytical requirements is crucial. Effect modification identified during exploratory analyses can shed light on important and previously unknown nuances of a scientific question; yet such findings should be interpreted with caution and distinguished from what in reality may be statistical artifacts.

Confounding and effect modification are important concepts that can undermine the validity of an otherwise well-designed and conducted study. As such they should be prominent in the minds of investigators during the conception and design of a study. Studies that do incorporate these concepts into their design should ensure that they are properly accounted for throughout, particularly during the analytical phase.

Key Points

- Observational (non-randomized) studies should address the role of confounding using study design or statistical analysis techniques.
- Associations between a dependent and primary independent variable may be modified by a third characteristic; failure to explore such relationships may preclude valuable insight.

Keywords: Analysis, Confounding, Effect Modification, Interaction, Statistics

Key Reading

Rothman, K. J., & Greenland, S. (1998). *Modern epidemiology* (2nd ed.). Philadelphia: Lippincott, Williams, & Wilkins.

References

Coffey, C. S., & Kairalla, J. A. (2008). Adaptive clinical trials: Progress and challengers. *Drugs in R & D, 9*(4), 229–242.

Concato, J., Peduzzi, P., Holford, T. R., & Feinstein, A. R. (1995). Importance of events per independent variable in proportional hazards analysis. I. Background, goals, and general strategy. *Journal of Clinical Epidemiology, 48*(12), 1495–1501.

Demidenko, E. (2007). Sample size determination for logistic regression revisited. *Statistics in Medicine, 26*(18), 3385–3397.

Erren, T. C., Jacobsen, M., & Piekarski, C. (1999). Synergy between asbestos and smoking on lung cancer risks. *Epidemiology, 10*(4), 405–411.

Friede, T., & Kieser, M. (2006). Sample size recalculation in internal pilot study designs: A review. *Biometrical Journal, 48*(4), 537–555.

Ghadirian, P., Lynch, H. T., & Krewski, D. (2003). Epidemiology of pancreatic cancer: An overview. *Cancer Detection and Prevention, 27*(2), 87–93.

Høidrup, S., Grønbaek, M., Pedersen, A. T., Lauritzen, J. B., Gottschau, A., & Schroll, M. (1999). Hormone replacement therapy and hip fracture risk: Effect modification by tobacco smoking, alcohol intake, physical activity, and body mass index. *American Journal of Epidemiology, 150*(10), 1085–1093.

Holt, G., Kay, M., McGrory, R., & Kumar, C. S. (2008). Emergency brake response time after first metatarsal osteotomy. *Journal of Bone and Joint Surgery, American Volume, 90*(8), 1660–1664.

Juniper, M., Hack, M. A., George, C. F., Davies, R. J., & Stradling, J. R. (2000). Steering simulation performance in patients with obstructive sleep apnoea and matched control subjects. *European Respiratory Journal, 15*(3), 590–595.

Kleinbaum, D. G., Kupper, L. L., & Morgenstern, H. (1982). *Epidemiologic research. Principles and quantitative methods.* New York: Van Nostrand Reinhold.

Lee, P. N. (2001). Relation between exposure to asbestos and smoking jointly and the risk of lung cancer. *Occupational and Environmental Medicine, 58*(3), 145–153.

Lenth, R. V. (2007). Statistical power calculations. *Journal of Animal Science, 85*(13S), E24–E29.

Lui, K. J. (1990). Sample size determination for case-control studies: The influence of the joint distribution of exposure and confounder. *Statistics in Medicine, 9*(12), 1485–1493.

MacMahon, B., Yen, S., Trichopoulos, D., Warren, K., & Nardi, G. (1981). Coffee and cancer of the pancreas. *New England Journal of Medicine, 304*(11), 630–633.

Miettinen, O. (1974). Confounding and effect-modification. *American Journal of Epidemiology, 100*(5), 350–353.

Peduzzi, P., Concato, J., Feinstein, A. R., & Holford, T. R. (1995). Importance of events per independent variable in proportional hazards regression analysis. II. Accuracy and precision of regression estimates. *Journal of Clinical Epidemiology, 48*(12), 1503–1510.

Peduzzi, P., Concato, J., Kemper, E., Holford, T. R., & Feinstein, A. R. (1996). A simulation study of the number of events per variable in logistic regression analysis. *Journal of Clinical Epidemiology, 49*(12), 1373–1379.

Riphaus, A., Gstettenbauer, T., Frenz, M. B., & Wehrmann, T. (2006). Quality of psychomotor recovery after propofol sedation for routine endoscopy: A randomized and controlled study. *Endoscopy, 38*(7), 677–683.

Rothman, K. J., & Greenland, S. (1998). *Modern epidemiology* (2nd ed.). Philadelphia, PA: Lippincott, Williams, & Wilkins.

Schafer, H., Timmesfeld, N., & Muller, H. H. (2006). An overview of statistical approaches for adaptive designs and design modifications. *Biometrical Journal, 48*(4), 507–520.

Smith, P. G., & Day, N. E. (1984). The design of case-control studies: The influence of confounding and interaction effects. *International Journal of Epidemiology, 13*(3), 356–365.

Swanson, J. A., Lee, J. W., & Hopp, J. W. (1994). Caffeine and nicotine: A review of their joint use and possible interactive effects in tobacco withdrawal. *Addictive Behaviors, 19*(3), 229–256.

Tavani, A. & La Vecchia, C. (2000). Coffee and cancer: A review of epidemiological studies, 1990–1999. *European Journal of Cancer Prevention, 9*(4), 241–256.

Thompson, W. D., Kelsey, J. L., & Walter, S. D. (1982). Cost and efficiency in the choice of matched and unmatched case-control study designs. *American Journal of Epidemiology, 116*(5), 840–851.

Uc, E. Y., Rizzo, M., Anderson, S. W., Shi, Q., & Dawson, J. D. (2006). Unsafe rear-end collision avoidance in Alzheimer's disease. *Journal of the Neurological Sciences, 251*(1–2), 35–43.

Vittinghoff, E., & McCulloch, C. E. (2007). Relaxing the rule of ten events per variable in logistic and Cox regression. *American Journal of Epidemiology, 165*(6), 710–718.

Wilson, S. R., & Gordon, I. (1986). Calculating sample sizes in the presence of confounding variables. *Applied Statistics, 35*, 207–213.

16

External Driver Distractions: The Effects of Video Billboards and Wind Farms on Driving Performance

Shaunna L. Milloy
University of Calgary

Jeff K. Caird
University of Calgary

Abstract

The Problem. External driver distractions are the largest category of distraction-related crashes. Two studies are reported that examined the effects of two new external distractions on driving performance, namely video billboards and wind farms. *Role of Driving Simulators.* Driving simulators allow potential external distractions to be examined in a safe, confound-controlled, and relatively cost-effective environment. The translation of certain bottom-up and top-down properties of these external distractions into a set of driving simulator scenarios provided a means to determine their impact on driving performance. *Scenarios and Dependent Variables.* Common to both evaluations, a lead vehicle braked in the presence of the wind turbines and digital video billboard. The video billboard was animated to show a sequence of advertising images that correspond with an operational billboard. A 500-meter-long row of wind turbines was animated to show differential responses to a virtual wind along a six-lane freeway that approximated a roadway southeast of Toronto. Perception response time, speed maintenance, and lane keeping were measured in the presence of each potential distraction and during corresponding baseline segments. ***Key Results of Driving Simulator Studies.*** In the wind turbine study, perception response time (PRT) to a lead vehicle that braked hard was not significantly different between baseline and wind turbine conditions. While passing the wind farms, drivers adopted slower speeds than without their presence. In the video billboard study, significantly more rear-end collisions occurred to the hard lead-vehicle braking event in the presence of video billboards than conventional billboard and control conditions. ***Platform Specificity and Equipment Limitations.*** External distractions can be brought into the simulator to obtain a more precise understanding of their relative distraction effects. Exact lighting replication, anticipation of probable co-occurring events and generality of results constrain the use of driving simulation.

16.1 Introduction

The use of driving simulators to investigate different kinds of driver distraction is well established. Specifically, driver distraction is defined as the momentary or transient redirection of attention from the task of driving to a thought, object, activity, event or person (Caird & Dewar, 2007, pg. 196). In this book, chap. 27 by Strayer, Cooper and Drews illustrates how driving simulators have been used to determine the effects of cell phones on driving distraction, but also see Caird, Willness, Steel and Scialfa (2008). The current chapter examines how driving simulators can be used to evaluate external distractions or those objects that draw the attention of the driver *outside* of the vehicle.

The relative contribution of different categories of driver distraction to crashes is described by Stutts, Reinfurt, Staplin and Rodgman (2001), Stutts et al. (2005) and is listed in Table 16.1. The category **outside person, object or event** row is highlighted. External distractions compose the single largest category of distraction-related crashes comprising about 23% and 29% of distraction crashes. Although external distractions are the most common contributor to crashes, the least is known about why these crashes occur (Caird & Dewar, 2007; Eby & Kostyniuk, 2003). The variety of people, objects and events available for

viewing by drivers may deter researchers from wading into this un-drainable empirical swamp. In contrast, in-vehicle technologies such as cell phones, iPods and navigation systems have been studied more extensively using driving simulators (Caird et al., 2008; Chisholm, Caird, & Lockhart, 2008; Tijerina, Johnson, Parmer, & Winterbottom, & Goodman, 2000, respectively). In general, a variety of external distractions do not readily collapse into distinguishable categories with some notable exceptions.

The studies reported in this chapter investigated two novel external technologies that are becoming part of rural and urban transportation landscapes; namely, wind turbines or wind farms, and video or electronic billboards. In general, investigations of driver distraction tend to focus where there are financial interests or liability exposure at stake. If alternative energies become a focus of sustainable economic growth, wind farms are likely to proliferate. Their presence in close proximity to roadways has the potential to affect traffic safety. In contrast, video billboards provide a means to reach consumers in their vehicles. Advertisers are willing to pay owners of video billboards in order to reach drivers. Determining whether it is safe to do so has financial repercussions to advertisers, billboard owners and drivers too. Within the literature reviews that follow, the practical and methodological issues of each kind of external distraction are elaborated.

TABLE 16.1　Driver Distraction Categories and Overall Percent (Standard Errors) for Each Category (Based on Stutts et al., 2001, p. 11 and Stutts et al., 2005, p. III-4)

Driver Distraction Category	1995 to 1999	2000 to 2003
Outside person, object, or event (e.g., vehicle, police, animal, novel events, people or objects in the road, etc.)	29.4 (2.4)	23.7[a]
Adjusting radio/cassette/CD	11.4 (3.7)	2.9
Other occupant (e.g., talking, yelling, fighting, child, infant)	10.9 (1.7)	20.8
Moving object in vehicle (e.g., insects, animals, objects)	4.3 (1.6)	3.7
Other device/object (e.g., purse, water bottle, etc.)	2.9 (0.8)	5.2[b]
Adjusting vehicle/climate controls	2.8 (0.6)	1.5
Eating/drinking (e.g., burger, tea, coffee, soda, alcohol, etc.)	1.7 (0.3)	2.8
Talking/listening/dialing cell phone (e.g., answer, initiate call)	1.5 (0.5)	3.6
Smoking-related (e.g., reaching for, lighting, dropping, etc.)	0.9 (0.2)	1.0
Other distraction (e.g., medical, other inside or outside events or objects, intoxicated, depressed, etc.)	25.6 (3.1)	34.8[c]
Unknown distraction	8.6 (2.7)	
Total	100.00	100.00

Source: Adapted From Caird, J. K., & Dewar, R. E. (2007). *Driver distraction.* In R. E. Dewar & R. Olson (Eds.), *Human factors in traffic safety* (2nd ed.) (pp. 463–485). Tucson, AZ: Lawyers & Judges Publishing (2007, p. 198).

[a] *Standard errors were not reported in Stutts et al. (2005).*

[b] Category was modified to using/reaching for object brought into the vehicle.

[c] Other and unknown categories were collapsed.

16.1.1 Video Billboards as a Potential External Distraction

Video billboards, electronic billboards, or video advertising signs represent a relatively new and unknown source of external distraction (Farbry, Wockinger, Shafer, Owens, & Nedzesky, 2001). At night, a walk through Times Square in New York City or Yonge-Dundas Square in Toronto affords an intense visual exploration of both electronic and conventional billboards (see Web Figure 16.1 on the *Handbook* web site). Video billboards are more vibrant (range of luminance and colors) and dynamic (contain movement and luminance changes) than static or conventional billboards, making them more conspicuous. Kline and Dewar (2004) define attention conspicuity as the extent to which a sign is sufficiently prominent in the driving scene to capture attention, and is a function of size, color, brightness, contrast relative to surroundings, and dynamic components such as movement and change. It follows that video billboards are more conspicuous than conventional billboards because of their dynamic and brightness properties.

In addition to bottom-up properties, a video billboard provides a stream of images and text that are available for apprehension and reading. The length of an advertisement, content and availability during an approach are some of the external factors that affect the extent to which information processing can occur during glances. Internal factors of the driver also affect the degree to which a driver will look at a video billboard. Familiarity with advertising content, willingness to be distracted in a particular context, and immediate threats to ongoing driving are some of the factors that are likely to affect visual interaction on any given approach. Thus, noticing and encoding of content from a video billboard are affected by numerous external and internal factors to the driver.

Despite a general concern about video billboards' potential to capture and divert attention, minimal research has been done to determine the impact of video signs on driver behavior and performance (Farbry et al., 2001). Reviews of existing literature up to 2001 find that the contribution of conventional and video billboards to increases in crash rate is difficult to determine due to a number of practical and methodological reasons. Crashes are often an infrequent occurrence and thus are not an ideal measure of relative safety. In addition, distraction may interact with other factors, such as weather and congestion, to produce a crash. A body of equivocal findings, where some studies find crash increases and others do not, was compiled (Farbry et al., 2001). From this review, one study that compared crash rates found a decrease in crashes at both an intersection after the installation of a video billboard and a control intersection. In another study, a lesser decrease in crash rates attributed to a sign outside Boston, compared to greater decreases elsewhere resulted in the removal of the sign. In yet another case, involving a three-vehicle crash, an airline owner of an electronic sign was found indirectly responsible for the crash and the sign was removed. A variety of mitigating factors were also identified across these cases including the context in which a given sign is placed: for example, straight freeway, curves (horizontal and vertical, intersections, work zones, etc.); properties of the sign itself (e.g., luminance, size, colors, message length, attention conspicuity, exposure time, etc.); and individual differences (e.g., driver and route experience, age, willingness to look, etc.).

Since 2001, several studies have examined eye movements to video signs. An on-road study on a raised six-lane expressway in Toronto with 25 drivers found that video billboards received significantly more glances per subject per sign ($M = 1.45$) and a greater proportion of longer glances (i.e., those greater than 0.75 seconds), compared to static billboards (Beijer, Smiley, & Eizenman, 2004). The mean duration of glances to video signs was 0.6 s. A number of researchers have argued that longer glances away from the road present a threat to traffic safety and can result in a higher likelihood of collisions (Green, 2007). In addition, billboards may contribute to the increasingly cluttered driving environment. A driver's ability to rapidly search visually dense scenes, extract critical information, and respond in a timely manner may be interfered with when a single source of information such as a video billboard draws the attention of the driver to it over information in other equally or more important signs (Ho, Scialfa, Caird, & Graw, 2001).

Smiley, Smahel, and Eizenman (2004) addressed whether the presence of video billboards affected glances to other signs, and related questions about the frequency and duration of glances to video billboards. Sixteen participants were fitted with an eye-tracking system and they drove a route with four video billboards; one on an expressway, and three in downtown Toronto. On average, participants looked at video billboards about 48% of the time. About one-quarter (23%) of glances were longer than 0.75 seconds. Loss of data from 29 of 112 sign approaches (Smiley et al., 2004), which represents a loss of about 26%, makes the statistical analyses and interpretations of the behavioral results somewhat problematic (Tabachnick & Fidell, 2006, chap. 4).

Smiley et al. (2005) summarize the previous eye movement study (i.e., Smiley et al., 2004) and report the results of a traffic conflict study, a before-and-after crash study, a before-and-after speed and headway study, and a public survey of the perceived safety of billboards. The same four locations in Toronto, which are listed above, were the focus of these studies. The traffic conflict study examined driver behaviors on the video billboard approach and non-video billboard approach to two downtown intersections. A higher incidence of braking without good cause was found in the video billboard approach for one of the two intersections. No differences in unwarranted lateral displacements or delayed start on a green light were found at either intersection.

The before-and-after crash study examined three downtown intersections before and after the installation of video billboards. When examined individually, two of the three intersections showed an increase in rear-end and total collisions on the video billboard approach, while the third intersection showed a decrease. The results of the before-and-after speed and headway study were inconclusive. Finally, the public survey indicated that 65% of the 152 drivers surveyed said that video billboards had a negative effect on driver attention to pedestrians and cyclists, and 86% of respondents supported restrictions on video billboards.

While these studies provide helpful information on the impact of video billboards on drivers, they lack the control and level of analysis that simulator studies provide. Specifically, before-and-after crash studies can be confounded by multiple other contextual issues such as roadway geometry, weather, and increases in traffic flow, while simulator studies control for these issues and provide more in-depth data regarding driver behaviors such as speed and lane position.

Using video clips while driving, Crundall, van Loon and Underwood (2006) investigated the impact of street-level and raised advertising on drivers' attention. Eye movement measures indicated that street-level advertising attracted attention to a greater degree than raised advertising because the street-level advertising was further from the line of sight. The extent to which a sign is in the periphery relative to the course of travel seems to affect both the frequency and duration of glances.

Shinar (2007) reports an unpublished study that examined the extent that a sample of experienced drivers in an on-road study looked right, left and at the roadway using a camera installed in the rear-view mirror. On the right hand side of one approach was a large static billboard with a topless male model in it. The opposite direction drive served as the comparison condition. When approaching the billboard, glances to the left (10%) and roadway (68%) were shifted to the billboard (23%). However, on the drive in the opposite direction (i.e., not facing the billboard) drivers looked to the right (10%), left (19%) and to the roadway (71%) accordingly. The explanation given for the results was that drivers allocated their spare attentional capacity to the billboard with little cost to the roadway or left-side glances. By extension, billboards are safe because drivers have spare visual attention to allocate. How spare is spare capacity, however, when an emergency event requires an immediate response? Driving simulators would allow the researchers to address this question and determine if allocating spare visual attention is safe when unexpected driving events such as a lead vehicle braking suddenly or a pedestrian emerging into the roadway occurs. These events are rare and difficult to measure in on-road studies.

To address this gap, a study was conducted to address three central questions: 1) Do video billboards cause an increase in response time to unexpected driving events such as a lead vehicle braking?; 2) Do video billboards interfere with visual search for traffic signs? 3) Are video billboards remembered more than static billboards because they attract more attention?

16.1.2 Wind Turbines as a Potential External Distraction

Wind turbines or wind farms are quickly multiplying along roadways in the U.S., Canada and Europe, but their impact on driver distraction is completely unknown. Many countries expect that wind turbines will provide a larger proportion of energy needs in the future. Wind turbines come in a wide variety of sizes and installation patterns, some of which appear along roadways (see Web Figure 16.2 on the *Handbook* web site).

An extensive web search and consultation with researchers from around the world found almost no research that attempted to determine the relative distractive potential of wind turbines. Typically, planning documents and opposition position papers cite driver distraction as a potential negative impact when considering new wind farm installations. Turbine visibility while driving and road proximity are also mentioned. However, evidence that establishes a firm link between crashes and turbine installations is lacking.

The only previous study on driver distraction and wind farms was performed by David George (2007) and was sponsored by Your Wind Energy Limited (U.K.). In this study four wind farms were investigated along roadways and before-and-after-crash frequencies were compared to matched control roads without farms. No differences in crash rates were found between the wind farm installations and comparison roadways. The body of available evidence on wind farms and driving distraction is insufficient to determine whether a given installation will cause crashes to occur on adjacent roadways.

An exploratory study was conducted to determine the impact of wind turbines placed along roadways on driver distraction. Specifically, do rear-end collisions occur when a lead vehicle brakes hard in the presence of the wind turbines? Are driving behaviors such as speed maintenance and lane keeping negatively impacted? Are older or younger drivers more or less likely to be distracted by wind turbines? Do wind turbines distract drivers from the task of driving?

16.2 Common Methods

16.2.1 University of Calgary Driving Simulator

A pair of experimental studies was conducted in the University of Calgary Driving Simulator (UCDS), which is described in detail elsewhere (Caird, Chisholm, Edwards, & Creaser, 2007; Chisholm et al., 2008). Briefly, a Saturn sedan is situated in front of three screens. Each projector displays onto a wrap-around screen, each of which measures 86.5" wide by 65" in height. The total projected forward field-of-view is 150 degrees. Traffic environments and experimental scenarios for the driving simulator were developed and run in HyperDrive (v.1.9.28), which also manages the data collection of a number of driving variables including perception response time (PRT), velocity, lane position, minimum headway, and collisions. An ASL 501 eye tracker was used in both studies (see Chisholm et al., 2008 for details), but the data was not analyzed for cost reasons.

16.2.2 Common Procedures

Participants in both studies were required to have a valid driver's license, to drive at least 10,000 kilometers per year, and to report not being susceptible to car sickness or being in an at-fault crash within the previous three years. Drivers with graduated or restricted licenses were excluded to control for driving experience. All participants were screened for visual acuity, contrast sensitivity and colour deficiencies.

Each participant drove a short practice drive to familiarize them with the visual properties of the simulated drives and the handling characteristics of the vehicle. Drivers were also told to drive as they normally would in their own vehicle and were reminded to adhere to the posted speed limit signs. Following the experimental drives and questionnaires, participants were debriefed and remunerated for their participation.

16.3 Methods—Video Billboard

16.3.1 Participants

A total of 21 participants (11 males and 10 females) volunteered for the Video Billboard study through the Department of Psychology's online experiment participation system. Their characteristics are summarized in Table 16.2. Participants were slightly older than the typical university student at 24.5 years of age on average (SD = 7.2 years).

16.3.2 Procedure

Participants drove three experimental drives with each one lasting approximately five minutes. Two static and two video billboards were present in each experimental drive (see Web Figure 16.3). The order of presentation of the three drives was counterbalanced across participants. In total, 12 billboards were encountered over the course of the three experimental drives. In the simulation, sequences of images were triggered at locations on the approach to a video billboard, somewhat like a flip-book. A new image was displayed as each trigger was passed while driving toward the billboard. Roadways consisted predominantly of two-lane urban highways with four-way and three-way intersections dispersed throughout. Oncoming traffic was present throughout the drives and cross traffic and operational stoplights occurred at intersections.

A lead vehicle was intermittently inserted ahead of the participant at certain times during the drives. The lead vehicle was coupled to the movement of the driver using a programmed script and maintained at least a minimum headway of one second. The lead vehicle would brake at 8 m/s² for 50 m and reaccelerate to the posted speed (see Figure 16.1 and Web Figure 16.4). The driver had to respond quickly to avoid a rear-end collision. Two

lead-vehicle braking (LVB) events occurred in each condition (i.e., video billboard, static billboard, and baseline).

The posted speed limit was 80 km/h throughout the experimental drives. However, drivers had to reduce their speed when the posted speed was changed to 70 km/h. This speed change task occurred once in the presence of a video billboard, once in the presence of a static billboard, and once with no billboard (i.e., baseline) (see Web Figure 16.5). Speed was measured before and after the 70 km/h sign.

Participant's memory of the static and video billboards was tested after the simulator drives were completed. Participants were shown twelve images of billboard advertisements on a computer and were asked to identify those that had been present in the experimental drives. The image set included three static billboards and three still images from video billboards presented during the drives, as well as six images that were not present during the drives.

16.4 Results—Video Billboard Study

16.4.1 Experimental Design and Analyses

A mixed-factor Analysis of Variance (ANOVA) with billboard condition (video, static, baseline) as the within-subjects factor, and order of presentation of drives (1, 2, 3) as the between-subjects factor, was used to analyze perception response time (PRT) and minimum headway distance (MHD). A significant order effect was found for the lead vehicle braking events. Thus, order was examined in each analysis.

A repeated measures ANOVA was used to analyze the effect of billboard condition for the speed change event and memory test. All follow-up analyses used the Bonferroni correction for multiple comparisons. If sphericity was violated, the Greenhouse-Geisser adjustment was used and is indicated with a G-G next to reported F tests. The frequency of collisions during the lead vehicle braking events were analyzed using Chi-square non-parametric tests. Drive order effects were not significant for the speed change event and memory test.

16.4.2 Perception Response Time

Perception response time (PRT) is defined as the elapsed time (in seconds) from when the lead vehicle brake lights were

TABLE 16.2 Video Billboard Study

	N	Mean Age (SD)	Average km/year	Crashes (Last Five Years)	Moving Violations (Last Five Years)	Visual Acuity, Left Eye	Visual Acuity, Right Eye
Male	11	24.18 (5.64)	18,175 (8,335)	0.36 (0.5)	1.09 (1.1)	20/21 (0.3)	20/21 (0.3)
Female	10	24.80 (8.90)	23,500 (12,331)	0.90 (1.1)	1.90 (2.5)	20/23 (0.3)	20/25 (0.4)
Total	21	24.48 (7.19)	20,710 (10,516)	0.62 (0.9)	1.48 (1.9)	20/22 (0.3)	20/24 (0.4)

FIGURE 16.1 **(See color insert)** Lead vehicle before (left) and during (right) a braking event in video billboard condition.

visible until the participant removes his or her foot from the accelerator pedal and begins to depress the brake pedal (Olson & Farber, 2003). PRT is used in road geometry and traffic control design guidelines and forensic investigations of crashes. There was no main effect of billboard condition on PRT, $F(2, 36) = 2.32$, $p > 0.05$. Means and standard errors were similar across conditions, video billboard ($M = 0.88$, $SE = 0.03$), static billboard ($M = 0.83$, $SE = 0.03$), and baseline ($M = 0.89$, $SE = 0.03$). Order of drive presentation was not significant, $F(2, 18) = 0.35$, $p < 0.05$. However, the two-way interaction between order and billboard condition was significant, $F(4, 36) = 6.00$, $p = 0.001$ (see Web Figure 16.6).

Follow-up analyses examined the effect of billboard condition within each order. Order 1 and order 2 did not differ significantly between conditions. However, order 3 follow-up tests revealed a significant effect of condition on PRT, $F(2, 12) = 13.426$, $p < 0.001$. Participants took significantly longer to respond to the lead vehicle braking in the video billboard condition ($M = 0.99$, $SE = 0.07$), compared with both static billboard ($M = 0.88$, $SE = 0.06$), and baseline ($M = 0.83$, $SE = 0.06$) conditions, $p < 0.05$.

16.4.3 Minimum Headway Distance

Minimum headway distance (MHD) is defined as the minimum distance in meters from the simulator's front bumper to the rear bumper of the lead vehicle during the braking event (Chisholm et al., 2008). Data from the onset of brake pressure until either velocity reached zero or brake pressure was removed were used. Billboard condition had a significant effect on MHD, $F(2,36) = 28.02$, $p < 0.001$. On average, participants traveled significantly closer to the lead vehicle when it braked in the video billboard condition ($M = 2.76$, $SE = 0.64$), than the static billboard ($M = 7.18$, $SE = 0.43$) and baseline ($M = 5.15$, $SE = 0.53$) conditions (see Figure 16.2 and Web Figure 16.7). A significant two-way interaction of condition and order was found $F(4, 36) = 4.984$, $p < 0.005$.

Follow-up analyses examined the effect of condition within each order presentation. Order 1 and order 2 revealed a

significant effect of billboard condition on MHD, $F(2, 10) = 9.41$, $p < 0.005$, and $F(2,16) = 5.01$, $p < 0.005$, respectively. Participants came much closer to the lead vehicle in the video billboard condition ($M = 3.51$, $SE = 1.31$) than in the static billboard condition ($M = 8.97$, $SE = 0.74$); however, no difference was found between video billboard and baseline. Order 3 follow-up tests also indicated an effect of billboard condition, $F(2, 10) = 17.008$, $p < 0.001$. Significantly shorter minimum headway distances were found in the video billboard condition ($M = 1.52$, $SE = 1.31$) than in either the static billboard ($M = 7.09$, $SE = 0.68$), $p < 0.05$, or baseline ($M = 6.96$, $SE = 1.22$), $p < 0.05$, conditions.

16.4.4 Lead Vehicle Braking Collisions

Lead vehicle-participant collisions were determined by a collision detection algorithm. Fifteen collisions occurred out of a total of 126 LVB events (6 events × 21 participants). Chi-square analyses found a significant effect of billboard condition on collision frequency ($\chi^2(2) = 12.71$, $p = 0.002$). More collisions occurred in the video billboard condition (11) than the static billboard condition (1) ($\chi^2(1) = 9.72$, $p < 0.05$). The video billboard also significantly differed from the baseline condition (3), ($\chi2(1) = 5.49$, p < 0.05).

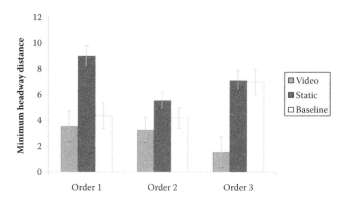

FIGURE 16.2 Minimum headway distance (m) to lead vehicle braking event by order and condition.

Thus, significantly more collisions occurred in the video billboard condition than in either the static billboard or baseline conditions.

16.4.5 Speed Change Event

The purpose of the speed change task was to determine whether the presence or absence of billboards caused drivers to delay or miss changes in the posted speed limit. Participant's velocity was recorded at the posted speed sign in km/h. Speed changes from 80 km/h to 70 km/h were present in all three conditions (video billboard, static billboard, baseline), but did not include those instances of lead vehicle braking. Analyses found no significant differences in velocity across the three conditions: video billboard ($M = 70.3$, $SE = 1.09$), static billboard ($M = 73.3$, $SE = 1.04$) and baseline ($M = 71.7$, $SE = 1.00$), $F(1, 21) = 1.014$, $p > 0.05$.

16.4.6 Memory Test

Participants were given a billboard recognition test at the conclusion of the drives. Twelve billboard pictures were shown sequentially on a computer. Six were present in the drives and six were not. Within each of these sets of six, three were video and three were conventional. A billboard was classified as being remembered when a participant identified if it was present in one of the experimental drives and whether it was a video or static billboard. Only correct responses for video and static billboards were analyzed. Overall, the main effect of billboard condition was not significant, $F(1,20) = 0.03$, $p > 0.05$. Overall, recognition of the video ($M = 0.35$, $SE = 0.08$), and static ($M = 0.33$, $SE = 0.06$) billboards occurred about one third of the time.

16.5 Discussion—Video Billboard Study

16.5.1 General Discussion

An experimental simulator study was conducted to determine the effects of video billboards on driver behavior and performance. The presence of video billboards affected a number of driver performance measures. Significantly more collisions occurred in the video billboard condition than in either the static billboard or baseline conditions. Perception response time (PRT) and minimum headway distance (MHD) measures were affected by the order of exposure to the video, static and baseline billboards. PRT results only found differences among billboard conditions in drive 3 where, as hypothesized, significant delays in PRT were found in the video billboard condition. In general, response times to the braking of the lead vehicle would be expected to decline (Chisholm et al., 2008) as participants anticipate the lead vehicle braking events in the presence of a billboard. Order 1 and 2 LVB events are interpretable in the context in which each occurred. In drive 3, the first critical incident occurred next to the video billboard. In drive 1 the first car braking occurred next to a video billboard. Learning may have occurred throughout the drives, and the presence of billboards may have been used as a cue to prepare to brake. Thus, because the first lead vehicle braking events in both drive 3 and drive 1 (order 3) were video billboard

conditions, and they were presented in that order, PRT in the video billboard condition may be higher than in the static or baseline conditions as learning had not yet occurred and participants were not cued by the presence of a billboard to anticipate a braking event.

When the LVB events occurred in the presence of the video, static and baseline conditions, the MHD results found that participants traveled significantly closer to the lead vehicle in the video billboard condition than the baseline condition, for all orders. Observing the video billboard most likely delayed the participant's response times to the lead vehicle braking. Taken together, the MHD and collisions results indicate that video billboards have a safety cost when certain events occur. Adjusting speed when a speed limit sign was present was not affected by the presence of video billboards. This result was based on discrete samples of speed at specified locations and did not examine speed adjustment delay over time.

The results of the simulator study, where a lead vehicle braked hard in the presence of the video billboard, found a *causal* relationship between the presence of the video billboard and collisions with, and delays in responding to, the lead vehicle. A realistic assumption of the performance of those in the simulator study is that their performance is near optimal. The luminance of the video billboards in this study was lower because of the brightness characteristics of the projectors. Thus, the capability to capture attention was not the same as video billboards typically found in traffic environments and therefore drivers' performances in the simulator would presumably be better than they would on the open road. Furthermore, the limited capability to evaluate night driving is available in some driving simulators (see Wood & Chaparro, this book, chap. 28). However, this obvious manipulation was not explored in the present study.

16.6 Methods—Wind Turbine Study

16.6.1 Participants

Participants for the Wind Turbine Study were recruited into the study using posters placed around the University of Calgary and nearby community centers. A total of 24 participants, stratified into the age groups of 18 to 25 (younger), 26 to 54 (middle-aged), and 55 and older (older), participated in the study. An equal number of men and women were in each age group. This sample of drivers was entirely different than the sample in the video billboard study.

The mean age of drivers in the younger age group was 22.4 ($SD = 1.8$), 30.9 ($SD = 4.5$) in the middle-aged group, and 65.5 ($SD = 7.1$) in the older age group. The characteristics of the participants in this study are summarized in Table 16.3.

16.6.2 Procedures

Participants drove two experimental drives each lasting approximately 12 minutes, which were counterbalanced across participants. The first experimental drive is illustrated in Web Figure 16.8.

TABLE 16.3   Wind Turbine Study

Group	N	Mean Age (SD)	Average km/year	Crashes (Last 5 years)	Moving Violations (Last 5 years)	Visual Acuity, Left Eye	Visual Acuity, Right Eye
Younger (18 to 25)	8	22.4 (1.8)	23,750 (20,234)	0.8 (0.9)	1.5 (1.9)	20/22 (0.2)	20/23 (0.4)
Middle-Aged (26 to 54)	8	30.9 (4.5)	21,550 (15,300)	0.4 (0.5)	1.1 (1.6)	20/23 (0.4)	20/24 (0.4)
Older (55 to 77)	8	65.5 (7.1)	14,375 (4,955)	0.3 (0.5)	1.1 (1.1)	20/26 (0.2)	20/24 (0.2)
Total	24	39.6 (19.6)	19,891 (14,834)	0.5 (0.7)	1.2 (1.5)	20/24 (0.3)	20/24 (0.3)

The second drive was similar but the order and location of events and billboards was varied. Each participant began in either a residential or industrial area with two lanes. They then drove through a suburban four lane road and entered a six lane freeway where the posted speed limit was 100 km/h. Drivers were told to stay in the right hand lane while on the freeway and were prompted to stay in that lane. The experimental decision to not allow drivers to change lanes to either the center or left lanes on the freeway ensured that measures of performance were uniformly taken from the lane closest to the wind turbines.

The freeway roadway was developed to resemble Highway 401 in Ontario (see Web Figure 16.9, left; **see color insert**). In particular, a barrier is present between the three lanes of traffic in either direction. Deciduous trees were placed on either side of the road to mimic the vegetation found along the 401. Static billboards are present along the route and serve as a comparison measure for participant's speed and lane variability in the presence of the wind farms.

Each modeled turbine was 5 meters in height, 12 meters from the road edge, and 15 meters from the next turbine in a line of turbines that extended for one-half kilometer along the roadway (see Web Figure 16.9, right; **see color insert**). The line of wind turbines incorporated several animation effects.

First, three blade speeds were modeled; namely: off, 60 rpm (slow) and 500 rpm (fast). (The fast speed produced a turbine that appeared to be moving at high speed, but not necessarily at 500 rpm due to frame rate limitations.) Within the line of turbines, two or three turbines (a group) were set to a given speed (e.g., slow). Adjacent groups of turbines were set to the next slower or faster speed so that groups of turbines followed a pattern of slow, fast, slow, off. Second, the turbines were rotated around the support structure or pole to simulate tracking the wind direction. A total of 33 turbines were installed; 17 of which tracked the wind as if it originated perpendicular to the roadway, and 16 as if the wind came parallel to the roadway and from behind the driver as they approached the turbines (see Figure 16.3 and Web Figure 16.10, right). Thus, the turbine blades faced the roadway or the approaching driver. The combined motion of all of the turbines in the wind farm, including the different blade speeds, was intended to collectively attract the attention of drivers.

The traffic or autonomous vehicles that participants encountered during the freeway drives mimicked the types of vehicles found on the 401 and was also similar to platooning behaviors typically found on freeways. For instance in light and moderate traffic flows, clusters or groups of vehicles often travel together

FIGURE 16.3   A close-up of a 15 kW wind turbine appears on the left. The line of turbine that was modeled in the UCDS is on the right.

while slower vehicles are overtaken and faster vehicles move on to other groups. A platoon or pod of 10 vehicles, including the lead vehicle in the right hand lane, was generated just prior to when the participant entered the freeway. A pod of vehicles would adhere to a number of rules including: moving to the right if between participant's vehicle and the lead vehicle (see below); and travel at 100 km/h if in the right lane, 105 km/h if in the center lane, and 110 km/h if in the left lane. At the interchanges, a number of autonomous vehicles exited the freeway and a similar number merged onto the freeway.

After several minutes of driving on the freeway, the lead vehicle, which was coupled to the velocity of the participant's vehicle at a 1.0 second time headway, braked at 8 m/s², which is a hard brake and required the participant or following driver to brake and/or steer to avoid a collision (see Web Figure 16.8, lower left image; **see color insert**). Drivers had to respond to the lead vehicle braking events both in the presence and absence of wind turbines. Two wind farms were present at different locations within each experimental drive (see Web Figure 16.8, lower right image; **see color insert**).

16.7 Results—Wind Turbine Study

16.7.1 Experimental Design and Analysis

The experimental design of the study was a 3 × 3 design with treatment (wind farm, billboards, baseline) as a within-subjects variable and age group (18–25, 26–54, 55+) a between-subjects factor. A repeated measures analysis of variance (RM ANOVA) was computed using SPSS (v. 14) using the General Linear Model (GLM). All follow-up analyses used the Bonferroni correction for multiple comparisons. If sphericity was violated, the Greenhouse-Geisser adjustment was used and is indicated with a G-G next to reported F tests. The analyses are organized into

lead vehicle braking perception response time, minimum headway distance, mean speed, location velocity, lane keeping, and debriefing questions.

16.7.2 Perception Response Time

Perception response time (PRT) was calculated in seconds from the onset of the lead vehicle braking until the participant responded by applying pressure to the brake pedal (Olson & Farber, 2003). Billboard was not a level of the treatment factor for the PRT analysis because the lead vehicle braking events only occurred at the wind farm and baseline levels. Braking events were limited to two conditions in order to avoid priming the participants from multiple braking events.

Figure 16.4 (Web Figure 16.11) illustrates that older drivers responded to the lead vehicle braking event slower than younger and middle-aged groups. Overall, age group was significant, $F (2, 21) = 6.3$, $p < 0.007$, but treatment ($p = 0.70$) and the treatment by age interaction ($p = 0.99$) were not. Simple effects tests indicated that age group was significant in the wind farms treatment, $F (2, 21) = 5.7$, $p < 0.011$ and specifically, the middle-aged group significantly differed from the older age group (p < 0.009). In the baseline condition, age group was not significant ($p = 0.12$), which is illustrated in part in Figure 16.4 by the overlap of the standard error bars.

16.7.3 Lead Vehicle Braking Collisions

A collision detection algorithm determined if the lead vehicle and the participant's vehicle collided. Out of 96 possible lead vehicle braking events, a single collision occurred in the presence of the wind farm. Analysis of the eye movements made by

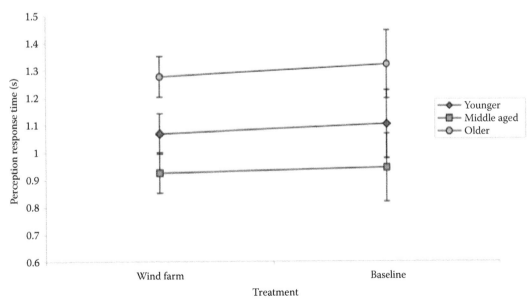

FIGURE 16.4 Perception response time (PRT) of drivers in the different age groups to the lead vehicle braking events in the presence of the wind farm and baseline treatment conditions. Standard error bars are shown.

the driver prior to the crash indicated that they were glancing at the speedometer just before the crash occurred. About six seconds prior to the crash, the same participant looked at the turbines and to their rear-view mirror.

16.7.4 Minimum Headway Distance

Minimum headway distance was measured in meters from the front bumper of the participants' vehicle to the lead vehicle rear bumper. The collection period for this variable started when the lead vehicle began to brake and ended when velocity reached zero or the participant removed his or her foot from the accelerator pedal. Minimum headway distance is indicative of how close a driver got to the lead vehicle when it braked. Billboard was not a level of the treatment factor for the minimum headway distance analysis because the lead vehicle braking events only occurred during the wind farm and baseline levels. The single collision identified in the PRT analysis was excluded from the analysis.

Overall, how close a driver got to the lead vehicle when it braked hard was significantly closer in the wind farm condition ($M = 13.5$ meters) than in the baseline condition ($M = 23.3$ meters), F (1, 21) = 41.7, $p < 0.000$. Age group ($p = 0.32$) and the age group by treatment interaction ($p = 0.082$) were not significant.

16.7.5 Mean Speed

Mean speed was collected in km/h during the presence of the wind farm, billboards and baseline treatment conditions. Data was collected during the one-half kilometer distance that the wind farm was located along the roadway. For the billboard, data was collected 100 meters before and after the sign. Baseline measures were collected for 200 meters at a number of matched locations.

Overall, the older driver group adopted slower average speeds ($M = 98.9$ km/h) in all three conditions compared to other age groups ($M = 99.6$, younger; $M = 103$, middle-aged) (see Figure 16.5 and Web Figure 16.12). Across treatment conditions, the slowest mean speed occurred in the wind farm condition ($M = 99.1$ km/h). The billboard ($M = 101.3$ km/h) and baseline ($M = 101.0$ km/h) conditions were similar.

The main effects of age group, F (2, 21) = 8.5, $p < 0.002$, and treatment, F (2, 42) = 4.0, $p < 0.027$, were significant, but the interaction between age and treatment was not ($p = 0.24$). Simple effects tests indicated that there were no age group differences in speed in the presence of the wind farms ($p = 0.12$). To determine if wind turbines are more distracting to older or younger drivers follow-up analyses were conducted. Significant effects between age groups were found for billboard, F (2, 21) = 6.8, $p < 0.005$, and baseline, F (2, 21) = 4.2, $p < 0.029$, conditions. For billboard, the middle-aged group differed significantly from the older age group ($p < 0.005$). This same pattern of significant group differences also held for the baseline condition ($p < 0.027$).

16.7.6 Location Velocity

Location velocity was collected in km/h at the 250 meter point of each wind farm, at the point perpendicular to the placement of the billboards and at matched baseline locations. Location velocity is similar to the analysis of mean speed, but is for a single location instead of over a distance surrounding an object of interest.

Age group, F (1, 21) = 9.03, $p < 0.001$, and treatment, F (2, 42) = 7.8, $p < 0.001$, were significant, but not the age group by treatment interaction ($p = 0.066$). Younger ($M = 98.4$ km/h) and older ($M = 98.0$ km/h) drivers had significantly lower speeds than middle-aged drivers ($M = 102.5$ km/h) ($p < 0.006$, $p < 0.003$, respectively). Drivers had significantly higher speeds

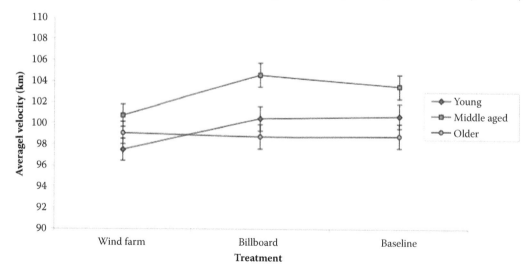

FIGURE 16.5 Age group differences in average velocity in km/h in the presence of the wind farms, billboards and baseline treatment conditions. Standard error bars are shown.

when they passed the billboard ($M = 101.4$ km/h) than when they passed the wind farm ($M = 98.0$ km/h) ($p < 0.002$), but not the baseline location ($M = 99.5$ km/h) ($p = 0.14$).

16.7.7 Lane Variability

Standard deviation of lane position (SDLP) in meters (m) was collected for the same treatment conditions as mean speed. SDLP is a measure that indicates how well a person maintains their lane position.

No significant differences were found for age group ($p = 0.50$), treatment ($p = 0.10$) or the age by treatment interaction ($p = 0.69$).

16.7.8 Debriefing Questions

Participants were interviewed when they debriefed at the close of the study. They were asked a number of questions including:

- Whether they recognized the road they had traveled on?
- Did they see the wind turbines?
- What they thought the purpose of the study was?

Drivers did not recognize the freeway that they had driven. Of those that guessed a highway ($N = 10$), most thought that it was Highway 1 near the mountains (i.e., TransCanada) or somewhere in Alberta (e.g., Edmonton). All participants said they saw the wind turbines while driving. However, none of the participants thought the study was about the potential impact of wind farms on driving distraction. Some said that the study was about reaction time or driver responses.

16.8 Discussion—Wind Turbine Study

16.8.1 Discussion

The purpose of this driving simulation study was to determine the effects of a wind farm on driving performance and is one of the first to do so. The discussion that follows is organized according to the objectives of the study.

The occurrence of a vehicle braking hard on a 100 km/h freeway is an unlikely event, but does represent a plausible worst case scenario. There were no differences in PRT to a lead vehicle braking whether the wind farms were present or not. However, drivers came significantly closer to the lead vehicle when the wind farm was present than when it was not (i.e., minimum headway distance). This may be because drivers were looking at the wind farm and following the lead vehicle too closely as a result. A single collision did occur in the presence of the wind farms when the lead vehicle braked. However, just prior to the collision, an analysis of eye movements found that the driver was looking inside the vehicle.

Drivers adopted slower speeds, probably due to the novelty of seeing the turbines, when driving by the wind farms. The

middle-aged group of drivers had the highest average speed in all conditions. The mean age of this group was relatively young ($M = 30.9$) compared to other samples and may account for the higher speed choice. Lane keeping, as measured by standard deviation of lane position (SDLP), was not affected by the presence of the wind farms.

Older drivers responded slower to the braking event and adopted slower speeds during freeway driving than other age groups. Results did not find an age by treatment interaction which would indicate that a particular age group was more or less affected by the presence of wind turbines.

The driving performance results of this study are similar to previous research on older drivers. As age increases, the speed adopted by drivers decreases in a variety of contexts (Caird et al., 2007) including freeway driving (Wasielewski, 1984). A slower response to a lead vehicle braking event was also found for older drivers by Strayer and Drews (2004), who were studying the impact of cell phone conversations on driver distraction. Older drivers respond to unexpected events slower than do other age groups (Olson & Sivak, 1986). In contrast to the lead vehicle braking, the one-half kilometer length of the wind farm evaluated in the study was not a surprising event. The turbines could be seen for about 20 seconds at about 100 km/h.

Drivers did look at the turbines when they passed by them. Overall, drivers slowed down slightly while doing so. Lane keeping in the right-hand lane next to the turbines was not affected. Not all drivers looked at the turbines when they drove by them. While some drivers momentarily looked at the wind turbines, the effect of doing so on a number of measures was minimal.

16.9 Conclusions

These reported studies are among the first to investigate novel external distractions such as video billboards and wind farms using driving simulation. Although these two studies were similar methodically, they produced somewhat different results. The video billboard study raised concerns regarding the safety of video billboards, specifically as they produced a higher incidence of rear-end collisions. This result was not found in the wind turbine study; however, minimum headway distance and speed reductions produced effects where the driver came closer to the lead vehicle when the wind farm was present and reductions in speed, which is considered an adaptation, respectively.

The lead vehicle braking events in both studies were programmed identically with the exception of the speed that participants were traveling at when the braking events occurred and the driving environment. In both studies the lead vehicle maintained a constant distance of 1 second and braked at 8 m/s^2 for 50 m. In the video billboard study the participant was traveling at 80 km/h through an industrial environment, while in the wind turbine study the participant was traveling at 100 km/h in a freeway environment. Participants in the wind turbine study traveling at 100 km/h had additional time to react to the braking event as the faster speed created more distance between the driver and lead vehicle.

Overall, the results of this study found minimal impact of wind farms on measures of driving performance. Compared to studies on cell phones and video billboards, small-scale wind farms placed on straight roadways affect driving performance to a lesser degree. The results obtained in the wind farm study are limited to small-scale turbines placed 12 meters from the edge of a straight freeway. The results do not generalize to large wind turbines (e.g., those 100 meters in height). One limitation of this study was that many drivers do not naturally stay in the right hand lane when they encounter slower vehicles. However, if lane change maneuvers were allowed a number of confounding variables would have been introduced into the study, thereby limiting the analysis and interpretation. Glances to the wind turbines from the center and left lanes are less likely because the turbines appear at a greater angle in the visual periphery.

The nature of the distraction in either study is somewhat different. Video billboards are designed to attract the attention of drivers. If looked at, video billboards provoke the driver to read and watch the changing images and text. Thus, video billboards can produce cognitive absorption for a brief period of time. A wind farm is present for a longer period of time and is interesting because of the movement of the turning blades. Not all drivers look at video billboards (Bejier et al., 2004; Smiley et al., 2004) and choosing to do so may redistribute glances from other sampled locations to where the billboard is located.

In theory, because drivers have spare attention, the frequency of glances to the roadway itself is preserved and the only shift of glances is from one side of the road to the other (Shinar, 2007). Thus, "spare" visual attention can be "re-allocated" to look at the billboard *without detriment to routine ongoing driving*. The question arises, however, what is "spare" and can attention be "re-allocated" when an emergency event occurs which requires the driver to respond quickly? If attention is directed to the billboard or wind farm at the same time that a lead vehicle brakes, how long will it take to detect the braking and can drivers respond adequately under the circumstances? In this kind of event-based test, attention is not assumed to be spare, or re-allocate-able at no cost to ongoing driving. Either the driver responds adequately or not. Further, glance behavior is considered merely descriptive. How long or how frequently a driver looks at a wind farm or billboard does not answer the question of whether looking is necessarily safe because no consequences necessarily follow. If the eyes are directed towards an external distraction and an emergency event cannot be adequately responded to, the external distraction is, by definition, not relatively safe. The value of the driving simulator, when used to test external distractions, is to determine relative safety when emergency events occur.

Future studies on external distractions may include methods from driving simulation, naturalistic studies and epidemiology. The extent that various distractions affect driving performance,

behavior and crash risk would be addressed. In particular, before-after studies (Hauer, 2002) that record crashes over time will hopefully resolve the relative safety of video billboards, wind turbines and wind farms. However, until more data on crashes, adverse behavior and performance decrements can be collected, the absolute safety of video billboards and wind farms has not been conclusively determined.

16.10 Recommendations

1. As a practical matter, the placement of video billboards or wind farms should be carefully considered to avoid locations where drivers need to attend to roadway geometry and traffic control devices. Particularly, placement at intersections, merges, curves (horizontal or vertical), and work zones is potentially dangerous. Roadways that have pre-existing accident rates and severity scores that are above norms should not be considered for video billboard installations (Farbry et al., 2001) or wind farms. In addition, these installations should not be visible to drivers while they are engaged in vehicle maneuvers or decision-making activities such as traffic sign reading.

2. More research is needed using valid before-and-after statistical methods (see, e.g., Hauer, 2002) that determine whether or not a relationship exists between crashes and the presence of wind farms and video billboards. The single study on wind farms by George (2007) and those on video billboards reviewed by Farbry et al. (2001) and found by Smiley et al. (2005) were equivocal and more research is needed.

3. Wind farms are currently novel and more likely to attract attention. Static billboards are more common and are frequently ignored by drivers. Additional research is needed to determine whether drivers exposed to wind farms over weeks and months continue to look or ignore them. Drivers may simply ignore wind farms with increased familiarity to them. Eventually, a mixture of drivers who have and have not seen the turbines will be driving by them. Additional research is needed to determine whether this is an issue.

Key Points

- The value of the driving simulator, when used to test external distractions, is to determine relative safety when emergency events occur.
- The absolute safety of video billboards and wind farms has not been determined. Additional research using driving simulation, instrumented vehicles, and injury epidemiology is still needed.
- As a practical matter, the placement of video billboards or wind farms should be carefully considered to avoid locations where drivers need to attend to roadway geometry and traffic control devices.

Keywords: External Distractions, Video Billboards, Wind Farms

Acknowledgments

The authors would like to thank SkyPower, and Charmaine Thompson, in particular, for funding the Wind Turbine study. Elise Teteris contributed to the experimental design, programming, running and analyzing the Wind Turbine study. We wish to also thank Amanda Ohlhauser who spent many hours reducing data for the Wind Turbine study. Daniel MacAulay and John Michalak helped to run the Billboard study. Bob Dewar provided valuable edits of the Billboard study. Don Fisher, Andrew Mayer and John Lee provided valuable critiques of previous versions of this chapter. Finally, thanks to Susan Chisholm for providing statistical analysis for both studies.

Glossary

Driver distraction: External distraction.
Driver distraction: The momentary or transient redirection of attention from the task of driving to a thought, object, activity, event or person (Caird & Dewar, 2007, p. 196).
External distraction: Those objects that draw the attention of the driver *outside* of the vehicle.

Web Resources

The *Handbook*'s web site contains supplemental materials for this chapter including a number of the printed figures in color, and additional web-only figures.

Web Figure 16.1: Times Square with a variety of street signs, conventional billboards and digital billboards.

Web Figure 16.2: Several rows of wind turbines located near Pincher Creek, Alberta, Canada.

Web Figure 16.3: Static billboard in driving scenario (left), video billboard (middle and right) in driving scenario.

Web Figure 16.4: Lead vehicle before (left) and during (right) a braking event in video billboard condition. (Color version of printed Figure 16.1).

Web Figure 16.5: Speed change sign with no billboard (left), and in the presence of a video billboard (right).

Web Figure 16.6: Perception response time(s) to lead vehicle braking event by order and condition.

Web Figure 16.7: Minimum headway distance (m) to lead vehicle braking event by order and condition. (Color version of printed Figure 16.2)

Web Figure 16.8 (see color insert): Experimental Drive 1 is shown from start to end with associated road types. The locations of the lead vehicle braking events, billboards, and wind farms are also indicated.

Web Figure 16.9 (see color insert): A picture of Highway 401 near the town of Puslinch southeast of Toronto (left). A similar section of freeway modeled in the University of Calgary Driving Simulator (UCDS) (right).

Web Figure 16.10: A close up of a 15 kW wind turbine appears on the left. The line of turbine that was modeled in the UCDS is on the right.

Web Figure 16.11: Perception response time (PRT) of drivers in the different age groups to the lead vehicle braking events in the presence of the wind farm and baseline treatment conditions. Standard error bars are shown. (Color version of Figure 16.4).

Web Figure 16.12: Age group differences in average velocity in km/hr in the presence of the wind farms, billboards and baseline treatment conditions. Standard error bars are shown. (Color version of Figure 16.5).

Key Readings

Caird, J. K., & Dewar, R. E. (2007). Driver distraction. In R. E. Dewar, & R. Olson (Eds.), *Human factors in traffic safety* (2nd ed., pp. 463–485). Tucson, AZ: Lawyers & Judges Publishing.

Farbry, J., Wockinger, K., Shafer, T., Owens, N., & Nedzesky, A. (2001). *Research review of potential safety effects of electronic billboards on driver attention and distraction.* Final Report. Washington, DC: Federal Highway Administration.

Smiley, A., Persaud, B., Bahar, G., Mollett, C., Lyon, C., Smahel, T., & Kelman, W. L. (2005). Traffic safety evaluation of video advertising signs. *Transportation Research Record, 1931,* 105–112.

References

Beijer, D. D., Smiley, A., & Eizenman, M. (2004). Observed driver glance behavior at roadside advertising. *Transportation Research Record, 1899,* 96–103.

Caird, J. K., Chisholm, S., Edwards, C., & Creaser, J. (2007). The effect of yellow light onset time on older and younger drivers' perception response time (PRT) and intersection behavior. *Transportation Research: Part F, 10*(5), 470–483.

Caird, J. K., & Dewar, R. E. (2007). Driver distraction. In R. E. Dewar & R. Olson (Eds.), *Human factors in traffic safety* (2nd ed., pp. 463–485). Tucson, AZ: Lawyers & Judges Publishing.

Caird, J. K., Willness, C., Steel, P., & Scialfa, C. (2008). A meta-analysis of cell phone use on driver performance. *Accident Analysis & Prevention, 40,* 1282–1293.

Chisholm, S. L., Caird, J. K., & Lockhart, J. (2008). The effects of practice with MP3 players on driving performance. *Accident Analysis & Prevention, 40,* 704–713.

Crundall, D., van Loon, E., & Underwood, G. (2006). Attraction and distraction of attention to roadside advertisements. *Accident Analysis & Prevention, 38*(4), 671–677.

Eby, D. W., & Kostyniuk, L. P. (2003). *Driver distraction and crashes: An assessment of crash databases and review of the literature* (Rep. UMTRI-2003-12). Ann Arbor, MI: University of Michigan Traffic Research Institute.

Farbry, J., Wockinger, K., Shafer, T., Owens, N., & Nedzesky, A. (2001). *Research review of potential safety effects of electronic billboards on driver attention and distraction* (Final Rep.). Washington, DC: Federal Highway Administration.

George, D. (2007, May 3). Claims that wind farms distract drivers are blown off course. *IHIE focus*, p. 20.

Green, P. (2007). Where do drivers look while driving (and for how long)? In R. E. Dewar & R. Olson (Eds.), *Human factors in traffic safety* (2nd ed., pp. 57–82). Tucson, AZ: Lawyers & Judges Publishing.

Hauer, E. (2002). *Observational before—after studies in road safety*. Amsterdam: Pergamon.

Ho, G., Scialfa, C. T., Caird, J. K., & Graw, T. (2001). Visual search for traffic signs: The effects of clutter, luminance, and aging. *Human Factors, 43*(2), 194–207.

Kline, D., & Dewar, R. (2004). The aging eye and transport signs. In C. Castro & T. Horberry (Eds.), *The human factors of transport signs* (pp. 25–48). New York: CRC Press.

Olson, P. L., & Farber, E. (2003). *Forensic aspects of perception and response* (2nd ed.). Tuscon, AZ: Lawyers & Judges Publishing.

Olson, P. L., & Sivak, M. (1986). Perception-response time to unexpected roadway hazards. *Human Factors, 28*, 91–96.

Shinar, D. (2007). *Traffic safety and human behavior*. Amsterdam: Elsevier.

Smiley, A., Persaud, B., Bahar, G., Mollett, C., Lyon, C., Smahel, T., & Kelman, W. L. (2005). Traffic safety evaluation of video advertising signs. *Transportation Research Record, 1931*, 105–112.

Smiley, A., Smahel, T., & Eizenman, M. (2004). The impact of video advertising on driver fixation patterns. *Transportation Research Record, 1899*, 76–83.

Strayer, D .L., & Drews, F. A. (2004). Profiles in driver distraction: Effects of cell phone conversations on younger and older drivers. *Human Factors, 46*(4), 640–649.

Stutts, J. C., Knipling, R. R., Pfefer, R., Neuman, R. T., Slack, K. L., & Hardy, K. K. (2005). *NCHRP Report 500: Guidance for implementation of the AASHTO Strategic Highway Safety Plan: Volume 14: A guide for reducing crashes involving drowsy and distracted drivers*. Washington, DC: National Cooperative Highway Research Program, Transportation Research Board.

Stutts, J. C., Reinfurt, D. W., Staplin, L., & Rodgman, E. A. (2001). *The role of driver distraction in traffic crashes*. Washington, DC: AAA Foundation for Traffic Safety.

Tabachnick, B. G., & Fidell, L. S. (2006). *Using multivariate statistics* (5th ed.). Boston, MA: Pearson, Allyn & Bacon.

Tijerina, L., Johnson, S., Parmer, E., & Winterbottom, M. D., & Goodman, M. (2000). *Driver distraction with wireless telecommunications and route guidance systems* (Rep. DOT-HS-809-069). Washington, DC: National Highway Transportation Safety Administration.

Wasielewski, P. (1984). Speed as a measure of driver risk: Observed speeds versus driver and vehicle characteristics. *Accident Analysis & Prevention, 16*, 89–103.

17

Measuring Physiology in Simulators

Karel A. Brookhuis
University of Groningen

Dick de Waard
University of Groningen

Abstract

The Problem. Almost all accidents can be traced back to human error or dysfunctioning. Many accidents are due to suboptimal condition of the operator. More specifically, drivers often conduct their driving task while they are not in adequate shape. Their mental condition may be inappropriate for the task condition, for instance, being stressed, fatigued, affected by external factors such alcohol or medicine, or simply distracted and/or not alert. It is difficult and dangerous, if not unethical, to study the circumstances and condition of drivers while they function inadequately in the actual driving environment. *Role of Driving Simulators*. Driving simulators are an ideal tool for the study of drivers in conditions of increasing accident likelihood because of poor mental condition. Driving circumstances of extreme task conditions can be created under strict experimental control on the one hand while meticulous measurement of drivers' physical and mental condition are feasible through various physiological parameters. Driving simulators are, in principle, no less than adequate, well-equipped laboratories, suited for almost any psychophysiological measurement. *Key Results of Driving Simulator Studies*. Driving simulator studies with concomitant psychophysiological measurement have been gaining popularity in recent years. Mental workload measures such as heart rate and heart-rate variability are now becoming reasonably common, mainly because of the relative ease of registration, enabling detection of circumstances that demand mental effort as instigated by the driving task and situational conditions. Registration of refined measures such as brain potentials enable close monitoring of information-processing capabilities upon relevant (and irrelevant!) signals in the driving environment. Gradually developing drowsiness is detected by means of frequency analysis of the electroencephalogram. Numerous studies have proved the fruitful combination of psychophysiological measurement in driving simulators. *Scenarios and Dependent Variables*. Often scenarios in this type of research involve some rather extreme task conditions. To measure drowsiness, for instance, extended driving trial times are included or, sometimes, nighttime driving. To measure stress, information overload, sudden unexpected events, dangerous maneuvers, etc. are quite common. Finally, effects of alcohol and drugs are almost necessarily tested in simulators, although sometimes trials on the road under strictly controlled conditions are conducted in some countries. Commonly, standard driving performance parameters in terms of longitudinal and lateral control are registered, alongside physiological measures and subjective measures to check whether drivers suspect what is going on regarding their constitution. *Platform Specificity and Equipment Limitations*. In principle no high demands on the platform and equipment are necessary for physiological measurements in simulators, except the physiological instrumentation itself. In a wider sense though, laboratory circumstances around the simulator, including light control and shielding against (electrical) noise, are necessary.

17.1 Introduction

Inadequate performance, labeled as human error by some, constitute the major cause of the majority of the accidents in the air, on the road, and on waterways (Smiley & Brookhuis, 1987), grossly because of imperfect perception, insufficient attention and inadequate information processing. A common factor in this field is the operator's mental condition; that is, (high or low) mental workload, (inadequate) mental activation or arousal, externally-driven mental conditions such as psycho-active substances, and emotion and motivation. Non-optimal mental workload, activation or condition are, in a broad sense, generally at the basis of poor performance, while an adequately workable relationship with accident causation is not easily established let alone measured in (traffic) research practice or even in (driving) simulators. De Waard and Brookhuis (1997) discriminated between underload and overload; with the former leading to reduced activation, and lowered alertness and attention, the latter to stress, distraction, diverted attention and insufficient time for adequate information processing. Both factors have been studied in relationship to driver impairment; however, the coupling to accident causation has not been established via a direct link (see also Brookhuis, De Waard, & Fairclough, 2003), mainly because of ethical concern, although simulators provide a perfect opportunity in that respect. Criteria for when impairment is below a certain threshold, leading to accidents, has to be established. Only then can accidents and mental activation and/or workload (high or low) be related to each other, in conjunction with their origins such as information overload, stress, underload, boredom, fatigue, or factors such as alcohol and drugs. The traffic environment and traffic itself will only gain in complexity, at least in the near future, with the rapid growth in numbers of automobiles and electronic applications. The study of the consequences of this forecast will largely concentrate in simulators because of their flexibility, certainly with respect to manipulating critical driving conditions and measuring (mental) workload and activation. Physiological measures are a natural type of index for mental activity and/or workload, since any work or activity, including mental work and mental activation, demand physiological activity by definition. The measurement of physiological parameters in simulators profits from the laboratory circumstances on the one hand while gaining in validity depending on the degree of reality of the type of simulator used on the other hand.

17.1.1 Measurement of Physiological Parameters

The various methods of measuring physiology have been copied from the medical field for human factors and ergonomics purposes in order to study operators in workplaces with respect to workload, or more specifically, mental workload. There are many reasons why the measurement of operators' mental workload earns great interest these days, and will increasingly enjoy this status in the near future because of the working conditions in the last few decades. Firstly, the nature of work has changed,

or has at least extended from physical (e.g., measured by muscle force exertion, this chapter) to cognitive (e.g., measured in brain activity), a trend that has not reached a ceiling yet. Secondly, accidents in workplaces of all sorts are numerous, costly and seemingly ineradicable, and in fact largely attributable to the victims themselves, human beings. Thirdly and most well-known, human errors related to mental workload in the sense of inadequate information processing are the major causes of the majority of accidents.

As long as 35 years ago, Kahneman (1973) defined mental workload as directly related to the proportion of mental capacity that an operator allocates to task performance. The measurement of mental workload is the specification of that proportion (O'Donnell & Eggemeier, 1986; De Waard & Brookhuis, 1997), in terms of the costs of the cognitive processing, which is also referred to as mental effort (Mulder, 1986). Mental effort is similar to what is commonly referred to as doing your best to achieve a certain target level, to even "trying hard" in the case of a strong cognitive processing demand, reflected in several physiological measures. The concomitant changes in effort will not show easily in work performance measures because operators are inclined to cope actively with changes in task demands, for instance in traffic, as drivers do by adapting their driving behavior to control safety (Cnossen, Brookhuis, & Meijman, 1997). However, the changes in effort will be apparent in self-reports of drivers and a fortiori in the changes in certain physiological measures, such as the activity in certain brain regions but also heart rate and heart rate variability (cf. De Waard, 1996).

Mulder (1986) discriminated between two types of mental effort; namely, the mental effort devoted to the processing of information in controlled mode (computational effort) and the mental effort needed when the operator's energetical state is affected (compensatory effort). Computational effort is exerted to keep task performance at an acceptable level; for instance, when task complexity level varies or secondary tasks are added to the primary task. In case of (ominous) overload, extra computational effort could forestall safety hazards in such a way. Compensatory effort takes care of performance decrement in the case of fatigue, for instance (up to a certain level). Underload by boredom, affecting the operator's capability to deal with the task demands, might be compensated for as well. In case effort is exerted, be it computational or compensatory, both task difficulty and mental workload will be increased. Effort is a voluntary process under control by the operator, while mental workload is determined by the interaction of operator and task. As an alternative to exerting effort, the operator might decide to change the (sub)goals of the task. Adapting driving behavior as a strategic solution is a well-known phenomenon. For example, overload because of an additional task, such as looking up telephone numbers while driving, is demonstrably reduced in a simulator by the lowering of vehicle speed (see De Waard, Van der Hulst, & Brookhuis, 1998, 1999).

The relationship between mental activation, mental workload and performance is depicted in Figure 17.1 (see Web Figure 17.1 on the *Handbook* web site for a color version), together with an

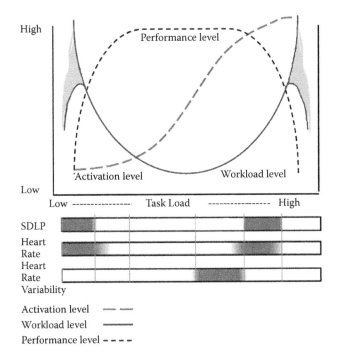

FIGURE 17.1 The relationship between activation level, workload and performance (upper part), and specific measures from the simulator computer and physiological activity derived from heart rate measures (lower part).

account of what some of the measurements that are feasible in simulators might indicate at various levels.

Physiological measures are the most natural type of workload index, since working, whether physical or mental, causes physiological activity. Physical (and also mental) workload has, for instance, a clear impact on heart rate and heart rate variability (Mulder, 1986, 1988, 1992; De Waard & Brookhuis, 1991), on galvanic skin response (Boucsein, 1992), blood pressure (Rau, 2001) and respiration (Mulder, 1992; Wientjes, 1992). Depending on the circumstances, mental workload might increase heart rate and decrease heart rate variability (HRV) at the same time (Mulder, De Waard, & Brookhuis, 2004). Other measures of major interest are event-related phenomena in the brain activity (Kramer, 1991; Kramer & Balopolsky, 2004; Noesselt et al., 2002) and environmental effects on certain facial, task irrelevant muscles (Jessurun, 1997).

17.1.2 Methodology of Physiological Measurement

In this chapter, the methodology of measuring a number of relevant physiological parameters, as applied in simulators, is elaborated to a limited extent. They are the cardiovascular parameters heart rate and heart rate variability, (electro) cortical parameters (i.e., frequency shifts in the electroencephalogram), event-related potentials and functional imaging, galvanic skin responses, blood pressure responses, respiration rate, eye and eyelid movements, and muscle activity.

The elaboration of measures starts in Section 17.2.1, with a short overview of a very old method, the measurement of electrical phenomena in the skin. The result comprises galvanic skin response (GSR), skin potential, peripheral autonomic surface potentials, and so forth, all with the aim to study electro-dermal activity. The latter can be regarded as a psychophysiological indicator of arousal, stress-strain processes, and emotion. The measurement of electro-dermal activity is used to investigate orienting responses and their habituation, for studying autonomic conditioning, for determining the amount of information-processing capacity needed during a task, and for determining arousal/stress levels, especially in situations evoking emotions. It has also been used for measuring workload, mental strain, and, specifically, for emotional strain; whereas increases in certain types of electro-dermal activity indicate readiness for action. The second section studies electromyography (EMG), that is, muscle function through the analysis of the electrical signals emanated during muscular contractions. EMG is commonly used in ergonomics and occupational health research for reasons of its non-invasiveness, allowing convenient measures of physical effort during movements, and non-invasively detecting physiological reactions caused by mentally-controlled processes. Specific muscles in the face, for instance, have been found to reflect aspects of workload in various conditions. In Section 17.2.3 heart rate is the central topic to be derived from the electrocardiogram (ECG), which reflects the (electrical) activity of the heart. For the assessment of mental effort it is not the ECG itself but the variation in time duration between heartbeats that provides the most interesting information (Mulder, 1992). During task performance, operators have to spend (mental) effort, which is usually reflected in increased heart rate and decreased heart rate variability, when compared to resting situations. The general cardiovascular response pattern that is found in many mental effort studies can be characterized by an increase of heart rate and blood pressure and a decrease of HRV and blood pressure variability in all frequency bands. This pattern is comparable with a defense reaction and is predominantly found in laboratory studies using short-lasting tasks requiring effortful mental operations in working memory.

Section 17.2.4 is on the evaluation of workload by means of ambulatory assessment of blood pressure. This type of ambulant technique permits the assessment of behavioral, emotional, and activation interaction with workload under real work conditions. Carryover effects of workload on activities, behavior, and strain after work, as well as on recovery during rest, can be measured. This implies an enhancement of the load-strain paradigm from short-term effects—such as fatigue, boredom, vigilance, and so forth—to long-term effects of work, such as disturbed recovery processes after work and cardiovascular health diseases, diabetes mellitus, depression, and so on. In Section 17.2.5 the measurement of respiration in applied research is studied. Respiratory measurement is potentially a powerful asset as it seems closely related to a variety of important functional psychological dimensions, such as response requirements and appraisal patterns. Respiratory measures may provide valuable (supplementary) information to alternatives such as subjective measures, (other) measures of operator workload such as heart rate, and to stressful or potentially hazardous aspects of task environments.

Section 17.2.6 is on mental chronometry, using event-related potentials (ERP), derived as a transient series of voltage oscillations in the brain, which can be recorded from the scalp, in response to discrete stimuli and responses. Some ERP components, usually defined in terms of polarity and latency with respect to discrete stimuli or responses, have been found to reflect a number of distinct perceptual, cognitive and motor processes thereby proving useful in decomposing the processing requirements of complex tasks. ERPs are being used to study aspects of cognition that are relevant for Human Factors and Ergonomics research such as vigilance, mental workload, fatigue, adaptive aiding, stressor effects on cognition, and automation.

The final section is about alertness monitoring by certain measures of ocular psychophysiology that have been identified for their potential to detect minute-to-minute changes in drowsiness and hypo-vigilance, associated with lapses of attention and diminishing alertness during performance. A measure of slow eyelid closure, referred to as percentage of closure (PERCLOS) correlates highly with visual vigilance performance lapses, and is now increasingly used to monitor operators in their working environment, such as (professional) drivers. (See also, this book, chap. 29 by Matthews, Saxby, Funke, Emo, & Desmond, "Driving in States of Fatigue or Stress".)

The selection criteria for inclusion in this chapter were non-intrusiveness and more or less regularly proven effects in relation to (mental) work conditions as studied in simulators. Measurement of some of the included physiological parameters is relatively easy, or at least feasible in the simulator environment; however, it is cumbersome for some measures of brain activity (event-related potentials in the electroencephalogram) or extremely difficult (magnetic resonance imaging within the cortex), at least for the time being. All of them, however, are relevant within the context of this research on simulators.

17.2 Physiological Measures

17.2.1 Galvanic Skin Response (GSR)

The most common term for all electrical phenomena measured on the skin is electro-dermal activity (EDA). EDA comprises the well-known galvanic skin response (GSR), skin potential, peripheral autonomic surface potentials, and so forth. It is easily measured with or without the application of external voltage to the skin, while a constant external voltage of about half a Volt is most widely used. Electro-dermal recording is actually a very old physiological method, dating back to the turn of the nineteenth century into the twentieth and has since then been used as a psychophysiological method (for an extensive

overview, see Boucsein, 1992, 2004). GSR is generated by sweat gland activity, originating from the sympathetic autonomous nervous system. GSR is considered by most researchers to be a psychophysiological indicator of arousal, stress-strain and emotion, independent of the parasympathetic autonomous nervous system. GSR has been measured to investigate orienting responses and following habituation, to study conditioning, to estimate information processing capacity used in a task, and to establish arousal or stress level, specifically in (negative) emotional situations (Boucsein, 1992). In the human factors and ergonomics area GSR has been used to monitor task load and (emotional) mental strain (Boucsein & Backs, 2000). In the area of task-related information processing, increases in phasic GSR reflect attention direction to a stimulus, whereas increases in tonic GSR indicate a high arousal level; that is,. readiness for action (Boucsein, 2004).

GSR is easy to measure but, depending on the study objectives, not always easy to interpret (for extensive guidelines see Boucsein, 1992). The major drawback for simulator use is, similar to the use in cars in reality, that the electrodes are fixed to one hand (or sometimes on the sole of one's foot), which is thereupon destined to be unusable. However, driving with one hand is feasible as long as the vehicle has automatic gears.

17.2.2 Electromyography (EMG)

Electromyography (EMG) is based on the analysis of electrical signals that are emanated during muscular contractions. EMG is used to detect muscle activity. Electrical potentials evoked during voluntary muscular contraction were observed already a long time ago but it required advanced semiconductor technology to reveal significant information from the complex and noisy electrode signal (for an overview see Goebel, 2004). Over the past several decades, sophisticated electronic instrumentation and powerful analysis methods allow the use of EMG in applied research environments such as simulators.

Muscle force is built by activity of a number of motor units, discharging alternating with different rates (5–50/s). Surface electrodes are mostly used that collect the sum of different motor unit potentials under the skin. The larger the electrode and the larger the distance between muscle fibres and electrode, the more motor units will be covered. The asynchronous activity of the different motor units sums to an interference pattern at the electrode, which can be described as a train of quasi randomly shaped spikes varying in amplitude and duration (Kramer, Kuchler, & Brauer, 1972). The average EMG signal amplitude increases with muscle contraction. Next to the average, sophisticated signal processing and pattern analysis may deliver more information about muscle activation and exertion.

Despite its inherent noisy character, relevant information can be processed from the EMG signal. In that sense it is important to realize that EMG measures voltage, representing local muscle recruitment, and does not measure force or joint position. In applications such as "cockpit" design, EMG may be used in simulators for design and scheduling of operator processes with respect to (static and dynamic) muscle load, muscle fatigue or overload, timing and coordination, and motor unit recruitment patterns, explaining low-level muscle fatigue and mentally-induced strain (Goebel, 2004). The activity of some selected facial muscles are indicative of task load (frontalis) while other facial muscles appear to be related to both workload and emotional effects (corrugator supercilii), as demonstrated by Jessurun, Steyvers and Brookhuis (1993), Van Boxtel and Jessurun (1993), and De Waard, Jessurun, Steyvers, Raggatt and Brookhuis (1995). An alternative method of monitoring emotions while driving in a simulator is online video analysis of facial expressions, which is not yet developed to its full potential but appears to be promising (Van Kuilenburg, Den Uyl, Israël, & Ivan, 2008).

EMG is additionally used as representing eye movements by placing electrodes directly next to and just above the eye(s) (Jessurun, 1997). Exact determination of a fixation point is not without problems with this method; for instance, frequent calibration will be necessary. However, (online) video-analysis of eyeball movement faces similar problems, though to a lesser degree (Mallis & Dinges, 2004).

17.2.3 Electrocardiogram (ECG)

The electrocardiogram (ECG) reflects the electrical activity of the heart as can be measured relatively easily through three electrodes attached to the human chest. Activity of the heart is based on the standard rhythm as instigated by the cardiac sino-arterial node, modulated by innervations from both the sympathetic and parasympathetic (vagal) activity of the autonomous nervous system, dependent on requirements because of physical and/or mental effort. Measuring physical effort is not opportune most of the times in simulators; for the assessment of mental effort the time duration between individual heartbeats provides the relevant information. The heart rate (HR) is the number of heartbeats within a fixed period of time (usually a minute); while mean heart rate or inter-beat interval (IBI) is the average time duration of the heartbeats in that period. Heartbeats have variable time durations with different oscillation patterns, leading to time series with source-characteristic patterns and frequency contents (e.g., Kramer, 1991), called heart rate variability (HRV). During task performance, when subjects have to exert mental effort, increased HR and decreased HRV, in particular in the 0.10 Hz frequency band, are usually clear in comparison to resting situations, effects dependent on amount and type of effort (for an extensive overview see Mulder, 1992; Mulder et al., 2004). In simulator environments HR and HRV are relatively easily measured by means of a few (AgAgCl)-electrodes and analyzed according to mental effort as exerted for task requirements. In Figure 17.2 an example is given of the effects of different task demands in simulated traffic environments, depending on situational events and challenges. The drivers clearly show effects of demanding tasks. In particular, the effect on heart rate of two gap-acceptance tasks can be seen: Judging when to cross a junction and when to accept a gap between meeting cars to turn left increased average heart rate by

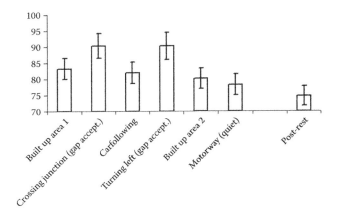

FIGURE 17.2 Average heart rate of 13 participants during a ride in the driving simulator. Data of the control group, also mentioned in Brookhuis et al. (2004), are depicted.

five beats per minute, a very large effect, similar to the effect of driving compared with the resting measurement. Similar effects of increased task demands were found on HRV (Brookhuis, De Waard, & Samyn, 2004).

17.2.4 Blood Pressure

Ambulatory blood pressure monitoring (ABPM) was originally developed for clinical purposes. For over 25 years it has served as an ambulatory assessment of strain or stress effects during and after work in relation to workload (for an extensive overview, see Pickering, 1991; Rau, 2001, 2004). The technique may well be used to investigate psychosocial work consequences with respect to cardiovascular diseases. The majority of work-related ABPM studies are based on the job demand and decision latitude model as proposed by Karasek and co-workers (Karasek, 1979; Karasek & Theorell, 1990). Theorell et al. (1991), Pickering (1991), Rau (1996), Schnall, Schwartz, Landsbergis, Warren and Pickering (1998), and Belkic, Emdad and Theorell (1998) report sympathetic nervous activity to be increased during exposure to high job strain, that is the result of high workload combined with low-decision latitude or control, which in turn relates to hypertension. ABPM is additionally linked to effects of social support, of perceived control, of effort-reward balances, and of sequential and hierarchical completeness of tasks (Rau, 2001, 2004). ABPM is conducted with the aid of portable recorders for automatic, repeated, non-invasive registration of arterial blood pressure; that is, the level of systolic and diastolic blood pressure, and sometimes also the arterial mean blood pressure. As such, it is fairly feasible to use ABPM in simulators, specifically when carrying out research on demanding driving tasks such as the extended driving of long haul freight vehicles.

17.2.5 Respiration

There are two parameters of respiratory assessment; namely, the establishment of the contribution to ventilation by depth and frequency of breathing and the measurement of parameters associated with gas exchange (see also Wientjes & Grossman, 2004). Depth of breathing is usually expressed in terms of tidal volume (i.e., the volume displaced per single breath), and frequency in terms of respiration rate (the number of breaths per minute). The volume of air displaced per minute is expressed as minute ventilation, which is the product of tidal volume and respiration rate, and typically reflects metabolic activity. In addition to these basic parameters, respiratory measurement often includes a more sophisticated evaluation of the breathing cycle in terms of the duration of the phases of the breathing cycle (inspiratory and expiratory time), total cycle time, mean inspiratory flow rate (tidal volume/inspiratory time) and duty cycle time. Measurement of gas exchange includes assessment of the volume, or quantity of oxygen (VO_2) consumed per time unit, and of the quantity of carbon dioxide (VCO_2) produced. These measures may be used to calculate the energy expenditure (for an extensive overview see Wientjes, 1992; Wientjes & Grossman, 2004). Specifically in high-demanding circumstances such as fighting situations in flight simulation, these measures grant the researcher considerable flexibility and measurement margin, whereas HR and HRV may run into ceiling effects.

17.2.6 Electroencephalogram (EEG) and Event-Related Potentials (ERP)

The analysis of the "raw" electroencephalogram, or background EEG (i.e., the collection of low-voltage oscillations between about 1 Hz and 30 Hz), is specifically useful for and indicative of level of activation of the brain. While the event-related potential (ERP) is a transient series of voltage oscillations in the brain and is to be discriminated from the background EEG in response to discrete stimuli and responses. EEG and ERPs are recorded from the scalp through (AgAgCl)-electrodes that, due to the low voltage, have to be amplified considerably, in the order of 1000×, which has far-reaching consequences for the measurement procedures and circumstances. If EEG is to be measured in driving simulators, the environment preferably has to be electrically shielded in order to avoid amplified noise of, for instance, the common 50 Hz (in Europe); otherwise filtering of the data is the only solution. Measuring EEG, even in laboratory circumstances is relatively demanding with respect to skills and facilities.

The content of the background EEG is usually subdivided in bins; from 1 to 5 Hz is called Delta-waves, 5–8 Hz Theta, 8–12 Hz Alpha, and above 12 Hz Beta, in various subcategories that are not reported here in the light of the restricted use for research in driving simulators. When Beta activity is predominant, the participant in the study is generally awake and alert, while the activity dropping to Alpha indicates developing drowsiness, and going further down into the Theta region may lead to falling asleep. Delta waves are normally an indication of various phases of actual sleep. The background EEG is by definition considered the most appropriate measure to monitor alertness c.q. vigilance state of operators in task situations (Åkerstedt, 2004).

Specific ERP components, usually defined in terms of polarity (P or N) and latency (in msec) with respect to a discrete stimulus or response, have been found to reflect a number of distinct perceptual, cognitive and motor processes, thereby providing useful information for decomposition of processing requirements in complex task situations (Fabiani, Gratton, & Coles, 2000), such as driving motor vehicles in various conditions. ERPs have been used to study effects on cognition and performance for more than 40 years now, successfully relating brain activity with operator performance under vigilance conditions, with variable mental workload, while fatigued, with adaptive driving support, and all kinds of stressors, and most importantly, automation (see Wilschut, 2009; Kramer & Belopolsky, 2004; Kramer & Weber, 2000; Byrne & Parasuraman, 1996; Brookhuis, 1989). For example, Brookhuis et al. (1981) found that increasing task load on a letter search task resulted in increases in the latency but in decreases in the amplitude of a late positive component of the ERP, the P300, alternatively labeled P3b. Kramer (1991) and Sirevaag, Kramer, Coles and Donchin (1987) found different effects in the amplitude of P300s elicited by primary and secondary tasks. Wilschut (2008) reported P300 effects of task load in a secondary task while participants in her driving simulator experiment performed lane-changing tasks on a simple highway. In Figure 17.3 the amplitude of the P300 is smaller at higher task loads, at least when the P300 is derived stimulus-locked (upper panel), while the response-locked derived P300 displays a completely different picture.

17.2.7 Eye and Eyelid Motions

The standard indicator for the level of alertness is the activity pattern in the EEG (see Lammers et al., 2005). Most applications in alertness or vigilance research note the changes that follow in the usual frequency bands of the EEG during wakefulness, such as Beta, Alpha and Theta. However, as mentioned before, measuring EEG, even in simulator environments, is relatively demanding and expensive. Facial muscle tone and activity can also serve to measure effort and emotional strain (Van Boxtel & Jessurun, 1993), whereas position and eyelid activity can also be used to detect sleepiness (see Åkerstedt, 2004). However, some easily-derived measures of ocular psychophysiology have the potential to accurately detect minute-to-minute changes in drowsiness and hypo-vigilance that are associated with lapses of attention during performance. Specifically, a measure of slow eyelid closure, labeled as percentage of closure or PERCLOS (Wierwille, Ellsworth, Wreggit, Fairbanks, & Kirn, 1994; Wierwille & Ellsworth, 1994), correlates highly with visual vigilance performance lapses (Dinges, Mallis, Maislin, & Powell, 1998; Mallis & Dinges, 2004). PERCLOS is operationalized by video-based scoring of slow eyelid closures, and correlates better to performance decrement than participant's ratings of their sleepiness (Mallis & Dinges, 2004). PERCLOS has the potential to detect fatigue-induced lapses of attention during task performance, certainly in the relatively protected simulator environment, in particular as soon as the PERCLOS

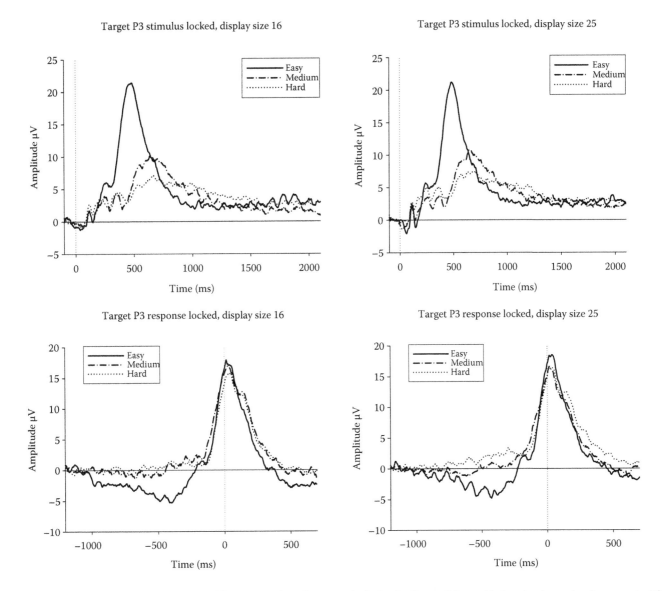

FIGURE 17.3 Grand average stimulus locked for target trials and response locked at Pz. Six conditions with three levels ranging from easy to difficult and two display sizes 16 versus 25 arrows. P3 at time >400 ms, positive up (Reprinted with permission from Wilschut, E. (2008). *The impact of in-vehicle information systems on simulated driving; Effects of age, timing and display type.* Doctoral dissertation, University of Groningen: Groningen, the Netherlands.)

scoring algorithm used by human observers in laboratory studies is automated in a computer algorithm, interfaced to provide informational feedback on alertness and drowsiness levels and scientifically validated in a controlled laboratory experiment.

17.3 Conclusions

Driving simulators, at least the advanced types that are used by demanding researchers, are in fact profitable to us in two ways: They enable us to study realistic conditions, without any objective risks, while at the same time they can function as a clean laboratory equivalent to many cognitive psychological laboratories. The preconditions that are necessary for human (psycho)

physiological measurements are relatively easily fulfilled. A major advantage of psychophysiological measures is that these can be taken continuously, without having to freeze or disturb task performance, as is required for self-reports. A number of delicate psychophysiological experiments have already been conducted in driving simulators. While neuro-psychological research such as fMRI (functional Magnetic Resonance Imaging) is difficult because of the measurement requirements, and is therefore still simplistic with respect to simulator equipment; the near future will undoubtedly bring it within reach. Our conviction is that the added value of being able to "look into the body" of a participant in driving simulator experiments is not common everywhere yet, but will quickly spread.

Key Points

- Driving simulators basically allow laboratory measurement environments.
- Physiological measurements in simulators enable close monitoring of operators.

Keywords: Driving Simulator, Mental Activation, Mental Condition, Mental Workload, Physiological Measures

Web Resource

The *Handbook* web site contains supplemental material for the chapter, including a color version of Figure 17.1.

Web Figure 17.1: The relationship between activation level, workload and performance (upper part), and specific measures from the simulator computer and physiological activity derived from heart rate measures (lower part) color version of Figure 17.1.

Key Readings

Backs, R. W., & Boucsein, W. (Eds.). (2000). *Engineering psychophysiology: Issues and applications.* Mahwah, NJ: Lawrence Erlbaum Associates.

Brookhuis, K. A. (2004). Psychophysiological methods. In N. A. Stanton, A. Hedge, K. A. Brookhuis, E. Salas, & H. W. Hendrick (Eds.), *Handbook of human factors and ergonomics methods* (Chapter 17). Boca Raton, FL: CRC Press.

De Waard, D., Jessurun, M., Steyvers, F. J. J. M., Raggatt, P. T. F., & Brookhuis, K. A. (1995). The effect of road layout and road environment on driving performance, drivers' physiology and road appreciation. *Ergonomics, 38,* 1395–1407.

References

Åkerstedt, T. (2004). Ambulatory EEG methods and sleepiness. In N. Stanton, A. Hedge, H. W. Hendrick, K. A. Brookhuis, & E. Salas (Eds.), *Handbook of ergonomics and human factors methods.* London: Taylor & Francis.

Belkic, K., Emdad, R., & Theorell, T. (1998). Occupational profile and cardiac risk: Possible mechanisms and implications for professional drivers. *International Journal of Occupational Medicine and Environmental Health, 11,* 37–57.

Boucsein, W. (1992). *Electrodermal activity.* New York: Plenum Press.

Boucsein, W. (2004). Electrodermal measurement. In N. Stanton, A. Hedge, H. W. Hendrick, K. A. Brookhuis, & E. Salas (Eds.), *Handbook of ergonomics and human factors methods.* London: Taylor & Francis.

Boucsein, W., & Backs, R. W. (2000). Engineering psychophysiology as a discipline: Historical and theoretical aspects. In R. W. Backs & W. Boucsein (Eds.), *Engineering psychophysiology: Issues and applications* (pp. 3–30). Mahwah, NJ: Lawrence Erlbaum Associates.

Brookhuis, K. A. (1989). *Event-related potentials and information processing.* Doctoral dissertation. University of Groningen, Groningen, the Netherlands.

Brookhuis, K. A., De Waard, D., & Fairclough, S. H. (2003). Criteria for driver impairment. *Ergonomics, 46,* 433–445.

Brookhuis, K. A., De Waard, D., & Samyn, N. (2004). Effects of MDMA (Ecstasy), and multiple drugs use on (simulated) driving performance and traffic safety. *Psychopharmacology, 173,* 440–445.

Brookhuis, K. A., Mulder, G., Mulder, L. J. M., Gloerich, A. B. M., Van Dellen, H. J., Van der Meere, J. J., & Ellermann, H. H. (1981). Late positive components and stimulus evaluation time. *Biological Psychology, 13,* 107–123.

Byrne, E. A., & Parasuraman, R. (1996). Psychophysiology and adaptive automation. *Biological Psychology, 42,* 249–268.

Cnossen, F., Brookhuis, K. A., & Meijman, T. (1997). The effects of in-car information systems on mental workload: A driving simulator study. In K. A. Brookhuis, D. de Waard, & C. Weikert (Eds.), *Simulators and traffic psychology* (pp. 151–163). Groningen, the Netherlands: Centre for Environmental and Traffic Psychology.

De Waard, D. (1996). *The measurement of drivers' mental workload.* Doctoral dissertation, University of Groningen, Haren, the Netherlands, Traffic Research Centre.

De Waard, D., & Brookhuis, K. A. (1991). Assessing driver status: A demonstration experiment on the road. *Accident Analysis & Prevention, 23,* 297–307.

De Waard, D., & Brookhuis, K. A. (1997). On the measurement of driver mental workload. In J. A. Rothengatter & E. Carbonell Vaya (Eds.), *Traffic and transport psychology* (pp. 161–171). Amsterdam, the Netherlands: Pergamon Press.

De Waard, D., Jessurun, M., Steyvers, F. J. J. M., Raggatt, P. T. F., & Brookhuis, K. A. (1995). The effect of road layout and road environment on driving performance, drivers' physiology and road appreciation. *Ergonomics, 38,* 1395–1407.

De Waard, D., van der Hulst, M., & Brookhuis, K. A. (1998). The detection of driver inattention and breakdown. In J. Santos, P. Albuquerque, A. Pires da Costa, & R. Rodrigues (Eds.), *Human factors in road traffic II: Traffic psychology and engineering.* Braga, Portugal: University of Minho.

De Waard, D., van der Hulst, M., & Brookhuis, K. A. (1999). Elderly and young drivers' reaction to an in-car enforcement and tutoring system. *Applied Ergonomics, 30,* 147–157.

Dinges, D. F., Mallis, M., Maislin, G., & Powell, J. W. (1998). *Evaluation of techniques for ocular measurement as an index of fatigue and the basis for alertness management* (DOT-HS-808-762). US Department of Transportation, National Highway Traffic Safety Administration.

Fabiani, M., Gratton, G., & Coles, M. G. H. (2000). Event-related brain potentials. In J. Cacioppo, L. Tassinary, & G. Bertson (Eds.), *Handbook of psychophysiology* (pp. 53–84). New York: Cambridge University Press.

Goebel, M. (2004). Electromyography (EMG). In N. Stanton, A. Hedge, H. W. Hendrick, K. A. Brookhuis, & E. Salas (Eds.),

Handbook of ergonomics and human factors methods. London: Taylor & Francis.

Jessurun, M. (1997). *Driving through a road environment.* Doctoral dissertation, University of Groningen, Groningen, the Netherlands.

Jessurun, M., Steyvers, F. J. J. M., & Brookhuis, K. A. (1993). Perception, activation and driving behavior during a ride on a motorway. In A. G. Gale, I. D. Brown, C. M. Haslegrave, H. W. Kruysse, & S. P. Taylor (Eds.), *Vision in Vehicles IV* (pp. 335–337). Amsterdam: Elsevier Science.

Kahneman, D. (1973). *Attention and effort.* Upper Saddle River, NJ: Prentice Hall.

Karasek, R. (1979). Job demands, job decision latitude, and mental strain: Implications for job redesign. *Administrative Science Quarterly, 24,* 285–307.

Karasek, R. A., & Theorell, T. (1990). *Healthy work.* New York: Basic Books.

Kramer, A. F. (1991). Physiological metrics of mental workload: A review of recent progress. In D. L. Damos (Ed.), *Multiple-task performance* (pp. 279–328). London: Taylor & Francis.

Kramer, A. F., & Belopolsky, A. (2004). Assessing brain function and mental chronometry with event-related potentials (ERP). In N. Stanton, A. Hedge, H. W. Hendrick, K. A. Brookhuis, & E. Salas (Eds.), *Handbook of ergonomics and human factors methods.* London: Taylor & Francis.

Kramer, A. F., & Weber, T. (2000). Application of psychophysiology to human factors. In J. Cacioppo, L. Tassinary, & G. Bertson (Eds.), *Handbook of psychophysiology* (pp. 794–814). New York: Cambridge University Press.

Kramer, H., Kuchler, G., & Brauer, D. (1972). Investigations of the potential distribution of activated skeletal muscles in man by means of surface electrodes. *Electromyography Clinical Neurophysiology, 12,* 19–24.

Lammers, G. J., Brookhuis, K. A., Declerck, A., Eling, P., Linskens, I., Overeem, S., . . . Coenen, A. (2005). Vigilance; evaluation and measurement. *NSWO, 16,* 21–35.

Mallis, M. M., & Dinges, D. F. (2004). Monitoring alertness by eyelid closure. In N. Stanton, A. Hedge, H. W. Hendrick, K. A. Brookhuis, & E. Salas (Eds.), *Handbook of ergonomics and human factors methods.* London: Taylor & Francis.

Mulder, G. (1986). The concept and measurement of mental effort. In G. R. J. Hockey, A. W. K. Gaillard, & M. G .H. Coles (Eds.), *Energetics and human information processing* (pp. 175–198). Dordrecht, the Netherlands: Martinus Nijhoff Publishers.

Mulder, L. J. M. (1988). *Assessment of cardiovascular reactivity by means of spectral analysis.* Doctoral dissertation, University of Groningen, Groningen, the Netherlands.

Mulder, L. J. M. (1992). Measurement and analysis methods of heart rate and respiration for use in applied environments. *Biological Psychology, 34,* 205–236.

Mulder, L. J. M., de Waard, D., & Brookhuis, K. A. (2004). Estimating mental effort using heart rate and heart rate variability. In N. Stanton, A. Hedge, H. W. Hendrick, K. A.

Brookhuis, & E. Salas (Eds.), *Handbook of ergonomics and human factors methods.* London: Taylor & Francis.

Noesselt, T., Hillyard, S. A., Woldorff, M. G., Schoenfeld, A., Hagner, T., Jäncke, L., . . . Heinze, H. J. (2002). Delayed striate cortical activation during spatial attention. *Neuron, 35,* 575–687.

O'Donnell, R. D., & Eggemeier, F. T. (1986). Workload assessment methodology. In K. R. Boff, L. Kaufman, & J. P. Thomas (Eds.), *Handbook of perception and human performance. Volume II: Cognitive processes and performance* (pp. 1–49). New York: Wiley Interscience.

Pickering, T. G. (1991). *Ambulatory monitoring and blood pressure variability.* London: Science Press.

Rau, R. (1996). Psychophysiological assessment of human reliability in a simulated complex system. *Biological Psychology, 42,* 287–300.

Rau, R. (2001). Objective characteristics of jobs affect blood pressure at work, after work and at night. In J. Fahrenberg & M. Myrtek (Eds.), *Progress in ambulatory assessment* (pp. 361–386). Seattle, WA: Hogrefe & Huber.

Rau, R. (2004). Ambulatory assessment of blood pressure to evaluate workload. In N. Stanton, A. Hedge, H. W. Hendrick, K. A. Brookhuis, & E. Salas (Eds.), *Handbook of ergonomics and human factors methods.* London: Taylor & Francis.

Schnall, P. L., Schwartz, J. E., Landsbergis, P. A., Warren, K., & Pickering, T. G. (1998). A longitudinal study of job strain and ambulatory blood pressure: Results from a three-year follow-up. *Psychosomatic Medicine, 60,* 697–706.

Sirevaag, E., Kramer, A. F., Coles, M. G. H., & Donchin, E. (1987). Resource reciprocity: An event-related brain potentials analysis. *Acta Psychologica, 70,* 77–90.

Smiley, A., & Brookhuis, K. A. (1987). Alcohol, drugs and traffic safety. In J. A. Rothengatter & R. A. de Bruin (Eds.), *Road users and traffic safety* (pp. 83–105). Assen, the Netherlands: Van Gorcum.

Theorell, T., De Faire, U., Johnson, J., Hall, E., Perski, A., & Stewart, W. (1991). Job strain and ambulatory blood pressure profiles. *Scandinavian Journal of Work and Environmental Health, 17,* 380–385.

van Boxtel, A., & Jessurun, M. (1993). Amplitude and bilateral coherency of facial and jaw-elevator EMG activity as an index of effort during a two-choice serial reaction task. *Psychophysiology, 30,* 589–604.

van Kuilenburg, H., et al., (2008). Advances in face and gesture analysis. In A. J. Spink, M. R. Ballintijn, N. D. Bogers, F. Grieco, L. W. S. Loijens, L. P. J. J. Noldus, G. Smit, & P. H. Zimmerman (Eds.), *Proceedings of measuring behavior 2008* (pp. 371–372). Wageningen, the Netherlands: Noldus Information Technology.

Wientjes, C. J. E. (1992). Respiration in psychophysiology: Measurement issues and applications. *Biological Psychology, 34,* 179–203.

Wientjes, C. J. E., & Grossman, P. (2004). Measurement of respiration in applied human factors and ergonomics research. In N. Stanton, A. Hedge, H. W. Hendrick, K. A. Brookhuis, &

E. Salas (Eds.), *Handbook of ergonomics and human factors methods.* London: Taylor & Francis.

Wierwille, W. W., & Ellsworth, L. A. (1994). Evaluation of driver drowsiness by trained raters. *Accident Analysis & Prevention, 26,* 571–581.

Wierwille, W. W., Ellsworth, L. A., Wreggit, S. S., Fairbanks, R. J., & Kirn, C. L. (1994). *Research on vehicle-based driver status/performance monitoring: Development,* validation, and refinement of algorithms for detection of driver drowsiness (DOT-HS-808-247). US Department of Transportation, National Highway Traffic Safety Administration.

Wilschut, E. (2008). *The impact of in-vehicle information systems on simulated driving: Effects of age, timing and display type.* Doctoral dissertation, University of Groningen, Groningen, the Netherlands.

18

Eye Behaviors: How Driving Simulators Can Expand Their Role in Science and Engineering

Donald L. Fisher
University of Massachusetts, Amherst

Alexander Pollatsek
University of Massachusetts, Amherst

William J. Horrey
Liberty Mutual Research Institute for Safety

Abstract

The Problem. It is almost a truism that what we don't see can threaten us as drivers. It is also becoming more and more apparent that even what we do see can threaten us if we are otherwise engaged. In either case, information about eye movements is essential in order to understand how differences in people (age, experience, disease), exposure to drugs and alcohol, the environment inside the vehicle (in-vehicle devices, texting, cellular phones), and the environment outside the vehicle (traffic, weather, night time, traffic control devices) impact driving. Much is known about what differences exist in the spatial and temporal characteristics of the eye behaviors among different groups of individuals. However, less is known about why these differences exist across the different groups of individuals. Nor is much known about how the spatial and temporal characteristics of eye behaviors change as a function of different environments inside and outside the cabin of the automobile, or for that matter why they change. *Role of Driving Simulators.* Eye movements are gathered in the field infrequently at best, even with today's relatively non-intrusive and inexpensive eye trackers, often because one cannot safely stage the sorts of events about which eye movements are so informative. As an alternative to gathering eye movements in the field, driving simulators have three distinct advantages: drivers can be deliberately placed in dangerous scenarios, the scenarios to which drivers are exposed can be rigorously controlled, and the frequency with which drivers encounter scenarios in which meaningful eye behaviors can be extracted can be greatly increased. *Key Results of Driving Simulator Studies.* Recent studies of eye movement behaviors on driving simulators have begun to help put together a picture of why there are large differences in the temporal and spatial distribution of the eye movements as a function of differences in drivers and the environment. Additionally, recent studies have used what is known about the factors that influence eye behaviors to change drivers or the environment in ways which reduce the most risky eye behaviors. *Scenarios and Dependent Variables.*

The types of scenarios used to capture information on eye movements depend critically on the questions that are being asked. It is made clear what the key characteristics of the scenarios are for each question. The types of dependent variables that are analyzed include characteristics of the spatial (e.g., horizontal and vertical standard deviation; range; looked versus did not look; first- and higher-order transition probabilities) and temporal (e.g., total time on task; average glance duration in a particular area or at a particular object; maximum glance duration away from the forward roadway) distribution. ***Platform Specificity and Equipment Limitations.*** Eye movement behaviors are almost always studied on simulators that have at least three screens. Gathering data on a single monitor presents special challenges. Equally important, there are limitations in all eye trackers that can make collecting and analyzing difficult. These more technical issues are not addressed in any detail in this chapter.

18.1 Overview

Eye movement research as we know it first goes back to the 1900s when photographic techniques were used to measure discontinuous eye movements (as cited in Wade & Tatler, 2005). Perhaps most is known about eye movements in reading (for a recent review see Rayner, Pollatsek, & Starr, 2003). However, there is a considerable body of literature on eye movements and driving, beginning in earnest in 1969 (Mourant, Rockwell, & Rackoff, 1969; also see Mourant & Rockwell, 1970, 1972; for a review of the impact of their research see Shinar, 2008). Not surprisingly, most of this research focuses on where drivers look and for how long. A recent, comprehensive review summarizes much of this information (Green, 2007). This includes a discussion of how eye behavior is influenced by the environment (e.g., curves, traffic, signs and signals), driver characteristics (experience, fatigue, alcohol and other drugs), vehicle characteristics (vehicle type, headlamp illumination), and in-vehicle devices.

In this chapter, we want to extend the discussion of eye movements to provide a more general understanding of what is known about them. We hope this aids the reader in designing experiments that can be used to identify the reasons for the differences in eye behaviors across individuals and situations and then use that understanding to create improvements in training or design. For example, consider the comparison of novice and experienced drivers. Not only would we like to know how the eye behaviors of novice and experienced drivers differ from one another, but, as scientists, we would also like to understand why these differences exist. Then, as engineers this would allow us to design interventions which might help the novice driver. Alternatively, consider different interfaces to in-vehicle music retrieval systems. We would like to know not only how the eye behaviors varied as a function of the differences in the music retrieval systems, but, as scientists, we would like to develop a general understanding of why one interface to a music retrieval system might cause drivers to spend longer with their eyes away from the forward roadway than some other music retrieval system. Then, as engineers, we would want to use the general principles to develop interfaces which are safer.

Below, we begin with a general discussion of what is known about eye movements, especially as that knowledge has developed in reading. Second, we show why an understanding of eye movements is such an important part of the larger puzzle of understanding why crashes occur, hopefully motivating

the reader to look further into the substance of this chapter. Third, we summarize very briefly what is known about where drivers look and for how long and then provide examples of studies which attempt to grow our understanding of why it is that differences exist in the eye behaviors of different groups of individuals or groups of individuals exposed to different environments. Because of the very large literature and the recent excellent surveys (e.g., Green, 2007), we have decided to focus the review in three different areas—individual differences, the environment inside the car, and the environment outside the car—selecting within each area those topics which we think are most timely. Finally, we discuss models of visual attention and their utility in describing and predicting driver scanning behavior.

18.2 Eye Movements

The importance of eye movements in understanding visual cognition is perhaps best documented by the fact that a review article by Rayner in 1998 entitled "Eye Movements in Reading and Comprehension: 20 Years of Research" has about 750 references, almost all on experiments using eye movements as the major dependent variable, and about half published since 1990. Although the majority of these articles are in reading, there is an increasing use of eye movements in the fields of scene and object recognition, where the findings are clearly relevant to the understanding of the role of eye movements in driving. The reason for this increasing interest is that the research that has been conducted makes it clear that what is fixated (i.e., the location at which the eyes are pointing) is very closely related to what is being encoded and processed. This is because acuity falls off quite rapidly from the fixation point (i.e., where the eyes are pointing) and the eyes usually need to get very close to what is processed to encode it well. One exception to this is the processing of motion, which can be done quite well even in the periphery. However, even in this case, the eyes are quickly drawn to the moving object.

18.2.1 Saccades, Fixations and Pursuit Movements

We should perhaps begin by describing the details of eye movement behavior. Standardized terms for eye movements in the context of vehicles are defined in SAE (Society of

Automotive Engineers, 2000) and ISO (International Standard Organizations, 2002) documents. We adhere to these definitions for the most part. There are basically three states for the eyes: fixations, saccades, and smooth pursuit movements. First consider static situations such as reading a sign while stopped at a signal. In these situations, one has only fixations and saccades; the eyes are in fixations most of the time, in which the eyes remain essentially still and these fixations are interspersed with rapid eye movements called *saccades*. During a saccade, no meaningful visual information is encoded (this is sometimes called *saccadic suppression*) and people are unaware of the blurry moving image on the retina during the saccade, largely because it is backwardly masked by the visual information from the fixation following the saccade. Thus, in static situations like this, the only meaningful information accrued is during fixations.

There is little of interest for our purposes in the details of saccades. The time to execute a saccade is basically a function of the distance of the saccade (longer saccades take more time). In contrast, as we will see, both the duration and location of fixations are of central interest as they both indicate that what is being fixated is being processed. The data below indicate that: (a) objects that are not fixated are unlikely to be encoded, and (b) the time spent fixating an object is closely related to the difficulty of processing that object.

In contrast to situations in which the image is static, *smooth pursuit movements* are an important component of eye movements in situations in which there are either moving objects or in which the observer is moving with respect to the environment. These are obviously relevant in driving. Smooth pursuit movements occur when observers are trying to keep their eyes pointing to an object that is moving with respect to them. These movements are much slower than saccades and, more importantly, the visual information coming in during such a movement is not suppressed. If the observer is not moving, such movements would occur when one attempts to follow a moving object. In driving, these would typically occur if someone is attempting to process an object that is not directly in front of the driver (such as attempting to read a speed limit sign). At a distance, such a stimulus would be close to straight ahead, but as the vehicle nears it, the stimulus would become more and more eccentric. Thus, if one wants to continue pointing the eyes at such an object, one needs to move the eyes further and further away (in terms of visual angle) from the center of the road. (The above is a slightly oversimplified version of eye behaviors, e.g., see Rayner et al., 2003.)

Typically, in static situations like reading, the data are collapsed into *fixation durations* and *fixation locations*. However, often greater emphasis is placed on a measure usually called *gaze duration*. For example, if one was interested in how long a specific target word was fixated, the measure usually relied on would be the gaze duration, which would be the sum of the fixation durations on the word when it was first fixated. Other measures that would include later fixations back to the word are often used; however, gaze duration is usually thought of as being the measure that best captures the time to encode the stimulus.

This is reinforced by the fact that when one manipulates variables like the frequency of a target word (i.e., the frequency in the language) or the predictability of the word from the prior text, one gets meaningful and significant effects of these variables on gaze durations. This is notable, since gaze durations on words in reading tend to be around 250–300 ms, so that it is quite a feat for the visual system to encode a word and use this information to plan a saccade to the next word in this short period of time. This speaks to the immediacy of eye movements.

As indicated above, in driving, encoding of objects may often not be in a fixation, but during a smooth pursuit eye movement. Our sense in reading some of the literature on driving is that these are counted as fixations, as in a functional way, they are: The eyes are being continually aimed at the same object. However, as with other aspects of eye movements, these are not perfect, and smooth pursuit eye movements are interspersed with short saccades that are attempting to get the object back into the center of fixation. Thus, it would seem like a measure analogous to gaze duration (e.g., the total time of the interval the eye is directed at or near an object of interest before moving elsewhere) is the most appropriate here for determining the time spent processing the object.

18.2.2 Eye Movements in Reading

Although reading is clearly a very different situation than driving, it is worth briefly summarizing a few relevant findings in reading, making it clear as we go how the studies are related to driving. First, when studying reading, one can use a far more accurate eye movement apparatus and thus be more confident about the details of the eye behavior, both in the temporal and spatial dimensions. In the most accurate Dual Purkinje apparatus, when the head is fixed by a bite bar, one can measure eye movements to within 1 ms and with an accuracy so that one is quite confident which letter is being fixated (letters typically subtend between 1/4 and 1/3 of a degree of visual angle). (The accuracy is somewhat less in the vertical dimension.) It is clear from these studies how rapidly acuity information falls off. For example, from studies involving displays that change contingent on where the eye is fixating, we know that people extract no useful information more than 14 characters from fixation (i.e., about five degrees of visual angle) in the horizontal dimension, and acuity falls off faster in the vertical dimension.

Obviously, the information needed to identify letters is finer-grained than much of what is used in driving, but these studies make clear that visual acuity falls off rapidly. Moreover, the reading research also makes clear that people are unaware of how little they are actually processing. One dramatic example comes from *moving window* experiments, in which the reader is presented with a window of normal text around the fixation point and then the letters outside the window are replaced by something like random letters. (Spaces between words, however, are preserved.) In conditions in which the window of normal text is the word that is fixated (for example, words other than the one fixated are just random letters until the reader fixates them), readers are almost always completely unaware that the text is changing even though they are reading at only something like 60% of their normal rate (Rayner, Well, Pollatsek, & Bertera, 1982). (The fact that readers are unaware of information upon which they are not focused might be explained today by the various theories of change blindness, e.g., Rensink, 2002). There are many implications for driving. For example, individuals making multiple sequential fixations on a sign to the side of the road are not apt to notice differences in the forward roadway that might signal a potential threat since the changes in the environment while one is glancing at the sign (the "text" outside the window) are all but impossible to notice. We will come back to this point several times in later portions of the chapter.

A second important finding from the eye movement research in reading is that the effects of cognitive processing show up quite immediately in the behavior of the eyes. For example, if the frequency of a target word is varied (with the length held constant), this produces an effect even on the first fixation duration on that word and a substantial effect on the gaze duration on that word. In addition, if a word or a sense of a word does not make sense given the prior sentence context, this also has a substantial effect on gaze duration on the word (Duffy, Morris, & Rayner, 1988; Rayner, Warren, Juhasz, & Liversedge, 2004). Although there is substantial evidence that processing of a word begins before it is fixated (the moving window data above is one of many such pieces of evidence), this still means that, not only is it clear that what is fixated is being processed, but that the processing of this information has quite immediate effects on eye behavior. Moreover, even though there are some "spillover" effects as well (e.g., the frequency of word n will affect fixation time on word $n+1$), they are generally substantially smaller and less reliable than the immediate effects. Thus, in reading, it is quite clear that the eyes are a reliable guide to what is being processed and that fixation times are a reliable indicator of the difficulty of processing the object (a word in this case). This adds weight to the arguments that have been made in the driving literature that the glance duration is an important indicator of the difficulty that drivers have processing a given scene or scenario (e.g., Chapman & Underwood, 1998, have shown that drivers have longer fixation durations during complex scenarios potentially leading to a crash).

A third important point from the reading studies is that what is being processed is not only a function of distance from the fixation point but of where covert attention is being directed (attention is covert when it is focused on an area of the scene which is not directly the locus of fixation). One finding supporting the existence of covert attention is that the functional window of what is being processed from the text is asymmetric. That is, although replacing the letters in the word to the right of the fixated word results in a sizeable decrement in reading speed, replacing all the letters in the words to the left of the fixated word has no measurable effect on reading. This is an attentional phenomenon, as the phenomenon is reversed for readers of Hebrew, where only information to the left of the fixated word is relevant. Moreover, removing relevant letter information from the line

below the line of text (which is quite close to fixation) had no measurable effect on reading. Thus, attention in reading appears to be directed to the word that is being processed and the word about to be processed. This is all accounted for in the E-Z Reader model (Pollatsek, Reichle, & Rayner, 2006b) in which attention is directed to one word at a time, and when processing is finished on the fixated word, attention shifts to the next word in advance of the eyes actually landing there. The implications for driving are less clear here simply because not enough is known. For example, one would like to know whether drivers reading a sign on the left side of the road can direct attention to the right (towards the center of the road) whereas drivers reading a sign on the right side of the road can direct attention to the left (again, towards the center of the road). Thus, ideally one would run the experiments required to understand just how malleable is the topography of the window of attention.

A fourth point that is worth commenting on is that saccadic eye movements are not completely accurate. In reading, it appears that the target of a saccade is the middle of the next word, but there is both random variation in targeting and a bias in which short programmed saccades tend to be overshot and long programmed saccades tend to be undershot (McConkie, Kerr, Reddix, & Zola, 1988; Pollatsek et al., 2006b). (This variation is consistent with the data on other motor movements.) Among other things, this predicts that short words may often be skipped due to programming error. However, there are also data indicating that short and/or highly frequent or highly predictable words are skipped because they are processed quickly before they are fixated. There are possible implications for driving from these word-skipping data. That is, perhaps frequent and/or easy to process stimuli such as stop signs (perhaps the equivalent of short words in driving) can be processed without being fixated. This has been looked at somewhat (Louma, 1988), but not in depth, and certainly needs to be pursued in more detail.

18.2.3 Scene and Object Perception

Processing objects in scenes is obviously a different situation than processing words in text. First, as indicated above, words are very special stimuli in which detailed information about letters is needed in order to make discriminations. Second, reading is a well-defined task in which one is basically trying to process the words in order from left to right in most alphabetic languages; in contrast, there is no well-defined task of "scene perception" and people can view scenes for various reasons. Thus, there is no reason to expect that the eye movement pattern will either be as orderly as in reading or as consistent over people. Usually, people who have studied object perception in scenes using eye movements have employed one of two tasks: (a) a visual search task (e.g., participants are given a target object to search for before the scene is presented and asked to press a button as soon as they locate it or decide it is not present in the scene), or (b) a recognition memory task (e.g., several seconds after the scene is presented and removed, participants are asked whether a dog was in the scene). In spite of these fundamental differences between reading and scene perception, however, the findings from the two types of situation are remarkably similar.

As indicated above, one area in which the scene perception and reading literatures agree strikingly is that little processing of a word or object occurs if it is not fixated. For example, Hollingsworth and Henderson (2002) found that if people looked at a scene and then were tested on whether a target object was present in the scene seconds after the scene was removed, performance was not significantly above chance if the object was not fixated. Another area in which the two literatures agree is that processing is quite immediate. For example, in a visual search task (Williams & Pollatsek, 2007) in which the target was an O in a background of Cs with varying gap sizes and the stimuli were in well-defined clusters, the fixation time on a cluster to determine whether an O was present (i.e., whether any of the characters in a cluster did not have a gap) was a function of the gap size for the Cs in the cluster. Moreover, the gap size of the Cs in cluster n had no effect on the fixation time on either the prior or succeeding cluster fixated. A follow-up experiment indicated that fixation time on a cluster perfectly mirrored response time on a cluster when the cluster was presented in isolation. That is, fixation time on a cluster mirrors processing time on the cluster indicating that fixation time on objects in non-linguistic displays can be taken seriously as indicating what is being processed at that moment. The implications for driving are straightforward here, as they were in reading. Where the eyes are pointing is a good indication of what is being processed and the duration of a glance is a good indication of the difficulty of the latent processing which transpires during the glance.

An obvious question about scene processing is what attracts attention and eye movements. It is clear that motion of a single object in an otherwise static display will attract attention and eye movements even when they are far from fixation. Note that this is detection of motion in one area of the retina versus no other motion present. However, less is known about when a moving object captures attention in a situation like driving. That is, when one is driving, there is motion being detected everywhere on the retina, so that detection of a moving object entails a more complex computation of relative motion. Intuitively, it would seem that an object moving horizontally across the retina would be easier to detect than one moving parallel to the motion of the driver's car because there would be a greater disparity between that motion and the "background" motion. However, as far as we know, this has not been determined.

One cue that has been reliably shown to attract attention and eye movements is an object that differs from all other objects in an array on a simple feature (e.g., a red object in a background of green objects or a square in a background of circles). The salience or conspicuity of a particular object or target has been a fundamental aspect of models of visual search (e.g., Itti & Koch, 2000). Highly salient objects often capture an observer's visual attention when engaged in a search for a visual target (e.g., Wolfe, 1994). This is a major reason that red is the only background color that can be used on stop signs. Unfortunately, salient yet irrelevant objects can easily become distractors for

more important task-relevant information. For example, within more urban settings, most red signs, even small ones, are probably not stop signs.

In contrast to the physical characteristics (salience) of an object, which is largely a bottom-up phenomenon, it is not clear that anything at a deeper (top-down) level does attract attention. For example, in a visual search paradigm, where one is looking for a lamp (cued by the word *lamp*) in an array of objects, the time elapsed before the target object is fixated did not differ between when the target was a prototypical lamp and a very idiosyncratic lamp (Castelhano, Pollatsek, & Cave, 2008). Thus, having a mental image of a lamp (which presumably is like a prototypical lamp) does not in any way make the image of the lamp stand out. On the other hand, the time elapsed between when the target is first fixated and the manual response indicating that the target is present was much less for the prototypical target objects. Thus, having a mental image of lamps similar to the prototype facilitates verifying that the object is a member of the category but does not help in attracting attention to it. However, finding a target object was significantly faster when it was cued by the image of the target object rather than being given a name. Thus, either having an exact image of an object or some significant feature of the object does help locate it (Jonides, 1981). Much work is clearly still needed to determine what kinds of events in driving automatically attract attention and eye movements. Perhaps more importantly, work is needed to catalogue the events in driving that people assume will automatically attract attention but which do not do so. So, for example, one would assume that a pedestrian entering a crosswalk would automatically attract the attention of the driver. But there is no evidence that such is the case, which is why the efforts to design safer crosswalks are still so intensive.

18.2.4 Spatial and Temporal Characteristics

18.2.4.1 Spatial Parameters

When the question is whether a given object in the scene or area in the scene has been processed, the fixation or gaze location is the most frequently reported measure. This is a local measure, in which there are planned scenarios on a drive in which there are signs, signals, and real or potential hazards (e.g., Pradhan et al., 2005; Romoser, Fisher, Mourant, Wachtel, & Sizov, 2005). For example, consider the measure of hazard anticipation. In the simulator, the potential hazard never appears because one does not want to make the driver unnaturally hyper-vigilant; the hazard virtually never occurs in road tests either. Thus, the measure here is very specific: Did the driver look away from straight ahead in the appropriate place (target zone) and at an appropriate time (launch zone) in order to anticipate the potential hazard? As we discuss below, such local measures can give important information about novice and older drivers' behaviors when one knows what an optimal fixation pattern should be.

Another type of measure that has been used with respect to eye location is a detailed analysis of the sequence of fixations. For example, such analyses typically partition the scene into areas of interest (AOI). Areas of interest (AOIs) are particularly useful in the study of driver eye movements. While drivers often must search and detect relevant targets in their environment (e.g., vehicles, pedestrians, traffic signs), the driving task is much more complicated than the visual search tasks described above. Rather than search for a static target at an uncertain location, drivers must monitor a series of dynamic processes at known locations (gauges, displays, roadway—each mapping onto a relevant AOI) in order to detect critical events or information at the appropriate time. Thus, the emphasis is on knowing both when and where to look. Models of supervisory control, discussed in a later section, are particularly relevant for the task of driving. Furthermore, because the outside traffic environment is dynamic (in contrast to the interior of the vehicle, which is fixed), it is often more practical to look at AOIs defined by the scenario rather than unvarying regions of space.

Analysis of eye behavior related to different AOIs will often utilize the proportion of glances in each region, which can be counted and compared across groups or conditions (Maltz & Shinar, 1999). In addition, one can analyze the actual scan paths of drivers to whatever level of detail one desires (Underwood, Phelps, Wright, van Loon, & Galpin, 2005). A scan path is defined quite simply as a sequence of gazes in different locations. The locations could be different AOIs. In this case, the description of a first order scan path would include the initial AOI and then the proportion of drivers gazing at each of the subsequent AOI. A description of a second order scan path would start with an initial AOI and then give the proportion of drivers gazing at each of the two succeeding AOIs, and so forth.

When the question is how well a driver is scanning a given scenario over time, the range (e.g., Mourant & Rockwell, 1972), the standard deviation in either the horizontal or vertical axes (e.g., Chapman & Underwood, 1998), and the product of the two standard deviations (Recarte & Nunes, 2003) are the measures most often used. These measures are most useful when there is no obvious hazard in the scene which would cause any other than very general strategies to be employed. When there is a particular hazard that should be anticipated, then more local analyses would appear to be preferred.

18.2.4.2 Temporal Characteristics

The other obvious question in making sense out of eye movement data is what useful information can be extracted from the durations of fixations. Analogous to Sections 2.2 and 2.3 above, one could obtain global measures such as the mean fixation duration and the standard deviation over an extended stretch of driving; however, it is not clear that anything meaningful will emerge from these very global measures. Instead, people have focused on how long fixations are on specific objects or areas of interest. As we argued in the prior section, it is also unlikely that the durations of individual fixations are going to be the measure of choice as there is likely to be a lot of "jitter" in a situation on whether the eyes remain in a fixation—or, more likely, in a smooth pursuit movement that is attempting to keep fixation on a particular target. Instead, we would favor using a measure

analogous to *gaze duration* in reading [i.e., the amount of time the person's fixation point is within a certain region (AOI) or in the area surrounding a predetermined target]. However, our experience is that, in the driving literature, there is considerable vagueness about how a fixation measure is defined; we think that it is likely that something like a gaze duration is being called a "fixation duration".

The other obvious question is what measures one wants to use indexing such duration measures. A standard measure is the mean, but because the histograms of such measures are virtually always skewed to the right (i.e., a long tail), people are skeptical about using the mean as a reliable measure of central tendency. In the driving literature it appears that it is assumed that the histogram has a log normal shape and thus the data is often transformed by using the logarithm, and then the mean of these values is taken as the measure of central tendency (often converted back to time). In the reading literature discussed above, the mean is often reported; however, there has been a recent trend to attempt to fit the histograms obtained with an "ex-Gaussian" distribution, which is a convolution of a normal distribution with an exponential distribution. Such analyses usually explicitly assume that there are two underlying processes, one giving a normal distribution and one producing an exponential. The focus of these analyses is typically whether a variable affects the mean of the assumed normal distribution (generally assumed to be the usual default underlying process) or whether it is merely affecting the exponential distribution (generally assumed to be some special process that occurs on a minority of the trials).

A point that we wish to stress is that it is far from clear what the real underlying form of the histogram is for fixation durations (or gaze durations). Although people have fit the distributions well assuming a log normal distribution or an ex-Gaussian distribution, a Gamma distribution also fits the data quite well. In any case, it is clear that the histogram is not symmetric and thus that one should be cautious about merely reporting the mean.

18.2.5 Eye Movements and Crashes

Inattention and distraction have long been classified as major contributors to *crashes*. (A crash is typically defined as any physical contact between the subject vehicle and another vehicle, fixed object, pedestrian, cyclist, animal, etc.) Ideally, in order to document the roles that inattention and distraction play, one would like to know immediately before a crash occurred where a driver is looking and for how long. Perhaps not surprisingly, early attempts to identify just how large a role inattention and distraction did play in the cause of crashes had to make do without such information. Estimates were often based on police crash reports, among others (e.g., Sabey & Staughton, 1975; Treat et al., 1979). In a recent, comprehensive meta-analysis of research on driver distraction, it is estimated that distraction is a cause of as few as 6% of crashes and as many as 25% of crashes, the different estimates being due both to competing definitions of what constitutes a distraction and to insufficient information about the crashes in the reports themselves (Caird & Dewar, 2007).

However, information available to researchers has changed dramatically over the past five or so years. In particular, eye behaviors have come to play an increasing role in helping researchers identify the cause of crashes due to distraction and inattention. In this section, we want to discuss those particular eye behaviors that have been linked, either directly or indirectly, to crashes. These are the eye behaviors one wants to monitor in order to determine whether differences between drivers or differences in the environment to which drivers are exposed were (or are going to be) the likely cause of crashes.

First, we want to describe the recent results which indicate that long glances away from the forward roadway that are not driving related (e.g., that are not related to looking at the mirror or speedometer or applying the emergency flashers) are associated with crashes. These results imply that anything that is not directly driving related and takes the driver's attention away from the forward roadway is a threat to safety. Eye movements have an obvious role to play here in letting us document the fact that the driver is not attending to the forward roadway for an extended period of time immediately before a crash occurs. Second, we want to describe the studies that show that drivers, especially those with a high crash rate, fail to look at locations on the roadway where likely hazards could emerge. Thus, arguably, anything which affects the driver's ability to anticipate hazards, even when the driver is actively scanning the roadway, is going to increase risk. Eye movements have a similar role to play here, letting us document that the driver fails to anticipate hazards (i.e., fails to scan the locations that are most likely to contain threats which could materialize suddenly). Third, we want to describe studies which show that a driver, even when looking directly at a threat, can fail to identify the threat and potentially crash into that threat. This is referred to in the more general psychological literature as *inattention blindness* (e.g., see Simons & Chabris, 1999). Cognitive distraction is a common cause of inattention blindness. Monitoring eye movements is critical here because it is only by documenting that a driver is looking at an emerging threat that one can conclude that a resulting crash may be due a cognitive distraction that causes inattention blindness.

18.2.5.1 Crashes: Glances Away from the Forward Roadway

Treat et al. (1979) estimate that some 5.6% to 9.0% of the crashes are due to internal distraction (i.e., a distracting event within the vehicle), the differing estimates being a function of how confident an investigator is that the crash would not have occurred had internal distraction not been present. This is similar to an estimate reported by Wang, Knipling and Goodman (1996). They found that 8.2% of the crashes could be linked to internal distraction.

Green was one of the first to put a hard number on the maximum time that a particular in-vehicle task should take to complete (Green, 1999a) in order to be considered safe, in this case a navigation and route guidance task. This number serves as the basis for SAE Recommended Practice J2364. Known as the 15-seconds rule, the rule specifies that an in-vehicle navigation task should take no more than 15 seconds in total to complete

(Society of Automotive Engineers, 2004). Green (2007) was very clear that this does not mean that it is safe for the driver to have his or her eyes off the road for 15 seconds continuously, but did not specify how long a glance (i.e., a gaze duration) should be, other than that it should not be too long.

Wierwille (1995) was one of the first to use eye movement data to predict actual crash involvement and to tie that prediction to the length of a glance. In particular, he was able to predict the number of vehicles involved in crashes in North Carolina per year due to glances to in-vehicle instruments from a knowledge of the mean glance duration at the instrument (m), the mean number of glances to the instrument in order to record its value (n), and the mean number of times per week a driver glanced at the instrument (f):

$$\text{Number of Involvements per Year} = -0.554 + [0.335 \times m^{1.4} \times n \times f]$$

It is clear that especially long glances are problematic as the mean glance duration, m, is raised to a power greater than 1.0. Green (1999b) generalized this formula so that it could predict the number of involvements per year throughout the entire United States.

Recently, two pieces of evidence, one in the field and one on a driving simulator, provided real advances in our understanding of the relation between eye movements away from the forward roadway and crashes. Not surprisingly, looking away for long periods of time is associated with an increase in crashes. What may be somewhat surprising is how dramatically the risk increases at the longer durations.

18.2.5.1.1 Naturalistic Field Study

In a naturalistic study of 100 instrumented cars reported in 2006, 241 drivers 18-years old and above were filmed inside their vehicles. The drivers logged over two million miles (Klauer, Dingus, Neale, Sudweeks, & Ramsey, 2006; also see Dingus et al., 2006). Detailed vehicle, event, environmental, driver state (eye behavior, drowsiness) and narrative data were gathered on *events* in the data base: *Crashes, near-crashes* and *incidents*. A *crash* was defined as above (any physical contact between the subject vehicle and another vehicle, fixed object, pedestrian, cyclist, animal, etc.) and was assessed by either lateral or longitudinal accelerometers located inside the driver's vehicle. A *near-crash* was defined as a conflict situation requiring a rapid, severe, evasive maneuver in order to avoid a crash. An *incident* was defined as a conflict requiring an evasive maneuver but of lesser magnitude than a near-crash. There were a total of 69 crashes, 761 near-crashes, and 8,295 incidents in the event database. A primary measure of driving related inattention for crashes and near-crashes was the total time the eyes were off the forward roadway in the interval between five seconds prior to and one second after a precipitating event (e.g., a lead vehicle braking).

As a control for the event database, a stratified random sample of 20,000 *baseline events* was also coded. Similar to the event database, each baseline event consisted of six seconds of information. The controls were stratified by the number of crashes, near-crashes and incidents in which each vehicle was involved (e.g., 200 baseline events would be randomly sampled for a vehicle involved in 1% of the crashes, near-crashes and incidents so that it contributed 1% of the 20,000 baseline events). Detailed vehicle and environmental data were recorded for each of the baseline events. In addition, driver state variables (eye glance behavior, drowsiness) were analyzed for 5,000 randomly sampled baseline events.

The *odds ratio* was then computed for various different glance durations. So for example, the odds ratio for glances greater than two seconds long was computed as the ratio of the A/B to C/D: A was the number of events (crashes and near-crashes) in which the driver's eyes were off the road for more than two seconds, B was the number of events in which the driver's eyes were off the road for two seconds or less, C was the number of baseline events in which the driver's eyes were off the road for more than two seconds, and D was the number of baseline events in which the driver's eyes were off the road for two seconds or less. Interestingly, it turns out that driving-related inattention to the forward roadway has a protective effect, both when the glances are shorter than two seconds (odds ratio 0.23) and when the glances are greater than two seconds [odds ratio 0.45, i.e., (A/B)/(C/D) = 0.45].

More generally, however, if the sum of the glance durations away from the forward roadway for more than two seconds during the six second window surrounding a crash was used as the measure, such behavior no longer had a protective effect; in fact, it was a contributing factor to crashes. (Note that the complete set of glances away from the forward roadway included all instances of driving-related inattention to the forward roadway and all instances of non-specific eye glances away from the forward roadway as well as a subset of the cases involving secondary task distraction and drowsiness.) While the odds ratio for eye glances away from the forward roadway for a total of less than 2.0 seconds was not significantly greater than 1.0, it was significantly greater than 1.0 for glances greater than a total of 2.0 seconds (odds ratio 2.19, Table 6.2, Klauer et al., 2006). Moreover, it was estimated that such glances away from the forward roadway for more than 2.0 seconds were the cause of more than 23% of the crashes and near-crashes (population attributable risk, Table 6.3, Klauer et al., 2006).

These data raise the question of why individual glances away from the forward roadway should have a protective effect. The most likely hypothesis is that some of these glances are related to noticing a potential hazard (such as a vehicle not stopping at a stop sign or a lead vehicle braking sharply) and thus being able to react accordingly. One likely special case of this is glances to the left, center and right rear-view mirrors. Indeed, as there are data that indicate that such glances are protective, it makes sense to remove such glances from the above calculations. When this is done, the odds ratio when the sum of the glances away from the forward roadway was greater than 2.0 seconds increased to 2.27 (Table 6.4, Klauer et al., 2006). The population attributable risk decreased to roughly 19%. In summary, it is clear that glances away from the forward roadway, for even a short period of time

when not related to the driving task itself, increase the odds of being in a crash significantly.

18.2.5.1.2 Simulator Study

The critical role that especially long glances play in the development of crashes has been clearly illustrated in a recent study on a driving simulator (Horrey & Wickens, 2007). Drivers had to navigate a single-lane road subject to small or large wind gusts, keeping the vehicle centered as best as possible. At the same time, drivers had to perform an in-vehicle task which required them to count the number of even digits in a string of digits 5 or 10 digits long (low- and high-complexity, respectively). The digit string was displayed inside the cabin of the vehicle so that the driver had to look away from the roadway to read the string. Finally, at various points in time, a pedestrian, animal or bicyclist emerged from behind an occluding object (e.g., parked car) into the path of the driver, creating the opportunity for a collision (there were a total of six opportunities for a collision over the drive which had three blocks which were each eight minutes long; collisions were identified by a separate algorithm than the one used by Klauer et al., 2006).

Neither the level of turbulence nor the load of the secondary task had an effect on drivers' mean glance durations, nor was there an interaction. However, when just the tails of the distribution were analyzed a very different picture emerged (Figure 18.1). The exact point at which one classifies the tail of

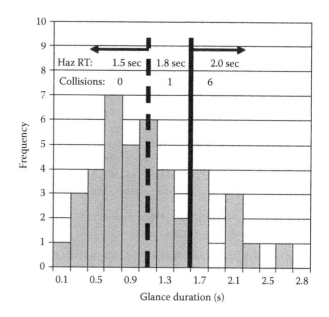

FIGURE 18.1 Distribution of glance durations immediately before hazard events. Overall mean glance duration (1.1 s) is marked by dashed line. The 1.6 seconds threshold is marked by the solid line. Mean hazard response time (Haz RT) and collision rate for different portions of curve are shown. (From Horrey, W. J., & Wickens, C. D. In-vehicle glance duration distributions, tails, and model of crash risk. *Transportation Research Record. Journal of the Transportation Research Board,* No. 2018, Figure 3, p. 25. Copyright, National Academy of Sciences, Washington, D.C., 2007. Reproduced with permission of the Transportation Research Board.)

the distribution has been the point of some disagreement. For example, Wierwille (1993) proposed a 1.6 seconds threshold for in-vehicle glances. The Alliance of Automobile Manufacturers recommends a 2.0 seconds threshold for in-vehicle glances (McGehee, 2001). However, one might reasonably argue, based on the data from Klauer et al. (2006), that a somewhat more complex criterion be used: The total time that a driver's eyes were off the road during any six second interval should be no more than 2.0 s. Regardless of the criterion used, the authors came to much the same conclusion. In particular, 86% of the crashes (six out of seven, Figure 18.1) were associated with the longest 22% of the glances (in this case the glances that were longer than 1.6 s). Moreover, the more complex in-vehicle task was implicated in six of the seven crashes. Interestingly, this was not because the drivers were more likely to be glancing down when the threat materialized in the complex in-vehicle task. The probabilities that the driver was looking away from the roadway at the onset of the hazard event were almost identical in the low- and high-complexity in-vehicle tasks. Instead, it was because there was a much longer tail in the glance duration in high-complexity tasks: 21% of all in-vehicle glances were longer than 1.6 seconds compared to 6% for the less complex task.

In summary, both the simulator and field studies point not only to the importance of understanding where the eyes are glancing when evaluating the crash risk of different groups but also to the importance of identifying the frequency of especially long glances away from the forward roadway. Anything that increases the likelihood of especially long glances is especially problematic.

18.2.6 Crashes: Anticipatory Glances

Crashes happen not only when the driver glances away from the forward roadway for too long. They also happen when the driver is looking at the forward roadway but fails to anticipate the materialization of a threat that is signaled by the road geometry, distribution of traffic, or status of existing traffic control devices. McKnight and McKnight (2003) estimate that the majority of crashes are caused by failures to scan the roadway adequately. It is not known what percent of these crashes is due to failures to anticipate a potential threat and make the appropriate anticipatory scanning maneuvers. However, it is known that experienced drivers are up to six times more likely to anticipate threats than novice drivers (Pradhan et al., 2005). Similarly, it is known that novice drivers during the first month of solo licensure are up to five times more likely to crash than novice drivers are during their sixth month of solo licensure (McCartt, Shabanova, & Leaf, 2003). Combined, these findings suggest that the failure of novice drivers to anticipate hazards is an important factor in their crashes. Eye movement data is some of the most, if not the most, direct evidence we can gather that potential hazards are being anticipated.

An example of a scenario in which drivers should be scanning for a potential threat is diagrammed in Figure 18.2. Here, a truck is parked in front of a marked midblock crosswalk. A pedestrian could step out in front of the truck into a driver's path. Thus, the

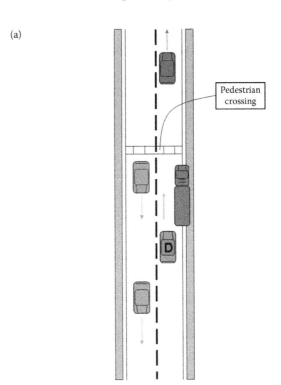

(a)

(b)

FIGURE 18.2　Truck Crosswalk Scenario. (a) Plan view; (b) Perspective view.

driver should anticipate this and glance to the right as he or she passes the front of the truck. A perspective view is given in Web Figure 18.1. In a recent study, the eye movements of novice drivers were monitored in the field in just such a scenario. The clear role that anticipatory eye movements have in avoiding crashes is illustrated by the videos that show, on the one hand, a driver who fails to make an anticipatory eye movement to the right as the driver travels across a marked midblock crosswalk that is obscured by a vehicle stopped in the parking lane just upstream of the crosswalk and, on the other hand, by a driver that does make such eye movements (Web Videos 18.1 and 18.2).

18.2.7 Crashes: Unseeing Glances at the Forward Roadway: Inattention Blindness

The percentage of crashes due to looking, but not seeing, has been estimated by various investigators from police crash reports. For example, Treat et al. (1979) report that "improper lookout" is responsible for between 17.6% and 23.1% of the crashes, again the difference being a function of how confident an investigator is that the crash would not have occurred had the factor (i.e., improper lookout in this case) not been present. Improper lookout consists both of "failed to look" and "looked but failed to see" (Treat, 1980). Instances of "looked but failed to see", or inattention blindness, are difficult if not impossible to identify without an eye tracker.

Instances of looking but failing to see are easy to identify with an eye tracker if there is a crash. But what if there is not a crash? The eye tracker will indicate that the driver's eyes are directed towards a vehicle, pedestrian, bicyclist or other potential danger, regardless of whether or not they are fully processing the object. So, in the absence of a crash, eye tracker evidence by itself is not enough to classify a driver as looking but not seeing. However, it is still possible to gather information on the occurrence of inattention blindness in the more general driving environment using an eye tracker. And one could argue that any increase in inattention blindness will increase the likelihood of crashes.

This is the approach taken in a recent study conducted on a driving simulator (Strayer, Drews, & Johnston, 2003). Participants drove through a virtual environment populated with various objects of relevance to safe driving while either talking or not talking on a cell phone. At the end of the drive, the participants were asked to recall several of the objects. Drivers were outfitted with an eye tracker. Only those trials on which the driver glanced at the object were included in the analysis. It is already known that drivers' eye movements while talking on a cell phone are constrained (Recarte & Nunes, 2003). Thus, it would not be surprising to find that they recalled fewer objects when on the cell phone because they probably glanced at fewer objects. However, Strayer et al. (2003) found that drivers recalled fewer objects when on the cell phone even when they glanced at those objects (a nearly 50% reduction in recall performance compared the same drivers not talking on a cell phone). Note that differences in the glance duration in the cell and no cell phone groups cannot explain the differences in recall since the analyses controlled for these differences.

In summary, the record of eye movements tells us much about how likely a driver is to crash (e.g., Scholl, Noles, Pasheva, & Sussman, 2003). In particular, the tails of the distribution of glance durations away from the forward roadway and where the driver looks in the forward roadway in anticipation of potential threats are both important indicators of the likelihood of a crash. In addition, at least some secondary tasks (e.g., conversation on a cell phone) affect the drivers' ability to process information even when the driver is fixating appropriately. We now want to move from a general consideration of the characteristics of eye behaviors which are most likely to provide us with information relevant to crashes to a discussion of the impact of individual differences, the environment inside the vehicle, and the environment outside the vehicle on eye behaviors and, indirectly, on crashes.

18.3 Individual Differences

There are important individual differences in eye behavior that appear to be related to crash rates. First, we examine the differences between novice and more experienced drivers and then we discuss next the differences between older and younger (middle aged) drivers. The older and novice groups are an obvious choice to study because of the greatly inflated risk for novice drivers during their first six months of solo licensure (McCartt et al., 2003) and older drivers after age seventy (Bryer, 2000). Within each section (novice and older drivers), we describe first what is known about the spatial and temporal characteristics of the eye behaviors. We then discuss several state-of-the-art studies that move beyond a simple description of the eye behaviors. These studies take one of several forms. Some have been used as a window into the differences in the functioning of the latent cognitive systems that govern the eye behaviors of drivers. Others have been used to create interfaces that alter the pattern of eye behaviors. Both efforts are critical to improvements in safety. We concentrate on those state-of-the-art studies that have been undertaken on driving simulators. However, where it is clear that a study not undertaken on a driving simulator has clear implications for what could be done on a simulator, we report that study as well.

18.3.1 Novice Drivers

18.3.1.1 Spatial and Temporal Profiles

One of the first studies to compare novice and experienced drivers' spatial patterns found that novice drivers scanned less broadly (horizontal and vertical range), looked less far downstream, and checked their mirrors less often (Mourant & Rockwell, 1972). A more recent study looked at the variability (variance) along the horizontal axis and found no difference between novice and experienced drivers, except on expressways, where the experienced drivers' scanning was more variable (Crundall & Underwood, 1998). Interestingly, the variability along the vertical axis was greater for novice drivers. This is consistent with a related study where the novice drivers had more glance sequences that included the road far ahead (Underwood, Chapman, Brocklehurst, Underwood, & Crundall, 2003). Taking a somewhat different approach, Pradhan et al. (2005; also see Pollatsek, Fisher, & Pradhan, 2006a) looked at anticipatory glances to areas of the roadway where a potential hazard might appear. Here, the differences were dramatic. Novice drivers can be some six times less likely to glance at an area of the roadway from which a potential hazard might appear, an example being given in the truck crosswalk scenario diagrammed in Figure 18.2.

As for the temporal characteristics of glance durations, most studies report the mean glance duration, and it is typically between 10 and 50 milliseconds shorter for experienced drivers (Laya, 1998; Crundall & Underwood, 1998). As discussed in Section 18.2 above, this is what one would expect if novice drivers were having more difficulty processing the information in each fixation. The difference in glance durations between experienced and novice drivers is magnified in situations where a hazard materializes (Chapman & Underwood, 1998). More recently, investigators have looked at the distribution of the glance durations inside the vehicle, both in the field (Wikman, Nieminen, & Summala, 1998) and on a driving simulator (Chan, Pradhan, Knodler, Pollatsek, & Fisher, 2008, 2010), for novice and experienced drivers. The results are disturbing. In the field, Wikman et al. (1998) found that glance durations inside the vehicle of at least 2.5 seconds occurred in 46% of the inexperienced drivers but only 13% of the experienced drivers. In the simulator the results are similar. Chan et al. found that for novice drivers glance durations inside the vehicle for at least 2.5 seconds occurred during the performance of 46% of the in-vehicle tasks whereas for experienced drivers glance durations of at least this length occurred during the performance only 12% of the in-vehicle tasks.

18.3.1.2 State-of-the-Art

As noted above, novice drivers' eye behaviors are different in a number of respects from those of more experienced drivers (e.g., Chapman & Underwood, 1998; Mourant & Rockwell, 1972; Pradhan et al., 2005). Here we want to focus on anticipatory eye movements and ask why it is that novice drivers make much fewer such eye movements than more experienced drivers. Obviously, there are some scenarios in which experience plays a big role. One might not expect a novice driver to have the knowledge required to make the correct anticipatory eye movement. But, there are other scenarios in which such knowledge is arguably available to almost every driver who has ridden in a car long before the driver gets his or her solo license or even before the driver practices driving with a parent during the learner's permit period. One such scenario was discussed above (Figure 18.2). Presumably all drivers, of whatever age and experience, realize that a pedestrian could be walking out from behind a vehicle stopped in a parking lane immediately in front of a marked midblock crosswalk and thereby pose a threat. Perhaps novice drivers simply fail to recognize that they need to look for the potential threat as they pass over the crosswalk. An alternative hypothesis is that novice drivers are overloaded with the demands of driving and so do not have the spare capacity left that it takes to make the prediction which would indicate that they need to launch the anticipatory eye movement.

These two hypotheses were evaluated in a recent experiment on a driving simulator using an eye tracker (Garay-Vega, Fisher, & Pollatsek, 2007). Novice and experienced drivers navigated across two different marked midblock crosswalks as they motored through a virtual world. In one scenario in the drive, no pedestrian emerged from behind the vehicle stopped in the parking lane immediately in front of the crosswalk. In a second scenario, there was a *cue* that a pedestrian might emerge: About three seconds before the driver crossed over the crosswalk a pedestrian crossed in front of the driver (but none emerged when the driver himself or herself drove over the crosswalk three seconds later). Assume that novice drivers are not looking for a hidden pedestrian because they are overloaded by the driving task, but would anticipate a pedestrian if they were not overloaded and take a precautionary glance. If so, one would predict that the novice drivers who did see the cue would now predict that a pedestrian could emerge and be likely to look to the right. Alternatively, it is possible that, even when cued, novice drivers do not realize that they need to take an anticipatory glance to the right in order to be safe. If so, there should be no increase in the percentage of anticipatory glances caused by the cue. In fact, it appears that both factors explain the differences between novice and experienced drivers. Novice drivers glanced to the right 61.4% of the time when cued but only 52.8% when not cued. Thus, load does appear to contribute somewhat to the depressed anticipatory glances for novice drivers. However, even when cued, the novice drivers were still 33.3% less likely to glance to the right than the experienced drivers (61.4% versus 94.7%). Thus, much of the difference between the two groups is that the novice drivers are not aware of the necessity of making such eye movements.

In summary, eye movements can play a critical role in determining why it is that novice drivers make so many fewer anticipatory glances than do experienced drivers. Without knowledge of eye movements in a test of the above hypotheses it would be impossible to determine whether the drivers had ever glanced at the cue and thus one could not analyze the data just from the subset of drivers who had looked at the cue. Obviously, one would not predict any effect of the cue if none of the drivers had glanced at it while approaching the crosswalk.

18.3.2 Older Drivers

18.3.2.1 Spatial and Temporal Profiles

It is well known that vision deteriorates with age and there are many studies of the effect of aging on vision (Klein, 1991; Kline et al., 1992). However, there are relatively fewer studies of either the spatial or temporal characteristics of the eye movements of older drivers (usually defined as 60 years old and older, though the crash risk does not increase appreciably until age 75). Consider first the spatial characteristics. In one study of strategic hazard anticipation (general scanning in the absence of a particular threat), older and younger drivers were asked to watch videos with and without hazards (Underwood et al., 2005). There was greater horizontal variability of the eye movements for both groups when there were hazards in the video, but there was no effect of age. In another study, a photograph of the view ahead was classified into areas of interest (Maltz & Shinar, 1999). Older adults were much more likely to concentrate their scan in particular areas of interest. As for tactical hazard anticipation, the two studies that have been reported come to opposite conclusions. In one study there was no difference in the frequency of the anticipatory eye glances of older and younger adults when compared on a driving simulator (DeRamus & Fisher, 2004); in the second study, there were large differences in the frequency of anticipatory glances both on a driving simulator (Romoser et al., 2005) and in the field (Romoser & Fisher, 2009). Perhaps these differences exist because the anticipatory glances did not necessarily require large head movements in the former study whereas they did in the latter study. The details of the latter study are discussed later in this chapter.

As for the temporal characteristics, very little has been published. In the one study available to these investigators, which compared the fixation durations of younger and older adults, there was no effect of age (Maltz & Shinar, 1999). However, the great weight of the general evidence on aging suggests that the glance durations on more difficult scenarios should be longer for older adults than they are for younger adults.

18.3.2.2 State-of-the-Art

Two examples are discussed here of how we can extend our understanding of older drivers' behaviors by examining their eye (or eye and head movements) in some detail. First, as noted in this book [chap. 26 by Trick and Caird, "Methodological Issues When Conducting Research on Older Drivers" and chap. 25 by Ball and Ackerman, "The Older Driver (Training and Assessment: Knowledge, Skills and Attitudes)"], older adults are particularly disadvantaged when turning into or across traffic. Furthermore, we know that one of the reasons they are so disadvantaged is because they fail to take a secondary look (Romoser et al., 2005). For example, a driver stopped at a T intersection waiting to take a right turn should look to the left for approaching traffic, look to the right for traffic in the opposing lane that may be turning or pedestrians or bicyclists that may be traveling in the lane into which the driver is turning, and then look again to the left for traffic as he or she is pulling out. The final look is referred to as a secondary look. It is particularly important for older adults to take this secondary look because they often accelerate slowly. What we want to understand here is why older adults are less likely to take secondary looks than younger adults.

There are several reasons older drivers are more likely to be involved in intersection crashes. In particular, they could fail to take a secondary look for opposing vehicles while navigating a turn, due either to changes in vision (which make them less able to abstract the information they need even if they do take a secondary look and therefore they do not do such; Klein, 1991), physical function (changes which make them unable to turn their head at such extreme angles; McPherson, Michael, Ostrow, & Shaffron, 1988), psychomotor coordination (which make them less able to steer as their head is turned; McGill, Yingling, & Peach, 1999), or cognition (which make them less likely to predict opponent vehicles encroaching during the turn; Caserta & Abrams, 2007). A recent experiment has addressed this question (Romoser & Fisher, 2009). In this experiment, older adults were asked to maneuver both T intersections and four-way intersections (signalized and unsignalized) on a driving simulator and in the field. Since all secondary looks require

eye movements greater than 15 degrees and such large movements are almost always accompanied by a head movement, head movements were used as a proxy for eye movements (Robinson, Erickson, Thurston, & Clark, 1972). Older adults were given tests of visual, physical, psychomotor and cognitive functions before driving in the simulator. Of the four factors, the cognitive measure (a combination of the scores on the Trail Making Test Trials 1 and 2, the Rey Auditory Verbal Learning Test, and the Rey-Osterreith Complex Figure Test) was the only significant predictor of the percentage of secondary glances for both the simulator and field drives. The psychomotor measure was not significant but did approach significance for both simulator and field drives. Neither the vision nor physical measure was a significant predictor in either drive.

In summary, we can use what we know about the larger context within which eye movements occur to predict why they may be less in evidence for older adults than they are for younger drivers. Armed with this knowledge, and in particular, the knowledge that a decrease in secondary looks is largely a function of cognitive status, one could then design training programs to remediate this decrease. In fact, such programs have been designed and appear to more than double the frequency of secondary looks on the simulator and in the field up to six weeks after training (Romoser & Fisher, 2009).

Second, consider an example provided by Maltz and Shinar (1999). It will be recalled that they found that older adults concentrated their glances in particular areas of interest when looking at a photograph of a traffic scene. The question one wants to ask at this point is whether that is because the older adults were moving their eyes about less frequently, spending large blocks of time within a given area of interest before moving to another area. Coupled with a concentration in particular areas of interest, this would suggest a reason why older adults' crash risk might increase so dramatically. In order to make this determination, Maltz and Shinar computed what they defined as a stability ratio: The ratio of the conditional probability that the next fixation in a given area of interest is within the area of interest to the unconditional probability that a fixation is within the area of interest. So, for example, if the area of interest is identified as x, the following *stability ratio* was computed:

$$P(\text{fixation } n+1 \text{ in } x \mid \text{fixation } n \text{ in } x) \, / \, P(\text{fixation } n \text{ in } x)$$

This ratio should be 1 if the probability that fixation $n + 1$ is in area of interest x is independent of where the driver was looking on fixation n. The larger it is than 1, the more likely the driver is to remain fixated in the same area of interest. Somewhat surprisingly, it is found that the stability ratio for younger drivers is 2.39 whereas the stability ratio for older drivers is 1.37. Thus older drivers are glancing more frequently between areas of interest than are younger drivers even though they are concentrating their attention on fewer areas of interest. In summary, an analysis of conditional probabilities can also provide a window on the scanning behaviors of older adults that otherwise would not be available.

18.4 The Environment Inside the Car

Distraction has long been recognized as a major contributor to automobile crashes among all drivers (Wang et al., 1996). While there are even problems when drivers interact with standard controls and displays (Klauer et al., 2006), the magnitude of the problem is likely to increase because of the growing popularity of in-vehicle tasks that require the driver to glance away from the forward roadway—most notably music retrieval operations (Chisholm, Caird, & Lockhart, 2008; Salvucci, Markley, Zuber, & Brumby, 2007) and text messaging with cell phones (Lerner & Boyd, 2004; Hosking, Young, & Regan, 2006; Reed & Robbins, 2008). Cell phones (even without text messaging) represent an enormous problem (see Strayer, Cooper, & Drews, this book, chap. 27). Recent comprehensive reviews have been devoted to driver distraction (e.g., Caird & Dewar, 2007) as has an entire book (Regan, Lee, & Young, 2008). Obviously, we can hope to cover only a small part of the entire topic.

18.4.1 In-Vehicle Controls, Gauges and Advanced Displays

18.4.1.1 Spatial and Temporal Profiles

Studies of drivers looking at controls, gauges and other displays inside the vehicle have generally included information on both the spatial and temporal characteristics of the eye glances. For example, an extensive list of the mean number of glances and glance times has been compiled for various in-vehicle tasks by Wierwille, Antin, Dingus and Hulse (1988). These include standard driver activities such as operating the fan (1.7 glances, 1.1 seconds mean glance duration) and tuning the radio (6.9 glances, 1.6 s), checking the speed (1.2 glances, 0.62 s) and time of day (1.2 glances, 0.83 s), and setting the stereo balance/volume (2.6 glances, 0.87 s) and cruise control (5.9 glances, 0.82 s). They also identified the mean glance durations associated with the identification of information on a then-current in-vehicle navigation device. The glance frequencies and times were much higher. For example, on average drivers glance 1.67 seconds (5.2 glances) at a cross street, 1.60 seconds (5.7 glances) at a road name, and 1.52 seconds at a road distance.

Tijerina, Barickman and Mazzae (2004) plotted the distribution of the glance durations to the forward roadway, the mirrors, and away from the forward roadway for drivers who were asked to navigate interstates, highways and city segments in an area of Columbus, Ohio. The distributions were all positively skewed and were categorized as log normal (though as we discussed above, it can be difficult to discriminate between log normal, ex-Gaussian and Gamma distributions). Here we are interested in the glance durations away from the forward roadway. The mean duration was 0.60 s, the standard deviation 0.46 s, and the 95th percentile 1.47 s. If we take the criteria used by Horrey and Wickens (2007) on the one hand (1.6 s) and Klauer et al. (2006) on the other hand (2.0 seconds in a six second window), from an ecological perspective, it does not appear that long glances

away from the forward roadway were especially problematic. However, in another study by Rockwell (1988), the mean glance duration was 1.1 seconds and there were a moderate number of glances greater than 2.38 seconds when just those glances at the radio were considered. Clearly, glances of such a long duration could create safety issues. No obvious explanation has been put forth for the differences in the two studies.

More detail is available on the temporal distribution from a study reported by Hada (1994). The drivers looked at information on the windshield, the instrument cluster, or the center console for as long as they felt was safe. They drove on an expressway, rural road or suburban street. Their mean glance durations were, respectively, 1.09 s, 1.00 seconds and 0.84 s. Perhaps more important, as we have learned above, the distribution of glance durations needs to be considered. The 95th percentile glance duration was 2.2 s. However, what was not explicitly reported was the percentage of glances longer than 1.6 s. Recall that Horrey and Wickens (2007) found that the 22% of the glances longer than 1.6 seconds were responsible for 86% of the crashes. By manually computing the percentage longer than 1.6 s, we find that approximately 23% were longer than 1.6 seconds in the Hada study. Thus, drivers on the open road are willing to make roughly the same proportion of long, high risk glances as drivers on a simulator. While the results are strikingly close to one another, as yet there is no evidence that the same percentage of crashes on the open road are due to these especially long glances.

18.4.1.2 State-of-the-Art

Two recent studies on driving simulators point directly to the impact on driving performance of interacting with advanced displays, in this case an iPod. In the first study (Salvucci et al., 2007), 17 drivers (no information on age was available) navigated a virtual roadway while selecting various media (music, podcast or video) on an iPod that was placed in a hands-free device holder. Each new request for an iPod task was made 30 seconds after the driver had completed the previous task, allowing for 30 seconds of control driving between secondary tasks. Our interest here is primarily in the song selection results. Selecting a song took an average of almost 32 seconds to complete. Furthermore, when drivers were selecting a song, the lateral deviation around lane center was larger than baseline. Involvement in a secondary task for such a long period of time is a clear threat to safe driving (Green, 1999a), as are increases in the lateral deviation around lane center. However, no direct measures were made of eye behaviors, so we do not know whether the glance durations away from the forward roadway were too long.

In the second study (Chisholm et al., 2008), eye behaviors were measured. Drivers between the ages of 18 and 22 (mean 19.1) were asked to navigate through a virtual world in which, among other things, a lead car braked suddenly, a pedestrian entered the roadway unexpectedly, and a vehicle pulled out into the roadway without warning. They drove a total of six different sessions, held on different days. In each session, the participants were asked to interact with an iPod, performing both easy retrieval tasks (2–3 steps, e.g., turning off the iPod) and more difficult ones (5–7 steps, turn on the iPod and find a particular song title). Eye movements were monitored throughout. Task completion times for the difficult iPod tasks did not differ from one another in the last three sessions and were around 28 s. Task completion times for the easy iPod tasks were much faster and were around 4 s. Perhaps not surprisingly, drivers engaged in a difficult iPod task performed less safely than did drivers engaged in no secondary task on all other measures. Of interest to us, the average glance durations inside the vehicle were longer when drivers were using the iPod. An in-vehicle area of interest (AOI) was defined which included any glances that were made into the vehicle, whether at the iPod device, center console, or speedometer. Mean glance durations made into the vehicle differed depending on secondary task. Longer glances were made in the difficult iPod condition ($M = 1.15$ s, $S.E. = 0.03$) compared to the baseline ($M = 0.54$ s, $S.E. = 0.04$) and easy iPod ($M = 0.66$ s, $S.E. = 0.03$) conditions, all of which differed significantly from each other. In the case of the difficult iPod condition, it can be estimated that 7.5% of the observations were over 1.6 s, the critical duration which several authors have argued leads to a greatly inflated risk of crashing (e.g., Horrey & Wickens, 2007).

Knowing that iPods create real problems, one can ask what can be done to reduce those demands. Of course, one could simply outlaw such systems. But music retrieval systems, including, of course, the radio, have long been a staple of driving and are likely to continue in that role. In order to reduce the demands, one needs to consider just what demands the iPod makes on the visual, voice, motor and memory systems of users and the effect that these demands have on eye glance behaviors such as hazard anticipation, the tail of the distribution of the glance durations, and the total time the driver spends with his or her eyes off the road.

First consider the demands. Clearly, the iPod requires multiple entries, makes demands on both the motor and visual systems (in order to see on which menu option the cursor is positioned), and requires attentional resources (in order to navigate the menu hierarchy). There are several obvious ways in which these demands could influence these measures of distraction: (1) multiple entries will lead to long task completion times, (2) the demands on the motor and visual systems will increase the frequency and duration of glances away from the forward roadway, and (3) the demands on attentional resources may influence hazard anticipation.

Perhaps the most obvious way to modify the touch interface is to use voice commands to enter the request. There are two types of voice interfaces one might consider evaluating. One is a multiple conversational turn voice interface that prompts the user at each step to enter a command appropriate to the level in the menu hierarchy. The second is a single turn voice interface that allows the user to give the entire request in one command. One would predict that, when compared with an interface which made touch and visual demands (e.g., the iPod), the two voice interfaces would reduce the number of glances away from the

forward roadway and the frequency of especially long glances because visual feedback was not necessary. Additionally, one would predict that all interfaces interfered with hazard anticipation since they make demands on the attentional resources of the drivers which could otherwise be used to scan the roadway and predict potential threats.

Recently, an experiment was run to test just these hypotheses (Garay-Vega et al., 2010). In this study, each of 17 participants between the ages of 18 and 30-years old was asked to use three different music-retrieval systems (one with a multiple entry touch interface, the iPod, one with a multiple turn voice interface and one with a single turn voice interface) while driving through a virtual world. Additionally, each participant had to complete a control drive in which no music retrieval system was operative. When compared with the touch interface, the voice interfaces reduced the total time drivers spent with their eyes off the forward roadway, especially the number of prolonged glances, as well as the total number of glances away from the forward roadway. Two indices computed were the number of short (less than 1.5 s) and long (greater than or equal to 1.5 s) glances away from the forward roadway. Both the average number of short glances per task and the average number of long glances per task were greater for the iPod than they were for either of the two voice interfaces. Finally, the iPod and single and multiple voice interfaces reduced the percentage of hazards that were anticipated below the baseline, although the difference was not significant.

18.4.2 Cell Phones

18.4.2.1 Spatial and Temporal Profiles

Perhaps the most ubiquitous task inside the car in which drivers are now engaged is conversations on a cell phone. Answering a call and dialing a phone take drivers' eyes away from the roadway ahead; talking on the phone removes a driver's attention from the roadway leading to inattention blindness (Caird & Dewar, 2007; Strayer & Drews, 2007). Texting is even worse. (See Strayer, Cooper, & Drews, this book, chap. 27.)

The effect a secondary task has on the spatial distribution of eye behaviors has been reported by a number of researchers. One of the earliest studies (Recartes & Nunes, 2000) did not use conversations per se, but instead used secondary verbal tasks (repeating words starting with a particular letter) and spatial tasks (cycling through the alphabet determining for each letter whether it had a particular geometry). A measure of variability used was the product of the standard deviations of the glance positions along the horizontal and vertical axis. There was a clear reduction in this variability measure for the various secondary tasks, compared with the variability when no secondary task was performed. The reductions ranged from 16% to 77% (the great majority of reductions were over 50%). Glances inside the car (at the speedometer and rearview mirror) were also reduced when the drivers were engaged in a secondary task.

Studies of the temporal distribution of eye behaviors while drivers are on the cell phone are more limited. We do know that dialing seven and ten digit numbers takes drivers' eyes away from the forward roadway for longer periods of time than is the case when no secondary task is present (Tijerina, Kiger, Rockwell, & Tornow, 1996). However, one would also like to know how talking on the phone affects the duration of individual glances inside the vehicle. One would expect that individual glances could be especially long because drivers, engaged in a conversation, may be less likely to notice that they are not glancing at the forward roadway. There are no such studies to the best of our knowledge, but it would be instructive to repeat the study reported above (Wierwille et al., 1988) with and without drivers engaged in a cell phone.

18.4.2.2 State-of-the-Art

In a recent study, an attempt was made to determine whether drivers engaged in a cell phone conversation can reduce some of the distracting potential of that conversation by focusing only those areas of the roadway that provide the most safety-relevant information (as described in Strayer & Drews, 2007). Participants were asked to navigate through a virtual world engaged or not engaged in a cell phone conversation and 30 target objects relevant to safe driving were placed along the roadway (e.g., pedestrians, cars, trucks, signs, billboards, etc.). At the end of the drive, the participants were asked to determine which of two objects had appeared in the drive. Each target object was paired with a foil that did not appear in the drive. Participants were not informed that they would be tested on their memory of the objects that they had seen until after the drive. The probability that a driver gave the correct answer, given that he or she had fixated the target object, was significantly smaller when the driver was engaged in a cell phone conversation. The difference was not influenced by how long the target object had been fixated. In order to understand whether the cell phone users focused on the most safety-relevant targets, the same participants were asked to rate the importance to safety of each object on a scale from 1 to 10 (10 being the most important). There was considerable variability in the ratings (the mean was 4.1 and they ranged from 1.5 to 8.0); however, there was no significant correlation between recognition memory performance and the safety-relevant ratings. Thus, there was no evidence that cell phone users were focusing primarily on safety-relevant targets.

18.5 The Environment Outside the Car

Eye movements are critical to understanding how drivers process information outside the car. This information directly impacts the behavior of the driver. On the one hand, there are the ever-present signs, signals and pavement markings. On the other hand, there are the new video billboards and dynamic message signs that may create real distractions for drivers.

18.5.1 Traffic Signs, Signals and Pavement Markings

18.5.1.1 Spatial and Temporal Profiles

Perhaps the closest researchers have come to a general study of the spatial distribution of eye behaviors on signs in general and traffic control devices in particular was reported by Mourant et al. (1969). They asked drivers either to scan the roadway looking for (a) all signs, (b) just the necessary signs, or (c) no signs—either on an open road or when following a lead vehicle. Consider just those trials where the driver looked for the necessary signs. As a percentage of the total glances when the driver was following a lead vehicle, 32.8% were directly ahead of the driver (but not on the lead or other vehicles), 40.4% on the lead (or other) vehicles, 4.3% on road and lane markers, 4.3% on road signs, 5.0% on bridges, and 13.8% out of view (not on any of the above). When the drivers were not following a lead vehicle, 54.2% of the glances were directly ahead of the driver, 4.0% on vehicles, 2.3% on road and lane markers, 6.2% on road signs, 8.1% on bridges, 25.2% out of view. What is striking about these results is how frequently the driver following a lead vehicle fixated that vehicle. These fixations came at the expense of fixations on objects which were out of view. Drivers not following a lead vehicle were almost twice as likely to scan the broader environment (25.2% versus 13.8%). This has implications for hazard anticipation, as discussed above. However, it is only infrequently that one is actually following a lead vehicle. What would be more interesting to know would be just how much increase in the volume of traffic or difficulty of negotiating a specific complex scenario reduces scanning of the broader environment.

There are also studies of the patterns of glances at particular signs (here we will combine the discussion of the spatial and temporal characteristics of the eye behaviors because they are so inextricably linked to one another). Typical of this research is a study by Zwahlen, Russ and Schnell (2003). They recorded the eye movements of drivers during nighttime while approaching six separate interchanges of a local or state road with a highway (the driver was on the local or state road). At each interchange a different pair of two identical ground-mounted diagrammatic signs, appeared 1/2 and 1/4 mile before the interchange. The overall median distance of the first look to the signs was found to be 125 m. On average, drivers glanced 2.9 times at the first sign and 2.0 times at the second sign. The average first glance durations were, respectively, 0.74 seconds and 0.66 seconds for the first and second signs, and the corresponding average last glance durations were respectively 0.68 seconds and 0.62 seconds. The average number of looks and look durations found in this study were also very close to the average number of looks and look durations for other signs reported in related studies (Zwahlen, 1981, 1987, 1988). In summary, diagrammatic signs, despite their visual complexity, do not appear to cause unnecessarily long glances. Unfortunately, this cannot be determined for certain since the distribution of eye glance durations was not reported.

18.5.1.2 State-of-the-Art

Perhaps the major use of eye movements when studying drivers' responses to different signs, signals and pavement markings is to verify that a traffic control device has indeed been fixated. Such knowledge is critical when trying to determine why a sign does or does not have an effect on drivers' behavior. For example, in one study done on a test track at night, the amount of reflective material varied on panels on the side of the track (Green & Olson, 1979). The panel contained target information which the driver needed to identify. As predicted, as the reflectivity increased so did the time that drivers glanced at the panel. Had there been no effect of the level of reflectivity, it would have been important to know whether the drivers were fixating the panels. In another study, drivers had to navigate a two-lane rural road with several curves (Zwahlen, 1987). For half of the drivers, speed advisory signs were included along with curve warning signs. The speed advisory signs were noticed by drivers, but the performance of the two sets of drivers did not differ. With the eye movement data, one can rule out the hypothesis that the speed advisory signs were ineffective simply because they were not noticed. Although these studies were not done on driving simulators, they illustrate how a driving simulator might be used here to good effect. For example, in the second study it would have been easy to vary a number of different factors associated with the geometry of the curve and the placement of the speed advisory sign that might have influenced eye movement and driving behavior.

There are two studies which have looked at anticipatory eye movements; both were on a driving simulator. One examined the location of eye movements of drivers taking a left turn at a four-way signalized intersection (Knodler & Noyce, 2005). Two sets of traffic lights were visible to drivers, one over the left lane and one over the right lane (a four-lane urban street, two lanes in each direction). Each driver negotiated three intersections with opposing traffic and three intersections without opposing traffic. The areas of interest included the protected/permissive left turn (PPLT) signal head over the left lane, the right lane signal head, opposing traffic, cross traffic, the path of travel (the cross street to the left), and the area where pedestrians would cross after the driver takes a left turn. All drivers, understandably, looked at opposing traffic at all intersections. When there was no opposing traffic, an anticipatory glance before turning for a pedestrian in the crosswalk in their intended path of travel was made on 24% of the intersection trials. When there was opposing traffic, such glances were made on only 3% of the intersection trials. Similarly, when there was no opposing traffic, an anticipatory glance for red light runners (cross traffic) was made on 21% of the intersection trials. When there was opposing traffic, such glances were made on only 3% of the trials. To the best of our knowledge, this is the first time that traffic in one direction has been shown to affect anticipatory glances in another direction.

Finally, there is one study that has looked at eye movements and pavement markings on a driving simulator (Garay-Vega, Fisher, & Knodler, 2008). The standard pavement markings included a stop line in front of the crosswalk. The alternative pavement markings included advance yield markings some 30 feet upstream of the crosswalk as well as a "Yield Here to Pedestrians" sign placed opposite the advance yield markings (Figure 18.3). This is known as a multiple threat scenario because the pedestrian, when crossing, is threatened both by vehicles stopped immediately adjacent to him or her as well as vehicles traveling in the adjacent lane. Advance yield markings increase the likelihood that a driver will yield for a pedestrian when there is an adequate sight distance (Van Houten, Malenfant, & McCusker, 2001). However, one would like to know whether drivers actually look more frequently for a pedestrian when the sight distance is not adequate. There is a clear role here for driving simulators because the multiple threat scenario cannot be staged safely in the field. With this as background, Garay-Vega asked drivers to navigate through a virtual world where either standard or alternative pavement markings were introduced at marked midblock crosswalks which covered four travel lanes, two in each direction. Participants in the advance yield marking group looked for pedestrians 69% of the time. Participants in the standard marking group looked for pedestrians 47% of the time. The differences were not statistically significant, though the results are in the direction one would expect.

18.5.2 External Distractions

18.5.2.1 Spatial and Temporal Profiles

Video billboards are perhaps the major external distraction of concern to appear in the last several years (see also Milloy & Caird, this book, chap. 16). In a recent study of the impact of video billboards on eye movements, it was found that drivers

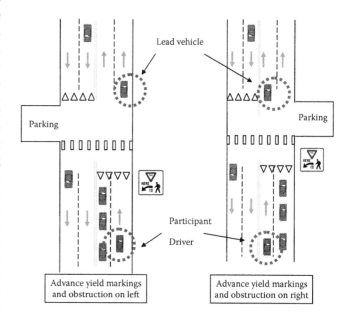

FIGURE 18.3 Multiple Threat Scenarios. (Threats consist of three vehicles stopped for pedestrian—in the left or right hand travel lanes—and participant driver's vehicle circled at the bottom of the figure in travel lane adjacent to pedestrian.)

in Toronto glanced more frequently (on average 1.45 glances per billboard) and glanced longer (defined here as glances greater than 0.75 s) at video billboards than they did at static billboards (Beijer, Smiley, & Eisenman, 2004). A second study provided more information, again on video billboards in the Toronto area. In particular, Smiley, Smahel and Eisenman (2003) reported that drivers glanced at video billboards 48% of the time with 23% of the glances being greater than 0.75 sec in duration.

18.5.2.2 State-of-the-Art

The above studies did not address a question central to the spatial distribution of eye movements that is also critical to an evaluation of their safety. In particular, one can ask whether the billboards are actually diverting drivers' attention from the roadway ahead and to the adjacent lane of traffic (typically the side opposite the billboard). Shinar (2007) reports an unpublished study which addresses this question. In this study, drivers' eye movements were monitored as they approached a video billboard located on the right hand side of the road that was either facing the driver or facing away from the driver (i.e., facing drivers approaching from the opposite direction). When the billboard was facing the driver, 10% of the glances were to the left, 68% were to the forward roadway and 23% were to the billboard on the right. When the billboard was facing away from the driver, 19% of the glances were to the left, 71% were to the forward roadway and 10% were to the right. It is interesting that the billboard had virtually no effect on the glances to the forward roadway. However, it is not clear whether the reduction in the glances to the left when the billboard face is visible will have an impact on safety.

In a second study described in this book (see Milloy & Caird, chap. 16), drivers' eyes were monitored on a driving simulator as they navigated past wind farms (a line of wind turbines which were rotating at different speeds). The wind farms were off to the right hand side of the road and the drivers were told to stay in the right hand lane on the freeway. There were two separate drives, each lasting about 12 minutes. A lead vehicle whose speed was linked to the speed of the participant driver would occasionally brake hard, once opposite a wind farm and once opposite open terrain in each drive (for a total of four lead vehicle braking events). The participant needed to brake hard and/or steer around the lead vehicle during these braking events. One can reasonably hypothesize that wind farms would prove especially attractive, thus making it difficult to see a lead vehicle braking in front of the driver when the driver was focused on the wind farm. Only the results of the eye movement analyses are discussed here. Out of the 96 total braking events, there was only one actual collision in the presence of a wind farm. Consistent with the analyses reported in Horrey and Wickens (2007) and Klauer et al. (2006), the driver was glancing away from the forward roadway (at the speedometer) just before the collision occurred. The driver had glanced at the wind farm six seconds prior to the crash. There is no obvious reason that the driver would glance at the speedometer soon after glancing at the wind farms. And even were that to

be the case, such glances clearly are important. Thus, it does not appear that the wind farm is involved, directly or indirectly, in this crash, counter to what one might have expected.

18.5.3 Modeling Drivers' Visual Scanning

While we know much about where drivers look, much less is known about why they scan the way they do. Models of visual attention, in particular those derived from supervisory control, might be useful in this respect. In this section, we briefly describe a few of these models and their underlying parameters. It is readily apparent that these concepts map onto many of the concepts discussed in this chapter, in particular, those described in Sections 4 and 5. We also add that driving simulation is a useful tool in the development and testing of different models of visual attention, though this should not be the ultimate proving ground for any such model.

Models of supervisory control tend to characterize the eye as a single server queue and visual scanning as the means of serving the queue (e.g., Senders, 1964; Moray, 1986). As noted earlier, these models tend to differ from visual search models in that the observer is monitoring a series of dynamic areas of interest rather than searching for a static target. Also, unlike search tasks, the pattern of scanning and the distribution of glances towards the different AOIs are the critical variable, more so than target detection times. Finally, the emphasis is on the observer's knowledge of when to look where as opposed to knowing simply where to look. This knowledge is a function of many factors, a few of which are described below.

In 1964, John Senders described a simple model of visual attention based on information bandwidth, which postulated that observers' visual sampling of locations (e.g., AOIs) is associated with the frequency with which relevant information occurs at that location (Senders, 1964). For example, drivers will seek out information regarding lane position much more frequently when driving on a curvy road, compared to a long, straight stretch with wide lanes. Other factors, such as speed, level of traffic, wind turbulence, condition of roadway, and weather can also impact the uncertainty in lane position and, hence, the relative information bandwidth. Note that knowledge of (or calibration to) information bandwidth is linked to experience—a function that might explain some of the discrepancies in the scanning of novice drivers, compared to more experienced ones (e.g., Underwood, Chapman, Bowden, & Crundall, 2002). The rate at which information is presented, and the resulting impact on scanning, is also an important concern for in-vehicle devices.

The relative value of the information is also an important determinant of visual scanning. As shown by Carbonell (1966), observers tend to try to maximize the benefits of correctly processing certain information (or minimize the costs of missing it) relative to other information, which may be less critical. The value of perceiving a vehicle that has cut across a driver's path, for example, is much greater than correctly perceiving the items on display in a storefront window.

Unfortunately, information bandwidth and value are not always associated positively. There are many cases, including the situation described above, in which highly critical information occurs very infrequently.

Wickens and colleagues subsequently have combined these factors in a model of visual scanning, the SEEV (Salience, Effort, Expectancy, Value) model (e.g., Wickens, Goh, Helleburg, Horrey, & Talleur, 2003; Horrey, Wickens, & Consalus, 2006). Expectancy (related to information bandwidth) and value are consistent with the above description. Salience refers to the conspicuity, and hence the attention-grabbing potential of information at a given location (e.g., Itti & Koch, 2000). A bright, flashy video billboard (as described in Section 5.2.1) is much more likely to draw the eyes (and for longer) than a static one. Finally, the amount of effort required to access information may inhibit visual scanning (e.g., Sheridan, 1970). For example, drivers may be less willing to adequately shoulder check (requiring a long re-fixation, including head and/or torso turn) compared to checking the mirrors, which are more proximally located. It is important to note that, in many cases, the value of the information should offset the amount of effort required; however, this is not always the case.

While a full description of the application of these models is beyond the scope of this chapter, preliminary work in simulated flight (Wickens et al., 2003) and driving (Horrey et al., 2006) suggest that these models may be useful in predicting scanning to different AOIs in a dynamic, multitask environment. The practical upside to using models of visual scanning is that they can help predict how drivers will allocate their attention under different circumstances and for various tasks, while pointing out potential vulnerabilities for missing critical safety-relevant information in the traffic environment. Ultimately, the validation of various models of visual attention will require testing and application in the real world. That being said, driving simulation is a very effective tool for early development and parameter testing.

18.6 Conclusions

The analysis of eye behaviors can provide real insights into why crashes occur and how the interface with the driver (signs, music retrieval systems, and so on) can be designed in ways which minimize the risk. At the outset, we spent some time discussing the research in reading. In our concluding remarks we want to come back to this research and talk briefly about the future of eye movement research in driving. The research in reading was instructive both for its content and its methodology. In terms of content, we learned that very little is seen outside the location which is fixated and that the gaze duration on a given object is a very finely tuned measure of the processing difficulty of the information which is being fixated. The former findings make clear the important role that eye location data plays in driving. What is not fixated is generally not processed. Thus, although in theory one might have critiqued the studies of anticipatory glances (e.g., Pradhan et al., 2005) because those studies considered only fixations on a very narrow area as indicating processing of the information in that area, the technique appears to be

justified. In this regard, it would be interesting to put together the sequence of fixations prior to those where an anticipatory glance is made. For example, one might ask whether the odds of glancing to the right in the truck crosswalk scenario (Figure 18.2) are increased greatly if the driver glances first at a sign indicating that a pedestrian crosswalk is ahead. And one could move backwards from there as well. The triggering events for anticipatory glances strike us as a very important use of eye movement behaviors in the future. The latter findings also have clear implications for driving research. It has been found in general that more complex situations require longer glance durations in driving. However, little is known about the factors that affect complexity. Using the length and distribution of the glance duration as an index of complexity could lead one to uncover just what and what does not make a given scenario increasingly complex, something for which we have little if any experimental evidence at the moment. Of course, these are only two of many different directions in which one could expand on the use of the above to general sets of results in driving.

As noted above, not only does the general content of the results of studies of eye movements in reading have implications for what research might be pursued in driving, but the general methodologies do as well. There are truly exciting possibilities here, especially with the recent extension of the moving window capabilities so long used in reading into the world of driving. For example, up until recently, it has been all but impossible to trigger roadway events based on where a driver is looking. However, one can now thread the vehicle and eye data in real time and determine when the driver is looking away from the forward roadway inside the vehicle, at a rear-view mirror, or wherever. One could then stage a lead vehicle braking event half of the time a driver was glancing inside the vehicle and half of the time the driver was glancing at the forward roadway. This would greatly increase the opportunities to understand in more detail just how dangerous it is to look away from the forward roadway because one is not relying on chance to synchronize looks away from the forward roadway with lead vehicle braking events.

Key Points

- Much is known about where individuals glance and for how long they glance as a function of individual differences, activities, and distractions inside the car, and events and stimuli outside the car.
- Where drivers glance is a very good indication of what they are processing; how long they glance is a good indication of the difficulty of what they are processing.
- Crashes can be traced to three factors in which knowledge of eye movements is critical: especially long glances away from the forward roadway, failures to anticipate hazards, and looking but not seeing.
- Novice drivers fail to anticipate hazards much more often than experienced drivers, both because they do not appreciate the need to make anticipatory glances and because they are overloaded with the demands of driving.

- Older drivers fail to anticipate hazards at T intersections largely because of cognitive declines, not because of physical, psychomotor, or visual loss.
- In-vehicle systems such as iPods lead to frequent and especially long glances away from the forward roadway, which can be partly mitigated by music retrieval systems with a single entry voice interface.
- Cell phone users are likely to look at safety-relevant information without fully processing it; they also fail to adjust their pattern of glances to focus on the most safety-relevant stimuli in the environment.
- Traffic signs and pavement markings can increase anticipatory eye movements.
- Digital billboards do not reduce the time the drivers spend with their eyes on the forward roadway, but they do reduce the time that drivers spend scanning the sides of the roadway.

Keywords: Crashes, Cell Phones, Digital Billboards, Novice Drivers, Eye Tracking, Eye Glances, External Distractions, In-vehicle Distractions, Older Drivers, Music Retrieval Systems, Traffic Control Devices

Acknowledgments

Portions of this research were funded by grants from the National Institutes of Health (1R01HD057153-01) and the National Science Foundation (Equipment Grant SBR 9413733 for the partial acquisition of the driving simulator) to Donald Fisher and Alexander Pollatsek.

Web Resources

The *Handbook* web site contains supplemental materials for this chapter including a color version of Figure 18.2 and two videos:

Web Figure 18.1: Truck Crosswalk Scenario Perspective View (color version of Figure 2).

Web Video 18.1: Video recording of a driver who fails to take an anticipatory look to the right for pedestrians entering a marked midblock crosswalk who are potentially obscured by a vehicle stopped in the parking lane.

Web Video 18.2: Video recording of a driver who does take an anticipatory look to the right for pedestrians entering a marked midblock crosswalk who are potentially obscured by a vehicle stopped in the parking lane.

Key Readings

Chapman, P. R., & Underwood, G. (1998). Visual search of driving situations: Danger and experience. *Perception, 27,* 951–964.

Horrey, W. J., & Wickens, C. D. (2007). In-vehicle glance duration distributions, tails, and model of crash risk. *Transportation Research Record, 2018,* 22–28.

Maltz, M., & Shinar, D. (1999). Eye movements of younger and older drivers. *Human Factors, 41,* 15–25.

Mourant, R. R., & Rockwell, T. H. (1972). Strategies of visual search by novice and experienced drivers. *Human Factors, 14,* 325–335.

Pradhan, A. K., Hammel, K. R., DeRamus, R., Pollatsek, A., Noyce, D. A., & Fisher, D. L. (2005). The use of eye movements to evaluate the effects of driver age on risk perception in an advanced driving simulator. *Human Factors, 47,* 840–852.

Strayer, D. L., Drews, F. A., & Johnston, W. A. (2003). Cell phone induced failures of visual attention during simulated driving. *Journal of Experimental Psychology: Applied, 9,* 23–52.

References

Beijer, D. D., Smiley, A., & Eizenman, M. (2004). Observed driver glance behavior at roadside advertising. *Transportation Research Record, 1899,* 96–103.

Bryer, T. (2000). Characteristics of motor vehicle crashes related to aging. In K. W. Schaie & M. Pietrucha (Eds.), *Mobility and transportation in the elderly* (pp. 157–206). New York: Springer Publishing Company.

Caird, J., & Dewar, R. (2007). Driver distraction. In R. Dewar & P. Olson (Eds.), *Human factors in traffic safety* (2nd ed., pp. 196–229). Tucson, AZ: Lawyers & Judges Publishing.

Carbonell, J. R. (1966). A queuing model of many-instrument visual sampling. *IEEE Transactions on Human Factors in Electronics, HFE-7,* 157–164.

Castelhano, M. S., Pollatsek, A., & Cave, K. R. (2008). Typicality aids search for an unspecified target, but only in identification, and not in attentional guidance. *Psychonomic Bulletin & Review, 15,* 795–801.

Chan, E., Pradhan, A. K., Knodler, M. A., Pollatsek, A., & Fisher, D. L. (2008). *Empirical evaluation on a driving simulator of the effect of distractions inside and outside the vehicle on drivers' eye behaviors.* Proceedings of the 87th Transportation Research Board Annual Meeting CD-ROM. Washington, DC: National Research Council.

Chan, E., Pradhan, A. K., Knodler, M. A., Pollatsek, A., & Fisher, D. L. (2010). Evaluation on a driving simulator of the effect on drivers' eye behaviors from distractions inside and outside the vehicle. *Transportation Research F, 13,* 343–353.

Chapman, P. R., & Underwood, G. (1998). Visual search of driving situations: Danger and experience. *Perception, 27,* 951–964.

Chisholm, S. L., Caird, J. K., & Lockhart, J. (2008). The effects of practice with MP3 players on driving performance. *Accident Analysis & Prevention, 40,* 704–713.

Crundall, D. E., & Underwood, G. (1998). Effects of experience and processing demands on visual information acquisition in drivers. *Ergonomics, 41,* 448–458.

DeRamus, R., & Fisher, D. L. (2004). The effect of driver age and experience on risk assessment and risk prediction. *Proceedings of the 48th human factors and ergonomics society's annual meeting* (pp. 2627–2631), New Orleans, LA. Santa Monica, CA: Human Factors and Ergonomics Society.

Dingus, T. A., Klauer, S. G., Neale, V. L., Petersen, A., Lee, S. E., Sudweeks, J., . . . Knipling, R. R. (2006). *The 100-Car Naturalistic Driving Study, Phase II—Results of the 100-Car Field Experiment* (Rep. DOT HS 810 593). Washington, DC: National Highway Traffic Safety Administration.

Duffy, S. A., Morris, R. K., & Rayner, K. (1988). Lexical ambiguity and fixation times in reading. *Journal of Memory and Language, 27*, 429–446.

Garay-Vega, L., Fisher, D. L., & Knodler, M. (2008). *Evaluation of drivers' performances in response to sight-limited crash scenarios at midblock crosswalks: Benefits of advance yield markings and symbolic signage.* Proceedings of the 52nd Annual Meeting of the Human Factors and Ergonomics Society, New York. Santa Monica, CA: Human Factors and Ergonomics Society.

Gary-Vega, L., Fisher, D. L., & Pollatsek, A. (2007). Hazard anticipation of novice and experienced drivers: Empirical evaluation on a driving simulator in daytime and nighttime conditions. *Transportation Research Record, 2009*, 1–7.

Garay-Vega, L., Pradhan, A. K., Weinberg, G., Schmidt-Nielsen, B., Harsham, B., Shen, Y., . . . Fisher, D. L. (2010). Evaluation of different voice and touch interfaces to in-vehicle music retrieval systems. *Accident Analysis & Prevention, 42*, 913–920.

Green, P. (2007). Where do drivers look while driving (and for how long)? In R. E. Dewar & R. Olson (Eds.), *Human factors in traffic safety* (2nd ed., pp. 57–82). Tucson, AZ: Lawyers & Judges Publishing.

Green, P. (1999a, April 19–22). *The 15 second rule for driver information systems.* Invited presentation at the ITS America Ninth Annual Meeting. Washington, DC.

Green, P. (1999b). *Visual and task demands of driver information systems* (Tech. Rep. 98–16). Ann Arbor, MI: University of Michigan Transportation Research Institute.

Green, P., & Olson, P. L. (1979). Eye fixations of drivers in response to various retroreflective treatments of a semitrailer. *HSRI Research Review, 10*(3), 19–24.

Hada, H. (1994). *Drivers' visual attention to in-vehicle displays: Effects of display location and road types* (Tech. Rep. UMTRI-94-9). Ann Arbor, MI: University of Michigan Transportation Research Institute.

Horrey, W. J., & Wickens, C. D. (2007). In-vehicle glance duration distributions, tails, and model of crash risk. *Transportation Research Record, 2018*, 22–28.

Horrey, W. J., Wickens, C. D., & Consalus, K. P. (2006). Modeling drivers' visual attention allocation while interacting with in-vehicle technologies. *Journal of Experimental Psychology: Applied, 12*(2), 67–78.

Hosking, S., Young, K., & Regan, M. (2006). *The effects of text messaging on young novice driver performance* (Report No. 246). Victoria, Australia: Monash University Accident Research Centre.

International Standards Organization. (2002). *Road vehicles—measurement of driver visual information with respect to transport information and control behavior—Part I: Definitions and standards* (ISO Committee Standard 15007-1). Geneva, Switzerland: International Standards Organization.

Itti, L., & Koch, C. (2000). A saliency-based search mechanism for overt and covert shifts of visual attention. *Vision Research, 40*, 1489–1506.

Jonides, J. (1981). Voluntary versus automatic control over the mind's eye's movement. In J. Long & A. Baddeley (Eds.), *Attention and performance IX* (pp. 187–203). Hillsdale, NJ: Lawrence Erlbaum Associates.

Klauer, S. G., Dingus, T. A., Neale, V. L., Sudweeks, J. D., & Ramsey, D. J. (2006). *The impact of driver inattention on near-crash/crash risk: An analysis using the 100-car naturalistic driving study data.* (Rep. DOT HS 810 594). Washington, DC: National Highway Traffic Safety Administration.

Klein, R. (1991). Age-related eye disease, visual impairment, and driving in the elderly. *Human Factors, 33*, 521–525.

Kline, D. W., Kline, T. J. B., Fozard, J. L., Kosnik, W., Schieber, F., & Sekuler, R. (1992). Vision, aging, and driving: The problems of older drivers. *Journal of Gerontology: Psychological Sciences, 47*, 27–34.

Knodler, M. A., & Noyce, D. A. (2005). *Tracking driver eye movements at permissive left-turns.* Proceedings of the Third International Driving Symposium on Human Factors in Driver Assessment, Training and Vehicle Design. Iowa City, IA: University of Iowa Public Policy Center.

Laya, O. (1992). Eye movements in actual and simulated curve negotiation tasks. *IATSS Research, 16*(1), 15–26.

Lerner, N., & Boyd, S. (2004). *Task report: On-road study of willingness to engage in distracting tasks.* Report jointly funded under National Highway Safety Administration (Contract DTNH22-99-D-07005), and National Institute of Child Health and Human Development (Contract GS-23F-8144H). Rockville, MD: Westat.

Louma, J. (1988). Drivers' eye fixations and perceptions. In A. G. Gale, M. H. Freeman, C. M. Haslegrave, P. Smith, & S. P. Taylor (Eds.), *Vision in Vehicles II* (pp. 231–237). Amsterdam, the Netherlands: Elsevier Science.

Maltz, M., & Shinar, D. (1999). Eye movements of younger and older drivers. *Human Factors, 41*, 15–25.

McCartt, A. T., Shabanova, V. I., & Leaf, W. A. (2003). Driving experience, crashes and traffic citations of teenage beginning drivers. *Accident Analysis & Prevention, 35*, 311–320.

McConkie, G. W., Kerr, P. W., Reddix, M. D., & Zola, D. (1988). Eye movement control during reading: I. The location of initial eye fixations on words. *Vision Research, 28*, 1107–1118.

McGehee, D. V. (2001). New design guidelines aim to reduce driver distraction. *Human Factors and Ergonomics Society Bulletin, 44*, 1–3.

McGill, S. M., Yingling, V. R., & Peach, J. P. (1999). Three-dimensional kinematics and trunk muscle myoelectric activity in the elderly spine: A database compared to young people. *Clinical Biomechanics, 14*, 389–395.

McKnight, J. A., & McKnight, S. A. (2003). Young novice drivers: Careless or clueless. *Accident Analysis & Prevention, 35*, 921–925.

McPherson, K., Michael, J., Ostrow, A., & Shaffron, P. (1988). *Physical fitness and the aging driver: Phase I.* Washington, DC: AAA Foundation for Traffic Safety.

Moray, N. (1986). Monitoring behavior and supervisory control. In K. R. Boff, L. Kaufman, & J. P. Thomas (Eds.), *Handbook of perception and human performance* (Vol. 2, pp. 40.1–40.51). New York: John Wiley & Sons.

Mourant, R. R., & Rockwell, T. H. (1970). Mapping eye-movement pattern to the visual scene in driving: An exploratory study. *Human Factors, 12*, 81–87.

Mourant, R. R., & Rockwell, T. H. (1972). Strategies of visual search by novice and experienced drivers. *Human Factors, 14*, 325–335.

Mourant, R. R., Rockwell, T. H., & Rackoff, N. J. (1969). Drivers' eye movements and visual workload. *Highway Research Record, 292*. Washington, DC: National Academy Press.

Pollatsek, A., Fisher, D. L., & Pradhan, A. K. (2006a). Identifying and remediating failures of selective attention in younger drivers. *Current Directions in Psychological Science, 15*, 255–259.

Pollatsek, A., Reichle, E. D., & Rayner, K. (2006b). Tests of the E-Z reader model: Exploring the interface between cognition and eye-movement control. *Cognitive Psychology, 52*, 1–56.

Pradhan, A. K., Hammel, K. R., DeRamus, R., Pollatsek, A., Noyce, D. A., & Fisher, D. L (2005). The use of eye movements to evaluate the effects of driver age on risk perception in an advanced driving simulator. *Human Factors, 47*, 840–852.

Rayner, K. (1998). Eye movements in reading and information processing: Twenty years of research, *Psychological Bulletin, 124*, 372–422.

Rayner, K., Pollatsek, A., & Starr, M. S. (2003). Reading. In A. F. Healy & R. W. Proctor (Eds.), *Handbook of psychology: Experimental psychology* (Vol. 4, pp. 549–574). New York: John Wiley & Sons.

Rayner, K., Warren, T., Juhasz, B. J., & Liversedge, S. P. (2004). The effect of plausibility on eye movements in reading. *Journal of Experimental Psychology: Learning, Memory and Cognition, 30*, 1290–1301.

Rayner, K., Well, A. D., Pollatsek, A., & Bertera, J. H. (1982). The availability of useful information to the right of fixation in reading. *Perception & Psychophysics, 31*, 537–550.

Recartes, R. A., & Nunes, L. M. (2003). Mental workload while driving: Effects on visual search, discrimination, and decision making. *Journal of Experimental Psychology: Applied, 9*(2), 119–137.

Reed, N., & Robbins, R. (2008, September). *The effect of text messaging on driver behavior: A simulator study* (Tech. Rep. 25-09-2008). Wokingham, Berkshire, England: Transport Research Laboratory (TRL).

Regan, M. A., Lee, J. D., & Young, K. (2008). *Driver distraction: Theory, effects, and mitigation.* Boca Raton, FL: CRC Press.

Rensink, R. A. (2002). Change detection. *Annual Review of Psychology, 53*, 245–277.

Rensink, R. A., O'Regan, J. K., & Clark, J. J. (2000). On the failure to detect changes in scenes across brief interruptions. *Visual Cognition, 7*(1/2/3), 127–145.

Robinson, G. H., Erikson D. J., Thurston, G. L., & Clark, R. L. (1972). Visual search by automobile drivers. *Human Factors, 14*(4), 315–323.

Rockwell, T. H. (1988). Spare visual capacity in driving-revisited. In A. G. Gale, M. H. Freeman, C. M. Haslegrave, P. Smith, & S. P. Taylor (Eds.), *Vision in Vehicles II* (pp. 317–324). Amsterdam, the Netherlands: Elsevier Science.

Romoser, M., & Fisher, D. L. (2009.) The effect of active versus passive training strategies on improving older drivers' scanning for hazards while negotiating intersections. *Human Factors, 51*(5), 652–668.

Romoser, M., Fisher, D. L., Mourant, R., Wachtel, J., & Sizov, K. (2005). The use of a driving simulator to assess senior driver performance: Increasing situational awareness through post-drive one-on-one advisement. *Proceedings of the 3rd international driving symposium on human factors in driver assessment, training and vehicle design* (pp. 456–463), Rockport, ME. Iowa City: University of Iowa Public Policy Center.

Sabey, B. E., & Staughton, G. C. (1975, September). *Interacting roles of road, environment, and road user in accidents.* Paper presented at the Fifth International Conference of the International Association for Accident and Traffic Medicine and the 3rd International Conference on Drug Abuse of the International Council on Alcohol and Addiction, London, England.

Salvucci, D. D., Markley, D., Zuber, M., & Brumby, D. P. (2007). iPod distraction: Effects of portable music-player use on driver performance. *Proceedings of the SIGCHI conference on human factors in computing system: CHI 2007* (pp. 243–250). New York: ACM Press.

Scholl, B. J., Noles, N. S., Pasheva, V., & Sussman, R. (2003). Talking on a cellular telephone dramatically increases "sustained inattentional blindness" [Abstract]. *Journal of Vision, 3*(9), Abstract 156, 156a, http://journalofvision.org/3/9/156/, doi:10.1167/3.9.156.

Senders, J. (1964). The human operator as a monitor and controller of multidegree of freedom systems. *IEEE Transactions on Human Factors in Electronics, HFE-5*, 2–6.

Sheridan, T. (1970). On how often the supervisor should sample. *IEEE Transactions on Systems Science and Cybernetics, SSC-6*, 140–145.

Shinar, D. (2007). *Traffic safety and human behavior.* Amsterdam, the Netherlands: Elsevier Science.

Shinar, D. (2008). Looks are (almost) everything: Where drivers look to get information. *Human Factors, 50*, 380–384.

Simons, D. J., & Chabris, C. F. (1999). Gorillas in our midst: Sustained inattentional blindness for dynamic events. *Perception, 28*(9), 1059–1074.

Smiley, A., Smahel, T., & Eisenman, M. (2004). The impact of video advertising on driver fixation patterns. *Transportation Research Record, 1899*, 76–83.

Society of Automotive Engineers. (2000). *Definitions and measures related to driver behavior using video based techniques* (SAE Recommended Practice J2396). Warrendale, PA: Society of Automotive Engineers.

Society of Automotive Engineers. (2004). *Navigation and route guidance function accessibility while driving* (SAE Recommended Practice J2364). Warrendale, PA: Author.

Strayer, D .L., & Drews, F. A. (2007). Cell-phone–induced driver distraction. *Current Directions in Psychological Science, 16*, 128–131.

Strayer, D. L., Drews, F. A., & Johnston, W. A. (2003). Cell phone induced failures of visual attention during simulated driving. *Journal of Experimental Psychology: Applied, 9*, 23–52.

Tijerina, L., Barickman, F. S., & Mazzae, E. M. (2004). *Driver eye glance behavior during car following.* Washington, D. C.: US Department of Transportation, National Highway Traffic Safety Administration.

Tijerina, L., Kiger, S., Rockwell, T., & Tornow, C. (1996). *Heavy vehicle driver workload assessment, Task 7A: In-car text message system and cellular phone use by heavy vehicle drivers on the road* (Rep. DOT HS 808 467 7A). Washington, DC: US Department of Transportation, National Highway Traffic Safety Administration.

Treat, J. R. (1980). A study of pre-crash factors involved in traffic accidents. *The HSRI, Research Review, 10*(1), 1–35.

Treat, J. R., Tumbas, N. S., McDonald, S. T., Shinar, D., Hume, R. D., Mayer, R. R., . . . Castallen, N. J. (1979). *Tri-level study of the causes of traffic accidents: Final report. Volume I: Causal factor tabulations and assessments* (Rep. DOT HS 805 085). Washington, DC: US Department of Transportation, National Highway Traffic Safety Administration.

Underwood, G., Chapman, P., Bowden, K., & Crundall, D. (2002). Visual search while driving: Skill awareness during inspection of the scene. *Transportation Research Part F: Traffic Psychology and Behaviour, 5*, 87–97.

Underwood, G., Chapman, P., Brocklehurst, N., Underwood, J., & Crundall, D. (2003). Visual attention while driving: Sequences of eye fixations made by experienced and novice drivers. *Ergonomics, 46*, 629–646.

Underwood, G., Phelps, N., Wright, C., van Loon, E., & Galpin, A. (2005). Eye fixation scan paths of younger and older drivers in a hazard perception task. *Ophthalmic & Physiological Optics, 25*, 346–356.

van Houten, R., Malenfant, J. E. L., & McCusker, D. (2001). Advance yield markings reduce motor vehicle/pedestrian conflicts at multilane crosswalks with an uncontrolled approach. *Transportation Research Record, 1773*, 69–74.

Wade, N., & Tatler, B. (2005). *The moving tablet of the eye: The origins of modern eye movement research.* Oxford, England: Oxford University Press.

Wang, J. S., Knipling, R. R., & Goodman, M. J. (1996). The role of driver inattention in crashes: New statistics from the 1995 Crashworthiness Data System. *Proceedings of the 40th annual meeting of the association for the advancement of automotive medicine* (pp. 377–392).

Wickens, C. D., Goh, J., Helleburg, J., Horrey, W. J., & Talleur, D. A. (2003). Attentional models of multi-task pilot performance using advanced display technology. *Human Factors, 45*(3), 360–380.

Wierwille, W. W. (1993). Visual and manual demands of in-car controls and displays. In B. Peacock & W. Karwowski (Eds.), *Automotive ergonomics* (pp. 299–320). Washington, DC: Taylor & Francis.

Wierwille, W. W. (1995). Development of an initial model relating driver in-vehicle visual demands to accident involvement. *Third annual mid-atlantic human factors conference proceedings* (pp. 1–7). Blacksburg, VA: Virginia Polytechnic and State University.

Wierwille, W. W., Antin, J. F., Dingus, T. A., & Hulse, M. C. (1988). Visual attention demand of an in-car navigation display system. In A. G. Gale, M. H. Freeman, C. M. Haselgrave, P. Smith, & S. P. Taylor (Eds.), *Vision in Vehicle II* (pp. 307–316). Amsterdam, the Netherlands: Elsevier Science.

Wierwille, W. W., & Tijerina. L. (1998). Modeling the relationship between driver in-vehicle visual demands and accident occurrence. In A. G. Gale, I. D. Brown, C. M. Haslegrave, & S. P. Taylor (Eds.), *Vision in vehicles—VI* (pp. 233–243). Amsterdam, the Netherlands: Elsevier Science.

Wikman, A., Nieminen, T., & Summala, H. (1998). Driving experience and time-sharing during in-car tasks on roads of different width. *Ergonomics, 41*(3), 358–372.

Williams, C. C. & Pollatsek, A. (2007). Searching for an O in an array of Cs: Eye movements track moment-to-moment processing in visual search. *Perception & Psychophysics, 69*, 372–381.

Wolfe, J. M. (1994). Guided search 2.0: A revised model of visual search. *Psychonomic Bulletin and Review, 1*, 202–238.

Zwahlen, H. T. (October 1981). Driver eye scanning of warning signs on rural highways. *Proceedings of the 25th annual meeting of the human factors and ergonomics society* (pp. 33–37).

Zwahlen, H. T. (1987). Advisory speed signs and curve signs and their effect on driver eye scanning and driving performance, *Transportation Research Record, 1111*, 110–120.

Zwahlen, H. T. (1988). Stop ahead and stop signs and their effect on driver eye scanning and driving performance, *Transportation Research Record, 1168*, 16–24.

Zwahlen, H. T., Russ, A., & Schnell, T. (2003). Driver eye scanning behavior while viewing ground-mounted diagrammatic guide signs before entrance ramps at night. *Transportation Research Record, 1843*, 61–69.

19

Situation Awareness in Driving

Leo Gugerty
Clemson University

Abstract

The Problem. This chapter focuses on three questions. First, what are the attentional component processes that drivers use to maintain situation awareness (SA)? Second, how can we measure drivers' ability to maintain SA, especially using driving simulators? Third, what is the empirical evidence, especially using simulators, that the particular component processes described here are key parts of maintaining SA and that these components affect driving performance? SA is defined here as the updated, meaningful knowledge of a changing, multifaceted situation that drivers use to guide choice and action. Regarding the first question, it is proposed in this chapter that SA involves component processes of focal vision (including attention allocation within tasks, event comprehension, and task management across concurrent tasks) as well as ambient vision processes (including attention capture by sudden peripheral events). *Role of Driving Simulators.* SA is a complex process that requires assessment by a variety of online (during driving) and offline (post-driving) measures. Driving simulators are used in many SA measures. *Key Results of Driving Simulator Studies.* Research using these measures shows that most of the above components of SA can be trained, improve with driving experience, and correlate positively with safe driving.

19.1 Introduction

Driving is a very attention-demanding task, isn't it? Either consciously or unconsciously, drivers attend to their route location, the shape of the road ahead, nearby traffic, signs and signals, unexpected hazards, the state of their vehicle, and how all these things are changing. As if that were not enough, drivers also attend to side tasks such as meals, radio shows, and conversations. In this chapter, I will make the argument, and provide evidence for it, that managing attention is the key to safe driving. Assuming for the moment that this claim is true, it becomes important to measure drivers' attention well. Good measures of attention will help both in increasing scientific understanding of

the important task of driving, and in evaluating the safety effects of changes in the driving process, such as use of cell phones and collision warning systems; and changes in driver state, such as fatigue and drunk driving. Driving simulators can be used effectively to measure drivers' attention. However, many simulator studies neglect attentional measures and focus mainly on psychomotor aspects of driving, such as steering and speed control.

In the first part of this chapter, I elaborate on the multifaceted construct I have been calling "attention" (which is similar to situation awareness, or SA), and discuss some psychological theories of the components of attention and SA as they have been applied to driving. Then I describe specific procedures that researchers can use to measure constructs related to attention and SA in their driving simulators. Finally, I review some empirical studies that applied these attentional and SA measures to the components of attention and SA identified earlier.

19.1.1 Definitions of Attention and Situation Awareness in Driving

Researchers studying human attention during real-time tasks such as driving and flying often use the related construct of SA. My definition of SA is: The updated, meaningful knowledge of an unpredictably-changing, multifaceted situation that operators use to guide choice and action when engaged in real-time multitasking. In psychological terms then, SA is a type of *knowledge*. For driving, the contents of this knowledge base are the items I listed above—the driver's route location, road curvature, location of nearby traffic and pedestrians, fuel level, and so on. Most cognitive psychologists assume that this knowledge is active in working memory during driving. Researchers have debated whether the knowledge an operator maintains is solely conscious or also unconscious. Endsley (1995) has argued that SA involves primarily conscious knowledge. Durso, Rawson and Girotto (2007) have argued that real-time knowledge can be conscious or unconscious; accordingly, they prefer the term *situation model* to refer to the knowledge aspect of SA. I agree with the latter group, but will still use the term SA.

However, in addition to focusing on SA as knowledge, researchers also need to understand the perceptual and cognitive *processes* that update and maintain this knowledge. Three levels of cognitive processing are probably involved in maintaining SA: 1) automatic, pre-attentive processes that occur unconsciously and place almost no demands on cognitive resources; 2) recognition-primed decision processes that may be conscious for brief periods (<1 s) and place few demands on cognitive resources; and 3) conscious, controlled processes that place heavy demands on cognitive resources. In terms of visual perception, the automated processes are probably using ambient vision; while the recognition-primed and controlled processes are probably using focal vision (Leibowitz & Owens, 1977).

Some have argued that for very experienced drivers, most driving subtasks are highly automated and therefore that analysis of attention and SA are not needed to understand expert driving performance (D. Norman, personal communication, 2006). I argue that even very experienced drivers regularly use all three of the above levels of processing. For example, vehicle control—perceiving optic flow and the changing shape of the road and using this information to control speed and heading—is probably an automated process. At the middle level, even experienced drivers may need to use recognition-primed processes involving some, albeit brief, conscious awareness when they make routine decisions about whether to change lanes, back up, or stop in response to a yellow light. At the other extreme, making navigational decisions in unfamiliar territory while avoiding hazards in heavy traffic probably engages controlled, conscious processes in a sustained fashion.

Horswill and McKenna (2004) reviewed studies demonstrating that experienced drivers used more cognitive resources for hazard perception than less experienced drivers, suggesting that hazard perception, a key process in SA, does not become automated with extensive experience, but instead remains a controlled process.

This three-level view of SA processes is at odds with the common view that maintaining SA involves recognition-primed and controlled processes, which use focal vision but not automated processes such as ambient vision. However, SA may involve processes other than focal vision. Later, I will examine evidence suggesting a broader view of SA as involving focal processes but also more automatic, ambient processes. Readers are free to adopt the narrow or broader view.

The automated, recognition-primed, and controlled levels of processing outlined above are conceptually different from Endsley's (1995) three levels of SA: 1. perception, 2. comprehension, and 3. projection. However, in practice, perceiving the elements of a situation (Endsley's Level 1 SA) is probably highly automated in most situations, while comprehension and projection (Levels 2 and 3) are more likely to use recognition-primed and controlled processes.

Since the output of the controlled processes, and sometimes the recognition-primed processes, is conscious recognition and comprehension of a meaningful event (e.g., "tailgater ahead"), the processes underlying event perception and comprehension are important ones for driving. Also, since the number of events that drivers may need to attend to using these processes is large (i.e., information overload), another important component process in driving is attention allocation. Attention allocation occurs both within tasks (e.g., when drivers determine their visual scan paths) and across tasks (e.g., when drivers multitask or divide attention among multiple tasks like driving, eating and conversing). These processes will be discussed further in the section on theories of attention and SA.

19.2 Theories of Attention and Situation Awareness in Driving

19.2.1 Focal and Ambient Vision

Schneider (1967) and others have distinguished between two modes of vision: Focal vision, which uses foveal input and serial processing to subserve object identification and conscious awareness; and ambient vision, which uses peripheral and foveal retinal input and parallel processing to subserve spatial localization and guidance of locomotion in an automated, unconscious manner. Leibowitz and Owens (1977) suggested that the main subtask of driving, vehicle control (or guidance), uses ambient vision; while other important driving subtasks, such as identifying hazards, use focal vision. They also hypothesized that, at night, focal vision degrades much more rapidly than ambient vision, so that drivers cannot identify hazards well but can follow the road easily. Empirical studies have supported these hypotheses (Owens & Tyrrell, 1999; Brooks, Tyrrell, & Frank, 2005). The problem with this selective degradation of the two visual modes is that drivers become overconfident in their ability to perform the overall task of driving at night because their ambient vision allows them to perform the main subtask of vehicle control well, and because they are unaware of the severe degradation of their focal vision. This overconfidence leads them to drive too fast and have more crashes. Using the terminology of this chapter, night drivers overestimate their SA, because ambient vision allows them to easily perform the vehicle control subtask of driving. Wickens and Horrey (2008) have suggested that the problem of overestimating SA in the face of degraded focal vision is more general than the situation of night driving. That is, drivers may overestimate SA when experiencing other factors that degrade focal vision, such as driver distraction or multitasking.

It is important to see ambient and focal vision as interrelated rather than separate systems. An important example of cross-talk between ambient and focal vision is the phenomena of attention capture. In attention capture, conspicuous events in the environment (e.g., sudden movements) are detected by stimulus-directed, parallel perceptual processes (similar to ambient vision) without the need for a prior, focal attention shift to the stimulus (Yantis & Jonides, 1984). The result is that focal vision is "captured" and attends to the stimulus, which often leads to comprehension and conscious awareness of the event. For example, a car looming ahead of a driver can capture attention and SA immediately, without waiting for the driver's normal scan pattern to detect it. The phenomenon of attention capture is a good example of how parallel ambient processes that are not usually considered as components of SA can be very important to maintaining SA.

Another example of focal-ambient interrelationships is the recent finding that abrupt visual onsets, which normally capture attention, no longer do so when cognitive load is elevated due to an auditory side task (Boot, Brockmole, & Simons, 2005). This finding that attention capture is diminished under cognitive load argues against the firm distinction between ambient processes as demanding little or no cognitive resources (i.e., automated processing) and focal processes as resource demanding (i.e., controlled processing). Instead, both focal and ambient processes are affected by cognitive load. These two examples of interrelationships between focal and ambient processes support the broader, three-level view of SA mentioned earlier.

19.2.2 Models of Allocating Focal Attention Within Driving Subtasks

Of the three main driving subtasks of vehicle control, monitoring and hazard avoidance, and navigation, the latter two require focal vision. The task of monitoring the roadway for hazards requires sequential shifting of focal attention to task-relevant parts of changing events (that is, attention allocation), as well as comprehension of the attended objects and the overall meaning of an extended event. Especially in cluttered, high-traffic situations, monitoring the roadway so that both routine traffic and hazards are detected is itself a complex task. My view is

that the process of attention allocation is a critical sub-skill in maintaining SA.

Wickens and his colleagues have recently developed the SEEV model, which stands for Salience, Effort, Expectancy and Value (Horrey, Wickens, & Consalus, 2006). SEEV is a stochastic model of how operators allocate their focal visual attention in real-time tasks such as driving. In other words, SEEV deals with how drivers make decisions (probably unconscious ones) about where to focus attention next. More specifically, SEEV estimates the probability of shifting focal attention to particular locations in an ongoing visual event. This probability is affected by the four SEEV factors according to the formula: $P = S - Ef + Ex + V$; where P is the probability of attending to a particular location, S is salience, Ef is effort, Ex is expectancy, and V is value. Salience refers to the visual conspicuity of objects in the visual field, and is affected by parallel perceptual processes such as attention capture. High contrast objects or sudden movements increase salience. Effort refers to the physical and mental effort to shift the visual focus to a particular new location, and is primarily affected by the visual angle between the current and new location. Thus, objects in the driver's blindspot that require large eye and head movements will be viewed relatively infrequently. Expectancy (also called bandwidth) refers to how frequently information is changing at a particular location. Thus, a vehicle moving erratically should be viewed relatively frequently. Value refers to the fact that operators seek task-relevant information, and is affected by the relevance of locations to particular driving subtasks (e.g., the rear-view mirror is highly relevant to the subtask of changing lanes) and the relative priority of subtasks within the overall task of driving (e.g., changing lanes safely is more important than navigating).

Based on eye tracking data, SEEV has been shown to accurately predict the frequency of looking at particular locations during driving and flying tasks (Horrey et al., 2006; Wickens, Goh, Helleburg, Horrey, & Talleur, 2003). These model validation efforts have focused mainly on the Expectancy and Value parameters of SEEV; the Salience and Effort parameters need further validation. One limitation of SEEV is that it does not deal with the cognitive product that is the goal of attention allocation, a mental representation of a meaningful event. The models in the next section deal with this.

19.2.3 Models of Comprehension Within Driving Subtasks

Durso et al. (2007) have sketched out how Kintsch's (1988) construction–integration model of text comprehension can be adapted to describe the perceptual and cognitive processes by which a driver builds up a meaningful model of an ongoing driving situation. The initial step in this process is the perception of objects in a particular scene, leading to surface-level scene and object representations with little meaning. In the second step, spreading activation among these representations and related representations in long-term memory leads to a more meaningful, interconnected representation of the scene. This process repeats cyclically across scenes, with the most important (i.e., most activated) information from each scene carried over to the next cycle, until an integrated, meaningful representation of a particular event (the "eventbase") is built up in working memory. The eventbase would contain knowledge like "those two cars ahead are on a collision path." In the final step, domain-specific knowledge and expertise (e.g., "maybe the Porsche can speed up enough to avoid a collision") is used to further elaborate the eventbase into a richer representation called the situation model. The situation model contains information about causal, temporal, and spatial relationships among the objects comprising an event that drivers use to anticipate future events and guide their actions.

A model of event and hazard comprehension is critical to understanding how people maintain SA; and Durso et al.'s adaptation of the construction–integration model to comprehending driving events seems promising. The main drawback of their model is that it has not been implemented in a specific computational model and validated against empirical driving data.

19.2.4 Models of Multitasking

Driving, even without side tasks, could be considered to involve multitasking, since it requires concurrent performance of at least three distinct subtasks: vehicle control, monitoring and hazard avoidance, and navigation. Side tasks such as conversation or eating only add to the multitasking load. Multitasking is made more difficult by the fact that in many multiple-task situations, tasks have unequal priorities and task priorities change over time. For example, drivers should not assign equal priority to maneuvering through heavy traffic and talking on their cell phone. Thus a key sub-skill during multitasking is task management (or task coordination), which can be thought of as cross-task attention allocation, and which includes setting subjective task priorities for a set of tasks that closely match changing objective priorities, and switching between tasks. In my view, effective task management during multitasking is another key sub-skill in maintaining SA.

Salvucci and Taatgen (2008) hypothesized that people have a general-purpose ability to multitask, distinct from their abilities to perform single tasks. They developed a process model of this general multitasking ability and integrated it into the Adaptive Control of Thought-Rational (ACT-R) cognitive architecture. The new model, which they call threaded cognition, models the processing of multiple tasks as multiple goals (threads) in working memory, and allows interleaving and switching among the multiple tasks as each one builds towards its next response. In three studies, Salvucci and Taatgen validated this model against human performance during simulated driving along with side tasks. Thus, this model seems to provide an accurate description of how people allocate attention and cognitive resources across driving subtasks and side tasks. However, this conclusion may only hold for the relatively simple driving subtasks considered in these validation studies. Also, the threaded cognition model uses a very simple task management rule that gives each subtask

equal priority. This rule may not work well in describing more complex and realistic driving situations involving unequal task priorities.

19.3 Measures of Attention and Situation Awareness in Driving

The prior discussion of the component processes of SA will, I hope, generate an interest in how to measure these components. In this section, I will describe a variety of SA measures and classify them as either online, where behavior is measured during a simulated driving scenario with little or no interruption, or offline, where behavior is measured when the driving scenario is not visible.

19.3.1 Online SA Measures

For almost 40 years, researchers have used eye trackers to monitor and record drivers' eye movements during real and simulated driving. Most researchers assume that overt eye movements (saccades) and fixations indicate the focus of attention most of the time, while understanding that focal attention can sometimes be shifted during a fixation without a saccade. The assumption that fixations track focal attention is especially safe for driving because drivers must gather information from about 270 degrees around them using head movements and large saccades. The most common eye movement variable in driving studies is the percentage of time fixating on particular locations (dwell time).

In performance-based SA measures, the researcher makes inferences about a drivers' SA based on their driving actions during a scenario. For example, in a low-fidelity simulator, Gugerty (1997) used drivers' driving actions to assess their awareness of cars about to collide with them and of cars in the blindspot.

In event detection SA measures, drivers report (e.g., verbally or by pressing a button) whenever they see predefined events during a simulated drive (e.g., a car swerving) (Greenberg et al., 2003). The driving scenario is not interrupted by reporting the event. In one important kind of event detection measure, called a hazard perception test, drivers view or drive through a scenario and report whenever they see any hazardous events (Horswill & McKenna, 2004). The most common variables are the speed and accuracy of event or hazard detection.

In the Situation Present Awareness Method (SPAM), an ongoing driving scenario in a simulator is paused at unpredictable times, but the scenario remains visible. Then the driver responds to one or two questions about the scenario (Durso, Bleckley, & Dattel, 2006). Response time is the main variable, since the answer to the question is visible and participants are very accurate.

The above SA measures are all probably best seen as assessing focal vision and attention processes. The Useful Field of View (UFOV; also called functional field of view) test was developed to assess ambient vision processes (Ball, Beard, & Roenker, 1988). It does this by requiring people to perform a central (focal) visual task along with a peripheral event detection task.

The peripheral stimuli are presented at varying eccentricities beyond 10 degree for a brief period that disallows saccades, thus emphasizing ambient vision. Frequency of reporting the peripheral stimuli is the main variable relevant to ambient vision. The central and peripheral visual stimuli in the original UFOV tests are not related to driving. However, Crundall, Underwood and Chapman (1999) have developed a version of the UFOV with a driving hazard perception task as the central task.

Subjective SA measures, in which participants rate their own SA, are not considered here due to lack of space.

19.3.2 Offline SA Measures

Endsley (1995) developed the Situation Awareness Global Assessment Technique (SAGAT) in which operators perform a simulated real-time task. At unpredictable times, the real-time scenario is interrupted and the simulator screen goes blank. Then, the operator is asked a series of questions about events in the scenario. Accuracy of responding to questions is the main variable in this test.

Strayer, Drews and Johnston (2003) used a post-drive memory test to assess drivers' SA during a simulated drive. In their post-drive test, participants saw pairs of driving scenes, with one scene in each pair from the previous drive and one not, and chose the one from the drive (i.e., a recognition memory test). Recognition accuracy was the main variable.

19.4 Using the SA Measures to Understand the Components of SA

In this section, research will be presented that uses these SA measures to understand the components of SA discussed earlier—within-task attention allocation, event comprehension, multitasking, ambient vision. The reader should not expect a mutually exclusive set of measures for each SA component, as many measures assess multiple components. This section will also present information about the validity of the SA measures by demonstrating, where possible, that they can detect expected skill differences between experts and novices, and that they correlate with driving performance (i.e., predictive validity). In addition to demonstrating the validity of the SA measures, expertise and training effects as well as correlations with real driving show the importance of these SA components to the overall task of driving.

19.4.1 Research on Within-Task Attention Allocation

Eye tracking studies comparing novice and experienced drivers are relevant to the question of whether attention allocation skill changes with experience. Some of these studies have found that experienced drivers look at their mirrors more than novices, look farther down the road than novices (who tend to focus close to the front of the vehicle), and have shorter fixations than novices (Chapman & Underwood, 1998; Mourant & Rockwell, 1972).

Eye tracking studies have also been used to show that eye movements and attention shifts are sensitive to characteristics of the driving environment, which is the main assumption behind attention allocation. The studies by Wickens and colleagues validating the SEEV model showed that eye movements are affected by characteristics such as the frequency of information change, the task relevance of information, and objects' conspicuity (Horrey et al., 2006). Chapman and Underwood (1998) showed that drivers who were faced with dangerous situations narrowed their scan pattern to focus on the danger; and that drivers used shorter fixations and longer saccades in complex urban environments than in simple rural driving. Finally, eye tracking studies have looked at how side tasks affect attention allocation to the roadway. Recarte and Nunes (2000) showed that a spatial side task lead to a narrowing of visual search across the forward roadway, less use of the mirrors, and longer fixations. These studies show that drivers' attention allocation develops with experience and is affected in expected ways by characteristics of the driving environment and by concurrent tasks.

Falzetta (2004) used an event detection task to assess how drivers' attention allocation was affected by the location and the type of events. Participants drove in a simulator while reporting (via a button press) swerves and sudden decelerations by traffic cars. The events occurred either ahead of the driver in the same or the oncoming lane, or behind the driver. As Figure 19.1 shows, participants detected forward events better than rear events, and detected swerves better than decelerations. The location effect is consistent with an attention allocation strategy that gives higher priority to the road ahead. By testing other locations and distances from the driver, this event allocation task could be used to map out the areas of high and low attention allocation.

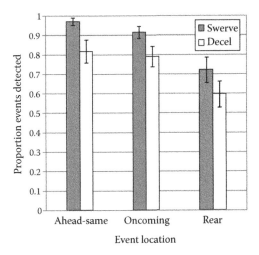

FIGURE 19.1 Proportion of swerve and deceleration events detected at three road locations, with standard error bars. (From Falzetta, M. (2004). *A comparison of driving performance for individuals with and without attention-deficit-hyperactivity disorder.* Unpublished Masters Thesis, Clemson University, Psychology Department, Clemson, SC).

19.4.2 Research on Event Comprehension

As mentioned above, drivers must do more than scan the road well (i.e., allocate attention), they must read the road well (i.e., comprehend the information scanned). Since driving is a risky activity, probably the most important aspect of event comprehension during driving is risk comprehension. The measures and studies reported here will focus on risk comprehension, which is also called hazard (or risk) perception (Horswill & McKenna, 2004).

Horswill and McKenna (2004) cited a number of studies supporting the conclusion that better hazard perception ability, as measured by their video-based test, is associated with fewer on-road crashes, and that hazard perception ability increases with driving experience. These authors also reviewed a number of studies showing that performance on the hazard perception test can be improved by explicit training—using techniques such as learner-generated and expert commentaries while viewing driving scenarios—and that this training transfers to hazard perception during simulated driving. In a similar vein, Pollatsek, Narayanaan, Pradhan and Fisher (2006) showed that PC-based training where drivers practiced identifying high-risk locations in plan-view (2D) driving scenes markedly increased the frequency of looking at risky locations during a simulated drive. In keeping with the theoretical approach taken here, Horswill and McKenna claim that their hazard perception test measures SA for hazardous situations, and that hazard perception is an effortful, non-automated process in which drivers maintain a mental model of the driving situation.

Research on expert-novice differences and predictive validity for the SAGAT and SPAM measures is not available for driving tasks, so aviation and air-traffic-control (ATC) tasks will be used here. Sohn and Doane (2004) found that scores on a modified SAGAT test related to instrument flight were higher for expert than for novice pilots. Also, individual differences in the novices' SAGAT scores were predicted only by their spatial working memory ability; while differences in the experts' scores were predicted most strongly by long-term working memory (i.e., short-term memory aided by learned flying knowledge). This study suggests that experts' greater SA (compared to novices) is based on learned domain knowledge which allows experts to quickly comprehend and remember real-time information. Durso et al. (2006) assessed the incremental predictive ability of the SAGAT and SPAM measures in the domain of air traffic control. In this study, the online SPAM measure predicted additional variance in performance on a simulated ATC task, over and above the variance accounted for by basic cognitive and personality attributes. However, the offline SAGAT test accounted for little incremental variance.

Gugerty (1997; Gugerty, Rakauskas, & Brooks, 2004) used SAGAT to assess SA in a low-fidelity driving simulator in which participants viewed driving scenarios and responded to driving hazards. Periodically, the scenario was blanked and participants then recalled traffic car locations or identified hazardous vehicles. The hazard identification test measured risk comprehension. In two studies assessing the effects of concurrent conversation tasks on driving SA, drivers' ability to recall car locations and identify hazards was strongly degraded by conversation.

These studies of event and risk comprehension show that drivers' comprehension ability improves with experience, training, and the build-up of driving knowledge, and that drivers' event comprehension is strongly degraded by multitasking.

19.4.3 Research on Multitasking

As mentioned above, effective task management during multitasking—including allocating attention to multiple tasks over time in keeping with their changing priorities—is a key sub-skill in maintaining SA. Ackerman, Schneider and Wickens (1984) have pointed out the difficulties in measuring task management skill as a separate component of performing a complex task. One of the measurement difficulties is separating general (i.e., cross-task) task management skill from skill at performing the individual tasks. To date, effective measurement of how individuals vary in general task management skill has proven difficult even with abstract laboratory tasks, and has not been done using more complex tasks like simulated driving.

Nevertheless, a number of studies using abstract laboratory tasks and more realistic tasks (e.g., flight training) have shown that people can improve their multitasking performance as a result of training in task management (Gopher, Weil, & Bareket, 1994; Gopher, Weil, & Siegel, 1989; Kramer, Larish, Weber, & Bardell, 1999). This training, often called variable-priority training, involves giving people explicit instructions to assign varying priorities to two concurrent tasks along with practice and explicit feedback on how well their performance reflects these priorities. The effectiveness of variable-priority training helps demonstrate the importance of task management skills in multiple task performance.

Horrey et al. (2006) investigated how people responded to differences in task priorities in multitasking situations during driving. They looked at how well people could allocate attention to a simulated driving task and an in-vehicle side task in response to changing conditions. The driving task was a steering task involving ambient vision; while the in-vehicle task was a focal task requiring reading, speaking, and comparing digits on a dashboard display. These studies showed that drivers could follow explicit instructions to prioritize either the driving or the in-vehicle task. Whichever task was prioritized received the most visual scanning and showed the highest performance.

However, as task priorities change during realistic driving, drivers do not usually receive explicit instructions to change their attention allocation. A more realistic test of whether drivers can manage task priorities well occurs when they have to detect changing priorities on their own, e.g., by noticing changes in task difficulties. Accordingly, Horrey et al., varied the difficulty of the steering task (via crosswind variations) and of the in-vehicle task. Drivers were able to adjust their attention allocation and task accuracy in response to difficulty variations; although some of these adjustments did not seem to increase safety. For example, as the difficulty of the in-vehicle task increased, scanning the road and steering performance decreased.

The Gugerty et al. (2004) study mentioned earlier also provided evidence that drivers could recognize when one side task (talking on a cell phone) was more difficult than another side task (talking to a passenger) and then allocate less attention to the more difficult side task in order to reduce its detrimental effect on a higher priority driving task. However, they also found that both types of conversation side tasks markedly degraded important driving tasks such as hazard identification, which suggests that people's ability to allocate attention away from conversation tasks during driving is limited.

These multitasking studies show that drivers have some ability to recognize and manage the varying priorities of their multiple tasks and this task management ability can improve with training. However, the studies also show that task management abilities during driving are quite limited. Performing side tasks like conversation while driving degrades multiple safety-critical driving subtasks.

19.4.3.1 Research on Ambient Vision

Under the broader view of SA mentioned earlier, ambient visual processes are important to maintaining SA. One important visual function performed by ambient vision is detecting salient peripheral events, such as sudden movements, and forcing focal vision to allocate attention to them (i.e., attention capture). A common measure of the ability to detect brief peripheral events that has been used in driving research is the UFOV test.

Individual differences studies have showed that reductions in the useful field of view, as measured by the abstract (non-driving) version of the test, correlate with increases in on-road crash involvement in older drivers (Owsley, Ball, Sloane, Roenker, & Bruni, 1991). Practice on the abstract UFOV test improves performance on the test itself as well as on some measures of simulated and on-road driving (Roenker, Cissell, & Ball, 2003). Using a driving-related UFOV test (where the central task involved perceiving hazards in a realistic driving video, and the peripheral task involved detecting peripheral visual onsets), experienced drivers detected the peripheral onsets more frequently than non-drivers (Crundall et al., 1999). These three studies show that the ambient-vision ability to detect sudden peripheral events while performing a focal task improves with experience and training and positively affects overall driving performance.

Crundall et al. (1999) also found that the frequency of detecting peripheral visual onsets decreased as the cognitive demand of the focal hazard-perception task increased. This supports the Boot et al. (2005) finding that ambient processes such as attention capture are degraded under cognitive load, and also supports the broader view of SA as including both ambient and focal processes. It seems important for tests of SA to document not only changes in focal attention but also when drivers' ability to detect peripheral events is changed.

19.5 Conclusion

In this article, I have defined SA as the updated, meaningful knowledge of an unpredictably-changing, multifaceted situation that operators use to guide choice and action when performing a real-time, multitasking task. Driving is a prime example of a real-time multitasking task. Also, I have fleshed out this definition, which focuses on SA as knowledge, by describing the key perceptual and cognitive processes used in maintaining SA. These consist of processes of focal vision and attention, including attention allocation, event comprehension, and task management (multitasking), and also, under a broader view of SA offered here, processes of ambient vision such as attention capture.

Given that SA involves maintaining rich knowledge of a dynamic environment by using attentional, comprehension, and executive control processes, it is not surprising that SA is difficult to measure. A variety of measures are used to understand SA, including: online measures such as eye tracking, performance measures, event detection, SPAM, and UFOV; and offline (memory-based) measures such as SAGAT and post-drive recognition tests. Most of these measures have been used extensively in conjunction with driving simulators, and all of them could be.

Although each of these SA measures can be used to study a variety of research questions related to driving, the following suggestions are offered. Eye tracking is useful for measuring how drivers allocate focal attention. Event detection measures (including focal events as assessed by hazard perception tests and peripheral events as assessed by the UFOV) are useful for measuring how SA for current (ongoing) events is affected by: 1) environmental factors such as the location, magnitude or type of an event; and 2) internal driver factors such as cognitive load, multitasking, distraction, or fatigue. Query techniques (including the offline SAGAT technique and the online SPAM measure) are useful for measuring how SA for past, current, and future events is affected by internal driver factors. Performance measures (and perhaps post-drive recognition memory tests) are useful for measuring the differences between implicit, unconscious and explicit, conscious aspects of SA.

Research using these SA measures has shown that people's ability to use focal processes related to SA including attention allocation, event and risk comprehension, and task management improves with training and with experience at a real-time task like driving. Also, ability at event and risk comprehension, a key component of SA, correlates positively with whole-task driving performance on the road or in a simulator. A similar pattern emerges for ambient processes that may be related to SA. People's ability to detect sudden peripheral events improves with training and with driving experience, and is positively correlated with on-road driving performance. This overall pattern suggests that both the focal and the ambient processes presented here as the component processes of SA are critical components of safe driving.

Key Points

- SA (situation awareness) can be defined as the updated, meaningful knowledge of an unpredictably changing, multifaceted situation that operators use to guide choice and action when performing a real-time, multitasking task.
- The key perceptual and cognitive processes used in maintaining SA consist of processes of focal vision and attention, including attention allocation, event comprehension,

and task management (multitasking), and also, under a broader view of SA offered here, processes of ambient vision such as attention capture.

- A variety of measures are used to understand SA, including: online measures such as eye tracking, performance measures, event detection, SPAM, and UFOV; and offline (memory-based) measures such as SAGAT and post-drive recognition tests.
- Research using SA measures has shown that people's ability to use focal processes related to SA including attention allocation, and event and risk comprehension, and task management improves with training and with experience at a real-time task like driving.
- Ability at event and risk comprehension, a key component of SA, correlates positively with whole-task driving performance on the road or in a simulator.
- Regarding ambient processes related to SA, people's ability to detect sudden peripheral events improves with training and with driving experience, and is positively correlated with on-road driving performance.
- This overall pattern suggests that both the focal and the ambient processes presented here as the component processes of SA are critical components of safe driving.

Keywords: Attention Allocation, Event Comprehension, Focal Vision, Situation Awareness, Task Management

Glossary

Attention allocation: Processes used to decide upon the sequence of focal-attention fixations.
Event comprehension: Understand the meaning of an event.
Focal vision: Controlled, effortful processes that focus visual attention on a small part of a scene and result in recognition of a meaningful event.
Situation awareness: The updated, meaningful knowledge of an unpredictably-changing, multifaceted situation that operators use to guide choice and action when performing a real-time, multitasking task.
Task management: Executive control processes used to organize lower level cognitive processes.

Key Readings

Crundall, D., Underwood, G., & Chapman, P. (1999). Driving experience and the functional field of view. *Perception, 29,* 1075–1087.

Durso, F., Rawson, K., & Girotto, S. (2007). Comprehension and situation awareness. In F. T. Durso, R. Nickerson, S. T. Dumais, S. Lewandowsky, & T. Perfect (Eds.), *The handbook of applied cognition* (2nd ed.). Chicester, England: John Wiley & Sons.

Horrey, W. J., Wickens, C. D., & Consalus, K. P. (2006). Modeling drivers' visual attention allocation while interacting with in-vehicle technologies. *Journal of Experimental Psychology: Applied, 12*(2), 67–78.

Horswill, M., & McKenna, F. (2004). Drivers' hazard perception ability: Situation awareness on the road. In S. Banbury & S. Tremblay (Eds.), *A cognitive approach to situation awareness: Theory, measures and application* (pp. 193–212). London: Ashgate Publishers.

Salvucci, D., & Taatgen, N. (2008). Threaded cognition: An integrated theory of concurrent multitasking. *Psychological Review, 115*(1), 101–130.

References

Ackerman, P., Schneider, W., & Wickens, C. (1984). Deciding the existence of a time-sharing ability: A combined methodological and theoretical approach. *Human Factors, 26*(1), 71–82.

Ball, K., Beard, B., & Roenker, D. (1988). Age and visual search: Expanding the useful field of view. *Journal of the Optical Society of America, A, Optics, Image & Science, 5*(12), 2210–2219.

Boot, W., Brockmole, J., & Simons, D. J. (2005). Attention capture is modulated in dual-task situations. *Psychonomic Bulletin and Review, 12*(4), 662–668.

Brooks, J. O., Tyrrell, R. A., & Frank, T. A. (2005). The effects of severe visual challenges on steering performance in visually healthy young drivers. *Optometry and Vision Science, 82*(8), 689–697.

Chapman, P., & Underwood, G. (1998). Visual search of driving situations: Danger and experience. *Perception, 27,* 951–964.

Crundall, D., Underwood, G., & Chapman, P. (1999). Driving experience and the functional field of view. *Perception, 29,* 1075–1087.

Durso, F., Rawson, K., & Girotto, S. (2007). Comprehension and situation awareness. In F. T. Durso, R. Nickerson, S. T. Dumais, S. Lewandowsky, & T. Perfect (Eds.), *The handbook of applied cognition* (2nd ed.). Chicester, England: Wiley.

Durso, F. T., Bleckley, M. K., & Dattel, A. R. (2006). Does SA add to the validity of cognitive tests? *Human Factors, 48,* 721–733.

Endsley, M. R. (1995). Towards a theory of situation awareness in dynamic systems. *Human Factors, 37*(1), 32–64.

Falzetta, M. (2004). *A comparison of driving performance for individuals with and without attention-deficit-hyperactivity disorder.* Unpublished Masters Thesis, Psychology Department, Clemson University, Clemson, SC.

Gopher, D., Weil, M., & Bareket, T. (1994). Transfer of skill from a computer game trainer to flight. *Human Factors, 36*(3), 387–405.

Gopher, D., Weil, M., & Siegel, D. (1989). Practice under changing priorities: An approach to the training of complex skills. *Acta Psychologica, 71*(1–3), 147–177.

Greenberg, J. A., Tijerina, L., Curry, R., Artz, B. A., Cathey, L., Curry, R., . . . Grant, P. (2003). *Transportation Research Record, 1843.*

Gugerty, L. (1997). Situation awareness during driving: Explicit and implicit knowledge in dynamic spatial memory. *Journal of Experimental Psychology: Applied, 3*(1), 42–66.

Gugerty, L., Rakauskas, M., & Brooks, J. (2004). Effects of remote and in-person verbal interactions on verbalization rates and attention to dynamic spatial scenes. *Accident Analysis & Prevention, 36*(6), 1029–1043.

Horrey, W. J., Wickens, C. D., & Consalus, K. P. (2006). Modeling drivers' visual attention allocation while interacting with in-vehicle technologies. *Journal of Experimental Psychology: Applied, 12*(2), 67–78.

Horswill, M., & McKenna, F. (2004). Drivers' hazard perception ability: Situation awareness on the road. In S. Banbury & S. Tremblay (Eds.), *A cognitive approach to situation awareness: Theory, measures and application* (pp. 193–212). London: Ashgate Publishers.

Kintsch, W. (1988). The role of knowledge in discourse comprehension: A construction-integration model. *Psychological Review, 95*(2), 163–182.

Kramer, A. F., Larish, J. L., Weber, T. A., & Bardell, L. (1999). Training for executive control: Task coordination strategies and aging. In D. Gopher & A. Koriat (Eds.), *Attention and performance XVII: Cognitive regulation of performance: Interaction of theory and application* (pp. 617–652). Cambridge, MA: The MIT Press.

Leibowitz, H. W., & Owens, D. A. (1977, July 29). Nighttime driving accidents and selective visual degradation. *Science, 197,* 422–423.

Mourant, R., & Rockwell, T. (1972). Strategies of visual search by novice and experienced drivers. *Human Factors, 14,* 325–335.

Owens, D. A., & Tyrrell, R. A. (1999). Effects of luminance, blur and age on nighttime visual guidance: A test of the selective degradation hypothesis. *Journal of Experimental Psychology: Applied, 5*(2), 115–128.

Owsley, C., Ball, K., Sloane, M., Roenker, D., & Bruni, J. (1991). Visual/cognitive correlates of vehicle accidents in older drivers. *Psychology and Aging, 6*(3), 403–415.

Pollatsek, A., Narayanaan, V., Pradhan, A., & Fisher, D. (2006). Using eye movements to evaluate a PC-based risk awareness and perception training program on a driving simulator. *Human Factors, 48*(3), 447–464.

Recarte, M., & Nunes, L. (2000). Effects of verbal and spatial-imagery tasks on eye fixations while driving. *Journal of Experimental Psychology: Applied, 6*(1), 31–43.

Roenker, D., Cissell, G., & Ball, K. (2003). Speed-of-processing and driving simulator training result in improved driving performance. *Human Factors, 45*(2), 218–233.

Salvucci, D., & Taatgen, N. (2008). Threaded cognition: An integrated theory of concurrent multitasking. *Psychological Review, 115*(1), 101–130.

Schneider, G. E. (1967). Contrasting visuomotor functions of tectura and cortex in the golden hamster. *Psychologische Forschung, 31,* 52–62.

Sohn, Y., & Doane, S. (2004). Memory processes of flight situation awareness: Interactive roles of working memory capacity, long-term working memory, and expertise. *Human Factors, 46*(3), 461–475.

Strayer, D. L., Drews, F. A., & Johnston, W. A. (2003). Cell phone-induced failures of visual attention during simulated driving. *Journal of Experimental Psychology: Applied, 9,* 23–32.

Wickens, C., & Horrey, W. (2008). Models of attention, distraction and highway hazard avoidance. In M. Regan, J. Lee, & K. Young (Eds.), *Driver distraction: Theory, effects and mitigation* (pp. 57–72). Boca Raton, FL: CRC Press.

Wickens, C. D., Goh, J., Helleburg, J., Horrey, W. J., & Talleur, D. A. (2003). Attentional models of multi-task pilot performance using advanced display technology. *Human Factors, 45*(3), 360–380.

Yantis, S., & Jonides, J. (1984). Abrupt visual onsets and selective attention: Evidence from visual search. *Journal of Experimental Psychology: Human Perception and Performance, 10*(5), 601–621.

20

Simulator Data Reduction

Michelle L. Reyes
The University of Iowa

John D. Lee
University of Wisconsin–Madison

Abstract

The Problem. Simulators produce a potentially overwhelming volume of data. The raw data generated by a simulator need to be parsed, aggregated, and combined to produce summary variables that relate to the theoretical constructs underlying the research questions that motivated the study. This process of transformation can be quite complex and error-prone, and manual reduction using a spreadsheet is often infeasible. ***Role of Driving Simulators.*** Driving simulator studies provide a more detailed account of human behavior than many other experimental approaches. This chapter provides suggestions for exploring this rich source of information through three basic steps: planning, writing, and testing. *Planning* emphasizes a focus on data reduction throughout the entire research process that begins with links to the theoretical constructs being measured and manipulated in the study. *Writing* describes a series of tips to avoid common frustrations in developing code for data reduction. *Testing* advocates a systematic plan that includes automatic checks for the bounds of the reduced variables and visualization to identify unexpected failures of the software. ***Key Results of Driving Simulator Studies.*** The data reduction demands of simulator studies repeatedly frustrate both novice and experienced researchers. Data reduction requires the power of software and researchers often find themselves victims of the many pitfalls associated with developing software. Undiscovered errors in the data reduction process have the potential to invalidate a research program and undermine the collective understanding of driver behavior. ***Scenarios and Dependent Variables.*** Future trends toward standardized scenarios and measures might avoid many data reduction challenges, but such standardizations make it more likely for researchers to blindly interpret outcome variables without careful consideration for how they relate to the theoretical constructs of interest. ***Platform Specificity and Equipment Limitations.*** The substantial differences between simulator platforms make it difficult for the content of this chapter to address the particular challenges any particular researcher will likely face. Data from different simulators reflect different underlying assumptions, definitions, and hardware configurations (e.g., eye tracker), but the general processes described in this chapter should help avoid the pitfalls commonly confronted when interpreting simulator data by increasing the opportunities for finding errors.

20.1 Introduction

Driving simulators can output dozens of variables at rates that typically range from 30 to 240 Hz. The enormous amounts of data that result from driving simulator experiments must be reduced into meaningful information that provides insight into driver behavior. Data reduction is the process of transforming raw data into summary dependent measures that are then subjected to analysis. (See in this book, chap. 21 by Boyle, "Analytical Tools".) Because research questions, experimental designs, dependent variables, and scenarios differ from one study to another, the data reduction process often needs to be tailored for each study. The amount of data and the complexity of data transformations make calculating summary measures of driver behavior using only spreadsheet-based calculations infeasible and impractical for most studies; programmatic approaches, that is, developing code or scripts, are needed to automate all or some of the data reduction process. In other words, driving simulator data reduction often involves some form of software development.

The data reduction process may seem to be a straightforward or even trivial task, but errors that can compromise the transformation from raw to summary data can easily be introduced during the development of the data reduction code. Reduction code containing simple programming mistakes can seem to process the data successfully and provide no obvious indicator of the error. For example, when calculating a mean reaction time across several events, if the event counter is initialized to 1 rather than the intended 0, when the total reaction time is divided by the inflated number of events, the calculated mean will be less than the actual mean. The aggregation of data can also sometimes mask problems with the raw data; for example, undetected high frequency noise in the brake pedal sensor can lead to inaccurate calculations of brake reaction time or time to maximum brake. Such problems are difficult to spot, but the consequences can be critical. The data reduction process is the crucial link between the raw data collected in the simulator and the data interpretation that underlies the conclusions of the study. At a conceptual level, assumptions concerning the links between the raw data and the concepts of interest can have profound consequences for the conclusions of the study. For these reasons, considerable effort should be made to maximize the opportunities for finding and correcting errors and ensure the quality of the data reduction process.

The earlier in the data reduction process an error is found, the less costly it is to correct. If the error is discovered by the person writing the code just after it was written, correcting the error may only take a minute. If the error is discovered several weeks later, it may take the coder a half an hour to re-familiarize him- or herself with the code to fix it. If an error is found after all of the data have been reduced and analyzed, it may take a hundred hours to fix the error and re-reduce and re-analyze the data. Finally, the costs of errors that are never discovered cannot be measured—such errors can invalidate a research program and undermine the collective understanding of driver behavior.

Formal software development practices cannot eliminate all errors in the data reduction process, but they can help to find errors and also prevent them in the first place. Unfortunately, researchers often have little formal software development training and may even have little experience with programming. Researchers who benefit from having dedicated programmers to write data reduction scripts are not immune to these problems. In these cases, the researcher must communicate the intended transformations to the programmer and, due to their different perspectives, this "meeting of the minds" can be fraught with difficulties. In addition, the researcher is partly responsible for the quality of the data reduction solution and needs to ensure that the solution is adequately tested. This chapter aims to help researchers address these problems so that the summary data provide a solid foundation for subsequent analysis.

Even though the development of data reduction code is a decidedly small-scale development effort compared to the major software development efforts that result in the applications that we use on our PCs every day, we describe how some of the formal practices of software development can be adapted for the development of data reduction code. The purpose of this chapter is to share with researchers at all levels of programming experience and various degrees of involvement in the data reduction process some practical advice about planning, writing, and testing data reduction procedures for driving simulator data.

To support this aim, three general practices inspired by formal software development procedures are presented. First, *plan* how the software code will be written and tested by considering the entire research process—specification should precede programming. Second, use the plan to *write* the data reduction software using good coding practices. Third, *test* the code throughout the writing process using visualization techniques to verify that it is performing the functions required to reduce and transform the data as intended. Certainly, no chapter can replace the effective teamwork, attention to detail, and internal reviews that are essential for conducting good science, or formal courses on structured programming and software development, but this chapter provides a starting point to reduce the risk of undiscovered errors in simulator data reduction that can undermine the conclusions of a research study.

20.2 Planning for Data Reduction

The first practice suggested by formal software development concerns *planning* the data reduction process. Formal software development efforts begin by specifying what the software is required to do: the inputs the software will receive, the outputs desired, and the functional processes necessary to obtain them. Software developers start writing code only after they have these "specifications." These requirements are analogous to a problem statement for an engineer or a statement of building users' needs for an architect. We have no doubt that many researchers are superb planners who would not begin data reduction without a plan in place. Our advice on this topic concerns *when* in the research process that planning takes place and how comprehensively it is carried out.

Some researchers consider data reduction to be a discrete step in the research process that follows after data collection. However, the planning of the data reduction process and the definition of the data reduction requirements are affected by nearly every phase of the research process. If planning occurs throughout the project rather than waiting until the data have been collected, adjustments and changes can be made to the other phases if needed. When considered within the research process, requirements for the data reduction code must be informed by the other phases of the research process. In fact, one could think of the research process as a wheel, the development of the data reduction process as the hub, and the requirements as the spokes, as shown in Figure 20.1 (Web Figure 20.1). Elements of the research process provide inputs to or receive outputs from the data reduction process. Some elements that seem remote from the details of data reduction, such as drawing conclusions from the study, may actually be more tightly coupled than many researchers recognize. The selection and specification of dependent variables can have a surprising influence on the conclusions. The remainder of this section describes how each phase in the research process contributes to specifications for the data reduction process.

20.2.1 Theoretical Constructs

Initial data reduction requirements should be grounded in the theoretical basis and the hypotheses that guide the experiment. Specifically, the variables and their definitions require a clear description of the underlying theoretical construct that a particular dependent variable is meant to capture. Ideally, these constructs would be linked to previous research and standard equations for calculating variables (e.g., see the discussions in

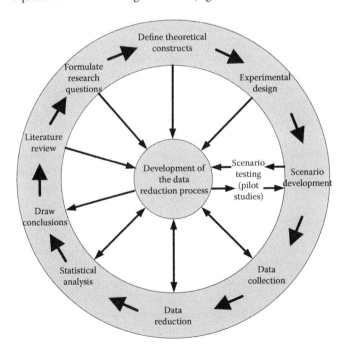

FIGURE 20.1 Research process implications for data reduction planning.

this book, chap. 17 by Brookhuis and de Waard, "Measuring Physiology in Simulators;" chap. 18 by Fisher, Pollatsek, & Horrey, "Eye Behaviors: How Driving Simulators Can Expand Their Role in Science and Engineering"; and chap. 19 by Gugerty, "Situation Awareness in Driving"). The literature review that guides the initial design of the experiment should consider the theoretical constructs and results that are relevant to addressing the research questions. Just as importantly, the literature review should consider how those constructs can be operationalized using variables reduced from the simulator data. The variables should reflect the research questions and the underlying theoretical constructs that are being manipulated and measured. Theoretical constructs of particular importance for describing driving behavior include: mental workload, expectation, attention, trust, situation awareness, and safety margins. Whether these constructs are useful or simply confusing often depends on how well researchers operationalize them in the variables they choose and the data reduction process they implement. The planning for the data reduction should consider the theoretical constructs and how the qualitative constructs will be quantified from the raw data for analysis. Clear definitions and documentation of formulas for calculating the variables are especially important if the data reduction is being done by a person who is not involved with formulating the theoretical constructs and designing the experiment.

Too often variables are selected without careful consideration for the behavior they are meant to capture. Dependent measures should not be selected simply because the simulator collects them, nor should a dependent measure be immediately rejected because it is not. By thinking about what dependent measures will be generated by the data reduction before the simulator scenarios are developed, it may be possible to add variables that are not currently collected by the simulator. If the measure cannot be acquired, then the limits of the simulator can be recognized and addressed at the beginning rather than at the end of the experimental design process.

20.2.2 Experimental Design

The experimental design specifies, among other things, the independent variables, number of subjects, number and ordering of trials, and statistical analyses. The final data set containing the summarized measures must also indicate the independent variables in a manner that is logical and unambiguous. For example, in a study that examined braking behavior for four different imminent collision situations, one column of the final data set contained the subject identifier; another column indicated whether the summary measures in that row were for the left incursion, right incursion, slowing vehicle, or stopped vehicle event; and a third column indicated whether the subject experienced that collision situation first, second, third, or fourth. Including the third column of information was critical for examining the effect of repeated exposure to the collision situations and would have been difficult to add after the data reduction had been completed.

The details of the experimental design help define the requirements for how the data reduction will handle the various experimental conditions as they are operationalized in the simulator scenarios. In the same study of braking, the reduction code that defined the start and end points of the collision situations had to be customized for each of the four situations; however, the code that calculated most of the dependent measures was generalized for all four situations.

20.2.3 Scenario Development and Validation

While scenario development is covered in detail in this book, chap. 6 by Kearney and Grechkin ("Scenario Authoring"), the interaction between scenario development and data reduction development is worth discussing in greater detail. Because both tasks involve software development and the simulator scenarios ultimately create the raw data that has to be reduced, scenario development and data reduction should be tightly coupled, as indicated in Figure 20.1.

Ideally, the interdependence between the scenario design and data reduction processes leads to a situation in which the scenarios and the data reduction procedures are developed in an iterative fashion. Ensuring that the scenario conforms to its specifications requires more than a qualitative inspection (i.e., the scenario looks and feels "right" when it is driven); a quantitative analysis in which specified scenario performance measures are calculated provides the systematic approach that is needed. For instance, verifying the timing of an event that presents drivers with a yellow light dilemma requires calculating the time to arrival at the intersection at the moment the stop light turns yellow. Effective data reduction can support the validation and refinement of scenarios before the data collection phase.

The initial scenario development can play a critical role in the development of the data reduction code. Frequently the simulator output data must contain customized variables that identify when certain events occurred (e.g., braking lead vehicle, changing stop light) or when the subject drove through a specific location (e.g., driver is approaching an interstate on-ramp). If the driver will encounter a disturbance (e.g., wind gusts) or a distraction task (e.g., radio tuning), the location or timing, duration, and magnitude must be known in order to evaluate the driver's response. All of these occurrences and the corresponding dependent measures must be defined before scenario development is complete, and the scenario developer and the data reductionist must work together to ensure that all of the information required to calculate the dependent measures is contained in the simulator output data. One strategy to enhance this collaboration is for the developers of the data reduction code to drive the scenarios as early pilot subjects. It is important for the data reductionist and analyst to be intimately familiar with the scenarios, driver responses, and associated data; participating as early pilot subjects and observing data collections can facilitate this intimacy.

The data from the pilot testing should also provide an early indicator as to whether data reduction code responds to the raw data properly. No one can account for all the potential variations of driver behavior. It is almost inevitable that a driver will do something unexpected that violates the underlying assumptions of the data reduction code. A simple example is a driver who changes lanes when the scenario and analysis assumes the driver will remain in the same lane. This unexpected behavior might result in data that are not meaningful and need to be eliminated from the analysis; therefore, part of data reduction involves identifying violations about assumed driver behavior. In some cases, the scenarios may need to be adjusted and in others the data reduction code may need to be changed to accommodate the behavior. It can also be useful for test drivers to intentionally violate assumptions and try to break scenarios by driving in unexpected ways. This can assess whether the scenarios are robust to unexpected behavior that drivers might exhibit during data collection.

20.2.4 Data Collection

The logistic issues of data collection need to be considered early on so they can be in place for the pilot testing, but these logistics have implications for the data reduction development as well. The contents and format of the raw data file produced by the simulator, along with the file naming and data storage conventions that will be used during data collection, determine the specifications for the code that will open and read the raw data. A particularly critical aspect of the data collection concerns the integration of data collected by any external systems (see Chapters 15 through 19, the subsection on Independent and Dependent Variables), such as data from an eye tracker or a device that presents secondary tasks to the participants. How do these extra variables affect the data reduction process? How will the data streams from the external systems be synchronized with the simulator data? Will data synchronization take place in real-time during data collection or be part of the data reduction process? Network delays (time it takes data from external systems to travel across the network to the simulator's data logger) and mismatched data collection rates (e.g., an eye tracker that runs at 58 Hz while the simulator runs at 60 Hz) may require a substantial effort to resolve. As such, researchers should consider the logistic and external system issues of data collection and their impact on the development of the data reduction scripts early in the research process.

20.2.5 Data Reduction

It is also critical to define how the actual reduction of the raw data will proceed. Reducing a single data file at a time is not very efficient; a reduction method that enables batch processing of the files is highly recommended. Most data analysis tools (e.g., R, MATLAB®, SAS®) have functions that make it possible to read and process all the files in a specified directory. The batch method should be flexible so that any number of files can be reduced at a particular time. It should also provide feedback about whether each file was successfully reduced and the data reduction code should be able to accommodate missing or corrupt files without failing and with appropriate feedback.

Another consideration related to batch reduction is how the output data will be written to a file. What will the file format be (it needs to be compatible with the statistical analysis software), and what method will be used to write the file? At what stage during the data reduction process will the data be written? One option is to write the data out as each variable or set of variables is calculated. Another is to write the data as each raw data file is reduced; while a third option is to store the output data in a temporary way while the entire batch runs and then write the entire data set at the end. The most obvious problem with the third option is that if an error occurs during the middle of the data reduction processing, all of the data reduced up to that point will likely be lost. Therefore, a method that writes the data incrementally is desirable.

20.2.6 Statistical Analysis

The statistical model(s) from the experimental design and the software that will be used to complete the statistical analyses dictate the structure and format of the final reduced data set. For example, some statistical analysis tools for conducting repeated measure analysis of variance may require that the final data set contain one row for each participant rather than one row per event or drive. This "wide" format contrasts with the "long" format that indicates the repetition explicitly and places the repeated measurements of a variable in a single column. This is an example of an important detail that can consume considerable effort and lead to a delay of the data analysis if it is not considered early on in the data reduction development process. The statistical software may also require that a specific character be used to indicate a missing value or that a column of data contains only text or numerals, but not both, in order to be read properly.

Beyond the independent variables defined in the experimental design and the dependent measures that reflect the theoretical constructs, covariates also need to be defined and recorded. These variables are essential in simulator experiments because the experimental conditions are only partially defined by the experimenter and depend, to some extent, on the behavior of the driver. As an example, in a car following situation, a driver might adopt different speeds and headways. These variables can have a powerful effect on the driver's brake reaction time when the lead vehicle decelerates and should be recorded for possible use as a covariate in the data analysis (Donmez, Boyle, & Lee, 2008). These quasi-independent variables can also serve as a manipulation check to ensure that the experimental conditions occurred as planned. For instance, the experimenter might assume that the driver was engaged in a secondary task, but the driver may have shed the task to accommodate the roadway demand. Recording secondary task activity would be a critical variable to ensure the experimental condition has been operationalized properly; that is, that the intended theoretical constructs were manipulated in the expected fashion.

The planning phase of data reduction should consider all phases of the research process. While this section has discussed many considerations, there are others not mentioned here and others that are study-specific. It is important to note that the planning phase also includes planning for testing the data reduction process and, where it is appropriate to do so, adding specifications to the data reduction plan to support testing. For example, the plan might specify that the data reduction code should generate plots to check whether drivers made anticipatory accelerator releases before a potential hazard became an actual threat. Testing provides opportunities for finding and correcting errors in the data reduction process and will be discussed in detail in an upcoming section.

20.3 Writing Code That Avoids Common Pitfalls

The purpose of this section is to offer tips, suggestions, and approaches to writing data reduction code in a manner that avoids errors and enables error discovery. Discovering errors early ultimately enhances efficiency because the sooner errors are found, the easier it is to fix them. This section is not intended to be a replacement for a formal course in programming; instead it provides an informal set of considerations that complement formal training. Because we most often use MATLAB from The MathWorks to write and test data reduction code and ultimately complete the data reduction, we will refer to it several times within this section; however, the same ideas can be applied for other software platforms, such as SAS, R, and SPSS®.

20.3.1 Become Familiar With the Data Reduction Software

The task of writing code for data reduction will be greatly assisted by familiarity with the data reduction software of choice. Though software packages have extensive help pages with examples and cross-references, sometimes it can be extremely difficult to find the specific information desired. Start out with some tutorial lessons in how to use the software. Seek other users—in your work group, in your university or company, or around the world. Finally, find a basic reference text for the language or software platform used to write the reduction code and keep it nearby. These books can be an invaluable resource for describing the data types and structures, coding conventions and syntax, and features of the development environment. For example, MATLAB treats all variables as matrices and provides extensive functions that make many matrix manipulations very easy. Not understanding the capabilities of MATLAB could lead to many lines of error-prone code which could be replaced with a single line that draws on the inherent capabilities of MATLAB. Similarly, not understanding that SAS is structured to process one line of data at a time (i.e., the data step) can make it difficult to capitalize on the strengths of the package.

MATLAB and R have many users who post solutions and toolboxes that support a wide range of data reduction and analysis techniques (e.g., data mining and signal processing). These can save hundreds of hours compared to developing the same tools from scratch. Such toolboxes and other online references can often be a very useful complement to the reference books. However, there is a danger in relying on toolboxes or code written by others without thoroughly verifying that they are doing what their authors claim that they do and that the toolbox or code fits your data and desired outcomes.

20.3.2 Write and Test the Code Incrementally

Developing the data reduction script by writing and testing in an incremental fashion increases the opportunities for error discovery. Some software packages are more amenable to incremental development than others. MATLAB is a particularly useful tool because it has a command window where code can be written and tested a few lines at a time and a workspace to hold variables. The transformation of the variables can be examined in a spreadsheet or visually in a plot. For example, if one of the dependent measures is maximum brake force, the calculation can be confirmed by plotting the brake force values from the raw data and then overlaying the calculated maximum. If the commands were successful, they can then be copied and pasted into the script in the code editor. This makes it very easy to incrementally test each line of code to ensure that it performs as expected.

Data visualization, discussed later in this chapter, plays a critical role in this process. The value of visualization is that it makes it possible to incrementally abstract the data so that the summary values (e.g., standard deviation and mean) can be seen relative to the underlying data and roadway environment. This incremental assessment makes it possible to identify aberrant data points and to assess whether the summary values accurately represent the underlying behavior.

20.3.3 Saving Your Work and Version Control

Frequently save your work. This seems obvious, but it is not uncommon to "get into the groove" and write many lines of code without saving. Many editors have an autosave feature that should be enabled. Become familiar with how the feature works, where the files are saved, and make sure that it is saving frequently enough.

A related consideration is to use "Save As." Frequently a piece of code may be working well, and the next set of changes will corrupt it. If the cause of the problem is not obvious, reverting to a previous version might be the easiest way to recover. Adopt a naming convention that will keep the various versions organized, for example, X_Y, where the first number X is incremented for large changes (e.g., adding a new section of code) and the second number Y for smaller changes (e.g., for incremental changes within that section). Avoid modifying code in multiple sections at a time and keep a table of changes so it is easy to determine which version to return to if it becomes necessary to revert.

Finally, make sure that the directories containing the data reduction scripts and the raw data are included in a regular data backup routine or that all of the data reduction scripts are frequently copied to at least one or two other locations. It is not a matter of "if" the hard drive or memory stick will fail but "when," and too often "when" comes when you can least afford it. Increasingly software is being developed to make backing up data very easy and inexpensive, and systems such as Dropbox and Apple's Time Machine® can help eliminate neglected backups.

20.3.4 Use Pseudocode

Successful and efficient writers often start by drafting outlines that define the scope and content of their written works. Software development can be aided by using a similar approach. When writing code, the outline is sometimes called "pseduocode." Pseudocode provides a simple description of the intended code that acts as a general structure to guide the implementation of the actual code. Just like an outline for writing, pseudocode can vary in level of detail. When writing the pseudocode, be sure that the planned program will satisfy the requirements that have been specified. With the pseudocode to guide the functionality and structure of the code, the actual writing of the code can proceed more efficiently. Here is a code excerpt from a MATLAB script for adjusting eye movement gaze data so the mode of the gaze distribution is located at (0,0). The pseudocode lines begin with "%" and were written first.

```
%round filtered gaze to nearest tenth of
degree
r_horz_gaze = RoundValues(orig_horz_
gaze,1);
r_vert_gaze = RoundValues(orig_vert_
gaze,1);


%use mode of front view to define the
center
horz_center = mode(r_horz_gaze(front));
vert_center = mode(r_vert_gaze(front));


%adjust gaze angle for center
horz_gaze = orig_horz_gaze - horz_center;
vert_gaze = orig_vert_gaze - vert_center;
```

Including pseudocode as commented text alongside the final code serves to document the intent of the code and makes it easier for others to understand the behavior of the final code.

20.3.5 Create Flexible and Easy-to-Modify Code

One approach to writing flexible and robust code is to avoid "hard coding" or specifying a fixed value for parameters within the code. Instead, create a new variable to hold the parameter's value. For example, rather than assuming a fixed sampling rate of 60 Hz when calculating the steering reversal rate, set a variable called "steer_sample_rate" equal to 60. Then use this variable throughout the code rather than the fixed value of 60. This provides greater flexibility, but more importantly it forces the person using the code to explicitly define the sampling rate, ensuring that the sampling rate used for the data reduction is consistent with the raw data. In addition, changing values which have been hard-coded requires a meticulous search to ensure the value change is carried throughout the data reduction code while leaving unrelated occurrences of the value unaltered.

Hard coding can also be problematic when defining the structure of input and output data arrays in the reduction code. For example, if the seventh column of the simulator data file contains speed data, rather than writing code that will read in the data contained in the seventh column and give it the name "speed", the code should search the column labels or header in the simulator data for "speed" and then assign the data in the associated column to the "speed" vector. By implementing this flexibility, if the structure of the simulator data changes through the addition or deletion of a variable, only a few lines of code have to be changed rather than changing the column numbers for all of the variables.

Similar care should be taken with the output file. If the column assignments have been hard coded, inserting a new dependent measure will require either putting the dependent measure at the end of the file, possibly away from other related measures, or modifying the hard coding for all of the columns following the insertion. This dilemma can be avoided by using a counter for column numbers to assign column names when the output data are defined. First initialize a counter to a value of 1, then set the dependent measure column label to the value of the counter, and finally increment the counter by 1. Repeat until all of the dependent variables have been assigned a column number. After the output file has been defined in this fashion, the label is used to specify the desired column in the output file rather than the column number. The code excerpt below illustrates how this approach might be implemented in MATLAB.

```
colCount = 1;    sp_avg_col = colCount;
colCount = colCount+1; sp_min_col =
colCount;
colCount = colCount+1; brk_max_col =
colCount;




.
.
.




dataOut(outputRow, brk_max_col) =
max(brake);
dataOut(outputRow, sp_avg_col) =
mean(speed);
dataOut(outputRow, sp_min_col) =
min(speed);
```

If a new dependent measure is needed in the output file, a new column name is inserted in the list and the column number is incremented. The next time the code is run, all of the column numbers will adjust accordingly. This approach, used in data reduction code for the National Advanced Driving Simulator (NADS) written by Yiannis Papelis, allows the flexibility of calculating dependent measures in any order and eliminates the need to explicitly match dependent measures with specific column numbers.

20.3.6 Become Thoroughly Familiar With the Simulator Data

Some setbacks encountered while writing data reduction code could be avoided with a solid understanding of the simulator data. A common pitfall in calculating even simple dependent measures, such as the average speed, is to neglect the units of measurement for both the raw data and the aggregate measures. The difference in units reflects the different frames of reference of the simulator developers and the researchers. Simulator developers might calculate and report speed in meters per second, but researchers might expect units of miles per hour or kilometers per hour. Some simulators generate raw data using a variety of units for similar or related variables and these differences must be resolved during data reduction.

Figure 20.2 (Web Figure 20.2) helps address this problem by overlaying a meaningful point of reference, the lane boundary, on the graph of lane position. This figure shows a simple plot of steering wheel angle and lane position. The light line represents the steering wheel angle scaled to fit the graph. The bold line represents the lane position in the context of the lane edge represented by the dotted lines. The horizontal line at the right indicates the mean lane position and the vertical line indicates +/− three standard deviations. Such plots force the analyst to consider the meaning of the data and the units. Although a seemingly trivial consideration, the failure of software developers at the National Aeronautics and Space Administration (NASA) to consider units has resulted in spectacular catastrophes (JPL Special Review Board, 2001; Mars Climate Orbiter Mishap Investigation Board, 1999). Similar problems can occur with researchers interpreting simulator data.

The frames of reference must also be considered when working with simulator data. For example, vehicle position may be represented in terms of a global coordinate system rather than in terms of distance along the route or position within the driving lane, and distances between vehicles may be represented in terms of the centers of gravity rather than distance from the vehicles' bumpers. Variables derived by simulators from primary data like position and speed, such as time-to-collision, whether a collision occurred, or lane position, can contain errors of varying degrees of severity. Unfortunately the method or algorithm used by the simulator for deriving these and other variables is not always disclosed so errors can be difficult to detect and quantify.

For a specific example, consider collision detection, which involves several transformations and subtle assumptions. The

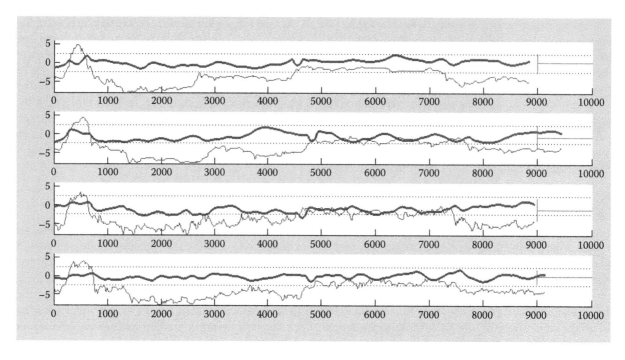

FIGURE 20.2 Visualization plot of raw, mean, and standard deviation of lane position as well as raw steering angle for four drivers.

typical convention is to give the coordinates of the subject vehicle's geometric center or perhaps the center of gravity. The boundary of the vehicle must be determined using the vehicle's width and length, or the location of the center of gravity relative to the vehicle boundary or geometric center, and the vehicle heading. The boundary of the hazard vehicle must be similarly determined. Once both vehicles are represented as polygons, intersection of the polygons at each time step must be assessed. If the simulator does not offer collision detection as an output measure, developing the code to extrapolate the edges of the vehicles from the center of gravity and then assessing overlap can involve many hours of coding. Likewise, if the simulator approximates curves with a series of short line segments, calculating measures related to lane position can be onerous (as we will demonstrate in a future example).

These examples illustrate how data reduction often involves the challenging process of translating from the perspective of simulator and scenario developers to the perspective of the researchers. This translation can only be successful if the data reductionist is thoroughly familiar with the simulator data.

20.3.7 Use Short Cuts With Caution

One common short cut is to modify previously-written code for use with a new data set. Do so with great care, especially if you did not write the original code, the source and purpose of the original code are not well documented, and it does not contain detailed comments. Reusing code without completely understanding its purpose and computational process makes errors more likely to occur and much more difficult to find. Another short cut that should be used with extreme caution is copying

code from one place and pasting it to a new location with the intention of modifying it. Just as written documents have an increasing number of such "cut and paste" errors, programming can suffer from the same problem. It is also tempting to use the "search and replace all" feature—for example, to change the name of a variable throughout the code—but this also has to be done with care. In the example where the sampling frequency was hard coded as 60, one might need to change all the occurrences of 60 to 120. Using "search and replace all" might successfully change all instances of 60 to 120, but it might also change 3600 to 31200. Saving a copy of the code before any large changes such as a global search and replace to change a hard coded variable can save substantial time recovering if something goes wrong.

20.3.8 Initialization

Loops are a common programming structure that is central to data reduction. Loops are used to repeat a calculation for each event, drive, or participant in a study. As a consequence, loops are a source of great efficiency in data reduction. They are also the source of many errors. Specifically, failing to "reset" or initialize variables each time the loop is processed will cause the data from one trial or participant to be combined with others. For example, if the count of lane deviations is not cleared as a loop processes each segment of a drive, the results might suggest that the later segments generate more lane deviations. The output data array (where summary data are stored until they are written to the output file) must also be *completely* cleared before each new event or drive. Do not depend on the newly calculated values to write over the old ones. In an experiment investigating driver response to potential hazards, for example, failing to completely clear the output data array might allow the braking response variables to remain in the array for the next driver so that they indicate that the driver responded when in fact he did not. The consequence of failing to initialize is not always obvious and can lead to insidious bugs that create erroneous summary data.

The advice for writing data reduction code presented here is by no means comprehensive. In addition some lessons can only be learned with experience, or through trial and error. For this reason, performing systematic testing concurrently with writing code is vital.

20.4 Testing With Systematic Planning and Visualization

Data reduction transforms raw data into meaningful measures of driver behavior. The most commonly used measures aggregate data with transformations such as the mean, standard deviation, median, minimum, and maximum, and often provide a reasonable description of driver behavior. However, aggregation involves assumptions and can mask the important features of driver behavior. Aggregation at the level of the individual driver should be assessed to ensure the aggregate measure is representative of the behavior of the individual. Likewise, aggregation over several individuals should be assessed to ensure the aggregate measure is representative of all the individuals. A driver with one slow lane deviation in each lateral direction may have the same mean lane position and standard deviation as a driver who made more frequent and abrupt lane deviations. Likewise, the mean may not reflect safety-relevant behavior. In some cases, such as with glance duration, the tails of a distribution can be more informative than the mean (Horrey & Wickens, 2007). Careful validation of the final aggregation is critical to ensure that the variables reflect the constructs of interest.

Data aggregation can also disguise problems with the data. For instance, if mean reaction time is derived from 20 events, then a spurious reaction time for one event might seriously distort the aggregate measure. A systematic testing plan combined with data visualization can help address these and other problems to ensure that the summary measures provide a meaningful description of the drivers' behavior.

20.4.1 Develop a Systematic Plan for Testing Data Reduction Software

Formal software testing consists of submitting pieces of software (it might be one line of code, an entire program, or any quantity in between) to test cases (a set of inputs and conditions) and evaluating whether the software responds as specified or expected. Ideally the data reduction code will be tested at several different stages of development. The test cases that are specified should be derived from the requirements and can utilize data collected during pilot testing, pilot data that has

been modified, or data that has been created for the specific test case. For example, a portion of pilot data might be modified to include a large lane deviation in order to test code for measuring lane departures.

There are two general types of software testing: black box and white box testing. Black box testing evaluates whether the code meets the specified requirements without concern for the details of the internal code. The conversion error that brought down the Mars Climate Orbiter should have been discovered during black box testing because the requirements called for units of Newtons, which could have been detected with a range check of the software output. White box testing considers the details of the internal code. Many different types of both black and white box testing exist (Hecht & Buettner, 2005), but we provide only a brief introduction here.

Both black and white box testing can be classified as nominal or negative. Nominal testing evaluates the software under normal, expected conditions; whereas negative testing considers situations that are likely to "break" the system (Hecht & Buettner, 2005). More rigorous negative testing might have discovered the oversensitive sensors that led to the loss of the Mars Polar Lander or the bugs in the Therac-25 software. Negative testing of data reduction procedures is important because simulator scenarios might not be exposed to a wide variety of driving behavior before the data collection process begins and drivers tend to behave unpredictably. For example, participants may manage to steer around an imminent collision situation that was designed to force the driver to brake. A lane departure may cause the simulated vehicle to miss the trigger that initiates a pullout vehicle. In a study of braking behavior, one might assume that the driver will release the accelerator before pressing the brake pedal, but some people drive with both feet and so they begin to brake before they release the accelerator. To address this issue, it is critical to explicitly state the assumptions of what nominal behavior is, including speed, navigation, and following distances (this list is certainly not exhaustive). The code must then be tested to determine accuracy when the assumptions are met and the ability to accommodate the exceptions when they are not. (Note that, as mentioned earlier, these assumptions about driver behavior can also be critical to development of the driving scenarios.)

The next level of testing evaluates sections of code or functions in the code. Much of this testing may be performed by the person writing the code; however, it is sometimes appropriate to begin a peer review process at this stage of coding. Submitting the code to review by at least one other person is a great way to find errors. This can be done at various stages throughout the code writing rather than leaving the reviewer(s) to wade through all of the code at once. Peer review also allows for the addition and clarification of the comments in the code so that others can quickly understand the code's purpose and logic. Finally, one other strategy that reviewers can employ is to try to "break" the code using extreme values that might not occur with a small sample of pilot data but might with a large data set.

Testing the overall reduction code can be a daunting task if the code was not tested as it was being written. If piece-wise testing of the code has been performed, the final testing should consist of making sure that all of the parts have come together correctly. Additionally, the data reduction code should be validated with spreadsheet calculations and data plots. No testing regime is perfect and it is virtually impossible to test a piece of software for every contingency. For example, if a loop has five possible paths and the loop is executed 20 times, complete testing would require over 100 trillion tests (Safeware Engineering, 2003). For this reason, testing procedures need to be carefully defined with the potential consequences of specific types of errors in mind.

When the code has been thoroughly tested, the summary measures produced by the reduction process need to be reviewed and validated. This should be done with the pilot data before data collection occurs and both during and after the final data reduction.

20.4.2 Verify Input Data With Automatic Data Checks and Plots

Verification of input data concerns the process of ensuring that the raw data accurately reflect the states of the vehicle, driver, and roadway that they intend to measure. Sensor noise, flaws in the road network database, unexpected driver behavior, and mistakes in the scenario authoring can all contribute to spurious raw data that must be resolved before they can be transformed into summary measures of driver behavior. In one instance this might be a file that is not written to disk and results in an empty dataset; in another instance it might be intermittent voltage spikes that corrupt the indicator of accelerator position. Several automatic data checks or visualizations can address these issues.

The simplest approach is to verify that all the variables in the raw data contain values and that the file is of the expected size. Two simple strategies can be used to identify problems. First, the files can be sorted by size and any dramatically smaller or larger files would suggest a problem. Second, and more precisely, each instance of each variable can be checked to ensure it contains a valid value and that there are an appropriate number of

instances in the data file. For example, if the dataset is meant to contain steering wheel angle and lane position sampled at 30 Hz for a 20-minute drive, there should be two numeric values for each of approximately 36,000 instances. These simple checks should be performed immediately after each drive to ensure that the basic data collection is successful.

A more thorough analysis assesses the integrity of each variable. The integrity of the variable depends on three factors. The first consideration is whether the values lie within the expected range of the variable. For example, depending on the lane configuration, lane deviation values should not exceed a range from −8 to +8 feet. Values outside this range are suspect and merit investigation. The second consideration is whether the values vary. A common failure mode is a sensor that stops indicating system state, but continues to produce a valid value. For example, a steering wheel angle of zero is valid, but if this value does not change for more than 10 seconds, the data should be investigated. This sort of error can be indicated if the minimum and maximum are the same and the standard deviation is zero. These metrics can be applied to the entire data stream or to specific segments. The third consideration is whether the variation in the values is reasonable. Sensor failures and artifacts in the database can cause discontinuities in the data which should be investigated. With lane position, physics limits the change in lateral position over time; if the car suddenly moves laterally at 100 miles per hour there is likely a problem with the data. Calculating the derivative of values (the difference between two sequential data points divided by the time between samples) and then calculating the minimum and maximum of this value should indicate discontinuities in the data. An important source of such discontinuities can occur when lane position is measured with reference to the center of the lane that the driver is currently in and the driver changes lanes. A discontinuity occurs when the driver moves from the far left of the right lane to the far right of the left lane, as shown in Figure 20.3. If such data

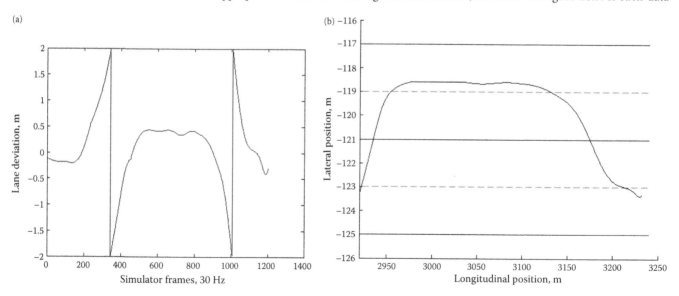

FIGURE 20.3 Discontinuities in lane deviation caused by the simulated vehicle moving from one lane to another.

were blindly analyzed by calculating the standard deviation of lane position, the results could be severely distorted. These three indicators—range, lack of variability, and discontinuities—can be automatically assessed or can be revealed by plotting the data.

Each of the three indicators of problems with the data depends on an implicit model of the driver, vehicle, and roadway. For example, the range of lane position depends on a simple model of lane width and permissible space in which the driver can travel. A more sophisticated approach involves screening the data relative to a more explicit and fully developed model. An example of such a model that could guide data verification uses the control laws that link steering wheel angle with lane position depending on the curvature of the roadway. If steering wheel angle, initial lane position, and road curvature are known, the subsequent lane position can be predicted. If the predications deviate from the reported lane position, the variables should be investigated. Such divergence can occur if the lane center or edge is defined by a series of line segments rather than a smooth curve (see Figure 20.4). As shown in Figure 20.5, the reported deviation from the lane center will include a fluctuating bias associated with the degree to which the edge defined by the line segments diverged from the smooth curve that defines the lane center. Such model-based comparisons can identify subtle flaws in the instrumentation and even the dynamics models that underlie the simulator that might not be revealed in the analysis of individual variables.

20.4.3 Validate Summary Measures

Validation concerns the process of ensuring that the summary measures accurately reflect the driver behavior of interest. Failures of summary variables to be representative of the underlying data can occur at two levels. First, measures based on raw data might not reflect important details of the actual time history of the data, such as when the standard deviation of lane position masks a lane departure because the data summarized also include a period of good lane keeping. Second, aggregating measures at the population level (either across events, drives, or

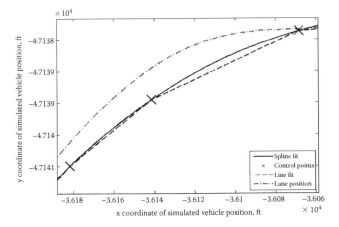

FIGURE 20.4 Lane center can be modeled with a line fit that contains discontinuities or a spline fit that does not.

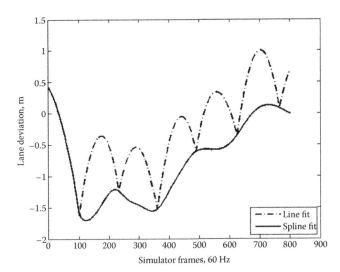

FIGURE 20.5 Lane deviation measured relative to the center of the lane when the center of the lane is defined by a series of short line segments (dotted line) and a spline fit (solid line).

people) might fail to reflect the underlying differences in behavior, such as when only some people respond to the treatment. Data visualization techniques provide useful tools for addressing both of these challenges to the validity of summary measures.

Visualization can superimpose the summary measure over the raw data and a brief inspection can assess whether the summary measure reflects the data sufficiently. For example, the time history of lane position can be augmented with summary measures, such as the minimum time to lane crossing and the standard deviation of lane position. Overlaying these variables along with the lane marker as a reference point makes it possible to roughly assess whether the underlying calculations are correct and whether the variables capture behavior. This could be particularly critical if one is developing a new variable that aims to distinguish between swerving, weaving, drifting, and lane straddling behavior. The standard deviation of lane position does not provide a good basis to differentiate between these behaviors and a new measure would need to be developed and validated against the actual behavior. A visual representation would ease the validation of these and more standard measures.

Figure 20.6 (Web Figure 20.3) shows lane position over time for six segments. Each pair has nearly identical measures of the standard deviation of lane position, and yet the underlying behavior is substantially different. The graph shows the lane position as the heavy line. The upper line represents a 40 seconds time filter of the data (straddling), the line below that represents a 20 seconds time window (drifting), and the one below that represents an 8 seconds time window (weaving). The line at the bottom represents the high frequency variation when the other signals have been subtracted (swerving). The horizontal bars represent the width of the corresponding time window used to filter the data. The standard deviation for each of the signals is represented by the vertical bars at the ends of the horizontal bars. The

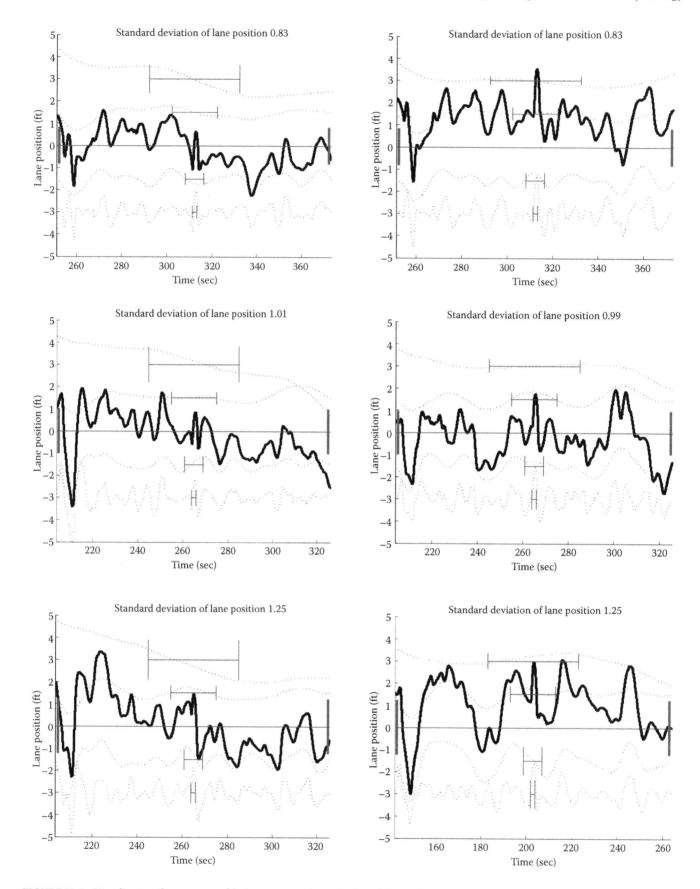

FIGURE 20.6 Visualization of components of the lane position. Identical values of the standard deviation of lane position do not reflect similar behavior.

plot clearly shows that the standard deviation of lane position fails to differentiate between quite different behaviors. The drives on the left have a substantial component of low frequency variation whereas the ones on the right do not, and yet the standard deviation for each pair is almost identical.

For aggregation of variables over a population of events, drives, or people, standard techniques from exploratory data analysis should be engaged (Behrens, 1997; Tukey, 1977). The underlying assumption of many statistical techniques, particularly the ANOVA, is that the summary variable is normally distributed, independent, and has homogenous variance over conditions. Violations of this assumption can compromise the statistical analysis. More importantly, the mean of a population might not provide a good representation of the underlying population, such as when the data follow a bimodal distribution. Plotting the distribution of variables using a boxplot or as a distribution of distributions as in a scatterplot matrix can help evaluate the assumption of normality, and these methods can also be used to describe the differences in behavior not characterized by the summary measures.

To illustrate these points with a concrete example, Figure 20.7 shows the information often used to summarize data. Typically, the lines extend as vertical bars to the horizontal axis and sometimes these include gratuititous three dimensional effects, but the basic information is often limited to the mean value for each condition. Such informationally impovershed plots can lead to incorrect conclusions. Specifically, comparing Events 3 and 4 in Figure 20.7 might lead one to conclude driver behavior in these situations is similar.

A boxplot shown in Figure 20.8 greatly expands the information contained in Figure 20.7 by placing the measure of central

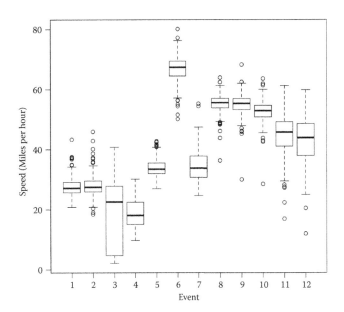

FIGURE 20.8 The boxplot shows the median in the context of the underlying distribution and data points that fall outside this distribution.

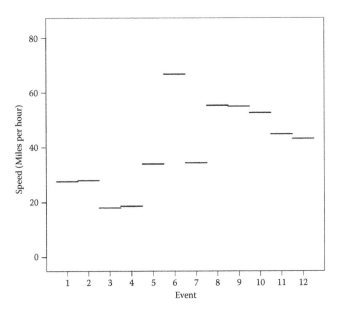

FIGURE 20.7 The basic information contained in the typical bar graph lacks any indication of how well the measure of central tendency represents the underlying data.

tendency, the mean value shown in Figure 20.7, in the context of the data needed to interpret its meaning. According to Tukey's definition of the boxplot, the bar in the center of the box represents the median, and the upper and lower elements of the box represent the 25th and 75th percentiles of the data—the interquartile range (Tukey, 1977). These provide a clear indication of the spread of the data, showing the degree to which the distributions that underlie each median overlap. The boxes also indicate whether the data generally conform to the assumptions that underlie the ANOVA: normality and equal variance. In order to conform to the assumptions, the boxes should be symmetrical and of similar size. Event 3 clearly violates the ANOVA assumptions as reflected by its oversized and asymmetrical box. Statistical tests can quantify these violations, but the boxplot places the violations in the context of the specific experimental conditions.

Although the ANOVA is generally robust to the violation of its assumptions, gross violations can distort the statistics and lead to erroneous conclusions. More importantly, violations of the assumptions may indicate instances where the conditions differ by more than just the standard measure of central tendency, such as the mean value reported in Figure 20.7. The asymmetry and size of the interquartile range in Figure 20.8 shows that the data in Event 3 might reflect a different speed control process compared to the other events.

The vertical lines that extend from the box represent the minima and maxima of the data, unless the data are beyond 1.5 times the interquartile range, in which case the lines show 1.5 times the interquartile range and the individual data points outside this range are plotted directly. These individually plotted data points represent instances that are not well represented by the median or mean value. These "outliers" merit special attention because these data may be a result of a different causal mechanism than that represented by the measure of central tendency. The participants that drive a speed of 15-20 mph in Event 12 seem to be responding in a qualitatively different manner than those who drive at 45 mph over the same stretch of road.

The boxplot reveals much more of the data underlying the summary measure than does the conventional bar graph, but the interquartile range summarizes data in a way that does not always reflect the details of the underlying data. The boxplot assumes unimodal distributions and can mask deviations from this assumption.

The beanplot offers an alternative to the boxplot in that it does not make such assumptions (Kampstra, 2008). Figure 20.9 shows the beanplot for the same data shown in Figure 20.7 and Figure 20.8. The dotted line through the center represents the grand mean across all participants. The short line shows the mean value for each event; the same value as in Figure 20.7. The black curve represents the kernel density function of the distribution. This density function must be interpreted with caution. Just as the elements of the boxplot make assumptions regarding the underlying data, the density function assumes that the underlying distribution is smooth and continuous. The tiny lines represent individual data points, which form an almost solid band when they are densely clustered. These lines make it possible to assess how well the density function and mean represent the data.

The beanplot of the data for Event 3 clearly shows a bimodal distribution underlying the mean value. The mean value resides between the two modes, and the beanplot clearly shows that this summary measure fails to represent any of the individual data points. This feature is completely masked by the traditional bar graph and is visible only in the breadth of the interquartile range

in the boxplot. Considering the details of Event 3, the reason for this bimodal distribution is clear: Event 3 presents drivers with a yellow light dilemma and the mean speed reflects whether they stopped at the light or not. This simple example shows the importance of going beyond simple graphs of single summary measures of the data distribution, such as mean values on a bar graph. In this case, Figure 20.7 might lead one to consider the behavior of drivers in Event 3 and 4 to be quite similar, but Figure 20.9 shows that such a conclusion is clearly wrong. Such plots can identify potential violations of the ANOVA, but more importantly, they can identify instances where summary measures fail to represent the underlying distribution. Here it shows that the mean speed for Event 3 can only be interpreted by separating those who stopped for the yellow light and those who did not.

20.5 Future Developments

The central challenge with simulator data reduction concerns extracting meaning from the huge volume of data that simulators produce. The future promises to make this process more effective, but also threatens to make it more challenging. One of the persistent challenges of data reduction involves the need to translate between the perspective of the simulator developers and the researchers. This challenge is beginning to diminish as simulators become more widespread and developers begin to respond to the needs of the simulator research community. Now that the basic scenario development and data reporting requirements have been met, the development community is likely to turn its attention to creating variables that are increasingly relevant to describing driver behavior. Fifteen years ago many driving simulators did not report lane position, and even now training simulators do not always provide such data. In the future, all simulators will report the distance between the closest point of contact between the driver's vehicle and the other vehicles in the environment. As simulators report variables that are more closely related to the driver behavior of interest, the difficulty and errors that stem from deriving these variables will diminish.

As the research community matures, the potential for greater standardization of scenarios and measures will increase. The development of standardized scenarios is an important topic (see this book, chap. 1) as is the development of standard measures (e.g., see in this book, chap. 17 by Brookhuis and de Waard, "Measuring Physiology in Simulators"). The combined progress on these issues makes it possible for simulators to directly produce the variables of interest for researchers and reduces the difficulty of developing the data reduction process for a given study.

Although promising, such standardization comes with a cost. Standardized measures embody many assumptions regarding the behavior of interest and may not pertain to the constructs the researcher intends. This threat to construct validity emerges as researchers come to rely on summary measures and neglect the underlying behavior. To the extent that critical insights depend on a deep understanding of the drivers' behavior, standardized scenarios and measures may distance

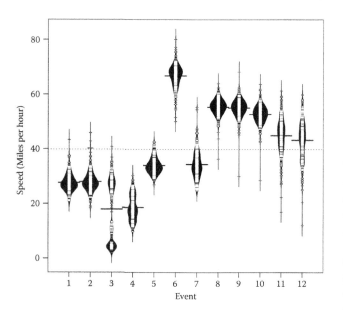

FIGURE 20.9 The beanplot shows the mean value in the context of the grand mean, as well as the individual data points and the probability density function.

researchers from the behavior and the chance for fundamental insights into driver behavior. In the case of a commonly used measure, such as the standard deviation of lane position, its clear and consistent application makes it easy to calculate and compare across experiments. Its rote application, however, also leaves us blind to some differences in lane keeping behavior, such as the drifting behavior shown in Figure 20.6. Such standardized scenarios and measures support evolutionary understanding of driver behavior but might interfere with more revolutionary developments in understanding behavior (Kuhn, 1970).

Careful design of simulators and analysis tools can avoid this trade-off between efficient research and deep understanding. Visualization tools can support researchers in detecting and understanding unexpected behavior. By visualizing the raw data in the context of the summary measures, researchers can quickly assess whether or not the summary measures provide a reasonable account of the behavior. Where the standard summary measures fail, new measures can be developed. Parallel to the interface developments supporting operators of complex processes, data reduction efforts could benefit from the advanced concepts of interface design and data representation (Tufte, 1990; Vicente, 2002; Vicente & Rasmussen, 1992), and even multimodal representations of data, such as sonification (Barrass & Kramer, 1999; Pang, Wittenbrink, & Lodha, 1997; Walker, 2002). Future simulator design should consider supporting researchers with an interface that enables insight into driving, just as process control systems consider supporting operators with interfaces that provide insight into the complex systems they control.

Key Points

- Data reduction is critical for interpreting the outcomes of a study and practices inspired by formal software development procedures can help researchers avoid undiscovered errors in the data reduction process.
- Planning the data reduction process should begin long before data collection, entails gathering specifications for data reduction from nearly every phase of the research process, and should also include plans for testing the data reduction process.
- Practices like writing and testing the code incrementally, using pseudocode, avoiding hard coding, becoming thoroughly familiar with the simulator data, and using shortcuts with caution can help avoid frustrations while writing the data reduction code.
- Testing the data reduction code calls for a systematic plan that includes verifying the raw data and using visualization to validate the summary measures.
- Although standardized scenarios and measures can simplify the data reduction process, they also embody assumptions about driver behavior that can make it more difficult to understand behavior that is not consistent with those assumptions.

Keywords: Data Analysis, Data Reduction, Data Verification, Pilot Testing, Research Process, Scenario Development, Software Development, Visualization

Glossary

Data reduction: The process of transforming raw simulator data into summary dependent measures which are then subjected to analysis.

Operationalize: The process of translating theoretical constructs into independent variables and relating the dependent measures to the theoretical constructs that were intended to measure.

Pseudocode: A description of the planned code that acts as a general structure to guide the implementation of the actual code.

Theoretical constructs: Conceptual elements that support the interpretation and generalization of measurable data.

Visualization: A graphical representation of data that makes it easier to perceive meaningful patterns.

Web Resources

The *Handbook* web site contains supplemental information for this chapter including color versions of three of the printed figures.

Web Figure 20.1: Research process implications for data reduction planning (Figure 20.1 in printed chapter).

Web Figure 20.2: Visualization plot of raw, mean, and standard deviation of lane position, as well as raw steering angle for four drivers (Figure 20.2 in printed chapter).

Web Figure 20.3: Visualization of components of the lane position. Identical values of the standard deviation of lane position do not reflect similar behavior (Figure 20.6 in printed chapter).

Key Readings

Behrens, J. T. (1997). Principles and procedures of exploratory data analysis. *Psychological Methods, 2*(2), 131–160.

DeMarco, T. (1979). *Structured analysis and system specification.* Upper Saddle River, NJ: Yourdon Press.

Tufte, E. R. (1990). *Envisioning information.* Cheshire, CT: Graphics Press.

Tukey, J. W. (1977). *Exploratory data analysis.* Reading, MA: Addison-Wesley.

References

Barrass, S., & Kramer, G. (1999). Using sonification. *Multimedia Systems, 7*(1), 23–31.

Behrens, J. T. (1997). Principles and procedures of exploratory data analysis. *Psychological Methods, 2*(2), 131–160.

Donmez, B., Boyle, L., & Lee, J. D. (2008). Accounting for the covariate effects in driving simulator studies. *Theoretical Issues in Ergonomics Science, 9*(3), 189–199.

Hecht, M., & Buettner, D. (2005). Software testing in space programs. *Crosslink, 6,* 31–35.

Horrey, W. J., & Wickens, C. D. (2007). In-vehicle glance duration: Distributions, tails and model of crash risk. *Transportation Research Record, 2018,* 22–28.

JPL Special Review Board. (2001). *Report on the loss of the Mars Polar Lander and Deep Space 2 Missions.* Retrieved from Jet Propulsion Laboratory, National Aeronautics and Space Administration Website: ftp://ftp.hq.nasa.gov/pub/pao/reports/2000/2000_mpl_report_1.pdf

Kampstra, P. (2008). Beanplot: A boxplot alternative for visual comparison of distributions. *Journal of Statistical Software, 8,* 1–8.

Kuhn, T. S. (1970). *The structure of scientific revolutions.* Chicago, IL: University of Chicago.

Mars Climate Orbiter Mishap Investigation Board. (1999). *Phase I Report.* Retrieved from NASA website: ftp://ftp.hq.nasa.gov/pub/pao/reports/1999/MCO_report.pdf

Pang, A. T., Wittenbrink, C. M., & Lodha, S. K. (1997). Approaches to uncertainty visualization. *Visual Computer, 13*(8), 370–390.

Safeware Engineering. (2003). *The risk of computers.* Retrieved November 30, 2007 from http://safeware-eng.com/Safety%20White%20Papers/Risk%20in%20Computers.htm

Tufte, E. R. (1990). *Envisioning information.* Cheshire, CT: Graphics Press.

Tukey, J. W. (1977). *Exploratory data analysis.* Reading, MA: Addison-Wesley.

Vicente, K. J. (2002). Ecological interface design: Progress and challenges. *Human Factors, 44*(1), 62–78.

Vicente, K. J., & Rasmussen, J. (1992). Ecological interface design: Theoretical foundations. *IEEE Transactions on Systems, Man, and Cybernetics, SCM-22*(4), 589–606.

Walker, B. N. (2002). Magnitude estimation of conceptual data dimensions for use in sonification. *Journal of Experimental Psychology: Applied, 8*(4), 211–221.

21

Analytical Tools

Linda Ng Boyle
University of Washington

Abstract

The Problem. High-fidelity driving simulators can generate a great deal of data. Traditional analytical techniques are useful and can provide answers across different treatment conditions. However, more complex analytical tools are needed to provide insights into variations in driver performance that occur within a treatment level or as drivers transition from one treatment level to the next. *Role of Driving Simulators.* Studies conducted in driving simulators provide greater control than studies in an on-road or naturalistic driving setting. Even if the data is collected with the same level of detail as naturalistic studies, the information related to drivers' performance is much greater given that the same safety situation under the same scenarios can be observed multiple times. *Key Results of Driving Simulator Studies.* Researchers who acquire knowledge related to various analytical tools will be able to compare and contrast differences such that the appropriate insights can be gained on crash causation or safety factors, rather than merely correlations and associations. *Scenarios and Dependent Variables.* This chapter provides information on how to analyze various dependent variables (including speed, time-to-collision, accelerator release time) across different scenarios and is therefore not specific to any dependent variable or scenario. *Platform Specificity and Equipment Limitations.* The tools described in this chapter can be used across all simulator platforms and are not restricted by limitations of the apparatus.

21.1 Introduction

The chapters in this *Handbook* demonstrate the impressive advancements in driving simulators as technology allows greater fidelity and degrees of motion. However, as fidelity and range of motion increase, so does the researcher's need and quest for more data and as Shakespeare's Hamlet says, "there's the rub." As simulators become more complex, so does the data that is being generated from them (see also, this book, chap. 20 by Reyes and Lee, "Simulator Data Reduction"). The ever-increasing complexity presents various analytical challenges that can influence the validity of the inferences drawn. If the correct statistical method is not employed, the results can be misleading and have serious implications for system designs and policies. For example, traditional ANOVA techniques may reveal no differences between two in-vehicle systems when summarized to a condition or treatment level. However, differences may actually exist within the condition but can only be observed using more advanced analytical tools that account for subtle changes over time. These subtle changes may also provide greater insights about driver behavior that may not be revealed otherwise.

This chapter provides an overview of the different tools and approaches that can be used to analyze data. The goal is not to teach statistics, but rather provide information on how best to use analytical tools within the context of simulator studies. Therefore, the assumptions prior to reading this chapter are that readers are familiar with hypothesis testing, Type 1 and Type 2 errors, degrees of freedom, some inferential statistics (F, t, and χ^2 statistics), and power analysis. With regards to power analysis, the analyst should also report the effect size along with the p-value because failure to find a significant effect may actually be related to the inability of the small sample size to discern a difference (Type 2 error). Reporting of effect sizes is good practice and should always be included in your manuscripts along with the reported F or t-tests.

The focus of this chapter centers on parametric statistics rather than non-parametric statistics. In other words, the outcomes, performance measures, or responses you gather in a driving simulator are assumed to follow some distribution. One does not transition to non-parametric statistics because the data is not normal, but because the data is distribution free. If the assumptions of normality do not appear to be met, the researcher should determine the distribution, conduct a normality transformation if possible, or consider a different technique best suited for the observed distribution.

Driving simulator studies require a great deal of ingenuity to ensure that the scenarios designed can have the potential to generate a significant change in participants' responses with some degree of certainty. The scenarios should also be controlled such that any differences observed are not confounded with other factors. This can only be examined if researchers take the time to conduct pilot studies to ensure any additional funds and resources are not expended unnecessarily. The selection of the appropriate analytical tools will be highly dependent on carefully-crafted problem statements and hypotheses. Each hypothesis will influence how the outcome will be manipulated, and hence, the variables to control (independent measures) and to observe (responses or dependent measures).

This chapter connects the appropriate analytical tools for data collected from various sampling strategies (as described in chap. 22 by Dawson) to reveal safety outcomes that can be measured from driving simulator studies (as identified in chap. 15 by McGwin). The chapter begins with the basics of analysis of variances (ANOVAs) techniques as observed in simulator studies to assess differences in driver performance across gender, age, scenarios, and systems. For example, are there significant differences in driver speed when using system X compared to system Y? From there, the chapter transitions to more complex

techniques needed to answer micro-level research questions such as "how does the variation in speed over repeated drives provide insights into a driver's propensity to speed?" The chapter then concludes with a brief discussion of techniques that have yet to be used for driving simulator data, but show great promise for providing insights on driving behavior.

21.2 Analysis of Variance: Some Basics

There are three major assumptions for conducting ANOVAs: (1) the dependent variable or outcome for each treatment level follows a normal distribution; (2) the variation across treatment conditions is homogenous (i.e., homogeneity of variance); and (3) the residuals (or error terms) are normally, identically, and independently distributed (n.i.i.d.) such that similar conclusions would be drawn if the experiment was repeated in other locations or by other researchers. If any of these assumptions are violated, the inferences drawn would be misleading.

The F statistic will be accurate and meaningful only if these assumptions are upheld. Although the ANOVAs are robust to violations of these assumptions and some researchers tend to make inferences even if the assumptions are violated, it is highly recommended that researchers not consider making inferences based on the latter unless the limitations with the outcomes are well understood and identified. There are transformations (e.g., log, inverse, square root, Box-Cox transformation) and corrections (e.g., Satterthwaite's) that can be applied to ensure that the data conform to these assumptions before embarking on the analysis. If readers are still unsure of whether assumptions for conducting an ANOVA are satisfied, Hicks (1993) and Box, Hunter and Hunter (2005) are excellent references.

ANOVA models can contain fixed or random effects, or both. These two terms are commonly used in the context of ANOVA and regression models. Fixed effect models assume that the data is purposely manipulated across several treatments and the error distribution for the residuals (i.e., variables that influence the model but were not controlled) comes from a normal population. For example, if the researcher is interested in observing how lane deviation may change with an auditory distraction as compared to a visual distraction, each level of distraction (auditory, visual) being manipulated is considered a fixed effect. Other researchers who use these same distraction conditions should theoretically generate similar outcomes.

In a random effects model, the researcher is making inferences about the entire population but only a segment of the population is collected. In driving simulator studies, the driver is considered a random effect since information on all types of drivers cannot be collected, but more than likely, only a subset of the entire population is sampled. A mixed effects model describes situations where both fixed and random effects are present. Thus, most driving simulator studies consists of mixed effects models. This is different than a mixed factorial design (based on the independent variables), which is described later in this chapter.

21.2.1 Completely Randomized Factorial Designs

Most human factors studies are experimental in nature and have been historically analyzed using a factorial design. Given that the majority of driving simulator studies are designed to examine human factors-related issues (e.g., system usability, operator performance, etc.), factorial designs would be a typical beginning point. Completely randomized factorial designs are set up such that each driver is randomly assigned to one treatment unit. This is sometimes referred to as a between-subjects factorial design, since the comparisons between different experimental groups are also a comparison between different participants (Presser, 2004). However, using between-subject factors does not necessarily mean you have a completely randomized design.

Between-subject factors are independent variables that do not repeat or experience all conditions. In driving simulator studies, the independent variable, "age group," is considered a between-subjects factor. Age groups can be segmented into several levels with the typical categories being younger, middle-aged, and older drivers (Szlyk, Seiple, & Viana 1995; see also, this book, chap. 25 by Ball & Ackerman, "The Older Driver"). The specific age range for each group can differ as exemplified in simulator studies conducted by Reed and Green (1999), and Briem and Hedman (1995). Readers who are interested in issues on age-related differences are referred to Chapters 24 through 26 of this *Handbook* (the subsection on Experience and Maturity). The purpose of discussing this independent variable here is merely to demonstrate why it is labeled a between-subjects factor. In other words, if a driver is in the "older age group", he or she will not be in the "middle-aged group". However, the use of this variable does not mean you have a completely randomized design since drivers cannot be randomly assigned to the older or middle-aged group. An example of a between-subject factor that can be randomly assigned to drivers would be the examination of different in-vehicle systems (e.g., system X, system Y). Issues may relate to the length of time it takes each driver to be tested in all conditions. Thus, some participants can be randomly placed in the treatment condition, system X while others are tested with system Y only.

The advantage of a completely randomized factorial design is the simplicity of the design—there is no need to account for restrictions on randomization in the model. The analysis is fairly straightforward and interpretation of results is very intuitive. There is no matching and each experimental unit receives only one treatment.

The disadvantage of a completely randomized factorial design for driving simulator studies is that the differences among drivers cannot always be controlled by random assignment of participants to treatment levels (as demonstrated in the age groupings). For complete randomization to be useful the participants also need to be relatively homogeneous or alternatively, a large number of participants need to be used. A large number of participants will also be needed if the analyst is interested in examining many different conditions. A full factorial design where one tests all combinations of all factor levels in a randomized fashion can have many experimental conditions and thus, can be quite expensive and time-consuming. Given that researchers are typically interested in examining multiple drivers over several driving conditions, this design is problematic. The more common driving simulator experiment is therefore, a mixed factorial design consisting of randomized trials repeated across drivers.

21.2.2 Mixed Factorial Design

A mixed factorial design is a combination of within- and between-subject variables. A within-subject factor implies that the participant will be examined across all levels of that factor. If the study was set up as a completely within-subject design, the age groups in our earlier example would be set up such that a younger driver between the ages of 18 and 25 could also be in the middle-aged driver category (between 35 and 55). Likewise, a single participant would have data collected under the male as well as female category. This is clearly not possible. More likely examples of within-subject factors are road type (curve, straight road) or distraction level (auditory, visual). These factors would be considered within-subject if the drivers encountered all levels as part of the experimental protocol. These treatment combinations (or levels) are typically randomized or counterbalanced across subjects and are also referred to as a "repeated measures factors" (Hicks, 1993) since drivers repeat all drives with different treatment combinations. When these factors are combined with between-subject factors, this is called a mixed-factorial design. A mixed-factorial design is the most likely scenario for a driving simulator study because researchers can have groups of individuals traverse through multiple drive scenarios.

The example in Figure 21.1 uses placement of traveler advisory information (four levels: in-vehicle, out-of-vehicle, both, none) as a between-subjects factor and the drive scenarios (fog level, snowplow level) as within-subjects factors. This study was designed to examine whether driver compliance to traveler advisory information would differ based on location and used 48 subjects randomly placed in one of four different system

System type	Fog		No fog	
	Snowplow	None	Snowplow	None
In-vehicle	*Subj 1-12*			⟶
Out-of-vehicle	*Subj 13-24*			⟶
Both	*Subj 25-36*			⟶
None	*Subj 37-48*			⟶

FIGURE 21.1 Example of mixed factorial design used for Boyle and Mannering (2004) study.

configurations (Boyle & Mannering, 2004). All drivers were asked to traverse a similar drive that included reduced visibility and slow moving vehicles that were placed in a counterbalanced order based on a Latin-Square design. The advantage of this design was the minimization of scenarios that needed to be developed and the ability to observe performance within as well as between drivers.

When a study includes within-subject factors, there will also be some autocorrelation since drivers tend to have less variability within themselves than across other drivers. In other words, drivers who tend to drive at higher speeds will most likely drive at higher speeds in all conditions when compared to another driver regardless of the scenario encountered. This autocorrelation violates the n.i.i.d. (normally, identically and independently distributed) assumptions of the basic ANOVA model.

The analyst will therefore need to account for the covariance structure for the effects of the within-subject factors in the ANOVA model. There are several covariance structures and the appropriate one will depend on the relationship among the observations collected. The more common covariance structures are compound symmetry and autoregressive (Figure 21.2). Compound symmetry assumes that all repeated observations have equal variance and covariances (i.e., the covariance between observations is due only to drivers). Autoregressive techniques assumes that repeated observations have greater correlations when observations are closer in time and less correlation as they get further apart. The more commonly observed autoregressive technique in driving simulator studies is the first-order autoregressive (AR(1)) where only one parameter, ρ, is needed to relate current values to previous values. Others, less commonly used, include second-order AR(2) where two parameters are needed, third-order AR(3) for three parameters, and so forth.

An unstructured covariance makes no assumption regarding equal variance or correlations. This structure is used when the data does not follow any pattern over time or space, or is unbalanced and the standard error estimates appear erratic. Thus, there are no assumptions made about the variances or covariances. This covariance structure is commonly used in longitudinal studies where the variances are rarely observed to be constant over time (Fitzmaurice, Laird, & Ware, 2004). This structure has also been used by Bloomfield et al. (1999) in a driving simulator study on antihistamines and alcohol to measure car following behavior with weeks being the repeated measures.

$$S^2\begin{bmatrix} 1 & \rho & \rho \\ \rho & 1 & \rho \\ \rho & \rho & 1 \end{bmatrix} \qquad \frac{S^2}{1-\rho^2}\begin{bmatrix} 1 & \rho & \rho^2 \\ \rho & 1 & \rho \\ \rho^2 & \rho & 1 \end{bmatrix}$$

Compound symmetry
(Equally correlated and equal variance)

Autoregressive (1st order)
(Correlation changes with each measurement)

FIGURE 21.2 Example of 3×3 pairwise covariance structure for compound symmetry and first order autoregressive.

21.2.3 Multiple Comparison Tests

Significant ANOVA findings (or a significant F-test) are typically followed-up with multiple comparison tests. Depending on the analyst's perspective, this can be based on an a priori or post hoc (a posteriori) hypothesis. Statisticians consider a priori hypothesis to be the best approach (Ruxton & Beauchamp, 2008) because it assumes that the researcher has carefully planned out the specific hypothesis to be tested and has some expectations on experimental outcomes. If an a priori hypothesis is developed, then the analyst does not have to compare every single interaction and main effect term, but rather focus on fewer pairwise comparisons. In these cases, orthogonal contrasts can be used where the comparison of means provides independent information. Consider a 2^8 design which can encompass 256 different runs or drive scenarios. In addition to the time and resources that will be used to design this driving simulator study, consider also the number of interaction effects that could possibly be examined. Although all eight main effects may be of interest, testing all possible interaction terms is probably unnecessary and will increase the risk for a Type 1 error. That is, by testing all interaction terms, the probability of observing significance when there is none is greatly enhanced.

Orthogonal contrasts provide a means of partitioning the sum of squares into individual degrees of freedom and are quite useful for focusing on the a priori hypothesis of interest. They have a further advantage of allowing the researcher to focus on specific comparisons even when there are missing treatment combinations. Let's say there were two independent variables, factors A and B. Factor A represents the distraction levels that were previously described: visual (A1) and auditory (A2). Factor B is the presentation of information within each distraction type: words (B1) and pictures (B2). Within the visual stimuli, the distraction can be presented as words or pictures; whereas with auditory stimuli, both levels are not possible (see Figure 21.3).

Based on this design, there are two null hypotheses for the main effect: no differences between visual (A1) and auditory (A2), and no differences between words (A1B1) and pictures (A1B2).

Main effects: $H_o : \mu_{A1} - \mu_{A2} = 0$ and $H_o : \mu_{A1B1} - \mu_{A1B2} = 0$
Interactions: $H_o : \mu_{A1B1} - \mu_{A2B2} = 0$

In this design, the hypothesis for the main effect of factor B cannot be tested with the auditory condition and any differences between words (B1) and pictures (B2) can only be identified within the visual condition. Alternatively, presentation type cannot be distinguished from the interaction of visual × words

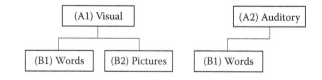

FIGURE 21.3 Incomplete (or unbalanced) factorial design.

(A1B1) and visual × pictures (A1B2). The only unique interaction that can be examined in this situation would be visual × words (A1B1) and auditory × words (A2B1). For this planned unbalanced design, orthogonal contrasts allow the ANOVA sum of squares to be partitioned into individual degrees of freedom such that the number of inferences made cannot be more than the total experimental degrees of freedom.

The disadvantage of orthogonal contrasts is that each hypothesis will need to be carefully crafted beforehand. Furthermore, even though sophisticated statistical packages exist, there will most likely be separate statements for each contrast since they are uniquely defined by the analyst. A very good example of how to program contrasts in SAS (Statistical Analysis Software) based on incomplete factorial designs is provided by Bergerud (1999) and similar examples are available for other statistical packages.

Post hoc hypothesis implies that there are no previous implications related to the differences across all of the conditions being examined and hence, every single combination of relationships will be compared (or contrasted). Examining all contrasts can be done with several types of post hoc pairwise comparisons. That being said, the following tests can also be conducted for a priori hypothesis that requires examination of all pairwise groups. The advantage of these procedures is the simplicity in programming. The majority are included as part of statistical software packages and require no more than one additional line of code. The disadvantage of these methods relates to a higher probability for a Type 1 error. The majority of multiple comparison tests described in this section have a correction factor included as part of their calculations to account for the fact that all possible contrasts are being conducted, but the correction factor can be quite different.

In driving simulator studies, several multiple comparison procedures have been used including Scheffe (Gawron & Ranney, 1988; Regan, Deery, & Triggs, 1998), Duncan (Boyle & Mannering, 2004), Tukey (Donmez, Boyle, & Lee, 2006), and Bonferroni (Lee, Lee, & Boyle, 2007). They all have the same purpose (to examine all possible pairwise combinations) and are available across many statistical packages. However, preferences for any procedure depends on how conservative or liberal the analyst is in reporting statistical findings, the need to include confidence intervals, and whether sample sizes across treatment groups are equal.

The Scheffe test is the most conservative and therefore the most robust to the F-test outcome (Klockars & Sax, 1986; Keppel & Zedeck, 1989). In other words, if the F-test is significant, the Scheffe test will also find at least one significant pair. If the F-test was not significant, the Scheffe test will not find any pair to be significant. The disadvantage is that the Scheffe test controls for the Type 1 error by requiring a large difference between groups before the null hypothesis is rejected. That is, the correction factor is quite large and lacks power and should never be used as a substitution for a small number of planned (or a priori) hypotheses.

A more powerful test is the Student Newman Keuls (or Newman-Keuls) procedure. This procedure has been used to identify significant differences in driving performance for those drivers using a cell phone (Liu, 2003) and for those drivers identified as drowsy (Eoh, Chung, & Kim, 2005). It is a stepwise approach based on the ordering of sample means from the smallest value to the largest. If the difference between the two closest means exceed some critical value, then the pair is considered to be significantly different. The downfall of this method is that the sample sizes for each mean being compared need to be the same. If they are not, this procedure will not be able to control for the overall family-wise error rate and the Type 1 error will most likely be greater than 0.05. Thus, although the test is more powerful, the risk of a Type 1 error is also much greater (Seaman, Levin, & Serlin, 1991). Another downfall of this procedure is that one cannot generate confidence intervals due to the very nature of the sequential calculations performed for this procedure (Muth, 2006). If sample sizes among treatment levels are different, the Tukey or Bonferroni procedure is recommended.

The Tukey (or Tukey-Kramer) procedure is not as powerful as the Student Newman Keuls (SNK) but more so than the Scheffe. The procedure for SNK is actually derived from the Tukey test (Glantz, 2005). The difference is that the Tukey procedure will keep alpha at 0.05 for all pairwise comparisons and enable confidence interval calculations. Thus, when compared to the SNK procedure, the Tukey is a more conservative test and will reveal less significantly different pairwise comparisons.

There are many other multiple comparison tests used in simulator studies, including the Duncan and Bonferroni as indicated earlier. The Bonferroni procedure can be used for multiple comparisons against a single control group (Glantz, 2005) but is not considered as powerful as the Tukey test (Keren & Lewis, 1993). More complete descriptions of these and other multiple comparisons are provided in several statistical textbooks including Glantz (2005) and Keren and Lewis (1993). As stated earlier, the disadvantage of a post-hoc comparison is that every combination will most likely need to be examined. In attempting to look at every possible main and interaction effect through post hoc comparisons, the consequences for Type 1 and Type 2 errors need to be considered carefully.

21.3 Relationships Among Dependent Variables

Driving simulator studies typically generate multiple dependent measures. This is one of the greatest advantages of simulator studies: The capability to examine the relations among various outcomes that are collected simultaneously including speed, acceleration, steering, and lane position. Simulators provide the sense of realism that many feel is necessary to explain the complex driving behavior that a simple reaction-time study or joystick-based study cannot accomplish. The disadvantage, however, is that there is a higher likelihood that the outcomes are correlated.

Although it is tempting to analyze these variables as separate univariate ANOVAs, this can again increase the risk of a Type 1 error. That is, you may find a significant difference even though it does not exist. An analyst may also consider combining the dependent variables into one variable if this appears reasonable. For example, rather than examining height and width, a new variable—area (which is the product of height and width)—may be more plausible. The analyst may not always find this manipulation of variables desirable since there may have been good reasons why both measures were collected.

One approach to account for interdependent variables is by using multivariate analysis of variance technique or MANOVA. Given the interdependencies of multiple dependent measures, one would naturally expect it to be relevant for driving simulator studies. Indeed, Staner et al. (2005) used MANOVA to account for the relationship among multiple subjective measures in their driving simulator study, and Rakauskas, Gugerty and Ward (2004) used it to account for the multiple speed maintenance measures. Thus, when similar measures are collected as is the case in driving simulator studies, MANOVA would be an appropriate tool.

In driving simulator studies, two measures of importance are steering behavior and lane deviation. As expected, these two variables are highly correlated as demonstrated in a study by Boyle, Tippin, Paul and Rizzo (2008) (see Figure 21.4). However, both measures were considered important since they have been examined in other driving simulator studies (Furukawa, Takei, Kobayashi, & Kawai, 1990; Risser, Ware, & Freeman, 2000; Summala, Nieminen, & Punto, 1996). In Boyle et al. (2008), the subpopulation of interest was sleepy drivers and again, both measures have been useful in finding insights on the driving performance of this group. In this case, the relationship of the driver to the vehicle, and the vehicle to the road were highly correlated but still provided unique insights which may not have been gained otherwise.

21.4 Including Covariates in the Model

The best design is one that adequately controls for all variables that can impact the outcome. However, in driving simulator studies, this may not always be feasible given the many interactions that can exist among several environmental, road, and traffic variables. For example, traffic is oftentimes depicted at random intervals on a simulated drive to give the participant a

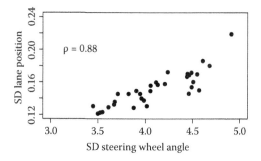

FIGURE 21.4 Correlation between standard deviation of lane position and standard deviation of steering wheel angle (based on data described in Boyle et al. (2008) study).

sense of realism when driving on an urban road or metropolitan area. However the randomness of the cars can have an indirect and unanticipated influence on the driver. Thus, an alternative but comparable approach when it is not possible to control for all specific variables is to set up the experimental model with one or more covariates treated as an independent or explanatory variable (i.e., Analysis of Covariance [ANCOVA]).

In a driving simulator study on driver distraction conducted by Donmez, Boyle and Lee (2008a), the inverse headway distance to a braking lead vehicle (which cannot be controlled) had a strong influence on accelerator release time. Accelerator release time was found to be influenced by the resulting magnitude of cues (i.e., the rate of change in the visual angle) suggesting a critical driver feedback loop. For that reason, inverse headway distance, representative of visual angle in that study, was included as a covariate in the model.

The difference between using MANOVA and incorporating a covariate in your model (ANCOVA) is the prior assumptions and the overall objective. In MANOVAs, the researchers are interested in examining the effect of the independent variables on several potentially correlated dependent measures. In ANCOVAs, researchers are controlling for unwanted sources of variance that may influence the dependent measure (Hartmann & George, 1999).

21.5 Linear Regression Models

Linear regression models are used to predict the linear relationships of a continuous dependent variable (or outcome) and the parameter estimates (or coefficients) from one or more predictor variables. Linear regression models were first developed by Karl Pearson in 1908 and are most likely the first model that readers were taught in a statistics class. Therefore, they are the most likely used model in any study and also, the most frequently misused model.

In developing regression models for simulator studies, it is important to understand how well the regression model fits the data being examined and whether the computed parameter estimates are unbiased, efficient, consistent, and asymptotically normal. The parameter estimates are computed from ordinary least squares (OLS) and any explanatory variables that may have been omitted is picked up in the residual (or error or disturbance term) which is also based on assumptions of n.i.i.d. (normally, identically and independently distributed). More specifically, the residuals should have a mean of zero and variances should be identical across observations as well as independent of other observations (i.e., homoscedasticity assumption).

Similar to design of experiment techniques, a linear regression model cannot be used if the assumptions are violated. For example, if the study design includes repeated (or time-series) measures or other within-subject effects, the residuals will not be independent or normally distributed and hence, the linear regression models will not be appropriate.

As the reader can probably discern, the outcomes of an ANOVA are actually a fitted linear regression model. The difference between fitted and predicted regression models relate to the philosophical approach of each. Design of experiment techniques were developed by Ronald Fisher, and although regression models were originally developed by Karl Pearson, each statistician had a different perspective on the use and value of this tool. See Aldrich (2008) for an interesting discussion on these two statisticians. Pearson's regression was based on the multivariate theory of correlation which included goodness of fit and inferences regarding the regression coefficient. Thus, it is more common to see predicted regression models for the examination of uncontrolled studies. Fisher reconceptualized the theory centered on a univariate approach, with the outcome conditioned to be normally distributed and the independent variable treated as categories that are fixed; as observed in a controlled study. That is, the explanatory variables (or independent variables) provide ancillary information about the regression coefficients (Aldrich, 2005).

The predictive linear regression model will allow several related variables to be examined within one model and thus, has been extended to include the controlled variables in the "generalized linear model" (GLM) which is becoming more common for the analysis of simulator studies (Meda et al., 2009; Horberry, Anderson, & Regan, 2005). The issue related to GLM is that it is highly dependent on the analyst specifying the right model to summarize the data. If the model is misspecified, the estimators are likely to be biased (or incorrect).

There are also alternatives if the observations (or responses) and disturbances are not normally distributed. In these cases, the preferred method for computing estimates for a regression model is not OLS, but rather maximum likelihood estimation (MLE), which is based on maximizing the probability of the sampled data. The advantage of this method is that the MLE for the variance is biased toward zero. That is, as the sample size gets larger, MLE estimates become more consistent. The disadvantage is that MLE requires the distribution of response variable to be specified a priori, whereas OLS does not. Thus, if the model is misspecified, the estimators will also be biased.

21.6 Categorical Outcomes

Most traditional designs of experiment techniques generate outcomes based on a continuous scale. However, there are situations when outcomes from discrete (or categorical) data is desired. Consider responses from a Likert scale (e.g., a preference scale that can range from strongly agree to strongly disagree), which are oftentimes collected along a five or seven-point scale. Although, researchers have tried to examine these variables using traditional ANOVA, this is incorrect for several reasons. Most importantly, the residuals would be highly non-normal. Responses from a Likert scale are typically bimodal (or even trimodal). People tend toward strongly agree or strongly disagree and rarely toward a mean response of moderately agree.

Within simulator studies, the outcome of interest may be the driver's decision regarding a chosen route (Lotan & Koutsopoulos, 1993). As stated earlier, a major assumption of using ANOVA is normality and route choice does not typically follow a normal distribution. Drivers do not randomly choose routes but tend to have preferences based on familiarity, perception of safety, and time on route. If this data were analyzed based on a normal distribution, the mean selection would be the route that was "moderately safe" or "moderately quick". This outcome masks all the distribution "tails" that are of greatest interest and therefore may not provide the insights that are anticipated. Thus, discrete choice models are designed to help researchers interpret this information in a more meaningful way.

A simple discrete choice response is binary (e.g., distracted or not, sleepy or not) and typically analyzed using a binary logit model. If there are more than two outcomes of interest (preference for system A, B or C), then a multinomial logit model can provide insights into whether or not drivers prefer using one system over two different ones; selection of alternate route 1 when compared to 2 and 3. The Likert scale example mentioned previously is an example of multiple discrete choices with an order effect. This outcome may be more appropriately modeled with an ordinal logit model. This section describes two discrete choice models observed in simulator studies: binary logit and ordered logit models

21.6.1 Binary Logistic Regression Model

Logistic regression is commonly used in studies related to crash and injury risks to predict the likelihood of being involved in a crash or severe injury (e.g., Janke, 2001; Lee, Lee, Cameron, & Li-Tsang, 2003; Hill & Boyle, 2006). These models are used to predict the probability of occurrence of an event as fitted by an S-shaped logistic curve and can be used for data collected from driving simulator studies if the sample size is appropriate. Guidelines for recommended sample sizes can be found in Peduzzi, Concato, Kemper, Holford and Feinstein (1996) and Demidenko (2007).

The logistic model is as follows:

$$\text{log odds of a event} = \ln\left(\frac{p}{1-p}\right) = \beta_0 + \beta\mathbf{X} + e \quad (21.1)$$

where p is the probability that an event will occur (e.g., serious injury in a crash), β_0 is the intercept, and β is the matrix of coefficient estimates for each respective predictor variable, $\mathbf{X}$ (e.g., vehicle position, gender). There is also a residual or random disturbance term, ε that follows a logistic distribution. This is included to reflect relevant variables omitted from the model.

For driving simulator studies, a discrete outcome of interest could be the likelihood of a driver deviating out of an intended lane (yes or no) or the likelihood of the driver going over the posted speed limit (yes or no). The binary nature of the dependent variable (or outcome measure) makes logistic regression a highly appropriate technique. They provide insights on the odds of an occurrence calculated directly by taking the inverse log of each side of Equation 21.1 as follows:

$$\text{odds of an event} = \left(\frac{p}{1-p}\right) = e^{\beta_o + \beta X}$$
$$= e^{\beta_o + \beta_1 X_1 + \dots + \beta_n X_n} \qquad (21.2)$$

This allows for the computation of the odds of a lane deviation as influenced by various explanatory factors. Perhaps more intuitive for readers is the odds ratio which is defined as the odds of a lane deviation in the presence of a set of conditions divided by the odds of a lane deviation in the absence of those conditions (Equation 21.3). Note that the independent variables can be coded such that $x_i = 0$ implies the absence of a condition of interest. The odds ratio can then be used to determine whether the presence of the set of conditions leads to an increase in the likelihood of a crash over the absence of the same set of conditions. Additional information regarding logistic regression models can be found in Fox (1997), Hosmer and Lemeshow (2000), and Washington, Karlaftis and Mannering (2003).

$$\text{Odds ratio} = \frac{e^{\beta_0 + \beta_1 x_1 + \dots + \beta_n x_n}}{e^{\beta_0}} \qquad (21.3)$$

Analysis of driving simulator studies have included logistic regression models but are typically complemented with data collected outside the simulator portion of the study. Turkington, Sircar, Allgar and Elliot (2001) used data from 150 drivers to set up a logistic regression model to predict the likelihood of a tracking error (i.e., whether or not they had good tracking ability). The binary outcome was based on driving simulator data and the explanatory variables included self-reported information such as number of near missed events in the previous three years. Other explanatory variables used in this study included age, gender, and alcohol use. The authors did two other logistic regression models with the binary outcomes being high number of off-road events, and good/bad reaction time. Others have used logistic regression techniques to compare a driver's crash history with their performance in a driving simulator (Lee et al., 2003; Rizzo, McGehee, Dawson, & Anderson, 2001).

21.6.2 Proportional Odds Model

A proportional odds or ordered logit model provides a strategy for examining the ordinal nature of data (McCullagh, 1980) whose constructs cannot be measured on an interval scale. That is, if the analyst was to assign numeric values to ordinal levels and perform a linear regression model assuming a continuous dependent variable with equal spacing between levels, the analysis would lead to improper conclusions.

For example, assessing levels of trust in an in-vehicle system can be more appropriately measured on a five-point ordinal scale that can range from "strongly agree" to "neutral" to "strongly disagree." This would then be represented with a set of equations as:

$$\ln\left[\frac{p_1}{1-p_1}\right] = \beta_{01} + \beta X_i + \varepsilon$$

$$\ln\left[\frac{p_1 + p_2}{1 - p_1 - p_2}\right] = \beta_{02} + \beta X_i + \varepsilon \qquad (21.4)$$

$$\vdots$$

$$\ln\left[\frac{p_1 + p_2 + \dots + p_n}{1 - p_1 - p_2 - \dots - p_n}\right] = \beta_{0n} + \beta X_i + \varepsilon$$

where $p_1 + p_2 + \dots + p_n = 1$. In this example, p_1 represents the probability of agreeing strongly in system trustworthiness, p_2 is agree in system trustworthiness, and so forth, to p_5 being strongly disagree. The odds ratio for each predictor is taken to be constant across all possible ways of collapsing the response variable and can be interpreted as a summary of the odds ratios from separate binary logit models. The odds ratios are then interpreted as the odds of being "lower" or "higher" on the outcome variable across the entire range of the outcome (Gameroff, 2005). Thus, the assumption for this modeling technique is that all equations are parallel to each other (Figure 21.5). If they are not, then the multinomial logit model would be more appropriate.

An example of this modeling technique may be observed in a study by Donmez, Boyle and Lee (2008b) that examined a five-point scale on trust based on different feedback types to mitigate distraction, adjusted for individual differences (e.g., age and gender) and usefulness measures collected after each simulator drive. The model examined the drivers' responses to four ordered responses rather than to five. In this study, none of the respondents strongly disagreed that the feedback timing was useful (that is, no one had great abhorrence toward it). Thus, the selection of "strongly disagree" was omitted. The intercept for model n was represented by β_{0n}, and β was the matrix of coefficient estimates for each respective predictor variable, X_i (e.g., age, gender, usefulness), and ε is the error term associated with parameters not included in the model. As with the binary logit model, the error term, ε, is logistically distributed. If the error term was observed to be normally distributed, then the model would convert to an ordered probit.

The parameter estimates in Table 21.1 depict the coefficients, B, in the ordered logit model. A positive parameter estimate indicates that the driver is more likely to agree strongly in

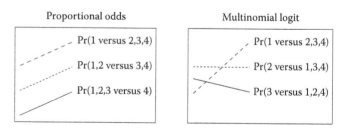

FIGURE 21.5 General assumption for the set of equations for the proportional odds model (parallel) compared to the multinomial logit model (not parallel).

TABLE 21.1 Proportional Odds Model of Trust Being Higher (Based on Results of Donmez et al. (2008b))

Parameter	Estimate	Standard Error	95% Confidence Int. (Lower, Upper)	Z	Pr > |Z|
β_{01} Intercept (Strongly Agree)	−2.90	1.26	(−5.37, −0.44)	−2.31	0.021
β_{02} Intercept (Agree)	1.14	1.10	(−1.02, 3.29)	1.03	0.301
β_{03} Intercept (Neutral)	4.03	1.26	(1.56, 6.50)	3.20	0.001
Usefulness	ns	ns	ns	ns	ns
Female	−3.78	1.16	(−6.05, −1.51)	−3.27	0.001
Drive Number	0.55	0.23	(0.11, 0.99)	2.43	0.015
Usefulness*Drive	−0.37	0.18	(−0.73, −0.01)	−2.03	0.042
Usefulness*Female	2.12	0.87	(0.41, 3.83)	2.43	0.015

system trustworthiness while a negative estimate reduces their likelihood of having strong trust. For example, the parameter estimate for female is −3.78, which indicates that the odds of "strongly agreeing" if you are female and male, respectively are:

$$P(Y = \text{strongly agree} \mid \text{female}) = e^{(-2.90-3.78)} = 0.0013$$
$$P(Y = \text{strongly agree} \mid \text{male}) = e^{(-2.90)} = 0.0550$$

Thus, the odds ratio of having strong trust in the system for females compared to males would be 0.0013/0.0550 = 0.0228. Given that this number is less than 1, this would indicate that females have lower levels of trust in the system when compared to males. However, there was an additive effect with perceived system usefulness such that if the female driver observed the system to be useful, this increased their odds of feeling strongly toward the system:

$$P(Y = \text{strongly agree} \mid \text{female}) = e^{(-2.90-3.78+2.12)} = 0.0105$$

The interaction, usefulness*female is included, but usefulness itself is not since it was not significant. These examples focused on only two parameters from the model as illustrations for the computation of odds and the odds ratio. The table provided is typical of how these parameter estimates are reported in engineering manuscripts. However, the reader can generate the ordered logit equations from the data in the table with 1 = strongly agree, 2 = agree, 3 = neutral as follows:

Logit $(P(Y = 1)) = -2.90 - 3.78$ Female $+ 0.55$ Drive $- 0.37$ Usefulness*Drive $+ 2.12$ Usefulness*Female

Logit $(P(Y \leq 2)) = 1.14 - 3.78$ Female $+ 0.55$ Drive $- 0.37$ Usefulness*Drive $+ 2.12$ Usefulness*Female

Logit $(P(Y \leq 3)) = 4.03 - 3.78$ Female $+ 0.55$ Drive $- 0.37$ Usefulness*Drive $+ 2.12$ Usefulness*Female

Notice that the coefficients for each estimated parameter are the same, but that the intercept changes (i.e., the equations are parallel).

A researcher may also be interested in repeating the ordered categories after each drive condition. This can be examined using a mulitlevel ordered logit model and readers interested in the calculations are referred to Agresti (1989) and Agresti and Lang (1993).

21.6.3 Other Discrete Choice Models

There are other models that have been traditionally used in the analysis of crash data which have not been used in conjunction with driving simulator data but could perhaps be considered in future analysis. These include nested logit, Poisson regression models, negative binomial models, and zero-inflated models. These models are useful when the distribution is skewed toward the left side as crash data typically is. That is, drivers have a higher probability of having zero, one, or two crashes in their lifetime. However, the number of drivers that have 5 to 10 crashes is less common, and it is even rarer to have more than 10 crashes.

The zero-inflated versions of these models account for overdispersion and excess zero values. In a driving simulator study, these models could account for drivers that had no lane tracking errors or missed pedestrian detections. These models assume that the data are actually a mixture of two separate distributions: one with drivers that generate only zeros; and the other follows either a Poisson or negative binomial distribution (Erdman, Jackson, & Sinko, 2008). These models typically require exposure information, which can include vehicle miles driven as observed in a crash analysis. In driving simulator studies, the exposure measure could be the time or distance driven in the simulator. In other words, these models can predict the likelihood of a driving error given the number of opportunities the driver had to make that error.

21.7 Multivariate Analyses

Multivariate techniques provide researchers with opportunities to explore the data and uncover relationships that may not have been realized a priori. These techniques are incredibly useful when additional insights are desired on groups of drivers who appear to have similar patterns of behavior or exhibit similar relationships among factors. Although there are many techniques that fall into this category (principal components analysis, discriminate analysis, multidimensional scaling, and even MANOVA), the two more common techniques used with simulator studies are described in this chapter. For further information on these and other multivariate techniques, Harris (2001), Hair, Black, Babin, Anderson and Tatham (2005), and Lattin, Carroll and Green (2002) are very good references.

21.7.1 Principal Components and Factor Analysis

Principal components and factor analysis are actually two distinct types of multivariate analysis. However, their major goals are somewhat similar—to find groupings in variables. Both are concerned with the variance-covariance structure of a set of *n* variables using *m* components. The objectives of both are to summarize patterns of correlation among observed variables, and reduce the complexity associated with large sets of variables into smaller, more meaningful sets of factors.

The difference between the two relates to the origin of its development. Factor analysis was developed by a psychologist (Charles Spearman) and principal components analysis (PCA) was invented by a statistician (Karl Pearson). Why is that relevant? With PCA, you typically get one outcome regardless of who is conducting the analysis since the major goal of PCA is to find the optimum number of subsets. However, with factor analysis, several analysts can get different numbers of factors with the same dataset. Factor analysis assumes that the variation associated with each variable is related to an underlying common factor and a unique factor that is associated with the measurement error in the experiment. That does not at all make factor analysis the wrong approach, but makes it necessary for the analyst to really understand the data being examined. In the driving domain, factor analysis has been used with survey data and in conjunction with other analytical techniques to provide insights on dimensions of driver stress (Hill & Boyle, 2006). However, they are quite useful for examining driving simulator data as well. For example, Takayama and Nass (2008) used principal components analysis (PCA) with driving simulator data (e.g., time on course, speeding incidents, number of pedestrians hit) to find a factor index that was correlated with unsafe driving behavior. Frank, Casali and Wierwille (1988) used PCA to generate two indices for vestibular disruption and degraded performance based on a combination of simulator data (steering, yaw, and seat movement) and subjective measures (questionnaires on simulator and motion sickness).

21.7.2 Cluster Analysis

Cluster analysis is an exploratory analysis tool for solving classification problems. It provides a way of sorting people, things, and events into groups that have some degree of association. It is a useful technique in driving studies given the diversity of drivers. Common traits may not always relate to gender and age, but to other behavior intrinsic within the driver. To find common traits among groups of drivers would help researchers and designers understand which drivers are more likely to trust a system, to use the system as designed, or to use a system at all. Understanding the common characteristics associated with these drivers can provide insights into what components can be improved as well as what features will be of greatest benefit.

This technique was developed by Robert Tryon in 1939 to isolate psychological constructs related to individual differences. Tryon viewed clustering like factor analysis but was not partial to the "traditional" factor analysis techniques which involved complicated math and assumptions that he felt were undesirable (Blashfield, 1980). Since then, others have developed mathematical procedures and different computation methods which have led to many different perspectives. Some clustering procedures are hierarchical (complete linkage, single linkage, Ward's method) while others are non-hierarchical (k-means clustering). The procedure used by an analyst will depend on the relationships that are being examined, the structure of the data, and the amount of data gathered. Thus, the number and interpretation of clusters can vary greatly from one analyst to another. Readers are referred to Blashfield (1980) for a more in-depth discussion on the evolution of cluster analysis.

Within the realm of driving simulators, they have been used to segment the driver population into more meaningful groups and have been applied prior to and after the simulator portion of the study. Deery and Fildes (1999) used cluster analysis to differentiate individuals of a population of younger drivers prior to conducting their simulator experiment. Their cluster analysis revealed the existence of five novice driver subtypes. They then used these subtypes to carefully select drivers for the simulator portion of their study. Alternatively, Donmez, Boyle and Lee (2010) clustered drivers that participated in a driving simulator study to reveal three groups of drivers that significantly differed based on eye glance behavior and driving performance: drivers with low-risk, moderate-risk, and high-risk behavior. Their study showed that those who were the riskiest drivers benefited the most from feedback on their distraction activities.

21.8 Time-Dependent Analysis

Data that comes from a driving simulator study are typically collected at many frames per second. Depending on the research question, there can be great benefits in looking at the micro as well as macro level. Research questions related to eyes off road time, or percent of visual sampling away from the roadway can be better examined using time-dependent analysis. Fluctuations in drivers' performances may change over time due to prolonged exposure to a distraction, adaptation to an in-vehicle system, or because of a physical or cognitive impairment. These techniques are not quite the same as a repeated measures design (or within-subjects study) as described earlier. Although data is examined within-subject and correlation does exist, the assumption in a repeated measure study is that time will not influence the outcomes of the results and the trials are randomized. In a time-dependent analysis, time greatly influences the outcome and needs to be accounted for such that insights can be gained on the relationship of each consecutive trial or drive. There are many techniques for examining time-dependent measures. Two techniques that are likely to be used with driving simulator studies are simultaneous equations and autoregressive moving average techniques.

21.9 Simultaneous Equations

Simultaneous equations involve analyzing multiple equations (typically sets of linear regression equations) simultaneously. The advantage of analyzing multiple regressions simultaneously rather than separately is that the outcomes of one variable can be simultaneously incorporated into the equation of another. As an example, if we are interested in observing how speed in one time period has an impact on the next or previous time period and how that relationship influences the variability of speed, an equation system can be written as,

$$u = \beta_1 + \alpha_1 X_1 + \phi_1 \sigma + \varepsilon_1$$
$$\sigma = \beta_2 + \alpha_2 X_2 + \phi_2 u + \varepsilon_2$$

(21.5)

where u is the mean speed for each driver in a (for example) one-kilometer segment, and σ is the standard deviation of the speed over the segment. In this model, mean speed and standard deviation of are interrelated endogenous variables while X_1 and X_2 are vectors of roadway, environmental, information, and subject characteristics. The variables β_1, β_2, ϕ_1, ϕ_2 are estimable scalars, α_1, and α_2 are estimable vectors, and ε_1, and ε_2 are normally distributed disturbance terms.

The advantage of this analytical technique for driving simulator studies is that we can take into account variation in one model that is highly dependent on the other. This technique was used in a simulator study by Boyle and Mannering (2004) to examine compensation behavior based on variable speed limit information using the equation system presented in Equation 21.5. A nested factorial model was used and ANOVA techniques showed no significant difference in overall mean speed when the outcomes were examined between conditions. A power analysis confirmed that the sample size was large enough to discern differences with a medium effect. A simultaneous equation model was then used to examine the speed variation within conditions. The outcome demonstrated that drivers did slow down when given information to do so only to speed up later to make up the time they lost earlier. Hence, the overall mean speed and variability was indeed the same across conditions, but the variability observed was due to drivers compensating for changes and information received along the route.

21.9.1 AutoRegressive Integrated Moving Average (ARIMA)

An interrupted time series analyses using an AutoRegressive Integrated Moving Average (ARIMA) model could be useful in examining whether the introduction of a system can provide any improvements in driver performance measures. Such models have been used in examining whether the introduction of the graduated driver licensing (GDL) program was effective in reducing crashes (Neyens, Donmez, & Boyle, 2008). Although they have not been used in driving simulator studies, there are clearly benefits in these models. An interruptive time series analysis allows one to identify whether a significant change in the slope of the time-based data series will occur while accounting for seasonal trends as well as those associated with the introduction of a new safety system.

21.10 Conclusions

The analytical tools presented in this chapter are not interchangeable. Rather, there is a right method for each study design and research question. However, some methods can be used in conjunction with others, such as clustering and ANOVAs, because they can contribute greater insights into the driver population being examined. Although there may be limitations with ANOVA techniques (see Vicente & Torenvliet, 2000, for a list), they do provide great insights for causality due to their ability to analyze outcomes that come from a highly controlled study. Highly controlled studies also require fewer data points, whereas observational studies that have no controlled variables require larger data sets to find insights into the event of interest. That said, it is important to understand how best to use ANOVA techniques and how best to supplement them with information that has already been collected.

This chapter described analytical tools for experimental setups traditionally used in driving simulator studies. However, there are many other more complex experimental designs (e.g., blocking, split plots, fractional factorials) that require advance knowledge of statistics but have been used in driving studies (Belz, Robinson, & Casali, 1999). Thus, the analyst should consider other analytical tools if the experimental data is not quite in line with any discussed in this chapter. Although there is limited space in this chapter, it is important to recognize that these more complex designs pose even greater restrictions on randomization and further limit interpretation based on confounding outcomes.

The merging of techniques traditionally used in experimental psychology with those used in epidemiology, engineering, and economics provide different perspectives that can be of great benefit to the analyst. For example, design scenarios (e.g., case-crossover, case-control) used in epidemiology can be used within the experimental framework of a simulator study. Paul, Boyle, Tippin and Rizzo (2005) set up an experiment that was originally set up with drowsy drivers and a within-subject factor that encompassed three identical but monotonous drives. The participants were observed to have microsleep episodes at different points in the experiment. These microsleep episodes were examined based on an epidemiologic-based case-crossover design (Maclure, 1991; Marshall, 1993). Case-crossover designs are used to compare the timeframe when a condition is present with another timeframe when the condition is absent, and have been successfully used in driving research to assess the associations between cellular-telephone calls and motor vehicle collisions (Redelmeier & Tibshirani, 1997). Paul et al. (2005) used this design to compare the drive lap when microsleep were present (the case) with the two other drive laps when they were absent (the crossover control). Thus, studies based on other experimental protocols can provide additional information on the use of these analytical tools and the reader is encouraged to examine these other designs since they may be well adapted to driving simulator studies.

As a final note, some behavioral scientists and engineers may not always feel it is necessary to understand all these various analytical tools. Some analysts may believe that everything can be analyzed using simple one-way ANOVAs, while others try to fit a linear regression model to everything. There are even some that try to avoid assumptions of normality completely by choosing nonparametric statistics for all studies. It is very common to use techniques that one is most comfortable with. However, designing a good driving simulator study also requires a good understanding of how the desired outcomes are to be achieved. As noted several times in this chapter, specifying the wrong assumptions for any modeling technique can have a negative impact on the outcome and lead the analyst to draw the wrong conclusions.

The recommendation that this author provides to all researchers is to "look and feel" the collected data before beginning any inferential statistics. Looking at the data entails developing the descriptive statistics first (e.g., means, median, standard deviations, box plots, scatter diagrams, etc.). Feeling the data implies absorbing the value of every data point and variable that was collected. Spending time on this aspect will reduce wasted resources later and generate a better understanding of which inferential statistics is best. Furthermore, it may force the analyst to think of an analytical tool that may not have been used before in driving simulators and hence provide ideas for the next version of this chapter.

Key Points

- All studies cannot be analyzed the same way.
- Understanding the differences in study designs can help in the identification of the best analytical technique, which can help in drawing appropriate conclusions.

Keywords: Analysis of Variance, Complex Designs, Regression Models, Multivariate Techniques

Acknowledgments

The author acknowledges the editors of this book and Birsen Donmez for their thoughtful comments and review time. The author also acknowledges all her graduate students who provided many comments throughout the years that contribute to the ideas presented in this chapter.

Key Readings

Box, G., Hunter, J. S., & Hunter, W. G. (2005). *Statistics for experimenters, design, innovation, and discovery* (2nd ed.). Hoboken, NJ: John Wiley & Sons.

Donmez, B., Boyle, L. N., & Lee, J. D. (2008a). Accounting for time-dependent covariates in driving simulator studies. *Theoretical Issues in Ergonomics Science, 9*(3), 189–199.

Washington, S., Karlaftis, M., & Mannering, F. (2003). *Statistical and econometric methods for transportation data analysis*. Boca Raton, FL: Chapman & Hall/CRC.

References

Aldrich, J. (2005). Fisher and regression. *Statistical Sciences, 20*(4), 401–417.

Aldrich, J. (2008). E. S. Pearson's reviews of R. A. Fisher's *Statistical methods for research workers*. Retrieved from http://www.economics.soton.ac.uk/staff/aldrich/fisherguide/esp.htm

Agresti, A. (1989). A survey of models for repeated ordered categorical response data. *Statistics in Medicine, 8*(10), 1209–1224.

Agresti, A., & Lang, J. (1993). A proportional odds model with subject-specific effects for repeated ordered categorical responses. *Biometrika, 80*(3), 527–534.

Belz, S. M., Robinson, G. S., & Casali, J. G. (1999). A new class of auditory warning signals for complex systems: Auditory icons. *Human Factors, 41*(4), 608–618.

Bergerud, W. (1999). ANOVA: Coefficient for contrasts and means of incomplete factorial designs, *Biometrics Information (Pamphlet No. 59)*. British Columbia Ministry of Forest and Range. Retrieved from http://www.for.gov.bc.ca/hre/biopamph/pamp59.pdf

Blashfield, R. K. (1980). The growth of cluster analysis: Tryon, Ward, and Johnson. *Multivariate Behavioral Research, 4*, 439–458.

Bloomfield, J., Weiler, J., Grant, A., Brown, T., Layton, T., Grant, P., . . . Young, M. (1999). Driving performance with first and second generation antihistamimes and alcohol. In M.A. Hanson, E.J. Lovesey, & S.A. Robertson, Eds. *Contemporary Ergonomics 1999. Proceedings of the Annual Conference of Contemporary Ergonomics* (pp. 63–67). Routledge: United Kingdom.

Box, G., Hunter, J. S., & Hunter, W. G. (2005). *Statistics for experimenters, design, innovation, and discovery* (2nd ed.). Hoboken, NJ: John Wiley & Sons.

Boyle, L., & Mannering, F. (2004). Impact of traveler advisory systems on driving speed: Some new evidence. *Transportation Research Part C: Emerging Technologies, 12*(1), 57–72.

Boyle, L. N., Tippin, J., Paul, A., & Rizzo, M. (2008). Driver performance in the moments surrounding a microsleep. *Transportation Research Part F: Traffic Psychology and Behavior, 11*(2), 126–136.

Briem, V., & Hedman, L. (1995). Behavioral effects of mobile telephone use during simulated driving. *Ergonomics, 38*(12), 2536–2562.

Deery, H. A., & Fildes, B. N. (1999). Young novice driver subtypes: Relationship to high-risk behavior, traffic accident record, and simulator driving performance. *Human Factors, 41*(4), 628–643.

Demidenko, E. (2007). Sample size determination for logistic regression revisited. *Statistics in Medicine, 26*, 3385–3397.

Donmez, B., Boyle, L. N., & Lee, J. (2006). The impact of driver distraction mitigation strategies on driving performance. *Human Factors, 48*(4), 785–804.

Donmez, B., Boyle, L. N., & Lee, J. D. (2008a). Accounting for time-dependent covariates in driving simulator studies. *Theoretical Issues in Ergonomics Science, 9*(3), 189–199.

Donmez, B., Boyle, L. N., & Lee, J. D. (2008b). *Trust influences real-time and post-trip feedback in mitigating driver distraction.* Proceedings of the 10th International Conference on Application of Advanced Technologies in Transportation, Athens, Greece.

Donmez, B., Boyle, L. N., & Lee, J. D. (2010). Differences in off-road glances: Effects on young drivers' performance. *ASCE Journal of Transportation, 136*(5), 403–410.

Eoh, H., Chung, M. K., & Kim, S. (2005). Electronencephalograhic study of drowsiness in simulated driving with sleep deprivation. *International Journal of Industrial Ergonomics, 35,* 307–320.

Erdman, D., Jackson, L., & Sinko, A. (2008). *Zero-inflated Poisson and zero-inflated negative binomial models using the COUNTREG Procedure* (Paper 322-2008). Proceedings of SAS Global Forum 2008. San Antonio, TX.

Fitzmaurice, G., Laird, N., & Ware J. (2004). *Applied longitudinal analysis.* Hoboken, NJ: John Wiley & Sons.

Fox, J. (1997). *Applied regression analysis, linear models, and related methods.* Thousand Oaks, CA: Sage Publications.

Frank, L. H., Casali, J. G., & Wierwille, W. W. (1988). Effects of visual display and motion system delays on operator performance and uneasiness in a driving simulator. *Human Factors, 30*(2), 201–217.

Furukawa, Y., Takei, A., Kobayashi, M., & Kawai, T. (1990, May 7-11). *Effects of active steering control on closed-loop control performance.* Paper Presented at the Eighteenth FISITA Congress, The Promise of New Technology in the Automotive Industry, Torino, Italy.

Gameroff, M. (2005). *Using the proportional odds model for health-related outcomes: Why, when, and how with various SAS procedures (Paper 205-30).* Proceedings of the Thirtieth Annual SAS® Users Group International Conference. Philadelphia, PA.

Gawron, V., & Ranney, T. (1988). The effects of alcohol dosing on driving performance on a closed course in a driving simulator. *Ergonomics, 31*(9), 1219–1244.

Glantz, S. (2005). *Primer of biostatistics* (6th ed.). New York: McGraw Hill, Medical Publishing division.

Hair, J., Black, B., Babin, B., Anderson, R., & Tatham, R. (2005). *Multivariate data analysis* (6th ed.). Upper Saddle River, NJ: Prentice Hall.

Harris, R. J. (2001). *A primer of multivariate statistics* (3rd ed.). Mahwah, NJ: Lawrence Erlbaum Associates.

Hartmann, D., & George, T. (1999). Design, measurement, and analysis in developmental Research. In M. Bornstein & M. Lamb (Eds.), *Developmental psychology: An advanced textbook* (4th ed.). Mahway, NJ: Lawrence Erlbaum Associates.

Hicks, C. R. (1993). *Fundamental concepts in the design of experiments* (4th ed.). New York: Oxford University Press.

Hill, J., & Boyle, L. (2006). Assessing the relative risk of severe injury in automotive crashes for older female occupants. *Accident Analysis & Prevention, 38,* 148–154.

Horberry, T., Anderson, J., & Regan, M. (2006). The possible safety benefits of enhanced road markings: A driving simulator evaluation. *Transportation Research Part F: Traffic Psychology and Behavior, 9*(1), 77–82.

Hosmer, D., & Lemeshow, S. (2000). *Applied logistic regression* (2nd ed.). Hoboken, NJ: John Wiley & Sons.

Janke, M. (2001). Assessing older drivers: Two studies. *Journal of Safety Research, 32,* 43–74.

Keppel, G., & Zedeck, S. (1989). *Data analysis for research designs: analysis of variance and multiple regression/correlation approaches.* New York: W. H. Freeman & Company.

Keren, G., & Lewis, C. (1993). *A handbook for data analysis in the behavioral sciences: Statistical issues.* Mahwah, NJ: Lawrence Erlbaum Associates.

Klockars, A., & Sax, G. (1986). *Multiple comparisons.* Newberry Park, CA: Sage Publications.

Lattin, J., Carroll, D., & Green, P. (2002). *Analyzing multivariate data.* Belmont, CA: Duxbury Press.

Lee, H. C., Lee, A. H., Cameron, D., & Li-Tsang, C. (2003). Using a driving simulator to identify older drivers at inflated risk of motor vehicle crashes. *Journal of Safety Research, 34*(4), 453–459.

Lee, Y., Lee, J., & Boyle, L. (2007). Visual attention in driving: The effects of cognitive load and visual disruption. *Human Factors, 49,* 721–733.

Liu, Y. (2003). Effects of Taiwan in-vehicle cellular audio phone system on driving performance. *Safety Science, 41*(6), 531–542.

Lotan, T., & Koutsopoulos, H. (1993). Models for route choice behavior in the presence of information using concepts from fuzzy set theory and approximate reasoning. *Transportation, 20*(2), 129–155.

Maclure, M. (1991). The case-crossover design—A method for studying transient effects on the risk of acute events. *American Journal of Epidemiology, 133*(2), 144–153.

Marshall, R. J. (1993). Analysis of case-crossover designs. *Statistics in Medicine, 12*(24), 2333–2341.

McCullagh, P. (1980). Regression models for ordinal data (with discussion). *Journal of the Royal Statistical Society, Series B, 42*(2), 109–142.

Meda, S., Calhoun, V., Astur, R., Turner, B., Ruopp, K., & Peralson, G. (2009). Alcohol dose effects on brain circuits during simulated driving: An fMRI study. *Human Brain Mapping, 30*(4), 1257–1270.

Muth, J. (2006). *Basic statistics and pharmaceutical statistical applications.* Boca Raton, FL: Chapman and Hall/CRC.

Neyens, D., Donmez, B., & Boyle, L. (2008). The Iowa graduated driver licensing program: Effectiveness of crash reduction. *Journal of Safety Research, 38*(4), 383–390.

Paul, A., Boyle, L. N., Tippin, J., & Rizzo, M. (2005). *Variability of driver performance during microsleeps*. Proceedings of Third International Driving Symposium on Human Factors in Driver Assessment, Training, and Vehicle Design. Rockport, ME.

Peduzzi, P., Concato J., Kemper E., Holford, T. R., & Feinstein, A. R. (1996). A simulation study of the number of events per variable in logistic regression analysis. *Journal of Clinical Epidemiology, 49*, 1373–1379.

Presser, S. (2004). *Methods for testing and evaluating survey questionnaires*. Hoboken, NJ: John Wiley & Sons.

Rakauskas, M. E., Gugerty, L. J., & Ward, N. J. (2004). Effects of naturalistic cell phone conversations on driving performance. *Journal of Safety Research, 35*(4), 453–464.

Redelmeier, D. A., & Tibshirani, R. J. (1997). Association between cellular-telephone calls and motor vehicle collisions. *New England Journal of Medicine, 336*(7), 453–458.

Reed, M., & Green, P. (1999). Comparison of driving performance on-road and in a low-cost simulator using a concurrent telephone dialing task. *Ergonomics, 42*(8), 1015–1037.

Regan, M., Deery, H., & Triggs, T. (1998). Training for attentional control in novice car drivers: A simulator study. *Proceedings of the 42nd annual meeting of the Human Factors and Ergonomics Society* (pp. 1452–1456). Chicago, IL. Santa Monica, CA: Human Factors and Ergonomics Society.

Risser, M. R., Ware, J. C., & Freeman, G. (2000). Driving simulation with EEG monitoring in normal and obstructive sleep apnea patients. *Sleep, 23*(3), 393–398.

Rizzo, M., McGehee, D., Dawson, J., & Anderson, S. (2001). Simulated car crashes at intersections in drivers with Alzheimer disease. *Alzheimer Disease and Associated Disorders, 15*, 10–20.

Ruxton, G., & Beauchamp, G. (2008). Time for some a priori thinking about post hoc testing. *Behavioral Ecology, 19*(3), 690–693.

Seaman, M. A., Levin, J. R., & Serlin, R. C. (1991). New Developments in pairwise multiple comparisons: Some powerful and practicable procedures. *Psychological Bulletin, 110*, 577–586.

Staner, L., Ertle, S., Boeijinga, P., Rinaudo, G., Arnal, M. A., Muzet, A., et al. (2005). Next-day residual effects of hypnotics in DSM-IV primary insomnia: A driving simulator study with simultaneous electroencephalogram monitoring. *Psychopharmacology (Berlin), 181*(4), 790–798.

Summala, H., Nieminen, T., & Punto, M. (1996). Maintaining lane position with peripheral vision during in-vehicle tasks. *Human Factors, 38*(3), 442–451.

Szlyk, J. P., Seiple, W., & Viana, M. (1995). Relative effects of age and compromised vision on driving performance. *Human Factors, 37*(2), 430–436.

Takayama, L., & Nass, C. (2008). Driver safety and information from afar: An experimental driving simulator study of wireless vs. in-car information services. *International Journal of Human-Computer Studies, 66*, 173–184.

Turkington, P., Sircar, M., Allgar, V., & Elliott, M. (2001). Relationship between obstructive sleep apnoea, driving simulator performance, and risk of road traffic accidents. *Thorax, 56*(10), 800–805. doi:10.1136/thorax.56.10.800.

Vicente, K. J., & Torenvliet, G. L. (2000). The Earth is spherical ($p < 0.05$): alternative methods of statistical inference. *Theoretical Issues in Ergonomics Science, 1*(3), 248–271.

Washington, S., Karlaftis, M., & Mannering, F. (2003). *Statistical and econometric methods for transportation data analysis*. Boca Raton, FL: Chapman & Hall/CRC.

22

Statistical Concepts

Jeffrey D. Dawson
The University of Iowa

Abstract

The Problem. Many researchers consider statistical aspects of the study only when analyzing the data. This can lead to study designs that do not address the intended research aims. *Role of Driving Simulators.* Driving simulator studies can address a variety of important research hypotheses and typically involve the collection, analysis, and interpretation of numerical data. *Key Implications of Statistical Concepts for Driving Simulator Studies.* This chapter emphasizes how statistical issues arise through many phases of the planning and implementation of simulator studies. This assertion is illustrated using a hypothetical study of an in-vehicle alerting device in a driving simulator. *Scenarios and Dependent Variables.* In our hypothetical study, the researchers want to test whether an alerting device in a two-way traffic scenario can increase safety. The researchers discuss potential definitions for dependent variables that could measure safety and other important experimental design issues are also discussed. *Generality.* Integration of statistical considerations into the entire research process is essential to the success of a driving simulator study.

22.1 Introduction

One research team was applying to a U.S. government agency for infrastructure funding for a large research center. The content of the application was to consist of descriptions of some of the central cores of the center, as well as summaries of several potential projects that would be pursued if the research center were to be funded. Each project summary was organized in the format of a brief grant application; that is, containing sections on specific aims, hypotheses, background, preliminary data, clinical methods, statistical methods, and potential ramifications of the proposed study. The first project summary had numerous details on the statistical analysis plan, including such items as the definitions of outcomes and predictors, a discussion of statistical tests that

would address the specific aims, and power calculations. In the summary for the second potential project, the statistical analysis plan consisted of only one sentence, paraphrased as follows: "We will give the data to the statistician, who will analyze the data." The disparate amount of information given by these two project leaders illustrates that some scientists are proactive in involving those with statistical expertise as part of their research teams, while others only see statistical analyses as afterthoughts—perhaps even nuisances.

This chapter emphasizes that statistical concepts, broadly conceived, should inform much more than just the formal analyses that are performed after the data are collected. Instead, statistical concepts should be considered throughout the entire research process. The formal statistical analyses are only a fraction of what is needed. This assertion will be illustrated with both hypothetical and real examples from simulator studies of driving behavior.

22.2 Steps of the Research Process

As an overview of the research process, consider the following steps:

1. Identify a research question;
2. Convert the research question into parameters to estimate and/or hypotheses to test;
3. Define quantifiable outcomes and predictors to address the hypotheses;
4. Design the study and develop the protocol;
5. Determine the appropriate statistical analyses to perform;
6. Choose a sample size based on power calculations;
7. Acquire resources to perform the study;
8. Begin the study;
9. Set up a secure database and collect data;
10. Perform descriptive analyses and formal statistical tests; and
11. Report and interpret the results, suggesting relevant future research questions.

These general steps have been written in somewhat chronological order; however, such order is not absolute, because as one particular step is being addressed, this often leads the research team to revisit one or more previous steps. Furthermore, many of these steps are closely linked and it can be difficult to say when one step ends and the next begins. Many of these general steps and their subcomponents are topics of individual chapters in this *Handbook*. We now discuss these steps in various amounts of detail, emphasizing the role of statistical concepts therein. Although some of the steps may not explicitly involve statistics, all implicitly are related to statistics due to their linked nature.

22.2.1 Identifying a Research Question

Most scientists, including driving simulator researchers, seem to be able to identify research questions quite easily. The challenge is to find research questions that are feasible to address, from both a technological and a funding aspect. As the subsequent steps are pursued, it often becomes clear whether a proposed research question can be addressed. Consider the question that will be used throughout this chapter: "Does an alerting device improve driving safety?" This particular question does not sound statistical in nature, but many statistical issues come into play when determining how to address it. There are also other potential research questions that are methodological in nature, such as, "What method of reducing time-series data obtained in a driving simulator is best for distinguishing between disease groups?" Such questions explicitly involve statistics, since the properties of specific statistical models are being investigated.

22.2.2 Converting the Research Question Into Specific Parameters and Hypotheses

Many simulator studies involve measuring a variable in multiple subjects in a way that will allow the estimation of one or more parameters. Parameters may be such quantities as means, variances, rates, proportions, and correlations in some population of drivers. Data are used to obtain point estimates of those parameters, as well as confidence intervals that reflect the uncertainty of the estimates due to the variability in the data sampling process. A specific hypothesis is often used to assess whether these parameter estimates are significantly different from some value (e.g., *is the population correlation, ρ, different than 0?*), or whether the parameters differ across different subgroups (e.g., *do drivers with epilepsy have the same mean response, μ, as those without?*)." Generally, an investigator states the null hypothesis as something that he/she hopes to disprove ($H_0: \mu_1 = \mu_2$) in favor of one or more alternative hypotheses ($H_A: \mu_1 > \mu_2$ or $\mu_1 < \mu_2$).

Researchers should avoid specific aims that are too vague, trivial, or self-fulfilling. For example, consider the following aim: "In this study of reaction times in the simulator, we aim to categorize the reaction times into three categories: slow, moderate, and fast." This aim is self-fulfilling, in that a researcher is almost guaranteed to be able to place the outcomes into three groups (unless there are only one or two distinct values of the reaction times). Some authors have suggested that even the traditional hypotheses described in the previous paragraph are trivial, claiming that few, if any, null hypotheses are likely to be absolutely true, and that failure to disprove them is just a function of a limited sample size (Cohen, 1994). This assertion underlines the importance of making the distinction between *statistical* and *practical* (or *clinical*) significance, which we will discuss later in this chapter.

The step of translating research questions into specifics can be illustrated by the following hypothetical conversation between the principal investigator (PI) of a potential study and a co-investigator who is the statistical expert (S) on the study team.

PI: I'd like to find out whether an alerting device can help people drive more safely. We have a front-mounted detector that will measure the speeds of all vehicles in a 20-degree field within a quarter mile range. If any oncoming vehicles are sensed with an approach speed greater than 5 mph, the detector sends a signal, causing a warning light on the dashboard to blink. We have an actual device that can do this in a real vehicle, but we first want to program and equip our simulator to act like this system, to see if it works in the driving lab before formally testing it on the road. I'm trying to write this up for a grant proposal. What kind of statistical method should I say we'll use to analyze the data?

S: Well, before we get to that, we need to work on specific hypotheses. What do you mean by "driving more safely"?

PI: Well, in the simulated drive, it is a two-lane highway, with oncoming traffic in the left lane. So, if somebody is driving safely, then they should not be too close to the center line.

S: *So, are you saying that "driving more safely" means maintaining a greater distance from the center line when an oncoming car passes?*

PI: *Yes. So maybe I could define a driver's outcome as the minimum lateral distance between the left wheels of the car and the center line, say, from the interval between the time four seconds prior to the front edges of the two cars being even with each other until 1 second after. If the driver's left front tire is on the right of the center line, then that is a positive value; if the tire just barely touches the center line, then that distance is 0; and if the left front tire overlaps with or straddles the center line, then that distance would be negative. So positive values are good—the higher the better—and I hope to show that the minimum distance to the center lane tends to be higher when an alerting device is being used.*

S: *That sounds reasonable, except it seems that if we had one driver who keeps the car in the exact center of the driving lane, and another driver whose front right tire is touching the shoulder line, then we would be claiming that the second driver is safer than the first.*

PI: *I see your point; driving too close to the shoulder is probably less safe than being in the middle of the lane, and driving on the shoulder should certainly be considered unsafe. However, since the alerting device is looking for oncoming vehicles, which are usually in the left lane, the critical issue should be the distance from the center line. So, I still like my original hypothesis, that minimum distance to the center line is higher when an alerting device is being used. Then, as a secondary hypothesis, maybe we could analyze whether the frequency of the car going onto the shoulder is increased when using the alerting device. This could help us look at whether drivers are overreacting to the alerts.*

S: *Sounds good.*

In this example, the minimum distance and the number of lane crossings were considered as bases for driving safety hypotheses. Many others could be considered, such as those based on lateral acceleration, variability in lane position, velocity control, steering wheel reversals, reaction times, and crashes. The above exchange is portrayed as a single conversation. In practice, such iterations towards a focused hypothesis can span days, weeks, or evens months.

22.2.3 Defining Quantifiable Outcomes and Predictors

Outcome variables are responses or "dependent" variables in statistical models. For example, if a researcher is testing whether a cognitive test is predictive of whether somebody crashes in a simulator scenario, the crash status is the outcome variable. If one finds that the cognitive test is predictive of crash status, then we may say that the probability of crashing "depends" on the cognitive test score. However, this is a statistical dependency, not necessarily a causal dependency. Generally, a cause-effect relationship between a predictor and an outcome can only be shown through a randomized experimental study. For non-randomized studies (i.e., either purely observational, or with a non-randomized experiment), such as assessing the relationship between cognition and driving, arguments for causality can only be made indirectly (Hennekens & Buring, 1987, pp. 39–43). Due to the complexity of simulator studies, a single research project may have elements of randomized experiments, non-randomized experiments, and observational studies.

In driving simulator studies, an outcome measure may be *dichotomous* (binary) in nature, such as when the presence/absence of a crash is noted, or it may be a numerical measurement, which could be *continuous* or *discrete* in its scale. There are also categorical outcomes, which may either be *nominal* (related to the word "name"; order is not important) or *ordinal* (meaning that order is important). An example of an ordinal outcome could be at a stop sign scenario, where the categories, ordered from best to worst, might be "stopped properly", "stopped improperly" (e.g., beyond the stop sign), "failed to stop but did not crash", and "crashed".

Driving simulator outcomes are objective when they are impartially derived from the digitized data. They may also be subjective in nature, such as a self-reported rating by the participant, or a rating of the participant's performance, as judged by one or more expert raters (perhaps based on video review). Subjective data are often analyzed in the same way as objective data, but one must be aware of potential sources of biases. For example, if expert raters know that a driver performed poorly on neuropsychological tests, they might assume that he/she will be less skilled at driving, and assign a pessimistic rating. In the case of testing for the effect of an alerting device, it would decrease the chance for bias if the reviewer was unaware to the presence/absence of the device. In order to decrease bias and increase precision of subjective expert ratings, it is also important to have standard definitions of outcomes, thorough training sessions, and investigation of between-rater reliability.

Predictor variables are those that are used as explanatory variables in statistical models. They may represent subject-level factors, such as demographics, medical diagnoses, or neuropsychological measures, or they may represent experimental factors such as devices, interventions, or conditions. Some textbooks refer to predictor variables as "independent" variables, because in the case of controlled experimental factors, an investigator may be able to set the levels of such variables "independently" of outside influence, or perhaps "independently" of each other. For example, in the hypothetical study, one could enroll 20 male drivers and 20 female drivers, and then stratify the randomization process so that 50% will use the alerting device and 50% will not. This would give a balanced factorial design (possibly analyzed using a two-way ANOVA), where gender (male versus female) and experimental setting (device versus control) are independent in the dataset. However, if gender was not forced to be balanced across the experimental groups, then gender and experimental setting could be correlated (and therefore not independent) with each other. Hence, calling them "independent" variables would be a misnomer.

Some predictor variables are linked to the hypotheses of a study, while others may be additional covariates that are being used to adjust for confounding in a statistical model. For example, suppose one wants to test whether the presence of a neurological disease is predictive of a driving outcome. If age is predictive of the outcomes, and if the average age of those with the disease is different than those without the disease, then the relationship between the disease status and the outcome is confounded or biased by the relationship between age and the outcomes. In this case, age would be an important predictor variable to consider, even though its relationship with the outcome is not of primary interest.

The variables measured in a simulator may be considered as outcomes or predictors, depending on the research question. In the hypothetical study of alerting devices above, the defined measure of safety would be the outcome variable, while the presence/absence of the alerting device would be the predictor variable. But one could also use the measure of safety as a predictor variable for other outcomes. For example, perhaps the safety measure at the beginning of the drive could be used for predicting unsafe events in later portions of the drive, or for predicting real-world crashes after the simulation study is performed.

Because of the complex nature of simulator experiments, it is often unclear what to use as an outcome variable, nor is it obvious whether to adjust for certain variables. In our alerting system example, one would anticipate that the vehicle speed would be associated with the safety outcome, so an analyst may consider including speed as a covariate. This would probably be fine if the speed covariate is obtained in a baseline drive or segment, *before* the first instance where the alerting device would start. If the speed covariate is measured during the instances where the outcome is measured (sometime *after* the first alert), then adjusting for speed may wash out the effect of the alerting device, since speed might be an intermediate outcome. Suppose that the unadjusted analysis shows that the alerting system helps, but an analysis that adjusts for post-alert speed shows no effect of the alerting system. The interpretation of this would likely be that the alerting system effectively improves safety by encouraging drivers to slow down. On the other hand, if the unadjusted analysis shows a large benefit of the alerting system and the adjusted analysis also shows benefit, but to a lesser magnitude, then the interpretation would likely be that the alerting device improves safety partly through reduced speed and partly through some other causal pathway (increased attention, perhaps).

An empirical comparison of unadjusted and adjusted analyses was made by Donmez, Boyle and Lee (2008). In that study, the outcome was accelerator release time in response to the braking of a lead vehicle, and the predictor was the presence and type of distraction. The authors' unadjusted analysis had the counterintuitive result of a *faster* mean reaction time in the presence of distractions compared to drives without distractions. However, when they adjusted for inverse headway distance the analysis suggested that distractions *slowed* reaction times, which seemed to make more sense. The reason for the discrepant results was that drivers tended to have shorter headway distances when they were distracted, so when the lead vehicle stopped they needed to react faster in order to avoid a collision. The authors inferred from the conflicting results in this study that it is important to adjust for time-varying covariates; however, this advice is inconsistent with the adage that adjustment should only be made for baseline factors (Friedman, Furberg, & DeMets, 1998, p. 298) in order to preserve the benefits of a randomized experiment. An alternative conclusion would be that the choice of outcome variables is very important, since outcomes that are too context-dependent may give counterintuitive results. So, it may have been better to define the primary outcome as the presence/absence of a crash, or perhaps the minimum headway distance. These measures seem to address the "bottom line" of safety better than accelerator release time, which is only crucial when there is insufficient headway distance.

Most of the discussion in this chapter has centered on research studies where the research hypothesis entails looking for an association between a predictor and an outcome. Not all studies follow this pattern. In some studies the goal is just to estimate some distribution or one of its parameters. For example, it might be of interest to describe the reaction times of subjects in a given scenario, perhaps for the purpose of proposing an initial setting of an intervention device. In this case the goal would be estimation and not hypothesis testing. Other times, the hypothesis test might involve comparing the value of an estimated parameter versus published normative values. In such a case, there would be an outcome but no predictor variables, and one would use one-sample statistical techniques. Finally, there may be times when the goal is to look for associations between two variables, without a clear choice as to which is the predictor variable and which is the outcome. Cross-sectional correlation studies and test-retest reliability studies fall into this category.

22.2.4 Designing the Study and Developing the Protocol

The design stage includes many aspects, such as defining the population of participants, developing a recruitment strategy, setting up inclusion and exclusion criteria, designing the driving scenarios, and determining the methods for measuring the predictors and outcomes. In a broad sense, it actually includes steps five and six of the research process, but these steps have been listed separately for emphasis. We can illustrate study design and protocol development by returning to the hypothesis example of the alerting device, considering several questions which would likely come up in the context of this study.

What will be the eligibility criteria for the study? Having very broad criteria will make the results of the study easier to generalize, and a certain amount of heterogeneity is desirable when assessing the association between predictors and outcomes.

However, when comparing subgroups, the statistical power will decrease as the within-group heterogeneity increases. Although there is no simple formula on how to balance these two aspects of heterogeneity, one strategy is to seek for high variability in the predictor variables, and limited variability in the response variable. For example, if a researcher was studying the relationship between cognitive ability and driving performance, then a broader range of cognitive scores might be found by recruiting participants from an entire community, rather than just recruiting on a university campus. At the same time, one could limit the study to those who had been licensed drivers for at least four years, hoping to have fewer extreme outliers due to novice drivers.

Will each driver drive under both conditions (e.g., with and without the alerting device), or will the drivers be randomly split up, so that half use the device and half don't use the device? In statistical terminology, this question is asking whether a study should be a *within-subject* design (also known as a *cross-over*, *block*, or *repeated measures* design) or a *between-subject* design (also known as a *nested* design). One advantage of within-subject designs is that each driver serves as his/her own control, which tends to increase statistical power. For example, suppose that in the alerting device example, there are data from five drives with the alerting device and from five drives without. Suppose also that the minimum distances from the center line (in *cm*) observed during the drives with the alerting device are 35, 33, 45, 57, and 46, while the distances observed without the device are 30, 25, 35, 50, and 40. The respective sample means (standard deviations) in the two groups would be 43.2 (9.7) and 36.0 (9.6). If these data had come from a between-subject design, the two sets of numbers would probably be compared using a two-sample t-test, which would have a test statistic of $t = 1.18$, a p-value of 0.271, and a mean (95% confidence interval) difference of 7.2 (−6.9, 21.3), showing no evidence of an effect of the alerting device. However, suppose these same data had come from a within-subject design, with the first driver having a value of 35 with the device, and 30 without; the second driver having respective values of 33 and 25, and so forth. With that design, the data would likely be analyzed based on a paired t-test, or a one-sample t-test on the within-subject differences (5, 8, 10, 7, and 6). In this case, the mean (standard deviation) of the paired differences would be 7.2 (1.9), the test statistic would be $t = 8.37$, the p-value would be 0.001, and the mean (95% confidence interval) would be 7.2 (4.8, 9.6). Hence, the between-subject design would fail to show a difference, while the within-subject design would show a consistent increased margin of safety of 5 to 10 cm. Furthermore, the within-subject design only used half of the subjects. By using each subject as his/her own control, the variability of the estimate of the effect of the alerting device decreased dramatically (from 9.6 to 1.9), because of the high correlation of within-subject outcomes (a Pearson correlation estimate of 0.98 for this dataset). But even with smaller amounts of correlation, within-subject designs tend to be more powerful.

So, why don't researchers always use within-subject designs? First of all, it may not be practical or ethical to change certain factors in individual drivers (gender, height, weight, drug use, etc.) midway through an experiment. Also, there can be carryover (or contamination) effects from one experimental condition to the next within a driver, which would dilute the between-condition differences of the outcomes, possibly offsetting the benefits gained from using the within-subject design. For example, if using an alerting device in one drive made a participant more alert in a subsequent drive when the device was not used, then there might be little benefit observed in that participant. In summary, between-subject designs are often simpler to perform and easier to interpret, but within-subject designs tend to have greater statistical power in situations where they are feasible and if there is negligible contamination across experimental conditions.

If each driver drives under both conditions, will this be in two separate drives, or will the use/non-use of the device be interspersed throughout a drive? There is no obvious answer to what should be done. A pilot study may be helpful in determining which approach is more feasible and promising. It may be simpler for the device to be completely present or absent during a given drive, and such a design might have less contamination, since more time would have elapsed between conditions compared to switching back and forth between conditions on the same drive. On the other hand, having the use/non-use of the device interspersed throughout a drive might increase the within-subject correlation of outcomes, which could improve the statistical power of the paired analysis.

Supposing that each driver drives two separate drives, with one drive per condition, how will the order of the drives be determined? The order should be randomized. Ideally, when a subject is enrolled, an identification (ID) number would be assigned, which would be linked to a previously prepared list of randomly chosen orders. Such orders can be chosen completely randomly with a computer-based flip of a fair coin, or this can be done with the constraint that exactly half of the subjects must do the drives starting with the device, and the other half staring without the device. This latter approach can be illustrated with 40 subjects as follows. First create a column in a database or spreadsheet with 20 entries of "Device 1st" and 20 entries of "Device 2nd". Then create a column of continuous random values between 0 and 1. Next, sort both columns in ascending order of the random values. Finally, enter in the potential ID numbers (say 01 to 40) into a third column, as a way of assigning subjects to order. This will ensure an exact balance so that order does not confound the results. However, even if order cannot bias the results, it is often helpful to test for the effect of order (e.g., by looking at residual errors versus order in a scatter plot, or by formal testing in a regression model). If order is important, then including it as a covariate in a regression or analysis of covariance model may decrease the residual variability and thereby increase statistical power.

If each driver only drives under one condition, how do we make that assignment? This should be randomized, and can be performed in a manner analogous to that described in the preceding paragraph. Also, it would be important collect the data so that the order is independent of time (e.g., in random order, or in the order in which subjects were enrolled). If the order were to be associated with time (e.g., doing all of the drives with the alerting device first, and then doing all of the drives without the device), it could invalidate the entire study because it would be impossible to know whether any between-group differences are due to the device or due to doing the drive on a different day. There are many factors which could vary from day to day, possibly causing confounding. The projection bulbs on the simulators might be brighter on one day, or maybe there is more ambient noise on one day, or maybe the research assistant administering the drives does something slightly different on one day. Even issues external to the driving simulator could have an effect on the outcomes. If it is snowing outside, some participants may be anxious about having to drive home in the snow. If the stock market is having a bad day, or if the local sports team won an important game the previous night, the mood and performance of the drivers could be affected. Randomization greatly reduces the risk that known and unknown factors will confound the experimental effects of the study, which is why randomized experiments are considered the gold standard of causal inference.

How long should the drive be, and how many instances of an oncoming car will there be in each drive? Here again, some pilot studies may be helpful. If the drive is too short, the drivers may not be acclimated to the simulator (McGehee, Lee, Rizzo, Dawson, & Bateman, 2004). If the drive is too long, the drivers may get fatigued. Regarding the second issue, having only one instance or trial of oncoming cars may not provide a very meaningful summary of the driver's performance on that drive; having several repeated trials in a drive could provide more robust information. However, with multiple trials per drive, or even across drives within subject, there can be large learning effects due to the driver knowing what to expect. Sometimes learning effects can be as large as the effect of the experimental factor (Lee, McGehee, Brown, & Reyes, 2002).

22.2.5 Determining the Appropriate Statistical Analyses to Perform

Specific statistical methods are covered in chapter 21 of this *Handbook*, as well as in dozens of textbooks. For biomedical researchers, Pagano and Gauvreau (2000) offer a concept-oriented introduction, while Rosner (2006) uses a more traditional approach, augmented with lots of real data examples. Zar (1999) provides at least two semesters worth of introductory material for audiences with strong technical skills, and includes excellent historical perspective and references. For regression, analysis of variance, and experimental design issues (including random effects and repeated measures), textbooks can be found at less technical (Kleinbaum, Kupper, Muller, & Nizam, 2008) or more technical levels (Kutner, Nachtsheim, Neter, & Li, 2005).

Generally, the design of the study should guide the choice of statistical test. In particular, it is important to accommodate random driver effects in the analyses. Returning to our hypothetical study, suppose that 40 drivers are randomly and evenly split between driving with and without the alerting device. Each driver does one drive each, with 10 instances of oncoming

vehicles in each drive where the measure of safety is obtained. The dataset might be organized with identification number in Column 1, condition (with or without device) in Column 2, oncoming vehicle number (from 1 to 10) in Column 3, and the safety measure in Column 4. With this layout, there would be 400 rows of data. A researcher might be tempted to simply perform a two-sample t-test to compare the 200 observations of the safety measure with the device versus the 200 without. However, this approach ignores the fact that observations within a driver will tend to be similar to each other (sometimes quantified by the *intra-class correlation*, or *ICC*, or the within-subject reliability), due to the random driver effects (i.e., some drivers will tend to be safer than others). Ignoring these random effects by doing this t-test (which would have $400 - 2 = 398$ degrees of freedom) would be inappropriate, in that risk of a Type I error would be greatly increased. For example, assuming a total standard deviation of 1 unit and a nominal ICC of 0.10, it can be shown through computer simulations that the real Type I error rate would be about 10% when the target rate is $\alpha = 0.05$. With a modest ICC of 0.50, the real Type I error rate increases to about 43%.

As a simple modification to address the issue of random effects in this example, one can reduce the data from 400 rows to 40 rows by replacing the individual safety measure observations with the observed mean across the 10 trials for the respective drives. At this point, a two-sample t-test (with $40 - 2 = 38$ degrees of freedom) comparing the mean safety measures would be valid, since there would only be one row per driver, causing each row to be independent of all of the others. Alternatively, one could fit a nested linear model, with trial nested within driver, and driver nested within driving condition (i.e., with versus without the device). In this linear model, the test statistic would be an *F* statistic, found by dividing the mean square of driving condition term by the mean square of the driver term. This *F* statistic would have 1 denominator degree of freedom and 38 numerator degrees of freedom, and would be the square of the 38-degree-of freedom *t* statistic mentioned above. Hence, these two approaches should give identical results. An advantage of the linear model's approach is that it permits estimation of between- and within-driver variance components (and hence, the ICC), which could aid in designing future studies. For example, if the within-driver variance component is high relative to the total variance (i.e., low ICC), then that suggests that it may be important to have several trials per driver in order to get precise estimates of the behavior of each driver. Conversely, if the within-driver variance component is relatively low, then one or two trials per driver may be sufficient. The linear model approach also has the advantage that the effect of "order" can be assessed by including the oncoming vehicle number as a covariate.

22.2.6 Choosing a Sample Size Based on Power Calculations

The statistical power is the probability of rejecting the null hypothesis, given the null hypothesis is false. For example, suppose that the outcome variable is normally distributed in two groups, where the population variance is the same in each group (i.e., $\sigma_1^2 = \sigma_2^2 = \sigma^2$), but the population means (μ_1 and μ_2) are potentially different. Then the following formula is a large sample approximation for the power of a two-sample, two-sided t-test of the null hypothesis of $H_0: \mu_1 = \mu_2$:

$$\Phi\left(\frac{|\mu_1 - \mu_2|}{\sigma\sqrt{1/n_1 + 1/n_2}} - \Phi^{-1}(1 - \alpha/2)\right),$$

where $\Phi(.)$ is the cumulative distribution function of the standard normal distribution. When using the traditional value of $\alpha = 0.05$ for the target Type I error rate, the value of $\Phi^{-1}(1 - \alpha/2)$ would be $\Phi^{-1}(0.975) = 1.96$. Since $\Phi(.)$ monotonically increases with its argument, the power function will increase if $|\mu_1 - \mu_2|$ increases, if σ decreases, or if the sample size (n_1 and/or n_2) increases. It will also increase if α is increased, but that is usually undesirable. To illustrate the calculation, suppose that we were doing our alerting device study as a between-subject design, and the outcome (minimum distance in cm) had a population standard deviation of $\sigma = 10$ in either group, and that the respective means for with and without device were $\mu_1 = 42$ and $\mu_2 = 30$ (1.2σ units apart). If we decided to enroll 20 subjects into each group, then the power when using $\alpha = 0.05$ would be approximately equal to

$$\Phi\left(\frac{1.2\sigma}{\sigma\sqrt{1/20 + 1/20}} - \Phi^{-1}(0.975)\right) = \Phi\left(\frac{1.2}{\sqrt{0.1}} - 1.96\right)$$

$$= \Phi(1.835) = 96.7\%.$$

Hence, if doing a two-group study with 20 subjects in each group, we would have about a 97% chance of finding a difference in the outcomes if the real difference is 1.2 standard deviations. If the difference were only, say, 0.6 standard deviations (e.g., $\mu_1 = 36$ and $\mu_2 = 30$), then the power would be $\Phi(-0.06) = 47.6\%$. When designing a study, statistical power of at least 80% or 90% is usually desirable.

The above formula assumes that we know the population variance in each group, so that a Z (standard normal) can be used as our test statistic. In practice, we usually do not know the variance, and we rely on a t-test for our test statistic; hence, the above formula is an approximation that is only accurate in very large samples (when the sample variance would be very similar to the population variance). Although formulas exist to improve upon that approximation in small and modest sample sizes, the calculation is more likely to be performed using a computer software program. In Proc Power of SAS 9.1.3 (SAS Institute, 2008), the following syntax is used:

    proc power;
    twosamplemeans test = diff meandiff = 1.2 stddev = 1 npergroup = 20 power = .; run;

Submitting these commands, one finds that the power is actually 95.9%—slightly lower than the approximation. Using the

same settings but with a sample size of 10, the precise power calculation from SAS is 71.8%, while the large sample approximation based on the normal formula is an inflated 76.5%. Note that the *power =.* statement indicates that we have entered in all of the other relevant information and want the power to be calculated. Alternatively, we could have put in a desired level of power (*power = 0.90*), and then asked for required sample size (*npergroup =.*) or the size of the difference (*meandiff =.*). Also note that many other software programs can perform power calculations, ranging from free internet-based software such as that found at http://www.stat.uiowa.edu/~rlenth/Power (Lenth, 2006), to very comprehensive commercial software such as PASS (Hintze, 2000).

The above discussion pertains to two-sample t-tests. Power calculations can be performed for virtually any statistical test. In most common situations, approximate formulas and computer software programs are available to facilitate calculations. In complicated situations, power can be accurately estimated by setting up a simulation study. For this approach, an investigator creates hundreds or thousands of datasets, each consisting of random draws from an assumed model that results in a specific sample size, and then calculates the test statistic (t, Z, F, chi-square, etc.) for each dataset. The percent of datasets wherein the null hypothesis is rejected is then the estimated power for that specific sample size and model setting.

The mechanics of power calculations are often fairly straightforward. The challenge is knowing what to use for the values of the parameters (especially the means and variances). Ironically, only after one does the proposed study on a very large sample can precise estimates of these parameters be obtained. So power calculations tend to be very speculative and approximate, based on information from a combination of historical normative values, studies from other participant populations, and small pilot studies. Nevertheless, they do serve the purpose of encouraging the research team to be very specific about the primary hypotheses and the outcome variables in a proposed study.

22.2.7 Acquiring Resources to Perform the Study

For many investigators, this step involves applying for grant funds. As explained in the introduction, it is important to show that the research team has integrated expertise from all important aspects of the study, such as psychological, medical, engineering, and statistical. Different funding agencies may expect different levels of statistical expertise in the study team, as well as different levels of statistical detail in a grant proposal. Generally, the research team should clearly indicate which member will have primary responsibility over statistical issues, and justify this choice through that person's degrees, coursework, and experience. If such a choice cannot be clearly justified, it may be important to add an appropriate statistical expert to the team.

Regardless of the eventual source of funds, it is essential to obtain proper clearance from the investigators' Institutional Review Board (IRB), which is charged with protecting human subjects used in research studies. In driving simulator studies, some common risks to the participants are dizziness, nausea, and headaches. Loss of confidentiality is another important risk, which we highlight below. More information on IRB issues can be found through online training programs at the NIH Office of Human Subject Research (http://ohsr.od.nih.gov) and the Collaborative Institutional Training Initiative (http://www.citiprogram.org).

22.2.8 Beginning the Study

It is important to obtain predictor and outcome data on pilot subjects in order to make sure that any technical or procedural problems are resolved in a timely manner. Examples of technical problems might be simulator program bugs, video projection failure, inappropriate audio levels, or incomplete data capture. Procedural problems might include unexpected driver fatigue (which might be addressed via an adjustment to the length of the drive) or confusion concerning what is being requested in the simulator tasks (which might be addressed via a better set of instructions). Clear and well-developed standard operating procedures are also crucial to increase the precision and to decrease the chance of confounding biases in the data collection process. For example, the instructions that a research staff member gives a driver could change the participant's driving strategy, which could impact the outcome variable. Hence, the script that is used to introduce the driver to the simulator test should be standardized for all subjects. Pilot testing can reveal a range of questions that participants are likely to have in response to the initial instructions, or perhaps during the drive. Based on the type of information obtained in the pilot phase, new instructions could be written, or an algorithm could be constructed on how to respond to potential questions.

22.2.9 Setting Up a Secure Database and Collecting Data

Data from driving simulators are often captured at high frequencies (e.g., 10 to 100 frames per second), so high volumes of data need to be captured and securely stored, with efficient back-up systems. Many studies also use demographic, medical, and/or psychological data, as well as data that are abstracted based on video review. It is important that these data are also securely stored and managed so that they can be merged with the driving simulator data to address the study aims. To help preserve confidentiality, a study ID number should be assigned to each subject upon enrollment, with the link between the ID and the subject's name being kept in a secure database that is not used in analyses. Similarly, government-issued numbers (e.g., Social Security numbers) should only be used for mandatory purposes (say, for documenting compensation payments to participants), and not for ID variables in analysis datasets.

Codebooks should be set up so that all members of the research team understand the definitions, format, scales, and interpretations of the data being collected. Data should be monitored for quality via within-variable range checks and across-variable logical checks. Range checks can be set up in some database programs either to force a certain range (e.g., participant weights must only be positive values) or to force the data entry person to double-check values outside certain ranges (e.g., weights outside of 30 to 200 kg). Range checks could also be done periodically by calculating the minimum and maximum observed values for various fields. Examples of across-variable logical checks include making sure that all gender-specific conditions (pregnancy, menopause, prostate cancer, etc.) are only occurring in participants of the corresponding gender; or that the age distribution of a diagnosis group is appropriate (e.g., no teenagers in the Alzheimer disease group). Such logical checks are more difficult to implement, but can be done with a little forethought and computer programming.

22.2.10 Performing Descriptive Analyses and Formal Statistical Tests

Descriptive analyses include numerical summaries (means, standard deviations, minimums, maximums, proportions, etc.) as well as visual displays. They are important tools to check the quality of the data and to illustrate the findings of a study. For simulator data, it is often beneficial to examine time-series plots of digitized information for portions of an individual drive, including input items such as accelerator position, brake position, and steering wheel position, as well as output information such as longitudinal speed, lateral acceleration, and lateral position. Displaying such data can help researchers give a better understanding of the study to those who are not driving specialists. However, sometimes such displays can be misleading. For example, note Figure 22.1, which was produced using commands in the statistical package *R* (R 2.0.1., 2004). This figure shows the lateral position of a driver with Alzheimer's disease over time, and has geographical markers plotted so that the entire plot resembles a roadway (Dawson, Cavanaugh, Zamba, & Rizzo, 2010). At first glance, this driver seems to be weaving in an extremely sharp manner. However, it is important to realize that since the vertical axis is distance (in meters) and the horizontal axis is time (in seconds), the "roadway" is not drawn to scale and tends to exaggerate the horizontal swerving. When the time-series data have been reduced to per-driver outcome measures, then standard graphical tools (box plots, scatter plots, etc.) can be applied to help check for data quality, choose appropriate parametric models, and illustrate the data.

To perform formal statistical tests, the ideal is to follow the original analysis plan. However, complications often arise, such as when the analysis plan assumed that the responses would be normally distributed, but the data suggest that this assumption is violated. To address the issue of non-normality, a numerical transformation of the data (e.g., logarithms, square roots, etc.) may be helpful. If not, a different parametric model could be considered, or a non-parametric or semi-parametric approach could be used.

Related to the issue of non-normality is the presence of outliers. If a few extreme outliers are seen, it can be helpful to perform the statistical analyses with the full dataset and with outliers temporarily removed, and then to compare the results.

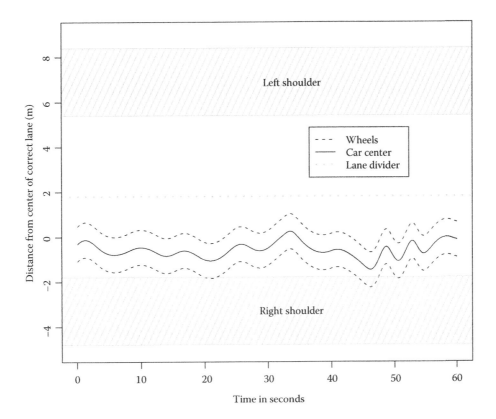

FIGURE 22.1 A time-series graph showing the lateral position of the center and edges of a simulated vehicle in a one-minute segment. Although it is drawn to resemble a road, the horizontal axis is time, not distance; hence, the lateral swerves are exaggerated.

If the results are very similar, then it is likely that the statistical results are not due to just one or two unexpected values. If they are quite different, then it may be useful to re-analyze the data in a non-parametric manner that is robust to outliers. However, throwing out data permanently just to "clean" the data should not be done, as that usually exaggerates the statistical significance of the results.

Also, sometimes technical problems result in missing data and an unbalanced design, which may require a more sophisticated analysis than originally planned. If data are missing for reasons unrelated to the outcome and predictors, one can usually analyze the available data as if that was the original design. If the data are missing for reasons likely associated with outcome and predictors, it is often unclear what to do. Little and Rubin (2002) discuss several potential techniques.

22.2.11 Reporting and Interpreting the Results, Suggesting Future Research Questions

The data analysis of a study often assesses whether certain trends are *statistically* significant. When interpreting such results, it is important to consider *practical* (or *clinical*) significance, as well. Confidence intervals can be helpful in this regard. For example, suppose that the alerting device in our example has a very miniscule benefit—say, reducing the crash risk in a simulator drive from 40% to 38%. With a large enough sample size, an investigator is likely to be able to conclude that the drivers who use

the device do "significantly" better than those who do not, but such a difference may not be clinically important. For example, suppose that a huge study was performed, and the drivers who use the alerting device have 3,450 crashes among 9,000 drivers (38.3%), while the control group (no device used) has 3,600 crashes among 9,000 (40.0%). The estimated reduction in crash rates due to intervention is 1.7% (95% confidence interval: 0.2%, 3.1%) a result that is statistically, but probably not practically, significant. On the other hand, if a much smaller study had 5/20 crashes in the device group and 10/20 crashes in the control group, the estimated difference is −25.0% (approximate confidence interval: −54.0% to 4.0%). It would be difficult to interpret this study, since the majority of the confidence interval seems to show an important clinical benefit to the intervention (up to a 54% reduction in crash risk), but the data are also consistent with no benefit (0% difference) and even a slight detriment of the intervention (4% increase). If a study had results showing 20/90 crashing in the intervention group versus 40/90 crashing in the control group, the estimated reduction would be 22.2% (confidence interval 8.8%, 35.6%), suggesting both statistical and practical significance.

22.2.12 Concluding Remarks

Altman (1998, p. 2670) attributed the following intriguing statement to Micheal Healy: "The difference between medical research and agricultural research is that medical research is done by doctors but agricultural research is not done by farmers." One interpretation of this statement is that physicians may bring many personal biases into their research in ways that agricultural researchers do not. This suggests that having multi-disciplinary research teams, with a variety of expertise, may provide a balance of perspective to minimize bias. Another interpretation is that being a good scientist involves more than just the expert knowledge of a content area, be it agriculture, medicine, engineering, psychology, or any other scientific field. Specific research skills are necessary, including an appropriate knowledge of statistical concepts, and the understanding that statistical reasoning applies to the entire research process and not just to the analysis stage.

Key Points

- Statistical issues should be considered throughout the stages of the study.
- It can be challenging to define or choose meaningful outcome variables that properly address the goals of the study.
- The data should be analyzed in a manner consistent with the study design.
- When there is more than one data point per study participant, the analyses need to accommodate random person effects.
- Confidence intervals should be presented when reporting study results.

Keywords: Hypothesis Testing, Power, Sample Size, Statistics, Study Design

Acknowledgments

The author has been supported by several NIH awards over the years, including AG17717, AG15071, NS044930, and AG026027. The author would like to thank all of the investigators and research staff in the Division of Neuroergonomics in the Department of Neurology at the University of Iowa. Special thanks go to the biostatistics graduate students who have worked with me on driving simulator studies for the past ten years: Kirk Bateman, Laura Stierman, Qian "Cicci" Shi, Yang Lei, JonDavid Sparks, Elizabeth Dastrup, Mijin Jang and Amy Johnson.

Glossary

Null hypothesis: The assumed setting of a model that an experimenter often hopes to disprove in order to show support for a research hypothesis.

Outcome variable: The dependent variable or the response in a statistical model.

Power: The probability of rejecting the null hypothesis in favor of an alternative (or research) hypothesis, given that the null hypothesis is false.

Predictor variable: The explanatory variable in a statistical model; sometimes called an independent variable.

Research hypothesis: The conjecture that a scientist is hoping to prove through a planned experiment. It is often the complement of a null hypothesis.

Type I error: The probability of rejecting the null hypothesis, given that the null hypothesis is true.

Key Readings

Altman, D. G. (1998). Statistical reviewing for medical journals. *Statistics in Medicine, 17,* 2661–2674.

Kleinbaum, D. G., Kupper, L. L., Muller, K. E., & Nizam, A. (2008). *Applied regression analysis and other multivariable methods* (4th ed.). Belmont, CA: Duxbury.

Kutner, M. H., Nachtsheim, C. J., Neter, J., & Li, W. (2005). *Applied linear statistical models* (5th ed.). New York: McGraw-Hill.

Rosner, B. (2006). *Fundamentals of biostatistics* (6th ed.). Belmont, CA: Thomson Brooks/Cole.

Zar, J. H. (1999). *Biostatistical analysis* (4th ed.). Upper Saddle River, NJ: Prentice Hall.

References

Altman, D. G. (1998). Statistical reviewing for medical journals. *Statistics in Medicine, 17,* 2661–2674.

Cohen, J., (1994). The earth is round ($p < .05$). *American Psychologist, 12,* 997–1003.

Collaborative Institutional Training Initiative. *Home Page.* Retrieved February 2, 2009 from http://citiprogram.org

Dawson, J. D., Cavanaugh J. E., Zamba, K .D., & Rizzo, M. (2010). Modeling lateral control in driving studies. *Accident Analysis & Prevention, 42*(3), 891–897.

Donmez, B., Boyle, L., & Lee, J. D. (2008). Accounting for the covariate effects in driving simulator studies. *Theoretical Issues in Ergonomics Science, 9*(3), 189–199.

Friedman, L. M., Furberg, C. D., & DeMets, D. L. (1998). *Fundamentals of clinical trials* (3rd ed.). New York: Springer.

Hennekens, C. H., & Buring, J. E. (1987). *Epidemiology in medicine.* Boston: Little, Brown and Co.

Hintze, J. (2006). *PASS 2000.* Kaysville, UT: Number Cruncher Statistical Systems.

Kleinbaum, D. G., Kupper, L. L., Muller, K. E., & Nizam, A. (2008). *Applied regression analysis and other multivariable methods* (4th ed.). Belmont, CA: Duxbury.

Kutner, M. H., Nachtsheim, C. J., Neter, J., & Li, W. (2005). *Applied linear statistical models* (5th ed.). New York: McGraw-Hill.

Lee, J. D., McGehee, D. V., Brown, T. L., & Reyes, M. L. (2002). Collision warning timing, driver distraction, and driver response to imminent rear end collisions in a high-fidelity driving simulator. *Human Factors, 44*(2), 314–334.

Lenth, R. V. (2006). *Java Applets for Power and Sample Size* [Computer software]. Retrieved November 24, 2008, from http://www.stat.uiowa.edu/~rlenth/Power

Little, R. J. A., & Rubin, D. B. (2002). *Statistical analysis with missing data.* New York: John Wiley & Sons.

McGehee, D. V., Lee, J. D., Rizzo, M., Dawson, J., & Bateman, K. (2004). Quantitative analysis of steering adaptation on a high performance driving simulator. *Transportation Research Part F: Traffic Psychology and Behavior, 7,* 181–196.

Office of Human Subject Research National Institutes of Health. (n.d.). *Home Page.* Retrieved February 2, 2009, from http://oshr.od.nih.gov

Pagano, M. P., & Gauvreau, K. (2000). *Principles of biostatistics* (2nd ed.). Pacific Grove, CA: Duxbury.

R Development Core Team. (2004). *R 2.0.1. A language and environment.* Vienna, Austria: R Foundation for Statistical Computing.

Rosner, B. (2006). *Fundamentals of biostatistics* (6th ed.). Belmont, CA: Thomson Brooks/Cole.

SAS Institute. (2008). *SAS 9.1.3.* Cary, NC.

Zar, J. H. (1999). *Biostatistical analysis* (4th ed.). Upper Saddle River, NJ: Prentice Hall.

The color figures are presented in order, beginning with those figures that provide a history and an overview of different types of simulators, continuing with those figures that depict the different layouts of simulator laboratories and a side by side comparison of real and simulated worlds as well as the scene and scenario development tools, and ending with scenes from studies in transportation engineering, distracted driving, and hazard anticipation.

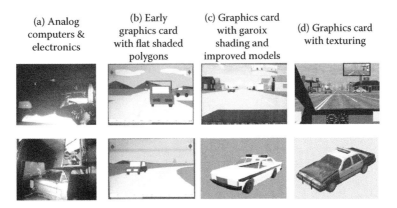

(a) Analog computers & electronics (b) Early graphics card with flat shaded polygons (c) Graphics card with garoix shading and improved models (d) Graphics card with texturing

FIGURE 2.2 Evolution of driving simulators and PC graphics.

TABLE 7.3 Driving Simulator Motion Systems

Type	Typical Example(s)	Notes
Small Stewart Platform	Moog 12 Inch Series	Electric actuator design. Displacements: X: +/− 0.25 m Y: +/− 0.25 m Z: +/− 0.18 m Roll: +/− 21 degree Pitch: +/−22 degree Yaw: +/−22 degree Payload: 1000 Kg
Large Stewart Platform	Ford VIRTTEX Simulator	Hydraulic actuator design. Displacements: X: +/−1.6 m Y: +/−1.6 m Z: +/− 1.0 m Roll: +/−20 degree Pitch: +/−20 degree Yaw: +/−40 degree Payload: ~6000 Kg

(continued)

TABLE 7.3 (Continued) Driving Simulator Motion Systems

Type	Typical Example(s)	Notes
Asymmetric Large Displacement Systems	VTI Driving Simulator III 	Electric actuator design: X or Y: +/−3.75 m Roll: −9 degree to +14 degree Pitch: +/−24 degree Vibration: Z: +/−0.06 m X: +/−0.06 m Roll: +/−6 degree Pitch: +/−3 degree
Symmetric Large Displacement Systems (Payload < 1500 Kg)	Renault Ultimate Driving Simulator 	Electric actuator design: X: 6.0 m Y: 6.0 m Z: +/−0.2 m Roll: +/−15 degree Pitch: +/−15 degree Yaw: +/−15 degree Payload: ~1,000 Kg
Symmetric Large Displacement Systems (Payload > 1500 Kg)	National Advanced Driving Simulator (NADS) 	Hydraulic/Electric actuator design: X: +/−9.75 m Y: +/−9.75 m Z: +/−0.6 m Roll: +/−25 degree Pitch: +/−25 degree Yaw: +/−330 degree There are also 3 vibration degrees of freedom. Payload: ~10,000 Kg
Asymmetric Large Dispalcement Systems (Payload > 1500 Kg)	Toyota Driving Simulator 	Hydraulic/Electric actuator design: X: +/− 17.5 m Y: +/− 10 m Z: +/− 0.6 m Roll: +/− 25 degree Pitch: +/− 25 degree Yaw: +/− 330 degree There are also 3 vibration degrees of freedom. Payload: ~10,000 Kg

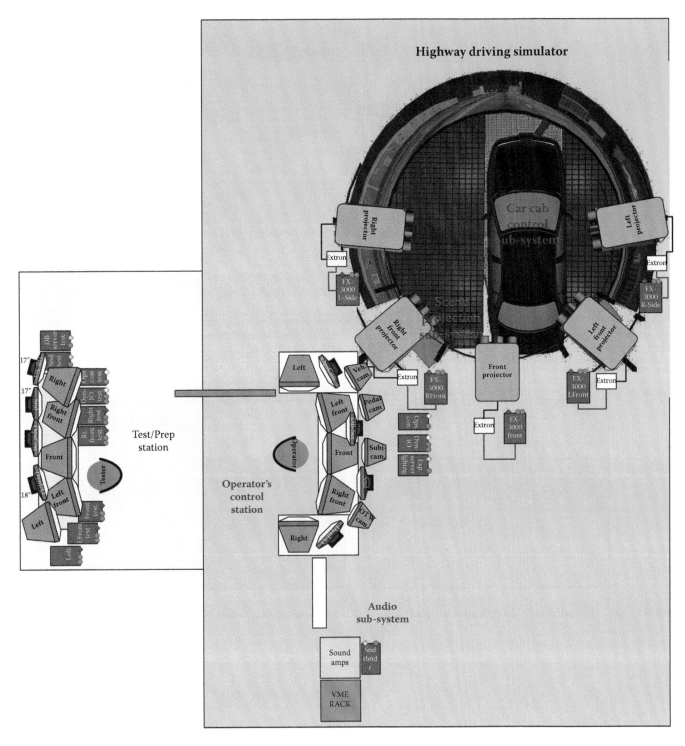

FIGURE 34.1 FHWA highway driving simulator is comprised of a number of networked computers.

WEB FIGURE 38.7 The experimenters' view of the UCDS with the Fallowfield highway-railway grade crossing shown. The displays from left to right are the quadraplex view (face, upper left; center column, upper right; pedals and feet, lower left; center view simulator, lower right), eye movement calibration and control monitor, Hyperdrive (v. 1.9.2) console and SimObserver recording monitor.

FIGURE 34.8 Photograph (left) and simulation (right) of Curve No. 2 in subsequent visibility studies.

WEB FIGURE 16.9 A picture of Highway 401 near the town of Puslinch southeast of Toronto (left). A similar section of freeway modeled in the University of Calgary Driving Simulator (UCDS) (right).

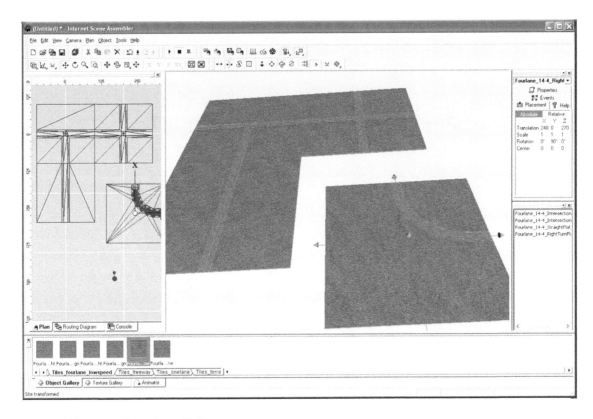

FIGURE 6.1   Assembling scene from tiles in SimVista.

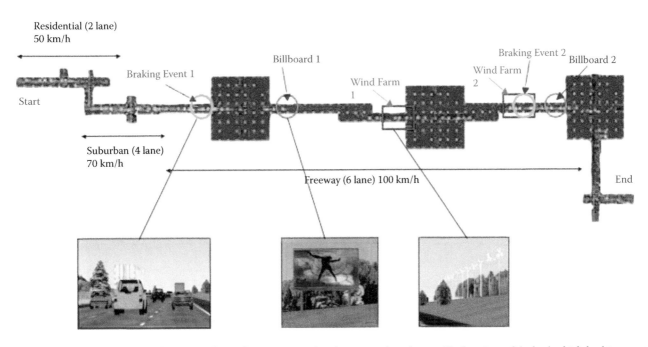

WEB FIGURE 16.8   Experimental Drive 1 is shown from start to end with associated road types. The locations of the lead vehicle braking events, billboards, and wind farms are also indicated.

FIGURE 34.9 In the round-about context, many drivers did not understand the meaning of the lane restriction sign (far right) and lane restriction markings (center).

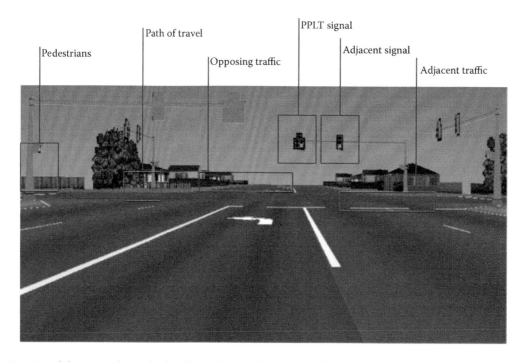

FIGURE 35.3 Partitioned driver simulation display. (From Knodler, M. A., Jr., & Noyce, D. A. (2005). Tracking driver eye movements at permissive left-turns. *Proceedings of the 3rd international driving symposium on human factors in driver assessment, training, and vehicle design* (pp. 134–142). Rockport, ME.)

FIGURE 36.9 Simulated changeable message sign created by inserting black box with scenario authoring software and using supplementary projector to project image of sign legend into black box.

FIGURE 36.11 Pavement marking shields over-projected onto the roadway scene.

FIGURE 16.1 Lead vehicle before (left) and during (right) a braking event in video billboard condition.

FIGURE 30.1 Adjacent truck left turn. (Cars in the opposing lane in front of the truck are obscured from the driver and can turn left into the path of the driver as he or she travels through the intersection.)

23

The Qualitative Interview

Jane Moeckli
The University of Iowa

Abstract

The Problem. Quantitative methods, including the use of subjective instruments such as questionnaires and structured interviews, provide data that can be easily analyzed using standard statistical methods. What questionnaires and structured interviews miss, however, are what driving means to different individuals and how these meanings reflect broader cultural values. These questions are better addressed using qualitative approaches. *Role of Driving Simulators.* Qualitative research methods can be woven into our simulation research designs through one of three methodological approaches that complement and/or build upon driving simulation research: the dominant-less dominant design, the multiple phase combined method design, or a more pure mixed method design. The degree to which the two are integrated depends on a number of factors. Primary among them is the research question, but pragmatic concerns, such as time to analyze data and access to appropriate resources, also influence the use of qualitative approaches. *Key Implications of Qualitative Methods for Driving Simulator Studies.* This chapter discusses the unique contribution qualitative research can make to our analyses. It focuses on the qualitative interview, detailing its strengths and limitations; question structure and type; the development and use of interview protocols; the recording, analyzing, and interpreting of data; and the writing up of the results. *Scenarios and Dependent Variables.* The qualitative interview is particularly useful in revealing drivers' perceptions of, feelings toward, and attitudes about driving, safety, and risk, as well as the cultural influences that shape these factors. While intervening variables such as trust, driver attitude, and self-confidence can be ascertained through quantitative analyses, qualitative methodologies often provide a deeper understanding of the affects of these variables, albeit at a smaller scale than descriptions engendered through quantitative analyses. *Generality.* Although variations on the interview are employed by many driving simulation researchers, the qualitative interview has a limited presence outside of verbal protocol analysis and focus groups. This chapter presents a step-by-step guide for determining the appropriateness of the qualitative interview for your study, and how to plan, conduct, and analyze interview data, using a method heavily influenced by grounded theory. Especially in the context of the "cultural turn" in traffic safety research, qualitative interviews provide a bridge between the research world of driving performance and that concerned with cultural values.

23.1 Introduction

"If you want to know how people understand their world and their lives, why not talk with them?" (Kvale & Brinkmann 2008, p. xvii)

Why include a chapter on qualitative interviewing in a *Handbook* on driving simulation? While quantitative data from our driving simulation experiments provide generalizable information about driver performance, the data fall short when attempting to understand how people think, behave, attribute meaning, develop understanding, make decisions, and solve problems. Qualitative approaches provide a complementary path to our ultimate goal of understanding driving phenomena, and qualitative interviews specifically generate insights and concepts relevant to answering *why* (as opposed to what or how) from an individual and collective lens.

The interview is not a new method for most human factors and driving simulation researchers. Interviews are used in parallel with other research methods, typically in either a dominant-less dominant (e.g., interviews conducted on a subset of driving simulation participants) or multiple-phase combined method design (e.g., focus group discussion to inform design decisions which are then evaluated in a simulation experiment), but occasionally they are conducted in combination with other methods in a more pure form of a mixed methodology design (e.g., verbal reports while driving in a simulated experiment) (Creswell, 1994). Different types of interviews are utilized for a variety of reasons, including focusing the development of a prototype, evaluating a design, capturing expert perspectives on an issue, analyzing driver attitudes, eliciting the type of knowledge accessed in task processing, and testing and improving survey instruments and our simulations. In these contexts, interviews draw out people's thoughts, opinions, and/or feedback; they help define future areas of interest while providing insight into past developments; and they serve to provide clarification.

While the interview is not a new technique for most readers, its use is rather restricted, and minimal reporting of interview findings masks the potential contributions interview data may have on developments in the field. The aim of this chapter is to provide the general reader with an account of the qualitative interview process in an effort to broaden the scope of how we think about, conduct, analyze, and ultimately use, qualitative interviews.

Take, for example, the following three interview passages. The excerpt presented in Table 23.1 demonstrates a fairly sophisticated understanding of the adaptive interface technology under study. Unique among other interpretations of the technology's function and operation, this participant recognized the broad reach that the in-vehicle system had to immediate and distant environments through flows of information transmitted through satellites, filtered through lasers and cameras, and displayed to the driver to discourage certain behaviors.

TABLE 23.1 Interview Excerpt Describing a Mental Model of an In-Vehicle Information System

Interviewer	*Can you tell me a little bit about your picture?*
Participant	Well, when I think of how it works, I think of technologically how all the information is being gathered and being put in here, and how it gets to me via the screen obviously which I've drawn down here as, well I'll get to that. Anyway, so the way I envision it is, the mapping system is probably transmitted to the car by satellite through GPS. The SMS or the text messaging is done through the cell phone system, which by the way I think is a bad idea. And of course the collision avoidance is real easy, I would imagine we're doing, I don't know, it's either a laser or some sort of sound bouncing off their laser/radar type of deal. That's easy enough to understand. I picture the lane change thing actually being able to read, you know, through a video camera, being able to see, you know, to key off of the... and I don't know whether that's important that the strips have reflective surface in them, you know, or what that is, whether it interacts with the road in any way or what. But you know it essentially takes that information and transmits it up to a light if you shouldn't be changing lanes...

The interview excerpt shown in Table 23.2 presents a more emotional response to the increasing proliferation of information technologies in our vehicles. The participant's own personal belief in the potential value of the hypothetical system is overshadowed by the dangerous connection that the participant draws among information, technology, regulation, and corporate interests. The last excerpt presented in Table 23.3

TABLE 23.2 Interview Excerpt About Potential Alcohol Impaired Driving Interventions

Interviewer	*Ok, now how about a system that could collect data in something like an airplane's black box which can be accessed by insurance companies or law enforcement after an accident or serious violation to detect if impairment was a factor?*
Participant	Yeah... I've read about things like that, and that gets back to your GPS and things like that, and that gets back, I understand what they're trying to do and personally that's a fine line, but what I don't like about that is what it opens the door for, and that's what I don't like about that.
Interviewer	*Ok, and what do...*
Participant	I don't want regardless with if I agree or not. I don't want...
Interviewer	*To, like, track your car or...*
Participant	Exactly. I don't want, I don't care what the hell I'm doing but I don't want somebody to, oh, you know, if we can use this for this then why don't we use this for [pause], so then the insurance companies [can say] "Oh, let's see, we can track, we can specifically say our drivers drive here, so these insurance rates, so he does 90% of his driving in 5 miles." I don't know. I don't like all the doors that that one opens. That opens up too many Big Brother, I don't know what the terminology is. I don't like that one at all.

TABLE 23.3 Interview Excerpt Characterizing Relationship With Vehicle

Interviewer	*Now can you talk about generally your relationship with your vehicle, with any vehicle it doesn't matter it could be our vehicle in there or it could be your vehicle. How do the two of you interact?*
Participant	I'd like to feel like we're talking to each other. I mean, if I listen to the radio, focus a little too much on the radio, but I'm changing stations a lot, if I don't like the song. I use the mirrors, I use both mirrors a lot. I just got an SUV, so I really need the mirrors a lot for all the blind spots. And the rear view mirror. So I feel like [I'm] on a team, I mean I don't, but yeah.

returns to the basic elements of the vehicle—its mirrors, the radio—but describes them again in terms of their communicative relationship with the driver. While each example illustrates individual understandings and attitudes toward the vehicle and its technologies, they also reflect an underlying cultural system that frames how people think about and interact with the world in terms of an interconnected network of information flows. The connection may be positive (e.g., my vehicle and I are a team) or threatening (e.g., information technologies open the door for unjustified invasion of privacy); it may be mediated through low-tech (e.g., mirrors) or high-tech means (e.g., GPS communicating with satellites to determine the vehicle's location). Nonetheless, our perceived relationship with the car has shifted from a mechanistic relationship to a communicative one. Such insights would not come from our driving data, and would be difficult to obtain from survey interviews or questionnaires.

The qualitative tradition is rich, expansive, and grounded in several disciplines. I come to the qualitative interview as a cultural geographer, so my perspective is less focused on the analysis of cognition or cognitive processes than those trained in the behavioral sciences, and rests instead on attempting to understand the social world through the meanings people bring to it. There are, of course, fruitful connections between these approaches, particularly as psychology has shifted from an analysis of internal cognitive processes to investigating cognition from an information processing perspective (e.g., Newell & Simon, 1972) or understanding the role of cultural and historical context on cognitive processes (Smagorinsky, 2001).

Accordingly, the process I outline here is one among many, and it is more heavily influenced by ethnographic field methods and grounded theory than by approaches more familiar to human factors researchers. Because there are many types of interview formats, which can occur in any number of settings, I have focused my discussion assuming that most of us would use interviews as part of a driving simulation study visit. And because interviews have an established history in design, I touch on some of the ways they can be used in that context as well. My goal in this chapter is to provide sufficient building blocks to

assist the novice interviewer, and to help refine the process for those with more experience.

23.2 Qualitative Research Design

Qualitative research methods are as varied as the topics investigated.* A black-and-white description of what does—and does not—constitute qualitative research design is offered by Strauss and Corbin (1998, pp. 10–11): "By the term 'qualitative research,' we mean any type of research that produces findings not arrived at by statistical procedures or other means of quantification".

Some methods, such as the interview, can fit equally well in qualitative or quantitative research designs, depending on the interview structure and how the resulting data are analyzed. Some quantitative data often appear in qualitative research, typically as background information to help contextualize the subject matter (e.g., census data, socio-economic factors of study sample). On the whole, however, qualitative research involves an interpretative approach that informs design decisions as well as to how the study is conducted and how the data are to be analyzed.

Qualitative approaches require a specific set of motivations that are borne out of the desire to conduct exploratory research, often where variables and a theory base are unknown (Creswell, 1994). In the context of driving research, qualitative inquiry is particularly relevant when seeking microanalyses of drivers' feelings, ways of thinking, beliefs, and emotions when they drive in congestion, with their families on vacation, on a date, or with their co-workers on a commute, for example; all aspects of social life which are not easily obtained through quantitative instruments (Strauss & Corbin, 1998).

Patton (1990, p. 12) provides a set of questions that help determine if qualitative research is an appropriate approach: Who will use the findings? What kinds of information are needed? How is the information to be used? When is the information needed? What resources are available to conduct the evaluation? If after answering these questions you find that you have the time, expertise, access to appropriate resources and funding agency support, then qualitative methods may be worth considering. If not, then look to other options. Studies requiring a large number of participants or seeking standardized responses for statistical analyses are typically better served by the more economically feasible questionnaire (Seashore, 1987).

23.2.1 The Interview

While many kinds of methods are used in qualitative research, the interview plays a significant role in discerning the meanings that people generate about the worlds they inhabit, information

---

* Wolcott (1990, 65) provides a partial list of some of the most prominent qualitative research approaches, including: ethnography, ethnomethodology, field study, participant observation, oral history, phenomenology, case study, criticism, investigative journalism, non-participant observation, human ethology, and natural history.

that would otherwise remain inaccessible (Peräkylä, 2005). Such insights complement our driving simulation data documenting interpretations of critical events, motivations behind the strategies employed while driving, as well as broader feelings about humans and automation or safety and risk. In our recent heavy truck electronic stability control (ESC) study, for example, drivers were asked to provide a verbal report of a critical event—in this case, an incursion from the right —as we watched the video of the drive, and then discuss their strategy in similar situations. Several participants noted that their decision to hit the vehicle instead of attempting to steer around it was motivated by employee training that balanced safety against fault. Simulator and video data captured drivers' performances, but interview data shed light on the fact that the benefits of new active safety technologies like ESC must be placed within the organizational context of carriers, risk management, and professional drivers. Without management support, while the system may be installed, its full potential may not be realized.

As with other methods, the interview's strengths are also the source of its limitations. First, Creswell (1994) notes that interviews provide indirect information because the data gathered are filtered not only through the participants' process of interpretation, but also through the social aspect of the interview itself, which may impose a particular bias on the response. Second, the indirect nature of interview data is a product of its detachment from the space and time of interest to the interviewer, as interviews typically occur outside of the participants' natural setting. Finally, not all participants are equally reflexive and communicative.

For driving, these limits are important because many of the skills utilized when driving cannot be verbalized. Particularly when attempting to ascertain participants' cognitive processes, for example, while responding to a safety-critical event, verbal reports rarely correlate with actual behavior (Nisbett & Wilson, 1977). Interviews do not provide direct access to participants' cognitive processes of evaluation, choices, or behavior, but rather when reporting how stimuli influence a response, we make plausible judgments about the causal relationship between the two based on cultural resources rather than introspectively accessing a memory of the mediating processes. Nisbett and Wilson suggest that "sometimes people tell more than they know" (1977, p. 247, referencing Polanyi, 1964), making assertions about mental events to which they have no access, but in the process communicating volumes about how culture mediates our understanding of our life experiences.

23.2.1.1 Interview as Social Interaction

Interviews can take many forms, from telephone interviews to focus groups. Regardless of its form (or rather because of it), the interview is a unique type of social interaction. Interviewees are not "vessels" of information from which the interviewer dips, but rather knowledge is "collaboratively constructed" through the exchange between the interviewer and interviewee (Warren & Karner, 2005, p. 156, referencing Holstein & Gubrium, 1995).

For many who practice quantitative research within the positivist tradition, it is precisely the interactive nature of qualitative research that challenges keystone concepts like objectivity, value-free knowledge, and singular reality. Because of the interactive nature of qualitative interviews, concern is voiced over the potential for the interviewer's perspective to dominate that of the participant's, or for the interviewer's agenda to weigh heavily on the knowledge he has a part in constructing. This is not an isolated issue for qualitative interviews; poorly designed and executed research—regardless of its epistemological origins or chosen methodology—can lead to skewed findings. Even well-conceived and executed studies carry the grander influence of the researcher in the assumptions underlying his or her research questions and methodological approaches.

While the majority of this chapter is dedicated to outlining techniques researchers can use which both facilitate the conduct of the interview and ensure the quality of the interview data, a few points can be made here about the management of interviewers and the interview process. Beyond the work typically conducted by researchers, the role of the interviewer is to ask questions, explore the participant's answers through additional probing, and record answers. Interviewers often are involved in defining the questions and the interview protocol, transcribing interviews, analyzing the data, and reporting findings. Less obviously, interviewers also have the important role of "training" participants to respond in a desirable manner. Fowler (2002) observes that interviewers have an important influence in shaping *how* participants respond to questions, not only through the content but also through the delivery of interview questions. If the pace of the interview is rushed, participants will provide quick answers; likewise deliberate, contemplative delivery suggests to the participant that their thoughtful answers will be appreciated. Expectations for the exchange are communicated through standardized instructions to participants about the interview's purpose (e.g., the goal of the interview and how it relates to other components of the study, if applicable), the interview process (e.g., provide answers in your own words to open-ended questions), and specific standards set for participants' answers (for example, priority setting statements that help participants understand that accuracy is favored over immediacy when providing responses) (Fowler, 2002).

While managing participants is important, so is the management of interviewers. The fundamental elements to minimizing the negative effects of interviewer influence are to develop good questions and a solid protocol (both discussed in more detail throughout the chapter), as well as to provide adequate interviewer training. For our purposes, interviewer training should address these components (adapted from Fowler, 2002, p. 124):

- Procedures for introducing the study;
- Conventions used in the design of the interview protocol with respect to wording and skipping instructions to ensure consistent and clear communication;
- Procedures for probing answers in a nondirective way;

- Procedures for recording answers to both open-ended and closed questions (depending on the protocol design); and
- Rules and guidelines for handling the interpersonal aspects of the interview in a nonbiased way.

Fowler observes that with increased training for interviewers, the accuracy of some reports has improved.

23.2.1.2 What Does Interview Data Tell Us?

As researchers we need to be cognizant of our epistemology vis-à-vis the interview, or what kind of knowledge we believe the interview can generate. Let's compare three approaches that use the interview as a data collection method to see how different epistemologies give interviewers different data: ethnographic methods, the unstructured or semi-structured interview, and the structured interview. Ethnographic research methods may include interviews as a complement to participant observation. Because of the researcher's long-standing involvement with the subculture studied, an emphasis is placed on the importance of context when analyzing interview data and interviews can provide additional insight to continued observations in the field. If we were studying issues associated with teen mobility, we could participate in ride-alongs to observe the driving and socializing behaviors of teenage drivers, recording field notes to be referenced at a later date when formally or informally asking questions about what car ownership or driving means in relationship to social status and identity construction. The qualitative interview, on the other hand, captures reports on participants' behavior, and can address past and present occurrences and future prospects, but data cannot be analyzed within a larger context of observed behavior (unless, of course, the object of study was the interview itself). At what point does the ability to drive for a newly-licensed teen shift from the extraordinary to the ordinary? Is this contextual (e.g., when driving facilitates sociability it remains exciting; but when it involves contributing to household labor, it is taken for granted or reviled)? Does the shift impact the decisions teens make when driving? Such questions are well-suited for an unstructured or semi-structured interview. The structured or survey interview, in contrast, collects data that can be more easily compared across participants but that restricts participants to pre-existing categories defined by researchers, potentially over-simplifying participant responses. Structured interviews, like questionnaires, may be better-suited to explore teens' use of cars and their travel preferences; for example, under what circumstances they drive (e.g., to school), how often they drive, if they drive alone, if they are the primary driver among their groups of friends, and so on.

Claims to knowledge are limited to *accounts* of behavior or of social life, and cannot be equated to actual observed "truths." Warren and Karner (2005, p. 158) nicely summarize this distinction: "If you want to find out what people *say* about what they do, what they *say* they believe or value, what they *say* their opinions are, then the interview method is ideal." Interview data, like many other types of evidence, will contain true and

false information. As intimated above, the questions shift from truth-finding to discovering sense-making and value-defining strategies:

Thus, Dean and Whyte (1958) argue that rather than asking, for example, "How do I know if the informant is telling the truth?" we should consider what the informant's statements reveal about his or her feelings and perceptions, and what inferences can be made from these about the actual environment or events he or she has experienced. The aim is not to gather "pure" data that are free from potential bias. There is no such thing. Rather, the goal must be to discover the correct manner of interpreting whatever data we have (Hammersley & Atkinson 1995, p. 131).

Qualitative interviews typically generate *narratives* that, in addition to the experimental context, also are embedded in a variety of social, economic, and political fields that inform the content and structure of the stories told. As noted above, participants rely on cultural resources, such as language, shared experiences, and common understandings, to communicate the stories they share during the interview. To this end, Warren and Karner (2005) remind us that interviews tell us something about an individual story and that singular story's reflection of a grander collective story. The interview excerpts in the introduction illustrate this point.

23.2.1.3 Interview Formats and Their Strengths and Limitations

Let's compare three formats that are either commonly used or could be attempted within driving simulation study designs. One-on-one, or dyadic, interviews involve one researcher interviewing one study participant in a face-to-face exchange, and are the most common form of interviews used at our laboratory. Interview protocols containing sensitive questions or using video review, for example, are ideal for the one-on-one format, where participant confidentiality can be best safe-guarded. But this format is also useful in other contexts. The singular focus on one participant's experience provides a unique opportunity to obtain his or her feedback, perspective, impressions, or opinion in a deep and meaningful manner. However, doing so can take a lot of time, which could affect the researcher's study schedule, and ultimately, their budget.

Triadic interviews (typically between one researcher and two respondents who know each other) provide two additional layers of information—that of the additional third person and of the shared meaning generated by the couple's interaction (Warren & Karner, 2005). This format could provide interesting data when examining teen-parent relationships and driving practices; or mobility issues specific to older couples; or between elders and their adult child caretakers. Importantly, triadic interviews necessarily involve the negotiation of "truths" within the context of complex social relationships, so the analysis of this and focus groups also requires recognition of this special form of speech act.

Focus groups involve a researcher engaging a group of participants in a discussion, often for market research but also for human factors and design research (Caplan, 1990). Again, the purpose of the interview will inform its structure—if the researcher is interested in assessing a program or product, the questions and format may be more structured and focused, but if the emphasis is on the participants' interpretations of a topic, then the format and questions may be more fluid. Focus groups enable a large amount of data to be collected in a relatively short amount of time, but the public nature of this format would make it an ill-fit for sensitive topics, and social dynamics within the group may affect who shares what, when, and how.

Choosing to conduct dyadic, triadic, or group interviews is ultimately a choice about the kind of conversation you hope to elicit (Warren & Karner, 2005). Talking with a newly-licensed teen driver independent of a parent will elicit, in most cases, a different kind of response than that expected in a triadic interview with the teen and his or her parent, or with a group of newly-licensed teens devoid of any parental authority figures.

23.3 Collecting Interview Data

The art and science of interviewing involves close attention to question structure, question type, interview preparation, the exchange itself, and ways in which we make sense of the copious amounts of data generated.

23.3.1 Question Structure

Differences in question structure and in the interviewing process can influence how questions are asked, and therefore, what data are collected. Hammersley and Atkinson (1995) characterize the difference in the structure of questions as either directive or non-directive in nature. Compare the following questions: "Describe how traffic poses a risk in your line of work?" and "What is the major risk associated with your line of work?" The first question assumes that traffic constitutes a major risk for, in this case, truck drivers, and therefore does not allow for alternative forms of risk—like poor eating habits, back pain, or corporate policies—to enter into the conversation. The first question is, as Dexter (1970, p. 55, in Hammersley & Atkinson, 1995, p. 152) describes, one "which sharply defines a particular area for discussion," whereas the latter question is designed to "trigger and stimulate" the participant to discuss in his or her own words what they believe constitute risk in their job.

Both question structures are useful. Directive questions address specific topics with narrow focus (e.g., "In what ways did your interaction with this particular interface affect your ability to safely perform your driving task?"), which may be necessary to test a developing hypothesis or to accommodate research design or time constraints; in the extreme, directive questions can be so limited *and* limiting, however, that another survey instrument, such as a questionnaire, may better serve the study design. Non-directive questions on the other hand, are essentially open-ended, allowing the interviewer and participant the space to discuss topics more broadly (e.g., "What

role does technology play in your car?"); but again, in the extreme, non-directive questions may elicit vague answers that lack the detail necessary to discern, for example, how interface design elements influence the driving task. Pure non-directive questions in the experimental environment have proven to be unexpected at best, and disconcerting at worst, for the majority of the participants we have interviewed. Although pure conjecture, I suspect the divide is an artifact of the research setting and the social expectations participants have for the study engagement (see Section 23.2.1.1).

Often the two structures are used in conjunction with each other. Because we are interested in the motivations, beliefs, and values that inform acceptance, annoyance, and so on, we often start with a mix of scripted directive (e.g., "Is your truck equipped with any advanced safety systems?") and non-directive questions (e.g., "What constitutes risk in your line of work?"), and a set of predetermined "steering" questions from which various interviewers are able to work (e.g., "You say traffic is the primary risk you face. I wonder if you could provide a recent example of how traffic posed a risk for you and what strategies you used to handle it?").

23.3.2 Types of Questions

Question structure determines the degree of focus your questions will have. Question type impacts the kind of information you hope to elicit. There are many types; anthropologist James Spradley (1979) has identified over 30, outlined in Table 23.4. I'll briefly discuss three.* As I mentioned before, interviews not only help us gain knowledge about participants' interactions with in-vehicle technologies or their perceptions about their driving performance when impaired—the typical kinds of questions we ask in our debriefing interviews—but they can also be used to gain an understanding of a particular culture in order to better understand the meanings generated out of human-machine interactions in the driving context. I'll touch more on this as we continue.

Descriptive questions provide samples of participants' language (Warren & Karner, 2005). Often procedural and meant to tap into everyday or typical actions or considerations (e.g., "Could you describe a typical driving event?", "Could you tell me what safety technologies you considered when purchasing your vehicle?"), descriptive questions also provide contextual information from which prompting questions can be asked for additional details.

Structural questions reveal information about domains ("the basic units in an informant's cultural knowledge" [Spradley, 1979, p. 60]), which in turn allow the researcher to discover how participants organize knowledge. Structural questions expose the linkages people in a particular cultural context make between things; it is these connections that say something about

---

* While written to provide basic instruction on the ethnographic interview, Spradley's (1979) detailed descriptions of question types will be useful for researchers new to the interviewing process.

TABLE 23.4   Ethnographic Interviewing Question Types, Definitions, and Examples

Question Type			Definition	Example
Descriptive	Grand Tour	Typical	Description of how things usually are. Asks participant to talk about a pattern of events.	"Could you describe a *typical* morning drive to work?" or "Could you tell me how you *usually* set your ACC in this driving context?"
		Specific	Asks about the most recent day, event, or the context best known to the participant.	"Could you describe what happened during this segment of your drive?"
		Guided	Asks participant to provide a tour of a place or an activity.	"Could you provide a tour of your vehicle?"
		Task-Related	Asks to demonstrate how something is done, then asks questions while the task is being performed.	"Could you explain how the IVIS works while using it?"
	Mini-Tour	Typical	Same as Grand Tour questions but on a smaller scale.	"Could you describe what goes into you setting your ACC in this context?"
		Specific		"Could you describe your initial response when seeing the left incursion?"
		Guided		"Could you describe your instrument panel to me?"
		Task-Related		"Could you draw me a picture of the IVIS display, and explain it to me?"
	Example		Asks for an example of an act or event identified by the participant.	Responding to "I was distracted that night," ask "Can you give me an example of how you were distracted?"
	Experience		Asks for experiences the participant has had in a particular setting.	"Could you tell me about some experiences you have had while driving to work?"
	Native-Language	Direct Language	Asks for examples of how people typically talk.	"How would you refer to driving around town with friends?"
		Hypothetical-Interaction	Provides typical situations and asks participant to describe how they communicate with others.	"If you were talking to a friend at school about your trip last night, would you say it that way?"
		Typical-Sentence	Asks for typical sentences that contain a word or phrase.	"What are some sentences that use the term *cruising*?"
Structural	Verification	Domain	Seeks to verify the existence of a domain (group, category) identified by the researcher.	"Are there different routes you use to travel to and from work?"
		Included Term	Seeks to verify whether one or more terms are included in a domain.	"Is the daycare route the kind of route you take on a typical day?"
		Semantic Relationship	Tests the appropriateness of the way a semantic relation is expressed.	"How would most of your friends say it, that aggressive drivers are a kind of group, or that aggressive drivers are one group?"
		Native-Language	Verifies whether a particular term is a *folk term* rather than a *translation*.	"Is this a term you would use to describe aggressive drivers?"
	Cover Term		Used to identify a domain.	"Are there different kinds of routes that you take?"
	Included Term		Included terms that may surface before a cover term or name for a domain is discovered.	"Are daycare, nighttime, bad weather, and fast all the same kind of thing?" "Yes, they're all kinds of routes I take to and from work." "Are there any other kinds of routes?"
	Substitution Frame		Remove one term from a statement uttered by the participant, and the participant is asked to *substitute* other similar terms.	"Can you think of any other terms that might go in that sentence?"
	Card Sorting Structural		Uses cards to help elicit, verify, and discuss a domain.	Routes are recorded on cards, and then participant is asked "Are these all the kinds of routes?"

(continued)

TABLE 23.4 (Continued)   Ethnographic Interviewing Question Types, Definitions, and Examples

Question Type	Definition	Example
Contrast Verification	Used after discovering a difference between two terms.	"Are these routes you drive to and from work? Are these routes you drive at another time?"
Directed	Begins with a known characteristic of one term and asks if other terms contrast with that characteristic.	"Your fast route takes the interstate. Does your nighttime route also include an interstate drive?"
Dyadic	Presents participants with two folk terms and asks if they see differences between them.	"What are the differences between your nighttime and fast routes?"
Triadic	Presents participants with three folk terms and asks if they see similarities and differences between them.	"Of these three routes, which two are alike and which one is different than the others?"
Contrast Set Sorting	Places all terms in a contrast set in relation to each other.	"Would you sort these cards into two or more piles in terms of how they are alike or different?" Afterwards the cards are placed in a single pile and the participant is asked to repeat the exercise.
Twenty Questions Game	Participant asks yes and no questions until they can guess the folk term that is part of a contrast set.	
Rating Questions	Used to discover the values placed on sets of symbols. Contrasts based on best, easiest, worst, most difficult, etc.	"Which alert type would you like to hear first, which ones next, and which one is last?"

(The left margin of the table is labeled "Contrast" spanning the rows.)

Source: Adapted from Spradley, J. P. (1979). *The ethnographic interview.* New York: Holt, Rinehart and Winston.

culture as members of particular cultures organize their knowledge of that culture in similar ways. For example, we could ask American car consumers "Which automobile manufacturers do you associate with safety?" and "What are all of your vehicle's safety features?" Safe automobile manufacturers and vehicle safety features are domains under which different companies and a variety of features, respectively, fall. French anthropologist G. Clotaire Rapaille (in Gladwell 2004, pp. 29–30) has observed that since the mid-1990s American consumers identify cup holders as safety features in vehicles. In different cultures, cup holders would not be linked with safety, but rather with danger because they enable distraction from beverage consumption while driving.

A third category of questions, contrast questions, is designed to discern the boundaries between objects, activities, people, and events in the participant's world. Meaning generation is, of course, relational—part of understanding what something *is* is to understand what it *is not*. By extension, then, similarities drawn between things also tell us something about how the participant's world is organized. In an interview for a study examining an adaptive interface technology, we asked study participants to compare the experience of driving with the test system to that of driving their personal vehicles (i.e., "How was your experience driving with the system similar to your experience driving your personal vehicle? How was your experience different?"). Most participants responded that the experiences were completely different—their vehicles did not flash lights or make sounds, referring to the visual and auditory alerts provided by the system during their study drives. Yet when additional question prompts were asked—"In what ways does your vehicle provide you with

information?"—many participants noted similarities between the study vehicle's system and the lights (e.g., dashboard icons) and sounds (e.g., "idiot" alerts when keys are left in the ignition or lights are left on) provided by their personal vehicles. Through this line of questioning we could see that while most participants characterized their interaction with the test system as dissimilar from their interactions with their personal vehicles, there was a familiarity with the "structure of the conversation" between the car and the driver (e.g., lights and sounds as alerts) that made these forms of communication more easily identified as warnings compared with the haptic seat, which did not fit the communicative framework with which our study participants were familiar.

23.3.3 Developing Interview Protocols

What should be recorded, and how should it be documented? Interview protocols provide a template of sorts for the questions to be asked, the responses that participants provide, and the researchers' reflections during the interview process. At a minimum, protocols encourage consistency when more than one researcher conducts interviews for any given study. When used to their full extent, the document itself is an extremely useful tool for analyzing and interpreting the interview data. The following are useful elements to include in an interview protocol: a heading, instructions to the interviewer, key research questions, probes to follow key questions, transition messages for the interviewer, space for recording comments, and space in which the researcher records reflective notes (Creswell, 1994). I advocate using a semi-structured interviewing style, which borrows elements from structured and reflexive interviews. A semi-structured interview is characterized by the scripting of some questions prior to the interview, similar to a structured interview, but when conducting the interview, there is flexibility to follow a line of questioning that is not scripted in order to obtain a deeper understanding of the topic studied, as well as to use different questioning styles to accommodate the research agenda. Your protocol should reflect the style of interview you plan to conduct.

23.3.4 Conducting the Interview and Recording Data

Note-taking during the interview can play a very productive role as you begin to conceptualize the data. When you are listening to the participant, write down words or phrases that stand out to you, either because they are similar to answers provided in previous interviews and therefore point to a pattern, or because they are different and therefore indicate a potentially meaningful departure. These notes do two things: they kick-start the analysis process by initially developing a framework that the researcher can test with each additional interview; and they "flag" the audio stream for sections of the interview or transcription that at the time of the interview seemed interesting. Importantly, the analytical frameworks developed at this stage must be rigorously challenged through an openness to the complexity of participants' responses so as not to shut off avenues of investigation so early in the analysis.

A balance should be struck between your attentiveness to the participant and jotting notes. Participants can become self-conscious if they see you are taking notes, and you may be so invested in recording that you fail to actively listen to the participant and therefore miss an opportunity to follow up on topics outside your initial purview (Hammersley & Atkinson, 1995). It is also important to keep your analytic notes separate from your records of what the participant said. Take care in formatting the protocol so that you can easily categorize your note-taking during the interview, perhaps by creating two columns so that analytic notes can be paired with the interview question and answer that triggered the thought or forged the connection. I'll return to iterative analysis in the next section.

It is unrealistic to assume that you will be able to accurately record participants' responses by hand during the interview, so audio recording the interview and transcribing at a later date is preferable. There are instances where video recording will be useful or even necessary, particularly when the body language people use provides important context for your analysis. Video recordings of focus group interviews also are essential. Imagine attempting to transcribe audio tapes without knowing to whom comments should be attributed! In both cases, I would suggest capturing an audio recording as well. Microphones on video cameras are generally not sensitive, cameras are usually positioned at some distance from the speaker in order to capture the desired image, and transcription machines are not designed to play video tapes. Keep in mind that video and audio recordings are two separate types of data, so you will want to make sure that both are included in your IRB application and informed consent document.

Because you will be interested in obtaining accurate, complete, and honest impressions as they relate to the research topic, establishing an open and equal exchange with the participant is critical. A few easy steps can be taken to cultivate rapport. Warren and Karner (2005, pp. 132–134) suggest that in all of your exchanges with the participant you are an active listener, which they characterize as empathetic, trustworthy, polite, non-judgmental, and professional. Importantly, professionalism here may mean something different than it does in other work-related contexts. Vigorously avoid presenting yourself as an authority figure. Instead adopt an "apprentice" role in which you seek the knowledge of the participant, or "master" (Beyer & Holtzblatt, 1998).

Try to blend in with your participants to make them feel more comfortable. Over the course of two studies examining safety devices in heavy trucks, our team realized that the truck drivers felt more at ease when we wore more casual clothing, and when we reinforced their position as field experts by including questions that allowed them to demonstrate their perceived driving expertise (e.g., "Based on your experience, do you think that having these technologies available promotes safe driving in your line of work? If no, in what ways do they hinder safety? If yes, in what ways do they promote safety?"). Also, indicate through verbal and physical cues that you are following what the participant is saying and are generally accepting of what he or she shares (Hammersley & Atkinson 1995). The goal is to encourage a conversation-style exchange within the constraints of the interviewing process, which may be counterintuitive to researchers trained in an approach that links "objective" with detachment.

The location of the interview addresses both ethical and practical considerations, as you are better able to protect the privacy of the participant by avoiding heavily-populated locations and your audio recording will be easier to transcribe with minimal background noise. Select a location for the interview that is safe, private, quiet, and convenient for the participant. Alternately, if you are conducting contextual inquiry as part of a design process (Beyer & Holtzblatt, 1998) or ethnographic research (Hammersly & Atkinson, 1995), asking questions while observing the participant as they engage in the specific tasks or cultures under analysis will yield a different, potentially richer account of the way the driver adapts his or her behavior to warnings issued by a system prototype, for example, or the difficulties older drivers have in processing information at busy intersections. In this case, audio recording may not be possible or preferable, so notes jotted during the exchange will serve as important triggers for interpretation sessions (for the contextual design process) or recording comprehensive field notes (for ethnographic interviews) afterward. In either case, let the participant know that he or she can choose not to answer any question without explanation. If using an audio recorder, show the participant how to pause or stop the recorder, and place it in an open space within easy reach of the participant. Invite participants to ask questions about the process before you begin, and remind them that they can ask for clarification at any point during the interview.

When you are ready to begin the interview, first ask for the participant's permission to begin recording, or at a minimum, notify the participant that you have turned on the recording device. In your introductory and transition remarks included in your protocol, briefly describe the themes you will be covering in the interview so that the participant knows what to expect.

Preparing the participant is important, but being prepared yourself is also critical for a successful interview. Know how to use the recording equipment, and test it before each interview to make sure it is working properly. Have extra batteries and tapes on hand and the electrical cord (if available) in case battery power does not work. Take an opportunity to ensure that background noise does not drown out your voices by making a short sample recording while showing the participant how to stop or pause the recording. Being prepared saves time and instills confidence in the participants that you are taking them and their contributions seriously.

If possible, purchase professional-grade equipment. Retailers such as J&R provide a good selection, and Andy Kolovos with the Vermont Folklife Center provides a thorough and continuously updated review of recording devices, including digital audio recording equipment (see Web Resources section for URL). As with analog audio cassette recorders, direct-to-laptop recordings, and the MiniDisc, many digital recording devices are now in their twilight years, but solid state memory card recorders hold some promise for quality and relative longevity.

23.3.5 Transcription

Frankly, transcription is the down-side to interview-based methods. Doing it undoubtedly assists in the methodical mental processing of the interview data, and often unexpected patterns will emerge just through listening again to interviews, particularly if you are listening to several at a time. But it is time-intensive, and it can be tedious. Although transcripts are generally considered verbatim accounts of the interview session, the content captured in the transcript can (and does) vary from the actual spoken exchange. Comprehensive transcriptions are most desirable, but they may not be necessary. While not recommended here, if you are certain about what does and does not constitute relevant data, audio tapes can be indexed using the counter, summarized, and only the essential sections transcribed, or at a minimum, create lists that document the primary themes, topics, and stories (if relevant) discussed. Time is saved, to be sure, but the documents that will come to stand in for the recordings themselves will be slim in content and narrow in focus. The level of detail in each transcription also is negotiable. Because most of us will not be engaging in discourse analysis, formal representation of speech patterns (e.g., speed of speech, pauses, interruptions, volume changes) is not necessary. The study goals and research questions should drive the level of detail captured in the transcription.

With that said, interview transcripts should do the following without exception. First, as already discussed, the interview is a speech event between at least two people. It is important to capture the interviewer's lines of questioning in the transcript, even if there is a designated script for the interview. Questions can be asked out of order, different words may be used than those specified on the script, and meanings can be shifted based on something as small as the emphasis on a particular word or the intonation the interviewer uses when asking questions. Second, it is strongly recommended that the interview transcripts include everything said by the participant, including responses that at the time of the interview or during its transcription you consider to be irrelevant. Relevance often shifts as the research focus matures, and what was once a digression can, in the end, be a critical insight (Warren & Karner, 2005). As most of us know, however, when people talk, digressions do occur. For many populations that we work with, being a participant is a social occasion. Life stories by older drivers and by more isolated populations, like truck drivers, are quite common and are one of the reasons that they participate in research.* During a recent debriefing interview for a heavy truck pilot study, a participant shared with me a tale of how he was shot, offering to show me the massive scar on his abdomen (I declined). When staff transcribed the interview, a bracketed note was added to the transcript summarizing the digression (e.g., "[participant recounted at length unrelated gunshot story]").

Third, steps should be put into place to guarantee the accuracy of the transcription. As with the first point, minor details like the placement of punctuation can radically alter meaning. Poland (2002, p. 632 in Warren & Karner, 2005, p. 154) provides this example to illustrate how important capturing conversation flow can be: "I hate it, you know. I do" carries a different meaning from "I hate it. You know I do." Also, words or phrases that are difficult to discern may be misinterpreted, transcribers may inadvertently alter the text to "fix" improper word usage or grammatical errors, or segments of the recording may be missing from the transcription altogether. A second reviewer should verify the interview data by listening to the recording while reading the typed transcript. As with verifying other data, discrepancies can then be addressed by a third party if needed. Fourth, all identifying information (names, home towns, workplace names) must be replaced with pseudonyms in order to protect the identity of your participants. We replace participants' names with their subject number in our transcriptions.

On a practical note, Hammersley and Atkinson (1995) suggest that the ratio between recorded time and transcription time may be five to one, if not higher, depending on the complexity of the interview. There are a few tools that can be used to speed up the process. A low-tech and widely available tool is the foot-pedal-operated transcription machine. Transcription machines reduce the need to move your hands from the keyboard, thus increasing efficiency. If you will be transcribing audio tapes, check what resources are available

---

* As an aside, oral history interviews may have a place when studying such populations, but it is more commonly used when studying a historical event, membership in a culture, or in documenting folklore.

at your home institution before you invest in a machine; transcription machines may be available for departments to rent on a monthly basis. Also, make sure that the tape size for your recorder is the same as the transcription machine you choose to use. Higher-tech solutions, such as digital recorders and speech-to-text software, cannot yet precisely capture interview conversation in textual form (although see Warren & Karner 2005, p. 153).

The audio and/or video recordings of your interviews should be archived in a safe and secure location, along with complete paper and/or digital copies of your transcribed interviews. These are your master copies, and should not be altered in any way. You should also have a participant log which lists, at a minimum, participant names and their codes. The log should be stored in a secure location as well (whether that is with or separate from your transcripts and other study materials is dictated by your IRB).

If you are able to work with a research partner(s), an alternative to transcription is the interpretation session developed by Beyer and Holtzblatt (1998). In the framework of their design process, after conducting a contextual inquiry (i.e., observation and interviews while trailing the customer during his or her typical interaction with the object or process under consideration), the interviewer is then interrogated by design partners to summarize the interview and record what was learned in a series of models that eventually assist the design team. In effect, the interviewer is interviewed about the exchange, relying on notes and memory to address their partners' questions. Online and paper notes and models generated during the interpretation session are then consolidated to create a view of the customer which helps define the market for designers. Similar techniques can be used for other kinds of qualitative interviews. Without a record of the exchange, however, there is increased potential for the interviewer's bias to cloud his or her recollection of the interaction. An explicit statement of how interviews were conducted, what constitutes the interview data, and how the information gathered was analyzed are of particular importance when relying solely on the interviewer's notes and memory.

23.4 Analyzing and Interpreting Interview Data

You will begin and end your analysis and interpretation of your interview data with the same process: that of "memoing." As I discussed in the previous section, the first stage of analysis can start as early as your first interview. As your analytic notes begin to accumulate, the next step is to develop these ideas in analytic memos. Described by Hammersley and Atkinson (1995, p. 191) as "occasional working notes whereby progress is assessed, emergent ideas are identified, research strategy is sketched out, and so on," analytic memos provide an opportunity to review and reflect on the data collected. Iterative reviews are important because qualitative methodologies typically hinge upon the idea that research problems evolve as the project

unfolds. This "progressive focusing" (Hammersley & Atkinson, 1995) obviously works better when the research design allows for the interview questions to also evolve over the duration of long-term study, but it can serve a similar function with the kinds of interviews we are more likely to conduct through providing a sharper focus to the follow-up questions we ask as the study proceeds. It might be useful to consider this process as one of model building in which the model is validated, ideally with the original participants.

After the data are reviewed and sorted and analytic descriptions are written, etic interpretations* can be documented through conceptual notes which articulate linkages between data in a "coherent and cohesive assemblage" (Warren & Karner 2005, p. 214). We will unpack this path from one form of memo to the other in the following sections; suffice it to say that both constitute thinking on paper, or in Hammersley and Atkinson's nomenclature, capturing on paper your internal dialogue about the research and data (1995, p. 192).

23.4.1 Splitting and Lumping

You will, no doubt, come to your interview data with a set of questions that originally provided guidance in defining your study's purpose. If you have followed some of the suggestions outlined thus far, you also may have analytic notes or analytic memos that document patterns you believe are emerging, as well as the evolution of your thinking vis-à-vis the research questions and the data you have collected. These, along with established theories and common-sense expectations, constitute deductive tools you can use to inform your analysis. The remainder of this section will focus on an inductive model for research, here paraphrased from Creswell (1994, p. 96): "researcher gathers information, researcher asks questions, researcher forms categories, researcher looks for patterns (theories), researcher develops a theory or compares pattern with other theories." Let's see how this process unfolds by turning to the data itself.

23.4.1.1 Inductive Coding

Once a comprehensive record of your interviews is preserved, use a second copy of the interview data to begin the "splitting and lumping" process. It is widely suggested that before you begin dissecting the transcripts, you first order them in a manner that fits your organizational scheme for the study. Did you interview teen drivers and their parents? Depending on the focus of your study, you can reorder your interviews by teens, then parents, in the chronological order of the interviews, or by teen-parent groupings in chronological order.

After organizing your data files, read and reread your transcripts, mark the texts' similarities, consistencies, or striking differences, and record general insights that arise from your review of the data. Warren and Karner (2005) opt for an unstructured approach to the data, remaining open to its variations and

---

* Etic interpretations refer to the researcher's understanding, compared to the participant's understandings, which are called emic interpretations.

recording all possible insights in an effort to sketch the "big picture" present in the data through a process they call "open coding," whereas Auerbach and Silverstein (2003) suggest a more focused approach, using your research concerns to cull potentially relevant text from its raw form. While the latter approach shows some deductive tendencies, especially in comparison with the former, both approaches achieve the similar end of identifying material of interest (although why it is interesting or relevant is yet to be determined).

When you are quite familiar with your data, possibly completing several sessions of open coding, you then begin to focus your analysis. Following the sage advice proffered by Wolcott (1990, p. 35), "the critical task in qualitative research is not to accumulate all the data you can, but to 'can' (i.e., get rid of) most of the data you accumulate." How you choose to focus will be driven in part by the research questions. Because time constraints typically are very real, the process of preliminary analysis (developing analytic notes and memos) allows for a set of relationships to be identified that can be employed when trying to define your focus. The endpoint is to establish a stable set of repeating ideas, and then themes, that you can use to systematically code all of the data.

Repeating ideas, or text-based categories, use the same or similar language to communicate a shared meaning (Auerbach & Silverstein, 2003). Two opposing ideas that surfaced in interview transcripts for a study we conducted on advanced warning systems provide an example: Several participants characterized their relationship with the warning system as a form of teamwork (e.g., the driver and the system working together), while other participants characterized the warnings as working against them (e.g., warnings were like having my mother in the passenger seat). We then could use the search function in our word-processing program to find specific words or phrases that capture similar meanings (although I would not want to solely depend on keyword searches to find similar instances that use different language). As you can imagine, the search would capture a wide range of responses that may or may not share the sentiment of our two example ideas (e.g., someone could use the term "mother" to denote protection instead of the overbearing connotation in the previous example). At this point we would just note what sentiment was attached to these keywords and then continue our search for other repeating ideas. Auerbach and Silverstein (2003, p. 60) note that any complete body of research materials are likely to render 40 to 80 repeating ideas.

From here the process is one of continually grouping similar repeating ideas into sensitizing concepts (Hammersley & Atkinson, 1995) or themes (Auerbach & Silverstein, 2003), and then to more abstract groupings in the form of identifying patterns (Creswell, 1994) or developing theoretical constructs (Auerbach & Silverstein, 2003). Developing themes is not easy. Auerbach and Silverstein appropriately summarize the perceived ambiguity of this task:

Several questions may be occurring to you as you do this exercise. How do you know when repeating ideas are sufficiently similar to express a common theme? How do you know what that theme is? There is no formula for answering these questions. (Auerbach & Silverstein, 2003, p. 64)

Some concepts will arise through use by a participant; others will be applied by the researcher, based on his or her common sense, personal experience, or common themes in various academic literatures. The themes you identify may be "mundane," to borrow from Hammersley and Atkinson, but that does not limit their subsequent development into more analytic constructs or patterns with further review and consideration, as shown in this example from Hammersley's analysis of teachers' talk:

…in his analysis of teachers' talk in a school staff room Hammersley developed categories that ranged from the very concrete (teacher talk about students, about teaching, about national political events, etc.) to rather more abstract and analytic ones (trading news about students, exchanging mutual reassurances, accounting for decline and crisis, defending teacher competence, etc.). (Hammersley & Atkinson 1995, 212)

Conceptual themes identified through open coding will not be fully formed elements of a complete analytical model, but rather they function as "germs of analysis" that suggest directions for how to approach your empirical data (Hammersley & Atkinson 1995, p. 205).

Sensitizing concepts or themes refer to an "implicit topic" around which groups of repeating ideas are organized (Auerbach & Silverstein 2003, p. 38). Along with participants using a teamwork analogy to describe their experience with the warning system, we, for example, could group other repeating ideas such as the system learning from and adapting to the drivers' behavior and notions that the driver, system, and vehicle need to "become one" for safe driving to occur as a theme of *partnerships with advanced driving technologies*. The repeating idea of the system working against the driver could be paired with other similar repeating ideas such as the contradictory nature of the system (simultaneously enabling the use of distracting devices while warning the driver that they are distracted) under the theme *conflict with advanced driving technologies*.

Auerbach and Silverstein (2003) use a comparative method when splitting and lumping their relevant text, taking one instance and comparing it against the running record of data in order to group like-minded instances together and then repeating for the next available instance. Other methods are perfectly acceptable as well: "coding the record" directly by demarcating segments of text that are then linked to descriptive or analytic codes in the document's margins, creating an analytical index of thematic headings under which segments of data would be placed, or physically sorting hard copies of segments in piles or folders, to name a few (Hammersley & Atkinson, 1995). Regardless of the approach taken, the coding scheme you choose is based on a shared underlying approach of segmenting and disaggregating your data. Each researcher will develop his or her own process: Warren and Karner note that the creativity and organizational schemes researchers bring to the data are part of qualitative research (such as Beyer & Holtzblatt's [1998] innovative use of models).

Pragmatically, if you are considering coding your transcripts (versus the comparative method advocated by Auerbach & Silverstein), Warren and Karner present both high-tech and low-tech methods for these early reviews. Warren uses colored pencils to color-code segments of text on hard copies of transcripts; while Karner works in (and on) electronic files, using editing tools in her word-processing program to add analytical comments in the margins or color-code segments of text using the highlight function. I've successfully used both methods. The benefit of Karner's approach is that these early codes are saved and stored on working electronic documents and researchers new to this process can conduct their analysis using tools at hand—there is no requirement to purchase and learn how to use a new software program designed to organize and assist in the analysis of interview data.

Several types of software packages can assist in textual analysis. Automated text mining approaches include software-aided content analysis and latent semantic analysis (LSA). These programs identify and classify words and phrases or extract concepts in the texts, respectively. LSA is based on the concept that the meaning of an utterance can be deciphered by analyzing its semantic structure. It assumes that words used in similar contexts have similar meanings, so queries to a data set analyzed using LSA will return words that reflect a shared concept, although the terms themselves may seem unrelated. Qualitative analysis programs, referred to as Computer-Aided Qualitative Data Analysis Software (CAQDAS), can be used to manage interview data and manually assign codes and notes to transcripts (see Tesch, 1990; and Koenig, n.d., for software reviews). Examples include NVivo, ATLAS.ti, and HyperRESEARCH. CAQDAS can also be used for more sophisticated analyses, including developing hierarchical coding trees and utilizing the software's more powerful search and theory-building applications. Text mining approaches and CAQDAS are useful tools, but they have steep learning curves, and more importantly, they cannot be used to learn how to analyze qualitative data. Researchers must develop analytic skills prior to using these programs.

More than likely, several of these techniques will be familiar to readers, only applied to different forms of data. Because we are not necessarily interested in counting or statistically analyzing our data, one important distinction that can be made between quantitative and qualitative approaches to coding is that there is no specific requirement that data be restrained to only one theme or category during the coding process (Hammersley & Atkinson, 1995). In fact, the linkages drawn between categories through the data can be quite compelling. And while this should apply more to open coding than to your final systematic coding, it is worth stating that the coding process is recurrent; new categories require that the entire data file be recoded.

23.4.1.2 Interpreting Results

Having identified themes within your data set, the next step is to clarify and make transparent the "mutual relationships and internal structures" (Hammersley & Atkinson, 1995, p. 213) of your themes, using the grounded theory method of "systematic shifting and comparison". At its basic level this is a task of delimiting the boundaries of what is and is not included in each theme, thereby refining the theme's definition. For example, the theme I called *conflict with advanced driving technologies* encapsulated the repeating ideas of the system working against the driver and the contradictory nature of the system. If, during further comparison between this theme and others, I note that the contradictory nature of the system actually does not constitute a conflict per se, but rather more accurately fits with other repeating ideas that characterize confusion over the system's operation, I have (a) refined the meaning of the conflict theme, and (b) created a new theme (or possible subtheme) that would then need to be examined against the corpus of the data. These kinds of ruptures may seem frustrating at the outset as you will need to recode or reconsider your data with each fracture in your themes, but in reality they suggest that you are developing a deeper and growing understanding of your data set.

Importantly, the task of sifting and comparing also involves sketching the links between themes, as well as the relations between your themes and those put forth in related research. By way of a quick example, our previously described themes could be grouped under the theoretical construct *user acceptance,* alongside other theoretical constructs which emerged through the "lumping" of additional repeating ideas and themes. Auerbach and Silverstein (2003, p. 67) describe this process as the development of theoretical constructs, which they define as "an abstract concept that organizes a group of themes by fitting them into a theoretical framework." In a sense, you have now effectively shifted your data set's initial organizational strategy based on descriptive categories (i.e., group status and time) to an analytic ordering that depends on logical linkages (Warren & Karner, 2005). At the end of this process you will have something akin to an outline that reflects your interpretation of how your data fit together into an accurate and comprehensive representation of your research, with theoretical constructs as your most general or abstract headings supported by increasingly specific subheadings and sub-subheadings developed as themes and then as repeating ideas.

Your interpretations may be completely new insights, grounded in the analysis of your data, but more than likely they will constitute a mix of personal insights derived from the data that are informed and enriched by the theories of others. Where a theme speaks to parts of theories, typologies, or models others have developed, models potentially can be expanded and/or tested through an exploration of the linkages between your theme and others in your data set. As Hammersley and Atkinson (1995) remark, most sociological models have yet to be sufficiently developed for extensive testing to occur. With this in mind, the authors observe that it is acceptable to stop at the point where your data describes or explains the phenomenon you are studying, and only pursue the development of systematic typologies or models when your analysis has the potential to apply to data derived from other situations. This, too, is the point where design recommendations can be made from insights gathered from the interview data. Those repeating ideas that comprise confusion over a system's operation, for example, provide critical information about user uncertainty, potential strategies or methods of adaptation used to compensate for a misunderstood or imperfect system, and the limitations of users' ability to adapt.

23.4.2 Reworking Validity and Reliability

Qualitative researchers, like their quantitative cohorts, support rigorous standards for evaluating research. Because the goal of most qualitative research is to develop a unique and insightful interpretation of the events (social processes, cultural exchanges, and so forth) that are studied, external validity (or generalizability) and reliability are less applicable in maintaining rigor in qualitative analyses. Auerbach and Silverstein (2003) do provide the useful concept of transferability as a replacement for generalizability, noting that abstract theoretical constructs may apply beyond the particulars of the case studied, but that the specificity of the patterns—that which is captured by repeating ideas and themes—remains bound to the (sub)culture studied. In this sense, limited generalizability is possible. When using the semi-structured interview format described in this chapter, publishing the detailed interview protocol provides a procedure, albeit incomplete, that could be replicated. Creswell (1994) also supports documenting the researcher's central assumptions, any known biases, as well as details about how informants were selected (in the case of ethnographic or case study methodologies), to assist in replicating the study design elsewhere.

Internal validity, on the other hand, does garner significant attention. For Janesick (1998, p. 50) the measure of validity can be succinctly defined as "Is the explanation credible?" Warren and Karner (2005, pp. 215–217) break down this query to three questions that assess the validity of your interpretations: Are your data robust enough to support the claims you make in your analysis (e.g., Does your gathered information address your specific questions? And is it comprehensive?)? Is there a "good fit" between your data and your analysis (e.g., Have you accounted for all possible and reasonable interpretations of the data in your analysis?)? And can your claims be supported externally through your research subjects (member checks), colleagues, or other data you have collected? Checking your analysis against other data collected as part of your research and data available through additional sources, which is known as triangulation, may be appropriate for interviews you conduct that can then be compared against questionnaire data, driving data, or findings presented in peer-reviewed literature.

23.5 Writing Up Qualitative Research

While results from qualitative interviews can be presented in a standard scientific report, they are more often presented in a descriptive, narrative form. The structure and content of the narrative can vary extensively (see Creswell, 1994, p. 160), which is fortunate for researchers in fields less acquainted with the qualitative narrative form. Auerbach and Silverstein (2003) outline an example manuscript that, for our purposes, is suggestive of a structure that should be familiar to most of us, as well as to our potential publication outlets: Introduction, Methodology (broken out into the following subsections: the qualitative research paradigm, research participants, focus group procedure, data analysis, reliability and validity), Results (organized by sensitizing concepts/themes and supporting text-based categories), Discussion, Theory-Driven Constructs, Implications for Further Research, Limitations of the Research, and the Conclusion.

Content decisions—which evidence to include and how much evidence is necessary—are another matter altogether. Wolcott (1990) provides some trenchant advice. In writing early drafts, include "an excess of illustrative material" (Wolcott, 1990, p. 68), as it is always easier to cut materials, and you may find that your needs change as your analysis evolves. Then, as the manuscript and your ideas mature, trim as required. One approach that has worked for me is to "[do] less more thoroughly" (Wolcott, 1990, p. 69). Is there a single instance that illustrates the point you want to make? Which pieces of evidence do you have that can stand for the whole? Use these to your advantage when reporting your results.

We will end with the question of when to begin. Wolcott (1990, p. 21) suggests that writing not only reflects thinking but that the act of writing itself constitutes a form of thinking. Why not write a preliminary draft of the study prior to beginning research? Much of this early (thinking and writing) work might already exist in your research proposal. Writing early and often can be an exercise that helps provide structure and flow to the document's content, a record of your assumptions, and a "systematic inventory of what you already know, what you need to know, and what you are looking for" (Wolcott 1990, p. 22). While this advice is particularly suited toward work in the field, it also provides insights into interview-based research.

23.6 Concluding Remarks

Qualitative methods operate from a different—but not necessarily adversarial—set of assumptions than do their quantitative counterparts: Qualitative inquiry is based on the understanding that reality is subjective and multiple; its focus is on meaning, or how people make sense of their lived experiences; data collection is mediated through the researcher instead of through other instruments; and engagement between the researchers and that which is researched is acknowledged, as is the influence of values and biases. Finally, research is an inductive process giving way to an emergent design that primarily is informed by the research participants themselves. Many of these tenets of the qualitative approach can be accommodated within a mixed-methodology research design that facilitates the pairing of different techniques and assumptions to attempt to develop a richer understanding of the driving phenomenon under study.

While not appropriate in all research contexts, qualitative methods—especially the qualitative interview—do provide access to drivers' thoughts, attitudes, beliefs, and values, garnering insights into both individual behaviors and the larger cultural systems in which they are positioned. Unlike the more widely-used survey interviews or questionnaires, qualitative interviews maintain a flexibility of structure that encourages deeper exploration while abiding by established standards of rigor. Qualitative interview methodology also values the multiplicity of meanings generated by participants, documenting and attempting to understand "extreme" cases that may fall outside of the majority of responses but that highlight potentially valuable insights.

Key Points

- While widely used to access measures such as driver acceptance or cognitive processes, qualitative methods also can provide valuable insight into the cultural meanings people have about driving and vehicles, as well as to the meaning-making process.
- Qualitative interviews are easily incorporated into a range of mixed-methodology research designs to complement driving simulation data.
- Interviews are social interactions, so managing the exchange is an important task that researchers must attend to. This is best facilitated through ample interviewer training.
- Dyadic, or one-on-one, interviews are underutilized in comparison to focus groups. While focus groups have a strong history in design and marketing work, dyadic interviews provide access to the individual's thoughts, perception, feedback, emotions, and the meanings that he or she ascribe to a technology or process, which provides another useful layer of understanding.
- Interview protocols provide a useful organizational structure to the exchange, and facilitate memoing and other note-taking procedures that aid in analyzing and interpreting interview data.
- While approaches to validity and reliability vary between quantitative and qualitative methods, qualitative researchers are committed to rigor and have established standards that work within a mix-methodology design.

Keywords: Culture, Inductive Data Analysis, Interview Protocols, Qualitative Interviews, Qualitative Research Design

Acknowledgments

I am indebted to the insights provided by and patience of John Lee, Don Fisher, and Jeff Caird, and to the assistance of our undergraduate research assistant, Tyler Weers. All errors remain mine alone.

Glossary

Culture: A complex term, but generally it consists of the beliefs, values, norms, and things people use, which guide their social interactions in everyday life. It also is dynamic, meaning it is a process by which society creates, reproduces, and justifies certain values and beliefs while suppressing others.

Inductive data analysis: A bottom-up technique that allows research findings to emerge from frequent, dominant, or significant themes represented in the raw data, in comparison to the top-down approach associated with deductive data analysis.

Qualitative interview: In-depth or intensive interviews that are typically focused on the meanings ascribed to experiences by the individuals being interviewed.

Reflexive interview: An interviewing technique that involves active listening and improvised questioning in a conversational style. While best used in naturalistic contexts, semi-structured interview protocols can reflect some of these traits.

Triangulation: A method used by qualitative researchers to ensure internal validity. It is based on the assumption that biases linked to a researcher, data source, or method are mitigated when used in conjunction with other researchers, data sources, or methods.

Web Resources

Vermont Folklife Center web site: http://www.vermontfolklifecenter.org/archive/res_audioequip.htm. Vermont Folklife Center provides up-to-date reviews of recording technologies.

Key Readings

Auerbach, C. F., & Silverstein, L. B. (2003). *Qualitative data: An introduction to coding and analysis.* New York: New York University Press. [Auerbach & Silverstein (2003) provide a detailed outline of qualitative data coding couched within the grounded theory approach. I would recommend reviewing Part III as a supplement to this chapter.]

Creswell, J. W. (1994). *Research design: Qualitative & quantitative approaches.* Thousand Oaks, CA: Sage Publications. [Creswell provides an accessible text that outlines quantitative, qualitative, and mix-methodology research design. It is particularly well-suited for undergraduate and graduate researchers.]

Fowler, F. J., Jr. (2002). *Survey research methods* (3rd ed.). Thousand Oaks, CA: Sage Publications. [Chapter 7 addresses how to manage survey interviewers, including what qualities to look for when hiring interviewers and how to supervise them under different conditions. While primarily focused on interviews within the post-positivist tradition, the issues raised are worth consideration.]

References

Auerbach, C. F., & Silverstein, L. B. (2003). *Qualitative data: An introduction to coding and analysis.* New York: New York University Press.

Beyer, H., & Holtzblatt, K. (1998). *Contextual design: Defining customer-centered systems.* San Diego, CA: Academic Press.

Caplan, S. (1990). Using focus groups methodology for ergonomic design. *Ergonomics, 33*(5), 527–533.

Creswell, J. W. (1994). *Research design: Qualitative & quantitative approaches.* Thousand Oaks: Sage Publications.

Fowler, F. J., Jr. (2002). *Survey research methods* (3rd ed.). Thousand Oaks, CA: Sage Publications.

Gladwell, M. (2004). Big and bad: How the S. U. V. ran over automotive safety. *The New Yorker,* 28–33. January 12.

Hamersley, M., & Atkinson, P. (1995). *Ethnography: Principles in practice* (2nd ed.). New York: Routledge.

Janesick, V. J. (1998). The dance of qualitative research design: Metaphor, methodolatry, and meaning. In N. K. Denzin & Y. S. Lincoln (Eds.), *Strategies of qualitative inquiry* (pp. 35–55). Thousand Oaks, CA: Sage Publications.

Koenig, T. (n. d.). *CAQDAS Comparison.* Retrieved July 23, 2009 from http://www.lboro.ac.uk/research/mmethods/research/software/caqdas_comparison.html

Kvale, S., & Brinkmann, S. (2008). *InterViews: Learning the craft of qualitative research interviewing.* Thousand Oaks, CA: Sage Publications.

Newell, A., & Simon, H. A. (1972). *Human problem solving.* Englewood Cliffs, NJ: Prentice-Hall.

Nisbett, R., & Wilson, T. (1977). Telling more than we can know: Verbal reports on mental processes. *Psychological Review, 84*(3), 231–259.

Patton, M. Q. (1990). *Qualitative evaluation and research methods* (2nd ed). Newbury Park, CA: Sage Publications.

Peräkylä, A. (2005). Analyzing talk and text. In N. K. Denzin & Y. S. Lincoln (Eds.), *The Sage handbook of qualitative research methods* (3rd ed., pp. 869–886). Thousand Oaks, CA: Sage Publications.

Seashore, S. E. (1987). Surveys in organizations. In G. Salvendy (Ed.), *Handbook of human factors* (pp. 313–328). New York: John Wiley & Sons.

Smagorinsky, P. (2001). Rethinking protocol analysis from a cultural perspective. *Annual Review of Applied Linguistics, 21,* 233–245.

Spradley, J. P. (1979). *The ethnographic interview.* New York: Holt, Rinehart & Winston.

Strauss, A., & Corbin, J. (1998). *Basics of qualitative research: Techniques and procedures for developing grounded theory* (2nd ed.). Thousand Oaks, CA: Sage Publications.

Tesch, R. (1990). *Qualitative research: Analysis types and software tools.* New York: Falmer.

Warren, C. A. B., & Karner, T. X. (2005). *Discovering qualitative methods: Field research, interviews, and analysis.* Los Angeles, CA: Roxbury Publishing Company.

Wolcott, H. F. (1990). *Writing up qualitative research.* Qualitative Research Methods Series 20. Newbury Park, CA: Sage Publications.

IV

Applications in Psychology

24

Understanding and Changing the Young Driver Problem: A Systematic Review of Randomized Controlled Trials Conducted With Driving Simulation

Marie Claude Ouimet
University of Sherbrooke

Caitlin W. Duffy
Northwestern University and National Institutes of Health

Bruce G. Simons-Morton
National Institutes of Health

Thomas G. Brown
Douglas Mental Health Research Institute

Donald L. Fisher
University of Massachusetts, Amherst

Abstract

The Problem. Younger drivers, from newly licensed teenagers to drivers in their early twenties, often lack driving skills and engage in driving behavior that may increase their crash risk. Clarification of the mechanisms underlying risk in young drivers and the development of new effective countermeasures to complement existing ones are needed. *Role of Driving Simulation.* Driving simulation can play a role in safely identifying factors associated with young drivers' crash risk and evaluating intervention programs to mitigate risk. It also offers the experimental control required for randomized control trials (RCTs), the gold standard for causal inference. The main goals of this chapter are to evaluate the present state of research in this area and to stimulate future research with young drivers combining RCTs and driving simulation. *Scenarios and Dependent Variables.* This chapter reviews RCTs with either parallel or factorial designs, organized with respect to their research topics, driving-related outcomes, and issues related to the internal and external validity of findings. Studies ($N = 37$) were identified using PsychINFO and MEDLINE (1950–2008), as well as other literature search methods. A systematic review was conducted on the studies' main topic (i.e., interventions to increase skills, chiefly risk or hazard perception; $n = 11$) and main driving-related outcome (i.e., speed; $n = 22$). Results indicated that interventions could increase short-term skill acquisition but left many other key issues unanswered, such as the long-term effects of these interventions on skill acquisition and crash reduction. Most studies (82%) reported significant results on speed in the hypothesized direction. Appraisal of study methodology and interpretation of results revealed widespread shortcomings in reporting practices with regards to randomization and blinding, as well as a frequent failure to adequately address the generalizability of results. This chapter also examines promising avenues for future research and the relevance of RCT guidelines from the Consolidated Standards of Reporting Trials (CONSORT) Statement to this research area. *Platform Specificity and Equipment Limitations.* Platforms varied greatly across studies, but there was no evidence that results in any study reviewed here were limited to the specific equipment used.

24.1 The Teenage and Young Driver Problem

Young drivers' crash risk is relatively low during the supervised driving period (Gregersen, Nyberg, & Berg, 2003; Mayhew, Simpson, & Pak, 2003). It increases dramatically during the first months of independent driving (e.g., Mayhew et al., 2003; McCartt, Shabanova, & Leaf, 2003; Sagberg, 1998) to up to five to seven times that of drivers from 30 to 69 years old (Insurance Institute for Highway Safety [IIHS], 2008a), and then decreases rapidly over the next six months. Additional but more gradual reductions continue into the mid-20s. Many factors have been associated with teenagers' high post-licensure crash risk, including young age at licensing, risky driving behavior like speeding (National Highway Traffic Safety Administration [NHTSA], 2007) and drink-driving (NHTSA, 2008), and exposure to risky driving conditions such as transporting teenage passengers and driving at night (Chen, Baker, Braver, & Li, 2000; Ouimet et al., 2010; Simons-Morton, Lerner, & Singer, 2005; Williams, 2003; Williams & Ferguson, 2002; Williams & Preusser, 1997). A lack of experience and deficiencies in skills, exemplified in the failure to adequately scan the roadway and maintain attention on the forward road, are also significantly associated with elevated crash rate in young drivers (McKnight & McKnight, 2003). Risk or hazard perception, a particular type of scanning, has been singled out as a critically important safety-related driving skill. For example, novice and young drivers have been found to be less likely to identify and slower to react to hidden or hard-to-predict hazards than older and more experienced drivers (Borwosky, Shinar, & Oron-Gilad, 2007; McKenna & Crick, 1991; Pradhan et al., 2005; Quimby & Watts, 1981). Preventing unsafe behaviors is essential to reducing crash risk.

Proposed countermeasures to reduce teenage and young driver crash risk include delaying the age of licensure, restricting driving under high-risk conditions, and increasing pre-license practice and skills attainment. There is strong evidence that delaying licensure and restricting high-risk driving conditions, which can also have an impact on exposure, inexperience, and risk-taking, are associated with reduced crash risk (e.g., Ferguson, Leaf, Williams, & Preusser, 1996; McKnight & Peck, 2002). Most graduated driver licensing programs in the United States address these risk factors (Chen, Baker, & Li, 2006; IIHS, 2008b) and parental collaboration is critical in supporting and enforcing these strategies (Simons-Morton, Ouimet, & Catalano, 2008). Evidence of post-licensure risk reduction is, however, either non-existent for driver education or weak for supervised driving (e.g., Mayhew & Simpson, 2002; McCartt et al., 2003; Simons-Morton et al., 2008). While delaying licensure and restricting driving under certain conditions can reduce some driving risks, these countermeasures fail to mitigate all the perils that teenage and young drivers inevitably face due to inexperience, lack of skills, involvement in risky driving behavior, and exposure to some risky driving conditions.

A broader range of interventions is needed for teenage and young drivers to make headway in further reducing their driving-related risk. Clarification of the mechanisms underlying teenage and young drivers' risky driving behaviors, the acquisition of driving skills, as well as the risks posed by various conditions (e.g., carrying a passenger, driving at night), is critical for informing innovations in evidence-based intervention. Unfortunately, the knowledge base in these areas is severely limited. One methodological reason for this lack of knowledge includes heavy reliance on study designs that use post-hoc analysis of citation and crash databases, and on descriptive studies that do not easily lend themselves to identifying distinct causal mechanisms. For example, it is not clear to what extent speeding reflects young drivers' inner propensity to engage in risky behaviors; their inclination to experiment with different behaviors and test limits; a lack of experience with certain driving conditions; their greater susceptibility to social or normative influence; an overconfidence in their driving abilities; or some combination of these as well as other factors. Similarly, multiple hypotheses are posited in the literature to explain the increased crash risk for young drivers when traveling with teenage passengers, including distraction, peer pressure, and social norms (e.g., Williams, 2003), though the research has yet to disentangle the relative contribution of these potential mediators of passenger risk. In sum, more rigorous experimental methodologies to identify specific mechanisms of risk, their amenability to change, and the impact of these changes on driving risk are needed in order to better understand the novice and young driver problem and how to effectively reduce it.

24.2 The Role of Randomized Controlled Trials Using Driving Simulation in the Teenage and Young Driver Problem

24.2.1 Randomized Controlled Trials

This section reviews the characteristics and benefits of randomized controlled trials (RCTs) and random assignment experiments related to driving simulation with young drivers. The demands of both, in terms of design and rigor, are practically the same. One distinction between RCTs and random assignment experiments, in this context, relates to the nature of the research question addressed. RCTs seek to improve skills and/or reduce risky behavior, and assignment is to an "intervention" or a "treatment." Random assignment experiments, in contrast, examine the impact of exposure to a condition (e.g., conversation on a cell phone) on some behavioral parameter, and assignment is to a "condition." For parsimony, the term RCT subsumes random assignment experiments, and the terms intervention and treatment subsume condition. For comprehensiveness, Table 24.1 makes these distinctions.

RCTs can clarify cause and effect relationships between interventions and outcomes. As described by Jadad and Enkin (2007)

TABLE 24.1 List of the Selected Studies by Design and Topic, Nature of Assigned Groups, Type of Simulation, Outcome and Outcome Measurement Tool, and Vote Counting Results for Speed ($N = 37$)

List of the Selected Studies by Design and Topic	Experimental or Control Condition/Intervention	Type of Simulation	Outcome (Measurement Tool)	Vote Counting (Speed: Experimental Versus Control Groups)
I. Random Assignment Experiments: Studies on Exposure to a Condition				
A. Talking and Listening While Driving				
-Gugerty et al., 2004 (Exp. 1)[a]	E: in-person conversation C: remote conversation	Sim tech	Crash (bp), situation awareness (q)	Na
-Gugerty et al., 2004 (Exp. 2)	E: in-person conversation C: remote conversation	Sim tech	Crash (bp), situation awareness (q)	Na
-Horswill & McKenna 1999b	E: prioritizing verbal task C: no prioritizing of verbal task	Sim tech	Gap acceptance (bp), hazard/risk perception (bp), headway (bp)	Na
-Hunton & Rose 2005	$E_{(1)}$: conversation with vehicle passenger $E_{(2)}$: conversation hands-free cell phone C: no conversation	Simulator	Crash (s), infraction (s)	Na
-Kass, Cole, & Stanny, 2007	E: conversation on cell phone C: no conversation	Simulator	Crash (s), lane position (s), running stop sign (s), speed (s), situation awareness (q)	+
-Kubose et al., 2006 (Exp. 1)	E: speech production C: speech comprehension	Simulator	Lane position (s), speed (s)	+
-Kubose et al., 2006 (Exp. 2)	E: speech production C: speech comprehension	Simulator	Headway (s), lane position (s)	Na
B. Driving Under the Influence of Alcohol				
-Bergeron et al., 1998	E: knowing drivers drank alcohol C: not knowing drivers drank alcohol	Sim tech	Hazard/risk perception (q)	Na
-Harrison & Fillmore 2005[a]	E: under influence of alcohol C: sober	Simulator	Lane position, speed (s)	Ns
-McMillen & Wells-Parker 1987	Alcohol expectancy/ consumption: $E_{(1)}$: high / high - $E_{(2)}$: high/ moderate - $E_{(3)}$: moderate/ high - $E_{(4)}$: moderate/ moderate $C_{(1)}$: moderate/ none - $C_{(2)}$: none/ none	Game	Overtaking (g), speed (ee)	+
-Oei & Kerschbaumer 1990[a]	E: confederate promoting advantages of drink-driving C: confederate stating disadvantages of drink-driving	Game	Attitude (q), crash (g), (crash, lane position, overtaking) (g), speed (g)	+
C. Other: psychological, perceptual, or physiological factors				
-Broughton, Switzer, & Scott, 2007	$E_{(1)}$: moderate fog; $E_{(2)}$: dense fog; C: clear visibility	Simulator	Headway (s), speed (s)	+
-Fischer et al., 2007 (Study 2)	E: racing game C: neutral game	Game	Affect (q), cognition (q)	Na

(continued)

TABLE 24.1 (Continued) List of the Selected Studies by Design and Topic, Nature of Assigned Groups, Type of Simulation, Outcome and Outcome Measurement Tool, and Vote Counting Results for Speed ($N = 37$)

List of the Selected Studies by Design and Topic	Experimental or Control Condition/Intervention	Type of Simulation	Outcome (Measurement Tool)	Vote Counting (Speed: Experimental Versus Control Groups)
-Fischer et al., 2007 (Study 3)[b]	E: racing game C: neutral game	Sim tech	Cognition (q), risky driving behavior (bp)	Na
-Funke, Matthews, Warm, & Emo, 2007	Stress/ level of vehicle automation $E_{(1)}$: induced/ automated driving – $E_{(2)}$: induced/ lead following – $E_{(3)}$: induced/ free driving – $E_{(4)}$: no induction/ automated driving – $C_{(1)}$: no induction/ lead following – $C_{(2)}$: no induction/ free driving	Simulator	Hazard/risk perception (bp), lane position (s), speed (s)	+
-Hennessy & Jakubowski 2007	E: perspective of at-fault driver C: perspective of driver following offending vehicle	Sim tech	Attribution of riskiness and responsibility about crash (q)	Na
-Horswill & McKenna 1999a (Study 2)	E: normal auditory feedback C: low auditory feedback	Sim tech	Speed (q)	+
-Horswill & McKenna 1999c	E: perceived control C: no perceived control	Sim tech	Gap acceptance (bp), headway (bp), overtaking (bp), speed (q)	+
-Lees & Lee 2007	$E_{(1)}$: 100% Accurate alarms; $E_{(2)}$: 29% accurate alarms with unnecessary alarms $E_{(3)}$: 29% accurate alarms with false alarms C: no alarms	Simulator	Acceleration (s), deceleration (s), lane position (s), speed (s)	+
-Matthews, Sparkes, & Bygrave, 1996	E: auditory reasoning task C: visual reasoning task	Simulator	Driving task related interference (q) Headway (s), lane position (s)	Na
-Rupp, Arnedt, Acebo, & Carskadon, 2004[a]	E: restricted sleep C: nonrestricted sleep	Simulator	Lane position (s), speed (s)	Ns
-Taubman Ben-Ari et al., 1999 (Study 3)[a]	E: mortality salience, positive feedback on performance C: emotionally neutral	Game	Speed (c)	+
-Taubman Ben-Ari et al., 1999 (Study 4)[a]	$E_{(1)}$: exposure to mortality salience, positive feedback on performance $E_{(2)}$: mortality salience, no feedback $E_{(3)}$: exposure to emotionally neutral task, positive feedback C: emotionally neutral, no feedback	Game	Speed (c)	+
-Taubman Ben-Ari et al., 2000 (Study 2)[a]	E: mortality salience C: emotionally neutral	Game	Speed (c)	+
-Ward & Beusmans 1998	E: display crash risk probability in terms of safety margin C: display crash risk severity in terms of kinetic energy	Simulator	Hazard/risk perception (q) Headway (s), speed (s)	+

II. Randomized Control Trials: Intervention Studies

A. Skill Training

-Chapman et al., 2002	E: commentary driving C: questionnaires	Sim tech	Eye glance (ee)	Na

Column 1	Column 2	Column 3	Column 4	Column 5
-Fisher et al., 2002	E: training similar to mediated instruction C: no training (intervention not described)	Simulator	Deceleration (s), lane position (s), speed (s)	+
-Fisher et al., 2007 (Exp. 2)	E: error training C: no training (intervention not described)	Simulator	Eye glance (ee)	Na
-Ivancic & Hesketh 2002 (Exp. 1)	E: error training C: errorless learning	Simulator	(Crash/infraction) (s), speed (s), confidence driving skills (q), situation awareness (q)	+
-Ivancic & Hesketh 2002 (Exp. 2)	E: guided error training C: errorless learning	Simulator	(Crash/infraction) (s), speed (s), confidence driving skills (q), situation awareness (q)	+
-McKenna et al., 2006 (Exp. 1)	E: commentary drive C: same drive as E but no commentary or instruction	Sim tech	Confidence driving skills (q), gap acceptance (bp), hazard/risk perception (bp), headway (bp), speed (q)	+
-McKenna et al., 2006 (Exp. 2)	E: commentary drive C: same drive as E but no commentary or instruction	Sim tech	Speed (q)	+
-Pollatsek et al., 2006[a]	E: error training C: no training (intervention not described)	Simulator	Eye glance (ee)	Na
-Pradhan et al., 2006	E: error training C: no training (intervention not described)	Simulator	Eye glance (ee)	Na
-Regan et al., 1998[a]	$E_{(1)}$: mediated instruction $E_{(2)}$: avoidance learning training C: similar elements to $E_{(1-2)}$ without putative active ingredients	Simulator	Confidence driving skills (q), lane position (s), speed (s)	-
-Regan et al., 1998[b]	E: Variable priority training (with feedback) C: Variable priority tasks (with no feedback)	Simulator	Acceleration (s), lane position (s), speed (s)	Ns

B. Driving Under the Influence of Alcohol

Column 1	Column 2	Column 3	Column 4	Column 5
-Howat et al., 1991	E: under influence alcohol C: questionnaires	Simulator	Driving under influence alcohol (q)	Na

Note: *Column 1:* Exp = experiment; [a]Studies identified in the PsychINFO and Medline search. [b]In this study, games were used to elicit behavior, which was measured with a simulation technique. *Column 2:* C = control group; E = experimental group. *Column 3:* Sim tech = simulation technique. *Column 4:* c = coded by research assistant; g = game; bp = button pressing; q = questionnaire; s = simulator. In classifying the outcomes, an effort was made to regroup similar concepts under similar terms. Some outcomes might be described with different terms in each article. In this classification, deceleration includes "braking." "Eye glance" includes movement of the eyes (sometimes inferring risk or hazard perception); for "hazard/risk perception" and "situation awareness" participants were probed. "Lane position" includes position of the vehicles, crossing the center or median lane, and steering. When more precision was given, such as driving completely off the road, the event was classified as a crash. "Risky behavior" is used when behaviors were pooled as one variable. Outcomes in brackets mean that behaviors were considered together. *Column 5:* '+' = speed effect higher or lower for the experimental group, as hypothesized; '-' = speed effect higher or lower for the control group, contrary to hypothesis; Na = not applicable, speed was not measured; Ns = between-group difference was not significant.

in their accessible overview: "… RCTs are quantitative comparative controlled experiments in which a group of investigators study two or more interventions in a series of individuals who are randomly 'allocated' (chosen) to receive them" (p. 2). The main feature of an RCT is random allocation of participants to groups where all participants have an equal chance of being assigned to any of the groups. The goal of randomizing is to control selection bias by randomly distributing potential confounds, including known factors, such as participants' characteristics, as well as unknown factors. Random allocation increases the likelihood of having comparable groups at baseline, and thus the possibility of observing the effect of an intervention, while decreasing the impact of other known and unknown factors that could influence outcome. When properly implemented, random allocation does not erase all group imbalances, but if there are imbalances, they are considered to be randomly distributed.

In a typical two-group parallel RCT, the outcome of the experimental group is compared to the outcome of a control (or comparator) group. This comparison between the experimental and control group explains why it is called a randomized "controlled" trial. In behavioral research, the control can be another standard or "usual care" intervention, a placebo intervention, or no intervention. The comparison of the experimental intervention with a placebo control intervention—that is, similar to the experimental intervention in terms of format and length, except for the putative active ingredient—would be to demonstrate the efficacy of the experimental intervention; the comparison to a standard or usual care treatment would be to demonstrate the superior efficacy of the experimental intervention over an existing intervention. Note that the comparison of the experimental intervention with no intervention does not rule out alternative explanations for findings, such as more attention being paid to the experimental group. When a study contains two or more groups to which participants have not been randomly assigned, this design can be described as quasi-random, and the study referred to as a quasi-experiment or a non-RCT.

It is important to establish a clear set of rules to create random allocation sequences and adhere precisely to them throughout the study. Flipping a coin, rolling a die, and using a table of random numbers are examples of simple ways to generate random allocation sequences. Even when sequences are random, however, there is always the possibility that group size and the distribution of participant characteristics will not be equal between groups. Strategies are available to assure better between-group balance. For example, blocking can be used to ensure equivalence of sample size. Stratification can minimize differences between selected characteristics of the participants (e.g., same number of males and females in each group), especially in small group studies where balanced allotment can increase statistical power (Kernan, Viscoli, Makuch, Brass, & Horwitz, 1999). When blocking or stratification are used, these strategies are referred to as "restricted randomization" (e.g., Altman et al., 2001). When randomization is properly implemented, any significant group differences are due to chance. Presenting demographic and baseline outcome data for each group is suggested

to help readers appraise if the results of a study could be relevant to other groups. Statistical between-group comparisons and the use as covariates of those variables on which groups differ are not necessary nor recommended unless these variables are expected to affect outcome (Altman et al., 2001; Assmann, Pocock, Enos, & Kasten, 2000)

Different types of RCTs were also identified by Jadad and Enkin (2007) according to how participants are exposed to interventions (parallel, factorial, or cross-over design) and whether researchers and participants know which intervention is being assessed (open versus blinded trials), among others. Regarding participants' exposure to interventions, the most common RCT design is the parallel design, briefly described earlier in this section, in which each group is exposed to only one intervention and the outcome of the experimental group is compared to the outcome of the control group. In the least complex factorial design, participants can be assigned to training *x*, training *y*, training *x+y*, or a control group. Each experimental intervention can be compared to the control intervention (e.g., training *x* versus control and training *y* versus control) and to the other experimental intervention (training *x* versus training *y*). Additionally, the interaction between interventions can be assessed (training in either *x* or *y* versus training *x+y*). In a cross-over design, participants serve as their own controls and each group is exposed to all interventions in random order. This design, often adopted by pharmacological studies, is powerful as it requires fewer participants. At the same time, it is inappropriate for studies where exposure to the experimental intervention is likely to result in a lasting change on some characteristic that could affect performance on subsequent trials (Nelson, Dumville, & Torgerson, 2009). As one of the focuses of this review was on interventions that have the potential to alter behavior over a long period of time, the vast majorities of cross-over design studies were not relevant and were therefore excluded as a category of research.

Blinding or masking of group assignment is meant to reduce the *ascertainment (or observation) bias* that can arise when individuals involved in the study are aware of group assignment. For example, participants could anticipate more benefit from their participation in one intervention as opposed to another, while researchers/research assistants and statisticians may react differently—even in subtle ways—towards participants or data based on group assignment, and may bias results in the hoped-for direction. There are different approaches to blinding: open (all parties are aware), single-blind (only participants are unaware), double-blind (participants and researchers/research assistants are unaware), triple-blind (participants, researchers, and statisticians are unaware), and so forth. Although double-blind trials are praised as being less vulnerable to bias (Schulz, Chalmers, & Altman, 1995), they may be easier to orchestrate in pharmacological investigations than in behavioral intervention research where group assignment may be more obvious to participants and/or researchers. Nevertheless, it may still be possible to blind researchers to the group assignment until the intervention is administered, and to blind statisticians until after the results are analyzed (i.e., by dummy coding of group assignment).

As summarized by Mercer and colleagues (2007), RCT design is considered optimal for controlling experimental bias, assessing the relative effects of two or more experimental interventions, and making causal inferences. RCTs are implemented in controlled environments and protect against selection bias, which is a serious threat to internal validity (i.e., the ability to make causal inferences about the effect of an experiment). Causal inferences are internally valid if differences between groups of participants can be attributed to the effect of the experimental design and to no other plausible cause. "… [RCT] with individual random assignment remains the gold standard for safeguarding internal validity" (Mercer, DeVinney, Fine, Green, & Dougherty, 2007, p. 139). Although establishing internal validity in experiments on teenage and young drivers is critical, conducting an RCT that effectively reduces threats to internal validity is challenging. The Consolidated Standards of Reporting Trials (CONSORT) Statement (Altman et al., 2001; Begg et al., 1996; Boutron, Moher, Altman, Schulz, & Ravaud, 2008; Moher, Schulz, & Altman, 2001) identifies 22 key elements pertaining to methodological quality in RCT research, including potential threats to the internal validity of findings. The CONSORT Statement also outlines descriptions that should be included in scientific reports to assist in appraising and synthesizing RCT-based evidence. The implications of the CONSORT Statement for both the design and reporting of RCT studies is considered in more depth in Section 24.4.

24.2.2 Driving Simulation Versus Other Research Tools

RCTs in driving research require safe experimental control. In this light, various research tools, including test tracks, road studies (i.e., naturalistic observations, observations by an in-vehicle research staff), crash or driving file databases, questionnaires, and driving simulation can be considered. Test track studies can offer well-controlled environments for experimental manipulation and observation. Some experiments using test tracks, however, such as studies on the effect of psychoactive substances or medications, might not be feasible due to potential risks for participants. In addition, some test track methodologies might affect observed results (e.g., required presence of an in-car passenger/experimenter). Test tracks are also not very accessible. Experiments on the road, on the other hand, are constrained by safety concerns and limited experimental control—two potential threats to internal validity. Databases and traffic observations are generally not well-suited for experimentation, and their use limits causal inferences. Driving simulation, which can involve a high level of safety and experimental control and is more accessible than test tracks, appears to represent the best research tool for conducting RCTs.

All research tools and designs represent a compromise between internal and external validity (or generalizability; namely, the ability to apply causal inferences to other populations or settings). Simulation presents a higher level of internal validity and a lower level of external validity while the converse is true for road studies. The potential lack of external validity is a common criticism of driving simulation studies. Consequently, many studies have attempted to demonstrate the correspondence between behaviors measured with a specific simulator and those assessed on the road (e.g., Blaauw, 1982; Riemersma, van der Horst, Hoekstra, Alink, & Otten, 1990; see also, this book, chap. 13 by Mullen, Charlton, Devlin, & Bédard, "Simulator Validity: Behaviors Observed on the Simulator and on the Road"). Consideration of potential threats to the external validity of findings is recommended by the CONSORT Statement (Altman et al., 2001; Begg et al., 1996; Boutron et al., 2008; Moher et al., 2001).

24.3 Systematic Review of the Literature

The main goal of this chapter is to summarize research in young drivers that employs both RCT methodology and driving simulation. In order to gauge the methodological quality of this literature, we also consider the internal and external validity of findings, presented in Section 24.4, with reference to the CONSORT Statement guidelines (Altman et al., 2001; Begg et al., 1996; Boutron et al., 2008; Moher et al., 2001).

24.3.1 Methods

24.3.1.1 Study Criteria and Identification and Selection of Manuscripts

For the sake of comprehensiveness, we initially used a broad set of search criteria with the intention of focusing them afterward if necessary. For this review, "teenage drivers" were defined as drivers with a maximum average age of 18 years, and "young drivers" referred to drivers with an average age of 25 years, thus subsuming teenage drivers. In order to include RCTs that used either parallel or factorial designs, random assignment of participants to interventions was the minimal inclusion criterion.

Three types of driving simulation were distinguished and guide our discussion below: driving simulator, simulation technique, and game. While all three types can create a complex simulated driving environment, they differ on three distinct dimensions: the degree of participants' interaction with the device; the variety of scenarios participants encounter; and the measurement of outcomes (e.g., simulation, external device). First, in a driving simulator and a game, participants can interact directly with the simulated environment to execute common driving maneuvers using, for example, the steering wheel and accelerator and brake pedals. With a simulation technique, the projection can involve either real or simulated driving scenes, and participants interact at specific moments by either pressing a key or a button or responding to questions. Second, with a driving simulator or a simulation technique, researchers can program different scenarios and measure multiple driving outcomes depending on the research questions. In a game, scenarios and driving outcomes are selected by the company that designed the game. Finally, a driving simulator can collect most of the driving outcomes; a game might collect a few outcomes and leave many to be coded by research staff (e.g., speed at a specific location); and a simulation technique relies almost exclusively on other means to collect data of interest. Note that given the above definitions, studies employing pursuit tasks that did not involve driving, such as following a moving target, were excluded as were studies in which driving simulation was employed but no driving-related outcomes were collected

A number of search strategies were used to find relevant studies. Publications were identified via a systematic search of two electronic databases: PsychINFO (1950–2008) and MEDLINE (1950–2008). English words used in the search were: 1) random, randomised, randomized; 2) adolescent(s), teen(s), teenage, teenager(s), young, youth; 3) driver(s), driving; and 4) game(s), video, simulated, simulation, simulator(s). Internet search engines and reference lists from relevant articles were also drawn upon. The grey literature was sampled, but sparingly. Only one published source was selected when multiple papers or conference proceedings emerged from the same study. When a paper summarized results from multiple studies, only findings from driving simulation RCTs were considered. When a paper reported on more than one RCT, all were summarized. When studies also included a non-randomized group, such as an older group of participants, the comparison between the randomized and non-randomized groups was not reported.

24.3.1.2 Data Analysis

In this systematic review, or qualitative synthesis of outcomes, the vote counting method (Bushman, 1994) was employed to answer research questions. The direction of the effect was identified as positive if hypothesized effects were supported, negative if the observed effects were in the opposite direction of what was hypothesized, and non-significant when no significant differences were found. When a variable was measured in different ways (e.g., mean speed, speed variability) but only one was significant, results were reported as being significant. All data extracted from selected studies were appraised by two coders. Inter-rater reliability was found to be above 80%, and all discrepancies in coding were discussed until a consensus was reached. To first categorize the selected studies, answers to two general research questions were sought with the goal of further refining research questions if deemed possible: What research topics were addressed? And what was the main driving-related outcome?

24.3.2 Findings

24.3.2.1 Study Characteristics

With the broad criteria discussed above, we found 39 studies from electronic databases, eight of which fulfilled the inclusion criteria, and 29 other studies were found by way of other search strategies described above, for a total of 37 studies. The list of the selected studies is presented in Table 24.1.

Age. In 16% of studies, the mean age was 18 years or younger, as these studies were conducted in countries that allow independent driving before age 18 (e.g., Australia, Canada, the U.S.). In 49% of studies, the mean age was between 19 and 21 years; and in 24% of the studies it was between 22 and 25 years. Two studies (5%) did not report any information on age but indicated that participants were either undergraduate students or newly-licensed drivers. Two others (5%) only reported age ranges exceeding 25 years but noted the recruitment of university students. It was assumed that participants in these four studies had a mean age of 25 years or younger. In many studies, participants' age range was important but age distribution was highly skewed in the direction of the minimum age of inclusion. For example, in the study by Gugerty, Rakauskas and Brooks (2004), the age range was 18 to 43 years old, but the mean age was 21 years. Table 24.2 presents additional information on participants' age, sex, and driving experience, as well as the country in which these studies took place.

TABLE 24.2 Participants Demographic Information and Countries in Which Studies Took Place (*N* = 37)

	N	%
Mean age		
≤ 18	6	16.2
19-21	18	48.6
22-25	9	24.3
No or minimal information	4	10.8
Age		
Mean and measure of dispersion	26	70.3
Only mean or measure of dispersion	9	24.3
No information	2	5.4
Sex		
Males and females included	29	78.4
Only males	8	21.6
Driving experience (in time)		
Mean and measure of dispersion	6	16.2
Only mean or measure of dispersion	8	21.6
Other information (e.g., minimum required)	18	48.6
No information	5	13.5
Driving experience (in distance)		
Mean and measure of dispersion	3	8.1
Only mean or measure of dispersion	5	13.5
Other information (e.g., minimum required)	5	13.5
No information	24	64.9
Countries in which studies were conducted		
United States	17	46
United Kingdom	8	22
Australia	6	16
Israel	3	8
Germany	2	5
Canada	1	3

Methodological Features. Detection of RCT studies was often challenging because participant randomization to interventions was rarely indicated (i.e., 11%) in the title or abstract as recommended by the CONSORT Statement (Item 1) (Altman et al., 2001; Begg et al., 1996; Boutron et al., 2008; Moher et al., 2001). Because of this widespread shortcoming in RCT reporting practices, it is possible that some pertinent studies were passed over in our search. Five studies included in our review failed to specify use of an RCT design in the abstract or in the text, although our access to other information sources led to their classification as an RCT. Four unspecified studies could be classified as RCTs based upon personal communication with an author, and one additional study was assumed to be an RCT based upon evidence that the same author had conducted a similarly designed study that was clearly labeled as an RCT (also reviewed here). As discussed in Subsection 24.2.1, Table 24.1 presents RCT studies classified as random assignment experiments if their goal was strictly to examine the impact of exposure to a condition on some behavioral parameter, usually driving skills or risky driving behavior (68%), or as RCTs (i.e., intervention studies) if their goal was explicitly to improve skills and/or reduce risky behavior (32%). Table 24.1 also shows the nature of the assigned group for each study, and Table 24.3 summarizes the methodological details of the studies reviewed.

Types of Simulation and Outcomes, and Outcome Measurement Tools. Table 24.1 shows types of simulation and outcomes, and how outcomes were measured. A simulator was used in 54% of the studies. Of the 30% ($n = 11$) of studies using a simulation technique, eight involved video projection of real driving scenes and two involved driving scenes captured with a simulator or a game (Bergeron, Laviolette, Perraton, & Joly, 1998; Hennessy & Jakubowski, 2007). One study measured outcomes (i.e., risky behavior) with a simulation technique after participants were first exposed to either a racing or a neutral game (Fischer, Kubitzki, Guter, & Frey, 2007). Games were employed in 16% of the studies. About 80% of the simulator studies used outcome data collected by the simulator, while only 33% of games did. Simulation techniques mostly used button pressing and questionnaires. External equipment to measure outcomes was distributed evenly between the three types of driving simulation.

24.3.2.2 Description of the Content of the Studies

The answers to the two general research questions were first sought: What research topics were addressed in the selected studies? What was their main driving-related outcome? Table 24.1 indicates the final classifications that emerged from this exercise. The research topic areas covered most frequently were the effects on driving behaviors and/or skills of: a) skill training (30%); (b) talking and listening while driving (i.e., operating a cell phone, engaging in a discussion with a passenger, responding to a verbal task) (19%); (c) driving under the influence of alcohol (13%); (d) other category including psychological, perceptual, or physiological factors (38%). Skill training, which represents 90% of the intervention studies and about a third of all studies, is the largest body of research employing RCT design and driving simulation

TABLE 24.3 Aspects of Internal and External Validity and Other Information on Study Methodology ($N = 37$)

	N	%
Mention of random assignment[a]		
In abstract and text	2	5.4
In abstract only	2	5.4
In text only	28	75.7
Unspecified	5	13.5
Adequate information on randomization and blinding procedures[b]		
Yes	1	2.7
No	36	97.3
Number of participants per group		
$n < 30$	19	51.4
$n \geq 30$	18	48.6
Control intervention[c]		
Comparable to experimental intervention (e.g., using simulator or simulation technique)	31	83.8
Partially comparable to experimental intervention (e.g., no simulator or simulation technique, but questionnaire)	2	5.4
No information	4	10.8
Scenarios counterbalanced or presented at random		
Yes	13	35.1
Partial	4	10.8
No	20	54.1
Baseline information on main outcomes for each group[d]		
Yes	20	54.1
No	17	45.9
Demographic information on each group[d]		
Yes or partial	20	54.1
No	17	45.9
External validity or generalizability of findings (mentioned)[e]		
Yes	21	56.8
No	16	43.2
External validity or generalizability of findings (tested)		
Yes	2	5.4
No	35	94.6

Note: Randomized controlled trial guidelines from the Consolidated Standards of Reporting Trials (CONSORT).
[a] Item 1.
[b] Items 8–11.
[c] Item 4.
[d] Item 15.
[e] Item 21.

in young drivers. Speed was the driving-related outcome most commonly examined (about 60% of all studies). Given that the number of studies on skill training was small and goals differed among studies measuring speed, no further research questions were asked and only skill training and speed studies were reviewed guided by key questions presented below.

24.3.2.2.1 Main Topic: Effect of Skill Training

Adequate skills acquisition prior to licensure represents an important safety strategy, and simulation skill training has been examined as a means to expeditiously achieve this aim. Guided by several questions, the findings of simulation skill training studies are summarized below using vote counting.

Which skills were subject to training? Eleven RCTs explored the impact on driving skills and behaviors of: 1) risk/hazard perception (or awareness) training (Chapman, Underwood, & Roberts, 2002; Fisher et al., 2002; Fisher, Pradhan, Pollatsek, & Knodler, 2007; McKenna, Horswill, & Alexander, 2006; Pollatsek, Narayanaan, Pradhan, & Fisher, 2006; Pradhan, Fisher, & Pollatsek, 2006; Regan, Deery, & Triggs, 1998a-b), and 2) other driving-related skills, including choosing appropriate speed and merging left (Ivancic & Hesketh, 2000). Training involved simulators, simulation techniques, or personal computers. Skill acquisition was most commonly measured using a simulator or simulation technique.

Did driving skills and/or safe driving behavior increase immediately after training? Overall, results showed that compared to untrained participants, those who were trained to perceive risks or hazards demonstrated behaviors indicative of enhanced perception or awareness of potential or real threats. Specifically, trained novice drivers: made more eye glances to areas where risks could materialize (e.g., Fisher et al., 2007; Pollatsek et al., 2006; Pradhan et al., 2006); had shorter durations of fixation and a wider spread of horizontal search while watching hazardous situations (Chapman et al., 2002); perceived hazards more quickly and gaps from oncoming traffic as more dangerous (McKenna et al., 2006); drove more slowly at safety-critical locations (Fisher et al., 2002; Ivancic & Hesketh, 2000) or chose a slower speed than the vehicle on the video they were watching (McKenna et al., 2006); had more appropriate increases in acceleration (Regan et al., 1998b) or decreases in acceleration before a potential hazard (Fisher et al., 2002); maintained a safer position on the road when approaching potentially dangerous situations (Fisher et al., 2002; Regan et al., 1998a); received fewer tickets; and were involved in fewer crashes (e.g., Ivancic & Hesketh, 2000). These learning effects were usually observed immediately after training or in a relatively short post-training period ranging from four days to one month (e.g., Fisher et al., 2002; Pradhan et al., 2006; Regan et al., 1998a). Some studies, however, failed to find significant differences between trained and untrained novice drivers on the following variables: chosen headway (McKenna et al., 2006); deviation in the lane position; reaction time to speed change (Regan et al., 1998b); number of recalled strategies to avoid errors (Ivancic & Hesketh, 2000); and speed (Regan et al., 1998a, b).

Were increases in driving skills and safe driving behavior long-lasting? Only one study examined longer-lasting effects of training on simulation performance during the three to six month follow-up period (Chapman et al., 2002). Compared to the untrained group, the trained group demonstrated shorter fixation durations and wider spread of search right after training, but only the latter at follow-up.

Did skills acquired in training generalize to other situations? Five studies examined the generalizability of skills learned when using scenarios both similar to those employed for training (i.e., near scenarios) and considerably different from those learned in training (i.e., far scenarios). Four studies found that compared to untrained participants, trainees had higher scores in both near and far scenarios (Fisher et al., 2007; Pollatsek et al., 2006; Pradhan et al., 2006; Regan et al., 1998a). One study found higher scores only in near scenarios (Ivancic & Hesketh, 2002).

Did skills acquired in training generalize to driving performance and outcome on the road? Generalization of simulation results to the road was examined in two studies. Fisher et al. (2007) found that compared to untrained participants, trainees had higher risk recognition scores in both near and far transfer scenarios in the simulator as well as on the road. Chapman et al. (2002) found longer-lasting training effects on the simulator during the three to six month follow-up period, but only short-term effects on the road. No study examined the effect of training on crash reduction.

What type of training was implemented, of the training approaches found to be effective, was one superior to the others? The training approaches used in these studies, which typically consisted of one session lasting no more than an hour and a half, were inspired by a variety of theories and approaches. Training included: variable priority training (i.e., asking participants to give different attentional priority to two tasks) (Regan et al., 1998b); mediated instruction (i.e., training on hazard perception and safe decision-making; practicing learned skills in a simulator; commentary driving on video replay, Regan et al., 1998a); avoidance learning training (i.e., watching materials on general information about driving; exposure to a near-miss situation in a simulator; replay of driving without comments, Regan et al., 1998a); error or direct training (i.e., drivers allowed to make mistakes and encouraged to learn from feedback, Fisher et al., 2002, 2007; Ivancic & Hesketh, 2000; Pollatsek et al., 2006; Pradhan et al., 2006); guided error or vicarious training (i.e., watching drivers making mistakes, Ivancic & Hesketh, 2000); and commentary driving (Chapman et al., 2002; McKenna et al., 2006). As described above, the review of these studies indicates that training was generally superior to a control intervention, but could not answer the question as to which training approach was superior to the others as only one study directly compared two experimental interventions (Regan et al., 1998a).

Is training associated with any negative side-effects? Self-confidence in driving skills was the negative side-effect most studied, as greater self-confidence is often believed to be associated with increased risk. Training did not appear to affect self-confidence in driving skills (e.g., McKenna et al., 2006; Regan et al., 1998a). One study showed a decrease in confidence, which could be considered a positive side-effect. Results obtained by Ivancic and Hesketh (2000) suggested that how training is conducted may be important, with direct training decreasing drivers' confidence in their own abilities in contrast to vicarious learning.

24.3.2.2.2 Main Driving-Related Outcome: Speed

Results were also summarized for the main driving-related outcome (i.e., speed). This subsection also addressed whether the type of simulation (i.e., simulator, game, simulation technique) influences measurement of speed and as a consequence, speed outcomes. Toward these ends, a systematic review of the extant research was conducted to answer the following questions: Have between-group differences been observed in study outcomes, and if so, were they in the hypothesized direction? Did studies employing different types of simulation generate similar results?

As summarized in Table 24.1, results of vote counting on the 22 retained studies indicated that 82% reported a significant result in the hypothesized direction. Examples are presented involving the three identified types of driving simulation. Kubose et al. (2006) used a simulator to investigate the impact of speech production and comprehension on speed. They found that the average mean speed was higher for participants randomly assigned to the speech production condition compared to those assigned to the speech comprehension condition. Oei and Kerschbaumer (1990) employed a video game to investigate the influence of peer pressure and blood alcohol level on speed. Participants were randomly assigned to a pro or con drink-driving condition in which confederates were instructed to convince each group to adopt a pro (i.e., confederate agreeing with drink driving) or con (i.e., confederate disagreeing with drink driving) drink-driving position. Results indicated that drivers in the pro condition drove faster as blood alcohol levels increased (i.e., 0.00%, 0.04%, 0.08%), while those in the con condition drove at about the same speed at all levels. Horswill and McKenna (1999a) used a simulation technique to examine the effect of auditory feedback on speed. Participants were randomly assigned to normal auditory (i.e., internal car noise) versus low auditory feedback conditions and were asked to rate the speed at which they would drive in the same situation as a vehicle they viewed in video footage. The investigators found that those in the normal auditory feedback condition reported they would drive slower than those in the low auditory feedback condition. Finally, speed was measured in 65% of simulator studies and 56% of simulation technique and game studies. Results were similar irrespective of the simulation type used.

Other Driving-Related Outcomes. Table 24.1 also describes all the driving-related outcomes and how each outcome was measured. After speed, the most common driving-related outcomes were, in descending order: lane position (e.g., position of vehicles, crossing the center or median lane, steering); hazard/risk perception and situation awareness based on probing participants, headway, crash (including driving completely off the road), eye glance (sometimes inferring risk or hazard perception), gap acceptance, overtaking, acceleration, and deceleration including braking patterns.

24.4 Internal and External Validity of Findings

In this section, studies are appraised for both internal and external validity with frequent reference to CONSORT Statement guidelines (Altman et al., 2001; Begg et al., 1996; Boutron et al., 2008; Moher et al., 2001). Table 24.3 summarizes some of the CONSORT criteria for methodological rigor and good reporting practices for RCT studies and the proportion of studies reviewed that were in compliance with each. Additional information on study methodology is also presented.

Random assignment of participants to interventions: This criterion pertains to whether a description was provided about how the random allocation sequence was generated (e.g., table of random numbers) and any restrictions employed (e.g., stratification) (CONSORT Statement, Item 8), whether the random allocation sequence was concealed until group assignment (CONSORT Statement, Item 9), and if details were provided about how it was implemented (i.e., who generated the sequence, enrolled participants, and assigned participants to groups?) (CONSORT Statement, Item 10). Of the 37 studies included in our review, only one investigation on driving under the influence of alcohol by McMillen and Wells-Parker (1987) adequately addressed random assignment procedures. Some studies provided partial information on some aspects of random assignment, such as stratification for sex (e.g., Bergeron et al., 1998; Fisher et al., 2007; Gugerty et al., 2004; Pradhan et al., 2006). In most cases, however, details were often not explicitly described. At times, outcomes suggestive of a restricted randomization procedure (e.g., exactly the same number of males and females in each group) were presented as if they were the (unlikely) result of simple randomization. Overall, the lack of information on randomization procedures made the validity of study results difficult to appraise.

Blinding in group assignment: This criterion refers to whether participants, researchers and research assistants administering the intervention, and statisticians were blind to group assignment, and how success in blinding was evaluated (CONSORT Statement, Item 11). Only the study by McMillen and Wells-Parker (1987) explicitly described its double-blind procedures. No other studies mentioned if participants, researchers, and statisticians were blind to group assignment. One study that did not report blinding procedures nevertheless tested the success of blinding (Oei & Kerschbaumer, 1990).

Type of control intervention administered: Item 4 of the CONSORT Statement indicates the need to describe in detail the interventions given to both treatment and control groups. A related issue involves the desirability of an equivalent intervention (e.g., in time, engagement, attention, contact with experimenters) employing driving simulation. Most studies (84%) had a comparable intervention. Four skill training studies (11%) did not describe the intervention to which the control group was assigned (Fisher et al., 2002, 2007; Pollatsek et al., 2006; Pradhan et al., 2006). Two studies (5%) had partially comparable interventions as the control groups completed questionnaires or interviews while the experimental group received training to improve skills or reduce risky driving behavior (Chapman et al., 2002; Howat, Robinson, Binns, Palmer, & Landauer, 1991).

Control for unwanted learning effects: One threat to internal validity, which has special meaning to simulation studies, stems from learning effects. Learning can arise from repeated exposure to the same or similar driving simulators, scenarios, or tasks. For example, performing one simulation task can systematically contaminate performance on subsequent tasks. To reduce the effect of this potential confound in simulation studies with repeated testing episodes, counterbalancing or random presentation of scenarios and tasks can be used. Only about one third of studies reported counterbalancing or randomly presenting multiple scenarios or tasks. In two other studies, either partial counterbalancing or varying the presentation order of training and testing was undertaken (Ivancic & Hesketh, 2000; Lees & Lee, 2007). Three studies that did not use counterbalancing or random presentation nevertheless attempted to retrospectively account for the potential impact of learning statistically (Taubman Ben-Ari, Florian, & Mikulincer, 1999, 2000).

Information on each group: In order to evaluate if the results of a study could be relevant to other groups, presenting demographic information and baseline data on outcome variables is recommended for each group (CONSORT Statement, Item 15). Table 24.3 shows that most studies did not report any demographic information for each group separately or only indicated that groups were equivalent on some characteristics. Concerning a description of the entire sample, Subsection 24.3.2.1 shows that information on age was sometimes incomplete. Table 24.2 highlights that sample information regarding driving experience in time and distance was reported in only approximately half of the studies. Less than 50% of the studies, including most of the skill training studies, did not collect baseline data on outcome variables.

External validity of findings: The CONSORT Statement (Item 21) recommends that authors discuss the external validity of their findings. In regards to driving simulation, we asked whether there was any discussion or presentation of data from other related research to support the external validity of results. More than half (57%) of the studies addressed the external validity by considering the relationship of driving simulation to actual road behavior. External validity of findings was directly assessed in two training studies by including a road task in the protocol (Chapman et al., 2002; Fisher et al., 2007). In some cases, investigators discussed external validity but for tasks not related to driving per se. At other times, only the concurrent validity (i.e., relationship between the measures and other previously validated measures) or predictive validity (i.e., how a performance score predicts a criterion measure such as traffic violations and crashes) of the data were addressed.

24.5 Discussion

Effective prevention strategies to reduce teenage drivers' risk have focused on delaying licensure and imposing post-licensure restrictions (e.g., Ferguson et al., 1996; McKnight & Peck, 2002; Simons-Morton et al., 2008). While these strategies can reduce some driving risks, they fail to mitigate all the perils that teenage and young drivers inevitably face due to factors such as inexperience, lack of skills, involvement in risky driving behavior, and exposure to risky driving conditions. This shortcoming may be due, in part, to our limited knowledge of the mechanisms underlying young drivers' risks. In this chapter, we argued that driving simulation, and its amenability to RCT methodology, makes it an essential tool to understand causal pathways to teenagers' risky behavior and to develop evidence-informed intervention programs. Our primary objective was to critically review the existing literature and in this way, promote high-quality research with young drivers utilizing RCT methodology coupled with driving simulation.

The use of RCT methodology and driving simulation seems promising, and there are many possibilities for future research. These include investigations into night driving and the influence of passengers, the operation of electronic devices, and the consumption of alcohol or drugs. The long-term impact of skill training on driving skills, safe driving behavior, and crash risk in real driving situations has yet to be demonstrated. In addition, other central questions concerning the training of novice drivers remain unaddressed. For example, is behavioral change produced by simulation training caused by the same mechanisms that mediate behavioral change under real driving conditions? What other critical driving skills besides risk or hazard perception can be acquired with simulation training? Are some types of training more efficient for developing specific skills and behaviors? Could the generalization of learning to new tasks not encountered during training be influenced by the type of training provided? More research on factors related to young drivers' increased risk and how to reduce it is needed.

This review also highlighted important inadequacies in RCT reporting practices. Even if most studies may have been rigorously designed and implemented, this shortcoming diminishes their credibility. For example, greater treatment effects have been found in studies with absent or inadequate reporting of key methodological elements related to internal validity (e.g., Schulz et al., 1995). This suggests that such studies may be more prone to bias. In contrast, the study by McMillen and Wells-Parker (1987) represents a good example of how RCT design and reporting should be orchestrated in driving-related investigations, especially in terms of randomization and blinding procedures. Good RCT design and reporting practices that comply with CONSORT Statement guidelines are possible in traffic injury prevention research, albeit with additional effort than currently appears to be the norm. We propose that the elements in RCT design and reporting outlined in the CONSORT Statement (Altman et al., 2001; Begg et al., 1996; Boutron et al., 2008; Moher et al., 2001) are indispensable for increasing the quality of driving simulation RCTs.

Despite adopting broad study inclusion criteria for our review, we found that the literature on driving simulation RCTs was not abundant. The number of studies on the same topic was small and goals differed among studies measuring the same outcome. Accordingly, a systematic review (i.e., a qualitative synthesis of the outcomes) was deemed more appropriate than a quantitative approach like meta-analysis (Engberg, 2008). Vote counting, which was employed to summarize the results, does not weigh results according to sample size or provide an estimate of effect size. More simulation RCTs and improved reporting practices will make other forms of review, such as meta-analysis, increasingly viable and effective in moving the field forward.

24.6 Conclusion

This review revealed that driving simulation RCTs with young drivers are rare. It also indicated many areas in which reporting in these studies could be improved to enhance their contribution to traffic injury prevention. Driving simulation, and its amenability to RCT methodology, can play a unique role in clarifying the mechanisms underlying young drivers' risky behavior and crash risk and to developing evidence-informed intervention programs.

Key Points

- Young drivers are over-represented in motor vehicle crashes and many factors have been associated with their increased crash risk including inadequate driving skills, engaging in risky driving behavior, and driving in risky conditions.
- Randomized controlled trials (RCTs) are generally considered the best design for establishing causal relationships. Implementing them in a safe and controlled driving simulation environment can help clarify the mechanisms underlying teenage drivers' risks and test the efficacy of interventions. To stimulate and guide future research with young drivers employing RCTs and driving simulation, a systematic review of selected studies was performed in addition to a consideration of potential threats to the internal and external validity of findings.
- Interventions to increase driving skills, especially risk or hazard perception, represent the only major body of research focusing on young drivers using the RCT design and driving simulation. Both RCT design and driving simulation could be utilized to address key research questions regarding the mechanisms that underlie young drivers' increased risk in dangerous driving conditions (e.g., the effect of driving at night, with passengers) and possible interventions to mitigate these risks (e.g., improving driving skills other than risk or hazard perception, reducing risk when driving with young passengers).
- All but one of the selected studies failed to document essential details of RCTs, especially randomization and blinding procedures that are vital to the appraisal of internal validity. Also, many studies did not adequately report demographic information, such as the mean and distribution of age and driving experience, so the specific relevance of the findings to either teenage or young drivers was sometimes unclear.

Keywords: Adolescent, Driving Simulation, Driving Skill, Novice Driver, Prevention, Random Assignment Experiment, Randomized Controlled Trial, Risky Behavior, Systematic Review, Young Adult

Acknowledgment

Support for this study was received from the Intramural Research Program of the Eunice Kennedy Shriver National Institute of Child Health and Human Development, National Institutes of Health (NIH). Portions of this research were funded by a grant from the NIH (1R01HD057153-01) to Donald Fisher.

Glossary

External validity (or generalizability): Ability to apply causal inferences to other different situations (e.g., location, population).
Internal validity: Ability to make causal inferences about the effect of an experiment.

Key Readings

Altman, D. G., Schulz, K. F., Moher, D., Egger, M., Davidoff, F., Elbourne, D., . . . Lang, T. (2001). The revised CONSORT Statement for reporting randomized trials: Explanation and elaboration. *Annals of Internal Medicine, 134,* 663–694.
Hunton, J., & Rose, J. M. (2005). Cellular telephones and driving performance: The effects of attentional demands on motor vehicle crash risk. *Risk Analysis, 25,* 855–866.
McKenna, F. P., Horswill, M. S., & Alexander, J. L. (2006). Does anticipation training affect drivers' risk-taking? *Journal of Experimental Psychology, 12,* 1–10.
Oei, T. P. S., & Kerschbaumer, D. M. (1990). Peer attitudes, sex, and the effects of alcohol on simulated driving performance. *American Journal of Drug and Alcohol Abuse, 16,* 135–146.
Taubman Ben-Ari, O., Florian, V., & Mikulincer, M. (1999). The impact of mortality salience on reckless driving: A test of terror management mechanisms. *Journal of Personality and Social Psychology, 76,* 35–45.

References

References preceded by an* are included in the systematic review.
Altman, D. G., Schulz, K. F., Moher, D., Egger, M., Davidoff, F., Elbourne, D., . . . Lang, T. (2001). The revised CONSORT Statement for reporting randomized trials: Explanation and elaboration. *Annals of Internal Medicine, 134,* 663–694.
Assmann, S. F., Pocock, S. J., Enos, L. E., & Kasten, L. E. (2000). Subgroup analysis and other (mis)uses of baseline data in clinical trials. *Lancet, 355,* 1064–1069.
Begg, C., Cho, M., Eastwood, S., Horton, R., Moher, D., Olkin, I., . . . Stroup, D. F. (1996). Improving the quality of reporting of randomized control trials: The CONSORT Statement. *The Journal of the American Medical Association, 276,* 637–639.

* Bergeron, J., Laviolette, E., Perraton, F., & Joly, P. (1998). The perception of risk involved in alcohol-impaired driving. In K. Brookhuis, D. de Waard, & C. Weikert (Eds.), *Simulators and Traffic Psychology* (pp. 115–124). Groningen, the Netherlands: Centre for Environmental and Traffic Psychology, University of Groningen.

Blaauw, G. J. (1982). Driving experience and task demands in simulator and instrumented car: A validation study. *Human Factors, 24*, 473–486.

Borwosky, A., Shinar, D., & Oron-Gilad, T. (2007). *Age, skill, and hazard perception in driving.* Proceedings of the Fourth International Driving Symposium on Human Factors in Driver Assessment, Training and Vehicle Design. Iowa City: University of Iowa.

Boutron, I., Moher, D., Altman, D. G., Schulz, K. F., & Ravaud, P. (2008). Extending the CONSORT Statement to randomized trials of nonpharmacologic treatment: Explanation and elaboration. *Annals of Internal Medicine, 148*, 295–309.

* Broughton, K. L. M., Switzer, F., & Scott, D. (2007). Car following decisions under three visibility conditions and two speeds tested with a driving simulator. *Accident Analysis & Prevention, 39*, 106–116.

Bushman, B. J. (1994). Vote-counting procedures in meta-analysis. In H. Cooper & L. V. Hedges (Eds.), *The handbook of research synthesis* (pp. 193–213). New York: Russell Sage Foundation.

* Chapman, P., Underwood, G., & Roberts, K. (2002). Visual search patterns in trained and untrained novice drivers. *Transportation Research Part F: Traffic Psychology and Behavior, 5*, 157–167.

Chen, L. H., Baker, S. P., Braver, E. R., & Li, G. (2000). Carrying passengers as a risk factor for crashes fatal to 16- and 17-year-old drivers. *The Journal of the American Medical Association, 283*, 1578–1582.

Chen, L. H., Baker, S. P., & Li, G. (2006). Graduated driver licensing programs and fatal crashes of 16-year-old drivers: A national evaluation. *Pediatrics, 118*, 56–62.

Engberg, S. (2008). Systematic reviews and meta-analysis: Studies of studies. *Journal of Wound, Ostomy, and Continence Nursing, 35*, 258–265.

Ferguson, S. A., Leaf, W. A., Williams, A. F., & Preusser, D. F. (1996). Differences in young driver crash involvement in states with varying licensure practices. *Accident Analysis & Prevention, 28*, 171–180.

* Fischer, P., Kubitzki, J., Guter, S., & Frey, D. (2007). Virtual driving and risk taking: Do racing games increase risk-taking cognitions, affect, and behaviors? *Journal of Experimental Psychology: Applied, 13*, 22–31.

* Fisher, D. L., Laurie, N. E., Glaser, R., Connerney, K., Pollatsek, A., Duffy, S. A., & Brock, J. (2002). Use of a fixed-base driving simulator to evaluate the effects of experience and PC-based risk awareness training on drivers' decisions. *Human Factors, 44*, 287–302.

* Fisher, D. L., Pradhan, A. K., Pollatsek, A., & Knodler, M. A. (2007). Empirical evaluation of hazard anticipation behaviors in the field and on driving simulator using eye tracker. *Transportation Research Record, 2018*, 80–86.

* Funke, G., Matthews, G., Warm, J. S., & Emo, A. K. (2007). Vehicle automation: A remedy for driver stress? *Ergonomics, 50*(8), 1302–1323.

Gregersen, N. P., Nyberg, A., & Berg, H. Y. (2003). Accident involvement among learner drivers—An analysis of the consequences of supervised practice. *Accident Analysis & Prevention, 35*, 725–730.

* Gugerty, L., Rakauskas, M., & Brooks, J. (2004). Effects of remote and in-person verbal interactions on verbalization rates and attention to dynamic spatial scenes. *Accident Analysis & Prevention, 36*, 1029–1043.

* Harrison, E. L. R., & Fillmore, M. T. (2005). Are bad drivers more impaired by alcohol? Sober driving precision predicts impairment from alcohol in a simulated driving task. *Accident Analysis & Prevention, 37*, 882–889.

* Hennessy, D. A., & Jakubowski, R. (2007). The impact of visual perspective and anger on the actor-observer bias among automobile drivers. *Traffic Injury Prevention, 8*, 115–122.

* Horswill, M. S., & McKenna, F. P. (1999a). The development, validation, and application of a video-based technique for measuring an everyday risk-taking behavior: Drivers' speed choice. *Journal of Applied Psychology, 84*, 977–985.

* Horswill, M. S., & McKenna, F. P. (1999b). The effect of interference on dynamic risk-taking judgments. *British Journal of Psychology, 90*, 189–199.

* Horswill, M. S., & McKenna, F. P. (1999c). The effect of perceived control on risk-taking. *Journal of Applied Social Psychology, 29*, 377–391.

* Howat, P., Robinson, S., Binns, C., Palmer, S., & Landauer, A. (1991). Educational biofeedback driving simulator as a drink-driving prevention strategy. *Journal of Alcohol & Drug Education, 37*, 7–14.

* Hunton, J., & Rose, J. M. (2005). Cellular telephones and driving performance: The effects of attentional demands on motor vehicle crash risk. *Risk Analysis, 25*, 855–866.

Insurance Institute for Highway Safety. (2008a). *Fatality Facts 2007: Older people.* Arlington, VA: Author.

Insurance Institute for Highway Safety. (2008b). *US licensing systems for young drivers.* Arlington, VA: Author.

* Ivancic, K., & Hesketh, B. (2000). Learning from errors in a driving simulation: Effects on driving skill and self-confidence. *Ergonomics, 43*, 1966–1984.

Jadad, A. R., & Enkin, M. W. (2007). *Randomized control trials: Questions, answers, and musings* (2nd ed.). Malden, MA: Blackwell Publishing.

* Kass, S. J., Cole, K. S., & Stanny, C. J. (2007). Effects of distraction and experience on situation awareness and simulated driving. *Transportation Research Part F: Traffic Psychology and Behavior, 10*, 321–329.

Kernan, W. N., Viscoli, C. M., Makuch, R. W., Brass, L. M., & Horwitz, R. I. (1999). Stratified randomization for clinical trials. *Journal of Clinical Epidemiology, 52,* 19–26.

* Kubose, T. T., Bock, K., Dell, G. S., Garnsey, S. M., Kramer, A. F., & Mayhugh, J. (2006). The effects of speech production and speech comprehension on simulated driving performance. *Applied Cognitive Psychology, 20,* 43–63.

* Lees, M. N., & Lee, J. D. (2007). The influence of distraction and driving context on driver response to imperfect collision warning systems. *Ergonomics, 50,* 1264–1286.

* Matthews, G., Sparkes, T. J., & Bygrave, H. M. (1996). Attentional overload, stress, and simulated driving performance. *Human Performance, 9*(1), 77–101.

Mayhew, D. R., & Simpson, H. M. (2002). The safety value of driver education and training. *Injury Prevention, 8,* ii3–ii7.

Mayhew, D. R., Simpson, H. M., & Pak, A. (2003). Changes in collision rates among novice drivers during the first months of driving. *Accident Analysis & Prevention, 35,* 683–691.

McCartt, A. T., Shabanova, V. I., & Leaf, W. A. (2003). Driving experience, crashes, and traffic citations of teenage beginning drivers. *Accident Analysis & Prevention, 35,* 311–320.

McKenna, F. P., & Crick, J. (1991). Experience and expertise in hazard perception. In G. B. Grayson & J. F. Lester (Eds.), *Behavioral Research in Road Safety* (pp. 39–45). Crowthorne, United Kingdom: Transportation Research Laboratory Limited.

* McKenna, F. P., Horswill, M. S., & Alexander, J. L. (2006). Does anticipation training affect drivers' risk-taking? *Journal of Experimental Psychology, 12,* 1–10.

McKnight, A. J., & McKnight, S. A. (2003). Young novice drivers: Careless or clueless? *Accident Analysis & Prevention, 35,* 921–925.

McKnight, A. J., & Peck, R. C. (2002). Graduated driver licensing: What works? *Injury Prevention, 8,* ii32–ii36.

* McMillen, D. L., & Wells-Parker, E. (1987). The effect of alcohol consumption on risk-taking while driving. *Addictive Behaviors, 12,* 241–247.

Mercer, S. L., DeVinney, B. J., Fine, L. J., Green, L. W., & Dougherty, D. (2007). Study designs for effectiveness and translation research: Identifying trade-offs. *American Journal of Preventive Medicine, 33,* 139–154.

Moher, D., Schulz, K. F., & Altman, D. G. (2001). The CONSORT Statement: Revised recommendations for improving the quality of reports of parallel-group randomized trials. *Lancet, 357,* 1191–1194.

National Highway Traffic Safety Administration. (2007). *Traffic safety facts: Speeding.* Washington, DC: Author.

National Highway Traffic Safety Administration. (2008). *Traffic safety facts 2007 Data Alcohol-Impaired Driving.* Washington, DC: Author.

Nelson, A., Dumville, J., & Torgerson, D. (2009). The research process in nursing. In A. L. K. Gerrish (Ed.), *The research process in nursing* (5th ed., pp. 239–259). Oxford, UK: Blackwell Publishing Ltd.

* Oei, T. P. S., & Kerschbaumer, D. M. (1990). Peer attitudes, sex, and the effects of alcohol on simulated driving performance. *American Journal of Drug and Alcohol Abuse, 16,* 135–146.

Ouimet, M. C., Simons-Morton, B. G., Zador, P. L., Lerner, N. D., Freedman, M., Duncan, G. D., & Wang, J. (2010). Using the U.S. National Household Travel Survey to estimate the impact of passenger characteristics on young drivers' relative risk of fatal crash involvement. *Accident Analysis & Prevention, 42*(2), 689–694.

* Pollatsek, A., Narayanaan, V., Pradhan, A., & Fisher, D. L. (2006). Using eye movements to evaluate a PC-based risk awareness and perception training program on a driving simulator. *Human Factors, 48,* 447–464.

* Pradhan, A. K., Fisher, D. L., & Pollatsek, A. (2006). Risk perception training for novice drivers: Evaluating duration of effects on training on a driving simulator. *Transportation Research Record, 1969,* 58–64.

Pradhan, A. K., Hammel, K. R., DeRamus, R., Pollatsek, A., Noyce, D. A., & Fisher, D. L. (2005). Using eye movements to evaluate effects of driver age on risk perception in a driving simulator. *Human Factors, 47,* 840–852.

Quimby, A. R., & Watts, G. R. (1981). *Human factors and driving performance* (Rep. LR 1004). Crowthorne, United Kingdom: Transportation and Road Research Laboratory.

* Regan, M. A., Deery, H. A., & Triggs, T. J. (1998a). A technique for enhancing risk perception in novice car drivers. *Proceedings of the road safety research, policing and education conference* (pp. 51–56). Wellington, New Zealand.

* Regan, M. A., Deery, H. A., & Triggs, T. J. (1998b). Training for attentional control in novice car drivers: A simulator study. *Proceedings of the 42nd annual meeting of the human factors and ergonomics society* (pp. 1452–1456). Chicago, IL.

Riemersma, J. B. J., van der Horst, A. R. A., Hoekstra, W., Alink, G. M. M., & Otten, N. (1990). The validity of a driving simulator in evaluating speed-reducing measures. *Traffic Engineering & Control, 31,* 416–420.

* Rupp, T., Arnedt, J. T., Acebo, C., & Carskadon, M. A. (2004). Performance on a dual driving simulation and substraction task following sleep restriction. *Perceptual and Motor Skills, 99,* 739–753.

Sagberg, F. (1998). *Month-by-month changes in accident risk among novice drivers.* Proceedings of the 24th International Conference of Applied Psychology. San Francisco, CA.

Schulz, K. F., Chalmers, I. H. R. J., & Altman, D. G. (1995). Empirical evidence of bias. Dimensions of methodological quality associated with estimates of treatment effects in controlled trials. *The Journal of the American Medical Association, 273,* 408–412.

Simons-Morton, B., Lerner, N., & Singer, J. (2005). The observed effects of teenage passengers on the risky driving behavior of teenage drivers. *Accident Analysis & Prevention, 37,* 973–982.

Simons-Morton, B. G., Ouimet, M. C., & Catalano, R. (2008). Parenting and the young driver problem. *American Journal of Preventive Medicine, 35,* S294–S303.

* Taubman Ben-Ari, O., Florian, V., & Mikulincer, M. (1999). The impact of mortality salience on reckless driving: A test of terror management mechanisms. *Journal of Personality and Social Psychology, 76,* 35–45.

* Taubman Ben-Ari, O., Florian, V., & Mikulincer, M. (2000). Does a threat appeal moderate reckless driving? A terror management theory perspective. *Accident Analysis & Prevention, 32,* 1–10.

* Ward, N. J., & Beusmans, J. (1998). Simulation of accident risk displays in motorway driving with traffic. *Ergonomics, 41,* 1478–1499.

Williams, A. F. (2003). Teenage drivers: Patterns of risk. *Journal of Safety Research, 34,* 5–15.

Williams, A. F., & Ferguson, S. A. (2002). Rationale for graduated licensing and the risks it should address. *Injury Prevention, 8,* ii9–ii14.

Williams, A. F., & Preusser, D. F. (1997). Night driving restrictions for youthful drivers: A literature review and commentary. *Journal of Public Health Policy, 18,* 334–345.

25

The Older Driver (Training and Assessment: Knowledge, Skills and Attitudes)

Karlene K. Ball
University of Alabama
at Birmingham

Michelle L. Ackerman
University of Alabama
at Birmingham

Abstract

The Problem. Normal age-related changes, as well as those related to medical conditions, may put older adults at risk while driving. Basic to the issue of driving competence, especially for older drivers, is the need to balance mobility and autonomy with personal and public safety. ***Role of Driving Simulators.*** Driving assessment tools and training methods are aimed at not only identifying at-risk older drivers, but developing effective methods to enhance driving competencies. Methods used to assess driving ability are discussed, including on-road driving tests, simulated driving tests, and batteries which evaluate driving-related skills. Promising methods to extend the safe mobility of older drivers are reviewed, including driving simulation training and training programs designed to sustain the skills needed for safe driving (e.g., cognitive training). ***Key Results of Driving Simulator Studies.*** Studies on driving simulator training suggest that such training may improve not only overall fitness to drive, but also specific skills such as visual scanning ability and right foot coordination. ***Limitations and Future Research.*** Conclusions are made regarding the need for standardization of assessment methods, and further investigation of the validity and reliability of on-road evaluations targeted to older drivers. Recommendations for future research are made.

25.1 Introduction

The safe mobility of older persons is a growing public health concern in the United States, as well as around the world. In the U.S., it is projected that by 2030, the population of adults over 65 years of age will grow from the current 12.6% to 20% (U.S. Census Bureau, 2008). As a result, the number of drivers over the age of 65 will also greatly increase. Although most older adults are safe drivers, this population is at greater risk for decline in functions that support safe driving (e.g., visual, cognitive, and motor functions), which are associated with aging, chronic disease, and medication use (Messinger-Raport & Rader, 2000). Therefore, this

projected increase in the number of older road users may lead to a corresponding rise in motor vehicle crashes (Lyman, Ferguson, Braver, & Williams, 2002). Research suggests that older drivers are also at increased risk of being killed or injured when in a motor vehicle crash (Dellinger, Kresnow, White, & Sehgal, 2004). These trends illustrate the need for research targeted toward maintaining the safe mobility of older drivers, especially in light of the important role that automobiles play in American society today (O'Neill, 2000). A majority of older Americans rely on their personal vehicles to perform necessary activities such as accessing healthcare (Collia, Sharp, & Giesbrecht, 2003; O'Neill, 2000), and the loss of one's private automobile may severely reduce personal mobility and independence (Freeman, Gange, Munoz, & West, 2006; Marottoli et al., 2000).

Basic to the issue of older driver safety is the need for older adults to remain autonomous for as long as possible, and independence is frequently associated with the ability to drive. Older driver assessment methods are aimed at not only identifying at-risk older drivers, but have also resulted in a concerted effort to develop effective methods to sustain the driving abilities of older drivers.

Recent thinking makes a distinction between driver "assessment" and driver "screening." According to consensus-based definitions developed at the North American Driver License Policies Workshop (Molnar & Eby, 2008), screening is not used to make licensing decisions, but to gauge driving-related skills and determine if further evaluation is warranted. For example, declining functional ability, as measured through screening, may trigger further testing (assessment). Assessment is a more in-depth examination of driving-related functional impairments, and can be used to determine the extent to which driving ability is impaired. Assessment provides a basis for identifying options for licensing recommendations and determining the possibility of remediation. By these definitions, this chapter will focus primarily on assessment methods, as well as driver training and training programs designed to sustain the underlying skills needed for safe driving.

Driving ability is measured in many ways, and there is currently no uniformly accepted strategy or protocol for determining when and how to evaluate driving competence. The gold standard of driving assessment continues to be on-road driving evaluations, but recent developments in simulator assessment have shown promise for assessment, and some driving assessment batteries of functional abilities have been validated relative to driving safety as well as to mobility (Ball et al., 2006; Vance et al., 2006).

Advances in driver training and cognitive training indicate that although some at-risk older drivers must restrict or cease their driving due to serious or permanent declines in driving-related functional abilities, there are many alternatives for those who wish to maintain or improve these abilities. There is emerging evidence to suggest that simulation training and cognitive training are effective in both safety outcomes as well as mobility outcomes (sustained independence) (Ball, McGwin, & Edwards, 2009; Edwards et al., 2009; Edwards, Delahunt, & Mahncke, 2009; Ross et al., under revision.)

25.2 Driving Assessment

25.2.1 On-Road Assessment

An exhaustive review of driving assessment measures and methods is beyond the scope of this chapter (see Ball, Wadley, & Edwards, 2002; Owsley, 2004). Instead, this section is designed to provide an overview of promising or commonly used assessments, as well as to provide information regarding the benefits and limitations of each. On-road driving evaluations are generally considered to be the gold standard method for determining

driving fitness (Odenheimer et al., 1994; Di Stefano & Macdonald, 2003). The on-road assessment has also conventionally been used as the measurement standard when evaluating driving simulations and validating off-road assessments of driving ability. Road tests quite obviously have better face validity than other current methods of assessing driving competency, which is important to individuals whose license may be under review. On-road testing also provides the opportunity to examine driver competency as drivers perform actual driving activities, and includes aspects of driving (including physiological stimulation, traffic interaction, and tactical planning) that may not be easily replicable by other testing means (Reimer, D'Ambrosio, Coughlin, Kafrissen, & Biederman, 2006). However, on-road testing is costly and can be dangerous when the driver is not competent. Results of on-road testing may also differ widely depending on traffic conditions and inconsistent driving demands from one drive to another. For example, a driver who is not competent may at times appear competent if the driving demands are minimal.

Despite, or perhaps as a result of, the face validity of on-road testing, few studies have been conducted to examine the reliability or validity of such assessments (Withaar, Brouwer, & Van Zomeren, 2000). Several general components have been recommended for valid and reliable road-testing. Researchers and clinicians generally agree that on-road assessment should be conducted in traffic, on a standardized route, in a vehicle with dual controls. A minimum of two assessors is needed to assume responsibility for either scoring or ensuring safety during the evaluation (Fox, Bowden, & Smith, 1998; Mazer, Gelinas, & Benoit, 2004; Withaar et al., 2000). A standardized route that includes specific road and traffic conditions is often used, but the elements of such routes are not always comparable to other such routes, and there is no real standardization for evaluator training across studies (Freund, Gravenstein, Ferris, & Shaheen, 2002). Criteria used to determine the outcome of the assessment is also extremely varied, ranging from a gestalt decision based on overall driving performance, to standardized observation and scoring procedures. Research has been done comparing scoring criteria to overall pass/fail status to determine which elements of scoring are most predictive of the overall assessment results, but even so, results are not consistent across studies.

There is evidence that some standardized scoring methods have good construct validity and inter-rater reliability (Kay, Bundy, Clemson, & Jolly, 2008). One study by Di Stefano and McDonald (2003) utilized data from an Australian licensing authority to examine driving errors made during on-road evaluations, and the relationship of types of error with test outcome. Participants in this study included 533 drivers aged 24 to 100 (mean 76.1 years), with no major physical or specific cognitive impairments, who were referred for driving evaluation because their driving competence was in question (63% referred by police). The study reported an overall test failure rate of 49%. The authors found that errors resulting in intervention by the test administrator were the strongest predictors of test failure. Other strongly-related errors included those involving intersection negotiations, maintenance of position and speed, and safety margin. Tester intervention was only performed if such intervention was necessary in order to maintain safety. Because of this criterion, and because tester intervention was the strongest indicator of test failure, the authors concluded that the on-road test was predictive of driver safety. The authors also concluded that some errors observed during road-testing (such as rolling past stop signs) result from habitual driving behaviors, and do not threaten safety. Consistent with other research, they recommend that flexibility be incorporated into scoring methods, placing more emphasis on errors that require tester intervention. However, there are some obvious flaws to this type of standardized scoring. The dual role of the test administrator to intervene for safety reasons, combined with their role to determine pass/fail status is not ideal. Scoring based on multiple raters or scoring where road-test errors could be validated against actual driving records could improve determination of whether errors observed threaten safety.

Currently, few studies have been done to determine whether or not on-road assessments are predictive of actual real-world driving ability, or road safety (McCarthy, 2005). A study examining national motor vehicle fatality information from the Fatality Analysis Reporting System (FARS; 1990–2000) found that the only state licensure policy associated with a lower driver fatality rate among drivers aged 85 years and older was in-person license renewal (Grabowski, Campbell, & Morrisey, 2004). Road-testing was not independently associated with reduced driver fatality. This is in contrast to earlier research with FARS data, which suggested that tests of visual acuity were associated with a lower fatal crash risk for drivers over 70 years old (Levy, Wernick, & Howard, 1995; Ship, 1998). This research examined only fatal crashes, however, and it is apparent that more research is needed to examine the relationship between pass/fail status on road-test measures and subsequent crash incidents as well as injury.

25.2.2 Simulator Assessment

Recently, substantial effort has been put forth to develop alternatives to on-road driving evaluations. One important goal of off-road driving assessment is to identify older drivers whose suspected or known impairments may put them at increased risk of crash or driving incidents without the expense or effort of administering an on-road evaluation. There are a variety of simulator types (see Greenberg & Blommer, this book, chap. 7), which can incorporate many of the benefits of on-road evaluations, and address some of the concerns. For example, the use of simulated driving assessments is more economical than on-road testing, while having better face validity than other off-road assessments of driving-related skills (Desmond & Mathews, 1997). Driving simulators also have the ability to incorporate a wide array of variables potentially impacting the driving performance of older adults, including aspects of road planning, and in-car modifications (see Trick & Caird, this book, chap. 26).

A major benefit to using simulated driving tasks is the ability to create driving situations that would be unethical or dangerous to experience in the real environment (Rizzo, McGehee, Dawson,

& Anderson, 2003). Important driving skills can be observed by programming deliberately challenging situations into the simulated environment which would be too dangerous for testing in an actual road environment. Simulators provide an avenue for evaluating driving skill without the possible risks associated with making mistakes in on-road testing. During an on-road evaluation, assessors intervene when safety seems to be at risk, while during simulator evaluations the driver can be allowed to proceed, giving them the opportunity to correct their actions.

Driving simulation testing can also address many of the reliability and validity concerns that challenge on-road driving routes. A simulated driving route allows for multiple observations of behavior under exactly replicated conditions, and negates concerns regarding tester training and ensuring test-retest reliability. However, many older drivers are unfamiliar with computer technology, and the use of driving simulations may be intimidating enough to impede their performance. Another criticism of simulated driving tasks is that they lack some of the complexity that exists in a real-world driving environment, which requires skills such as interacting with traffic and unexpected road hazards. Some researchers have noted that actions in a simulated driving environment have no real consequences (McCarthy, 2005; see also Ranney, this book, chap. 9). Another major concern that arises with simulator use is simulator adaptation syndrome (also known as "simulator sickness"; see Stoner, Fisher, & Mollenhauer, this book, chap. 14). This phenomenon results from the discrepancy between complex visual motion and static vestibular feedback in a simulated driving environment. The incidence of simulator adaptation syndrome may be decreased by the use of more advanced moving-base simulators versus fixed-base simulators, and there are measures available to screen for individuals that may be particularly vulnerable. Even so, this syndrome can be a deterrent to the use of simulator assessment.

As with on-road testing, there is no standardized simulation available for comparison across studies (Messinger-Rapport, 2003). Furthermore, it is difficult to compare across studies of driving simulation (as with on-road testing) because there is no standardized set of individual difference measures. Comparability would be significantly enhanced across studies if all investigators were to report key characteristics of the sample being investigated. Common measures that would assist in comparing participant samples include: Age (mean and range), gender, race, mental status (preferably MMSE for consistency), cognitive function (Useful Field of View clock test [UFOV], Trails, Motor Free Visual Perception Test [MVPT]), visual function (visual acuity and contrast sensitivity), and physical function (walk time). Information should also be included as to where the sample was obtained. If this information was uniformly reported across studies, researchers would be able to better explain conflicting findings with respect to standardized simulator scenarios.

As with on-road driving assessment, limited research has been done to examine simulator performance as a predictor of actual driving ability or driving safety. However, some evidence suggests that simulated driving performance is associated with driving-related cognitive functioning (Lee, Drake, &

Cameron, 2002), and on-road driving assessment performance. A pilot study by Freund and colleagues (Freund et al., 2002) examined the relationship between simulated driving performance and on-road driving performance. The authors found that among nine older adults (four of whom were cognitively impaired), simulated driving performance was strongly associated with on-road performance in both cognitively impaired and unimpaired adults. Similarly, Lee, Cameron and Lee (2003) administered a driving simulation and on-road driving test to 129 licensed drivers between the ages of 60 and 88 years old. Principal component analysis was used to create a scoring index for simulation performance and on-road performance. The authors found that more than 65% of the variability in the Road Assessment Index could be explained by the Simulated Driving Index, after accounting for gender and age. Another study by Lee and colleagues (Lee, Lee, Cameron, & Li-Tsang, 2003) using the same sample of 129 licensed drivers administered a 45-minute driving simulation that included 10 reliable assessment criteria. Better performance on driving tasks that involved working memory, decision and judgment, and speed compliance were found to be associated with lower self-report of crash occurrence for the previous year.

25.2.3 Assessment of Driving-Related Skills

The International Older Driver Consensus Conference on Assessment, Remediation and Counseling for Transportation Alternatives recommended in 2003 (Stephens et al., 2005) that driver assessment batteries should include cognitive, sensory and motor domains. Although methodological challenges to conducting such research exist (see Trick & Caird, this book, chap. 26), substantial efforts have been made to translate knowledge about the cognitive, visual, and motor skills associated with driving competence (i.e. Ball et al., 2006; Marottoli, Cooney, Wagner, Doucetter, & Tinetti, 1994; Mazer, Korner-Bitensky, & Sofer, 1998) into driving assessment measures. Although some have argued that the use of any single off-road screening test of driving fitness to determine safe driving on a pass/fail basis is not justified (Langford, 2008), this research has resulted in the development and evaluation of several batteries designed to identify at-risk drivers. A study by Wood and colleagues (Wood, Anstey, Kerr, Lacherez, & Lord, 2008) found that a battery measuring vision, cognition, and motor skills, in addition to self-reported driving exposure, was able to accurately predict on-road driving safety among a sample of older drivers aged from 70 to 88 years old. Performance on three measures (motion sensitivity, color choice reaction time, and postural sway) combined with self-reported driving exposure, was predictive of on-road driving performance with 91% sensitivity (correctly identifying drivers with unsafe performance on the driving task) and 70% specificity (correctly identifying drivers with safe performance on the driving task).

Another assessment battery, the Driving Health Inventory (DHI; Edwards et al., 2008) was developed based on a body of research that aimed to translate knowledge about functional abilities required for safe driving into appropriate assessments. Specifically, a study by Ball and colleagues (Ball et al., 2006) was conducted to evaluate the relationship between performance-based measures and subsequent at-fault crash involvement among drivers aged 55 and greater. Participants were older adults presenting to renew their driver's license at three Motor Vehicle Administration sites in Maryland who agreed to participate in an additional series of assessments after completing their license renewal. The battery of tests administered included physical measures (such as rapid walk, foot tap, arm reach, and head/neck rotation), cognitive measures (cued and delayed recall, symbol scan, Motor Free Visual Perception Test [MVPT], Trails A and B), Useful Field of View [UFOV]® sub-test two, and a self-reported mobility questionnaire. Results from this study indicated that participants with poor performance on the MVPT Visual Closure sub-test were more than twice as likely to crash over the next three years as those with fewer errors, and those with poor performance on UFOV® sub-test two were also twice as likely to be involved in an at-fault crash compared to participants with faster speed of processing.

This battery, refined through repeated evaluation in the Maryland Motor Vehicle Administration field sites, now includes measures of visual acuity, motor skills (particularly leg strength), and cognitive skills (visual information processing speed, visualization of missing information, divided visual search, memory, executive function). One recent study examined 258 licensed drivers ranging in age from 18 to 87 years who were administered the DHI battery (Edwards et al., 2008). Results suggested that older drivers performed significantly worse than younger drivers, and older drivers with crash involvement in the previous two years performed worse than older drivers without recent crashes.

Other batteries combine measures of driving-related functional abilities with an on-road evaluation. The American Medical Association developed The Assessment of Driving-Related Skills (ADReS), which was published as part of the *Physicians' Guide to Assessing and Counseling Older Drivers* (Wang, Kosinski, Schwartzberg, & Shanklin, 2003), to assist physicians in identifying patients who may be at risk for unsafe driving. This battery includes measures of vision (visual fields, visual acuity), motor ability (leg strength, range of motion testing, manual muscle test), and cognition (executive function), as well as an on-road driving evaluation. McCarthy and Mann (2006) evaluated the ADReS among 50 licensed, older drivers who completed an on-road driving evaluation, a clinical examination during which the ADReS was administered, and a telephone interview. All participants that received a failing score on the road-test were identified by the ADReS as needing an intervention; however, the authors reported that the assessment had poor specificity. Seventy percent of the sample was recommended for intervention; however, 35% of those recommended for intervention received a passing score on the road-test. This battery has not been validated relative to crash records.

It is apparent that many important steps have been taken towards identification of at-risk drivers; however, more research is needed to improve the specificity of such instruments.

In addition to identifying drivers who are no longer safe to drive, it is vital to determine if training or rehabilitation can be utilized to regain or sustain driving competency. Another key step in maintaining safe mobility in older adults is to facilitate the continued driving of those older adults who are not at risk. Training and rehabilitation methods may also prove to be important tools for these older drivers if they can prevent or delay the loss of important functional abilities. Extension of driving can be achieved in a variety of ways. Improving the driver's abilities (including health and functioning) may increase safety. When the capacity and skills of the driver cannot be enhanced, the driving environment can be adapted to be less challenging. This second option includes modifications to the vehicle and the driving environment, including driver self-regulation to avoid challenging driving situations. The next section of this chapter will focus its discussion on improvement of the driver's abilities.

25.3 Driver Training and Remediation

Advances in driver training and protocols for improving the functional abilities needed to drive safely have led to the conclusion that some "at-risk" older drivers have options other than to cease driving permanently. Depending on the severity of the health, functional, or cognitive deficit, which has resulted in "at-risk" status, interventions may be offered to improve driving ability among some drivers deemed unsafe, allowing them to continue or resume safe driving. Researchers have suggested that interventions to improve older adults' driving competence should focus primarily on the maintenance or improvement of the complex skills necessary for safe driving performance (Parker, McDonald, Rabbitt, & Sutcliffe, 2000). Many varieties of intervention have been investigated, and one promising area includes training on driving tasks specifically. Driver training that takes place in an on-road setting is largely impractical, due to the at-risk nature of drivers in need of such training (as opposed to a novice driver, for example). Despite adaptations installed in training cars to ensure safety, individuals who have been identified as unsafe drivers may still be at risk of a crash or other adverse outcome (Akinwuntan et al., 2005). Off-road driver training methods, including those that target improvement of functional abilities such as vision or cognition, offer a higher level of safety than on-road training, and there is emerging evidence that such methods are effective in improving the driving skills and safety of older drivers.

Educational driver training programs have been very popular in the past years, and remain so despite a lack of compelling evidence to suggest that they improve driving safety. Evidence suggests that educational interventions improve driving awareness and driving behavior, but do not necessarily reduce crashes (Kua, Korner-Bitensky, Desrosiers, Man-Son-Hing, & Marshall, 2007). Owsley and colleagues (Owsley, Stalvey, & Phillips, 2003) found that visually impaired drivers who received an educational intervention regarding their visual abilities also reported more frequent self-regulatory practices. However, participants did not exhibit reduced crash risk in the following two years (Owsley et al.,

2003). Some support has been found for the use of educational interventions in combination with on-road training (Bédard et al., 2008). Participants in a study by Marottoli and colleagues (Marottoli et al., 2007) included 126 drivers aged 70 or older who were randomized either to an intervention or control group. The educational intervention included eight hours of classroom training and two hours of on-road instruction that addressed common driving errors in general, and specific errors made by the driver during baseline on-road testing. Control group participants received information about in-home and vehicle safety. Post-training evaluation results indicated that after eight weeks, the intervention group scored significantly higher on a knowledge test, and performed better on the road-test. Conclusions about the efficacy of the educational component of such combined interventions are difficult to make, however, without longitudinal follow-up and collection of other safety outcomes. In general, findings indicate that educational interventions are not generally efficacious in improving driving safety, and the addition of an on-road training component confounds the ability to attribute improvements in driving performance to the education component, the on-road component, or the combination.

25.3.1 Driving Simulator Training

Studies suggest that simulator training may be useful for improving or maintaining a variety of driving-related skills. In addition to improving performance on simulated driving tasks, simulators may be used to retrain specific skills such as right foot coordination, or visual scanning ability (Stephens et al., 2005). Researchers have also suggested that training conducted with a driving simulator may increase older drivers' self-awareness of their capacities and limitations (Lee et al., 2003), thus allowing them to make appropriate decisions for self-regulation.

There is some evidence that driving simulators have been effective in improving fitness to drive in populations such as stroke patients. Akinwuntan and colleagues (Akinwuntan et al., 2005) examined 83 stroke patients to determine the efficacy of simulator-based training in improving driving ability. Participants were randomized into a five-week simulator training (experimental group) or driving-related cognitive skills training (control group) program. The experimental group received two to three hours of familiarization with the simulator, followed by a pre-training assessment. Experimental subjects then received 15 hours of training (three hours per week over five weeks) on common traffic demands such as speed control and road sign recognition as well as responding to complex traffic-related situations. Control subjects also received a schedule of 15 hours of training over five weeks, performing driving-related cognitive tasks such as flash-card recognition of roads and traffic signs, route-finding on a map, and pattern and memory training. The authors reported that post-training, both the experimental and control groups showed significant improvement on a visual task as well as a variety of neuropsychological tasks. Both groups also showed significant improvement from pre- to post-training for performance during the study-administered on-road assessment; however, significantly more experimental subjects (73%) passed the official pre-driving assessment (administered six to nine months post-stroke), and were determined fit to drive than subjects in the control group (42%).

Simulator training may also improve driver fitness by programming the simulator to elicit errors from participants during the simulation training, then providing feedback regarding such errors. One such study utilized a simulated driving task that was designed to elicit errors during five key events (Ivancic & Hesketh, 2000). Baseline performance on the simulated task was observed, then participants received simulator training in one of two groups: (1) The error training group received feedback where errors on any of the five key events resulted in either a collision or a police ticket, indicated by a loud siren; (2) the errorless training group received no feedback for any errors made (no collisions or citations occurred). A simulated post-test was then given, which incorporated the same elements as the initial key events. The authors reported that participants in the error training group made significantly fewer errors on the post-test than the errorless training group. These findings are consistent with those typically found in less applied training studies indicating that feedback is a key element for improving performance on any task.

25.3.2 Improvement of Functional Abilities

It is typically assumed that declining functions (e.g., poorer vision, motor, or cognitive function) are the reason for any driving difficulties experienced by older adults. However, there are decades of research illustrating how difficult it is to find assessments that differentiate an "at-risk" older driver from the majority who are not at risk. For those few abilities where data indicate improvement of function might result in improved driving competence, interventions have been evaluated.

Declining cognitive and physical abilities have been found to be associated with poor driving performance, or increased risk of crashes. Training methods targeting these abilities have been developed and investigated to examine their efficacy in improving both driving-related skills and overall driving performance (usually assessed through on-road testing). Past research by Marottoli and colleagues (Marottoli et al., 1994; 1998) reported that several specific physical problems (such as axial and extremity flexibility, coordination, and speed of movement) were predictive of driving safety. A recent study conducted by Marottoli and colleagues (2007) examined the impact of a physical conditioning program in increasing safe driving among 178 drivers aged 70 or older who were determined to be at-risk drivers due to physical impairments (but not visual or cognitive impairment). Participants were randomized to either a physical conditioning (experimental), or in-home education (control) group. The physical conditioning program incorporated exercises which became progressively more difficult to improve axial/extremity conditioning; upper extremity coordination, dexterity and hand strength; and foot abnormalities. Performances on pre- and post-training on-road assessments were compared, and the results indicated that the intervention group maintained driving

safety across three months, while driving performance of the control group declined.

A measure of visual processing speed, the useful field of view or more specifically the UFOV® test, has been shown to be associated with crash risk in older adults (Ball et al., 2006; Clay et al., 2005; Goode et al., 1998; Owsley et al., 1998; Rizzo, Reinach, McGehee, & Dawson, 1997), including the probability of passing an on-road driving evaluation (Myers, Ball, Kalina, Roth, & Goode, 2000). Older adults with poor UFOV® scores have been found to be twice as likely to incur an at-fault crash in large population-based studies of licensed drivers (Ball et. al, 2006; Rubin, Bandeen-Roche, Keyl, Freeman & West, 2007). Studies examining several populations, including stroke patients, have shown that speed of processing (SOP) training is effective in increasing driving ability. Mazer and colleagues (2003) conducted a randomized clinical trial that included 97 participants who were referred for a driving evaluation following stroke. Participants received either 20 hours of SOP training, or traditional computerized visuo-perception retraining. The authors found that participants with right-sided lesions who received UFOV® training demonstrated a doubled on-road evaluation pass-rate.

One study by Roenker and colleagues (Roenker, Cissell, Ball, Wadley, & Edwards, 2003) compared driving ability among 104 older drivers who received simulator training, speed-of-processing training, or served as a low-risk reference group. The simulator training group served as an at-risk control group, while the low-risk reference group served as an age-matched no-intervention control group. The authors found that both the speed of processing and simulator training groups improved their driving skills; however, the nature of the improvement varied. Simulator training improved on two specific driving maneuvering skills, although this improvement largely disappeared by 18-month follow-up. Speed-of-processing training improved a specific measure of useful field of view, which transferred to driving and resulted in fewer dangerous maneuvers during the on-road driving assessment; these improvements (in decreasing the degree of useful field of view reduction) largely persisted through the 18-month follow-up.

A recent study by Edwards and colleagues (Edwards et al., in press) sought to replicate and extend the findings of Roenker by examining data from the Staying Keen in Later Life (SKILL) study. Five hundred older drivers who participated in the baseline phase of the study, were current drivers at baseline, and completed a three year follow-up interview, were included. Participants in this study with poor baseline UFOV® test performance were randomized to receive speed-of-processing training or to receive training in how to navigate the internet, which served as a social contact control group. Participants who did not meet criteria to participate in a training group at screening were classified as a low risk reference group. Results of this study indicated that participants with poor speed of processing who did not receive SOP training (social contact control group) experienced steeper declines in driving mobility across the three-year period relative to the reference group, as indicated by increased driving difficulty and decreased driving exposure and driving space (the distance to which an individual drives within their environment; i.e., whether the individual drives to places beyond their neighborhood, more than 10 miles from their home, more than 25 miles from their home, outside their state, etc.). Those who did receive SOP training experienced some increased difficulty driving (while driving alone, making lane changes, and making left-hand turns across traffic) when compared to the low risk reference group, but did not differ across time from the reference group in driving exposure, driving space, or driving difficulty during high traffic, at night, in the rain, during rush hour, or when merging with traffic.

The Advanced Cognitive Training for Independent and Vital Elderly (ACTIVE) program was developed to evaluate the effects of three cognitive training interventions on improving daily functioning, including mobility and driving outcomes (Jobe et al., 2001). This study is a longitudinal multi-site single-blind randomized clinical trial which involved 2832 adults aged 65 years and older, possibly at risk for loss of independence from functional or cognitive decline. Eligibility criteria and further details of this study are reported elsewhere (Jobe et al., 2001). Participants were divided into four groups: speed of processing training, reasoning training, memory training, and a no-contact control group. Additional subsamples of compliant trained participants from the three training groups were randomly assigned to receive booster training of up to four one-hour sessions at annual visits one and three. Follow-up data was collected annually at one, two, three, and five years.

Ross and colleagues (Ross et al., under revision) used longitudinal modeling methods to examine the impact of SOP training on self-reported driving over a five-year period among ACTIVE participants. The authors found that participants with poor initial processing speed who received SOP training maintained their driving frequency, and those who further received booster training reported greater maintenance of driving frequency as well as greater driving exposure (specific situations encountered during typical driving). Edwards et al. (2009) investigated a combined sample from the SKILL and ACTIVE studies to investigate the effectiveness of cognitive training in delaying driving cessation. SOP-trained individuals from the ACTIVE study were only included if they met SKILL inclusion criteria for training (i.e., poor baseline UFOV® performance). Results indicated that older drivers who completed speed of processing training were 40% less likely to cease driving over the subsequent three years as compared to the controls.

25.4 Conclusion

As the population ages, the study of older drivers and an understanding of the impact, both to the individual and to society as a whole, of maintaining safe driving continues to grow in importance. This importance stems primarily from the links between driving, and outcomes such as loss of mobility, increased depression, and risk of nursing-home entry. Fortunately, training studies indicate that at least some driving-related abilities can be improved in older adults who are experiencing declining function, and that improved abilities are associated with improved driving performance in older adults. There is also some evidence to suggest that such improved performance endures for a period of years. This provides hope to those who are experiencing deficits that impact their driving safety. Although some older drivers must restrict their driving because of a progressive or permanent decline in function and associated driving-related skills, there are alternatives for those who are motivated and capable of maintaining or improving those skills.

With respect to driving assessments, future research should explore the validity of on-road driving assessments in predicting subsequent driving performance. Not only are on-road evaluations the favored method for determining driving safety in licensing settings, on-road assessments remain the primary criterion measure for developing off-road driving assessments as well as investigating the efficacy of interventions designed to improve driver competence. It is vital that research be done to investigate the validity of such on-road evaluations in predicting subsequent driving ability (i.e., sensitivity and specificity—correctly determining not only who is at-risk, but also who is still safe to drive).

More effort is needed to arrive at a standardized driving assessment battery. The diverse assessment methods currently in use constitute a major stumbling block when making comparisons across studies and samples, or in drawing conclusions about the efficacy of driving interventions. Driving simulator use as part of an assessment battery seems promising, although more research is needed to investigate its association with real-word driving performance and driver safety outcomes. The combination of driving-related skills' evaluation with simulated driving performance would provide an abundance of information about the driver's current abilities, as well as suggest possibilities for the most effective intervention method. Simulators are also useful to encourage driver self-awareness, and to encourage appropriate self-regulation. The use of simulated driving tasks is extremely useful for assessment before and after targeted interventions, as well as for training.

Further research in this area will also need to focus on making worthwhile interventions both affordable and accessible to older adults. Along these lines, an ongoing project is currently investigating the effectiveness of training software that can be self-administered in a variety of settings (in-home, libraries, senior centers, etc.). Although the software is available, further research is needed to determine the impact on future driving safety.

Key Points

- Although most older drivers are safe drivers, this population is at greater risk for decline in functions that support safe driving (e.g., visual, cognitive, and motor functioning), which are associated with aging, chronic disease, and medication use.
- More research is needed to determine the predictive ability of on-road or simulated driving assessments in relation to road safety measures.

- There is emerging evidence that simulation training and cognitive training are effective in sustaining the driving skills and safety of older drivers.
- Future research in this area will need to focus on making effective interventions both affordable and accessible to older adults.

Keywords: Cognitive Training, Driving Assessment, Older Drivers, Simulator Training

Acknowledgments

Preparation of this chapter was supported by the UAB Roybal Center for Translational Research on Aging and Mobility (Edward R. Roybal Center grant 5 P30 AG022838). Karlene Ball owns stock in the Visual Awareness Research Group, Inc. (formerly Visual Awareness, Inc.), and Posit Science, the companies that market the Useful Field of View Test and speed of processing training software. Dr. Ball continues to work on the design and evaluation of these assessment and training programs within several Roybal Center projects, and as a member of the Posit Science Scientific Advisory Board. The other author has no financial disclosure or conflict of interest.

Glossary

Cognitive training: A strategy that seeks to sustain, improve, or restore an individual's cognitive functioning. The overall purpose is often to decrease the everyday problems faced by individuals with cognitive difficulties, thereby improving the quality of their lives.

Driver screening: A more basic method of gauging driving-related skills to determine if further evaluation is warranted. Screening is not an appropriate basis for making licensing decisions.

Driving assessment: An in-depth examination of driving-related functional impairments that can be used to determine the extent to which driving ability is impaired. Assessment provides a basis for identifying options for licensing recommendations and determining the possibility of remediation.

Key Readings

Ball, K., Roenker, D. L., Wadley, V. G., Edwards, J. D., Roth, D. L., McGwin, G., . . . Dube, T. (2006). Can high risk older drivers be identified through performance-based measures in a department of motor vehicles setting? *Journal of the American Geriatrics Society, 54,* 77–84.

Kay, L., Bundy, A., Clemson, L., & Jolly, N. (2008). Validity and reliability of the on-road driving assessment with senior drivers. *Accident Analysis & Prevention, 40,* 751–759.

Kua, A., Korner-Bitensky, N., Desrosiers, J., Man-Son-Hing, M., & Marshall, S. (2007). Older driver retraining: A systematic review of evidence of effectiveness. *Journal of Safety Research, 38,* 81–90.

McCarthy, D. P. (2005). Approaches to improving elders' safe driving abilities. *Physical and Occupational Therapy in Geriatrics, 23*(2/3), 25–42.

References

Akinwuntan, A. E., De Weerdt, W., Feys, H., Pauwels, J., Baten, G., Arno, P., & Kiekens, C. (2005). Effect of simulator training on driving after stroke. *Neurology, 65,* 843–850.

Ball, K., McGwin, J., & Edwards, J. D. (2009). The effects of training on driving competence—crash risk. *Transportation Research Board Annual Meeting CD-ROM.* Washington, DC: Transportation Research Board of the National Academies.

Ball, K., Roenker, D. L., Wadley, V. G., Edwards, J. D., Roth, D. L., McGwin, G., . . . Dube, T. (2006). Can high risk older drivers be identified through performance-based measures in a department of motor vehicles setting? *Journal of the American Geriatrics Society, 54,* 77–84.

Ball, K., Wadley, V., & Edwards, J. (2002). Advances in technology used to assess and retrain older drivers. *Gerontechnology, 1*(4), 251–261.

Bédard, M., Porter, M., Marshall, S., Isherwood, I., Riendeau, J., Weaver, B., . . . Miller-Polgar, J. (2008). The combination of two training approaches to improve older adults' driving safety. *Traffic Injury Prevention, 9*(1), 70–76.

Clay, O. J., Wadley, V. G., Edwards, J. D., Roth, D. L., Roenker, D. L., & Ball, K. K. (2005). Cumulative meta-analysis of the relationship between useful field of view and driving performance in older adults: Current and future implications. *Optometry and Vision Science, 82*(8), 724–731.

Collia, D. V., Sharp, J., & Giesbrecht, L. (2003). The 2001 national household travel survey: A look into the travel patterns of older Americans. *Journal of Safety Research, 34,* 461–470.

Dellinger, A., Kresnow, M., White, D., & Sehgal, M. (2004). Risk to self versus risk to others: How do older drivers compare to others on the road? *American Journal of Preventative Medicine, 26*(3), 217–222.

Desmond, P. A., & Mathews, G. (1997). Implications of task-induced fatigue effects for in-vehicle countermeasures to driver fatigue. *Accident Analysis & Prevention, 29*(4), 515–523.

Di Stefano, M., & McDonald, W. (2003). Assessment of older drivers: Relationships among on-road errors, medical conditions and test outcome. *Journal of Safety Research, 34,* 415–429.

Edwards, J. D., Delahunt, P., & Mahncke, H. W. (2009). Speed of processing training delays driving cessation. *Journals of Gerontology: Series A, Biological Sciences and Medical Sciences, 64*(12), 1262–1267.

Edwards, J. D., Leonard, K. M., Lunsman, M., Dodson, J., Bradley, S., Myers, C. A., & Hubble, B. (2008). Acceptability and validity of older driver screening with the Driving Health® Inventory. *Accident Analysis & Prevention, 40,* 1157–1163.

Edwards, J. D., Myers, C., Ross, L. A., Roenker, D. L., Cissell, G. M., McLaughlin, A. M., & Ball, K. (2009). The longitudinal impact of cognitive speed of processing training on driving mobility. *The Gerontologist, 49*(4), 485–494.

Fox, G. K., Bowden, S. C., & Smith, D. S. (1998). On-road assessment of driving competence after brain impairment: Review of current practice and recommendations for a standardized examination. *Archives of Physical Medicine and Rehabilitation, 79*(10), 1288–1296.

Freeman, E. E., Gange, S. J., Munoz, B., & West, S. K. (2006). Driving status and risk of entry into long-term care in older adults. *American Journal of Public Health, 96*(7), 1245–1259.

Freund, B., Gravenstein, S., Ferris, R., & Shaheen, E. (2002). Evaluating driving performance of cognitively impaired and healthy older adults: A pilot study comparing on-road testing and driving simulation. Letter to the Editor. *Journal of the American Geriatrics Society, 50*, 1309–1315.

Goode, K. T., Ball, K. K., Sloane, M., Roenker, D. L., Roth, D. L., Myers, R. S., & Owsley, C. (1998). Useful field of view and other neurocognitive indicators of crash risk in older adults. *Journal of Clinical Psychology in Medical Settings, 5*, 425–440.

Grabowski, D. C., Campbell, C. M., & Morrisey, M. A. (2004). Elderly licensure laws and motor vehicle fatalities. *Journal of the American Medical Association (JAMA), 291*(23), 2840–2846.

Ivancic, K., & Hesketh, B. (2000). Learning from errors in a driving simulation: Effects on driving skill and self-confidence. *Ergonomics, 43*(12), 1966–1984.

Jobe, J. B., Smith, D. M., Ball, K., Tennstedt, S. L., Marsiske, M., Willis, S. L., . . . Kleinman, K. (2001). ACTIVE: A cognitive intervention trial to promote independence in older adults. *Controlled Clinical Trials, 22*(4), 453–479.

Kay, L., Bundy, A., Clemson, L., & Jolly, N. (2008). Validity and reliability of the on-road driving assessment with senior drivers. *Accident Analysis & Prevention, 40*, 751–759.

Kua, A., Korner-Bitensky, N., Desrosiers, J., Man-Son-Hing, M., & Marshall, S. (2007). Older driver retraining: A systematic review of evidence of effectiveness. *Journal of Safety Research, 38*, 81–90.

Langford, J. (2008). Usefulness of off-road screening tests to licensing authorities when assessing older driver fitness to drive. *Traffic Injury Prevention, 9*(4), 328–335.

Lee, H. C., Cameron, D., & Lee, A. H. (2003). Assessing the driving performance of older adult driver: On-road versus simulated driving. *Accident Analysis & Prevention, 35*, 797–803.

Lee, H. C., Drake, V., & Cameron, D. (2002). Identification of appropriate assessment criteria to measure older adults' driving performance in simulated driving. *Australian Occupational Therapy Journal, 49*, 138–145.

Lee, H. C., Lee, A. H., Cameron, D., & Li-Tsang, C. (2003). Using a driving simulator to identify older drivers at inflated risk for motor vehicle crashes. *Journal of Safety Research, 34*, 453–459.

Levy, D. T., Wernick, J. S., & Howard, K. A. (1995). Relationship between driver's license renewal policies and fatal crashes involving drivers 70 years or older. *JAMA, 274*, 1026–1030.

Lyman, S., Ferguson, S. A., Braver, E. R., & Williams, A. F. (2002). Older driver involvement in police reported crashes and fatal crashes: Trends and projections. *Injury Prevention, 8*, 116–120.

Marottoli, R. A., Allore, H., Araujo, K. L. B., Iannone, L. P., Acampora, D., Gottschalk, M., . . . Peduzzi, P. (2007). A randomized trial of a physical conditioning program to enhance the driving performance of older persons. *Journal of Geriatric Internal Medicine, 22*, 590–597.

Marottoli, R. A., Cooney, L. M., Wagner, D. R., Doucetter, J., & Tinetti, M. E. (1994). Predictors of automobile crashes and moving violations among elderly drivers. *Annals of Internal Medicine, 121*, 842–846.

Marottoli, R. A., Mendes de Leon, C. F., Glass, T. A., Williams, C. S., Cooney, L. M., & Berkman, L. F. (2000). Consequences of driving cessation: Decreased out-of-home activity levels. *Journals of Gerontology Series B: Psychological Sciences and Social Sciences, 55*, S334–340.

Marottoli, R. A., Richardson, E. D., Stowe, M., Miller, E. G., Brass, L. M., Cooney, L. M., Jr., & Tinetti, M. E. (1998). Development of a test battery to identify older drivers at risk for self-reported adverse driving events. *Journal of the American Geriatric Society, 46*, 562–568.

Marottoli, R. A., Van Ness, P. H., Araujo, K. L. B., Iannone, L. P., Acampora, D., Charpentier, P., & Peduzzi, P. (2007). A randomized trial of an education program to enhance older driver performance. *The Journals of Gerontology, 62A*(10), 1113–1119.

Mazer, B. L., Gelinas, I., & Benoit, D. (2004). Evaluating and retraining driver performance in clients with disabilities. *Critical Reviews in Physical and Rehabilitation Medicine, 16*(4), 291–326.

Mazer, B. L., Korner-Bitensky, N., & Sofer, S. (1998). Predicting ability to driver after stroke. *Archives of Physical Medicine and Rehabilitation, 79*(7), 743–750.

Mazer, B. L., Sofer, S., Korner-Bitensky, N., Gelinas, I, Hanley, J., & Wood-Dauphinee, S. (2003). Effectiveness of a visual attention retraining program on the driving performance of clients with stroke. *Archives of Physical Medicine and Rehabilitation, 84*(4), 541–550.

McCarthy, D. P. (2005). Approaches to improving elders' safe driving abilities. *Physical and Occupational Therapy in Geriatrics, 23*(2/3), 25–42.

McCarthy, D. P., & Mann, W. C. (2006). Sensitivity and specificity of the assessment of driving-related skills older driver screening tool. *Topics in Geriatric Rehabilitation, 22*(2), 139–152.

Messinger-Rapport, B. (2003). Assessment and counseling of older drivers. *Geriatrics, 58*(12), 16–24.

Messinger-Rapport, B. J., & Rader, E. (2000). High risk on the highway. *Geriatrics, 55*(10), 32–40.

Molnar, L. J., & Eby, D. W. (2008). *Consensus-based recommendations from the North American License Policies Workshop.* Washington, DC: American Automobile Association Foundation for Traffic Safety.

Myers, R. S., Ball, K. K., Kalina, T. D., Roth, D. L., & Good, K. T. (2000). Relationship of useful field of view and other screening tests to on-road driving performance. *Perceptual and Motor Skills, 91*, 279–290.

Odenheimer, G. L., Beaudet, M., Jette, A. M., Albert, M. S., Grande, L., & Minaker, A. L. (1994). Performance-based driver evaluation of the elderly driver: Safety, reliability, and validity. *Journal of Gerontology: Medical Sciences, 49*(4), M153–M159.

O'Neill, D. (2000). Safe mobility for older people. *Reviews in Clinical Gerontology, 10*, 181–191.

Owsley, C. (2004). Driver capabilities. *Proceedings of transportation in an aging society: A decade of experience* (pp. 44–55). Washington, DC: Transportation Research Board of the National Academies.

Owsley, C., Ball, K., McGwin G., Jr., Sloane, M. E., Roenker, D. L., White, M. F., & Overley, E. T. (1998). Visual processing impairment and risk of motor vehicle crash among older adults. *JAMA, 279*(14), 1083–1088.

Owsley, C., Stalvey, B. T., & Phillips, J. M. (2003). The efficacy of an educational intervention in promoting self-regulation among high-risk older drivers. *Accident Analysis & Prevention, 35*, 393–400.

Parker, D., McDonald, L., Rabbitt, P., & Sutcliffe, P. (2000). Elderly drivers and their accidents: The aging driver questionnaire. *Accident Analysis & Prevention, 32*, 751–759.

Reimer, B., D'Ambrosio, L. A., Coughlin, J. F., Kafrissen, M. E., & Biederman, J. (2006). Using self-reported data to assess the validity of driving simulation data. *Behavior Research Methods, 38*(2), 314–324.

Rizzo, M., McGehee, D. V., Dawson, J. D., & Anderson, S. N. (2003). Simulated car crashes at intersections in drivers with Alzheimer disease. *Alzheimer Disease and Associated Disorders, 15*(1), 10–20.

Rizzo, M., Reinach, S., McGehee, D., & Dawson, J. (1997). Simulated car crashes and crash predictors in drivers with Alzheimer's disease. *Archives of Neurology, 54*, 545–553.

Roenker, D. L., Cissell, G. M., Ball, K. K., Wadley, V. G., & Edwards, J. D. (2003). Speed of processing and driving simulator training result in improved driving performance. *Human Factors, 45*(2), 218–233.

Ross, L. A., Edwards, J. D., Ball, K., Roth, D. L., & Wadley, V. G., & Vance, D. (under revision). Can driving be maintained with cognitive training in older adults at-risk for mobility loss? *Psychology and Aging.*

Rubin, G. S., Bardeen-Roche, K., Keyl, P. M., Freeman, E. E., & West, S. K. (2007). A prospective population-based study of the role of visual impairment in motor vehicle crashes among older drivers: The SEE study. *Investigative Ophthalmology and Visual Science, 48*(4), 1483–1491.

Shipp, M. D. (1998). Potential human and economic cost-savings attributable to vision testing policies for driver license renewal. *Optometry and Vision Science, 75*, 103–118.

Stephens, B. W., McCarthy, D. P., Marsiske, M., Shechtman, O., Classen, S., Justiss, M., & Mann, W. C. (2005). International older driver consensus conference on assessment, remediation and counseling for transportation alternatives: Summary and recommendations. *Physical and Occupational Therapy in Geriatrics, 23*(2/3), 103–121.

U.S. Census Bureau. (2008). *2008 National population projections.* Retrieved November 4, 2008 from http://www.census.gov/population/www/projections/summarytables.html

Vance, D. E., Roenker, D. L., Cissell, G. M., Edwards, J. D., Wadley, V. G., & Ball, K. K. (2006). Predictors of driving exposure and avoidance in a field study of older drivers from the state of Maryland. *Accident Analysis & Prevention, 38*, 823–831.

Wang, C. C., Kosinski, C. J., Schwartzberg, J. G., & Shanklin, A. V. (2003). *Physician's guide to assessing and counseling older drivers.* (Rep. DOT HS 809 647) Washington, DC: US Department of Transportation, National Highway Traffic Safety Administration.

Withaar, F. R., Brouwer, W. H., & Van Zomeren, A. H. (2000). Fitness to drive in older drivers with cognitive impairment. *Journal of the International Neuropsychological Society, 6*, 480–490.

Wood, J. M., Anstey, K. J., Kerr, G. K., Lacherez, P. F., & Lord, S. (2008). A multidomain approach for predicting older driver safety under in-traffic road conditions. *Journal of the American Gerontological Society, 56*, 986–993.

26

Methodological Issues When Conducting Research on Older Drivers

Lana M. Trick
University of Guelph

Jeff K. Caird
University of Calgary

Abstract

The Problem. Older drivers are an important segment of the driving population and they are frequently included in studies that use driving simulation. There is a pressing need for research on this growing segment of the population to aid policymakers and traffic safety professionals. *Role of Driving Simulators.* Older drivers are disproportionately at risk for certain types of collisions. It is essential to understand the factors that endanger older drivers if their collision rate is to be reduced. Driving simulators enable researchers to reproduce the conditions associated with collisions in a safe environment. Increasingly, simulators have been used to help identify older drivers at risk or evaluate the effectiveness of rehabilitation programs for these drivers. *Key Results of Driving Simulator Studies.* There are a variety of methodological issues to consider when conducting research on older drivers. Foremost are those that originate from the heightened incidence of simulator sickness. Moreover, in any study with older drivers, it is necessary to contend with factors confounded with age that also affect performance. Three factors are discussed in depth: driving exposure, age-related health issues, and medications. *Scenarios and Dependent Variables.* A number of scenarios have been investigated with older drivers, including car-following, way-finding, left turns, late yellow lights, and highway-railway grade crossings. Performance measures fall into these categories: longitudinal control (e.g., speed, time headway), lateral control (e.g., SD of lane position), hazard response time variants (e.g., perception response time), and eye movement measures (e.g., total fixation duration, average fixation frequency, etc.). *Platform Specificity and Equipment Limitations.* Minimizing the number of turns and stops can reduce simulator sickness, but this restricts the scope of research questions that can be addressed. High dropout rates due to simulator sickness limit the extent that results can be generalized.

26.1 Introduction

With the demographic shift towards ever-increasing numbers of older drivers on the road, research on older drivers becomes more and more necessary. Older drivers have a disproportionate risk of collision per kilometer of exposure (McGwin & Brown, 1999), particularly drivers in excess of 75 years of age. Though the implications of this finding have been debated (e.g., Hakamies-Blomqvist, Raitenen, & O'Neil, 2003), it is clear that older drivers are more likely to die or sustain serious injury as a result of collision (Evans, 2004; Hauer, 1988). Older drivers are especially at risk for certain types of collision: multi-vehicle collisions that take place at intersections or when merging (McGwin & Brown, 1999; Preusser, Williams, Ferguson, Ulmer, & Weinstein, 1998). There are age-related declines in attention (Ball & Owsley, 1991; Brouwer, Waterink, Van Wolffelaar, & Rothengatter, 1991; Caird, Edwards, Creaser, & Horrey, 2005; Staplin & Lyles, 1991), perception, motor processes, and memory (Craik & Salthouse, 2000; Perfect & Maylor, 2000) that may explain these risks. The study of driving can contribute to the theory of lifespan development insofar as it provides a way of assessing the impact of these transitions in a real-life task (Groeger, 2000).

Driving simulators are often used in this research because they make it possible to control moment-to-moment factors that affect driving performance. More importantly, they permit study of the situations that result in collisions without putting lives at risk. However, there are methodological challenges to working with older drivers. The purpose of this chapter is to alert researchers new to the field to issues that we have encountered. By way of example, we hope that others will not repeat our mistakes or, at the very least, they will acknowledge the need to carefully interpret their results given these problems. In what follows, the first section lists the different categories of research that involve older drivers; the second focuses on considerations in choosing dependent measures to best assess performance in older drivers; and the third highlights methodological challenges.

26.2 Research on the Older Driver

An exhaustive review of the literature on older drivers is beyond the scope of this chapter (for reviews, see Dewar & Olson, 2007; Evans, 2004; Shinar, 2007). Instead, this section is designed to provide an overview of the types of research that involve older drivers tested in driving simulators. Although epidemiological studies of collisions involving older drivers are a useful starting point, current theories of lifespan development influence the research questions given emphasis and the experimental paradigms employed to answer those questions. Consequently we will begin with a brief description of the theoretical context in which research on older drivers takes place. Normal aging and age-related disorders compromise long-term memory (Craik & Salthouse, 2000), vision (e.g., Scialfa & Kline, 2007), hearing (e.g., Willott, Chisolm, & Lister, 2001), motor flexibility, speed, and strength (e.g., Spirduso, Francis, & MacRae, 2005) but theories that stress attention and/or executive working-memory are

dominant in the study of older drivers. Selective attention determines which perceptual information reaches awareness, which information is stored and retrieved from long-term memory, and which actions are chosen from the response repertoire (for a discussion, see Trick, Enns, Mills, & Varnik, 2004). Executive working memory determines how these various processes are timed, coordinated, and prioritized (Baddeley, 1993; Shallice & Burgess, 1993).

A variety of theories have been proposed to explain age differences in attention and working memory. For example, Salthouse (1996) suggests that the neural changes that occur with age result in general cognitive slowing and this slowing is the source of age-related declines in performance. Hasher, Stoltzfus, Zacks and Rypma (1991) propose that specific deficits in inhibitory processing make it more difficult for older adults to inhibit distracting stimuli and irrelevant responses. Craik and Bialystok (2006) contend that age-related deficits emerge because cognitive control declines as a result of age-related changes in the brain, particularly changes in the frontal lobes.

Nonetheless, it is resource theories that are the most influential in the driving literature (e.g., Resource theory: Kahneman, 1973; Multiple-resource theory: Wickens, 2002). According to these theories, it is not possible to fully perceive and respond to everything at once because cognitive resources are limited. These resources are necessary for controlled processing (Shiffrin & Schneider, 1977). Controlled processes are conscious, deliberate, slow, effortful, attention-demanding processes, and because it is not possible to attend to everything at once, performing two similar controlled processes at one time creates interference, which is to say that the speed and accuracy of one or both suffers. Moreover, controlled processing is necessary whenever the task is challenging, novel, or unexpected and when there is a need to modify the process on-line, in response to feedback from the environment. Driving, by its nature, requires carrying out a number of different tasks at once, including hazard detection, steering and speed control, way-finding, planning, and decision-making (Michon, 1985), and many of these require controlled processing. From the perspective of resource theory, older drivers experience deficits in performance because they have fewer resources for controlled processing, either simply as a result of aging, or because age-related sensory and motor deficits force older adults to use cognitive resources to carry out operations that are normally carried out automatically (without controlled processing), thus reducing the resources available for other types of deliberate, attention-demanding process (Trick et al., 2004; see also Sensory-Cognitive interaction theory, Baldwin, 2002). Regardless of the source of this limitation, if older adults have fewer resources for controlled processing this means they would be especially at risk in complex unfamiliar visual environments and when they have to perform several attention-demanding operations at once.

Research on older drivers can be divided into three broad categories. The first investigates differences in driving behavior between older and younger drivers. The second evaluates new technologies and driving environments with samples that include older drivers. The third focuses on identifying older drivers at risk of collision (see also Ball & Ackerman, this book, chap. 25). In the following sections, we briefly describe these three types of research.

26.2.1 Measuring Age Differences in Driving Performance

The most straightforward type of study simply involves comparing performance in older and younger drivers. There are a number of studies that investigate how performance of older drivers varies as a function of cognitive load. Cognitive load is typically increased by having participants drive while performing some type of secondary task, such as having participants monitor displays for a specific letter or number, answer questions of various types, and so forth. Most of these studies manipulate the complexity of the secondary task to assess the impact on driving performance but some use the converse approach, manipulating the complexity of the drive and assessing the effects on secondary task performance (Makishita & Matsunaga, 2008; Shinar, Tractinsky, & Compton, 2005). Overall, these studies suggest that older drivers have more problems dual-tasking than younger drivers, particularly when the secondary task involves controlled processing, which suggests that they have fewer resources for controlled processing (see Riby, Perfect, & Stollery, 2004, for a meta-analysis and review).

Although these studies are informative, most manipulate complexity by combining driving with another task that does not normally go with it (e.g., mental arithmetic is a common task). It is possible that older drivers have special difficulties with unusual combinations of tasks. Another approach is to manipulate complexity by investigating how performance varies with different types of challenge within the driving task (Trick, Lochner, Toxopeus, & Wilson, 2009). There were three forms of challenge: visibility (driving in clear weather versus fog), traffic density (one car every 1500 meters as opposed to one car every 150 meters), and way-finding. The way-finding manipulation involved comparing performance where drivers simply had to follow the road to arrive at their destination to a situation where drivers had to memorize the name of a specific destination and then find their way to that town, making use of direction signs and landmarks. Two age groups were compared (M ages = 18 and 71 years, respectively). The older drivers missed significantly more turns because they failed to react appropriately to the directional signs and landmarks, but otherwise their performance was as good as or better than that of younger drivers. This may be because older drivers engaged in compensatory slowing in the face of challenges. Older drivers had significantly fewer collisions but they also reduced their speed more in the face of each type of challenge. In contrast, the younger adults maintained the same speed, regardless of the visibility, traffic density, or way-finding challenge.

With attentional load there is decreased processing in the periphery (Owseley, Ball, Sloane, Roenker, & Bruni, 1991) and intersections are attentionally-demanding locations insofar as drivers have to keep track of signs, lights, and the activity of road

users coming up from the left, right, ahead, and occasionally even from behind the vehicle. Epidemiological studies indicate collision risk is especially high at intersections for older drivers (McGwin & Brown, 1999; Preusser et al., 1998), and this suggests deficits in hazard anticipation. One way to measure hazard anticipation is to study eye movements. DeRamus and Fisher (2004) measured hazard anticipation at intersections in a driving simulator. Four age groups were tested: middle-age (40–50), young-old (65–69), middle-old (70–74), and old-old (75–79). This study showed no evidence that tactical hazard anticipation skills deteriorated, even in the old-old group. However, a more recent study investigated T-intersections, where drivers needed to take a second glance to the right or left after the first to ensure safe turning to the left or right (Romoser, Fisher, Mourant, Wachtel, & Sizov, 2005). Eye and head movements were assessed. This study showed that drivers over the age of 70 were less likely to take that second glance than drivers between the ages of 25 and 55. In fact, older drivers failed to take secondary looks almost three times more often than younger drivers (it is not known yet whether this is due to physical decline or attentional deficits).

Fatigue and the use of alcohol might also be understood as factors that reduce the resources available for controlled processing: fatigue because individuals withdraw resources for effortful processing, and alcohol because it depresses neural activity in general and this especially affects controlled processes because they are the most time-consuming (see Trick et al., 2004 for a review). If fatigue and alcohol reduce resources available for controlled processing, there should be interactive effects, such that fatigue and alcohol may have even stronger effects on older than on younger drivers. The interactive effects of fatigue and alcohol consumption have been investigated on older drivers in studies using driving simulators (e.g., fatigue: Campagne, Pebayle, & Muzet, 2004; Reimer, D'Ambrosio, & Coughlin, 2007, alcohol: Quillian, Cox, Kovatchev, & Phillips, 1999). The effects of alcohol and fatigue are reviewed in this book, chap. 44 by Creaser, Ward, & Rakauskas, and chap. 29 by Matthews, Saxby, Funke, Emo, & Desmond respectively.

26.2.2 Evaluating the Impacts of Road Design or In-Vehicle Technologies on Older Drivers

Research on how older adults respond to different road geometries, signage, and pavement markings is critical for policy-makers and highway engineers alike, and the effects of these factors have been assessed using driving simulators (e.g., Caird, Milloy et al., 2008; Lerner, 1994; Staplin, Lococo, & Byington, 1998; Shechtman et al., 2007). Simulators have also been used to test older drivers' intersection performance during left turns (Staplin, 1995), at late yellow lights (Caird, Chisholm, Edwards, & Creaser, 2007), and at highway-railway grade crossings (see Caird, Smiley, Fern, & Robinson, this book, chap. 38).

Similarly, for those involved in the design and evaluation of in-vehicle devices, it is essential to understand how these technologies affect older drivers. One such technology is the cellular telephone, which has now been shown to have a deleterious effect on driving performance in a variety of different age groups (e.g., Caird, Willness, Steel, & Scialfa, 2008; Shinar et al., 2005; Strayer & Drews, 2004). Then there are the "intelligent" in-vehicle technologies or telematics; that is, technologies designed to assist or even take over some functions of the driver. Navigation, night vision, and collision warning systems fall into this category. Caird (2004), Davidson (2007), and Green (2001) all review the research on the impact of these systems on older drivers, citing simulator and on-road studies. In general, older drivers may not accept these systems as readily as other drivers if they see no clear advantage to using the technology. Older drivers may also experience greater detriments to performance as a result of design flaws that limit the usability of a given device. Furthermore, although some of these systems are explicitly designed to help older drivers, they may ultimately compromise their safety. Older adults who have given up nighttime driving may install night vision systems and once again drive at night because the systems give them additional confidence (Caird, Horrey, & Edwards, 2001). This puts them at increased risk if the system fails or if it is used in a situation for which it was not intended. Often the balance between the mobility of the individual and the safety of society is not given adequate consideration when new technologies are developed.

26.2.3 Identifying the Collision-Prone Older Driver

Another type of research focuses on finding ways to identify older drivers at risk of collision—information critical for those involved in driver assessment and rehabilitation (see also Ball & Ackerman, this book, chap. 25). Epidemiological approaches have been used to identify the demographic characteristics of high-risk groups (McGwin & Brown, 1999; Waller, 1985) and psychological tests or batteries of tests have also been tried (e.g. Ball, Owsley, Sloane, Roenker, & Bruni, 1993; Janke & Eberhard, 1998; Marshall et al., 2007; Molnar, Patel, & Marshall, 2006; Staplin, Gish, & Wagner, 2003). For example, one of the most popular attentional measures is the *Useful Field of View* (UFOV: Owsley, Ball, Sloane, Roenker, & Bruni, 1991; see Hoffman, McDowd, Atchley, & Dubinsky, 2006 for a critical review). Simulators are sometimes used to validate these measures, and test scores can be correlated across simulator and on-road driving as well (e.g., Rizzo, Jermeland, & Severson, 2002). In fact, there has been a recent move to use driving simulators per se in order to identify older drivers at risk as a cost-effective alternative to on-road testing (e.g., Lee, Cameron, & Lee, 2003; Lee, Lee, Cameron, & Li-Tsang, 2003). There is also a rapidly growing literature on the impact of a variety of age-related disorders on driving (Frittelli et al., 2008; McGwin, Sims, Pulley, & Roseman, 2000; Rizzo, Reinach, McGehee, & Dawson, 1997; see also Rizzo, this book, chap. 46 for a review). The effectiveness of driver rehabilitation programs for older drivers at-risk has also been assessed, and some of these studies involve

simulators (see Ball & Ackerman, this book, chap. 25; Hunt & Arbesman, 2008).

26.3 Considerations in Choosing Dependent Measures for Simulator Studies on Older Drivers

A variety of different dependent measures can be used to assess driver performance and the choice of variables will in part be dictated by the research question. Investigators should be wary of ceiling and floor effects when testing different age groups no matter what dependent measure they choose. Pilot testing can help ensure that the task is neither too easy nor too difficult to reveal age differences. Dependent measures can be classified into the following general categories: lateral control, longitudinal control, response times, and eye movements. Considerations in using these measures with older drivers are outlined below.

26.3.1 Lateral Control Measures

Lateral control measures assess how well drivers maintain vehicle position within a lane. These include standard deviation of lane position, lane excursions, and deviations within a lane to increase clearance of other vehicles and road users (for example, moving left to avoid a cyclist). Lateral control measures can be sensitive to eyes off the road from distractions, perceptual-motor declines, and some cognitive declines. However, lateral control measures are also affected by the handling characteristics of the driving simulator, and the simulator vehicle may differ markedly from the one that the participant normally drives. Older drivers may have more problems adapting to these differences in handling, and this may be especially problematic when frequent right and left turns are required. Consequently, it is vital that older participants be given adequate practice so that they can get used to how the simulator vehicle handles.

26.3.2 Longitudinal Control Measures

Longitudinal control measures assess speed and headway. Speed maintenance is an essential driving skill and it can be related to tactical decisions (Michon, 1989). For older drivers, the most common pattern is to adopt slower speeds to increase available response time (Chu, 1994). Older drivers may use this strategy in order to exert some control over their circumstances and compensate for age-related increases in response time. When age comparisons are made, driving speeds are typically more variable within a given drive in samples of older drivers. In general, driving speeds are about three to five kilometers per hour slower in older drivers, but this may change depending on the posted speed limit of a roadway (e.g., see this book, chap. 16 by Milloy & Caird describing external distractions, and the chap. 38 by Caird et al., describing highway-railway crossings).

Because fewer perceptual cues to depth are available in a simulator (see also Andersen, this book, chap. 8), drivers may have difficulty calibrating their speed, a problem which may be exacerbated by reductions in image contrast (Horswill & Plooy, 2008). Also, older drivers have a smaller pupil diameter (limiting the amount of light reaching the retina) and reduced contrast sensitivity. Scialfa, Adams and Giovanetto (1991) argue that older drivers are generally less sensitive to how fast they are traveling, but this effect has not been replicated in a simulator to this point. Nonetheless, given that depth cues are less available in the simulator and given that these cues may be even less effective for older drivers than young, before testing begins it is essential to provide participants with opportunities to learn how to calibrate their speed appropriately in the simulator. In particular, it is vital to have a warm-up period where drivers are encouraged to repeatedly check the speedometer and then the roadway. Because participants may lose track of how fast they are going when they become preoccupied with other things, during this warm-up period it is a good idea to provide occasional verbal prompts to remind drivers to monitor their speed.

Another measure of longitudinal control is headway. Headway can be defined in terms of the distance to a lead vehicle or the time-to-contact. The distribution of headways for a given driver may reflect following preferences and the need to respond to surrounding traffic. Older drivers who maintain a greater headway may have others pull into their headway gap. Certain drivers attempt to block others from pulling into a gap ahead, though at this point, there has never been a scenario designed to assess this behavior. Furthermore, there are no simulations that measure how drivers accommodate to other drivers changing lanes, though it seems very plausible that such a scenario would reveal age differences. Research on this topic is a logical extension of automated vehicle algorithms (see Kearney & Grechkin, this book, chap. 6) but it has not been the focus of study.

Overall, longitudinal measures assess how well drivers achieve or maintain a certain target speed. However, it is pertinent to keep in mind that compared to younger drivers, older drivers take longer to accelerate to the posted speed limit (Strayer & Drews, 2004) and brake over longer distances (Caird et al., 2007). There is evidence that older drivers try to stop with more precision than younger drivers and, in general, older drivers may be more inclined to value accuracy over speed. It is important to consider this possibility when designing instructions and, if possible, it may be beneficial to try to assess whether participants are putting differential emphasis on speed or accuracy—perhaps with a post-test questionnaire.

26.3.3 Reaction Time Measures

An obvious way to measure differences in driving performance is to look for age-differences in collisions, but collisions are relatively rare and, as a result, *perception response times* are often measured (see also Caird & Horrey, this book, chap. 5). There are a variety of different response time measures, and different issues emerge depending on the event that gives rise to the response. For example, one common scenario requires drivers to respond to lead-vehicle braking (e.g., Strayer & Drews, 2004). Complications emerge when this scenario is used with

older drivers. Because older drivers often adopt larger following distances than younger drivers, they may avoid a collision even though they take longer to brake in response to lead-vehicle braking (see this book, chap. 16 by Milloy & Caird, "External Driver Distractions: The Effects of Video Billboards and Wind Farms on Driving Performance"). Another way to measure performance on this task is to assess minimum headway: how close the driver comes to a lead vehicle when it brakes (Caird, Chisholm, & Lockhart, 2008).

There are also studies that measure response times to sudden hazards that emerge from the periphery: pedestrians, cyclists, or vehicles that travel into the path of the driver. Response times can be measured in different ways. For instance, if a pedestrian walks into the roadway, detection time could be defined as the time from the appearance until the eyes of the driver land on the pedestrian (i.e., fixation) followed by an adequate avoidance of the pedestrian. More commonly, *perception response time* (PRT) is defined as the time from first appearance of the hazard until the driver places his or her foot on the brake. Perception and response time can be further fractionated. *Perception time* is the time from appearance of the hazard until the driver removes his or her foot from the accelerator and *response time* is the length of time from leaving the brake until placing it on the brake (Olson & Farber, 2003). PRT is used to determine whether drivers can respond adequately to traffic control devices, signs, and road geometry, and it is an essential component of many design assumptions (e.g., see Staplin et al., 1998). Accident reconstructionists use PRT to determine if a driver responded with due care and attention when a crash has occurred. Estimated PRT falling outside of certain distributions are used to infer proportional blame (Olson & Farber, 2003).

A variety of other reaction time measures require drivers to respond to the appearance of a probe stimulus, for example, the sudden illumination of a light-emitting diode within the vehicle (Lamble, Kauranen, Laakso, & Summala, 1999). In this type of study, participants are required to do two things at once: drive (the primary task) and respond to the probe (the secondary task). The assumption is that attention is shared between tasks. If there are more attentional resources to be shared, or if the driving task demands fewer resources for a given individual, then response to the secondary task should be more efficient. Thus, these probe response time tasks are an index of the amount of "spare" attention available for secondary tasks. A number of reviews have addressed a range of common response time measures (Green, 2000; Summala, 2000). Older drivers typically take longer to respond to these stimuli (Olson & Sivak, 1986). This may indicate that they have less "spare" attention, but it may also show that older drivers are more cautious and place a greater emphasis on driving than carrying out secondary tasks.

26.3.4 Eye Movement Measures

It has become increasingly common to use eye movement systems in driving simulator studies, but there are a number of pragmatic considerations that make it difficult to measure eye movements when testing older drivers. First, the eye tracking system must be calibrated to each driver to accommodate his or her glasses. Our experience is that about three-quarters of older drivers can be calibrated, but this is dependent on the system that is used and the acquisition of calibration expertise by research assistants (see Caird et al., this book, chap. 38). Reflections of the infrared beam from eyeglass lenses and frames may interfere with obtaining a reliable corneal reflection; loss of eye movement data can result. In general, head-mounted systems may restrict head movements because of the cordsa that exit the mounting band on the back of the head. Restriction of head movements due to age-related loss of range of motion may be confused with this imposed restriction. This is a problem in studies that measure head movements, such as those investigating eye-head relationships when participants perform lane change maneuvers. Sorting out the various baseline restrictions imposed by head-mounted systems would require a separate study. Dash-mounted eye movement systems also have limitations in calibration and reliability over the course of an experiment. For instance, detecting the position of the eye and gaze relative to the environment may be difficult because many older drivers wear corrective lenses. Loss of data and unbalanced cells must be properly analyzed (e.g., see Tabachnick & Fidell, 2006).

A large number of different eye movement measures can be collected. Often they require time-consuming data reduction and extraction. (See this book, chap. 38 by Caird et al., describing highway-railway crossings for a more complete description of the processes involved in collecting eye movements with older drivers; and chap. 30 by Pollatsek, Vlakveld, Kappé, Pradhan, & Fisher, for a more thorough review of eye movement measures.) Given the difficulties inherent in collecting eye movement data from older drivers and the amount of work necessary in extracting the results, it is especially important for researchers to think carefully before including eye movement measures in their study. What will knowing the total fixation duration or the total length of time extracting information from a location necessarily tell you about an older driver, a disease, a roadway modification or technology? Are there proxies of eye movements that can provide insight into the same question without incurring the same time costs? For example, when testing the impact of using a new in-vehicle interface, it may be as effective to measure time to task completion (time to use the interface) or an aspect of driving performance (e.g., stopping accuracy).

If a researcher is not deterred by our warnings, a variety of fixation measures can be collected. Variations on fixation frequency and duration are common. Definitions of these measures can be found in Green (2007). Older drivers typically look at an object longer and more frequently to extract the same information from it as younger drivers (e.g., Ho, Scialfa, Caird, & Graw, 2001). In addition, older drivers look at a sign later (when they are in closer proximity) given restrictions to legibility—an effect that may be even more pronounced when testing is conducted in a driving simulator (see, Kline & Dewar, 2004; Caird et al., this book, chap. 38).

26.4 Methodological Challenges

In this section we will first focus on general issues of reliability and validity when testing older drivers, and then discuss complications that emerge due to the higher incidence of simulator adaptation syndrome (colloquially known as "simulator sickness") among older drivers.

26.4.1 Reliability

When conducting research it is important to strive for reliable measurement techniques: ones that yield consistent results across repeated assessments of the same individual. However, the performance of older drivers is more variable across different times than younger drivers (Waller, 1991). This means that an older driver may perform very well on one occasion and much worse on another, even if the measurement technique is deemed reliable when used on younger populations. This increased variability over time may originate from several sources. Age-related health conditions may produce marked day-to-day fluctuations in how older participants feel. In addition, many older adults experience periodic problems getting a good night's sleep, and fragmented sleep has been shown to have deleterious effects on cognitive abilities (Oosterman, Van Someren, Vogels, & Scherder, 2009).

This increased variability across time in older drivers is a problem if a researcher wants to use a given driver's performance on one day to predict his or her performance at some later time—as occurs in studies that endeavor to identify collision-prone drivers (Waller, 1991)—because it undermines the correlations between time one and time two performances. Depending on how the study is designed, it can also be a problem in studies of age differences or studies of the effects of road design or in-vehicle technologies on the older driver. Many of these studies involve one or more within-subject (repeated measures) manipulations in addition to the age comparison. In a repeated measures manipulation, each participant experiences every condition of the independent variable, and thus the time required per participant increases with the complexity of the design (the number of conditions and the number of repeated measures variables). With a complex enough design, this may necessitate multiple testing sessions. However, in a within-subjects design, the variability of a given individual across time contributes to error variance. There is a danger that this error variance may obscure the effects of the manipulations—especially when the sample size is relatively small. If multi-session testing is necessary, it may be necessary to obtain a larger sample to counteract the effects of increased variability across time in the performance of older adults. As tempting as it is to add conditions and variables, it is best to keep the design as simple as possible when testing older drivers.

Overall, if consistency across time is crucial, it may be a good idea to find out how older adults feel and how they slept before testing begins (ideally in enough time that the session can be rescheduled if necessary). If a driver must be tested over multiple sessions, it is better to schedule tests at the similar times so that the different tests fall in approximately the same phase of their circadian rhythm (older adults "peak" earlier in the day than younger adults: Zilli, Giganti, & Uga, 2008).

26.4.2 Validity

When age-differences are found, it is often unclear whether these differences are the result of age per se or of variables confounded with age. One solution to this problem is to test younger participants who match the older drivers in terms of the confounding variable(s). A second is to screen out drivers if their scores do not fall within a certain range on the confounding variable(s). However, it is frequently difficult to find younger participants to match older drivers and screening increases time and costs incurred in acquiring samples. Either solution reduces the representativeness of the resultant samples and thus limits the external validity of the research. A third solution involves using statistical controls such as regression or analysis of covariance to partial out the effects of confounding variables. Failing that, at the very least, the confounding variables should be reported and conclusions qualified accordingly. Regardless of the solution chosen, the first step is to measure the confounding variables. In the following section we describe three classes of variables that are a cause for concern.

The first, and arguably the most troublesome, is differential exposure. Older drivers do not drive as often or as far as younger drivers and they are more likely to avoid driving in challenging conditions (Evans, 2004; Unsworth, Wells, Browning, Thomas, & Hal, 2007). It has been argued that differential exposure underpins age-differences in collision rates (Hakamies-Blomqvist et al., 2003; Langford, Methorst, & Hakamies-Blomqvist, 2006). Also, in previous generations, men did the majority of driving (Hakamies-Blomqvist, 1994; Waller, 1991). Some older women do not start driving on a regular basis until their husbands no longer drive. These women may be novice drivers though they have been licensed for many years. Driving is a complex task that demands coordinating multiple tasks, including steering, controlling velocity, detecting hazards, way-finding, and responding to signs, traffic control devices, and driving challenges such as slippery roads and bad weather. With regular practice, some of these tasks can become automatic, or "effortless", and can be performed without interfering with other ongoing processes (Shiffrin & Schneider, 1977). Some components of driving may take up to five years of practice to become fully automatic (e.g., Groeger, 2000; Shinar, Meir, & Ben-Shoham, 1998). However, processes gain or lose automaticity based on the frequency and recency of practice (Trick et al., 2004), and if older drivers do not drive on a regular basis, they may have to use effortful deliberate (controlled) processes to carry out tasks that were once automatic.

As a result, it is useful to gather detailed information about the frequency, recency, and the regularity with which drivers use their cars before making strong conclusions about age-related effects. Self-report measures are often used to assess exposure, such as the number of kilometers driven over a certain time interval (a week, a month, or a year). Unfortunately, self-reports of driving behavior may be inaccurate (Arthur et al., 2005; Owsley, 2004). Drivers may not have a clear idea of how far they have traveled. Some studies use GPS technologies or electronic devices such as the CarChip to assess exposure (e.g., Huebner, Porter, & Marshall, 2006; Marshall et al., 2007) but in many cases this is not feasible and thus self-report measures may be the only practical alternative. Some investigators require participants to keep detailed driving diaries over a period of time, though this increases the time and costs of carrying out the research. Research is needed to find the optimal way to design self-report measures to ensure the most accurate estimates for the amount and type of driving exposure. A minimum yearly travel distance should be considered as a screening criterion, unless distance traveled per year is the basis of a central research question.

A second problem originates from difficulties in distinguishing the effects of normal age-related changes from those of age-related disorders. There are a variety of such disorders, including cataracts, macular degeneration, glaucoma, diabetic retinopathy, heart disease and stroke, sensori-neural deafness, dementias of various kinds, and arthritis (see Rizzo, this book, chap. 46). At the very least, researchers should include a questionnaire to assess general health. Unfortunately, participants are not always aware that they have an age-related disorder which compromises their ability to perceive and react to the challenges of the driving environment. For this reason, researchers often include measures of acuity (Casson & Recette, 2000) such as the *ETDRS* (Early Treatment of Diabetic Retinopathy Scale: Ferris, Kassoff, Bresnick, & Bailey, 1982), and contrast sensitivity, such as the *Pelli-Robson* (Pelli, Robson, & Wilkins, 1988) or *Vistech* tests (Evans & Ginsburg, 1985; Gilmore, 2002). Many perceptual disorders affect color vision, and deficits in color vision may compromise a driver's ability to detect signs or react to hazards. Given that it is often necessary to distinguish between problems that originate from sensory processes and those that originate in attention, it may be useful to assess color vision, using measures such as the *Ishihara color plates* (Ishihara, 1993). Cognitive status is often assessed using tests such as *the Mini-Mental State Exam* (MMSE: Folstein, Folstein, & McHugh, 1975). However, for more comprehensive testing, Eby, Molnar, Shope and Dellinger (2007) suggest a battery of 17 measures that might be useful when conducting longitudinal studies of older drivers, where it is critical to evaluate changes that might affect driving performance. The battery takes about an hour to administer, and it measures vision, attention, motor performance, and cognitive status (see also Dawson, Anderson, Uc, Dastrup, & Rizzo, 2009).

The third problem originates from the fact that many older adults take one or more prescription drugs which may impair driving (e.g., Rapoport & Banina, 2007; Lococo & Staplin, 2006). It may be useful to have the participants bring a list of the medications that they are using to the testing session. Participants may be excluded if they are taking a drug that is known to adversely affect driving, such as cyclic antidepressants (Wilkinson & Moskowitz, 2001). Doctors and pharmacists do not always inform patients of side-effects. However, in many cases the effects of drugs and drug interactions on driving performance are unknown.

When reporting the characteristics of the sample, it is also a good idea to include information on age (mean and standard deviation), gender, education, types of license, region (rural or urban), and whether the sample contains commercial drivers. Moreover, it is useful to measure crash history and moving violations, though for these variables there may be biases in self-reports (Arthur et al., 2005; Owsley, 2004). Information about crashes and moving violations may be obtained from insurance companies and government licensing agencies but, when using these sources, it is important to recognize that many crashes and violations are not reported. There may also be institutional biases among the authorities judging who is at fault. Access to databases involves overcoming privacy and bureaucracy issues.

Overall, it is essential to remember that older drivers are not a uniform group and notable changes occur between the ages of 60 and 90 years, in transition from the "young-old" to "old-old". As well, there is more biological diversity among samples of the same chronological age for older than younger adults (Waller, 1991). In any given study, there is a danger that the sample may not be truly representative of the general population of older drivers because of how the sample was obtained. That is why it is a good idea to specify how the sample was recruited (from newspaper ads, booths at seniors' recreational centers, senior's residences, hospitals, etc.). Selection bias may be a factor insofar as older volunteers who agree to be tested in a driving simulator may be especially good drivers, or ones who are exceptionally confident about their driving. Even then, we have observed that many older drivers are very concerned about having their driving performance evaluated and before they agree to participate they may require additional assurances that information about their performance will not be shared with licensing agencies.

26.4.3 Simulator Adaptation Syndrome

There are many types of simulator, ranging from desktop models to those that involve car bodies and moving bases that support six-degrees-of-freedom motion (see Greenberg & Blommer, this book, chap. 7) but when simulators are used, there is the risk of simulator adaptation syndrome, colloquially known as "simulator sickness". When a participant has simulator adaptation syndrome they experience one or more of the following symptoms: disorientation, dizziness, headache, stomach discomfort and nausea. If this occurs, testing may have to be halted for ethical reasons.

The exact prevalence of simulator sickness is unknown because dropout rates are seldom reported (see Stoner, Fisher, & Mollenhauer, this book, chap. 14). Nonetheless, there are indications that the incidence is especially high among older adults. Several years ago, the second author contacted a variety of institutions, including the University of Minnesota, INRETS, the Texas Transportation Institute (TTI) and the University of Iowa, asking about dropout rates for older drivers due to simulator

adaptation syndrome. Estimates ranged from between 35% and 75%, with the average at around 40%. Dropout rates varied substantially from study to study for a number of reasons, but older drivers and females seem to be especially at risk. For example, Caird et al. (2007) found that drivers over the age of 65 comprised 56% of dropouts (19% were aged 55 to 64, and 25% were aged 18 to 35). Of the dropouts, 87.5% were women. In the following sections we discuss issues that arise due to the incidence of simulator adaptation syndrome.

26.4.3.1 Screening and Monitoring

Screening is a crucial first stage when conducting research on older drivers using driving simulators. To begin with, it is a good idea to screen for cognitive status because it is critical that participants fully understand the risks of simulator testing before they consent to participate in the study. In community samples, where participants give their own consent, we require MMSE scores of 26 (or higher) out of 30 to participate in a simulator study. (Most volunteers easily surpass this criterion.) Then there are questionnaires designed to identify and screen out individuals at risk of simulator sickness (the Simulator Sickness Questionnaire: Kennedy, Lane, Berbaum, & Lilienthal, 1993; Test of Postural Stability: Stoffregen, Pagulayan, Bardy, & Hettinger, 2000). Careful screening can help reduce the number of participants who experience simulator sickness in a given study, but these tests do not have perfect predictive validity. Consequently, there will be people who pass the screening tests and nonetheless develop simulator adaptation syndrome.

Thus, some of the recruited sample will be excluded in the screening process. Others will be excluded when they develop simulator sickness during testing. There is a danger that the remaining sample will no longer be representative of the population of older drivers (if it ever was). In fact, there is a possibility that the excluded participants are the very ones that most need to be studied. For example, crash risk is related to falls and vestibular disturbances (Vance et al., 2006). Vestibular disturbances can produce postural instability and one common screening test screens out participants based on postural instability (the Test of Postural Stability: Stoffregen et al., 2000). This does not negate the importance of research on the remainder of the participants, but it means that researchers should acknowledge that there are drivers whose problems they cannot address. Because high dropout rates limit the generality of the findings, it is important to document the incidence of simulator adaptation syndrome. In fact, if progress is to be made in developing better screens to identify those at risk of simulator adaptation syndrome, in addition to recording characteristics of the sample, it is critical to keep records on the characteristics of those that were dropped from the sample. Information about predictors of simulator adaptation syndrome is necessary to understanding the adaptation processes used in coordinating and re-weighting different types of sensory information for purposes of perceptual-motor control and balance (Reed-Jones, Vallis, Reed-Jones, & Trick, 2008).

Given the prevalence of simulator adaptation syndrome in older adults, research personnel should be trained to identify early symptoms. Furthermore, it would be a good idea for them to rehearse what they will do in advance because symptoms may emerge suddenly. Early symptoms include repeated head movements, excessive swallowing, skin pallor, sweating, and slowed verbal response. Researchers should query participants about how they feel on multiple occasions, especially after turns and stops (the maneuvers most likely to produce discomfort). Some drivers will not admit that they are feeling ill. This is where online physiological monitoring could be particularly valuable (see Stoner et al., this book, chap. 14, for a review of online techniques). Ultimately, when participants deny discomfort, the decision to terminate the session is up to the researcher. For both of our laboratories, the policy is to err on the side of caution, and to stop the study at the first indication of symptoms even if the participant denies feeling uncomfortable. This makes it necessary to run additional participants to achieve the required number per condition and results in higher costs for participant payments, recruiting, and simulator time.

26.4.3.2 Mitigating the Risk of Simulator Sickness Through Scenario Design

A variety of techniques have been used to prevent simulator adaptation syndrome, including reducing the room temperature, using a fan to circulate fresh air, decreasing the luminance of the projectors, providing a horizon line for participants to look at, having participants wear bracelets on pressure points on the wrists, and giving them more time to acclimate to the simulator (see Stoner et al., this book, chap. 14, for a comprehensive review). However, some of the most effective interventions involve changing the nature of the simulated drive: that is, limiting the number of stops and turns and reducing the field of view.

Unfortunately, restricting the drive in these ways may make it more difficult to observe age differences in performance. Crash records reveal that older adults are especially at risk at intersections (Preusser et al., 1998), but it is hard to test drivers at intersections without giving them a wide field of view and opportunities to stop and turn. There are issues of reliability in measurement when the numbers of stops and turns are minimized. Perception response time is one of the most common indices of driving performance (Olson & Farber, 2003) and most response time measures require braking. In the laboratory research on attention, it is standard practice to require at least 10 replications per data point to ensure reliable response time data. Therefore, reliable response time data requires multiple braking events but simulator studies cannot afford to have a large number of braking events in a short period of time for fear of increasing the risk of simulator adaptation syndrome to unacceptable levels. (There also is the danger that drivers may become sensitized if there are too many events per drive [Caird, Chisholm, & Lockhart, 2008].) Similarly, to study way-finding (finding the way to a desired destination using signs and landmarks, e.g., Trick et al., 2009), it is necessary to give drivers choice points where they have opportunities to turn. It is also important to consider the corrections that will be necessary if drivers miss their designated turn. To obtain reliable way-finding

data it may be necessary to provide drivers with a number of such decision points, but each turn (required or committed in error) increases the probability of simulator adaptation syndrome.

Thus, in driving simulation studies, increases in reliability are bought at the expense of augmented risks of simulator adaptation syndrome. It is possible to reduce these risks by extending testing over several days (rather than concentrating it in one long session), but this adds to the cost of testing, and with multi-session studies there is always a danger of attrition, and variability across time (see section on reliability). When testing older drivers with simulators, it is often necessary to wrestle with a variety of conflicting demands. There are compromises that can be made: ways to minimize the risk of simulator adaptation and yet keep a reasonably wide field of view; ways to study intersections and yet reduce the number stops and turns (see Stoner et al., this book, chap. 14). Researchers should familiarize themselves with these techniques and use them when they can. However, before they undertake an investigation, it is critical that they ensure that their research question is of adequate importance and their project sufficiently well planned to justify the enhanced risk of simulator sickness for older participants. This means a thorough literature review and careful pilot-testing.

26.5 Conclusions

Simulator studies have yielded valuable insights into the performance of older drivers and this research has significant implications for both theory and practice. However, there are methodological challenges inherent in testing older drivers and additional complexities emerge when these studies involve driving simulators. This research requires patience, logistic support, and a research plan. Investigations inevitably take longer than expected. It may take many months to gather an adequate sample given the dropout rate and the number lost to screening and simulator adaptation syndrome. Furthermore, each individual testing session may take longer than anticipated because many older drivers like to talk before, during, and after the study. It has been our experience that older drivers are very concerned with driver safety and they have many valuable insights about problems they encounter or see in other drivers. As a result, it is especially important that researchers listen and observe. There is much to be learned.

Key Points

- Research on older drivers is essential to policymakers, professionals involved in testing and rehabilitating older drivers, and those involved in designing and evaluating new driving environments and in-vehicle technologies.
- A variety of factors threaten the reliability and validity of research on older drivers and there are a number of issues to be taken into consideration when choosing dependent measures to best assess their driving performance. The variability among individuals of the same chronological

age is higher among older than younger drivers. It is important to record participant characteristics, including health, medications, visual acuity, cognitive status, and driving exposure.

- The prevalence of simulator sickness is higher among older than younger adults. Measures can be taken to reduce simulator sickness but nonetheless it is sometimes necessary to drop participants from the study because they are experiencing symptoms. The remaining sample may not be typical of the general population of older drivers.

Keywords: Aging, Age-Related Decrements, Lifespan Driving, Older Drivers, Simulator Sickness

Acknowledgments

Many of the ideas expressed by the authors benefited from presentations and feedback at many meetings at the Transportation Research Board and Driving Assessment conferences. Both authors received funding from the Canadian Foundation for Innovation (CFI) and AUTO21 Network Centers of Excellence (NCE). The second author also received funding from the Transport Canada, Transportation Development Center. We would like to thank Ryan Toxopeus and Andrew Mayer for their help in preparing the manuscript.

Key Readings

Evans, L. (2004). *Traffic safety*. Bloomington Hills: Science Serving Society.

Transportation Research Board. (2004). *Transportation in an aging society: A decade of experience*. Washington, DC: Transportation Research Board.

References

Arthur, W., Bell, S. T., Edwards, B. D., Day, E. A., Tubre, T. C., & Tubre, A. H. (2005). Convergence of self-report and archival crash involvement data: A two-year longitudinal follow-up. *Human Factors, 47*(2), 303–313.

Baddeley, A. (1993). Working memory or working attention? In A. Baddeley & L. Weiscranz (Eds.), *Attention: Selection, awareness, and control. A tribute to Donald Norman* (pp. 152–170). New York, NY: Clarendon Press/Oxford University Press.

Baldwin, C. L. (2002). Designing in-vehicle technologies for older drivers: Applications of sensory-cognitive interaction theory. *Theoretical Issues in Ergonomic Science, 3*(4), 307–329.

Ball, K., & Owsley, C. (1991). Identifying correlates of accident involvement for the older driver. *Human Factors, 33*(5), 583–595.

Ball, K., Owsley, C., Sloane, M. E., Roenker, D. L., & Bruni, J. R. (1993). Visual attention problems as a predictor of vehicle crashes in older adults. *Investigative Ophthalmology and Visual Science, 34*(11), 3110–3123.

Brouwer, W. B., Waterink, W., Van Wolffelaar, P. C., & Rothengatter, T. (1991). Divided attention in experienced younger and older drivers: Lane tracking and visual analysis in a dynamic driving simulator. *Human Factors, 33*(5), 573–582.

Caird, J. K. (2004). Intelligent transportation systems (ITS) and older drivers' safety and mobility. *Transportation in an aging society: A decade of experience*. Washington, DC: Transportation Research Board.

Caird, J. K., Chisholm, S., Edwards, C., & Creaser, J. (2007). The effect of yellow light onset time on older and younger drivers' perception response time (PRT) and intersection behavior. *Transportation Research: Part F, 10*(5), 383–396.

Caird, J. K., Chisholm, S., & Lochhart, J. (2008). The effect of in-vehicle advanced signs on older and younger drivers' intersection performance. *International Journal of Human Computer Studies, 66*(3), 132–144.

Caird, J. K., Edwards, C. J., Creaser, J. I., & Horrey, W. J. (2005). Older driver failures of attention at intersections: Using change blindness methods to assess turn decision accuracy. *Human Factors, 47*(2), 235–249.

Caird, J. K., Horrey, W. J., & Edwards, C. J. (2001). Effects of conformal and non-conformal vision enhancement systems on older driver performance. *Transportation Research Record, 1759*, 38–45.

Caird, J. K., Milloy, S., Ohlhauser, A., Jacobson, M., Skene, M., & Morrall, J. (2008). *Evaluation of four bicycle lane treatments using driving simulation: Comprehension and driving performance results*. Annual Conference of Canadian Road Safety Professionals. Whistler, British Columbia.

Caird, J. K., Willness, C., Steel, P., & Scialfa, C. (2008). A meta-analysis of cell phone use on driver performance. *Accident Analysis & Prevention, 40*, 1282–1293.

Campagne, A., Pebayle, T., & Muzet, A. (2004). Correlation between driving errors and vigilance level: Influence of the driver's age. *Physiology & Behavior, 80*, 515–524.

Casson, E. J., & Recette, L. (2000). Vision standards for driving in Canada and the United States. A review for the Canadian Ophthalmological Society. *Canadian Journal of Ophthalmology, 35*, 192–203.

Chu, X. (1994). *The effect of age on driving habits of the elderly: Evidence from the 1990 national personal transportation study* (Rep. DOT-T-95-12). Washington, DC: US Department of Transportation.

Craik, F. I. M., & Bialystok, E. (2006). Cognition through the lifespan: Mechanisms of change. *Trends in Cognitive Sciences, 10*(3), 131–138.

Craik, F. I. M., & Salthouse, T. A. (2000). *Handbook of aging and cognition* (2nd ed.). Mahwah, NJ: Lawrence Erlbaum Associates.

Davidson, R. (2007). *Assisting the older driver: Intersection design and in-car devices to improve the safety of the older driver*. Leidschendam: The Netherlands. SWOV.

Dawson, J. D., Anderson, S. W., Uc, E. C., Dastrup, E., & Rizzo, M. (2009). Predictors of driving safety in early Alzheimer disease. *Neurology, 72*, 521–527.

DeRamus, R., & Fisher, D. (2004). The effect of driver age and experience on risk assessment and risk prediction. *Proceedings of the human factors and ergonomics society annual meeting* (pp. 2627–2631). New Orleans, LA.

Dewar, R. E., & Olson, P. (2007). *Human factors in traffic safety* (2nd ed.). Tucson, AZ: Lawyers & Judges Publishing.

Eby, D. W., Molnar, L. J., Shope, J. T., & Dellinger, A. M. (2007). Development and pilot testing of an assessment battery for older drivers. *Journal of Safety Research, 38*, 535–543.

Evans, L. (2004). *Traffic safety*. Bloomington, Hills: Science Serving Society.

Evans, D. W., & Ginsburg, A. P. (1985). Contrast sensitivity predicts age-related differences in highway-sign discriminability. *Human Factors, 27*(6), 637–642.

Ferris, F. L., Kassoff, A., Bresnick, G. H., & Bailey, I. (1982). New visual acuity charts for clinical research. *American Journal of Ophthalmology, 94*, 91–96.

Folstein, M. F., Folstein, S. E., & McHugh, P. R. (1975). Mini-Mental State: A practical method for grading the cognitive status of patients for the clinician. *Journal of Psychiatric Research, 12*, 189–198.

Frittelli, C., Borghetti, D., Iuduce, G., Bonanni, E., Maestri, M., Tognoni, G., ... Iuduce, A. (2008). Effects of Alzheimer's disease and mild cognitive impairment on driving ability: A controlled clinical study by simulated driving test. *International Journal of Geriatric Psychiatry, 24*(3), 232–238. doi: 10.1002/gps.2095

Gilmore, G. C. (2002). Scoring of contrast sensitivity on Vistech charts. *Journal of the Neurological Sciences, 205*, 85.

Green, M. (2000). "How long does it take to stop?" Methodological analysis of driver perception-brake times. *Transportation Human Factors, 2*(3), 195–216.

Green, P. (2001). *Variations in task performance between younger and older drivers: UMTRI research on telematics*. Association for the Advancement of Automotive Medicine Conference on Aging and Driving, Southfield, MI.

Green, P. (2007). Where do drivers look while driving (and for how long)? In R. E. Dewar & R. Olson (Eds.), *Human factors in traffic safety* (2nd ed., pp. 57–82). Tucson, AZ: Lawyers & Judges Publishing.

Groeger, J. A. (2000). *Understanding driving*. Philadelphia, PA: Taylor and Francis.

Hakamies-Blomqvist, L. (1994). Aging and fatal accidents in male and female drivers. *Journal of Gerontology: Social Sciences, 49*(6), S286–S290.

Hakamies-Blomqvist, L., Raitanen, T., & O'Neill, D. (2003). Driver aging does not cause higher accident rates per km. *Transportation Research Part F, 5*, 271–274.

Hasher, L., Stoltzfus, E. R., Zacks, R. T., & Rypma, B. (1991). Age and inhibition. *Journal of Experimental Psychology: Learning, Memory, & Cognition, 17*, 163–169.

Hauer, E. (1988). The safety of older persons at intersections. In *Transportation in an aging society* (Vol. 2, pp. 195–252). Washington, DC: Transportation Research Board.

Ho, G., Scialfa, C. T., Caird, J. K., & Graw, T. (2001). Traffic sign conspicuity: The effects of clutter, luminance, and age. *Human Factors, 43*(2), 194–207.

Hoffman, L., McDowd, J. M., Atchley, P., & Dubinsky, R. (2005). The role of visual attention in predicting driving impairment in older adults. *Psychology and Aging, 20*(4), 610–622.

Horswill, M. S., & Plooy, A. M. (2008). Reducing contrast makes speeds in a video-based driving simulator harder to discriminate as well as making them appear slower. *Perception, 37*, 1269–1275.

Huebner, K. D., Porter, M. M., & Marshall, S. C. (2006). Validation of an electronic device for measuring driving exposure. *Traffic Injury Prevention, 7*, 76–80.

Hunt, L. A., & Arbesman, J. (2008). Evidence-based perspective of effective interventions for older clients that remedy or support improved driving performance. *American Journal of Occupational Therapy, 62*(2), 136–148.

Ishihara, S. (1993). *Ishihara's test for color-blindness*. Tokyo: Kanehara.

Janke, M. K., & Eberhard, J. W. (1998). Assessing medically impaired older drivers in a licensing setting. *Accident Analysis & Prevention, 3*, 347–361.

Kahneman, D. (1973). *Attention and human performance*. Pacific Palisades, CA: Goodyear.

Kennedy, R. S., Lane, N. E., Berbaum, K. S., & Lilienthal, M. G. (1993). Simulator sickness questionnaire: An enhanced method for quantifying simulator sickness. *International Journal of Aviation Psychology, 3*(3), 203–220.

Kline, D., & Dewar, R. (2004). The aging eye and transport signs. In C. Castro & T. Horberry (Eds.), *The human factors of transport signs* (pp. 115–134). New York: CRC Press.

Lamble, D., Kauranen, T., Laakso, M., & Summala, H. (1999). Cognitive load and detection thresholds in car following situations: Safety implications for using mobile telephone (cellular) telephones while driving. *Accident Analysis & Prevention, 31*, 617–623.

Langford, J., Methorst, R. M., & Hakamies-Blomqvist, L. (2006). Older drivers do not have a high crash risk – A replication of the low mileage bias. *Accident Analysis & Prevention, 38*(3), 574–578.

Lee, H. C., Cameron, D., & Lee, A. H. (2003). Assessing the driving performance of older adult drivers: On-road versus simulated driving. *Accident Analysis & Prevention, 35*, 797–803.

Lee, H. C., Lee, A. H., Cameron, D., & Li-Tsang, C. (2003). Using driving simulators to identify older drivers at inflated risk of motor vehicle crashes. *Journal of Safety Research, 34*, 453–459.

Lerner, N. (1994). Giving the older driver enough perception-reaction time. *Experimental Aging Research, 20*, 225–233.

Lococo, K. H., & Staplin, L. (2006). *Literature review of polypharmacy and older drivers: Identifying strategies to collect drug usage and driver functioning among older drivers* (Rep. DOT HS 810 558). Washington, DC: National Highway Traffic Safety Administration.

Makishita, H., & Matsunaga, K. (2008). Differences of drivers' reactions times according to age and mental workload. *Accident Analysis & Prevention, 40,* 567–575.

Marshall, S. C., Wilson, K. G., Molnar, F. J., Man-Son-Hing, M., Stiell, I., & Porter, M. M. (2007). Measurement of driving patterns of older adults using data logging devices with and without global positioning system capability. *Traffic Injury Prevention, 8*(3), 260–266.

McGwin, G., Jr., & Brown, D. B. (1999). Characteristics of traffic crashes among young, middle-aged, and older drivers. *Accident Analysis & Prevention, 31,* 181–198.

McGwin, G., Jr., Sims, R. V., Pulley, L., & Roseman, J. M. (2000). Relations among chronic medical conditions, medications, and automobile crashes in the elderly: A population-based case-control study. *American Journal of Epidemiology, 152*(2), 424–431.

Michon, J. (1985). A critical view of driver behavior models: What do we know, what should we do. In L. Evans, & R. Schwing (Eds.), *Human behavior and traffic safety.* New York: Plenum Press.

Michon, J. (1989). Explanatory pitfalls and rule-based driver models. *Accident Analysis & Prevention, 21*(4), 341–353.

Molnar, F. J., Patel, A., & Marshall, S. C. (2006). Clinical utility of office-based predictors of fitness to drive in persons with dementia: A systematic review. *Journal of the American Geriatric Society, 54,* 1809–1824.

Olson, P. L., & Farber, E. (2003). *Forensic aspects of driver perception and response* (2nd ed.). Lawyers & Judges, Tucson, AZ.

Olson, P. L., & Sivak, M. (1986). Perception-response time to unexpected roadway hazards. *Human Factors, 28,* 91–96.

Oosterman, J. M., Van Someren, E. U. W., Vogels, R. L. C., & Scherder, E. J. A. (2009). Fragmentation of the rest-activity rhythm correlates with age-related cognitive deficits. *Journal of Sleep Research, 18,* 129–135.

Owsley, C. (2004). Driver capabilities. *Transportation in an aging society: A decade of experience* (pp. 44–55). Washington, DC: Transportation Research Board.

Owsley, C., Ball, K., Sloane, M. E., Roenker, D. L., & Bruni, J. R. (1991). Visual/cognitive correlates of vehicle accidents in older drivers. *Psychology and Aging, 6*(3), 403–415.

Pelli, D. G., Robson, J. G., & Wilkins, A. J. (1988). The design of a new letter chart for measuring contrast sensitivity. *Clinical Vision Sciences, 2,* 187–199.

Perfect, T. J., & Maylor, E. A. (2000). *Models of cognitive aging.* Toronto: Oxford University Press.

Preusser, D. F., Williams, A. F., Ferguson, S. A., Ulmer, R. G., & Weinstein, H. B. (1998). Fatal crash risk of older drivers at intersections. *Accident Analysis & Prevention, 30*(2), 151–159.

Quillian, W. C., Cox, D. J., Kovatchev, B. P., & Phillips, C. (1999). The effects of age and alcohol intoxication on simulated driving performance, awareness, and self-restraint. *Age and Aging, 28,* 59–66.

Rapoport, M. J., & Banina, M. C. (2007). The effects of psychotropic medications on simulated driving: A critical review. *CNS Drugs, 21*(6), 503–519.

Reed-Jones, R. J., Vallis, L. A., Reed-Jones, J. G., & Trick, L. M. (2008). The relationship between postural stability and virtual environment adaptation. *Neuroscience Letters, 435*(3), 204–209.

Reimer, B., D'Ambrosio, L., & Coughlin, J. F. (2007). Secondary analysis of time of day on simulated driving performance. *Journal of Safety Research, 38,* 563–570.

Riby, L. M., Perfect, T. J., & Stollery, B. T. (2004). The effects of age and task domain on dual task performance: A meta-analysis. *European Journal of Cognitive Psychology, 16*(6), 863–891.

Rizzo, M., Jermeland, J., & Severson, J. (2002). Instrumented vehicles and driving simulators. *Gerontechnology, 1*(4), 291–296.

Rizzo, M., Reinach, S., McGehee, D., & Dawson, J. (1997). Simulated car crashes and crash predictors in drivers with Alzheimer's disease. *Archives of Neurology, 54,* 545–553.

Romoser, M., Fisher, D. L., Mourant, R., Wachtel, J., & Sizov, K. (2005). The use of a driving simulator to assess senior driver performance: Increasing situational awareness through post-drive one-on-one advisement. *Proceedings of the third international driving symposium on human factors in driver assessment, training and vehicle design* (pp. 456–463). Iowa City, IA: University of Iowa Public Policy Center.

Salthouse, T. A. (1996). The processing-speed theory of adult age differences in cognition. *Psychological Review, 103,* 403–428.

Scialfa, C., & Kline, D. (2007). Vision. In *Encyclopedia of gerontology: Age, aging, and the aged* (2nd ed., pp. 653–660). New York, NY: Academic Press.

Scialfa, C. T., Adams, E., & Giovanetto, M. (1991). Reliability of the Vistech Contrast Test System in a life-span sample. *Optometry and Vision Science, 68,* 270–274.

Shallice, T., & Burgess, P. (1993). Supervisory control of action and thought selection. In A. Baddeley & L. Weiscranz (Eds). *Attention: Selection, awareness, and control. A tribute to Donald Norman* (pp. 171–186). New York: Clarendon Press/Oxford University Press.

Shechtman, O., Classen, S., Stephens, B., Bendixen, R., Belchior, P., Sandhu, M., Davis, E. (2007). The impact of intersection design on simulated driving performance of young and senior adults. *Traffic Injury Prevention, 8,* 78–86.

Shiffrin, R., & Schneider, W. (1977). Controlled and automatic human information processing: II. Perceptual learning, automatic attending, and a general theory. *Psychological Review, 84,* 127–190.

Shinar, D. (2007). *Traffic safety and human behavior.* Amsterdam: Elsevier.

Shinar, D., Meir, M., & Ben-Shoham, I. (1998). How automatic is manual gear shifting? *Human Factors, 40,* 647–654.

Shinar, D., Tractinsky, N., & Compton, R. (2005). Effects of practice, age, and task demands, on interference from a phone task while driving. *Accident Analysis & Prevention, 37,* 315–326.

Spirduso, W. W., Francis, K. L., & MacRae, P. G. (2005). *Physical dimensions of aging.* Champaign, IL: Human Kinetics.

Staplin, L. (1995). Simulator and field measures of driver age differences in left-turn gap judgments. *Transportation Research Record, 1485*, 49–55.

Staplin, L., Gish, K. W., & Wagner, E. K. (2003). MaryPODS updated: Updated crash analysis and implications for screening program implementation. *Journal of Safety Research, 34*, 389–397.

Staplin, L., Lococo, K., & Byington, S. (1998). *Older driver design handbook* (Rep. FHWA-RD-97-135). McLean, VA: Federal Highway Administration. Retrieved November 10, 2008, from http://ntl.bts.gov/DOCS/older/rec/index_m.html

Staplin, L., & Lyles, R. W. (1991). Age differences in motion perception and specific traffic maneuver problems. *Transportation Research Record, 1325*, 23–33.

Stoffregen, T., Pagulayan, R., Bardy, B., & Hettinger, L. (2000). Modulating postural control to facilitate visual performance. *Human Movement Science, 19*(2), 203–220.

Strayer, D. L., & Drews, F. A. (2004). Profiles in driver distraction: Effects of cell phone conversations on younger and older drivers. *Human Factors, 46*(4), 640–669.

Summala, H. (2000). Brake reaction times and driver behavior analysis. *Transportation Human Factors, 2*(3), 217–226.

Tabachnick, B. G., & Fidell, L. S. (2006). *Using multivariate statistics* (5th ed.). Boston, MA: Pearson, Allyn & Bacon.

Transportation Research Board. (2004). *Transportation in an aging society: A decade of experience.* Washington, DC: Transportation Research Board.

Trick, L. M., Enns, J. T., Mills, J., & Vavrik, J. (2004). Paying attention behind the wheel: A framework for studying the role of selective attention in driving. *Theoretical Issues in Ergonomic Science, 5*(5), 385–424.

Trick, L.M., Lochner, M., Toxopeus, R., & Wilson, D. (2009). Manipulating drive characteristics to study mental load and its older and younger drivers. *Proceedings of the 5th international symposium on human factors in driving assessment, training, and vehicle design* (pp. 363–369). Iowa City, IA: University of Iowa Public Policy Center.

Unsworth, C. A., Wells, Y., Browning, C., Thomas, S., & Hal, K. (2007). To continue, to modify, or relinquish driving: Findings from a longitudinal study of healthy ageing. *Gerontology, 53*(6), 423–431.

Vance, D. E., Ball, K. K., Roenker, D. L., Wadley, V. G., Edwards, J. D., & Cissell, G. M. (2006). Predictors of falling in older Maryland drivers: A structural-equation model. *Journal of Aging and Physical Activity, 14*(3), 254–269.

Waller, J. A. (1985). The older driver: Can technology decrease the risks. *Generations, Fall*, 36–37.

Waller, P. (1991). The older driver. *Human Factors, 33*(5), 499–505.

Wickens, C. (2002). Multiple resources and performance prediction. *Theoretical Issues in Ergonomic Science, 3*, 159–177.

Wilkinson, C., & Moskowitz, H. (2001). *Polypharmacy and older drivers: Literature review.* Unpublished manuscript. Los Angeles, CA: Southern California Research Institute.

Willott, J. F., Chisolm, T. H., & Lister, J. J. (2001). Modulation of presbycusis: Current status and future directions. *Audiology & Neurotology, 6*(5), 231–249.

Zilli, I., Giganti, F., & Uga, V. (2008). Yawning and subjective sleepiness in the elderly. *Journal of Sleep Research, 17*, 303–308.

27
Profiles in Cell Phone-Induced Driver Distraction

David L. Strayer
University of Utah

Joel Cooper
Texas Transportation Institute

Frank A. Drews
University of Utah

Abstract

The Problem. Driver distraction is a leading cause of motor vehicle accidents, accounting for up to 25% of crashes on the roadway. Multitasking activities, such as concurrent use of a cell phone while driving, are increasingly becoming a significant source of such distraction. ***Role of Driving Simulators.*** Driving simulators provide an important research tool for systematically and safely studying driver distraction. Carefully constructed studies can be designed to establish a causal link between a multitasking activity (e.g., cell phone conversations, text messaging, passenger conversations) and driving impairment. ***Key Results of Driving Simulator Studies.*** We used a car-following paradigm to examine the effects of cell-phone conversations on simulated driving. We found that driving performance was negatively influenced by both hand-held and hands-free cell-phone conversations and that the impairment was equivalent for these two modes of communication. Compared to non-distracted driving conditions, the cell-phone driver's brake reaction times were slower and they took longer to recover their speed that was lost following braking. When drivers were conversing on a cell phone, they adopted a compensatory strategy to attempt to counteract their delayed brake reactions; however, accident rates were still higher than when drivers were not conversing on the phone. This pattern of driving behavior differs qualitatively from that of the driver conversing with a passenger in the vehicle and from that of the driver using the phone for text messaging. ***Scenarios and Dependent Variables.*** We examined a number of dependent variables related to the safe operation of a motor vehicle (e.g., speed, following distance, reaction time, etc.). Quantitative analysis of the distribution of reaction times provides insight into the mechanisms underlying the cell-phone-induced dual-task interference. We also discuss a number of other scenarios and dependent measures that have been used to study distracted driving. ***Platform Specificity and Equipment Limitations.*** Finally, we address what we believe are some of the critical hardware and software issues relevant to the use of driving simulators to study the effects of multitasking on driving performance.

27.1 Introduction

Driver distraction is a leading cause of motor vehicle accidents, accounting for up to 25% of crashes on the roadway (National Highway Transportation Safety Administration, 2001). There are many sources of driver distraction. Some "old standards" include talking to passengers, eating, drinking, lighting a cigarette, applying make-up, and listening to the radio (cf. Stutts et al., 2003). However, the last decade has seen an explosion of new wireless devices (e.g., surfing the Internet, sending and receiving e-mail or fax, communicating via cellular device, etc.) that have made their way into the automobile. It is likely that these new sources of distraction are more impairing than the old standards because they are more cognitively engaging and are often performed over more sustained periods of time.

This chapter focuses on how driving simulators can be used to evaluate the impact of cellular communication on driving performance because this is one of the most prevalent exemplars of this new class of multi-tasking activity. Indeed, the National Highway Transportation Safety Administration estimates that 8% of drivers on the roadway at any given daylight moment are using their cell phone (Glassbrenner, 2005). It is now well-established that cell phone use significantly impairs driving performance (e.g., McEvoy et al., 2005; Redelmeier & Tibshirani, 1997; Strayer & Johnston, 2001; Strayer, Drews, & Johnston, 2003; see Drews & Strayer, 2009 for a comprehensive review). Taken together, the data from these studies support an inattention-blindness interpretation wherein the disruptive effects of cell phone conversations on driving are due in large part to the diversion of attention from driving to the phone conversation (Strayer & Drews, 2007).

Reported herein is a driving simulator-based study using a car-following paradigm (see also Alm & Nilsson, 1995; Anderson & Sauer, 2007; Lee, Vaven, Haake, & Brown, 2001) to determine how driving performance is altered by conversations over a cell phone (see the Discussion part of this chapter for a consideration of other techniques for using driving simulators to study the impact of cellular communication on driving). Car following is an important requirement for the safe operation of a motor vehicle. In fact, failures in car following account for approximately 30% of police-reported accidents (National Highway Transportation Safety Administration, 2001). In our study, the performance of a non-distracted driver was contrasted with the performance of that same driver when they were conversing on either a hand-held or hands-free cell phone. We were particularly interested in examining the differences in driving performance of the hand-held cell

phone driver with that of the hands-free cell phone driver because there are several states that prohibit the former, while allowing the latter form of cellular communication (e.g., chap. 69 of the Laws of 2001, Section 1225c State of New York). Our analyses will show that the performance of drivers engaged in a cell phone conversation differs significantly from the non-distracted driver and that there is no safety advantage for hands-free over hand-held cell phones. We also will discuss why cell phone conversations differ from in-vehicle conversations with a passenger and show how another source of interaction with a cell phone, namely text messaging, produces a different, and potentially more dangerous, source of driver distraction.

27.2 Method

27.2.1 Participants

Forty adults (25 male and 15 female), recruited via advertisements in local newspapers and paid $10 per hour, participated in the study. Participants ranged in age from 22 to 34, with an average age of 25. All had normal or corrected-to-normal vision, normal color vision (Ishihara, 1993) and a valid driver's license with an average of eight years of driving experience. Seventy-eight percent of the participants owned a cell phone and 87% of the cell phone owners reported that they use their cell phone while driving.

27.2.2 Stimuli and Apparatus

A PatrolSim driving simulator, illustrated in Figure 27.1, was used in the study. The simulator incorporates proprietary vehicle

FIGURE 27.1   A participant using a hands-free cell phone while driving in the PatrolSim driving simulator.

dynamics, traffic scenario, and road surface software to provide realistic scenes and traffic conditions. The dashboard instrumentation, steering wheel, gas, and brake pedal were taken from a Ford Crown Victoria® sedan with an automatic transmission.

The simulator used a freeway road database simulating a 24-mile multi-lane highway with on and off-ramps, overpasses, and two and three-lane traffic in each direction. A pace car, programmed to travel in the right-hand lane, braked intermittently throughout the scenario. Distractor vehicles were programmed to drive between 5% and 10% faster than the pace car in the left lane, providing the impression of a steady flow of traffic. Four unique driving scenarios, counterbalanced across participants, were used for each of the conditions in the study. In each scenario, the pace car was programmed to brake at 32 randomly distributed locations. Measures of real-time driving performance, including driving speed, distance from other vehicles, and brake inputs, were sampled at 30 Hz and stored for later analysis.

Cellular service was provided by Sprint PCS. The cell phone was manufactured by LG Electronics Inc. (model TP1100). For hands-free conditions, a Plantronics M135 headset (with ear piece and boom microphone) was attached to the cell phone (see Figure 27.1).

27.2.3 Procedure

When participants arrived for the experiment, they completed a questionnaire assessing their interest in potential topics of cell phone conversation. Participants were then familiarized with the driving simulator using a standardized 20-minute adaptation sequence. Participants then drove four 10-mile runs on a multi-lane highway. The duration of each run was approximately 10 minutes, but varied as a function of the driving speed of each participant. The first and last runs in the study were used to assess performance in the single-task (i.e., non-distracted) driving condition and the second and third runs were used in the dual-task conditions (i.e., driving and conversing on a hand-held or hands-free cell phone). The order of hand-held and hands-free cell phone conditions was counterbalanced across participants.

The participant's task was to follow a periodically braking pace car that was driving in the right-hand lane of the highway. When the participant stepped on the brake pedal in response to the braking pace car, the pace car released its brake and accelerated to normal highway speed. If the participant failed to depress the brake, they would eventually collide with the pace car. That is, like real highway stop-and-go traffic, the participants were required to react in a timely and appropriate manner to vehicles slowing in front of them.

Figure 27.2 presents a typical sequence of events in the car-following paradigm. Initially, both the participant's car (solid line) and the pace car (long-dashed line) were driving at about 62 mph with a following distance of 40 meters (dotted line). At some point in the sequence, the pace car's brake lights illuminated for 750 msec (short-dashed line) and the pace car

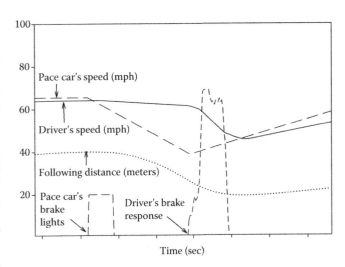

FIGURE 27.2   An example of the sequence of events occurring in the car-following paradigm.

began to decelerate at a steady rate. As the pace car decelerated, the following distance decreased. At a later point in time, the participant responded to the decelerating pace car by pressing the brake pedal. The time interval between the onset of the pace car's brake lights and the onset of the participant's brake response defines the brake reaction time. Once the participant depressed the brake, the pace car began to accelerate, at which point the participant removed his or her foot from the brake and applied pressure to the gas pedal. Note that in this example, the following distance decreased by about 50% during the braking event.

The dual-task conditions involved naturalistic conversation on a hand-held or hands-free cell phone with a research assistant. The participant and the research assistant discussed topics that were identified in the questionnaire as being of interest to the participant. These naturalistic conversations were unique to each participant and the research assistant was instructed to maintain a dialog in which the participant spoke and listened in approximately equal proportions. The phone call was initiated before the drive in order to eliminate any interference associated with dialing or answering the phone.

27.2.4 Dependent Measures

We examined four parameters associated with the participant's reaction to the braking pace car. *Brake reaction time* is the time interval between the onset of the pace car's brake lights and the onset of the participant's braking response (i.e., a 1% depression of the brake pedal). *Following distance* is the distance between the rear bumper of the pace car and the front bumper of the participant's car. *Speed* is the average driving speed of the participant's vehicle. *Recovery time* is the time it took participants to recover 50% of the speed that was lost during a braking episode.

27.2.5 Design and Statistical Analysis

We performed univariate analyses on each of the dependent measures using both Analysis of Variance (ANOVA) and a series of pair-wise t-tests. A significance level of $p < 0.05$ was adopted for all inferential tests.

27.3 Results

In order to better understand the changes in driving performance with cell phone use, we examined driver performance profiles in response to the braking pace car. Driving profiles were created by extracting 10-second epochs of driving performance which were time-locked to the onset of the pace car's brake lights. That is, each time that the pace car's brake lights were illuminated, the data for the ensuing 10 seconds were extracted and entered into a 32×300 data matrix (i.e., on the jth occasion that the pace car brake lights were illuminated, data from the 1st 2nd, 3rd, ..., and 300th observations following the onset of the pace car's brake lights were entered into the matrix $X_{[j,1]}, X_{[j,2]}, X_{[j,3]}...X_{[j,300]}$; where j ranges from 1 to 32 reflecting the 32 occasions in which the participant reacted to the braking pace car). Each driving profile was created by averaging across j for each of the 300 time points in a manner identical to that used to extract Event-Related Brain Potentials (ERPs) from time-locked electroencephalography measures (Donchin, Ritter, & McCallum, 1978). Here we report driving profiles of the participant's speed and following distance.

Figure 27.3 presents the driving speed profile, averaged across participants, time-locked to the onset of the pace car's brake lights, for the three conditions in the experiment. Over the 10-second epoch, participants in the single-task condition drove at a greater speed than when they were conversing on a cell phone, $F(2,78) = 3.3$, $p < 0.05$; however, vehicle speed during the pre-breaking interval did not differ significantly between conditions. Driving speed reached nadir between two and three seconds after the onset of the pace car's brake lights, whereupon the participant's vehicle reaccelerated towards pre-braking speed. The difference in overall speed was primarily determined by the time it took participants to recover the speed lost during braking. In particular, the recovery time was significantly shorter in the single-task condition than the hand-held or the hands-free cell phone conditions, $F(2,78) = 4.4$, $p < 0.01$. Subsidiary pair-wise t-tests indicated that single-task recovery was significantly faster than either the hand-held or the hands-free cell phone conditions and that the rate of recovery time did not differ for the two cell phone conditions. This sluggish behavior appears to be a key characteristic of the driver distracted by a cell phone conversation and such a pattern of driving is likely to have an adverse impact on the overall flow of dense highway traffic (e.g., Vladisavljevic & Martin, submitted).

Figure 27.4 presents the average following distance profile, time-locked to the onset of the pace car's brake lights, for the three conditions in the experiment. Over the 10-second epoch, participants in the single-task condition had a smaller following distance than when they were conversing on a cell phone, $F(2,78) = 3.7$, $p < 0.05$; however, following distance during the interval prior to the participant's brake response did not differ significantly between conditions. Inspection of the figure indicates that following distance decreased as the pace car began to decelerate and then increased as the participant applied their brakes and the pace car reaccelerated.

Figure 27.5 cross-plots driving speed and following distance to illustrate the relationship between these two variables over the braking episode. In the figure, the beginning of the epoch is indicated by an arrow and the relevant symbol (circle, triangle, or square) is plotted every third of a second in the time series. The distance between the symbols provides an indication of how each function changes over time (i.e., on a given function, symbols closer together indicate a slower change over time than symbols

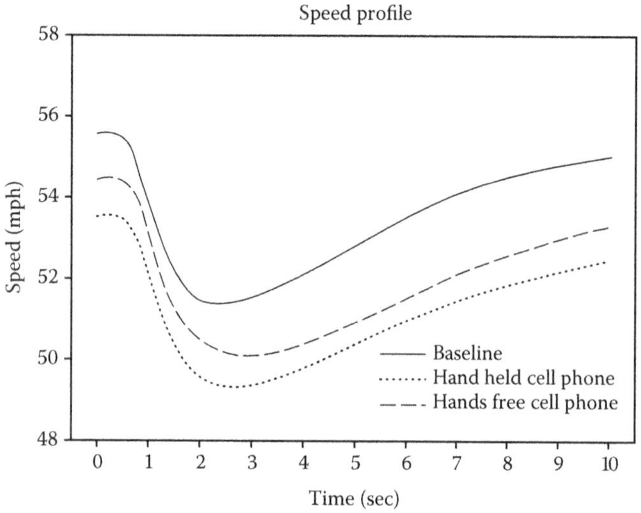

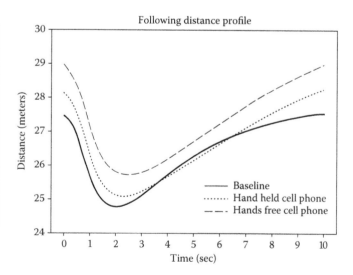

FIGURE 27.3 Participant's time-locked driving speed profile in response to the braking pace car.

FIGURE 27.4 Participant's time-locked following distance profile in response to the braking pace car.

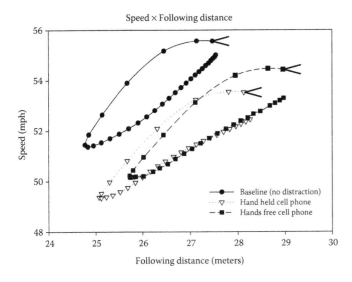

FIGURE 27.5 A cross-plot of driving speed and following distance. The "V" indicates the beginning of each plot.

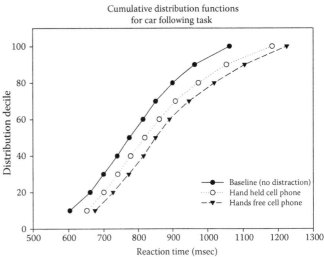

FIGURE 27.6 The reaction rime (RT) cumulative distribution functions for each of the conditions of the study.

farther apart). The figure clearly illustrates that the relationship between driving speed and following distance is virtually identical for the driver distracted by either a hand-held or hands-free cell phone. By contrast, the performance of the participant in single-task conditions provides a qualitatively different pattern than what is seen in the dual-task conditions. In particular, the functions representing the dual-task conditions are displaced towards the lower right quadrant, indicative of a driver operating the vehicle more conservatively (i.e., somewhat slower and greater following distance from the pace car) than in single-task conditions.

Figure 27.5 also illustrates the dynamic stability of driving performance following a braking episode. From a dynamic systems perspective, driving performance in single-and dual-task conditions can be characterized as operating in different speed-following distance basins of attraction with performance returning to equilibrium following each braking perturbation. Note also that the curves in Figure 27.5 for the non-distracted driver and the driver conversing on a cell phone did not intersect. This suggests that the basin of attraction created with either the hand-held or hands-free cell phone conversations was sufficiently "deep" that participants returned to their respective pre-braking set points after a braking episode had perturbed their position in the speed/following distance space.

We also examined the brake reaction time of participants as they reacted to the pace car's brake lights. The braking reaction time Vincentized cumulative distribution functions (CDFs) are presented in Figure 27.6. In the figure, the reaction time at each decile of the distribution is plotted and it is evident that the functions for the hand-held and hands-free cell phone conditions are displaced to the right, indicating slower reactions, compared to the single-task condition. Analysis indicated that reaction time (RT) in each of the dual-task conditions differed significantly from the single-task condition at each decile of the distribution, whereas the distributions for hand-held and hands-free conditions did not differ significantly across the deciles. A

companion analysis of median brake reaction time found that braking reactions were significantly slower in dual- than single-task conditions, $F(2,78) = 13.0$, $p < 0.01$. Subsidiary pair-wise t-tests indicated that the single-task condition differed significantly from the hand-held and hands-free cell phone conditions and the difference between hand-held and hands-free conditions was not significant.

The brake reaction times were also analyzed by fitting the CDFs for each subject and condition to the Ex-Gaussian distribution. The Ex-Gaussian, a convolution of the Gaussian and Exponential distributions, has been shown to do an excellent job of fitting reaction time data (see Pashler, 2002; Schmiedek, Oberauer, Wilhelm, Süß, & Wittmann, 2007) and is determined by three parameters, μ, σ, and τ. μ and σ represent the mean and standard deviation of the Gaussian distribution. τ parameterizes the Exponential distribution and is sensitive to the positive skew that typifies RT distributions (i.e., a τ of 0 would result in a symmetrical distribution). The mean of the total reaction time distribution is the sum of μ and τ and the overall variance of the reaction time distribution is the sum of σ^2 and τ^2. Importantly, the parameter τ has been linked to fluctuations in attention and cognitive control (e.g., Schmiedek et al., 2007), with greater values of τ reflecting greater attentional fluctuation. Recent research has shown that fluctuations in the τ parameter are associated with individual differences in frontal lobe functioning associated with executive attention (e.g., Watson & Strayer, 2010). Thus, attention-based interpretations of cell phone-induced driver distraction predict that τ would be greater in dual-task than in single-task conditions and that there would be little difference in the τ parameter between hand-held and hands-free dual-task conditions.

The CDFs for five of the participants were not well characterized by the Ex-Gaussian, resulting in a total of 35 participants' data to be fitted. For baseline conditions, the Ex-Gaussian parameters were 682, 69, and 55 for μ, σ, and τ, respectively. For the hand-held cell phone condition, the Ex-Gaussian parameters were 713, 72, and 94 for μ, σ, and τ, respectively. For the hands-free cell phone condition, the Ex-Gaussian parameters were 742, 58, and 96 for μ, σ, and τ, respectively. A Multivariate Analysis of Variance (MANOVA) found a significant effect of condition, $F(6,29) = 3.4$, $p < 0.01$. Subsidiary analyses found that both the μ and σ parameters significantly differentiated between hands-free and hand-held cell phone conditions, but the differences in μ and σ between single- and dual-task conditions were not reliable. By contrast, the parameter τ was significantly greater for dual- than single-task conditions whereas the difference in τ between the hand-held and hands-free dual-task conditions was not significant. This τ effect indicates that the dual-task RT CDFs were more positively skewed than the single-task condition. In effect, the differences between single- and dual-task conditions became greater as the reaction time interval increased (but note that the differences between single- and dual-task conditions were significant across the entire CDF). As noted above, the τ effect is suggestive of greater attentional fluctuations in the dual-task conditions; an interpretation that is consistent with the attention-based interpretations of cell phone-induced driver distraction proposed by Strayer and Drews (2007; see also Strayer, Drews, & Johnston, 2003; Strayer & Johnston, 2001). Moreover, ongoing research has found that participants with greater working memory capacity tend to show smaller τ–based effects than participants with low working memory capacity (e.g., Watson & Strayer, 2010). These findings suggest that there are important individual differences in multi-tasking ability that are related to the executive attentional system.

27.4 Discussion

Taken together, the data demonstrate that conversing on a cell phone impaired driving performance and that the distracting effects of cell phone conversations were not significantly different for hand-held and hands-free devices. Compared to single-task conditions, cell phone drivers' brake reaction times were slower and they took longer to recover the speed that was lost following braking. The cross-plot of speed and following distance showed that drivers conversing on a cell phone tended to have a more cautious driving profile, which may be indicative of a compensatory strategy to counteract the delayed brake reaction time. However, the compensation would appear to be inadequate because reaction times were longer and accident rates were higher in the dual-task than in the single-task conditions (three versus zero accidents, respectively). Elsewhere, Brown, Lee and McGehee (2001) found that sluggish brake reactions, such as the ones described herein, can increase the likelihood and severity of motor vehicle collisions.

One may be tempted to argue that increasing following distance and driving slower is desirable because this would tend to decrease the likelihood and severity of accidents. While this may hold for the non-distracted driver, the argument does not hold if drivers adopt this strategy so that they can take on more risk by engaging in additional multi-tasking activities. Furthermore, the compensatory changes in following distance and speed are

likely to have a direct impact on other motorists on the highway. Indeed, reaction time, speed, and following distance are all considered key variables in describing the stability and flow of traffic. From this systemic perspective, what may seem to be strategic compensation for an individual driver can also be viewed as a burden to the traffic system as a whole, increasing congestion and adding impediments to traffic flow. Models of highway traffic have, in fact, shown that traffic congestion increases in direct proportion to the number of drivers distracted by a cell phone on the highway (e.g., Vladisavljevic & Martin, submitted).

It is also important to note that performance decrements for both hands-free and hand-held cell phone conversations were obtained even when there was no possible contribution from dialing or answering the cell phone, thus calling into question legislation that restricts hand-held devices but permit hands-free devices. It is noteworthy that epidemiological-based studies comparing hand-held and hands-free cell phones have also found no safety advantage for hands-free cell phones over hand-held units (e.g., McEvoy et al., 2005; Redelmeier & Tibshirani, 1997). In particular, laws that prohibit the use of hand-held cell phones while allowing the use of hands-free cell phones are not likely to eliminate the problems associated with using cell phones while driving because these problems can be attributed in large part to the distracting effects of the phone conversations themselves (see also Strayer & Drews, 2007).

It is interesting to contrast cell phone conversations with another type of conversation commonly engaged in while driving: conversation with a passenger seated next to the driver in the vehicle. On the one hand, given that both dual-tasks involve a conversation, one possibility is that there would be no difference between these two forms of communication. On the other hand, these conversation types tend to differ because passengers can adjust their conversation based on driving difficulty; often helping the driver to navigate and identify hazards on the roadway and pausing the conversations during difficult sections of the drive. This real-time adjustment based upon traffic demands is not possible with cell phone conversations.

In fact, we recently conducted a study in which participants were required to drive on a multi-lane freeway and exit at a rest stop approximately eight miles down the road (Drews, Pasupathi, & Strayer, in press). We found that the majority of drivers (88%) who were conversing with a passenger successfully completed the task of navigating to the rest area, whereas only 50% of the drivers talking on a cell phone successfully navigated to the rest area. Analysis of the video recordings indicated that a primary difference between these two modes of communication was that the passenger helped the driver in the navigation task by reminding them to exit at the rest stop. Moreover, our analysis of the content of the conversation indicated that references to traffic conditions were more likely with passenger conversations than with cell phone conversations.

Additional evidence of differences between cell phone conversations and passenger conversations comes from epidemiological studies showing differences in the relative risk of being in a motor vehicle accident when concurrently engaging in one of these conversations. As noted above, for cell phone drivers (either hand-held of hands-free), the relative risk of being in a motor vehicle accident increases by a factor of four (e.g. Redelmeier & Tibshiraini, 1997; McEvoy et al., 2005). By contrast, epidemiological studies (Rueda-Domingo et al., 2004; see also Vollrath, Meilinger, & Krueger, 2002) find a strikingly different pattern for situations where a passenger is in the vehicle. In particular, when drivers have a passenger in the vehicle, the relative risk of a motor vehicle accident is lower than when the driver drives by him or herself (i.e., the odds ratio of an accident with a passenger in the vehicle is 0.7). Thus, there is clear evidence from both simulator and epidemiological studies that passenger and cell phone conversations have very different effects on driving performance (see Ouimet, Duffy, Simons-Morton, Brown, & Fisher, this book, chap.24 on the effects of passenger conversations on teen drivers).

Another mode of cell phone use that is gaining popularity, particularly among teen drivers, is text messaging while driving (Drews, Yazdani, Godfrey, Cooper, & Strayer, 2009). We recently completed a study where we examined the effects of text messaging on driving performance using the car-following paradigm described in this chapter. Overall, we found that the relative risk of being involved in a crash increased by a factor of six when drivers were texting. Moreover, the brake reaction time CDFs yielded a very different pattern from what was observed with cell phone conversations (Figure 27.6). In particular, the text messaging CDFs did not differ from the non-distracted baseline in the first five deciles of the distribution and then significantly diverged, with slower reactions in dual-task conditions. That is, the text messaging CDFs indicated that about 50% of the trials showed no dual-task cost and the remainder exhibited a significant lengthening of brake reaction time (by contrast, recall that the CDF at each decile differed significantly from the non-distracted baseline with cell phone conversations). This pattern suggests that drivers switched in and out of an impaired state when text messaging (exhibiting the greatest impairment when typing in a message) whereas drivers remained in an impaired state throughout the duration of a cell phone conversation (i.e., while both talking and listening). This observation highlights the utility of examining the reaction time distributions, because two very distinct patterns of dual-task performance were revealed for cell phone conversations and text messaging.

Car following is a very useful paradigm for studying driver distraction because it provides diagnostic dependent measures (e.g., brake reaction time, driving speed, following distance, etc.) that can help to identify the changes in driving behavior with different kinds of distraction. The version of the car-following paradigm reported in this chapter included discrete braking events that required the driver to react with alacrity to avoid a collision (thereby providing useful measures of brake reaction time). However, Anderson and Sauer (2007) reported a version of the car-following paradigm that did not include discrete braking events, but rather used a linear combination of sinusoidal functions to vary the speed of the lead vehicle. In this version of car-following, measures of following distance and coherence can be used to assess driver distraction. There

are a number of other driving scenarios that have been used to study driver distraction in the simulator. For example, Cooper, Vladisavljevic, Strayer and Martin (in press) varied traffic density on an interstate highway and compared lane changing behavior of drivers who were talking on a hands-free cell phone with those same drivers when they were not distracted. A similar scenario was used by Drews, Pasupathi and Strayer (2008) in which participants were required to drive until they came to a rest area. As noted above, the use of a cell phone significantly increased the likelihood of missing the rest area exit, whereas this was not the case for in-vehicle conversations. Strayer, Drews and Johnston (2003; see also Strayer & Drews, 2007) used an eye tracker to determine what information in the driving scene was processed by the distracted driver (compared to a non-distracted baseline). In these studies, objects that varied in relevance to safe driving were placed along the interstate and participants were subsequently probed for their recognition of the objects that they encountered. With the use of eye-tracking measures, the assessment of recognition memory was based on just those instances where the participant actually fixated upon the objects during the driving portion of the study. The analyses indicated that drivers talking on a hands-free cell phone had a 50% reduction in recognition memory relative to their non-distracted baseline performance. Furthermore, when Event-Related Brain Potentials (ERPs) were recorded to the onset of braking events in a car-following paradigm identical to that reported in this chapter, Strayer and Drews (2007) found that the amplitude of the attention-sensitive P300 component was reduced by 50%. This partial list indicates the diversity of driving simulator scenarios that have been used to study cell phone-induced driver distraction and that each paradigm provides the opportunity to collect important dependent measures that allow the researcher to draw inferences about the nature and source of driver distraction. For a more comprehensive analysis of the methods used to study cell phone-induced driver distraction, see Drews and Strayer (2009).

We believe that the most important characteristic of driving simulation is that it provides sufficient functional realism to mimic the phenomena of interest (see Kantowitz's discussion of microworlds, this book, chap. 3). As heretical as it may seem, we have not found that motion-based simulators are necessary to study driver distraction and in some cases a simple pursuit-tracking task can shed light on the important mechanisms of dual-task interference better than a high-fidelity simulator The majority of our research, however, has been performed using the high-fidelity fixed-base driving simulator built by L3 communications (formerly I-SIM, see Figure 27.1), and we believe that this level of simulation is necessary and sufficient for most studies of driver distraction. For successful driving simulation, it is important to have complete control of the driving environment so that creative scenarios can be repeated in a carefully controlled and counterbalanced study. Perhaps more important is the requirement to be able to collect meaningful dependent measures from the simulation. In our studies, we collect a large volume of real-time data from the simulation for off-line analysis (in our studies, this required the development of customized software for data acquisition and data analysis). It is also essential to be able to interface peripheral devices with the simulator (e.g., eye trackers, ERP recording equipment, etc.) to compliment the data recorded from simulation.

A final comment concerns the nature of the cell phone conversations in our study. Unlike earlier research using working memory tasks (Alm & Nilsson, 1995; Briem & Hedman, 1995), mental arithmetic tasks (McKnight & McKnight, 1993; Harbluk, Noy, & Eizenmann, 2002), reasoning tasks (Brown, Tickner, & Simmonds, 1969) or simple word generation (Strayer & Johnston, 2001), the conversations in our current study were designed to be naturalistic casual conversations centering on topics of interest to the participant. However, one limitation with naturalistic conversation is the lack of adequate methods for measuring the cognitive load imposed in the dual-task conditions. That is, there is a tradeoff between the fidelity of the conversation and the quality of the dependent measures that can be used to quantify the nature of the conversation. In some situations, it is desirable to make the conversation as naturalistic as possible (e.g., Drews, Pasupathi, & Strayer, 2008), whereas in other situations it is important to be able to precisely quantify the parameters of the conversation surrogate (e.g., Watson & Strayer, 2010).

In sum, driving simulators have helped to provide converging evidence into the nature of driver distraction, elucidating the causal mechanisms underlying the interference. The strength of diving simulation is that it allows for rigorous experimental control that can determine the underlying nature of patterns of association obtained with observational and epidemiological studies. In the specific case of concurrent use of cell phones while driving, the data indicate that cell phones cause impairments to driving by diverting attention from the processing of information necessary for the safe operation of a motor vehicle and that this impairment is similar for both hand-held and hands-free devices.

Key Points

- Driving simulators provide a valuable tool for the scientific study of driving behavior and driver distraction.
- Laboratory studies using driving simulators can establish causal relations between different activities performed in the vehicle and driving impairment.
- Conversing on a cell phone significantly impairs driving performance.
- The proximal cause for cell-phone-induced driver distraction is inattention blindness.
- Hand-held and hands-free cell phones cause identical patterns of driving impairment.

Keywords: Attention, Car Following, Cell Phones, Driver Distraction, Inattention Blindness, Reaction Time

Acknowledgments

Portions of the data reported in this study were published in aggregated form in Strayer, Drews and Crouch (2006). This research was funded in part by a grant from the Federal Aviation Administration.

Glossary

Driver distraction: An impairment to driving performance caused by a concurrent activity that is unrelated to the task of driving (e.g., talking on a cell phone, eating, drinking, etc.).

Inattention blindness: A decrement in seeing or detecting objects caused by a diversion of attention from the processing of information in the visual field.

Key Readings

McEvoy, S. P., Stevenson, M. R., McCartt, A. T., Woodward, M., Haworth, C., Palamara, P., & Cercarelli, R. (2005). Role of mobile phones in motor vehicle crashes resulting in hospital attendance: A case-crossover study. *British Medical Journal, 331,* 428–433.

Redelmeier, D. A., & Tibshirani, R. J. (1997). Association between cellular-telephone calls and motor vehicle collisions. *The New England Journal of Medicine, 336,* 453–458.

Strayer, D. L., & Drews, F. A. (2007). Cell phone-induced inattention blindness. *Current Directions in Psychological Science, 16,* 128–131.

Strayer, D. L., Drews, F. A., & Crouch, D. J. (2006). Comparing the cell phone driver and the drunk driver. *Human Factors, 48,* 381–391.

Strayer, D. L., & Johnston, W. A. (2001). Driven to distraction: Dual-task studies of simulated driving and conversing on a cellular phone. *Psychological Science, 12,* 462–466.

References

Alm, H., & Nilsson, L. (1995). The effects of a mobile telephone task on driver behavior in a car following situation. *Accident Analysis & Prevention, 27*(5), 707–715.

Anderson, G. J., & Sauer, C. W. (2007). Optical information for car following: The driving by visual angle (DVA) model. *Human Factors, 49,* 878–896.

Briem, V., & Hedman, L. R. (1995). Behavioral effects of mobile telephone use during simulated driving. *Ergonomics, 38*(12), 2536–2562.

Brown, I. D., Tickner, A. H., & Simmonds, D. C. V. (1969). Interference between concurrent tasks of driving and telephoning. *Journal of Applied Psychology, 53*(5), 419–424.

Brown, T. L., Lee, J. D., & McGehee, D. V. (2001). Human performance models and rear-end collision avoidance algorithms. *Human Factors, 43,* 462–482.

Cooper, J. M., Vladisavljevic, I., Medeiros-Ward, N., Martin, P. T., & Strayer, D. L. (2009). Near the tipping point of traffic stability: An investigation of driving while conversing on a cell phone in simulated highway traffic of varying densities. *Human Factors, 51,* 261–268.

Donchin, E., Ritter, W., & McCallum, C. (1978). Cognitive psychophysiology: The endogenous components of the ERP. In E. Callaway, P. Tueting, & S. Koslow (Eds.), *Brain event-related potentials in man* (pp. 371–387). New York: Academic Press.

Drews, F. A., Pasupathi, M., & Strayer, D. L. (2008). Passenger and cell phone conversation during simulated driving. *Journal of Experimental Psychology: Applied, 14*(4), 392–400.

Drews, F. A., & Strayer, D. L. (2009). Cellular phones and driver distraction. In M. A. Regan, J. D. Lee, & K. Young (Eds.), *Driver distraction: Theory, effects, and mitigation.* Boca Raton, FL: CRC Press.

Drews, F. A., Yazdani, H., Godfrey, C., Cooper, J. M., & Strayer, D. L. (2009). Text messaging during simulated driving. *Human Factors, 51,* 762–770.

Glassbrenner, D. (2005). Driver cell phone use in 2004 – Overall results. *Traffic safety facts. Research note* (DOT HS 809 847).

Harbluk, J. L., Noy, Y. I., & Eizenmanm, M. (2002, January). *Impact of cognitive distraction on driver visual behavior and vehicle control.* Paper presented at the 81st Annual Meeting of the Transportation Research Board, Washington, DC.

Ishihara, S. (1993). *Ishihara's test for color-blindness.* Tokyo: Kanehara.

Lee, J. D., Vaven, B., Haake, S., & Brown, T. L. (2001). Speech-based interaction with in-vehicle computers: The effects of speech-based e-mail on drivers' attention to the roadway. *Human Factors, 43,* 631–640.

McEvoy, S. P., Stevenson, M. R., McCartt, A. T., Woodward, M., Haworth, C., Palamara, P., & Cercarelli, R. (2005). Role of mobile phones in motor vehicle crashes resulting in hospital attendance: A case-crossover study. *British Medical Journal, 331,* 428–433.

McKnight, A. J., & McKnight, A. S. (1993). The effect of cellular phone use upon driver attention. *Accident Analysis & Prevention, 25*(3), 259–265.

National Highway Transportation Safety Administration. (2001). *Traffic safety facts – 2001* (Rep. DOT 809 484). Washington, DC: US Department of Transportation.

Pashler, H. (2002). Analysis of response time distributions. In H. Pashler, S. Yantis, D. Medin, R. Gallistal, & J. Wixted (Eds.), *Steven's handbook of experimental psychology* (3rd ed., p. 472). Hoboken, NJ: John Wiley & Sons.

Redelmeier, D. A., & Tibshirani, R. J. (1997). Association between cellular-telephone calls and motor vehicle collisions. *The New England Journal of Medicine, 336,* 453–458.

Rueda-Domingo, T., Lardelli-Claret, P., Luna-del-Castillo, J., Jiméénez-Moleón, J., Garcia-Martin, M., & Bueno-Cavanillas, A. (2004). The influence of passengers on the risk of the driver causing a car collision in Spain: Analysis of collisions from 1990 to 1999. *Accident Analysis & Prevention, 36,* 481–489.

Schmiedek, F., Oberauer, K., Wilhelm, O., SüB, H.-M., & Wittmann, W. W. (2007). Individual differences in components of reaction time distributions and their relations to working memory and intelligence. *Journal of Experimental Psychology: General, 136,* 414–429.

Strayer, D. L., & Drews, F. A. (2007). Cell phone-induced inattention blindness. *Current Directions in Psychological Science, 16,* 128–131.

Strayer, D. L., Drews, F. A., & Crouch, D. J. (2006). Comparing the cell phone driver and the drunk driver. *Human Factors, 48,* 381–391.

Strayer, D. L., Drews, F. A., & Johnston, W. A. (2003). Cell phone-induced failures of visual attention during simulated driving. *Journal of Experimental Psychology: Applied, 9,* 23–52.

Strayer, D. L., & Johnston, W. A. (2001). Driven to distraction: Dual-task studies of simulated driving and conversing on a cellular phone. *Psychological Science, 12,* 462–466.

Stutts, J., Feaganes, J., Rodman, E., Hamlet, C., Meadows, T., Rinfurt, D., . . . Staplin, L. (2003). *Distractions in everyday driving.* AAA Foundation for Traffic Safety. Available from the AAA Foundation Web site: http://www.aaafoundation.org/pdf/distractionsineverydaydriving.pdf

Vladisavljevic, I., & Martin, P. (submitted). *Impact of cell phone conversation while driving on car-following behavior.*

Vollrath, M., Meilinger, T., & Krüger, H.-P. (2002). How the presence of passengers influences the risk of a collision with another vehicle. *Accident Analysis & Prevention, 34,* 649–654.

Watson, J. M., & Strayer, D. L. (2010). Supertaskers: Profiles in extraordinary multi-tasking ability. *Psychonomic Bulletin and Review, 17*(4), 479–485.

28

Night Driving: How Low Illumination Affects Driving and the Challenges of Simulation

Joanne Wood
*Queensland University
of Technology*

Alex Chaparro
Wichita State University

Abstract

The Problem. Night-time driving can be dangerous. When adjusted for mileage, the fatality rate at night is two to four times higher than during the daytime. The risks are markedly elevated for pedestrians, who are up to seven times more vulnerable to a fatal collision at night. Although multiple factors, including alcohol and fatigue, contribute to the increased fatality rate at night, poor visibility is a leading cause of collisions with pedestrians, cyclists, and other low-contrast obstacles, and these effects are exacerbated for older drivers. *Role of Driving Simulators.* Driving simulators can assist in understanding the problems of night-time driving and have a role in evaluating interventions developed to improve night-time driving safety. However, there are a number of factors that are very different when driving at night compared to the daytime which have important implications for the design of simulator scenarios that seek to represent the night-time driving situation. *Differences between Day and Night-time Driving.* These include reduced ambient illumination levels and headlamps (both those of the driver and of oncoming vehicles), which can have a dramatic impact on the visual performance of the driver, as well as the fatigue and alcohol levels of drivers in the traffic mix. This chapter will emphasize the impact of reduced illumination levels on driving performance. *Scenarios and Dependent Variables.* Examples of the sorts of scenarios that have been used to study night-time driving both in simulators and the real world will be discussed, together with an examination of the key differences in driving performance between day and night-time. The limitations and difficulties in replicating appropriate scenarios will be outlined.

28.1 Introduction

Crash statistics indicate that driving at night-time is more dangerous than during the daytime. While only 25% of all traffic is present at night-time, and the distances traveled are shorter,

the number of crashes are equal under both day and night-time conditions (Rumar, 2002). When adjusted for mileage, the night-time fatality rate is two to four times higher than that for daytime (National Safety Council, 1999–2004). Recent data also suggest that night-time crashes are more severe than those occurring during the day (Plainis, Murray, & Pallikaris, 2006). The night-time elevation in road safety risk has been shown to be even greater for pedestrians, who are up to seven times more vulnerable to a fatal collision at night than in the day (Sullivan & Flannagan, 2002).

Although multiple factors, including alcohol and fatigue, may contribute to the elevated night-time fatality rate, the basic difference between night and daytime driving is the reduction in illumination at night. The increase in crash rates, particularly collisions with pedestrians, cyclists, and other low-contrast obstacles during night-time driving, can be attributed largely to poor visibility (Owens & Sivak, 1996). Recent findings show that even the low illumination levels available from a full moon can have a positive effect on pedestrian fatalities. For example, Sivak, Schoettle and Tsimhoni (2007) found that pedestrian fatalities were 22% lower on nights with a full moon than on moonless nights. The problems of poor visibility are further compounded by drivers' misjudgement of their visual limitations at night, where drivers' confidence appears to be largely based upon their lane-keeping ability at night, which is relatively unimpaired compared to the significant decrement in "focal" vision found under low-light conditions (Owens, 2003; Owens & Tyrrell, 1999). Despite reduced visibility under night-time conditions, studies of general traffic flow report little or no difference between average speeds under day and night conditions (Herd, Agent, & Rizenbergs, 1980), demonstrating that drivers do not slow down to compensate for the decrease in their visual performance.

An important challenge for researchers is to better understand the problems involved in night-time driving; yet there has been only limited research in this area, despite the risk to road safety that it incurs. In addition, there has been a widespread emergence of night vision systems and other devices which purport to assist in night-time driving, including near and far infrared night vision systems (Tsimhoni, Bargman, Minoda, & Flannagan, 2004) and other devices (e.g. adaptive headlamps), which require proper evaluation and validation under as realistic conditions as possible to ensure that they actually do provide safety benefits. While driving simulators potentially have a role in investigations of night-time driving safety issues, it is imperative that researchers undertaking such studies are aware of the differences involved in day and night-time driving, and the limitations that simulators have in accurately replicating the night-time driving environment.

28.2　What Are the Key Differences Between Driving Under Day and Night-Time Conditions?

There are a number of differences in the driving environment encountered under day and night-time conditions which need to be taken into account when simulating night driving. Of central concern is the ability to replicate, as nearly as possible, the levels of illumination derived from the headlamps of the driver's vehicle and that of oncoming vehicles, as well as those from ambient lighting sources (e.g. street lights, moonlight). These illumination conditions (both in terms of absolute values and variations in these levels) will have a significant impact on the visual performance of the driver. Failure to replicate these illumination levels, or take them into account, may invalidate the outcomes of results and any potential inferences drawn from them.

28.2.1　Luminance

28.2.1.1　Street Lighting

The level of street lighting differs by road type, traffic density and a range of other factors (Kaptein, Hogema, & Folles, 1997). The standard illumination levels of busy, well lit roads at night range from 0.5–1.5 cd/m², whereas those of less-busy roads, which have a lower traffic density, range from 0.35–0.5 cd/m² (Joint Technical Committee LG-002, 2005a). These values represent luminance levels measured at 1.5 m above the carriageway from a point halfway across the driver's section of the carriageway (Joint Technical Committee LG-002, 2005b). While the regions beside the roadway are not normally covered by lighting recommendations, they are important for the detection of potential hazards, such as animals or pedestrians, (Rea, 2001) and should be accounted for in simulation research.

Analysis of fatal crash data indicates that pedestrian fatality rates are particularly sensitive to ambient light levels. Whereas single vehicle run-off-road crashes show little difference between dark and light periods, pedestrians are up to seven times more at risk at night than in the day (Sullivan & Flannagan, 2002). Importantly, the number of night-time crashes involving vehicles and pedestrians has been shown to decrease following the installation of overhead lighting (Elvik, 1995), and when the days are longer during the summer months in the Northern Hemisphere (Owens & Sivak, 1996). Recent studies have also shown that the severity of crashes at night is worse than in the daytime and that injury severity at night is reduced in the presence of street lighting by a factor of three (Plainis et al., 2006).

The effects of inclement weather conditions are also likely to have a greater effect at night than under daytime conditions given that wet road surfaces, unlike dry surfaces, produce specular reflection at night-time, increasing the effects of disability glare (Figure 28.1). Indeed, Gordon (1977) found that observers reported poorer visibility of roadside targets, including pedestrians wearing reflective clothing, when road surfaces were wet. These effects are likely to be exacerbated for older drivers, especially those with cataracts.

28.2.1.2　Headlamps

Vehicle headlamps provide a three-dimensional illumination of the roadway that varies with the type and configuration of headlamps, as well as whether the headlamps are used at a low- or high-beam setting. Interestingly, field studies indicate that most drivers continue to use their low-beam headlamps under

FIGURE 28.1 Photograph of a wet roadway at night showing the reflections and disability glare resulting from oncoming vehicle headlamps. (Courtesy of Trent Carberry.)

roadway conditions that require the use of high-beams (Sullivan, Adachi, Mefford, & Flannagan, 2004), suggesting that they are not concerned by the limited visibility provided by the illumination from low-beam headlamps (Andre & Owens, 2001). A roadside survey of motorists during the long Scandinavian winter nights found that most drivers did not recognize the adverse effects of their dirty headlamps until luminous intensity had decreased by 60% or more (Rumar, 1974).

The importance of replicating the lighting effects of headlamp beams is particularly critical given the body of evidence advocating the importance of different forms of headlamps in reducing night-time vehicle and pedestrian crashes. In field studies, supplementary headlamps emitting ultraviolet light (UVA) have been shown to have significant benefits for the detection and recognition of pedestrians when used in conjunction with normal low-beam headlamps (Lestina, Miller, Langston, Knoblauch, & Nitzburg, 2002). Changing from low- to high-beams has been shown to increase pedestrian recognition distances, particularly for darkly clad pedestrians (Mortimer & Olson, 1974; Shinar, 1984; Wood, Tyrrell, & Carberry, 2005), with one study finding that recognition distances increased by a factor of 3.5× for high- compared to low-beams (Wood, Tyrrell, & Carberry, 2005).

Sivak, Flannagan, Kojima and Traube (1997) reported an extensive photometric analysis of low-beam headlamp patterns for a sample of different vehicles and found significant variations between vehicles. Building upon this research, Andre and Owens (2001) took a user-centered approach to describe the area of visibility provided to the driver by a range of different headlamp designs. Rather than defining the beam patterns using conventional photometric approaches, they characterized the headlamp beams in terms of the three-dimensional area of useful illumination in the road environment, known as the twilight envelope. The envelope boundaries were based upon general limitations of visual function at the fovea (the central and most sensitive portion of the retina), using a limit of 3.3 lux as the criterion illuminance. They found large inter-vehicle differences in the twilight envelope

(see Figure 28.2) which they hypothesized arose from differences in headlamp size, optical characteristics of the reflectors, lenses, filament position, aiming and variations in voltage.

It is important to note that the reported light levels for headlamps, or those of driving scenes, are based on lux or luminance meters that use the photopic spectral luminosity efficiency function—this describes the variation in sensitivity of the visual system to light of different wavelengths. This function, however, is correct only for the photopic or cone mediated vision at luminance levels greater than approximately 3 cd/m^2 (Alferdinck, 2006), which is higher than the recommended European and American standards for average roadway illumination (Viikari, Ekrias, Eloholma, & Halonen, 2008). Figure 28.3 shows the spectral luminosity function for both low luminance, rod-mediated scotopic, and high luminance, cone-mediated photopic visual systems.

The specification of mesopic light levels, or the effective illuminance produced by lights or target stimuli under real or simulated night-time conditions, is hampered by the absence of an accepted spectral luminosity function for mesopic light levels where both photopic (cones) and scotopic (rods) contribute to vision. Indeed, at present there is still no widely accepted method for calculating the sensitivity of the human visual system at mesopic luminance levels, or predicting the relative visibility of target stimuli that might be used in simulations.

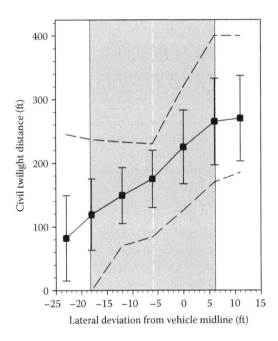

FIGURE 28.2 Mean (solid line) of the civil twilight envelope measured at ground level from a sample of 13 different vehicles. The dashed black lines represent the shortest and longest illumination patterns from the sample of 13 vehicles. The error bars indicate ±1 SD. The grey background illustrates the dimensions of a two-lane roadway. (Adapted from Andre, J., & Owens, D. A. (2001). The twilight envelope: A user-centered approach to describing roadway illumination at night. *Human Factors, 43*(4), 620–630. With permission.)

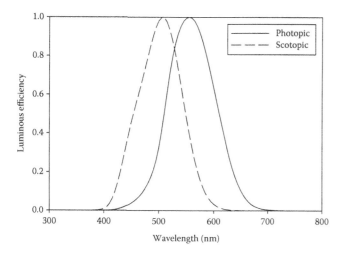

FIGURE 28.3 Spectral luminosity function of the human visual system under photopic (high) and scotopic (low) light levels.

Collectively, the studies described in this section serve to underscore the importance of including consideration of headlamp beams and roadway illumination in studies of night-time driving and road object visibility. While many simulator studies have attempted to address some of these issues (see Section 28.4.1) they still have important limitations which must be acknowledged when interpreting study findings. See also Andersen, this book (chap. 8) for a review of other sensory and perceptual considerations relevant to the design of simulation displays.

28.3 How Does Reduced Illumination Impact on Drivers' Visual Performance

Under typical night-time driving conditions the driver's adaptation levels tend to fall within the lower end of the mesopic luminance region (Plainis, Chauhan, Murray, & Charman, 1999), with a recent study reporting that both photopic (cone) and scotopic (rod) mechanisms mediate visual function during night driving in dark rural environments (Plainis, Murray, & Charman, 2005). Most aspects of visual function are significantly impaired under dim conditions because vision is mediated to a greater extent by the rod photoreceptors, which have relatively poor spatial and temporal processing abilities. The other photoreceptor type, the cones, operate at high or photopic luminance levels and are

responsible for mediating fine detailed vision, color vision and rapid motion perception. Hence, under night-time driving conditions, the visual environment is relatively impoverished and this has an important impact upon performance under real-world night driving conditions. Figure 28.4 shows the range of scotopic, mesopic and photopic vision and corresponding luminance values expressed in cd/m^2.

28.3.1 Spatial Vision

As luminance levels decrease and vision is mediated to a greater extent by rods, which have coarser spatial processing characteristics than cones, the spatial resolution of the visual system is decreased (Barlow, 1965). Visual acuity is thus reduced centrally (Arumi, Chauhan, & Charman, 1997; Johnson & Casson, 1995; Sturr, Kline, & Taub, 1990; Wood & Owens, 2005) and peripherally (Bedell, 1987). There are also reductions in the contrast sensitivity function (CSF), which describes the lowest contrast level at which objects of different sizes (i.e., spatial frequency) can be detected. At lower luminance levels the high spatial frequency cut-off (the finest detail that can be seen at 100% contrast) moves to lower frequencies, in line with the reduction in visual acuity, and peak sensitivity is reduced and shifts to progressively lower spatial frequencies (Woodhouse & Barlow, 1982). At very low luminance levels, the CSF changes from a bandpass to a low-pass function. There is also an additive effect of decreased luminance and contrast with the amount of refractive error (degree of short- or long-sightedness) of the individual (Johnson & Casson, 1995).

Importantly, the link between visual changes and illumination levels is not linear and different aspects of visual function are affected to a greater or lesser extent by changes in illuminance. This was demonstrated in driving simulator studies which assessed the differential effects of reduced luminance (through the use of neutral density filters mounted before the participants' eyes) on those tasks considered to be "focal", such as visual acuity and object recognition, and those of "ambient" vision, including steering, maintenance of lane position, heading and speed (Brooks, Tyrrell, & Frank, 2005; Owens & Tyrrell, 1999). These studies supported a key element of the so called selective degradation hypothesis advocated by Leibowitz and colleagues, which suggests that "ambient" vision, which mediates steering abilities and the ability to maintain lane position, is preserved at night even when the ability to recognize objects and hazards mediated by "focal" vision is not. Drivers are generally

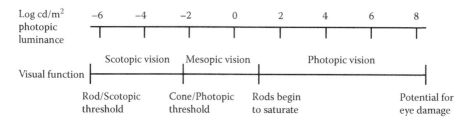

FIGURE 28.4 Relationship between luminance and visual function. (Adapted from Hood, D. C., & Finkelstein, M. A. (1986). Sensitivity to light. In K. R. Boff, L. Kaufman, & J. P. Thomas (Eds.), *Handbook of perception and human performance*. New York, NY: John Wiley & Sons.)

unaware of this phenomenon because many objects in the night-time environment, such as road signs, are engineered to be highly visible at night and drivers experience little difficulty in steering.

Disparities in the spatial resolution of simulations and the visual system of the observers can also pose a challenge for researchers interested in studying night-time sign legibility or drivers responses to road signs using simulators. The spatial resolution of these displays is lower than the spatial resolution of the eye, which means that signs rendered on the display may not be as legible as signs viewed from the same distance in the real-world (see, Andersen, this book, chap. 8). This, combined with the fact that the reflective properties of signs are not well modeled by simulations, means that researchers should employ caution in the interpretation of their findings.

28.3.2 Temporal Processing and Motion Sensitivity

Temporal processing and motion perception are both impaired under night-time conditions because of the poor temporal processing characteristics of the rods. Hence when driving at night, because of the involvement of the rod pathways, there is a possibility of underestimating the speed of other vehicles given that psychophysical studies report that velocity perception decreases by as much as 20% under scotopic compared to with photopic conditions (Gegenfurtner, Mayser, & Sharpe, 1999; Gegenfurtner, Mayser, & Sharpe, 2000). These findings raise the possibility that, in the absence of speedometer information, drivers may underestimate their actual speed at night. Contrary to this hypothesis, however, Triggs and Berenyi (1982) found that speed estimates made by passengers while riding on a freeway were underestimated in daylight and were generally more accurate at night.

Considerations of the temporal processing of the rod and cone systems have important implications for the acceptable refresh rates for the visual displays that are used in simulations. Because of the lower temporal resolution of the rod visual system monitor, refresh rates can be lower without significantly impacting on the observers' perception of smooth motion. This is not the case for higher mesopic and photopic illumination levels where the cone visual system supports higher temporal resolution (see this book, chap 8).

28.3.3 Glare Sensitivity

Glare from light scattered from the headlamps of oncoming vehicles is a particular problem at night and can result in discomfort glare (the sensation of discomfort in response to the glare source) or disability glare (the reduction in visual function resulting from the light scattered from the glare source) which can substantially impair a driver's vision (Pulling, Wolf, Sturgis, Vaillancourt, & Dolliver, 1980). Glare originates from light scatter in the optical media of the eye, which results in a veil of stray light over the retina that reduces the contrast of the retinal image. Stray light does not change significantly for pupil diameters between two and eight mm (Franssen, Tabernero, Coppens, & van den Berg, 2007), which led these authors to conclude that stray light or glare sensitivity measured with photopic pupils can be used to predict stray light for mesopic and scotopic pupils, such as those encountered under night-time driving. Glare from oncoming headlamps has been shown to exacerbate the effects of simulated mild lens opacities, such that dynamic visual acuity, the ability to resolve details of a moving object, is reduced by a factor of six in the presence of glare from vehicle headlamps on low- compared to high-beam (Anderson & Holliday, 1995).

While the potential effect of glare on driving performance is frequently alluded to for both day and night-time driving, there is little evidence to support these assertions. This failure has been attributed to methodological difficulties in defining "glare" and measuring it, as well as to a poor understanding of what drivers mean when they report glare problems (Owsley & McGwin, 1999). Indeed, studies of drivers with cataracts who are widely reported to suffer with glare problems, have failed to find an association between glare sensitivity and unsafe driving performance (Owsley, Stalvey, Wells, Sloane, & McGwin, 2001; Wood & Carberry, 2006).

Researchers have sought to simulate the effects of glare in a variety of ways, including mounting extraneous light sources to simulate the glare of oncoming head lamps (see Section 28.4.1) (e.g., Featherstone et al., 1999; Horberry, Anderson, & Regan, 2006). This does at least replicate some aspects of the stray light from headlamps experienced under night-time driving conditions, but does not reflect the dynamic nature of the illumination at night, nor reflections from the roadway.

28.3.4 Perception of Depth

Relative to photopic levels, the variability of inter-individual distance and depth estimations is increased under mesopic and scotopic levels (Bourdy, Cottin, & Monot, 1991). This may reflect changes in the availability and effectiveness of monocular and binocular depth cues under low illumination that, in normal daytime viewing conditions, aid in judging the relative distance of objects. Some of the monocular depth cues, such as texture density and relative size, may only be available in the near region illuminated by the vehicle headlamps, offering the driver little time to identify the relative distance of a potential hazard. Under stereoscopic viewing conditions observers can use binocular disparity and oculomotor cues, including accommodation and convergence eye movements. These cues offer information about distance but with relatively poor accuracy and are effective over short distances typically less than approximately six meters (20 ft) (Leibowitz & Moore, 1966; Wallach & Floor, 1971). Studies of steering and lane keeping show that these behaviors are well-maintained even under degraded visual conditions lacking stereoscopic cues (Brooks, Tyrrell, & Frank, 2005; Owens & Tyrrell, 1999). Stereoscopic cues, however, appear more important for time-to-contact judgments that become less accurate when disparity cues are not present (Gray & Regan, 1998).

Under impoverished viewing conditions like those at night, drivers may instead attempt to use changes in the visual angle of objects, such as the size or separation of vehicle headlamps or taillights, to judge the relative distance of other vehicles. However, this cue is also only effective for near distances. Figure 28.5 shows the changes in the visual angle subtended by 12-, 6- and 3-inch headlamps at various distances; it is clear that the visual angle does not change appreciably with distance until the vehicle is within approximately 50 ft.

28.3.5 Color Perception

Color vision is mediated by the cone photoreceptors which function at mesopic and photopic light levels. Signals originating in the rod photoreceptors may also influence color perception through retinal rod-cone interactions (Stockman & Sharpe, 2006). The rendering of color for low luminance conditions may be important for visual search in clutter, where color may enhance an object's salience, facilitate tracking of objects, or facilitate identification of one's own or another's vehicle.

The cone photoreceptors that mediate color vision function best at high photopic illumination levels. Color appearance and chromatic sensitivity both change at low photopic or mesopic illumination levels. Studies of color appearance show that color saturation and color gamut gradually decline as illumination levels are reduced. Chromatic sensitivity also deteriorates, especially at luminances below 3 cd/m^2, with observers showing a preferential loss of blue-yellow sensitivity at mesopic levels. This is consistent with the findings of Pokorny, Lutze, Cao and Zele (2006) who investigated hue perception of paper color samples across a range of illuminance levels between 10 and 0.0003 lux. Participants were instructed to sort the color samples into groups that they could categorize by a specific color term (i.e., red, pink, orange, yellow, green, blue, purple and gray). They

reported that color sorting of the samples remained consistent for illuminance levels between 10 and 0.32 lux. However, at lower illuminance levels (i.e., 0.1–0.01 lux) categorization of reds and oranges remained consistent, but other color samples tended to be categorized as either black or green or blue-green depending on their reflectance.

28.3.6 Interaction With Driver Age and Eye Disease

Age-related changes in vision are also likely to exacerbate the hazards of night driving. Unlike younger adults, older drivers commonly report difficulties with night vision, and some are reluctant to drive at night (Kosnik, Sekuler, & Kline, 1990; Schieber, 1994; Shinar & Schieber, 1991). However, many older drivers continue to drive at night, with crash data indicating that the proportion of drivers who strike pedestrians at night increases progressively after middle age (Owens & Brooks, 1995).

Many of the problems of night-time driving experienced by older drivers are likely to result from the visual changes that occur as we age. Normal ageing is associated with a number of optical changes in the eye that reduce the amount of light reaching the retina. In particular, the eye lens gradually loses its transparency (Said & Weale, 1959), while the pupil (the aperture of the eye) becomes more constricted and less able to enlarge under low light levels. Together these changes reduce the amount of light reaching the retina of a normal 60-year old to approximately one-third of that reaching the retina of a 20-year old (Weale, 1992). Thus older people need much more light to achieve the same amount of retinal illumination as a younger person and have more problems under low luminance conditions, particularly for night driving (Kosnik et al., 1990). The eye lens of an older person also scatters more light and absorbs the shorter wavelengths of light (the blue end of the light spectrum), reducing contrast sensitivity as well as the accuracy of color perception. Changes also occur at retinal and neural levels, where alterations become evident in the integrity of pigment in the central retina (macular region), the retinal nerve fiber layer and the visual pathways (Lovasik, Kergoat, Justino, & Kergoat, 2003; Spear, 1993; Weale, 1992). Together, these changes lead to decreased light and contrast sensitivity, increased glare sensitivity, reductions in visual acuity and visual fields, and prolonged dark adaptation (see Weale, 1992, and Haegerstrom-Portnoy & Morgan, 2007, for an overview). The effects of reduced illumination on visual function are also greater for older compared to younger individuals (Sturgis & Osgood, 1982).

The presence of eye disease is more common in older populations (Attebo, Mitchell, & Smith, 1996; Klein, Klein, Linton, & De Mets, 1991) and exacerbate the problems of night-time driving for older individuals. While there has been little objective data documenting how commonly-occurring eye diseases affect night driving, there are a number of studies of self-reported problems in night-time driving in these populations.

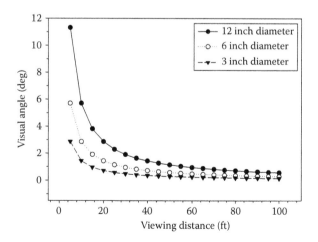

FIGURE 28.5 Comparison of the visual angle subtended by three different sized targets as a function of viewing distance.

Individuals with early age-related maculopathy report difficulty with night driving (as derived from the Activities of Daily Vision Scale [ADVS]) compared with age-matched controls (Scilley et al., 2002), and these difficulties were associated with measures of scotopic sensitivity. Similarly, individuals with cataracts report more difficulties with night-time driving compared to controls, (Owsley, Stalvey, Wells, & Sloane, 1999) and self-reported improvements in the night driving subscale of the ADVS have been shown following cataract surgery (McGwin, Scilley, Brown, & Owsley, 2003). In a recent closed-road study (Wood, Chaparro, Carberry, & Chu, 2010), the effects of simulated cataracts were shown to have a significant effect on driving performance under night-time conditions, over and above that of refractive blur, which reduced visual acuity to the same amount (Figure 28.6). Importantly, these effects were greater at night relative to day, where Figure 28.6 also shows comparable data collected in a separate study under daytime conditions.

28.3.7 Night Myopia

Night myopia is a physiological phenomenon that has been identified for a number of centuries and describes a condition observed under low illumination, where normal observers become myopic (short-sighted) in the absence of a strong visual stimulus to drive the accommodative response (Leibowitz & Owens, 1975). Thus, in dim lighting, near objects may be in focus whereas distant objects will be blurred. While there have been some reports of increased night-time crashes for those with night myopia (Cohen et al., 2007), the evidence is limited and as Arumi et al. (1997) have demonstrated, night myopia only becomes significant at light levels below that of 0.03 cd/m^2, which is lower than that normally encountered under night-time driving conditions.

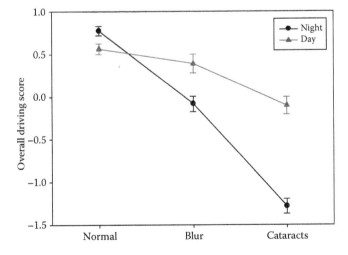

FIGURE 28.6 Effect of simulated cataracts and optical blur on overall driving scores measured at a closed-road circuit under day and night-time conditions.

28.4 Studies of Night-Time Driving (Derived From a Range of Driving Outcome Measures)

28.4.1 Driving Simulators

A number of studies have used driving simulators to investigate driving performance under night-time conditions. While many of these have developed night-time scenarios specifically for the assessment of night driving (Campagne, Pebayle, & Muzet, 2004; Featherstone et al., 1999; Gillberg, Kecklund, & Åkerstedt, 1996; Horberry et al., 2006; Szalmas, Bodrogi, & Sik-Lanyi, 2006), there have been only limited attempts to ensure that the lighting and environmental conditions are representative of the night-time driving situation. For example, some researchers have incorporated extraneous light sources in an attempt to simulate the effect of headlamp glare (Featherstone et al., 1999; Horberry et al., 2006), however, it is unclear how accurately these reflect the effects of oncoming headlamps on either vision or driving performance. Featherstone et al. (1999) used scenarios in the Iowa Driving Simulator designed to replicate the low contrast conditions of night-time driving and incorporated simulated road signs developed to match the color, contrast and luminance values of real-world traffic signs. While the authors state that a field study was conducted to characterize an automobile headlight as a glare source at night, it is unclear from the methodology how this was achieved and how representative the effect was. Horberry et al. (2006) similarly used driving scenarios which aimed to replicate wet night-time driving conditions based on rural night-time scenes. However, the authors do not explicitly state how the night-time scene was recreated and while the effects of oncoming headlamps were simulated using a two-LED light source, mounted on the simulator vehicle bonnet and directed towards the driver, its ability to replicate the effects of oncoming head lamps is less clear (Horberry et al., 2006). Simulations of this sort require judicious choices regarding the positioning and orientation of the light beam and selection of a light source that approximates the spectral, light distribution and intensity profile of a vehicle headlamp. The relative importance of these considerations for effective night-time simulations seems to be largely unknown.

The majority of night-time driving simulator studies have been undertaken using conventional driving simulator systems, with the drivers wearing neutral density goggles to reduce the luminance of the viewed scene (Alferdinck, 2006; Brooks et al., 2005; Bullough & Rea, 2000; Owens & Tyrrell, 1999). Filters, however, uniformly decrease the illumination of the entire rendered scene. This is very different from the non-uniform illumination typical of night-time driving, where the features of the near scene illuminated by the headlamps are more visible than those outside the region illuminated by the headlamps. This heterogeneous pattern of illumination also produces differing states of light adaptation across the retina, which have consequent effects on visual abilities including visual acuity, contrast sensitivity and motion perception. An observer's ability to perceive environmental features and their ability to detect, resolve and attend to task-relevant information in the scene is likely to differ. These limitations may account in part for the range of discrepancies in results from night-time simulator studies and those undertaken in the real world.

Owens and Tyrrell (1999) investigated the effects of age and visual degradation on driving performance in a driving simulator. They found that reduced luminance (resulting from viewing the stimulus display through ND filters) had little or no effect on steering accuracy; while tunnel vision (such as that which results from eye diseases such as glaucoma or retinitis pigmentosa) affected steering ability (this effect was only measured at photopic and not scotopic luminance levels); these findings were replicated subsequently using a more sophisticated simulator (Brooks et al., 2005). However, a closed-road study by Owens, Wood and Owens (2007), which manipulated headlamp intensity, showed that lane keeping and hence steering were impaired when the luminance of the vehicle headlamps was reduced; driver speeds were also slower under these conditions (Owens et al., 2007). The driving recognition ability of both young and older participants was also impaired when headlamp beam intensity was reduced, impairments in performance that were better predicted by contrast sensitivity and low luminance visual acuity than standard measures of high contrast visual acuity (Wood & Owens, 2005). Owens et al. (2007) suggested that the finding of differences in steering performance on the road for reduced illumination conditions as compared with simulators can be attributed to three factors: speed control, task familiarity, and potential crash risk, all of which raise problems for generalizing results from simulators to driving in the real world. Firstly, speed is fixed in some simulator studies, preventing drivers from adjusting their speed to compensate for task difficulty, as they would under real driving conditions—particularly in the case of older drivers (Wood, 2002). Secondly, driving a simulator is an unfamiliar experience for most participants, again, particularly for older individuals. The fixed-base simulators, which have been used in many of the previous night driving studies, lack inertial and proprioceptive variations present in a real vehicle which can make steering a simulator feel strange and more demanding of attention than actual driving. Under normal driving conditions, lane keeping is a relatively automatic task; and is strongly influenced by an individual's habitual driving style and their speed choices. Finally, driving on a real road involves safety risks that are not encountered in a simulator. While the risk of a crash on a closed-road course is relatively low, it is still present as it is under normal driving conditions, and is likely to influence driver behavior (Evans, 2004); drivers are likely to adopt more cautious strategies than they might otherwise in a simulator. In addition, care should be applied when generalising research findings from studies that simulate night-time driving by having participants view scenes through neutral density filters. These simulations do not capture the pattern of illumination produced by a vehicle's headlamps, nor the large differences in the relative visibility of scene features lying inside and outside the regions of the environment that are illuminated. It is possible that performance under the simulated

conditions may be easier, since the whole scene is visible but reduced uniformly in luminance, whereas under real night-time conditions a driver may rely to a greater degree on limited proximal areas of the road scene illuminated by the vehicle.

28.4.2 Closed and On-Road Driving Studies

There have been only a limited number of studies which have reported information about night-time driving using closed-road circuits or on secluded areas of real-world road systems. In many of these closed-road studies (Owens et al., 2007; Tyrrell, Wood, & Carberry, 2004; Wood & Owens, 2005; Wood et al., 2005), researchers have utilized headlamps linked to sensors, so that drivers passing through them initiate the onset of the headlamp beam to simulate the effects of oncoming headlamps, given that most of these studies are undertaken in locations which lack other vehicles for standardisation and safety reasons.

Indeed, many of the studies of night-time pedestrian visibility have been undertaken under real-world driving conditions, such as closed-road circuits, rather than filming road situations and subsequently presenting them in the laboratory or simulators (see Kwan & Mapstone, 2004; Langham & Moberly, 2003, for comprehensive reviews). An important rationale for this is that the filming and projection of road scenes results in degradation of visual information, mainly due to the limited resolution of video film, resulting in the loss of contrast and contour information and projection of the images which results in further loss of definition. Hughes and Cole (1986) compared a video-based methodology with a field study and observed that search conspicuities for objects in the laboratory experiments were around one-third less than those observed in a field study. While it was suggested by these authors that real-world conspicuities could be predicted from laboratory or simulator experiments if a scaling factor was applied, there are other factors that need to be taken into account. As stated in other sections of this review, the interactive effects of illumination from the driver's own headlamps and that of oncoming headlamps are challenging to replicate in a simulator environment and thus the glare experienced due to stray light cannot be accounted for in the visibility measures. Importantly, the amount of glare experienced will also be increased as a function of the age of the driver and in the presence of cataracts (Wood & Carberry, 2006).

28.5 Implications for the Design of Appropriate Scenarios for Night-Time Driving Simulator Studies

As described in the previous sections, there are many important characteristics of night-time driving that make it quite different, particularly visually, from the driving conditions that the majority of simulators are designed to replicate. One of the major problems faced by researchers attempting to simulate night-time driving conditions is that of replicating the effects of the headlamps of both oncoming vehicles as well as those of the vehicle being driven by an individual participant. There are extensive data, as described in the previous sections, that document the detrimental effect of the glare of oncoming vehicles in detecting low contrast obstacles, such as pedestrians, and these effects are exacerbated for older drivers and those with eye diseases, such as cataracts. Driving simulator scenarios cannot properly represent the dynamic range of luminance at night—ranging from complete darkness (rural roads) to the intense brightness of oncoming headlamps—nor the interactions with the road surface, glare sources and the driver.

Another important limitation is that the lack of a photometric model of mesopic vision makes it impossible to accurately predict the relative luminosity, and hence the visibility, of stimuli of the different spectral compositions that researchers seek to simulate and participants may need to detect. These issues are relevant to whatever system of night-time driving simulation researchers choose to use. While simulations of night driving may support the investigation of reduced visibility on driving more generally, it is more problematic in conditions where researchers are interested in issues specifically related to roadway illumination and the effects of glare. In addition, in order to enhance the generalizability of their findings, simulator researchers should include driver participants who vary widely in experience and age, as well as in levels of general and visual health. Finally, researchers need to recognize the limitations of simulation in terms of representing the lighting or environmental conditions, vehicle or pedestrian motion, or the driver workload encountered in real-world night-time driving tasks and to qualify any conclusions based on the experimental findings accordingly.

Key Points

- Characteristics of night driving are different to those in the day: fatality rates are higher, particularly for pedestrians, and it is believed that reduced illumination is a key factor.
- Illumination of the road scene is affected by a number of factors including street lighting and vehicle headlamps.
- Reduced illumination at night-time has an important impact on visual performance; these changes include reduced visual acuity, contrast sensitivity, motion sensitivity, and color perception. There is also increased glare from oncoming headlamps.
- The effects of reduced illumination on visual performance at night-time are exacerbated in older drivers, particularly those with eye diseases such as cataracts.
- A number of studies have used driving simulators to investigate night-time driving, but there are many limitations with these approaches, which mostly fail to capture many of the critical visual aspects of the night-time driving scene. These include the effects of oncoming headlamps, as well as those of the vehicle itself, and the variation in illumination levels commonly encountered at night.

Keywords: Night Driving, Older Drivers, Reduced Illumination, Visual Function

Acknowledgments

The authors would like to acknowledge the helpful comments of Ralph Marszalek on a previous version of this chapter and Trent Carberry for his photograph used in Figure 28.1.

Key Readings

Owens, D. A., & Sivak, M. (1996). Differentiation of visibility and alcohol as contributors to twilight road fatalities. *Human Factors, 38*(4), 680–689.

Plainis, S., Murray, I. J., & Charman, W. N. (2005). The role of retinal adaptation in night driving. *Optometry & Vision Science, 82*(8), 682–688.

Pokorny, J., Lutze, M., Cao, D., & Zele, A. J. (2006). The color of night: Surface color perception under dim illuminations. *Visual Neuroscience, 23*(3–4), 525–530.

Stockman, A., & Sharpe, L. T. (2006). Into the twilight zone: The complexities of mesopic vision and luminous efficiency. *Ophthalmic & Physiological Optics, 26,* 225–239.

Wood, J. M., & Owens, D. A. (2005). Standard measures of visual acuity do not predict drivers' recognition performance under day or night conditions. *Optometry & Vision Science, 82*(8), 698–705.

References

Alferdinck, J. W. (2006). Target detection and driving behavior measurements in a driving simulator at mesopic light levels. *Ophthalmic & Physiological Optics, 26,* 264–280.

Anderson, S. J., & Holliday, I. E. (1995). Night driving: Effects of glare from vehicle headlights on motion perception. *Ophthalmic & Physiological Optics, 15*(6), 545–551.

Andre, J., & Owens, D. A. (2001). The twilight envelope: A user-centered approach to describing roadway illumination at night. *Human Factors, 43*(4), 620–630.

Arumi, P., Chauhan, K., & Charman, W. N. (1997). Accommodation and acuity under night-driving illumination levels. *Ophthalmic & Physiological Optics, 17,* 291–299.

Attebo, K., Mitchell, P., & Smith, W. (1996). Visual acuity and the causes of visual loss in Australia: The Blue Mountains Eye Study. *Ophthalmology, 103*(3), 357–364.

Barlow, H. B. (1965). Visual resolution and the diffraction limit. *Science, 149,* 553–555.

Bedell, H. E. (1987). *Eccentric regard, task and optical blur as factors influencing visual acuity at low luminances.* Paper presented at Night Vision, Current Research and Future Directions. National Research Council Symposium Proceedings.

Bourdy, C., Cottin, F., & Monot, A. (1991). Errors in distance appreciation and binocular night vision. *Ophthalmic & Physiological Optics, 11*(4), 340–349.

Brooks, J. O., Tyrrell, R. A., & Frank, T. A. (2005). The effects of severe visual challenges on steering performance in visually healthy younger drivers. *Optometry & Vision Science, 82*(8), 689–697.

Bullough, J. D., & Rea, M. R. (2000). Simulated driving performance and peripheral detection at mesopic and low photopic light levels. *Lighting Research & Technology, 32*(4), 194–198.

Campagne, A., Pebayle, T., & Muzet, A. (2004). Correlation between driving errors and vigilance level: Influence of the driver's age. *Physiology & Behavior, 80*(4), 515–524.

Cohen, Y., Zadok, D., Barkana, Y., Shochat, Z., Ashkenazi, I., Avni, I., et al. (2007). Relationship between night myopia and night-time motor vehicle accidents. *Acta Ophthalmologica Scandinavica, 85*(4), 367–370.

Elvik, R. (1995). A meta-analysis of evaluation of public lighting as an accident countermeasure. *Transportation Research Record, 1485,* 112–123.

Evans, L. (2004). *Traffic Safety.* Bloomfield Hills, MI: Science Serving Society.

Featherstone, K. A., Bloomfield, J. R., Lang, A. J., Miller-Meeks, M. J., Woodworth, G., & Steinert, R. F. (1999). Driving simulation study: Bilateral array multifocal versus bilateral AMO monofocal intraocular lenses. *Journal of Cataract & Refractive Surgery, 25,* 1254–1262.

Franssen, L., Tabernero, J., Coppens, J. E., & van den Berg, T. J. T. P. (2007). Pupil size and retinal straylight in the normal eye. *Investigative Ophthalmology & Visual Science, 48*(5), 2375–2382.

Gegenfurtner, K. R., Mayser, H., & Sharpe, L. T. (1999). Seeing movement in the dark. *Nature, 398* (6727), 475–476.

Gegenfurtner, K. R., Mayser, H. M., & Sharpe, L. T. (2000). Motion perception at scotopic light levels. *Journal of the Optical Society of America, 17*(9), 1505–1515.

Gillberg, M., Kecklund, G., & Åkerstedt, T. (1996). Sleepiness and performance of professional drivers in a truck simulator—Comparisons between day and night driving. *Journal of Sleep Research, 5*(1), 12–15.

Gordon, P. (1977). Appraisal of visibility on lighted dry and wet roads. *Lighting Research & Technology, 9,* 177–188.

Gray, R., & Regan, D. (1998). Accuracy of estimating time to collision using binocular and monocular information. *Vision Research, 38,* 499–512.

Haegerstrom-Portnoy, G., & Morgan, M. W. (2007). Normal age-related vision changes. In A. A. Rosenbloom (Ed.), *Rosenbloom & Morgans's vision and aging* (pp. 31–48). St. Louis: Butterwoth Heinemann.

Herd, D. R., Agent, K. R., & Rizenbergs, R. L. (1980). Traffic accidents: Day versus night. *Transportation Research Record, 753,* 25–30.

Hood, D. C., & Finkelstein, M. A. (1986). Sensitivity to light. In K. R. Boff, L. Kaufman, & J. P. Thomas (Eds.), *Handbook of perception and human performance.* New York, NY: John Wiley & Sons.

Horberry, T., Anderson, J., & Regan, M. A. (2006). The possible safety benefits of enhanced road markings: A driving simulator evaluation. *Transportation Research Part F: Traffic Psychology & Behavior, 9*(1), 77–87.

Hughes, P. K., & Cole, B. L. (1986). Can the conspicuity of objects be predicted from laboratory experiments? *Ergonomics, 29*(9), 1097–1111.

Johnson, C. A., & Casson, E. J. (1995). Effects of luminance, contrast and blur on visual acuity. *Optometry & Vision Science, 72*(12), 864–869.

Joint Technical Committee LG-002. (2005a). Australia/New Zealand Standard: Lighting for roads and public spaces. AS/NZS 1158.2:2005. *Standards Australia*: Sydney, Australia.

Joint Technical Committee LG-002. (2005b). *Australia/New Zealand Standard: Lighting for roads and public spaces.* AS/NZS 1158.1.1:2005. Standards Australia: Sydney, Australia.

Kaptein, N. A. Hogema, J. H., & Folles, E. (1997). *Dynamic public light (DYNO).* Paper presented at the 8th European Lighting Conference. Lux Europa, Amsterdam, The Netherlands.

Klein, R., Klein, B. E. K., Linton, K. L. P., & De Mets, D. L. (1991). The Beaver Dam Eye Study: Visual acuity. *Ophthalmology, 98,* 1310–1315.

Kosnik, W. D., Sekuler, R., & Kline, D. W. (1990). Self-reported visual problems of older drivers. *Human Factors, 32,* 597–608.

Kwan, I., & Mapstone, J. (2004). Visibility aids for pedestrians and cyclists: A systematic review of randomized controlled trials. *Accident Analysis & Prevention, 36*(3), 305–312.

Langham, M., & Moberly, N. (2003). Pedestrian conspicuity research: A review. *Ergonomics, 46*(4), 345–363.

Leibowitz, H. W., & Moore, D. (1966). Role of changes in accommodation and convergence in the perception of size. *Journal of the Optical Society of America, 56*(8), 1120–1123.

Leibowitz, H. W., & Owens, D. A. (1975). Night myopia and the intermediate dark focus of accommodation. *Journal of the Optical Society of America, 65*(10), 1121–1128.

Lestina, D. C., Miller, T. R., Langston, E. A., Knoblauch, R., & Nitzburg, M. (2002). Benefits and costs of ultraviolet fluorescent lighting. *Traffic Injury Prevention, 3,* 209–215.

Lovasik, J. V., Kergoat, M. J., Justino, L., & Kergoat, H. (2003). Neuroretinal basis of visual impairment in the very elderly. *Graefe's Archive for Clinical & Experimental Ophthalmology, 241,* 48–55.

McGwin, G., Scilley, K., Brown, J., & Owsley, C. (2003). Impact of cataract surgery on self-reported visual difficulties: Comparison with a no-surgery reference group. *Journal of Cataract & Refractive Surgery, 29*(5), 941–948.

Mortimer, R. G., & Olson, P. L. (1974). *Evaluation of meeting beams by field tests and computer simulation.* Ann Arbor, MI: University of Michigan Highway Safety Research Institute.

National Safety Council. (1999–2004). *Injury facts.* Chicago

Owens, D. A. (2003). Twilight vision and road safety: Seeing more than we notice but less than we think. In J. Andre, D. A. Owens, & L. O. Harvey (Eds.), *Visual perception: The influence of H. W. Leibowitz.* Washington, D.C: American Psychological Association.

Owens, D. A., & Brooks, J. C. (1995). *Drivers' vision, age, and gender as factors in twilight road fatalities.* Ann Arbor, MI: The University of Michigan Transportation Research Institute.

Owens, D. A., & Sivak, M. (1996). Differentiation of visibility and alcohol as contributors to twilight road fatalities. *Human Factors, 38*(4), 680–689.

Owens, D. A., & Tyrrell, R. A. (1999). Effects of luminance, blur, and age on night-time visual guidance: A test of the selective degradation hypothesis. *Journal of Experimental Psychology: Applied, 5*(2), 115–128.

Owens, D. A., Wood, J. M., & Owens, J. M. (2007). Effects of age and illumination on night driving: A road test. *Human Factors 49*(6), 1115–1131.

Owsley, C., & McGwin, G. (1999). Vision impairment and driving. *Survey of Ophthalmology, 43*(6), 535–550.

Owsley, C., Stalvey, B., Wells, J., & Sloane, M. E. (1999). Older drivers and cataract: Driving habits and crash risk. *Journal of Gerontology: Biological Science & Medical Sciences, 54*(4), M203–211.

Owsley, C., Stalvey, B. T., Wells, J., Sloane, M. E., & McGwin, G., Jr. (2001). Visual risk factors for crash involvement in older drivers with cataract. *Archives of Ophthalmology, 119*(6), 881–887.

Plainis, S., Chauhan, K., Murray, I. J., & Charman, W. N. (1999). Retinal adaptation under night-time driving conditions. *Proceedings of vision in vehicles VII,* (pp. 61-70).

Plainis, S., Murray, I. J., & Charman, W. N. (2005). The role of retinal adaptation in night driving. *Optometry & Vision Science, 82*(8), 682–688.

Plainis, S., Murray, I. J., & Pallikaris, I. G. (2006). Road traffic casualties: Understanding the night-time death toll. *Injury Prevention, 12*(2), 125–138.

Pokorny, J., Lutze, M., Cao, D., & Zele, A. J. (2006). The color of night: Surface color perception under dim illuminations. *Visual Neuroscience, 23*(3–4), 525–530.

Pulling, N. H., Wolf, E., Sturgis, S. P., Vaillancourt, D. R., & Dolliver, J. J. (1980). Headlight glare resistance and driver age. *Human Factors, 22*(1), 103–112.

Rea, M. S. (2001). The road not taken. *The Lighting Journal, 66,* 18-25.

Rumar, K. (1974). Dirty headlights: Frequency and visibility effects. *Ergonomics, 17*(4), 529–533.

Rumar, K. (2002). *Night driving accident in an international perspective.* Paper presented at the First International Congress Vehicle and Infrastructure Safety Improvement in Adverse Conditions and Night Driving.

Said, F. S., & Weale, R. A. (1959). The variation with age of the spectral transmissivity of the living human crystalline lens. *Gerontologica, 3,* 213–231.

Schieber, F. (1994). High-priority research and development needs for maintaining the safety and mobility of older drivers. *Experimental Aging Research, 20,* 35–43.

Scilley, K., Jackson, G. R., Cideciyan, A. V., Maguire, M. G., Jacobson, S. G., & Owsley, C. (2002). Early age-related maculopathy and self-reported visual difficulty in daily life. *Ophthalmology, 109*(7), 1235–1242.

Shinar, D. (1984). Actual versus estimated night-time pedestrian visibility. *Ergonomics, 27*(8), 863–871.

Shinar, D., & Schieber, F. (1991). Visual requirements for safety and mobility of older drivers. *Human Factors, 33,* 507–519.

Sivak, M., Flannagan, M. J., Kojima, S., & Traube, E. C. (1997). *A market-weighted description of low-beam headlighting patterns in the US.* Ann Arbor, MI: The University of Michigan - Transportation Research Institute.

Sivak, M., Schoettle, B., & Tsimhoni, O. (2007). *Moon phases and night-time road crashes involving pedestrians.* Ann Arbor, MI: University of Michigan Transportation Research Institute.

Spear, P. (1993). Neural bases of visual deficits during aging. *Vision Research, 33,* 2589–2609.

Stockman, A., & Sharpe, L. T. (2006). Into the twilight zone: The complexities of mesopic vision and luminous efficiency. *Ophthalmic & Physiological Optics, 26,* 225–239.

Sturgis, S. P., & Osgood, D. J. (1982). Effects of glare and background luminance on visual acuity and contrast sensitivity: Implications for driver night vision testing. *Human Factors, 24*(3), 347–360.

Sturr, J. F., Kline, G. E., & Taub, H. A. (1990). Performance of young and older drivers on a static acuity test under photopic and mesopic luminance conditions. *Human Factors, 32,* 1–8.

Sullivan, J. M., Adachi, G., Mefford, M. L., & Flannagan, M. J. (2004). High-beam headlamp usage on unlighted rural roadways. *Lighting Research & Technology, 36*(1), 59–65.

Sullivan, J. M., & Flannagan, M. J. (2002). The role of ambient light level in fatal crashes: Inferences from daylight saving time transitions. *Accident Analysis & Prevention, 34*(4), 487–498.

Szalmas, A., Bodrogi, P., & Sik-Lanyi, C. (2006). Characterizing luminous efficiency functions for a simulated mesopic night driving task based on reaction time. *Ophthalmic & Physiological Optics, 26*(3), 281–287.

Triggs, T. J., & Berenyi, J. S. (1982). Estimation of automobile speed under day and night conditions. *Human Factors, 24*(1), 111–114.

Tsimhoni, O., Bargman, J., Minoda, T., & Flannagan, M. J. (2004). *Pedestrian detection with near and far infrared night vision enhancement.* Ann Arbor, MI: University of Michigan Transportation Research Institute.

Tyrrell, R. A., Wood, J. M., & Carberry, T. P. (2004). On-road measures of pedestrians' estimates of their own night-time conspicuity. *Journal of Safety Research, 35*(5), 483–490.

Viikari, M., Ekrias, A., Eloholma, M., & Halonen, L. (2008). Modeling spectral sensitivity at low light levels based on mesopic visual performance. *Clinical Ophthalmology, 2*(1), 1–13.

Wallach, H., & Floor, L. (1971). The use of size matching to demonstrate the effectiveness of accommodation and convergence as cues for distance. *Perception & Psychophysics, 10,* 423–428.

Weale, R. A. (1992). *The senescence of human vision.* Oxford: Oxford University Press.

Wood, J. M. (2002). Age and visual impairment decrease driving performance as measured on a closed-road circuit. *Human Factors, 44*(3), 482–494.

Wood, J. M., & Carberry, T. P. (2006). Bilateral cataract surgery and driving performance. *British Journal of Ophthalmology, 90*(10), 1277–1280.

Wood, J. M., Chaparro, A. Carberry, T. P., & Chu B. S. (2010). Effect of simulated visual impairment on day and night-time driving performance. Unpublished raw data.

Wood, J. M., & Owens, D. A. (2005). Standard measures of visual acuity do not predict drivers' recognition performance under day or night conditions. *Optometry & Vision Science, 82*(8), 698–705.

Wood, J. M., Tyrrell, R. A., & Carberry, T. P. (2005). Limitations in drivers' ability to recognize pedestrians at night. *Human Factors, 47*(3), 644–653.

Woodhouse, J. M., & Barlow, H. B. (1982). Spatial and temporal resolution and analysis. In H. B. Barlow (Ed.), *The senses* (pp. 152–162). Cambridge: Cambridge University Press.

29

Driving in States of Fatigue or Stress

Gerald Matthews
University of Cincinnati

Dyani J. Saxby
University of Cincinnati

Gregory J. Funke
University of Cincinnati

Amanda K. Emo
Federal Highway Administration

Paula A. Desmond
Southwestern University

Abstract

The Problem. Fatigue is a serious safety problem for drivers. There are several separate sources of fatigue including sleep loss, circadian rhythm effects, and the intrinsic demands and workload of the driving task. It is important to understand both the processes that generate fatigue in the driver and the impact of fatigue states on information processing, performance, and safety. Fatigue effects involve a number of separate mechanisms and so a clear theoretical understanding of these mechanisms is needed in order to guide interventions to enhance safety. This chapter reviews methods for inducing fatigue during simulated driving, theoretical frameworks for interpreting fatigue data and practical applications. *Role of Driving Simulators.* The driving simulator affords the means for controlled experiments on the interplay between fatigue and driver performance. Fatigue states are readily induced on the simulator. They may be evaluated using subjective and psychophysiological measures. The simulator also affords measurement of fatigue-induced changes in vehicle control, attention, and risk-taking behaviors. Simulator methods are also pivotal for assessment of individual differences in fatigue vulnerability, investigation of clinical fatigue conditions, and evaluation of countermeasures for fatigue. *Key Results of Driving Simulator Studies.* One important contribution of simulator studies is to map how environmental factors such as monotony and workload elicit fatigue responses and to track the development of fatigue states over time. Fatigue responses may be understood within the transactional theory of stress; fatigue is the outcome of a self-regulative process of managing personal discomfort and tiredness. Simulator studies have also proved valuable in testing theories of performance decrement. Studies have explored how depletion of attentional resources and breakdown of effort-regulation may contribute to driver performance impairments. *Scenarios and Dependent Variables.* Scenarios for inducing fatigue typically involve some combination of monotony and extended drive duration. High workload is not directly tied to fatigue response, but may also contribute to fatigue. Several types of dependent variables may be assessed, including measures of subjective state, psychophysiological responses, and performance data logged by the simulator.

29.1 Introduction

Fatigue refers to a cluster of disturbances of psychological and physiological states including tiredness, sleepiness, loss of alertness, reduced task-directed effort and cortical de-arousal (Hitchcock & Matthews, 2005). Various factors including sleep loss and sleep disturbance, prolonged high workload and time pressure, and driving patterns that run counter to circadian rhythms all contribute to fatigue. Fatigue is widely-recognized as a serious driver safety problem, especially for commercial motor vehicle drivers. Morrow and Crum (2004) cite National Transportation Safety Board statistics indicating that 31% of truck driver fatalities are attributable to fatigue.

Different groups of drivers may be more or less sensitive to fatigue-related impairments (Lee, 2006). Factors that increase sleepiness such as sleep apnea, long and/or irregular work shifts, and use of alcohol and drugs may all increase vulnerability. Younger and inexperienced drivers also seem to be more at risk (Pack et al., 1995). A recent large-scale prospective study in France (Nabi et al., 2007) found various factors relating to experiencing episodes of sleepy driving, including being male, younger (within a restricted range), working nights or overtime, using alcohol or drugs for emotional disorders, and having an untreated sleep disorder. However, it may be difficult to differentiate group differences in vulnerability to fatigue from group differences in exposure to fatiguing driving conditions. For example, younger drivers may be more likely than older drivers to drive during the night-time hours when circadian rhythms contribute to fatigue. The advantage of simulator studies is that the vulnerability of groups to fatigue can be observed under controlled conditions, thereby separating vulnerability and exposure effects.

Simulator methods are critical for understanding driver fatigue, and the allied issue of driver stress. This chapter assumes a bidirectional association between the demands of the driving task and fatigue responses (Matthews, 2002). That is, task demands elicit fatigue and stress responses, which in turn feed back to influence the driver's coping efforts with those demands. One objective for simulator research is to manipulate task factors such as workload, drive duration and monotony, and evaluate their impact on fatigue response. A second objective is to investigate how fatigue responses relate to objective indices of driving performance and safety, using either experimental or correlational methods. Thus, fatigue may be understood within the theoretical framework of the transactional theory of stress (Lazarus, 1999), which relates stress and fatigue to the ongoing, dynamic interaction between the task environment and the driver's coping efforts (Matthews, 2002; Matthews & Desmond, 2002).

Simulation provides an imperfect correspondence with reality, in that the driver is at no risk of physical harm. Caution is needed in drawing conclusions from simulator studies about real-life fatigue and its implications for safety. Where simulation is useful is in identifying psychological mechanisms and processes that may operate in real driving also. Such processes may include loss of attention, reluctance to apply effort to the task, and impairments in judgment and decision-making (Fairclough, 2001; Matthews, Davies, Westerman, & Stammers, 2000).

Simulator research provides a number of methodological and theoretical challenges. Perhaps the most pressing methodological issue is the multidimensional nature of fatigue, and its overlap with stress (Matthews et al., 2002). Fatigue may be expressed as a range of subjective states, behavioral responses (e.g., falling asleep), and physiological changes. These various responses do not necessarily coincide, and may have differing impacts on driving performance. For example, although subjective tiredness is a central feature of fatigue states, one tired driver may be fighting to maintain active involvement with the task, whereas a second tired driver enters a more passive, "mindless" state. Different task characteristics of the drive may profoundly influence the pattern of subjective and objective fatigue responses that ensue.

It is emphasized that the study of fatigue is considerably more than the study of sleepiness. Prolonged high-workload driving, which involves little probability of falling asleep, nevertheless induces fatigue (and stress) states with consequences for safety (Matthews & Desmond, 2002). Some researchers have distinguished fatigue as a consequence of high workload from sleepiness as a consequence of sleep deprivation (Philip et al., 2005).

Simulation also affords measurement of different components of performance that may be differentially sensitive to fatigue, such as attention, psychomotor control, and the processing of risk. In driving, as in other operational settings, fatigue effects on objective performance are often elusive, and dependent on task and contextual factors (Matthews et al., 2000). Does fatigue influence specific component processes, or some more general attentional "resource"? Alternatively, are the effects of fatigue primarily strategic, for example, in influencing how much effort the driver is willing to expend on the task (Fairclough, 2001)? Simulator studies must incorporate the methods and principles of cognitive psychology that allow for testing different information-processing mechanisms that mediate observed performance change.

Another theme here is individual differences in vulnerability to fatigue and performance impairment. Following both real and simulated drives, individuals differ markedly in the extent and patterning of fatigue responses that ensue (e.g., Desmond, Matthews, & Bush, 2001). Identifying fatigue-prone drivers and monitoring the build-up of fatigue over time is important for applied research. Fatigue vulnerability is a stable *trait* that changes little over time. During a given drive, researchers may also need to assess the transient *state* that may change rapidly as the driver progresses; again, various instruments have been developed for this purpose.

The last research issue covered here is the applied utility of simulator research on fatigue and stress. The transactional model emphasizes that stress and fatigue states derive from the interplay of external and environmental factors, and the individual's coping with external demands. Thus, countermeasures for fatigue may be focused on either external factors (e.g., work schedules) or on the individual (e.g., training more effective coping). Simulator studies support various applications of these kinds, including evaluation of alerting devices, investigation of the impact of clinical disorders related to fatigue, and analysis of the role of new technology such as vehicle automation in alleviating (or exacerbating) fatigue states.

29.2 Simulator Methods for Driver Fatigue Research

29.2.1 Fatigue Inductions: Methodological Issues

At its simplest, fatigue induction requires only a prolonged simulator drive. Indeed, much research has done just this, with little attention to the role of the characteristics of the driving task. However, understanding of induced fatigue states requires attention to the workload factors of the driving task. Careful design of simulator manipulations is important in order to control both the magnitude and the qualitative patterning of fatigue symptoms. Environmental factors that are likely to induce fatigue include length of drive, lack of variability in stimulation, and various driver workload factors (Oron-Gilad & Hancock, 2005). Drive duration may also be manipulated to study the development of fatigue over differing time intervals (Saxby, Matthews, & Hitchcock, 2007).

Workload manipulations are effective in accelerating the development of fatigue on the simulator. They may be either intrinsic to the drive, such as level of interaction with traffic, or extrinsic, such as an additional secondary task (Matthews & Desmond, 2002). However, fatigue develops in low- as well as high-workload conditions. Desmond and Hancock (2001) describe a passive fatigue state, in which workload is low and fatigue derives from monotony and boredom. A real-life instance might be driving long-distance on a straight freeway through featureless terrain. Thiffault and Bergeron (2003) confirmed that monotony is a critical factor for fatigue, by varying background scenery on the simulator. By contrast, Desmond and Hancock's (2001) active fatigue state derives from high workload, such as negotiating heavy, fast-moving traffic. As further discussed below, driving simulators may be configured to impose either active or passive fatigue on the driver, by manipulating the nature of the demands of the simulator environment (Saxby et al., 2007, 2008).

Studies of sleepiness require special care. Occasionally, initially wakeful drivers will fall asleep during the course of fatigue inductions. In general, though, such research requires methods for sleep deprivation, entailing overnight monitoring of sleep loss, control for circadian rhythms and attention to possible practice effects (e.g., Baulk, Biggs, Reid, van den Heuvel, & Dawson, 2008). Sleep loss studies commonly monitor for the episodic "microsleeps" that may have dire safety consequences (Boyle, Tippin, Paul, & Rizzo, 2008). These authors defined microsleeps as 3–14 second episodes during which 4–7 Hz (theta) activity replaced the waking 8–13 Hz (alpha) background rhythm. Drivers showed greater variability in maintaining lane position during microsleep episodes.

A final design issue is that, ideally, fatigue studies require a control condition in which the driver does not experience fatigue. The problem is that routine simulator driving tends to induce fatigue even without any overt induction of fatigue. Saxby et al. (2007) showed that in a "control" condition, task engagement was maintained through 10 minutes of driving, but showed a moderate drop relative to initial baseline (−0.4 SD) after 30 minutes and a large decline (−1.4 SD) after 50 minutes of driving. Thiffault and Bergeron (2003) have shown that the monotony or interest of background scenery—a factor that is often not controlled explicitly in research—plays a critical role in fatigue response. Similarly, Saxby et al. (2007) conducted a second study that suggested that adding background scenery slowed, but did not eliminate, the build-up of fatigue.

It is recommended that fatigue studies use either a short drive as a control condition, or a longer drive with varied scenery and/or traffic conditions. In either case, explicit evaluation and comparison of driver response is necessary.

29.2.2 Evaluation of Fatigue Response

A major difficulty for fatigue assessment is the multi-faceted nature of the construct. Figure 29.1 shows an outline taxonomy for different aspects of fatigue (Hitchcock & Matthews, 2005). Typically, simulator studies are concerned with inductions of acute fatigue, although an evaluation of chronic fatigue

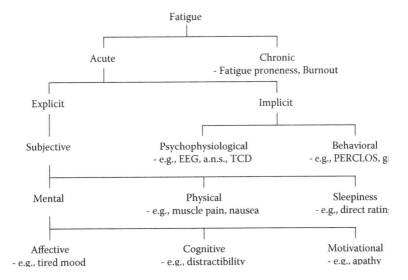

FIGURE 29.1 A provisional taxonomy of fatigue constructs.

symptoms is also feasible. A basic distinction is between explicit elements of fatigue, which the person can report verbally, and implicit fatigue components, which are unconscious but expressed in psychophysiological and/or behavioral response. Next, we briefly review some of the main psychophysiological and subjective fatigue indicators used in simulator studies.

29.2.2.1 Psychophysiological Assessment

Traditionally, much research was based on the attribution of fatigue to loss of general cortical arousal. The various central and autonomic nervous system indices used to monitor general arousal are well-known and need no discussion here (see Lal & Craig, 2001, for a review). Slow-wave activity in the EEG (electroencephalogram: delta and theta bands) appears to be one of the more reliable indices of fatigue; other popular arousal indices such as EEG alpha are less consistent as markers for fatigue (Wijesuriya, Tran & Craig, 2007). There is some overlap between psychophysiological and subjective indices, but simulator evidence suggests that drivers have limited insight into their neurophysiological alertness (Moller, Kayumov, Bulmash, Nhan, & Shapiro, 2006).

Certainly, use of electroencephalographic, cardiovascular, electrodermal and other measures may be useful in depicting the fatigue state that may develop during simulated driving. However, these traditional measures also have limitations. First, general arousal theory is largely discredited as an explanatory theory in studies of stress and performance (Matthews et al., 2000). Second, loss of autonomic arousal may not map into changes in information-processing in any simple way, limiting its relevance to performance change. Third, as discussed in the next section, fatigue inductions may also be stressful and liable to increase some forms of arousal (e.g., tense arousal), so that general arousal indices become equivocal.

Psychophysiological assessment is most useful when the index of choice can be shown to have some functional significance for performance. For example, in sleepiness research the EEG may be used to identify potentially hazardous microsleeps during simulated driving (Boyle et al., 2008; Moller et al., 2006). Oculomotor indices such as frequency and duration of eye closures are also promising (e.g., Bergasa, Nuevo, Sotelo, Barea, & López, 2006), as further discussed below. Recent work in our laboratory (Warm, Matthews, & Parasuraman, 2009) suggests a further promising diagnostic index: measurement of cerebral blood flow velocity (CBFV) in the medial arteries, using transcranial Doppler sonography (TCD). Declines in CBFV are generally diagnostic of performance decrement on a range of tasks requiring sustained attention, including monitoring visual displays and prolonged use of working memory. Reinerman, Warm, Matthews and Langheim (2008) found that CBFV declined during a monotonous 36-minute simulated driving task that required drivers to monitor pedestrian hazards during the task. Variability in lateral position increased concurrently with loss of CBFV during the drive. Importantly, CBFV does not decline during passive viewing of task stimuli, indicating that declining CBFV is functionally related to the workload and attentional demands of the task rather than some general loss of cortical arousal (Warm et al., 2009).

29.2.2.2 A Multidimensional Framework for Subjective Fatigue

Although single scales or even single ratings are often used to assess fatigue and sleepiness (e.g., Fairclough, 2001; Philip et al., 2005), it is preferable to adopt multidimensional measurement frameworks to assess the various components of fatigue states. Hierarchical models may reconcile various dimensional models of fatigue that have been proposed. Although such models are often used in other areas of assessment, such as intelligence testing, little attention has been allocated to them in fatigue assessment. The Dundee Stress State Questionnaire (DSSQ; Matthews et al., 2002) is a 96-item measure designed to assess the full spectrum of transient states associated with stress, arousal, and fatigue. It represents a hierarchical model in which 10 primary factors support three broader, higher-order factors: task engagement, distress, and worry. State factors correlate moderately with psychophysiological indices (Fairclough & Venables, 2006), including TCD (Matthews et al., in press). Changes in state are expressed in SD units derived from a large normative sample, affording a standard scale for comparison of fatigue effects across different studies (Matthews et al., 2002). Loss of task engagement is critical to fatigue. At the primary-factor level, loss of engagement corresponds to tiredness, de-motivation and distractibility. Both simulator and field studies have shown substantial decreases (>1 SD) in task engagement during protracted driving (Desmond et al., 2001; Matthews & Desmond, 2002).

While general mental fatigue and tiredness (corresponding to low-task engagement) is readily assessed, additional fatigue components should also be evaluated. In a study of simulated driving, 256 subjects completed the Task-Induced Fatigue Scale (TIFS; Matthews & Desmond, 1998) and the DSSQ, before and after a fatiguing drive. A factor analysis discriminated physical fatigue symptoms, including muscular and visual fatigue, in addition to mental fatigue associated with boredom. The TIFS scales also correlated with general state measures, such as mood and cognitive interference, but a factor analysis showed that mental and physical fatigue could be operationally distinguished from general distress. A simulator study showed that substantial increases in both forms of fatigue (c. 1 SD) were produced by a high-workload simulated drive. Saxby et al. (2007) validated a new Driver Fatigue Scale comprising seven correlated factors, including scales related to mental and physical fatigue, awareness of performance deterioration and use of coping strategies intended to counter fatigue.

29.2.3 Individual Differences in Fatigue Susceptibility

Variability in the individual's susceptibility to fatigue is evident in both real-life and in simulator studies. General traits such as extraversion-introversion have been implicated as sources of variability in fatigue response (Lal & Craig, 2001; Wijesuriya et al., 2007). The Driver Stress Inventory (DSI; Matthews, Desmond, Joynes, & Carcary, 1997) was specifically developed to assess five stable traits associated with driving: aggression, dislike of driving, hazard monitoring, fatigue-proneness, and thrill-seeking. The DSI is more predictive of driver stress and performance than are more traditional measures of personality (Matthews, 2002). The most relevant trait in this context is fatigue-proneness, referring to vulnerability to a spectrum of mental and behavioral fatigue symptoms.

Matthews and Desmond (1998) verified that drivers high in fatigue-proneness on the DSI experienced both lower levels of task engagement and also greater physical fatigue during simulated driving. Similar findings have been reported in real-life drives performed by commercial drivers and the general public (Desmond et al., 2001). In fact, fatigue-prone drivers experienced elevated fatigue symptoms even prior to driving in these studies, but DSI fatigue-proneness predicts loss of engagement during driving with initial, baseline fatigue controlled. Other DSI scales relate to further disturbances of mental state; dislike of driving predicts distress and worry, whereas aggression relates to anger. Thus, individual differences may introduce considerable variability in stress and fatigue responses, but use of dispositional scales geared to the driving context may help to explain this variability.

29.3 Theoretical Issues

29.3.1 Transactional Theory as a Framework for Understanding Fatigue

In the driving context, the transactional theory of stress suggests that stress reactions (including fatigue) are the product of the dynamic interaction between the driver and the traffic environment, as the driver attempts to manage environmental demands (Matthews, 2001). The transactional model emphasizes the key role of cognitive appraisals of task demands and of personal competence in moderating the impact of the task environment. As Gawron, French and Funke (2001) state, "individuals choose and modify coping strategies and motivational factors based on their fatigue levels and the changing demands of the tasks".

29.3.1.1 A Transactional Perspective on Active and Passive Fatigue

Fatigue may derive from prolonged high task loads ("active") or from boredom and monotony ("passive": Desmond & Hancock, 2001). From a transactional perspective, the difference is that active fatigue affords the driver scope for maintaining task-focused coping, whereas passive fatigue offers little opportunity for exerting control over the task. Hence, the focal fatigue symptoms of loss of task engagement (i.e., tiredness, de-motivation, distractibility) should be more prevalent when the simulator is configured to produce passive fatigue.

Saxby et al. (2007) tested the prediction. Drivers in the active fatigue condition were exposed to random simulated wind gusts, while drivers in the passive fatigue condition were placed in a supervisory role over a fully-automated vehicle system. Drivers in the control condition drove normally. Drivers in the passive fatigue condition experienced a substantial loss of task engagement, while drivers in the active fatigue condition did not (see Figure 29.2). Nevertheless, drivers in both the active and passive condition experienced a significant increase in distress. The results suggest that loss of task engagement (e.g., boredom, tiredness) can occur simultaneously with distress (e.g., negative mood) depending on task demands.

29.3.1.2 Overlap Between Fatigue and Stress

There is considerable overlap between fatigue and stress, which is often neglected in the fatigue literature; signs of emotional distress can often be seen in fatigued drivers. Within the transactional model, stress is a sign that the driver is over-taxed by demands of the driving task (Matthews, 2002). Because fatigue threatens the driver's coping abilities, it is not surprising that stress symptoms commonly accompany fatigue in simulator studies. However "stress" is a rather vague term, and so it is useful to discriminate different dimensions of stress, including the subjective state factors of (loss of) task engagement, distress and worry defined in Section 29.3.2.2. Different manipulations may induce different multidimensional patterns of state change, as assessed by the DSSQ (Matthews et al., 2002). For example, both passive and active fatigue manipulations tend to increase distress, but only passive fatigue reliably lowers task engagement (see the Saxby et al., 2007, data depicted in Figure 29.2).

Although it is common for distress to accompany fatigue in simulator studies, the converse does not apply. A "winter drive" manipulation, in which the driver was exposed to episodic loss of vehicle control (attributed to black ice), was effective in elevating distress and worry, without any effect on subjective task engagement, measured with the DSSQ (Matthews, 2002). In this case, cognitive appraisals of lack of control drive the stress response, especially in those drivers high in DSI (Dislike of Driving). Thus, different forms

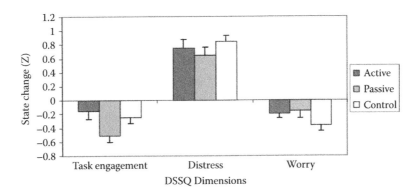

FIGURE 29.2 Subjective state responses to active and passive fatigue inductions, assessed by standardized change scores on the DSSQ.

of challenge may be simulated, eliciting different constellations of subjective symptoms. We can loosely distinguish fatigue and stress responses, corresponding to different appraisals and coping preferences, but both states are best represented as multidimensional state patterns of the kind shown in Figure 29.2, which overlap in some respects.

29.3.2 Information-Processing Mechanisms for Performance Impairment

Performance commonly deteriorates during prolonged driving, depending on the performance index. A methodological difficulty of the field is the wide variety of performance indices used. Analyses of driving as a multi-component skill (e.g., Angell et al., 2006; Lee, 2006) differentiate a wide range of cognitive functions required for the task, as well as a variety of specific performance measures that may be used to index those functions. Key components include visual perception and attention, psychomotor skills, and judgment of risk, as well as executive control processes supporting active search for hazards, anticipation of events, multi-tasking and monitoring personal competence and safety. Any given function of these kinds may be assessed on the simulator with a multiplicity of behavioral measures. For example, visual attention may be measured through: (1) naturalistic indices such as response time to a lead vehicle braking; (2) inclusion of sub-tasks requiring explicit response to hazards or secondary task stimuli; (3) assessment of distractor effects on driver performance; (4) metrics for eyeglance behavior; or (5) composite indices such as useful field of view (see Angell et al., 2006; Lee, 2006; Saxby et al., 2008). Similarly, vehicle control may be assessed with a variety of indices of lateral and longitudinal tracking, or performance during execution of specific maneuvers such as passing (e.g., Emo, Funke, Matthews, & Warm, 2004).

Simulator research has not yet provided any standard performance metrics for evaluation of fatigue effects, although some indices may be more sensitive than others. Sleep deprivation produces the most consistent performance impairments, of which lane drift appears to be the most reliable (Baulk et al., 2008; Fairclough, 2001). However, fatigue-related performance decrements show considerable variability both across different drivers and across different traffic environments. For example, van der Hulst, Meijman and Rothengatter (2001) found that steering, measured by SD of lateral position, became impaired over time, although fatigue had no effect on responses to decelerations of a lead vehicle, taken as an index of hazard avoidance. The authors suggest that collision avoidance is the higher-priority element of performance, and thus is preserved better under fatigue. By contrast, Matthews and Desmond (2002) found that task-induced fatigue had stronger effects on lateral control of the vehicle than on visual attention. In two studies, they found that fatigue increased heading error and also reduced frequencies of small-magnitude steering movements. Visual attention was assessed by having drivers press a button in response to moving pedestrians. Fatigue effects on perceptual sensitivity on this task were more fragile than those found for measures of lateral control, appearing only when the attentional task was relatively easy. Across three studies, Oron-Gilad and Hancock (2005) showed that, although a performance decrement was found in each one, the performance index subject to fatigue varied across different types of drive (e.g., winding versus straight road). For example, on a winding road fatigue was expressed as increased speed, whereas on a straight road impairments in vehicle control without change in speed were observed. Drivers appeared to focus on maintaining the safety-critical elements of driving for each road type.

Matthews (2001) suggests that stress and fatigue effects on performance may be divided into those that are biocognitive, reflecting changes in basic information-processing parameters (e.g., attentional resource availability), and those that are cognitive-adaptive, reflecting changes in voluntary strategy choice (e.g., prioritization of task goals). The challenge posed by fatigue effects is that they may reflect both changes in the efficiency of multiple processing components, as well as strategy changes such as reluctance to apply effort (Matthews & Desmond, 2002). Use of the driving simulator affords the control over task demands that is necessary to test for the information-processing mechanisms sensitive to fatigue. Next, we illustrate the use of such methods, in relation to two leading theories that attribute fatigue effects to loss of attentional resources (Kahneman, 1973), and to disruption of effort-regulation (Hancock & Warm, 1989).

29.3.2.1 Resource Theory

Attentional resource theories propose that attentional capacity or resources exists in a fixed quantity (Matthews et al., 2000). Resources are believed to facilitate information processing, such that insufficiency of resources may cause performance decrement. Workload may be defined as the proportion of resources required to meet the demands of a task (e.g., Wickens & Hollands, 1999). Problems with resource theory include the lack of a precise definition, and the possibility that resources are multiple rather than unitary (Wickens, 2004). Nevertheless, the construct is useful as a means for formalizing loss of performance resulting from increasing workload (Funke, Matthews, Warm, & Emo, 2007).

Fatigue produced by prolonged performance may deplete resources (Warm, Matthews, & Finomore, 2008). Hence, performance deterioration over time should increase as task load increases. The prediction is amply supported for laboratory signal detection tasks (Warm et al., 2008), but it appears to be less successful in driver fatigue studies. Matthews and Desmond (2002) found that fatigue effects on driving performance were more pronounced when workload was low (driving a straight road) than when workload was high (driving curves), contrary to prediction from resource theory. Resource theory may be more relevant as an explanation for individual differences in performance in the fatigued driver. The task engagement subjective state dimension described in Section 29.2.2.2 correlates reliably with performance of vigilance and other attentional tasks (Warm et al., 2008). Funke et al. (2007) showed that engagement correlated with superior steering control, although generally, studies in our laboratory suggest that driver performance is less sensitive to variation in engagement than is vigilance.

29.3.2.2 Effort-Regulation Theory

Effort-regulation accounts attribute performance impairments to a failure to match effort to environmental task demands (Hancock & Warm, 1989). Fatigued drivers may misperceive task demands or fail to apply appropriate effort to the task (Matthews & Desmond, 2002). The effort-regulation model of driver performance may explain the paradoxical finding that fatigue is more damaging to low-workload driving. Fatigued drivers may cope with discomfort by withdrawing effort selectively when the task load is relatively low, as evidenced by reduced steering activity in low-workload conditions (Matthews & Desmond, 2002). Even when driving is routine, some effort is necessary to maintain effective vehicle control and attention, and the fatigued driver may reduce effort to a level below the modest investment required to maintain safety. By contrast, when perceived workload is high, drivers must mobilize additional effort to maintaining performance, attentional focus, and to mitigating the effects of fatigue. Fairclough (2001) makes the similar suggestion that fatigued drivers switch from external goals of road safety to more internal goals such as maintaining personal comfort, leading to reduced task-directed effort.

29.4 Applications

Simulator methods are widely used for evaluation of various factors that may accentuate or mitigate driver fatigue. Studies may be directed towards either systems or people. In the former category, a focus for future research is on the impact of in-vehicle technology on fatigue, such as the role of automated systems (Funke et al., 2007). Much work has also been done to evaluate countermeasures to fatigue, such as pharmacological agents, and the effects of rest breaks and napping (e.g., Macchi, Boulos, Ranney, Simmons, & Campbell, 2002). Technological countermeasures depend on detecting fatigue and providing a warning, as discussed in Section 29.4.3 below. In addition, simulator studies have addressed which people are most sensitive to fatigue, in relation to both individual difference factors in the general population, and clinical conditions characterized by fatigue symptoms.

29.4.1 Assessment of Fatigue Vulnerability

As previously discussed, drivers differ considerably in vulnerability to fatigue. Several strategies, of increasing context-specificity, may be used to identify fatigue-prone drivers. One approach is to measure fatigue-proneness as a general personality dimension using a validated questionnaire. Traits including extraversion and sensation-seeking may be useful for this purpose (Thiffault & Bergeron, 2003), although instruments designed specifically for the driving context may have greater validity (Matthews & Desmond, 1998). Another possibility is to use the simulator to provide a "work-sample" test. Drivers who show large fatigue *state* responses to the simulated drive as assessed by the DSSQ or other scales may be deemed fatigue-prone. Further work is needed to establish the test-retest stability of fatigue response across multiple drives.

The simulator may also be used to identify fatigue-prone drivers on the basis of objective measures of performance deterioration, rather than subjective response. For example, Philip et al. (2003) administered a simulated drive at a freeway rest area, and found that long-distance driving was related to poorer steering control (but not divided attention). Simulated performance provides diagnostic information that cannot be supplied by simple information-processing tasks, such as psychomotor vigilance (Baulk et al., 2008). Again, further research is needed to identify driver performance metrics that uniquely identify fatigue; decreased steering activity provides a possible index of loss of task-directed effort (Matthews & Desmond, 2002).

29.4.2 Investigation of Clinical Fatigue Conditions

Simulator methods have often been used with sleep disorder patients. In addition, many clinical groups (with both psychological and physiological) conditions experience fatigue as a side-effect either from the illness itself or from medications used to treat the illness. In clinical settings, questionnaires and other assessment tools are often used to assess level of functioning and fatigue. Driving simulators offer a unique and ecologically valid way to assess performance impairments related to fatigue in clinical populations. Sudden onset of sleep (SOS) is a scarcely understood phenomenon of Parkinson's disease (PD), which has elicited some concern given the potential consequences that could ensue should SOS occur during activities such as driving. The safety consequences of PD and other neurological disorders may be investigated using the risk-free environment of the simulator (e.g., Zesiewicz et al., 2002).

Research studies with clinical populations using driving simulators may also be useful in training coping mechanisms to help clients safely address fatigue-related concerns. For example, studies using driving simulators could be aimed at helping clients recognize signs of fatigue, when it is no longer safe to drive, and what conditions are most hazardous for them to drive in. Driving simulators provide realistic, yet carefully controlled scenarios that allow for the therapeutic process to happen as clients experience their fears. Indeed, some studies have shown that virtual-reality graded exposure therapy helps relieve symptoms related to driving phobia and post-traumatic stress disorder (Wiederhold & Wiederhold, 2005). Similarly, virtual driving systems might be used with PD patients, for example, to explore which of their fears about falling asleep during driving are realistic and which are unrealistic. Therapy could then be directed towards the avoidance of situations that might lead to sleep episodes and towards increasing confidence in situations in which falling asleep is not a danger.

29.4.3 Evaluation of Countermeasures for Fatigue

The goal of in-vehicle fatigue devices is two-fold: First, the device must be able to accurately detect fatigue in a timely fashion (e.g., before a crash or uncorrectable driver error); and second, the device must be able to trigger some type of warning mechanism that alerts the driver of the potential fatigue state. Driver simulator studies play a key role in validating fatigue detection devices and exploring their operational effectiveness. Psychophysiological techniques for detecting fatigue include the use of EEG and electrocardiac activity as described in Section 29.2.2.1. For example, Lal, Craig, Boord, Kirkup and Nguyen (2003) used a simulator to validate an algorithm for detecting fatigue based on EEG changes in all frequency bands. However, changes in EEG patterns in studies of driver fatigue have shown inter-individual variability; different tasks and different methods of inducing fatigue may also elicit somewhat different EEG responses (Lal & Craig, 2001). EEG alone appears to be of limited practical use in detecting driver fatigue, although it may contribute to fatigue monitoring as one of multiple physiological measures (Rimini-Doering, Manstetten, Altmueller, Ladstaetter, & Mahler, 2001). It may be most productive to monitor psychophysiological indices that are directly related to the cognitive demands of the driving task, such as heart rate variability (Lal & Craig, 2001) or through monitoring cerebral blood flow via TCD, as described previously (Reinerman et al., 2008). However, in many practical contexts, fitting sensors for physiological recording to the driver may be impractical.

Currently, the most widely utilized in-vehicle fatigue countermeasure devices are designed to detect ocular activity. Measures of eye movement, eyelid closure, and head movement appear to signal transitions from alertness to fatigue that seem to occur in the majority of participants (e.g., Lal & Craig, 2001). The traditional electro-oculogram (EOG) requires the use of electrodes. A less invasive measure is the PERCLOS (percentage of eyelid closure) index, which can be obtained using video cameras and machine vision technology. Simulator and field studies concur in suggesting that the PERCLOS index is quite reliable in detecting driver fatigue (Williamson & Chamberlain, 2005), and it has been favorably evaluated by NHTSA (Dinges, Mallis, Maislin, & Powell, 1998). One of the difficulties of the PERCLOS method is that it may be difficult to decide on the criterion percentage of eyelid closure that is used to determine an unsafe level of fatigue. Indeed, drivers may experience potentially hazardous loss of alertness before eyelid closures are apparent (Williamson & Chamberlain, 2005). Schleicher, Galley, Briest and Galley (2008) suggest that oculomotor measures may distinguish "driving without awareness", with open eyes, from sleepiness. Loss of interest in the environment may be indicated by increasing blink rate, whereas sleepiness relates to increasing blink duration. There are also considerable inter-individual differences in eye closure responses which may imply that criteria for fatigue should be set individually for each driver (Schleicher et al., 2008; Svensson, 2004). Another difficulty (Williamson & Chamberlain, 2005) is determining an effective warning message to follow detection of fatigue. A fundamental difficulty is that the appropriate response for the driver may depend on the type of fatigue. The only safe remedy for sleepiness is to sleep, whereas relatively minor fatigue induced by the driving task may be alleviated with rest periods and/or caffeine.

Simulator studies may be useful for addressing issues of these kinds. For example, Schleicher et al. (2008) tracked changes in oculomotor indices and subjective alertness over the course of a monotonous two-hour drive. They were able to identify oculomotor parameters that appeared to define different levels of alertness. Other simulator studies have investigated multiple fatigue indices. A typical study was conducted by Heitmann, Guttkuhn, Aguirre, Trutschel and Moore-Ede (2001), who evaluated several fatigue monitoring devices, including a head-position sensor, an eye-gaze system and pupillometric indices, using a driving simulator. The authors concluded that all these technologies showed promise, but information from multiple sensors should be integrated for optimal results. Although these devices are promising, there is currently no single detection device available which can accurately and reliably detect fatigue the majority of the time (Heitmann et al., 2001). One promising approach may be to combine oculomotor indices with other behavioral measures: Bergasa et al. (2006) report on an algorithm that combines PERCLOS with other measures obtained from facial behaviors such as face direction and fixity of gaze. Since automobile manufacturers are showing increasing interest in engineering devices for detecting fatigue and inattention, there is likely to be an important role for simulator studies in testing the effectiveness of such devices in a range of driving conditions.

29.5 Conclusions

Fatigue is a multi-faceted driver safety problem. Driving simulators contribute to safety research, by allowing experimental control over the multiple sources of fatigue, including sleepiness, prolonged driving and circadian rhythms. Simulators also allow ready measurement of behavioral and psychophysiological outcomes of fatigue, including multiple performance indices. We have suggested in this chapter that simulator research should be guided by a functional understanding of driver fatigue that discriminates between different forms of fatigue, including active and passive fatigue. Subjective and psychophysiological measures may be used to monitor distinct patterns of fatigue response, and to evaluate the utility of monitoring multiple fatigue indices in delivering warnings to the driver. We have proposed also that behavioral expressions of fatigue may vary with task demands and workload. Simulator studies allow the impact of fatigue to be studied in a wide range of different driving scenarios, including those characterized by monotony and those that overload the driver. Level of risk may also be manipulated, including the onset of safety-critical events that may be difficult to study in the field. The normal caveats about transferability of results from the laboratory to real-life driving should be respected. However, simulator research on fatigue appears to have been quite successful in eliciting and elucidating some of the physiological and psychological responses that may be pivotal for driver safety.

Key Points

- Fatigue is a serious safety issue for drivers; simulator studies make an important contribution to investigating driver fatigue and stress.
- Driver simulators may be configured to induce qualitatively different fatigue states including sleepiness, passive fatigue (related to monotony and underload), and active fatigue (related to overload).
- Multidimensional assessments of fatigue response that may include both subjective and psychophysiological indices are recommended.
- Detrimental effects of fatigue may be observed in relation to a variety of performance indices, including measures of attention and psychomotor control, but different studies may show differing performance effects.
- Fatigue effects on driver performance may be moderated by workload factors; the passive fatigue that develops in underload conditions may be especially dangerous.
- Simulators may be used to investigate individual differences in fatigue vulnerability and clinical conditions associated with fatigue that may compromise safety.
- Simulators afford the systematic testing of countermeasures to fatigue and the evaluation of methods for detecting unsafe levels of fatigue, such as monitoring oculomotor behaviors.

Keywords: Behavior, Driver Human Performance, Fatigue, Stress, Psychophysiology

Acknowledgments

The first author gratefully acknowledges support from the U.S. Army Medical Research and Materiel Command under Contract No. DAMD17-04-C-0002. The views, opinions and/or findings contained in this report are those of the author(s) and should not be construed as an official Department of the Army position, policy or decision unless so designated by other documentation. In the conduct of research where humans are the subjects, the investigator(s) adhered to the policies regarding the protection of human subjects as prescribed by 45 CFR 46 and 32 CFR 219 (Protection of Human Subjects). The contents of this report reflect the views of the authors, who are responsible for the facts and accuracy of the data presented herein. Any opinions, options, findings, conclusions, or recommendations expressed herein are those of the authors and do not reflect the official policy or position of the U.S. Department of Transportation, or the U.S. Government. This report does not constitute a standard, specification, or regulation.

Glossary

Attentional resources: One or more reservoirs of a metaphorical "energy" that may be required for certain demanding processing activities. The concept is useful for understanding variation of performance with mental workload.

Driver fatigue: A set of responses to prolonged driving and/or sleep loss including tiredness, sleepiness and loss of cortical arousal.

Fatigue countermeasures (technological): A collection of techniques for detecting fatigue during driving from behavioral or psychophysiological responses, and delivering an appropriate warning message to the driver.

Mental workload: The total level of demands on the operator imposed by task performance in relation to available processing resources or capacity.

Oculomotor indices: Measures of eye closure behaviors that may signal fatigue or the onset of sleep, such as closures of long duration.

Task engagement: A subjective state characterized by high energy, motivation and concentration; the opposite of fatigue.

Key Readings

Baulk, S. D., Biggs, S. N., Reid, K. J., van den Heuvel, C. J., & Dawson, D. (2008). Chasing the silver bullet: Measuring driver fatigue using simple and complex tasks. *Accident Analysis & Prevention, 40,* 396–402.

Desmond, P. A., & Hancock, P. A. (2001). Active and passive fatigue states. In P. A. Hancock & P. A. Desmond (Eds.), *Stress, workload, and fatigue* (pp. 455–465). Mahwah, NJ: Lawrence Erlbaum.

Fairclough, S. H. (2001). Mental effort regulation and the functional impairment of the driver. In P. A. Hancock & P. A. Desmond (Eds.), *Stress, workload, and fatigue* (pp. 479–502). Mahwah, NJ: Lawrence Erlbaum.

Lal, S. K. L., & Craig, A. (2001). A critical review of the psychophysiology of driver fatigue. *Biological Psychology, 55,* 173–194.

Matthews, G. (2002). Towards a transactional ergonomics for driver stress and fatigue. *Theoretical Issues in Ergonomics Science, 3,* 195–211.

References

Angell, L., Auflick, J., Austria, P. A., Kochhar, D., Tijerina, L., Biever, W., . . . Kiger, S. (2006). *Driver workload metrics task 2 final report* (Report for National Highway Traffic Safety Administration contract #DTFH61-01-X-00014.)

Baulk, S. D., Biggs, S. N., Reid, K. J., van den Heuvel, C. J., & Dawson, D. (2008). Chasing the silver bullet: Measuring driver fatigue using simple and complex tasks. *Accident Analysis & Prevention, 40,* 396–402.

Bergasa, L. M., Nuevo, J., Sotelo, M. A., Barea, R., & López, E. (2006). Real-time system for monitoring driver vigilance. *IEEE Transactions on Intelligent Transportation Systems, 7,* 63–77.

Boyle, L. N., Tippin, J., Paul, A., & Rizzo, M. (2008). Driver performance in the moments surrounding a microsleep. *Transportation Research Part F: Traffic Psychology and Behavior, 11,* 126–136.

Desmond, P. A., & Hancock, P. A. (2001). Active and passive fatigue states. In P. A. Hancock & P. A. Desmond (Eds.), *Stress, workload, and fatigue* (pp. 455–465). Mahwah, NJ: Lawrence Erlbaum.

Desmond, P. A., Matthews, G., & Bush, J. (2001). Individual differences in fatigue and stress states in two field studies of driving. *Proceedings of the 45th annual meeting human factors and ergonomics society* (pp. 1571–1575). Santa Monica, CA: Human Factors and Ergonomics Society.

Dinges, D. F., Mallis, M. M., Maislin, G., & Powell, J. W. (1998). *Evaluation of techniques for ocular measurement as an index of fatigue and the basis for alertness management* (DOT HS 808 762), April 1998.

Emo, A. K., Funke, G., Matthews, G., & Warm, J. S. (2004). Stress vulnerability, coping, and risk-taking behaviors during simulated driving. *Proceedings of the 48th annual meeting of the human factors and ergonomics society* (pp. 1228–1232). Santa Monica, CA: Human Factors and Ergonomics Society.

Fairclough, S. H. (2001). Mental effort regulation and the functional impairment of the driver. In P. A. Hancock & P. A. Desmond (Eds.), *Stress, workload, and fatigue* (pp. 479–502). Mahwah, NJ: Lawrence Erlbaum.

Fairclough, S. H., & Venables, L. (2006). Prediction of subjective states from psychophysiology: A multivariate approach. *Biological Psychology, 71,* 100–110.

Funke, G. J., Matthews, G., Warm, J. S., & Emo, A. (2007). Vehicle automation: A remedy for driver stress? *Ergonomics, 50,* 1302–1323.

Gawron, V. J., French, J., & Funke, D. (2001). An overview of fatigue. In P. A. Hancock & P. A. Desmond (Eds.), *Stress, workload, and fatigue* (pp. 581–595). Mahwah, NJ: Lawrence Erlbaum.

Hancock, P. A., & Warm, J. S. (1989). A dynamic model of stress and sustained attention. *Human Factors, 31,* 519–537.

Heitmann, A., Guttkuhn, R., Aguirre, A., Trutschel, U., & Moore-Ede, M. (2001). Technologies for the monitoring and prevention of driver fatigue. *Proceedings of the first international driving symposium on human factors in driver assessment, training and vehicle design* (pp. 81–86).

Hitchcock, E. M., & Matthews, G. (2005). *Multidimensional assessment of fatigue: A review and recommendations.* Proceedings of the International Conference on Fatigue Management in Transportation Operations.

Kahneman, D. (1973). *Attention and effort.* Englewood Cliffs, NJ: Prentice Hall.

Lal, S. K. L., & Craig, A. (2001). A critical review of the psychophysiology of driver fatigue. *Biological Psychology, 55,* 173–194.

Lal, S. K. L., Craig, A., Boord, P., Kirkup, L., & Nguyen, H. (2003). Development of an algorithm for an EEG-based driver fatigue countermeasure. *Journal of Safety Research, 34,* 321–328.

Lazarus, R. S. (1999). *Stress and emotion: A new synthesis.* New York: Springer.

Lee, J. D. (2006). Driving safety. In R. S. Nickerson (Ed.), *Review of human factors* (pp. 172–218). Santa Monica, CA: Human Factors and Ergonomics Society.

Macchi, M. M., Boulos, Z., Ranney, T., Simmons, L., & Campbell, S. S. (2002). Effects of an afternoon nap on night-time alertness and performance in long-haul drivers. *Accident Analysis & Prevention, 34,* 825–834.

Matthews, G. (2001). Levels of transaction: A cognitive science framework for operator stress. In P. A. Hancock & P. A. Desmond (Eds.), *Stress, workload and fatigue* (pp. 5–33). Mahwah, NJ: Lawrence Erlbaum.

Matthews, G. (2002). Towards a transactional ergonomics for driver stress and fatigue. *Theoretical Issues in Ergonomics Science, 3,* 195–211.

Matthews, G., Campbell, S. E., Falconer, S., Joyner, L., Huggins, J., Gilliland, K., . . . Warm, J. S. (2002). Fundamental dimensions of subjective state in performance settings: Task engagement, distress and worry. *Emotion, 2,* 315–340.

Matthews, G., Davies, D. R., Westerman, S. J., & Stammers, R. B. (2000). *Human performance: Cognition, stress and individual differences.* London: Psychology Press.

Matthews, G., & Desmond, P. A. (1998). Personality and multiple dimensions of task-induced fatigue: A study of simulated driving. *Personality and Individual Differences, 25,* 443–458.

Matthews, G., & Desmond, P. A. (2002). Task-induced fatigue states and simulated driving performance. *Quarterly Journal of Experimental Psychology: Human Experimental Psychology, 55,* 659–686.

Matthews, G., Desmond, P. A., Joyner, L. A., & Carcary, B. (1997). A comprehensive questionnaire measure of driver stress and affect. In E. Carbonell Vaya & J. A. Rothengatter (Eds.), *Traffic and transport psychology: Theory and application* (pp. 317–324). Amsterdam, the Netherlands: Pergamon.

Matthews, G., Warm, J. S., Reinerrman, L. E., Langheim, L., Washburn, D. A., & Tripp, L. (in press). Task engagement, cerebral blood flow velocity, and diagnostic monitoring for sustained attention. *Journal of Experimental Psychology: Applied.*

Moller, H. J., Kayumov, L., Bulmash, E. L., Nhan, J., & Shapiro, C. M. (2006). Simulator performance, microsleep episodes, and subjective sleepiness: Normative data using convergent methodologies to assess driver drowsiness. *Journal of Psychosomatic Research, 61,* 335–342.

Morrow, P. C., & Crum, R. (2004). Antecedents of fatigue, close calls, and crashes among commercial motor-vehicle drivers. *Journal of Safety Research, 35,* 59–69.

Nabi, H., Salmi, L. R., Lafont, S., Chiron, M., Zins, M., & Lagarde, E. (2007). Attitudes associated with behavioral predictors of serious road traffic crashes: Results from the GAZEL cohort. *Injury Prevention, 13,* 26–31.

Oron-Gilad, T., & Hancock, P. A. (2005). Road environment and driver fatigue. *Proceedings of the third driving assessment symposium* (pp. 318–324).

Pack, A. I., Pack, A. M., Rodgman, E., Cucchiara, A., Dinges, D. F., & Schwab, C. W. (1995). Characteristics of crashes attributed to the driver having fallen asleep. *Accident Analysis & Prevention, 27*, 769–775.

Philip, P., Sagaspe, P., Moore, N., Taillard, J., Charles, A., Guilleminault, C. et al. (2005). Fatigue, sleep restriction and driving performance. *Accident Analysis & Prevention, 37*, 473–478.

Philip, P., Taillard, J., Klein, E., Sagaspe, P., Charles, A., Davies, W. L. et al. (2003). Effect of fatigue on performance measured by a driving simulator in automobile drivers. *Journal of Psychosomatic Research, 55*, 197–200.

Reinerman, L. E., Warm, J. S., Matthews, G., & Langheim, L. K. (2008). Cerebral blood flow velocity and subjective state as indices of resource utilization during sustained driving. *Proceedings of the 52nd annual meeting of the human factors and ergonomics society* (pp. 1252–1256). Santa Monica, CA: Human Factors and Ergonomics Society.

Rimini-Doering, M., Manstetten, D., Altmueller, T., Ladstaetter, U., & Mahler, M. (2001). Monitoring driver drowsiness and stress in a driving simulator. *Proceedings of first international driving symposium on human factors in driver assessment, training and vehicle design* (pp. 58–63).

Saxby, D. J., Matthews, G., Hitchcock, E. M., Warm, J. S., Funke, G. J., & Gantzer, T. (2008). Effects of active and passive fatigue on performance using a driving simulator. *Proceedings of the 52nd annual meeting of the human factors and ergonomics society* (pp. 1252–1256). Santa Monica, CA: Human Factors and Ergonomics Society.

Saxby, D. J., Matthews, G., & Hitchcock, T. (2007). *Fatigue states are multidimensional: Evidence from studies of simulated driving.* Proceedings of the Driving Simulation Conference North America 2007. Iowa City, IA: University of Iowa.

Schleicher, R., Galley, N., Briest, S., & Galley, L. (2008). Blinks and saccades as indicators of fatigue in sleepiness warnings: Looking tired? *Ergonomics, 51*, 982–1010.

Svensson, U. (2004). *Blink behavior based drowsiness detection - method development and validation.* Unpublished M.Sc. Thesis, University of Linköping.

Thiffault, P., & Bergeron, J. (2003). Monotony of road environment and driver fatigue: A simulator study. *Accident Analysis & Prevention, 35*, 381–391.

van der Hulst, M., Meijman, T., & Rothengatter, T. (2001). Maintaining task set under fatigue: A study of time-on-task effects in simulated driving. *Transportation Research Part F: Traffic Psychology and Behavior, 4*, 103–118.

Warm, J. S., Matthews, G., & Finomore, V. S. (2008) Workload and stress in sustained attention. In P. A. Hancock & J. L. Szalma (Eds.), *Performance under stress* (pp. 115–141). Aldershot, UK: Ashgate Publishing.

Warm, J. S., & Matthews, G., & Parasuraman, R. (2009). Cerebral hemodynamics and vigilance performance. *Military Psychology, 21*(S1), S75–S100.

Wickens, C. D. (2004). Multiple resource time sharing model. In N. A. Stanton, E. Salas, H. W. Hendrick, A. Hedge, & K. Brookhuis (Eds.), *Handbook of human factors and ergonomics methods* (pp. 40.1–40.7). Boca Raton, FL: CRC Press.

Wickens, C. D., & Hollands, J. G. (1999). *Engineering psychology and human performance* (3rd ed.). Upper Saddle River, NJ: Prentice-Hall.

Wiederhold, B. K., & Wiederhold, M. D. (2005). Fear of driving. In B. K. Wiederhold & M. D. Wiederhold (Eds.), *Virtual reality therapy for anxiety disorders: Advances in evaluation and treatment* (pp. 147–155). Washington, DC: American Psychological Association.

Wijesuriya, N., Tran, Y., & Craig, A. (2007). The psychophysiological determinants of fatigue. *International Journal of Psychophysiology, 63*, 77–86.

Williamson, A., & Chamberlain, T. (2005). *Review of on-road driver fatigue monitoring devices.* Technical Report, NSW Injury Risk Management Research Centre, University of New South Wales.

Zesiewicz, T. A., Cimino, C. R., Malek, A. R., Gardner, N., Leaverton, P. L., Dunne, P. B. et al. (2002). Driving safety in Parkinson's disease. *Neurology, 59*, 1787–1788.

30

Driving Simulators as Training and Evaluation Tools: Novice Drivers

Alexander Pollatsek
University of Massachusetts, Amherst

Willem Vlakveld
SWOV Institute for Road Safety Research

Bart Kappé
Netherlands Organization for Applied Scientific Research, TNO

Anuj K. Pradhan
National Institutes of Health

Donald L. Fisher
University of Massachusetts, Amherst

Abstract

The Problem. We know that newly licensed 16- and 17-year-old drivers during their first six months on the road with a restricted license are at a greatly inflated risk of crashing. This inflated crash rate has not changed over the last 50 years. *The Question.* The question we address is whether there are training techniques that show some promise of reducing these high crash rates. *Role of Driving Simulators.* Driving simulators represent an important tool for evaluating the efficacy of training programs in situations that would be too unsafe to study on the open road; they may also be of value in training, although their widespread use may be limited by their cost. *Key Results of Driving Simulator Studies.* Studies using driving simulators and the open road have revealed that newly licensed drivers can be trained to anticipate specific hazards, to scan more broadly within the general driving environment, to prioritize their attention, and to maneuver their vehicle more safely, all without becoming overconfident. *Scenarios and Dependent Variables.* Examples of the sorts of simulator scenarios that are used to study scanning, attention maintenance, and vehicle management skills are discussed in detail. Difficulties that attend the development of such scenarios are described. Examples of the dependent variables used to differentiate between trained and untrained novice drivers are discussed, as are the procedures that are needed to reduce the data to meaningful summary measures. *Limitations.* Although the studies so far have shown that programs that are effective on a driving simulator are also effective on the open road, one cannot assume that this is always true.

30.1 Introduction

The goals of driver education programs have varied over time. In the United States, the very first programs simply attempted to teach the most basic skills of maneuvering a vehicle (Butler, 1982). This has evolved considerably. In 1973 at the Fifth National Conference on Driver Education, the stated purpose of driver education was: "to develop safer and more efficient highway users who understand the essential components of the highway transportation system in a manner that will enhance the effectiveness of such components (Aaron & Strasser, 1977)." Most recently, in a 1994 report to the United States Congress, the National Highway Traffic Safety Administration (NHTSA) defined driver education as follows: "Driver education is a training program of organized learning and practice designed to provide the basic knowledge, attitudes and skills needed to drive safely, and to provide the advanced knowledge and skills needed for safe driving performance under special circumstances" (NHTSA, 1994, p. 1).

If the goals of driver education are to provide the skills, knowledge, and attitudes necessary for safe driving, then one measure of the success of such programs would be a reduction in the crashes of drivers who were exposed to those driver education programs. Although the first known driver education program in the United States was established in 1916 [NHTSA 1994, p. 3], it was not until 1976 that a full-scale controlled evaluation of driver education was undertaken in a suburb of Atlanta, Georgia. Teen drivers were randomly assigned to three groups: one received extensive training, (the Safe Performance Curriculum, SPC) one received more limited training (the Pre-Driver Licensing Curriculum, PDL), and the third group was used as a control. The results were disappointing. The SPC (National Highway Traffic Safety Administration, 1974) contained a component referred to as simulation, but it was a film that was not responsive to any control input and thus different from what we mean by simulation today. Although there was about a 15% reduction in the number of crashes per licensed driver in the groups that received training during the first six months of driving (i.e., the SPC, PDL, and control groups had 0.105, 0.107, and 0.122, crashes per licensed driver; Stock, Weaver, Ray, Brink & Sadoff, 1983), there were no differences between the trained and control groups when evaluated two (Stock et al.) and four years (Smith & Blatt, 1987) after training had ended. Twenty years later, Mayhew and Simpson (1996) reviewed 30 studies from several countries that evaluated the effect of driver training programs on crashes, and found very little support for the claim that formal driver education decreased crash involvement. More recently, a number of literature reviews of the effectiveness of standard driver education programs have been conducted (standard driver education programs typically have 30 hours of classroom instruction and six hours of behind-the-wheel instruction). The reviews have spanned the globe, including ones undertaken in Australia (Woolley, 2000), Britain (Roberts & Kwan, 2002), Canada (Mayhew & Simpson, 2002), Sweden (Engström, Gregersen, Hernetkoski, Keskinen, & Nyberg, 2003) and the United States (Vernick, Li, Ogaitis, MacKenzie, Baker, & Gielen, 1999; Nichols, 2003), and most recently a comprehensive, international review sponsored by the AAA Foundation for Traffic Safety (Clinton & Lonero 2006; also see Lonero, 2007). These reviews are uniform in concluding that in the great majority of experimental evaluations of standard driver education programs, no reduction in the crash rates among newly-licensed drivers is observed. In fact, some reviewers concluded that standard driver education programs may actually increase the crash rates, both by reducing the age at which solo driving is allowed and by teaching novice drivers skills such as skid control that may increase a novice driver's willingness to take risks (Nichols, 2003). Importantly, it is not clear from the reviews whether any of the standard programs required training on a driving simulator where the driver's path through the virtual world was determined by the driver's manipulation of the vehicle controls. However, it seems somewhat doubtful that any did, since fully articulated simulator training programs have not been reported in the published literature until very recently (e.g., Allen, Park, Cook, & Viirre, 2003).

In this context, one can ask three major related questions: (1) What are the behaviors of novice drivers that are causing them to crash? (2) Are standard driver education programs addressing those behaviors? (3) And, if not, are there any existing alternative driver education programs that address these behaviors? The focus here is on the use of driving simulators both to evaluate behaviors that have been inferred to be a contributing cause of crashes and to train drivers to avoid crashes. However, several of these key studies in which simulators were not involved—either in training or in the evaluation—are discussed as well because of their clear relevance to the topic at hand. At the end of the chapter, we discuss more general issues relevant to the evaluation and training of novice drivers on simulators.

30.1.1 Causes of Newly-Licensed Driver Crashes

To the general public, alcohol and high speeds are perhaps thought to be the two major reasons for the high crash rate for newly-licensed drivers. However, during the first six months, the percentage of newly-licensed drivers who crash while under the influence of alcohol (NHTSA, 2006) or while traveling at very high speeds (McKnight & McKnight, 2003) is relatively small. Instead, analyses of police crash reports (McKnight & McKnight) indicate that failures of (a) *visual scanning* (ahead, to the sides, and to the rear), (b) *attention maintenance* (distribution of attention between the forward roadway and other locations inside and outside the automobile cabin), and (c) *speed management*, are responsible, respectively, for 43.6%, 23.0% and 20.8% of the crashes (the causes overlap) among drivers between the ages of 16 and 19 years old. In addition, although the absolute number of crashes decreased as a function of increases in the experience of the young drivers, these three percentages did not change appreciably. Thus, if overall crashes per licensed driver are decreasing rapidly during the first six months, unless there is a not-yet-identified factor causing the decrease, it seems likely that newly-licensed drivers are improving in all three areas. Moreover, there is evidence from laboratory studies and from naturalistic and experimental studies in the field that young drivers do differ considerably from more experienced drivers in all of the three areas identified by McKnight and McKnight as causing problems for teen drivers (2003).

Visual scanning: Hazard anticipation. There is now a large body of literature that indicates that young drivers perform more poorly on hazard anticipation tests that are administered outside of an actual driving situation (Horswill & McKenna, 2004). However, when the hazard is easily detected (e.g., by motion), differences between younger drivers and more experienced drivers are less clear-cut (Sagberg & Bjørnskau, 2006). These differences in hazard anticipation skills also appear when people are on a driving simulator: Young drivers are much less likely than experienced drivers to scan for potential hazards when these hazards are difficult to detect, such as a pedestrian that might emerge suddenly from behind a vehicle stopped in front of a midblock crosswalk (Pollatsek, Narayanaan, Pradhan, & Fisher,

2006). In addition, these differences between groups appear even when the possible risk is "foreshadowed" (e.g., a pedestrian is seen crossing a considerable amount of time before the car enters the area, Garay-Vega, Fisher, & Pollatsek, 2007). The above differences in *tactical hazard anticipation scanning* (i.e., the scanning pattern observed when a feature in the environment suggests that a hidden threat is especially likely to materialize at a particular location and time in a scenario) coexist with differences in *strategic hazard anticipation scanning* (i.e., the scanning pattern observed when there is no such key feature). Specifically, as indicated by studies of strategic scanning on the open road young drivers: a) scan less broadly from side to side, especially when changing lanes (Mourant & Rockwell, 1972); b) have, on average, less widely spaced eye movements as measured along the horizontal axis (Crundall & Underwood, 1998); and c) are less likely to make consecutive fixations on objects in the periphery (Underwood, Chapman, Brocklehurst, Underwood, & Crundall, 2003). It is worth mentioning that these differences between age groups in tactical and strategic hazard anticipation scanning in a driving simulator and on the open road appear even though the drivers know they are being tested, and thus are, in some sense, "on their best behavior".

Attention Maintenance. Studies on the open road—both controlled (Wikman, Nieminen, & Summala, 1998) and naturalistic (Klauer, Dingus, Neale, Sudweeks, & Ramsey, 2006)—and in the laboratory (Chan, Pradhan, Knodler, Pollatsek, & Fisher, 2010) indicate that young drivers are much more likely to gaze continuously for longer than two seconds inside the vehicle when searching for something inside the vehicle (i.e., are more likely not to attend to the roadway). (A common measure employed here is *glance duration*, which is the duration of a period when fixations are in a particular region—in this case, on locations inside the vehicle.) For example, in a controlled study on the open road (Wikman et al., 1998), it was found that only 13% of the experienced drivers had a *glance duration* inside the vehicle of at least 2.5 seconds during an episode of searching for something inside the vehicle whereas 46% of inexperienced drivers had a glance duration of at least this length during such an episode (an *episode* begins when the driver is asked to perform the in-vehicle task and ends when the driver completes the task; there may be several glances inside and outside of the vehicle during any one episode).

Vehicle Management: Hazard Response. Speed management was identified by McKnight and McKnight (2003) as the third most prevalent cause of crashes among young drivers. This included adjusting the speed of the vehicle to traffic/road conditions, slowing on curves, and slowing on slick surfaces. It is clear that speed management is part of a more general category of *vehicle management* which includes behaviors such as maintaining the proper space between vehicles, adjusting the vehicle lane position in response to traffic and road conditions, and responding when a hazard requires an evasive maneuver. *Tactical* vehicle management is particularly important when potential hazards might materialize. A study of the differences in the tactical hazard anticipation vehicle management skills of young and experienced drivers indicates that the differences are pronounced in situations where hazards are difficult to detect (Fisher et al., 2002). So, for example, consider a driver who is approaching an intersection and who intends to drive straight through it (the Truck Left Turn scenario, **Figure 30.1;** for a color version, see the insert pages or Web Figure 30.1 on the *Handbook*

FIGURE 30.1 **(See color insert)** Adjacent truck left turn. (Cars in the opposing lane in front of the truck are obscured from the driver and can turn left into the path of the driver as he or she travels through the intersection.)

web site). Imagine that the driver is passing a truck in a left turn lane that is blocking the driver's view of traffic in the opposing lane which could be taking a left turn across the driver's path. The aware driver should slow down as he or she passes the truck. In fact, experienced drivers are more likely than novice drivers to apply their brakes as they travel past the truck on the left.

30.1.2 Driver Education Programs and Critical Skills

Until recently, there has been little empirical evidence to support the development of a specific content for the training of crash-reduction behaviors in driver education programs. It is true that the recommendation has been made that driver education programs should focus on hazard perception and risk assessment (Mayhew & Simpson, 1995, 1996). But this recommendation is too broad to be of much assistance to driver education instructors, and thus the coverage of the tactical or strategic behaviors in scenarios which would decrease novice drivers' risk has been minimal in driver education (Mayhew & Simpson, 2002). Clearly something is missing from the current driver education programs as a safety countermeasure, since such programs have little impact on crash rates. New driver training programs have been developed which target the critical driving skills to which reference has been made above: Hazard anticipation, attention maintenance, and hazard response. Some of these training programs make use of driving simulators.

30.1.3 Simulators as Training Devices

Simulators are increasingly being used in driver education programs around the world (Vlakveld, 2006). In the Netherlands alone, there are over 100 simulators in operation in different driving schools (Kappé & van Emmerik, 2005). Despite their increasing use, there has been (and probably will continue to be) vigorous debate about the utility of driving simulators as training devices. Most researchers agree, however, that simulators are the only way in which one can safely learn how to respond in many emergency situations, and most would also agree that simulators are poor substitutes for automobiles for training the psychomotor skills associated with basic driving maneuvers.

In contrast, there is real disagreement about whether simulator-based training programs can be used to train higher-order cognitive skills (e.g., situation awareness, hazard anticipation, attention maintenance) which then transfer to the field. For example, Groeger and Banks (2007) argue that driving simulators are unlikely to have much of a role to play in the training of higher-order cognitive skills because positive transfer of any skills from the simulator would usually require evaluation of a situation that drivers had not experienced before and they would have to apply the skills in less than a second (see also Christie & Harrison, 2003). Those who argue that driving simulators can and should be used to train these higher-order cognitive skills (e.g., Wheeler & Triggs, 1996) note that—unlike more basic psychomotor skills—they do not depend on feedback in a dynamic

environment, but are conscious and usually planned (well) ahead of time. A good example is the eye movement that a driver should make in front of the truck stopped before a midblock crosswalk in order to see a pedestrian who might be hidden by the truck (see Figure 30.2). Often, two to three seconds elapse in a situation like this—even for skilled drivers—between the recognition that the situation is one in which a potential hazard can materialize and the resulting response (moving the eyes), but this is still sufficient for the eyes to focus on the potential hazard in time.

Are higher-order skills, such as hazard anticipation, when used on the open road, radically different from these skills when exercised in training? Groeger and Banks (2007), drawing on the work of Barnett and Ceci (2002), have provided a useful framework for discussing the transfer of training. Briefly, transfer is a function of the *content* (of what might transfer from prior experience) and the *circumstances* in which the transfer will occur (the relation

(a)

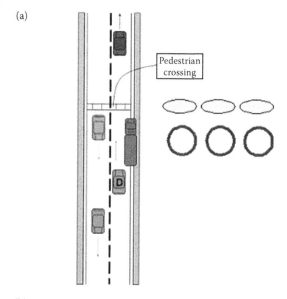

(b)

FIGURE 30.2 Truck crosswalk scenario. (a) Plan view. (b) Perspective view.

between the context of the prior experience and the context in which the transfer is expected to occur). In terms of content, transfer will occur most easily if the skill is well-practiced, and if the initiation of the skill is prompted externally so that the learner does not have to remember to execute it. In the hazard anticipation scenario discussed above, eye movements are well-practiced, and the speed and accuracy of the eye movements demanded in the field are no different than they would be on a driving simulator. Although a driver on the road is not automatically reminded that a particular hazard must be scanned (and, in fact, this takes some mental effort, Horswill & McKenna, 2004), the recognition of a potential hazard is often simple (e.g., a marked midblock crosswalk signals the driver of a potential hazard each and every time it appears). Thus, from the standpoint of content, it is clear that training of the recognition of this hazard should involve repeated practice of recognizing the hazard; this is easily done with simulators or computer-controlled training programs. However, it would appear that when transfer is at an abstract—conceptual—level, at least when hazard anticipation is involved (Pollatsek, Fisher, & Pradhan, 2006), training is required on only one instance of a scenario in order to achieve generalization, not a large number of instances.

Then there are the "circumstances" of transfer, or the relation between the context of training and the situation to be transferred. Much can be made the same between the simulator and the actual road experience (e.g., cabin noise, sound of traffic, and lighting) but without motion it is difficult to mimic important cues that might help in vehicle management such as cornering (although there is no clear experimental evidence to suggest that motion in training is important). This suggests that only certain types of skills will transfer or will transfer well. In addition, although transfer would likely decline for larger delays between the training and on-road driving, there is little reason to believe that the rate of decline is much different for higher-order driving skills than for other skills. Moreover, because the first couple of months on the road are especially critical, it is not necessary that the effects of training have to last for a long period of time, certainly no more than the first couple of months. Of course, the question of whether some skills can be trained on a driving simulator and then transfer to real driving is ultimately an empirical one, and we turn now to a discussion of the effectiveness of various training programs, some of which use a driving simulator and others do not.

30.2 Alternative Driver Education Programs

30.2.1 Australia

Researchers at the Monash University Accident Research Center (MUARC) have engaged in extensive development (Regan, Triggs, & Wallace, 1999) and evaluation (Regan, Triggs, & Godley, 2000) of a novice driver training program. The program that was eventually developed, DriveSmart, combined CD-ROM and simulator training. A total of 14 different content areas in

driving were identified as requiring emphasis (e.g., hazard anticipation, attention maintenance). These content areas were then taught in one of three different driving contexts (rural, freeway, urban) using either digitized real-world videos or a three-dimensional virtual world. The combination of content, context, and instructional modes led to a product with six training modules (Table 30.1). In the Scanning module, video-clips were used to provide the background for the training of hazard anticipation skills. Each of the 67 clips lasted between 20 and 30 seconds. The video was stopped at a given point and the driver was asked, for example, to click on likely risks. In the Keeping Ahead and Playing Safe module, the video was paused and the driver was asked to indicate what the driver should do next. In the Concentration module, three-dimensional virtual words were used to train prioritization of attention. As a drive through a simulated world unfolded, either the participants were asked to keep a constant distance between their car and a car ahead whose velocity varied sinusoidally or they were asked to monitor areas of the periphery in which numbers would suddenly appear and then to perform operations on those numbers. In the fifth and sixth video modules, "On the Road—Urban Driving" and "On the Road—Country Driving", participants were again asked to both scan the roadway and plan ahead. The drives that were presented in these modules exposed the participants a number of times to *near transfer* scenarios (i.e., scenarios similar to ones upon which they would be evaluated after training on a driving simulator) and a limited number of times to *far transfer* scenarios. An example of a near transfer scenario in which novice drivers were trained on the PC and then evaluated on a driving simulator is a right turn at an intersection against opposing traffic (left turn in the United States). An example of a far transfer scenario that was not trained, but which was evaluated, is the driver's behavior in a scenario where it suddenly becomes foggy.

In order to evaluate DriveSmart, 103 learner drivers between the ages of 16 years and 11 months, and 17 years and 10 months were randomly assigned to the treatment (training with DriveSmart) or control (training with a Microsoft flight simulator, which had no obvious relevance) groups. Training was undertaken for both the experimental and control groups in four separate sessions. Immediately after training and then four weeks later, participants were evaluated on a driving simulator in each of four 5-minute *risk perception* drives and three 5-minute

TABLE 30.1 Elements of the DriveSmart Training Program

Module	Medium	
	Digitized Real-World Video	3D Virtual World Simulation
1. Introduction	X	
2. Scanning	X	
3. Keep Ahead and Play Safe	X	
4. Concentration		X (near/far)
5. Urban	X (near/far)	
6. Country	X (near/far)	

attentional control drives. There were a total of four scenarios in each risk perception drive in which it was possible to score vehicle behaviors indicative of what we have defined as tactical hazard perception (and thus a total of 16 scenarios across all four drives). One or more dependent variables were used to index the participant's safe driving performance in each scenario. Two of the scenarios in each drive mirrored what had been trained in DriveSmart (near transfer); two were different (far transfer). Immediately after training, the treatment group was significantly more likely than the control group (at the 10% level) to detect a hazard in four of the near transfer and four of the far transfer scenarios (as indicated by at least one of the measures). There was no statistically significant difference in the remaining eight scenarios. Similar results were observed four weeks later, with the treatment group performing better in three of the eight near transfer, and four of the eight far transfer scenarios. In each of the three attention drives, the speed limit changed in six different places. During each attention drive, the participants had to listen to a series of two-digit numbers (for example, "83"), and then report aloud the absolute value of the difference between the first and last digit, (i.e, "5"). In the evaluation immediately after training the treatment group reached the speed limit more quickly and drove closer to the posted limit than did the control group, all the while performing the arithmetic task equally accurately. In the evaluation undertaken four weeks after training, the treatment group still reached the speed limit more quickly. In part due to this work, passing a hazard perception test is now part of the Victoria, Australia licensing requirement. That test can be downloaded from: http://www.vicroads.vic.gov.au/Home/Licensing/LicenceTests/AboutLicenceTests/HazardPerceptionTest.htm. Unfortunately, the dependent measures were not reported for each scenario so it is difficult to know exactly how large a practical effect training had. Additionally, one cannot know what effect the training might have on actual crash rates.

30.2.2 England

Researchers in England, although not using a driving simulator in the training or evaluation of novice drivers, have focused on a behavior (strategic hazard anticipation) which has been identified as critical to reducing novice driver crashes; thus some discussion is warranted (Chapman, Underwood, & Roberts, 2002). Specifically, hazards were presented in video clips to novice drivers. The training took approximately one hour. No clip that appeared in the training phase appeared later in the evaluation phase. There were five phases in training. In the first, drivers saw four clips, and commented on what they saw and were supposed to press a button whenever they saw a hazard. In the second phase, drivers viewed five clips run at half speed twice; on the first run, they commented on areas that were circled in blue (general areas of interest) and on areas that were circled in red (specific hazards) and then listened to experts' comments on why the different areas were circled. In the third phase, five new clips were played and paused at critical points and the

participant's task was to state aloud what will happen next that could pose a threat. The clip was then restarted and an expert explained why there was an impending potential hazard. In the fourth phase, the 10 clips from the two previous phases were played at full speed with the red and blue circles overlaid appropriately. The driver was supposed to comment on the clip and to anticipate the hazards. In the fifth phase, the driver watched four new clips and provided commentary on potential hazards and other driving-relevant information as well as pressing a button at each potential hazard.

The trained group was evaluated three times: three months before training (soon after passing their driving test), immediately after training, and three to six months after training. A control group that was not trained was evaluated at the same times. One evaluation was conducted in the field on a predetermined course on the open road at 18 selected points. The full drive included three different road types (urban, rural, and dual carriageway) and four different speed limits (30, 40, 60 and 70 mph). The eye movements of the participants were evaluated during these drives. Additionally, on each of these testing occasions, after completing the drives the eye movements of the participants were monitored while they watched video clips and pushed a button when they detected a hazard (there were 13 different clips in each phase). In the analyses of the effects of training on driving on the open road course, several different indices of these effects were used—notably time headways, which were computed in situations where a vehicle was directly in front of the driver, and eye movements. For time headway (which was set equal to the distance headway divided by the velocity), there were no differences in the time headways between the untrained and trained novice drivers either immediately after training or three months post-training. However, there were effects of training on the variability of the search; this was computed on three sections of roadway which were selected, one from each roadway type, of roughly 45 seconds in duration that had relatively little traffic and few traffic signals.*

Chapman et al. (2002) concentrated on the variability of fixations along the horizontal axis, which presumably reflects searching areas of potential risk. The standard deviation of the trained drivers increased from 6.18 degree before training to 7.30 degree after training, whereas the standard deviation actually decreased for the untrained drivers from the first to the second test (6.34 degree to 6.08 degree). However, there were no significant effects of training on the third test which occurred three to six months after training. The eye movement patterns from the video clips were similar. The video clips were divided into dangerous sections and not dangerous sections. The horizontal variability of the eye movements was less in all three phases of testing in the dangerous sections. More importantly, the horizontal variability in the dangerous sections was significantly greater for the trained than for the untrained novice drivers both immediately after training (2.34 degree versus 2.05 degree)

---

* Chapman et al. (2002), reported variances, but we have converted them to standard deviations to simplify exposition.

and three to six months after training (2.28 degree versus 2.00 degree), but not before training (1.93 degree versus 2.08 degree). Finally, it was possible to compare the eye movement behavior of the trained and untrained novice drivers in the video clips in this experiment (Chapman et al., 2002) with experienced and untrained novice drivers seeing the same clips in another experiment (Chapman & Underwood, 1998). In the clips identified as hazardous in both studies, the experienced drivers in the Chapman and Underwood study had a larger search variance than then untrained novice drivers, just as the trained novice drivers in the Chapman, Underwood and Roberts study had a larger search variance than the untrained novice drivers. However, in the scenarios identified as nonhazardous in both studies, the variance of the trained novice drivers' scan continued to be large (Chapman et al. 2002), unlike the experienced drivers (Chapman & Underwood, 1998), suggesting that the training is not helping the newly-licensed driver discriminate hazardous from nonhazardous situations. Note that Chapman et al. did not determine whether the trained drivers were actually looking at areas of the roadway which might reduce their likelihood of a crash; they used only global measures of the amount of eye (and head) scanning behavior. Thus, although the training of Chapman et al., and Underwood and Roberts (2002) focused on knowledge, scanning, and anticipation, rather than on training eye movements per se, without more specific information than the horizontal variance of the search it is difficult to know whether the training is changing much other than the observable scanning patterns.

30.2.3 Sweden

From 2000 to 2004 the European research project TRAINER took place. The aim of this project, which was funded by the European Commission and in which various research institutes participated, was the development of new methods for driver training in which Computer-Based Training (CBT) and simulator training are key elements (TRAINER, 2002). The objectives for the simulator training and CBT were derived from the GDE (Goals of Driver Education) framework, which was the result of a literature review of the causes of the high crash rate of young novice drivers in an earlier European research project called GADGET (Hatakka, Keskinen, Hernetkoski, Gregersen, & Glad, 2003). An abridged version of the GDE-framework is presented in Table 30.2. The last three rows in Table 30.2 are the three levels of skills and control of the driving task which were proposed by Michon (1985): the operational level (vehicle maneuvering), the tactical level (mastery of traffic situations), and the strategic level (driving goals & context). The first row indicates a more global level: "goals for life and the skills for living", which deals with personality, lifestyle and norms and values and how these affect driving behavior. The first column indicates relevant domains of knowledge and skills, the second column indicates the factors that increase risk for young novice drivers, and the third column indicates the required calibration (i.e., balancing task demands and capabilities on the basis of an assessment of the complexity of the task and an assessment of one's own capabilities). As the GADGET project stated that only the lower left boxes are normally addressed in initial driver training, the aim of the TRAINER project was to address all boxes with the aid of simulators and computers.

The CBT that was developed in the TRAINER program differed from DriveSmart (the Australian CBT that is mentioned in Section 2.1) and Driver ZED (the CBT that was developed in the United States, mentioned in Section 2.5.1). In contrast with DriveSmart and Driver ZED, only a small proportion in TRAINER-CBT is about hazard anticipation. This part contains video clips that freeze after about 15 seconds. The learner driver has to click on

TABLE 30.2 Sweden's Goals of Driver Education (GDE) Framework

Essential Contents (Examples)	Knowledge & Skills	Risk-Increasing Factors	Self-Evaluation
Hierarchical levels of behavior			
Goals for life & skills for living (general)	Knowledge about & control over how life goals & personal tendencies affect driving behavior Motivation	Risky tendencies Acceptance of risk Self-enhancement through driving Use of alcohol & drugs	Self-evaluation: Personal skills for impulse control Risky tendencies
Driving goals & context (journey related)	Knowledge & skills concerning: Effects of journey goals on driving Effects of social pressure inside the car	Risks connected with: Driver's condition (mood, blood alcohol concentration, etc.) Driving environment (e.g., urban/rural)	Self-evaluation: Personal planning skills Typical driving goals
Mastery of traffic situations	Knowledge & skills concerning: Traffic regulations Speed adjustment Communication	Risks caused by: Wrong expectations Risk increasing driving style (aggression) Vulnerable road-users	Self-evaluation: Strong & weak points of basic traffic skills Personal driving style
Vehicle maneuvering	Knowledge & skills concerning: Control of direction & position Tire grip & friction	Risks connected with: Insufficient automatism or skills Unsuitable speed adjustment	Awareness of: Strong & weak points of basic maneuvering skills Realistic self-evaluation

the spot in the frozen picture where a potential risk is visible (e.g., a pedestrian in the distance on the sidewalk). All potential risks are explicit and there are no hidden risks. The rest of the CD-ROM contains questions and answers about participant behavior such as drunk-driving, peer group pressure, and fatigue.

The simulator training that was developed had a total of 31 scenarios divided into five training blocks. The blocks were: (1) *basic knowledge* (application of rules of the road and vehicle control), (2) *manoeuvring and safety, divided attention* (car following and overtaking), (3) *manoeuvring and safety, hazard perception* (search strategies, gap acceptance, hazard anticipation), (4) *particular situations with higher risk* (road and weather conditions, darkness) and (5) *particular situations: New technology and personality aspects* (driving with ABS, ecological driving, distraction and attention, motives for driving).

Falkmer and Gregersen (2003) tested whether the hazard perception skills of learner drivers improved when the CBT and the simulator training of the TRAINER project was part of the regular initial driver training in Sweden. Two training simulators, a so-called Low Cost Simulator (LCS) and a so-called Mean Cost Simulator (MCS) were used. The LCS consisted of a driver chair, pedals, a gear lever, a steering wheel, a dashboard, only one monitor (40 degrees field of view horizontally) right in front of the driver, and a sound generator. The pedals and steering wheel had force feedback, but there was no motion system. The MCS had the same configuration but with three monitors (a field of view of about 120 degrees horizontally) and had a simple motion and vibration system. All the participants were learner drivers from a driving school. They had professional driver training (on the road with an instructor) but had not yet taken the driving test. The participants were divided into three groups. The first group initially did CBT and after that received the simulator training on an MCS, the second also did CBT first but received the simulator training on an LCS, and the third (control) group neither did CBT nor simulator training.

To test the acquired hazard perception skills, a high-end research simulator with a moving base was used to present six scenarios to the participants. In the first, the participants drove in an urban environment. At a junction, a bus approached from the right hand side, and according to the rules of the road, the bus has no right-of-way; however, it did not stop. This situation demands early detection and immediate reaction in order to avoid a crash. In the second scenario, the participants drove on a rural road with forest all around. A moose suddenly crossed the road and, after a few seconds, two calves followed. In this condition, situation awareness is required for early detection. In the third scenario, the participant receives an SMS (short message service) on his/her mobile phone when driving on a rural road, and directly after the phone signal, the participant passes a traffic sign indicating a lower speed limit (it was 90 km/h and the sign says 50 km/h). Of key interest is whether participants are distracted by the phone signal or whether they recognize the change in speed limit in time (by reducing speed). Scenarios four to six are actually one long scenario divided into three parts. In the first part, fog gradually reduced the visibility to 100 metres.

(Do participants notice gradual decreases in visibility in time and do they adjust their speed?) In the second, the fog disappears and then the driver enters a second fog bank in which a van appears that is driving in the same direction. If the participant is driving too fast, a rear-end collision will occur. In the third, the fog disappears, but then the van (which was directly in front of the participant driver) starts to accelerate (over the speed limit). Will the participant also start to accelerate over the speed limit?

For each scenario, Falkmer and Gregersen (2003) used a different set of dependent variables. All of them dealt with car performance such as speed, onset of braking, following distance, time-to-collision, lateral position, and so forth, so that the drivers' eye behavior was not measured directly (e.g., the scanning behavior was not measured with an eye tracker). There were no significant differences on all dependent variables between the three groups with regard to the bus, moose and the third fog (car following) scenarios. There was a significant difference on one of the dependent variables in the mobile phone scenario and the first fog scenario, and there were significant differences on two dependent variables in the second fog scenario. With regard to the mobile phone scenario, the participants that were trained in the MCS stayed on course better (measured by lateral position) during the secondary task with the mobile phone than either the control group or the group that was trained in the LCS. With regard to the first fog scenario, the average of the speeds of the MCS and LCS groups taken together and the average speed of MCS group by itself were significantly slower than the average speed of the control group when the visibility gradually decreased. In the second fog scenario there was a significantly larger minimum time-to-collision and a larger distance to the van in the MCS group compared to the LCS group, but not compared to the control group.

In summary, the simulator training and the CBT had some positive effects on the driving performance of learner drivers and the group that was trained on the MCS did slightly better than the group that was trained on the LCS. However, there was no improvement when the testing scenarios were considerably different in their appearance from the training scenarios even though there was the same underlying principle in both. That is to say, there was some near transfer but there was no far transfer.

30.2.4 The Netherlands

In the Netherlands, driving simulators are used both for initial and advanced driver training, with more than 100 systems currently in use (Kappé & van Emmerik, 2005). Using driving simulators for driver training requires that cost and effectiveness are carefully balanced. Even though driver training in the Netherlands is relatively expensive (taking about 35–40 hours at €35–40 an hour to pass the practical driving test) driving schools operate a business with small margins. Thus these driving simulators need to be cheap and effective training devices.

The cost-effectiveness of driving simulators was one of the issues in the ELSTAR (European Low-cost Simulators for the Training of ARmed forces) project (see Korteling, Helsdingen,

& von Baeyer, 2000). In this project, aimed at a method for developing cost-effective training simulators, the driving task was dissected into 20 "elementary driving tasks" for each of which the required training cost (in hours) and the required simulator hardware (display, motion base, traffic, etc.) was determined. The resulting cost-effectiveness analysis showed that a cost-effective driver trainer is fixed-base, has a wide field of view (180 degrees horizontal, or wider), a traffic model with a targeted set of traffic scenarios, and a focus on didactics. This allows students to practice a large portion of the curriculum, including procedural aspects of vehicle operation (starting, stopping, changing gears, etc.), and traffic participation (negotiating intersections, roundabouts, etc.) in a relatively cheap simulator. The subtleties of operating the clutch, (hard) braking, curve negotiation and skidding make up only a small portion of the curriculum and rely on motion cues that require an expensive moving base. Those aspects are trained more economically in a real car.

In any training simulator, cost is associated with hardware, software, creating and maintaining the lessons that run on the system, and the instructor operating the system. One of the most important factors in the return on investment is obviously the effectiveness of the training on the simulator. Here the didactical quality of the simulator system, the number of simulators operated by a single instructor and the integration of the simulator in the training curriculum play an important role (Farmer, van Rooij, Riemersma, Jorna, & Moraal, 1999). Since the traffic can be controlled by the simulator system, driving simulators allow an instructive traffic scenario to be presented on demand. Thus, students can, at least in theory, learn faster in the simulator than they can in a practical driving lesson, since they can be programmed to encounter more instructive situations per period of time. Most driving simulators are equipped with a "virtual driving instructor" that guides the student through the lesson, and may provide real-time instruction and feedback as well as briefing and debriefing. With such a system, the role of the human instructor is reduced, allowing multiple simulators to be operated by a single instructor. Some systems claim that they can be used without any human intervention at all, providing "stand-alone" virtual instruction. Even the perfect simulator can fail, however, if its lessons do not fit the training curriculum of the school. Driving schools have to carefully examine how the simulator lessons are embedded in their training. This can be achieved by taking the entire curriculum into account and determining where simulator lessons are best-suited in respect to the other training means (see MASTER, Farmer et al., 1999).

In 2002 ANWB (Algemene Nederlandse Wielrijders Bond; the Dutch Automobile Association) started using simulators in driver training that were developed using ELSTAR and MASTER principles. With a PC-based simulator using a car mock-up, a wide 180-degree three channel projection system, and a sophisticated, scripted, traffic system with virtual instructor, they provided 18 lessons, each of 20 minutes' duration, in which students learned how to master basic vehicle operation

and traffic participation. The simulators are located in regional centers, with clusters of two to four simulators operated by a single instructor. Each simulator lesson is focused on a specific topic (e.g. highway driving, negotiating complex intersections, negotiating roundabouts). These specific topics were also treated in the theory book (the homework for that day), in Computer-Based Training (CBT) prior to the simulator lessons, and as the focus of the practical driving lessons that day. An evaluation of the effectiveness of the system showed that practical driving instructors rated the performance of simulator students above average when compared to students that did not receive simulator training. Being freed from explaining and training the relatively boring basic principles of traffic participation, practical driving instructors commented that they had more time to teach students the more cognitive "higher-order" aspects of driving. One ANWB simulator instructor estimated that students learned three times faster in the simulator than in a practical driving course. Based on this evaluation, ANWB driving school decided to proceed with the introduction of the simulator in their curriculum. They now use 30 simulators in their schools, all with a supervising simulator instructor. In 2003, other Dutch driving schools started using driving simulators, but they tended to use simpler driving simulators that were operated in clusters or as stand-alone machines (depending on the size of the school). These simulators did not present specific scenarios to the student, but instead manipulated more general traffic parameters (traffic density, percentage of "aggressive" drivers, etc) to deliver lessons of varying difficulty levels. These schools generally offer the simulator lessons at a reduced cost as an "extra", prior to the start of the regular practical driver training.

Apart from the evaluation that was conducted at ANWB driving schools, the Dutch driving simulators have not been subjected to a classical validation study. Such a study would compare the performance of randomly assigned groups of students trained with the normal practical driving curriculum with that of a group trained with a simulator curriculum. In practice, validation studies are very difficult to perform. They are hampered by their practical setting (at a school, and not in a laboratory) and are sensitive both to differences in the student population (simulators may or may not attract specific subgroups) and to the way the simulator is used (as an "extra to" or as a "replacement for" practical driving lessons). Also, the results of a classical validation study are only valid for a specific simulator in a specific training curriculum, and thus they are difficult to generalize. In addition, the cost and the time involved in a classical validation study have been a hurdle for the Dutch driving schools. They need information on the validity of simulator training at an early stage to help them make a proper decision on the introduction of driving simulators. At that stage, there is often just one prototype driving simulator available with a limited number of students (at best) who can do a small set of lessons that are not fully embedded in the school's curriculum yet. As the effectiveness of training depends primarily on the quality of the simulator lessons and their integration into the curriculum, a classical validation would not be fair at that stage. When the driving simulators are in use, the schools generally have a good sense of the validity of the lessons and are no longer inclined to spend money on a full-blown validation of their system.

Apart from the positive responses of the driving schools, there is some circumstantial evidence that driving simulators like these can be effective trainers. For example, as discussed in more detail in the next section, Allen, Park, Cook and Fiorentino (2007) found that simulator training in the United States can lower novice driver crash rates (compared with the general U.S. novice driver population). In the Netherlands, De Winter, Wieringa, Kuipers, Mulder and Mulder (2007) reported that students that performed well during their driving simulator training have a higher chance of passing their driving test the first time (a correlation of 0.18 in a regression analysis), and that students that required fewer practical driving lessons made fewer errors in the simulator training (predictive correlation 0.45). The curriculum was based on Dutch driver training and consisted of 15 lessons of 27 minutes each. Thus, students that performed well in the simulator also performed well during their subsequent practical driving lessons and on the test, which gives some indication of the validity of the driving simulator that was used.

Although these results indicate that driving simulators can play a role in initial driver training, there are several qualifying observations that can be made (see also, in this book, chap. 5 by Caird & Horrey, "Twelve Practical and Useful Questions About Driving Simulations"). First, simulated environments present a relatively poor abstraction of reality; they display relatively clean, predictable virtual environments with little clutter and without subtle visual cues (e.g., those that signal the characteristics and intentions of other road-users, and potentially hazardous situations). Furthermore, simulated traffic tends to behave very stereotypically, and looks clumsy and erratic from time-to-time (depending on the quality of the simulated traffic). Pedestrians and cyclists also generally look and behave like robots.

An additional concern is the quality of the virtual instruction. Virtual instruction often lacks the subtle student-teacher interactions that characterize practical driver training. This is a relatively minor problem for the first phases of teaching, where instructors guide the student through the "scripts" (i.e., more-or-less standardized patterns) that are required to negotiate different situations in traffic. It is a major problem when attempting to teach higher-order, more cognitive aspects of driving with a virtual instructor. Human instructors tend to know why an error is made (and will ask if they are not sure) and will give appropriate feedback. Virtual instructors generally do not have a clue why an error is made, and will not give the right feedback. Also, virtual instructors cannot measure all the relevant behavior of the student. For example, scanning behavior is difficult and expensive to measure, and there are currently no commercial driving simulators equipped to measure scanning behavior. However, as proper scanning is an important learning goal in driver education, instruction and feedback on scanning behavior need to be provided otherwise. This requires that a human instructor is present to provide instructions and feedback on

scanning behavior. Although the driving instructor can not always follow the gaze of the participants, it seems unacceptable to allow a student to learn such aspects in a stand-alone simulator, as they may learn inappropriate scanning motor patterns which would need to be "deprogrammed" in the following practical driving lessons.

As with any driving simulator, the novice driver trainees exhibit some simulator sickness, though with drastically reduced rates compared to experienced drivers. Novice drivers, who have not steered a vehicle, seldom become sick (less than 5%), whereas 50% of the experienced drivers (e.g., driving instructors) experience some symptoms of sickness, in the same simulator driving the same lessons. Reasons for these differences are discussed elsewhere (e.g., see in this book, chap. 14 by Stoner, Fisher & Mollenhauer, "Simulator and Scenario Factors Influencing Simulator Sickness").

All things considered, it seems that driving simulators are best-suited for training the basics of vehicle operation and traffic negotiation. This is exactly how they are used for novice driver training in the Netherlands. However, this does not mean that driving simulators cannot be used to teach expert drivers. The Dutch police driving school has been using six driving simulators for their advanced driving courses. These simulators are used for improving the general driving skills of cadets, as well as for driving in emergency situations. These simulators have the same basic characteristics and shortcomings described above. However, their didactical use is entirely different. Instead of having a single student in the simulator, the students are trained in groups of six. One student is in the simulator, the others are in an adjacent room looking at the performance of their colleague on a separate three-channel display. These students have a response button which they are required to press when an error is made or when there is room for improvement. The teacher uses these responses to trigger discussions on cognitive aspects of the driving task (e.g., What did he or she do wrong? Why? What would you have done?). According to the school's instructors, 60% of the didactical value is in these discussions, and only 40% is in the simulator. (It should be noted that Dutch police always train students in small groups, and that these groups are comfortable in expressing and receiving critique.) This result is in line with Gregersen, Brehmer and Morén (1996) who compared the effectiveness of different types of training on accident involvement rates, and found that having group discussions outperformed books, CBT and skidding courses. It remains to be determined whether these methods for teaching experts might have value for training novice drivers as well.

As with any training medium, it is the didactical setting for simulators that is most relevant in determining its learning value. Driving simulators are well-suited for novice driver training, but may, under the right conditions, be used for experienced drivers too. Given the relatively slow progress in simulator hardware and software, driving schools should have a focus on simulator didactics. That is the single aspect that is most under their control, is most directly related to the validity of simulator training, and is most directly related to making a profit using the simulator.

30.2.5 United States

There have been several attempts in the United States over the past 10 years or so to introduce training that is more focused on hazard anticipation, attention maintenance, and vehicle management. Most evaluations have been conducted only on a driving simulator. One recent evaluation has been carried out both on a driving simulator and in the field.

30.2.5.1 Driver ZED

About 10 years ago, the American Automobile Association (AAA) developed a risk awareness training program called Driver ZED (Zero Errors Driving). Participants sat in front of a PC watching a total of 80 different scenarios filmed in city, town and rural settings (Willis, 1998). The scenarios contained views filmed from the cabin in a moving vehicle—both of the roadway ahead and the roadway as seen in the side- and rear-view mirrors. The participant needed to take one of several actions sometime during or after each scenario, with the action depending on the mode of presentation. There were four modes: scan, spot, act and drive. In the *scan* mode, the driver needed to answer questions at the end of a scenario which assessed how well the driver was paying attention to everything in the scenario (e.g., the driver might be asked whether there was a vehicle approaching in the rear-view mirror). In the *spot* mode, the scenario was stopped at the last frame. The driver was asked to use the mouse to click the cursor on each risky element in the scenario (e.g., a child playing with a ball on the sidewalk). In the *act* mode, the driver was asked what action he or she would take midway through a scenario (e.g., the driver might be asked whether he or she should speed up or slow down at an intersection where the traffic signal was displaying a yellow globe). Finally, in the *drive* mode, the driver needed to click on the mouse at the point in a scenario when they would take an action that could potentially avoid a crash (e.g., the participant might need to brake suddenly when approaching a driveway out of which a car was backing). (Note that Driver ZED has evolved. The most current version, Driver ZED 3.0, can be downloaded at http://www.driverzed.org/home/.) The risky scenarios in Driver ZED were selected specifically because they were ones in which younger drivers frequently crashed (Lonero, Clinton, Black, Brock, & Wilde, 1995).

Fisher et al. (2002) evaluated the effects of Driver ZED training on a driving simulator. Specifically, the vehicle behaviors of three groups of drivers on the driving simulator were recorded: Novice drivers (high school students with a learner's permit) who were trained to recognize risks using Driver ZED; novice drivers who were trained only through standard drivers' education programs; and more experienced drivers (qualified college students who were driving buses for the university in which they were enrolled). Each group drove through a number of potentially risky scenarios on the driving simulator and vehicle behavior was recorded throughout each drive. The 12 scenarios Fisher et al., chose to evaluate on the driving simulator included examples not only from categories of crashes in which younger adults are frequently involved, such as proceeding straight ahead

through an intersection, but also examples from categories in which younger adults are only infrequently involved but which are often fatal, such as passing other vehicles (Aizenberg & McKenzie, 1997, p. 53). Fisher et al. (2002) examined the simulator data and found that, overall, there was not a single dependent variable based on vehicle behavior that captured the efficacy of training across all 12 scenarios. However, some clear effects emerged when they evaluated the effect of training on a subset of six of these scenarios using different dependent variables as the outcome measures. Moreover, the behavior of the trained novice drivers closely resembled that of the experienced drivers. For example, in the Truck Left Turn scenario displayed in Figure 30.1, the experienced drivers braked opposite the truck more frequently than the untrained novice drivers, indicating that they recognized the risk that an obscured vehicle in the opposing lane could be in the process of turning across their path; the more frequent braking was also characteristic of the ZED-trained novice drivers. Similar differences in driving behavior between the ZED-trained and untrained novice drivers were reported in other scenarios. However, from these data it could not be determined how much of the differences between the untrained novice drivers on the one hand, and the trained novice and experienced drivers on the other, were due to differences in hazard anticipation skills (predicting that a car might intrude) and a combination of hazard and speed control skills (actually braking). Perhaps all drivers predicted the presence of a potential hazard, but only the trained and experienced drivers were able to act on this information.

30.2.5.2 Driver Assessment and Training System (DATS)

In a recent study, over 500 novice drivers were recruited in the state of California to participate in a study of how effective driving simulators were at reducing crash rates among this population of drivers (Allen et al., 2007). The drivers were assigned to one of three simulator training modes: A single-monitor desktop simulator, a three-monitor desktop simulator, or a wide-screen vehicle cab simulator (135 degrees horizontal). One group of drivers was recruited from the Department of Motor Vehicles and assigned either to the three-monitor desktop or wide-screen vehicle cab simulators. Training occurred in a laboratory setting. A second group of drivers was recruited from high school driver education courses. Training in the high schools was on the single-monitor desktop simulator. Crash data was obtained from the California DMV for up to two years post-training.

In the simulator training that was part of a larger Driver Assessment and Training System (DATS), the students navigated six 12–15 minute drives (Park, Cook, Allen, & Fiorentino, 2006). The temporal and spatial relations between the driver's vehicle and other traffic, traffic signals, and pedestrians were controlled in each of the hazardous situations that a participant encountered in a simulated drive. The scenarios were designed to require and train skills, including situation awareness, hazard perception, and decision-making under time pressure. The order of hazard encounters was varied in the scenarios prepared for training so that students could not anticipate upcoming events. A summary of the critical events that occurred in each scenario is contained in Table 30.3 (Parke et al., 2006). For example, there were six pedestrian events, two coming from the right and four from the left, where the trigger times for pedestrians moving in front of the participants were designed to present the drivers with critical decisions that had to be made in a very short period of time. A participant's performance was scored on each drive, including use of turn signals, time-to-collision, and number of crashes, road edge incursions, traffic signal violations, and posted speeds which were exceeded by the driver. Earlier work suggested that performance on each of these indices reached asymptote after six trials for most participants (e.g., Allen, Park, Cook, & Viirre, 2003). Accordingly, participants could graduate on the sixth trial if they met performance criteria including no crashes, no more than one ticket, nominal values for lane and speed deviations, and reasonable use of the turn signal indicators (Allen et al., 2007; Allen, Park, Cook, Fiorentino, & Viirre, 2003). Otherwise, they were allowed to improve their score on up to another three drives. Almost 80% of the participants graduated.

The crash rates per licensed driver were then computed for each of the three simulator groups over a two-year period. At the end of two years, 17% of the one-monitor desktop simulator participants had a crash, 14% of the three-monitor desktop simulator participants had a crash, but only 7% of the wide-screen vehicle cab simulator participants had a crash (the percentages are estimates because they are taken from a figure). As a control group, the authors compared the cumulative crash rates of drivers in California and Canada over the course of the first two years after obtaining their license. Crash rates were regressed on the time since licensure. The linear increases in cumulative crash rates of drivers in California and Canada were indistinguishable from each other. Both were clearly higher than increases in the cumulative crash rates of the participants in the wide-screen vehicle cab simulator but about the same as the cumulative crash rates of the other two groups. Unfortunately, assignment of the

participants to the experimental groups was not random nor, obviously, was assignment of the participants to experimental and control groups, so it is difficult to be certain that the lower crash rates of the wide-screen vehicle cab participants would be replicated in a properly controlled study. Nonetheless, the results are encouraging for the use of the wide-screen vehicle cab simulator (presumably because it mimics what one sees on the road most closely).

30.2.5.3 Risk Awareness and Perception Training (RAPT) Program

A program to train novice drivers to anticipate hazards has been developed and then evaluated; it was motivated by the finding that younger drivers are less likely to anticipate hazards than more experienced drivers (Pradhan et al., 2005). In the simplest version of this Risk Awareness and Perception Training (RAPT) program, participants were presented a plan view of a hazardous situation. For example, imagine a driver is approaching a marked midblock crosswalk. There are two travel lanes in each direction and a parking lane (see Figure 30.2; Web Figure 30.2). A truck or other large vehicle is stopped in front of the crosswalk in the parking lane that (from the perspective of the driver) is potentially obscuring a pedestrian who might enter the crosswalk. (The driver is the gray vehicle labeled with a "D" on the roof). Six markers (three yellow ovals and three red circles) were positioned off to the side. Participants had to drag the yellow ovals to areas on the plan view where a potential threat might be located (in front of the truck) and the red circles to locations on the plan view where they would expect the threat to materialize (to the left and front of the truck). The number of markers needed in each plan view varied between one and three. Participants were then told the correct locations of the ovals and circles (Pollatsek, Narayanaan, et al., 2006). The complete set of plan views used in one of the evaluations of training (Diete, 2007) is presented in Web Figure 30.2 to Web Figure 30.10. (The complete training program can also be seen by going to http://www.ecs.umass.edu/hpl, clicking on "Younger Drivers" and then clicking on "RAPT-2".). In other versions of RAPT, sequences of photographs of hazardous situations were used. The sequence was generated by taking a picture every second or two as one approached and passed beyond an actual hazardous situation. The participant was required to use the mouse to move a cursor to locations in each photograph where a risk might appear but was not visible. Again, feedback was provided. (This sequence is viewable by going to http://www.ecs.umass.edu/hpl, clicking on "Younger Drivers" and then clicking on "RAPT-3").

In a series of studies, it has been shown (principally using eye movement measures) that RAPT increases the likelihood that newly-licensed drivers anticipate hazards, to the point that their performance is not much different than much more experienced drivers. These effects of training were clear both immediately after exposure to the training program (Pollatsek, Narayanaan, et al., 2006) and for up to one week later (Pradhan, Fisher, & Pollatsek, 2006). They were present both on a driving simulator and in the field (Pradhan, Pollatsek, Knodler, & Fisher, 2009), and they were

TABLE 30.3 Hazardous Events in STI Simulator Drivers

1 head-on collision event

6 pedestrian walkout events:

    a) 2 from right
    b) 4 from left

10 intersections

    a) 1 green signal light event
    b) 2 red signal light events
    c) 4 yellow light decision making events
    d) 2 cross-traffic stop signs – driver did not have to stop at intersection
    e) 1 with-traffic stop signs – driver had to stop at intersection
    f) 1 right and 1 left turn

2 obscured vehicle pullouts from right

2 overtaking vehicle events

    2-lane, dashed line road, vehicle approaches from behind, moves into opposing traffic lane, overtakes driver, returns in front of driver vehicle

1 construction zone environment

evident both in scenarios which are similar to ones that were trained and ones that were quite dissimilar (Pollatsek, Fisher, et al., 2006). For example, consider one study done on the driving simulator and one in the field where the same version of RAPT was used (Fisher, Pradhan, Pollatsek, & Knodler, 2007). The principal dependent variable was whether participants fixated in a certain region in a certain time window, indicating that they were looking for a specific potential hazard (e.g., fixating around the front of the truck in the example above). There was an overall training effect of 37.4 percentage points on the driving simulator: The trained group fixated the critical region 77.4% of the time, whereas the control group fixated the critical region only 40.0%. The training effect was 41.7 percentage points for the nine near transfer tests (77.4% versus 35.7%) and 32.6 percentage points for the nine far transfer tests (76.8% versus 44.2%). The overall training effects observed on the driving simulator were somewhat larger than the overall averages observed in the field study. In the field study the overall training effect was 27.1 percentage points (64.4% versus 37.3%), 38.8 percentage points (79.2% versus 40.4%) for near transfer and 20.1 percentage points (58.3% versus 38.2%) for far transfer. Videos of trained and untrained drivers scanning behavior in the field and on the driving simulator in the Truck Crosswalk Scenario are available on the *Handbook* web site (Web Videos 30.1–30.4). The three major limitations of this work are that it remains unknown just how long the training effects will last, it remains unclear whether drivers will be motivated to apply what they have learned in training when they are out on the open road by themselves, and it is not certain that there will be a reduction in crashes.

An interesting variation of the training program was recently evaluated. In that variation, a head-mounted driving simulator was used to train tactical hazard anticipation scanning skills (SIMRAPT) in addition to the training they received with RAPT (Diete, 2007). The driver sat in a real car in which inputs from the wheels, brake, and accelerator were sent to a computer which then displayed movements through the virtual world on the head-mounted display consistent with these inputs. Participants who did not make a head movement indicating that they recognized a potential hazard had to repeat the drive. At the end of the combined RAPT and SIMRAPT training, both the trained participants and a group of participants who had not been trained were evaluated on a more advanced driving simulator. The training effect here was no different than the training effect with just RAPT alone. It is not entirely clear why the additional simulator training did not provide a larger benefit, although perhaps the critical hazard anticipation behaviors learned with the head-mounted display (head turns) are sufficiently different from the behaviors that are typically used to anticipate hazards (eye movements and head turns).

30.3 General Discussion

It is clear from the above discussion that training programs for novice drivers are now at the point where novice drivers' hazard anticipation scanning skills, attention prioritization skills, and hazard anticipation vehicle handling skills can all be improved, perhaps even to the point where there is an impact on crash rates. Continued progress in this area will depend on researchers' ability to improve simulation as a tool for both training and evaluation. Above, we have discussed in general terms those aspects of training which are likely to affect the success of training transferring to on-road driving (Groeger & Banks, 2007; Barnett & Ceci, 2002). Below, we discuss what, in general, will be needed to advance our understanding and what limitations there are to using simulation to train and evaluate newly-licensed drivers.

30.3.1 Driving Simulators: Training and Evaluation

Individuals who want to use driving simulators for training and or evaluating novice drivers' scanning, attention maintenance and vehicle management skills need to understand something about the general development of scenarios which have proved to be useful and the equipment (in addition to simulators) and analyses which it is important to be able to undertake.

30.3.1.1 Hazard Anticipation: Scanning and Vehicle Handling

The scenarios that need to be developed in order to evaluate tactical scanning and vehicle management hazard anticipation skills do not generally require the coordination of the movement of other vehicles and pedestrians with the movement of the participant's car. This is because it is not necessary to materialize the potential hazard, as the driver's scanning and vehicle management responses should be the same whether or not the hazard actually appears. This greatly simplifies the construction of the scenarios. There are generally two categories of scenarios used to evaluate scanning and vehicle management hazard anticipation skills. In the first category are scenarios where a threat is obscured by a built object (e.g., a truck in Figure 30.2) or natural object (e.g., a bush in Web Figure 30.3) in the environment. In order for a driver to anticipate the hidden threat, it must be cued in advance, such cues including pavement markings (the crosswalk stripes in Figure 30.2), signs (e.g., the left fork sign in Web Figure 30.6), or the geometry and placement of objects in a given environment (e.g., the truck in the left turn lane of a four-way signalized intersection as shown in Figure 30.1). In the second category are scenarios where the threat is visible, but latent. In this case, the threat is often cued by the geometry and knowledge of other drivers' intentions. For example, a line of cars in the left of two travel lanes waiting to turn left constitutes a potential hazard for a driver in the right travel lane because one or more of the drivers in the left travel lane could pull out suddenly into the right travel lane, perhaps deciding that they actually wanted to go straight. No one currently knows what other categories of scenarios are representative of hazards which need to be anticipated or even how many examples of hazards within each category actually need to be trained in order to get good generalization to the entire category. But, at least currently, it does not seem like there are any issues particular to constructing scenarios used for the simulator training or evaluation of tactical hazard anticipation skills that are holding back progress.

Measures of tactical hazard anticipation skills typically include vehicle parameters (e.g., speed, lane position), information that is collected automatically by all simulators, and eye behavior (e.g., glance locations, range), information which requires specialized eye tracking devices. Analysis of the vehicle information is usually straightforward. Analysis of the eye tracker information can require much more work, especially when the issue is tactical hazard anticipation scanning skills. Here, for each hazardous scenario, a *launch zone* (the location in which the eye movement can be initiated) and *target zone* (the location in which the eye movement must land) must be defined. The eye tracker record must then be analyzed one frame at a time to determine whether there was an eye movement that meets the criteria of fixating appropriately. Fortunately, most scenarios used to evaluate tactical hazard anticipation scanning skills are so constructed that the eccentricity of the eye movements needed to identify the location at which a potential hazard might be, or from which the potential hazard might materialize, is so large that the scoring of the eye tracker videotape is almost always straightforward. Nine out of the 10 near transfer scenarios in the evaluation of RAPT (Pollatsek, Narayanaan, et al., 2006) required eye movements greater than 10 degree in the near transfer scenarios, with the eye movements that were scored as meeting the criteria ranging from as small as 5.4 degree to as large as 27.8 degree (see Web Table 30.1).

30.3.1.2 Attention Maintenance

There has been much less research on programs that deal explicitly with attention maintenance. The one notable exception in this regard is DriveSmart (Regan, Triggs, & Godley, 2000; Regan, Triggs, & Wallace, 1999), where attention prioritization in a combined primary driving and secondary auditory task has been the focus of the training efforts. There is nothing about this training program, or the evaluation of the training program, that could not be implemented on any of the existing simulators. Nor is special equipment required to record the response in these scenarios or advanced algorithms to analyze the data. However, when attention prioritization in a combined primary driving and secondary visual in-vehicle task becomes the focus of a training program, then some type of head and/or eye tracking device will be needed in order to determine how long the continuous glances are in which the driver is not fixating the forward roadway (e.g., when looking for a map on the front seat; see Chan et al., 2008).

30.3.2 Limitations

Newly-licensed drivers, almost to an individual, are not scanning as well as more experienced drivers, are not maintaining attention as well as more experienced drivers, and are not managing their vehicle as well as more experienced drivers. The discussion of the training programs above suggests that notable advances

can be made among newly-licensed drivers in each of these skills. However, it would be naïve to assume that skill training alone would be the only influence on newly-licensed drivers' performance. If not motivated to put into practice the skills they have learned, newly-licensed drivers will continue to crash at high rates (Christie, 2001). Moreover, if they actively still choose to behave in a risky fashion, something which is well-documented among teens (Beirness, 1996), the crash rates will remain high. Finally, there is the problem that skills training can sometimes lead to overconfidence, which has a net negative impact on crash rates (Mayhew & Simpson, 1996).

Key Points

- Police crash reports indicate that novice drivers are at a greatly increased risk of crashing during the first six months of solo licensure due to poor hazard anticipation, attention maintenance, and speed management skills.
- Driving simulators have been used to compare novice drivers' eye and vehicle behaviors with those of more experienced drivers in each of the above three skill areas; such comparisons are consistent with the crash reports, showing real deficiencies among novice drivers in their ability to anticipate hazards, maintain attention, and manage speed.
- PC-based programs have been developed that train novice drivers to anticipate hazards and maintain attention.
- The effects of these programs generalize from the PC to both the driving simulator and the open road, from scenarios which are similar to those trained on the PC to those which are far removed, and from evaluations undertaken immediately after training to evaluations taken up to a week after training.
- Although these training programs have been shown to have an effect on driver behavior in fielded observations, it is still not clear that they will reduce crashes.

Keywords: Attention Maintenance, Eye Movements, Hazard Anticipation, Novice Drivers, Simulator Evaluation, Simulator Training

Acknowledgments

A number of studies reported in this chapter in which the first author was involved were run on a driving simulator purchased with funds from a National Science Foundation major research instrumentation grant (SBR-9413733). Portions of the research for these same studies were supported by a grant from the National Highway Traffic Safety Administration (Cooperative Agreement Number: DTNH22-05-H-01421). Support for writing this chapter was provided in part by the aforementioned grant and a grant from the National Institutes of Health (1R01HD057153-01) and in part by a subcontract from Dunlap and Associates, Inc. (under Task Order Number 4 of contract No. DTNH22-05-D-35043 from the National Highway Traffic Safety Administration). Special thanks go to Richard Blomberg and Dennis Thomas for their careful editing of this manuscript.

Glossary

Attention maintenance: There are two varieties of attention maintenance, broadly speaking. First, drivers need to distribute their physical resources (eye movements) appropriately inside and outside the vehicle in order to maintain attention to the roadway. Second, while looking at the forward roadway, drivers still need to concentrate on their driving to the exclusion of talking on the cell phone, interacting with the music retrieval system, or doing anything in general which reduces their capacity to process events in the forward roadway.

Eye behaviors: These typically include only the sequences of fixations, though in some studies they might include pupil diameter and the percentage of eyelid closure.

Tactical and strategic hazard anticipation scanning: *Tactical hazard anticipation scanning* is the scanning pattern observed when a feature in the environment suggests that a hidden threat is especially likely to materialize at a particular location and time in a scenario whereas *strategic hazard anticipation scanning* is the scanning pattern observed when there is no such key feature.

Vehicle behaviors: These include such measures as velocity, acceleration, steering wheel angle and brake pressure.

Web Resources

The *Handbook*'s web site contains supplemental materials for this chapter, including:

Web Table 30.1: Scoring of Near Transfer Scenario in RAPT.

Web Figure 30.1: Adjacent Truck Left Turn scenario (color version of Figure 30.1).

Web Figure 30.2: Truck Crosswalk scenario. (a) Plan View. (b) Perspective View.

Web Figure 30.3: Amity-Lincoln scenario. (a) Plan View. (b) Perspective View.

Web Figure 30.4: Adjacent Truck Left Turn scenario. (a) Plan View. (b) Perspective View.

Web Figure 30.5: T-Intersection scenario. (a) Plan View. (b) Perspective View.

Web Figure 30.6: Left Fork scenario. (a) Plan View. (b) Perspective View.

Web Figure 30.7: Opposing Truck Left Turn scenario. (a) Plan View. (b) Perspective View.

Web Figure 30.8: Blind Drive scenario. (a) Plan View. (b) Perspective View.

Web Figure 30.9: Pedestrian on Left scenario. (a) Plan View. (b) Perspective View.

Web Figure 30.10: Mullins Center scenario. (a) Plan View. (b) Perspective View.

Web Video 30.1: Truck Crosswalk Scenario on Simulator: Untrained.

Web Video 30.2: Truck Crosswalk Scenario on Simulator: Trained.

Web Video 30.3: Truck Crosswalk Scenario on Open Road: Untrained.

Web Video 30.4: Truck Crosswalk Scenario on Open Road: Trained.

Key Readings

Allen, R. W., Park, G. D., Cook, M. L., & Fiorentino, D. (2007). The effect of driving simulator fidelity on training effectiveness. *Proceedings of Driving Simulation Conference (DSC) 2007 North America.* Retrieved January 11, 2008, from http://www.nads-sc.uiowa.edu/dscna07/DSCNA07CD/main.htm

Mayhew, D. R., & Simpson, H. M. (2002). The safety value of driver education and training. *Injury Prevention, 8* (Suppl. II), ii3–ii8.

McKnight, J. A., & McKnight, S. A. (2003). Young novice drivers: Careless or clueless. *Accident Analysis & Prevention, 35,* 921–925.

Pradhan, A. K., Pollatsek, A., Knodler, M., & Fisher, D. L. (2009). Can younger drivers be trained to scan for information that will reduce their risk in roadway traffic scenarios that are hard to identify as hazardous? *Ergonomics, 52*(6), 657–673.

Regan, M. A., Triggs, T. J., & Wallace, P. R. (1999). *DriveSmart: A CD ROM Skills training product for novice car drivers.* Proceedings of the Traffic Safety on Two Continents Conference, September 20–22, 1999, Malmo, Sweden.

References

Aaron, J. E., & Strasser, M. K. (1977). *Driver and traffic safety education* (2nd ed.). New York: Macmillan Publishing.

Aizenberg, R., & McKenzie, D. M. (1997). *Teen and senior drivers* (CAL-DMV-RSS-97-168). Sacramento, CA: California Department of Motor Vehicles.

Allen, R. W., Park, G. D., Cook, M. L., & Fiorentino, D. (2007). The effect of driving simulator fidelity on training effectiveness. *Proceedings of driving simulation conference (DSC) 2007 North America.* Iowa City, IA: University of Iowa. Retrieved January 11, 2008, from http://www.nads-sc.uiowa.edu/dscna07/DSCNA07CD/main.htm

Allen, R. W., Park, G. D., Cook, M. L., Fiorentino, D., & Viirre, E. (2003). Experience with a low-cost, PC- based system for young driver training. In L. Dorn (Ed.), *Driver behavior and training,* Proceedings of the First International Conference on Driver Behavior and Training (pp. 349–358). Stratford-upon-Avon, England: Ashgate, Aldershot, England.

Allen, R. W., Park, G. D., Cook, M. L., & Viirre, E. (2003). Novice driver training results and experience with a PC-based simulator. *Proceedings of the second international driving symposium on human factors in driver assessment, training and vehicle design.* Iowa City, IA: University of Iowa.

Barnett, S. M., & Ceci, S. J. (2002). When and where do we apply what we learn? A taxonomy for far transfer. *Psychological Bulletin, 128*, 612–637.

Beirness, D. J. (1996). The relationship between lifestyle factors and collisions involving young drivers. In H. Simpson (Ed.), *New to the road: Reducing the risks for young motorists, proceedings of the first annual international symposium of the youth enhancement service* (pp. 71–77). Los Angeles, CA: Youth Enhancement Service, Brain Information Service, University of California Los Angeles.

Butler, G. T. (1982). *Effectiveness and efficiency in driver education programs* (DOT-HS-806-135). Washington, DC: National Highway Traffic Safety Administration.

Chan, E., Pradhan, A. K., Knodler, M. A, Pollatsek, A., & Fisher, D. L. (2010). Empirical evaluation on a driving simulator of the effect of distractions inside and outside the vehicle on drivers' eye behaviors. *Transportation Research Part F: Traffic Psychology and Behavior, 13*, 343–353.

Chapman, P. R., & Underwood, G. (1998). Visual search of driving situations: Danger and experience. *Perception, 27*, 951–964.

Chapman, P., Underwood, G., & Roberts, K. (2002). Visual search patterns in trained and untrained novice drivers. *Transportation Research Part F: Traffic Psychology and Behavior, 5*, 157–167.

Christie, R. (2001). *The effectiveness of driver training as a road safety measure: A review of the literature*. Noble Park, Victoria, Australia: Royal Automobile Club of Victoria Ltd.

Christie, R., & Harrison, W. (2003). *Driver training and education programs of the future* (Rep. 03/03). Melbourne, Australia: Royal Automobile Club of Victoria, Ltd.

Clinton, K., & Lonero, L. (2006). *Evaluation of driver education: Comprehensive guidelines*. Washington, DC: AAA Foundation for Traffic Safety.

Crundall, D. E., & Underwood, G. (1998). Effects of experience and processing demands on visual information acquisition in drivers. *Ergonomics, 41*, 448–458.

De Winter, J. C. F., Wieringa, P. A., Kuipers, J., Mulder, J. A., & Mulder, M. (2007). Violations and errors during simulator-based driver training. *Ergonomics, 50*, 138–158.

Diete, F. (2007). *Evaluation of a simulator based, novice driver risk awareness training program*. Unpublished Master's thesis, Graduate School of the University of Massachusetts at Amherst, Amherst, MA.

Engström, I., Gregersen, N. P. K., Hernetkoski, K., Keskinen, E., & Nyberg, A. (2003). *Young novice drivers, driver education and training: Literature review* (VTI 491A). Swedish National Road and Transport Research Institute.

Falkmer, T., & Gregersen, N. P. (2003). The TRAINER project—The evaluation of a new simulator-based driver training methodology. In L. Dorn (Ed.), *Driver behavior and training*, Proceedings of the First International Conference on Driver Behavior and Training (pp. 317–330). Stratford-upon-Avon, England: Ashgate, Aldershot, England.

Farmer, E., van Rooij, J., Riemersma, J., Jorna, P., & Moraal, J. (1999). *Handbook of simulator-based training*. Aldershot, Hampshire, England: Ashgate.

Fisher, D. L., Laurie, N. E., Glaser, R., Connerney, K., Pollatsek, A., Duffy, S. A., & Brock, J. (2002). The use of an advanced driving simulator to evaluate the effects of training and experience on drivers' behavior in risky traffic scenarios. *Human Factors, 44*, 287–302.

Fisher, D. L., Pradhan, A. K., Pollatsek, A., & Knodler, M. A., Jr. (2007). Empirical evaluation of hazard anticipation behaviors in the field and on a driving simulator using an eye tracker. *Transportation Research Record, 2018*, 80–86.

Garay-Vega, L., Fisher, D. L., & Pollatsek, A. (2007). Hazard anticipation of novice and experienced drivers: Empirical evaluation on a driving simulator in daytime and nighttime conditions. *Transportation Research Record, 2009*, 1–7.

Gregersen, N. P., Brehmer, B., & Morén, B. (1996). Road safety improvement in large companies. An experimental comparison of different measures. *Accident Analysis & Prevention, 8*(3), 297–306.

Groeger, J. A., & Banks, A. P. (2007). Anticipating the content and circumstances of skill transfer: Unrealistic expectations of driver training and graduated licensing? *Ergonomics, 50*, 1250–1263.

Hatakka, M., Keskinen, E., Hernetkoski, K., Gregersen, N. P., & Glad, A. (2003). Goals and contents of driver education. In L. Dorn (Ed.), *Driver behavior and training*, Proceedings of the First International Conference on Driver Behavior and Training (pp. 309–315). Stratford-upon-Avon, England: Ashgate, Aldershot, England.

Horswill, M. S., & McKenna, F. P. (2004). Drivers' hazard perception ability: Situation awareness on the road. In S. Banbury & S. Tremblay (Eds.), *A cognitive approach to situation awareness* (pp. 155–175). Aldershot, England: Ashgate.

Kappé, B., & van Emmerik, M. L. (2005). *Mogelijkheden rijsimulatoren in de rijopleiding en het rijexamen* [Possibilities of driving simulators in driver training and driving tests], (Report TNO-DV3 2005 C114) (in Dutch). Soesterberg, the Netherlands: TNO Human Factors.

Klauer, S. G., Dingus, T. A., Neale, V. L., Sudweeks, J. D., & Ramsey, D. J. (2006). *The impact of driver inattention on near-crash/crash risk: An analysis using the 100-car naturalistic driving study data* (Rep. DOT HS 810 594). Washington, DC: National Highway Traffic Safety Administration.

Korteling, J. E., Helsdingen, A., & von Baeyer, A. (2000). *ELSTAR handbook low-cost simulators* (Rep:11.8/ ELS-DEL/5-HB), the Netherlands: European Co-operation for the Long Term in Defence (EUCLID).

Lonero, L. (2007). *Trends in driver education, evaluation and development*. Retrieved January 9, 2008, from http://drivers.com/article/941/

Lonero, L., Clinton, K. M., Laurie, I., Black, D., Brock, J., & Wilde, G. (1995). *Novice driver education model curriculum outline*. Washington, DC: AAA Foundation for Traffic Safety.

Mayhew, D. R., & Simpson, H. M. (1995). *The role of driving experience: Implications for the training and licensing of new drivers*. Toronto, Ontario: Insurance Bureau of Canada.

Mayhew, D. R., & Simpson, H. M. (1996). *Effectiveness and role of driver education and training in a graduated licensing system*. Ottawa, Ontario: Traffic Injury Research Foundation.

Mayhew, D. R., & Simpson, H. M. (2002). The safety value of driver education and training. *Injury Prevention, 8*(Suppl. II), ii3–ii8.

McKnight, J. A., & McKnight, S. A. (2003). Young novice drivers: Careless or clueless. *Accident Analysis & Prevention, 35*, 921–925.

Michon, J. A. (1985). A critical view of driver behavior models: What do we know, what should we do? In: L. Evans & R. C. Schwing (Eds.), *Human behavior and traffic safety* (pp. 485–520). New York: Plenum Press.

Mourant, R. R., & Rockwell, T. H. (1972). Strategies of visual search by novice and experienced drivers. *Human Factors, 14*, 325–335.

National Highway Traffic Safety Administration. (1974). *Driver education curriculums for secondary schools: User guidelines*. Safe Performance Curriculum and Pre-Driver Licensing Course (Final Rep. DOT-HS-003-2-427). Washington, DC: Office of Driver and Pedestrian Research, US Department of Transportation.

National Highway Traffic Safety Administration. (1994). *Research agenda for an improved novice driver education program*. Report to Congress, May 31, 1994 (DOT-HS-808-161). Washington, DC: US Department of Transportation.

National Highway Traffic Safety Administration. (2006). *Beginning teenage drivers* (DOT-HS-810-651). Washington, DC: National Highway Traffic Safety Administration.

Nichols, J. L. (2003). *A review of the history and effectiveness of driver education and training as a traffic safety program*. Washington: National Transportation Safety Board.

Park, G. D., Cook, M. L., Allen, R. W., & Fiorentino, D. (2006). Automated assessment and training of novice drivers. *Advances in Transportation Studies: An International Journal,* Special Issue, 87–96.

Pollatsek, A. Fisher, D. L., & Pradhan, A. K. (2006). Identifying and remediating failures of selective attention in younger drivers. *Current Directions in Psychological Science, 15*, 255–259.

Pollatsek, A., Narayanaan, V., Pradhan, A., & Fisher, D. L. (2006). The use of eye movements to evaluate the effect of PC-based risk awareness training on an advanced driving simulator. *Human Factors, 48*, 447–464.

Pradhan, A. K., Fisher, D. L., & Pollatsek, A. (2006). Risk perception training for novice drivers: Evaluating duration of effects on a driving simulator. *Transportation Research Record, 1969*, 58–64.

Pradhan, A. K., Hammel, K. R., DeRamus, R., Pollatsek, A., Noyce, D. A., & Fisher, D. L. (2005). The use of eye movements to evaluate the effects of driver age on risk perception in an advanced driving simulator. *Human Factors, 47*, 840–852.

Pradhan, A. K., Pollatsek, A., Knodler, M., & Fisher, D. L. (2009). Can younger drivers be trained to scan for information that will reduce their risk in roadway traffic scenarios that are hard to identify as hazardous? *Ergonomics, 52*(6), 657–673.

Regan, M. A., Triggs, T. J., & Godley, S. T. (2000). *Simulator-based evaluation of the DriveSmart novice driver CD-ROM training product*. Proceedings of the Road Safety Research, Policing and Education Conference. Brisbane, Australia.

Regan, M. A., Triggs, T. J., & Wallace, P. R. (1999). *DriveSmart: A CD-ROM Skills training product for novice car drivers*. Proceedings of the Traffic Safety on Two Continents Conference, September 20–22, 1999, Malmo, Sweden.

Roberts, I., & Kwan, I. (2002). School-based driver education for the prevention of traffic crashes (Cochrane Review). *Cochrane Library, Issue 1*. Oxford, England.

Sagberg, F., & Bjørnskau, T. (2006). Hazard perception and driving experience among novice drivers. *Accident Analysis & Prevention, 3*, 407–414.

Smith, M. F., & Blatt, J. (1987). *Follow-up evaluation—Safe performance curriculum driver education project: Summary of preliminary results*. Paper presented at the American Driver and Traffic Safety Education Association, Research Division Annual Conference, Spokane, WA.

Stock, J. R., Weaver, J. K., Ray, H. W., Brink, J. R., & Sadoff, M. G. (1983). *Evaluation of safe performance secondary school driver education curriculum demonstration project*. (DOT-HS-806-568.) Washington, DC: National Highway Traffic Safety Administration.

TRAINER. (2002). *Deliverable 5.1: TRAINER assessment criteria and methodology*. Retrieved February 1, 2008, from http://ec.europa.eu/transport/roadsafety/publications/projectfiles/trainer_en.htm

Underwood, G., Chapman, P., Brocklehurst, N., Underwood, J., & Crundall, D. (2003). Visual attention while driving: Sequences of eye fixations made by experienced and novice drivers. *Ergonomics, 46*, 629–646.

Vernick, J., Li, G., Ogaitis, S., MacKenzie, E., Baker, S., & Gielen, A. (1999). Effects of high school driver education on motor vehicle crashes, violations and licensure. *American Journal of Preventive Medicine, 16*, 40–46.

Vlakveld, W. (2006). Will simulator training in basic driver education help to enhance road safety? *Proceedings of the HUMAN centered design for information society technologies* (Task Force G: Workshop, organized by UPM, European Guidelines for the Application of New Technologies for Driver Training and Education) Retrieved April 24–25, 2008, from http://www.noehumanist.org/workshop-madrid_presentations.php

Wheeler, W. A., & Triggs, T. J. (1996). A task analytical view of simulator based training for drivers. *Proceedings of the road safety research and enforcement conference "effective partnerships"* (pp. 217–221), Coogee Beach, New South Wales, Australia.

Wikman, A., Nieminen, T., & Summala, H. (1998). Driving experience and time-sharing during in-car tasks on roads of different width. *Ergonomics, 41*(3), 358–372.

Willis, D. K. (1998). The impetus for the development of a new risk management training program for teen drivers. *Proceedings of the human factors and ergonomics society 42nd annual meeting* (pp. 1394–1395). Santa Monica, CA: Human Factors and Ergonomics Society.

Woolley, J. (2000). *In-car driver training at high schools: A literature review*. Walkerville, South Australia: Safety Strategy, Transport SA.

31

The Commercial Driver

Myra Blanco
Virginia Tech Transportation Institute

Jeffrey S. Hickman
Virginia Tech Transportation Institute

Richard J. Hanowski
Virginia Tech Transportation Institute

Justin F. Morgan
Virginia Tech Transportation Institute

Abstract

The Problem. The use of driving simulators is becoming increasingly frequent in Commercial Vehicle Operations (CVO). However, the full extent of their use for training and maintaining the skills of commercial vehicle drivers has not been completely examined. *Role of Driving Simulators.* Driving simulators address many current problem areas and driver training needs for commercial motor vehicle (CMV) drivers, both before and following licensure (i.e., entry-level training, screening, evaluation, retraining). *Key Results of Driving Simulator Studies.* Previous research and companies implementing commercial vehicle simulators have indicated cost and safety benefits from simulator-based commercial driver training. Further investigation of the long-term effect of CMV simulator-based training, especially in regard to driver safety, is warranted. *Scenarios and Dependent Variables.* Entry-level simulator-based driver training presents an alternative to, or enhancement of, traditional behind-the-wheel (BTW) training by allowing drivers to gain basic skills and driving time in a low-risk environment. In addition, commercial vehicle simulation allows drivers to be introduced to hazardous or dangerous driving conditions (such as weather and traffic conditions) without the risk associated with BTW training. *Platform Specificity and Equipment Limitations.* Although a range of commercial vehicle simulators exist, the range of training possible with each level of simulator varies. In order to train higher-level rules and knowledge type tasks, simulators with greater fidelity of visual and kinesthetic modeling are required.

31.1 What Is a Commercial Motor Vehicle Driver?

Commercial motor vehicle (CMV) drivers operate large motor vehicles, hauling either goods or passengers across the nation. Operating a CMV requires specialized knowledge and skills. However, in the United States of America (USA) prior to the Commercial Motor Vehicle Safety Act of 1986, the commercial driver's license (CDL) testing and licensing program was non-existent and specialized CMV knowledge and skills tests were not required in all states. Before the Commercial Motor Vehicle

Safety Act, a person with an automobile license could drive a CMV (i.e., tractor-trailer or bus) without the need to provide qualifications to handle such a vehicle. However, there was an implementation delay of several years from the development of the Safety Act to actual licensing enforcement. It was not until 1 April 1992 that all individuals interested in becoming a CMV driver were actually required to have a CDL prior to becoming a CMV driver.

There are three different classes of CDLs that a CMV driver can hold: Class A, B, or C. These three classes refer to the type of CMV that can be operated. The Federal Motor Carrier Safety

Administration (FMCSA, 2007a) defines these three CDL classes as follows:

- *Class A.* Any combination of vehicles with a gross vehicle weight rating (GVWR) of 26,001 or more pounds (11,794 kg) provided the GVWR of the vehicle(s) being towed is in excess of 10,000 pounds (4,536 kg).
- *Class B.* Any single vehicle with a GVWR of 26,001 or more pounds (11,794 kg), or any such vehicle towing a vehicle not in excess of 10,000 pounds (4,536 kg) GVWR.
- *Class C.* Any single vehicle, or combination of vehicles, that does not meet the definition of Class A or Class B, but is designed to transport 16 or more passengers (including the driver) or is placarded for hazardous materials.

According to FMCSA, the goal of the Commercial Motor Vehicle Safety Act of 1986 was to prohibit unqualified drivers from operating CMVs and to remove any drivers who, since obtaining a CDL, became unqualified. The Act requires states to meet minimum national testing standards when licensing drivers, and maintain special information-sharing system requirements in order to ensure "one driver, one record, one license."

Ensuring accurate tracking of drivers is critical since, of the over 8 million drivers with a CDL, approximately 11% of CDL drivers have lost their CDL privilege at least once (FMCSA, 2007a). Moreover, the Large Truck Crash Facts report (FMCSA, 2007b) suggests that large trucks (i.e., trucks with a GVWR greater than 10,000 pounds) account for 7% of all vehicle miles traveled and 3% of all registered vehicles in the USA. As part of the motor vehicle crash statistics, large trucks represent 8% of vehicle fatal crashes, 3% of injury crashes, and 5% of crashes that result only in property damage. Over 5,000 fatalities and 114,000 people injured result from crashes involving a large truck (FMCSA, 2007b). However, this problem not only speaks to the need for accurate driver records, but also suggests the need for both entry-level and experienced CMV driver training. Because of this problem, many have viewed implementations of simulation-based CMV driver training as a potential solution.

31.1.1 Basic Skills

CMV drivers are responsible for the mass movement of goods and services. The large size and truck-trailer articulation of these vehicles, as well as the complex driving conditions they operate within, necessitate that these drivers possess special qualifications beyond those required to operate a light vehicle. These specialized qualifications involve detailed knowledge and skills to ensure CMV driver safety and the safety of other motorists that share the roadway with them. The main purpose of the CDL program is to ensure that the required knowledge and skills are in place before the operation of a CMV. The licensing program serves as a critical safety gate to help ensure CMV drivers are well prepared for operating a heavy vehicle on the nation's roadways. Although there are no current nationwide standards for the training of CMV drivers within the USA, some independent organizations (e.g., the Professional Truck Driver

Institute; PTDI) have developed minimum knowledge and skills that CMV drivers should possess. Specifically, PTDI (1999) states that a well-prepared CMV driver should have knowledge and skills developed in the following areas:

- Basic operation of the vehicle (e.g., basic control systems, backing and docking)
- Safe operating practices (e.g., visual search, space management)
- Advanced operating procedures (e.g., night operation, hazard perception)
- Vehicle systems and reporting malfunctions (e.g., identification, diagnosing)
- Non-vehicle activities (e.g., hours of service regulations, trip planning)

The FMCSA has begun investigating the use of commercial vehicle simulations to train CMV drivers for these knowledge and skill sets (Pierowicz et al., 2002), and ongoing research should help determine the extent to which simulation-based training may be used in CMV driver training.

31.1.2 Driver Finishing Programs

Obtaining a CDL is the first step in becoming a safe CMV driver. Safety-conscious companies have identified a number of areas benefiting from additional training. These additional areas, such as defensive driving and sleep hygiene, require newly licensed drivers to take part in "driver-finishing programs." Those programs can take different forms, including pairing novice drivers with experienced drivers before driving solo. However, in the USA, not all companies have such programs. Moreover, these finishing programs do not have standardized or certified curriculums, and their implementation in terms of hours driven (behind the wheel of either a real vehicle or a simulator) are not required to be logged or reported to any external institution or agency for review or approval.

Yet crash statistics suggest appropriate training and assessment of CMV drivers is essential for ensuring safe drivers. The findings from Knipling, Hickman and Bergoffen (2003) that suggest insufficient training is an area of concern for fleet managers echo the crash data, such as the finding that 17% of single-vehicle fatal crashes were caused by driving too fast for the conditions (FMCSA, 2007b). Knipling et al. (2003) also identify several other problem areas in commercial vehicle operations (CVOs), many of which commercial vehicle simulation is potentially well-suited to address:

1. Insufficient training:
   a. Lack of basic driving skills
   b. Poor knowledge of federal, state, and/or company rules
2. At-risk driving behaviors (e.g., speeding, tailgating)
3. Aggressive driving
4. Lack of defensive driving skills (e.g., space management around vehicle)

5. Driver fatigue/drowsiness
6. Delays associated with loading and unloading (e.g., resulting in long working hours, tight schedules, and fatigue)
7. Alcohol and/or illicit drug abuse
8. Driver health and wellness problems, specifically:
    a. Lifestyle/general health-related (e.g., poor diet, smoking)
    b. Sleep apnea
    c. Cardiovascular illness/heart disease
    d. Prescription drug side-effects (e.g., drowsiness)
    e. Mental illness (e.g., depression, anxiety, mood disorders)
9. Poor attitude and morale, loneliness, alienation, unhappiness
10. Driver turnover resulting in unstable workforce
11. Drivers unfamiliar with routes
12. Neglect of vehicle maintenance (e.g., brakes, ties)
13. Failure to inspect vehicle (e.g., pre-/post-trip)
14. Unsecured loads
15. High-risk drivers (all causes combined; i.e., the degree to which managers should focus on the worst 10–20% of their drivers)

These 15 items provide a broad scope of areas that may be considered a part of a comprehensive driver-finishing program curriculum and, as stated earlier, simulation provides a convenient route for addressing many of these aspects. Furthermore, the existence of such programs provides a basis for the concept that obtaining a CDL is only the starting point for a driver becoming a safe driver; additional instruction, assessment, and training is required to ensure a driver's safety.

The information provided in the manuals for entry-level CMV drivers varies from state to state and it is not comprehensive. These driving manuals provide a general knowledge of some of the topics delineated above. However, a number of topics (such as delays associated with loading and unloading, company rules, health and wellness, attitude, high-risk drivers, unfamiliar routes, and turnover) are not covered in many manuals. This requires CMV driver knowledge and skills to be supplemented by private training and/or a finishing program in order to fulfill these needs. Simulation provides a platform for training and assessment that may prove beneficial in a variety of settings, and allows for flexibility in implementation.

31.2 CMV Training, Assessment, and Driving Simulators

31.2.1 The Evaluation of Current Commercial Driver Training Programs

When assessing the adequacy of CMV training in the USA, Dueker (1995) found that entry-level CMV drivers were generally not receiving adequate training to drive heavy trucks, motor coaches, and school buses. Based on anecdotal data, there seems to be a consensus in the CMV industry that passing the CDL skills test generally does not mean a driver is ready for solo-operational

driving. As noted, many safety-conscious fleets have driver-finishing programs for new CDL holders and carrier-based training for all new hires (Knipling et al., 2003).

In contrast to the USA, Europe as a whole has strict training standards. European countries typically use a comprehensive vocational training approach; this is different from the minimum knowledge and skill approach used in the USA, and makes greater use of simulators in their CMV driver training. However, the regulatory systems are narrow and based on the societal and industrial structures of each European country. Both public and private commercial driver entities have established standardized curriculums that include simulators, as well as internet access, classroom lecture, and behind-the-wheel (BTW) training (Hartman et al., 2000).

The combination of an expanding industry and shifting driver demographics will spotlight CMV driver shortages and training needs in the coming decades. As noted, the PTDI provides the only USA certification of CMV driver training programs. Certification is at the course level, as opposed to the broader school level or the narrower instructional unit level. Furthermore, no studies have empirically demonstrated the effectiveness of CMV driver training. However, it has been hypothesized that formal training, either BTW or in a simulator, will have long-term effects in the driving performance of the CMV driver (Robin et al., 2005). Moreover, insufficient driver training is usually highlighted as a need and potential for improvement when commercial vehicle operation (CVO) management techniques are evaluated (Knipling et al., 2003).

Unfortunately, in the period leading up to 2008, the demand for CMV drivers has outpaced what driver training schools are currently producing. This has left CMV fleets to hire inexperienced and untrained drivers. Given the increasing demands on CMV fleets to deliver goods and services, coupled with new technological innovations and increasingly congested roadways, effective training and education programs are critical elements in a fleet's safety management system. Efforts such as the FMCSA-sponsored Commercial Motor Vehicle Driving Simulator Validation Study (SimVal) being performed by the Virginia Tech Transportation Institute (VTTI) seeks to examine multiple aspects of CMV driver training. This research study includes a comparison of drivers that obtain training versus drivers that do not (as no minimum training level is currently required to obtain a CDL). The comparisons include the examination of a driver's Department of Motor Vehicle (DMV) records (i.e., DMV test scores, crashes, violations, citations), performance in both real-truck and commercial vehicle simulator versions of road and range driving tests, fleet company evaluations, and a CDL road and range driving retest at four months after employment as a CMV driver.

31.2.2 Commercial Driver Training Programs: Delivery and Content

31.2.2.1 Standard Driver Training Programs

Driver shortages coupled with poorly trained entry-level drivers have forced many fleets to become more dependent on their in-house training and education programs (also called driver-finishing programs). Of those fleets surveyed in a study performed by the American Trucking Associations Foundation (ATAF, 1999a) only 14% relied on certified driving schools for driver training and education, while 85% maintained their own in-house driver training programs. Most fleets (91%) reported hiring new drivers in a probationary status and requiring newly-hired drivers to train with either a driver trainer or senior driver. Similarly, Stock (2001) examined practices of safety managers and found that 75% of fleets required new drivers to train with an experienced driver before driving solo, with 23% requiring attendance at defensive driving courses. These studies highlight the need for well-trained entry-level drivers and emphasize the importance of quality training programs. In fact, 83% of respondents in Stock's (2001) study of safety management rated their in-house training programs as important to carrier safety. Typically, in-house training programs for new hires focus on topics including:

- Administrative policies and procedures
- Equipment loading and operation
- Customer relations
- Driving safety and skills training, including:
  - Defensive driving
  - Fatigue management techniques
  - Equipment inspections
  - Driver health and wellness

Although such in-house training programs often include safe driving instructions and training, the extent to which they specifically address the training needs outlined previously (PTDI, 1999; Knipling et al., 2003) is unknown. In a survey of fleet safety managers, Hickman et al. (2007) reported on both driving and non-driving behaviors taught to new drivers during in-house training programs. The two most prevalent driving behaviors in these training programs were driving inattentively (54.8%) and speeding (54.8%), while the two most prevalent non-driving behaviors for new drivers in training programs were pre- and post-trip inspection (75.8%) and completing paperwork (25.8%). The primary teaching techniques employed by in-house programs to train and educate entry-level CMV drivers may be grouped into three major categories:

- *Classroom.* Knowledgeable professionals or former drivers instruct new hires on the rules of the road, usually in a lecture/discussion format (Horn & Tardif, 1999). Increasingly, interactive computer teaching programs (e.g., CD-ROMs and Internet-based programs) are replacing classroom lectures. These programs are very cost-efficient and present the information using a multimedia experience (Ryder, 2000).
- *Practice range.* Experienced drivers instruct new hires how to handle a truck and allow them to experience driving the vehicle on a closed-off driving course/range. The training vehicle usually has three to four extra seats in the sleeper cab so the trainer can teach several other new

hires through feedback, observation, and commentaries (Horn & Tardif, 1999). Some aspects of many commercial vehicle simulators, such as the ability to provide the driver with an overhead bird's-eye view of his/her position, make simulation-based training useful for training certain range backing maneuvers. For example, trainees may practice these maneuvers and learn the relationship between steering inputs and movements of the articulated vehicle with the ability to instantly view the results from above.

- *On-the-road driving.* This is similar to the practice range, except new hires are driving on the road with other vehicles. This may include the instructor and new hire going on long-haul trips (Horn & Tardif, 1999). This is one area in which simulation-based training may prove beneficial for commercial vehicle drivers. Newly hired drivers may be introduced to infrequently occurring situations, such as steering tire blowouts, in a simulator scenario with a minimal level of risk. This allows drivers to be exposed to the situation and learn appropriate responses before ever encountering the situation in the real world.

Simulation-based training has been used in the two latter categories (practice range and on-the-road driving). Additionally, the use of a CMV driving simulator in these training categories is being examined in the VTTI SimVal study.

31.2.2.2 General Characteristics of Simulator Training Programs

While still in its infancy and (at the present time) too expensive for widespread adoption, virtual reality training simulators offer a tantalizing glimpse into the future. Simulators are able to monitor an individual driver's performance and create a database to help classify their progression against a desired skill set, all occurring in training sessions that can be standardized and repeated. Further, simulators allow drivers to be suddenly and unexpectedly placed in dangerous driving situations (such as poor weather or road conditions) where extremely precise driving maneuvers are required, all in a risk-free environment. As suggested by Robin et al. (2005), the many advantages of the use of advanced training simulators include:

- Safety: Practice of otherwise dangerous maneuvers;
- Scenario versatility: Creation of operational situations that may otherwise be unavailable (e.g., weather or roadway environments);
- Standardization: Scenarios developed for specific instructional objectives, and organized to ensure that all students are exposed to each learning activity;
- Repeatability: Lessons or tests can be replayed to permit extra practice and skill mastery;
- Improved perspectives: Provision of overhead or other visual perspectives;
- Sophisticated performance measurement: Recording and analysis of student performance in more precise and quantitative ways; and

- Efficiency: More training events can occur in a given time period.

In addition to the benefits described by Robin et al. (2005) and Hartman et al. (2000) cites French and Swedish claims regarding the positive cost-benefits of simulation training, although no large-scale, controlled evaluative studies were cited. USA and foreign militaries are the biggest users of truck driver training simulators (Emery, Robin, Knipling, Finn, & Fleger, 1999), followed by European CMV driver training organizations (Lester, Rehm, & Vallint, 2003). USA CMV driver training trails these other training venues in the use of driving simulators, and truck driving training technology in general.

Simulation-based training can be viewed through the Skills, Rules, Knowledge (SRK) model of behavior (Rasmussen, 1983). The SRK model is based on the level of cognitive processing required of an individual driver during any task. This SRK model includes training oriented towards skill-, rule-, and knowledge-based behaviors. Skill-based behaviors are low in cognitive loading and demand, such as a driver's dynamic use of mirrors. Rule-based behaviors are those requiring the driver to apply a set of known pre-defined rules to a specific situation, such as how to pull up to a bus stop. Knowledge-based behaviors are those in which the driver does not have existing applicable rules; these scenarios require the driver to complete the task by applying his or her own knowledge, such as "reading the road."

Although most implementations of CMV simulators accommodate training at the skill- and rule-based level, training for knowledge-based tasks and behaviors typically requires a more sophisticated simulation facility that accommodates the behavior targeted for training (e.g., understanding traffic interactions). Additionally, training for higher-level rule- and knowledge-based behaviors may require simulators with a higher degree of precision in visual fidelity, motion cueing, and network interaction. The range of CMV simulation requirements and training abilities is summarized in Table 31.1. (Some examples of different simulators are shown in chap. 2 by Allen, Rosenthal, & Cook, "A Short History of Driving Simulation", and chap. 12 by Jamson, "Cross-Platform Validation Issues", both in this book.)

Using simulators for training offers many potential advantages and benefits. Playback of trainee performance, including a review of errors made, is probably the most important part of simulator-based driver training (Robin et al., 2005). Learning is more effective if trainees are allowed to make errors and then receive corrective feedback on those errors, as opposed to simply being instructed on or shown correct driving. Figure 31.1 (Web Figure 31.1 for color version) shows the similarities between real-truck (a) and simulator (b) test conditions for the rule-based task of backing a tractor-trailer.

Truck simulators allow drivers to learn from their mistakes without the property damage and injuries associated with real-world mistakes in a similar driving scenario. Thus, a newly licensed CMV driver can learn how to interact with traffic in an intersection, drive a double-trailer on an eight percent grade

TABLE 31.1 Simulator Requirements and Capabilities

Behavioral Level	Simulator Requirements	Examples of Possible Training
Skills	Part-Task	Basic Vehicle Controls Basic Visual Scanning Skills Shifting Simple Perception-Decision-Reaction Time
Rules	Mid-Range, Some Part-Task	*Skill Level Training Items,* and: Avoidance Maneuvers Backing Basic Defensive Driving Grades and Slopes Intersections/Turning Lane Change/Merges Load Shifting Parking Signaling and Mirror Use Space Management
Knowledge	Mid-Range, High-End	*Rules Level Training Items,* and: Advanced Defensive Driving Traffic Interactions

roadway, or manage a tank-trailer during a heavy snowstorm with slick roadways and black ice Figure 31.2 (Web Figure 31.2), all before actually experiencing these events in the real world. Simulators allow instructors to design tasks which mimic those that their students will encounter in a range of conditions, from CDL range test type tasks to other more extreme conditions unique to the advanced capabilities offered by a simulator (such as the ability to replicate a steering tire blowout). Selecting the appropriate simulator, by deciding on the appropriate levels of perceptual fidelity, for the training of interest should be the first step in this process. The reader is referred to the chapters on simulator selection in this book.

As mentioned, the demand for CMV drivers has outpaced the training school output. That, coupled with the lack of training standardization, has moved companies to take a proactive approach and create in-house training and driver-finishing programs. These programs are a combination of classroom, practice range, on-the-road driving, and—in some instances—advanced simulator capabilities, in order to cover all the driver training needs.

31.2.3 The History of Commercial Driver Simulator Training Programs and Research

Beginning in 1986, minimum national standards for testing were developed by states for licensing CMV drivers. Previously, the qualifications for receiving a CDL varied widely between states. The current (post-1986) CDL program places a more standardized set of requirements on the CMV driver, motor carriers, and the states. Organizations such as the PTDI and the Canadian Trucking Human Resources Council (CTHRC) have developed such standards and guidelines for CDL training. The

FIGURE 31.1 Large truck backing task in real truck versus truck simulator. Views starting from the top left quadrant in a clockwise fashion show the driver's face, forward view from the truck, driver's side of the truck and trailer, and passenger's side of the truck and trailer.

aforementioned PTDI (1999) provides certification for entry-level truck driver training and driver-finishing programs in the USA, while the CTHRC is an accreditation and certification agency for Canadian institutes and programs. However, the true effect of such training programs remains to be seen. Neither these agencies, nor the government, provide a standardized method to test driver performance and measure the safety benefits of pre-license training on CMV driver performance.

31.2.3.1 Simulator Training Programs

Although driving simulators have been around for many decades, it was not until the last decade that the technology progressed to the point where it was both affordable enough and held sufficient realism to mimic real-world driving scenarios. High-fidelity driving simulators can help CMV drivers by enhancing training and licensing, and aiding the evaluation of new technologies, roadway design, and other potential safety improvements (Keith et al., 2005; Pierowicz et al., 2002).

The increased availability of high-fidelity driving simulators has led to their increasing adoption by more safety-conscious fleet operations. Some fleet companies, such as Schneider National,

FIGURE 31.2 CMV simulator training scenarios. From top to bottom: city intersection, 8% grade, snow-covered road.

Inc., are currently using full-mission and part-task simulators in their entry-level and finishing programs. Schneider National, Inc. (2006) claims to have experienced extremely positive results from simulator-based training. They suggest that their simulator-based training has lowered new driver dropout and termination rates by approximately 10%, and lowered accident frequency and severity by as much as 20%. Based on the cost-benefit analysis, the implementation of simulation-based training has saved Schneider millions of dollars.

The New York City transit system, which integrated a bus simulator into its driver training program, is an example of how simulators can be integrated as a core training component (Vidal & Borkoski, 2000). This was accomplished through cooperation between a simulator manufacturer and New York City transit drivers that created a simulator dedicated to their operational needs, prior to attempting development of specific training exercises. Many other training areas can also benefit from simulators. For example, carriers can screen candidate drivers' knowledge, skills, and abilities without having to put them in a real vehicle.

31.2.3.2 Simulator Training Research

Simulator technology may play an important role in developing such testing and measurement standardization. The FMCSA-sponsored SimVal study being performed by VTTI is examining if simulator technology can enhance tractor-trailer driver training and on-the-job performance (Robin et al., 2005). The principal purpose of this study is to examine if simulator technology, compared to conventional CMV driver training methods, can facilitate and enhance tractor-trailer driver training, long-term driving performance, and long-term driving safety. Additionally, the study will examine the implications for simulator-based CDL testing. This study will ascertain the safety benefits of simulation training, and of training itself, by comparing the research results of these four different types of training methods:

- *Long certified conventional BTW training.* An eight-week entry-level training course, certified by PTDI, where all the driving practice is performed in a real tractor-trailer;
- *Long certified simulator-based training.* An eight-week entry-level training course, certified by PTDI, where approximately 60% of the driving practice is performed in a tractor-trailer simulator and 40% in a real tractor-trailer;
- *Informal training.* No formalized entry-level training is obtained. Friends or family members train the driver; and
- *Short, CDL-focused truck driver training school.* A two- to three-week entry-level training course focused on the topics covered by the CDL test.

Favorable results from this study, for the simulator-based entry-level training group, may result in certification agencies allowing schools using simulators to certify students graduating from their programs where a proportion of the required driving hours are performed in a simulator. Moreover, if the SimVal study results reveal that more structured training (i.e., long certified or CDL-focused) leads to improved driving performance and safety, it would be important for regulatory agencies to consider entry-level training as a potential requirement for obtaining a CDL.

Finally, the SimVal study highlights the advanced capabilities of a truck simulator to replicate emergency and evasive maneuvers and provide effective assessment of drivers performing these maneuvers. These maneuvers are being examined in van-, tanker-, and doubles-trailer truck configurations. This

assessment will also provide more information regarding the appropriate levels of visual and kinesthetic fidelity for simulations of emergency maneuvers and extreme conditions. Very little information exists on the appropriate levels of fidelity with regard to training CMV drivers during high-demand scenarios. How experienced drivers respond to the simulation is not only informative of the general level of fidelity achieved by the simulation, but also indicative of the visual and kinesthetic requirements for simulating such events. Feedback from these experienced drivers will better inform future simulation efforts with respect to appropriate modeling of physical interactions in the simulated environment, as well as the increased fidelity of kinesthetic and visual cues provided to the driver in these special scenarios. These results have important implications for the future of commercial vehicle simulation design, as knowledge of the various aspects of simulator fidelity increases.

The results from the SimVal study, together with the suggested performance assessment method by Brock, McFann, Inderbitzen and Bergoffen (2007), will benefit both practitioners and researchers by identifying main areas of interest in training and CDL licensing that impact safety. Implications from the study could include training requirements prior to obtaining a CDL. Additionally, the results from the SimVal study could potentially influence entry-level training certification by allowing simulator-based driving hours to count towards total training hours (regulations similar to this are in place in Europe; Parkes, 2003).

31.2.4 The Future of Driving Simulators and Commercial Operations: Screening, Evaluation, and Retraining

As simulators become more readily available, they can be a key tool to enhance CVO safety and reduce cost due to driving-related incidents which could be avoided through appropriate training and assessment. Although CMV fleets have started to benefit from driving simulators at different levels, simulation is still primarily contained within the training arena. However, simulators have the potential to couple with screening and on-the-job evaluation and retraining techniques to further improve driving safety.

31.2.4.1 Driver Screening

Driver selection has important implications for both safe and economical operation of a commercial vehicle. The ATAF (1999a) simply states "...starting with the right people is key to overall safety performance." Specific practices that are recommended include requiring in-person applications, personal interviews, screening for stable employment history, maximum point limits for moving violations, minimum years of experience, driving tests, a physical examination, and reviewing the past financial performance (e.g., credit rating) of drivers (ATAF, 1999a).

The *Truck Driver Risk Assessment Guide* (ATAF, 1999b) summarizes effective screening and recruiting methods necessary to target safe drivers. One important aspect of this is recruitment ads that include:

- Company practices relevant to safety (e.g., regular schedules, well-maintained equipment);
- Specific safety-related criteria for employment (e.g., minimum age, years of experience, driving record);
- Specific "do's and don'ts" for employment application forms and structured interviews; and
- Federal regulations on required background checks as well as additional recommended background checks.

In addition, this guide recommends the use of commercial services providing employment-related databases and personality inventories or other psychological tests purported to predict driver safety.

Stock (2001) examined the hiring practices of fleet safety managers. Overall, these safety managers considered safety-related screening criteria to be most important when hiring drivers. More than 90% of respondents required in-person interviews, called past employers to review employment histories, tested for alcohol and drugs during screening, and conducted on-road driving tests before hiring. Other screening practices reported by fleet safety managers included the use of third party services to review driver histories (36%), requiring a minimum number of years driving experience (56%), specific maximum number of points/crashes/violations (82%), and requiring a written test on U.S. Department of Transportation (U.S. DOT) regulations (41%).

Corsi and Barnard (2003) surveyed motor carriers with very high safety ratings regarding their management practices. They identified a number of common driver characteristics considered to be most important in making hiring decisions for company drivers. These screening techniques included lack of prior dismissals for alcohol and drug abuse, lack of past chargeable crashes, driving experience with other carriers, no prior traffic violations, solo driving experience, recommendations from other carriers, and being age 25 or older. Essential driver personality traits identified by this select group of carriers included reliability, honesty, self-discipline, self-motivation, and patience.

Although the aforementioned traits and criteria are widely valued in the industry, simulation affords only a limited opportunity to test for their presence or magnitude. However, in addition to these important aspects, a screening drive can be used as an evaluation tool of the driver's knowledge and skills during a comprehensive set of staged scenarios (Allen & Tarr, 2005). However, in terms of logistics, insurance, liability, and other constraints, this step may be more suited for a truck driving simulator than a real truck. Exposure to other drivers, risk from driving on public-access roadways, and potential for damage to company equipment is minimized by (at least initially) screening drivers with a simulator drive. In these instances, truck driving simulators provide the ability to examine an individual driver's knowledge and skill level in a controlled, safe, and economical manner. Additionally, such unambiguous assessments of the driver's starting level reveals areas in which the driver may benefit from specialized finishing programs or other forms of remedial training.

Driving simulators are not a panacea for driver screening and selection. Driver history, characteristics, and traits all play important roles in the long-term success or failure of an individual driver (ATAF, 1999b; Corsi & Barnard, 2003). These aspects of the driver are not easily, if at all, accessible through simulator-based assessment. However, when used in conjunction with other selection techniques, simulation can provide additional information that is critical to the hiring decision-making process. Although the body of literature provides several techniques for use in driver screening, the ultimate purpose is to ensure that the knowledge and skill level of the new hires is known. Driving simulation provides an addition to these techniques and, as such, standard driving routes and scenarios of interest could easily be integrated into the screening process. Such integration of simulator-based screening promises more reliability in the driver screening process.

31.2.4.2 On-the-Job Evaluation

In-service performance evaluation is a way for fleets to measure their risks, assess the need for and provide countermeasures, maintain performance expectations, and promote meaningful safety-focused communication. Observation and feedback, in their various forms, are key processes in continuously maintaining and promoting driver safety. Many elements of driver performance can be monitored, including: driving skills, driving habits, hours-on-duty, miles driven, moving violations, crashes, cargo loss, vehicle inspection and maintenance, and non-driving activities such as loading/unloading practices. Tools used to gather this information include performance standards by driving task, driver evaluation forms, employee appraisal forms, crash follow-up procedures, direct observation, and feedback of driving and non-driving work behaviors (ATAF, 1999b).

In Stock's (2001) examination of fleet safety managers, "driver monitoring" was considered important to safety by more than 90% of respondents. Almost all respondents continuously monitored violations, citations, and crash reports. About one-third of respondents used some form of in-vehicle recording to observe drivers while they were in-service making their normal, revenue-producing deliveries. This technique was more likely to be observed in larger fleets rather than smaller fleets (Stock).

Tracking of fleet safety statistics helps to assess overall fleet safety performance, identify risk factors, and evaluate the effects of safety programs implemented in the fleet (Corsi & Barnard, 2003). Of course, once drivers are hired, their in-service safety performance must be continuously monitored and evaluated. Continuous tracking of driver crashes, incidents, and violations was practiced by virtually all (99%) of the safety manager respondents in a survey on effective safety management techniques in CMVs (Knipling et al., 2003). Many safety managers felt that non-DOT reportable incidents, such as damage to cargo or loading areas, are important events to track along with crashes, traffic violations, and inspection violations as a number of small incidents are likely to lead to a major incident.

"Periodic observation of driving" was the second most-practiced driver evaluation method reported by Knipling et al.

(2003). Such observations can be in the truck ("ride-along"), or can be from outside the truck (from a "shadow vehicle"). Less than half of the respondents employed managers and/or senior drivers to perform the ride-along. Nevertheless, ride-along observations can be a way of providing one-on-one instruction and behavioral counseling to drivers. They should include explicit feedback on driving behaviors. Limitations of the ride-along method and other driving observations include the fact that they are time-consuming for managers and they may not provide accurate appraisals of drivers' actual on-road behavior if the conditions where the driver might have deficiencies never occur during the ride-along. The information gained from these observational practices can be beneficial in identifying undesirable driver characteristics, and may help better target training (including simulator-based training) towards such behaviors.

Two other driver evaluation practices addressed by Knipling et al. (2003) were "How's My Driving?" placards and on-board safety monitoring devices. The use of safety placards has some advantages and several disadvantages. It is a method for identifying risky driving behavior by drivers before these behaviors result in a crash. Corrective management actions (e.g., reprimands, counseling, retraining) can follow reports from the public about a driver's unsafe driving. Third-party companies providing placards and receiving 800-number calls from the public claim their use results in fleet crash rate reductions. The primary disadvantage is the reliance on other motorists to call the toll-free number to report an event; that is, there is no guarantee that risky driving will be reported. Callers may or may not describe the incident accurately, leading to possible disputes with drivers. Most of the phone calls from the public, and therefore the feedback received by drivers, will be negative, perhaps leading drivers to resent the method and feel that they have been more unlucky than unsafe.

On-board safety monitoring devices would seem to be a technique with tremendous potential to assess commercial driver safety performance and particularly to identify potential unsafe driving performance. Interest in these technologies is increasing and an ongoing study by VTTI is evaluating the safety benefits of a low-cost driver behavior management system in CVO. Technology is at a point where many safety-critical measures (e.g., speed, acceleration, brake use, driving times associated with hours-of-service compliance) can be continuously monitored electronically. Emerging technologies can measure forward headway (to detect tailgating), rollover risk on curves, lane tracking, lateral encroachments toward adjacent vehicles (e.g., during lane changes), and even driver alertness and attention (Knipling et al., 2003). Such technologies may provide safety performance feedback, to both drivers and their managers, in addition to providing collision warnings (Knipling & Olsgard, 2000). On-board safety management devices can serve as the basis for short- and long-term safety performance feedback, counseling to drivers, data source to identify if retraining is needed, and can be employed in support of fleet safety management techniques.

In spite of this safety potential, commercial drivers and fleet safety managers have not, as of yet, widely embraced the use of on-board safety management devices. A major barrier to the widespread use of these systems is driver acceptance. Penn and Schoen Associates (1995) found that on-board safety monitoring devices were not well-accepted by commercial drivers because they perceived it as an invasion of privacy and/or as a sign of disrespect for their professionalism as drivers. Ironically, drivers in this study acknowledged the potential safety benefits of these devices. A recent commercial driver focus group study conducted by Roetting, Huang, McDevitt and Melton (2003) found that drivers were willing to be monitored and receive feedback from on-board technologies, but only if the feedback was specific, constructive, individualized, and implemented within a positive and supportive management environment. Therefore, positive reinforcement of appropriate driving behaviors where safety managers take advantage of this type of technology to develop safety-related rewards and incentives should be part of any successful fleet safety program.

On-road monitoring can provide valuable information to fleet management. As noted, on-the-job evaluations can be conducted in several different ways, including monitoring and feedback in order to promote and maintain driver safety. One possible use of the information gained from on-road monitoring is targeted training, including simulation-based targeted training. Situations the driver commonly encounters, or difficult situations that the driver has struggled with, are becoming easier to identify, while the ability to replicate these situations in a simulated environment has also increased. The findings from on-the-job evaluations provide areas that can be focused on during retraining or coaching sessions. The ability to integrate these two methods and provide a reasonable simulacrum of a real-world scenario with which the driver is struggling is an incredible strength of simulation-based targeted training.

31.2.4.3 Retraining

Due to a shortage of commercial drivers, the trucking industry is experiencing difficulty finding and retaining qualified drivers (Min & Lambert, 2002). Therefore, fleet companies have started investing in various driver development tools. This investment may reduce turnover rates, crash rates, and operating costs. Although such driver development programs are increasing in popularity and provide a way for companies to ensure drivers are adequately trained, there is still the possibility that bad driving habits may develop over time, with or without the driver's knowledge. On-the-job monitoring provides methods of identifying these behaviors, and allows for retraining to be more precisely focused.

The goal of retraining is not to punish the driver, but rather to help improve driving behavior. It is estimated that retraining marginal drivers can save fleets $5,000 to $6,000 in recruitment, drug testing, and CDL qualification costs per driver (Siegel, 1992). Several fleets are currently investing in the training\retraining approach. Acquiring part-task and/or full-mission simulators in order to retrain for scenarios or conditions in which drivers are considered deficient is becoming more common for fleet companies as part of their safety initiative (Lockheed Martin, 2001; MPRI, 2005).

Retraining should be seen as an investment and integrated with any process of driver monitoring. Due to the driver shortage, retraining is very important and encouraged. In order to foster driver acceptance of retraining, it should never be treated or portrayed as a punishment. Instead, retraining should be formulated in a positive atmosphere where drivers can self-request retraining based on their own observations, periodic on-the-job evaluations, on-the-job monitoring, or at the recommendation of an evaluator. Retraining should be focused on the specific areas identified, and the reasoning should be presented to the driver in order to avoid creating additional resistance to retraining. Simulators present a unique opportunity to offer driver retraining in a safe and controlled environment that cannot always be achieved through other methods.

31.3 Conclusions

The Commercial Motor Vehicle Safety Act of 1986 put in place a number of key safety initiatives to improve motor carrier safety. Driver training was a positive outcome of this initiative, and has become a key focus area to help ensure safe driving performance of CMV drivers. Moreover, as many of the safety-conscious fleets have determined, there is an increasing recognition that driver training does not end with licensing. Many fleets are using driver training as a continual learning process that drivers engage in over the course of their careers.

Although simulators hold the promise of aiding not only training, but also screening and retraining, their implementation remains limited. At present, commercial vehicle simulators are typically used (if they are present at all) as training devices in large fleet operations. Simulators have been successfully integrated into a number of CVOs with favorable results; however, few investigations of simulator-based training and screening for commercial drivers have occurred. Recent and ongoing studies will help fill these lacunae. CVOs wishing to implement simulators are finding increasingly capable configurations of simulators available (see, this book, chap. 2 by Allen, Rosenthal, & Cook or chap. 12 by Jamson for examples), and an increasingly wide knowledge base on developing instructional and screening programs for their commercial drivers. Simulation should, over the coming decade, become a frequently used tool in the selection of drivers due to its strength in creating and measuring scenarios that the driver is likely to encounter in the real world. Likewise, simulation will only play a larger role in the retraining of marginal drivers within CVOs. Commercial vehicle simulation is likely to never fully subsume these operations. However, the complementary role commercial vehicle simulators play within them will only become larger.

Driver training is a concept that has evolved and improved over the years. Curriculums have changed and, with the advancement of technology, the method of administering training protocols is evolving as well. Truck driver training is presently at an exciting stage as new driving simulation technology is introduced and its efficacy is tested. Positive results from these investigations may lead to a significant change in training

protocol; namely, the use of driving simulators for driver training both pre- and post-CDL. In line with the Commercial Motor Vehicle Safety Act of 1986, as training improves, it would be expected that safety for CDL drivers and the general motoring public would also improve. Through innovations in commercial driving simulator technology, the safety benefits gained in scenarios practiced within the virtual world will ideally manifest as safety improvements in the real world.

Key Points

- The process of operating a commercial vehicle is fundamentally dissimilar from the operation of an automobile. Accordingly, the simulation needs for commercial drivers are different from those of automobile drivers.
- Insufficient training is a major contributor to commercial motor vehicle (CMV) crashes. Training of CMV drivers is a continuous process, beginning with pre-CDL training and continuing throughout a driver's career. Many fleets recognize simulation as a valuable tool for training CMV drivers.
- In addition to training, commercial vehicle simulators may be used to support the pre-hire screening process and targeted training (retraining) of commercial drivers. Simulation allows for these processes to occur in safe and controlled environments, minimizing both the risk and costs associated with screening and retraining drivers.
- Commercial vehicle simulators are sufficiently advanced to allow training of a wide variety of commercial driving behaviors.
- Simulators have demonstrated safety and cost benefits in some limited testing. Ongoing studies will help further define what benefits may be obtained from use of a training simulator for training novice commercial vehicle drivers.

Keywords: Commercial Driver License, Commercial Vehicle Simulator, Driver Evaluation, Driver Screening, Driver Training

Acknowledgments

This chapter presents a current research effort funded by the U.S. DOT Federal Motor Carrier Safety Administration (FMCSA), *Commercial Motor Vehicle (CMV) Driving Simulator Validation Study (SimVal): Phase II* (DTNH22-05-01019, Task Order #9).

Glossary

Commercial driver's License (CDL): A class of license allowing a driver to operate a commercial vehicle (e.g., tractor-trailers, motor coaches, tankers). Obtaining a CDL in the USA requires meeting minimum testing standards and passing a medical exam.

Commercial vehicle simulator: A wide range of simulators (including partial task simulators) which provide a simulation of one or more commercial vehicles. Typical features of truck simulators include controls such as air brakes and shift towers, and physics modeling allowing for the accurate simulation of articulated vehicle maneuvering.

Driver screening: The process of pre-hire review, including a review of driving history, criminal history, drug or alcohol abuse, and driving ability. Simulation-based screening may be integrated into this process.

On-the-job evaluation: The process of measuring risk associated with the behaviors and skill levels of individual drivers. This is commonly accomplished through driver monitoring (such as in-vehicle recorders and ride-alongs) and can be used to improve driver retraining.

Retraining: The process of targeting training to improve marginal driver skills. Simulators allow for the creation of such targeted training in a safe, measurable, and replicable environment.

Web Resources

The *Handbook* web site contains color version of the chapter's figures.

Web Figure 31.1: Large Truck Backing Task in Real Truck versus Truck Simulator (color version of Figure 31.1).

Web Figure 31.2: Specialized Training Scenarios (color version of Figure 31.2).

Key Readings

Allen, R. W., & Stein, A. C. (1990). *The use of simulation in truck safety research, driver training and proficiency testing* (Report SAE 902271). Warrendale, PA: Society of Automotive Engineers.

Brock, J. F., Jacobs, C., Van Cott, H., McCauley, M., & Norstrom, D. M. (2001). *Simulators and bus safety: Guidelines for acquiring and using transit bus operator driving simulators* (Report TCRP 72). Washington, DC: Transportation Research Board.

Pierowicz, J., Robin, J., Gawron, V., Watson, G., Nestor, B., & Murphree, W. (2002). *Commercial truck simulators reassessment and evaluation* (Publication FMCSA RT-03-008). Washington, DC: Federal Motor Carrier Safety Administration, US Department of Transportation.

Robin, J. L., Knipling, R. R., Derrickson, M. L., Antonik, C., Tidwell, S. A., & McFann, J. (2005). Truck simulator validation ("SimVal") training effectiveness study. *Proceedings of the 2005 truck & bus safety & security symposium* (pp. 475–483). Alexandria, VA: National Safety Council.

Strayer, D. L., & Drews, F. A. (2003). Simulator training improves driver efficiency: Transfer from the simulator to the real world. *Proceedings of the second international driving symposium on human factors in driver assessment, training and vehicle design* (pp. 190–193). Iowa City, IA: University of Iowa.

References

Allen, T., & Tarr, R. (2005). Driving simulator for commercial truck drivers—Human in the loop. *Proceedings of the third international driving symposium on human factors in driver assessment, training, and vehicle design* (pp. 335–341). Iowa City, IA: University of Iowa.

American Trucking Associations Foundation (ATAF). (1999a). *Truck driver risk assessment guide and effective countermeasures: Recommended management practices.* Arlington, VA: Author.

American Trucking Associations Foundation (ATAF). (1999b). *Safe returns: A compendium of injury reduction and safety management practices of award-winning carriers* (Publication No. C0938). Arlington, VA: Author.

Brock, J. F., McFann, J., Inderbitzen, R. E., & Bergoffen, G. (2007). *Effectiveness of commercial motor vehicle driver training curricula and delivery methods* (Commercial Truck and Bus Safety Synthesis Program, Synthesis 13). Washington, DC: Transportation Research Board.

Corsi, T. M., & Barnard, R. E. (2003). *Best highway safety practices: A survey of the safest motor carriers about safety management practices, final report* (FMCSA Contract DTFH61-98-X-00006). Washington, DC: Federal Motor Carrier Safety Administration, US Department of Transportation.

Dueker, R. L. (1995). *Assessing the adequacy of commercial motor vehicle driver training: Final report* (Publication FHWA-MC-96-011). Washington, DC: Federal Highway Administration Office of Motor Carriers, US Department of Transportation.

Emery, C., Robin, J., Knipling, R., Finn, R., & Fleger, S. (1999). *Research design: Validation of simulation technology in the training, testing and licensing of tractor-trailer drivers. Final report* (Publication FHWA-MC-99-060). Washington, DC: Federal Highway Administration, US Department of Transportation.

Federal Motor Carrier Safety Administration. (2007a). *Commercial drivers' license program.* Retrieved September 24, 2007, from the FMCSA Web site: http://www.fmcsa.dot.gov/registration-licensing/cdl/cdl.htm

Federal Motor Carrier Safety Administration. (2007b). *Large truck crash facts 2005* (Publication FMCSA-RI-07-046). Washington, DC: Author.

Hartman, K., Pritchard, R., Jennings, K., Johnston, J., Knipling, R. R., MacGowan, J., et al. (2000). *Commercial vehicle safety: Technology and practice in Europe* (Publication FHWA-PL-00-010). Washington, DC: Federal Highway Administration Office of International Programs, US Department of Transportation.

Hickman, J. S., Knipling, R. R., Inderbitzen, R. E., Wiegand, D. M., Bergoffen, G., & Hanowski, R. J. (2007). *Impact of behavior-based safety techniques on commercial motor vehicle drivers* (Commercial Truck and Bus Synthesis Program, Synthesis 11). Washington DC: Transportation Research Board.

Horn, B. E., & Tardif, L. -P. (1999). Licensing and training of truck drivers. *IATSS Research, 23*(1), 16–25.

Keith, K., Trentacoste, M., DePue, L., Granda, T., Huckaby, E., Ibarguen, B., et al. (2005). *Roadway human factors and behavioral safety in Europe* (Publication FHWA-PL-05-005). Washington, DC: Federal Highway Administration, US Department of Transportation.

Knipling, R. R., Hickman, J. S., & Bergoffen, G. (2003). *Effective commercial truck and bus safety management techniques.* Commercial Truck and Bus Safety Synthesis Program, Synthesis 1. Washington, DC: Transportation Research Board.

Knipling, R. R., & Olsgard, P. J. (2000, May). *Prospectus: The behavioral power of on-board safety monitoring feedback.* Proceedings of the 10th Annual Meeting of the Intelligent Transportation Society of America (ITS America). Boston, MA.

Lester, T., Rehm, L., & Vallint, J. (2003). *Research for the development and implementation of a purpose built truck simulator for the UK: Phase I* (Publication PR/T/122/02). Berkshire, UK: Transportation Research Laboratory.

Lockheed Martin. (2001, May 2). *High-tech total driver training system enhances training and safety in the commercial trucking industry* [Press Release]. Retrieved from http://www.lockheedmartin.com/news/LockheedMartin DeliversTruckDriverTr.html

Min, H., & Lambert, T. (2002). Truck driver shortage revisited. *Transportation Journal, 42*(2), 5–17.

MPRI. (2005, October 11). *Nation's largest trucking company gives driver training program a makeover* [Press Release]. Retrieved from http://www.mpri.com/main/newsroom/ NationsLargestTruckingComp.html

Parkes, A. M. (2003). Truck driver training using simulation in England. *Proceedings of the second international driving symposium on human factors in driver assessment, training and vehicle design (pp. 59–63).* Iowa City, IA: University of Iowa.

Penn & Schoen Associates, Inc. (1995). *User acceptance of commercial vehicle operations (CVO) services; Task B: Critical issues relating to acceptance of CVO services by interstate truck and bus drivers. Final report* (Contract DTFH61-94-R-00182). Washington, DC: Research and Innovative Technology Administration, US Department of Transportation.

Pierowicz, J., Robin, J., Gawron, V., Watson, G. Nestor, B., & Murphree, W. (2002). *Commercial truck simulators reassessment and evaluation* (Publication FMCSA-RT-03-008). Washington, DC: Federal Motor Carrier Safety Administration, US Department of Transportation.

Professional Truck Driver Institute (PTDI). (1999). *Curriculum standard guidelines for entry-level tractor-trailer driver courses.* Retrieved from http://www.ptdi.org/errata/ CURRICULUM_STANDARDS.pdf

Rasmussen, J. (1983). Skills, rules, and knowledge: Signals, signs, and symbols, and other distinctions in human performance models. *IEEE Transactions on Systems, Man, and Cybernetics, SMC-13*(3), 257–266.

Robin, J. L., Knipling, R. R., Derrickson, M. L, Antonik, C., Tidwell, S. A., & McFann, J. (2005). Truck simulator validation ("SimVal") training effectiveness study. *Proceedings of the 2005 truck & bus safety & security symposium (pp. 475–483).* Alexandria, VA: National Safety Council.

Roetting, M., Huang, Y-H., McDevitt, J. R., & Melton, D. (2003). When technology tells you how to drive: Truck drivers' attitudes toward feedback by technology. *Transportation Research Part F: Traffic Psychology and Behavior, 6*(4), 275–287.

Ryder, A. (2000). A smarter way to train. *Heavy Duty Trucking, 79*(11), 60–62.

Schneider National Inc. (2006, July 10). *Schneider National's simulation-based driver training program receives national training and development Excellence in Practice award* [Press Release]. Retrieved from http://www.schneider.com/news/ Final_ASTD_Release.html

Siegel, S. (1992). Driving safely: Your best investment. *Fleet Owner, 87*(5), 56–62.

Stock, D. (2001). *I-95 corridor coalition field operational test 10: Coordinated safety management. Volume I: Best practices in motor carrier safety management, final report.* Rockville, MD: I-95 Corridor Coalition.

Vidal, S., & Borkoski, V. (2000). *MTA New York City transit's bus operator simulator: A successful public/private partnership.* Paper presented at the 2000 American Public Transportation Association Bus & Paratransit Conference. Retrieved from: http://www.faac.com/pdfs/nyct2.pdf

Driving Rehabilitation as Delivered by Driving Simulation

Harsimran Singh
University of Virginia Health System

Brent M. Barbour
University of Virginia Health System

Daniel J. Cox
University of Virginia Health System

Abstract

The Problem. The ability to drive safely constitutes an important aspect of individual freedom in today's society. Partial or complete loss of driving ability can seriously impair an individual's daily functioning, well-being, and quality of life. Assessment of driving abilities and delivering driving rehabilitation are complex procedures, especially when dealing with clinical populations (e.g., patients who have suffered a traumatic brain injury). Traditional measures such as neuropsychological tests and computerized tests may offer information on specific cognitive, motor, and perceptual skills that facilitate driving; however, they fall considerably short of providing a comprehensive picture in terms of real-life, complex, and demanding driving situations. *Role of Driving Simulators.* Virtual reality driving simulation (VRDS) offers tremendous advantages over traditional assessment tools in that it has the potential to evaluate both specific driving behaviors and general driving skills objectively, reliably, and safely in an assimilated real-life experience using a variety of challenging, but standardized and monitored, settings. Use of VRDS in driving rehabilitation, although still an emerging field, shows a lot of promise. It offers an opportunity to incorporate real-life driving situations (without compromising safety of participants) that can be manipulated to train people in a systematic manner. *Key Results of Driving Simulator Studies.* The literature strongly supports the use of VRDS as a sensitive and an effective tool for driving assessment and it can be hoped that researchers, in the near future, can work towards standardization of driving evaluations (in terms of simulators and the software used) across various clinical and non-clinical conditions. The use of VRDS in driving rehabilitation is, however, surprisingly limited. The few studies that have been published in this area offer compelling evidence highlighting their potential effectiveness in providing driving retraining to those who have suffered a neurological and/or physical compromise. *Conclusions and Way Forward.* In the face of encouraging evidence indicating the value of driving simulators in driving rehabilitation, it is important to appreciate that this is a relatively new field of application for driving simulators. Therefore, a few practical and technological concerns (e.g., standardization, simulation adaptation syndrome) will need to be addressed before driving simulators are accepted as the tool of choice to deliver and assess driving rehabilitation.

32.1 Introduction

In today's heavily mobile-dependent society, the ability to drive safely is considered an important expression of individual freedom which contributes significantly towards a person's well-being (Fonda, Wallace, & Herzog, 2001), feelings of self-worth and aspects of their quality of life (DeCarlo, Scilley, Wells, & Owsley, 2003). Therefore, it becomes essential to not only understand the extent and impact of loss of adequate driving skills in people who have suffered physical or neurological compromise but also, where possible, help individuals get appropriate re-training in skills that can enable them to resume independent and safe driving.

The assessment of driving abilities and delivery of driving rehabilitation are often lengthy and complicated procedures, especially when dealing with clinical populations (Hopewell, 2002). Clinical populations are themselves quite varied, including those with conditions that may recover with time and exercise including traumatic brain injury, stroke, brain surgery, and driving phobias as well as those with conditions that can be progressive, such as Alzheimer's disease, Parkinson's disease, and Huntington's disease, among others (assessment and rehabilitation in the last three conditions are discussed by Uc and Rizzo, this book, chap. 48). Even within the same condition (for example, traumatic brain injury [TBI]), patients vary in their degrees of cognitive, motor, and perceptual impairments (Brouwer, Withaar, Tant, & van Zomeren, 2002) and rates of recovery. Thus, someone with TBI who is currently certified as not fit to drive may soon show significant improvement in their condition which could allow them to consider driving rehabilitation. On the other hand, an aging patient with progressing visual deficits may currently be able to drive sufficiently well but will be unable to engage in safe driving sometime in the future. It therefore becomes essential to have available clinically standardized, reliable and valid assessment techniques and rehabilitation opportunities that can provide sensitive and objective assessments of patients' abilities to continue or to resume driving without jeopardizing their safety or that of others on the road (Korteling & Kaptein, 1996; Fox, Bowden, & Smith, 1998).

Traditional measures such as neuropsychological tests and computerized tasks have been employed very commonly to study the relationship between cognitive impairment and functional performance. These tools may offer information on specific cognitive, motor, and perceptual skills that have been compromised in an individual; however, they fall considerably short of providing a comprehensive picture in terms of assessing individual behavior in real-life, complex and demanding situations such as driving (Wilson, 1993). Neuropsychological tests have also shown poor face, ecological and predictive validity (Wilson, 1993). The use of Virtual Reality Driving Simulators (VRDS) in driving assessment and rehabilitation offers various advantages over existing tools as it allows for the use of virtual reality to enable patients to experience and react to varying levels of interactive "real-life" driving situations, employing driving-relevant responses, without posing a risk to themselves or others on the road.

32.2 Use of Driving Simulators in Driving Assessment

Schultheis and Mourant (2001) have usefully summarized the benefits of VRDS as an assessment tool and discussed how the application of VRDS in the area of driving assessment has helped to overcome some of the major limitations that exist with the currently used assessment methods, for example, focusing

on certain individual skills, such as reaction time or speed-of-processing on an independent computer task, but not providing an overall assessment of the multitask driving behaviors. The driving experience involves a very dynamic and complex environment which is difficult to capture other than by driving itself (Wald, Liu, & Reil, 2000). However, the use of VRDS in driving assessment offers tremendous opportunities to assimilate the "real experience" of driving through space and time while negotiating traffic and traffic signals in a standardized and a safe setting for objective monitoring. The VRDS technology offers an ecologically valid assessment measure which allows the individual to react to highly interactive and real-life driving scenarios which can be manipulated to introduce varying levels of complexities in driving situations.

A major challenge to developing a standardized VRDS assessment is to incorporate a representative sample of driving demands reflective of safe driving. A possible analogy to a comprehensive, reliable, and accurate assessment of driving competency is that of assessing intelligence. Tests for IQ (Intelligence Quotient) do not assess all domains of intellectual abilities, but instead select accepted key domains, such as information, vocabulary, and verbal memory, identify discriminating and reliable items for each, convert performance on these domains (or subtests) to z scores to be summed into an IQ, and then demonstrate reliability of the IQ score, its ability to discriminate known groups, and its ability to predict future outcomes such as school achievement. Similarly, VRDS must first select essential domains of driving safety, such as the use of rear-view mirrors to merge or pass slow lead vehicles, speed control to properly accelerate or decelerate around anticipated and unanticipated stops, steering control to negotiate curvy roads and turns at intersections, and then develop representative probes of each selected domain. VRDS further need to demonstrate reliability of performance on these probes as well as both discriminative and predictive validity of the resulting "DQ" or Driving Quotient. As with the IQ tests, which establish a cut-off as a threshold for mental retardation; the DQ will have to establish a minimal threshold to determine driving competency.

An early study involving the use of VRDS as an assessment tool was conducted in Canada on patients recovering from traumatic brain injury (TBI) (Liu, Miyazaki, & Watson, 1999). The researchers used a simple PC-based interactive driving simulator ("DriVR") including a head-mounted display with a 30-degree horizontal field of view to compare the performances of 17 adults with TBI with those of 17 matched healthy controls. DriVR's measures were able to discriminate between participants who had head injuries from their healthy counterparts and provided support of discriminative validity for the simulator. Wald, Liu and Reil (2000) (Table 32.1) in their attempt to further validate the DriVR, compared measures of DriVR performance with other indicators of driving performance—for example, on-road tests, cognitive and visual-perceptual tests, and driving video tests on a sample of 28 adults who had had TBI.

TABLE 32.1 Published Studies Showing the Use of VRDS in Assessing Driving Performance in Patients

Author	Sample	Simulator	Study Design	Main Results
1. Biederman et al. (2007), USA	20 adults with ADHD & 21 controls	An instrumented 2001 Volkswagen Beetle with a 40-degree visual angle. STISEM Drive and STISIM Open Module simulation software was used to compute graphics	All participants drove the simulator for 1 hour. This driving period included a training period (10 min), a high stimulus testing period (urban road with high traffic demands) & a low stimulus testing period (rural road with "monotonous" driving)	Compared to the controls, ADHD adults were more likely to collide with an obstacle during the monotonous segment of driving
2. Muhlberger et al. (2007), Germany	15 people with high fear of driving in tunnels & 15 matched controls	Graphics computer (ONYX2 3-pipe Infinite Reality, SGI, CA) that projected on a curved screen. Visual angle of 180-degree horizontally & 55-degree vertically. Three virtual driving scenarios of 8.9 km each	Verbal & physiological fear reactions were assessed in 3 virtual driving scenarios—open environment, a partially open tunnel & a closed tunnel	Participants with high tunnel fear reported significantly greater fear while driving in the tunnel and the gallery
3. Lew et al. (2005), USA	11 patients with TBI & 16 healthy controls. Both groups matched on age	STI driving simulator program on a PC, 21-inch monitor screen, 30–35 min test route	Time-1: TBI patients completed baseline evaluation on driving simulator & in-car road test. Healthy controls provided normative data on driving simulator. Time-2: Family members observed patients' driving over 4 weeks & rate performance	• TBI patients showed poorer performance on simulator compared to healthy controls • Relationship between patients' driving simulator performance & long-term driving performance Driving simulators provide unique info. regarding patients' driving skills (e.g., fluctuations in acceleration)
4. Kotterba et al. (2004), Germany	13 patients with Narcolepsy & 10 healthy controls	Computer-Aided Risk Simulator (Dr. Ing. R. Forest, Gummersbac, Germany)	Vigilance, alertness & divided attention assessed using neuropsychological tests. All participants drove on a highway in the simulator for 60 min during different weather & daytime conditions. Five patients reinvestigated after optimized medication therapy	• No correlation between driving performance & neuropsychological test results • After therapy: No change in neuropsychological tests but accident rates & concentration lapses as indicated by the simulator declined significantly
5. Lengenfelder et al. (2002), USA	3 patients with TBI & 3 healthy controls (matched on age, gender & education)	VR-Driving Simulator computer program, PC with a 21-inch monitor screen, 1.75 mile long driving route	Participants administered neuropsychological measures & Useful Field Of View (UFOV) with two tasks: Primary (VR driving task) & Secondary (divided-attention task)	• No difference in driving speeds between TBI patients & controls • TBI patients showed greater number of errors on the secondary task
6. Wald et al. (2000), Canada	28 patients with brain injury (mean age = 40 yrs)	DriVR simulator software on a PC, Virtual I/O I-glasses, 30-degree horizontal field of view, 15 min test route	Compare DriVR measures to other indicators of driving ability e.g. on-road, cognitive, visual-perceptual & driving video tests	Concurrent validity coefficients identified between DriVR measures & other measures of driving ability
7. Lundqvist et al. (1997), Sweden	Patients with brain lesion n = 29 & 29 controls (matched on age, gender, education & driving experience)	VTI driving simulator, wide-angle (120-degree) visual screen & 5 subsystems, 80 km test route with diff driving situations	All participants assessed on: neuropsychological test battery, simulator driving & on-road driving	Patients & controls showed similar performance in predictable situations on the simulator; however, in unpredictable situations, patients showed longer reaction times & safety margins

The strongest correlations were found between DriVR variables of "maintaining lane position" and "lane tracking" with on-road test results ($r = 0.56$ and 0.50, respectively). The driving video tests consisted of two main tests: The Driver Performance Test II, which evaluated participants' driving-related knowledge and behavior while watching a driving video (higher scores indicating better performance), and the Driver Risk Index II, which measured a driver's willingness to take risks (higher scores indicating greater risk-taking behavior). Positive correlations were revealed between DriVR variables of "Driveway Choice" (i.e., pulling into the driveway of the house that has a car parked in front of it) and Driver Performance Test II (0.46), and "lane tracking" and Driver Risk Index II (0.45). DriVR measures also correlated with the cognitive and visual-perceptual tests; however, these relationships were less substantial (all less than 0.40) compared to others.

Although, initial evidence of concurrent validity of the DriVR was established as study findings showed correlations between performance measures of the simulator and other driving indicators, replication of these findings is warranted. Wald, Liu and Reil (2000) also recommended measuring psychological variables such as anxiety in participants who are tested using a VRDS as non-familiarity with the VR technology might be a significant contributor to differences on VRDS performance measures. Someone who is acquainted with or used to VR simulations (e.g., VR-based video games) may find it easier to understand this technology and may quickly accommodate to the research task at hand. On the other hand, failure to accommodate to VRDS may reflect unfamiliarity with the virtual reality technology, or it may reflect the limitations brought on by cognitive impairments in question.

In a more recent study, Lew et al. (2005) attempted to seek evidence for the predictive validity of driving simulators using 11 patients with moderate to severe TBI and 16 healthy participants (Table 32.1). At baseline, participants were measured on 12 driving parameters (categorized into speed control and direction control) on a PC-based Systems Technology Incorporated (STI version 8.16) driving simulator. Ten months later, patients were observed while driving by family members who also completed a Driving Performance Inventory (DPI). TBI patients performed significantly worse on the simulator compared to the healthy participants—the TBI patients had greater difficulty in controlling the speed and direction of their vehicle. They had more problems performing on divided-attention tasks, multitasking while driving, and were also five times more likely to commit driving violations using the virtual vehicle compared to their healthy counterparts. One major finding of this study was that patients' performance on the driving simulator correlated significantly ($r = 0.66$, $p = 0.01$) with their long-term real-road driving performance as rated by their family member on the DPI 10 months later. It is, however, interesting and useful to note that the driving simulator proved more valuable in capturing reliably subtle aspects of driving performance (e.g., fluctuation in acceleration and capacity to divide attention between tasks) compared to significant others who had simply assessed patients' driving through observation. These variables may not be assessed reliably using other assessment tools including road-test evaluations, which are designed to minimize driving-related threats for patients and their evaluators and are therefore unable to offer a wide variety of testing environments.

In an earlier study, Lengenfelder, Schultheis, Al-Shihabi, Mourant and DeLuca (2002) had attempted to evaluate the effect of divided attention on the driving performance of TBI and healthy participants using a VRDS simulator computer program that was presented on a PC computer (21-inch screen) connected to a steering wheel and gas/brake pedals (Table 32.1). The study involved a small sample of three TBI patients matched for age, gender, and education to three healthy participants. Results provided evidence in favor of the use of VRDS in that it can be employed to achieve a specific, objective, and direct evaluation of the relationship between complex driving-relevant, cognitive, motor and perceptual tasks that result in divided attention, and driving performance, which could previously not be studied using traditional methods of assessment. TBI patients committed a greater number of errors on the secondary task (i.e., number recall) compared to the healthy participants as part of the divided-attention driving task on the simulator.

Similar evidence in favor of using VRDS (over other assessment tools) to reliably assess cognitive functions and their impact on driving ability has been provided by researchers working with other clinical populations, such as Narcolepsy patients (a neurological condition marked by excessive daytime sleepiness). Kotterba et al. (2004) compared driving simulator performance and neuropsychological test results in 13 patients with Narcolepsy and 10 healthy controls (Table 32.1). Overall, patients had significantly more crashes on the simulator compared to controls ($p < 0.01$). Study findings indicated that, compared to the neurological tests, driving simulator investigations allowed for a more comprehensive patient assessment (including recording any concentration lapses) and were more sensitive to any improvements in driving performances. Simulated scenarios were a very close reflection of real-life driving situations (e.g., varying traffic conditions, daytime driving versus night-time driving, driving under different weather conditions), thereby offering greater face validity to VRDS tests.

Although, an on-road evaluation is accepted as an ultimate test of an individual's driving performance, such tests may fall short while assessing patients with Narcolepsy as evaluators or researchers accompanying the patient driver may actually enhance alertness in the patient, thus compromising patient data. Assessing the effects of sleep deprivation on driving performance in real-life driving can be potentially very unsafe. Furthermore, it is extremely difficult to develop a reliable on-road testing procedure since it is nearly impossible to ensure the same traffic patterns, road and weather conditions from one examination to another. Narcolepsy and other similar clinical problems that can impact upon patients' driving ability can be examined both safely and reliably in a virtual reality driving simulator, which closely simulates real-life driving situations (Rizzo & Kim, 2005; see Andersen, this book, chap. 8), allows for interaction between the "driver" and these scenarios, while maintaining high experimental control at all times. For example, VRDS have been used to investigate the extent and quality of phobic fear in people who are scared of driving through tunnels by having them drive through simulated scenarios that induce fear progressively (Muhlberger, Bulthoff, Wiedemann, & Pauli, 2007).

A recent study by Biederman et al. (2007), evaluated driving performance in adults with Attention-Deficit Hyperactivity Disorder (ADHD) using a driving simulator (full cab 2001 Volkswagen Beetle) with a front projection screen with a 40-degree view of virtual roadway. Participants were assessed in a variety of driving situations that were modeled on impulsivity, hyperactivity, and inattention while driving (taking account of the clinical features of the population under study). Results highlighted "inattention" while driving, due to a low-stimulus driving environment as the key contributing factor to the impaired driving performance in people with ADHD. In another study, Cox and colleagues (2006) assessed the influence of manual versus automatic transmission on driving in adolescents with ADHD. Participants were assessed on 10 specific driving performance variables, including steering control, speed control, driving off the road and braking using an Atari Research driving simulator (Atari Inc.) featuring three 25-inch computer screens and a 160-degree visual field. Results of the study were in favor of manual transmissions, suggesting that young drivers with ADHD showed an improved driving performance when manually controlling the clutch, gearshift, and so forth.

Use of a driving simulator with people with diabetes has helped to identify the risky blood sugar levels (<3.8 mmol/l) at which these patients become susceptible to driving mishaps and the related impairment in cognitive-motor functions (Cox, Clarke, Gonder-Frederick, & Kovatchev, 2001). These and other similar studies evaluating driving capabilities in clinically vulnerable populations underline the suitability of VRDS as a better assessment tool compared to other existing measures in that virtual reality offers opportunities to systematically manipulate driving test environments to "mimic" real-life situations, thereby enabling objective, accurate and safe assessments of driving skills.

Given the strong evidence in favor of VRDS being a sensitive and an effective tool for assessing driving skills it can be hoped that researchers, in the near future, can work towards the standardization of driving evaluations across various clinical (e.g., patients with TBI, ADHD, Narcolepsy and driving phobias) and non-clinical conditions (e.g., enhancing specific driving skills in novice drivers). These standardization procedures (both in terms of machines and the software used) would not only help in establishing driving simulators as reliable and valid measures of driving assessment, but also make VRDS technology more cost-effective. Researchers and healthcare professionals would be able to use and share with each other specific standardized VRDS scenarios that have already been established to assess their patients without having to develop simulated scenarios from start, which can take substantial time, effort and finances.

32.3 Use of Driving Simulators in Driving Rehabilitation

The high level of user interaction offered by the VRDS along with the increased flexibility of providing different driving environments and the quality of driving performance variables assessed together make VRDS a tool suitable for delivering driving rehabilitation. In a detailed analysis of the application of virtual reality in the field of rehabilitation medicine, Rizzo and Kim (2005) highlighted the strengths of virtual reality rehabilitation (as translated to VRDS):

- It offers an opportunity to incorporate real-life driving situations (without compromising the safety of the trainer and the trainee) which could be manipulated to test and train people in a systematic manner. This feature can be valuable while rehabilitating people who have suffered some serious cognitive impairment as a result of brain injury, spinal cord injury, stroke, and so forth. On-road tests may prove quite dangerous in such cases, as patients may not necessarily be equipped with sufficient skills to drive the car, thereby posing a risk to themselves and/or their evaluators (Ku et al., 2002).

- It allows for systematic and controlled delivery of driving-related stimuli (Rizzo, Schultheis, Kerns, & Mateer, 2004). This is particularly useful during rehabilitation as the same driving scenario can be presented repeatedly (Schultheis et al., 2006) or in a hierarchical manner for training purposes (e.g., exposure therapy delivered using a VRDS to help a patient manage their driving-related anxieties and resume safe driving (Thacker, 2003)).

- It allows performance to be captured in detail and later evaluated objectively (by both the user and the provider) to better target "problem" driving behaviors during the rehabilitation process.

- It allows for real-time performance feedback while "interacting" with a driving environment. For example, if in a particular VRDS session a "driver" chooses to drive closer to an emergency ambulance on duty, the driver could be made to hear its siren loud and clear. However, if the driver chooses to slow down and let the ambulance pass or if he or she changes the route then the sound of the siren could be appropriately weakened.

- It makes it possible to include driving-related gaming tasks (e.g., a NASCAR drive course) that would help to maintain or enhance motivational levels in patients by making driving re-training therapy sessions more interesting and engaging. This is useful because driving rehabilitation training can prove to be lengthy and boring for a few patients who may then find it difficult to continue with their therapy.

32.3.1 Stroke Patients

Akinwuntan and colleagues (2005) examined the driving performances of 83 stroke patients from a rehabilitation unit in Belgium (Table 32.2). Participants were randomized to either the simulator-based training group (i.e., experimental group) or the control group where participants were asked to conduct driving-related cognitive tasks, such as route-finding on a map, memory training with numbers, and recognition of road and traffic signs. Both groups participated in routine hospital rehabilitation programs and, in addition, received extended

training in a 15-hour driving-related series of exercises. As part of this extra training:

a. Patients in the experimental group were trained on VRDS which incorporated a Ford Fiesta. The VRDS was powered by a STISIM Drive System (version 1.03) and had a projection 'screen with a 45-degree visual angle. The driving course included a 13.5 km scenario which took approximately 25 minutes to complete. This scenario advanced from a relatively simple driving course (two-lane road with urban traffic) to a more complex driving situation (four-lane highway

TABLE 32.2 Published Studies Showing the Use of VRDS in Successfully Delivering Driving Rehabilitation

Author	Sample	Simulator	Study Design	Main Results
1. Cox et al. (2010), USA	11 military personnel from a rehabilitation center with TBI	Model T³ providing 180-degree field of view with rear and side view "mirror" images, optional 5-speed manual transmission, turn signal, gas and brake pedals, steering wheel & air conditioner for temperature control. Each driving scenario involves a 12-mile course	Experimental: $n = 6$, Control: $n = 5$. Experimental group received 4–6, 60 to 90 minute rehabilitation training sessions involving practicing progressively more complex driving skills through progressively more demanding traffic	• Significant improvement in driving performance in experimental group along with significant reduction in road rage and risky driving at post assessment
2. Beck et al. (2007), USA	6 patients experiencing post-traumatic stress disorder as a result of a serious motor accident	An SGI Origin 3400 & a Moog 6-degree of freedom motion base (fitted with a steering wheel, gas pedal & a brake pedal) were used for the computational & graphics system. Simulated scenes were projected on a 10ft by 8 ft screen	All patients completed 10 virtual reality exposure therapy (VRET) sessions. Patients were assessed pre-and post-treatment (1 month after the last therapy session)	• Significant reductions in PTSD symptoms • High treatment satisfaction reported by patients
3. Akinwuntan et al. (2005) & Devos et al. (2009), Belgium	Post-stroke patients ($N = 83$), < 75 yrs	STISIM Drive system (v 1.03), 25-minute scenario, visual angle of 45-degree	Experimental: $n = 42$, Control: $n = 41$ / 5 week, 15 hour training	Pre- versus post-training differences between groups: • Experimental group significantly better on road sign recognition • Most patients who improved from "unfit to drive" to "fit to drive" were from experimental group • 73% experimental patients resumed driving versus 42% control patients
4. Mazer et al. (2005), Canada	42 patients from a rehabilitative center with TBI	(Information not available)	Patients randomized to experimental group (simulator training twice a week for 8 weeks) or control group (no intervention)	Improvement in passing on-road evaluations for the experimental group compared to the control participants (57% versus 29%)
5. Tomasevic et al. (2000), Australia	A patient reporting a driving-related phobia	Simulator consisted of 2 computer systems, a 3D audio system, a projection system (with a 180-degree field of view forward & a 60-degree field of view to the rear) & a vehicle control system	Patient was systematically de-sensitized (over a broken 3-hour period) using 11 driving scenarios that ranged from least anxiety-provoking to the most anxiety-provoking (for the patient)	The patient was effectively treated and had maintained treatment effects at an 11 month follow-up assessment

with a possibility of overtaking other cars) before terminating on a two-lane road in a rural setting. Participants in the experimental group were first given an opportunity to become acquainted with the simulator (over 2–3 hours in a week) following which they took part in a pre-training assessment where they drove the 13.5-km scenario on the simulator and were given feedback on their performance. The actual training sessions involved participants driving through various 5-km training scenarios with common traffic demands and complexities (e.g., speed control, unexpected multitasking situations). Finally, in the post-training assessment (fifth week of simulator training), all participants drove through the 13.5 km driving scenario again on consecutive trials. These trials, however, were designed so that participants could not predict their driving tasks on a particular trial based on their experience of a previous trial.

  b. The control group on the other hand, as part of their extended driving training, received only standardized training, which included performing driving-related cognitive tasks (e.g., route-finding on a paper map, recognition of road and traffic signs using cards with different traffic situation).

Patients from both the experimental and the control groups underwent visual and neuropsychological evaluations along with an on-road driving examination both before and after their respective training assessments. At the end of their on-road test, participants were placed in one of three categories which constituted the primary outcome measures at post-training: (1) fit to drive, (2) temporarily unfit to drive, or (3) unfit to drive. After completing their post-training assessments, participants were encouraged to take an official driving assessment examination (after first six months of having a stroke) with the Department of Belgian Road Safety Institute in Brussels. Primary outcome measures at follow-up constituted the results from the on-road test at post-training (i.e., "fit to drive", "temporarily unfit to drive" or "unfit to drive") and the outcome of the official driving assessment.

Results showed that both the experimental and the control group had improved on the visual and the neuropsychological tests after completing their training assessments. No significant differences were revealed between the two groups from pre- to post-training except on the road sign recognition test where the simulator-trained participants performed significantly better than the controls ($p = 0.007$). Concerning the on-road driving test, more participants from the experimental group progressed from the "unfit to drive" category to the "fit to drive" category compared to the controls.

At follow-up again, results were in favor of the simulator-trained group as significantly more experimental patients (73%) could legally resume driving (as per the official driving assessment) compared to only 42% of the control patients ($p < 0.05$). Compared to the controls, the experimental participants also performed significantly better on (1) anticipation and perception of road signs ($p = 0.001$), (2) visual behavior and communication ($p = 0.002$), (3) quality of traffic participation ($p < 0.05$), and (4) turning left ($p < 0.05$) at the follow-up assessment (Devos et al.,

2009). It is encouraging to note that the effects of driving training which had been completed at 11–15 weeks post-stroke were sustained at the follow-up assessment which was held at 24–36 weeks post-stroke. This opens new doors for designing effective driving training interventions using simulators for other conditions that may neurologically compromise the individual, and for ensuring that they are introduced in the patients' rehabilitation programs at an appropriate time during their recovery for maximum benefit.

32.3.2 Patients With Traumatic Brain Injury

Mazer et al. (2005) reported on a Canadian study evaluating the effectiveness of VRDS rehabilitation employing a randomized controlled trial (Table 32.2). Forty-two residents in a rehabilitation center with non-degenerative acquired TBI who did not successfully pass a functional driving evaluation were randomized to either the simulator retraining group or a control group. The experimental group received an intervention using a driving simulator two times a week for eight weeks. The intervention approach was standardized. Those in the control group did not receive any additional intervention. Overall results indicated an improvement in the rate of passing the on-road evaluation (57% versus 29%) after simulator training.

We recently conducted a pilot study with 11 military personnel recovering from mild or moderate TBI. Participants were randomly assigned to the control group involving residential rehabilitation ($n = 5$), or to the residential rehabilitation plus virtual reality driving simulation rehabilitation training group ($n = 6$). All participants underwent pre- and post-assessments that included simulator driving and completing questionnaires assessing road rage and risky driving. Between assessments, experimental participants received four–six rehabilitation training sessions each lasting between 60–90 minutes. The training sessions involved practicing progressively more complex driving skills (lane position, speed control, turn signal usage, etc.) through increasingly more demanding traffic conditions. Virtual reality training was well received, considered realistic and effective by trainees, with no reported simulation sickness. Driving performance improved significantly in the experimental group only ($p < 0.01$). Experimental participants also demonstrated a significant reduction in road rage ($p = 0.01$). A similar trend was noticed for risky driving ($p = 0.06$) at post-assessment in this group (Cox, Singh, & Cox, 2010).

32.3.3 Patients With Driving-Related Phobias

Driving rehabilitation as delivered by a VRDS is specially suited to patients who have specific driving-related phobias. Virtual reality can help generate driving scenarios based on real-life driving situations that can be tailored to a patient's needs and presented systematically (i.e., progressively invoking fear) for a controlled period as part of patient's treatment (e.g., exposure therapy) (Wald & Taylor, 2001). Tomasevic, Regan, Duncan and Desland (2000) presented a case study wherein a patient who had an extreme fear of overtaking trucks while driving could successfully resume normal driving after receiving a broken three-hour period exposure therapy session on a VRDS. The session included 11 driving scenarios ranging from the least anxiety-provoking to the most anxiety-provoking situation for the patient. Follow-up assessments showed that therapy effects had remarkably been maintained even at 11 months post-session. Similarly, the application of VRDS to treatment of patients diagnosed with post-traumatic stress disorder (PTSD) related to a motor vehicle accident has produced encouraging results. Beck, Palyo, Winer, Schwagler and Ang (2007) used a VRDS to administer individual exposure therapy sessions (*n* = 10) to six participants who had been diagnosed with motor vehicle related PTSD. Therapy sessions constituted a mix of four driving scenes (highway, urban, suburban, and rural) that employed varying driving environments (e.g., amount of traffic, changes in the weather, etc.). Patients were encouraged to continue with a specific virtual driving environment until their anxiety had been reduced by 50% on the subjective units of distress (SUDS) scale. Overall, patients reported high treatment satisfaction with the exposure therapy as delivered by using virtual reality and clinically significant improvements were seen in PTSD symptoms (such as reduction in avoidance and emotional numbing) one month post-treatment. Due to the lack of any long-term follow-up data in this study, the extent to which these therapy effects can be maintained over time can only be speculated over. Future studies will understandably be expected to provide not only impressive post-treatment results but also offer evidence in favor of long-term effectiveness of virtual reality led driving therapies before they are established as effective forms of treatment for people with driving-related phobias. It has been argued that virtual reality exposure therapy may prove to be more effective and show longer-sustaining effects if it is paired with real-life driving at an appropriate time during treatment (Wald, 2004).

32.3.4 Individual Factors Affecting Rehabilitation's Effectiveness

Akinwuntan and associates (2005) usefully note that a few individual characteristics may make patients more receptive to simulator-based training which consequently helps their therapy adherence and ability to resume real-life driving. In their study, patients with high academic qualifications, a limited level of disability and a left-sided brain lesion compared to those with a lesion on the right side of their brain showed greater improvement in their driving skills within the experimental group. Interestingly, "level of education" was not related to improvements in driving performance of the control group. Furthermore, it is likely that patients who are very motivated to resume driving and "get their life back" experience greater benefits from the simulator training programs compared to their less hopeful counterparts.

Other than individual factors that could define the value of VRDS as a rehabilitation tool, the usability factor of VRDS and its effect on the user should also be considered while evaluating the success or failure of re-training results. Usability has been defined as, "the quality of a tool that makes it easy to learn, easy to use, easy to remember, makes it error tolerant and subjectively pleasing" (Schultheis, Rebimbas, Mourant, &

Millis, 2007, page 2). A VRDS which has a low-usability factor and does not address other provider and user perspectives (e.g., not being cost-effective, has considerable side effects) would probably be less likely to establish itself as an effective tool for delivering rehabilitation. Age (of the patient/user), however, is an important moderator that could potentially influence the usability factor of a VR-based driving simulator and consequently the training process (Schultheis, Rebimbas, Mourant, & Millis, 2007). Younger people may be more accommodating and enthusiastic regarding new information technologies (such as virtual reality) and may feel more comfortable using the VRDS for rehabilitation purposes compared to their older counterparts who may not have had much experience with this technology (Morris & Venkatesh, 2000).

In addition to usability, VRDS must be standardized in terms of type of equipment used (e.g., 40 versus 200-degree field of view, use of rear-view mirrors), driving scenarios (e.g., length of drive, urban, rural highway roads, number and type of driving challenges and required maneuvers), and performance variables (e.g., quantification of speed, brake and steering control). To paraphrase Henry Ford, you can get a model T in any color you like, as long as it is black. Similarly, driving science will not progress until we use a common instrument to quantify and train driving performance. Once an agreed upon metric is accepted, then normative data needs to be generated against which an individual patient can be compared to quantify his/her particular deficit. However, the specific elements of this normative data that define safe driving from which one predicts future accidents will also have to be identified. In addition to having standardization that discriminates high-risk individuals, this system must also be relatively inexpensive. Again, taking a cue from Henry Ford, like the Model T that standardized routine driving, the Model T simulator must be mass-produced in order to lower costs and encourage widespread use. Of course, broad acceptance of such standardization will be very controversial, since acceptance of one system will mean rejection of many other VRDS systems, as is the case in all industries.

32.3.5 Practical Concerns

Use of VRDS in rehabilitation medicine is still a developing field. Despite evidence of significant advantages that VRDS enjoys over other assessment and training tools (Rizzo, Schultheis, Kerns, & Mateer, 2004), we were able to identify only three randomized control studies that have actually evaluated its rehabilitative element in the area of driving. Although there are various rehabilitation units and occupational therapy centers across the USA and internationally that impart driving rehabilitation services using VRDS (e.g., Palo Alto Polytrauma Rehabilitation Center [USA], Sister Kenny Rehabilitation Institute [Minneapolis, USA], St. Cloud Hospital [Minnesota, USA]), they may not necessarily be actively involved in publishing their driving rehabilitation research. Researchers have recognized a few challenging concerns which would need to be addressed to strengthen the case to be made for employing VRDS in helping people to resume driving. Setting-up and maintaining a sophisticated and an effective VRDS could be a time-consuming and an expensive process (Rizzo & Kim, 2005). This may put off researchers from exploring the VRDS route for rehabilitation or they may decide to compromise on the materials and programs used as part of the virtual reality simulator, which could affect the validity of results. Additionally, the problem of side effects of using VRDS would need to be addressed to increase its usability for both patients as well as providers (Schultheis, 2005).

Simulator Adaptation Syndrome (SAS) is one of the leading methodological problems in driving simulation research. It refers to feelings of nausea, disorientation, headache, and problems with focusing that are sometimes experienced while operating the simulator and sometimes even several hours after participation. Although, these symptoms could vary in intensity between individuals, for a few people they could be severe enough to significantly impair their driving performance on the simulator, which would result in their driving data being dropped from analyses. Our research has demonstrated that among senior drivers, of ages above 60 years, SAS is more common among women and those who report a history of motion and/or sea-sickness (Cox, Singh, Guerrier, & George, unpublished manuscript). Furthermore, use of the ReliefBand®, which provides mild electrical stimulation to the radial nerve at the wrist, can significantly reduce SAS symptoms as compared to a placebo ($p = 0.003$) (Cox et al., 2010).

It remains unclear why SAS actually occurs and why only a few people and not others experience it (Stanney, Mourant, & Kennedy, 1998). However, considering the adverse effects of SAS, it is recommended that participants both be screened for their susceptibility to SAS—for example, by asking participants (before they initiate their simulator trial) if they have any history of motion sickness, nausea, or headache while driving, and so forth; and that participants be also monitored during their performance on the driving simulator at various intervals—for example, have examinees rate their SAS on a 0 ("I feel fine") to 4 ("I feel so bad that I have to stop") scale every five minutes throughout the simulated drive. These efforts would be useful not only to protect individuals from the adverse effects associated with SAS but also because experiencing SAS could reduce or negate the potential rehabilitation effects that would be offered by VRDS under normal circumstances. (Readers may also want to look this book, chap. 14 by Stoner, Fisher, & Mollenhauer which directly addresses SAS.)

32.3.6 Process of Assessment and Rehabilitation

In Figure 32.1, we depict a possible assessment and rehabilitation process that may be feasible in the future for patients using a VRDS. A VRDS could assess both general and specific driving

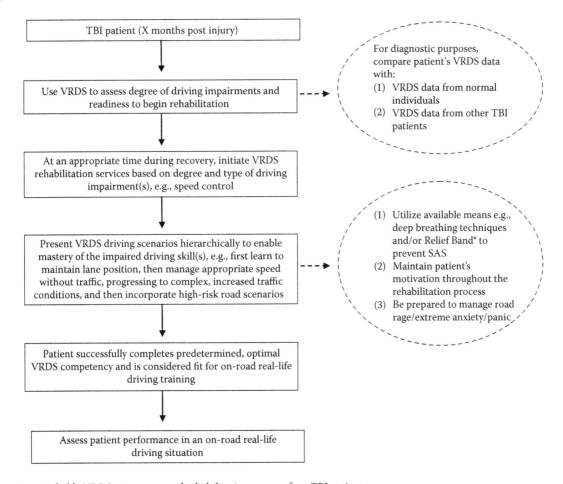

FIGURE 32.1 A probable VRDS assessment and rehabilitation process for a TBI patient.

impairments. This information could then be evaluated in relation to normative data, which would help achieve a much more comprehensive understanding of a patient's driving deficits. At an appropriate time during the patient's recovery period, she/he can be introduced to a driving rehabilitation process offered using a VRDS. The VRDS will offer a safe environment for the patient to practice and master his or her impaired driving skills while being evaluated objectively by clinicians. During this stage, clinicians will make efforts to protect the patient against possible side effects of the VRDS and to ensure that the patient's motivation and interest in the rehabilitation process are maintained (e.g., using NASCAR driving scenarios). Once patients have achieved the pre-determined rehabilitation goal using the VRDS, they can be introduced to an on-road driving situation, which could serve as their ultimate test before they resume driving.

32.4 Conclusion and Way Forward

It would be reasonable to expect that with further research and development in the area of virtual reality and its application to driving simulation, VRDS would be more widely-recognized as a suitable tool to deliver driving assessment and rehabilitation. Strong multi-disciplinary collaborations (e.g., VR specialists, engineers, healthcare professionals) targeted towards developing VR technology in driving rehabilitation may both help increase our general understanding of its potential and help overcome the few limitations (including cost reductions) that have been highlighted for this field, making the technology more sensitive and acceptable to insurance providers as well as users.

Key Points

- Neuropsychological tests, although they offer a good indication of the relationship between cognitive impairments and their impact on functional performance, fall short of providing a comprehensive prediction of driving performance in people who have neurological impairments.
- The use of virtual reality driving simulators (VRDS) in evaluating driving performance in clinical and non-clinical populations offers opportunities to give individuals the 'real experience' of driving through space and time while negotiating simple and complex road situations in a standardized and safe setting for objective monitoring.
- VRDS has proven to be an effective assessment tool that can be used to provide a detailed evaluation of an individual's general and specific driving skills in both clinical and healthy populations.
- VRDS is especially suitable as a driving rehabilitation tool as it offers the opportunity to generate, manipulate, and present real-life driving scenarios to target individual needs in a safe and objective setting without having to wait for an appropriate driving situation in real life, which could potentially involve an impractical amount of

resources in terms of time and finances as well as compromise the safety of both patients and their evaluators.
- It is important to note that like most assessment and treatment procedures, the use of VRDS may not be suitable for everyone. Various individual factors (e.g., age, previous experience with virtual reality environments, severity of the clinical condition, susceptibility to experience simulation adaptation syndrome) have been recognized that can potentially limit the effectiveness of VRDS for certain people.

Keywords: Driving Assessment, Driving Performance, Driving Rehabilitation, Simulation Adaptation Syndrome, Virtual Reality Driving Simulator

Glossary

Concurrent validity: Discriminative validity, driving rehabilitation, neuropsychological tests.

Concurrent validity: The ability of a measure to vary directly with a measure of the same construct or indirectly with a measure of an opposite construct. It shows that a test is valid by comparing it with an already valid test.

Discriminative validity: The ability of a measure to adequately differentiate or not differentiate between groups that should differ or not differ based on theoretical reasons or previous research.

Driving quotient (DQ): A measure of a person's driving performance as indicated by a driving test on a virtual reality driving simulator.

Driving rehabilitation: The process of restoring an individual's driving abilities to their former capacity.

Neuropsychological tests: Testing techniques through which neuropsychologists can acquire data about an individual's cognitive, motor, behavioral, language, and executive functioning.

Key Readings

Akinwuntan, A. E., Weerdt, W. D., Feys, H., Pauwels, J., Baten, G., Arno, P., et al. (2005). Effect of simulator training on driving after stroke: A randomized controlled trial. *Neurology, 65*, 843.

Lew, H. L., Poole, J. H., Lee, E. H., Jaffee, D. L., Huang, H. C., & Brodd, E. (2005). Predictive validity of driving-simulator assessments following traumatic brain injury: A preliminary study. *Brain Injury, 19*(3), 177.

Rizzo, A. S., & Kim, G. J. (2005). A SWOT analysis of the field of VR rehabilitation and therapy. *Presence, 14*(2), 119.

Schultheis, M. T., & Mourant, R. R. (2001). Virtual reality and driving: The road to better assessments for cognitively impaired populations. *Presence, 10*(4), 431.

Schultheis, M. T., Rebimbas, J., Mourant, R., & Millis, S. R. (2007). Examining the usability of a virtual reality driving simulator. *Assistive Technology, 19*, 1.

References

Akinwuntan, A. E., Weerdt, W. D., Feys, H., Pauwels, J., Baten, G., Arno, P., et al. (2005). Effect of simulator training on driving after stroke: A randomized controlled trial. *Neurology, 65,* 843–850.

Beck, J. G., Palyo, S. A., Winer, E. H., Schwagler, B. E., & Ang, E. J. (2007). Virtual reality exposure therapy for PTSD symptoms after a road accident: An uncontrolled case series. *Behavior Therapy, 38,* 39–48.

Biederman, J., Fried, R., Monuteaux, M. C., Reimer, B., Coughlin, J. F., Surman, C. B., . . . Faraone, S. V. (2007). A laboratory driving simulation for assessment of driving behavior in adults with ADHD: a controlled study. *Annals of General Psychiatry, 6*(4).

Brouwer, W. H., Withaar, F. K., Tant, M. L. M., & van Zomeren, A. H. (2002). Attention and driving in traumatic brain injury: A question of coping with time-pressure. *Journal of Head Rehabilitation, 17*(1), 1–15.

Cox, D. J., Davies, M., Singh, H., Barber, B., Nidiffer, F. D., Trudel, T., . . . Moncrief, R. (2010). Driving rehabilitation for military personnel recovering from traumatic brain injury using virtual reality driving simulation: A feasibility study. *Military Medicine, 175*(6), 411–416.

Cox, D. J., Gonder-Frederick, L. A., Kovatchev, B., & Clarke, W. L. (2001). Self-treatment of hypoglycemia while driving. *Diabetes Research and Clinical Practice, 54*(1), 17–27.

Cox, D. J., Punja, M., Powers, K., Merkel, R. L., Burket, R., Moore, M., . . . Kovatchev, B. (2006). Manual transmission enhances attention and driving performance of ADHD adolescent males: Pilot study. *Journal of Attention Disorders, 10*(2), 212–216.

Cox, D. J., Singh, H., & Cox, D. M. (2010, October 19). *Effectiveness of acupressure and acustimulation in minimizing driving simulation adaptation syndrome.* Poster presented at the Association for the Advancement of Automotive Medicine Annual Scienticfic Conference, Las Vegas, USA.

Cox, D. J., Singh, H., Guerrier, J., & George, C. (2008). *Driving simulator induced sickness in older adults.* Unpublished manuscript.

DeCarlo, D. K., Scilley, K., Wells, J., & Owsley, C. (2003). Driving habits and health-related quality of life in patients with age-related maculopathy. *Optometry and Vision Science, 80*(3), 207–213.

Devos, H., Akinwuntan, A. E., Nieuwboer, A., Tant, M., Truijen, S., De Wit, L., . . . Weerdt, W. D. (2009). Comparison of the effect of two driving retraining programs on on-road performance after stroke. *Neurorehabilitation and Neural Repair, 23*(7), 699–705. doi: 10.1177/1545968309334208

Fonda, S. J., Wallace, R. B., & Herzog, A. R. (2001). Changes in driving patterns and worsening depressive symptoms among older adults. *The Journals of Gerontology Series B: Psychological Sciences and Social Sciences, 56,* S343–S351.

Fox, G. K., Bowden, S. C., & Smith, D. S. (1998). On-road assessment of driving competence after brain impairment: Review of current practice and recommendations for a standardized examination. *Archives of Physical Medicine and Rehabilitation, 79,* 1288–1296.

Hopewell, A. (2002). Driving assessment issues for practicing clinicians. *Journal of Head Rehabilitation, 17*(1), 48–61.

Korteling, J. E., & Kaptein, N. A. (1996). Neuropsychological driving fitness tests for brain-damaged subjects. *Archives of Physical Medicine and Rehabilitation, 77,* 138–146.

Kotterba, S., Mueller, N., Leidag, M., Widdig, W., Rasche, K., Malin, J. P. et al. (2004). Comparison of driving simulator performance and neuropsychological testing in narcolepsy. *Clinical Neurology and Neurosurgery, 106,* 275–289.

Ku, J. H., Jang, D. P., Lee, B. M., Lee, J. H., Kim, Y., & Kim, S. I. (2002). Development and validation of virtual driving simulator for the spinal injury patient. *Cyberpsychology & Behavior, 5*(2), 151–156.

Lengenfelder, J., Schultheis, M. T., Al-Shihabi, T., Mourant, R., & DeLuca, J. (2002). Divided attention and driving: A pilot study using virtual reality technology. *Journal of Head Rehabilitation, 17*(1), 26–37.

Lew, H. L., Poole, J. H., Lee, E. H., Jaffee, D. L., Huang, H. C., & Brodd, E. (2005). Predictive validity of driving-simulator assessments following traumatic brain injury: A preliminary study. *Brain Injury, 19*(3), 177–188.

Liu, L., Miyazaki, M., & Watson, B. (1999). Norms and validity of the "DriVR" — A virtual reality driving assessment for persons with head injury. *Cyberpsychology & Behavior, 2*(1), 53–67.

Mazer, B., Gelinas, I., Vanier, M., Duquette, J., Rainville, C., & Hanley, J. (2005). Effectiveness of retraining using a driving simulator on the driving performance of clients with a neurologic impairment [Abstract]. *Archives of Physical Medicine and Rehabilitation, 86,* E20.

Morris, M. G. & Venkatesh, V. (2000). Age differences in technology adoption decision: Implications for a changing work force. *Personnel Psychology, 53,* 375–403.

Muhlberger, A., Bulthoff, H.H., Wiedemann, G., & Pauli, P. (2007). Virtual Reality for the psychophysiological assessment of phobic fear: Responses during virtual tunnel driving. *Psychological Assessment, 19*(3), 340–346.

Rizzo, A. S. & Kim, G. J. (2005). A SWOT analysis of the field of VR rehabilitation and therapy. *Presence: Teleoperators & Virtual Environments, 14*(2), 119–146.

Rizzo, A. S., Schultheis, M. T., Kerns, K. A., & Mateer, C. (2004). Analysis of assets for virtual reality applications in neuropsychology. *Neuropsychological Rehabilitation, 14*(1/2), 207–239.

Schultheis, M. T. (2005). "Virtual driving" versus "real driving": Pros and cons of using a VR driving simulator in rehabilitation. In K. Stanney (Ed.), *Advances in virtual environments technology: Musings on design, evaluation, & applications.* Mahwah, NJ: Lawrence Erlbaum Associates.

Schultheis, M. T., & Mourant, R. R. (2001). Virtual reality and driving: The road to better assessments for cognitively impaired populations. *Presence: Teleoperators & Virtual Environments, 10*(4), 431–439.

Schultheis, M. T., Rebimbas, J., Mourant, R., & Millis, S.R. (2007). Examining the usability of a virtual reality driving simulator. *Assistive Technology, 19,* 1–8.

Schultheis, M. T., Simone, L. K., Roseman, E., Nead, R., Rebimas, J., & Mourant, R. (2006). *Stopping behavior in a VR driving simulator: A new clinical measure for the assessment of driving.* Proceedings of the 28th IEEE EMBS Annual International Conference, New York.

Stanney, K. M., Mourant, R. R., & Kennedy, R. S. (1998). Human factors issues in virtual environments: A review of the literature. *Presence: Teleoperators & Virtual Environment, 7*(4), 327–351

Thacker, P. D. (2003). Fake worlds offer real medicine: Virtual reality finding a role in treatment and training. *JAMA, 290*(16), 2107–2112.

Tomasevic, N., Regan, M.A., Duncan, C.C., & Desland, M.L. (2000, February 29). *Use of advanced simulation to treat a driving-related phobia.* Paper presented at the SimTecT Conference and Exhibition. Sydney, Australia.

Wald, J. (2004). Efficacy of virtual reality exposure therapy for driving phobia: A multiple baseline across-subjects design. *Behavior Therapy, 35,* 621–635.

Wald, J. L., Liu, L., & Reil, S. (2000). Concurrent validity of a virtual reality driving assessment for persons with brain injury. *Cyberpsychology & Behavior, 3*(4), 643–654.

Wald, J. & Taylor, S. (2001). Efficacy of virtual reality exposure therapy to treat driving phobia: A case report. *Journal of Behavior Therapy and Experimental Psychiatry, 31,* 249–257.

Wilson, B. (1993). Ecological validity of neuropsychological assessment: Do neuropsychological indexes predict performance in everyday activities? *Applied & Preventive Psychology, 2*(4), 209–215.

V

Applications in Engineering

33

The Importance of Proper Roadway Design in Virtual Environments

Douglas F. Evans
DriveSafety, Inc.

Abstract

The Problem. An important factor in the usefulness of any given driving simulation is the fidelity of the roadway models presented in the virtual environment. The effects of a roadway's design on vehicle performance and driver comfort and safety are dramatic on both real and simulated roads. Simulators that do not provide adequate roadway model fidelity may yield driving experiences that do not correspond well to the real world. *Driving Simulators: Software Specificity and Equipment Limitations.* There are practical limitations to how accurately real-world roads can be represented in a simulator. These are discussed, along with some strategies that are used to optimize simulated roadways. *Driving Simulators: Increasing Fidelity.* A brief history of the development of driving simulator roadway modeling tools is presented, including a discussion of four types of tools that have emerged. Finally, we discuss underlying roadway design standards. Real-world roadways are designed and constructed according to rigorous guidelines developed over decades by transportation officials and highway engineers seeking to optimize drivability, comfort, and safety. Compliance with these standards requires care by developers in at least four principal areas. These are (1) roadway classification, (2) junction design, (3) curve design, and (4) traffic control devices. Key concepts in each of these areas are presented, not in an attempt to provide a comprehensive design methodology, which is better done by rigorous extant policy and engineering texts, but only to make simulator developers and users aware of the key issues involved.

33.1 Introduction

An important factor in the usefulness of any given driving simulation is the fidelity of the roadway models used in the virtual environment. Real-world roadways are designed and constructed according to rigorous guidelines developed over decades by transportation officials and highway engineers seeking to optimize drivability, comfort and safety (Barnett, 1940a, b; American Association of State Highway and Transportation Officials, 2004). There is a lot that goes into a well-designed road. For example, real roads employ complex 3D pavement surfaces

effecting alignment changes through vertical and horizontal curves and transitions, as well as controlling lane widths and cross-sections. The design of these surfaces, along with roadway markings, signs and traffic control devices are all regulated by guidelines and standards (Federal Highway Administration, 2009).

The effects of a roadway's design on vehicle performance and driver comfort and safety are dramatic on both real and simulated roads. They govern safe travel speeds, sight-distances on curves and hills, steering side-forces required to negotiate curves, time and space allowed to accomplish merges and weaves, and more

(Ogden, 1996). Driving simulations that do not model the roadways with adequate fidelity may yield driving experiences that do not correspond well to the real world.

Common pitfalls in simulations that have unrealistic models include traffic lanes that are either too narrow or too wide, either of which can have untoward effects on lane-tracking performance measures such as edge crossings and standard deviation of lane position. Other problems observed by the author in some simulations are intersection corners that are unrealistically sharp, making turn navigation awkward or even impossible at such junctions. Real urban intersection "corners" actually have fairly generous radii, often 10 meters or more, to facilitate safe, comfortable navigation without encroachment into adjacent lanes (Federal Highway Administration, 2009). Yet some simulators present intersection corners with tiny radii that are completely unrealistic. Also of concern are freeway entrance and exit ramps that are too short, as well as horizontal curves that are flat (lacking appropriate superelevation.) Such model failings make it much more difficult to drive the resulting roads and compromise correlation to real driving experiences.

In the United States, roadways are generally designed according to a publication of the American Association of State Highway and Transportation Officials (AASHTO) called "A Policy on Geometric Design of Highways and Streets." This thick green book is euphemistically known as the "Green Book" by highway engineers (AASHTO, 2004). Signage, traffic control devices and roadway markings on U.S. roads are regulated by the Manual on Uniform Traffic Control Devices for Streets and Highways (MUTCD: Federal Highway Administration, 2009). Similar guidelines are used in other countries, and there are sometimes significant differences between countries, such as which side of the road vehicles travel on; while others are more subtle, such as the details of the signage and the road markings that are used. All of these are important and care should be given that simulator content is appropriate for the region in which it is to be used.

33.2 Practical Limitations to Realism

There are practical limitations, of course, to how accurately real-world roads can be represented in a simulator. These limits arise principally from cost and performance constraints on simulator computation and cueing systems. For example, the geometric detail given to the representation of roadway surfaces is limited by the rendering capacity of the simulator's graphics subsystems. Increasingly complex models may also carry higher development costs. Those who design simulator applications or experiments need to ensure the models used provide sufficient fidelity for the needs of their applications. Generally, roadway model geometry, appearance and function should be consistent with applicable highway design regulations.

Graphical models built for realtime simulators are quite different from those built for highway design or visualization, where it does not much matter how long it takes to render images of the model. Highway design CAD models may employ hundreds of thousands of polygons to represent the geometry of a

relatively short section of highway or even a single intersection. Realtime models built for simulators do not have the luxury of such data density. In a 60 Hz simulator, complete updated renderings of the visual environment are required every 16 milliseconds. Even with the great advances made in graphics hardware performance, there are limits to the amount of visual detail that can be processed at such rates. Hence, a number of visual effects and load-management strategies are employed by simulation system developers to create realistic-looking scenes that can be rendered efficiently. These include the use of simplified models with range-dependent levels of detail, photo and algorithmic textures with MIP-mapping, backdrops and facades, and other such devices.

While such strategies have clearly elevated the quality of visual scenes that can be rendered in realtime, there are challenges and artifacts that can show up as well. A classic example of this is the "popping" in and out of environment objects on the horizon or in the distance as a driver proceeds down a roadway. This can result from over-aggressive level-of-detail range settings, rendering system overloads, or some systems' lack of "fade LOD" capabilities, or combinations thereof. Another problem exhibited by some simulators has been the blurring of roadway stripes caused by sub-optimal texture MIP-Mapping techniques. Visual database development and rendering strategies are truly something of an art, and care must be given by simulator manufacturers to achieve attractive, consistent visual renderings.

Note that the question about realism goes much deeper than we can discuss here. That is, it goes beyond issues of attractive visual renderings which are consistent with existing guidelines. In essence, the goal in simulator studies is to obtain the same behavior in the simulator as one would in the real world. As many chapters in this *Handbook* indicate, most simulators cannot in principle provide drivers with fully realistic visual and physical cues, even when the roadway design in the simulator is a true one (e.g., see chap. 7 by Greenberg & Blommer, "Physical Fidelity of Driving Simulators", chap. 8 by Andersen, "Sensory and Perceptual Factors in the Design of Driving Simulation Displays", and chap. 36 by Chrysler & Nelson, "Design and Evaluation of Signs and Pavement Markings Using Driving Simulators"). This means that, on occasion, one may actually need to construct the simulated roads, signs, signals and pavement markings in ways that do not comply with current design standards in order to provide drivers with the cues that are comparable to those they would receive were they driving on real roads. An example of this is the need in many simulators (because of their limited display resolutions) to present oversized roadside signage, so that the sign detection and legibility distances for drivers are similar to those on real roads.

33.3 Roadway Modeling Tools

When early driving simulators were developed in the 1980s and 1990s, there were no commercially available tools for creating realtime models of roadways. Modeling tool development efforts over several years by independent parties led to a number of tools in the industry for making simulator roadway models (see also, this book, chap. 6 by Kearney & Grechkin, "Scenario Authoring"). Some of the earliest work was done by simulation engineers at Daimler-Benz, BMW, Toyota and at the University of Iowa. The Iowa work was adopted by MultiGen for use in its Creator software as the MultiGen RoadTools—the first commercially available solution for creating high-fidelity roadway models specifically for realtime simulation. The RoadTools excelled at building roadways with all kinds of cross-sections and curves and hills, but did not automate construction of roadway junctions, which had to be modeled "by hand." This would involve the low-level creation of geometry for intersection curbs, islands, pavement markings; and most notably, the definition of the logical roadway network connectivity and attribution within intersections for use in traffic simulation. Subsequent developments by several groups have made the modeling of whole road networks, including intersections, much easier.

Most currently available simulator roadway modeling tools can be broken down into four principal types, outlined in the following sections.

33.3.1 Generic Low-Level Database Creation Tools

There are tools such as MultiGen (Presagis) Creator and its RoadTools (http://www.presagis.com/products/database/creator) that allow simulation engineers to create driving environments from scratch using an interactive graphical modeling system. Models are built based on source data such as maps, digital terrain and roadway data, roadway design parameters, digital photos and video sources, possibly including satellite imagery, and so forth. The developer must also create all required correlated datasets and build the models in compliance with the system data requirements of the target simulation system, which can be a complex task. Driving simulators often have three or more correlated models of the driving environment, only one of which is the visual database rendered in the graphics system. The other correlated models or datasets represent other non-visual aspects of the environment that are important to other simulator sub-systems. These typically include: 1) analytic surface models used by the tire-dynamics model as the vehicle traverses drivable surfaces; and 2) logical roadway network models used in traffic simulation and scenario control. Low-level database creation tools give maximum flexibility to build any kind of model desired but can be demanding and are generally used by simulation system developers, as opposed to simulator users or customers.

33.3.2 Script-Based Linear Progression Models

This type of proprietary solution lends itself well to research or applications on a specific target simulator and where the desired simulation experience is best described as a sequence of roadway elements and traffic events. An excellent example of such a modeling solution is the Scenario Definition Language

(SDL) developed by Systems Technology of Hawthorne, CA, for use on its STISIM products (Allen et al., 2001) (http://www.stisimdrive.com).

33.3.3 Tile-Based Authoring Tools

This type of proprietary solution allows users to drag and drop database "tiles" (pre-defined driving database regions) into their simulated world to assemble a roadway network and its surrounding environs. This can be a powerful solution for users who do not require geo-specific roadway models, but want to be able to quickly build highly customized roadway networks and environment models. The first such tile-based modeling tools were created at the University of Iowa in the early 1990s in support of the Iowa Driving Simulator. Solutions of this type include the NADS modeling suite (Tile Mosaic Tool, Interactive Scenario Authoring Tool) (National Advanced Driving Simulator, 2010); the SimVista authoring tools from Realtime Technologies (http://www.simcreator.com/simvista/simvista.htm), and the HyperDrive authoring tools from DriveSafety (http://www.drivesafety.com/Products/Software/HyperDrive.aspx).

33.3.4 Geo-Specific Modeling Tools

This type of modeling solution allows for the importing of digital terrain and roadway data from real-world locations to support the automatic or semi-automatic generation of simulation models of specific places and real road networks. This type of tool has been used for many years in urban planning and transportation project visualization applications that do not require the specialized database structures and performance optimization methods required by realtime interactive visual simulation. There have been, however, a number of simulator developers who have made powerful geo-specific modeling tools, and this remains a very active development area in the field. One of the earliest solutions of this type was developed in the mid 1990s by Bodo Randt of the German simulation company STN-Atlas in Bremen (START: Randt, 1996). His modeling system not only supported automated generation of extensive real-world road networks, but excelled even at representations of the complex irregular intersections typical of European cities. Similar work has been done by French developers at Renault and INRETS simulation facilities, and others. The SCANeR Evariste software (http://www.scanersimulation.com) is a good example. There remains a need, in the U.S. especially, for the development of robust geo-specific modeling tools that support the automated creation of high-quality realtime models from imported GIS and CAD data models of existing roadways. Most tile-based modeling tools, discussed in the previous section, allow some level of import of geometry from GIS and CAD models, but the challenge is to convert such data-intensive engineering models into efficient realtime rendering models. Automatic geometry decimation and level-of-detail methods continue to be areas of research to better support such import.

33.4 Modeling Guidelines

Building realistic roadway models from scratch requires essentially the same skill set and reference materials as highway design engineering. Fortunately for those lacking that full skill set there are high level modeling tools available for many simulation systems, as previously outlined. For those who are required to actually design and construct realistic roadway models, it is strongly recommended to become acquainted with the AASHTO *Green Book* (2004) and the MUTCD guide (2009) for U.S. roadways, or the comparable regulations for other countries. This discussion does not seek to provide similar coverage, but only to make developers aware of the key issues involved.

Compliance with roadway design standards requires care by developers in at least four principal areas. These are: 1) Roadway Classifications, 2) Junction Design, 3) Curve Design, and 4) Traffic Control Devices. Key concepts in each of these areas are presented in the following sections.

33.4.1 Roadway Classifications

While there are many varieties of roads, most fall into a small set of types or functional classifications. For each of these there are useful standards of construction and geometric design that are helpful in assuring their safe and efficient function. In the U.S., the Federal Highway Administration (FHWA) defines three functional classifications: arterials, collectors and local roads. Some transportation organizations choose to define expressways also, as distinct from Arterials.

Specifically, the following definitions are used in the FWHA publication, *Flexibility for Highway Design* (2010):

> "Typically, travelers will use a combination of arterial, collector, and local roads for their trips. Each type of road has a specific purpose or function. Some provide land access to serve each end of the trip. Others provide travel mobility at varying levels, which is needed en route. There is a basic relationship between functionally classified highway systems in serving traffic mobility and land access…. Arterials provide a high level of mobility and a greater degree of access control, while local facilities provide a high level of access to adjacent properties but a low level of mobility. Collector roadways provide a balance between mobility and land access" chap. 3.

When modeling different roadway types it is important to use class-appropriate design speeds, lane widths, shoulder and median treatments, sight distances, markings and signage.

33.4.2 Junction Design

Junction design is defined broadly here to include the design of all types of connections between roadways. Junctions may be built between roads of similar or disparate functional classifications. For example, on an interstate freeway there may be

junctions with other freeways as well as with surface streets—or even an unmarked gravel road, such as at a rural ranch exit. Junctions should be designed to support the expected traffic type and volume for each roadway and to provide safe crossings and connections between them.

An unfortunate error that simulator developers have sometimes made is to ignore basic rules of intersection design either in ignorance or in attempting to simplify their models. An example of this is using unrealistically sharp corners at intersections, making it impossible to navigate a turn without encroaching into adjacent lanes. Another is using unrealistically short gore, merge or weave sections on highway interchange models. Road geometry and traffic channelization at junctions must provide adequate maneuvering room for the expected traffic speeds and vehicle types.

33.4.3 Curve Design

Curve Design refers to the geometric design standards governing horizontal and vertical curves on roadways. These standards have also sometimes been violated by developers of driving simulators to the detriment of their users. Examples would include curves that are too sharp for a given roadway's design speed and curves that have inadequate (or no) transition section or superelevation, making curve navigation difficult and unrealistic. Vehicle performance and driver safety are both dramatically affected by the geometry of roadway curves. In general, horizontal curves are not flat, but are superelevated (or banked) to assist in the generation of the steering side forces required to negotiate the curve at the design speed of the road. In most well-designed horizontal curves, both the degree of curvature and the superelevation of the road surface are introduced gradually for drivers as they come into the curve through what are called clothoid or Euler spiral transitions. This allows natural steering dynamics into and out of curves. Vertical curves (sags and crests) on roads are generally formed using equal-tangent parabolic profiles. Equal-tangent implies that the horizontal distance from the center of the vertical parabolic curve is equal in both directions. Many real-world roads have the added complexity of simultaneous horizontal and vertical curves. All of the key design parameters of roadway curves are governed by the design speed of the road. Speed dictates what are acceptable transition lengths, degrees of curvature and superelevation, crest and sag lengths (which effect daylight and night-time sight distances), and so on. Excellent guidelines for all of these parameters are provided in AASHTO (2004).

33.4.4 Traffic Control Devices

Traffic Control Devices are broadly defined here to include all roadway signs, markings and traffic signals. The design and use of traffic control devices are governed in the U.S. by the *Manual on Uniform Traffic Control Devices for Streets and Highways* (MUTCD) (2009). Most developed countries have analogous guidelines.

Care should be given by simulator developers to adhere to the appropriate guidelines for the driving environments they seek to represent. Some simulators provide their users with extensive libraries of standard signs, with varieties of regulation, warning, route guidance, other driver information and commercial signing. In the case of traffic safety research simulators, it is useful for the system to allow researchers access to and manipulation of traffic control device logic and timing parameters in order to allow experimentation.

A brief discussion of sign size in simulators is appropriate here. The MUTCD provides standard ranges of size for each type of official roadway sign. These ranges can vary depending on the functional classification and design speed of the road. Sign sizes are derived by recognition and legibility distances. In most driving simulators the display technologies used yield lower visual resolution than what drivers experience in the real world. Hence, sign recognition and legibility distances for a given sign size may be shorter in the simulator. To achieve similar sign recognition and legibility distances in the simulator, developers may need to use larger-sized signs than in the real world. How much larger the signs should be depends on the display solutions being used, including the per-channel rendered resolution and field of view (Carpenter, 2002; see this book, chap. 2 by Allen, Rosenthal, & Cook, "A Short History of Driving Simulation"). The effectiveness of the texturing methods for sign graphics can also have a significant effect on sign recognition and legibility.

33.5 Conclusion

A driving simulation's correspondence to real-world driving is significantly affected by the quality of the roadway models used in the virtual environment. Developers should give care to adhere to proper roadway design guidelines, including applicable standards for geometric design of curves and intersections, for roadway signs and markings and traffic control devices. Negligence in any of these areas could lead to untoward effects on vehicle performance and driver behavior in the simulation. Having said this, given the limitation of today's driving simulators and their cueing systems, it is not enough just to make sure that the design standards are followed rigorously. Ultimately, the goal is to make sure that the results on driving simulators generalize to the real world and this may require generating designs that do not follow the particular guidelines faithfully.

Key Points

- The fidelity of a simulator's roadway model effects how well the driving experience corresponds to the real world.
- There are limits to the realism of environment models in simulators.
- Several different methods and types of tools can be used to model driving environments.
- Engineering standards govern proper roadway design and should be followed in best practice for simulator model development.

Keywords: Modeling, Realism, Roadway, Standards

Acknowledgments

The author would like to acknowledge the editors, Professor John D. Lee (University of Wisconsin–Madison) and Professor Donald L. Fisher (University of Massachusetts, Amherst) for their very helpful review, suggestions and assistance in the development of this chapter; as well as Professor Johnell R. Brooks and Vijay Bendigeri of Clemson University for their review and research assistance.

Key Readings

American Association of State Highway and Transportation Officials (AASHTO). (2004). *The "Green Book"—A policy on geometric design of highways and streets* (5th ed.). Washington, DC: Author.

Federal Highway Administration. (2009). *Manual on uniform traffic control devices for streets and highways.* Washington, DC: US Department of Transportation.

Federal Highway Administration. (2010). *Flexibility in highway design.* Washington, DC: Department of Transportation. Retrieved February 24, 2010 from http://www.fhwa.dot.gov/environment/flex/ch03.htm

Ogden, K. W. (1996). *Safer roads: A guide to road safety engineering.* Cambridge, MA: University Press.

References

Allen, R. W., Rosenthal, T. J., Aponso, B., Parseghian, Z., Cook, M., & Markham, S. (2001, June 29). *A scenario definition language for developing driving simulator courses.* Paper presented at DSC2001 Driving Simulator Conference (Tech. Rep. 580), Nice, France. Hawthorne, CA: Systems Technology, Inc.

American Association of State Highway and Transportation Officials (AASHTO). (2004). *The "Green Book"—A Policy on geometric design of highways and streets* (5th ed.). Washington, DC: Author.

Barnett, J. (1940a). Safe side friction factors and superelevation design. *Highway Research Board, 16,* 69–80.

Barnett, J. (1940b). *Transition curves for highways.* Washington, DC: Federal Works Agency, Public Roads Administration.

Carpenter, R. (2002). *Factors affecting legibility distance in an advanced, fixed-base driving simulator* (a Master's thesis submitted in partial fulfillment of a Master's degree). Amherst, MA: Department of Mechanical and Industrial Engineering, University of Massachusetts.

DriveSafety. (n.d.). *HyperDrive driving simulation software* web site. Accessed February 24, 2010, at http://www.drivesafety.com/Products/Software/HyperDrive.aspx

Federal Highway Administration. (2009). *Manual on uniform traffic control devices for streets and highways.* Washington, DC: Federal Highway Administration, US Department of Transportation.

Federal Highway Administration. (2010). *Flexibility in highway design.* Washington, DC: Federal Highway

Administration, US Department of Transportation. Retrieved February 24, 2010, from http://www.fhwa.dot .gov/environment/flex/ch03.htm

National Advanced Driving Simulator (NADS). (2010). *National Advanced Driving Simulator overview.* Retrieved February 24, 2010, from http://www.nads-sc.uiowa.edu/press/pdf/ NADS_Overview.pdf

Ogden, K. W. (1996). *Safer roads: A guide to road safety engineering.* Cambridge, MA: University Press.

OKTAL. (n.d.). *Evariste SCANeR simulation software* web site. Accessed March 3, 2010, at http://www.scanersimulation.com

Presagis. (n.d.). *Creator* web site. Accessed March 8, 2010, at http://www.presagis.com/products/database/creator/

Randt, B. (December 6–7, 1996). *START: A toolset for generation of large city and rural driving simulation databases.* Workshop on Scenario and Traffic Generation for Driving Simulation, Orlando, FL.

Realtime Technologies. (n.d.). *RTI Simvista* web site. Accessed February 24, 2010, at http://www.simcreator.com/simvista/ simvista.htm

System Technology. (n.d.). *STISIM drive* web site. Accessed March 8, 2010, at http://www.stisimdrive.com

34

The Use of High-Fidelity Real-Time Driving Simulators for Geometric Design

Thomas M. Granda*
Federal Highway Administration

Gregory W. Davis
Federal Highway Administration

Vaughan W. Inman
Science Applications International Corporation

John A. Molino†
Science Applications International Corporation

Abstract

The Problem. Roadway geometrics play an important role in road safety and mobility. The success or failure of a geometric design is directly related to whether or not the physical manifestation of the design elicits the appropriate responses from the driver. So roadways must be designed with the driver in mind. *The Role of Driving Simulators.* Driving simulators can play an important role in helping roadway engineers design and evaluate alternative roadway configurations that conform to driver expectations and maximize the probability that drivers react appropriately. This is especially true in the case of unique roadways that exist only in the design or early development stage. *The Requirements of Geometric Design-Related Driving Simulation.* To help define the parameters of driving simulators that are used for geometric research, the physical and functional requirements of such driving simulators are discussed in terms of mechanical set-up, computer hardware and software requirements, and rendition of visual imagery. *Examples of the Use of Driving Simulators in Geometric Design.* This is followed by a number of applied research examples in a variety of geometric areas such as intersections and interchanges, geometric design safety measure evaluations, countermeasures to problematic geometric elements, visibility, and speed management. *Related Issues.* These research examples are followed by a discussion of selected issues related to the use of driving simulators for evaluating highway geometrics. These examples include advantages and disadvantages of driving simulators, validation of driving simulators, and simulator sickness.

---

* Retired from Federal Highway Administration.
† Present affiliation Tech-U-Fit Corporation.

34.1 Introduction

When roadway design engineers use the term "geometrics", they primarily refer to the physical aspects of the roadway. That is, they are largely concerned with topics such as horizontal and vertical alignment, road curvature, road width, and grades. Associated issues include sight distance, signing and marking, traffic signals, landscape, and lighting (American Association of State Highway and Transportation Officials, 2001). However, roadway engineers clearly understand that the physical system, and its related components, is not an end in itself. The success or failure of a geometric design is directly related to whether or not the physical manifestation of the design elicits the appropriate responses from drivers.

Roadway geometrics and appropriate driver responses come together in the concept of self-organizing roads (Keith et al., 2005). With self-organizing roads the physical features of the road (e.g., curvature, lane width, grading, and landscaping) implicitly tell, or provide information to, the driver on how fast to drive, when to slow, and when curves are approaching. But, of course, a self-organizing road is an ideal situation, and whether or not the road functions in a self-organizing way is an empirical question. For most roads, this goal remains unrealized. Therefore, understanding how drivers react to the roadway geometrics and to the total roadway system is important. This understanding is especially significant when designing or redesigning a roadway because implementing a poor roadway design can result in fatalities and crashes as well as an expensive retrofit. This is where a driving simulator can be valuable.

To provide a coherent perspective on the connection between driving simulator technology and the study of roadway geometrics, the remainder of the paper will focus on responding to three fundamental questions. The first question is "What kind of driving simulator do you need to study roadway geometrics and what are the physical and functional requirements for such a simulator?" Discussions on the mechanical set-up of the simulator, the computer hardware and software requirements, and the rendering of visual imagery are intended to address this question. These topics are important because they reveal some of the complexities in addressing geometric design using driving simulation.

The second question is "What are the examples of applied research studies that have used driving simulators to examine roadway geometrics?" The answer to this question involves reporting on research conducted on intersections, both at-grade and grade separated interchanges, the validation of measures used to assess the safety of geometric designs, countermeasures to problematic geometric elements, visibility, and speed management. The chief significance of this information is that it shows that driving simulators can be an effective tool in the investigation of roadway geometrics and their effects on drivers.

The third question is "What are some of the issues that should be considered when using driving simulators for investigating roadway geometric matters?" The issues discussed in responding to this question involve the advantages and disadvantages of using driving simulators to study roadway geometrics, validation of driving simulator findings, and simulator sickness. These issues are critical because failure to address them may easily result in expensive and nonproductive research efforts.

34.2 Physical and Functional Requirements for Driving Simulators When Used for Evaluating Highway Geometrics

The success of a driving simulation is greatly dependent on technology. The rapid growth of computer graphics technology, which is pushed in large part by the gaming industry, has resulted in an almost continuous increase in the fidelity and realism of driving simulations. However, such increases in graphics fidelity require the latest in hardware designs and software releases, which can be very expensive. Furthermore, high levels of fidelity may not be required to obtain reliable and valid data. Depending on the variables to be investigated, less expensive simulators that offer a lower level of graphics, motion and sound fidelity may still provide researchers with valid results. However, for the investigation of geometric design, a professional staff of programmers and graphics designers is required to develop project-specific roadway scenarios. Some of the issues that drive the need for a professional staff to operate the simulator are discussed by Williams et al. (2003).

The following sections discuss some of the considerations in matching hardware, software, and personnel requirements to the research requirements for exploring geometric design issues.

34.2.1 Hardware Configuration

Table 34.1 provides a brief summary of matching simulator requirements with research needs. For a basic desktop PC-based driving simulator, the requirements are relatively minimal. A desk and comfortable office chair would suffice. Additional equipment would be a steering wheel and foot pedals, which can be purchased from most merchants that sell PC gaming devices. A larger mid-level simulator can be arranged in two ways: with video monitors in front of the vehicle cockpit, or a projector set-up with two to four projection screens. These simulators require more room and physical space. However, there are vendors that sell self-contained simulators that are ready to use. For a high-fidelity simulator, the space requirements can be extensive and sometimes require a dedicated building. Components of high-fidelity simulators include racks for housing computer hardware, ceiling-mounted projectors (e.g., CRT or LCD), a complete car cab or self-contained pod, and a hydraulic or electro-mechanical motion base.

34.2.2 Computer Hardware and Software Requirements

A desktop-based driving simulator can utilize PCs and popular operating systems. High-end or hobbyist PCs designed for gaming can, however, offer superior performance due to the inclusion of more and faster memory, more sophisticated graphics

TABLE 34.1 Research Requirements (Column 1) Versus Recommended Driving Simulator Configurations

Research Variables to be Studied	Desktop PC Simulator	Driving Buck with Projectors or Multiple Monitors	High-Fidelity Simulator
Evaluation of basic Roadway Geometry	X		
Sign Placement	X		
Advanced Roadway Geometry		X	
Sign and Signal Visibility/ Comprehension		X	X
Complex Geometric Design		X	X
New Geometric Designs and Prototyping			X
Night-time Visibility studies			X[a]

[a] Night-visibility studies are best performed using CRT-type projectors, as they offer the highest contrast ratios when compared to LCD type projectors.

processors, and faster input-output data buses. Software to operate these machines is available from a variety of vendors that specialize in training and research simulation. Keep in mind, that for optimal performance and mitigating simulator sickness, a frame rate of 30–60 fps (frames per second) is considered minimal, and faster frame rates are desirable. Frame rates of 20 fps or less are noticeable to most people and the choppy image that results can be uncomfortable. Also, low frame rates can increase disturbing lags between driver inputs and simulator response.

Mid-level simulators may rely on one or more PCs. These PCs are typically of the high-end variety. These simulators may be purchased as self-contained systems, but may also accommodate user customization. Vehicle cockpit, screens, projectors, and software are often included in a self-contained module. Technical support is often available, sometimes at an additional cost.

A high-fidelity simulator requires the latest in hardware and software technology. An optimal configuration would be a series of computers, each dedicated to a specific task. For example, one computer would be dedicated to data collection, another for operating the motion base, another for computing vehicle dynamics, and one to render scenes for each projector. The process of linking all the computers together to operate as one system is known as clustering. This type of hardware configuration allows each processor in each of the PC cases to be dedicated for performing one task. The end result is the decreased likelihood of encountering a slower frame rate due to a job overload. These machines, as mentioned previously, require the latest in PC technology including multi-core processors and the latest graphics cards. Highly-trained programmers and modelers who are versed in real-time

graphics technology often program these simulators. Often the operating system and application software used are open source licensed such as Linux and Open GL® (open graphics language), so that scenario development has greater flexibility. These simulators are well suited for testing novel roadway infrastructure designs that can be modeled directly in graphics applications or by importing engineering drawings from computer-aided design applications. Dynamic interactive traffic, special effects such as weather, or specialized items such as headlight patterns can be incorporated into the test scenario. This type of simulator, while offering the greatest flexibility for scenario design, is also the most expensive type of simulator configuration.

34.2.3 Rendering of Visual Imagery

Cluster rendering (or distributed rendering) is the most common and efficient method for displaying images in a real-time environment. Cluster rendering utilizes more than one computer or video card simultaneously to render a scene. Multiple views can be rendered at the same time, each on a different computer. Thus, what the driver sees directly ahead may be rendered on one computer while other computers render views to the right and left, and still others generate perspectives for each of the rear-view mirrors. The driving simulator at the U.S. Federal Highway Administration (FHWA) utilizes this methodology. Figure 34.1 (Web Figure 34.1 is the color version) shows the configuration for the FHWA Highway Driving Simulator. A separate computer is dedicated to rendering the view projected from each of the five projectors. Not shown in Figure 34.1 are the three computers that generate the three rear-view mirror views.

34.3 Applied Research

We are aware of only a few instances where driving simulators have been used to directly address geometric design issues. Here we describe several studies, three of which were done in the Federal Highway Administration (FHWA) Highway Driving Simulator, that directly relate to geometric design. The first two of these studies were conducted to address safety concerns regarding new intersection designs. In both of these cases, intersections were proposed that briefly shift drivers from driving on the right side of the road (which is typical in the United States) to driving on the left side of the road. The concern was that drivers might bear to the right out of habit and create the risk of head-on collisions. The third example was an attempt to validate a driving simulator for conducting safety evaluations of intersection geometric designs. Finally, we present examples of the evaluations of signs and markings to mitigate challenges posed by specific geometric designs.

When creating an experimental scenario, care should be taken when placing or modeling visual objects (e.g. trees, buildings, parked cars) that are not part of the experiment. Objects added to create more visual complexity, and thus further the sense of

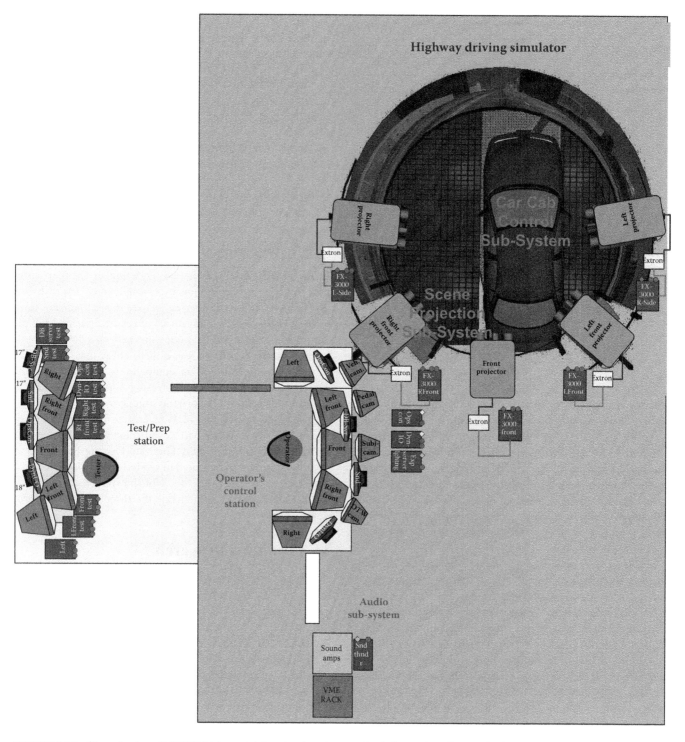

FIGURE 34.1   **(See color insert)** FHWA highway driving simulator is comprised of a number of networked computers.

realism, have the potential of becoming confounding variables. For example, when designing a curve recognition study, it may be easiest to run the experiment by building models by which the computer can present to the subject in a random order; with one model presenting a left curve scenario, and another model presenting a right curve scenario. If the left curve model includes a barn in the scene, and the "right turn" has a ranch house instead of a barn, an unintentional cue exists which participants can use to predict the direction of upcoming curves. This is an example of how confounding variables can unexpectedly enter into an experiment when modeling a scenario and why care must be taken in building driving simulation experiments.

34.3.1 Exploring New Intersection Designs

34.3.1.1 Diverging Diamond Interchange

The diverging diamond interchange design accommodates left-turning movements at signalized, grade-separated intersections of arterials and limited-access highways while eliminating the need for left-turn signal phasing as well as the need for a left-turn lane. On the arterial, traffic crosses over to the left side of the roadway between the ramp terminals on either side of the interchange. Two-phase traffic signals are installed at the crossovers. Once on the left side of the arterial roadway, vehicles can turn left onto limited-access ramps without stopping and without conflicting with through traffic. Figure 34.2 (Web Figure 34.2) provides a bird's-eye view of the diverging diamond interchange that was used in an FHWA simulation.

Because this design eliminates conflicts between left-turn movements and opposing through movements, it is expected to produce a safety benefit. Additional benefit might come from the lower speeds the design would induce. However, the potential safety benefit could be illusory if drivers go the wrong way at the crossover. When the Missouri Department of Transportation was considering the construction of a diverging diamond interchange, the FHWA offered the use of its driving simulator to evaluate the potential for these wrong-way errors.

The design proposed by Missouri was implemented in the FHWA driving simulator. The implementation included faithful simulation of the proposed horizontal geometric design, signs, markings, and traffic signals. In addition, median barriers and glare guards were included in the simulation. The resulting driver perspective of the visual scene at the eastbound crossover can be seen in Figure 34.3 (Web Figure 34.3).

The diverging diamond interchange safety evaluation had three stages. The first stage was a visualization exercise in which Missouri engineers drove through the simulated interchange and made adjustments in their design. As a result of this visualization, near-side traffic signals were added to address a sight-line problem. The near-side signals can be seen in Figure 34.4 (Web Figure 34.4). In addition, a visual path was evident to the left (oncoming)

FIGURE 34.3 Diverging diamond interchange crossover as it appeared in the driving simulator.

traffic lanes from one of the left-turns off of the freeway off-ramp. In Figure 34.5, this path can be seen from a driver's perspective at the end of the freeway off-ramp. The arterial median between interchange nodes was extended to eliminate that potential path to driver error. Additional examples of the use of driving simulators for visualization may be found in other chapters in this book, including Manore and Papelis's chapter (chap. 39), which focuses on driving simulation and visualization.

In the next stage of the evaluation, three alternative interchange designs were constructed in the simulator. Two of these designs were versions of the Missouri diverging diamond, and differed from each other only in the number of redundant signs and markings that were provided to guide drivers through the interchange. These alternatives were used to evaluate how well the interchange geometry directed drivers when minimal signs and markings were provided. The other alternative was a traditional diamond interchange design that featured the addition of protected left-turn lanes on the arterial instead of a crossover. The purpose of this alternative was to provide a source of comparison with driver behavior in the novel diverging diamond

FIGURE 34.2 Diverging diamond interchange traffic flow.

FIGURE 34.4 Near-side traffic signals were added to the design as a result of sight-line problems detected during the visualization.

FIGURE 34.5 A path around the median to a wrong-way violation could be seen from the end of one off-ramp.

designs. Figure 34.6 (Web Figure 34.5) shows a bird's-eye view of the west side of that traditional diamond interchange.

The final stage of the evaluation was an experiment in which driver performance in each of the three interchanges was observed. Seventy-two drivers drove through each of the three interchanges. The order in which drivers experienced the three interchanges was counterbalanced, with each driver assigned to one of three groups of 24 drivers that experienced the interchanges in a different order. The chief safety concern with the diverging diamond design was that drivers would bear right at the crossovers and travel against the intended traffic flow. The major finding of the study was that wrong-way maneuvers at the crossovers never happened, although the experiment afforded 12 opportunities for each of 72 drivers to make this error. Wrong-way maneuvers were observed in the experiment, but none of these were associated with the unique aspects of the diverging diamond interchange. In four cases drivers turned from the arterial into freeway off-ramps, but corrected their errors during or shortly after making

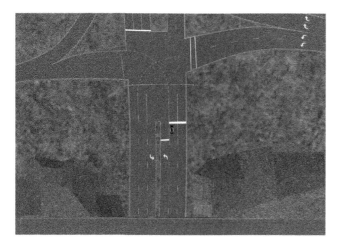

FIGURE 34.6 Bird's-eye view of one side of the conventional diamond interchange that was used to provide a baseline against which driver performance in the diverging diamond interchange could be judged.

these erroneous left-turns, two of which occurred with the conventional diamond design. The fifth wrong-way error could only have occurred in the conventional design—one driver completed a left-turn off of a freeway ramp by driving into the oncoming left-turn lane.

Another important finding from the evaluation was that the reverse curvature on approaches to the diverging diamond crossovers was successful in slowing drivers. Speeds through green signals at the diverging diamond intersections were 10 mph less than arterial speeds through the green signals in the conventional diamond design. Speed averaged 24 mph in both versions of the diverging diamond, and 34 mph in the conventional diamond interchange. Additional information on the diverging diamond interchange evaluation may be found in a Federal Highway Administration (FHWA) TechBrief (FHWA, 2007).

34.3.1.2 Continuous Flow Intersection

The continuous flow intersection (CFI) design eliminates the need for left-turn signal phasing at a main intersection by crossing left-turn traffic to the left side of the road upstream of the main intersection. An example of a CFI is shown in Figure 34.7. This allows left-turning traffic and through traffic to make their respective movements at the same time on the same green signal phase. There are already a few examples of CFIs in the U.S., and wrong-way drivers have not been a problem at those. Therefore, the research focus for the CFI simulation addressed the type and amount of signing required to enable drivers encountering a CFI for the first time to safely maneuver into the correct lane. The simulation enabled implementations of CFI designs that were identical except for variations in signs and markings. Additional information on the CFI simulations study may be found in an FHWA Summary Report (2008), *Evaluation of Sign and Marking Alternatives for Displaced Left-Turn Intersections.*

34.3.1.3 Geometric Design Safety Measure Validation

Investigators at the University of Central Florida (Yan, Abdel-Aty, Radwan, Wang, & Chilakapati, 2008) simulated an existing high crash-rate intersection in Orlando, FL. These researchers then observed driver behaviors in the high-fidelity simulator to determine whether some behaviors in the simulator might predict crashes in the real high crash-rate intersection. Thus, these researchers were not evaluating the geometric design of the simulated intersection, but rather they were validating crash surrogates that might be used in evaluating future geometric designs. The challenge in the investigation of high crash locations in a simulator is that crashes are rare events, possibly as rare in a high-fidelity simulator as they are in the real world. Surrogates for crashes are events that occur more frequently than crashes, but whose frequency is strongly correlated with the frequency of crashes of a particular type.

Because the surrogates that Yan et al. (2008) identified have not yet been validated prospectively, the surrogates they identified retrospectively should be viewed as reasonable candidates for further prospective validation. Nonetheless, several face-valid surrogates were identified. For instance, on corners where the stop line

FIGURE 34.7 CFI intersection in Baton Rouge, LA, with left-turn movements traced with red arrows. (From Inman, VW. (2009). *Evaluation of signs and markings for partial continuous flow intersection.* Transportation Research Board. With permission.)

was closer to the intersection, right-turns on red without stopping were more frequent in the simulator, and in the real intersection more collisions involving right-turns were recorded at those corners. Also, on intersection approaches where high rates of deceleration were observed in the simulator, higher rear-end crash rates have been observed at the real intersection. The results suggest that high rates of deceleration may serve as simulator surrogates for rear-end crashes, and right-turn on red violations may serve as surrogates for right-turn across path crashes.

34.3.2 Safety Countermeasures for Problematic Geometric Elements

Curvy two-lane rural roads tend to have high crash rates. In 1999 there were 8,901 single-vehicle run-off-road fatal crashes. Half of those fatal crashes occurred on rural two-lane roads (Neuman et al., 2003). Recently, FHWA coordinated with the Pennsylvania Department of Transportation to evaluate pavement-marking countermeasures to crashes on a short segment of rural Pennsylvania road that was experiencing about 16 non-fatal crashes per year (Molino, Katz, Donnell, Opiela, & Moyer, 2008). The evaluation included parallel data collection in the field and in the FHWA Highway Driving Simulator. The field data collection took place at night on a temporarily closed road. The geometric data (horizontal and vertical curvature

and road width), markings, roadway furniture, and other cultural features were imported into the simulation. This methodology enabled validation of measures that could be used in future simulation studies which would not require the expense and risk of field data collection. The investigators reported that drivers' subjective ratings of countermeasures were similar in the two environments, which was taken to mean that future studies of run-off-road signing and marking that used subjective measures could be done in the driving simulator without the need for field data collection. Figure 34.8 (see insert or Web Figure 34.6 for color version) shows a daytime comparison of one of the Pennsylvania curves and the simulation of that same curve.

Another simulation study (Molino, Inman, & Katz, 2007) addressed signing and marking issues at double-lane roundabouts. That study showed that, regardless of marking styles (that is, fishhooks or standard arrows) or whether lane restrictions were indicated on regulatory signs or on navigation guide signs, less than half of the drivers tested understood situations in which either the left or right lane was appropriate for their intended destination. In addition, participants chose to use lanes that were not legal for proceeding to their intended destination about 10% of the time. For instance, in the example shown in Figure 34.9 (see insert or Web Figure 34.7 for color version), drivers who intended to go straight ahead on the far side of the roundabout

FIGURE 34.8   (See color insert) Photograph (left) and simulation (right) of Curve No. 2 in subsequent visibility studies.

were equally likely to use the right or left lanes in spite of the conspicuous lane signs and markings that indicated the right lane must use the roundabout exit to the right. The FHWA is continuing to use its driving simulator to explore additional signing and marking strategies to increase comprehension of lane-use restrictions and use of correct lanes at multilane roundabouts.

34.3.2.1 Visibility

34.3.2.2 Curve Recognition

Since 2001, the FHWA has conducted a number of studies that address countermeasures to run-off-road crashes on rural curves at night. These studies specifically examine factors that affect the visibility of these curves. Early FHWA visibility studies employed simple driving simulator scenarios and explored the ability of drivers to recognize curves from a distance (Molino, Opiela, Andersen, & Moyer, 2003; Molino, Opiela, Andersen, & Moyer, 2004). These early experiments recorded participants' curve recognition distances as a function of the relative retroreflectivity/luminance of pavement markings (PMs), the relative luminous intensity/luminance of retroreflective raised

FIGURE 34.9   (See color insert) In the roundabout context, many drivers did not understand the meaning of the lane restriction sign (far right) and lane restriction markings (center).

pavement markers (RRPMs), and combinations of PMs with RRPMs. In these early simulation experiments the roadway geometry was simple compared to the geometries in later experiments which were described in the preceding section (Molino et al., 2008). The technical challenge was not in portraying the roadway geometry, but in adequately modeling the photometric parameters of the visual stimuli. Three challenges had to be met in this regard. The first was to achieve absolute luminances for modeled PMs and equivalent simulated luminances for modeled RRPMs that would portray what was typically measured in the field. The second was to achieve as low an absolute background luminance as possible (true black) in the visual display to simulate real night-time viewing. The third was to achieve the extreme visual dynamic range which is required to faithfully simulate bright RRPMs on a dark roadway. At the time that the experiments were conducted only the cathode ray tube (CRT) projector technology could provide a satisfactory balance for these requirements. The CRT projector technology could approach true black and retain a comparatively large visual dynamic range. In the end, by using the CRT projection technology, the absolute luminances of PMs could be simulated adequately, but the luminous intensity of RRPMs was difficult to simulate. A relatively low background luminance could simulate darkness sufficiently well, but the extreme visual dynamic range required to simulate bright, new RRPMs on a dark roadway could not be achieved.

Despite the simplicity of the simulated roadway geometry, these experiments succeeded in answering practical engineering questions of importance to FHWA. Two simulator experiments were conducted in the FHWA Highway Driving Simulator (HDS); and a third field validation study was conducted using an FHWA instrumented vehicle (Molino, Katz, Duke et al., 2004). Sixty-two research participants completed 11,572 trials with various combinations of RRPMs and PMs in the three experiments. With the exception of certain precautions with regard to RRPMs, the correspondence between the simulator and the field data was within expected tolerances for the various combinations of PMs employed in the two simulator experiments. The matrices of data relating curve recognition distance to retroreflectivity showed considerable correspondence between the laboratory and the field measurements. The mean curve recognition distances were somewhat greater in the field than in the laboratory. However, multiplying the mean curve recognition distance from the laboratory by 1.8 produced a reasonable correspondence with the mean recognition distance from the field.

One goal of the study was to compute "trading ratios" between the relative retroreflective/luminance of pavement markings and the retroreflective/luminance of RRPMs. That is, how much more could the PMs be allowed to degrade if RRPMs were installed to maintain the same curve recognition distance? The trading ratio is the proportion by which the retroreflectivity of the painted pavement markings can be allowed to be reduced if RRPMs are in place. In the case of retroreflectivity/luminance trading ratios, no scale factor was needed to translate between laboratory and field data because such ratios are relative. The results showed that retroreflectivity/luminance trading ratios determined in the field and in the laboratory had identical values of 0.41. Thus, pavement markings could be allowed to be degraded in recognition distance by 41% if RRPMs were in place and result in no overall reduction in driver curve recognition distance. Such results enhanced confidence in utilizing a simulator to investigate driver responses to novel patterns and combinations of PMs of differing retroreflectivity.

34.3.2.3 Curve Negotiation

The early visibility investigations only studied the effects of roadway delineation treatments on curve recognition, and did not observe the effects of these treatments on driver performance negotiating the curves. A subsequent pair of studies was conducted that employed more complex roadway geometries and measured performance while drivers negotiated curves on a rural roadway at night (Molino, Dietrich, Donnell, Opiela, & Moyer, 2006). These roadways contained a complex mix of both horizontal and vertical curvature. Some of the geometric complexity in these subsequent experiments, as well as the fidelity and realism of the simulation, can be seen in Figure 34.9 (Web Figure 34.7). Vehicle speed, lane position, and subjective rating responses to roadway delineation effectiveness were recorded as dependent variables. The application of various combinations of PMs and RRPMs to the roadway were the primary independent variables, to which wider edge lines and horizontal signing were also added. The FHWA Highway Driving Simulator had a limited three-degrees of freedom motion base, which was helpful in achieving a realistic simulation of an extremely curvy and hilly roadway. For simulations involving severe horizontal and vertical roadway curvature, some sort of motion base should be considered.

The first experiment was a field experiment conducted with 16 research participants who drove an instrumented FHWA test vehicle. The participants were recruited from rural Pennsylvania. On eight consecutive nights they drove in both directions a 3.5-mile (5.6-km) stretch of curvy rural Pennsylvania roadway. Each night various curves were treated with different combinations of pavement markings, pavement markers, and horizontal signage.

The second experiment was a laboratory simulation of the field study. The simulation was conducted in the highway driving simulator with 18 research participants recruited from the Washington, DC, metropolitan area. Although the simulator has the advantage of more easily enabling controlled and counterbalanced experimental designs that might be nearly impossible in the field, both experiments employed the same procedures and sequences of experimental sessions and conditions. This was done so that the FHWA Highway Driving Simulator could be more readily validated by comparison of field and simulator results.

Three of the dependent variables evaluated—vehicle speed, lane position and lane excursions—revealed important distinctions in comparing field and simulator data for the same driving task. Clear similarities were observed in the shapes of the speed and lane position profiles, and consistent indications of curve lengthening (i.e., driving closer to the right edge line for right-hand curves and closer to the centerline for left-hand curves, thus taking the fastest path through the curve) appeared on sharper curves in both testing environments. The quantitative data often required a mathematical transform or scale factor to make the absolute values of the data correspond between the field and the laboratory, but the same patterns in the data seemed to appear in both testing environments. Although the same patterns may occur in both environments, these patterns tended to be more pronounced and more variable in the simulator than in the field. The main feature was a tendency toward curve lengthening. Also evident was increased variability in lane position, which was evidenced by more pronounced curve lengthening in the simulator. These results imply that the same underlying driving behaviors were being measured both in the field and in the simulator. Such an outcome helped to validate the FHWA Highway Driving Simulator for future research in predicting driving behaviors on rural two-lane roads at night. However, there are several caveats to this validation. First, the simulator produces more variable driving behavior, which must be compensated for in comparing field and simulator data. When negotiating sharp curves, for example, driving performance measures of speed, lane position or lane excursions as a function of distance, may show similar overall behavioral patterns for mean measurements across research participants both in the field and in the simulator, but the variability of these means will be greater in the simulator than in the field. Second, the simulator may require a larger number of research participants, and possibly more repeated experimental trials, in order to show equivalent effects. Third, the simulator data may require transforms or scale factors in order to predict driver responses in the field. Based on the data from the present experiments, transforms need to be developed to accommodate the following observed differences: 1) Lane excursions were about 13 times more frequent in the simulator (mostly due to more pronounced curve lengthening); 2) Speed was an average of about 5.8 mph (9.3 km/h) faster in the simulator; 3) Speed variability was greater in the simulator, increasing the range by about 11 mph (17.7 km/h); 4) Lane position variability on tangent sections of roadway was somewhat greater in the simulator, increasing the range by about 1.0 ft (0.30 m). This increase in lane position variability represented a substantial portion of the total lane width of 11 ft (3.4 m).

In both the early and subsequent visibility experiments, research participants rated the effectiveness of the roadway delineation on each curve using a seven-point scale (Molino et al., 2008). An important methodological question was: Can such a subjective rating scale be effectively used in a driving simulator to reveal real-world stimulus-response relationships? The mean effectiveness ratings (MERs) were similarly sensitive to improved roadway delineation both in the simulator and in the field. The MERs increased with increasing luminance/retro-reflectivity of the PMs and RRPMs, and increased when multiple curve negotiation visual cues were present at the same time. These relationships were indicative of an internally consistent additive model of curve safety countermeasures for night-time driving. In general, the subjective rating scale proved to be effective in predicting the strength of both individual and combined stimulus effects, both in the simulator and in the field, and were often more effective than the objective metrics of vehicle speed, lane position, lane excursions, and so forth. The objective measures, especially lane position and lane excursions, were sometimes more variable, and more difficult to relate to stimulus parameters. Despite the overall superiority of the subjective metric, the rating scale tended to produce more variability and weaker overall results in the simulator relative to the field. The simulator may require a larger sample of research participants and more repeated experimental trials to produce equally strong effects.

34.3.3 Speed Management

Roadway geometry can be used to modify motor vehicle speeds and assist in meeting speed management goals. A recent FHWA simulator experiment studied low-cost safety improvements for rural curves and towns in the highway driving simulator (Molino, Katz, Hermosillo, & Dagnall, 2009). The rural curve portion of this experiment was directed at investigating pavement markings, post-mounted delineators (PMDs), and PMDs enhanced by LED lamps that produce a streaming pattern of light. These

low-cost safety improvements for curves were designed to modify vehicle entrance speeds to sharp curves. The town portion of this experiment was directed at investigating bulb-outs and chicanes as means to slow vehicle speeds in small rural towns. Bulb-outs are a traffic-calming measure consisting of curb extensions at intersections that widen the sidewalk and narrow the street. This speed control measure guards the pedestrian while they wait to cross, reduces roadway walking distance, and enhances pedestrian visibility. At the same time it reduces lane width, which is intended to slow traffic. Chicanes are traffic-calming measures that consist of curb extensions which alternate from side-to-side of the roadway, and thus create serpentine curves (S-shaped curves) in the roadway. These curb extensions are usually located away from intersections, and the S-shaped driving path tends to slow traffic.

34.4 Selected Issues Related to the Use of Driving Simulators for Evaluating Highway Geometrics

34.4.1.1 Advantages and Disadvantages of Driving Simulators

As with all research tools, driving simulators have their advantages and disadvantages when it comes to investigating driver reactions to roadway geometrics. The advantages and disadvantages, as shown in Table 34.2, are characterized. While the items in the table will not be discussed in this section, some are addressed directly or indirectly throughout the chapter.

34.4.1.2 Validation

Because driving simulators are essentially measuring instruments (e.g., measuring driver responses to geometric design alternatives) they must undergo validation. Two general aspects of validation should be mentioned here, validation type and validation procedure. In terms of validation type Tornros (1998) mentions absolute and relative validity. Absolute validity means that the driver performance data obtained in the simulator are identical, or virtually identical, with the data obtained in the real

TABLE 34.2   The Advantages and Disadvantages to Using Driving Simulators for Investigating Driver Reactions to Roadway Geometrics

Advantages:
- Experimenter controls road geometry and roadway environment
  - Experimental control over the roadway scenario
  - Experimental control over treatment and control conditions
- Unique and/or non-existent roadways can be created, developed, driven, and evaluated interactively
- Drivers can be subjected to countermeasures that are more expensive to test on real roads
- Drivers can be subjected to countermeasures that might be dangerous on real roads
- Roadway designers can visualize and test alternative roadway geometries before construction
- Reusable scenarios can be cost-effective in the long run

Disadvantages:
- The vehicle dynamics associated with complex roadway geometries cannot be fully experienced
- The manipulation of roadway geometries, especially complex ones, can induce simulator sickness
- Scenario development can be initially costly
- Simulator results need to be validated on a real road

world. The authors are unaware of any driving simulator studies in which absolute validity has been claimed. Relative validity means that the data obtained in the driving simulator and the real world are not the same but are consistent in regard to the relative magnitude and ordinal relations to the independent variables. It is the opinion of the authors that driving simulators are most often used to gauge the relative effect of various geometric treatments and therefore relative validity is usually sufficient to answer applied research questions.

There are three basic classes of driving simulator validity: face validity, validity for a specific area of research, and study-specific validity. Face validity assumes that driver performance in the simulator will be relatively similar to on-road performance because of the physical similarities of the simulation to the real world (Fildes, Godley, Triggs, & Jarvis, 1997). To establish face validity, demonstrating these physical similarities would be sufficient.

In validating a driving simulator for a specific area of research, experiments are conducted that encompass the type of tasks and measures to be investigated, such as the effect of independent variables on driver speed (Bella, 2005; Godley, Triggs, & Fildes, 2002). In this type of validation it may be assumed that once absolute or relative validity between real-world and simulation tasks is established, further validation is not necessary, or it may be periodically confirmed.

In the study-specific approach to validation, it is assumed that any simulation study requires real-world validation. When each experiment deals with different research topic (e.g., a novel intersection, new signage approaches, new road marking treatment), or contains novel driver tasks, it may not be assumed that the validation can be established across a series of studies even if they are in the same general area of research.

34.4.1.3 Simulator Sickness

The topic of simulator sickness is also addressed by Stoner, Fisher and Mollenhauer, this book, chap. 14. However, geometric design studies pose challenges that are not amenable to the usual countermeasures to simulator sickness.

Simulator sickness is related to motion sickness. The most common symptoms of simulator sickness resemble motion sickness: general discomfort, apathy, drowsiness, headache, disorientation, fatigue, pallor, sweating, salivation, stomach awareness, nausea, vomiting, retching, postural instability, and flashbacks (Kolasinski, Goldberg, & Hiller, 1995). There are two prominent theories for explaining the phenomenon of simulator sickness: the sensory conflict theory (Reason & Brand, 1975) and the postural instability theory (Riccio & Stoffregen, 1991). Both theories predict increases in simulator sickness where there is frequent or severe horizontal and vertical curvature, or intersections where stops and turns are required. There are two ways to combat such increases: reduce the complexity of the simulated roadways; or provide graduated adaptation and training for the research participants. In the case of geometric design research, the geometry of the roadway is the primary factor being investigated. Simplifying the roadway geometry

would defeat the purpose of the study. Therefore, graduated training regimes are often the only option for minimizing the risk of sickness.

The determinants of simulator sickness fall into three major categories: individual factors, simulator factors, and task factors. Potential individual factors include age, gender, ethnicity, experience, health, and personality. Simulator factors include color, contrast, field of view, time lag, resolution, motion platform, scene content and frame rate. Task factors include degree of control, duration, head movements, speed, vection, and difficult or unusual maneuvers (Kolasinski, Goldberg, & Hiller, 1995). As concerns roadway infrastructure and design, two important factors are the simulator and task factors. For example, Mourant and Thattacherry (2000) found that participants who drove at faster speeds had more sickness symptoms than those who drove at slower speeds. Also Johnson (2005) found that maneuver intensity, such as aggressive or violent actions, can be implicated in increased simulator sickness. This was confirmed by Mourant, Rengarajan, Cox, Lin and Jaeger (2007), who found that scenarios with curves and turns resulted in more simulator sickness than scenarios without these roadway features. Other factors are the age and gender of the test drivers. Wesley, Sayer and Tengler (2005) found more simulator sickness among older research participants. Mourant and Thattacherry (2000) found more simulator sickness among female participants; and Park, Allen, Fiorentino, Rosenthal and Cook (2007) found more symptom incidence for older drivers, particularly older female drivers.

Because roadway geometrics are at the core of many roadway and infrastructure design issues, and because some of these geometrics are by their nature complex, simulator sickness becomes a major consideration in the simulation of geometric designs. To control for simulator sickness, formal simulator sickness procedures have been implemented in the FHWA Highway Driving Simulator. These procedures have been reviewed and approved by an Institutional Review Board (IRB). The procedures are presented here as an example of one institution's approach to the screening, training, monitoring, identification and recovery associated with managing simulator sickness. Other institutions have different procedures.

All staff members who collect data are trained to monitor for the symptoms of simulator sickness. This training covers such topics as background, history, theories, biochemical mechanisms, minimization techniques, screening, symptoms, treatment and pre- and post-session tests. Staff members are trained to identify simulator sickness.

In the process of recruiting research participants, persons who have participated in earlier highway driving simulator experiments and not experienced more than mild symptoms are contacted first. New recruits who have not participated in previous driving simulations are screened for susceptibility to motion and simulator sickness. This telephone screening interview process probes five different types of sickness: sickness induced by an airplane, boat, car, simulator or other device.

Two tests are administered to detect and monitor simulator sickness: the Simulator Sickness Questionnaire (SSQ) (Kennedy,

Lane, Berbaum, & Lilienthal, 1993), and a sway magnetometry test for postural stability (Cobb & Nichols, 1998). These tests are given before the participant drives in the simulator, during breaks, and at the end of the simulator session. These tests are also administered if a participant becomes ill, and periodically thereafter to monitor recovery. In the event of simulator sickness, the baseline measurements taken before beginning the simulation are used to assess recovery.

Simulation sessions are stopped whenever the onset of significant symptoms of simulator sickness is observed. A camera mounted in the simulator and directed toward the participant's face is used to enable researchers to monitor for such symptoms. Signs that alert researchers to a potential problem include:

- Changes in facial expression (good or bad, any changes from the norm usually indicate there is a change in the participant's comfort level)
- Hand-to-face contact
- Sweating or removal of outer garments
- Verbal cues/sounds ("whoa", groaning or grunting)
- Increased swallowing—usually indicative of increased salvation or possible nausea
- Looking away from the projection screen
- Burping
- Blinking as if to clear or refocus their vision
- Other signs of anxiety or discomfort

When illness occurs, the participant is guided to a comfortable chair and asked to sit with her/his feet elevated and eyes closed. A motion sickness bag is provided if needed. The participant is encouraged to sit in this position until symptoms begin to subside. The researcher talks with the participant periodically to determine if sufficient recovery has occurred to justify administering the SSQ and the sway magnetometry tests. As necessary, these tests are administered every 20 minutes until the participant approaches baseline performance, or until two hours have elapsed. The participant is released when SSQ and sway magnetometry scores approximate baseline. If the affected participant does not recover within the two hours, the researcher offers to pay for a taxi fare home, and back the next day to retrieve her/his car. The researcher later calls the participant at home to monitor the recovery. If an affected participant insists on being released, a waiver form is provided for signature.

34.5 Summary

In this chapter we have endeavored to show that geometric design issues can be successfully addressed in driving simulator studies. However, we have also endeavored to show that geometric design studies pose a number of challenges. First, the resources required in terms of hardware and software need to be addressed. For studies where intersections must be modeled or where drivers need to negotiate large radius curves, a wide horizontal field of view (e.g., at least 180 degrees) is required.

Where signs and markings are used to assist drivers in comprehending the roadway, high-resolution graphics are required. In our experience, the current state of the art is insufficient to display signs and markings without compromise. Generally signs must be oversized to be read at distances at which they could be easily read in the real world. Special attention needs to be paid to sign recognition distances and the relationship of these distances to driver performance. We have shown that retroreflectivity, luminance, and sight distances for signs and markings can be successfully addressed in driving simulations, but because of limitations in display technology, the ranges of contrast and luminance that can be achieved in a driving simulator are considerably less than in the real world. These limitations require extra care in validating the use of the simulator for visibility applications. Finally, simulator sickness is an especially daunting challenge when exploring geometric design issues. Reducing speeds, limiting changes in speed, and using only mild curves are ways to reduce simulator sickness. However, these countermeasures to inducing sickness are often impractical in geometric design studies.

Key Points

- When roadway design engineers use the term geometrics, they primarily refer to the physical aspects of the roadway. That is, they are largely concerned with topics such as horizontal and vertical alignment, road curvature, road width, and grades. Associated issues include such items as sight distance, signing and marking, traffic signals, landscape, and lighting.
- Essentially, driving simulators can be classified into three different categories: (1) desktop PCs with a single monitor, steering wheel, and pedals, (2) mid-level driving simulators that utilize a non-motion-based driving buck and projectors, and (3) high-fidelity simulators that use a full car cab on a motion base, a wide field of view using projectors, and software that is specifically programmed on an experiment-by-experiment basis by trained computer graphics professionals. Obviously, the capabilities within these three categories can vary greatly.
- The physical and functional requirements for driving simulators that are used for evaluating highway geometrics can be organized into the following categories: mechanical set-up, computer hardware and software requirements, rendition of visual imagery, and integration of these elements to achieve the overall project objectives.
- Driving simulators can be used to study such geometric topics as new intersection designs, geometric design safety measure validation, evaluating countermeasures for problematic geometric elements, visibility of roadway delineation treatments, and speed management.
- While there are advantages to using driving simulators for studying roadway geometrics, there also are disadvantages.

- Using a driving simulator can lead to simulator sickness. But there are ways to predict and/or mitigate these effects.
- There are various ways to validate driving simulators used for studying the effects of geometric design on drivers.

Keywords: Driving Simulator Research, Geometric Design, Geometric Evaluation, Roadway Geometrics, Roadway Simulation, Simulator Sickness, Simulator Validation

Acknowledgments

The authors would like to thank Dr. John Farbry and Mr. Steve Fleger for reviewing, commenting, and offering technical recommendations to this chapter. Without their generous assistance, this chapter would not have been possible.

Glossary

Continuous flow intersection: An at-grade signalized intersection where traffic that intends to turn left is shifted to the left side of the roadway before it reaches the main intersection.

Curve recognition distance: The distance (usually in feet or meters) before the point of curvature at which a driver can identify whether a curve goes to the right or to the left.

Diverging diamond interchange: A grade separated interchange of an arterial with a limited access roadway. On the arterial, traffic on the arterial travels on the left side of the roadway between intersections on either side of the interchange. This configuration allows both left and right turns from the arterial to be made without the need for additional left or right turn traffic signal phases, thus greatly improving arterial throughput without adding lanes.

Point of curvature: The point on a road where a straight section of road (i.e., tangent) intersects a curve.

Roadway geometry: Physical aspects of the roadway that includes horizontal and vertical alignment, road curvature, road width, and grades.

Trading ratio: The proportion of new pavement marking reflectivity that can be allowed to remain after degradation if retroreflective pavement markings are in place.

Web Resources

The *Handbook* web site contains supplemental material for the chapter, including color versions of a number of the figures.

Web Figure 34.1: FHWA Highway Driving Simulator is comprised of a number of networked computers (color version of Figure 34.1).

Web Figure 34.2: Diverging diamond interchange traffic flow (color version of Figure 34.2).

Web Figure 34.3: Diverging diamond interchange crossover as it appeared in the driving simulator (color version of Figure 34.3).

Web Figure 34.4: Near-side traffic signals were added to the design as a result of sight-line problems detected during the visualization (color version of Figure 34.4).

Web Figure 34.5: Bird's-eye view of one side of the conventional diamond interchange that was used to provide a baseline against which driver performance in the diverging diamond interchange could be judged (color version of Figure 34.6).

Web Figure 34.6: Photograph (left) and simulation (right) of Curve No. 2 in subsequent visibility studies. (Figure 34.8).

Web Figure 34.7: In the roundabout context, many drivers did not understand the meaning of the lane restriction sign (far right) and lane restriction markings (center) (Figure 34.9).

Key Readings

Federal Highway Administration. (2007). *Drivers' evaluation of the diverging diamond interchange.* FHWA TechBrief. Author. Retrieved November 19, 2008, from http://www.tfhrc.gov/safety/pubs/07048/07048.pdf

Federal Highway Administration. (2008). *Evaluation of sign and marking alternatives for displaced left-turn intersections.* Summary Report (Rep. FHWA-HRT-08-071). Author.

Molino, J. A., Opiela, K. S., Andersen, C. K., & Moyer, M. J. (2003). Relative luminance of retroreflective raised pavement markers and pavement marking stripes on simulated rural two-lane roads. *Transportation Research Record, 1844,* 45–51.

Tornros, Jan. (1998). Driving behavior in a real and a simulated road tunnel—a validation study. *Accident Analysis & Prevention, 30*(4), 497–503.

Yan, X., Abdel-Aty, M., Radwan, E., Wang, X., & Chilakapati, P. (2008). Validating a driving simulator using surrogate safety measures. *Accident Analysis & Prevention, 40,* 274–288.

References

American Association of State Highway and Transportation Officials. (2001). *A policy on geometric design of highways and streets—2001.* Author.

Bella, F. (2005). Validation of a driving simulator for work zone design. *Transportation Research Record, 1937,* 136–144.

Cobb, S. V., & Nichols, S. C. (1998). Static posture tests for the assessment of postural instability after virtual environment use. *Brain Research Bulletin, 47*(5), 459–464.

Federal Highway Administration. (2007). *Drivers' evaluation of the diverging diamond interchange.* FHWA TechBrief. Author. Retrieved November 19, 2008, from http://www.tfhrc.gov/safety/pubs/07048/07048.pdf

Federal Highway Administration. (2008). *Evaluation of sign and marking alternatives for displaced left-turn intersections.* Summary Report (Rep. FHWA-HRT-08-071). Author.

Fildes, B., Godley, S., Triggs, T., & Jarvis, J. (1997). *Perceptual countermeasures: Simulator validation study.* Monash University. Retrieved June 13, 2007, from http://www.monash.edu.au/muarc/reports/atsb169.html

Godley, S. T., Triggs, T. J., & Fildes, B. N. (2002). Driving simulator validation for speed research. *Accident Analysis & Prevention, 34*, 589–600.

Johnson, D. M. (2005). *Introduction to and review of simulator sickness research* (Rep. 1832). Arlington, VA: US Army Research Institute for the Behavioral and Social Sciences.

Keith, K., Trentacoste, M., Depue, L., Granda, T., Huckaby, E., Ibarguen, B. et al. (2005). *Roadway human factors and behavioral safety in Europe* (Rep. FHWA-PL-05-005). Washington, DC: Federal Highway Administration. Retrieved November 19, 2008, from http://www.international.fhwa.dot.gov/human-factors/contents.htm

Kennedy, R. S., Lane, N. E., Berbaum, K. S., & Lilienthal, M. G. (1993). Simulator sickness questionnaire: An enhanced method for quantifying simulator sickness. *The International Journal of Aviation Psychology, 3*(3), 203–220.

Kolasinski, E. M., Goldberg, S. L., & Hiller, J. H. (1995). *Simulator sickness in virtual environments*. Alexandria, VA: US Army Research Institute for the Behavioral and Social Sciences. Retrieved November 19, 2008, from http://handle.dtic.mil/100.2/ADA295861

Molino, J. A., Dietrich, S. R., Donnell, E. T., Opiela, K. S., & Moyer, M. J. (2006). *Nighttime driving behavior on curves with different roadway delineation: Preliminary comparison of field and simulator results*. Unpublished technical report.

Molino, J. A., Inman, V. W., & Katz, B. J. (2007). *Lane restriction marking and signing for double-lane roundabouts: results of an FHWA simulator experiment*. Proceedings Institute of Transportation Engineers (ITE) Annual Conference, San Diego, CA.

Molino, J. A., Katz, B. J., Donnell, E. T., Opiela, K. S., & Moyer, M. J. (2008). *Using a subjective scale in a driving simulator to predict real-world stimulus-response relationships concerning nighttime delineation for curves*. 87th Annual Meeting of the Transportation Research Board Papers (CD-ROM). Washington, DC: Transportation Research Board.

Molino, J. A., Katz, B. J., Duke, D., Opiela, K. S., Andersen, C. K., & Moyer, M. J. (2004). *Field validation for the relative effectiveness of combinations of pavement markings and RPMs in recognizing curves at night*. 83rd Annual Meeting of the Transportation Research Board Papers (CD-ROM). Washington, DC: Transportation Research Board.

Molino, J. A., Katz, B. J., Hermosillo, M. B., & Dagnall, E. E. (2009). *Simulator evaluation of low-cost safety improvements for rural curves and small towns*. Manuscript in preparation.

Molino, J. A., Opiela, K. S., Andersen, C. K., & Moyer, M. J. (2004). *Relative luminance of center line vs. edge line pavement markings on simulated rural two-lane roads at night*. 83rd Annual Meeting of the Transportation Research Board Papers (CD-ROM). Washington, DC.

Molino, J. A., Opiela, K. S., Andersen, C. K., & Moyer, M. J. (2003). Relative luminance of retroreflective raised pavement markers and pavement marking stripes on simulated rural two-lane roads. *Transportation Research Record, 1844*, 45–51.

Mourant, R. R., Rengarajan, P., Cox, D., Lin, Y., & Jaeger, B. (2007). *The effect of driving environments on simulator sickness*. Proceedings of the Human Factors and Ergonomics Society 51st Annual Meeting, Baltimore, MD. Santa Monica, CA: Human Factors and Ergonomics Society.

Mourant, R. R., & Thattacherry, T. R. (2000). Simulator sickness in a virtual environments driving simulator. *Proceedings of international ergonomics association/human factors and ergonomic society* (IEA/HFES) 2000 Congress (pp. 534–537). San Diego, CA.

Neuman, T. R., Pfefer, R., Slack, K. L., Hardy, K. K., McGee, H., Prothe, L., et al. (2003). *Volume 6: A guide for addressing run-off-road collisions.* (NCHRP Report 500). Washington, DC: Transportation Research Board, National Cooperative Highway Research Program.

Park, G. D., Allen, R. W., Fiorentino, D., Rosenthal, T. J., & Cook, M. L. (2007). *Simulator sickness scores according to symptom susceptibility, age and gender for an older driver assessment study*. Proceedings of the Human Factors and Ergonomics Society, 51st Annual Meeting, Baltimore, MD. Santa Monica, CA: Human Factors and Ergonomics Society.

Reason, J. T., & Brand, J. J. (1975). *Motion sickness*. London: Academic Press.

Riccio, G. E., & Stoffregen, T. A. (1991). An ecological theory of motion sickness and postural instability. *Ecological Psychology, 3*(3), 195–237.

Tornros, J. (1998). Driving behavior in a real and a simulated road tunnel—a validation study. *Accident Analysis & Prevention, 30*(4), 497–503.

Wesley, A. D., Sayer, T. B., & Tengler, S. (2005). *Can sea bands be used to mitigate simulator sickness?* Proceedings of the Third International Driving Symposium on Human Factors in Driving Assessment, Training, and Vehicle Design. Rockport, ME.

Williams, J. R., Liao, D., Molino, J. A., Inman, V. W., Davis, G. W., & Wink, J. M. (2003). *Event driven dynamic interactive Experiments (EDDIE)*. Proceedings of the Driving Simulation Conference North America 2003. Retrieved from http://www.dsc-na.org/

Yan, X., Abdel-Aty, M., Radwan, E., Wang, X., & Chilakapati, P. (2008). Validating a driving simulator using surrogate safety measures. *Accident Analysis & Prevention, 40*, 274–288.

35

Traffic Signals

David A. Noyce
University of Wisconsin–Madison

Michael A. Knodler, Jr.
*University of Massachusetts,
Amherst*

Jeremy R. Chapman
University of Wisconsin–Madison

Donald L. Fisher
*University of Massachusetts,
Amherst*

Alexander Pollatsek
*University of Massachusetts,
Amherst*

Abstract

The Problem. Drivers rely on traffic control devices for regulatory, warning, and guidance information while traveling. This chapter addresses the use of driving simulation in the design of one type of traffic control device: traffic signal systems at controlled intersections. *Role of Driving Simulators.* Driving simulators have allowed traffic signal researchers to more cost-effectively develop repeatable study designs and look specifically at traffic signal head design (three-section, four-section, five-section), signal head placement, signal phasing schemes (protected-only, permissive-only, protected/permissive), and signal indications, including the flashing yellow arrow permissive left-turn option. Driving simulators provide tremendous flexibility allowing experimentation in traffic signal design and operational concepts that could not be evaluated in real-time field conditions because of cost, safety, liability, and repeatability. One of the largest case study applications for driving simulator technology with regards to traffic signals is the use of flashing yellow arrows for left-turning vehicles. *Scenarios and Dependent Variables.* Virtually any geometry and traffic signal combination of an at-grade signalized intersection can be effectively evaluated in a driving simulator environment. Variable levels of simultaneous traffic and non-motorized users of the intersection can be included in urban and rural settings. The impact of specific geometric features within signalized at-grade intersections and their impact on traffic operations and safety can also be considered. Dependent variables commonly include driver comprehension, driver understanding, compliance, lane keeping, speed, perception-reaction time, and several subjective preference measures. *Key Results of Driving Simulator Studies.* Driving simulator studies have led to the development and implementation of the flashing yellow arrow as an indication used for permissive left-turn operations. Driving simulators have also been used to evaluate driver understanding and comprehension of traffic signals including traffic signal head placement, signal display type, lane configurations, and opposing traffic gap acceptance. Combining driving simulators with eye trackers allows experimenters to also consider how drivers search for information as an independent variable in signal display design. *Platform Specificity and Equipment Limitations.* As described in other chapters, graphics resolution limits the ability of most driving simulators to render the necessary detail to effectively evaluate realistic size and viewing distances as well as signal head placement and simultaneous visual cues. Specific issues related to the use of driving simulators in traffic signal research include location bias, proper selection of subjects, accommodation of all modes, and the potential of simulator sickness. Traffic operational aspects related to acceleration, deceleration, and turning forces cannot be included with most driving simulators.

35.1 Introduction

35.1.1 Problem: Traffic Signal Comprehension and Uniformity

Accommodating all movements and modes at signalized intersections provides a challenge to traffic engineers. Effective communication is essential to accommodate all intersection users in a safe and efficient manner. Furthermore, the methods of communication must be comprehensible to all users and lead to the desired behavior. For example, in order to complete a left-turn at a signalized intersection, the driver must either find an appropriate gap in the opposing traffic stream or have all conflicting traffic stopped while the appropriate signal indication is displayed. Due to these conflicting movements, the successful accommodation of left-turn and opposing through-movement vehicles is critical to the safe and efficient operation of signalized intersections. Additionally, effectively communicating this desired operation in a uniform and easily understood manner is the seemingly simple challenge that the engineering community faces.

Research has shown that many traffic control devices used in our current transportation system are not universally understood. Traffic signals are no exception. The United States created the *Manual on Uniform Traffic Control Devices* (MUTCD) in an effort to help resolve problems of comprehension through the uniform application of traffic signal devices (Federal Highway Administration, 2009). Specifically, the purpose of the MUTCD is to provide national guidelines for the implementation of traffic control devices used on U.S. roadways. However, due to the limited guidance provided in the MUTCD regarding traffic signals, a variety of displays have been developed and used in ways that are far from uniform. A tremendous challenge exists in developing a system that maximizes human performance by minimizing the errors in comprehension common to traffic signals today.

35.1.2 Difficulties in Creative Technological Development

A search of traffic signal system literature identifies numerous research studies focused on the evaluation of user comprehension of traffic signals and the residual behavior. Historically, driver comprehension studies of traffic signals would include a pen-and-paper analysis showing sketches of traffic signals and asking participants to respond to a series of questions. Technological growth and the computer era allowed researchers to use a two-dimensional framework of intersection photographs or animated traffic signal systems to improve upon the realism of the signal indications or displays being evaluated. Difficulties

in these analysis methods were readily identified, and researchers understood that the dynamics and competing information associated with real-time traffic signal comprehension was not being captured. The creative researcher went to closed-course environments and created a "demo" intersection that used real traffic signals but did not have live traffic operations or other real-time conditions. Although useful findings were developed, limitations on the impact of competing cues and the inability to consider multiple signal display scenarios minimized the value of this experimentation methodology. It was well understood that the only way to effectively research traffic signals was to implement the desired study scenarios at an active signalized intersection. Numerous problems prevented and still inhibit this method of research, including the substantial time and costs associated with technology development along with the implementation and traffic impacts necessary to create the experimental design. More significantly, the safety and potential liability of changing the traffic signals under live traffic flow has understandably limited extensive use of this experimental method. The limitations of past study designs highlight the challenges faced by researchers interested in traffic control devices and traffic signal systems and the inability to push the technological envelope in traffic signal systems research.

35.1.3 Driving Simulation as a Research Tool

Driving simulation has recently evolved as an effective mechanism for overcoming the difficulties mentioned above. Beyond the obvious strengths of the ability to explore limitless numbers of geometric, traffic signal operations, and traffic signal display scenarios, research is conducted in a controlled and safe environment, costs are reduced, and the ability to communicate with research subjects adds to the body of knowledge.

35.1.4 Challenges

Although driving simulators have allowed traffic signal researchers to more effectively develop study designs, driving simulators are not without challenges. Concerns have been raised about research result bias due to the single location and non-transportable nature of most simulators. For example, we do not know whether drivers in Massachusetts drive and inherently interpret traffic signal information the same way as drivers in Wisconsin, Texas, or California. There are also regional differences in the driving culture, commonly identified through behavioral differences. For example, left-turn jumper and sneaker behavior is not consistent throughout the U.S. Therefore, research conducted in Massachusetts using the local population as subjects may have limited applicability and acceptance in other locations. Consistent driving simulator technologies and visual world transferability in some of the newer simulators allow the same study design to be conducted in different regions to overcome these geographical concerns (e.g., see this book, chap. 6 by Kearney & Grechkin, "Scenario Authoring").

There are still other challenges in the use of driving simulators, including: the selection of subjects that are representative of the driving population; flexibility in dynamic scenario development for all traffic situations; accommodating all modes in the simulation process; and file size limitations preventing the ability to fully populate a congested traffic scenario, although recent advances in computer hardware have improved the latter. Furthermore, the fact that the visual worlds are developed in an animated format rather than using actual video limits the realism of the images observed by subjects.

An additional concern with simulator experiments is the potential of simulator sickness, which has been likened to motion sickness (see also, on simulator sickness by Stoner, Fisher, & Mollenhauer, this book, chap. 14). The U.S. Army Research Institute for the Behavioral and Social Sciences (ARI) conducts research to improve the effectiveness of training simulators and simulations used in training soldiers in in-vehicle fighting techniques. Kolasinski (1995) completed an extensive literature review on simulator sickness, identifying trends and causes. Although she identified a number of factors that may trigger the onset of simulator sickness, she believes that a primary cause of simulator sickness is "inconsistent information about body orientation and motion received by the different senses, known as the cue conflict theory. For example, the visual system may perceive that the body is moving rapidly, while the vestibular system perceives that the body is stationary," (p. vii).

ARI has also evaluated the effects of several demographic factors, which may be important when using simulation in experiments at signalized intersections. ARI reports that for participants between the ages of 12 and 50 years, age has no statistically significant effect on the likelihood of simulator sickness. The literature also suggests that females might be more susceptible to simulator sickness than males. Ongoing research seeks to determine the specific factors responsible for simulator sickness. A solution to this potential problem is highly desirable due to the benefits of driving simulation studies.

Other challenges bearing on simulator sickness and related to the evaluation of traffic signals in driving simulator environments are described in other chapters throughout this book (e.g., see in this book, chap. 8 by Andersen, "Sensory and Perceptual Factors in the Design of Driving Simulation Displays", and chap. 26 by Trick & Caird, "Methodological Issues When Conducting Research on Older Drivers"). In any event, the desirability of a solution is critically important to driving simulator studies focused on traffic signals and intersections, given the typically large number of turning movements and other maneuvers commonly associated with simulator sickness.

35.1.5 Traffic Signal Scenarios

35.1.5.1 Typical Applications

The strength of using driving simulators in the evaluation of signalized intersection operations and communication methods is that there is no typical application. Rather, nearly any experimental variable of interest can be effectively evaluated. Nevertheless, typical applications to date have focused on the "human factors" elements of traffic signal displays and signal phasing operations primarily related to driver comprehension and behavior. Although driving simulators can be effectively used to consider geometric and traffic operations features of intersections, some of the challenges cited above have limited experimentation in this area. Driving simulators are also being considered as a tool for design evaluations and the proper placement of traffic signal and other traffic control devices before the construction phase of new projects, especially at complex intersections and interchanges (see this book, chap. 39 by Manore & Papelis, "Roadway Visualization").

35.1.5.2 Protocol

Protocols for traffic signal and signal intersection scenarios are consistent with most other study designs. The intent is always to create a virtual driving environment that seamlessly represents the actual driving environment and effectively accomplishes the experimental objective. The following paragraphs provide a summary of a typical protocol used in evaluating traffic signal indications. The intent is to provide a working example, understanding that other protocols have been successfully implemented.

Most driving simulator-based experiments focused on traffic signals will use a variety of signalized intersection scenarios with the variable of interest appropriately counterbalanced. For example, one intersection approach was created for each of the experimental signal display/indication combinations recently completed in a study of left-turn signal operations (Knodler, Noyce, Kacir, & Brehmer, 2005). Specifically, variables evaluated included changes from the permissive left-turn indication to other aspects of the signal indications including changes in:

- Traffic Signal Mode of Operation
  - Steady indication to steady indication
  - Flashing indication to steady indication
- Location of Traffic Signal Indication
  - Bimodal (i.e., green and yellow arrow indications in same signal section)
  - Section change for each indication
- Traffic Signal Display Configuration
  - Three-Section Vertical
  - Four-Section Vertical/horizontal
  - Five-Section Cluster
- Left-Turn Indication Shape
  - Arrow to circular

Figure 35.1 (Web Figure 35.1 for color version) depicts a typical signalized intersection used in this research. Additionally, several intersections that require the driver to evaluate different information and/or turn right, to proceed straight, or to turn left on a protected green arrow are included as part of the visual world. The additional movements are included to provide experimental variability and counterbalance, and reduce the probability of the subject driver keying into the nature of the evaluation. Variability in intersections also reduces the learning curve effects.

Additional variability in the experimental protocol can be provided through the creation of multiple driving modules and starting positions. In the left-turn experiment, two modules were developed, each presenting the experimental display/indications in a different order. Further, the module orders are varied to provide counterbalancing. Drivers observed each of the experimental displays only once by traversing the modules. Additionally, starting points for each drive are designed to provide further variability across subject drivers. Flexibility in module development, including reversing the intersection orders, along with different starting positions within each module, creates numerous unique module order and starting point combinations.

Protocols with traffic signal display systems require some starting point in the signal phasing sequence being presented. In the left-turn study, all experimental signal displays within the simulation started in red (all circular red indications). Signal displays were changed to the *test* indications as the driver approached the intersection. For example, between 30 to 120 meters prior to the intersection stop bar, the signal display was "triggered" and changed from a circular red indication to the selected permissive left-turn (or protected) indication. Additionally, the through-movement indication also changes from a circular red to a circular green indication in scenarios with a green through movement. A second trigger was placed at selected intersections to release queued opposing traffic and/or to change the signal indication from the protected or permissive left-turn indication to another signal indication. The location of these trigger points can vary by intersection scenario.

Protocols should include as many of the dynamic aspects of the signalized intersection as possible. One of the significant issues with the evaluation of permissive left-turns is the opposing traffic operations. In most protocols, it is important to have opposing traffic in each scenario, yet critically important to control the opposing traffic such that gap selection or related variables do not become uncontrolled variables in the experiment. In the left-turn experiment, the introduction of opposing traffic required drivers to simultaneously evaluate the signal display and indication, traffic movement, and opposing gaps to complete a safe permissive left-turn maneuver. As already mentioned, this methodology replicates the decision process required during actual operation of a motorized vehicle within the roadway system.

At experimental intersections in which comprehension of left-turn indication was evaluated, all gaps in opposing traffic were consistently applied. At least two vehicles were always positioned at the stop bar in the two through lanes opposing the left-turn driver. The remaining opposing vehicles were positioned further

FIGURE 35.1 Screen capture of typical simulated intersection. (From Knodler, M. A., Jr., & Noyce, D. A. (2005). Tracking driver eye movements at permissive left-turns. *Proceedings of the 3rd international driving symposium on human factors in driver assessment, training, and vehicle design* (pp. 134–142). Rockport, ME.)

upstream in a specified gap sequence. Past research suggested gap sequences of three and seven seconds in a series of 7-3-7-7; therefore, opposing vehicles crossed the intersection 7, 10, 17, and 24 seconds behind the initially queued opposing vehicles. Specifically, these gap sizes were derived from the theory associated with the critical gap concept. The Transportation Research Board's Highway Capacity Manual (2000) suggests a critical gap value of approximately 5.5 seconds for permitted left-turn maneuvers in the design of a four-lane roadway. Therefore, a three-second gap was selected because it is small enough that a majority of drivers will not accept it, and a seven-second gap was selected because it is generally acceptable to most drivers. Selecting a consistent sequence of three and seven-second gaps can prevent variables such as gap selection from being a significant variable in the evaluation.

After creating the simulator scenarios, but prior to conducting the experiment, protocols must include a beta test of the simulator visual environment. The purpose of the beta test is to identify necessary modifications that can only become apparent after administering the experiment. The number of subjects invited to participate is a function of the statistical sample size and the availability of subjects, along with experimental budgets. Demographic factors, possession of a valid driver license, and vision testing are common screening and selection variables. All demographic data are recorded as assigned to the subject.

Although the procedural protocol will vary across study types and locations, it is critical to adhere to a consistent protocol across participants. Consider again the left-turn study for which a generalized procedure is outlined below. When drivers arrived to participate in the simulator experiment, they were provided with an overview of the experimental procedure. Subject drivers were then seated in the driving simulator and given a specific set of instructions regarding the procedure. Drivers were asked to fasten their seatbelt, adjust mirrors, and adjust the radio as they would in their own vehicle. Again, the idea is to replicate their normal driving environment to the extent possible. Engine noise, vehicle vibration and wind through the driver side-window are all included.

The driving portion of the experiment began with a practice module that provided the opportunity for drivers to traverse a virtual network and familiarize themselves with the operational characteristics of the driving simulator. Drivers were asked to *drive* the simulator vehicle as they would drive their own vehicle. At this stage of the experiment, the driver's well-being is always closely observed for early signs of simulator sickness. Drivers who successfully completed the practice course, free of simulator sickness, are most often permitted to continue with the simulator experiment.

Following the practice course, subjects drive through the experimental modules. As previously noted, drivers start from different positions within the driving module and the order of experimental scenarios varies from module to module. To minimize the need for verbal communication during the evaluation, subjects were navigated through the modules by guide signs provided on each intersection approach. In addition, drivers were asked to observe speed limit signs, providing a higher level of realism and speed control in the virtual network. The driving portion of the experiment, including the practice module, required an average of 15 to 20 minutes.

Driver responses to each signal display scenario are recorded both manually and electronically. Correct responses were recorded accordingly and incorrect responses were categorized by the nature of the failure. A video record is completed with each experiment such that it can be reviewed to confirm driver responses. Additionally, the simulator computer system records selected data used to determine perception/reaction time, lane position, and related control variables.

Throughout the experiment, subject drivers are asked to *think out loud* and verbally express their thoughts about anything they observed. Research team members are present to record the results of the simulation, including the responses at each intersection and other driving-related factors such as indecision, unnecessary braking, or any pertinent verbal comments made. After completing the driving simulator evaluation, drivers are compensated for their participation. The amount of compensation will vary. Compensation may also be used as a tool to attract more participants in the research.

The above protocol is only one example of many potential protocols that can be used in signalized intersection experimentation using driving simulators. Nevertheless, this protocol has proven successful over 15 years of experimentation on the subject.

35.1.5.3 Challenges

In addition to the challenges previously cited, the ability to get experimental subjects to drive "as they normally would" and operate in a typical fashion while being observed in an experimental environment has the potential for bias (see this book, chap. 9 by Ranney, "Psychological Fidelity: Perception of Risk"). Additionally, attracting the proper cross-section of drivers to participate in the experiments is always a challenge.

35.2 Traffic Signal Requirements

Driving simulator experiments involving traffic signals and signalized intersections must specifically follow the guidelines of the *Manual on Uniform Traffic Control Devices* (Federal Highway Administration (FHWA), 2009), known by its acronym MUTCD, and the individual state supplements that may exist in the states in which the study is being conducted. Inaccurate placement and design of the experimental traffic signals will lead to invalid and unsupported results.

Since 1935 the MUTCD has provided guidance with regards to the meaning and application of traffic control devices. Recognized under the Code of Federal Regulations, the MUTCD provides national standards that must be adhered to when employing traffic control devices. Part 4 of the MUTCD is dedicated to providing current requirements and guidance regarding the application of traffic signals.

By definition, the MUTCD considers a traffic control signal to be "any highway traffic signal by which traffic is alternately directed to stop and permitted to proceed", where traffic is defined as "pedestrians, bicyclists, ridden or herded animals, vehicles, streetcars, and other conveyances either singularly or together while using for purposes of travel any highway or private road open to public travel" (chap. 1A, Section 1A.13, MUTCD). The broad definition of a traffic control signal within the MUTCD has led to the inclusion of a significant number of devices. Specifically, the MUTCD includes the following applications and types of traffic control signals: traffic signals; pedestrian signals; hybrid beacons; emergency-vehicle signals; traffic control signals for one-lane, two-way facilities; traffic control signals for freeway entrance ramps; traffic control signals for movable bridges; toll plaza traffic signals; flashing beacons; lane-use control signals; and in-roadway lights (chap. 4A, Section 4A.01, MUTCD). Each of these devices can be successfully evaluated in a driving simulator environment.

35.2.1 MUTCD Criteria

Traffic signal indications are limited to the following:

Operational mode—steady (i.e., solid) or flashing;
Color—green, yellow, or red; and
Shape—circular or arrow

The primary meaning of the traffic signal indications are further summarized in Table 35.1 below.

Equally as important to the meaning of traffic signal indications are the applications. Specifically, the situational setting and specific sequencing under which each indication is utilized is of paramount importance in the design and operation of traffic control signals. A complete description of the application for traffic signal indications is provided in Sections 4D.04 through 4D.31 of the MUTCD. Some of the more critical application elements that effect traffic signal experimentation in driving simulators include (but are not necessarily limited to) the following:

TABLE 35.1 Summary of Meanings for Vehicular Signal Indications

Indication Attribute	Description
Steady Green Indications	• Vehicular traffic, except pedestrians, facing a CIRCULAR GREEN signal indication is permitted to proceed straight through or turn right or left or make a U-turn movement except as such movement is modified by lane-use signs, turn prohibition signs, lane markings, roadway design, separate turn signal indications, or other traffic control devices. Such vehicular traffic, including vehicles turning right or left or making a U-turn movement, shall yield the right-of-way to pedestrians lawfully within an associated crosswalk and other vehicles lawfully within the intersection. • Vehicular traffic, except pedestrians, facing a GREEN ARROW signal indication, displayed alone or in combination with another signal indication, is permitted to cautiously enter the intersection only to make the movement indicated by such arrow, or such other movement as is permitted by other signal indications displayed at the same time.
Steady Yellow Indications	• Vehicular traffic, except pedestrians, facing a steady CIRCULAR YELLOW or YELLOW ARROW signal indication is thereby warned that the related green movement or the related flashing arrow movement is being terminated or that a steady red signal indication will be displayed immediately thereafter when vehicular traffic shall not enter the intersection.
Steady Red Indications	• Vehicular traffic facing a steady CIRCULAR RED signal indication [alone]… shall stop at a clearly marked stop line…and shall remain stopped until a signal indication to proceed is shown, or as provided below. • Vehicular traffic facing a steady RED ARROW signal indication shall not enter the intersection to make the movement indicated by the arrow and… shall remain stopped until a signal indication or other traffic control device permitting the movement indicated by such RED ARROW is displayed.
Flashing Yellow Indications	• Flashing yellow—vehicular traffic, on an approach to an intersection, facing a flashing CIRCULAR YELLOW signal indication is permitted to cautiously enter the intersection to proceed straight through or turn right or left or make a U-turn except as such movement is modified by lane-use signs, turn prohibition signs, lane markings, roadway design, separate turn signal indications, or other traffic control devices. • Vehicular traffic, on an approach to an intersection, facing a flashing YELLOW ARROW signal indication, displayed alone or in combination with another signal indication, is permitted to cautiously enter the intersection only to make the movement indicated by such arrow, or other such movement as is permitted by other signal indications displayed at the same time. • Such vehicular traffic, including vehicles turning right or left or making a U-turn, shall yield the right-of-way to pedestrians lawfully within an associated crosswalk and other vehicles lawfully within the intersection. In addition, vehicular traffic turning left or making a U-turn to the left shall yield the right-of-way to other vehicles approaching from the opposite direction so closely as to constitute an immediate hazard during the time when such turning vehicle is moving across or within the intersection.
Flashing Red Indications	• Flashing red—vehicular traffic, on an approach to an intersection, facing a flashing CIRCULAR RED signal indication shall stop at a clearly marked stop line; but if there is no stop line, before entering the crosswalk on the near side of the intersection; or if there is no crosswalk, at the point nearest the intersecting roadway where the driver has a view of approaching traffic on the intersecting roadway before entering the intersection. The right to proceed shall be subject to the rules applicable after making a stop at a STOP sign. • Vehicular traffic, on an approach to an intersection, facing a flashing RED ARROW signal indication if intending to turn in the direction indicated by the arrow shall stop at a clearly marked stop line; but if there is no stop line, before entering the crosswalk on the near side of the intersection; or if there is no crosswalk, at the point nearest the intersecting roadway where the driver has a view of approaching traffic on the intersecting roadway before entering the intersection. The right to proceed with the turn shall be limited to the direction indicated by the arrow and shall be subject to the rules applicable after making a stop at a STOP sign.

Source: From Federal Highway Administration. (2009). *Manual of uniform traffic control devices* (2009 ed.). Washington, DC: U.S. Department of Transportation, Federal Highway Administration.

- Application of steady signal indications
- Design, illumination, color and shape of signal indications
- Size of vehicular signal indications
- Positions of signal indications within a signal face (horizontal or vertical)
- Number of signal faces per approach
- Visibility, aiming and shielding of signal faces
- Lateral and longitudinal positioning of signal faces
- Mounting height of signal faces
- Yellow change and red clearance intervals
- Flashing operation of traffic signals

Much more information pertaining to traffic signal and signalized intersections can be found in the MUTCD. Perhaps the one issue of critical importance that should be re-emphasized here is signal display arrangement and placement within the visual world. Traffic signal displays can be three-, four-, or five-section arranged in a vertical, horizontal, or cluster arrangement. Placing the correct number of required traffic signal displays at the proper location, and in the required arrangement within each scenario, is significant to the research design proposed. Each of these variables has been found to influence driver comprehension and behavior related to traffic signals.

The MUTCD is a dynamic document that will be periodically updated. Researchers should consult the current version of the MUTCD when developing their experimental design and research protocol.

35.2.2 Issues Specific to Left-Turn Operations with Traffic Signals

The left-turn maneuver is widely recognized by both drivers and traffic engineers as one of the most difficult to safely execute on U.S. roadways. Driver confusion in and around the more than 300,000 signalized intersections in the United States is directly responsible for an increase in delay and heightened crash potential. The safe and efficient accommodation of left-turning vehicles at signalized intersections is a primary source of concern for traffic engineers, and this concern has led to the use of several unique design and operational practices. While dedicated turn lanes and protected left-turn phases have improved intersection operation and safety, they have often done so at the expense of intersection efficiency, as the time provided for an exclusive left-turn phase must be taken away from other critical movements at the intersection. To minimize this problem, protected/permissive left-turn (PPLT) signal phasing was developed.

PPLT signal phasing provides an exclusive, or protected, phase for left-turns as well as a permissive phase during which left-turns can be made if gaps in opposing through traffic allow, all within the same signal cycle (Noyce, Bergh, & Chapman, 2007). The theory of PPLT signal phasing is to minimize the exclusive left-turn phase time requirements while increasing the opportunity for left-turn maneuvers. Use of PPLT phasing can lead to increased left-turn capacity and reduced delay, thus improving the operational efficiency of the intersection.

Because previous editions of the MUTCD provided limited guidance for PPLT applications, a variety of adaptations of PPLT arrangements were established throughout the country (see Section 4D.20 and specifically Figure 35.4D-11 of the 2009 MUTCD for examples). The variability in PPLT arrangements contributed to the lack of a uniform national standard for PPLT control, which the MUTCD has only now begun to address.

As PPLT signal phasing was introduced throughout the country, many research efforts were undertaken to evaluate the potential impacts of protected phasing versus PPLT phasing. Agent (1979) compared protected-only and PPLT signal phasing in Kentucky. Using before-and-after trial data, the researcher discovered a 50% reduction in delay with PPLT phasing. Additionally, PPLT phasing was initially associated with a higher crash potential but the number of crashes decreased as drivers became familiar with the signal phasing. A public opinion survey conducted as part of the research showed that 90% of drivers preferred PPLT phasing and an economic analysis showed that PPLT phasing almost always generated a benefit-cost ratio of greater than one.

In a second research project, Agent (1985) completed an evaluation of crash data on intersections converted from protected-only to PPLT signal phasing. Research findings indicated that the introduction of PPLT phasing resulted in a reduction in vehicle delay, but resulted in higher crash rates. The researcher formulated a series of guidelines and criteria for applications where any form of permitted phasing should not be introduced because of the potential safety implications, including: high speed applications; three or more opposing lanes to cross; and intersections where left-turning vehicles have limited sight distance.

A driver opinion poll conducted by the Washington Section of the Institute of Transportation Engineers reported that Seattle-area drivers supported PPLT signal phasing. Results showed that 85% of all respondents felt traffic delay was reduced with PPLT signal phasing, 73% felt traffic safety was not compromised with PPLT signal phasing, and 73% also responded that they were in favor of PPLT signal phasing and would like PPLT signal phasing applied at more intersections (Institute of Transportation Engineers, 1985). By contrast, when participants of the 1988 Indiana State Fair were asked which left-turn signal phasing they preferred among protected-only, permitted-only, and PPLT, an overwhelming 98% of drivers opted for the protected-only phasing (Hummer, Montgomery, & Sinha, 1993).

Asante, Ardekani and Williams (1993) evaluated protected-only, PPLT, and permitted-only phasing using both field studies and static driver surveys in Texas. The field studies were conducted at over 100 locations and included both accident and conflict rates. The researchers found protected-only phasing had both lower accident and conflict rates than PPLT phasing. Permitted-only phasing resulted in the lowest levels of delay, but a significantly higher conflict rate than both protected-only phasing and PPLT signal phasing.

Perhaps the most significant impact to traffic signals and signal operations made to date has come through the use of driving simulators to study left-turn signalization especially related to

PPLT signal phasing. Specifically, research has considered the safety and operational impacts of protected and permissive left-turn signal indications and displays. Some of this research is discussed later in this chapter.

35.2.3 Traffic Signal Challenges

Although the potential benefits of traffic signals have been identified, they can only be achieved when the information is correctly presented to and comprehended by the driver. As has been noted, traffic signal information is presented to the driver through the illumination of circular and arrow-shaped indications within a traffic signal display. The meaning of all signal indications is transmitted through a combination of color, shape, orientation, and position of the signal display. Many challenges exist with regards to the successful application of a traffic signal, including but not limited to: driver error; dual meanings of indications; a general lack of uniformity; overall signal complexity; the aging driver population; and existing levels of congestion and delay.

Driver error is often associated with breakdowns in driver comprehension of the signal display message being presented. One of the well understood examples of problems with driver comprehension is the dual meaning of the circular green indication in traffic signals. Although it appears simplistic on the surface, the fact that a circular green indication has different meanings for through movement (go) and permissive left-turn movements (yield) has led to concerns over driver error and safety problems.

Uniform traffic signal displays aid the road user by simplifying recognition and understanding. Uniform displays also help road users, police and enforcement personnel, and traffic courts to interpret appropriate driver behavior. Uniform traffic control displays are also economical because they enable using consistent manufacturing, installation, maintenance, and administration processes.

Complexity in traffic signals and the associated aspects of the signalized intersection are growing as traffic volumes and the number of drivers on the road increase. Traffic congestion and delay are a common element in most roadway systems. Traffic signal operations are becoming more complex and include more dynamic phase options that can violate the expectancy of some drivers. Furthermore, the population continues to age, with the number of drivers over the age of 65 increasing yearly. Human factors issues common to older drivers are an important consideration.

35.3 Traditional Methods of Traffic Signal Research

35.3.1 Research Protocols

35.3.1.1 Surveys

Traditionally, pen-and-paper comprehension tests in which the driver simply marks what he/she believes to be the correct answer have been used. The critique of this methodology has focused on the belief that the study responses provided might not be consistent with decisions made in the actual driving environment.

35.3.1.2 Static Photos

To add more realism to driver behavior experiments, computer technology has been employed to provide static photos of actual driving environments and superimposing PPLT signal displays within them. Although this technology is believed to be a major step forward in experimentation, the static nature and lack of dynamic cues may still lead drivers through a different decision process, one which is inconsistent with the actual driving process.

35.3.1.3 Computer-Based Investigations

The ability to show subject drivers photographs or videos of specific intersection or traffic signal scenarios and the growth in computer technology has led to the increase in computer-based investigations. Computer programs can be easily written to create specific queries and to automate the data collection process. A computer-based method has been a common component of the flashing yellow arrow (FYA) left-turn research as described later in this chapter.

35.3.1.4 Test Tracks

Research facilities that have closed-course test tracks available allow for signalized intersections to be created and evaluated outside of the traveling public. Subjects can use their own vehicles in the experiment which has obvious advantages in driver comfort and familiarity. Clearly, there can be significant costs associated with the experimental design and getting subjects to the test track location can be a challenge. The use of test tracks is further discussed in chap. 13 in this book (Mullen, Charlton, Devlin, & Bédard, "Simulator Validity: Behaviors Observed on the Simulator and on the Road"). In the United States, one of the major tests tracks is located at the Virginia Tech Transportation Institute (http://www.vtti.vt.edu/virginiasmartroad.html).

35.3.1.5 Others

At times, field implementation of traffic signals can be used as the experimental platform. For example, field implementation of an experimental flashing yellow arrow display was conducted at several locations to field test a display that had shown promise in safety and driver comprehension in previous task activities. In conjunction with use of the experimental display (which required permission from Federal Highway Administration to operate), "before" and "after" studies were completed at each intersection where the flashing yellow arrow display was installed and at nearby control sites. These "before" and "after" studies allowed researchers to quantify the impact of the changeover from the MUTCD circular green indication to the flashing yellow arrow indication.

35.4 Driving Simulators as a Traffic Signal Research Tool

The literature contains a number of examples of how driving simulators have been used in traffic signal research. This section provides a number of examples of how driving simulators have been effectively used.

In a study of left-turn gap selection for drivers traveling at either 30 or 60 mph, Staplin and Fisk (1991) used driving simulation techniques to evaluate the willingness of drivers to select gaps of different durations. The study recorded driver information using a 20-inch monitor, a large-screen video projector, and a large-screen cinematic display. Additionally, gap selection field data were obtained for comparative purposes. Only the large-screen cinematic display corroborated what was occurring in the field, for which the minimum gap length increased as the speed increased. Staplin and Fisk (1991) reported that higher levels of realism, especially as it relates to the visual field and traffic signals, provided more accurate results, providing an early demonstration of the benefits of driving simulation in traffic signal experimentation.

In a study of driver comprehension of left-turn displays, Szymkowiak, Fisher and Connerney used a full-scale driving simulator to test drivers' reaction to a set of 40 different left-turn signal display/sign combinations (Szymkowiak et al., 1997). Thirty-two drivers were tested in 40 signalized intersection scenarios (32 test scenarios) in a realistic driving environment which included dynamic opposing traffic. The results indicated better driver comprehension than the Staplin research, which included only static opposing traffic, leading the researchers to conclude that drivers may "benefit from an even more realistic, that is dynamic, environment."

Smith and Noyce (2000) used a driving simulator in a study of drivers' comprehension of various traffic signal displays. Smith reported differences between the completed static surveys and the simulation experiment he conducted, finding a higher level of comprehension related to the circular green permissive indication. The added level of realism provided in the simulation experiment appeared to provide drivers with the usual cues which may be found on the roadway and was likely responsible for the increased comprehension. Recommendations included the length of time drivers should be tested and the need for added realism. Smith reported that the experiment required one hour of driving and a half-hour of follow-up questioning to complete each subject, yet he concluded that drivers should reasonably be expected to drive for a period of only 30–45 minutes (or shorter) if feasible. Smith also concluded that the driving simulator was effective in the evaluation of driver comprehension of traffic signal displays.

35.4.1 Using Full-Scale Driving Simulation for Traffic Signal Analysis

The inclusion of traffic signals within the framework of a driving simulation must consider several key elements related to traffic signal design. Every effort should be made to design traffic signals that present a realistic scenario to simulator study participants.

35.4.2 Driver Eye-Movement Evaluation

Although NCHRP Report 493 (Brehmer, Kacir, Noyce, & Manser, 2003) concluded that the flashing yellow arrow (FYA) is well comprehended by drivers, knowledge of the sources of information

that are most important to permissive left-turning drivers can be used to further improve driver comprehension. Knodler and Noyce (2005) researched the eye movements of 11 drivers at six virtual intersections with permissive left-turn phases to identify what driver information is commonly used while executing the maneuver. The study resulted in 66 evaluations using the full-scale driving simulator in the Human Performance Laboratory at the University of Massachusetts, Amherst. A detailed visual world was created for drivers to navigate through real-world driving situations. While drivers navigated the virtual roadway environment in the simulator's four-door Saturn Sedan, an Applied Science Laboratories (ASL) Series 5000 eye tracker with head-mounted optics was used to monitor eye movement. The ASL unit converted the eye position to an external point of gaze by superimposing crosshairs on a projected video screen.

The experiment evaluated three permissive indications installed at six intersections, as shown in Figure 35.2 (Web Figure 35.2). Opposing vehicles were present at three of the permissive left-turning intersections, and absent at the remaining three intersections. To provide additional experimental variability, drivers were required to complete movements at eight intersections that were not permissive left-turns.

Analysis of the experimental intersections included driver response and driver eye movements. Driver eye movement was found to be most valuable. In order to evaluate driver eye movements, the driver display was partitioned into multiple "areas of interest." Each area of interest coincided with a potential cue that drivers may use to complete the permissive left-turn. Partitions included: the PPLT signal display; the adjacent through signal; the cross traffic; opposing vehicles; and the location where pedestrians may be in the crosswalk to the left of the driver. Figure 35.3 (Web Figure 35.3 for color version) shows the partitions used in the evaluation.

Several of the major trends identified through a limited number of driver eye-movement evaluations included:

- Drivers used more sources for driver information when no opposing vehicles were present;
- The driver's eye primarily "rested" on the opposing traffic;
- Glancing away from the eye resting point, drivers fixated on other sources of information for less than one second at a time; and
- Drivers scanning multiple sources of information tended to scan from the right side of the intersection to the left.

PPLT signal configuration and permissive indication	G		G

FIGURE 35.2 Permissive left-turn displays evaluated. (From Knodler, M. A., Jr., & Noyce, D. A. (2005). Tracking driver eye movements at permissive left-turns. *Proceedings of the 3rd international driving symposium on human factors in driver assessment, training, and vehicle design* (pp. 134–142). Rockport, ME.)

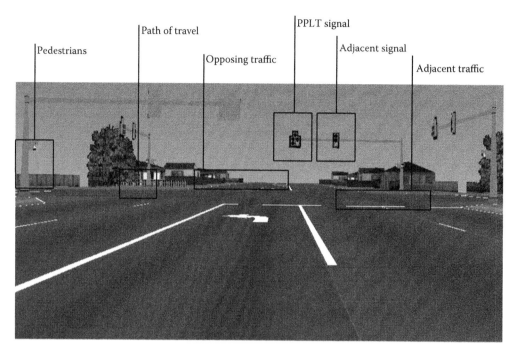

FIGURE 35.3 **(See color insert)** Partitioned driver simulation display. (From Knodler, M. A., Jr., & Noyce, D. A. (2005). Tracking driver eye movements at permissive left-turns. *Proceedings of the 3rd international driving symposium on human factors in driver assessment, training, and vehicle design* (pp. 134–142). Rockport, ME.)

It was interesting to note that drivers often did not use the traffic signal display and signal indication as the primary source of left-turn information. Specifically, drivers often used opposing traffic flow and the availability of gaps in the opposing traffic as the decision variable for permissive left-turns with secondary glances at the signal itself. The effectiveness of driving simulators in this type of research was further highlighted.

35.4.3 Simultaneous Permissive Left-Turn Indications

The PPLT left-turn movement has been displayed in multiple configurations throughout the United States. Requirements outlined in Section 4D.11 of the MUTCD state that:

> …a minimum of two primary signal faces shall be provided for the through movement. If a signalized through movement does not exist on an approach, a minimum of two primary signal faces shall be provided for the signalized turning movement that is considered to be the major movement from the approach.

A common scenario that meets this requirement is to utilize a five-section cluster signal for the left-turn and through movements and an adjacent three-section head for through and right-turn movements (see Figure 35.3, PPLT and adjacent signal, respectively).

Implementation of the FYA will require an interim retrofit display which will be displayed simultaneously with either a circular green (CG), circular yellow, or circular red indication in the adjacent lanes. Knodler et al. (2005) evaluated driver comprehension of the FYA indication when used simultaneously with the adjacent through-movement signal indication in a shared five-section cluster signal head. Additional research evaluated the simultaneous use of the FYA indication and various adjacent through-movement displays. Driver comprehension under various display scenarios was compared. Figure 35.4 (Web Figure 35.4) shows the seven-signal display scenarios evaluated.

Driver evaluation involved two groups of participants (Knodler et al., 2005). The first group participated in a dynamic driving simulator and follow-up computer-based static evaluation in Massachusetts. The second group only conducted the static evaluation in either Massachusetts or Wisconsin. Using a full-scale driving simulator, experimental drivers completed a test course and two study modules in the simulated environment. Each module included 14 intersections, 8 of which were permissive left-turn movements. Each driver that completed the driver simulator also completed a follow-up computer-based static evaluation that included 29 scenarios. Independent of the driving simulator, a diverse group of 210 drivers completed an independent static evaluation which resulted in a total of 2,310 scenarios being evaluated.

Drivers in all evaluations encountered several intersection scenarios and were asked to respond to each dual-display scenario with go, yield, stop, or stop and wait. The responses to the permissive scenarios were classified for evaluation as follows:

- Go. Considered an incorrect response, the driver chose to turn assuming they had right-of-way. A crash or narrowly-avoided crash occurred.

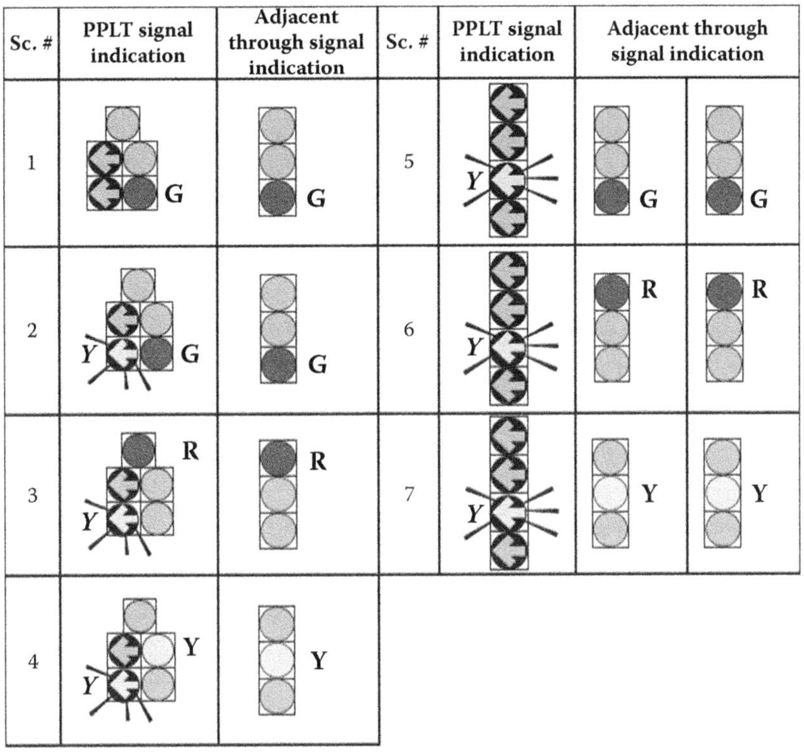

FIGURE 35.4 Signal display scenarios evaluated. (From Knodler, M. A., Jr., Noyce, D. A., Kacir, K. C., & Brehmer, C. L. (2005). Evaluation of flashing yellow arrow in traffic signal displays with simultaneous permissive indications. *Transportation Research Record, 1918*, 46–55.)

- Yield. Considered a correct response, the driver completed a left-turn after yielding to opposing traffic.
- Stop first. A conservative response, the driver stopped at the stop bar prior to completing a left-turn after finding an appropriate gap in opposing through traffic.
- Stop and wait. The driver waited at the stop bar even after opposing through traffic had cleared the intersection. The driver was considered to be waiting for the signal to change prior to proceeding.

There were no statistically significant differences in the responses across the presented signal display scenarios in the driving simulator. The research did find statistically significant differences in the distribution of yield responses across the five displays resulting from the follow-up static evaluation. The five-section cluster arrangement displaying a flashing yellow arrow/circular green (an FYA/CG) permissive indication had a significantly higher percentage of yield responses than all other displays, except the FYA in a four-section vertical configuration when the adjacent through-movement indication was CG.

The researchers concluded that the simultaneous display of the FYA and the adjacent through-movement indication did not affect driver understanding of the permissive indication. Specifically, the researchers concluded through a follow-up static evaluation that the retrofit five-section display improved driver comprehension of the permissive indication compared

to comprehension of the existing circular green (CG) display at a 95% level of confidence. This result, along with others determined from the study, led researchers to recommend the five-section cluster signal display to be acceptable for interim use while converting signals to the recommended four-section vertical protected/permissive left-turn (PPLT) display.

35.4.4 Two Allowable Permissive Left-Turn Indications

The FYA creates an alternative permissive indication that will result in two acceptable left-turn displays, both intended to indicate the same message to the driver. A logical concern is that driver understanding of the CG PPLT indication will change when an alternative PPLT indication, the FYA, is concurrently installed. For example, when drivers become more familiar with the FYA and identify it as a permissive left-turn indication, it is plausible that the CG indication would be interpreted as a protected left-turn.

Knodler, Noyce and Fisher (2007) conducted research using both static evaluations and a dynamic driving simulator to determine driver response to the CG indication after having been exposed to the FYA indication. The hypothesis of their research was that "drivers are more likely to interpret the CG permissive indication to indicate a right-of-way situation if the FYA is gradually implemented at a number of intersections and drivers comprehend the FYA indication (p. 55)."

A total of 25 drivers used a computer-based training program to learn about the FYA indication prior to participating in the driving simulation. The simulation environment included 14 intersections, half of which involved left-turn maneuvers with an FYA indication, a protected green indication, or a CG indication. Drivers were first exposed to the FYA, and then encountered an intersection display with the CG indication that was used for evaluation. The follow-up static evaluation and the independent static evaluation maintained the same consistent approach used in the simulated environment.

The researchers provided statistical evidence that the implementation of the FYA may not impact drivers' understanding of the CG indication during a short period after implementation. In fact, the follow-up static evaluation showed that drivers who were familiar with the FYA responded with a higher percentage of *yield* (correct) responses to the CG indication, compared to those unfamiliar with the FYA. Driver responses in the dynamic driver simulator were shown to be independent of exposure to the FYA indication. The independent static evaluation did not provide any significant findings which have not already been discussed in previously cited literature.

35.4.5 Driver Comprehension of Solid Yellow Indication

Traffic engineers have voiced concern about how to effectively terminate a signal phase once the FYA permissive indication is implemented on a broad scale. The concern is based on the fundamental understanding that, traditionally, when drivers see a yellow indication, it means that a signal phase is being terminated. By implementing the FYA as a second yellow indication that does not imply termination of another phase, driver understanding of the circular yellow (CY) indication may be impacted. Additional concern is related to the identification of an effective way to indicate termination of the FYA permissive phase.

Traditional three-section signal heads include two variables: color and location. Together they provide redundancy in the display shown to drivers. Redundancy improves driver comprehension and minimizes error. Driver error is hypothesized to be increased by a lack of redundancy caused by the use of two subsequent displays using the color yellow: the first being the CY change and the second being the FYA.

Knodler et al. (2007) used a computer-based static evaluation to determine the impact that exposure to and comprehension of the FYA indication may have on driver comprehension of the solid yellow arrow (SYA) indication. A total of 212 drivers completed sequential evaluations in Madison, Wisconsin and Amherst, Massachusetts. The experiment was sequential in evaluating the SYA before drivers were exposed to the FYA operations. Subsequent evaluation of the SYA after exposure to the FYA was focused on any change in driver response to the SYA. The researchers concluded that there was no evidence to suggest that the FYA permissive indication may negatively affect drivers' understanding of the SYA indication.

35.4.6 Applications of the FYA in Separated Left-Turn Lanes

Intersection geometry is a major factor with the potential to impact the probability that drivers erroneously comprehend the CG permissive left-turn indication and assume that they have right-of-way. Wide intersections with left-turn lanes that are separated from the adjacent through and right-turn lanes are one example of geometric design configurations that have this potential. In this scenario multiple jurisdictions have installed a flashing red arrow (FRA) to indicate a permissive left-turn. When the FYA indication was recommended in NCHRP Report 493, this particular geometric configuration was not thoroughly considered. Knodler, Noyce, Kacir and Brehmer (2006a) conducted follow-up research in an effort to identify the effectiveness of the FYA indication compared to the FRA indication at wide intersections with left-turn geometry as described.

The evaluation included three components: a dynamic driving simulator experiment, a follow-up static evaluation, and an independent static evaluation. Four permissive signal displays were evaluated, as shown in Figure 35.5 (Web Figure 35.5). The research was conducted in Massachusetts and Wisconsin—both states that use the CG for permissive left-turns and protected-only left-turn phasing at wide intersections. Therefore, it was assumed that participants were unfamiliar with the FYA and the FRA permissive indications.

The results of this research were summarized as follows:

- In the driving simulator, 10 fail-critical (go) responses were observed at the two scenarios with FYA permissive indications. However, all but one of these responses occurred on the first observation of a FYA display by each driver. Alternatively, no fail-critical responses were observed at the FRA scenarios. These results were statistically significant.
- In the follow-up static evaluation, there was no statistically significant difference between the correct response for the FYA and the FRA scenarios. The yield response occurred for the FYA in a similar proportion to the stop first response to the FRA.
- In the independent static evaluation, there was a statistically significant higher percentage of go responses for the

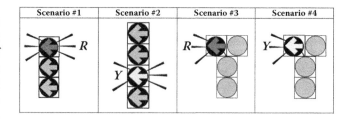

FIGURE 35.5 Permissive displays evaluated at wide median applications. (From Knodler, M. A., Jr., Noyce, D. A., Kacir, K. C., & Brehmer, C. L. (2006). Potential application of the flashing yellow arrow permissive indication in separated left-turn lanes. *Transportation Research Record, 1973*, 10–17.)

FYA scenarios than the FRA scenarios. Alternatively, there was also a statistically higher percentage of stop and wait responses to the FRA scenarios than for the FYA scenarios.

At wide intersections where the left-turn lanes are separated from the through and right-turn lanes, and the left-turn driver cannot see the through-movement indication, it was recommended that the FYA not be installed until it becomes widely implemented and more drivers are familiar with the indication. At wide intersections where protected-only left-turn phasing is not desirable, the use of the FYA or FRA permissive indications can be used.

35.4.7 Driver and Pedestrian Comprehension of Requirements for Permissive Left-Turns

In order to provide effective traffic operations, pedestrian phases are commonly operated with the parallel traffic movement. This phasing leads to conflict when a permissive left-turn is run concurrently with the pedestrian movement parallel to the through traffic. To address this concern, Knodler, Noyce, Kacir and Brehmer (2006b) conducted evaluations of driver and pedestrian comprehension of operational requirements of permissive left-turn applications. Both static and dynamic evaluations were administered to understand comprehension of the FYA from pedestrian and driver perspectives.

A dynamic driving simulator evaluation was administered to consider driver comprehension of the need to yield to pedestrians in addition to oncoming vehicles. Five signal display scenarios,

as shown in Figure 35.6 (Web Figure 35.6), were evaluated by 36 drivers. Several factors were varied within the driving simulator, including: presence or absence of pedestrians crossing the intersection; three-leg or four-leg intersection geometry; and driver maneuver made at the intersection (straight, right-turn, left-turn).

An additional 103 participants completed the static evaluation, which resulted in a total of 139 driver evaluations of the static displays. Each driver completed 25 scenarios in the static evaluation, which included an additional variable not evaluated in the dynamic simulator. The type of pedestrian varied among no pedestrian, a pedestrian waiting to cross, and a visually-impaired pedestrian (with guide-dog) waiting to cross. An example scenario from the computer-based static survey is shown in Figure 35.7 (Web Figure 35.7).

Pedestrian comprehension of the FYA indication was evaluated by a computer-based static evaluation. The evaluation presented the user with a pedestrian-crossing situation and the user was asked how they would respond under the given conditions. Seven permissive left-turn signal displays were evaluated—with and without a pedestrian signal present—to determine pedestrian comprehension of appropriate crossing opportunities. The evaluation also presented combinations of signal indications including CR, CY, SYA, and variations with pedestrian signal head scenarios.

The researchers comprehensively found that the study supports the use of the FYA as a safe and effective device for left-turn traffic. Additional conclusions are summarized below.

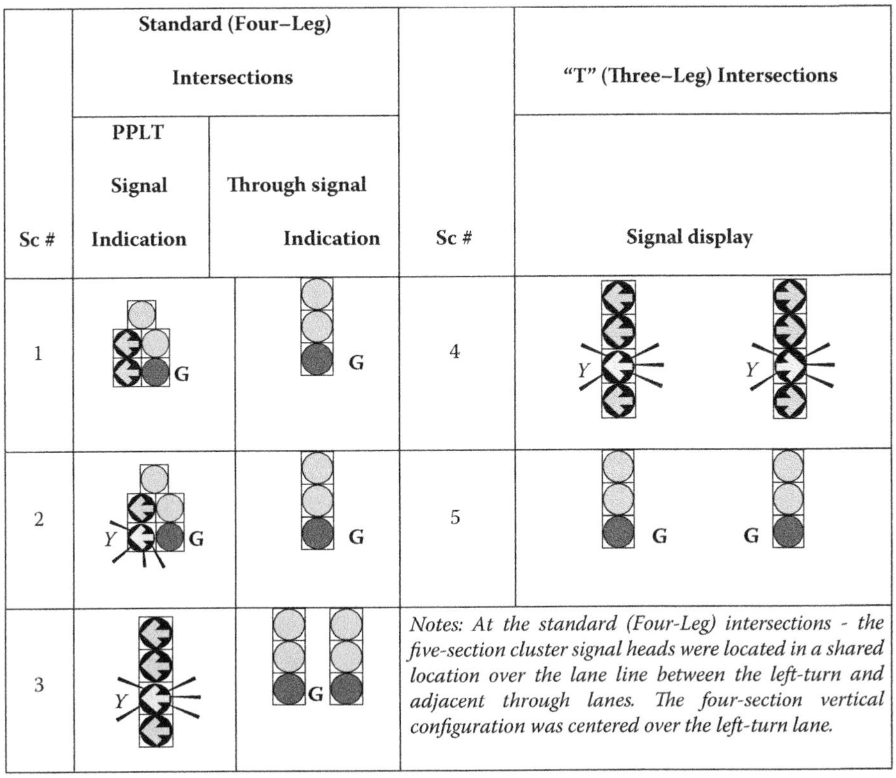

| Sc # | Standard (Four–Leg) Intersections | | Sc # | "T" (Three–Leg) Intersections |
	PPLT Signal Indication	Through signal Indication		Signal display
1			4	
2			5	
3			*Notes: At the standard (Four-Leg) intersections - the five-section cluster signal heads were located in a shared location over the lane line between the left-turn and adjacent through lanes. The four-section vertical configuration was centered over the left-turn lane.*	

FIGURE 35.6 Permissive scenarios evaluated in driving simulator. (From Knodler, M. A., Jr., Noyce, D. A., Kacir, K. C., & Brehmer, C. L. (2006). An analysis of driver and pedestrian comprehension for permissive left-turn application. *Transportation Research Record, 1982,* 65–75.)

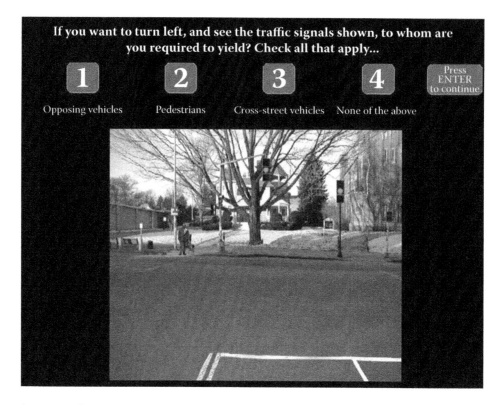

FIGURE 35.7 Sample scenario of computer-based static evaluation. (From Knodler, M. A., Jr., Noyce, D. A., Kacir, K. C., & Brehmer, C. L. (2006). Potential application of the flashing yellow arrow permissive indication in seperated left-turn lanes. *Transporation Research Record, 1973*, 10–17.)

35.4.7.1 Driver Simulation

Drivers exhibited low comprehension of the requirement to yield to pedestrians who were legally within the crosswalk. This was determined by combining the results of each permissive scenario evaluated by 36 drivers to find that the percentage of *yield* (correct) responses was lower than the percentage of fail-safe responses (the driver indicates that he or she has to stop before initiating a turn). Unfortunately, similar results can be found at most signalized intersections with pedestrian crosswalks in the U.S.

35.4.7.2 Static Driver Evaluation

At a "T" intersection, drivers were statistically more likely to respond correctly to the FYA scenario than the CG scenario. Drivers observing the FYA were statistically more likely to respond with a yield (correct) or stop and wait (fail-safe) response than the standard three-section head with a CG display. The CG display scenario was statistically more likely to result in a driver response of go (incorrect).

35.4.7.3 Static Pedestrian Evaluation

Less than half of the participants in the static evaluation of pedestrian comprehension understood correct crossing procedures, which in the absence of dedicated pedestrian signals is to cross parallel and simultaneous with traffic. Furthermore, when no pedestrian signal heads were available, pedestrians correctly identified crossing opportunities more often when viewing the CG signal display scenario than the FYA display.

35.4.8 Summary

Multiple studies have included full-scale driving simulators to evaluate driver comprehension of traffic signals including the FYA indication. Many made use of a computer-based static evaluation to supplement the driving simulator results. Key findings from both driving simulator and static evaluation studies include:

- Drivers are expected to comprehend the FYA indication while it is being implemented.
  - Driver comprehension of the FYA permissive indication remains high.
  - There is no evidence from driver studies that suggests use of the FYA will reduce comprehension of the CYA indication in use as a clearance interval.
  - The use of the FYA is not expected to reduce driver comprehension of the CG indication in use as a permitted phase at neighboring intersections. Use of the FYA may improve comprehension of the CG indication.
- Major geometric features influence driver comprehension of the FYA display.
  - The use of permissive left-turn phasing at intersections with wide medians should be considered only

after the FYA indication has been widely used and drivers have become more familiar with its use.

- Drivers were statistically more likely to respond correctly to the FYA scenario than the CG scenario at "T" intersections.
- Drivers exhibited low comprehension of the requirement to yield to pedestrians legally within the crosswalk.
  - Although drivers exhibit a cautious (fail-safe behavior) there is little pure comprehension of the need to yield the right-of-way to pedestrians.

35.5 The Future

The use of driving simulators for traffic signal experimentation is growing and will continue to grow as more driving simulators are added to transportation research laboratories. As described earlier in this and other chapters (e.g., see this book, chap. 36 by Chrysler & Nelson, "Design and Evaluation of Signs and Pavement Markings Using Driving Simulators"), the next generation of driving simulators must consider the limitations in graphic resolution in current simulators that can reduce the ability to render the necessary detail to effectively evaluate realistic size and viewing distances as well as traffic signal head placement and simultaneous visual cues. Comprehensive analyses of traffic signal systems require a detailed and accurate visual world. There remains a need to improve the quality of the visual worlds used, leading towards the implementation of video-based depictions of actual driving environments. Future generations of driving simulators will also be better able to commonly consider traffic operational aspects related to acceleration, deceleration, and turning forces not included with most driving simulators today (e.g., see this book, chap. 4 by Hancock & Sheridan, "The Future of Driving Simulation").

Key Points

- Limitations in the size, resolution, and brightness of most visual display systems restrict the ability of driving simulators to effectively evaluate detailed placement of traffic signal displays.
- Driving simulators can be very effectively used to test drivers' comprehension of traffic signal indications and displays, along with the associated driver behavior.
- The ability to replicate as much of the driving environment as possible (sound, wind, vibration, visual cues, seat belt use, and other in-vehicle attributes) is key to successful experimental use of driving simulators.
- Using more than one driving simulator in experimentation helps overcome the 'local' bias of subject drivers from one geographical area.
- Comprehensive analyses of traffic signal systems require a detailed and accurate visual world. There remains a need to improve the quality of the visual worlds used, leading towards the implementation of video-based depictions of actual driving environments.

Keywords: Driver Behavior, Driver Comprehension, Flashing Yellow Arrows, Indications, MUTCD, Permissive Left-Turn, Traffic Signals

Acknowledgments

The authors wish to acknowledge their respective laboratories and colleagues who contributed to this chapter. Specifically, we would like to thank the researchers and staff at the Traffic Operations and Safety (TOPS) Laboratory at the University of Wisconsin–Madison and the Human Performance Laboratory at the University of Massachusetts, Amherst. Many of these individuals have contributed to the traffic signal research described in this chapter. We would also like to acknowledge our research sponsors and supporters, including the National Cooperative Highway Research Program, the Federal Highway Administration, the American Association of State Highway and Transportation Officials (AASHTO), the National Committee on Uniform Traffic Control Devices, the Massachusetts Department of Transportation, the Wisconsin Department of Transportation, the University of Massachusetts, Amherst, and the University of Wisconsin–Madison.

Glossary

Permissive mode: A mode of traffic control signal operation in which left or right turns are permitted to be made after yielding to pedestrians, if any, and/or opposing traffic, if any. When a CIRCULAR GREEN signal indication is displayed, both left and right turns are permitted unless otherwise prohibited by another traffic control device. When a flashing YELLOW ARROW or flashing RED ARROW signal indication is displayed, the turn indicated by the arrow is permitted.

Protected mode: A mode of traffic control signal operation in which left or right turns are permitted to be made when a left or right GREEN ARROW signal indication is displayed.

Signal head: An assembly of one or more signal faces that is provided for controlling traffic movements on one or more approaches.

Signal indication: The illumination of a signal lens or equivalent device.

Signal phase: The right-of-way, yellow change, and red clearance intervals in a cycle that are assigned to an independent traffic movement or combination of movements.

Traffic control device: Sign, signal, marking, or other device used to regulate, warn, or guide traffic, placed on, over, or adjacent to a street, highway, private road open to public travel, pedestrian facility, or shared-use path by authority of a public agency or official having jurisdiction, or, in the case of a private road open to public travel, by authority of the private owner or private official having jurisdiction.

Traffic control signal (Traffic signal): Any highway traffic signal by which traffic is alternately directed to stop and permitted to proceed.

Yellow change interval: The first interval following the green or flashing arrow interval during which the steady yellow signal indication is displayed.

Web Resources

The *Handbook* web site contains supplemental material for this chapter, including color versions of all of the printed figures.

Web Figure 35.1: Screen Capture of Typical Simulated Intersection (Color version of Figure 35.1).

Web Figure 35.2: Permissive Left-Turn Displays Evaluated by Knodler & Noyce (2005) (Color Version of Figure 35.2).

Web Figure 35.3: Partitioned Driver Simulation Display (Color Version of Figure 35.3).

Web Figure 35.4: Signal Display Scenarios Evaluated (Figure 35.4).

Web Figure 35.5: Permissive Displays Evaluated at Wide Median Applications (Figure 35.5).

Web Figure 35.6: Permissive Scenarios Evaluated in Driving Simulator (Figure 35.6).

Web Figure 35.7: Sample Scenario of Computer-based Static Evaluation (Figure 35.7).

Key Readings

Brehmer, C. L., Kacir, K. C., Noyce, D. A., & Manser, M. P. (2003). *Evaluation of traffic signal displays for protected/permissive left turn control* (NCHRP Report 493), Washington, DC: Transportation Research Board, National Cooperative Highway Research Program.

Federal Highway Administration. (2009). *Manual of uniform traffic control devices* (2009 ed.). Washington, DC: U.S. Department of Transportation.

Knodler, M. A., Jr., Noyce, D. A., Kacir, K. C., & Brehmer, C. L. (2005). Evaluation of flashing yellow arrow in traffic signal displays with simultaneous permissive indications. *Transportation Research Record, 1918*, 46–55.

Noyce, D. A., Bergh, C. R., & Chapman, J. R. (2007). *Evaluation of the flashing yellow arrow permissive-only left-turn indication field implementation* (NCHRP Web-only Document 123). Washington, DC: National Research Council, Transportation Research Board. Available at: http://onlinepubs.trb.org/onlinepubs/nchrp/nchrp_w123.pdf

References

Agent, K. R. (1979). *An evaluation of permissive left-turn phasing* (Research Rep. 519), Kentucky Department of Transportation.

Agent, K. R. (1985). *Guidelines for the use of protected/permitted left-turn phasing* (Research Rep. UKTRP-85-19), Kentucky Transportation Research Program.

Asante, S. A., Ardekani, S. A., & Williams, J. C. (1993). *Selection criteria for left-turn phasing, indication sequence, and auxiliary sign* (Rep. 1256-1F). Arlington, TX: Civil Engineering Department, University of Texas at Arlington.

Brehmer, C. L., Kacir, K. C., Noyce, D. A., & Manser, M. P. (2003). *Evaluation of traffic signal displays for protected/permissive left turn control* (NCHRP Report 493), Washington, DC: Transportation Research Board, National Cooperative Highway Research Program.

Federal Highway Administration. (2009). *Manual of uniform traffic control devices* (2009 ed.). Washington, DC: U.S. Department of Transportation, Federal Highway Administration.

Hummer, J. E., Montgomery, R. E., & Sinha, K. C. (1993). Driver understanding of and preferences for left-turn signals. *Transportation Research Record, 1421*, 1–10.

Institute of Transportation Engineers (ITE). (1985). *Flashing yellow protected-permissive signal evaluation*. Final Report. ITE Washington Section of District 6.

Kolasinski, E. M. (1995). *Simulator sickness in virtual environments* (Tech. Rep. 1027; Army Project Number 2O262785A79 —Education and Training Technology). Alexandria, VA: US Army Research Institute for the Behavioral and Social Sciences. Retrieved January 26, 2010, from http://handle.dtic.mil/100.2/ADA295861

Knodler, M. A., Jr., & Noyce, D. A. (2005). Tracking driver eye movements at permissive left-turns. *Proceedings of the 3rd international driving symposium on human factors in driver assessment, training, and vehicle design* (pp. 134–142). Rockport, ME.

Knodler, M. A., Jr., Noyce, D. A., & Fisher, D. L. (2007). Evaluating effect of two allowable permissive left-turn indications. *Transportation Research Record, 2018*, 53–62.

Knodler, M. A., Jr., Noyce, D. A., Kacir, K. C., & Brehmer, C. L. (2005). Evaluation of flashing yellow arrow in traffic signal displays with simultaneous permissive indications. *Transportation Research Record, 1918*, 46–55.

Knodler, M. A., Jr., Noyce, D. A., Kacir, K. C., & Brehmer, C. L. (2006a). Potential application of the flashing yellow arrow permissive indication in separated left-turn lanes. *Transportation Research Record, 1973*, 10–17.

Knodler, M. A., Jr., Noyce, D. A., Kacir, K. C., & Brehmer, C. L. (2006b). An analysis of driver and pedestrian comprehension for permissive left-turn applications. *Transportation Research Record, 1982*, 65–75.

Noyce, D. A., Bergh, C. R., & Chapman, J. R. (2007). *Evaluation of the flashing yellow arrow permissive-only left-turn indication field implementation* (NCHRP Web-only Document 123). Washington, DC: National Research Council, Transportation Research Board. Available at: http://onlinepubs.trb.org/onlinepubs/nchrp/nchrp_w123.pdf

Smith, C. R., & Noyce, D. A. (2000). *An evaluation of five-section protected/permitted left-turn signal displays using driving simulator technology*. Compendium of Technical Papers for the 70th Annual Meeting of the Institute for Transportation Engineers. Nashville, TN.

Staplin, L., & Fisk, A. D. (1991). A cognitive engineering approach to improving signalized left turn intersections. *Human Factors, 33*(5), 559–571.

Szymkowiak, A., Fisher, D. L., & Connerney, K. A. (1997). False yield and false go decisions at signalized left-turn intersections: A driving simulator study. In K.-P. Holzhausen (ed.) *Advances in multimedia and simulation. Human-machine-interface implications,* Proceedings of the Human Factors and Ergonomics Society Annual Conference in Bochum, November 1997 (pp. 226–235). Bochum, Germany: Fachhochschule Bochum, University of Applied Sciences.

Transportation Research Board. (2000). *Highway capacity manual.* Washington, DC: National Research Council: Transportation Research Board.

Design and Evaluation of Signs and Pavement Markings Using Driving Simulators

Susan T. Chrysler
Texas Transportation Institute

Alicia A. Nelson
Texas Transportation Institute

Abstract

The Problem. Drivers rely on traffic control devices for guidance, regulatory, and warning information while traveling. These devices, primarily signs and pavement markings, must be easily seen and comprehended in order to be most effective. *Role of Driving Simulators.* Driving simulators offer a safe and low-cost environment in which to test the effectiveness of traffic control devices in a variety of roadway conditions. They provide repeatability for experimental control and flexibility to quickly evaluate numerous alternative designs. *Key Results of Driving Simulator Studies.* Driving simulator studies have provided useful evaluations of dynamic message sign cycle time, guide sign designs, pavement marking designs, and work zone devices. These results have been implemented through traffic engineering guidelines and manuals. *Scenarios and Dependent Variables.* Typical traffic control device scenarios include normal driving, with and without other traffic present, in rural and urban scenes. Dependent variables include lane keeping, speed, reading time, comprehension, navigation to a specified destination, and subjective preference measures. *Platform Specificity and Equipment Limitations.* Graphics resolution limits the ability of many simulators to render the fine detail of some traffic signs at realistic sizes and viewing distances. The dynamic range of projection and computer screens limits the ability of all simulators to accurately re-create a night-time driving environment and the luminance changes which occur with vehicle headlamps and retroreflective traffic control device materials.

36.1 Introduction

Traffic control devices (TCDs) are an important part of the roadway communication system. Traffic signs and pavement markings are often tested in closed-course and on-road studies of detection, legibility, and comprehension. These methods

provide the necessary range of luminance and contrast to evaluate photometric properties of traffic control devices, particularly at night. Closed-course studies of traffic sign properties such as conspicuity and legibility are challenging and expensive. They require signs to be fabricated, often a work crew to change signs, an instrumented vehicle, good weather, and careful control over lighting conditions. Tests of pavement markings are even more difficult because of the labor involved with changing stimulus features, such as brightness or width of the markings, during a single experimental session. While these methods provide the best replication of the luminance of signs in real-world driving conditions, the driving task itself is often not very demanding. The lack of other traffic, the simple roadway design of most closed-courses, and the focus on the sign reading task can create an artificial situation that may produce results that would not be replicated on actual roads.

Laboratory methods can also be used to evaluate traffic control devices. These approaches are particularly well-suited for tests of comprehension and sign design. Tachistoscopic displays of traffic sign images can be used to duplicate the limited viewing time of a sign present on the roadway. These laboratory methods offer the opportunity to test multiple versions of a sign design inexpensively and with proper experimental design and control.

Driving simulators can offer a convenient and safe method of evaluating the effectiveness of traffic control devices. Simulation can provide a challenging driving task that requires normal vehicle control while the experimental subject is also completing some task-related to the sign or marking. Simulation's main drawback for traffic control device lies in the resolution and dynamic brightness range of display systems.

This chapter will describe several measures of effectiveness which can used to evaluate traffic control devices. Within each method, traditional laboratory and in-vehicle evaluation methods will be reviewed as well as studies using driving simulators. The focus in this chapter is on the methodology rather than the results of the research. We hope to demonstrate the advantages and disadvantages of using driving simulators to evaluate traffic control devices. We also present some solutions to methodological problems that we have employed at the Texas Transportation Institute.

36.2 Measures of Effectiveness Used in Non-Simulator Studies

The measures of effectiveness for traffic control devices range from low-level perceptual tests of detection to higher level cognitive tests of comprehension. For traffic engineers, the ultimate test of effectiveness is driver compliance with the device. In this section, we present classic studies of traffic control devices representing a range of methodologies and measures of effectiveness.

36.2.1 Detection Distance

The first thing a driver must do with a traffic control device is to notice it. In a cluttered driving scene, traffic signs, pavement markings, delineators, and work zone devices may not attract the visual attention of drivers. During the daytime, it is primarily color and shape contrast that identifies traffic signs, though luminance does play a role (Jenkins & Cole, 1985; Brekke & Jenssen, 1997). Night-time detection is largely controlled by luminance contrast with the surrounding visual scene. The

vast majority of passive (i.e., not internally illuminated) traffic control devices utilize retroreflective materials in their fabrication. These materials reflect the light of the vehicle's headlamps back toward the driver. The device itself can be thought of as a filter of this light, changing its color, intensity, and direction as a function of the properties of the retroreflective material.

Conspicuity is the ability of an object to attract attention to itself. Conspicuity in traffic control device research is most often operationalized by detection distance or speed. The primary measure of effectiveness for detection is distance from the target. Distance can be measured dynamically in a vehicle by having the subject press a button directly or speak out loud when a target is detected. A verbal response can trigger a voice key connected directly to a computer, or more typically, an in-vehicle experimenter presses a response button upon hearing the verbal response. The response button may be linked to a distance-measurement instrument wired to the vehicle odometer or a global positioning system (GPS) within the vehicle. The target sign's location, which must be marked along the course using the same device and detection distance, is derived through subtraction. Research subjects can be drivers or passengers in the vehicle.

This basic methodology has been used to evaluate the daytime and night-time conspicuity of traffic signs (Jenkins & Cole, 1997; Zwahlen, 1989) in closed-course and open-road testing. It has also been used to evaluate pavement marking brightness (Zwahlen & Schnell, 1999) and pedestrians (Chrysler, Danielson, & Kirby, 1996). In addition to simple detection, color and shape recognition distances have also been tested in the same way (Brekke & Jennsen, 1997). Zwahlen (1989) illustrates that many roadway targets, such as signs and parked vehicles, initially appear 15 degrees or more off the driver's direct line of sight. This level of eccentricity may pose problems for some driving simulation systems.

An alternative way to measure detection distance is to have the subjects stationary and to measure the probability of detection at a series of fixed distances. In this way a cumulative probability distribution can be derived. The advantage to this method is that it allows more precise measurement because of the removal of error due to the moving vehicle. Methodologically, the stationary observer can in fact be obtained by moving the target successively closer to the subject or by moving the subject in discrete steps toward the target. Researchers at the Pennsylvania Transportation Institute used this method to evaluate pavement markings and post-mounted delineators at night (Pietrucha, Hostetter, Staplin, & Obermeyer, 1995). Conspicuity can also be measured in a laboratory setting. Research methods here include search time for brief displays of sign stimuli using occlusion devices (Scheiber, 2002).

36.2.2 Legibility Distance

For traffic signs, both static and dynamic, the next task for a driver is to read the words or symbols on the sign. The earliest work on traffic sign typefaces and design was conducted by Forbes in the 1930s using full-size signs in the field (Forbes & Holmes, 1939).

Later work by Olson and Sivak again used field conditions to assess legibility (Sivak & Olson, 1985). Further test track work was completed using full-sized signs and instrumented vehicles to test new fonts and retroreflective sheeting materials (Chrysler, Carlson, & Hawkins, 2002). On-road studies of legibility similarly used actual signs and distance-measuring instruments while observers drove actual roadways (Chrysler, Stackhouse, Tranchida, & Arthur, 2004). On-road studies of changeable message signs have also been completed (Ullman, Ullman, Dudek, & Ramirez, 2005). Again, the main measure of these field studies is the distance at which observers can accurately read words or identify symbols. In some cases, actual words that might appear on traffic signs are used, while in others non-traffic words, random letter combinations, or Landolt-C rings were used to assess legibility as a function of some traffic sign feature such as brightness or color. Landolt-C rings are standard optotypes consisting of a circle with a gap, equal to the stroke width of the circle, which can be rotated. Visibility is assessed by having the observer identify which direction the gap is facing. Standard optotypes such as this remove inherent bias in the familiarity of target words, can be used with subjects of different English reading abilities, and correspond directly to optometric principles of visual angle of resolution.

Laboratory studies of legibility have been executed to evaluate font type, brightness, color contrast, and other sign design features. These studies have used computer monitors located down long hallways (Garvey, Pietrucha, & Meeker, 1997). In these methods, the stimulus is often displayed briefly while the observer is stationary to determine a probability of correct legibility as a function of distance. Other laboratory methods have used "sign simulators", which are essentially zoom slide projectors wherein a high-resolution image of a sign is slowly enlarged until the observer correctly reads it (Katz, Hawkins, Kennedy, & Howard, 2008; Dewar, Kline, Schieber, & Swanson, 1994).

36.2.3 Comprehension

After drivers detect and read a sign, they must understand the intended message in order for it to be effective. Comprehension is often measured through surveys using line drawings of signs in isolation, or pictured in roadway scenes. Research participants may answer open-ended questions asking them to describe their understanding of the sign (Dudek, Ullman, Trout, Finley, & Ullman, 2006b) or answer multiple choice questions (Picha, Hawkins, & Womack, 1995). These surveys often permit unlimited viewing time of the sign which may inflate the comprehension scores. Some surveys, using computer or slideshow presentations, may limit the amount of time the stimulus is viewed. These methods have been used for comprehension of pavement markings and changeable message sign text as well (Parham, Womack, & Hawkins, 2003; Stackhouse & Burrus, 1996).

One study directly compared these methods to driving simulation and found that a limited viewing time presentation of sign images in roadway scenes produced comprehension scores equivalent to those seen in a driving simulator (Chrysler, Wright, & Williams, 2004).

36.2.4 Compliance

Traffic engineering studies which observe driver's compliance with traffic control devices are the ultimate measure of effectiveness. The U.S. Department of Transportation (USDOT) Manual on Uniform Traffic Control Devices (MUTCD) is the document, empowered in the Code of Federal Regulations, which governs the design and application of traffic signs, signals, and pavement markings on all roadways open to public travel (USDOT, 2009). It is the "signing bible" used by traffic engineers throughout the U.S. The MUTCD requires an evaluation of new devices by an agency wishing to experiment with a new sign, marking, or warning device. These evaluations typically involve traffic counters to measure speed, stopping, lane keeping, or some other traffic feature. Surveillance, either directly on-site or through cameras, is often used to observe vehicular and pedestrian traffic patterns at a site before and after the new device is deployed. An evaluation of crash and violation history in a "before-and-after" study is another method.

In our experience with working with the traffic engineers responsible for implementing changes to the MUTCD, the desire for actual traffic performance data is great—a field operational test that measures driver compliance with a new TCD is the gold standard. All of the other methods described in this section, including driving simulation studies, often do not satisfy the traffic engineer's criterion for the effectiveness of a traffic control device. These laboratory and field methods often are used to screen candidate designs in order to deploy the most promising into the field for a traffic study.

36.3 Driving Simulators and Traffic Control Devices

36.3.1 Limitations of Driving Simulation

All driving simulators are limited by their display systems in terms of image resolution, color accuracy, and luminance range. These limitations prevent accurate rendering of traffic signs and pavement markings, particularly when viewed at a distance.* This is a particular problem for studies of detection and legibility. The limitations of the display systems are described in more detail in this section. In addition to limitations of graphics rendering by display systems, driving simulator scene authoring and scenario authoring present additional limitations for TCD testing. These limitations are more specific to the simulator system being used. Solutions to some of these problems are presented at the end of the chapter.

36.3.1.1 Dynamic Range

All display systems output light. Even when a black object is displayed, there is still light being output. For this reason there are

---

* The comments regarding limitations of simulation are based on the authors' experience with simulators made by Systems Technology Inc. during the mid-1990s (single CRT screen) and DriveSafety between 2001 and 2008 (three screens, ceiling LCD projectors, software version 1.6).

no "black blacks" when using a computer screen or projection system. Similarly, the brightest white is limited by the illumination system used. The dynamic range of these light outputs (i.e., the range of values dark to light) in no way duplicates the real world where the difference between the darkest night and the brightest day is orders of magnitude. Even for daytime scenes, simulators typically do a poor job of accurately reproducing the luminance range available in a typical driving scene.

36.3.1.2 Constant Luminance

In outdoor scenes the luminance of objects depends on their reflective characteristics and the quality and direction of their illumination. In the daytime, shadows and imperfections of surfaces create mottled appearances and changes in brightness. At night, the luminance of retroreflective signs and markings depends on their particular material properties and the distance and direction of the illumination source. The brightness of a sign, for instance, can change ten-fold while a vehicle approaches it. Indeed, much research on detection has been focused on the sign material and vehicle illumination properties. The detection threshold depends on luminance contrast and size of the target. In simulators that only render a constant luminance of a target, the sign often appears artificially bright at far and near distances. This could be overcome by assigning a dynamic brightness to the sign target which follows, in a relative way, the luminance curves measured and modeled in photometric research (Uding, 1993; Chrysler, 1999).

36.3.1.3 Resolution and Refresh Rate

Another limitation of display systems is their resolution. This is a particular problem for traffic control device research because renderings of signs and markings at a far distance down the road often are pixilated and appear to shimmer on the screen. This shimmering continues until the image becomes large enough to resolve itself. In research in our laboratory, subjects have reported that the shimmering is distracting and causes them to watch the sign in an unnatural fashion.

Resolution and refresh rate also affect changeable message signs that may have more than one phase (i.e., alternate between two message sets). Many driving simulators are not equipped to have objects in the scene that change their appearance dynamically in this way. This is an example of a limitation due to scenario authoring (see Kearney & Grechkin, this book, chap. 6 for additional discussion of scenario authoring).

36.3.1.4 Display Size

The legibility of signs depends on the critical features surpassing some critical visual angle threshold in order for the eye to resolve the image and identify the component parts. In simulators, the researcher often fails to adjust the sign image size or viewing distance to control for the visual angle of the target traffic control device. The eccentricity (i.e., how far off the straight line of sight) of the target is often not controlled either, this is a particular problem for single channel systems. The viewing time of an overhead sign may be cut short because the

image disappears at the top of the screen sooner than it would on a real roadway.

36.3.1.5 Verisimilitude of Traffic Control Device Images

The stock signs and markings available in many driving simulation systems do not use the typefaces, colors, and symbol designs of the MUTCD. Color rendering in particular is often quite inaccurate and could contribute to slight delays in drivers' recognition of familiar signs. Pavement marking appearance is another area of great variation between simulation systems. This is true for both line width and the patterns of skip lines. Changeable message signs, if available at all in a simulation library, are often displayed with solid line characters rather than the pixels present on the majority of LED signs on roadways. These are all examples of limitations due to the scene authoring, or object development, software used by the simulator system. While these may seem like minor complaints, these slight variations all contribute to the artificiality of the simulation for the drivers. In addition, traffic engineers visiting our laboratory often fixate on these minor discrepancies and dismiss simulation wholesale because of them.

36.3.2 Detection and Legibility

As described in the previous section, the limitations of simulator display systems are most restrictive for studies of detection and legibility of traffic control devices. For these reasons, not many simulator studies have been done using these measures of effectiveness. Some work on pavement markings and raised pavement markers (road studs) have been done, but the authors acknowledge the issues of dynamic range and constant luminance as affecting their conclusions. Two early studies of pavement marking line width and brightness used low-fidelity driving simulators (McKnight, McKnight, & Tippetts, 1998; Allen & O'Hanlon, 1979) to assess driver lane keeping as a function of marking features. Other studies have examined raised pavement marking brightness and patterns and have found the dynamic range issues of the displays to limit the generalizability of their results (Bartelme, Watson, Dingus, & Stoner, 1995; Molino, Opiela, Andersen, & Moyer, 2003). A few studies of sign legibility have been completed, primarily in the area of changeable message signs (Guerrier & Wachtel, 2002).

36.3.3 Comprehension

In driving simulator studies, the line between using comprehension and compliance as a measure of effectiveness is a fine one. Drivers can demonstrate their comprehension of a device by heeding its direction or advice through some driving behavior such as speed modification or lane changes. In this section, we view those studies that use driving behavior metrics as measuring comprehension rather than compliance. These are presented in the next section.

Some studies have directly measured comprehension of signs and pavement markings in a driving simulation context. Often the simulator is used essentially as an elaborate display system and task-loading device. Driving performance metrics are never examined. Instead, comprehension is tested in the same way as the laboratory and survey studies described earlier. Test subjects encounter test signs while driving a simulator scenario and the experimenter interrupts the scene to ask multiple-choice or open-ended comprehension questions either verbally or on paper or on a computer screen. Several studies of changeable message sign abbreviations, message phasing, and graphic features have been conducted using these methods (Ullman, Ullman, Dudek, Nelson, & Pesti, 2005; Dudek, Schrock, Ullman, & Chrysler, 2006; Wang, Collyer, & Hesar, 2007). Evaluations of the comprehension of different symbol sign designs and understanding of warning sign colors were also completed using early driving simulators (Lum, Roberts, DiMarco, & Allen, 1983; Allen, Parseghian, & Van Valkenburgh, 1980). While the addition of a driving task may more accurately represent the workload of real-world driving, the repeated comprehension questions do cue the participant to pay particular attention to the signs so they can answer the questions correctly.

36.3.4 Compliance

Tests of compliance are where driving simulators really offer a great value to traffic control device researchers. Simulators can offer a low-cost substitute for the "gold standard" field study of driving behavior that traffic engineers desire. Simulation studies allow the researcher to present traffic control devices as naturally-occurring roadway features and observe driving behavior, just as in a real-world field study. One problem with all of the measures of effectiveness discussed up to this point is that the research subject usually figures out that it is the devices that are the topic of study. They may then increase the amount of attention they pay to signs and markings in order to please the experimenter. Simulator studies of compliance allow the experimenter to develop a cover story or to assign a natural driving task, such as car following or wayfinding, in order to minimize the emphasis on the different devices being evaluated (see Ranney, this book, chap. 9 for more information on the importance of the psychological fidelity of simulator scenarios). Compliance has been assessed through evaluation of many driving behaviors and each is presented in the following sections.

36.3.4.1 Lane Change

Many signs and pavement markings are installed on the roadway in order to tell drivers which lane they should be in. Simulators offer the opportunity to see how soon after seeing a sign or pavement marking these lane changes are initiated by drivers. This measure of effectiveness has been used in evaluating guide signs in advance of freeway interchanges in several studies (Chrysler, Nelson, Funkhouser, Holick, & Brewer, 2007; Upchurch, Fisher, & Waraich, 2005). It has also been used to evaluate advance exit signs in tunnels (Upchurch, Fisher, Carpenter, & Dutta, 2002), variable message signs (Dutta, Fisher, & Noyce, 2005), and one-way sign

designs (Laurie et al., 2004). The same measure of effectiveness was used by Ullman et al. (2008) to evaluate pavement marking route symbols and text in advance of work zones and exits.

One challenge in using this metric is determining the exact point at which a driver's steering inputs indicate lane change intent. Normal steering behavior has oscillations of the steering wheel to maintain lane position. Data reduction strategies that rely only on searching for steering wheel reversals to indicate lane change initiation may inaccurately classify a lane maintenance steering input for a lane change start. Methods which use lane crossing criteria or continuous steering input in one direction are more reliable.

In all of these studies, the display issues associated with traffic control devices are mitigated because the measure is taken *after* the sign or marking is passed by the driver. A candidate sign or marking is deemed to be successful because drivers made the desired lane change soon after passing the sign, or did not make any unnecessary lane changes.

36.3.4.2 Lane Placement

Position within a lane has been used in evaluations of treatments designed to improve lane placement, such as rumble strips and lane markings (Stanley, 2006; Chrysler, Williams, & Park, 2006). Metrics include absolute lane position, changes to lane position over time, standard deviation of lane position and number of excursions.

In conducting a study on rural road lane width and the presence of edgelines, the authors discovered great inconsistency in field and simulator tests with regard to the way in which lane excursions were scored (Chrysler & Williams, 2005). Figure 36.1 illustrates the different ways in which lane excursions may be defined and measured. Some simulators measure lane placement from the theoretical center of gravity of the vehicle and edgeline position may also be measured from the center of the lane. When comparing simulator results to field or closed-course studies, these differences of two to eight inches matter because

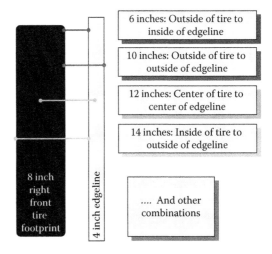

FIGURE 36.1 Illustration of variability sources in calculating lane position and lane excursion status.

most field studies of the effects of markings on lane placement may show differences of only six to ten inches.

36.3.4.3 Speed

Speed mitigation is the goal of many traffic control devices. Speed is easily measured in simulators. In addition to vehicle velocity, driver input to throttle and brake pedals can also be used as measures of speed control. Speed has been used as a measure of effectiveness in many studies, including assessments of fog-warning sign messages (Hogema & van der Horst, 1994). Speed has also been used to assess curve warning devices and treatments such as tunnel wall treatments and internally-illuminated barrier treatments (Manser & Hancock, 2007; Vercruyssen et al., 1995).

It should be noted that in all of these studies, the measure was actually speed in the treatment condition compared to some baseline condition or some change in speed following the device. Because drivers' perception of speed in a simulator is typically not realistic, direct generalizations from simulator to roadway speeds are not recommended (though note that sometimes the perception of speed in the two environments can be surprisingly similar; e.g., Hurwitz, Knodler, & Dulaski, 2005).

36.3.4.4 Headway

In addition to speed, another measure of longitudinal vehicle control is headway to a lead vehicle. Dudek, Schrock, Ullman and Chrysler (2006a) evaluated the workload of different changeable message sign designs by asking drivers to maintain a constant headway behind a lead vehicle. At certain times, the simulated lead vehicle began changing its speed unpredictably and without the activation of brake lights. Dudek, Schrock and Ullman (2007) found that drivers were slower to notice these lead vehicle speed changes (and thus showed greater headway variation) when the changeable message signs contained complex messages.

36.4 Tricks of the Trade

During the authors' 15 years of using driving simulators to evaluate traffic control devices of all kinds, we and our colleagues have created workarounds for many of the limitations of the simulation systems with which we have worked. This section presents some of our "tricks of the trade" which readers may find helpful in designing their own studies.

36.4.1 Creating Sign Images

We have found that the best way to create accurate MUTCD compliant sign images is by using commercially-available sign design software such as SignCad® (SignCad Systems, Inc.). This software includes standard fonts, arrow designs, and route markers as well as layout grids. The images can be exported as bitmaps which can typically be mapped to a blank sign in the simulator.

36.4.1.1 Aspect Ratio

For simulator systems, such as DriveSafety, which have limited aspect ratios available for customizable sign objects, the SignCad

drawing template may have to be adjusted to match the simulator sign dimensions in order to avoid any skewing when imported. For other systems which allow the use of re-sizing handles or direct entry of dimensions, care must be taken to maintain the aspect ratio specified in the MUTCD and the Standard Highway Signs *Handbook* (USDOT, 2004).

36.4.1.2 Shape

Some sign shapes such as octagons, triangles, and diamonds do not correspond directly to a polygon in the scene authoring software. In addition, sign drawing software such as SignCad typically exports rectangular files with white space surrounding non-rectangular portions of the image. So, for images such as the typical freeway sign shown in Figure 36.2, the main portion of the sign is one rectangular object and the exit number plaque is a separate rectangular object. These two objects are then placed in the scene to be contiguous. This extra step of placing the two objects can be error prone and, once placed, re-sizing and moving the two objects is troublesome but it avoids having blank or white space at the top of the sign next to the exit number plaque.

36.4.1.3 Color

The colors created in sign design packages, such as SignCad, may not match the color of stock sign images included in the simulator system. Images of new signs may have to be edited in graphics software in order to dull the images to match the overall apparent scene brightness in the simulator. Additional color adjustment and texturing may also be needed to match new signs to stock signs in the scene authoring software. The MUTCD does reference color specifications but these are for daytime appearance of retroreflective materials. The daytime appearance of signs in the real world is affected by sun angle, cloud cover, and other ambient factors. Iterative adjustments of

FIGURE 36.2 A standard freeway sign exported from SignCad. The main sign rectangle and the exit plaque rectangle may have to be treated as separate objects in the scene authoring software. The font shown is ClearviewHWY®, a FHWA-approved option to Highway Gothic.

FIGURE 36.3 Large, multi-lane overhead guide sign created by splicing two customizable signs together in order to maintain correct aspect ratio.

color and texture appearance of novel sign faces must be completed using the researcher's judgment about how veridical the new signs appear.

36.4.2 Creating Large Guide Signs

For large guide signs which may extend over multiple roadway lanes the dimensions and aspect ratio of the customizable sign objects provided in the simulator software may not be adequate. In these cases, we have drawn multiple panels in SignCad which are then used to create multiple signs objects of the simulator-determined size (see Figure 36.3). These multiple panels are then placed side-by-side in the scenario to create one large sign. Care must be taken to avoid visible seams. When viewed at a distance the shimmering due to pixilation may be visible along these seams.

36.4.3 Stopping Pixilation

For custom signs and changeable message sign panels, the simulator often produces a shimmering pixilated image. We have resolved this issue by creating one sign image where the words are replaced by large rectangles the same color as the text. These large blocks do not suffer the same shimmering at a distance. Through trial and error, the distance at which the shimmering stops is determined. A trigger is then inserted into the scenario such that at that distance the colored rectangle image is replaced by the actual sign text image.

Another trick to avoid shimmering is the use of anti-aliasing; most graphic programs allow you to save your images with anti-aliasing. When you look at an image closely, you often see a jagged, stair-step effect along the edges of colors. This is referred to as aliasing. Anti-aliasing is a process whereby the software creates

blends of the adjacent colors for a smoother color transition along the edges. The result is a more aesthetically appealing image with fewer hard edges that appear to shimmer in the simulator (see Figure 36.4). This is extremely important when you are dealing with sign legibility.

36.4.4 Creating Two-Phase Changeable Message Signs

This same technique of replacing one sign face image with another has been used to create two-phase changeable message signs. In this case a time-based trigger is inserted into the scenario which replaces the Phase 1 image with the Phase 2 image on a regular schedule, much like an actual CMS.

FIGURE 36.4 Illustration of anti-aliasing. Bottom image was saved using anti-alias settings. Note the differences in jagged appearance on rounded portions of the image.

36.4.5 Creating Sign-Mounting Structures

In the simulator systems we have used, the sign poles are often not separate objects from the stock signs. When creating a customized sign image, there is often not a structure available on which to place it. We have overcome this limitation in a number of ways. For ground-mounted signs we have used a stock sign and then placed the customized sign face immediately in front of it such that the pole from the stock sign is visible and appears to be attached to the new sign face. For overhead signs, we have used street light objects and placed a sign face in front of the luminaire of the street light (see Figure 36.5, Web Figure 36.1 for color version). Again, the pole is visible and approximately the right size and height to appear to be supporting the sign. For full-span overhead signs, we have used a street light on each side of the road. For systems—such as Realtime Technologies, Inc.—that allow the importing of VRML objects sign structures can be drawn in VRML editors and imported directly.

For trailer-mounted portable changeable message signs typically found in advance of work zones, sign face images can be placed in front of other trailer types which exist in the scene authoring software library (see Figure 36.6, Web Figure 36.2 for color version).

36.4.6 Adjusting Size to Provide Adequate Reading Time

For some studies, we have intentionally made the custom sign image larger than scale in order to provide adequate reading time (see Figure 36.7, Web Figure 36.3). This is an iterative process of trial and error in which experimenters judge the point at which a sign becomes legible in various sizes. This technique

FIGURE 36.5 Illustration of a light pole object from the scene authoring software with a sign face placed "floating" above the roadway to give the appearance to a driver of a cantilever sign support structure.

FIGURE 36.6 Portable changeable message sign typically used in work zones. It was created by placing a customizable rectangular text sign in front of a construction arrow trailer found in the scene authoring library.

works best when higher speeds are involved as the subject is less likely to notice the larger sign when passing by at high speed.

36.4.7 Punching a Hole in the Scene

In order to get the highest resolution of a sign face, we have created black boxes in the simulator scene in fixed positions, roughly located on the side of the road where a sign would appear. A second projector (see Figure 36.8) is then used to display high-resolution images of changeable message sign images into that black box (Dudek, Schrock, Ullman, & Chrysler 2006a). This allows realistic color and LED pixel representation of CMS sign faces. Note that this technique only works for signs with black backgrounds.

When using this technique, we embedded a text cue to the experimenter in the lower corner of the scene which was out of view of the driver (due to the vehicle door). When this cue appeared the experimenter activated a second laptop which projected the image into the black box (see Figure 36.9, Web Figure 36.4). One drawback to this technique is that the black box remains in the

FIGURE 36.7 Changeable message sign out of roadway scale to provide adequate reading distance.

FIGURE 36.8 Supplemental projector used to show sign face images in "black box" in simulator scene.

scene at all times, or is triggered to appear at a certain point which will cue the drivers that a message is about to appear.

36.4.8 Over-Projecting

Another use for a secondary projector is to display an image on top of the simulator-generated scene. These secondary images often are lighter than the simulator scene and do not move as the scene changes. In a study of pavement marking text and symbols, we were able to project images onto the roadway surface and control their screen position by placing the second projector on a swivel base (see Figure 36.10). An experimenter then stood next to the projector during the experiment and made adjustments in the lateral position of the marking as the driver changed lanes. A simple Powerpoint animation was used to make the image drop down the

FIGURE 36.10 Supplementary projector mounted on a swivel base to allow the over-projecting and repositioning of images on the roadway surface.

screen and increase in size as the driver approached it. The resulting scene gave the appearance of pavement marking shields coming into view and gave the impression of the car passing over them (Ullman et al., 2008) (see Figure 36.11, Web Figure 36.5).

36.5 Conclusion

Driving simulators provide a safe, climate-controlled environment in which to test driver understanding of traffic control devices. The limitations of the scene authoring software and display technologies may prevent evaluations on aspects of traffic signs that rely on visual cues, such as tests of legibility or conspicuity. With some ingenuity, many of these limitations can be overcome by creative use of scene authoring objects, the creation of new objects, and the use of supplementary projectors.

FIGURE 36.9 (See color insert) Simulated changeable message sign created by inserting black box with scenario authoring software and using supplementary projector to project image of sign legend into black box.

FIGURE 36.11 (See color insert) Pavement marking shields over-projected onto the roadway scene.

Key Points

- Limitations in the size, resolution, and brightness of display systems limits the applicability of driving simulators to tests of detection and legibility.
- Driving simulators are best applied to studies of traffic control device compliance.
- Researchers must create unconventional solutions to some limitations of common driving simulation software.

Keywords: Comprehension, Legibility, Luminance, Pavement Markings, Traffic Signs

Acknowledgments

The authors wish to thank the many members of the Bryan/College Station community who have volunteered as research participants in our laboratory. The authors wish to acknowledge the Center for Transportation Safety at the Texas Transportation Institute (TTI) for its continuing support of the driving simulation laboratory. We also thank TTI staff, former and present, who have contributed to our driving simulator traffic control device research: Steve Schrock, Brooke Ullman, Jerry Ullman, Nada Trout and Conrad Dudek. These individuals have been creative problem-solvers in overcoming the limitations of our simulator and in developing new methodologies and measures of effectiveness. We also acknowledge our research sponsors including the Texas Department of Transportation, the Federal Highway Administration, and the Southwest University Transportation Center.

Web Resources

The *Handbook*'s web site contains supplemental materials for this chapter including videos of driving simulator runs illustrating some of the tricks described in the chapter and color versions of Figures 36.5, 36.6, 36.7, 36.9 and 36.11.

Web Figure 36.1: Illustration of a light pole object from the scene authoring software with a sign face placed "floating" above the roadway to give the appearance to a driver of a cantilever sign support structure (color version of Figure 36.5 in print chapter).

Web Figure 36.2: Portable Changeable Message Sign typically used in work zones. It was created by placing a customizable rectangular text sign in front of a construction arrow trailer found in the scene authoring library (color version of Figure 36.6).

Web Figure 36.3: Changeable Message Sign out of roadway scale to provide adequate reading distance (color version of Figure 36.7).

Web Figure 36.4: Simulated Changeable Message Sign created by inserting black box with scenario authoring software and using supplementary projector to project image of sign legend into black box (color version of Figure 36.9).

Web Figure 36.5: Pavement marking shields over-projected onto the roadway scene (color version of Figure 36.11).

Web Video 36.1: Use of auxiliary projector to display pavement markings on roadway surface in DriveSafety Ver 1.6.

Web Video 36.2: Use of novel channelizing devices which were drawn in Google SketchUp in Realtime Technologies system.

Key Readings

Chrysler, S. T., Williams, A. A., & Park, E. S. (2006). *Traffic engineering applications of driving simulation.* Southwest University Transportation Center, Texas A&M University (Rep. 167142-1). Available from http://swutc.tamu.edu/publications/technicalreports/167142-1.pdf

Dudek, C. L., Schrock, S. D., Ullman, G. L., & Chrysler, S. T. (2006). Flashing message features on changeable message signs. *Transportation Research Record, 1959,* 122–129. Available from http://trb.metapress.com/content/v3368577l76w5527/fulltext.pdf

Ullman, G. L., Finley, M. D., Pike, A. M., Knapp, K. K., Songchitruksa, P., & Nelson, A. A. (2008). *Studies to improve temporary traffic control at urban freeway interchanges and pavement marking material selection in work zones* (Rep. 0-5238-2). College Station, TX: Texas Transportation Institute Report. Available from http://tti.tamu.edu/documents/0-5238-2.pdf

Ullman, G. L., Ullman, B. R., Dudek, C. L., Nelson, A. A., & Pesti, G. (2005). *Advanced notification messages and use of sequential portable changeable message signs in work zones* (Rep. 0-4748-1). College Station, TX: Texas Transportation Institute. Available from http://tti.tamu.edu/documents/0-4748-1.pdf

Wang, J., Collyer, C., & Hesar, S. (2007). Adding graphics to dynamic message sign messages. *Transportation Research Record, 2018,* 63–71. Available from http://trb.metapress.com/content/7524034278251223/

References

Allen, R. W., & O'Hanlon, J. F. (1979). Effects of roadway delineation and visibility conditions on driver steering performance. *Transportation Research Record, 739,* 5–8.

Allen, R. W., Parseghian, Z., & Van Valkenburgh, P. G. (1980). *Age effects on symbol sign recognition* (Rep. FHWA-RD-80-126). Washington, DC: Federal Highway Administration.

Bartelme, M. J., Watson, G. S., Dingus, T. A., & Stoner, J. W. (1995). Driving simulation studies of raised pavement markers for roadway delineation. *Proceedings of the 65th annual meeting of the institute of transportation engineers* (pp. 609–613). Washington, DC.

Brekke, B., & Jenssen, G. D. (1997). A comparison of the visibility of fluorescent and standard retroreflective traffic signs day and night. *Proceedings of PAL '97—Progress in automobile lighting* (pp. 108–118). Darmstadt University of Technology, Darmstadt, Germany.

Chrysler, S. T. (1999). *Luminance: A systems approach to the visibility of traffic control devices.* Paper presented at the 69th Annual Meeting of the Institute of Transportation Engineers, Washington, DC.

Chrysler, S. T., Carlson, P. J., & Hawkins, H. G. (2002). *Nighttime legibility of ground-mounted traffic signs as a function of font, color, and retroreflective sheeting type* (Rep. FHWA/

TX-03/1796-2). College Station, TX: Texas Transportation Institute. Available from http://tti.tamu.edu/publications/catalog/record_detail.htm?id=2979

Chrysler, S. T., Danielson, S. M., & Kirby, V. M. (1996). Age differences in visual abilities in night-time driving field conditions. *Proceedings of the human factors and ergonomics society annual meeting* (Vol. 2, pp. 923–927). Santa Monica, CA: Human Factors and Ergonomics Society.

Chrysler, S. T., Nelson, A. A., Funkhouser, D. S., Holick, A. J., & Brewer, M. A. (2007). *Driver comprehension of diagrammatic freeway guide signs* (Rep. 0-5147-1). College Station, TX: Texas Transportation Institute. Available from http://tti.tamu.edu/documents/0-5147-1.pdf

Chrysler, S. T., Stackhouse, S., Tranchida, D., & Arthur, E. (2004). Improving street name sign legibility for older drivers. *Proceedings of the human factors and ergonomics society annual meeting* (pp. 1597–1601). Santa Monica, CA.

Chrysler, S. T., & Williams, A. (2005, June 27–30). *Driving performance in a simulator as a function of pavement and shoulder width, edgeline presence, and oncoming traffic.* Paper presented at the 3rd International Driving Symposium on Human Factors in Driver Assessment, Training, and Vehicle Design, Rockport, ME.

Chrysler, S. T., Williams, A. A., & Park, E. S. (2006). *Traffic engineering applications of driving simulation* (Rep. 167142-1). Southwest University Transportation Center, Texas A&M University. Available from http://swutc.tamu.edu/publications/technicalreports/167142-1.pdf

Chrysler, S. T., Wright, J., & Williams, A. (2004). *3D visualization as a tool to evaluate sign comprehension* (Rep. 167721-1). Southwest University Transportation Center, Texas A&M University. Available from http://swutc.tamu.edu/publications/technicalreports/473700-00021-1.pdf

Dewar, R., Kline, D., Schieber, F., & Swanson, S. (1994). *Symbol signing design for older drivers* (Rep. FHWA-RD-94-069). McLean, VA: Federal Highway Administration.

Dudek, C. L., Schrock, S. D., & Ullman, B. R. (2007). License plate and telephone numbers in changeable message sign AMBER alert messages. *Transportation Research Record, 1959*, 64–71. Available from http://trb.metapress.com/index/16VKV54401179H03.pdf

Dudek, C. L., Schrock, S. D., Ullman, G. L., & Chrysler, S. T. (2006a). Flashing message features on changeable message signs. *Transportation Research Record, 1959*, 122–129. Available from http://trb.metapress.com/content/v3368577l76w5527/fulltext.pdf

Dudek, C. L., Ullman, B. R., Trout, N. D., Finley, M. D., & Ullman, G. L. (2006b). *Effective message design for dynamic message signs* (Rep. 0-4023-5). College Station, TX: Texas Transportation Institute. Available from http://tti.tamu.edu/documents/0-4023-5.pdf

Dutta, A., Fisher, D. L., & Noyce, D. A. (2005). Use of a driving simulator to evaluate and optimize factors affecting understandability of variable message signs. *Transportation Research Part F: Traffic Psychology and Behavior, 7*, 209–227.

Forbes, T. W., & Holmes, R. S. (1939). Legibility distances of highway destination signs in relation to letter height, letter width, and reflectorization. *Highway Research Board Proceedings, 19*, 321–335. Washington, DC: Transportation Research Board.

Garvey, P. M., Pietrucha, M. T., & Meeker, D. (1997). Effects of font and capitalization on legibility of guide signs. *Transportation Research Record, 1605*, 73–79. Available from http://trb.metapress.com/content/v8161k7254410184/fulltext.pdf

Guerrier, J. H., & Wachtel, J. (2002). Drivers' response to changeable message signs of differing message length and traffic conditions. In P. T. McCabe (Ed.), *Contemporary ergonomics* (pp. 223–228). London: Taylor & Francis.

Hogema, J. H., & van der Horst, A. R. A. (1994). Driver behavior under adverse visibility conditions. *Proceedings of the world congress on applications of transport telematics and intelligent vehicle-highway systems. Towards an intelligent transport system* (Vol. 4, pp. 1623–1630). Boston, MA: Artech House.

Hurwitz, D., Knodler, M., Jr., & Dulaski, D. (2005). *Speed perception fidelity in a driving simulator environment.* Proceedings of the Driving Simulation Conference 2005 North America, Orlando, FL.

Jenkins, S. E., & Cole, B. L. (1985). Daytime conspicuity of road traffic control devices. *Transportation research circular no. 297* (pp. 14–15). Washington, DC: Transportation Research Board.

Katz, B., Hawkins, H. G., Kennedy, J., & Howard, H. R. (2008). Design and evaluation of selected symbol signs. Traffic Control Devices Pooled Fund Study Report (Rep. TPF-5(065)). McLean, VA: Federal Highway Administration. Available from http://www.pooledfund.org/documents/TPF-5_065/symbol_sign_report_final.pdf

Laurie, N. L., Zhang, S., Mundoli, R., Duffy, S. A., Collura, J., & Fisher, D. L. (2004). Evaluation of alternative DO NOT ENTER signs: Failures of attention. *Transportation Research Part F: Traffic Psychology and Behavior, 7*, 151–166.

Lum, H. S., Roberts, K. M., DiMarco, R. J., & Allen, R. W. (1983). *A highway simulator analysis of background colors for advance warning signs* (HS-036 594). Washington, DC: Federal Highway Administration.

Manser, M. P., & Hancock, P. A. (2007). The influence of perceptual speed regulation on speed perception, choice, and control: Tunnel wall characteristics and influences. *Accident Analysis & Prevention, 39*(1), 69–78.

McKnight, A. S., McKinght, A. J., & Tippetts, A. S. (1998). The effect of line width and contrast upon lane keeping. *Accident Analysis & Prevention, 30*(5), 617–624.

Molino, J. A., Opiela, K. S., Andersen, C. K., & Moyer, M. J. (2003). Relative luminance of retroreflective raised pavement markers and pavement marking stripes on simulated rural two-lane roads. *Transportation Research Record, 1844*, 45–51. Available from http://www.ltrc.lsu.edu/TRB_82/TRB2003-002008.pdf

Picha, D. L., Hawkins, H. G., & Womack, K. N. (1995). *Motorist understanding of alternative traffic signs* (Rep. FHWA/

TX-96/1261-5F). Washington, DC: US Department of Transportation.

Pietrucha, M. T., Hostetter, R. S., Staplin, L., & Obermeyer, M. (1995). *Pavement markings and delineation for older drivers. Volume I: Final Report* (Rep. FHWA-RD-95-118). State College, PA: Pennsylvania Transportation Institute.

Parham, A. H., Womack, K. N., & Hawkins, H. G. (2003). Driver understanding of pavement marking colors and patterns. *Transportation Research Record, 1844,* 35–44. Available from http://trb.metapress.com/content/82wx451g816p37mn/fulltext.pdf

Scheiber, F. (2002). *Searching for fluorescent colored highway signs: Bottom-up versus top-down mechanisms.* Paper presented at the Transportation Research Board's 16th Biennial Symposium on Visibility University of Iowa, Iowa City, IA. Available from http://www.usd.edu/~schieber/pdf/Schieber2002.pdf

SignCad software. SignCad Systems, Inc. Minneapolis, MN. http://www.signcad.com/

Sivak, M., & Olson, P. L. (1985). Optimal and minimal luminance characteristics for retroreflective highway signs. *Transportation Research Record, 1027,* 53–57.

Stackhouse, S. P., & Burrus, M. (1996). *Human factors evaluation of driver multitasking and genesis message formats. Final Report* (Rep. MN/RC-96/13). Minneapolis, MN: University of Minnesota Center for Transportation Studies.

Stanley, L. M. (2006). *Haptic and auditory interfaces as a collision avoidance technique during roadway departures and driver perception of these modalities.* Bozeman, MT: Western Transportation Institute, College of Engineering, Montana State University. Available from http://www.coe.montana.edu/wti/wti/pdf/4W0767_Final_Report.pdf

Uding, K. D. (1993). Exact road geometry output program for retroreflective road sign performance. *Transportation Research Record, 1421,* 61–68.

Ullman, B. R., Ullman, G. L., Dudek, C. L., & Ramirez, E. A. (2005). Legibility distances of smaller letters in changeable message signs with light-emitting diodes. *Transportation Research Record, 1918,* 56–62. Available from http://trb.metapress.com/content/95720vl84mg52306/fulltext.pdf

Ullman, G. L., Finley, M. D., Pike, A. M., Knapp, K. K., Songchitruksa, P., & Nelson, A. A. (2008). *Studies to improve temporary traffic control at urban freeway interchanges and*

pavement marking material selection in work zones (Rep. 0-5238-2). College Station, TX: Texas Transportation Institute Report. Available from http://tti.tamu.edu/documents/0-5238-2.pdf

Ullman, G. L., Ullman, B. R., Dudek, C. L., Nelson, A. A., Pesti, G. (2005). *Advanced notification messages and use of sequential portable changeable message signs in work zones* (Tech. Rep. 0-4748-1). Texas Transportation Institute, College Station, TX. Available from http://tti.tamu.edu/documents/0-4748-1.pdf

U.S. Department of Transportation. (2009). *Manual on uniform traffic control devices.* Federal Highway Administration. Available from http://mutcd.fhwa.dot.gov/

U.S. Department of Transportation. (2004). *Standard highway signs handbook.* Federal Highway Administration. Available from http://mutcd.fhwa.dot.gov/ser-shs_millennium.htm

Upchurch, J., Fisher, D. L., Carpenter, R. A., & Dutta, A. (2002). Freeway guide sign design with driving simulator for Central Artery Tunnel: Boston, Massachusetts. *Transportation Research Record, 1801,* 9–17. Available from http://trb.metapress.com/content/880h534617x10755/

Upchurch, J., Fisher, D. L., & Waraich, B. (2005). Guide signing for two-lane exits with an option lane: Evaluation of human factors. *Transportation Research Record, 1918,* 35–45. Available from http://trb.metapress.com/content/g630862h080726w1/fulltext.pdf

Vercruyssen, M., Hancock, P. A., Knecht, W., Olofinboba, O., Chrysler, S. T., Williams, G., . . . Newell, R. (1995). Lighted guidance devices: Environment modulation of driving behavior through work zones. *Intelligent Transportation Systems, 3,* 118–127.

Wang, J., Collyer, C., & Hesar, S. (2007). Adding graphics to dynamic message sign messages. *Transportation Research Record, 2018,* 63–71. Available from http://trb.metapress.com/content/7524034278251223/

Zwahlen, H. T. (1989). Conspicuity of suprathreshold reflective targets in a driver's peripheral visual field at night. *Transportation Research Record, 1218,* 28–40.

Zwhalen, H. T., & Schnell, T. (1999). Driver preview distances at night based on driver eye scanning recordings as a function of pavement marking retroreflectivities. *Transportation Research Record, 1692,* 129–141. http://dx.doi.org/10.3141/1692-14

37

Advanced Guide Signs and Behavioral Decision Theory

Konstantinos V.
Katsikopoulos
*Max Planck Institute for
Human Development*

Abstract

The Problem. Advanced guide signs can display information that affects driver behavior. Thus they can be used to relieve traffic and parking congestion. To use this opportunity effectively, we need a theory of the effects of travel and parking information on driver decisions. *Role of Driving Simulators.* Driving simulators can provide quality data in controlled experiments that investigate the effects of travel and parking information on decision-making. *Key Results of Driving Simulator Studies.* There is evidence that people employ simple heuristics, often involving one-piece-at-a-time processing of information and the use of thresholds, for deciding whether to choose a travel route or a parking spot. *Scenarios and Dependent Variables.* Examples of the simulation scenarios used include driving on a highway and having to choose whether to divert to an alternative route or not, as well as driving on a city street and having to choose among several parking lots. *Platform Specificity and Equipment Limitations.* Not all simulators have the scenario development tools required to implement the scenarios discussed above because there are requirements on the quality of visual images.

37.1 Introduction

37.1.1 Opportunity: Advanced Guide Signs and Congestion

It is not possible for drivers to easily ignore guide signs. In their basic form, guide signs are sheets made of aluminum, posted on the right side of the road, with white text on a green background that say something such as "Downtown Boston 10". In general, they display information such as the distance to a destination, which is useful for the drivers to plan and execute their trip. In their more advanced form, signs can display more up-to-date information about, say, the travel time on a route to a popular destination, or the number of parking spots available in a busy lot. Guide signs can also be located inside the vehicle, but I will not consider such signs in this chapter.

Advanced guide signs can affect driver behavior (Kantowitz, Hanowski, & Kantowitz, 1997). Thus they also represent an opportunity for alleviating traffic and parking congestion. For example, if we knew the percentage of drivers who, after reading a certain piece of travel time information, decide to divert to the surface streets (as opposed to staying on the highway), we could intelligently switch the sign on and off to control the rate of traffic. Do we know the effect of information displayed on guide signs on driver decision-making?

37.1.2 The Role of Behavioral Decision Theory

The study of the decisions people make is called *behavioral decision theory*. It is an interdisciplinary study with contributions from fields such as economics, engineering, operations research, and psychology. The most commonly-used model of the decisions that people are faced with is the following: Each possible decision leads to a number of outcomes where each outcome is obtained with a probability. For example, for a particular driver, a route may have two possible travel times, 70 minutes and 120 minutes, each occurring with a probability ½. One may criticize this model because, for example, the possible decisions are assumed fixed before the trip, while in reality at least some possible decisions are constructed during the trip. I agree with this criticism but I am not aware of any decision models that are immune to it and are widely applicable.

A typical solution to this model comes from economics and operations research. It can be mathematically proven (von Neumann & Morgenstern, 1947) that accepting some logical axioms is equivalent to assigning utilities to outcomes and picking a decision with optimum expected utility. For example, consider the above example where a route can take one of two times, 70 minutes or 120 minutes. If the utility of 70 minutes driving equals, for a driver, ¾ and that of 120 minutes equals ¼ (the utility scale runs from 0 to 1; we do not discuss how utilities are assessed here), and if the route is equally likely to take one time as the other, then the expected utility for this route equals ½. More formally, if the travel time of a route R equals r_i minutes with probability p_i and the utility of driving for r minutes equals $u(r)$, the expected utility of route R equals $u(R) = \Sigma_i \, p_i u(r_i)$.

The utility approach has been developed further in transportation engineering, resulting in sophisticated mathematical models (Ben-Akiva & Lerman, 1997; Avineri & Prashker, 2005). For example, it may be assumed that drivers do not always pick a route with optimum expected utility. Instead, it is possible that a driver chooses any route where the probability of choosing a route depends on how the expected utility of the route compares with the expected utilities of all routes that are being considered. More formally, if the expected utility of route R^j equals $u(R^j)$, the probability that R^j is chosen equals $Pr(R^j) = u(R^j)/\Sigma_i u(R^i)$. The choice probability function can also be more complicated where, for example, utilities are exponents, and parameters are also used.

If you are wondering how it can be that people would want and be able to determine utilities and probabilities, combine them in the expected-utility way, and also use the expected utilities to come up with choice probabilities, you are not alone. Nobel prizes in economics have been awarded to researchers (Simon, H.; Kahneman, D.) who have spelled out these difficulties and have provided empirical evidence that people do not seem to perform all these calculations, at least not under all conditions.

It can and has been said, however, that utility theory remains the only good theory we have for predicting human rational choice (many Nobel prizes have been awarded to workers in utility theory too). According to this argument, even if people do not explicitly perform in their heads all calculations of expected utility theory, the theory can still account for a large chunk of people's decisions; especially in situations where there is pressure to be "rational". For example, when drivers make route choices, they would, according to the argument, suffer the consequences of choosing routes with slow travel times, and thus they may eventually decide approximately "as if" they optimize their expected utility due to travel time.

Even when the "as if" interpretation is granted, a problem with expected utility as a behavioral theory is that it no longer

aims at describing the processes (cognitive, neural, or hormonal) underlying a decision but just at predicting the final decision. This narrowing of the scope may be acceptable, depending on how much of human behavior one is satisfied with understanding. It is an empirical question as to how helpful a theory can be that does not show how decisions are made to consumers of behavioral decision theory. To the extent that a transportation policy-maker finds expected utility modeling to lack transparency, one may doubt how much she/he would trust a practical recommendation that is based on it.

An alternative approach holds that people use decision *heuristics* (Gigerenzer & Selten, 2001; Reinhard Selten is an economics Nobel laureate). Models of heuristics specify the underlying cognitive processes and also make quantitative predictions. Some of the studies reviewed here show that the heuristics can account for behavioral data that pose a challenge for utility theory.

37.1.3 The Role of Driving Simulators

Driving simulators represent a controlled experimental environment for studying the effects of travel and parking information on decision-making. The information can be manipulated systematically, it can be more or less ensured that drivers attend to it, and behavior can be easily recorded.

It is of course possible to use other methodologies such as paper-and-pencil devices or telephone interviews. There are, however, good reasons for preferring simulators: It is intuitive that using all controls in a car and attending to all information present in a high-fidelity virtual driving world can lead to the realistic cognitive processing demands that occur in a driving situation but not in a survey. In fact, it has been shown that drivers behave differently inside and outside of the car (Szymkowiak, Fisher, & Connerney, 1997). Additionally, the results obtained in driving simulators match closely the results obtained in the field for route and parking choice (Bonsall, Firmin, Anderson, Palmer, & Balmforth, 1997; Mahmassani & Jou, 2000; Bonsall & Palmer, 2004). For example, the route choices of drivers when a road network was simulated closely mimicked their real choices; this was also true for drivers with varying degrees of familiarity with the network (Bonsall et al., 1997). As Bonsall and Palmer (2004, p. 323) summarize: "The most complex/expensive simulators have a manifestly high degree of ecological validity and their use is appropriate when considering man–machine interface (MMI) issues, particularly those related to safety". There are, of course, always issues with simulating the effects of some of the variables that impact upon decision-making such as stress, fatigue, or boredom, or with simulating the effects of decision-making per se, such as the time spent waiting for a parking spot. In some of the studies that are reviewed here, efforts were made to simulate these effects.

37.2 The Research

There is quite a lot of research on route and parking choice in driving simulator environments. Most of it looks at the data using the lens of the utility approach. As mentioned above, it can be questioned how transparent this approach is. It may be fruitful to also try alternative approaches. I review a few studies where heuristics were used to understand driver decision-making.

37.2.1 Route Choice

In a series of experiments run on a mid-level driving simulator (Katsikopoulos, Duse-Anthony, Fisher, & Duffy, 2000, 2002), we investigated the choices that highway drivers made between two routes to a common destination when information about travel time on the routes was given, as shown in Figure 37.1.

The experimental instructions emphasized that I–93 was the default route in the sense that participants were currently driving on it. Thus the choice was framed as staying on the default route or diverting to the alternative, Route 28, a common real-world situation. Signs with such travel time information operate in highways in metropolitan areas in the United States (e.g., Atlanta, Saint Louis).

In a first experiment, the default route always had a certain travel time (100 minutes) while the alternative route had an interval of travel times (in the Figure 37.1 example, from 70–120 minutes). The average travel time of the alternative was taken to be the midpoint of the interval (in the Figure 37.1 example (70 + 120)/2 = 95 minutes). The range of this interval is 120 – 70 = 50 minutes. By crossing three levels of average travel time (95, 100, 105) with five levels of range (20, 30, 40, 50, 60), we get 15 route choice scenarios that we tested. To increase the data quality, we applied controls such as counterbalancing the presentation order of scenarios within participants.

We used the University of Massachusetts driving simulator (see Figure 37.2) that consists of a car (Saturn 1995 SL 1) connected to an Onyx Infinite Reality Engine 2 and an Indy computer (both manufactured by Silicon Graphics, Inc.) The images on the screen subtend 60 degree horizontally and 30 degree vertically, may be artificial or natural, and are developed using Designer's Workbench (by Centric). The movement of other cars on the road is controlled by Real Drive Scenario Builder (Monterey Technologies, Inc.). The system was assembled by Illusion Technologies.

Work on route choice has attempted to determine whether drivers are risk-averse (Abdel-Aty, Kitamura, & Jovanis, 1997). A

FIGURE 37.2 Driving simulator located in the Human Performance Laboratory at the University of Massachusetts, Amherst.

risk-averse driver is less likely to divert to the alternative route as the range increases (while the average remains the same). For example, she/he is less likely to choose Route 28 in the example in Figure 37.1 where the travel time ranged from 70–120 minutes than she/he would be to choose a route whose travel time ranged from 80–110 minutes (the average, in both cases, equals 95 minutes). A risk-seeking driver is more likely to choose the 70–120 minutes route.

It has been claimed that drivers are risk-averse. This is reminiscent of the early claim of economics that people are risk-averse (with the explanation being that they use a concave utility function). The psychological literature, however, suggests that people are risk-averse when the choice is framed as belonging to *gains* but risk-seeking when the choice is framed as belonging to *losses* (Tversky & Kahneman, 1981). In route choice, scenarios in which the alternative has a shorter average time than the default belong to the domain of gains. The scenario in Figure 37.1 is framed as a choice in the domain of gains because the average equals 95 minutes (the default takes 100 minutes). If the alternative route ranged from 90–120 minutes, the choice would be in the domain of losses because the average would equal 105 minutes.

Before running the experiment, one may expect that drivers are risk-averse in the domain of gains because diverting is, on the average, an improvement over the default and there is no need to risk more by preferring routes with larger range. On the other hand, one may also expect that drivers are risk-seeking in the domain of losses because diverting is, on the average, not an improvement over the default and larger ranges have a better chance of leading to an improvement. Past research on route choice had only tested scenarios in the domain of gains. We also tested route choices framed as losses. What did we find?

In Figure 37.3, we graph the number of participants (out of 30) that diverted, as a function of range. There is a decreasing trend for gains indicating risk aversion, a roughly flat line for the case where the alternative route has the default average (indicating

Estimated travel time to downtown Boston	Route
100 min	I–93
70–120 min	Route 28

FIGURE 37.1 Sign with travel time information, to guide route choice. (From Katsikopoulos, K. V., Duse-Anthony, Y., Fisher, D. L., & Duffy, S. A. (2000). The framing of drivers' route choices when travel time information is provided under varying degrees of cognitive load, *Human Factors, 42*(3), 470–481. With permission.)

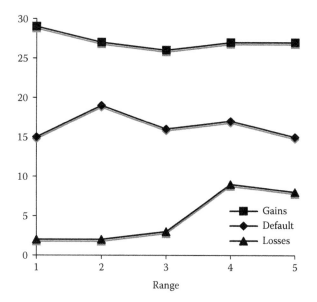

FIGURE 37.3 Percentage diverting to an alternative route as a function of the range of travel times. When choosing routes, drivers are risk-averse for gains and risk-seeking for losses. (From Katsikopoulos, K. V., Duse-Anthony, Y., Fisher, D. L., & Duffy, S. A. (2000). The framing of drivers' route choices when travel time information is provided under varying degrees of cognitive load, *Human Factors, 42*(3), 470–481. With permission.)

risk neutrality), and an increasing trend for losses indicating risk-seeking-ness. (For details on statistical tests on the significance of these trends, see Katsikopoulos et al., 2000).

There is a commonly-used model of framing effects in the intersection of behavioral economics and psychology, prospect theory (Kahneman & Tversky, 1979; and its cumulative version, Tversky & Kahneman, 1992), which is a modification of expected utility theory. In expected utility theory, the utility of route R (which takes r_i minutes with probability p_i) equals $u(R) = \Sigma_i p_i u(r_i)$. In prospect theory, the possible outcomes (r_i) and their associated probabilities (p_i) are both transformed before the expected utility formula is used. Each travel time r_i is mapped to the difference, $r_i - r$, where r is a reference travel time that equals the driver's apriori expectation about the magnitude of the travel time. For example, r could equal the average travel time along the default route. And each probability p_i is mapped to a probability weight $w(p_i)$ that expresses the decision-maker's subjective interpretation of the objective probability. The utility function u is often modeled by $u(r) = r^a$ for $r \geq 0$ and $u(r) = -k(-r)^b$ for $r < 0$, which has three parameters a, b, and k. The probability weighting function w is often modeled by $w(p) = pc/[pc + (1 - p)c]^{1/c}$ with another parameter c. Technically, prospect theory applies when there are only two possible outcomes, while cumulative prospect theory is its generalization for any number of outcomes.

Consider the example of Figure 37.1 where there is a choice to be made between a route with $r_1 = 70$ and $r_2 = 120$ and the default route with $r = 100$. As no information is given about probabilities, it seems reasonable to assume that $p_1 = p_2 = \frac{1}{2}$. Prospect theory

would determine the probability of diversion by calculating the quantities $w(\frac{1}{2})u(-30) + w(\frac{1}{2})u(20)$ and $u(0)$. To make these calculations, and all other calculations in our experiment, we would need to estimate the four parameters a, b, c, and k in the functions w and u, for each single driver or an "average" driver. These numerical estimates summarize the information the consumer of prospect theory has about a driver's decision-making. As said above, the problem is that prospect theory does not explain the process by which the driver decides to divert or not.

We developed a heuristic that postulates a number of steps that lead to the choice of a route and is also computationally simpler than prospect theory. In this heuristic, the driver takes a single sample to estimate the travel time along the alternative route (where she/he perceives that the travel time of the alternative route is following a normal probability distribution over the interval of possible travel times), and chooses to divert if this estimate is less than the reference travel time. This modeling incorporates a common theme of behavioral decision theories, that the driver is tuned to the probabilistic nature of decision-making (a normal distribution is used because many quantities in the real world are distributed normally and because it is mathematically convenient). The heuristic also incorporates a second emerging theme of behavioral theories—that human decision-making can be modeled by the processing of just one piece of information, here the driver's single-sample travel time estimate (Katsikopoulos & Gigerenzer, 2008).

More formally, the estimate of travel time along an alternative route with possible travel times in the interval from r_1 to r_2 is a linear combination of the midpoint and the range of the interval, $(r_1 + r_2)/2 + k(r_2 - r_1)$, where $k \sim N(\mu, \sigma^2)$. The estimate is centered on the midpoint when $\mu = 0$, but otherwise it is centered on another point in the interval. The estimated travel time may be rewritten as $(\frac{1}{2} - k) r_1 + (\frac{1}{2} + k) r_2$, which has the form of an expected utility according to prospect theory with $u(r) = r$ and the probability weighting function involving a random component. That is, prospect theory can mimic the predictions of the heuristic for the decision outcome; of course, as noted above, prospect theory cannot make process predictions. The probability of diversion equals $Pr[k < (r - (r_1 + r_2)/2)/(r_2 - r_1)]$.

In the example of Figure 37.1, where $r_1 = 70$, $r_2 = 120$, and $r = 100$, this probability equals $Pr[k < 0.1]$. This prediction, and all other predictions of the heuristic, requires the estimation of two parameters, μ and σ^2. In this experiment, the estimates of μ and σ^2 equaled -0.01 and 0.12 respectively, and the sum of squared errors equaled 72.87 (there were 15 data points additional to the 15 data points in Figure 37.1).

This simple heuristic accounts well for the framing effects on route choice (see Katsikopoulos et al., 2000, for more details). In a subsequent study, the heuristic was tested further (Katsikopoulos et al., 2002). In that driving simulator experiment, the default route (I-93) had a range of travel times. This manipulation complicated the results, in a way that generalized Figure 37.3. Specifically, the same pattern of risk attitudes with Figure 37.3 was observed when the default range was smaller than the alternative range, but the pattern was reversed when the range of the default route was larger. The heuristic can predict both patterns of results. The reversed pattern cannot be predicted by cumulative prospect theory because the predictions of this theory do not depend on the range of the reference point, only on its average value. Also, the standard versions of utility theory that do not use a reference point cannot account for any pattern of framing effects, either those found in the first simulation experiment or in the second one. For other non-utility models of framing effects in transportation engineering, see also Fujii and Takemura (2003).

37.2.2 Parking Choice

Choosing a route seems to be explained by simple heuristics. One may argue that this is no great surprise because there is so much going on cognitively while having to choose whether to take an exit or not that the drivers have to keep it simple. In the choice of a parking lot, however, the argument would go and one may expect different results because the speed is lower on urban streets than it is along the highway and thus there may be more time to make a decision. How do people decide where to park?

Among many studies of parking, there is one that pitted expected utility against heuristics (Hester, Fisher, & Collura, 2002). Their experiment also used the University of Massachusetts at Amherst driving simulator. To see what happened in a trial, refer to Figure 37.4.

FIGURE 37.4 A depiction of a parking choice scenario. (Top: Schematic; Middle: Destination; Bottom: Open Spaces).

Participants started a parking choice scenario by going through the stop sign (denoted with an asterisk, "*") in the left of Figure 37.4 (top panel). Soon after that, they encountered a sign (denoted with a pound sign, "#") that informed them of their destination for this scenario, which was one of the four rectangles marked A, B, C, or D. After the destination sign came a VMS (variable message sign; another term used for advanced guide signs) that provided information about the number of parking spots available in each one of the four lots, A, B, C, and D (as seen on Figure 37.4, top panel, lot A is closest to destination A, and so on). Examples of signs, as appeared to the driver, are shown in the middle and bottom panels of Figure 37.4.

The participants were instructed to park in a lot so as to minimize the expected travel time to their destination. They were told that this travel time was the sum of their driving time to the parking lot plus walking time from the parking lot to the destination, possibly plus any waiting time for a parking lot that ended up being full when they arrived (it was explained that by the time they got to a lot, the number of available parking spots could have changed, as would happen in the real world). They were told how to precisely compute these times: Each link in Figure 37.4 took one minute to drive (i.e., two minutes to drive from the start to lot A) and three minutes to walk (i.e., six minutes to walk from lot B to destination A); also waiting time for any full lot took five minutes.

How to specify the expected utility model? The utility of travel time t, $u(t)$, needs to be determined, as does the probability, $p(k)$, that a lot with k available spots when the sign is read will have at least one empty spot when the driver has arrived at the parking lot. In Hester et al. (2002), $u(t) = t$ and $p(k) = \frac{1}{2} + \frac{1}{2}[(a^{k-b} - a^{b-k})/(a^{k-b} + a^{b-k})]$.

As a heuristic alternative, Hester et al. (2002) suggested that drivers consider parking lots one at a time (starting from the destination lot) and choose to park at the lot where the number of available spots k exceeds some threshold d. That is, there is one parameter, d. The heuristic incorporates the two emerging themes in behavioral decision theory discussed above, one-at-a-time processing of pieces of information and the use of thresholds (Katsikopoulos & Gigerenzer, 2008).

Among many parking scenarios tested, Hester et al. (2002) considered six parking scenarios where the utility model and the heuristic made different predictions. It was found that parking choices were more consistent with the heuristic than with the utility model (for details, see Hester et al., 2002). For example, only 10% of the participants made all six parking choices consistent with the utility model and 35% of the participants made all six choices inconsistent with the utility model. On the other hand, data was fit well by the heuristic, as measured by a chi-square statistic.

In an as-yet unpublished study (Katsikopoulos, Hutchinson, Todd, & Gula, 2003), we used the University of Massachusetts driving simulator to further test parking heuristics. This experiment studied parking choice within a single lot, where spots were arranged sequentially. Four of the scenarios tested are shown on Figure 37.5.

Participants started driving towards the destination, looking for an empty spot that would minimize their total travel time.

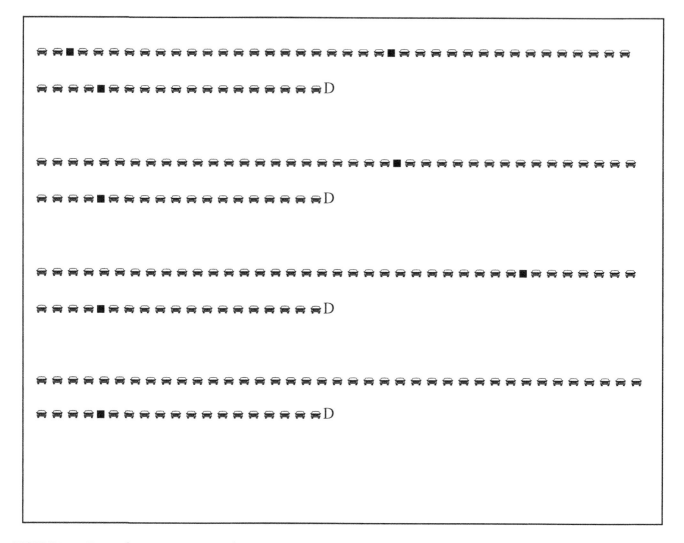

FIGURE 37.5 Four parking scenarios; 🚗 is a taken spot, ▪ is an empty spot, and D is the destination.

They were told that their total travel time equaled driving time to the spot plus walking time to destination, and the rules for calculating those times were given. An incentive structure was established in that participants were paid according to their incurred travel times. Additionally, they had to wait, after each trial, inside the car for an amount of time equal to their walking time. If the destination was passed, participants had to drive back to the start and repeat the process. The geometry and visuals of the simulation were such that participants could not see beyond the next (closest to the destination) spot.

It can be mathematically proven that, under some reasonable probabilistic assumptions, a threshold-based heuristic minimizes expected travel time (Ferguson, 2007). This heuristic has the form: Take the first available spot after d spots have been passed. Game-theoretic analyses have also shown that heuristics that use thresholds lead to low travel times (Hutchinson, Fanselow, & Todd, in press).

This experiment cannot discriminate between expected utility models and threshold-based heuristics because, as mentioned, expected travel time is minimized when a threshold is used. It can, however, discriminate between different threshold-based heuristics. For example, in the scenarios of Figure 37.5, consider a driver who has arrived at the last empty spot before the destination (she/he does not know that this is the last empty spot). If the threshold that she/he is using depends only on the number of spots passed, the driver would be equally likely to decide to take this spot in all four scenarios. If, however, the number of empty spots until this point also matters, the driver would not be equally likely to accept the spot in the four scenarios. Consistent with the general idea that drivers make simple decisions, we found that, in the four scenarios, an approximately equal percentage of drivers who made it to the last empty spot, decided to park.

37.2.3 The Bigger Picture

To use advanced guide signs effectively for alleviating congestion, transportation engineers need to be able to understand how drivers make decisions such as diverting to an alternative travel route or taking an empty parking spot. Driving simulators can be used to investigate these decisions. This chapter reviewed two route choice studies and two parking choice studies, all performed at the University of Massachusetts, Amherst mid-level driving simulator.

Even though a large part of the transportation literature uses utility-based approaches to model driver decisions, this chapter focused on models of simple heuristics. The starting point for arguing for heuristics is that they model the cognitive processes by which decisions are arrived at, whereas it can be argued that the utility approach does not specify this process. In addition, the heuristics are computationally simpler. As such, one may also argue that the heuristics approach is more transparent than the utility approach and could be more acceptable to the consumers of transportation decision modeling (e.g., policy-makers). In the studies discussed here, route and parking choices are not summarized by parameter values but by simple rules of thumb. These rules rely on the processing of single pieces of information (such as one sample from the travel time distribution), and on the use of thresholds (such as the number of parking spots passed). Note that the rules that were presented here referred to the statistical structure of the decision problem but not to other aspects of the problem which can also affect the decision such as habits, stress, or the physical characteristics of the vehicle. Making this extension is a good topic for future research.

Interestingly, some of the behavioral data, such as the more complex framing effects in route choice, cannot be accounted by standard expected utility theory or by its modifications such as prospect theory. It is possible that the more powerful approach that mixes utility with econometric modeling to study dynamic decision-making (Gao, Frejinger, & Ben-Akiva, 2008) can do this. Avineri and Prashker (2005) also discuss a number of models for route choice, including some models that were discussed in this chapter. However, as Gao et al. (2008) point out, an issue with this modeling is that it is not easy to estimate the multiple parameters reliably. How economically and how well can it be done in practice is an empirical question. On the other hand, even if data is scarce or expensive, simple rules of thumb offer ready insights into driver decision-making. Quality data from well-designed driving simulator studies coupled with a good understanding of the basic rules of driver decision-making can help toward using guide signs more effectively.

Key Points

- Driving simulation can be used to study people's route and parking choices.
- Traffic and parking congestion may be relieved if people's choices can be described mathematically.
- Route and parking choices seem to be described as resulting from the use of simple heuristics.
- The heuristics people seem to employ involve the sequential processing of information and the use of thresholds.

Keywords: Guide Signs, Heuristics, Parking Choice, Route Choice, Utility Theory

Glossary

Guide signs: Physical objects that display information, such as distance to a destination, which is useful for drivers to plan and execute a trip.

Theory of heuristics: The claim that people make decisions by using only some of the probabilities and utilities of only some of the possible outcomes of the different decision options.

Utility theory: The claim that people make decisions by performing computations that involve all probabilities and utilities of all possible outcomes of the different decision options.

Key Readings

Ben-Akiva, M., & Lerman, S. R. (1997). *Discrete choice analysis.* Cambridge, MA: MIT Press.

Gigerenzer, G., & Selten, R. (Eds.). (2001). *Bounded rationality: The adaptive toolbox.* Cambridge, MA: MIT Press.

Hester, A. E., Fisher, D. L., & Collura, J. (2002). Drivers' parking decisions: Advanced parking management systems. *Journal of Transportation Engineering, 128*(1), 49–57.

Katsikopoulos, K. V., Duse-Anthony, Y., Fisher, D. L., & Duffy, S. A. (2002). Risk attitude reversals in drivers' route choice when range of travel time information is provided, *Human Factors, 44*(3), 466–473.

References

Abdel-Aty, M., Kitamura, R., & Jovanis, P. (1997). Using stated preference data for studying the effect of advanced traffic information on drivers' route choice. *Transportation Research Part C: Emerging Technologies, 5*(1), 39–50.

Avineri, E., & Prashker, J. N. (2005). Sensitivity to travel time variability: Traveler's learning perspective. *Transportation Research C: Emerging Technologies, 13,* 157–183.

Ben-Akiva, M., & Lerman, S. R. (1997). *Discrete choice analysis.* Cambridge, MA: MIT Press.

Bonsall, P., Firmin, P., Anderson, M., Palmer, I., & Balmforth, P. (1997). Validating the results of a route choice simulator. *Transportation Research C: Emerging Technologies, 5,* 371–387.

Bonsall, P., & Palmer, I. (2004). Modeling drivers' car parking behavior using data from a travel choice simulator. *Transportation Research C: Emerging Technologies, 12,* 321–347.

Ferguson, T. S. (2007). *Optimal stopping and applications.* Retrieved August 20, 2007, from http://www.math.ucla.edu/~tom/

Fujii, S., & Takemura, K. (2003). *Contingent focus model of decision framing under risk.* Technology. Rep. 67, Tokyo Institute of Technology (pp. 51–67).

Gao, S., Frejinger, E., & Ben-Akiva, M. E. (2008). Adaptive route choice models in stochastic time-dependent networks. *Transportation Research Record, 2085,* 136–143.

Gigerenzer, G., & Selten, R. (Eds.). (2001). *Bounded rationality: The adaptive toolbox*. Cambridge, MA: MIT Press.

Hester, A. E., Fisher, D. L., & Collura, J. (2002). Drivers' parking decisions: Advanced parking management systems. *Journal of Transportation Engineering, 128*(1), 49–57.

Hutchinson, J., Fanselow, C., & Todd, P. M. (in press). Car parking as a game between simple heuristics. In P. M. Todd, G. Gigerenzer, & the ABC Research Group, *Ecological rationality: Intelligence in the world*. New York: Oxford University Press.

Kahneman, D., & Tversky, A. (1979). Prospect theory: An analysis of decision under risk. *Econometrica, 47*(2), 263–291.

Katsikopoulos, K. V., Duse-Anthony, Y., Fisher, D. L., & Duffy, S. A. (2002). Risk attitude reversals in drivers' route choice when range of travel time information is provided. *Human Factors, 44*(3), 466–473.

Katsikopoulos, K. V., Duse-Anthony, Y., Fisher, D. L., & Duffy, S. A. (2000). The framing of drivers' route choices when travel time information is provided under varying degrees of cognitive load. *Human Factors, 42*(3), 470–481.

Katsikopoulos, K. V., & Gigerenzer, G. (2008). One-reason decision-making: Modeling violations of expected utility theory. *Journal of Risk and Uncertainty, 37*(1), 35–56.

Katsikopoulos, K. V., Hutchinson, J. M. C., Todd, P. M., & Gula, B. (2003). *Supporting automobile drivers' parking decisions*. Poster presented at the Europe Chapter of the Human Factors and Ergonomics Society Annual Meeting, Lund, Sweden.

Kantowitz, B. H., Hanowski, R. J., & Kantowitz, S. C. (1997). Driver acceptance of unreliable traffic information in familiar and unfamiliar settings. *Human Factors, 39*(2), 164–177.

Mahmassani, H. S., & Jou, R. (2000). Transferring insights into consumer behavior dynamics from laboratory experiments to field surveys. *Transportation Research A, 34*, 243–260.

Szymkowiak, A., Fisher, D. L., & Connerney, K. A. (1997). False yield and false go decisions at signalized left-turn intersections: A driving simulator study. In K.-P. Holzhausen (ed.) *Advances in multimedia and simulation. Human-machine-interface implications,* Proceedings of the Human Factors and Ergonomics Society Annual Conference in Bochum, November 1997 (pp. 226–235). Bochum, Germany: Fachhochschule Bochum, University of Applied Sciences.

Tversky, A., & Kahneman, D. (1992). Advances in prospect theory: Cumulative representation of uncertainty. *Journal of Risk and Uncertainty, 5*, 297–323.

Tversky, A., & Kahneman, D. (1981). The framing of decisions and the psychology of choice. *Science, 211*, 453–458.

von Neumann, J., & Morgenstern, O. (1947). *Theory of games and economic behavior*. Princeton, NJ: Princeton University Press.

38

Driving Simulation Design and Evaluation of Highway–Railway Grade and Transit Crossings

Jeff K. Caird
University of Calgary

Alison Smiley
Human Factors North

Lisa Fern
*San Jose State University
Research Foundation*

John Robinson
Delphi-MRC

Abstract

The Problem. This chapter addresses the design and evaluation of a section of roadway using driving simulation. *Role of Driving Simulators.* A driving simulator was used to model a section of roadway which crossed a rail crossing and two at-grade intersections, one of which was a bus transitway, all in close proximity. The purpose of the simulator was to optimize the design through visual simulation as well as through testing with a sample of drivers. Specifically, plans for the section of roadway, on Fallowfield Road, north of Ottawa, Canada, were computer modeled in extensive detail and integrated into the University of Calgary Driving Simulator (UCDS), which was a research first in Canada. After driving the simulated Fallowfield Roadway tile, numerous elements of the design were changed and modified based on expert input from the organizations involved in the project. Approximately 20 design changes were incorporated into the final simulation model. *Scenarios and Dependent Variables.* To further evaluate the design elements, 47 participants, stratified into the age groups of 18–24, 25–55, and 55 and older, drove the Fallowfield simulation model in both directions, and on selected runs were challenged by two traffic events, namely a late yellow light and a stalled truck just past the railroad tracks. *Key Results.* Results showed that older drivers had significantly lower speeds at the rail crossing in both eastbound and westbound directions and had lower comprehension of a number of signs than other age groups. Eye movement analyses indicated that several signs were not fixated by the majority of participants and these signs had still fewer fixations when traffic was present. Driver responses to the two challenging traffic events in these contexts were similar to those observed in real-world situations. *Platform Specificity and Equipment Limitations.* A set of recommendations is made with respect to signs and signals identified in the evaluation phase. The utility of high-fidelity driving simulation models to visualize and problem-solve complex highway engineering designs and evaluate resulting solutions had a number of positive safety and design benefits, which are discussed.

38.1 Background and Objectives

Driving simulators have been used to evaluate roadway designs for about 15 years. More recently, the number of projects undertaken worldwide has risen. For example, interest in the area has increased to where the First International Conference on Road Safety and Simulation was held in Rome in 2007. The reasons for using driving simulation prior to, during, or after construction are often associated with a lack of understanding about the response of drivers to specific treatments and the relative safety therein (Smiley, 2008). Design changes that are safe and benefit drivers are much easier to make before the concrete is poured, so to speak.

For instance, over the span of about five years, Upchurch, Fisher, Carpenter and Dutta (2003) modeled approximately 163 miles of the Central Artery Tunnel in Boston, including associated over- and underpasses, to optimize the effectiveness of the sign design and placement. Using the University of Massachusetts driving simulator, the tunnel geometry was replicated, including the horizontal and vertical roadway curvature and ceiling height, which restricted the placement of signs. Fourteen target destinations with associated signs and alternatives were evaluated using a series of steps. Step one involved highway engineers and human factors experts driving the tunnel to identify problematic situations including misalignment of signs with horizontal road curvature. Step two had 24 naïve participants drive a specific route to determine the number of participants who missed the exit (3 out of 24). A second experiment with 18 participants examined lane change behavior in advance of important exits. Steps three and four repeated steps one and two to further improve practical solutions and check that drivers responded as expected.

Other projects using driving simulators have investigated the use of perceptual speed regulatory treatments (Denton, 1980) to decrease the speed of drivers in tunnels (Manser & Hancock, 2007). Specifically, stripes on the tunnel walls that decreased in width caused drivers to initially reduce their speed. Van der Horst and de Ridder (2007) also looked at speed and lateral vehicle position with respect to roadside features including trees, guardrails, barriers and the like in the driving simulator at the Netherlands Organization for Applied Scientific Research (TNO). In general, drivers moved their vehicles away from these roadway features and slowed slightly. Still others (Caird et al., 2008) used driving simulation to evaluate four bicycle lane treatments (white dashed, blue dashed, blue solid and sharrows [i.e., a bicycle icon with arrows]) and whether familiarization affected comprehension or reduced conflict events. The sharrows treatment had the highest level of comprehension and was adopted by the Transportation Association of Canada as a standard. These selected examples provide useful information about the relationships between drivers and infrastructure as well as methods for design and evaluation.

Against this background, the case study described herein was part of a larger effort by Delphi-MRC to provide risk management and due diligence safety review to the City of Ottawa. This chapter addresses the design and evaluation of a section of roadway that was undergoing construction (see Figure 38.1; Web Figure 38.1 for the color version). A number of safety concerns were raised by the City of Ottawa regarding the proximity of the transit and grade crossing intersections to one another. Based on a preliminary safety evaluation by Delphi-MRC, it was decided that the University of Calgary Driving Simulator (UCDS) would be used to evaluate the design of the crossings and to identify further design issues when evaluated by drivers using the simulator.

The evaluation of the effectiveness of certain highway–railway grade crossing elements should include a number of types of research. Ideally, a multi-method approach will yield convergent evidence about the effectiveness of various elements (Caird,

FIGURE 38.1   Aerial view of Fallowfield Road with CN Rail line and transitway extensions shown.

Creaser, Edwards, & Dewar, 2002). This is the approach used here. Two primary processes were engaged in to examine and evaluate the Fallowfield Road Rail Crossing. First, Fallowfield was computer modeled, based on drawings and other documents, and then driven by a panel of highway and human factors engineers to determine the effectiveness of a range of sign, signal and roadway treatments. Second, the final design, which was arrived at through the first design process, was evaluated by having a representative sample of drivers drive the modeled Fallowfield Roadway.

The specific goals of this project were to:

- Iteratively revise the proposed roadway design based on expert input using the UCDS to visualize specific changes;
- Use the UCDS to examine likely driver behavior at a proposed rail crossing in a complex road environment;
- Compare driver performance by younger, middle-aged and older driver age groups across the Fallowfield Roadway; and
- Based on the design process and evaluation results identify desirable features for the crossing.

38.1.1 Site Description

Fallowfield Road is a divided arterial road with two lanes in each direction (see Figure 38.1 and Figure 38.2). In the eastbound direction, drivers encounter advance signs for a railway crossing, a signalized intersection within which lies the railway crossing and a bus transitway, followed within a few seconds' travel by a second signalized intersection. The traffic signals are timed so that they turn red well in advance of the railway gates coming down. The traffic signals for the second intersection are designed so that they are not visible until the driver is part way through the first intersection. The traffic signals and railway flashers and gates are timed so that the driver either stops at the first intersection, or travels through that intersection over the rail line, across the bus transitway and across the following, close proximity intersection. In the westbound direction, the railway crossing is not contained within the intersection with the bus transitway but follows it. Thus driving task demands are somewhat reduced.

38.1.2 Design Activities

DriveSafety was contracted to model the Fallowfield Roadway based on drawings, video and photographs. Below, we refer to this model as the *Fallowfield tile*. Extensive input from engineers from the City of Ottawa, Delphi-MRC, Human Factors North, Inc., and researchers from the Cognitive Ergonomics Research Laboratory (CERL) further refined the model to achieve an exceptionally high level of visual fidelity. (The precision of all visual features in location, size and function, and correspondence with the built environment is construed as visual fidelity.) A large number of specific changes (Web Figure 38.7; **see color insert**) to the Fallowfield model were made to improve the precision of the signage and traffic control placement, to modify the way that

certain elements looked or should look, and to include elements that were thought to be essential to the fidelity of the simulated model. After a full-day simulator session with all parties, a number of digital videos of updated drives through the Fallowfield tile were provided to obtain additional input. Once the contractors and City of Ottawa-Carleton personnel were able to view the baseline simulation, a number of modifications were made to improve the baseline design. These modifications included:

- Removal of traffic signal advance signs on eastbound (EB) approach to reduce sign clutter;
- Relocation of advance rail signal flashers (i.e., Advance Warning Signals or AWS) and accompanying markings to the former location of the traffic signal advance signs on EB approach to improve advance warning of the rail crossing; and
- The height of the STOP LINE sign was reduced and the location on the side of the road was moved to the right so that the rail flashers would be visible to approaching drivers (see Web Figure 38.2).

A decision was taken to include the red light camera (RLC) sign for evaluation purposes (Web Figure 38.3), but not to inform subjects that there is a penalty for running the red light. The form of the RLC sign was extensively debated. Bilingual requirements and resolution limits of the UCDS were issues. A bilingual sign with two icons (see Web Figure 38.3, left) was dropped in favor of a simpler English-only sign that is used in the Calgary area (see Web Figure 38.3, right).

Cantilevered signals for east- and westbound approaches were considered (see Web Figure 38.4, left). However, when they were simulated, they were found to create a visually confusing scene on the eastbound approach, raising the likelihood that drivers would miss the traffic signal, and so were dropped from the simulation model. NO TURNS signs, instead of the iconic NO RIGHT and NO LEFT signs, were also considered for the transit crossing. However, these would need to be bilingual in the Ottawa area and so the iconic ones were retained (see Web Figure 38.8).

A uniform white concrete treatment between the train tracks and the transit intersection was considered, but not adopted (see Web Figure 38.4, right). The rationale for this decision included discussion of the cost, the pavement stopping characteristics, and the desirability of creating a visible link between the intersections. A visible link between the intersections has the potential for misinterpretation by drivers. Drivers tend to be literal interpreters of simulations so that they may have thought snow had fallen only in that location; that the area between the two intersections was special (but not signed as such); or that the two intersections were functionally one big intersection. In combination, the potential for misinterpretation and increased cost of the real-world implementation resulted in not including the white concrete treatment in the design or evaluation. The white treatment was applied to the rail bed to highlight the importance of the rail crossing (e.g., see Web Figure 38.5, left).

To improve the realism of the simulation, a number of dynamic traffic movements were implemented, including:

- A bicyclist headed eastbound, and another one headed westbound, near the transitway intersection (see Web Figure 38.5, left);
- Pedestrians walking parallel to the eastbound flow on Fallowfield Road in the vicinity of the Barhaven Mall (see Web Figure 38.5, right);
- Oncoming traffic flow set at a constant rate of approximately 2.21s between vehicles of varied types, which was intended to reflect peak traffic flow;
- Traffic ahead of driver, including trucks, had the same average 2.21s headway; and
- Buses waiting or crossing, depending on the signal status, at the transitway (see Web Figure 38.6, left), and trains entering the railway crossing (Web Figure 38.6, right).

Our initial plan was to compare two test scenarios—a baseline and an alternative design. The baseline, or original design, was to be the plans as proposed, according to all applicable standards, with minor adjustments made to ensure signs and signals were not obstructed by each other and with modifications decided as a result of viewing the simulation. The alternative design was to be a modification of the baseline design and based on interactions with it and application of various human factors principles such as positive guidance (Alexander & Lunenfeld, 1986) in which alterations are made to improve the conspicuity of critical information and distribute information to reduce driver workload. After extensive modifications to the original design, with parties being in agreement about the changes made, it was decided to evaluate only the alternative design or final design.

Reflecting on the design process, changes that could be made to objects such as a sign—if they were part of the database—were relatively easy to make. Changes to the road geometry which required additional modeling (and outsourcing) were more costly, and required time and planning. Dynamic animations, such as the gates raising and lowering with appropriate flashing lights, required some modeling as well as complicated coding and scripting. Attention to engineering details and managing the lists of model changes was important to the success of the project. Working with all parties in the simulator for at least a full day and dissemination of digital videos with new design changes were really important means of communication during the project.

38.2 Experimental Methods

38.2.1 Research Questions

Based on the extensive iterative redesign of the Fallowfield model, a single finalized design was tested to determine:

- Where subjects looked and the speed they adopted on westbound (WB) and EB approaches;
- Where drivers stopped on EB approach with advanced warning signals (AWS) providing sufficient warning;

- Whether being required to wait for the entire crossing signal cycle on the EB approach changes driver willingness to run a late yellow light;
- Whether drivers start forward when the railway gates rise even though the traffic signal remains red, giving priority to the bus transitway;
- Whether drivers found any signs or signals confusing based on driver actions, eye movements and structured interviews; and
- What drivers did if they were trapped behind a truck that stopped abruptly just past the tracks with a train approaching.

38.2.2 Participants

One-third of the sample was drivers over the age of 55, one-third under age 25, and one-third was between 25 and 55 years of age. Half of the participants in each age group were female and half, male. The 47 volunteers for the study were required to hold a valid driver's licence, be 18 years of age or older, drive a minimum of 10,000 kilometers per year, and have no previous simulator experience.

38.2.3 UCDS, Eye Movement and Digital Video Systems

The University of Calgary Driving Simulator (UCDS) is described in detail elsewhere (Caird, Chisholm, Edwards, & Creaser, 2007). Applied Science Laboratory (ASL) 501 system was used to measure participants' eye movements. Eye movements are sampled at a rate of 60 Hz (ASL Manual Version 2.0, 2001) (see Chisholm, Caird, & Lockhart, 2008, for a full description). A set of crosshairs is superimposed onto the scene video display that is integrated into a SimObserver video (see below) via the multiplexer. Fusion of simulation and eye movement data is performed using a custom C# program.

SimObserver records the camera multiplexed views of the driver, the feet, the forward field of view and eye movements to DVD in mpeg-2 format, along with raw simulator and eye movement data for later review and analysis. A purpose-specific workstation was configured to handle the transfer of digital video data, performance, and eye movement variables from the simulator. A custom software application, called Data Distillery, was developed by Realtime Technologies, Inc., to assist trained research assistants to extract the location fixated by the eye. The spatial accuracy of the digital video with crosshairs introduces experimenter error when researchers classify glances to specific targets and an unknown but limited amount of error comes from the ASL system. The timing error is one frame or 33 ms.

38.2.4 Procedure

Participants were screened for visual acuity, contrast sensitivity, color deficiencies, annual mileage, health, and simulator sickness. Participants in the 18–24 and 25–55 age groups were not allowed to participate if they wore glasses because glasses make calibration more difficult. Contact lenses were permitted. Those 55 and older were allowed to participate if they wore corrective lenses or glasses (eight participants did). This allowance was necessary to recruit a sufficient number of older drivers into the study.

Participants completed about a five-minute practice drive to familiarize themselves with the simulator and to get used to traveling at the posted speed limits. Throughout the drives participants were reminded to travel at the posted speed limits. The experimental portion of the study was composed of two drives that were predominately on suburban, rural and countryside roadway. These roadways were chosen to resemble the Fallowfield area. Each participant approached the Fallowfield crossing four times: three times in the EB direction and once in the WB direction during the experimental portion of the study. The essential verbal protocol for practice and experimental drives was to "drive as you normally would in your own vehicle; remember to obey all the rules of the road, such as speed limit, lights and signs."

Crossing approaches, because each was different, were designated A, B, C, D and E. A description and purpose for each approach follows. WB-A and EB-A were drives through the crossing without traffic in the indicated direction of travel. The reason for these drives was to obtain baseline data for speed and glance behavior to the signals and signs. In EB-B, drivers were faced with a late yellow light (2.21 sec) and had to decide whether to stop or go. In EB-C, drivers encountered the activated AWS and the gates down at the crossing. They then had to wait for 94 sec until the train had passed and the controls had cycled (see Web Figure 38.6, right). The train was only visible when it passed in front of participants who were stopped at the crossing. All participants experienced four of the crossing approaches in this order: WB-A, EB-A, EB-B, and EB-C. Half of the participants then encountered another late yellow light (EB-D) at 2.2 sec from the railway stop line, and half experienced a truck perform an emergency stop just past the train tracks (EB-E). Participants were randomly assigned to one of these two approaches (i.e., EB-D or EB-E).

The purpose of the late yellow light event was to determine whether having to wait for the entire grade crossing light cycle would cause drivers to change their decision criteria and run the subsequent traffic light (i.e., EB-B versus EB-D). The truck braking event was designed to explore potential responses by drivers. First, as the truck went through the intersection it blocked a light change from green to yellow. Would participants follow the truck through the intersection and over the tracks? Second, when the truck braked at 6 m/s², which was emergency braking, would drivers respond appropriately? Prior to the lead truck braking, a pace car, coupled by speed and distance to the participant's vehicle, blocked the route to the left throughout the approach to the crossing.

At the completion of the experimental drives, participants were asked a series of questions about how well they understood the various signs and signals that they had just encountered. Four pictures were used in association with specific questions to heighten the recall and recognition of drivers. Finally, participants filled out a post-simulator sickness questionnaire, were debriefed about the purpose of the study, and were remunerated $40 for completing the study session.

38.3 Results

38.3.1 Participant Characteristics

Three middle-aged and four older drivers withdrew from the study after experiencing simulator discomfort and sickness symptoms and all but one were replaced to given $n = 47$. Based on vision testing and the Driver Experience Questionnaire, the characteristics of those who participated in this study can be found in Table 38.1.

38.3.2 Dependent Measures

Three types of measures were collected during the course of the study; namely, driver performance in the simulator, questionnaire and interview items, and eye movement measures. Driving simulation measures that were collected included velocity in both east- and westbound directions, stopping accuracy at the rail crossing stop line, and responses to the late yellow light and

TABLE 38.1 Age Group Categories, Number of Participants in Each Group, Mean Participant Age (Standard Deviation), Average Kilometers Driven per Year, Total Number of Crashes Since Licensure, Moving Violations, and Visual Acuity (Minimum Angle of Resolution, Mar) With Correction

Age Group	N	Mean Age (SD)	Km/Year (SD)	Crashes, Last Five Years (SD)	Moving Violations, Last Five Years (SD)	Visual Acuity, Left (SD), Right (SD)
18 to 24 years	16	21.25 (1.6)	23,525 (12,472)	0.38 (0.62)	0.69 (1.1)	1.25 (0.43) 1.03 (0.30)
25 to 55 years	16	33.81 (8.5)	30,412 (14,566)	0.70 (0.80)	1.1 (1.0)	1.33 (0.46) 1.30 (0.42)
55+ years	15	61.4 (8.5)	27,693 (16,401)	0.33 (0.62)	0.80 (1.0)	1.14 (0.33) 1.20 (0.42)
Total/Means (SD)	47	38.3 (17.7)	27,200 (14,497)	0.47 (0.69)	0.87 (1.0)	1.24 (0.41) 1.17 (0.39)

truck braking events. Perception response time (PRT) and categorization of responses (e.g., crashed, stopped and backed up, etc.) were useful measures for these latter events (also see Caird et al., 2007).

Participants were asked a series of questions about how well they understood the various signs and signals that they had just encountered when they exited the simulator at the conclusion of the experimental drives. A series of images were used to heighten the recall and recognition of specific traffic elements of interest.

A number of eye movement-dependent measures were collected over the Fallowfield intersections. In accord with glance measure standards (ISO, 2002a; ISO, 2002b), a *glance* is defined as two or more successive frames, with each frame corresponding to 33 ms. A glance measures the time that the eye lands on an object, as indicated by the crosshairs in the digital video, until it leaves that target for another. It does not include the transition or saccadic time. A glance includes multiple fixations on a sign but excludes saccades into and out from the glance. In total, 3389 glances at road markings, signs and signals were extracted from approximately 18,000 fixations made during the experimental drives.

Using the integrated simulator data and eye tracking video from SimObserver, and the Data Distillery software applications, detection (yes or no) of a sign or signal was determined; the distance at which each sign or signal was first glanced was identified; the distance at which the last glance was made; and the total number of glances to each sign or signal was extracted from the digital video and eye movement data. Two research assistants coded each variable using Data Distillery, which provided some automation support to the process. The two research assistants worked together to categorize the variables for the first half of the drives that were analyzed. After working together to establish classification criteria, each researcher classified another 25% of the drives on their own. Issues of precise timing were resolved by moving repeatedly forward and backward through the digital video together until agreement on frame (i.e., time or distance) could be made. Expansions of the time frame or consideration of larger portions of drives often provided sufficient additional context with which to make a determination. In a number of cases, object tracking was used to determine the length of time an object was fixated.

Distances to a sign or signal were determined by extracting the "X" coordinate for each and comparing it to the location where a glance was made. The "X" dimension represents the distance ahead of the driver down the center of the roadway. *First glance distance* is defined as the distance from the center of the vehicle to an element in the "X" dimension in meters when it is first fixated. Experimenter error was likely higher in this measure because the signs and signals at first glance were close to the line of sight where most glances were made. Thus, a look along the line of sight was sometimes difficult to distinguish from a glance at a sign or signal near the line of sight. This is reflected in the variability of the first glance distance. *Last glance distance* was determined the same way except that the location corresponded to the last fixation to the sign or signal. The *total glance duration* was equal to the sum of the separate glance times to a sign or signal over the course of a drive.

Specific signs and signals were of interest in the analyses. In the eastbound direction, the advanced warning sign (AWS), the "X" on the roadway, the red light camera sign (RLC), the crossbucks and flashing train lights, the train grade crossing, the DO NOT STOP ON TRACKS sign, the distance at which drivers first looked at 3M lights before they were visible, and the 3M lights when they were visible were extracted (see Web Figure 38.8). The westbound elements included the AWS, "X", RLC sign, the DO NOT STOP ON TRACKS sign, and the traffic lights at the transit crossing.

The six dependent measures listed above and the nine signs and signals (also above) were analyzed using a repeated-measures MANOVA with age group (18–24, 25–54, 55+) as a between-subjects variable. Adjustments to the degrees of freedom and *p* values were made, where necessary, for violations of sphericity using the Greenhouse-Geisser correction. Post hoc comparisons were made using the Bonferroni adjustment. The results are organized by area of travel on the eastbound drive; namely, advanced signage (i.e., AWS, "X", red light camera), rail crossing information (i.e., traffic lights, grade crossing, stop line, crossbucks) and transit traffic lights.

The presentation of results in subsequent sections is organized by the sequence of drives experienced by participants (i.e., from A to E). Thus, free flow in east and west directions is presented first (EB-A, WB-A). An analysis of each sign or signal is accompanied by results from the driving simulation, interviews and eye movements. Results for stopping for the train, the late yellow light and truck events follow.

38.3.3 East- and Westbound Free Flow

Velocity was calculated as the participant's speed, in km/h, at various locations across the Fallowfield intersections. The selection of locations for speed measurement in the east- and westbound directions (see Figure 38.2; Web Figure 38.9) was based on where sign information and road geometry may have influenced speed regulation (Summala & Hietamaki, 1984). When participants were allowed to select a speed in the eastbound direction and the lights were green throughout the rail and transit crossings (EB-A), velocity was significantly different over the six measurement locations ($F (5, 40) = 18.01$, $p < 0.0001$, adjusted for sphericity violations, see Tabachnick & Fidell, 2001). As evident in Figure 38.3 (Web Figure 38.10), drivers tended to slow as they approached the railway crossing and then resumed their speed by the time they reached the transit crossing. Several researchers have found the same pattern of results in field observations (Shinar & Raz, 1992; Ward & Wilde, 1995). Specific to this study, the absence of lead traffic in the free-flow drives (EB-A & WB-A) is not a situation in which most drivers would find themselves except, perhaps, during early morning or late night drives.

Age approached significance ($p = 0.085$): The older age group tended to drive more slowly as measured at the six locations, whereas younger and middle-aged drivers drove faster and were quite similar in their speed profiles across the Fallowfield Roadway. Specifically, the mean speed reduction was about

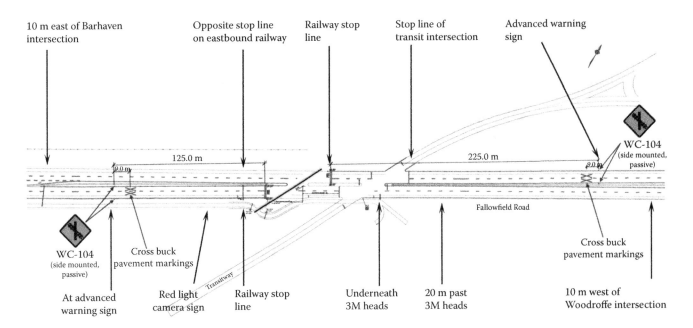

FIGURE 38.2 Fallowfield plan view with specific velocity measurement locations indicated. The eastbound locations appear across the bottom, whereas the westbound locations are shown across the top.

8 km/h at the rail stop line for the older age group and about 4 km/h for the other age groups.

The westbound velocity profile can be seen in Figure 38.4 (Web Figure 38.11). Drivers slowed as they approached the tracks and

resumed their speed afterwards. The main effect of velocity, that is, over the points where velocity was measured, was significant, $F(1, 234) = 5.95$, $p < 0.0001$. Post hoc Bonferroni tests indicated a pattern of specific differences where the rail track stop line and

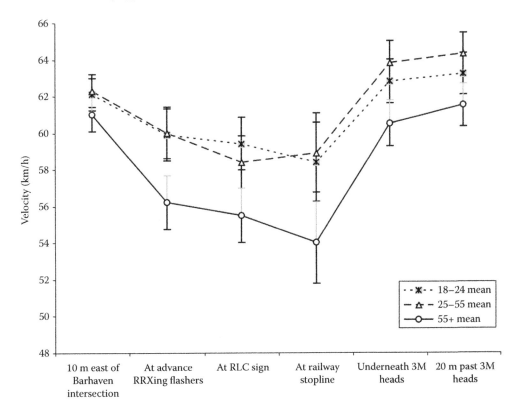

FIGURE 38.3 Eastbound velocity profile by age group with standard error bars is shown. The locations listed on the x axis are indicated on the plan view of Figure 28.2.

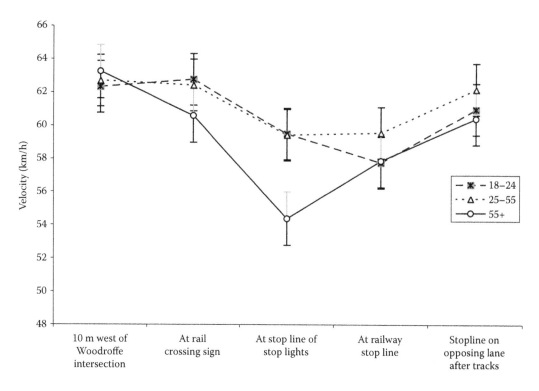

FIGURE 38.4 Westbound velocity profile by age group with standard error bars is shown. The locations listed on the x axis are also shown on the plan view of Figure 28.2.

intersection stop line were significantly slower than the other locations wherein drivers were traveling at a higher speed. Age and gender were not significant.

38.3.4 Eastbound Active Advanced Warning Sign (AWS)

Percent looked was calculated based on the total opportunities to see the sign or signal (i.e., four) multiplied by the total number of participants (N = 44). (Three older drivers could not be calibrated by the eye movement system.) From Table 38.2, the percentage of drivers who glanced at the AWS (on either side of

the road) in all eastbound drives was quite high, ranging from 100% when the flashing light was on (EB-C) (Web Figure 38.12) to a low of 91.3% during the truck event (EB-E).

Far fewer drivers looked at the "X" on the roadway over the various drives, from a high of 66% when the AWS was active (EB-C, see Web Figure 38.12) to a low of 30% when following the truck (EB-E). The high of 66% of drivers was accompanied by a jump in fixation frequency and duration (see Table 38.3), which most likely indicates that drivers were trying to determine what the flashing light and "X" meant in combination. The distance at which drivers made their first and last glances at the "X" was closer to the vehicle than was the case for other

TABLE 38.2 Percentage of Drivers Who Glanced at Each Object During Each Drive

Object> Drive	AWS	X on Roadway	RLC Sign	Right Traffic Light	Center and Left Lights	Do Not Stop on Tracks	Cross- Bucks and Flashers	3M Visible
WB-A	93.2	72.7	93.2	n/a	86.4	68.6	70.5	n/a
EB-A	95.5	38.6	65.9	45.5	77.3	50.0	65.9	65.9
EB-B	97.7	53.5	74.4	62.8	93.0	53.5	48.8	86.0
EB-C	100.0	66.7	73.8	66.7	92.9	59.5	76.2	80.5
EB-D	95.2	38.1	52.4	61.9	95.2	47.6	28.6	76.2
EB-E	91.3	30.4	60.9	60.9	82.6	43.5	47.8	76.9

Note: WB-A and EB-A were free flow in either direction. EB-B and EB-D contained the first and second late yellow light events. In EB-C, drivers came to a stop at the rail crossing. The truck stopping past the tracks event occurred in EB-E. See text for additional descriptions.

TABLE 38.3 Glance Percent, Duration and Distance

Sign or Signal	Mean Duration			Mean Glance Distance	
	% Glanced	Per Glance	Total	First	Last
AWS	100.0	0.39 (0.17)	1.4 (0.92)	112.8 (36.1)	30.7 (20.2)
X on Roadway	66.7	0.34 (0.20)	0.72 (0.46)	37.1 (12.6)	16.9 (7.0)
RLC Sign	73.8	0.30 (0.16)	0.44 (0.34)	57.2 (29.3)	39.0 (15.6)
Near Traffic Light	66.7	0.26 (0.13)	0.93 (0.81)	168.2 (74.5)	73.7 (53.5)
Center and Left Lights	92.9	0.26 (0.11)	1.6 (1.2)	188.1 (82.4)	39.1 (35.6)
Do Not Stop on Tracks	59.5	0.54 (0.31)	1.6 (1.5)	70.6 (36.2)	36.7 (33.5)
Crossbucks and Flashers	76.2	0.26 (0.24)	0.72 (0.60)	84.6 (42.4)	40.7 (26.6)
3M Visible	80.5	0.30 (0.19)	0.53 (0.42)	69.0 (11.3)	51.7 (15.9)

Note: Values are given for each sign or signal during the drive which required drivers to stop at crossing (EB-C). Measures were only collected when the vehicle was in motion, not while drivers waited at the crossing.

sources of information. This, in part, reflects the requirement to look down at the roadway to see the "X". Main effects of glance frequency, $F(5, 17) = 5.1$, $p < 0.000$, and last glance distance, $F(5, 17) = 5.32$, $p < 0.000$, across drives were significant. Post hoc comparisons indicated that significantly more glances were made to the "X" on the roadway ($M = 2.3$) when the AWS was flashing (EB-C) than not ($M = 1.3$) ($p < 0.001$), which probably indicates that drivers were looking for a connection between the "X" and the AWS. The last glance distance to the "X" on the roadway was significantly closer when the AWS was active ($M = 17.3$ m) than on eastbound drive A ($M = 30.9$ m) ($p < 0.01$).

A small number of participants misinterpreted the meaning of the "X" on the roadway in the eastbound direction when the advanced warning sign (AWS) was flashing in EB-C (see Web Figure 38.12). The "X" appears on the roadway just past the advanced warning flashers. Three out of 47 drivers stopped prematurely, well in advance of the grade crossing (approx. 100 m). In interviews after driving the simulator, comprehension of the meaning of the active AWS was only 55% for all age groups, although comprehension was slightly higher for the oldest age group (70%). Comprehension of advance active signs has been found to range from 68–91% when tested with a questionnaire (Lerner, Llaneras, McGee, & Stephens, 2002, p. 38).

The Xs in advance of Railroad (RR) and light rail transit (LRT) crossings in Calgary appear immediately before the crossing and not in advance of it. However, the confusions observed only occurred when the AWS was active and not in other encounters with it during other drives. From a design consistency principle, east- and westbound AWSs should be the same. The absence of vehicles in front of the drivers in the first drives (i.e., EB-A,

WB-A) forced them to self-regulate their speed and decide how to respond to active AWS. When other vehicles are present additional cues for action can be acquired by drivers.

38.3.5 Red Light Camera (RLC) Sign

The RLC sign was located prior to the rail crossing and a number of signs and signals in the same location also attracted attention. The main effect of age was significant for last glance distance, $F(2, 17) = 8.9$, $p < 0.000$, and average glance duration, $F(2, 17) = 3.26$, $p < 0.043$. Younger drivers were significantly further away from the red light camera (RLC) sign when they made their last glance than either the middle or older age groups ($p < 0.01$) (see Web Figure 38.3). Older drivers had significantly longer ($M = 0.42$ sec) average glance durations to the RLC sign than the younger drivers ($M = 0.31$ sec) ($p < 0.05$). The percentage of drivers who glanced at the RLC sign ranged from a high of 74% in the first late yellow light drive (EB-B), to a low of 52% in the second yellow light event (EB-D) (see Table 38.1). Total glance duration or the total amount of time looking at it was relatively low and the distance at which it was last glanced was relatively close (see Table 38.2). In interviews, most drivers said that they noticed the RLC sign, but only 29% of older drivers (55 years and above) knew what it meant compared with 69% of those aged 18–24 and 75% of drivers aged 25–54.

In previous research, right-angle collisions were estimated to decrease by 26% upon implementation of red light cameras (Flannery & Maccubbin, 2003). However, rear-end crashes increased by 40% according to another study (Golob, Cho, Curry, & Golob, 2003). In contrast, since there is no cross traffic at the railway grade crossing, these studies would suggest that there is no potential to decrease angle crashes; only to increase rear-end crashes. Consequently, red light cameras are not appropriate at this location because placement is likely to increase rear-end crashes.

38.3.6 Rail Crossing Signs and Signals

The redundant signs and signals at the rail crossing are intended to protect the crossing from unsafe driver behavior and ensure that sufficient information is available to the driver under a variety of circumstances. When the rail crossing signal was activated, the information presented to drivers is quite redundant. Figure 38.5 (Web Figure 38.13) shows the rail crossing has six signs: three are bilingual; one has two tabs and one has one tab. (The red light camera sign is not shown.) In addition, eight red lights, four of which are oscillating in the horizontal plane, are visible. (More are visible if the westbound flashers are counted.) All drivers stopped at the activated crossing without exception.

When the rail crossing is not active, the same number of signs and signals are available, although the gates are in the up position and the train flashers are not active. In this situation, at the rail crossing, the highest percent of glances were to the center and left traffic lights (see Table 38.3, Figure 38.5, Web Figure 38.13). Although drivers were in the right lane, the right traffic light and crossbucks with flashers—which provide redundant information—had lower levels of glances than the left and center lights. The percentage of those who looked at the crossbucks and flashers when the yellow light and truck events occurred was lower than during the free-flow drive.

Table 38.2 shows looking behaviour associated with various traffic control devices at the crossing. With the exception of the center and left lights, the probability of looking at the RLC sign, the DO NOT STOP ON TRACKS sign and crossbucks declined from EB-B to EB-D, which may, in part, reflect increasing familiarity with the crossing as well as the increase in those who ran the light in D (83%) as compared to B (75%). When more time was available to view the crossing during the deceleration to a stop at the crossing (EB-C), signs and lights at the crossing were more likely to be looked at and the number of fixations to each also increased when compared to drives where less time was available (e.g., EB-B or EB-D).

At the train intersection lights, main effects for age were significant for last glance distance, $F(2, 14) = 3.6$, $p < 0.042$, and glance frequency, $F(2, 14) = 3.6, p < 0.03$. Older drivers made significantly more glances at the right hand or near traffic light ($M = 3.2$) than those in the middle-aged group ($M = 2.3$) ($p < 0.05$). Older drivers made their last glance to the far traffic lights (i.e., center and left hand) at a significantly closer distance than the younger age group ($p < 0.05$). A main effect of drive was significant for glance frequency, $F(2, 17) = 2.9$, $p < 0.05$. When drivers had to stop at the rail crossing (i.e., EB-C), they made significantly more glances ($M = 6.2$) to the DO NOT STOP ON TRACKS sign than when they did not stop (EB-A) ($M = 3.3$) ($p < 0.001$). Younger drivers were significantly further away from the sign when they made their last glance than older drivers ($p < 0.05$). The main effect of drive was significant for last glance distance, $F(5, 16) = 5.4$, $p < 0.000$. The last glance to the crossbucks and flashers (right or left) was significantly closer in the drive where participants had to stop at the crossing (EB-C) than in either the free-flow drive ($p < 0.001$) or first late yellow light event (EB-B) ($p < 0.05$).

FIGURE 38.5 Red light, crossing flashers and gates down when train imminent. The sign on the far right indicates DANGER, DO NOT STOP ON TRACKS (in French NE PAS ARRETER SUR LA VOIE PERVEE). The signs immediately to either side of the stop line show STOP LINE (LIGNE D'ARRET).

Glances to the crossbucks and flashers included fixations to the stop line sign for several reasons (see Figure 38.5, Web Figure 38.13). The stop line sign was not visible during a portion of the approach because the DO NOT STOP ON TRACKS sign blocked it. Once it was visible, glances to the stop line sign could not be reliably distinguished from glances to the crossbucks with flashers because they were in the same region of space. In the design visualization process, the height of the sign was already reduced so that it did not block the flashers.

Stop/Go Probability is the observed likelihood of participants stopping or proceeding through an intersection. Here it is expressed as a percent of those who either stopped or ran the late yellow light. The purpose of choosing a challenging time to stop line (TSL) value, or a location deep in the dilemma zone, was an experimental choice. In another study (Caird et al., 2007), only 30.5% of drivers stopped at the yellow light at 2.21 sec TSL at 70 km/h. In the present study, the posted speed was 60 km/h but other time constraints favored running the light. Traffic lights are not typically present at rail crossings. This study is one of the first to describe drivers' responses in this context.

When participants were faced with a late yellow light (2.21 sec in advance of the rail crossing stop line at 60 km/h) most (74.5%) continued through the intersection, whereas 25.5% stopped. Many slowed but could not come to a stop and so they went ahead and drove through both intersections. Younger drivers (18–24) stopped more frequently (37.5%) than did middle-aged (18.8%, 25–54) and older (20.0%, 55-plus) drivers. After being required to wait for a train for a full 94 sec on a subsequent drive and then encountering the same late yellow light (2.21 sec), the percentage of drivers who stopped for the light decreased (17.4%) and most ran it (82.6%).

A comparison of glance percent during the yellow light events can be made by examining Table 38.1. With the exception of the center and left lights, the probability of looking at the RLC sign, the DO NOT STOP ON TRACKS sign and crossbucks declined from EB-B to EB-D, which may, in part, reflect increasing familiarity with the crossing as well as the increase in those who ran the light in D (83%) as compared to B (75%). When more time was available to view the crossing during the deceleration to a stop at the crossing (EB-C), signs and lights at the crossing were more likely to be looked at and the number of fixations to each also increased when compared to drives where less time was available (e.g., EB-B or EB-D).

When required to wait for the train to pass (see Web Figure 38.13), there was no evidence of jumping the light (i.e., starting forward prior to a green light after the gates rose) after waiting for the train to pass by any participant for any drive. Although this behavior was not observed in this study, it does not mean that it will not occur.

Stopping Accuracy was measured in meters by using the y coordinates of the first stop line of the intersection crossing. Stopping accuracy had a number of outlying data points caused by drivers stopping in advance of the railway stop line (EB-C).

With these exceptions removed, stopping accuracy did not differ across the age groups ($p = 0.71$) and most drivers stopped slightly before the stop line.

38.3.7 Transit Crossing Signs and Signals

The 3M traffic heads were a focus of evaluation. These signals were the second set of signals encountered by eastbound drivers. They showed a green arrow, permitting through movements only through the rail crossing/bus transitway intersection. The green arrow became visible just after the railway crossing (see Web Figure 38.14). Fixations to the 3M heads and transit intersection, prior to their being visible, were also analyzed. Most drivers looked ahead to the intersection in advance of the rail crossing. However, distinguishing which of those glances were to the lights or just to the road was more difficult.

The mean percentage of drivers who looked at the transit intersection in advance of the 3M heads being visible was 89%. In the first yellow light event, this mean was 75% but jumped to 95% in the second yellow light event where more drivers ran the traffic light. Drivers may have gained more experience as to where to look for this information, although it was not necessarily visible when they actually looked.

After the 3M heads were visible, the percentage of drivers who made a glance at them increased from 66% in EB-A to 86% in the first yellow light event (EB-B see Table 38.2, last column). The distance at which drivers first looked at the 3M heads averaged about 70.0 meters and varied little over different drives. For last glance distance, a main effect of age was significant, $F(2, 14) = 4.6, p < 0.012$. Older drivers looked at the 3M lights significantly closer than the middle-aged group ($p < 0.01$).

The signs for NO TURNS right and left, were understood by only 48% of drivers in post-simulation interviews. For example, four older drivers thought that the signs indicated turning lanes (an opposite interpretation). The size and resolution of these signs in the simulation may have contributed to this misunderstanding by participants. Monitoring of erroneous turns into the transitway is recommended.

38.3.8 Westbound Transit Intersection

Where to stop was identified as a potential problem by participants who were asked if others would have any difficulties with the westbound transit intersection at the transit crossing. The width of the intersection, given the angle of the crossing, may have contributed to this concern expressed by participants. Snow has the potential to cover up the stop line at this location, but this is mitigated by STOP HERE signs which bracket the stop line (see Web Figure 38.15).

The percentage of glances to analogous signs and signals differed in the westbound (WB-A) and eastbound (EB-A) drives (see Table 38.2). The percentage of glances to the AWS in either direction was similar. However, the percent of glances to the "X" differed, with eastbound being significantly lower (79% versus 39%). Perhaps once the "X" was seen on the first drive (WB-A), participants may have filtered it out as redundant information on subsequent drives (cf., Macdonald & Hoffmann, 1991). The RLC sign and traffic lights were in different locations in either direction. The higher percentage of glances to the DO NOT STOP ON TRACKS sign and crossbucks in the westbound direction (67%) than the eastbound drive (50%) is most likely due to there being less sign and signal information at the westbound rail crossing (i.e., no traffic lights).

38.3.9 Truck Braking Event

A challenging scenario was created to test drivers' responses in an emergency situation that required an appropriate response to remain out of harm's way. As participants followed the truck (EB-E), they were less likely to look at many of the signs at the rail crossing (see Table 38.2, last row). The pattern of looking most likely shifted to the truck while drivers followed it, which has also been found by other researchers (Crundall, Sharton, & Underwood, 2004; Tijerina, Barickman, & Mazzae, 2004).

The truck braking event was designed to evoke several responses by drivers. First, as the truck went through the intersection it blocked a participant from seeing a light change from green to yellow. Would participants follow the truck through the intersection and over the tracks? Second, once the truck drove over the tracks it braked at 6 m/s², which was emergency braking. Would drivers respond appropriately given the actions of the truck?

What drivers did when unexpectedly forced to stop on or near the tracks was not surprising. Of the 59.2% of drivers who responded appropriately to the rapid braking of a lead truck, 25% stopped at the stop line prior to the crossing and 12.5% drove around the truck to either the right or left. Of those who stopped at the grade crossing stop line, none were in the older group of drivers, whereas 50% were in each of the younger and middle-aged groups. Some drivers (8.3%) were trapped between the truck and the rail tracks and 12.5% stopped and reversed back to the stop line to escape the approaching train (see Web Figure 38.16, right). None of these participants were told that the simulator functioned in reverse. Only one driver steered into the pull out refuge to the right of the truck (see Web Figure 38.16, left). The visibility of the refuge from behind the truck was limited. When the lead truck braked rapidly (see Figure 38.16, left), 41.7% of drivers crashed into it. Of those who collided with the truck, 50% were older drivers, whereas 30% were younger drivers and 20% were in the middle-aged group.

The circumstances that coincide to create a crash-likely event are complex and difficult to simulate. The event chosen by all parties—the truck braking event—produced a range of interesting driver behaviors, some of which were not sufficient to remove the driver from a crash (also see Strayer, Drews, & Johnston, 2003). The responses to the emergency braking of the truck were not necessarily dependent on the design of the crossing elements.

38.4 Discussion

38.4.1 Design Implications

After driving the Fallowfield tile, numerous elements of the design were changed and modified based on expert input from the organizations involved in the project. Approximately 20 design changes were incorporated into the final simulation model. These changes included:

1. Moved AWS with flashers and pavement "X" forward away from the rail crossing to a location where it would more likely be seen and understood;
2. Included a red light camera (RLC) sign for evaluation;
3. Decided to use white concrete treatment on tracks only, instead of the region between the tracks and the transit intersection;
4. Reduced the STOP AT LINE sign height to make flashing crossbuck signals visible;
5. Considered removal of DANGER DO NOT STOP ON TRACKS sign to reduce visual clutter;
6. Used iconic NO RIGHT or LEFT signs on transit mast head; and
7. Considered cantilevered crossing signals for east- and westbound railway crossing approaches.

The ability of high-fidelity simulation models to visualize and problem-solve complex highway engineering designs and evaluate resulting solutions had a number of positive safety and design benefits. In addition to the design changes that were made during the visualization phase, another seven design issues were identified during the evaluation phase, which are described below in the Evaluation Results section. In the absence of these two phases, the system safety at Fallowfield would likely be significantly less than it is today. In hindsight, the use of driving simulation to problem-solve complex designs for safety issues is common sense because once the infrastructure is built, reconsideration of safety issues is often cost-prohibitive.

The process of design visualization and driver evaluation should be considered in cases where complex sign, signal and road geometries raise safety concerns. In many cases, the cost-effectiveness of the process must be considered. Not all the phases that were engaged in during the course of this process proceeded as planned. The creation and expert evaluation of the Fallowfield model was relatively quick. However, the process of conducting and analyzing the evaluation was time consuming—especially the analyses of eye movements.

38.4.2 Evaluation Results

Forty-seven participants, stratified into the age groups of 18–24, 25–55, and 55 and older, drove the Fallowfield simulation model in both directions when a number of traffic events occurred to determine if any additional design issues could be identified.

The speed measures taken without other vehicles provide a baseline indication of driver velocity without interaction with other vehicles. In both the eastbound and westbound directions at the rail crossing, speed reductions were 4 km/h for younger drivers and 8 km/h for the older age group. Older drivers tended to adopt slower speeds when approaching intersections (Caird et al., 2007; Caird & Hancock, 2007). In general, few studies have examined speed changes to signs (Shinar & Raz, 1982; Summala & Hietamaki, 1984; Ward & Wilde, 1995) and the speed reductions observed are typically in the range of 2–3 km/h or less. However, attribution of this speed reduction to a specific information source in this evaluation would be unwarranted. The reduction in speed may have resulted from drivers' habitual responses to traffic lights. Reductions in speed at the eastbound crossing are likely to be the result of the combination of signs, grade crossing contextual cues and driver experience. Common to both speed reductions in either direction was the presence of traffic lights.

Traffic lights are not typically present at rail crossings. Thus, these results are one of the first to describe drivers' responses in this context. When less time was available to look at all the sign and signal information at the crossing, the center and left traffic lights continued to be looked at, whereas the likelihood of glancing at the crossbucks and flashers was somewhat less. The late yellow light results, which showed a high percentage of drivers who ran the yellow light, are consistent with previous simulator and field studies (e.g., Chang et al., 1985; Van der Horst & Wilmink, 1986). For example, using the same time to stop line (TSL) of 2.2 sec and a posted speed of 70 km/h, Caird et al. (2007) found that 69.5% of drivers ran the light.

Guidelines for reducing sign clutter or providing just enough redundant information to guide drivers is not completely understood by researchers (Drory & Shinar, 1982; Green, 2007; Ho, Scialfa, Caird, & Graw, 2001; Hughes & Cole, 1986; Luoma, 1988; Olson, Battle, & Aoki, 1989; Serafin, 1994). For example, eye movement analyses indicated that several signs were not fixated by the majority of participants and these signs had still fewer fixations when traffic was present or the time to act was limited. Does this mean that those signs should not be used? Given the amount of sign and signal information at the rail crossing, how much is too much redundant information? Prior eye movements studies (e.g., Mourant, Rockwell, & Rackoff, 1969) have recorded the percentage of time that drivers looked at signs while driving. This number is dependent on the quantity, density and importance of signs on the route measured. For example, Olson et al. (1989) found that drivers' glances to signs were 6% of fixations and the mean glance duration to a sign was approximately 0.3 sec. Mean glance durations to signs tend to be somewhat longer if more information must be read from the sign (Luoma, 1988). Martens and Fox (2007) examined whether eye movements to signs changed with repeated exposures on-road. As in the present study, drivers reduced the number of fixations to signs that were seen over days of exposure.

Previous studies have not combined the multi-measure approach in the present study nor have prior studies attempted to make specific recommendations about sign location or design based on eye movement measures, with some exceptions (Crundall, van Loon, & Underwood, 2006; Smiley, Smahel,

& Eizenman, 2004; Zwalen, 1987). Real-time recording of eye movements to the signs and signals in Fallowfield intersections produced new insights into clutter and the relationships between visual behavior and driver performance.

38.4.3 Recommendations for Design of the Fallowfield Roadway

The recommendations that follow are based on the results of this study, other traffic safety literature, and professional human factors judgment. A number of signs and signals were identified in the evaluation phase as requiring further design consideration. These include:

- Stopping in advance of the crossing, confirming statements from the interviews and eye movement data indicate that a proportion of drivers did not understand how to respond when faced with the combination of AWS and "X" on the roadway. Based on these results, we recommended that the flashing lights should be removed from the sign and the "X" on the roadway should be reconsidered.
- The placement of the AWS and "X" together, further from the Barhaven Mall than prescribed, is an exception to current design practice. Based on the eye movement analysis, moving the AWS from near the Barhaven intersection to this location, where no other information is available, appears to have been a good decision and we recommended that this placement be maintained. The probability that the AWS would have been seen just past the Barhaven intersection would have been less than observed in its present location.
- We recommended removing RLC cameras and signs in the east- and westbound directions. There is evidence that the incidence of rear-end crashes increases both at intersections with red light cameras and at rail crossings (Flannery & Maccubbin, 2002). While RLC cameras are associated with a decrease in angle crashes, there is little opportunity for these at this location due to the limited amount of traffic on the bus transitway. Comprehension of the RLC sign by older drivers was considerably lower than for other age groups. Older drivers, in general, tend to have lower comprehension for many types of traffic symbols (Dewar, Kline, & Swanson, 1994).
- The DO NOT STOP ON TRACKS sign should be removed in the eastbound direction, but not the westbound, because sign clutter at the rail crossing should be reduced (see Figure 38.5, Web Figure 38.13). When time is limited, less than half of drivers looked at it and it blocks the drivers' sightline to the stop line sign for a portion of the approach. The gates and stop line roadway marking provide redundant information about where to stop. The white pavement marking also indicates the location of the tracks. Under the time constraints imposed on drivers at the eastbound rail crossing, signs that were less salient had a lower probability of fixation.

- Monitoring exceptional driver behavior at the rail and transit crossings such as red light running, driving under descending gates and stopping between the crossing, and entering the transitway is highly recommended. The use of loop detectors and incidence detection systems between the rail and transit crossings is suggested (and was implemented).
- The use of the 3M heads in their current configuration prevents confusion as to which lights apply to which intersection. Drivers seem to look at them more frequently with exposure. Therefore, the current use of the 3M heads is recommended.
- Drivers do not necessarily expect enforcement at grade crossings (Tidwell & Humphreys, 1981). A program focused on enforcement of violations of signs and signals at this location should be put in place should monitoring indicate problems.

38.4.4 Advantages and Disadvantages of Driving Simulation

Driving simulation has a number of advantages and disadvantages that should be weighed against cost, time, and potential to address the research questions of central interest.

The advantages of driving simulation include the following:

- Experimental control over extraneous variables;
- Precise collection of eye movement and driving simulation variables;
- Removal of participants from harm's way (i.e., trains, other traffic); and
- Cost-efficiency to evaluate a range of crossing elements (e.g., generation of trains, placement of crossing elements, placement of drives over multiple crossings, re-creation of known problems such as two trains, pedestrians, lack of visibility, etc.).

The disadvantages of driving simulation include:

- The range of ambient illumination and light reflection properties is not the same in the simulator as in the real world. For example, the UCDS cannot simulate bright daylight conditions;
- The resolution of signs presented to drivers in real-time is less than in other testing methods. The net effect of this limitation would be to constrain sign reading, depending on the size of the text (see, e.g., Carpenter, 2002). As a result, drivers may have to wait until they are closer to a sign to read it than they would in a field setting;
- Driver behavior, in the absence of consequences such as hitting a train, may not be the same as it is in the real world;
- Ordinarily, the frequency that an individual driver encounters a highway–railway grade crossing is limited. To reduce the number crossings encountered by each

participant, the number of volunteers required to evaluate crossing elements rises accordingly;
- Many of the highway–railway grade crossings and elements will need to be modeled using drafting software before they can be evaluated at an added cost;
- Not all worst-case events can or should be included in a simulation study. Practical constraints from the experimental design prohibit it. For example, the following behaviors were also of interest, but are beyond the scope of the experimental approach because they are likely to be quite rare:
  - For drivers who missed their destination (e.g., Barhaven Mall), do they attempt to make a U-turn at either the crossing or transitway openings in the center barrier?
  - When a school bus and truck stops and starts at the crossing how does this affect the behavior of others and the traffic flow?

The results of this case study reveal considerable benefits of driving simulation for a complex situation at an at-grade rail crossing in close proximity to two intersections, one of which is a bus transitway. Numerous design changes were made as a result of being able to visualize the location in three dimensions, as well as being able to observe drivers negotiating it. Clearly other road design reconstruction projects could also benefit from the same process.

Key Points

- Numerous design changes were made as a result of being able to visualize the highway–railway grade crossing in three dimensions, as well as being able to observe drivers negotiating it.
- A number of signs and signals were identified in the evaluation phase as requiring further design consideration.
- Clearly other road design reconstruction projects that have complex safety concerns could also benefit from the same process.

Keywords: Driving Performance, Eye Movements, Grade Crossings, Highway–Railway, Roadway Design

Acknowledgments

The City of Ottawa funded this unique study and we thank Paul LeMay for his support. Geoff Millen at Delphi-MRC was integral to all aspects of this study. Don Fisher, John Lee and Neil Lerner provided valuable editorial suggestions. We would like to thank Mike Boyle for the real-time programming and data reduction solutions that he provided. Susan Chisholm assisted with the statistical analyses. Stephanie Kramer helped to run participants, and Cale White scripted many of the scenarios that were used. Sondra Iverson, Ken Melnrick and Aaron Miller at DriveSafety helped to resolve innumerable Fallowfield model and programming solutions. Mike Mollenhauer, from RealTime, carefully worked with us to integrate and develop

SimObserver and Data Distiller to analyze the eye movement data. A grant from AUTO21 Network of Centers of Excellence (NCE) to the first author provided support for the eye movement software, hardware and statistical analyses. The Canadian Foundation for Innovation (CFI), Alberta Science and Innovation and the University of Calgary provided the base funding from which the UCDS was created. An abstract of the design portion of this study was presented at the International Conference on Road Safety and Simulation in Rome, Italy (2007). The eye movement data was presented at Vision in Vehicles XI in Dublin, Ireland (2006).

Web Resources

The *Handbook* web site contains supplemental material for this chapter, including color versions of some of the figures in the print version of the chapter, and additional web site-only figures.

Web Figure 38.1: Aerial view of Fallowfield Road with CN Rail line and transitway extensions shown (color version of Figure 38.1 in printed chapter).

Web Figure 38.2: Location of STOP LINE sign (French LIGNE D'ARRET) (left) and with height reduction (right) so that the left rail crossing flashing light was no longer obstructed. The DANGER, DO NOT STOP ON TRACKS signs (NE PAS ARRETER SUR LA VOIE PERVEE) were moved to the right and the $45.00 fine tab was deleted.

Web Figure 38.3: Proposed red light camera (RLC) sign (left) and Alberta specified sign that was tested (right).

Web Figure 38.4: Proposed cantilevered rail crossing lights (left) and pavement treatment between rail and transit crossings (right) shown from the westbound direction.

Web Figure 38.5: Route of bicyclist after the railway crossing (left) and pedestrian activity at the Barhaven Mall intersection (right).

Web Figure 38.6: Westbound transitway crossing illustrating the travel of buses through this intersection (left). Westbound train present in the railway crossing (right).

Web Figure 38.7 (see color insert): The experimenters' view of the UCDS with the Fallowfield highway–railway grade crossing shown. The displays from left to right are the quadraplex view (face, upper left; center column, upper right; pedals and feet, lower left; center view simulator, lower right), eye movement calibration and control monitor, Hyperdrive (v. 1.9.2) console and SimObserver recording monitor.

Web Figure 38.8: Fallowfield plan view with associated signing.

Web Figure 38.9: Fallowfield plan view with velocity measurement locations. Eastbound locations used to generate the velocity profile are indicated across the bottom, whereas westbound locations are shown across the top from right to left (Figure 38.2 in print chapter).

Web Figure 38.10: Eastbound velocity profile by age group with standard error bars is shown. The locations listed on the "X" axis are indicated on the plan view of Web Figure 38.9 (Figure 38.3 in print chapter).

Web Figure 38.11: Westbound velocity profile by age group with standard error bars is shown. The locations listed on the "X" axis are also shown on the plan view of Web Figure 38.9. (Figure 38.4).

Web Figure 38.12: The advance warning sign (AWS), which indicates PREPARE TO STOP WHEN FLASHING (PREPAREZ VOUS A ARRETER QUAND LE PUE LIGNOTE), and "X" roadway treatment on the eastbound approach to Fallowfield.

Web Figure 38.13: Red light, crossing flashers and gates down when train imminent. The sign on the far right indicates DANGER, DO NOT STOP ON TRACKS (in French NE PAS ARRETER SUR LA VOIE PERVEE). The signs immediately to either side of the stop line show STOP LINE (LIGNE D'ARRET) (Figure 38.5 in print chapter).

Web Figure 38.14: The 3M directional sign heads at the transit crossing with right and left NO TURNS signs on light masts.

Web Figure 38.15: Westbound transit intersection showing stop line and STOP HERE (LIGNE D'ARRETE) signs.

Web Figure 38.16: Truck position once it braked to a stop just past tracks (left) and the position of a driver once he or she backed up so as not to be struck by the train (right). Notice the pull out to the right.

Key Readings

Caird, J. K., Creaser, J. I., Edwards, C., & Dewar, R. E. (2002). *A human factors analysis of highway–railway grade crossing accidents in Canada* (Rep. No. TP 13938E). Montreal, Quebec: Transport Development Center, Transport Canada.

Green, P. (2007). Where do drivers look while driving (and for how long)? In R. E. Dewar & R. Olson (Eds.), *Human factors in traffic safety* (pp. 77–110). Tucson, AZ: Lawyers & Judges Publishing.

Lerner, N., Llaneras, E., McGee, H. W., & Stephens, D. E. (2002). *Traffic-control devices for passive railroad-highway grade crossings* (NCHRP Rep. No. Project 3–57). Washington, DC: Transportation Research Board.

Upchurch, J., Fisher, D., Carpenter, R. A., & Dutta, A. (2003). Freeway guide sign design and driving simulator for central artery-tunnel. *Transportation Research Record, 1801*, 9–17.

References

Alexander, G. J., & Lunenfeld, H. (1986). *Driver expectancy in highway design and traffic operations*. Washington, DC: Federal Highway Administration.

Applied Sciences Laboratory. (2001). *Eyenal manual (v. 1.59)*. Bedford, MA: Author.

Caird, J. K., Chisholm, S., Edwards, C., & Creaser, J. (2007). The effect of yellow light onset time on older and younger drivers' perception response time (PRT) and intersection behavior. *Transportation Research Part F: Traffic Psychology and Behavior, 10*(5), 383–396.

Caird, J. K., Creaser, J. I., Edwards, C., & Dewar, R. E. (2002). *A human factors analysis of highway–railway grade crossing accidents in Canada* (Rep. No. TP 13938E). Montreal, Quebec: Transport Development Center, Transport Canada.

Caird, J. K., & Hancock, P. A. (2007). Left turn and gap acceptance crashes. In R. E. Dewar & R. Olson (Eds.), *Human factors in traffic safety* (pp. 591–630). Tucson, AZ: Lawyers & Judges Publishing.

Caird, J. K., Milloy, S., Ohlhauser, A., Jacobson, M., Skene, M., & Morrall, J. (2008). *Evaluation of four bicycle lane treatments using driving simulation: Comprehension and driving performance results.* Annual Conference of Canadian Road Safety Professionals, Whistler, British Columbia.

Carpenter, R. (2002). *Static and dynamic legibility distances of simulated signs.* Unpublished masters thesis, Department of Operations and Industrial Engineering. University of Massachusetts, Amherst.

Chang, M.-S., Messer, C. J., & Santiago, A. J. (1985). Timing traffic signal change intervals based on driver behavior. *Transportation Research Record, 1027,* 20–30.

Chisholm, S., Caird, J. K., & Lockhart, J. (2008). The effects of practice with MP3 players on driving performance. *Accident Analysis & Prevention, 40,* 704–713.

Crundall, D., Sharton, C., & Underwood, G. (2004). Eye movements during intentional car following. *Perception, 33,* 971–986.

Crundall, D., van Loon, E., & Underwood, G. (2006). Attraction and distraction of attention to roadside advertisements. *Accident Analysis & Prevention, 38,* 671–677.

Denton, G. G. (1980). The influence of visual pattern on perceived speed. *Perception, 9,* 392–402.

Dewar, R. E., Kline, D. W., & Swanson, H. A. (1994). Age differences in comprehension of traffic sign symbols. *Transportation Research Record, 1456,* 1–10.

Drory, A., & Shinar, D. (1982). The effects of roadway environment and fatigue on sign perception. *Journal of Safety Research, 13,* 25–32.

Flannery, A., & Maccubbin, R. (2002). *Using meta-analysis techniques to assess the safety effect of red light running cameras.* Transportation Research Board 2003 Annual Meeting CD-ROM, Washington, DC: Transportation Research Board.

Golob, J. M., Cho, S., Curry, P. E., & Golob, T. F. (2002). *Impacts of the San Diego photo red light enforcement on traffic safety.* Transportation Research Board 2003 Annual Meeting *CD-ROM,* Washington, DC: Transportation Research Board.

Green, P. (2007). Where do drivers look while driving (and for how long)? In R. E. Dewar & R. Olson (Eds.), *Human factors in traffic safety* (pp. 77–110). Tucson, AZ: Lawyers & Judges Publishing.

Ho, G., Scialfa, C. T., Caird, J. K., & Graw, T. (2001). Traffic sign conspicuity: The effects of clutter, luminance, and age. *Human Factors, 43*(2), 194–207.

Hughes, P. K., & Cole, B. L. (1986). What attracts attention when driving? *Ergonomics, 3*(29), 377–391.

International Standards Organization (ISO). (2002a). *International Standard 15007-1: Road vehicles—measurement of driver visual behavior with respect to transport information and control systems—Part 1: Definitions and parameters.*

International Standards Organization. (2002b). *International Standard 15007-1: Road vehicles—measurement of driver visual behavior with respect to transport information and control systems—Part 2: Equipment and procedures.*

Lerner, N., Llaneras, E., McGee, H. W., & Stephens, D. E. (2002). *Traffic-control devices for passive railroad-highway grade crossings* (NCHRP Rep. No. Project 3–57). Washington, DC: Transportation Research Board.

Luoma, J. (1988). Drivers' eye fixations and perceptions. In A. G. Gale, M. H. Freeman, C. M. Haslegrave, P. Smith, & S. P. Taylor (Eds.), *Vision and Vehicles II* (pp. 231–237). Amsterdam: Elsevier Science.

Macdonald, W. A., & Hoffmann, E. R. (1991). Drivers' awareness of traffic sign information. *Ergonomics, 5*(34), 585–612.

Manser, M. P., & Hancock, P. A. (2007). The influence of perceptual speed regulation on speed perception, choice, and control: Tunnel wall characteristics and influences. *Accident Analysis & Prevention, 39*(1), 69–78.

Martens, M. H., & Fox, M. (2007). Does road familiarity change eye fixations? A comparison between watching a video and real driving. *Transportation Research Part F: Traffic Psychology and Behavior, 10,* 33–47.

Mourant, R. R., Rockwell, T. H., & Rackoff, N. J. (1969). Drivers' eye movements and visual workload. *Highway Research Record, 292,* 1–10.

Olson, P. L., Battle, D. S., & Aoki, T. (1989). *Driver eye fixations under different operating conditions* (Rep. UMTRI-89-3). Ann Arbor, MI: University of Michigan Transportation Research Institute.

Richards, S. H., & Heathington, K. W. (1988). Motorist understanding of railroad-highway grade crossing traffic control devices and associated traffic laws. *Transportation Research Record, 1160,* 52–59.

Serafin, C. (1994). *Driver eye fixations on rural roads: Insight into safe driving behavior* (Rep. UMTRI-94-21). Ann Arbor, MI: University of Michigan Transportation Research Institute.

Shinar, D., & Raz, S. (1982). Driver response to different railroad crossing protection systems. *Ergonomics, 25*(9), 801–808.

Smiley, A. (2008). Point of view: Driver behavior—A moving target. *Transportation Research News, 254,* 19–24.

Smiley, A., Smahel, T., & Eizenman, M. (2004). The impact of video advertising on driver fixation patterns. *Transportation Research Record, 1899,* 76–83.

Strayer, D. L., Drews, F. A., & Johnston, W. A. (2003). Cell phone-induced failures of attention during simulated driving. *Journal of Experimental Psychology: Applied, 9*(1), 23–32.

Summala, H., & Hietamaki, J. (1984). Drivers' immediate responses to traffic signs. *Ergonomics, 27*(2), 205–216.

Tabachnick, B. G., & Fidell, L. S. (2001). *Using multivariate statistics* (4th ed.). Boston, MA: Allyn and Bacon.

Tijerina, L., Barickman, F. S., & Mazzae, E. N. (2004). *Driver eye glance behavior during car following* (Rep. DOT HS 809 723). East Liberty, OH: National Highway Traffic Safety Administration.

Upchurch, J., Fisher, D., Carpenter, R. A., & Dutta, A. (2003). Freeway guide sign design and driving simulator for central artery-tunnel. *Transportation Research Record, 1801*, 9–17.

Van der Horst, R., & de Ridder, S. (2007). Influence of roadside infrastructure on driving behavior: Driving simulator study. *Transportation Research Record, 2018*, 36–44.

Van der Horst, R., & Wilmink, A. (1986). Drivers' decision-making at signalized intersections: An optimisation of the yellow timing. *Traffic Engineering and Control, December*, 615–622.

Ward, N. J., & Wilde, G. J. S. (1995). A comparison of vehicular approach speed and braking between day and nighttime periods at an automated railway crossing. *Safety Science, 19*, 31–44.

Zwalen, H. T. (1987). Advisory speed signs and curve signs and their effect on driver eye scanning and driving performance. *Transportation Research Record, 1111*, 110–120.

39

Roadway Visualization

Michael A.
Manore
*Vispective Management
Consulting, LLC*

Yiannis Papelis
Old Dominion University

Abstract

The Opportunity. This chapter describes the potential for incorporating driving simulators as an integral part of highway geometric design (including work-zone layout) and public involvement. ***Role and Benefit of Driving Simulators.*** Visualization is directly related to driving simulation in that both depend on the development of virtual three-dimensional (3D) information in a computerized environment for the purpose of prompting some form of human response. Whether it is awareness, a decision, pattern recognition, stress, psychomotor response, or any of a multitude of other human reactions, they both present visual information that affects human emotion, thought (conscious or otherwise), and reactions. With the growing interest in the use of visualization for public involvement, highway geometric design, and 3D modeling for construction, along with concurrent interest in incorporating driving simulators for human-in-the-loop assessments of design and work-zone alternatives, there is an opportunity to step back and consider the collective elements of a design alternative, including, especially, driving simulation, as part of a more visually enhanced and immersive project development environment. ***Influencing Factors.*** At least four factors must be present to realize the potential that driving simulation has for the design community. First, the interoperable exchange of 3D and related visual data between different systems (i.e., those used for design and those used for driving simulation) will need to be improved. Second, U.S. transportation agencies need both to understand the processes and guidelines for implementing systems that allow both 3D design visualization and driving simulation, and to be convinced of the return on investment by doing so. Third, 3D models should become the new "deliverable" from which visualizations could be readily created and 2D plans could be extracted as needed. This is part of a new process looking to be adopted from the vertical construction industry called virtual design and construction (or VDC) where 3D models and visualization are an integral part of a multidisciplined analysis, performance, and business process, in which driving simulators stand to play a valuable and integral role. This will, however, pose professional liability concerns for engineers, especially when it comes to the historical inconsistencies between simulator data and on-road data (Campbell et al., 1997; Hoffman, Brown, Lee, & McGehee, 2002). More discussion is provided on this topic throughout this chapter. And finally, relationships between agencies and those research institutions having simulators will need some rethinking to consider alternative business models, operational logistics, liability issues, and overall expectations for integrating this form of "project-specific/human-in-loop analysis" into highway geometric design.

39.1 Introduction

Visualization in the highway industry has evolved to imply many things, from simple rendering and animation of 3D models to large-scale virtual cities used as part of regional planning efforts. The momentum for computer-based visualization in highway design began in the early 1990s, primarily as a tool for public and stakeholder involvement. More recently, there has been growing interest in enhancing the engineer's ability to perform more "human-in-the-loop" analysis of proposed designs, and in incorporating the use of driving simulators as part of the design process.

This is a considerable opportunity, as highway engineers have had limited tools to evaluate directly the potential "human" impacts of their proposed design alternatives. Traditionally, engineers rely on professional journals, manuals, design standards, and equations to assess safety and performance, with the effects of their final design on driver behavior only coming forward after the project is completed and open to users. Today, highway engineers have more access to 3D design through existing functionality available in their geometric design software (typically Bentley or Autodesk in the U.S.), but are not "actively" designing in 3D, which, if they did, would set a more solid foundation to begin leveraging driving simulators. This relative absence of 3D design is in part due to generational mindsets, current institutional processes, business and contractual models, education and training, and most often, a simple lack of the awareness of the capabilities, measurable benefits, or professional liabilities in undertaking a 3D design. Improvements in these areas will certainly build a working appreciation for driving simulators, but another shift in the industry may provide an additional boost, and that is the industry's current trend to shift from "criteria-based" analysis of designs, to one that is more "performance-based".

In his presentation at the 2009 TRB Joint Session 786, *Beyond 3D: Progressive Visualization for Geometric Design*, Brian Ray describes the potential value that 3D design will bring in combination with a new performance-based mindset. He noted current practices prompt the design of roadway elements to be undertaken in relative isolation (structures, roadway alignment, drainage, landscaping, signalization & signing, etc.) and the overall products of the process (2D plans, specifications, and engineering estimates) are "aimed toward the contractor and not the end user". He notes the convergence of visualization, performance-based design, and driving simulators will assist geometric designers to more "iteratively" juggle the trade-offs between design standards, cost, facility performance, end-user performance, and overall safety as they strive to enhance the quality of the infrastructure they deliver.

Visualization and driving simulation are interdependent at two levels:

1. Technical: They both depend on a 3D-modeled representation of the facility; the 3D model in design facilitates decisions regarding position, form, relationship, and "functional" performance (e.g., structural performance of a bridge), and the 3D model in the simulator facilitates design decisions based on end-user behavioral responses to the design and the surrounding viewscape.

2. Institutional: They both depend on a conscious shift from a current practice of delivering plan sets to one that provides a more collaborative environment for delivering infrastructure that is optimized and balanced regarding community needs, functional requirements, operational performance, together with end-user performance and safety. Visualization and driving simulation can be performed independent of each other; however, the efficiencies and benefits to be gained will be minimal relative to the more integrated and collaborative approach just described.

39.1.1 Bridging 3D Design With Simulators

The full integration (technical and institutional) between 3D roadway design and driving simulators is yet to be seamless. In fact, the challenge today is three-fold since it is necessary to:

1. Improve the ability for highway engineers to intuitively work and design in 3D;
2. Enable a more streamlined hand-off between the 3D roadway data created in design and delivered to the simulators; and
3. Establish the professional metrics by which highway engineers employ simulators.

39.1.2 Progress in Highway Visualization

There are a number of departments of transportation in the United States today using visualization but primarily for public involvement according to Landphair and Larsen (1996) and Hixon (2006), and leaving the creation of the 3D models (Figure 39.1) to the typically small team of visualization experts either internal to or external (e.g., visualization consultants) to the organization. Within the transportation consulting community, visualization is still used primarily to deliver products to their clients for public involvement and also to help win projects, leaving the 3D model development to the visualization teams that are separate from the highway engineers who actually design the roadway.

Recognizing the potential value in design, however, and with a large interest in the construction community for 3D subgrade models used in stake-less construction and machine control, this stands to change in the near future. Since roughly the late 1990s, the highway construction industry has been taking advantage of the advancements in GPS technologies and on-board computer systems that, in effect, transform large earthwork machinery into self-navigating robots. Traditional methods required that wooden stakes be driven into the ground to mark the proposed roadway alignment, shoulders, slope catch-points, along with markings to indicate the amount of cut or fill at any given point along the alignment. Machine operators would then use these to remove or add earth until the subgrade designed elevations and alignment were achieved. With a 3D model available, most of this effort can be replaced with earthwork machines that know where they are

FIGURE 39.1 Image of US1/1-95 interchange in Alexandria, VA. Part of the Woodrow Wilson Bridge project: (Image Courtesy of Jeff Coleman, URS Corporation.)

on the project site, as well as what direction they are moving and how much fill or cut they must make at any given location. Survey crews that traditionally used to set the cut and fill stakes can use the same 3D models and GPS systems to check progress and construction accuracy. Contractors have observed upwards of 15% reductions in cost by leveraging 3D models and GPS systems, which can amount to considerable returns on large construction projects.

Also, in 2005, the transportation reauthorization bill SAFETEA-LU specifically called for state, regional, and local planners to incorporate visualization as part of the short- and long-term planning strategies. The Federal Highway Administration subsequently enacted a formal "rule-making" process that incorporated visualization as part of short- and long-term planning. The final ruling was published in 2007 regarding the use of visualization technologies and methods to improve communication with the public. The ruling was kept general in nature, allowing state and local agencies the freedom to implement the tools and processes that best supported their specific needs. In short, the ruling called for the use of visualization tools and methods to improve communication and participation with the public and stakeholders in the development of state, regional, and local transportation improvement plans.

39.1.3 Progress in Simulators for Highway Design

In the mid-1990s, as part of the Minnesota Department of Transportation's (MnDOT) Visualization Assessment and Development Project (Manore and Larson, 1996), exploratory efforts were initiated to take the 3D models that were being created for design, and supply them to the University of Minnesota Human Factors Lab to see if there was value in such a process. Subsequently in 1998, Scallen and Carmody, through the University of Minnesota Center for Transportation Studies, performed a research project titled "Investigating the Effects of Roadway Design on Driver Behavior: Applications for the Minnesota Highway System", Report 1999-10(4), which employed the use of an immersive driving simulator to assess various roadside treatments on an active MnDOT design to reduce speeding on a rural highway. The results of the simulator evaluations, balanced with other constraints such as aesthetics, safety, and maintainability, pointed to a unique solution that eventually made its way into the final contract specifications and, subsequently, to the construction of the site. Though the project was deemed a success, and MnDOT reflected positively on the use of the simulator because it allowed them to better understand and compare the impacts of various roadside treatments (landscaping, striping configurations, and contrasting pavement colors on the shoulders and median), no study was performed during post-construction to validate the performance in the simulator. Despite not coming full circle, and despite the fact that the 3D models had to be re-created for the simulator due to data inoperability, this was the first time that a simulator was used to directly facilitate design decisions for one of MnDOT's active highway projects.

In 2005, the Federal Highway Administration, under the International Technology Exchange Program, performed a SCAN of European countries (Keith et al., 2005). Six countries were visited to review research involving human factors to improve safety as part of design, many of which utilized driving simulator technology. Of all the findings published in the report, two stood out most significantly:

1. There is value in using driving simulators for roadway design and visualization. Two examples cited therein include:
   a. A team of highway designers visited the Swedish National Road and Transport Research Institute (abbr. VTI) to perform an informal review of their alternate road plans in the VTI simulator. Planning to make this one-day drive-through activity, they decided to extend their stay an additional two days making "several design changes based on their simulator experience".
   b. The Foundation of Scientific and Industrial Research at the Norwegian Institute of Technology (abbr. SINTEF) held a more formal design evaluation of the world's longest tunnel (the Laerdal project) in western Norway to assess lighting strategies to address potential driver anxiety. This evaluation demonstrated that by using certain lighting arrangements of blue, yellow, and green lights, driver safety and comfort were increased while anxiety was decreased when changes to these configurations were incorporated every 2 km. Driver performance since the tunnel's opening has verified these results, citing high comfort ratings from users and no crashes recorded up to the date of the report (2005).
2. Driving simulators can provide for a more human-centered roadway analysis and design. Or to quote the report (Keith et al., 2005, p. xiii), "Human-centered design starts with the limitations and preferences of the driver, and then derives appropriate technology from these human principles." The SINTEF example cited in 1b above, is relevant here in that the study "anticipated and minimized driver anxiety and boredom in the tunnel", the results of which directly impacted design decisions which were ultimately proven correct under operational conditions.

At the TRB 5th International Geometric Design Symposium, there were a number of papers touching on the use of 3D models to evaluate roadway design, and one research paper (Lamm, Cafiso, LaCava, & Beck, 2005) stood out as a successful attempt to integrate human stress measurements while driving the simulator. The research demonstrated the potential of evaluating the "human" impacts of a highway design on the driver by incorporating sensors to evaluate the eye movement and heart rate of test subjects as they navigated through roadway geometry in a simulator. The measured human response data were then correlated to specific locations along the path of the simulated roadway. This research revealed a potential to measure the stress levels of humans as they navigate through a proposed design alternative. If refined, this type of simulator-based assessment (combined with other strategies) would provide a valuable insight to engineers as they look to expose potentially unsafe elements.

Finally, in November, 2007, the International Conference on Road Safety and Simulation (RSS 2007) was held in Rome. Here, professionals from around the world convened and spoke on the various applications of driving simulation technology. Though much of the work focused on basic research issues, there were several papers that demonstrated the benefits of driving simulation to assess innovative designs (Aarts & Davidse, 2007) and even the impacts of alternative signage as a retrofit to improve safety and mobility of in-place geometry (Calvi & Amico, 2007). In one, Trentacoste (2007) described efforts at the Turner-Fairbank Highway Research Center (TF) to assess a progressive Diverging Diamond Interchange (DDI) design (first developed in France) as a way to improve safety by reducing the number of conflict points compared to traditional designs, and reduce costs from land acquisition and construction. A team of geometric designers from Missouri DOT, considering the DDI for Kansas City, MO, had prepared a set of plans but needed some way to test it before proceeding with construction. Their design was brought to the TF facility and initially evaluated by the team exposing sight distance issues along with needed adjustments to signing and signalization. TF then developed a study on the design, developing the appropriate evaluation measures followed by an assessment of 75 participants' potential navigation issues and speed tendencies. This study suggested that concerns over inappropriate driver behaviors (wanting to exit right when the design was requiring a non-typical left exit) were unwarranted. Since this study, Missouri moved ahead with its construction, and other states have either implemented or made plans to do so, including Michigan, Oregon, Utah, and Ohio, to name but a few.

Other related research presented at the RSS 2007 conference included a study by Reed, Luke and Parkes (2007) which explored the potential impacts of new operational strategies (or "novel schemes") on an existing roadway. Here, a simulator was used in the U.K. to measure comprehension and compliance to a proposed implementation of a high occupancy vehicle (HOV) lane, together with changeable message (or "Active Traffic Management") signing and marking strategies. The results demonstrated a high level of "comprehension and compliance" for normal operating conditions, but with measurable confusion when emergency operational conditions (accidents, evacuation) were in effect. The study demonstrated that simulators could provide valuable insight for engineers and managers designing operational modifications for existing facilities.

In another operational example, Upchurch, Fisher, Carpenter and Dutta (2002) used a simulator to evaluate the large number of different and novel freeway guide signs in Boston's Central Artery/Tunnel project. Alternative signs needed to be constructed for many reasons, including a restriction on the height of such signs to three feet, making the usual diagrammatic signs all but impossible. The total number of signs in the tunnel also had to be reduced in the surface section because of limited sight distances and possible sign locations. The full facility was recreated in the simulator, and a study was performed, both to expose specific problems with proposed sign placements, and to evaluate potential remedies providing sign designers with possible solutions for optimal sign designs and the effective placement thereof.

Finally, chap. 38 herein (Caird, Smiley, Fern, & Robinson) offers a prominent example of design decisions for optimal signing and signaling of an at-grade rail crossing to optimize safety by assessing driver behaviors specific to that crossing. The study offered many recommended physical location adjustments to some signs, while recommending the complete removal of others in order to address clutter and driver overload. This study is referenced for two reasons: first, to contrast the diversity in project scope and complexity for which driving simulators can support project-specific design decisions; and second to note that professional experience, design manuals, and human factors guides can get us close in our designs, but they cannot expose potential anomalies in the context of the entire viewable environment as seen through the eyes of a driver. Driving simulators—and the appropriate use thereof—provide designers with added insight into the potential consequences of their decisions.

We believe that the awareness and interest in the professional community to leverage this technology as part of the highway design will continue to grow, finally providing the practicing engineer with an innovative tool to more effectively bring the human into the process.

39.2 Establishing the Foundation

This chapter has so far explored the trends and opportunities regarding visualization and driving simulators, as well as factors influencing the industry's ability to move to the next level. This section presents suggestions and considerations that would support a more effective integration of driving simulation into the highway project development process, leveraging integrated visual design practices (visualization), and will also touch on potential legal and organizational concerns.

39.2.1 How Simulators Can Leverage Visualization

As noted earlier, visualization's primary role in highway project development today is to support more effective public involvement during planning and design, and very little 3D modeling is being done to facilitate highway design. However, this will change quickly with the advent of virtual design and construction (or VDC), where 3D modeling and visualization are at the core, and become fully integrated with multi-disciplined analysis, performance, and business processes throughout project development and on into operations. Because of the implications VDC will have on the industry, the TRB Visualization in Transportation Committee (ABJ95) initiated a joint subcommittee at the beginning of 2010, which includes participation from the Users Performance Section (AND00), to begin looking at research on a subject which will fundamentally shift the way highway projects are developed and delivered. For the near-term, and until VDC takes root, one method for visualization to support the use of driving simulation is to approach the opportunity from three directions:

1. Approach driving simulators as part of a more visually-centric project development process: In other words, instead of looking at simulators as a stand-alone investment requiring the full re-creation of design models, position them to leverage some of the same 3D data being generated to support other visually-supported decision processes in other parts of the project;

2. Assess, develop, and organize more in-house and consultant-based use of visualization as part of the normal design process: An organization is better positioned to realize and incorporate the benefits of simulators if they develop the staff and the means to create and utilize 3D/visualization data; and

3. Use driving simulation (by leveraging parts of the visualizations used for public information/involvement) as part of the public involvement process: Most stakeholders become a more productively-engaging resource for a project if they feel that: a) they are being heard; b) they are invited to be part of the process; and c) they feel they have contributed in a productive manner to its success. Inviting stakeholders to drive through an immersive version of what they saw the previous week at a public meeting stands to generate a more cooperative atmosphere for a project, and can provide valuable information to engineers if performance measurements are taken while these end-users drive through the various design alternatives. The benefit and validity of this concept has yet to be tested and scrutinized. The value of professional decisions to be made as a result of having driver performance data would need to be weighed against the cost and effort of capturing them in the first place. If anything, such an effort might expose potential problem areas in a proposed design where more structured simulation studies might be focused at a later date. To some extent, this concept of merging simulators with public information and involvement has already been attempted through Upchurch et al. (2002) on the Boston Central Artery/Tunnel project (Selingo, 2002), where simulator drive-throughs were recorded and posted on the internet (http://www.ecs.umass.edu/hpl; see Big Dig) to assist drivers in developing a level of familiarity when it comes to navigating this facility.

With visualization capabilities in place, design organizations can leverage the benefits of driving simulation in a few areas. First, as early as the *30% completion* stage (or earlier if necessary), a first review can expose major design anomalies, omissions or data errors, especially those that may require alignment adjustments effecting changes in right-of-way acquisition. Decisions can be made relating to sight distance, combined vertical and horizontal alignments, positioning of major structure elements (i.e., bridge piers), entrance and exit ramps, merge/diverge lanes, and other major geometric elements that may greatly affect driver behavior and/or performance. Adjustment decisions to these major design elements made as a result of a simulator assessment at this stage would have a direct impact on secondary and tertiary design elements such as drainage structures, utilities, major signing, signalization, major landscaping, and so forth. Exposing and addressing major issues directly related to driver performance early in a design process can save considerable re-working and project delays further downstream.

The second simulator review at the *60% completion* stage can be performed to validate previous adjustments to geometry and perform checks on traffic control devices once the geometry has been defined. It should be noted that advancements in "micro-simulation" (modeled output of traffic performance analysis software) will eventually find its way into the simulator so human factors assessments can be performed under more accurate operational conditions. This will be important as human stress and performance levels can be quite different when driving on an open road versus congested conditions. It is also at this stage that, depending on the nature of the project, simulators might be introduced as part of a public involvement initiative. Enough complexity to the 3D design model should be available at this point where stakeholders can begin to develop an appreciation for the major design decisions (some probably controversial) that have gone into the project. Instead of burdening the public with professional speak and analyses, driving simulators can provide them the opportunity to "experience" and "observe" professional decisions in a way they might appreciate more. Also, at this stage in the project, they will probably have seen a wealth of visualized design alternatives at previous public meetings, on the local news, or via the web. This opportunity would enable stakeholders to be in tune with the process, and may remove significant barriers to moving the project to the next phase. Preliminary signing, striping, and delineation would need to be incorporated at this point in order to maintain an acceptable level of realism and "relevance" to the design.

Depending on the complexity of the project, a subsequent *90% completion* check might be performed to "tweak" the final design to measure optimal performance and evaluate the impacts of traffic control devices, lighting, landscaping, and other roadside features. Of equal importance is the fact that simulators can and should be used to evaluate the safety and efficiency of "work-zone" staging layouts. These evaluations could be conducted to include human performance through work zones in the context of construction operations (i.e., the movement of materials, personnel, and heavy equipment). Contractors are currently using visualization and project simulation software (i.e. NavisWorks® by AutoDesk, and Navigator® by Bentley) to optimize staging operations for safety and efficiency; however this does not guarantee the same for stressed and impatient drivers trying to navigate a complex work zone. Bella (2004) conducted a simulator study assessing driver speeds and maneuvering comfort levels in relation to a new "Technical regulation regarding systems of traffic signs, differentiated according to the type of road" issued by the Italian Ministry of Infrastructure and Transport in 2002. The goal of the regulation was to require work zone signing configurations advising approaching drivers of an upcoming work zone, reduced speeds, and lane change adjustments. The goal of the study was to determine whether the new regulation would actually have any desired effect on driver behavior by testing the "regulated" configuration and alternative configurations in a driving simulator for a four-lane divided highway. The results of the study demonstrated that signing alone had no effect on reducing driver approach speeds to and through the work zone. It went on to demonstrate that the regulated configuration actually created a potential "zone of conflict" at the median crossover point because participants had to do so from the left-most lane. An alternative configuration was proposed and tested where traffic was merged to the right-most lane, and then diverted left at a more gradual rate over the median, which demonstrated an increase in driver comfort as participants passed through this "zone of conflict". In neither case were speeds reduced through the work zone, demonstrating that regulatory signing alone would have the desired effect. Though this example focused on a new regulation as apposed to an actual roadway project, it does demonstrate the ability of driving simulators to evaluate driver performance in the context of work zones and the potential value they might bring to optimizing complex work zones.

39.2.2 Getting the Organization Started in Visualization

Though variations exist, all highway design organizations incorporate some form of computer aided design and drafting (CADD) and roadway design technology to generate the necessary plan sheets that are ultimately provided to a contractor for construction. Most of these same tools already contain the 3D modeling and rendering capabilities to create both still and animated visualization output. The challenge for working in 3D is that 2D plans and specifications are still considered the documents of record for any project, and 3D visualization is looked upon primarily as a necessity for large-scale and controversial projects. Additionally, today's workflows, software expertise, business and contractual models, and data management practices do not fully support working in 3D, and there are legitimate concerns in the professional engineering practice regarding the liability of signing and sealing of 3D models.

These organizational and institutional concerns, however, are the target of focus for the vertical building industry as part of their progression towards building information modeling, or BIM. In short, BIM involves an integrated series of technologies that not only include the three-dimensional "geometry" (form and aesthetics) of a proposed facility, but also uses databases (integrated with 3D models) in order to "simulate" how the facility will operate once it is complete. Everything from structural performance, utilities, heating and cooling, security, fire suppression, human traffic movements, and so on, can be effectively evaluated and optimized. Efficiencies in design, construction, materials fabrication, and ultimately operations and maintenance, can leverage BIM to reduce costs throughout the facility's life cycle, and improve its overall quality.

According to Ashcraft (2008), the returns described above are a result of BIM's ability to enhance collaboration between the owner/operator, engineers, architects, and contractors. The core nature of this enhanced collaboration is to break down the barriers thereto for the overall benefit of the "project". This enhanced collaboration, however, conflicts with traditional business and contractual models and, together with the desired delivery of a 3D BIM dataset instead of traditional 2D plan sets to the contractor and subsequently the owner/operator, has the industry rethinking the distribution of both equity and liability (from the organization to the engineer-in-charge) on any project.

More on this comparison with BIM and its relationship with visualization and simulation for highways is covered later in the chapter, and the highway industry will stand to gain much by leveraging the progresses in BIM. In the meantime, there are steps that organizations and professionals can take to enable the foundations for effectively incorporating visualization and driving simulators.

39.2.2.1 Assessing Need and Current Resources

It is a fact that transportation agencies throughout the United States vary greatly in how much design they actually perform in-house (prompting full-scale design teams), and how much is farmed out to the consulting community. Regardless of their approach, they are still tasked with delivering safe and effective infrastructure for public and commercial use. That said, all organizations that have successfully incorporated visualization have had two things in common: a) a visionary who is willing to champion its development, and b) an executive willing to support him or her.

Due to the institutional challenges mentioned earlier, most internal visualization development has been performed through pilot initiatives, crawling first, and then evolving to the commitment that best suits the organization's needs, comfort-level, and available resources. Once an exploratory commitment is made and a supported visionary is found, the following approach can be used to assess an organization's ability to develop its visualization capabilities:

1. Assessing the Need for Visualization:
   a. Consider the quantity and complexity of change orders during construction to address design errors, omissions, or improper construction due to the complexities of 2D plans that could potentially be remediated with 3D plans and simulation;
   b. Assess the crash rates and causes on projects that went operational in the past three to five years;
   c. Determine the average project delays (in years) due to project or Environmental Impact Study complexities and/or public resistance (consider delays in acquiring the needed right-of-way);
   d. Consider the complexity and duration of design review milestones, and how visualization might streamline those stages especially if reviewers are distributed geographically;
   e. Assess how constructability reviews are being delayed due to design and staging complexities; and
   f. Inquire about other organizations that have initiated and incorporated visualization successfully, in the process learning from their experiences.

2. Assessing the Organization's Capability:
   a. Survey your CADD expertise and find out who might have dabbled in 3D, either on projects, casually, or at home off the job;
   b. Survey your more "creative" expertise, typically found in your Public Communication Office and/or Landscaping Units. In many cases they have already created some forms of visualization;
   c. Determine their availability to be part of a small start-up team to further evaluate visualization;
   d. Assess the consulting firms that are used for design as many of them carry some form of visualization expertise that can be drawn on for both services and knowledge for start-up efforts;
   e. Assess whether the organization's IT infrastructure (networks, servers, data storage, workstations, design software, etc.) can handle the larger storage, processing, and bandwidth requirements;
   f. Evaluate the organization's workflows and determine where basic 3D visualization might be best served with minimal investments in training, technology, start-up time, and disruption to current operations;
   g. Assess the current data being made available from the Survey Section as this will be necessary to develop "credible" 3D models. Note that not all visualization requires a detailed survey. The needed accuracy depends greatly upon what decision is trying to be made and how far along the project is. Earlier decisions may not require as much geometric precision;
   h. Explore readily-available driving simulators which are typically found in research institutions. Determine data transfer requirements, and map out what it might take to improve interoperability. Also discuss start-up requirements such as research initiatives to validate processes and expected "value";

i. Explore "semi-immersive" desktop driving simulators, which are becoming more readily available, and allow for preliminary human factors assessments without much investment.

It is important to note that simply connecting visualization capabilities in a highway agency to desktop simulators or driving simulators and performing human factors assessments is not enough, and engineers will need more than training on the simulators themselves. They will need a new level of human factors knowledge, including a better understanding of human subject studies and how all of these elements stand to affect their professional design decisions. An organized training and collaboration effort will need to be developed between engineers and human factors experts to make this link effective. This can and should be made an integral part of activities outlined in items "h" and "i" above.

39.2.2.2 Enabling the Visualization Team and the Organization

Once the above assessments are complete, and a "go" decision is reached, it has typically been the approach to start small and grow into a complexity that supports the identified needs of the organization while balancing available resources, talent, and comfort levels of professional practice.

1. Initiate a pilot team with the following constraints:
   a. Set a reasonable duration to exist as such (typically six months to one year);
   b. Develop the team's skills and talents on real projects and plan to set binding commitments for deliverables to project managers. Together with these managers, identify a diversity of projects varying in size, complexity, and project-specific issues;
   c. Consider revising your CADD standards to account for 3D;
   d. *Most Critical*: Establish a set of "measures" for the team such as start-up time, monies/time saved versus monies and time invested, and client satisfaction surveys in order to establish a tangible anticipation for Return on Investment (ROI). This does two things: 1) This enables credibility and accountability to the pilot visualization team; and 2) It facilitates more sound and supportable decisions at the executive level when determining future scope and monetary investments for visualization and simulation.
2. Prepare legitimate "strategies" and methods for the development of project visualization plans. Such plans look to assist the pilot team with:
   a. Interviewing strategies for scoping a proposed project;
   b. Determining project-specific needs which will drive what types of visualization will be required at what stage of a project, and define the data requirements necessary to support the needed visualizations;
   c. Validating whether visualization is needed in the first place, or if it will require visualization that is beyond the means of the team;
   d. Helping to establish levels of commitment, timelines, and deadlines;
   e. Exposing any training needs to enable the design teams to employ the visualization effectively.
3. Make plans to assemble identified talent into an integrated team. With today's networking capabilities, the team members can reside in different offices and still function effectively.
4. Define the visualization level of complexities you wish to start on, and determine holes in training, available tools, and other related needs.
5. Develop a budget accounting for items such as: a) technology; b) training; c) supplies; d) travel and e) communication.
6. Find and secure funding.
7. Engage your organization's consultant agreements Office and initiate (if not in place) a form of umbrella contract for qualified consultants to compliment your Pilot Team in the development and delivery of visualization services to the organization.
8. Engage your human resources office and evaluate current position descriptions for how they might be enhanced/modified to define specific skills needed for a visualization team; and consider the development of non-existent position descriptions in anticipation of hiring from outside the organization should long-term commitment to visualization arise.
9. Establish an "exit strategy" for completion of the pilot phase. This may include continuation expansion of a centralized team, centralized balanced with district or departmental complimentary strategies, or total disbandment.

It cannot be overemphasized that the ultimate goal of any pilot initiative is to gather new knowledge, awareness, and supporting data regarding the "value" that visualization may bring to the organization. Executives deciding on further expansion and investment will need to have a clear understanding of how the new tools and processes will help them deliver a transportation program that is safer, timelier, of higher quality, and more cost-effective than current tools and practices allow. For this reason, measures (as outlined in Part 1-d. above) demand considerable thought in their development, relevance to the organization, follow-through in their use, and clarity in their delivery if executives are to consider future investment.

39.2.2.3 Bridging Visualization Start-up Efforts With Driving Simulation

As discussed by Ashcraft (2008) regarding BIM, we need to consider visualization and driving simulation (and public involvement, and digital construction, and virtual operations, etc.) for highways not by the gains that they produce in separate arenas, but by the larger gains attained by integrating them into a collective mindset. In other words, as an organization begins piloting a visualization start-up, take care in considering all related processes that stand to connect both technically and as part of the project delivery process. Because decisions made in driving simulators can have direct and more immediate impacts on the design process, a visualization start-up could gain value and relevance if conducted in tandem with a simulator pilot initiative.

39.2.3 Moving 3D Design Data Into Driving Simulators

In part, for highway designers to see value in driving simulators they will need to know that the critical 3D data elements they create in CADD can be readily and accurately exported into the simulators. This is an issue of cost, efficiency, and even touches on professional liability. This section attempts to provide a few insights relative to process and the more technical issues that will need to be considered.

39.2.3.1 In the Context of Process

As previously noted, the incorporation of driving simulators in the design process may progress successfully without the need for an organization to develop its visualization capabilities. However, considering the benefits of leveraging data already created as part of 3D design and/or visualization practices, not to mention the "comfort level" of knowing what to specify regarding visualization-related deliverables, the organization would be best served by developing its capabilities in front of, or in tandem with, implementing driving simulation as part of design.

That said, as most highway design organizations use either Bentley® or Autodesk® software to perform their designs, consideration will need to be made regarding data translation into the simulators. One of the main differences between CADD environments and simulator environments is the way in which they store and display model information. In general, for every CADD package currently on the market, there is a unique and proprietary way to define a line, a surface, a solid, and so on. Additionally, there are a number of translation programs that exist just to translate models from one format to the next. Since 3D CADD files are typically viewed in a less dynamic environment as compared to driving simulators, rendering and texture mapping can be more complex because refresh rates are less burdensome to the graphics processor.

On the other hand, driving simulators must, by nature, support a more constant and rapid movement through a 3D design, thereby placing greater demands on rendering speeds and refresh rates, thus prompting the descriptions of the model in a "database"-centric environment. In short, the complexity and "realism" in a driving simulator are typically much less than that of a highly realistic CADD model. Considering that one is moving through a model in a simulator at much faster speeds than while designing at a desktop, the human eye does not require the level of complexity and fidelity than may be required for a public display of a visualization where different decisions are being made. At the same time, visualization through a driving simulation provides a uniquely useful perspective that reflects a realistic context by which a design can be evaluated in terms of usability and user comfort. Adjustments to the perspective can also be used to provide vehicle-specific vantage points (i.e., low for passenger cars, higher for trucks), providing further benefits to the early review process.

The quandary that remains, however, is how does one translate 3D CADD data from design into the database-centric environment of a simulator without pulling in all the design complexities while maintaining a reasonable level of precision, and minimizing the amount of model reconstruction that will be required to reproduce the design in the simulator?

39.2.3.2 Technically Speaking

This problem can be addressed at various levels. At the low level, geometry information must be converted from the CADD model into the "real-time" database representations typically used by simulator image generators (IGs). At a higher level, the procedural or parameterized description of the roadway as used for generating geometry at the CADD level can be re-used in order to generate the simulator specific format. Various issues associated with each of these approaches are discussed below.

39.2.3.2.1 Translating Primitives

At the lower level, the problem of converting between CADD and simulators can be further subdivided into two components, data interchange issues and semantic representation issues. The data interchange issue may be the simplest one to solve and is the area where most efforts have been focused to date. Even though this problem should be relatively straightforward, practical experience and anecdotal evidence indicates that more often than not, this process is more challenging than anticipated.

In general terms, the problem can be simply stated as converting geometry primitives from one representation to the other. As discussed earlier, CADD packages store road information by using a variety of primitives, including polygons, surfaces, lines, and so on. Each primitive is further attributed with surface properties such as color and texture. To the degree that both the CADD system and simulator IG support similar primitives and attributes, such conversion should be straightforward. However, when primitives from one system do not have a direct equivalent to the other system, additional effort is necessary in order to properly map the primitives. As an example, if the CADD program is utilizing an analytical surface representation such as NURBS, but the IG only supports polygons, there is the need to convert the analytical representation into polygons that can be rendered. Even when the representations have a straightforward mapping (i.e., triangle to triangle), nuances can create visible artifacts that require further processing. As discussed earlier, one of the key requirements of simulator IGs is real-time performance.

39.2.3.2.2 The Issue of Rendering

Before the recent explosive growth in personal computer graphics processing, simulator IGs utilized specialized data structures to enable consistent real-time rendering. In a typical real-time

IG format, such as OpenFlight®, graphics primitives are organized in data structures which allow the renderer to quickly eliminate primitives that are not visible. Other structure primitives dynamically change the detail of rendered features based on the distance of the eye point from a feature; for example, a level-of-detail (LOD) node is used to force this evaluation, following which the system effectively renders objects that have less detail, and thus improves performance.

Under these conditions, converting a road network representation into a simulator IG format requires not only mapping between primitives in the CADD program and the simulator, it also requires developing a good enough structure for these primitives to maintain real-time rendering performance. Depending on the performance of the IG, this can be anywhere between trivial and prohibitive. Typically, the growth in graphics computing power has diminished the need for extensive structuring of the data, although peculiarities in graphics chip architectures still require domain-specific knowledge in order to avoid problems. To better understand this issue, consider the following comparison.

One of the common image generators used in real-time simulation up to the late-1980s, utilized an architecture that was bound by triangle processing capacity; that is, it could only process up to a certain number of triangles per rendering step while maintaining real-time rendering. That number of triangles varied but was in the order of low thousands. The amount of data structures, such as level-of-detail nodes and amount of texturing applied on each triangle did not further load the system because there was enough hardware to perform such calculations in parallel. Experienced modelers for that IG were aware that using an LOD node to save as few as eight triangles was worth it, because of the small overall triangle budget and the ability of the architecture to process the LOD notes with little penalty. As a result, visual databases for such IGs utilized complicated data structures containing numerous LOD nodes to minimize the number of rendered triangles.

Contrast this approach to the load management in a modern graphics chip's architecture, which can render millions of triangles per frame. These chips are optimized for rendering triangles, each with multiple textures and other attributions. When using such architecture, excessive data structures can slow down rendering because processing of the data structures happens in the CPU of the host PC, as opposed to rendering of the triangles that happens in the graphics chip; in this case, using an LOD node may only be worth it if can save in excess of tens of thousands of triangles, as opposed to just a few. Also consider that despite their triangle rendering capacity, modern chips remain relatively slow when having to render triangles that utilize different textures.

So, for example, having a data structure that forces the IG to render 1000 groups of 100 triangles, with each group having a different texture applied to it, is much slower than rendering 100,000 triangles all of which utilize the same texture. Addressing such peculiarities is important because it is often the case that a straight conversion fails, either because of poor image quality or because of poor performance. In summary, even a straightforward conversion between CADD and simulator IG poses numerous issues that must be addressed, and most importantly, requires specific technical knowledge which is typically not available within transportation agencies.

39.2.3.2.3 Some Options

On a more practical level, it could prove useful to develop common representations that encapsulate the data as well as semantic information about that data. One path may lie in the realm of XML (Extensible Markup Language), which allows the user to define their own elements and data structure. Roadway design vendors are currently developing and releasing their software with XML in mind. Due to the variation of driving simulator providers and civil design providers, each organization will need to account for this if they choose to make their respective CADD and simulator environments more interoperable. Because of this implementation-specific uniqueness, and the constraints of this guide, no specific resolution can be supplied or recommended.

Another consideration for improving smoother data conversion between CADD and simulator IGs is the use of a higher-level description for features of interest. When first starting out, a highway engineer will utilize high-level descriptions of the roadway, which includes constraints as well as design parameters. Examples of such fundamental road characteristics include curve radius, super-elevation, entry and exit spiral lengths, and so forth. The CADD program will then operationalize such features by generating a geometrical description of the designed roadway.

The earlier portion of this section focused on the issues associated with converting the resultant description into a simulator IG. However, recent advances in the architectures used for representing virtual databases in simulator IGs are making it possible to consider a conversion at the road specification level, as opposed to the graphic primitive level. Specifically, there has been a recent trend of changing the traditional model of a visual database from a "static" representation of structured graphics primitives into a dynamically-generated representation that expands a core set of data with user-specified layers imported from a variety of domains.

Perhaps the best example of this shift is the Synthetic Environment Core (SECORE) project sponsored by the U.S. Army program executive office for simulation training and instrumentation (PEO STRI). One of the initiatives within SECORE is the Database Virtual Environment Development (DVED). The model used by this project and its initiatives is to utilize a master database of features that are augmented by multiple data sources and associated filters. The feature relevant to the highway design process is that data sources are uniquely identified with filters which have to be validated before acceptance. As an example, a master database may include medium-resolution elevation data for a particular region. Typically, additional data sources may include higher-resolution elevation data and high-resolution imagery.

However, consider the possibility of developing a filter that can import the original highway design parameters and constraints into the core database. Once validated and accepted, this filter can act as a universal converter for highway designs into a format usable by any IG that is compatible with the particular architecture. On the commercial front, several vendors are following the same architectural pattern for the representation of visual data. For example, the database generation system from the Rockwell Collins® image generators provides built-in terrain elevation and land coverage data for the entire glob at a 30 arc-second resolution. It can dynamically import DTED or DEM data for terrain elevation, ESRI shape files for terrain features, and various other formats for specific models.

Similarly, the GenesisRT™ visualization system from Diamond Visionics provides a real-time rendering environment that will build a 3D scene by importing real-time data from various sources such as ESRI shape files, imagery and other features. It provides facilities for building filters that can implement custom processing of input data before display. The system already includes a rudimentary road generation system which utilizes centerline shape files and produces roadways that trace the centerline of the line feature. Ongoing research at Old Dominion University is seeking to develop custom filters that will utilize roadway design features (such as radius of curvature, entry and exit spirals) and produce a high-quality rendering of the highway properly integrated with surrounding terrain elevation and imagery.

Such efforts, if successful, will provide a turnkey solution to support the conversion of highway designs into high-quality 3D visualization environments that do not have to deal with the low-level details of the conversion process, and thus eliminate the need for specialized skills. Additionally, such a solution will begin to address the business and professional reservations that are preventing the full use of driving simulators as part of the highway design process.

39.2.4 Research Center and Transportation Agencies—A New Relationship

This section describes how many of the opportunities and potential benefits of simulators can be made a more integral part of the highway design process. The intent is to balance potential benefit with potential business, professional, and contractual needs. Part of the discussion below leverages similar concerns and progress described by Ashcraft (2008) in the context of BIM, but still have direct relevance here.

39.2.4.1 A Summary of the Opportunities

It is important to recap the opportunities and potential benefits of simulators described throughout this chapter in order to place the following narrative in proper context. Overall, the opportunities and benefits include (but are not limited to):

- Exposing complex design anomalies unrecognizable in 2D plans or 3D models;
- Exposing specific and potentially dangerous human performance issues unique to the project;
- Identifying human behavior issues in work zones that may impact construction staging;

- Evaluating driver performance in the context of the "total environment" for a proposed project;
- Evaluating proposed operational and emergency modifications on existing roadways;
- Evaluating innovative geometric design strategies or roadside treatments (i.e., diverging diamond interchange); and
- Supporting a structured public involvement process on complex or controversial projects

39.2.4.2 Center/Client Relationships (Current and Potential)

Typical human factors research centers (and their simulators) reside in university settings where student and professional resources may be drawn on to complete formal studies. Some centers receive annual grants in support of general operations, with direct research-related revenue streams coming from a diversified client portfolio which typically include: public (as from transportation agencies), private (as from automobile or drug manufacturers), and defense (i.e., armed forces) sectors. Such revenue streams may result from successful responses to structured requests for proposal (RFPs), unsolicited proposals, or a direct approach from any of the aforementioned sources. Once contracted, the work is carried out, a report is filed, and spin-off work may or may not follow.

As driving simulation evolves to become an integral part of the highway design process, both transportation agencies and research centers will be placed in a position to reconsider their professional, business, and contractual models in order to facilitate a sound working environment for conducting effective project-specific human factors evaluations. This new working relationship will place both parties in relatively different roles and will raise some legitimate questions:

- How will state and federal roadway projects be tagged for human factors assessments? In other words, what criteria should be put in play? Complexity may not be the sole consideration as the University of Minnesota study demonstrated (Scallen & Carmody, 1999), and there is value in using simulators on less complex rural two-lane road projects.
- Is there an efficient method for scoping those projects that are tagged?
- What contract language will be needed to define specific "deliverables" when the specifics of human factors issues on any project are yet to be discovered?
- How does an agency effectively budget for these forms of human factors studies?
- How (if at all) does professional liability play into this (see expanded discussion—Section 2.4.3 herein)?
- Is there need for a "warranty" clause with respect to driver performance once the facility is completed and fully operational? Perhaps not, however, it is worth noting.

These are some obvious examples, and certainly others will arise as momentum builds for these professional services, which is what this new relationship stands to be.

Elaborating further, we can contrast the role that research centers play today with the role they would look to play as part of a highway project team. Traditionally a human factors problem statement is identified and defined, which drives a study resulting in some form of a findings report reviewed by researchers and engineers alike, followed by further evaluations, studies or prototypes which may or may not become part of an agency's formal design standard. With simulator-based human factors assessments integrated with the design process, professional engineering decisions will be made based on the very results of those assessments.

The potential benefits, however, (improved operational safety, overall facility and driver performance, reductions in retrofits, and accelerations to design schedules due to improved decision-making) stand to outweigh the intrinsic reservations.

39.2.4.3 Professional Liabilities and Risk

Revisiting the larger picture for a moment, simulator-supported human factors assessments for active projects are suggested as a part of an overall virtual design, construction, and operational mindset with links to the public involvement process. Each element by itself will have a varying benefit and return on investment; but collectively, we enable a highly-collaborative and more efficient process that converges to support the delivery of a quality piece of highway infrastructure. This vision for highways does bring forward specific considerations. To put them in some relevant perspective, we again reference Ashcraft's (2008) discussions on BIM for vertical construction by quoting the National Institute of Building Sciences, and their official definition thereof:

A BIM is a digital representation of physical and functional characteristics of a facility. As such it serves as a shared knowledge resource for information about a facility forming a reliable basis for decisions during its lifecycle from inception onward.

Assuming, for a moment, the highway industry has developed a comparable definition for virtual project development, a discussion can be raised regarding the potential considerations by looking at the following example. Assume that a DOT has committed to a virtual design and construction approach for a complex urban highway project, complete with bridges (which will be outsourced). 3D bridge models will be integrated with 3D terrain, roadway, roadside features, and drainage elements; subsurface strata and utilities will be located via 3D as well. The nature of the project convincingly merits the use of simulators to assess driver behavior and performance within the "collective" environment of the project. In its simplest form, an Engineer of Record for the overall project will be in the position of certifying the pending bid package, with a structural Engineer of Record (certifying the bridges) included therein.

As a human factors assessment milestone is reached during the design, and in the spirit of a fully virtual process, elements of both roadway and bridge models are transmitted to the simulator research center for arduous study, which exposes and verifies prominent human factors issues, prompting alignment

changes in both bridges and roadway, which are incorporated by the engineers. These changes address the issues as verified by subsequent evaluation in the simulator, the design is then finalized, awarded and constructed leveraging the final/certified 3D roadway and bridge models. The resulting considerations from this simplified example include the following:

1. Maintaining Model Completeness and Precision—The simulator team will not require the entirety of the bridge and roadway models-of-record to conduct their assessment. For example, they will not require any subsurface utility data, or 3D bridge rebar data. However, the 3D model data that are delivered to the simulator team will need to maintain completeness and precision at a level that does not adversely influence the outcomes of the human factors assessment itself. Equally relevant is the modeled environment residing outside of the project models, but that is within the driver's field of view (i.e., rivers, buildings, billboards, etc.), which might be added by the simulator team, all of which may influence driver behavior, and subsequently the assessment findings from which the engineering team based its design modification decisions.

2. Validation/Certification of Simulator Results—In the typical research environment, a simulator is "validated" by comparing human performance therein with human behavior in a corresponding real-world environment. Doing this for a specific roadway which has yet to be built will require that protocols be in place to ensure the validity of any assessment results in order to hold any pertinent value to the engineers. Whether this merits a requirement that the simulator team be in a position to "certify" such assessment results is a question yet to be explored. A possible solution may simply require appropriate documentation of the simulator technology used, the participants, the results, and a copy of the modeled environment used as part of the assessment. Proper archiving of this information should also be considered.

3. Liability, Risk, Ethics, and Standard of Care (Due Diligence–as it is commonly known)—Professional Engineers (PEs) are bound by rules of practice and corresponding codes of ethics and conduct which are defined and set forth by the respective governing boards of each state in which they are licensed to practice. These boards also require that they engage in, and complete, a pre-defined number of hours for continuing education (including recurrence of ethics and a code of conduct training) as part of their professional development, or undertake certain professional activities in order to remain up-to-date within their respective areas of practice as a condition of their licensure. There are a number of proactive organizations that support them in this cause, such as (in cooperation with the state boards) the National Society of Professional Engineers and the American Society of Civil Engineers. These organizations provide training courses and guidance regarding professional liability and risk. The participation of organizations such as the Transportation

Research Board (TRB), the Institute of Transportation Engineers (ITE) and the American Association of State Highway Transportation Officials (AASHTO) must also be recognized in supporting continuing education.

The pertinence of the above can be summarized thus: As driving simulators become part of standard practice, engineers will find themselves making professional-level decisions based on their "direct" application in the design process. The same holds true for the 3D modeling and visualization tools they will be using as part of virtual design. In other words, the new insights and benefits that these tools stand to provide can and will affect the engineer's perception of the problems and the remedies, which will feed into a professional decision that may heighten their exposure and liability.

Ethics and liability may also come into play if simulators are used as part of any public involvement process. Such has been the case already for simpler forms of visualization (still photo-simulations and animations). Agencies that have employed them have needed to incorporate disclaimers when showing proposed design alternatives due, in part, to the high levels of realism that the technologies have been able to deliver, and the implied expectations that they conveyed. There has also been some backlash from the public when engineers are perceived to be using visualization to market an unpopular alternative or pitch a preconceived agenda.

There are a couple of ways to address this. First, the research centers already savvy in simulator use can begin proactive efforts in educating the engineers, the organizations that employ them, and the organizations that oversee their professional practice and conduct regarding potential benefits and implications. Syntheses can be performed to gather an understanding of how professional engineers are addressing similar concerns in other industries such as aerospace and automotive design. Applied research may also be formulated and introduced to develop usable insights specific to the needs of highway engineers in order to support more effective education, training, and the development of guidelines.

39.2.4.4 Business Models and Staffing

Through this new opportunity (simulators becoming directly part of geometric highway design), highway research centers will have the opportunity to advance beyond their current business models indicative of the research environment, and become active contributors to highway infrastructure delivery. Again, with this opportunity comes some consideration.

Research centers already compete for research opportunities, and it can be safely assumed they will need to compete for project funds in a similar way to the consultants and contractors who bid on highway projects. This brings up the issue of "teaming". Many sizable highway projects today are contracted to project-specific enterprises made up of comprehensive teams of design consultants and contractors (as in the case of Design/Build projects). Assuming comprehensive human factors assessments are budgeted for the project, research centers might find themselves partnering with these teams. A preferred alternative

could position the research center to perform their work as part of a pre-established agreement via the contracting agency.

How might the center need to adjust staffing? The timeline demands of an active highway project may place new forms of staffing availability requirements to the center, possibly requiring a professional engineer trained in the proper use of the simulator. Perhaps the project consultant or the contracting agency will supply the certified engineer. The research center will also need to consider its potential exposure and liability in the context of professional practice, definition of deliverables, and even data management and security.

39.2.5 Driving Simulation as Part of Public Involvement

Up to the date of this publication, this chapter's lead author is unaware of any published research or reported best practices describing the use of driving simulators as part of Public Involvement. This does not imply that there are no benefits in doing so, and there are industrial and societal factors that merit further consideration of this concept:

1. The growing awareness of visualization and simulation in society: Small and Vorgan (2008) discuss the day-to-day spectrum of technologies that are impacting both the physical brains, and the resultant minds of humans throughout society. Email, instant messaging, computer-generated graphics in movies and television, interactive 3D graphics in education, video games, and all available via desktop and mobile, are contributing to these changes. Roberts, Hoehr and Rideout (2005) completed a Kaiser Family Foundation Study on the extent to which children aged 8–18 are exposed to these technologies, citing notable statistics such as the following: 83% of this group of children have at least one video game console at home with 31% having three or more. This group is exposed to eight and a half hours of digital and video sensory stimulation each day (including four hours of "passive" participation for television or video, almost two hours of listening to music, and almost two hours of "active" participation playing video games or using a computer).

The relevance this has to driving simulation and visualization is this: Our instantaneous access to global information and a diversity of experiences will have an affect on stakeholder expectations for a greater "active" involvement with the highway project development process. A case in point is the following. Bailey et al. (2007) employed a Structured Public Involvement process that combined both visualization of bridge alternatives and digital polling technology (similar to that used during the 2008 U.S. presidential debates to measure focus group likes and dislikes for candidate responses to questions). Engineers then adjusted designs based on the computer-tabulated data and the stakeholders reacted most positively to the process because the visualizations helped them to understand the problems and their solution alternatives, and the polling tools gave them a structured means of interacting with the engineers. Passive observations of engineering teams presenting before and after photos-simulations or animations will give way to more comprehensive forms of Structured Public Involvement. The 8–18 year-olds of today will become the "stakeholders" of tomorrow. And having grown up learning and playing with simulation and video game technologies, they will have little patience for passive observation of visualizations or 2D plans in public meetings.

2. The strong push to improve how transportation agencies engage and involve the public: Policy-makers have already cracked the ceiling for both visualization and more structured public involvement through their requirements set forth as part of SAFETEA-LU in 2005. Though this law and the subsequent ruling set forth by FHWA which focuses primarily on local, regional, and statewide transportation "planning", momentum will certainly spread over into latter phases of the project development process (design and construction). Simulators properly integrated with public involvement would build on this policy-level shift towards improved public participation, and leverage the ongoing societal shifts in thinking.

3. The evolving and ever more available cost-effective visualization and simulation technologies: This simply implies that traditional limitations due to cost, complexity, and portability relating to driving simulation technologies will be "positively" influenced by Moore's Law. Transportation organizations will also be in a position to reconsider their use. This is not to suggest that research centers will lose their importance and place in the design process. In fact, it should have just the opposite effect as agencies and consultants seek guidance on their professional use.

Considering, for a moment, that this concept of simulators in public involvement takes root, structured research (scientific and applied) will be needed to validate its worth together with potential fallout and processes. For example, everyone in a project community, given the chance, may want to drive a proposed design in the simulator, but simulator sickness may become an issue prompting subject-screening protocols (see this book, chap. 14 by Stoner, Fisher, & Mollenhauer on simulator sickness). Putting this in context, if the intent is to improve public involvement and develop buy-in for a design alternative, a nauseous stakeholder in a simulator or neighbor who is screened out due to a propensity for simulator sickness will not build much of a relationship much less any confidence in a proposed alternative. Also, any proposed research should look to:

- Shed light on any legal implications to practicing professionals;
- Strive to provide reliable and defendable best practices; and

- Consider leveraging the opportunity of engaging the public this way by incorporating methods to measure their performance as part of the process.

More importantly, research must be undertaken to measure the real value to the project development process, and the implications to the professional engineer, or in the spirit of BIM, the project as a whole. This aside, the extension of driving simulators as a part of public involvement stands to improve the whole of the highway project development process.

39.2.6 Chapter Summary

Visualization is an integral part of driving simulators as applied to 3D geometric design and potentially public involvement. The success of this integration can be enhanced by the professional and organizational commitment to enable a more visually enhanced and immersive project development environment. As with all changes in practice, there are risks; but with regards to implementing visualization, those risks can be readily minimized and the benefits far outweigh the same. The growing needs of society and the transportation infrastructure that supports it require that innovative tools and methods be provided to the professionals tasked with delivering safer and ever-progressive transportation systems faster and more cost-effectively than ever before. Visualization, in concert with driving simulation, can make this a reality.

Key Points

- 3D design/virtual design and construction will become the new environment for geometric highway design.
- Driving simulators should be used as an integral part of this new 3D design practice for human-in-the-loop assessments of geometric design to improve safety and performance.
- There are organizational opportunities and considerations for transportation agencies, research centers, and consulting firms in order to make points 1 and 2 a reality.

Keywords: 3D, Design, Human, Simulation, VDC, Virtual, Visualization

Web Resources

(See also those web sites listed below.)

About the National BIM Standard®. (2010). Building Smart Alliance®, a council of the National Institute of Building Science. Retrieved from http://www.buildingsmartalliance. org/index.php/nbims/about/

Allen, W., Rosenthal, T., & Park, G. (2006, October 23–26). *Efficient means for prototyping and reviewing roadway designs through visualization.* Presentation at the 5th International Visualization, in Transportation Symposium and Workshop. Denver, CO. [Powerpoint video]. Retrieved from http://www.teachamerica.com/VIZ/10_Allen/index.htm

Granda, T. (2006, October 23–26). *Efficient means for prototyping and reviewing roadway designs through visualization.* Presentation at the 5th International Visualization, in Transportation Symposium and Workshop. Denver, CO. [Powerpoint video]. Retrieved from http://www.teachamerica.com/VIZ/10_Granda/index.htm

Hannon, J. J. (2007). *Emerging technologies for construction delivery.* NCHRP Synthesis 372, Chapter 5, Washington, D.C., Transportation Research Board, NCHRP. Retrieved from: http://onlinepubs.trb.org/onlinepubs/nchrp/nchrp_syn_372.pdf

Hixon III, C. (2007). *Visualization for project development,* NCHRP Synthesis of Highway Practice 361. Washington, DC. Transportation Research Board, National Cooperative Highway Research Program. Retrieved from http://www.teachamerica.com/VIZpdf/nchrp_syn_361.pdf

Kunz, J., Fischer, M. (2009). *Virtual design and construction: Themes, case studies and implementation suggestions.* Center for Integrated Facility Engineering – Stanford University, CIFE Working Paper #097, Version 10: October 2009. Retrieved from: http://www.stanford.edu/group/CIFE/online.publications/WP097.pdf

Transportation Research Board, Visualization in Transportation Committee web site: http://www.trbvis.org

Transportation Research Board. (2009). *Annual Meeting.* Proceedings of Session 786 - Beyond 3D: Progressive Visualization for Geometric Design. Retrieved from http://www.trbvis.org/MAIN/786-Proceedings/786-Proceedings.html

Science Blog. (2002, December). UMass professor invites travelers to test-drive Boston's big dig—before hitting the road—Web-based project aimed at familiarizing drivers with new routes. *Science Blog.* Retrieved from http://www.scienceblog.com/community/older/2002/B/20026172.html

Wikipedia (n.d). Virtual design and construction. Accessible at: http://en.wikipedia.org/wiki/Virtual_Design_and_Construction

Key Readings

Ashcraft, H. W. (2008). Building information modeling: A framework for collaboration. *Construction Lawyer, 28.* Retrieved from http://www.hansonbridgett.com/docs/articles/bim_building_information_modeling_a_framework_for_collaboration.pdf

Keith, K., Trentacoste, M., Depue, L., Granda, T., Huckaby, E., Ibarquen, B., . . . Wilson, T. (2005). *Roadway human factors and behavioral safety in Europe* (Rep. FHWA-PL-05-005). Washington, DC: International Technology Exchange Program, Federal Highway Administration, US Department of Transportation.

Manore, M. A. (2005). *Enabling a more visual and immersive engineering environment for the highway transportation industry.* IMAGE 2005 Conference, Scottsdale, AZ: Image Society.

Taylor, M., & Moler, S. (2010). Visualization's next frontier. *Public Roads Magazine, 73*(4; January/February 2010). Retrieved from http://www.tfhrc.gov/pubrds/10janfeb/02.htm

References

Aarts, L. T., & Davidse, R. J. (2007). *Behavioral effects of predictable rural road design: A driving simulator study.* Road Safety and Simulation Conference 2007. Rome, Italy.

Bailey, K., Grossardt, T., Ripy, J., Toole, L., Williams, J. B., & Dietrick, J. (2007). Structured public involvement in context-sensitive large bridge design using CAsewise Visual Evaluation (CAVE): Case study, Section 2, of the Ohio River Bridges Project. *Transportation Research Record, 2028,* 19–27.

Bella, F. (2004). Safety in work zones: Experiences with driving simulator. *Institute of Transportation Engineers (ITE) 2004 Annual Meeting and Exhibit* (Paper 340, Publication ID AB04H593).

Calvi, A., & Amico, F. (Eds.). (2007). *Abstracts booklet, International Conference on Road Safety and Simulation, RSS 2007* (pp. 15, 17). Rome, Italy, November 7–9.

Campbell, J. L., Moyer, M. J., Granda, T. M., Kantowitz, B. H., Hooey, B. L., & Lee, J. D. (1997). Applying human factors research tools for ITS (SAE Paper 972670). *Series: Advances in intelligent transportation system design (SP-1285).* Warrendale, PA: Society of Automotive Engineers.

Fukui, J., & Shirato, M. (2005). *Performance-based specifications for Japanese highway bridges.* Rotterdam: Millpress.

Hoffman, J. D., Brown, T. L., Lee, J. D., & McGehee, D. V. (2002). Comparisons of braking in a high-fidelity simulator to braking on a test track. *Transportation Research Record, 1803,* 59–65.

Keith, K., Trentacoste, M., Depue, L., Granda, T., Huckaby, E., Ibarquen, B., . . . Wilson, T. (2005). *Roadway human factors and behavioral safety in Europe* (Rep. FHWA-PL-05-005). Washington, DC: International Technology Exchange Program, Federal Highway Administration, US Department of Transportation.

Lamm, R., Cafiso, S., LaCava, G., & Beck, A. (2005). *To what extent the human being is so far regarded in modern highway geometric design: An international review and a personal outlook.* Transportation Research Board Third International Symposium on Highway Geometric Design Compendium of Papers (CD-ROM). Chicago, IL: National Academies.

Landphair, H. C., Larsen, T. R. (1996). *Applications of 3-D and 4-D visualization in transportation* (NCHRP Synthesis of Highway Practice 229). Washington, DC: Transportation Research Board, National Cooperative Highway Research Program.

Manore, M. A. (2005). *Enabling a more visual and immersive engineering environment for the highway transportation industry.* IMAGE 2005 Conference. Scottsdale, AZ: Image Society.

Manore, M. A. (2007). It's about decisions: Improving transportation project development with visualization technologies. *TR-NEWS Magazine of the Transportation Research Board, 248.*

Manore, M. A., & Larson, D. (1996). *MnDOT's 3D visualization assessment and development project initial report.* St. Paul, MN: Minnesota Department of Transportation.

Manore, M. A. et al. (2007). Visualization in transportation: Empowering innovation. *TR-NEWS Magazine of the Transportation Research Board, 252.*

Ray, B. (2009). *Beyond 3D: Progressive visualization for geometric design.* Opening Presentation at the 2009 Transportation Research Board Annual Meeting, Session 786. Washington, DC.

Reed, N., Luke, T., & Parkes, A. (2007). *Driving simulator study of a motorway share lane: Lessons for the design and implementation of novel schemes.* Road Safety and Simulation Conference 2007 (RSS 2007). Rome, Italy.

Roberts, D. F., Foehr, U. G., & Rideout, V. (2005). *Generation M: Media in the lives of 8–18 year-olds. A Kaiser family foundation study.* The Henry J. Kaiser Family Foundation.

Scallen, P. A., & Carmody, J. T. (1999). *Investigating the effects of roadway design on driver behavior: Applications for Minnesota highway systems* (Final Report 1999-10). St. Paul, MN: University of Minnesota Center for Transportation Studies.

Selingo, J. (2002, December 19). Taking a virtual tour of Boston's new roads. *New York Times*, p. G-4.

Small, G., & Vorgan, G. (2008). *iBrain: Surviving the technological alteration of the modern mind.* New York: Harper Collins.

Trentacoste, M. F. (2007). *Integrating actual road design into highway driving simulators for design. Research, and consumer applications.* Road Safety and Simulation Conference 2007. Rome, Italy.

Upchurch, J., Fisher, D., Carpenter, R., & Dutta, A., (2002). Freeway guide sign design with driving simulator for the Central Artery tunnel. *Transportation Research Record: Traffic Control Devices, Visibility, and Rail-Highway Grade Crossings* (Highway Operations, Capacity, and Traffic Control), *1801,* 9–17.

40

Advanced Warning Technologies: Collision, Intersection Incursion

Michael P. Manser
University of Minnesota

Abstract

The Problem. Advanced technologies that provide safety critical warnings to drivers about collision and intersection incursion events have been shown through driving simulation studies to influence driving behaviors. Changes in driving behaviors consistent with the goals of the advanced technology can potentially reduce rates of collisions and unwanted intersection incursions. However, there has not been a systematic summary of the factors and variables that need to be considered by researchers and engineers when conducting these studies in driving simulators. *Role of Driving Simulators.* The usefulness of driving simulators for examining collision and intersection warning technologies cannot be underestimated due to their ease of use; their relatively low cost compared to on-road research; their ability to examine a variety of design iterations in a relatively short period of time; and most importantly, their ability to examine the utility of safety critical systems without risking driver safety. *Key Results of Driving Simulator Studies.* Studies examining the usefulness of these technologies have indicated significant performance gains as a result of the technology alerting and indicating the criticality of an event to a driver. Recently developed technologies build on this success by actively controlling portions of the driving task to further reduce the propensity of a collision. *Scenarios and Dependent Variables.* A wide range of dependent variables are available when examining these technologies. For example, time headway, time-to-contact, time-to-line crossing, accepted and rejected safety margins, and reaction time can provide a basic and robust understanding of driver behavior. A useful and quickly administered metric that provides an indication of workload associated with the use of these technologies is the Rating Scale Mental Effort. Technology usability can be assessed through the use of standard usability questionnaires and through the use of acceptance and satisfaction rating scales. When selecting dependent variables researchers must consider the relationship among dependent variables, scenario selection, and the advanced warning technology to be evaluated in order to conduct studies that allow for greater generalization. *Platform Specificity and Equipment Limitations.* While driving simulators are useful tools, researchers and engineers need to be cognizant of the potential cost, effort, and complexity associated with installing and maintaining advanced warning technologies within driving simulators.

40.1 Introduction

Collisions between motor vehicles on roadways or at intersections are a significant issue facing the motoring public. According to the Fatality Accident Reporting System (National Highway Transportation Safety Administration, 2007), in the United States, in 2007, 37% of the approximately 41,000 fatalities were related to multiple vehicle crashes while approximately 15% of all fatalities occurred at intersections. In response to this public health issue researchers and product designers have developed and evaluated technologies that can warn drivers of imminent collisions and can manipulate vehicle control systems to reduce the possibility and severity of a collision. While a primary tool employed in this research has been driving simulators no summary exists that critically details their use for examining these technologies. This chapter provides an introductory examination of the role of driving simulators in land transportation research for the evaluation of in-vehicle technologies designed to warn drivers of potential crashes on roadways and at intersections. Particular emphasis is directed toward identifying types of collision and intersection incursion warning technologies (CIIWT), appropriate dependent variables, driving scenarios that may be employed in simulation-based evaluations of CIIWT, and technological issues that should be addressed when installing and maintaining these systems in driving simulators.

To facilitate our understanding of these in-vehicle technologies it is necessary to first identify basic classification frameworks that define their operational characteristics. Technologies that are applied to the problems in surface transportation are classified as 'intelligent transportation systems' (ITS). These technologies can be designed for "...advanced information processing, communications, sensing, and computer control technologies to the problems of surface transportation." (Mast, 1998). Due to their ubiquitous nature, ITS technologies can exist both in- and outside of vehicles (i.e., as part of the infrastructure). For example, ramp metering technology has been implemented at on-ramps to reduce travel times on highways by controlling the number of vehicles that can enter a highway at any single time and reducing congestion (see Levinson & Zhang, 2006, for a recent review of the effectiveness of ramp metering).

Within the domain of ITS are those in-vehicle systems, commonly referred to as driver support systems (DSS), that directly support driver behavior, cognition, and perception for the purpose of improving levels of driving safety and comfort. As an example, adaptive cruise control (ACC) systems function similarly to traditional cruise control systems in that they regulate vehicle speed according to a criterion speed determined by the driver. However, ACC systems also have the capability to adapt by slowing and then maintaining a criterion time headway to a slower moving lead vehicle. In a driving simulator-based evaluation of ACC Ma and Kabar (2005) presented 18 participants with a car following task in which they were prompted to change speeds, change lanes, and navigate curves. Fifty percent of the participants were required to respond to and interact with

a cell phone while all participants experienced driving with and without the use of an ACC. Results of their work indicated that ACC use was associated with improved lane keeping abilities in curves and reduced mental effort as provided by ratings on a uni-dimensional rating scale. Results of their work suggest that this DSS can be beneficial to performance and workload because, presumably, when a driver's need to monitor the environment for slower vehicles is reduced, it may free cognitive resources. These newly-freed resources can then be reinvested into lateral vehicle control. As another example, consider anti-lock brake systems (ABS) that facilitate steering and braking capabilities in inclement weather conditions by modulating brake pressure to prevent wheel lock (Evans & Gerrish, 1996; Rompe, Schindler, & Wallrich, 1987).*

In addition to reducing the behavioral, cognitive, or perceptual needs of drivers, DSS can also be designed to augment drivers' behavioral, cognitive, or perceptual capabilities. Recent research examining the utility of a haptic accelerator pedal supports this contention. In a driving simulator, participants were instructed to follow a lead vehicle at a constant distance at all times while performing no-DSS (baseline) or DSS drives. Participants in the DSS conditions experienced a haptic accelerator pedal that provided increasing or decreasing accelerator pedal resistance relative to the speed and distance between their vehicle and a lead vehicle. The changing resistance provided drivers with a source of information relative to the gap between vehicles, in addition to that normally available with vision alone. Results of their research indicated that the distances maintained in the DSS conditions were more consistent (Kuge, Yamamura, Boer, Ward, & Manser, 2006; Manser, Ward, Kuge, & Boer, 2005) and, when the lead vehicle slowed suddenly, driver responses to the event were faster as compared to the no-DSS baseline condition (Manser, Ward, Kuge, & Boer, 2004). See also Hjälmdahl and Várhelyi (2004) and Várhelyi, Hjälmdahl, Hydén and Droskóczy (2004) for allied research examining the utility of haptic accelerator pedals.

40.1.1 Collision and Intersection Incursion Technologies

There exist within DSS a subclass of devices, termed here as collision and intersection incursion warning technologies (CIIWT). These devices purport to reduce the frequency and severity of motor vehicle crashes. They are designed to accomplish this goal by providing drivers with a warning within or outside a vehicle cockpit regarding an imminent collision (i.e., collision warning technology). These devices can also accomplish this goal by signaling to drivers that they have entered an intersection illegally (e.g., violating a red light) or legally (e.g., stop at a stop sign and

---

* Electronic stability control systems that control wheel speed can also facilitate vehicle dynamics capabilities that may result in a decreased propensity for collision. However, these systems are not addressed here because they are primarily designed to mitigate vehicle loss of control (e.g., skidding).

then proceed) but that the situation is unsafe due to, for example, an approaching vehicle (i.e., intersection incursion warning technology). This simple description trivializes the complexity of CIIWT conceptualization and operation. Conceptually, these systems are consistent with the notion of a field of safe travel (see Gibson & Crooks, 1938) in that they notify a driver of lateral and longitudinal incursions into this field from objects that approach a driver's vehicle (e.g., vehicles, pedestrians) or from objects that a driver approaches. Conveying the degree of an imminent collision or intersection incursion is often accomplished by scaling the warning signal such that the frequency, tone, or intensity of warning increases as a collision becomes more certain. These systems have been designed to operate in roadway and intersection environments, urban and rural areas, and can often warn drivers of both vehicle-to-vehicle (e.g., sideswipe, forward collision, or rear collision) as well as vehicle to non-vehicle collisions (e.g., pedestrians). Operationally, these systems can employ sophisticated laser, radar, or acoustic sensors (Bellomo-McGee, 2003) that are integrated within a vehicle to detect the distance and speed of surrounding objects and to determine if a collision may occur. Similar sensors can be integrated within the roadway infrastructure to detect the presence, location, and speed of vehicles. When vehicle or infrastructure-based sources of information are coupled with algorithms to calculate the possibility and timing of a collision or intersection incursion, they can inform a driver with warnings via audition (e.g., tones, instructions), haptics (e.g., active accelerator pedals, vibrotactile seats), or vision (e.g., text or icons on a heads-up display, on dashboards, and on signs placed in the environment) (see Pierowicz, Jocoy, Lloyd, Bittner, & Pirson, 2000; Campbell, Richard, Brown, & McCallum, 2007 for a review of modalities and associated warning characteristics).

There exist several types of CIIWT that can be differentiated according to the primary location of the technology. The technology employed in these systems is based within a vehicle, within the transportation infrastructure, or, more recently, is a combination of vehicle and infrastructure. Table 40.1 provides a brief description of common CIIWT, relative to the location of the system. An example of a vehicle-based CIIWT would include forward collision warning systems (FCW). Commonly, a FCW consists of a radar sensor embedded in the front bumper of a vehicle that detects the distance to a lead vehicle. The distance information can then be combined with vehicle speed information to generate a value that reflects the criticality level (e.g., dangerousness, potential for a collision) between the two vehicles. The FCW can then provide a simple auditory warning (e.g., beep) or illuminated icon when a driver's vehicle detects that the proximity of the lead vehicle is too close (Pierowicz et al., 2000). Rear collision warning (RCW), side collision warning (SCW), and lane change warning (LCW) systems operate in a similar manner.

Transportation infrastructure-based systems consist of sensing and warning equipment that resides in the environment, typically installed along roadways or at intersections. These

TABLE 40.1 Primary Types of Collision and Intersection Incursion Warning Technologies

Technology Location	CIIWT	Purpose	Collision Mitigation
Vehicle-based	Forward Collision Warning (FCW)	Warn a driver the gap between their vehicle and a lead vehicle is below a safety threshold.	Collision to the front of driver's vehicle.
Vehicle-based	Rear Collision Warning (RCW)	Warn a driver the gap between their vehicle and a following vehicle is below a safety threshold.	Collision to the rear of driver's vehicle.
Vehicle-based Vehicle and Infrastructure-based	Lane Departure Warning (LDW)	Warn a driver they are exceeding their lane boundary.	Sideswipe collision to each side of vehicle.
Vehicle-based Vehicle and Infrastructure-based	Roadway Departure Warning (RDW)	Warn a driver they are exceeding the roadway boundary.	Single vehicle roadway departure.
Vehicle-based	Lane Change Warning (LCW)	Warn a driver when a vehicle is occupying the lane into which they want to change.	Sideswipe collision to each side of vehicle.
Vehicle-based	Side Collision Warning (SCW)	Warn a driver when the gap between their vehicle and a vehicle approaching from the side is below a safety threshold.	Collision to the left or right side of a driver's vehicle.
Infrastructure-based	Intersection Collision Warning (ICW)	Warn a driver who is approaching an intersection about the presence of vehicles on the intersecting roadway.	Left turn across path.
Vehicle and Infrastructure-based		Warn a driver at an intersection about the presence of approaching vehicles.	Entry without adequate gap.
			Violation of traffic control signal.
			Violation of traffic control signal.

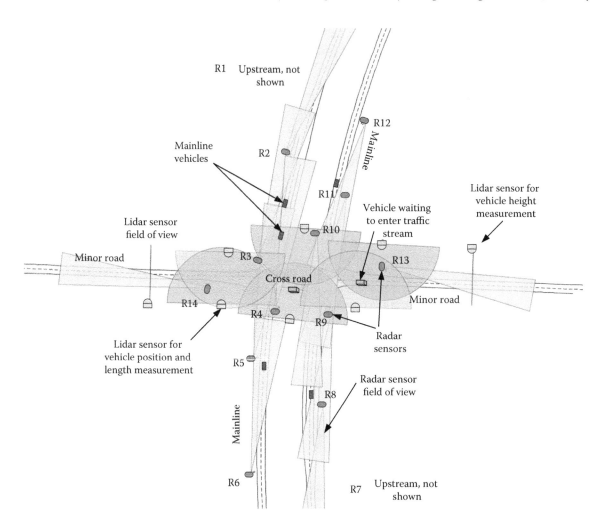

FIGURE 40.1 Plan view of an instrumented rural expressway intersection. Sensors on the mainline roadway (north and south in the figure) are radar which provides range, altitude, rate, and azimuth of vehicles relative to the sensor. Sensors on the minor roadway are lidar (i.e., light detection and ranging), which provide range to vehicles.

systems can help improve levels of safety by providing information or warnings to drivers to help them make better driving decisions. For example, previous research has indicated that a factor contributing to high rates of crashes at some rural thru-stop intersections* is the ability of drivers on a minor roadway, who are controlled by a stop sign, to estimate accurately the gap between approaching vehicles on a major intersecting roadway (i.e., highway) (Chovan, Tijerina, Pierowicz, & Hendricks, 1994; Najm, Koopmann, & Smith, 2001). An inability to accurately determine approaching vehicle gaps may lead drivers to select gaps that do not support safe intersection crossing behaviors. In an effort to address this situation researchers have developed an infrastructure-based system, termed "intersection decision

support system" (IDS) (see Creaser, Rakauskas, Ward, Laberge, & Donath, 2007; and Laberge, Creaser, Rakauskas, & Ward, 2006, for an extended presentation of IDS technology), that can inform drivers on a minor roadway of the gaps between vehicles on a major roadway. The system consists of a series of sensors placed along a section of a major roadway (i.e., highway) that detect the presence and speed of vehicles. See Figure 40.1 (Web Figure 40.1 for color version) for a depiction of the sensor layout at an example intersection. This information is then forwarded to a central computer near the intersection that can provide information to a driver via an illuminated variable message sign. See Figure 40.2 (Web Figure 40.2) for a depiction of an example variable message sign presenting unsafe gap information. This is known as a cooperative intersection collision avoidance system - stop sign assist (CICAS-SSA). Creaser et al. (2007) examined the utility of four IDS conditions for both younger and older drivers during day- and night-time conditions in a driving simulator. In their study, participants crossed a simulated rural thru-stop intersection twice in each of five conditions: a stop-sign condition which

---

* A rural thru-stop intersection is characterized here as a highway segment (e.g., two lanes in each direction with a separating median) that is intersected by a minor roadway (e.g., two-lane roadway). Traffic on the highway segment has the right-of-way whereas traffic on the minor roadway is controlled by a stop sign.

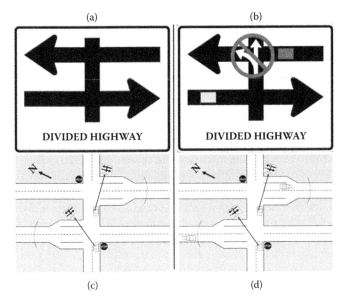

FIGURE 40.2 Two CICAS-SSA variable message signs presenting gap and warning information to a driver positioned at a stop sign. The sign would be presented to drivers on the minor roadway only. In the example, a driver would be attempting to cross a highway with two lanes of traffic in each direction separated by a median. The sign presented in Figure 40.2a indicates to the driver on the minor road that no vehicles are present on the roadway, as depicted in Figure 40.2b. The white icon (presented as yellow in experiments) in Figure 40.2c indicates that, for the near lanes of traffic, an approaching vehicle has entered a zone where the gap should be accepted with caution (zones are marked by a curved line). The same sign indicates that, for the far lanes of traffic, an approaching vehicle has entered a zone where the gap should not be accepted. This is depicted as a white circle and diagonal line (presented as red in experiments).

served as a baseline against which the other conditions were compared, and four IDS conditions that presented signs to drivers waiting to cross on a minor roadway with differing levels of dynamic traffic information about vehicles on a major roadway (each of the four sign conditions was presented with the baseline stop sign condition). Results of their work indicated that the use of all tested IDS signs was associated with drivers accepting gaps that were significantly larger than the stop sign only condition. However, IDS signs that provided gap (e.g., time-to-arrival and gap warnings) and advisory information were associated with the best performance as compared to the signs that did not provide this information.

Vehicle and transportation infrastructure-based CIIWT can share information between in-vehicle and infrastructure-based systems. For example, research efforts are underway to develop and evaluate a cooperative intersection collision avoidance system that provides an in-vehicle warning to a driver if there is a potential they will violate a traffic control signal or stop sign. This CIIWT collects information from the infrastructure to determine if a signal is in a prohibited phase and collects information from a vehicle to determine if it is turning or moving through an intersection. If the maneuver is prohibited, the

CIIWT wirelessly transmits a signal to an in-vehicle display to warn a driver of a prohibited intersection incursion. While information regarding the utility of this system is not available at this time, it is hoped that this technology can reduce the rate of collisions at intersections due to prohibited incursions.

As suggested by the current discussion, CIIWT are intended to reduce the frequency and severity of motor-vehicle crashes or intersection incursions by providing drivers with warnings within or outside a vehicle. CIIWT are identified here as one type of a broader class of systems that are intended to support drivers within the domain of ITS technology.

40.1.2 Collision and Intersection Incursion Scenarios

In light of the notion that many CIIWT evaluated in driving simulators may be deployed in real-world situations, it is necessary to optimize the ability to generalize results between real-world and driving simulator settings. One method for increasing generalizability is to include in driving simulator evaluations those real-world scenarios in which CIIWT are expected to operate. The purpose of this section is to identify and describe the primary roadway and intersection driving scenarios contributing to high crash rates that prototypical and existing CIIWT have been designed to address. Previous research efforts (Keifer et al., 1999; Keifer et al., 2003; Najm & Smith, 2007; Pierowicz et al., 2000) have collectively identified six basic driving scenarios which include rear-end (RE), lane departure (LD), left turn across path (LTAP), entry without adequate gap (EWAG), violation of traffic control (VOTC), and violation of traffic control signal (VOTS). The rear-end scenario occurs most often in urban environments and is characterized by a subject vehicle (SV) that crashes into the rear of a lead vehicle (LV) due to a variety of reasons including inattentive driving, distracted driving, poor visibility, aggressive driving habits, tailgating, and cut-in behaviors by a vehicle from an adjacent lane (Keifer et al., 1999). Mitigation of crashes within this scenario is primarily addressed by FCW systems that warn a driver of events that may occur in front of their vehicle (see Figure 40.3, location 1 for a depiction of this scenario). The lane change scenario is characterized by an SV that steers into an adjacent lane which is occupied by another vehicle (see Figure 40.3, location 2). This scenario is common to both urban and rural environments in which there exist two lanes of traffic moving in the same direction. CIIWT technologies such as lane departure warning (LDW) are best suited to address these crash scenarios due to their ability to identify and warn against events occurring laterally to the SV. The final four scenarios are indicative of intersection incursions (see Pierowicz et al., 2000 for full descriptions of these scenarios). LTAP is characterized by an SV that has the right-of-way (i.e., green-light) at a signalized intersection and is presented with an opposing vehicle turning left that has not yielded (see Figure 40.3, location 3). EWAG is depicted in Figure 40.2, location 4 and occurs when an opposing vehicle stops at the traffic control device (stop sign) and then proceeds through the intersection (to turn left, right, or proceed

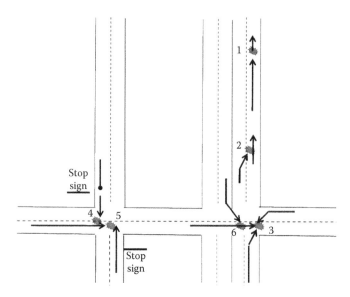

FIGURE 40.3 Roadway and intersection crash scenarios. The six individual crash scenarios are identified by numerals 1–6.

straight) without allowing an adequate gap between their vehicle and the SV. VOTC occurs when an opposing vehicle violates a traffic control device (e.g., runs a stop-sign or a red light) and is struck by the SV that has the right-of-way from the left or right (see Figure 40.3, location 5), while VOTS is the opposite (i.e., SV violates a traffic control device and strikes opposing vehicle) (see Figure 40.3, location 6). Note, that while these scenarios occur frequently in urban environments they are also not uncommon (especially LTAP, EWAG, and VOTC) in rural environments that contain low-cost traffic control devices (i.e., stop signs) or no devices at all. FCW are one type of CIIWT to deploy at intersections to address LTAP, EWAG, VOTC, and VOTS scenarios due to the fact that events primarily occur in front of the SV. However, more advanced intersection collision warning systems such as those that track individual vehicle movements, typically through the use of vehicle-based global positioning satellite technology, also present an effective method to reduce collisions at intersections.

40.2 Characteristics of Design

The overall design of a CIIWT depends predominantly on its intended purpose. However, the operation can also be defined based on underlying design characteristics that are exhibited to a greater or lesser extent for most CIIWT. It is important for driving simulator users to be aware of these design characteristics in order to integrate these systems into driving simulators adequately and to be prepared for the resources required for integration. Note that while all CIIWT presented earlier can be integrated into most driving simulators, the design characteristics of each system will allow for easier or more difficult integration. In addition, users should carefully consider each of the design characteristics because their manipulation (i.e., they are independent variables) can significantly influence CIIWT

effectiveness. Given that each characteristic can have a greater or lesser influence on CIIWT effectiveness, each exists along a continuum. The following section summarizes each of the primary characteristics of design along with a description of the ease of integration into a driving simulator.

A characteristic of all CIIWT is system action/initiation, which refers to the degree to which a system will act or initiate warnings based on information it has received (e.g., vehicle presence, distance to vehicle) or generated (e.g., time headway, time to lane crossing). On one end of the continuum a CIIWT can warn when it detects an imminent collision such as when a FCW provides short tones to warn a driver they are approaching a slower vehicle (Keifer et al., 1999) or a LDW that warns a driver when they are exceeding lane boundaries (Houser, Pierowicz, & Fuglewicz, 2005). These warnings simply indicate to a driver they should act and are the easiest to integrate into a driving simulator due to their simple technological features (e.g., distance-detection device and a system for producing a warning sound). The middle of the action/initiation continuum represents those CIIWT that are increasingly powerful and omnipresent. Versions of CIIWT have begun to combine warnings along with information regarding the criticality of the situation, such as a FCW that indicates the time in seconds to a lead vehicle (Alkim, Bootsma, & Looman, 2007). The provision of warnings and information allow a driver to better assess the situation and select an appropriate response, essentially indicating to the driver that: a) they should act, and b) how quickly they should act. These CIIWT can also be easy to integrate into a driving simulator due to the relatively simple technology required. At the end of the system action/initiation continuum are those most recently developed CIIWT that incorporate automated vehicle control. While these systems often employ warnings and information from the middle and the opposite end of the system action/initiation continuum they can also unilaterally initiate vehicle responses, augment driver behaviors, or attenuate driver behaviors. For example, some current model automobiles are equipped with brake assist systems that can detect a critical scenario developing in front of their vehicle through radar sensors using FCW. These can warn and inform the driver of the urgency of the scenario through a series of increasingly frequent tones, and can initiate actions to reduce or eliminate the criticality of the situation with or without driver awareness by either slowing a vehicle through engine braking or, in highly critical situations, through near-threshold braking. As evidenced by these examples CIIWT can range from simple warnings to warnings that are coupled with rich information and extensive vehicle actions.

The varying level of complexity within this continuum raises important issues that must be considered by users of driving simulators. First, as CIIWT system complexity increases along the system action/initiation continuum the effort and associated costs for installation, programming, and long-term support will also increase. For example, CIIWT on the latter end of the system action/initiation continuum can be difficult to integrate into a driving simulator due to the need for equipment that can implement actions. In the example presented above it would be easy to "program" a simulator to detect a critical scenario and it would be easy to output a warning signal to a speaker system. However, to initiate an action such as near-threshold braking, a simulator would most likely need a modest motion-base system to accurately depict vehicle dynamics and subsequent braking.* The effort and cost associated with the installation and use of a partial or full motion-base simulator are not insignificant. As a result, the simulation user must determine if adequate resources are available to support a desired experimental evaluation. Second, as CIIWT system complexity increases there will be a corresponding increase in the complexity of experimental parameters to be addressed. These may include the type of experimental design, range of dependent variables and, due to longer periods of time to perform an evaluation, physical fatigue, visual fatigue, and boredom will also need to be addressed. The system action/initiation continuum and the examples presented provide tentative support for the notion that when designing a driving simulator-based experiment that will employ a CIIWT, or when integrating a CIIWT into a driving simulator, it will be necessary to identify where on the system action/initiation continuum the CIIWT system exists. This will allow driving simulator users to more accurately determine the true effort and cost associated with CIIWT system installation and operation and to account for essential experimental variables.

Temporal/distance system activation (TSA) represents the time and/or distance at which a CIIWT initiates a warning or action after event detection. The endpoints of this continuum are simple from a conceptual standpoint; one end is essentially "zero time" in which a system is activated simultaneously with, or just previous to, an event while the opposite end represents system action/initiation presented significantly before an event (see Figure 40.4 for a depiction of this continuum relative to time and distance). Practically, the endpoints can be difficult to assess. For example, the ability of a CIIWT to determine when a collision will occur can be influenced by the capabilities of a sensor (e.g., sensors that detect vehicles, intersections, pedestrians), sensor type (e.g., radar, laser, lidar, acoustic), and additional factors that may introduce non-constant variables into sensor capabilities or into the algorithms employed to calculate when an event would occur (e.g., wireless transmission delay between sensor and vehicle, weather conditions that may interfere with sensor sensitivity). In addition, while oftentimes a determination of the "moment" an event may occur (e.g., collision, intersection incursion) is calculated with precision, some algorithms include a confidence variable which considers the degree to which the information gained from the sensors may be accurate or reliable. If the information is deemed to be of low accuracy or reliability, a system may delay a warning or action until the confidence value exceeds a predetermined threshold (i.e., accuracy or reliability has improved). In addition, if one type of distance sensor can

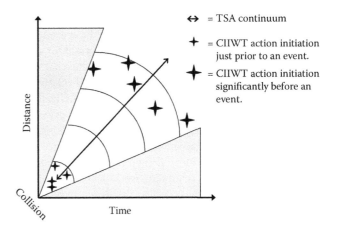

FIGURE 40.4 Temporal system activation continuum for CIIWT which depicts when a system initiates an action after event detection. Note that the shaded areas represent time and distance combinations that are normally not experienced in typical driving situations.

detect objects at greater range than another it is reasonable to expect that a warning could be provided to drivers at an earlier stage.

Within this continuum researchers must also be cognizant of the collision warning timing presented to drivers due to the notion that inappropriately-timed warnings may be associated with a decrease in driving safety (see McGehee & Brown, 1998). Research conducted in a driving simulator has indicated that early warnings are more beneficial than late warnings (or no warnings) for both distracted and undistracted drivers (Lee, McGehee, Brown, & Reyes, 2002). However, as indicated by Weise and Lee (2004), great care must be taken when alerts are paired temporally. Their research found that when a non-collision alert (email notification) was presented 1000 ms previous to a collision warning, driver responses were enhanced while the same warning presented only 300 ms in advance of the collision warning degraded driver response, presumably due to dual-task interference. To incorporate into driving simulation experiments CIIWT systems that are truly representative of actual systems, driving simulator users must assiduously define the operational characteristics of a CIIWT system's endpoints along the TSA continuum and, if a system is to be deployed in the real-world, researchers must consider carefully the warning timing functions (see Campbell et al., 2007 for a review of multiple warning signal prioritization and integration).

CIIWT systems along the TSA continuum are equally easy to implement into a driving simulator technologically because the underlying equipment that will provide a warning immediately prior to a collision, and which will provide a warning in advance of collision, will be identical, or quite similar. The difficulty in implementing systems with regard to the TSA continuum is the increased amount of programming effort that may be required to develop advanced algorithms on which the warnings are based.

The physical location in which a system has been designed to operate is termed here as geospatial system activation (GSA), with urban and rural environments composing opposite ends

---

* Note that braking dynamics can be emulated in a driving simulator without a partial or full motion-base system. However, to promote driver behaviors that are similar or identical to real-world situations it may be necessary to allow drivers to experience realistic sensations of braking.

of the associated continuum. Given that CIIWT are typically designed to address specific events and that these events typically occur in specific environments, it is important for driving simulator users to couple events and environmental characteristics to accurately assess the true utility of a CIIWT relative to its intended design. As an example, recent work has indicated that thru-stop crashes, which are those crashes in which drivers on a minor road attempt to cross a median of a major road (e.g., highway) and misjudge the gap afforded by approaching vehicles, are a significant problem in rural environments (Chovan et al., 1994; Najm et al., 2001). It is not surprising that CIIWT designed to mitigate thru-stop crashes consider the factors specific to this event (e.g., speed, sight distance, power connections for remote data collection along the major road, etc.) and that these factors are defined by the location (see Figure 40.3) in which the crash type exists. This notion supports the contention that in order to impartially evaluate a system, driving simulators must account for the environment in which the CIIWT will be deployed and account for the primary factors specific to the crash type to be mitigated. Implementing CIIWT in a driving simulator that is specific to each end of the GSA continuum will require similar levels of effort and cost due to the similarity in the underlying technology and programming requirements.

The characteristics identified previously related to CIIWT functionality; however, it is also important for driving simulator users to consider the physical aspects that are associated with the integration between a CIIWT and driving simulator. One aspect relates to the location of the CIIWT hardware and software components. A vehicle-based CIIWT may require minimal effort and hardware to integrate into a driving simulator. For example, a simple FCW implemented in a driving simulator could consist of gathering time headway data to a lead vehicle via software, programming logic to determine the time headway threshold at which a warning will be activated, a digital to analog converter to generate an analog warning signal, and a buzzer that can receive a signal and provide a warning. Integrating this type of CIIWT into a driving simulator should be relatively easy. In contrast, there are CIIWT that are physically complex and, as a result, may require significant efforts to replicate in driving simulators. For example, the infrastructure-based CICAS CIIWT described previously requires that a) the position and speed of all vehicles traveling through an area surrounding an intersection be identified; b) that gaps between all vehicles and between the vehicles and the intersection be calculated; and c) that this information be presented on a simulated variable message sign. While each of these three activities within a driving simulator is possible, they can require substantial programming efforts and, as a result, will be significantly more challenging to implement.

A second aspect relates to the level of CIIWT integration within a vehicle, environment, or infrastructure. A system that requires a greater degree of integration will be associated with an increase in driving simulator programming and hardware installation effort. For example, the integration of a vehicle-based rear collision warning system (RCW) may include a simple distance sensor and speaker to provide a warning. However,

more complex systems that combine FCW, RCW, LDW, and intersection incursion warnings into a unified vehicle and infrastructure-based CIIWT will require extensive intra-system, vehicle, environment, and infrastructure coordination and integration.

The physical mechanism that will deliver a warning to a driver is a final aspect that should be considered during installation and continued operation of a CIIWT within a driving simulator because of the potential technical hurdles and costs that may need to be addressed. A mechanism that can provide a discrete auditory or visual warning (e.g., a tone or a flashing light) will require minimal integration and cost while a system that provides feedback via a vibrotactile seat (Ho, Reed, & Spence, 2006) or heads-up display (Caird, Chisholm, & Lockhart, 2008) will require significant costs due to the required equipment purchases and integration costs.

The characteristics of functionality and physical aspects can vary greatly between differing types of CIIWT. For those CIIWT systems that are functionally and physically uncomplicated the development and installation of these devices into driving simulators will be tractable. However, CIIWT that are complicated in terms of warning initiation type, warning initiation timing, geographic location of system components, and physical design may present formidable obstacles to the implementation, continued operation, and use of CIIWT within driving simulators unless simulation users and engineers anticipate these issues.

40.3 Simulation and Experimental Designs for CIIWT Research

In order for an examination of the utility of CIIWT within driving simulators to yield valid, reliable, and veridical results, researchers should consider a range of simulation and experimentation issues. The purpose of this section is to identify several of the prominent issues to be considered when examining CIIWT. However, the reader is encouraged to expand the scope of their understanding by reviewing allied chapters within this book that address scenario selection, independent and dependent variables (e.g., this book, chap. 15 by McGwin), and various aspects of fidelity (e.g., simulation, traffic, physical, perceptual, and behavioral; see in this book, chap. 7 by Greenberg & Blommer, chap. 9 by Ranney, chap. 8 by Andersen, and chap. 10 by Angell) because of the significant role they may also play in conducting CIIWT research within a simulation environment.

40.3.1 System Selection

Generalization (Chapanis, 1988; Schmidt & Lee, 2005) is arguably one of the more prominent considerations when employing a driving simulator for CIIWT. Generalization is the degree to which research results can be expected to extrapolate to real-world scenarios or, relative to the current topic, how well research results obtained from studies employing driving simulators that examine the efficacy of CIIWT can predict the utility of the same technologies when they are deployed in real-world environments. Global methods to increase generalization of research results are provided elsewhere in this book (e.g., see this book, chap. 4 by Hancock & Sheridan). Specific to CIIWT, the most promising method is to equate the psychological processes between the testing and real-world environments (Kantowitz, 1992; Schiff & Arnone, 1995). However, as mentioned previously (Manser & Hancock, 1996), the nature and extent of the primary psychological processes in real-world environments has yet to be established adequately. How then, can users of driving simulators who are examining CIIWT promote generalizability? Due to the notion that a highly realistic environment can be expected to contain a greater number of veridical psychological processes than a non-realistic environment, a primary technique that can be employed is to increase realism. Realism is the degree to which testing environments represent the real world. Increasing the realism of driving environment simulation-based research can be accomplished by several methods that can be described according to vehicle and environmental constructs. Realism elements within the vehicle construct should include accurate representations of vehicle dynamics, (e.g., steering, acceleration, and deceleration), highly functional vehicle cab, and a full range of CIIWT characteristics to be investigated. Relative to the environment, realism elements should include high-resolution images, wide field of view to promote an accurate sense of speed, presentation of driving scenarios that are based on comparable real-world scenarios (e.g., collision, intersection, vehicle incursion), and depiction of realistic traffic flows (e.g., vehicle size, traffic density, gap presentation, and vehicle speed) that are representative of those appearing in real-world environments under consideration. Although technically difficult, an attempt should also be made to include salient real-world elements specific to the scenario and/or CIIWT to be examined (e.g., pedestrians, buildings, traffic signaling, night/day conditions).

40.3.2 Dependent Variables

A determination of the veridical utility of a CIIWT begins with an identification of dependent variables that are supported by a sound rational. Previous work in an allied area that examined in-vehicle information systems (Green, 1994) provides a sound rational for the selection of dependent variables within driving simulator-based evaluations (traditional dependent variables employed in driving simulator work can be found elsewhere in this book; see the chapters mentioned above). Adapted from that work, the following criteria for dependent variable selection relative to the examination of CIIWT are suggested: (1) Dependent variables should be directly related to the underlying research hypothesis being considered; (2) dependent variables should be of sufficient sensitivity to detect differences due to the use or inclusion of a CIIWT; (3) dependent variables that increase risk or embarrassment for drivers should not be included; (4) dependent variables that add to the cost and complexity of experiments should be assessed to determine both if they can be included and the degree of their worth; (5) dependent variables should be of sufficient simplicity so that they can be analyzed effectively; (6) metrics must exhibit repeatability in that they can reliably detect performance

differences; (7) metrics should have been validated in real-world and driving simulation settings previous to their inclusion in driving environment simulation experimentation; and (8) metrics should be appropriate to the driving scenario under examination (e.g., lane deviation metrics should be employed when examining an in-vehicle lane departure warning system).

In light of these "guidelines" governing the selection of dependent variables, what then are the dependent variables that should be employed when examining the utility of CIIWT? The answer to this question is not straightforward. The utility of a CIIWT can be measured in terms of safety, which is dictated by changes in human performance. How a CIIWT may influence driver stress is also an important consideration given the potential relationship between stress and performance. In addition, the degree to which a system is usable can significantly impact drivers' perceptions of a CIIWT and subsequent system use. The following sections identify some potential dependent variables within the constructs of performance, stress, and usability that may be necessary in order to gain a broad understanding of the true utility of a CIIWT.

40.3.2.1 Performance Construct Dependent Variables

Dependent variables within the performance construct relate to a vehicle's *input* devices (i.e., steering, braking, and acceleration) that reflect direct driver control of a vehicle and relate to resultant vehicle *output* (i.e., often measured in safety margins such as time-to-contact or time headway) as a result of CIIWT use. These variables can provide an understanding of the utility of a CIIWT with respect to driver performance. Within the input and output classifications, dependent variables can be further categorized according to task type (adopted from Green, 1994) which include: 1) primary task (i.e., the main task of the driver which is to control a vehicle), 2) CIIWT task (i.e., the task of interacting with the CIIWT and associated performance changes), and 3) overall task which will include performance metrics not associated with direct vehicle or CIIWT control (e.g., eye gaze behaviors). Input and output classification and task type can serve as the basis from which dependent variables for CIIWT research in driving simulators can be selected. A list of potential input and output dependent variables is presented in Tables 40.2 and 40.3, respectively.

Many of the input and output dependent variables that can be employed to assess the utility of a CIIWT have been identified previously (Green, 1994; Keifer et al., 2003); however, several variables identified in Tables 40.1 and 40.2 have been developed and employed successfully more recently and deserve further consideration. Consistent with the selection criteria that recommends dependent variables exhibit sensitivity and in light of the significant impact a CIIWT can have on event detection and response, researchers have employed input dependent variables that identify stages of perception-response times (see Olson, 2001 for an extended description of perception-response stages) that are indicative of CIIWT efficacy. The four perception-response stages thought to exist, and as they apply to CIIWT, include detection (i.e., first recognition of an object or event such as an approaching vehicle), identification (i.e., first determination that an object or event is a threat), decision (i.e., the moment an

TABLE 40.2   Performance Dependent Variables for Driver Input Controls

Task Type	Control Direction	Dependent Variables
Primary (i.e., driver control of vehicle)	Lateral	– steering wheel, movements per unit of time – steering wheel, reversals – steering wheel, reversal rate – steering wheel, action rate – angular wheel, position change average – angular wheel, position change standard deviation
	Longitudinal	– throttle position, average – throttle position, standard deviation – brake application, count – brake pressure/position, average – brake pressure/position, standard deviation
CIIWT (i.e., driver control and response to non-vehicle tasks)	System Utility (detection performance)	– accelerator reaction time – accelerator movement response time – accelerator movement time headway – brake response time – brake response time headway – max brake pedal pressure/position – 85th percentile brake pedal pressure/position – transition time between accelerator pedal release and brake pedal activation
Overall	Vision	Frequency and duration of glances to the: – road – mirrors, HVAC, display cluster – CIIWT (if visible and requiring driver interaction)

Source: Adapted from Green, P. (1994). Rep. FHWA-RD-94-058. McLean, VA: Federal Highway Administration.

appropriate response to the threat is identified such as deciding to apply a vehicle's brakes), and response (i.e., the moment the appropriate response is executed such as initiating movement of the lower leg to begin the braking response). An understanding of perception-response stages and their application to CIIWT evaluations offers unique insights into event detection and identification, into the selection of responses that are appropriate for addressing a developing situation, and into whether these potential benefits result in faster responses and decreased potential for crashes. For example, research that examined the utility of haptic feedback (i.e., accelerator pedal force changes) as an information source to drivers (i.e., employed for detecting changes in the gap between a driver's vehicle and a lead vehicle) (Manser et al., 2004) decomposed the primary driver response of pedal movements into their principle components consistent with human motor control measures of time and speed (e.g., reaction time, response time, transition time, movement time) (Schmidt & Lee, 2005). These principle components then served as input dependent variables that identified driver "decisions" and "responses". In this work, participants drove through a simulated environment, once with the use of the haptic accelerator pedal and once without, and were tasked with following a lead vehicle at a constant, close, and safe distance at all times. Throughout the drive the speed of the lead vehicle gradually increased and decreased; however, while an in-vehicle secondary device distracted drivers, the lead vehicle slowed suddenly. Results of their work indicated the use of a haptic accelerator pedal could facilitate event detection as indicated by significantly faster reaction times to the sudden braking situations. Although this technology is not a collision warning system the utility of these metrics for

understanding perception-response time and associated information processing can be applied easily to CIIWT evaluations.

Traditional output dependent variables for CIIWT research have included metrics that indicate the degree to which safety margins (e.g., time-to-contact) (Keifer et al., 2003) are violated relative to the use or non-use of a CIIWT. Non-traditional output dependent variables such as coherence, amplification, and delay (Boer, Ward, Manser, & Kuge, 2006; Brookhuis, De Waard, & Mulder, 1994; Ward, Manser, De Waard, Kuge, & Boer, 2003) can also be useful for identifying performance changes as a result of system use. Coherence represents a correlation between a lead and following vehicle's speed profiles that is reflective of the accuracy of drivers' speed adaptations (i.e., drivers' ability to maintain a safe and constant distance between their vehicle and a lead vehicle). Modulus represents the degree to which a participant over- or underestimates the maximum and minimum speed of a lead vehicle. The postponement in time between a lead vehicle and following vehicle speed change is termed delay. Research that employed a car following task in a driving simulator found that coherence, modulus, and delay (Ward et al., 2003) are sensitive to changes in driver performance due to the use of a continuous information source (via a haptic accelerator pedal identified earlier) that provides information relative to the gap between a driver's vehicle and a lead vehicle.

40.3.2.2 Stress Construct Dependent Variables

Decreased levels of driver stress due to positive interactions with a CIIWT may greatly facilitate an overall positive impression of a system and promote continued use. The selection of dependent variables that describe an association between stress and CIIWT use can be difficult due to the notion that changes in stress levels may

TABLE 40.3 Performance Dependent Variables for Vehicle Output Controls

Task Type	Control Direction	Dependent Variables
Primary (i.e., vehicle output)	Lateral	– lateral, deviation mean
		– lateral, deviation mean absolute
		– lateral, deviation standard deviation
		– lateral, acceleration
		– lateral, yaw rate
		– lateral, yaw angle
		– lateral, yaw acceleration
		– lane deviation count
		– percent and total time outside lane
		– steering entropy
	Longitudinal	– speed, mean
		– speed, standard deviation
		– acceleration/deceleration, average
		– accelerations/decelerations, exceeding a g criterion
		– headway mean, to lead vehicle
		– headway standard deviation, to lead vehicle
CIIWT (i.e., driver control and response to non-vehicle tasks)	System Utility (detection performance)	– time to lane crossing
		– time to contact to lead vehicle
		– time to arrival at intersection
		– coherence to a lead vehicle
		– modulus to a lead vehicle
		– delay to a lead vehicle
Overall		– crashes
		– near misses

Source: Adapted from Green, P. (1994). Rep. FHWA-RD-94-058. McLean, VA: Federal Highway Administration.

occur over long periods of time as compared to the relatively succinct support offered by a CIIWT. This section summarizes metrics of stress that are common to evaluations of CIIWT and that could be employed in future evaluations. As a result, there is the potential that measures of stress may reflect changes due to factors not associated with system use. However, traditional stress measures may provide utility if researchers bear in mind the potentially imperfect ability of these metrics. One method for evaluating driver stress is through the use of questionnaires. Mental workload, a surrogate measure of overall mental stress, can be measured relatively quickly with the Rating Scale Mental Effort (RSME) developed by Zijlstra (1993). The RSME consists of a univariate scale on which drivers indicate their perceived level of mental effort associated with the driving task. It has been shown to be a good measure of mental effort in cases where a secondary task is presented and may prove helpful in determining general increases/decreases in mental effort due to CIIWT use. The Mood Adjective Checklist (MACL) can be employed to measure driver affective mood state (mood stress) (Matthews, Jones, & Chamberlain, 1990). In this self-report measure drivers indicate their level of feeling for each of 29 mood adjectives on a four-point scale that are then combined to assess the composite mood factors of hedonic tone (cheerful, contented, satisfied), tension (anxious, jittery, tense, nervous), and vigor (active, energetic, alert, vigorous). These measures can also be combined with more traditional measures, such as the National Aeronautics Space Administration – Task Load Index (NASA-TLX) (Hart & Staveland, 1988), to provide confirmatory evidence of changes in mental effort.

Psycho-physiological metrics of driver stress are involuntary in that they respond to stress without conscious control

or without subjective bias on behalf of participants. Galvanic skin response (GSR) is a metric of arousal/effort that measures changes in skin conductance levels as a result of perspiration generated from increased stress levels. GSR is typically collected via sensors placed on the first and second fingers of a driver's hand. Heart rate variability (HRV) is quantified in terms of beats per minute. 10 Hz sinus arrhythmia measures changes in the profile of heart beats in response to workload/effort (Brookhuis & De Waard, 2001).

40.3.2.3 Usability Construct Dependent Variables

Designers of CIIWT should be aware of the "usability" of their systems because poor usability could lead to dissatisfaction and non-use, which would negate the potential benefits afforded by a system. This section summarizes common metrics that could be employed to evaluate system usability. Usability is the degree to which drivers accept, understand, learn, and interpret a system. The intangible factors of usability are made tangible through answers provided on questionnaires exploring a range of diverse issues such as how drivers perceive the usefulness of a system; how satisfied they are with a system; to what degree do drivers trust a system to operate correctly; whether the system promotes improved performance; and how easy a system is to learn and understand. While the use of many questionnaires can provide a robust indication of usability it is often sufficient to administer only a few to gain a reliable understanding of usability perceptions. One such questionnaire is the usability scale (developed by Van der Laan, Heino, & De Waard, 1997), which has routinely been employed to assess driver perceptions of telematic and DSS systems. This measure requires drivers to rate their perceptions on a number of bipolar adjective scales that are then summed to produce separate scores for the level of perceived satisfaction and usefulness. The scores can be positive or negative to reflect satisfaction/dissatisfaction and usefulness/non-usefulness.

The degree of trust drivers attribute to a particular CIIWT is a critical component of overall usability due to the notion that drivers attributing low degrees of trust to a particular system are likely to disengage or not utilize the beneficial information provided to avoid a collision, a situation which defeats the purpose of the system. Drivers can be provided with a trust metric which assesses their understanding and perceived reliability of the CIIWT system (Lee & Moray, 1992). This is a nine-point scale that rates the degree of agreement with a series of statements regarding system functioning. The scale can be aggregated to produce an average trust score and a number of additional subcomponents indicative of trust.

40.4 Conclusions

This chapter provided a review of the role of driving simulators in research examining the utility of CIIWT within land-based transportation environments. CIIWT are a subclass of driver support systems that provide in-vehicle collision warnings to drivers so they can respond appropriately to avoid collision-likely events by making informed decisions. An identification of CIIWT characteristics consisting of system actions, temporal activation, geospatial activation, hardware and software components, system integration with a vehicle or infrastructure, and the physical mechanism to deliver collision warnings helps to illuminate the complexity of these systems and the range of potential difficulties that one may be faced with when incorporating them into driving simulators for research purposes. The complexity of CIIWT technology evaluations increases still further when one considers the range of potential dependent variables within performance, workload, and usability constructs and the driving scenarios that should be paired with specific types of CIIWT technologies in order to conduct examinations that allow for a high degree of generalization of driving environment simulation results to real-world situations. The potential complexity of these evaluations should not dissuade scientific endeavors that seek a better understanding of the utility of these systems. Instead, it is hoped that the complexity of these evaluations will intrigue scientists and challenge them to conduct exceptional research which will highlight the veridical potential of these devices to improve transportation safety.

Key Points

- Driving simulators are a useful tool for evaluating the effectiveness of collision and intersection incursion warning technologies (CIIWT) due to their capacity to provide real-world scenarios in a risk-free environment at a relatively low cost (compared to real-world experiments).

- Improving the generalizability of driving simulator-based research that examines CIIWT can be facilitated by employing driving scenarios that are representative of those in which CIIWT have been designed to operate.

- The design characteristics of CIIWT can greatly influence the effort and finances required to integrate these systems into driving simulators. In addition, design characteristics such as system action/initiation, temporal/distance system activation, physical location, and level of system integration relative to each CIIWT to be evaluated must be understood and integrated into driving simulator-based evaluations in order to produce valid and objective research results.

- To capture the totality of the influence of a CIIWT it is necessary to select dependent variables that reflect changes in constructs related to performance, stress (workload and comfort), and usability.

- Non-traditional dependent variables that examine continuous task performance (e.g., coherence, modulus, and delay) and discrete task performance (e.g., measures of perception response stages) offer new and insightful views of the efficacy of CIIWT.

Keywords: Collision Avoidance, Collision Warnings, Driver Support Systems, Driving Simulator, Intelligent Transportation Systems

Acknowledgment

The author acknowledges the support of the Intelligent Transportation Systems Institute and the Intelligent Vehicles Laboratory at the University of Minnesota; their support is particularly appreciated.

Glossary

Adaptive cruise control (ACC): Vehicle-based system that maintains a criterion speed and, when following a vehicle that is traveling slower than the criterion speed, will maintain a predetermined time headway.

Anti-lock brake system (ABS): A vehicle-based system that prevents the wheels from locking/skidding.

Collision and intersection incursion warning technology (CIIWT): Technology within the vehicle and/or within the transportation infrastructure that can provide warnings from either within or outside a vehicle to a driver of an impending collision or dangerous incursion into an intersection.

Cooperative intersection collision avoidance system - Stop sign assist (CICAS-SSA): Intersection-based technology that can provide gap information to a driver on a minor roadway (who is attempting to cross over a major roadway) relative to traffic on a major roadway.

Driver support system (DSS): A general term referring to a class of products that intend to support driver intentions, increase driver safety, or increase driver comfort.

Forward collision warning system (FCW): A radar or laser-based in-vehicle system that warns a driver if a lead vehicle is within a predetermined time headway.

Lane departure warning (LDW): An in-vehicle system that warns a driver if their vehicle begins to move out of lane.

Web Resources

The *Handbook* web site contains supplemental material for the chapter, including color versions of two figures.

Web Figure 40.1: Plan view of an instrumented rural expressway intersection (color version of Figure 40.1).

Web Figure 40.2: Two CICAS-SSA variable message signs presenting gap and warning information to a driver positioned at a stop sign (color version of Figure 40.2).

Key Readings

Kuge, N., Yamamura, T., Boer, E. R., Ward, N. J., & Manser, M. P. (2006). *Study on drivers' car following abilities based on an active haptic support function* (SAE Tech. Paper 2006-01-0344). Warrendale, PA: Society of Automotive Engineers.

Laberge, J. C., Creaser, J. I., Rakauskas, M. E., & Ward, N. J. (2006). Design of an intersection decision support (IDS) interface to reduce crashes at rural stop-controlled intersections. *Transportation Research Part C: Emerging Technologies, 14,* 39–56.

McGehee, D. V., & Brown, T. L. (1998). *Examination of drivers' collision avoidance behavior in a lead vehicle stopped scenario using a front-to-rear-end collision warning system* (DTNH22-93-C-07326). Washington, DC: National Highway Traffic Safety Administration.

Najm, W. G., & Smith, J. D. (2007). *Development of crash imminent test scenarios for integrated vehicle-based safety systems* (Rep. DOT HS 810 757). Washington, DC: National Highway Traffic Safety Administration.

Pierowicz, J., Jocoy, E., Lloyd, M., Bittner, A., & Pirson, B. (2000). *Intersection collision avoidance using ITS countermeasures, task 9 final report* (Rep. DOT HS 809 171). Washington, DC: National Highway Traffic Safety Administration.

References

Alkim, T., Bootsma, G., & Looman, P. (2007). *The assisted driver.* Delft, the Netherlands: Ministry of Transport, Public Works and Water Management.

American Association of State Highway and Transportation Officials. (1997). *Highway safety design and operations guide.* Washington, DC: AASHTO.

Bellomo-McGee. (2003). *Intersection collision avoidance study. Final report.* McLean, VA: Federal Highway Administration. Retrieved from http://www.itsdocs.fhwa.dot.gov/JPODOCS/REPTS_TE/14105.htm

Boer, E. R., Ward, N .J., Manser, M. P., & Kuge, N. (2006). Driver-model-based assessment of behavioral adaptation. *Transaction of Society of Automotive Engineers of Japan, 37*(4), 21–26.

Brookhuis, K. A., & De Waard, D. (2001). Assessment of drivers' workload: Performance and subjective and physiological indexes. In P. A. Hancock & P. A. Desmond (Eds.), *Stress, workload, and fatigue* (pp. 321–333). Mahwah, NJ: Lawrence Erlbaum Associates.

Brookhuis, K. A., De Waard, D., & Mulder, L. J. M. (1994). Measuring driving performance by car-following in traffic. *Ergonomics, 37,* 427–434.

Caird, J. K., Chisholm, S. L., & Lockhart, J. (2008). Do in-vehicle advanced signs enhance older and younger drivers' intersection performance? Driving simulation and eye movement results. *International Journal of Human-Computer Studies, 66*(3), 132–144.

Campbell, J. L., Richard, C. M., Brown, J. L., & McCallum, M. (2007). *Crash warning system interfaces: Human factors insights and lessons learned* (Rep. DOT HS 810 697). Washington, DC: National Highway Traffic Safety Administration.

Chapanis, A. (1988). Some generalizations about generalization. *Human Factors, 33,* 253–267.

Chovan, J. D., Tijerina, L., Pierowicz, J. A., & Hendricks, D. L. (1994). *Examination of non-signalized intersection, straight crossing path crashes and potential IVHS countermeasures* (Rep. DOT HS 808 152). Cambridge, MA: National Highway Traffic Safety Administration.

Creaser, J. I., Rakauskas, M. E., Ward, N. J., Laberge, J. C., & Donath, M. (2007). Concept evaluation of intersection decision support system (IDS) interfaces to support drivers' gap acceptance decisions at rural stop-controlled intersections. *Transportation Research Part F: Traffic Psychology and Behavior, 10*(3), 208–228.

Dingus, T. A., Gellatly, A. W., & Reinach, S. J. (1997). Human computer interaction applications for intelligent transportation systems. In M. G. Helander, T. K. Landauer, & P. V. Prabhu (Eds.), *Handbook of human-computer interaction* (pp. 1259–1282). Amsterdam: North-Holland Publishing.

Evans, L., & Gerrish, P. H. (1996). Antilock brakes and risk of front and rear impact in two-vehicle crashes. *Accident Analysis & Prevention, 28*(3), 315–323.

Gibson, J. J., & Crooks, L. E. (1938). A theoretical field-analysis of automobile-driving. *The American Journal of Psychology, 1*(3), 453–471.

Green, P. (1994). *Measures and methods used to asses the safety and usability of driver information systems.* Final report (Rep. FHWA-RD-94-058). McLean, VA: Federal Highway Administration.

Hart, S. G., & Staveland, L. E. (1988). Development of NASA-TLX (Task Load Index): Results of empirical and theoretical research. In P. A. Hancock & N. Meshkati (Eds.), *Human mental workload* (pp. 139–184). Amsterdam: Elsevier Science.

Hjälmdahl, M., & Várhelyi, A. (2004). Speed regulation by in-car active accelerator pedal: Effects on driver behavior. *Transportation Research Part F: Traffic Psychology and Behavior, 7*, 77–94.

Ho, C., Reed, N., & Spence, C. (2006). Assessing the effectiveness of "intuitive" vibrotactile warning signals in preventing front-to-rear-end collisions in a driving simulator. *Accident Analysis & Prevention, 38*(5), 988–996.

Houser, A., Pierowicz, J., & Fuglewicz, D. (2005). *Concept of operations and voluntary operational requirements for lane departure warning system (LDWS) on-board commercial motor vehicles. Final Report* (Rep. FMCSA-MCRR-05-005). Washington, DC: Federal Motor Carrier Safety Administration.

Kantowitz, B. H. (1992). Selecting measures for human factors research. *Human Factors, 34*, 387–398.

Keifer, R. J., Casser, M. T., Flannagan, C. A., LeBlanc, D. J., Palmer, M. D., Deering, R. K., & Shulman, M. A. (2003). *Forward collision warning requirements project: Refining the CAMP crash alert timing approach by examining "last-second" braking and lane change maneuvers under various kinematic conditions* (Rep. DOT HS 809 574). Washington, DC: National Highway Traffic Safety Administration.

Keifer, R., LeBlanc, D., Palmer, M., Salinger, J., Deering, R., & Shulman, M. (1999). *Development and validation of functional definitions and evaluation procedures for collision warning/avoidance systems* (Rep. DOT HS 808 964). Washington, DC: National Highway Traffic Safety Administration.

Kuge, N., Yamamura, T., Boer, E. R., Ward, N. J., & Manser, M. P. (2006). *Study on driver's car following abilities based on an active haptic support function* (SAE Tech. Paper, 2006-01-0344). Warrendale, PA: Society of Automotive Engineers.

Laberge, J. C., Creaser, J. I., Rakauskas, M. E., Ward, N. J. (2006). Design of an intersection decision support (IDS) interface to reduce crashes at rural stop-controlled intersections. *Transportation Research Part C: Traffic Psychology and Behavior, 14*, 39–56.

Lee, J. D., McGehee, D. V., Brown, T. L., & Reyes, M. L. (2002). Collision warning timing, driver distraction, and driver response to imminent rear-end collisions in a high-fidelity driving simulator. *Human Factors, 44*, 314–334.

Lee, J., & Moray, N. (1992). Trust, control strategies and allocation of function in human-machine systems. *Ergonomics, 35*(10), 1243–1270.

Levinson, D., & Zhang, L. (2006). Ramp meters on trial: Evidence from the Twin Cities metering holiday. *Transportation Research Part A: Policy and Practice, 40*(10), 810–828.

Ma, R., & Kabar, D. B. (2005). Situation awareness and workload in driving while using adaptive cruise control and a cell phone. *International Journal of Industrial Ergonomics, 35*, 939–953.

Manser, M. P., & Hancock, P. A. (1996). Influence of approach angle on estimates of time-to-contact. *Ecological Psychology, 8*(1), 71–99.

Manser, M. P., Ward, N. J., Kuge, N., & Boer, E. R. (2004). Influence of a driver support system on situation awareness in information processing in response to lead vehicle braking. *Proceedings of the human factors and ergonomics society 48th annual meeting* (pp. 2359–2363). Santa Monica, CA: Human Factors and Ergonomics Society.

Manser, M.P., Ward, N.J., Kuge, N., & Boer, E.R. (2005). Driver behavioral adaptation in response to a novel haptic driver support system. *Proceedings of the human factors and ergonomics society 49th annual meeting*, USA, 1950–1954.

Mast, T. (1998). Introduction to ITS. In W. Barfield & T. A. Dingus (Eds.), *Human factors in intelligent transportation systems* (pp. xv–xxi). Mahwah, NJ: Lawrence Erlbaum Associates.

Matthews, G., Jones, D. M., & Chamberlain, A. G. (1990). Refining the measure of mood: The UWIST Mood Adjective Checklist. *British Journal of Psychology, 81*, 17–42.

McGehee, D. V., & Brown, T. L. (1998). *Examination of drivers' collision avoidance behavior in a lead vehicle stopped scenario using a front-to-rear-end collision warning system* (Rep. DTNH22-93-C-07326). Washington, DC: National Highway Traffic Safety Administration.

Najm, W. J., Koopmann, J. A., & Smith, D. L. (2001). *Analysis of crossing path crash countermeasure systems*. Proceedings of the 17th International Technical Conference on the Enhanced Safety of Vehicles, Amsterdam, the Netherlands.

Najm, W. G., & Smith, J. D. (2007). *Development of crash imminent test scenarios for integrated vehicle-based safety systems* (Rep. DOT HS 810 757). Washington, DC: National Highway Traffic Safety Administration.

National Highway Transportation Safety Administration. (2007). *Fatality Accident Reporting System*. Accessed at http://www-fars.nhtsa.dot.gov/Crashes/CrashesLocation.aspx

Olson, P. L. (2001). Driver perception-response time. In R. E. Dewar & P. L. Olson (Eds.), *Human factors in traffic safety* (pp. 43–76). Tucson, AZ: Lawyers & Judges Publishing Company.

Pierowicz, J., Jocoy, E., Lloyd, M., Bittner, A., & Pirson, B. (2000). *Intersection collision avoidance using ITS countermeasures, task 9 final report.* (Rep. DOT HS 809 171). Washington, DC: National Highway Traffic Safety Administration.

Rompe, K., Schindler, A., & Wallrich, M. (1987). *Advantages of an anti-wheel-lock system (ABS) for the average driver in difficult driving situations*. Proceedings of the 11th International Technical Conference on Experimental Safety Vehicles (pp. 442–448) (Rep. DOT HS 807 223). Washington, DC: National Highway Traffic Safety Administration.

Schiff, W., & Arnone, W. (1995). Perceiving and driving: Where parallel roads meet. In P. A. Hancock, J. Flach, J. Caird, & K. Vicente (Eds.), *Local applications of the ecological approach to human-machine systems* (Vol. 2, pp. 1–35). Hillsdale, NJ: Lawrence Erlbaum Associates.

Schmidt, R. A., & Lee, T. D. (2005). *Motor control and learning: A behavioral emphasis* (4th ed.). Champaign, IL: Human Kinetics Publishers.

Van der Laan, J. D., Heino, A., & De Waard, D. (1997). A simple procedure for the assessment of acceptance of advanced transport telematics. *Transportation Research C: Traffic Psychology and Behavior, 5*(1), 1–10.

Várhelyi, A., Hjälmdahl, M., Hydén, C., & Droskóczy, M. (2004). Effects of an active accelerator pedal on driver behavior and traffic safety after long-term use in urban areas. *Accident Analysis & Prevention, 36,* 729–737.

Ward, N. J., Manser, M. P., De Waard, D., Kuge, N., & Boer, E. (2003). Quantifying car following performance as a metric for primary and secondary (distraction) task load: Part A—Modification of task parameters. *Proceedings of the human factors and ergonomics society 47th annual meeting* (pp. 1870–1874). Santa Monica, CA: Human Factors and Ergonomics Society.

Weise, E. E., & Lee, J. D. (2004). Auditory alerts for in-vehicle information systems: The effects of temporal conflict and sound parameters on driver attitudes and performance. *Ergonomics, 47*(9), 965–986.

Zijlstra, F. R. H. (1993). *Efficiency in work behavior.* Unpublished doctoral dissertation, Technical University, Delft, The Netherlands.

41

Adaptive Behavior in the Simulator: Implications for Active Safety System Evaluation

Johan Engström
Volvo Technology Corporation

Mikael Ljung Aust
Volvo Car Corporation

Abstract

The Problem. Driving is, most of the time, a self-paced task where drivers proactively control the driving situation, based on their expectations of how things will develop in the near future. Crashes are typically associated with unexpected events where this type of proactive adaptation failed in one way or another. These types of scenarios are the main targets for active safety systems. In evaluation studies, drivers' responses to expected events may be qualitatively different from responses to similar, but unexpected, events. Hence, creating artificial active safety evaluation scenarios that truly represent the targeted real-world scenarios is a difficult challenge. ***Role of Driving Simulators.*** Driving simulators offer great possibilities to test active safety systems with real drivers in specific target scenarios under tight experimental control. However, in simulator studies, experimental control generally has to be traded against realism. The objective of this chapter is to address some key problems related to driver expectancy and associated adaptive behavior in the context of simulator-based active safety system evaluation. ***Key Results of Driving Simulator Studies.*** The chapter briefly reviews common types of adaptive driver strategies found in the literature and proposes a general conceptual framework for describing adaptive driver behavior. Based on this framework, some key challenges in dealing with these types of issues in simulator studies are identified and potential solutions discussed. ***Scenarios and Dependent Variables.*** Key variables representing adaptive driver behavior include the selection of speed, headway, and lane position as well as the allocation of attention and effort. It will never be possible to create artificial simulator scenarios for active safety evaluation that perfectly match their real-world counterparts, but there are several means that could be used to reduce the discrepancy. Problems with expectancy and resulting adaptive behavior may at least be partly overcome by various means to "trick" drivers into critical situations, several of which are addressed in the chapter. ***Platform Specificity and Equipment Limitations.*** The issues discussed in this chapter should apply across all types of driving simulator platforms. However, some of the proposed methods for tricking drivers into critical situations may require specific simulator features, such as a motion base.

41.1 Introduction

Active safety systems provide functions that intervene prior to a potential crash; for example, by alerting the driver to potential hazards or by partly automating some aspects of the driving task (e.g., headway maintenance, steering or braking). Several systems have already entered the market or are close to introduction, including forward collision warning, collision mitigation by braking, lane keeping aid, and drowsiness warning (there is yet little agreement on terminology in the field). A key challenge in active safety system development is the assessment of system performance. Because the most common purpose of active safety systems is to enhance driving performance in critical situations, the key issue here, beyond technical reliability, is to what extent a system succeeds in doing this. For example, will a forward collision warning (FCW) system improve the driver's braking performance if a lead vehicle suddenly brakes hard in front of him/her, and is that improvement sufficient to either prevent a crash or to mitigate its consequences? Moreover, while an active safety system may be beneficial in some targeted scenarios, it may also have adverse consequences in the very same scenarios or in other, non-targeted, scenarios. Such unintended effects may result from a wide range of factors including abuse of, or over-reliance on the system or an inadequate understanding of system functionality.

In the context of active safety system evaluation, driving simulators offer great possibilities to test systems with real drivers in specific target situations under tight experimental control. However, this requires the creation of driving scenarios which are sufficiently similar to the real-world target scenarios that they are intended to represent. This is a difficult challenge that is generally subject to a trade-off between experimental control and realism. Driving is, most of the time, a self-paced task where drivers proactively control the evolution of the driving situation based on their expectations of how things will develop in the near future (Näätänen & Summala, 1976; see Ranney, this book, chap. 9). Thus, if the driver somehow expects that a situation may turn critical, she/he will normally take action to avoid this, for example, by slowing down, increasing headway, focusing attention and/or mobilizing more effort.

Real-world crashes are typically associated with unexpected pre-crash events where this type of proactive adaptation failed in one way or another. This may be due to a sudden occurrence of a highly improbable event, but may also be related to inadequate perception and/or situation assessment by the driver, perhaps driven by erroneous expectancies (e.g., van Elslande & Faucher-Alberton, 1997; Räsänen & Summala, 1998). The recent 100-car naturalistic driving study indicates that many accidents, in particular rear-end collisions, may occur due to an "unfortunate" combination of an off-road eye glance and an unexpected event, such as a glance to the speedometer when a lead vehicle suddenly starts braking (Dingus et al., 2006; Klauer, Dingus, Neale, Sudweeks, & Ramsey, 2006).

These types of real-world scenarios, where expectancy-based adaptation fails, are the key target scenarios for many active safety systems. If the simulator scenario used for evaluation

does not truly represent the relevant aspects of the corresponding real-world scenario, the validity of the results could be called into question. For example, consider a simulator study set up to assess a forward collision warning (FCW) system's influence on brake response time (BRT) in a scenario where a lead vehicle suddenly brakes hard in front of the subject vehicle. If the lead vehicle braking event is anticipated by the test subjects, the measured BRTs may represent something else than BRTs for truly unexpected and urgent situations where emergency braking is needed to avoid the crash. First, BRTs to expected events are generally significantly shorter than for surprise events (Olson & Sivak, 1986; see the review in Green, 2000). Second, subjects' adaptive behavior, such as reduced speed, increased headway and increased attention and effort, may create a situation that is qualitatively different from the targeted real-world situation (Summala, 2000). Most importantly for the present discussion, the effect of the FCW measured in the simulator may be different from the effect that it would have had in the targeted real-world scenario. For example, if a subject participating in an experimental evaluation study suspects that a lead vehicle will brake and therefore proactively allocates attention to cope with the anticipated situation, a warning which would have made a critical difference in a truly unexpected scenario may be ignored, or even be perceived as a nuisance by the driver.

The objective of this chapter is to address some key problems related to driver expectancy and associated proactive adaptive behavior in the context of simulator-based active safety system evaluation. The following section reviews different types of adaptive driver behavior typically found in the literature. In Section 41.2, a general framework for conceptualizing these effects is proposed. Based on this framework, some key challenges in dealing with these issues are identified in Section 41.3, and potential solutions discussed. Finally, Section 41.4 provides some general conclusions and recommendations for further research.

41.2 Varieties of Adaptive Driver Behavior in the Simulator

A common general form of adaptive driver behavior is the adjustment of kinematic parameters such as speed, headway and lane position in response to variations in driving task demand. For example, it has been found that increasing lane and shoulder width, or improving road surface generally results in greater speeds (Organization for Economic Cooperation and Development [OECD], 1990).

Drivers also adapt to changes in performance characteristics of the vehicle. For example, Rumar, Berggrund, Jernberg and Ytterbom (1976) found that cars with studded tires were driven faster than cars with non-studded tires. Kinematic adaptation strategies are also commonly found in response to secondary task demand. For example, while operating a visually-demanding in-vehicle information system (IVIS), drivers tend to reduce speed (Curry, Hieatt, & Wilde, 1975; Antin, Dingus, Hulse, & Wierwille, 1990; Jamson & Merat, 2005; Engström, Johansson, & Östlund, 2005), presumably to compensate for the reduced visual input of the road scene. Rauch, Gradenegger and Krüger (2008) also demonstrated that drivers tend to slow down in *anticipation* of secondary task engagement and potential traffic conflicts.

In managing driving and non-driving-related tasks, drivers also employ various attention and task allocation strategies. Early evidence for this can be found in the classic occlusion study by Senders, Kristofferson, Levison, Dietrich and Ward (1967). The authors equipped their subjects with a visor mounted on a football helmet which occluded the subjects' vision. While driving, the subjects could open the visor for half a second whenever they considered it necessary and the self-chosen viewing time was assumed to represent the drivers' need to allocate visual attention to the road. The results demonstrated that drivers' viewing time strongly correlated with driving demand, represented, for example, in terms of vehicle speed. More recently, similar results have been obtained using direct eye movement measurement, where visual attention allocation can be represented in terms of the frequency and duration of glances. For example, in a study by Victor, Harbluk and Engström (2005), subjects drove a simulated rural road with both straight and curve sections while performing a visually-demanding secondary task. During curve negotiation (representing high driving demand), the mean duration of glances to the display was significantly reduced compared to straight road driving sections (representing low driving demand), indicating that relatively more attentional resources were allocated to the road when driving demand increased. Similar results have been found in the field, for example by Wikman, Nieminen and Summala (1998). The latter study also demonstrated that efficient visual time-sharing strategies develop with driving experience. In particular, although mean glance durations did not differ between experienced and novice drivers, it was found that overly long glances (>2 seconds) were much more common for novices. This result has been replicated in a recent simulator study by Chan, Pradhan, Pollatsek, Knodeler and Fisher (2008) for newly licensed U.S. drivers (16–17 year olds).

Drivers also selectively focus their visual attention in anticipation of critical events. In a simulator study by Lee, McGehee, Brown and Reyes (2002), with the aim of investigating the effects of a forward collision warning system in a rear-end collision scenario, subjects were repeatedly exposed to the same lead-vehicle braking scenario. It was found that accelerator pedal release times (the time from the lead vehicle brake onset until the accelerator pedal is released) for the second exposure were, on average, 430 ms faster than for the first exposure. A further analysis of eye movements indicated that this response enhancement was associated with an increase of visual attention to the forward road following the first exposure, presumably in anticipation of similar events.

Rauch et al. (2008) analyzed different strategies adopted by drivers to match secondary task performance to situational demands. In this study, the participants were *offered* to initiate the task (menu navigation in a simulated IVIS) at different points along a simulated route. This experimental paradigm could be expected to be more similar to real-world situations than the more common approach where subjects are instructed to perform secondary tasks at pre-defined locations along the route. The results showed that the subjects often adapted to potential conflicts either by rejecting the task altogether or by delaying task initiation. These strategies were found to be more effective in protecting against critical conflicts than strategies involving concurrent performance of the secondary task.

A key factor driving these types of proactive task- and attention-allocation strategies is *expectancy*, which in turn depends partly on the *predictability* of stimuli/events. For example, if the driver can predict with a high degree of certainty when and where a bicyclist will cross the road, visual attention can be temporally and spatially pre-allocated and there is a better chance that the bicyclist will be detected in time to avoid collision. However, if expectancies turn out to be wrong—for example if the bicyclist appears at another location than the expected (see e.g. Räsänen & Summala, 1998)—performance may even be worse than for "neutral" events which involve no specific expectations (as demonstrated in numerous laboratory studies, see for example Posner, 1980). Victor, Engström and Harbluk (2008) proposed the concept of a continuum ranging from fully predictable (e.g., frequently repeated artificial signals) to entirely unpredictable stimuli/events (e.g., an airplane emergency landing on a public road hitting a car).*

41.3 A Conceptual Framework for Understanding Adaptive Driver Behavior in Driving Simulators

In order to be able to describe and reason more systematically about adaptive driver behavior, a conceptual framework is needed. In this section, such a framework is proposed, largely based on the zero-risk theory by Näätänen and Summala (1976) and more recent accounts that also incorporate emotion as a key motivational determinant of adaptive driver behavior, in particular Summala (2007), Vaa (2007) and Fuller (2007). A more general application of this framework in the context of active safety requirement specification and evaluation is outlined in Ljung Aust and Engström (2010).

41.3.1 Adaptivity

Adaptivity is a hallmark of biological systems. From a general perspective, adaptivity can be viewed as a consequence of the need of an organism to sustain itself over longer periods of time in a continuously-changing environment (Pfeiffer & Scheier, 2000). Pfeiffer and Scheier (2000, based on Meyer & Guillot, 1991) proposed the general conceptualization of adaptivity illustrated in Figure 41.1 (see also the early cybernetics models of Wiener, 1948, and Ashby, 1956). The state of an organism and

---

* On February 7, 2008, a small DA40 Diamond Star had to perform an emergency landing on Riksväg 45 just outside of Gothenburg, Sweden. The landing gear hit the roof a car that was driving on the road. No one was critically injured.

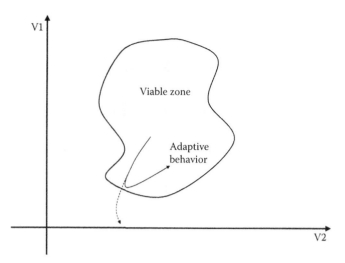

FIGURE 41.1 General conceptualization of adaptive behavior. (Adapted from Pfeifer, R., & Scheier, C. (2000). *Understanding Intelligence.* Cambridge, MA: MIT Press.)

its environment can be described in terms of a space defined by variables relevant for its survival (for example, blood sugar and body fluid). Within this space there is a *viable zone* that the organism must stay within to function properly and, ultimately, to survive. If the state approaches the boundary of the viable zone (e.g., the blood sugar level drops), actions need to be taken to remain within the zone (e.g., find food), which results in adaptive behavior. Two key pre-requisites for this type of self-regulatory mechanism to function efficiently are: (1) a proper *anticipation* and/or *perception* of the critical state, and (2) a *motivation* to do something about it. We would starve if we did not feel hunger and, if hunger was not an unpleasant feeling (and thus something we are motivated to avoid), we would be at greater risk of leaving the viable zone and starve to death. As suggested below, similar principles can be applied to the understanding of adaptive behavior in driving.

41.3.2 Adaptive Behavior in Driving

Driving is, in most situations, a self-paced activity. By analogy to the biological example above, adaptive behavior in driving can be viewed as a consequence of the need to maintain vehicle *control* in a continuously-changing driving environment. The key concept of control can be generally defined as the ability to direct and manage the development of events (Hollnagel & Woods, 2005) or, more specifically, the achievement and maintenance of a *goal state* in face of disturbances (Engström & Hollnagel, 2007). Critical control variables that need to be kept stable in basic vehicle handling (operational control in the terminology of Michon, 1985) include contact with the road surface, rate of optical flow information and available response time (Fuller, 2007; see also, Flach, Jagacinski, Smith, & McKenna, this book, chap. 43). However, driving also involves higher-level control tasks such as tactical maneuvering and strategic route selection (Michon, 1985; Alexander & Lunenfeldt, 1986).

In the hypothetical biological control system in Figure 41.1, the general goal state of the organism is to stay within the viable zone, that is, to keep blood sugar and body fluid with the boundaries required for survival. In the driving domain, a corresponding zone could be defined within a hypothetical space of driver, vehicle and environment (DVE) variables relevant for controlling the different sub-tasks involved in driving (steering, braking, navigating, etc.). Partly based on ideas developed over 70 years ago by Gibson and Crooks (1938), Summala (2007) suggests the concept of a *safety zone*, the boundary of which represents the limits of possible action, for example, the speed at which a vehicle starts to skid or roll over in a curve, or the minimum time-to-collision for which it is possible to avoid a crash by braking. The safety zone boundary thus separates driving situations into two categories: those which result in successful outcomes (i.e., maintaining or regaining control of the vehicle), and those which do not (see also Fajen, 2005). The safety zone is, on the one hand, determined by the properties of the environment (e.g., road surface conditions, curvature, lane width, etc.). However, it is also a function of the *action capabilities* of the agent. In driving, action capabilities depend on both driver and the vehicle properties. For example, the capability for overtaking depends partly on driver skills but is also stronger if one drives a Ferrari compared to a heavy truck. Thus, the "agent" in this case is best conceptualized as the joint driver-vehicle system (JDVS) rather than the driver or the vehicle in isolation (Vaa, 2007; Hollnagel & Woods, 2005).

While the safety zone is an objectively measurable property of the joint driver-vehicle system relative to the environment, the driver's control behavior is governed by the subjective experience of this relation. For example, the safety zone required for a following JDVS to stay clear of a braking lead vehicle is objectively determined by a number of parameters, such as the vehicles' braking capacity, the road surface conditions, the driver's level of attentiveness, and so on. However, the following driver's positioning and braking behavior depends on the his/her existing knowledge and/or subjective perception of relevant information; for example, optically specified time-to-collision (Lee, 1976), texture patterns providing cues about the road surface condition, knowledge about the vehicle's braking capability and/or estimation of one's own attentional state. More generally, the zero-risk theory, originally developed by Näätänen and Summala (1976), states that drivers normally control their behavior in order to maintain subjectively perceived *risk* at a zero level. Vaa (2007) has further developed this general idea, incorporating Damasio's (1994) concept of somatic markers. This model states that adaptive driver behavior is to a large extent governed by physiological reactions to threatening situations; that is, *emotions*, experienced by the driver in terms of unpleasant *feelings*. Somatic markers are emotional signals that represent the (positive or negative) value of actions and their outcomes. According to the model, somatic markers are mainly established through learned associations between stimuli and the effect of acting upon them (see also Fuller, 2007), but there may also be universal, innate, somatic markers such as reactions to looming stimuli and other basic reflexes (as indicated, for example, by the results of Schiff, Caviness and Gibson (1962) who found strong avoidance responses to looming stimuli in infant Rhesus monkeys).

The key idea behind Vaa's model is that, in the same way as organisms in general strive to avoid extreme hunger, drivers seek to avoid unpleasant feelings, associated, for example, with risk or threat. This is not to say that driver behavior is solely determined by emotions. In many situations, drivers' choice of action can still be understood as a product of more "cognitive" reasoning; for example, the application of explicit rules such as "keep a three-second headway to the vehicle in front" or a decision to change routes to reduce travel time in rush hour traffic. However, Vaa's model suggests that emotional mechanisms play a much more important role in shaping driver behavior than has traditionally been acknowledged. It may even be argued that the traditional distinction between emotion and cognition, or between "hot" and "cold" cognition (Abelson, 1963), is essentially an artefact deriving from the prevailing information processing metaphor (Engström & Hollnagel, 2007).

In line with Vaa's model, Summala (2007) proposes a generalization of the zero-risk model where the concept of risk is substituted for *discomfort* (similar ideas are also developed in Fuller, 2007). According to this model, the driver strives to maintain a state of zero discomfort. Discomfort includes feelings of immediate risk or threat (e.g., in a critical traffic situation), but may also be associated with the mobilization of *effort* to cope with excessive task demands (Hockey, 1997). The notion of zero discomfort can be conceptualized in terms of a *comfort zone* that the driver prefers to stay within. In order to ensure safe driving, the comfort zone should lie within the safety zone. The distance between the (subjective) comfort and the (objective) safety zone boundaries is the *safety margin* selected by the driver. Safety margins may be operationalized in terms of time-to-contact variables such as time-to-collision (van der Horst, 1990) and time-to-line crossing (Godthelp, Milgram, & Blaauw, 1984). The concepts of the safety zone, comfort zone and safety margin within a state space of DVE parameters are illustrated in Figure 41.2.

When the safety margin is violated, and this is perceived by the driver, a feeling of discomfort will be experienced, resulting in corrective actions—that is, adaptive behavior. As suggested by Fuller (2007), this will also influence future responses in similar situations by means of associative learning of somatic markers—strengthened associations between stimuli related to the particular situation and emotional reactions experienced as feelings of discomfort. Hence, to the extent that this learning is successful, the next time the driver experiences a similar situation, she/he will behave proactively to prevent the safety margin violation and maintain a state of zero-discomfort. In this way, proactive adaptive behavior is grounded in real-time interaction with the environment. The associations between situations and their potential outcomes established through

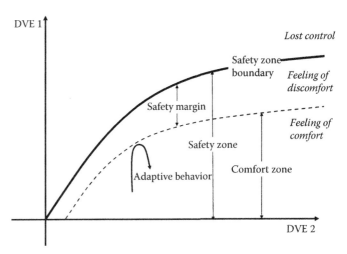

FIGURE 41.2 Illustration of the (objective) safety zone, the (subjective) comfort zone and the resulting safety margin between them in a space defined by two Driver-Vehicle Environment (DVE) parameters.

real-life driving will be brought into the simulator and shape the behavior of the test subjects, resulting in individually different adaptive strategies.

An important difference between the simulator and the real world is the lack of actual risk in the latter case as well as the different social context in the laboratory (Näätänen & Summala, 1976; see Mullen, Charlton, Devlin, & Bédard, this book, chap. 13 on psychological fidelity). However, as is further discussed below, simulated critical events may still be efficient in activating learned or innate somatic markers, thus yielding authentic initial avoidance responses, although these initial reactions may be subsequently suppressed by "intellectual" processes.

41.3.3 Optimizing Versus Satisficing

The key idea developed so far is that, as suggested by Näätänen and Summala (1976), driver behavior is driven by goals and motives which are balanced by inhibitory forces, conceptualized here in terms of discomfort (following Summala, 2007). It could be further suggested that the goals and motives could be satisfied by exploiting *opportunities for actions* afforded by the environment but also actively created by the driver. For example, if the driver is in a hurry and there is a large enough gap to oncoming traffic in the left lane (an opportunity for action), she/he can exploit this to satisfy the goal of overtaking the vehicle ahead. In terms of the present framework, the objective opportunities for actions are defined by the safety zone. The actual behavior is then determined by: (1) the *perceived* opportunities for actions, and (2) the degree of motivation to exploit these opportunities. The perceived opportunities for action are determined by the predicted discomfort associated with the action. Normally, if a possible action is predicted to violate safety margins or require excessive effort (and hence induce discomfort), it will not be performed unless there is a strong motivation for it or the driver has no other choice (in forced-paced conditions such as panic braking). If the driver is strongly motivated towards a specific goal, she/he would invest effort to stay as close to the comfort zone boundary as possible. For example, before overtaking, she/he would maintain the smallest headway to the lead vehicle that she/he is comfortable with. However, if there is little motivation to overtake, the driver may be quite indifferent to the exact headway adopted, as long as it does not violate his/her safety margins. The former case is an example of *optimizing* while the latter is an instance of *satisficing*. Optimizing refers to the case where one is actively attempting to optimize performance according to a specific criterion, which often (but not necessarily) requires effort. In control theoretic terms, this corresponds to minimizing error—that is, the difference between the actual and a target (or reference) state. A typical example of optimizing is the performance of classical manual tracking tasks where subjects are normally instructed to minimize the distance between the target and a manually controlled cursor.

By contrast, satisficing involves maintaining performance at a satisfactory ("good enough") level rather than minimizing error. For example, when driving in a wide lane at moderate speed, there is normally little reason to minimize deviations from the lane center. In such situations drivers could rather be expected to adopt a satisficing criterion, such as "don't get too close to the edge", while being less concerned about the exact lane position "error". Satisficing can be regarded as the normal mode of operation in everyday driving (Boer, 1999; Summala, 2007) as well as in human decision-making in general (Simon, 1955; Goodrich, Stirling, & Boer, 2000). In driving, optimizing normally takes place on specific occasions when the driver operates near or beyond his/her safety margins, such as when driving in adverse conditions, racing, multitasking or when trapped in a critical, forced-paced situation. It should be noted that satisficing and optimizing are not intended as all-or-none categories but rather as endpoints on a continuum.

In controlled experiments (both in simulators and in the real world), optimizing may also take place due to factors relating to the experimental situation. Even if not explicitly instructed to optimize performance, subjects may be motivated to invest additional effort in the driving- and/or secondary tasks in order to prove their abilities and/or to comply with what they think is the purpose of the experiment. For example, in the field study by Olson and Sivak (1986), brake response times for an unexpected event (a foam obstacle appearing while cresting a hill) were compared to brake response times in situations where the subjects knew that the obstacle would appear behind the crest. In the latter case, the subjects may have realized that the purpose of the study was to measure their response performance, thus attempting to respond as quickly as possible when detecting the obstacle (i.e., optimizing). However, in a corresponding real-life situation, it is possible that subjects would have waited to press the brake pedal until they felt it was necessary, that is, when encountering their comfort zone boundaries (i.e., satisficing). To obtain as "natural" behavior as possible in controlled experiments, these types of effects need to be taken into account.

In particular, the instructions given to the subjects, such as those in terms of task prioritization, are of key importance (see also, this book, chap. 13).

41.3.4 Applying the Framework to Conceptualize Adaptive Behavior

An example of the application of the proposed framework is given in Figure 41.3 and Figure 41.4. The trajectories in these figures represent the hypothetical temporal evolution of two DVE variables, *time headway to lead vehicle* and *attention to road*, in a potential rear-end collision scenario. In this case, the safety zone boundary represents the minimum time headway for which a collision could be avoided as a function of the level of attention to the forward roadway (based on the assumptions that brake response time increases with reduced attention and that both vehicles brake to maximum capacity). Figure 41.3 illustrates an example of successful proactive adaptive behavior in such a scenario. A driver is following a vehicle on a rural road and wants to enter a new destination on a navigation system. Based on perception of the environment and previous experience, the driver assumes that the lead vehicle will continue driving at the current speed and direction for the next few seconds (1). Thus, there is both motivation and perceived environmental opportunities for temporarily sharing attention between the road and the navigation system. In order to safely remain within the comfort zone (i.e., maintain zero discomfort), the driver increases the safety margin (headway) to compensate for the reduced attention to the road and protect against potential unexpected variance in the lead car's behavior. At (2) she notices a toll booth ahead, anticipates that the lead vehicle will brake, interrupts the destination entry task and redirects full attention to the road. At (3), the lead vehicle begins to slow down and our driver can brake smoothly and avoid collision without violating his/her safety margins, thus remaining in the comfort zone.

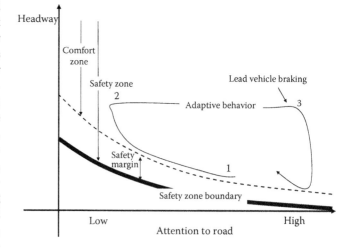

FIGURE 41.3 Example of successful adaptive behavior (see text for explanation).

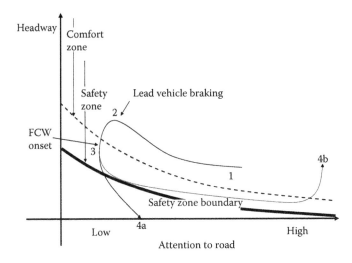

FIGURE 41.4 Examples of unsuccessful adaptation leading to crash and near-crash scenarios (see text for explanation).

Figure 41.4 illustrates a similar scenario, but with a different outcome. Here, the lead vehicle brakes hard and unexpectedly at (2) (e.g., due to an animal suddenly appearing on the road) while our driver looks towards the display. Since she/he is not prepared, and has attention directed elsewhere, the adopted safety margin is insufficient to avoid the rear-end collision at 4a.* However, in an alternative scenario, the driver is given a forward collision warning at (3). The warning redirects the driver's attention to the hazard and she/he is able to take corrective action (by emergency braking) just in time to stay within the safety zone and avoid the accident (4b). This represents a switch from a satisficing to an optimizing control mode; that is, a change in performance criterion from maintaining sufficient safety margins to getting back to the comfort zone as quickly as possible. The driver's safety margins were strongly violated, which produced a strong emotional response, experienced as feelings of discomfort. According to the somatic marker model outlined above, this will strengthen the associations between stimuli associated with this scenario and emotional signals (somatic markers), leading to different proactive behaviors (e.g., increased attention allocation to the road and/or the adoption of greater safety margins) the next time a similar situation is encountered. As outlined in the remainder of this chapter, this general conceptualization of adaptive behavior can be useful in the context of simulator-based evaluation studies, in particular for characterizing evaluation scenarios for active safety systems. In terms of the proposed framework, a key challenge in creating representative crash or near crash scenarios for simulator-based active safety evaluation is to trick drivers out of their comfort zones at precise moments in time (where leaving the comfort zone leads to a crash or a near-crash). Different concrete ways of achieving this are outlined in Section 41.4 below.

41.3.5 Summary of the Argument So Far

At this point, a brief summary of the argument so far may be useful. Adaptive strategies are commonly reported in the driver behavior literature. Such strategies are the result of, on the one hand, the active exploitation of opportunities for actions afforded by the environment, and/or created by the driver, to satisfy goals and motives and, on the other hand, the desire to maintain a state of zero discomfort. Adaptive strategies are generally proactive, driven by expectations on how driving scenarios will unfold, and include modification of driving situation kinematics to reduce driving demand and increase safety margins (e.g., by slowing down and increasing headway) as well as different attention and task allocation strategies (e.g., focusing attention on the road ahead when a critical event is expected).

The safety zone defines the objective opportunities for action and its boundary separates situations in two classes: those that result in successful outcomes and those that do not. In most conditions, drivers seek to maintain a feeling of zero discomfort. This may be conceptualized as a region in Driver-Vehicle-Environment state space, the comfort zone, which should lie within the safety zone to ensure safe driving. The difference between the zones is the safety margin chosen by the driver, which can be operationalized, for example, in terms of time-to-contact variables. The maintenance of safety margins relies on perception of relevant information, such as optically-specified expansion rates or road surface textures. However, it can also be based on more general knowledge, obtained, for example, through traffic messages warning for slippery roads or animals on the road ahead. In normal driving conditions, drivers generally operate in a satisficing mode, only investing as much effort that is needed to remain in the comfort zone. Excitatory forces, in terms of specific goals and motives, push the driver towards the comfort zone boundary. If these excitatory forces are strong, or the situation is forced-paced, operation shifts to an optimizing control mode. The mobilization of strong effort itself induces discomfort in terms of strain or fatigue and is thus generally avoided unless there is strong motivation for it or if the task is forced paced.

Failures in this adaptation process, for example, due to erroneous estimation of the safety zone boundary (including overestimation of one's own, or the vehicle's, action capabilities relative to the environment) and/or the sudden occurrence of an infrequent and unexpected event are key mechanisms behind real-world accidents. Hence, it is these types of situations that need to be re-created in simulator-based active safety evaluation. This challenge is addressed from a more practical perspective in the following section.

41.4 Issues for the Implementation of Critical Scenarios in the Simulator

How, then, can we create simulated evaluation scenarios for active safety systems that yield behavioral responses from subjects that are representative of responses in real crash and

---

* As mentioned above, this combination of inattention and unexpected events appears to be a common mechanism behind crashes in general, and rear-end crashes in particular, as suggested by the results from the 100-car study (Dingus et al., 2006).

near-crash situations? This section discusses different aspects of this problem and offers some concrete solutions, based on the conceptual framework outlined above.

41.4.1 Recreating Kinematic Conditions

The first challenge in evaluating an active safety system with respect to its ability to prevent a certain type of crash is to re-create the typical kinematic conditions of the real-world (target) pre-crash scenario. Conceptually, this amounts to obtaining certain values of the relevant V (vehicle) and E (environment) parameters in the DVE state space (e.g., a small headway to the lead vehicle combined with a certain speed on a certain type of road for a vehicle with certain braking characteristics).

A common approach for obtaining the targeted scenario kinematics is to couple target and subject vehicle behaviors. For example, consider a scenario where a target vehicle in the opposing lane suddenly turns left across the driver's path at an intersection where the driver has right-of-way. If the driver does not expect this to happen, she/he will not properly adjust safety margins and, thus, his/her comfort zone is likely to exceed the safety zone boundary when the target vehicle starts to turn. If the scenario succeeds, the driver has thus been tricked out of his/her comfort zone at the critical moment and has to perform an emergency maneuver to avoid collision (similar to the scenario illustrated in Figure 41.4, above). In order to create this kind of conflict, the target vehicle's speed can be controlled based on the subject's vehicle's approach speed so that the subject's vehicle always appears at the same distance (or time-to-contact) from the target vehicle when the latter begins to turn. However, coupling target and subject vehicle behaviors can become problematic if the desired conditions represent a too strong violation of the driver's safety margins; in other words, if it lies far beyond the comfort zone boundary. For example, subjects may end up in a "spiral" of mutual adaptation where compensatory speed reductions are responded to by a further speed reduction of the programed target vehicle ending in a stand-still of both vehicles and a very frustrated subject. Even worse, the coupling itself may become transparent to the subject. In early piloting of the intersection scenario just described (for a yet unpublished study involving the present authors) several subjects noticed that as they slowed down, so did the target vehicle. Discovery of this mirroring behavior of the target vehicle gave drivers a reported "eerie" feeling, and made them very cautious in their intersection approach which precluded the conflict from occurring. In this particular case, the problem was resolved by increasing the distance that the target vehicle had to travel before arriving at the intersection, hence effectively increasing the target vehicle's approach speed. This reduced the mirroring effect sufficiently to prevent the subjects from detecting the coupling. Another approach, employed by Lee et al. (2002), is to use a gradual coupling so that the target vehicle behavior does not become fully linked to the subject's behavior until just before the critical event. How this is best implemented naturally depends on the scenario at hand.

The general problem illustrated by these examples is that the desired target conditions often represent a state beyond most drivers' normal comfort zone boundary. If the target condition is somehow predictable, drivers typically adopt proactive strategies such as allocating more attention, slowing down and/or increasing headway to remain within their comfort zones (as illustrated in Figure 41.3 above). It may be suggested that this problem may be overcome by *forcing* drivers out of their comfort zones. For example, if a certain headway is targeted in a lead vehicle braking scenario, an adaptive cruise control (ACC) function can be used to exactly set the headway of the subject vehicle (as long as the driver cannot override the ACC settings). However, the validity of this approach can be strongly questioned. For example, if a subject is feeling uncomfortably close to a lead vehicle but is unable to slow down, the subject will most likely allocate extra attentional resources to cope with the situation, thus increasing the general response readiness. In addition, because the ACC relieves the driver of the longitudinal control task, further attentional resources become available, which can be re-allocated to the monitoring of potential hazards. Hence, despite precisely replicating the targeted kinematic conditions (E- and V-dimensions), this type of scenario must be regarded as fundamentally different from any realistic real-life scenario. More specifically, what is lacking is the consideration of driver state, that is, the D-dimension of the general DVE state space.

In particular, factors that govern the formation of expectations need to be taken into account in order to design representative target scenarios for active safety. In this sense, the task of the experimenter is similar to that of an illusionist, that is, to exploit human fallacies of perception and judgment to fool the drivers out of their comfort zones when a critical event occurs. The next section discusses some possible methods that can be used to achieve this.

41.4.2 Tricking Drivers out of Their Comfort Zones

The simulator experimenter can exploit a variety of means to manipulate drivers' expectancies and trick them into critical situations. This section discusses a number of approaches and their respective pros and cons.

41.4.2.1 Continuation Expectancy

One approach to trick drivers into critical conditions is to exploit the phenomenon known as *continuation expectancy*, which is a basic, and largely automated, tendency of humans to expect the continuation of the perceived events or states of affairs in the immediate future (Näätänen & Summala, 1976). This phenomenon was exploited in a recent simulator study with the main objective being to investigate the effects of repeated exposure to critical lead vehicle braking events on drivers' braking responses, and how this may interact with cognitive load (Engström, Ljung Aust, & Viström, in press). The study employed a cut-in scenario where a passing target vehicle, after cutting in at a very short time headway (0.9 seconds), continued to accelerate away from the subject vehicle before suddenly braking at a pre-determined time headway of 1.5 seconds. Driving was self-paced and subjects were instructed to maintain a speed of around 70 km/h (however, speeds above 80 km/h were prevented by means of a speed limiter). The relatively small initial time headway of 1.5 seconds could be expected to lie outside most drivers' comfort zone in a normal car following situation. The scenario was repeated six times and interleaved with "fake" scenarios (catch trials) where the target vehicle did not brake.

The analysis revealed that there was virtually no anticipatory speed compensation for the first scenario exposure; the average initial speed was very close to the instructed speed of 70 km/h and the difference between the speed just prior to scenario initiation and the speed at the moment of lead vehicle brake onset was close to zero. For the five remaining exposures, there was a small but significant reduction of speed as a function of repeated exposure to the scenario, with an average speed difference of 3.9 km/h between the first and the sixth exposure. This speed reduction is still surprisingly small, given the large number of exposures. In fact, several drivers did not exhibit any speed adaptation at all and some even increased speed just before the repeated event. Moreover, there were no strong signs of anticipatory braking responses. For example, the accelerator pedal was only released prior to the moment of POV brake onset for a few of the total 240 scenario exposures.

Hence, this cut-in scenario was thus quite successful in re-creating the desired initial kinematic conditions of the critical braking event, despite the fact that it was repeated and entirely self-paced. A possible explanation could be that drivers did not experience the short headway as threatening—that is, as intruding into their comfort zone, due to the fact that the lead vehicle was initially moving away from the subject vehicle and the drivers (unconsciously) expected it to continue that way. However, further analysis of maximum deceleration values (not reported in Engström et al., in press) revealed that the driver's braking response changed qualitatively with exposure, from "slamming the brake" in the first scenario to smoother deceleration patterns, with larger individual differences, in the later exposures. One factor behind this effect was probably an increased allocation of visual attention towards the lead vehicle as a result of expectancies developed during the repeated exposures. As also found by Lee et al. (2002), this significantly reduces response times, which gives the subjects more "room" for exploiting alternative braking strategies. However, the effect may also be due to perceptual-motor learning and calibration to simulator-specific characteristics such as visual cues for detecting the lead vehicle braking and the dynamic properties of the simulated vehicle (e.g., its braking capacity; see Fajen, 2005, for a theoretical account of perceptual learning and calibration). This, again, shows that the re-creation of kinematic conditions may not be sufficient to generate truly critical scenarios if the drivers have some degree of expectation and/or previous experience of the scenarios. Thus, the cut-in scenario was only fully successful for the first exposure despite the fact that the targeted kinematic conditions (in terms of initial speed and headway) were generally obtained for all exposures. The influence of repeated scenario exposure on driver response is further discussed below.

41.4.2.2 "Kinemagics"

In a simulated environment, it is possible to violate the laws of physics in order to create unexpected critical situations. An example of such "kinemagics" is the approach used by Ljung et al. (2007) originally described in Blommer, Curry, Kozak, Greenberg and Artz (2006). This study used a motion-base driving simulator to investigate differences between various lane departure warning (LDW) modalities, such as sound, light and haptics. The challenge was to get the subject vehicle to cross the lane boundary to trigger a warning. To achieve this, forced lane departures were generated while drivers were looking away from the road (performing a visual distraction task), by adding a small yaw deviation to the visual vehicle dynamics but not sending it to the motion platform. The vehicle thus started drifting out of lane, but drivers did not receive the normal motion feedback associated with lane drifting. Consequently, drivers only discovered the lane drift when looking back to the road which was generally too late to avoid the lane departure. The method proved successful in reliably generating authentic lane departures which would trigger warnings. A related method, employed in the same study, was used to create critical lead vehicle braking scenarios on a two-lane, median-divided highway. At the target headway, while the driver was performing a distraction task, a passing vehicle in the left lane suddenly "jumped" over to the right lane (i.e., performed an instantaneous lane change), and then braked in front of the subject vehicle. Neither of these two "kinemagic" tricks were discovered by any of the test subjects.

An explanation of why these types of "kinemagic" tricks seem to pass unnoticed by the subjects can be found in the literature on change blindness, where it has been demonstrated that people are surprisingly blind to changes when the transient is sufficiently masked (e.g., Rensink, O'Regan, & Clark, 1997). Hence, "kinemagic" tricks could be expected to work best when the transient change occurs outside the subject's field of view or when it is masked by other types of visual disruptions.

41.4.2.3 Distraction

As mentioned in the previous section, distraction can be used to move the driver's attention away from artificial pre-crash events such as "kinemagic" tricks. However, it can also be used in its own right to create critical scenarios for active safety evaluation. In this context, a main distinction could be made between *endogenously* (top-down, proactive) and *exogenously* (bottom-up, reactive) controlled distraction (Posner, 1980). The former includes cases where the driver is instructed to perform a secondary task, such as operating an IVIS, prior to the critical event. In this case, the allocation of attention is under the proactive control of the driver, who can adopt various strategies, for example, he/she can allocate more attention to the road if the driving demand and/or the anticipated risk increase (as illustrated in Figure 41.3 and demonstrated empirically by Victor et al., 2005; Wikman et al., 1998).

As reviewed above, performance of visually-demanding secondary tasks is also commonly associated with kinematic control strategies, such as slowing down (Curry et al., 1975; Antin et al., 1990; Jamson & Merat, 2005; Engström et al., 2005). Hence, it may be difficult to obtain a critical target critical scenario under conditions of endogenous attention control. Moreover, forcing subjects to perform secondary tasks in demanding situations may not yield behavior representative of real-world conditions where drivers generally decide themselves when to engage in distracting activities (Rauch et al., 2008).*

An important issue when using endogenously-controlled distraction tasks to generate critical events is to avoid the tasks becoming cues for the events. For example, if a repeated braking scenario is always preceded by a distraction task, the driver is likely to form strong specific expectancies for the scenario when the pattern is discovered. Thus, in order to prevent that specific expectancies are created by the secondary task, it is important to ensure that the two events are "decoupled" so that the distraction task occurs with similar probability in other, non-critical, scenarios.

Exogenously-controlled distraction, by contrast, refers to cases where the driver's attention is captured by an external, highly salient, event. Exogenous attention capture was employed in a simulator study with the purpose to evaluate forward collision warning strategies (reported in Lind, 2007). In this scenario, a helicopter flew across the landscape at low altitude just prior to a lead vehicle braking situation. This stimulus was highly successful in capturing the driver's attention, thus reliably yielding a truly critical situation. However, this type of scenario can obviously only be used one time per subject.

41.4.2.4 Using a Ruse to Reduce General Expectancies

A more general problem is that, when entering the laboratory, subjects are normally very interested in the purpose of the experiment (see Ranney, this book, chap. 9). One way to reduce general expectancies on what will happen in a study is to use some kind of ruse in the pre-test instructions. One can, for example, tell the subjects that the goal of the study is something unrelated to the critical situations of interest, such as data collection to be used for system development (e.g., Ljung et al., 2007; Engström et al., in press). The exact story naturally depends on the study at hand, and the ruse should, of course, be mentioned in the debriefing.

41.4.3 The Problem With Repeated Exposure

In order to obtain sufficient statistical power while keeping the number of subjects at a reasonable level, within-subject repetition of critical events is often deemed to be necessary in active safety evaluation studies. Indeed, the great majority of published work

in this area (e.g., Lee et al., 2002; Ljung et al., 2007; Cummings, Kilgore, Wang, Tijerina, & Kochhar, 2007; Scott & Gray, 2008) employ repeated events. However, as already touched upon, repeated exposure usually has strong effects on expectancy and resulting adaptive behavior. First, response times decrease radically with repeated exposure (e.g., Lee et al., 2002). Moreover, as found in the preliminary analysis of the cut-in braking scenario described above (Engström et al., in press), the qualitative nature of the braking response patterns changed with increased exposure, with drivers adopting increasingly smoother braking strategies as the expectancy for the events increased. The critical question for the present discussion is whether it is appropriate to use these repeated events in, for example, the evaluation of an FCW system.

Theoretically, the only way to obtain truly unexpected scenarios in simulator studies is to use data from a single exposure per subject. Thus there is always a trade-off between validity and cost/labor of the experiment. In this trade-off, the "accepted" level of expectancy (and resulting adaptive behavior) will depend on the specific system to be tested. If repeated events are to be used, a ground rule is that the target scenarios should at least be mixed with "fake" events (catch trials) and implemented in different settings, thus at least reducing expectancy from being specific—"this car will soon brake"—to a general increase in awareness that "cars in this study may sometimes brake". However, even if this may help reduce the adoption of proactive adaptive strategies, it does not guarantee that the responses to the repeated events are in any way representative of genuinely rare events.

It may be suggested that counterbalancing can be used as one way to get around the problem with repeated exposure. However, while counterbalancing may eliminate the effect of exposure between experimental conditions, it does not solve the more fundamental problem related to validity; that is, that responses to expected events may be *qualitatively* different from responses to unexpected events. Thus, for example, if the effect of a forward collision warning system is measured as the average response time for a series of repeated, and counterbalanced, rear-end collision scenarios, the result may not be ecologically valid if only the first exposure generates a response that is truly representative of the targeted real-world scenario.

It should also be noted that the severity of this problem depends on the purpose of the study. If the goal of the study is to assess specific aspects such as the acceptance of different types of warnings or the efficiency of different stimulus-response combinations in generating fast responses, the use of repeated events may be less damaging than if the study has the more general goal to assess the effects on driving performance in critical situations.

41.4.4 Differences in Actual Risk and Motivation Between Simulators and the Real World

Another important issue, only briefly addressed so far, is the basic problem that simulated driving does not impose any real risk to the subjects. As also suggested in this book, chap. 13

---

* However, the actual level of self-pacing may vary somewhat between different types of secondary tasks. For example, answering the phone is somewhat less self-paced than operating the radio and many people find it hard to resist answering an incoming phone call, even in very demanding driving situations (Green, 2004).

by Mullen et al., this may lead subjects to exhibit more risky behaviors in the simulator than in the real world. For example, Victor et al. (2005) found longer off-road glance durations during visual time-sharing with a secondary task while driving on a simulated motorway compared to a comparable real motorway. In line with this, Engström et al. (2005), in an analysis of the same motorway data, found degraded lane keeping performance during visual time-sharing in a fixed-base simulator but no effect in the field (the effect was marginally significant in a moving-base simulator). In terms of the present framework, this could be explained in terms of the adoption of greater safety margins (i.e., smaller comfort zones) in the field condition. However, other studies did not find any such differences between the simulator and in the field. For example, in two studies with the purpose to evaluate the effectiveness of a training program to improve drivers' anticipatory visual scanning for hazards, the probability of fixating an area containing a hidden potential hazard was almost identical between the simulator (Garay-Vega, Fisher, & Pollatsek, 2007) and the field (Pradhan, Pollatsek, Knodler, & Fisher, 2009). However, a key difference between the visual time-sharing study (Victor et al., 2005) and the scanning behavior studies (Garay-Vega et al., 2007; Pradhan et al., 2009) is that, in the former case, drivers could be expected to have operated close to their comfort zone boundaries, thus adopting a more optimizing control mode, especially in the field. By contrast, in the visual scanning studies, there was no immediate threat and thus the subjects probably operated in a more satisficing mode, further from the comfort zone boundary. Hence, even if the subjects in the simulator visual scanning study (Garay-Vega et al., 2007) indeed adopted smaller safety margins than the subjects in the field study (Pradhan et al., 2009), this difference would not show up in their scanning behavior. Thus, it could be predicted that performance differences between simulator and field conditions should mainly occur in more demanding or critical scenarios where drivers are "balancing" on their comfort zone boundary (i.e., optimizing). However, further empirical work is needed to further test this hypothesis.

The general issue with potential differences in motivation and perceived risk between simulators and the real world has led some authors to suggest that driving simulators are generally of limited value in driver behavior research (e.g., Summala, 2007). If, as suggested here, driver behavior is driven by, on the one hand, motivations and, on the other, feelings of risk/discomfort, it could be argued that behavioral data obtained in simulators could never be representative of what drivers would do and not do in the real world. This is certainly true to some extent. While motivations can be manipulated to some extent by means of incentives (see e.g., Stein, Allen, & Schwartz, 1978; see Ranney, this book, chap. 9), driving simulators can never recreate the entire motivational or emotional context of real driving situations, or the social situations occurring inside vehicles (see Ranney, this book, chap. 9; Näätänen & Summala, 1976).

However, the ability of artificial stimuli to evoke real emotional responses should not be underestimated. For example,

Bottomore (1999) cites a number of reports from the early days of cinema on the so called "train effect", which refers to "an anxious or panicky reaction to films of approaching vehicles" (p. 177). For example, consider the following report from 1896 quoted by Botttomore (translated from French and originally appearing in Le Mémorial de Vosges, 25 June 1896, quoted in Blaise Aurora, Histoire du Cinema en Lorraine: du cinematographe au cinema forain, 1896–1914):

> The train is the most stirring spectacle you can image. On the platform the porters and passengers are waiting. It's here! You see it arrive, get bigger, gain speed; it seems to come out of the screen, to race towards you. Real flesh-and-blood women recoil in horror, the reality is so striking! The train is the star of the show, if a train can be a star (p. 187).

Although, as pointed out by Bottomore (1999), these stories are probably somewhat exaggerated, the multitude of reports on the train effect suggests that strong emotional reactions did indeed occur. Today, the cinema audience does not react this strongly to approaching trains, presumably due to a combination of habituation to movie stimuli and "intellectual suppression". However, everybody who has tried a virtual rollercoaster knows that the only way to suppress the feeling in your stomach when going down the slopes is to close your eyes, despite knowing, intellectually, that you are sitting on a chair in a theater. It could thus be argued that the difference between reality and virtual reality is, in a sense, an intellectual one, not directly linked to "lower level", automated, perceptual and emotional processes that govern driver responses in critical situations. Artificial stimuli, such as looming objects, may thus initially trigger learned (or innate) avoidance reactions and somatic markers, even if these reactions are "intellectually suppressed" in the next moment. Interestingly, there is some evidence that such "intellectual suppression" of looming responses is less effective under conditions of distraction (King, Dykeman, Redgrave, & Dean, 1992). Hence, especially in the types of unexpected, near crash, situations that are the focus in active safety evaluation, initial behavioral responses to critical events obtained in simulators may be expected to be at least qualitatively similar to their real-world counterparts (given that the events are truly surprising). In order to investigate these issues further, naturalistic driving data, for example, from the 100-car study (Dingus et al., 2006), could be useful. More specifically, drivers' responses obtained in simulated unexpected pre-crash situations could be compared to responses to unexpected events captured on video in naturalistic driving, to see if they are qualitatively similar or if they differ in some fundamental way.

41.4.5 Indicators of Criticality

How do we know that an experimental simulator scenario is truly critical? One obvious way is to look at the actual outcomes of the drivers' avoidance responses, such as the number of crashes that occurred. However, a further indicator of whether the scenario succeeded in tricking the driver out of his comfort zone is to look at the between-subject behavioral variance. Low variance in self-paced conditions is indicative of a critical scenario where the majority of the drivers were forced to adopt an optimizing strategy to regain control once they realized the threat. By contrast, a large between-subject variance indicates that subjects did not experience the scenario as particularly critical, and thus found "room" for adopting individual strategies, that is, operating at least partly in a satisficing mode. An example of the latter was found in the analysis of lead vehicle braking study cited above (Engström et al., in review), where, after repeated exposure, some subjects began to brake smoother while others continued to perform the same emergency braking maneuver as for the first exposure, leading to a large between-subject variance in maximum deceleration values for the later scenario exposures. However, it should be noted that behavioral variance may also be high in certain critical scenarios where drivers have different avoidance options (e.g., steering or braking). Hence this principle may not be universally applicable and should be used with caution.

41.5 Conclusions

Critical traffic situations typically occur when adaptation fails—for example, due to infrequent events or misperception/misjudgement of the driving situation. These types of situations are the key targets for active safety systems. While the goal of the road safety professional is to prevent such adaptation failures, the challenge for the active safety evaluation experimentalist is quite the opposite; that is, to make these failures happen in the laboratory. In terms of the present framework, this entails tricking the drivers out of their comfort zones without them realizing it until a precise moment in time. To this end the experimenter, like the illusionist, can exploit different psychological phenomena such as continuation expectancy, change blindness and distraction, as well as utilize the possibilities to violate the laws of physics, that is, using "kinemagic" tricks in the simulator.

It will never be possible to create critical simulator scenarios that perfectly match their real-world counterparts, especially if repeated exposure has to be used. However, the issues of expectancy and resulting adaptive behavior need to be accounted for as far as possible when designing scenarios for active safety evaluation as well as when interpreting the results. What constitutes an "acceptable" level of expectancy depends on the nature of the study. Further empirical research, using, for instance, naturalistic driving data, is needed to determine the extent to which behavioral responses obtained in specific simulated scenarios could be said to adequately represent real-world behavior.

Scenarios for simulator-based active safety evaluation also need to be grounded in the analysis of real accidents. Even if a simulated scenario reliably tricks drivers out of their comfort zones, the ecological validity of the results may still be questioned if the scenario cannot be related to a real-world pre-crash scenario known to produce accidents. For example, the cut-in scenario described in Section 41.3 (Engström et al., in review), while being highly successful in generating critical events [at least for the first exposure], does not appear to be frequently represented in crash statistics (e.g. Najm, Smith, & Yanagisawa, 2007). Pre-crash typologies such as the "37 crashes" identified by Najm et al. (2007) based on statistical accident data, provide a useful starting point here. However, in order to obtain a more detailed understanding of the accident-producing mechanisms in the pre-crash phase, other methods, such as driver-centered methods for in-depth accident analysis (e.g., Ljung, Fagerlind, Lövsund, & Sandin, 2007) and large-scale naturalistic driving studies (Dingus et al., 2006), are needed. An important future research task is to link the results from such studies to the development of methodologies for simulator-based evaluation. Furthermore, although simulator scenarios always have to be adapted to the specific study at hand, there is a strong need for the development of common approaches and definitions of "standardized", validated, scenarios for active safety evaluation. In particular, this is needed to ensure that critical parameters, such as expectancy, are taken into account so that the results from different simulator studies can be adequately compared.

A further way to address the problems with expectancy and adaptive behavior is to complement empirical studies with computer simulations in which the human driver is replaced by a simulated driver. At least in theory, simulations of driver behavior in critical target scenarios can be parameterized to represent the case of an unprepared driver. Since the simulated driver will have no "memory" of previous exposures unless this is explicitly included in the driver model, any scenario can be run infinitely many times to test, for example, different system parameter settings (see e.g., Lee et al., 2002, for initial work in this direction). However, this requires sophisticated driver models which are not yet in place and further research is needed before computer simulations will be able to faithfully represent driver behavior in critical driving situations.

Finally, the proposed conceptual framework, in particular the somatic marker model proposed in different versions by Vaa (2007), Summala (2007) and Fuller (2007), also has several implications beyond scenario definition in driving simulators. For example, if proactive avoidance behavior is largely governed by emotional associations established through real-world encounters with critical events, this would have implications for driver training. In particular, the explicit knowledge gained through regular driving lessons, or specific training programs such as the Risk Awareness and Perception Training (RAPT) program (e.g., Pradhan et al., 2009), may serve the important purpose of providing novice drivers with useful heuristics for safe driving (e.g., "adopt a three second headway" or "always check for hazards at blind intersections"). However, according to the somatic marker model, this explicit knowledge needs to be grounded in real-world sensory-motor, and emotional, experience in order to persist in the long-term and become a natural part of drivers' "behavioral repertoire". More research is clearly needed to test how this and other predictions of the somatic marker model hold up against empirical evidence.

Key Points

- Driving simulators offer great possibilities to test active safety systems with real drivers in specific target scenarios under tight experimental control.
- Proactive adaptive driver behavior resulting from the development of expectancies poses challenges for the design of truly representative active safety evaluation scenarios in driving simulators.
- Adaptive driver behavior can be understood in terms of a desire to maintain certain safety margins, which could be represented as the difference between an objective safety zone and a subjectively defined comfort zone.
- Drivers may operate as optimizers (minimizing error) or satisficers (striving to achieve "good enough" performance), depending on, for example, the criticality of the driving situation and/or other motives to invest additional effort in the driving task (e.g., experimental instructions).
- Re-creating critical target scenarios in driving simulators entails tricking drivers out of their comfort zones at particular moments in time. To this end, the experimenter, like an illusionist, can exploit different psychological phenomena such as continuation expectancy, change blindness, and distraction. The experimenter can also utilize "kinemagics", i.e., the possibility of violating the laws of physics in the simulator.
- Perhaps the most difficult challenge in the context of active safety evaluation is driver expectation induced by repeated exposure to critical events.
- Further work is needed for a better understanding of possible quantitative and qualitative differences between simulated and real-world scenarios, possibly based on naturalistic driving data.
- Computer simulation, with a mathematical driver model replacing the real driver, is a promising future complement to empirical methods for active safety evaluation.

Keywords: Active Safety Systems, Adaptive Driver Behavior, Evaluation, Expectancy, Simulator Scenario Design

Acknowledgments

The authors would like to thank Hans-Erik Pettersson, John Lee, Don Fisher and Jeff Caird for commenting on draft versions of this chapter. We would also like to thank Trent Victor and Marco Dozza for fruitful discussions on some of the key issues addressed by the chapter.

Glossary

Comfort zone: The subjectively defined region in Driver-Vehicle-Environment state space where the driver experiences no discomfort.

Control: Safety zone, Comfort zone, Safety margin, Optimizing, Satisficing.

Control: The achievement and maintenance of a goal state in the face of disturbances.

Optimizing: Minimizing performance error relative to a target state.

Safety margin: The distance in Driver-Vehicle-Environment state space between the safety zone boundary and the comfort zone boundary.

Safety zone: The region in the Driver-Vehicle-Environment state space that objectively defines the limits of safe action.

Satisficing: Maintaining performance within acceptable boundaries.

Key Readings

Fajen, B. R. (2005). Perceiving possibilities for action: On the necessity of calibration and perceptual learning for the visual guidance of action. *Perception, 34,* 717–740.

Fuller, R. (2007). Motivational determinants of control in the driving task. In P. C. Cacciabue (Ed.), *Modeling driver behavior in automotive environments: Critical issues in driver interactions with intelligent transport systems* (pp. 165–188). London: Springer.

Pfeifer, R., & Scheier, C. (2000). *Understanding intelligence.* Cambridge, MA: MIT Press.

Summala, H. (2007). Towards understanding motivational and emotional factors in driver behavior: Comfort through satisficing. In P. C. Cacciabue (Ed.), *Modeling driver behavior in automotive environments: Critical issues in driver interactions with intelligent transport systems.* London: Springer.

Vaa, T. (2007). Modeling driver behavior on the basis of emotions and feelings. In P. C. Cacciabue (Ed.), *Modeling driver behavior in automotive environments: Critical issues in driver interactions with intelligent transport systems.* London: Springer.

References

Abelson, R. P. (1963). Computer simulation of "hot" cognition. In S. S. Tomkins & S. Messick (Eds.), *Computer simulation of personality.* New York: John Wiley & Sons.

Alexander, G. J., & Lunenfeld, H. (1986). *Driver expectancy in highway design and traffic operations* (Rep. FHWA-TO-86-1). US Department of Transportation.

Antin, J. F., Dingus, T. A., Hulse, M. C., & Wierwille, W. W. (1990). An evaluation of the effectiveness of an automobile moving-map navigation display. *International Journal of Man Machine Studies, 33,* 581–594.

Ashby, W. (1956). *An introduction to cybernetics.* New York: John Wiley & Sons.

Blommer, M., Curry, R., Kozak, K., Greenberg, J., & Artz, B. (2006). *Implementation of controlled lane departures and analysis of simulator sickness for a drowsy driver study.* Proceedings of Driving Simulation Conference (DSC) 2006, Paris, France.

Boer, E. R. (1999). Car following from the driver's perspective. *Transportation Research Part F: Traffic Psychology and Behavior, 2*(4), 201–206.

Bottomore, S. (1999). The panicking audience?: Early cinema and the "train effect". *Historical Journal of Film, Radio and Television, 19*(2), 177–216.

Chan, E., Pradhan, A. K., Pollatsek, A., Knodler, M. A., & Fisher, D. L. (2008). *Evaluation on a driving simulator of the effect of distractions inside and outside the vehicle on drivers' eye movements*. Technical report, Human Performance Laboratory, Department of Mechanical and Industrial Engineering, University of Massachusetts, Amherst.

Cummings, M. L., Kilgore, R. M., Wang, E., Tijerina, L., & Kochhar, D. S. (2007). Effects of single versus multiple warnings on driver performance. *Human Factors, 49*(6), 1097–1106.

Curry, G. A., Hieatt, D. J., & Wilde, G. J. S. (1975). *Task load in the motor vehicle operator: A comparative study of assessment procedures*. Ottawa, Ontario: Ministry of Transport, Road and Motor Vehicle Traffic Safety Branch.

Damasio, A. R. (1994). *Descarte's error: Emotion, reason and the human brain*. New York: Groset/Putnam.

Dingus, T. A., Klauer, S. G., Neale, V. L., Petersen, A., Lee, S. E., Sudweeks, J., . . . Knipling, R. R. (2006). *The 100-car naturalistic driving study phase II—Results of the 100-car field experiment* (Rep. HS 810 593). US Department of Transportation, National Highway Traffic Safety Administration.

Engström, J., & Hollnagel, E. (2007). A general conceptual framework for modeling behavioral effects of driver support functions. In P. C. Cacciabue (Ed.), *Modeling driver behavior in automotive environments: Critical issues in driver interactions with intelligent transport systems*. London, Springer.

Engström, J., Johansson, E., & Östlund, J. (2005). Effects of visual and cognitive load in real and simulated motorway driving. *Transportation Research Part F, 8*(2), 97–120. doi:10.1016/j.trf.2005.04.012

Engström, J., Ljung Aust, M., & Viström, M. (in press). Effects of working memory load and repeated scenario exposure on emergency braking performance. To appear in *Human Factors*.

Fajen, B. R. (2005). Perceiving possibilities for action: On the necessity of calibration and perceptual learning for the visual guidance of action. *Perception, 34*, 717–740.

Fuller, R. (2007). Motivational determinants of control in the driving task. In P. C. Cacciabue (Ed.), *Modeling driver behavior in automotive environments: Critical issues in driver interactions with intelligent transport systems* (pp. 165–188). London, Springer.

Garay-Vega, L., Fisher, D. L., & Pollatsek, A. (2007, January). *Hazard anticipation of novice and experienced drivers: Empirical evaluation on a driving simulator in daytime and nighttime conditions*. Paper presented at the 86th Transportation Research Board (TRB) Annual Meeting. Washington, DC: TRB, National Research Council.

Gibson, J. J., & Crooks, L. E. (1938). A theoretical field-analysis of automobile-driving. *The American Journal of Psychology, 51*(3), 453–471.

Godthelp, H., Milgram, P., & Blaauw, J. (1984). The development of a time-related measure to describe driving strategy. *Human Factors, 26*(3), 257–268.

Goodrich, M. A., Stirling, W. C., & Boer, E. R. (2000). Satisficing revisited. *Minds and Machines, 10*(1), 79–109.

Green, M. (2000). "How long does it take to stop?" Methodological analysis of driver perception-brake times. *Transportation Human Factors, 2*(3), 195–216.

Green, P. (2004). *Driver distraction, telematics design, and workload managers: Safety issues and solutions* (SAE Paper 2004-21-0022). Warrendale, PA: Society of Automotive Engineers.

Hockey, G. R. J. (1997). Compensatory control in the regulation of human performance under stress and high workload: A cognitive energetical framework. *Biological Psychology, 45*, 73–93.

Hollnagel, E., & Woods, D. D. (2005). *Joint cognitive systems: Foundations of cognitive systems engineering*. Boca Raton, FL: Taylor & Francis/CRC Press.

Jamson, H., & Merat, N. (2005). Surrogate in-vehicle information systems and driver behavior: Effects of visual and cognitive load in simulated rural road driving. *Transportation Research Part F: Traffic Psychology and Behavior, 8*(2), 79–96. doi:10.1016/j.trf.2005.04.002

King, S. M., Dykeman, C., Redgrave, P., & Dean, P. (1992). Use of a distracting task to obtain defensive head movements to looming visual stimuli by human adults in a laboratory setting. *Perception, 21*(2), 245–259.

Klauer, S. G., Dingus, T. A., Neale, V. L., Sudweeks, J. D., & Ramsey, D. J. (2006). *The impact of driver inattention on near-crash/crash risk: An analysis using the 100-car naturalistic driving study data* (Rep: DOT HS 810 59). US Department of Transportation.

Lee, D. N. (1976). A theory of visual control of braking based on information about time-to-collision. *Perception, 5*, 437–459.

Lee, J. D., McGehee, D. V., Brown, T., & Reyes, M. L. (2002). Collision warning, driver distraction, and driver response to rear-end collisions in a high-fidelity driving simulator. *Human Factors, 44*(2), 314–334.

Lind, H. (2007). *An efficient visual forward collision warning display for vehicles* (Paper 2007-01-1105). Society of Automotive Engineers World Congress. Detroit, MI.

Ljung, M., Blommer, M., Curry, R., Artz, B., Greenberg, J., Kochhar, D., . . . Jakobsson, L. (2007). *The influence of study design in results in HMI testing for active safety* (Paper 07-0383). Enhanced Safety Vehicles (ESV) Conference, Lyon, France.

Ljung, M., Fagerlind, H., Lövsund, P., & Sandin, J. (2007). Accident investigations for active safety at CHALMERS—new demands require new methodologies. *Vehicle System Dynamics, 45*(10), 881–894.

Ljung Aust, M., & Engström, J. (2010). A conceptual framework for requirement specification and evaluation of active safety functions. *Theoretical Issues in Ergonomics Science*. First published on: 15 June 2010 (iFirst).

Meyer, J.-A., & Guillot, A. (1991). Simulation of adaptive behavior in animats: Review and prospects. In J.-A. Meyer & S. W. Wilson (Eds.), From animals to animats: *Proceedings of the first international conference on simulation of adaptive behavior* (pp. 2–14). Cambridge, MA: MIT Press.

Michon, J. A. (1985). A critical review of driver behavior models: What do we know? What should we do? In L. A. Evans & R. C. Schwing (Eds.), *Human behavior and traffic safety* (pp. 487–525). New York: Plenum Press.

Näätänen, R., & Summala, H. (1976). *Road user behavior and traffic accidents*. Amsterdam: North Holland.

Najm, W. G., Smith, J. D., & Yanagisawa, M. (2007). *Pre-crash scenario typology for crash avoidance research* (Rep. DOT HS 810 767). US Department of Transportation, National Highway Traffic Safety Administration.

Olson, P. L., & Sivak, M. (1986). Perception-response time to unexpected roadway hazards. *Human Factors, 28*(1), 91–96.

Organization of Economic Cooperation and Development. (1990). *Behavioral adaptations to changes in the road transport system.* Prepared by an OECD Expert Group, Road Transport Research Program.

Pfeifer, R., & Scheier, C. (2000). *Understanding intelligence.* Cambridge, MA: MIT Press.

Posner, M. I. (1980). Orienting of attention. *Quarterly Journal of Experimental Psychology, 32*, 3–25.

Pradhan, A. K., Pollatsek, A., Knodler, M., & Fisher, D. L. (2009). Can younger drivers scan for information that will reduce their risk in roadway traffic scenarios that are hard to identify as hazardous? *Ergonomics, 52*(6), 657–673.

Rauch, N., Gradenegger, B., & Krüger, H-P. (2008). *Identifying user strategies for interaction with in-vehicle information systems while driving in a simulator.* Proceedings of the European Conference on Human Centered Design for Intelligent Transport Systems, Lyon, France.

Räsänen, M., & Summala, H. (1998). Attention and expectation problems in bicycle–car collisions: An in-depth study. *Accident Analysis & Prevention, 30*, 657–666.

Rensink, R. A., O'Regan, J. K., & Clark, J. J. (1997). To see or not to see: The need for attention to perceive changes in scenes. *Psychological Science, 8*, 368–73.

Rumar, K., Berggrund, U., Jernberg, P., & Ytterbom, U. (1976). Driver reaction to a technical safety measure—Studded tires. *Human Factors, 18*, 443–454.

Schiff, W., Caviness, J. A., & Gibson, J. J. (1962). Persistent fear responses in rhesus monkeys to the optical stimulus of "looming". *Science, 136*, 982–963.

Scott, J. J., & Gray, R. A. (2008). Comparison of tactile, visual, and auditory warnings for rear-end collision prevention in simulated driving. *Human Factors, 50*(2), 264–275.

Senders, J. W., Kristofferson, A. B., Levison, W. H., Dietrich, C. W., & Ward, J. L. (1967). The attentional demand of automobile driving. *Highway Research Record, 195*, 15–33.

Simon, H. A. (1955). A behavioral model of rational choice. *The Quarterly Journal of Economics, LXIX*, 99–118.

Stein, A. C., Allen, R. W., & Schwartz, S. H. (1978). Use of reward-penalty structures in human experimentation. *14th Annual conference on manual control* (pp. 167–178). NASA Ames Research Center.

Summala, H. (2000). Brake reaction times and driver behavior analysis. *Transportation Human Factors, 2*(3), 217–226.

Summala, H. (2007). Towards understanding motivational and emotional factors in driver behavior: Comfort through satisficing. In P. C. Cacciabue (Ed.), *Modeling driver behavior in automotive environments: Critical issues in driver interactions with intelligent transport systems.* London: Springer.

Vaa, T. (2007). Modeling driver behavior on the basis of emotions and feelings. In P. C. Cacciabue (Ed.), *Modeling driver behavior in automotive environments: Critical issues in driver interactions with intelligent transport systems.* London: Springer.

van der Horst, R. (1990). *A time-based analysis of road user behavior in normal and critical encounters.* Doctoral dissertation, Delft University of Technology, The Netherlands.

van Elslande, P., & Faucher-Alberton, L. (1997). When expectancies become certainties: A potential adverse effect of experience. In T. Rothengatter, & E. Vanya (Eds.), *Traffic and transport psychology.* Tarrytown, NY: Pergamon Press.

Victor, T., Engström, J., & Harbluk, J. (2008). Distraction assessment methods based on visual behavior and event detection. In M. Regan, J. Lee, & K. Young (Eds.), *Driver distraction: Theory, effects and mitigation.* Boca Raton, FL: CRC Press.

Victor, T. W., Harbluk, J. L., & Engström, J. (2005). Sensitivity of eye-movement measures to in-vehicle task difficulty. *Transportation Research Part F: Psychology and Behavior, 8*(2), 167–190.

Wiener, N. (1948). *Cybernetics, or, control and communication in the animal and the machine.* New York: Wiley.

Wikman, A., Nieminen, T., & Summala, H. (1998). Driving Experience and time-sharing during in-car tasks on roads of different width. *Ergonomics, 42*(3), 358–372.

42

Cognitive Architectures for Modeling Driver Behavior

Dario D. Salvucci

Drexel University

Abstract

The Problem. The driving task involves a host of relevant cognitive, perceptual, and motor abilities. Cognitive architectures—frameworks for modeling human cognition and behavior—have evolved to integrate known theories of human cognition into unified theories that account for a wide range of cognitive phenomena. *Role of Driving Simulators.* Models of driver behavior based on cognitive architectures can be linked into driving simulators to act as virtual drivers, using simulated perceptual and motor processes to drive in the simulator environment. Typically, the output of these virtual drivers is analogous to that for human drivers, thus facilitating comparison between model and human data. *Key Results of Driving Simulator Studies.* This chapter highlights one model in particular developed in the ACT-R cognitive architecture. This driver model has been shown to account for several important aspects of driver behavior, particularly in the context of curve negotiation and lane changing. *Scenarios and Dependent Variables.* The ACT-R driver model can be placed in various environments with driving as the only task or with driving while performing another secondary task (such as dialing a cellular phone). The model's simulation output includes predictions of both driver behavior as well as behavior on the secondary task (e.g., total dialing time). *Platform Specificity and Equipment Limitations.* The ACT-R driver model currently runs on a Macintosh platform and LISP environment. However, a recent Java reimplementation of the model within the Distract-R prototyping system has made the model more widely available on a range of platforms, in addition to facilitating its use by non-modeling designers and practitioners.

42.1 What Is a Cognitive Architecture?

Allen Newell (1973) famously described the state of psychology as fragmented into many disparate areas of research, listing 59 psychological phenomena that researchers have studied in fascinating detail but in largely separate efforts. His idea to address this problem evolved into the notion of a *cognitive architecture* (see Newell, 1990), which attempts to unify our understanding and theories of disparate psychological domains into a single large-scale effort. There are many varieties of cognitive architecture in use today, all of which rely on computation and simulation as a methodology for generating and predicting behavior (see, e.g., Gluck & Pew, 2005; Gray, 2007). Generalizing broadly,

the architectures are sometimes characterized as symbolic (manipulating symbolic units of information) or connectionist (involving highly-connected networks of computational units). For example, production systems (see, e.g., Anderson, 2007; Meyer & Kieras, 1997) that represent knowledge as condition-action rules can act upon symbolic facts (e.g., "3 + 4 = 7"), whereas neural networks or related constructs (see, e.g., O'Reilly & Frank, 2006) can represent such knowledge as patterns of activation across networked nodes. Most architectures today are not purely symbolic or connectionist, but rather include aspects of both approaches—for example, by linking symbols (e.g., "3", "4", "3 + 4 = 7") into a connectionist-like network of information with activations on nodes and spreading across network links.

An explicit goal of a cognitive architecture is to be integrative (see Byrne, 2007; Gray, 2007): The architecture should unify the many cognitive, perceptual, and motor aspects of the human system into a single framework. In computational terms, a cognitive architecture typically includes computational modules that make theories of these various aspects explicit. For instance, Anderson's (2007) theory of human memory provides a set of mathematical equations that govern memory processes for this knowledge. As one example, Anderson's theory defines a quantity *base-level activation* for each fact that defines how easily that fact can be retrieved. This activation changes over time, based on the usage of fact—for instance, through repeated recall when rehearsing a fact. The theory expresses base-level activation mathematically as follows:

$$B_i = \ln\left(\sum_{k=1}^{n} t_k^{-d}\right)$$

The base-level activation B_i of fact i depends on each usage t_k, where t_k represents the time elapsed between the usage and the current time. The decay factor d causes each usage to contribute less to the base-level activation over time. This straightforward formulation accounts for many known phenomena in human memory, such as the fact that more rehearsals and/or more recent rehearsals facilitate recall. Similarly, a cognitive architecture typically incorporates embedded theories of the perceptual and motor systems that serve to make behaviors more plausible and human-like; for example, an architecture may include a computational model of eye movements (Salvucci, 2001b) or computer mouse movements (Byrne, 2001) that abides by known constraints of these systems and produces plausible predictions of perceptual and motor performance.

Simulation is an essential ingredient of a cognitive architecture, especially when accounting for behavior in complex dynamic domains, such as air-traffic control (Taatgen & Lee, 2003) or fighter piloting (Jones et al., 1999). For some simpler laboratory domains (e.g., a simple choice reaction task), one may be able to examine the core workings of the architecture and predict behavior without running simulations (Schweickert, Fisher, & Proctor, 2003). However, for driving and similar domains, the complexities of both the human behavior—including cognitive, perceptual, and motor—and the environment with which the driver interacts—including the roadway, other vehicles, even vehicle dynamics—all make such off-line analysis difficult, if not impossible. Instead, models developed in a cognitive architecture can tie directly into a driving simulation, essentially allowing an experimenter to substitute a human driver with the computational model. The behavior exhibited by the model, then, generates exactly the same output as that of a human driver; namely, a simulation protocol that includes whatever data are appropriate for the given environment. Thus, model and human behavior can be directly compared using well-known measures of driver performance, providing a straightforward method for testing the accuracy and plausibility of computational driver models.

Besides the ability to simulate human behavior, cognitive architectures offer the significant benefit of an integrative community effort with respect to theory development. In contrast

to independently-developed models of behavior, any model developed in the context of a cognitive architecture immediately inherits the theoretical predictions of the architecture itself; for example, a model of behavior on a computer interface inherits the architecture's predictions about visual attention to screen icons, mouse movements on the interface, and potential memory constraints during task behavior. Moreover, as individual components of the architecture are further improved and account for new phenomena, the entire modeling community using that architecture benefits from the improved predictions of that architectural component. In fact, theoretical and applied work related to the cognitive architecture are mutually beneficial: Theoretical work on the architecture itself provides applied work in particular domains with increased fidelity in accounting for basic human processes; and applied research on individual domains such as driving helps to validate the architecture and challenge it to broaden the scope of its predictions.

42.2 Simulating Driver Behavior Using a Cognitive Architecture

There have been a variety of attempts to model and simulate driver behavior using cognitive architectures, including Aasman's (1995) model of intersection negotiation developed in the Soar architecture (Laird, Newell, & Rosenbloom, 1987), Tsimhoni and Liu's (2003) model of steering in the QN-MHP architecture (Liu, Feyen, & Tsimhoni, 2005), and my own integrated driver model (Salvucci, 2006) in the Adaptive Control of Thought-Rational (ACT-R) architecture (Anderson, 2007); Levison and Cramer's (1995) integrated driver model was also developed in the same spirit, incorporating cognitive, perceptual, and motor processes in a single model. We will highlight the ACT-R driver model in the exposition below and, in particular, two applications of the driver model: accounting for behavior during normal highway driving, and accounting for behavior while performing secondary tasks such as interacting with a cellular phone.

42.2.1 The ACT-R Integrated Driver Model

The ACT-R cognitive architecture (Anderson, 2007) posits that declarative fact-based knowledge can be represented as a network of associated *chunks*, and procedural skill-based knowledge can be represented as condition-action *production rules* that act upon these chunks. The chunks and production rules have symbolic properties in that they utilize symbols as basic units of information (e.g., *three* and *four*). At the same time, chunks and rules have continuous-valued properties (e.g., chunk activation that represents the facility with which the chunk can be recalled) that can change over time, and chunks are linked together through associations with other chunks (e.g., a chunk *three + four* might link *three* and *four* into an addition fact).

ACT-R incorporates various modules for cognitive, perceptual, and motor processing, as shown in Figure 42.1. Cognitive processing occurs centrally as the instantiation, or *firing*, of production rules in the central procedural resource. Cognitive

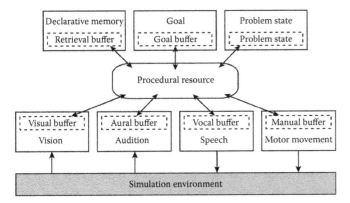

FIGURE 42.1 The ACT-R cognitive architecture: cognitive and perceptual-motor components, including interaction with the external simulation environment.

processing also includes declarative processing by which memory chunks can be recalled. Perceptual processing includes mechanisms for visual attention and eye movements (see, e.g., Salvucci, 2001b), as well as aural attention. Motor processing includes two hands (and feet in some scenarios) through which interaction with virtual environments occurs. As indicated in the figure, all communication occurs between the procedural resource and *buffers* within each module: In essence, a production rule in the procedural resource collects information from the buffers to check the conditions under which the rule fires (e.g., check for a visual stimulus), and sends commands to the modules through the buffers (e.g., press a key) as specified by the rule that has fired.

The ACT-R integrated driver model (Salvucci, 2006) centers on a simple set of core rules for basic control—that is, steering and speed control. The steering rules are based on work by Salvucci and Gray (2004), who posited that steering could be enacted as a control law based on two salient visual points as shown in Figure 42.2: a *near point* immediately in front of the vehicle to help with centering, and a *far point* well ahead of the vehicle to help with anticipation of the upcoming roadway (see also Donges, 1978); the far point can be the vanishing point of a straight roadway, the tangent point of a curved roadway, or even the position of a lead car (see Salvucci & Gray, 2004). Specifically, let us define variables θ_{near} and θ_{far} as the visual angles to the near and far points, respectively, and $\Delta\theta_{near}$ and $\Delta\theta_{far}$ as the changes in those values over a period of time Δt. The steering law then updates steering angle φ_{steer} by the amount $\Delta\varphi_{steer}$ based on the following equation:

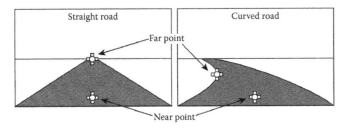

FIGURE 42.2 Near and far points on straight and curved roads.

$$\Delta\varphi_{steer} = k_{vf}(\Delta\theta_{far}) + k_{vn}(\Delta\theta_{near}) + k_n(\theta_{near})\Delta t$$

with constants scaling the individual terms. The straightforward interpretation of this equation is that the control law strives to: (1) maintain a stable far point (such that $\Delta\theta_{far} = 0$), (2) maintain a stable near point (such that $\theta_{near} = 0$), and (3) maintain a near point close to the center of the roadway (such that $\theta_{near} = 0$).

With this control law as a starting point, the ACT-R driver model dictates how cognitive and visual processing utilize this law in a more plausible behavioral model. The model comprises four production rules that perform the following steps:

1. Find the near point. The first rule initiates the movement of visual attention to the near point of the current lane.
2. Find the far point. After noting the location of the near point, the second rule initiates the movement of visual attention to the far point of the current lane.
3. Update steering and acceleration. After noting the location of the far point, the third rule uses the above control law to update steering angle and accelerator/brake pedal depression. For steering, this step uses the values gotten in the previous iteration to compute the changes in visual angle for the near and far points. The scheme for pedal depression follows a similar control law (see Salvucci, 2006). This production rule also initiates an eye movement to the far point.
4. Check and iterate. The fourth rule checks for the stability of the model with respect to the changes in visual angle; specifically, it checks to ensure that the vehicle's lateral velocity and position are within some threshold of acceptability, and if so, the rule restarts this four-step process. (If the situation is not stable, the model does not allow for other tasks to intercede—an important criteria when driving while performing a secondary task.)

Note that these four rules place the mathematical control law in the context of the ACT-R cognitive architecture, forcing behavior to abide by the cognitive, perceptual, and motor constraints of the architecture. One of the most important constraints is that, in ACT-R, each production rule requires 50 ms to execute, thus predicting that one update to steering and acceleration (i.e., one iteration of the model's four rules) requires approximately 200 ms (i.e., roughly five times per second).

In addition to using the control law for lane keeping and curve negotiation, we also incorporated lane changing into the model through a straightforward generalization: When the model wants to change lanes, it simply switches from using the near and far points of the current lane to using the near and far points of the destination lane. The change in near-point position allows the control law to steer in the direction of the destination lane; at the same time, the stability constraints in the control law ensure that the movement is performed in a smooth manner. Nevertheless, some drivers may wish to change lanes more slowly because of safety concerns, and thus the model uses a modified control law (Salvucci, 2006) that limits the contribution of the near-point position in the steering calculation:

$$\Delta\varphi_{steer} = k_{vf}(\Delta\theta_{far}) + k_{vn}(\Delta\theta_{near}) + k_n \min(\theta_{near}, \theta_{nmax})\Delta t$$

Here, the visual angle to the near point is limited to the value θ_{nmax}, which ensures that the change in steering angle does not get too large as the lane change is being enacted.

Besides the aspects of control described above, the driver model incorporates, albeit in a limited way, a model of monitoring for maintaining situation awareness. In particular, with some probability, the model periodically checks the forward or backward views and notes the presence of any vehicles in that location, along with their position and velocity. This information is stored in ACT-R declarative memory for later retrieval. When a lane change is desired, the model uses this knowledge to determine whether another car is present in a danger zone around the driver's vehicle. When the danger zone is clear, the model can initiate a lane change. All these steps are implemented in production rules in a fashion similar to the four control-related production rules.

42.2.2 Running the Model in Simulation

When models developed in cognitive architectures interact with simulation environments there are a number of possible methods for allowing the model to hook in and control the simulation. If the cognitive architecture and the simulation are developed in the same programming language and/or development framework, integration can be as straightforward as adding some communication code that glues the two components together. Typically, however, they do not share a common development framework, and furthermore, the source code may not be available for one or both components. Thus, some creativity is usually required to link up the model and the simulation.

In our experiences with the ACT-R driver model, we have found no "magic bullet" solution, but rather several solutions that have distinct advantages and disadvantages. The ACT-R cognitive architecture has been developed in the LISP programming language, and ACT-R models generally incorporate bits of LISP code in addition to standard model code. However, our (original) driving simulation was developed in C using OpenGL graphics routines, and thus there was no way to incorporate one body of code into the other. We briefly explored using network communication channels (i.e., sockets) between the components, but eventually followed a different approach. First, we implemented a minimal-graphics version of the driving simulator in our LISP implementation, with the same vehicle dynamics as the original but with extremely minimal graphics. Second, we ran the model and had it directly control the LISP simulation to generate a behavioral protocol equivalent to that of the primary C simulation environment; this protocol comprised

tab-separated columns of data including global data (e.g., timestamp), environmental data (e.g., locations of all vehicles in space), data particular to the driver's vehicle (e.g., steering-wheel angle, pedal depression), and other behavioral data (e.g., driver eye movements). We could then replay the model-generated protocols in the full-graphics environment to visualize its behavior, if desired. This solution may not be feasible for some situations; for instance, some vehicle-dynamic components of full-fledged simulators may be too complex to make re-implementation in another programming language practical, even without graphics. Nevertheless, for our situation, the solution avoided networking issues and produced protocols equivalent to what human drivers would generate in the full simulator.

42.2.3 Comparing Model Behavior to Human Behavior

Given that the same data protocols were generated by human drivers in the full simulator and by the ACT-R driver model with its simulation, we collected data from each and attempted to validate the model with respect to human behavior. Our efforts aimed at two broad categories of behavior: driving on a multi-lane highway, and driving while performing secondary tasks. The first environment comprised a multi-lane highway environment with moderate traffic generated by automated vehicles. The environment included a four-lane highway, with two lanes in each direction, and a number of automated vehicles which travel at varying desired speeds and can pass other vehicles (including the driver's car) when necessary. We used this environment to test the basic aspects of the model's behavior, focusing on the two tasks of lane keeping (including curve negotiation) and lane changing. In particular, over a number of simulation runs, we collected data from the driver model navigating this environment, driving among the automated vehicles and changing lanes when it encounters slower vehicles. We also collected data from human drivers driving as naturally as possible in the same environment, providing us with two data sets with which we can compare model and human data. The model included four parameters (k_{vf}, k_{vn}, k_n, θ_{nmax}) whose values were estimated to provide the best fit to the human data presented below.

Figure 42.3 shows the time course of two sets of measures for driver behavior during curve negotiation, namely steering angle and lateral position. Because the curved segments used for these results varied in length, the data were redistributed into 10 equal time segments between the start and end of the curve, each indicated by a vertical line. These segments were extrapolated before and after the curve, and then all the curve data were averaged together to produce the graphs. In the human data (solid lines), we see that the human drivers began steering just before entering the curve, maintained a steady steering angle throughout the curve, and then gradually reduced the steering angle back to zero around the end of the curve. The lateral-position results show that, except for a slight deviation at the start of the curve, drivers maintained a fairly central lane position throughout the curve. The model data (dashed lines) reproduces these aspects of the human

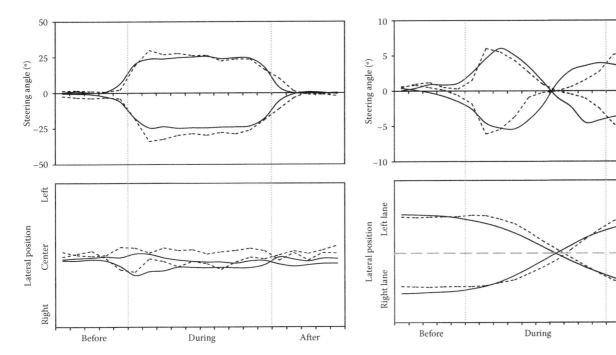

FIGURE 42.3 Steering angle and lateral position during curve negotiation, human data (solid lines) and model data (dashed lines). (Adapted from Salvucci, D. D. (2006). Modeling driver behavior in a cognitive architecture. *Human Factors, 48,* 362–380.)

FIGURE 42.4 Steering angle and lateral position while lane changing, human data (solid lines) and model data (dashed lines). (Adapted from Salvucci, D. D. (2006). Modeling driver behavior in a cognitive architecture. *Human Factors, 48,* 362–380.)

data, although it began changing steering angle somewhat later in its approach to the curve. Interestingly, the model produces very similar behavior at the end of the curve with respect to steering angle: Like human drivers, the model begins reducing the steering angle before the curve ends, and follows through in reducing the angle to zero gradually after the curve ends.

Figure 42.4 shows the same measures for lane changes, this time with the vertical lines representing the start and end of the lane change. (For the human drivers, the start and end were noted by the drivers themselves through verbal protocol; for the model, the start and end were also "verbally reported" in simulation.) The human-driver steering angle (solid lines) shows the characteristic steering pattern for a lane change (see also Salvucci & Liu, 2002), with a steering action in the direction of the destination lane followed by a corrective action to straighten the vehicle in the new lane. The lateral-position results show how these steering changes direct the vehicle smoothly into the new lane. Again, the model (dashed lines) reproduces the basic aspects of the human drivers' behavior. The most notable discrepancy between model and human behavior is the model's somewhat flattened steering angle in the middle of the lane change (in contrast to the human drivers' smoother transition in this area). Nevertheless, the model does a reasonable job in accounting for the qualitative and quantitative aspects of human driver performance.

In addition to modeling normal highway driving, we have also completed a number of studies examining how the driver model can be integrated with models of secondary tasks to account for performance under distraction. For example, in one recent study (Salvucci & Taatgen, 2008), we created a straightforward ACT-R

model of cellular-phone dialing and, through integration of this model with the driver model, ran simulations of the dual-task model to predict driver behavior while dialing. In fact, the dialing model could dial a phone using any of four distinct interactions: *full-manual* dialing of the entire number, *speed-manual* dialing of a single "speed number," *full-voice* dialing by speaking the entire number, and *speed-voice* dialing by speaking a representative phrase (such as "home" or "office"). Figure 42.5 shows the human from the original empirical study (Salvucci, 2001a) along with the model's behavior; the results include measures of reaction time in baseline (no driving) and driving conditions, and of lateral velocity as a measure of driver performance. The human drivers exhibited only a very small increase in total dialing time while driving compared to dialing only. They also exhibited a significant increase in lateral velocity for the manual-dialing conditions, and a lack of significant effect for lateral velocity in the voice-dialing conditions. For the model, the visual demands of the manual-dialing conditions steals time away from the driver model, thus decreasing the frequency with which it can update steering, and thus impacting its performance. At the same time, the model can reasonably well interleave dialing and driving, thus exhibiting only a small increase in dialing time in the driving conditions.

We have done other studies of phone dialing (e.g., Salvucci, 2005; Salvucci & Macuga, 2002), in addition to studies of other secondary tasks such as radio-tuning (Salvucci, 2005). The distraction from these tasks primarily arises from the competition for the visual resource between the secondary task and the driving task. However, because of cognitive limitations in the ACT-R cognitive architecture—most importantly the constraint that

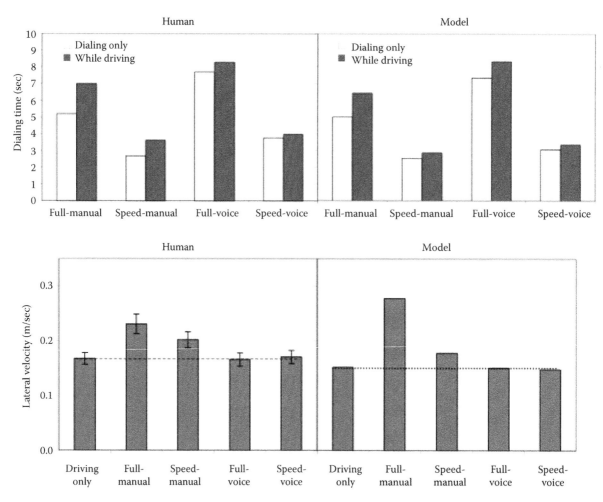

FIGURE 42.5   Reaction time and lateral velocity while phone dialing, human and model data. (From Salvucci, D. D., & Taatgen, N. A. (2008). Threaded cognition: An integrated theory of concurrent multitasking. *Psychological Review, 115,* 101–130.)

only one cognitive production rule can fire at a time—the driver model can also account for distraction from primarily cognitive tasks. For instance, we (Salvucci & Taatgen, 2008) have integrated the driver model with a model that performs a "sentence-span" task in which a person must respond to sentences while remembering the last word of a set of sentences, all while driving. In our exploration of how "cognitive distraction" can arise from such a task, we found that the model's behavior somewhat nicely corresponded with empirical results of human drivers in the same dual-task scenario (Alm & Nilsson, 1995).

42.3 Looking Ahead: Predictive Models and Model-Based Applications

The development of the ACT-R driver model is, in part, a theoretical effort aiming to better understand driver behavior and performance. At the same time, from an engineering standpoint, the model has good potential for use in the practical development of model-based applications. One such avenue of research explores how such a model can be used in a predictive way, with the primary purpose of evaluating new in-vehicle interfaces for their

potential for driver distraction. To this end, we have developed a working system called Distract-R (Salvucci, 2009), a standalone application and internet-based Java applet that allows for rapid prototyping and evaluation of in-vehicle interfaces.

Distract-R provides an interface with which a user can quickly piece together a prototype in-vehicle device, such as the radio device shown in Figure 42.6. The user then specifies tasks on that interface by demonstration: He or she simply clicks on the interface components and thus demonstrates the steps with which a driver would perform the secondary task. (Note that many interfaces, such as that of a cell phone or radio, may be associated with a number of possible tasks.) Distract-R then runs the specified interfaces in simulation with the ACT-R driver model to generate a set of behavioral predictions much like those described in the previous section, which can be viewed in tabular or graphical form (the graphical form included in Figure 42.6). Distract-R incorporates a highly-optimized (Java) implementation of the (LISP) ACT-R driver model, and runs simulations approximately 1000 times faster than real time—thus allowing in-vehicle device designers to rapidly iterate through ideas and revisions to identify the most promising concepts for further empirical testing.

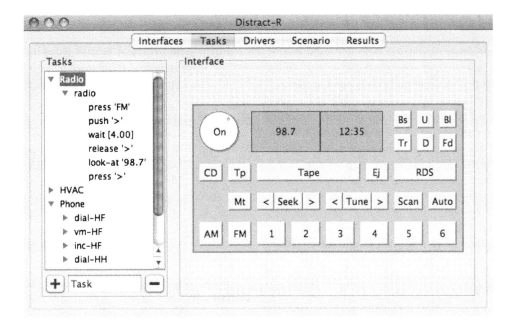

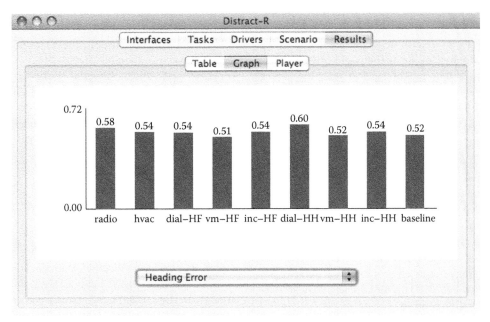

FIGURE 42.6 Sample Distract-R screens showing rapid prototyping and task specification (top panel) and graphical presentation of simulation results (bottom panel).

Interestingly, a more engineering-centered system like Distract-R requires close bootstrapping with relevant theoretical work. For example, specifying driver parameters led us to investigate the effects of driver age on performance (Salvucci, Chavez, & Lee, 2004). In this effort, we borrowed results from modeling efforts in the EPIC cognitive architecture (Meyer & Kieras, 1997) in order to computationally specify a major cognitive difference between a "younger" and "older" model (specifically a slowdown in cognitive processing time). In turn, this theoretical development allowed us to incorporate at least some account of age in the Distract-R system, thus allowing users to simulate and analyze

predicted differences in distraction between younger and older drivers (Salvucci, 2009).

While the fairly lower-level description of behavior in the ACT-R driver model provides a more detailed understanding of driver behavior, engineering approaches that utilize the model may not need descriptions at this level. We have found it useful to develop and specify our models at different levels of abstraction in order to satisfy the varying requirements of different applications. As an example, in one effort to detect lane changes using the driver model (Salvucci, Mandalia, Kuge, & Yamamura, 2007), we began by utilizing the full ACT-R driver model in a *model tracing* system that matches the behavior of models at the different levels of abstraction with observed driver performance. Our initial tests, however, showed that a simplified driver model actually worked better in detecting lane changes; specifically, a stripped-down version of the model that incorporated only aspects of the control law worked better than the full rule-based model. As another example, the Distract-R system incorporates a higher-level specification of ACT-R models called ACT-Simple (Salvucci & Lee, 2003) which allows users wanting to generate a model to do so much more easily. For example, instead of directly writing ACT-R production rules (which requires on the order of days to weeks of training), users can specify behaviors by using simple commands such as "look-at", "move-hand", and "press-key". This facilitation of model development and ability to model behavior at different levels of abstraction should benefit both theoretical and engineering developments in models of driver behavior.

Finally, we should note the relationship between models of driver behavior and surrogate methods and metrics described this book, chap. 10 by Angell. Almost all of the surrogate methods involve tasks that are simpler than driving. There is good reason to believe that, as surrogate methods gain in acceptance, models could be developed in cognitive architectures to predict behavior in related surrogate tasks. Such developments would further increase the utility of the surrogate methods and driving simulations more generally, as they could further increase the number of prototypes that could be evaluated and revised over multiple design cycles.

Key Points

- Cognitive architectures provide a unified psychological theory and associated computational framework for building models of driver behavior.
- Steering control can be represented as a two-level control law based on salient near and far visual features of the roadway.
- Based on this control law, the ACT-R driver model describes driving as centered on a four-step iterative process requiring roughly 200 ms for a single update to steering and acceleration controls.
- The ACT-R driver model has been shown to account for various aspects of driver behavior both for driving in multilane highway environments and for driving while performing a secondary task.
- Driver models based on cognitive architectures can be incorporated into engineering design systems that allow for rapid prototyping and evaluation of in-vehicle interfaces and tasks.

Keywords:　ACT-R, Cognitive Architectures, Driver Distraction, Rapid Prototyping and Devaluation

Acknowledgment

This work was supported by ONR Grant #N00014-09-1-0096 and NSF Grant #IIS-0426674.

Glossary

Cognitive architecture: A psychological theory and associated computational framework intended to enable simulation of human cognitive and behavior.

Key Readings

Anderson, J. R. (2007). *How can the human mind occur in the physical universe?* New York: Oxford University Press.

Gray, W. D. (Ed.). (2007). *Integrated models of cognitive systems.* Oxford, UK: Oxford University Press.

Newell, A. (1990). *Unified theories of cognition.* Cambridge, MA: Harvard University Press.

Salvucci, D. D. (2006). Modeling driver behavior in a cognitive architecture. *Human Factors, 48,* 362–380.

References

Aasman, J. (1995). *Modelling driver behavior in SOAR.* Leidschendam, The Netherlands: KPN Research.

Alm, H., & Nilsson, L. (1995). The effects of a mobile telephone task on driver behavior in a car following situation. *Accident Analysis & Prevention, 27,* 707–715.

Anderson, J. R. (2007). *How can the human mind occur in the physical universe?* New York: Oxford University Press.

Byrne, M. D. (2001). ACT-R/PM and menu selection: Applying a cognitive architecture to HCI. *International Journal of Human-Computer Studies, 55,* 41–84.

Byrne, M. D. (2007). Cognitive architectures. In A. Sears & J. Jacko (Eds.), *The human-computer interaction handbook* (2nd ed., pp. 93–113). Mahwah, NJ: Lawrence Erlbaum Associates.

Donges, E. (1978). A two-level model of driver steering behavior. *Human Factors, 20,* 691–707.

Gluck, K. A., & Pew, R. W. (Eds.). (2005). *Modeling human behavior with integrated cognitive architectures: Comparison, evaluation, and validation.* Mahwah, NJ: Lawrence Erlbaum Associates.

Gray, W. D. (Ed.). (2007). *Integrated models of cognitive systems.* Oxford, UK: Oxford University Press.

Jones, R. M., Laird, J. E., Nielsen, P. E., Coulter, K., Kenny, P., & Koss, F. (1999). Automated intelligent pilots for combat flight simulation. *AI Magazine, 20,* 27–42.

Laird, J. E., Newell, A., & Rosenbloom, P. S. (1987). Soar: An architecture for general intelligence. *Artificial Intelligence, 33,* 1–64.

Levison, W. H., & Cramer, N. L. (1995). *Description of the integrated driver model* (Rep. FHWA-RD-94-092). McLean, VA: Federal Highway Administration.

Liu, Y., Feyen, R., & Tsimhoni, O. (2005). Queueing Network-Model Human Processor (QN-MHP): A computational architecture for multi-task performance in human-machine systems. *ACM Transactions on Computer-Human Interaction, 13,* 37–70.

Meyer, D. E., & Kieras, D. E. (1997). A computational theory of executive cognitive processes and multiple-task performance: Part 1. Basic mechanisms. *Psychological Review, 104,* 3–65.

Newell, A. (1973). You can't play 20 questions with nature and win: Projective comments on the papers of this symposium. In W. G. Chase (Ed.), *Visual information processing* (pp. 283–308). New York: Academic Press.

Newell, A. (1990). *Unified theories of cognition.* Cambridge, MA: Harvard University Press.

O'Reilly, R. C., & Frank, M. J. (2006). Making working memory work: A computational model of learning in the prefrontal cortex and basal ganglia. *Neural Computation, 18,* 283–328.

Salvucci, D. D. (2001a). Predicting the effects of in-car interface use on driver performance: An integrated model approach. *International Journal of Human-Computer Studies, 55,* 85–107.

Salvucci, D. D. (2001b). An integrated model of eye movements and visual encoding. *Cognitive Systems Research, 1,* 201–220.

Salvucci, D. D. (2005). A multitasking general executive for compound continuous tasks. *Cognitive Science, 29,* 457–492.

Salvucci, D. D. (2006). Modeling driver behavior in a cognitive architecture. *Human Factors, 48,* 362–380.

Salvucci, D. D. (2009). Rapid prototyping and evaluation of invehicle interfaces. *ACM Transactions on Human-Computer Interaction.*

Salvucci, D. D., Chavez, A. K., & Lee, F. J. (2004). Modeling effects of age in complex tasks: A case study in driving. *Proceedings of the 26th annual conference of the cognitive science society* (pp. 1197–1202). Mahwah, NJ: Lawrence Erlbaum Associates.

Salvucci, D. D., & Gray, R. (2004). A two-point visual control model of steering. *Perception, 33,* 1233–1248.

Salvucci, D. D., & Lee, F. J. (2003). Simple cognitive modeling in a complex cognitive architecture. *Human factors in computing systems: CHI 2003 conference proceedings* (pp. 265–272). New York: ACM Press.

Salvucci, D. D., & Liu, A. (2002). The time course of a lane change: Driver control and eye-movement behavior. *Transportation Research Part F: Traffic Psychology and Behavior, 5,* 123–132.

Salvucci, D. D., & Macuga, K. L. (2002). Predicting the effects of cellular-phone dialing on driver performance. *Cognitive Systems Research, 3,* 95–102.

Salvucci, D. D., Mandalia, H. M., Kuge, N., & Yamamura, T. (2007). Lane-change detection using a computational driver model. *Human Factors 43,* 532–542.

Salvucci, D. D., & Taatgen, N. A. (2008). Threaded cognition: An integrated theory of concurrent multitasking. *Psychological Review, 115,* 101–130.

Schweickert, R., Fisher, D. L., & Proctor, R. (2003). Steps towards building mathematical and computer models from cognitive task analysis. *Human Factors, 45,* 77–103.

Taatgen, N. A., & Lee, F. J. (2003). Production compilation: A simple mechanism to model complex skill acquisition. *Human Factors, 45,* 61–76.

Tsimhoni, O., & Liu, Y. (2003). Modeling steering using the Queuing Network - Model Human Processor (QN-MHP). *Human factors and ergonomics society annual meeting proceedings, surface transportation* (pp. 1875–1879).

43

Combining Perception, Action, Intention, and Value: A Control Theoretic Approach to Driving Performance

John M. Flach
Wright State University

Richard J. Jagacinski
The Ohio State University

Matthew R. H. Smith
Delphi Electronics & Safety

Brian P. McKenna
Resilient Cognitive Solutions

Abstract

The Problem. Driving is a prototypical example of a closed-loop control problem. This chapter presents a tutorial introduction to the logic of closed-loop systems and the implications for theory and research on driver performance. The human-vehicle system is parsed into three coupled components: (1) the problem constraints; (2) the observer constraints; and (3) the control constraints. We hope to illustrate the importance of each component to a full appreciation of the closed-loop dynamic and to consider how the coupling across these components will determine the emergent stability of human-vehicle systems. We hope that this tutorial will help to provide common ground between behavioral scientists and engineers, so that each can better appreciate how the human factor, the vehicle, and the driving ecology interact to shape system performance. We hope that this will provide a theoretical context for interdisciplinary driver research using simulators. *Role of Driving Simulators*. There is a growing appreciation among those who study human performance that context matters. Thus, increasingly it becomes important to be able to evaluate performance under conditions that are representative of natural life experiences. Driving simulators provide a unique bridge between the complexity of natural environments and the demands for controlled observation in order to test hypotheses about human performance. This is equally important for addressing practical issues associated with training and design, as well as for basic issues associated with understanding adaptive cognitive systems.

43.1 Introduction

This chapter is designed to illustrate how a control theoretic approach might inform research on driving performance and simulation. In doing this, the focus will not be on specific control theoretic models, but on the motivation and rationale behind the models and the implications for research programs utilizing simulation to evaluate and improve driving performance. We believe that the driving problem in particular, and human-vehicle systems in general, are important instances where a closed-loop system architecture is

required for deep understanding of the relations among component behavioral elements. Frankly, we believe that this generalization can be made to most of the field of human performance and cognition (e.g., Flach, Dekker, & Stappers, 2008). The closed-loop architecture captures the mutually causal relationships between perception and action as well as emergent systemic properties such as stability and varieties of convergence toward behavioral goals.

This chapter will first consider three components of any control problem: 1) the *problem specification*; 2) the *observer problem*; and 3) the *control problem*. An additional section will consider

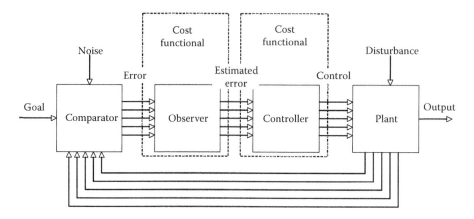

FIGURE 43.1   This abstract model illustrates a generic control system.

what Weinberg and Weinberg (1979) call the *Fundamental Regulator Paradox*. The final section will summarize key points and suggest how the control theoretic orientation might contribute toward a general approach to human-machine systems.

Figure 43.1 provides an abstract view of a generic control system. It is abstract in that the components do not necessarily correspond to any single mechanism or object. Rather, these components reflect logical aspects of a control problem (e.g., to move through a problem space from an initial state to a goal state) and of the system that might solve this problem (e.g., operations that determine movement in the space). To a certain extent, two of the boxes in Figure 43.1 represent constraints of the control problem (the Plant and the Comparator) and two of the boxes represent means for solving this problem (the Observer and the Controller).

The plant represents the dynamic process that is to be controlled. In the case of driving, this would typically be the vehicle dynamics. It is typical to describe this process in terms of state variables. The state variables reflect the condition of the system at any specific time t, such that knowledge of this state, together with knowledge of the input (control actions and disturbances) from t forward, is sufficient to predict the performance (output) of the system from t forward. For example, for a simple inertial system the state variables could be position and velocity. Thus, if you know position and velocity at time t and the forces acting on the body from t forward, then it is possible to specify the path of the body from time t forward.

Figure 43.1 shows multiple feedback paths from the plant to the comparator. Each path represents a measure of a state variable of the plant. Typically, there would be at least one measure for each state variable. The comparator represents comparisons between the current states of the plant and the goal states. To the extent that all measures are in a common metric, the comparator can be visualized as a subtraction to yield an error measure (i.e., the difference between the current state and the target state). For example, the target state might be the center of the lane on a curving road; the relevant states would be the vehicle's position and heading angle; and the errors would be the lateral position error (center of lane—vehicle position) and the heading angle error (road lane angle—vehicle heading angle).

The role of the observer is to estimate error at time $t + \Delta$ from the potentially noisy observations at time t (and perhaps earlier

times). Note that the observer function involves both estimation and prediction. Figure 43.2 (upper labels) illustrates the estimation process. The observation process typically involves weighting and integrating information over time. For example, the estimate for any particular state variable may correspond to a weighted function of all observations at time t and of previous estimates. Again, consider the simple inertial system; position at $t + \Delta$ might be estimated as a weighted function of the position and velocity observed at time t and previous estimates of these quantities. In a designed observer the weightings would be chosen to reflect knowledge about the process being observed. That is, there is at least an implicit internal model of the plant guiding the observation process. See Jagacinski and Flach (2003, Chapter 18) for some examples of simple observers.

The observation process in Figure 43.1 is shown in the context of a "cost functional". The point of this is to make it clear that the observation process will typically require a trade-off among multiple criteria or preferences. This trade-off is seen in Signal Detection Theory in terms of the choice of a decision criterion (β). The decision criterion can be chosen to maximize the

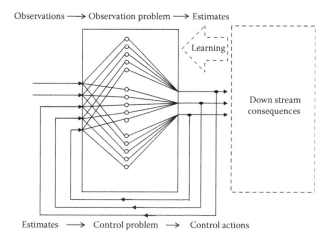

FIGURE 43.2   This figure illustrates observation (upper labels) and/or control (lower labels) as a process of weighting and possibly integrating information. The small open circles represent gains or multiplicative weightings of the various signals. The convergence of lines to a single point represents a summation of signals. This figure highlights parallels between the observation and control problems.

expected value associated with a specific payoff matrix (e.g.,value of hits and correct negatives and costs of misses and false alarms). In filtering, on the other hand, the trade-off is typically reflected in the gain of the filter (Jagacinski & Flach, 2003). This can be thought of as the relative weight given to the current observation relative to the expectation based on the integration of previous observations and/or an internal model. If the gains on new observations are high, the filter will be biased in favor of the current observation. This system would be very sensitive to a changing signal, but would also have a tendency to chase the noise. If the gains on new observations are low, the filter will be biased in favor of the expectation. This system would be more effective at filtering out noise, but would be sluggish to respond to real changes in the signal. Ideal Observer Theory provides analytical tools for specifying an optimal value for the filter gains, given a specific set of criteria (e.g., Gelb, 1974). Thus, the major point of the observer cost functional box in Figure 43.1 is to emphasize that evaluation of an observation process (i.e., determining whether it is optimal or satisfactory) involves consideration of an extrinsic payoff matrix or cost functional regarding the importance of new observations and prior estimates.

Like the observation process, the control process can also be envisioned as a weighting process based on some, at least implicit, model of the process (Figure 43.2, lower labels). In the case of control, the inputs are the state estimates coming from the observer process and the outputs would be control actions (i.e., manipulations of the control surfaces). In general, the weights can be thought to reflect the gains or sensitivities to the various estimated errors. As with the observation process, there typically will be optimal gains. If the gains are below optimal levels, the control system will be slow to eliminate error (i.e., it will be sluggish). However, if the gains are very much above optimal levels, then the control system can become unstable. That is, rather than reducing error, the control actions will tend to increase error (e.g., pilot induced oscillations—see National Research Council, 1997). If there are effective time-delays in the observer, controller, or the controlled process (e.g., the vehicle dynamics), then there will always be a stability limit on the gains. Choosing the right gains for the controller also requires consideration of a cost functional or payoff matrix. Typically, in control systems one is balancing the cost of control action (or effort) versus the cost of error. For example, the Optimal Control Model of the human tracker typically uses the velocity of control action as a measure of effort. The model assumes that the human is minimizing the integral of mean-squared error and mean-squared control velocity (Kleinman, Baron, & Levison, 1971). Optimal control theory provides analytic tools for specifying the optimal values for control gains (e.g., Kirk, 1970).

Typically, novice drivers show high degrees of effort and a tendency to weave within their lane. This behavior could be modeled as high gains on either the observer (i.e., tracking the noise) or the controller (i.e., over-correcting or pilot-induced oscillations). With the development of more proficiency, driving seems to take much less effort and drivers become better at tracking within their lane. This progression could be modeled as tuning toward or discovering more "optimal" gains for the observer or the controller, or both. This might be described as tuning of the internal control model.

Note that in the design of automatic control systems, an internal model together with the cost functional or payoff matrix guides the choices of the designers (e.g., the gain settings for the observer and the controller), but it is not always an explicit component of the control system. For biological systems, however, which are essentially "self-designing" or "self-organizing", some degree of knowledge/experience with the action, information, and value constraints will be fundamental components of the system. Thus, for researchers who are trying to model and/or explain skill development, this component needs to be more explicit. Much of the empirical research on control theoretic models has finessed the issue of these experiential and value constraints by exclusively modeling only well-trained and skilled operators. In essence, these skilled operators have become tuned to the task demands and their internal models and cost functions can be assumed to result in near-optimal weightings of the relevant task criteria (e.g., see Kleinman et al. 1971; or Pew & Barron, 1978). A major obstacle to applying optimal control theory to modeling novices is an inability to specify this knowledge and/or the cost functional. This will be explored in a later section in the context of the regulator.

Figure 43.3 provides a more detailed illustration of control in the context of driving. Figure 43.3 parses the control system to reflect the individual physical components of the system. Note that the human, as a component of both the observer and the controller, is at the heart of the control problem solution. This suggests that the attunement or skill level of the human will be critical to the effectiveness of the overall control solution. Figure 43.3 also includes a box to reflect the ecology. Classically, control problems are framed in terms of the vehicle being controlled. However, in framing the larger problem of driving safety, and particularly in considering the use of simulation, it may be very important to explicitly consider the ecology as part of the total system. This ecology reflects the driving surfaces, information displays (e.g., traffic signals), and obstacles (including pedestrians and other vehicles). In a simulator, this ecology must be explicitly specified. This specification will have important implications for how well performance in the simulator will generalize to natural driving contexts.

Figure 43.3 will be examined more carefully in the following sections as we delve deeper into four aspects of a control theoretic approach: the problem specification, the observer problem, the control problem, and the adaptive control problem.

43.2 The Problem Specification: The Situation Demands

Kirk (1970) cites the old adage that "a problem well put is a problem half solved." This certainly is consistent with the experiences of early workers in artificial intelligence. For example, Newell and Simon (1972) made the following observation:

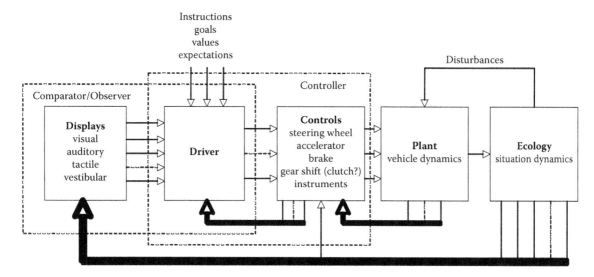

FIGURE 43.3 This figure shows the components of the human/vehicle control system.

It would be perfectly possible for the psychologist to follow the route of the economist: to construct a theory of concept formation that depended on no characteristic of the subject other than his being motivated to perform well. It would be a theory of how perfectly rational man would behave in that task environment—hence, not a psychological theory but a theory of the structure of the task environment. (p. 54)

In this sense, a control model is in essence a theory of the task environment or the situation demands. Ashby (1956) states this in terms of his *Law of Requisite Variety,* noting that only variety in the controller can reduce variety due to the environment (i.e., the disturbances). In other words, if the control has too few degrees of freedom, there may be aspects of the environment that cannot be reached or disturbances that cannot be attenuated of Weinberg and Weinberg (1979) formalize this as the Parallel Principle:

> A regulator must be "like" the environment it regulates. (p. 206)

From a research perspective, the idea of problem specification or situation demands is related to ecological validity (i.e., the degree to which the research context is representative of the natural phenomena of interest). In the context of simulation, this issue is typically framed in terms of fidelity. It might be valuable to consider three aspects of fidelity:

1. *Vehicle fidelity*—to what extent do the control dynamics of the simulation correspond to the dynamics of the vehicle? (e.g., this book, chaps. 8, 9, & 10)
2. *Situation fidelity*—to what extent do the scenarios and tasks presented in the simulation represent situations that would be encountered in natural driving contexts? (e.g., Chap. 16, 17, 18, & 19)
3. *Experiential fidelity*—to what extent do the experiences of the driver in the simulator correspond with experiences in natural driving contexts? (e.g., Chap. 11, 12, 13, & 14)

These three aspects of fidelity are addressed in greater depth in earlier chapters, as noted above. For this chapter we simply want to emphasize that in using simulation as a research tool for modeling or predicting driving performance, it is important to realize that the generality of models and predictions will be impacted by how well the simulation represents the problem. If the model of the vehicle is inaccurate, the situation contexts naïve, and/or the experience of the participants is biased by demand characteristics, then the generality of any models or predictions based on performance in that simulation will be suspect. Similarly, if the simulation is being used for training, then the degree of positive transfer to natural environments will depend to a large extent on how well the simulation represents the control problems of the natural environment.

Thus, the first step to effective modeling of human performance is to clearly specify the nature of the problem that the human is being asked to solve. Again, we quote Newell and Simon (1972):

> ... we shall often distinguish two aspects of the theory of problem solving as (1) demands of the task environment and (2) psychology of the subject. These shorthand expressions should never seduce the reader into thinking that as a psychologist he should be interested only in the psychology of the subject. The two aspects are in fact like figure and ground—although which is which depends on the momentary viewpoint. (p. 55).

Neisser (1987) made the case even more explicitly when he wrote:

> If we do not have a good account of the information that perceivers are actually using, our hypothetical models of their 'information processing' are almost sure to be wrong. If we do have an account, however, such models may turn out to be almost unnecessary. (p. 11)

If care is taken to ensure that the problems presented to participants in the simulator are good (ecologically valid) representations of the driving problem, then the control models of this performance can be important elements in an ecological approach to human performance (e.g., Brunswik, 1956). However, if the simulator is used to recreate abstract control problems (e.g., stationary compensatory tracking, simple discrete decision tasks), then the simulator can become simply a demonstration device that reifies well-known principles of information and control systems using humans. In a task designed to be accomplished by a simple servomechanism, the well-trained and well-motivated human can be expected to behave much like a simple servomechanism. A million such studies do not "prove" the relevance of control theory for modeling everyday human performance. They simply confirm the internal consistency of control theory.

Thus, we would propose an inverse to the Weinbergs' Parallel Principle as cited above. *We propose that to build an accurate model of a regulator (or control system), one must begin with an accurate model of the environmental relationships it regulates.*

43.3 The Observer Problem: The Awareness Demands

In classical models of vehicle control, the state variables are typically specified in terms of Euclidean distances and their derivatives consistent with Newtonian models of motion. In these models, both goals and state variables are presumed to be measured in common units. Thus, negative feedback is accomplished by the subtraction of a measure of the current state of the system from a goal or target state for that variable. This approach fits well with classical approaches to "space" perception that assume that the function of perception is to construct an internal analog of the world consistent with the "truth" as reflected in the models associated with Newtonian physics and Euclidean space. Thus, a classical psychophysics program focuses on the perception of (e.g., thresholds and differential sensitivity to) Euclidean distances (e.g., depth) and their derivatives (e.g., velocity). In both views, the functional feedback channels are defined in terms of Euclidean dimensions.

However, for a number of reasons, many people are beginning to suspect that for biological systems the observation problem is either far more complex or far simpler, depending on your perspective. One piece of evidence that suggests this is research using a paradigm of "blind walking" (e.g., Loomis, Da Silva, Fuijita, & Fukusima, 1992). In this task, people are shown a position on the ground in front of them and are then asked to close their eyes and walk to the position. Although classical research on depth-perception shows systematic biases in human judgments of depth, when people are asked to walk to a position visually specified in front of them without visual feedback (blind), they show no systematic biases. Thus, while classical psychophysical research suggests a flawed internal model of depth, when tested in an action context, humans prove surprisingly capable. While judgment of depth seems difficult for humans, control of action relative to depth seems fairly simple.

Gibson and Crooks (1938) were the first to imagine that the states relevant for controlling human locomotion may be different than the classical states suggested by Newtonian models. They introduced the construct of a "safe field of travel". It should be clear that in describing the safe field of travel, Gibson and Crooks are making a statement about the dimensions of the driver's state space. They define this space as "the field of possible paths which the car may take unimpeded" (p. 120). That this field is relevant to the observation problem, and that it is different from classical Newtonian measures of motion, is made clear in the following statement about the minimum stopping zone: "The driver's awareness of how fast he is going does not consist of any estimate in miles per hour; instead, he is aware, among other things, of this distance within which he could stop" (p. 123).

Thus, Gibson and Crooks (1938) are suggesting that animals do not construct an internal model of a Euclidean space. Rather, people "see" the environment relative to action capabilities (e.g., minimum stopping distance) and consequences (e.g., imminence of collision). The concept of safe field of travel was a precursor of Gibson's (1979/1984) more general construct of "affordance". The idea of direct perception of affordances can be understood in control theoretic terms as the idea that there is no need for an internal model of Euclidean space to mediate between the observation and control processes. To the extent that there is an internal model, an ecological approach suggests that the model is framed in the language of perception and action, not the language of classical physics.

While the paper by Gibson and Crooks is important to help people imagine how the states of motion for humans may be different from the classical physical dimensions, they did not provide any suggestions about how this information might be fed back for observation. However, Gibson, Olum and Rosenblatt (1955) laid the foundation for modeling structure in optical flow fields that would provide a way to directly measure the states specified by Gibson and Crooks. The connections and the implications for control of animal locomotion were first described in Gibson (1958), where he provided explicit recipes for how motion can be controlled via structure in optical flow. For example, here is Gibson's postulate about steering:

> The center of the flow pattern during forward movement of the animal is the direction of movement. More exactly, the part of the structure of the array from which the flow radiates corresponds to that part of the solid environment toward which he is moving. If the direction of his movement changes, the center of flow shifts across the array, that is, the flow becomes centered on another element of the array corresponding to another part of the solid environment. ... To aim locomotion at an object is to keep the center of flow of the optic array as close as possible to the form which the object projects. (p. 155)

The first empirical demonstrations of the plausibility of direct coupling of action to optical structures were inspired by Lee's (1976) analysis of the optical control of braking. Lee (1976) showed how a control system based on optical variables could work. Lee focused on the optical variable *tau* and its derivative *tau-dot*. *Tau* is the ratio of projected optical size of an object (i.e., optical angle) to the expansion rate (i.e., looming or rate of change of optical angle). Since Lee's analysis, a large amount of empirical data has been published that is consistent with the idea that collision actions (e.g., braking to avoid collisions, or swinging a bat to create collisions) are directly contingent on a *tau* criterion (e.g., Hecht & Salvelsbergh, 2004). That is, people (and other animals) tend to initiate action (braking or bat swing) at a constant value of *tau*, independent of distance, speed, and object size.

More recently, Flach, Smith, Stanard and Dittman (2004) have suggested that the observation process may involve measuring and weighting optical angle (θ) and optical expansion rate ($\dot{\theta}$) as independent observables as illustrated in Figure 43.4. Additionally, this model provides a better fit to performance early in practice, where people appear to give disproportionate weight to expansion rate. This model is consistent with errors observed in many of the empirical studies of collision control. Research consistently shows that when object size is manipulated, people tend to respond early to the largest objects; and that when object speed is manipulated, people tend to respond early to the slowest objects (e.g. Smith, Flach, Dittman, & Stanard, 2001). Both these trends can be accounted for by a disproportional weight on expansion rate (relative to the fixed proportional weighting assumed in Lee's original model). Note that our model and Lee's model both assume a linear function of angle and expansion rate. However, Lee's model assumes a zero intercept (a one-parameter line, allowing only the slope to vary), and our model loosens this constraint (a two-parameter line, allowing both the intercept and slope to vary).

It is important to note that with practice in the control of collisions, people tend toward a proportional weighting of angle and expansion rate as predicted by Lee's early model. Thus, our model is not a contradiction of Lee, but rather a refinement that allows Lee's intuitions to be extended over a wider range of skill levels. The major conflict with Lee's original hypothesis is that there is no need to include *tau-dot* as a feedback variable.

In addition to feedback of optical angle (θ) and expansion rate ($\dot{\theta}$), Figure 43.4 also shows feedback of global optical flow rate (GOFR), which is a function of the overall (e.g., average) angular speed of ground texture elements (Warren, 1982). Whereas θ and $\dot{\theta}$ specify relative rates associated with a looming obstacle (e.g., a stopped or slowing lead vehicle), GOFR specifies the observer's speed relative to the ground surface and fixed objects on that surface. GOFR for a flat ground plane is directly proportional to movement speed and inversely proportional to the distance of the eyes above the surface (i.e., eye height). Since eye height is constant in most driving situations, GOFR can be a reliable source of information about the surface speed of the vehicle (i.e., speed as measured by a speedometer). However, changes in eye height associated with changing vehicle sizes (e.g., going from a sports car to a large SUV) might create some negative transfer (e.g., Owen, 1984).

Another potential source of information for vehicle speed is edge rate (Denton, 1980). For regularly spaced textures, such as the lane markers on a highway, the rate at which these texture elements pass through a fixed field of view will be proportional to vehicle speed. There is a growing literature exploring the various optical structures and the role that they play for the perception and control of self-motion (e.g., Fajen, 2005; Flach & Warren, 1995; Warren & Wertheim, 1990). This work is consistent with Gibson's (1979/1984) concept of ecological optics. The essential idea is that structure in dynamic perspective resulting from a moving point of observation, can specify states of self-motion and that animals can use this information to skillfully control that motion.

Note that the various optical variables together span the same space as the classical state variables (distance and speed). And they do so in a way that there are common components associated with the classical variables—for controlling inertial systems these common components might be the "signal". Additionally, each variable has sources of variability associated with other

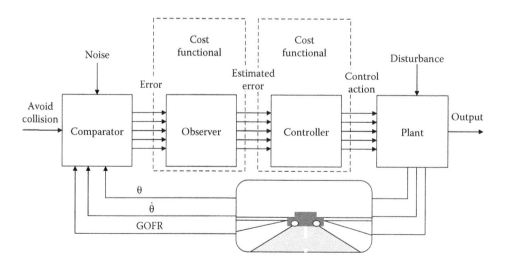

FIGURE 43.4 It is important to consider the possibility that input to the comparator and observer processes may be direct input of perceptual primitives (e.g., optical angles and angular rates), rather than Newtonian primitives (i.e., Euclidean distances and derivatives).

properties that might be considered "noise" with regard to the problem of controlling motion (e.g., object size and distribution of optical texture on surfaces).

Although the emphasis here has been on ecological optics, we believe that Gibson's (1958) intuitions about dynamic optical perspective and animal motion can be generalized to other modalities (e.g., auditory, vestibular, and tactile). We believe that an important step in modeling the observer components in any biological control system is to consider the possible sources of information feedback and their relations to the states of motion. In general, we expect that skilled drivers will learn to take advantage of all or most of the information available to them. Thus, knowledge of the limits of this information will provide valuable intuitions about the limits of human performance. Also, knowledge of how information is specified in natural contexts can suggest more effective ways to configure information in engineered displays. For example, Smith (2002) observed faster driver responses for forward collision warning (FCW) displays that featured a looming stimulus (an expanding vehicle angle) when compared with displays that featured only a unidimensional scale presentation. The data from these driving simulator studies led Delphi to select the "looming" display for the large-scale field operational test (ACAS FOT) that evaluated both adaptive cruise control (ACC) and FCW.

These assumptions suggest both unique opportunities and challenges for the use of simulators in research and training. Driving simulators provide a unique opportunity to control and manipulate information structures such as optical flow fields and to measure the implications for driving performance. This general program of research has been called "active psychophysics" (Warren & McMillan, 1984). In the active psychophysics research programs on visual control of locomotion, the independent variables are typically motivated by ecological optics, while the tasks (e.g., tracking), the measures (e.g., frequency response), and analyses (e.g., Fourier or Bode analysis) are motivated by control theory (e.g., Jagacinski & Flach, 2003, Chapter 22).

An important challenge in the use of simulators to study and to train drivers is that it is important to represent the natural information constraints as well as possible. Inferences about and transfer to natural situations will generally be constrained by the level of perceptual fidelity that is possible, as noted earlier. One of the benefits of ecological optics is that it has increased our capacity to analytically specify optical fidelity beyond the engineering or simple psychophysical constraints (e.g., screen resolution, frame rate). However, it is clear that there are some aspects of optical fidelity that are difficult to simulate, such as the unfolding of nested texture that is associated with approach to an object in natural situations. For example, from a distance a cornfield appears as a single element of texture, but as an observer gets closer, the field disappears as a texture unit and individual stalks of corn become the texture elements. Other modalities can provide even greater challenges (e.g., vestibular fidelity and spatial audio fidelity). Thus, when the risks allow, instrumented cars operating on test tracks or in natural environments can be a valuable supplement to research using simulators.

In our view, identifying the potential sources of information is a prerequisite to studying the unique limitations associated with human information processing. Again, consider the Neisser quote, cited earlier. However, it should be clear that the observer problem involves many considerations above and beyond the ecological analysis of information sources. In addition to general psychophysical limits, it should be clear that the ability to utilize potential information will be constrained by experience (e.g., Gibson, 1969) and by situational demands. For example, a growing safety concern is the proliferation of electronic media (e.g., cell phones and MP3 players) with the potential to distract drivers (Caird, Willness, Steel, & Scialfa, 2008; Chisholm, Caird, & Lockhart, 2007; Strayer & Johnston, 2001). Also, while many control models assume continuous feedback and proportional control, it is likely that, due to the spatial distribution of information (e.g., windows, side mirrors, rear-view mirrors, dashboard) and the presence of distractions, the human driver is more likely to function as a discrete control device (intermittently sampling information and using discrete stereotypic/automatic control responses over short time periods, e.g., see Senders, 1964).

The main point for this section is that the problem of how humans and other animals solve the observer problem may be more complex than assumed by classical models. Researchers should not assume that the dimensions which are conventionally used in physical models of space and motion are the feedback dimensions for biological control systems. Thus, it is dangerous to uncritically generalize from engineering analyses and engineered control solutions to biological systems. Nevertheless, the methodologies and logic of control theory as reflected in "active psychophysics" will play an important role in exploring biological control systems to identify the information that is available to specify the relevant states of motion.

43.4 The Control Problem: The Action Demands

Given that the driver can solve the estimation problem so that there are good estimates of the states of motion relative to a slowing line of traffic or relative to a winding road, how does the driver translate these estimates into appropriate actions? That is, at what point should the driver release the accelerator? At what point should the driver initiate braking? How hard should the driver press the brake? How much should the driver turn the wheel? The control problem involves choosing actions that bring the system closer to the goal states. In other words, the controller (driver) needs to choose the action that best nulls the error.

The overall control problem can be parsed into two subproblems which we will call the *degrees of freedom problem* (Bernstein, 1967) and the *gain problem*. The *degrees of freedom problem* is associated with the numbers of possibilities. This can mean that there may be many different actions (or means) that lead to the same goal (or ends), but also can mean that there are many paths to failure. Since there are many possibilities—the control problem can be very complex—which approach is best? This is well-illustrated by the problem of the golf swing (e.g.,

Cochran & Stobbs, 1968). The human body has many degrees of freedom as a result of the various joints involved in the golf swing. Of all the possible combinations, only a very few will yield the results desired by the golfer. Thus, a golfer seeking to eliminate a persistent slice has many possibilities to consider. Is it my head position or my stance, or my right elbow, or my grip, or some combination of these? Vehicle control systems tend to be relatively low degree of freedom systems; that is, the options are limited. If the traffic in front of you is slowing, you have two basic options to avoid a collision, to brake or to change lanes to pass the slowed traffic. Thus, we will focus on the gain problem.

The *gain problem* typically involves the timing and the magnitude of action relative to the state observations. This applies both for proportional control where the gain determines the magnitude of a proportional response and to discrete control actions where the gain may reflect the amplitude of a stereotypic response pattern (e.g., a bang-bang control). For example, given that you choose to brake (because there is no room to pass), at what point do you initiate the braking, and how hard do you press the brake? Note that while the ratio of visual angle-to-expansion rate may specify the time-to-collision, it does not specify the right time-to-brake. The right time-to-brake depends not only on the optics, but also on the state of the braking system (and perhaps the road). With poor brakes or slippery surfaces, it would be prudent to begin braking much earlier than with better brakes or drier surfaces.

This is another facet of Gibson's (1979/1984) affordance construct. The information has to be seen relative to the action capabilities. Many of the limits on the field of safe travel are limits on the control dynamics. In control theory, these limits are often visualized as boundaries in state space. Figure 43.5 is an example where presenting human performance data relative to control boundaries in a state space can provide insight into skill development.

The data in Figure 43.5 show the performance for one subject in an experiment conducted by McKenna (2004). The experiment was conducted using a desktop computer simulation. The simulation started with the car positioned on a road at one of five distances from an obstacle (at the zero position (left boundary) of the state space). The task was to drive from the initial stopped position to a stopped position just in front of the obstacle. Participants were asked to do this as quickly as possible while avoiding collisions with the obstacle. The data show the state of motion (position and velocity) when the driver released the accelerator. The open symbols show performance early in training (Block 1) and the filled symbols show performance late in training (Block 25). Each block involved 25 trials, five at each of five initial distances (in meters) from the obstacle (100 [squares], 150 [circles], 200 [diamonds], 250 [triangles], and 300 [circles]).

The first thing to note in Figure 43.5 is the dimension of the space. The dimensions in this case reflect the two state variables (position and velocity) for motion of an inertial system.

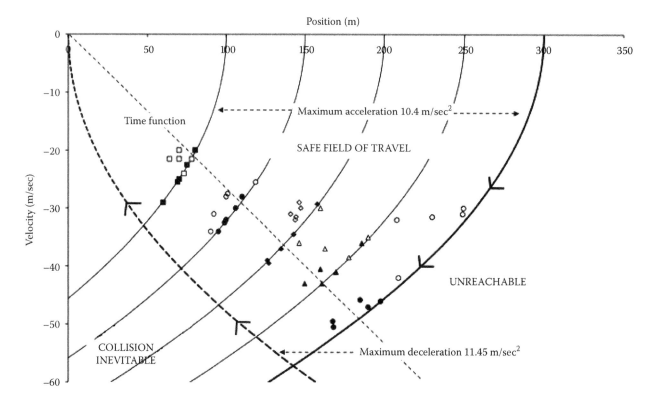

FIGURE 43.5 A state-space (position × velocity) for a braking control task employed by McKenna (2004) is illustrated. The data indicate the motion state of the vehicle when the driver released the accelerator in order to stop in front of an obstacle at the zero position (left boundary of the state space). The open symbols show performance in the first block of trials. The filled symbols show performance in the twenty-fifth block of trials. Five trials from each block are shown for each of five starting positions.

The quadratic curves represent control limits for each of the five starting positions. The curves extending from the initial vehicle positions to the left (i.e., toward the velocity axis) show maximum acceleration curves for the simulation; that is, the acceleration that would result from full deflection of the accelerator pedal. The dashed curve extending to the origin from the right shows the maximum deceleration for the simulation (i.e., the deceleration that would result from full deflection of the brake pedal).

The state space, together with the control limits, provides one way to analytically represent constraints within the field of travel. We will focus on the darkened curve for the farthest position (300 m). However, the same logic can be applied to the curves associated with any of the five initial conditions. The region enclosed by the acceleration curve and the two coordinate axes represents the field of possible travel. It is possible for the driver to reach any state inside this region. The region below and to the right of this curve represents the states that are unreachable in the simulation. Given that the car is constrained to only move forward and there is a constraint on maximum acceleration, there is no possible way to reach states in that region. The fastest path to the position of the collision object (vertical axis) is along the maximum acceleration path. However, this path leads to a collision at high velocity (over 60 m/sec).

The region enclosed by the maximum acceleration curve, the dashed maximum deceleration curve, and the horizontal coordinate axis represents the safe portion of the possible field of travel. From within this region it will be possible to stop prior to collision. The region below the dashed deceleration curve represents collision conditions. That is, for all combinations of position and speed in this region, it is impossible to reach zero velocity prior to reaching zero distance—a collision is inevitable. If a driver enters this region, he/she has "overdriven his/her brakes". Thus, this region is outside the safe field of travel.

The path along the edge of the safe field of travel, moving from the initial position along the maximum-acceleration curve until it intersects with the maximum-deceleration curve and then following the deceleration curve to the origin, represents the minimum-time path through the state space that satisfies the goal of stopping prior to collision. That is, any other path through the safe field of travel will take longer to travel to the same position. This path would be achieved through a bang-bang style of control—full acceleration to the point of intercepting the maximum-deceleration curve, then full braking to the point of zero velocity.

This driver from McKenna's (2004) study tended to converge toward a bang-bang style of control with practice. By the twenty-fifth block he/she tended to move along the maximum-acceleration curve, releasing the accelerator prior to intersecting the maximum-deceleration curve. The dashed line extending from the origin represents a *tau-like* function (constant time-to-collision, assuming a constant velocity of motion) fit to the data for Block 25. In fact, when computed based on optical state variables (visual angle and expansion rate) a linear function consistent with a *tau* strategy accounted for 80% of the performance variance for Block 1 and 89% of the variance for

Block 25. Using the physical states (position and velocity) less variance is accounted for (approximately 42% of the variance). Thus, this particular driver appears to be using a bang-bang style of control and the switch from acceleration to braking seems to be triggered by a criterion that is consistent with Lee's optical *tau* function.

If you observe the progression from Block 1 (open symbols) to Block 25 (filled symbols), it suggests that the driver is learning or tuning to the control limits. That is, the point of accelerator release tends to be closer to the edge of the field of safe travel in the later block. This is consistent with the task instructions to get to the target as quickly as possible. One might hypothesize that the driver is learning to "see" his action capabilities in terms of the optical variable *tau*. He is beginning to "see" the optimal time to release the accelerator and initiate hard braking for the dynamic constraints (braking limits) of the simulation. Or it might be said that the driver is developing an accurate internal model of the simulation dynamics (e.g., Jagacinski & Miller, 1978).

Figure 43.5 and the associated experiment by McKenna (2004) illustrate one facet of the field of travel (or affordances): the action constraints or control limits. However, a potentially important facet of the natural field of travel is not well-represented in either the diagram or the experiment. That is the consequences which are associated with the actions. In the simulation, while the participants were instructed to avoid collisions, the actual consequences of stopping a little too early or a little too late were essentially the same. In the natural driving environment this is not the case. The consequences of braking too late are dramatically different (a collision with the potential for costly damage to the vehicle and the driver), than the consequences of braking too early (need for corrective action and increased time to target).

Another important consequence of action in natural driving conditions that was not experienced by the participants in McKenna's study was g-force. With the desktop simulation, the task was purely visual (kinematic), whereas in the natural driving situation there is a kinetic (g-force) component. It would be unusual to see drivers adopt a bang-bang style of control to accomplish tasks similar to those in McKenna's study (e.g., pulling into a parking spot against a building). If a driver did adopt this strategy, his/her passengers would not be happy. Figure 43.6 provides another hypothetical model of the safe field of travel with respect to braking. The boundaries of the field of travel represent acceptable levels of g-force. Normal braking typically results in forces on the order of 0.3 g and hard braking results in forces on the order of 0.6 g. A driver who was concerned about his/her own comfort (and that of their passengers) would typically strive to keep within the comfort zone region of state space. They would try to avoid hard accelerations and situations that might require hard braking.

Thus, in the driver's control solution, it is important to consider both attunement to the control limits (action constraints) and the value criteria reflected in the cost functional that the driver is attempting to satisfy (consequences). Again, the choice of control magnitude (i.e., gain) in solving the control problem depends both on the dynamics of the vehicle (e.g., the status of

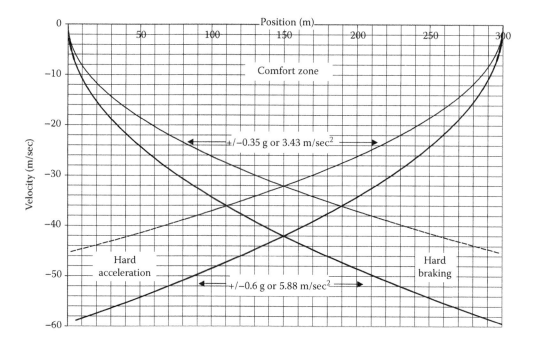

FIGURE 43.6 This figure illustrates a hypothetical model of the comfort zone for accelerating and braking in which the limits are specified in terms of g-forces. The zero position (left boundary) of this state space represents a potential collision object or obstacle, and the axes represent position (distance from this obstacle) and velocity toward the obstacle.

the brakes) and the criteria for successful control as typically reflected in a cost functional. Note that the cost functional is not typically represented in control diagrams. However, this function is an important consideration to the engineer who is choosing the control gains or the control algorithms.

A major point for this section is the need to consider both the constraints on action and the value or cost constraints that guide the choice from among the options for action. These constraints are not typically represented in control diagrams. However, they are critical to the logic for choosing one control solution relative to another. We suggest that state spaces showing the action and comfort/success boundaries can be an important tool for researchers who are interested in understanding the logic of biological control systems. In this context, the normative models for optimal control are used to partition the state or problem space into qualitative regions that provide clear benchmarks for evaluating and interpreting the behaviors of biological control systems.

It is also important to note any potential differences between the constraints in the simulator environment and constraints in the environments to which either training is intended to transfer, or to which models are intended to generalize. Thus, for example, the lack of g-forces in a simulator may lead to very different control behavior in that context (e.g., more aggressive bang-bang strategies).

43.5 The Fundamental Regulator Paradox

The task of a regulator is to eliminate variation, but this variation is the ultimate source of information about the quality of work. Therefore, the better job a regulator

does, the less information it gets about how to improve. (Weinberg & Weinberg, 1979, p. 250)

This quote is Weinberg and Weinberg's definition of the Regulator Paradox. It is interesting in the context of this book that they go on to illustrate this paradox with a driving example:

This lesson is easiest to see in terms of an experience common to anyone who has ever driven on an icy road. The driver is trying to keep the car from skidding. To know how much steering is required, she must have some inkling of the road's slickness. But if she succeeds in completely preventing skids, she has no idea how slippery the road is.

Good drivers, experienced on icy roads, will intentionally test the steering from time to time by 'jiggling' to cause a small amount of skidding. By this technique, they intentionally sacrifice the perfect regulation they know they cannot attain in any case. In return, they receive information that will enable them to do a more reliable, though less perfect job. . .

Without the jiggling . . . the driver . . . doesn't have any way of knowing if success comes from skill, or from the momentarily benign environmental conditions. The jiggling ... tactic is an investment in model improvement. Not only does it refine the model of the transitory environment, but also, even more important, it refines the model of the system's own regulatory ability . . . (p. 251).

Simple automatic control systems are typically designed off-line. That is, the jiggling needed to build a "model" of the control problem is done by the designer, not by the control system. In the off-line design of simple control systems, engineers have found it to be convenient to treat the observer and controller processes independently. Thus, engineers and engineering courses often focus on one of these problems (ideal observers or optimal controllers) independent of the other. However, biological systems typically design their control solutions through trial and error. That is, they have to tune their control solutions while actively engaged in solving the control problem. This is generally referred to as the problem of adaptive control. It requires a kind of meta-cognition in which drivers monitor system performance and adjust their effective *control laws* in order to improve performance (i.e., satisfy some value index). This adjustment could be a change of goals, priorities, and values. It could involve shifts of attention or attunement of the observation process to monitor states that are discovered to be relevant. It can also involve changing the mapping from state to control. In other words, while simple, automatic systems are designed by an extrinsic agent, biological control systems are self-organizing.

Figure 43.7 adds outer loops to the control diagram to reflect the meta-cognitive processes associated with evaluation and adaptation. The different loops are analogous to some of the tactics that have been used by control systems designers to solve the adaptive control problem. These tactics include *gain scheduling, model reference adaptive control,* and *self-tuning.* In each case, the simple controller is augmented by an outer-loop that monitors and adjusts inner-loop performance. *Gain scheduling* reflects a kind of context sensitivity. This control system monitors some aspect of the control context and it chooses one of several pre-stored control algorithms that best fits the context. For example, in aviation systems, the control system chooses different gains for different ranges of altitude. In a driving context, a driver may choose a different style of control for driving in different road conditions (e.g., dry, wet, snowy, or icy).

A second tactic is *model reference adaptive control.* With this tactic, the actual response of the vehicle is compared to an "internal model" or expectation of what the "normal" response should be. For example, a driver may have expectations about how brakes should respond. When the expectations are not met (i.e., the driver is surprised), the control process is adjusted in a direction intended to bring the actual response into closer alignment with the expected normal response. For example, if the response to braking is less deceleration than expected due to poor brakes or surface conditions, the driver might increase the braking force, until the deceleration matches expectations.

A third tactic for monitoring and adapting control processes is *self-tuning.* With this tactic, the input-output relations are being continuously observed to directly or indirectly (i.e., via revisions to an internal model of the plant) adjust the controller. This continuous process is typically supported by active tests (e.g., exploratory actions) of hypotheses about the plant. For example, after driving through high water a driver might tap the brakes to see if the braking dynamics have changed. The jiggling in the Weinbergs' example above illustrates a technique called dithering, where continuous small amplitude, high-frequency movements are made together with normal control actions. The small amplitude movements are designed to have minimum impact on the target response (e.g., lane maintenance), while providing continuous information about changes in the plant dynamic (e.g., road slickness).

In the self-organization process, the coupling of observation, control, and adaptation may be much more intimate than suggested by the partitioning of the observation and control problems as suggested by Figure 43.1, Figure 43.2, and Figure 43.7. Whereas, observation, control, and adaptation tend to be separable operations in engineered systems, in biological systems these processes are more likely to be tightly coupled processes. That is, every action may function to simultaneously reduce error in service of control and to generate information in service of observation of both the situation and the self. Gibson (1963) recognized this coupling with the terms *performatory action* (control function) and *exploratory action* (observer function). Figure 43.8 represents our attempt to make this intimate coupling salient.

The circles in Figure 43.8 represent three sets of constraints similar to those of Peirce's triadic model of semiotics [see Hawthorne, Weiss and Burks (Eds.) 1931–1966] and Neisser's (1976) perception-action cycle. Beginning from the left, reading the non-italicized labels, Figure 43.8 illustrates a feedback control process in which <u>intentions</u> originating from experience are compared with <u>consequences</u> being fed back from situations as represented in an information medium. The differences resulting from this comparison (i.e., <u>errors</u>) can be reduced either through <u>performatory actions</u> on the situation or through change of the intention. Simultaneously, an observation process is also happening in which *expectations* are compared with *results* stimulated by *exploratory actions.* The comparison can result in *surprise,* which can lead to both change in expectations and further exploration. Any action (e.g., depressing the brake) will have both performatory (e.g., decelerating the car) and exploratory (e.g., updating expectations about brake dynamics or road surfaces) implications. These actions can change beliefs (experiential constraints) as well as the physical circumstances of those beliefs (situational constraints). The link that allows the simultaneous coordination of control and observation is the informational constraints.

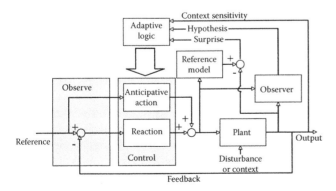

FIGURE 43.7 Three styles of outer-loop adjustments are shown as possible components in an adaptive control process.

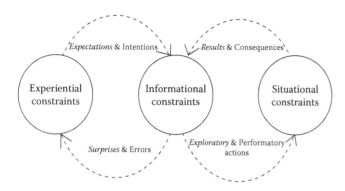

FIGURE 43.8 This diagram attempts to represent the intimate coupling between control (action) and observation (perception), where these processes are operating in parallel.

The adaptive processes associated with parallel control and observation reflects a process of "learning by doing" that is consistent in many ways with Peirce's concept of *abduction*. Peirce introduced abduction as a pragmatic alternative to classical logical models of human thinking (i.e., deduction and induction). An abductive process is driven by surprise. In such a system, belief is tentative. It reflects hypotheses about situations. These hypotheses are tested, not in terms of abstract truth or normative logical processes, but in terms of success (i.e., the consequences of acting on them). Beliefs that lead to successful actions are maintained, while beliefs that lead to unsuccessful actions are revised and modified. In an abductive system, beliefs are shaping actions, while simultaneously being shaped by the consequences of those actions.

The simplicity of Figure 43.8 compared to Figure 43.1, Figure 43.2, and Figure 43.7 is deceiving. In the earlier diagrams the complexity is parsed in a logical way that reflects the modular organization of designed systems and of the associated academic disciplines. It would be nice if we could divide the system so that the psychologist could model the human, while the mechanical and automotive engineers modeled the vehicle, the computer scientists and electrical engineers modeled the information displays, and the civil engineers modeled the highway environments. It would be nice if we could neatly separate the control problem from the observer problem in ways that were compatible with the normative models of control and observation. Unfortunately, the coupling of perception and action (observation and control, performance and exploration) are both richer and more complex than suggested by many of the normative control models or designed control systems. As Juarrero (2002) notes:

> … complex adaptive systems are typically characterized by … feedback processes in which the product of the process is necessary for the process itself. Contrary to Aristotle, this circular type of causality is a form of self-cause. Second, when parts interact to produce wholes, and the resulting distributed wholes in turn affect the behavior of their parts, interlevel causality is at work. Interactions among certain dynamical processes can create a systems-level

organization with new properties that are not the simple sum of the components that constitute the higher level. In turn, the overall dynamics of the emergent distributed system not only determine which parts will be allowed into the system: the global dynamics also regulate and constrain the behavior of the lower level components. (p. 5–6)

Thus, we are suggesting caution when using engineered control systems as models for adaptive human-in-the-loop systems. We strongly believe that intuitions from working with simple closed-loop systems and the methodologies of control engineering will be essential for unpacking the complexity of biological control systems. However, when simple control systems are used as a metaphor, we fear that they can hide as much as they reveal. The result can sometimes be a trivialization of the phenomena, rather than a deepening understanding. We believe that simulators can be critical components in research programs that have the goal of converging on that deeper understanding. They allow simultaneous measurement at multiple levels of detail under controlled conditions. For example, in evaluating steering behavior as a function of risk and optical flow information (e.g., comparing performance on the open road versus performance in a construction zone with an elevated cement barrier on one edge of the lane) it may be that gross measures of lane excursions or tracking error may show no performance differences. However, time histories of the steering activity or of the car path may show differential properties (e.g., more high-frequency power in the barrier condition) that would suggest subtle changes in control strategy (e.g., higher gain).

43.6 Control Theory: More Than Meets the Eye

It is important to differentiate between control theory, as a guide to research and design, and the specific analytic tools and solutions that have been motivated by this approach. As systems become very complex, the application of many of the analytic tools may be of limited value or may be completely infeasible, and any specific engineered control mechanism is likely to be a relatively weak metaphor. Nevertheless, we feel that the intuitions of control theory can be an important guide. In fact, the theory is very important even where the analytic tools and simple metaphors fail. At this point, a mindless application of the tools will not be useful and theoretically informed judgment and hypothesis (i.e., abduction) become essential. For those interested in the analytic techniques themselves, we recommend Jagacinski and Flach (2003) as a tutorial introduction to these techniques and the qualitative insights they provide for dynamic behavior.

We hope it is becoming clear at this point that many of the constraints that shape the solutions to control and observation problems are not well-represented in the block diagram images used to illustrate control mechanisms. It is important to differentiate between the control problem and the control mechanism. For any control problem there will be many potential mechanisms that might yield satisfactory or stable performance. For example, a cost functional reflects constraints on the control problem, but it is not an explicit component of the control mechanism. Thus, there is a need for some way, other than the diagram of a control mechanism, to summarize the problem constraints. This fact is an important motivation for Rasmussen's (1986) Abstraction Hierarchy. The Abstraction Hierarchy is a nested hierarchy of means-ends relations that are used to model the problem space for a complex control system.

Table 43.1 is an attempt to map the levels in the Abstraction Hierarchy to constraints of control problems. The left column shows Rasmussen's levels, and the right column is our view of how these levels are reflected in control theoretical terms. At the highest level, *Functional Purpose*, the focus is on the goals and resources associated with the control problem. The priorities that specify how goals are traded-off against each other (e.g., lane maintenance might compete with obstacle avoidance) or how the goals are traded-off against resources (e.g., precision versus effort) are reflected in cost functions. Note that the cost functions or value constraints that are guiding a biological control system will not always be obvious or clearly specified. Thus, in analyzing the control problem, one should consider what the effective cost functional might be. Normative models (e.g., the optimal control model) might be very useful to provide benchmarks associated with specific cost functions (e.g., the quadratic cost functional or a minimal time solution). However, in using the abstraction hierarchy as a component in a formative program of work analysis as described by Vicente (1999), the goal would be to consider possible or plausible cost functions. What are the goals and resources? What range of trade-offs would be reasonable or likely? How might different training, display, or control innovations impact the priorities or trade-offs (e.g., reduce the resource costs or workload)?

At the second level, *Abstract Function*, attention turns to the space of possibilities in its broadest sense. This space should obviously include the goal states and the potential paths of movement through the problem space. The dimensions of this space should reflect the states of the problem, reflecting physical, information, resource, and regulatory constraints on motion in that space. In other words, the dimensions of this space should reflect any variable that might be required to specify functional distance from a goal or that might be needed to differentiate various paths to the goal. For example, for a vehicle control task the state space should

at least include functions of position and velocity for each axis of motion. The space might also take into account information constraints. For example, Smith et al. (2001) found that using an optical state space (visual angle and expansion rate) helped in identifying the decision rules for a collision control task. This space reflects the joint constraints of the inertial properties of motion and the optics of flow fields. Again, this space is not a property of the control mechanism, but a property of the problem context.

At the third level of the Abstraction Hierarchy, *General Function*, the classical functional flow diagram such as reflected in Figure 43.1 provides a useful way to visualize the constraints associated with the partitioning of function and flow of information. At this stage, consideration is given to the demands created by the problem and the general types of processes that might meet those demands. Classically, this has been the focus for information-processing models inspired by the Cybernetic Hypothesis (Weiner, 1948). While this level of abstraction offers its own insights into the dynamic constraints associated with specific system architectures (e.g., open- versus closed-loop dynamics), these insights are often missed because of tendencies to emphasize open-loop, causal relations between pairs of stages, rather than the more global constraints (e.g., stability limits) of the overall system organization (e.g., closed-loop dynamics). As Jagacinski and Flach (2003) have noted, the information-processing modelers have often included closed-loops in their diagrams, but have rarely addressed the circular flow of mutual causality that such diagrams imply.

At the fourth level of the Abstraction Hierarchy, *Physical Function*, consideration is given to the particular types of systems that might satisfy the functions identified at the General Function level. For example, considerations of whether braking is accomplished by a human driver or by an automated braking system become important. Choices at this level can have important implications for considerations at all the other stages. In modeling an automated braking system, the dimensions of the state space (the Abstract Function level) should reflect the dimensions assumed in the design of the automated system (e.g., functions of distance and velocity). However, in modeling manual braking, it may be important to consider optical constraints in addition to the inertial constraints in choosing the dimensions of the state space. Individual differences among drivers may be an important consideration here, since the driver is the physical

TABLE 43.1   Control Problem Constraints Organized According to Level of Abstraction

Rasmussen's Levels (Rasmussen, 1986)	Control System Constraints
Functional Purpose	Goals or forcing functions; resources; cost functions or value systems.
Abstract Function	The physical (e.g., laws of motion), information (e.g., ecological optics), and regulatory (e.g., speed limits) constraints that guide the choice of state space dimensions.
General Function	This reflects the general functional organization as reflected in block diagrams of generic functions such as Figure 43.1.
Physical Function	This level reflects the allocation of function among different physical systems (e.g., which functions are accomplished by humans as illustrated in Figure 43.3 or possibly the different modalities of information).
Physical Form	This level considers the physical details (e.g., the layout of controls and displays and or the detailed geometry of information sources such as in an optical flow field).

system that satisfies important control functions. What should the performance criteria be for deciding whether a person is capable of functioning safely? Can simulators be used to test driver capabilities? For example, Caird, Chisholm, Edwards and Creaser (2007) examined how age impacts perceptual reaction time relative to the response to a yellow light.

At the lowest level of abstraction, *Physical Form*, the physical details of specific systems become important to the success of the control system. A significant example of the importance of such details for control solutions is the case of unintended accelerations (Schmidt, 1989). A physical detail such as the rotation of the head over the right shoulder while initiating a backing-up maneuver may bias the foot trajectory to the right, away from the brake and toward the accelerator. This physical detail can be the difference between an action that satisfies the intention and one that leads to potentially catastrophic failure. As an aside, unintended accelerations can be an example where at least for a brief time, human drivers function as effectively open-loop, automatic systems that are unable to make use of the error feedback indicating that they are depressing the accelerator while intending to brake.

The key point is that a research program to understand driving *must* consider multiple levels of abstraction as represented in Rasmussen's Abstraction Hierarchy. It is important to differentiate between the use of a specific control system as a metaphor for human performance, and a serious program to identify the constraints shaping driver performance. Understanding the system as a control system means to understand the loose couplings across these different levels of constraints, for example, to differentiate when drivers are functioning as a closed-loop system (e.g., adjusting on the basis of feedback) and when drivers are functioning open-loop (e.g., applying automatic routines that have been reinforced by consistent experiences).

It is important to appreciate that no single level of abstraction is privileged (e.g., as ultimate cause or as ground truth) with respect to the others; and that each level has the potential to significantly affect the emergent behaviors of the total system. Also, different forms of representation (e.g., block diagrams or state spaces) may be better for different levels of abstraction. As with the levels, no single representation is sufficient. Each representation reveals some facets of the problem and hides others.

Simulators can be critical tools for exploring the couplings across levels of abstraction. The simulation can allow researchers to manipulate any of the five levels of abstraction and to measure the implications at all other levels. For example:

- The level of Functional Purpose can be manipulated through instructions and payoff matrices.
- The level of Abstract Function can be manipulated through adjusting the vehicle/surface dynamics or information constraints.
- The levels of General Function and Physical Function can be manipulated by controlling access to certain types of information or by constraining/automating or uncoupling some of the functions (e.g., occluding the speedometer to force the person to estimate speed from optic flow).

- The Physical Form (e.g., position of the pedals) can be directly manipulated.

In many human social technical systems (e.g., nuclear power plants, air traffic control, or high performance aircraft) the simulator may be the only context where such manipulations and measurements are feasible. In driving, instrumented cars can also provide access to multiple levels of measurement and opportunities to sample across many contexts that reflect important differences associated with various levels of abstraction.

The US Department of Transportation has funded field operation tests (FOTs) to investigate the human interaction with emerging active-safety technologies (e.g., forward collision warning and lane departure warnings). These studies have cost tens of millions of dollars, and in collecting driver behavior for hundreds of thousands of miles, have contributed greatly to our understanding of driver acceptance of active safety systems. However, despite the grand scale of these field operational tests, they still provide an insufficient basis to make solid conclusions about the ability of active safety systems to minimize collisions. Although on global or national scales automotive crashes occur far too frequently and have the potential to produce devastating consequences for the people involved, on the scale of any given mile of travel, the chances of collision are low. Police-reported crashes in the United States occur on average only two times every million miles or once every 32 years of driving (derived from NHTSA, 2006; or see Evans, 1991). The approach of waiting for collisions to occur naturally, although extremely informative (see Dingus et al., 2006), is clearly impractical for most research applications and research time-scales.

For the purposes of human subject testing, test tracks represent a compromise in many respects between driving simulator research and on-road testing. However, test tracks are also "simulations" in that the situations faced may be quite distinct from those in natural driving contexts (e.g., limited traffic). Because of the level of control, test track research may provide an opportunity to examine collision minimization more efficiently than field testing. In some circumstances, methods can be developed on the test track where the driver believes that they are at risk when in reality there is actually little risk to the driver. One of the difficulties with human subject research on test tracks is that drivers may not feel as completely at ease as they might in a well-designed driving simulator test or in the field. The need to communicate with traffic control and other vehicles, the responsibility of driving in a novel environment, and the accompanying experimenter can sometimes overwhelm participants, preventing them from relaxing or adopting realistic driving behaviors. As a result, it may be quite difficult to distract drivers to a large extent. Test-track research is also usually less efficient than driving simulator research and often involves logistical issues that can become expensive and time-consuming. Nevertheless, test track research can be a valuable tool for validating driving simulator testing, conducting system-verification tests, and for rapidly educating subjects about the system capabilities. Due to the respective advantages and disadvantages of the three venues (field, test-track, and driving simulation), a recent ongoing government-sponsored program utilized a combination of venues; an efficient between-subject driving simulator methodology to focus on collision minimization, field-testing to focus on driver acceptance, and test-track work to rapidly expose drivers to the system prior to field testing (Smith, Bakowski, & Witt, 2007). In this context, control theory may help to identify the long threads that connect levels of abstraction that, in turn, allows integration of insights obtained across many different research venues.

In conclusion, we argue that it is very clear that driving is a control problem. That is, behaviors are being adapted at multiple time scales to satisfy various intentions. As such, the intuitions of control theory will be invaluable for a research program aimed at driver performance, training systems design, or highway safety. However, it is important not to confuse control theory with any specific control mechanism. Control theory provides tools for a search for explanations of how drivers might behave. It is not the answer; but it can provide important intuitions for unpacking the very complex dynamics of an adaptive, self-organizing biological control system. Control theory is not the end, but one means for attacking this very important and interesting phenomena of driving.

Key Points

- The human-vehicle and the human-simulator systems are complex control systems.
- Control theory can provide an important framework when using simulators for performance evaluation to meet the applied goals of performance evaluation and training design.
- Control theory can provide an important framework for integrating simulators into basic research programs to evaluate humans as adaptive cognitive systems.

Keywords: Abstraction Hierarchy, Control Theory, Observer, Optimal Control Model, Regulator Paradox, Stability

Key Readings

Ashby, W. R. (1956). *An introduction to cybernetics.* London: Chapman & Hall.

Gibson, J. J. (1958). Visually controlled locomotion and visual orientation in animals. *British Journal of Psychology, 49,* 182–194. Reprinted in Reed, E., & Jones, R. (Eds.). (1982). *Reasons for Realism* (pp. 148–163). Hillsdale, NJ: Lawrence Erlbaum Associates.

Jagacinski, R. J., & Flach, J. M. (2003). *Control theory for humans.* Mahwah, NJ: Lawrence Erlbaum Associates.

References

Ashby, W. R. (1956). *An introduction to cybernetics.* London: Chapman & Hall.

Bernstein, N. (1967). *Coordination and regulation of movement.* New York: Pergamon Press.

Brunswik, E. (1956). *Perception and the representative design of psychological experiments.* Berkeley, CA: University of California Press.

Caird, J. K., Chisholm, S. L., Edwards, C. J., & Creaser, J. I. (2007). The effect of yellow light onset time on older and younger drivers' perception response time (PRT) and intersection behavior. *Transportation Research Part F: Traffic Psychology and Behavior, 10,* 383–396.

Caird, J. K., Willness, C. R. Steel, P., & Scialfa, C. (2008). A meta-analysis of the effects of cell phones on driver performance. *Accident Analysis & Prevention, 40*(4), 1282–1293. doi:10.1016/j.aap.2008.01.009

Chisholm, S. L., Caird, J. K., & Lochhart, J. (2007). The effects of practice with MP3 players on driving performance. *Accident Analysis & Prevention, 40*(2), 704–713. doi:10.1016/j.aap.2007.09.014

Cochran, A., & Stobbs, J. (1968). *The search for the perfect swing.* Philadelphia, PA: J. B. Lippincott.

Denton, G. G. (1980). The influence of pattern on perceived speed. *Perception, 9,* 393–402.

Dingus, T. A., Klauer, S. G., Neale, V. L., Petersen, A., Lee, S. E., Sudweeks, J, . . . Knipling, R. R. (2006). *The 100-car naturalistic driving study: Phase II—Results of the 100-car field experiment* (NHTSA DTNH22-00-C-07007). US Department of Transportation, National Highway Traffic Safety Administration. Retrieved from http://www-nrd.nhtsa.dot.gov/departments/nrd-13/driver-distraction/PDF/100CarMain.pdf

Evans, L. (1991). *Traffic safety and the driver.* New York: Van Nostrand Reinhold.

Fajen, B. R. (2005). Calibration, information, and control strategies for braking to avoid collision. *Journal of Experimental Psychology: Human Perception and Performance, 31*(3), 480–501.

Flach, J. M., Dekker, S., & Stappers, P. J. (2008). Playing twenty questions with nature (The surprise version): Reflections on the dynamics of experience. *Theoretical Issues in Ergonomics Science, 9*(2), 125–145.

Flach, J. M., Smith, M. R. H., Stanard, T., & Dittman, S. M. (2004). Collisions: Getting them under control. In H. Hecht & G. J. P. Salvelsbergh (Eds.), *Time-to-contact.* (pp. 67–91). Amsterdam: Elsevier Science.

Flach, J. M., & Warren, R. (1995). Low-altitude flight. In P. Hancock, J. Flach, J. Caird, & K. Vicente (Eds.), *Local applications of the ecological approach to human-machine systems* (pp. 65–103). Hillsdale, NJ: Lawrence Erlbaum Associates.

Gelb, A. (Ed.). (1974). *Applied optimal estimation.* Cambridge, MA: MIT Press.

Gibson, E. J. (1969). *Principles of perceptual learning and development.* New York: Appleton-Century-Crofts.

Gibson, J. J. (1958). Visually controlled locomotion and visual orientation in animals. *British Journal of Psychology, 49,* 182–194. Reprinted in Reed, E., & Jones, R. (Eds.). (1982). *Reasons for Realism* (pp. 148–163). Hillsdale, NJ: Lawrence Erlbaum Associates.

Gibson, J. J. (1963). The useful dimensions of sensitivity. *American Psychologist, 18*, 1–15.

Gibson, J. J. (1979/1984). *The ecological approach to visual perception.* Hillsdale, NJ: Erlbaum.

Gibson, J. J., & Crooks, L. E. (1938). A theoretical field-analysis of automobile driving. *American Journal of Psychology, 51*, 435–471. Reprinted in Reed, E., & Jones, R. (Eds.). (1982). *Reasons for Realism* (pp. 119–147). Hillsdale, NJ: Lawrence Erlbaum Associates.

Gibson, J. J., Olum, P., & Rosenblatt, F. (1955). Parallax and perspective during landings. *American Journal of Psychology, 68*, 372–385.

Hawthorne, C., Weiss, P., & Burks, A. (Eds.). (1931–1966). *Collected papers of Charles Sanders Peirce.* 8 volumes. Cambridge, MA: Harvard University Press.

Hecht, H., & Salvelsbergh, G. J. P. (Eds.). (2004). *Time-to-contact.* Amsterdam: Elsevier Science.

Jagacinski, R. J., & Flach, J. M. (2003). *Control theory for humans.* Mahwah, NJ: Lawrence Erlbaum Associates.

Jagacinski, R. J., & Miller, R. A. (1978). Describing the human operator's internal model of a dynamic system. *Human Factors, 20*, 425–433.

Juarrero, A. (2002). *Dynamics in action. Intentional behavior as a complex system.* Cambridge, MA: MIT Press.

Kirk, D. E. (1970). *Optimal control theory.* Englewood Cliffs, NJ: Prentice-Hall.

Kleinman, D. L., Baron, S., & Levison, W. H. (1971). A control theoretic approach to manned-vehicle systems analysis. *IEEE Transaction on Automatic Control, AC-16*, 824–832.

Lee, D. N. (1976). A theory of visual control of braking based on information about time-to-collision. *Perception, 5*, 437–459.

Loomis, J. M., Da Silva, J. A., Fujita, N., & Fukusima, S. S. (1992). Visual space perception and visually directed action. *Journal of Experimental Psychology: Human Perception and Performance, 18*, 906–921.

McKenna, B. P. (2004). *Role of optical angle and expansion rate in a braking action.* Unpublished Masters thesis. Wright State University, Dayton, OH.

National Highway Transportation Safety Administration (NHTSA). (2006). *Traffic safety facts* (Rep. DOT HS 810 623). Washington, DC: US Department of Transportation. Retrieved from http://www-nrd.nhtsa.dot.gov/Pubs/810623.pdf

National Research Council. (1997). *Aviation safety and pilot control: Understanding and preventing unfavorable pilot-vehicle interactions.* Washington, DC: National Academy Press.

Neisser, U. (1987). From direct perception to conceptual structure. In U. Neisser (Ed.), *Concepts and conceptual development: Ecological and intellectual factors in categorization* (pp. 11–24). Cambridge, MA: Cambridge University Press.

Newell, A., & Simon, H. A. (1972). *Human problem solving.* Englewood Cliffs, NJ: Prentice-Hall.

Owen, D. H. (1984). *Optical flow and texture variables useful in detecting decelerating and accelerating self-motion*

(AFRHL-TP-84-4, AD-1248 718). Williams Air Force Base, AZ: Operations Training Division, Air Force Research Laboratory.

Pew, R. W., & Baron, S. (1978). The components of an information processing theory of skilled performance based on an optimal control perspective. In G. E. Stelmach (Ed.), *Information processing in motor control and learning.* Cambridge, MA: MIT Press.

Rasmussen, J. (1986). *Information processing and human-machine interaction: An approach to cognitive engineering.* New York: North-Holland.

Schmidt, R. A. (1989). Unintended acceleration: A review of human factors contributions. *Human Factors, 31*(3), 345–364.

Senders, J. (1964). The human operator as a monitor and controller of multidegree of freedom systems. *IEEE Transactions on Human Factors in Electronics, HFE-5*, 2–6.

Smith, M. R. H. (2002). *Automotive collision avoidance system field operational test. Warning cue implementation summary report.* US Department of Transportation, Federal Highway Administration. Retrieved from: http://www.itsdocs.fhwa.dot.gov//JPODOCS/REPTS_TE/13674.pdf

Smith, M. R. H., Bakowski, D. L., & Witt, G. J. (2007). *The utilization of different testing venues for the evaluation of safety and acceptance in the SAVE-IT program.* Proceedings of the Driving Simulation Conference North America. Iowa City, IA. Available from: http://www.dsc-na.org/

Smith, M. R. H., Flach, J. M., Dittman, S. M., & Stanard, T. (2001). Monocular optical constraints on collision control. *Journal of Experimental Psychology: Human Perception, and Performance, 27*, 395–410.

Strayer, D. L., & Johnston, W. A. (2001). Driven to distraction: Dual task studies of simulated driving and conversing on a cellular phone. *Psychological Science, 12*(6), 462–466.

Vicente, K. J. (1999). *Cognitive work analysis.* Mahwah, NJ: Lawrence Erlbaum Associates.

Warren, R. (1982). *Optical transformation during movement: Review of the optical concomitants of egomotion* (Final Tech. Rep. for Grant N. AFOSR-81-0108). Columbus, OH: Ohio State University. Also appears as Tech. Rep. AFOSR-TR-82-1028, Bolling Air Force Base, DC: Air Force Office of Scientific Research. (NTIS no. AD-A122 275).

Warren, R., & McMillan, G. R. (1984). Altitude control using action-demanding interactive displays: Toward an active psychophysics. *Proceedings of the 1984 IMAGE III conference* (pp. 37–51). Phoenix, AZ: Air Force Human Resources Laboratory.

Warren, R., & Wertheim, A. (Eds.). (1990). *Perception & control of self-motion.* Hillsdale, NJ: Erlbaum.

Weinberg, G. M., & Weinberg, D. (1979). *On the design of stable systems.* New York: John Wiley & Sons.

Weiner, N. (1948). *Cybernetics or control and communication in the animal and machine.* Cambridge, MA: The MIT Press.

VI

Applications in Medicine

Acute Alcohol Impairment Research in Driving Simulators

Janet I. Creaser
University of Minnesota

Nicholas J. Ward
University of Montana

Michael E. Rakauskas
University of Minnesota

Abstract

The Problem. Acute alcohol impairment is an important research topic for traffic safety as alcohol is a factor in a significant portion of traffic crashes. This chapter discusses how to identify the types of impairment due to different levels of blood alcohol concentration using a driving simulator. *Role of Driving Simulators.* Driving simulators have been used for almost 40 years to test the effect of alcohol on driving performance. Driving simulation provides the scenario control and ethical management of risk to support experimentation of impairment effects across levels of alcohol. *Key Results of Driving Simulator Studies.* Research using driving simulation has identified impairment of steering and braking performance, problems with maintaining vehicle position and speed, and impairment of reaction time to hazards. In general, driving simulators were able to identify many of the impairment effects seen in closed-course testing using a real vehicle. *Scenarios and Dependent Variables.* A standard set of driving scenarios does not exist for alcohol-impaired driving as the goal of previous research has been to identify all areas of driving skill that are impaired by alcohol. In general, scenarios can include common driving situations used in most types of driving simulation research, such as car following, responding to emergency events, and general driving on a variety of road types. Measures that are observable in both test track and simulator studies may hold the most validity for alcohol research and these include speed variability, lateral position deviations, and responses to hazards in the driving environment. However, not all measures have been tested in both settings; therefore, it is important to select measures that reflect the research question or for which impaired performance may lead to a safety critical situation. *Platform Specificity and Equipment Limitations.* Research shows that driving performance impairment occurs in simulated driving and similar measures can be observed from lower- and higher-complexity simulator systems. However, certain measures may not be sensitive to alcohol effects in lower-complexity simulators due to the lower task demand imposed by a low-complexity system that does not require attention management between multiple sensory components analogous to the real world.

44.1 Introduction

Alcohol is a drug that is recognized as a risk factor for many categories of unintentional injury (Smith, Branas, & Miller, 1999; Skog, 2001). In 2006, 17,602 people were killed in the United States (U.S.) in alcohol-related crashes where at least one driver, pedestrian or cyclist had a blood alcohol content (BAC) of 0.01 g/dl or over (National Highway Traffic Safety Administration (NHTSA, 2007). This represents 41.3% of all motor-vehicle crash fatalities for 2006. The number of people killed in alcohol-related crashes where at least one driver or motorcycle operator had a BAC over the U.S. legal limit of 0.08 g/dl was 13,470 (NHTSA, 2008). Of those 13,470 fatalities, 64% were the impaired drivers, 18% were passengers riding with the impaired driver, 11.9% were occupants of other vehicles, and 6.1% were non-occupants (NHTSA, 2008). In 2000, alcohol-involved crashes accounted for $50.9 billion (22%) of the $230.6 billion cost associated with motor vehicle crashes (Blincoe et al., 2002). Given the high incidence and associated cost of alcohol crashes, NHTSA has identified the reduction of alcohol-related traffic fatalities as a priority for improving traffic safety (NHTSA, 2001).

44.1.1 Alcohol and Driving Impairment

Changes in physiological functioning caused by alcohol can be expected to affect psychological functioning of the intoxicated individual, including their capacity for skilled performance. Elemental skills refer to the basic sensory, perceptual, cognitive, and motor control components that are combined to produce complex behaviors such as operating a motor vehicle in traffic. These skills are often assessed with simplified and abstract laboratory tests. For example, the Critical Flicker Fusion test is used as a measure of visual processing speed (perceptual skill), the Digit Substitution Test is used as a measure of information processing (cognitive skill), and Compensatory Tracking is used as a measure of motor control steadiness (motor skill) (Holloway, 1995). The purpose of these tests is to examine the effect of alcohol on rudimentary psychological functions that result from deviations from an optimal physiological state. In driving, impairment of elemental sensory, perceptual, cognitive, and motor skills results in the impairment of the more complex skills which are required to safely operate a vehicle.

Moskowitz and Fiorentino (2000) recently conducted a review of 112 studies published between 1981 and 1997 that examined alcohol impairment tests of elemental skills related to driving. Table 44.1 presents the lowest BAC level that produced significant impairment in a specific test. Table 44.1 also presents the BAC level above which 50% of the studies that included that test demonstrated a significant effect. The results of this review are generally consistent with their conclusions that tasks involving controlled skills requiring attention to multiple information sources and effort to process complex information at different levels are most sensitive to the impairment effects of alcohol.

From these analyses, it is reasonable to postulate that one of the primary loci of alcohol impairment is the disruption of the ability (or willingness) to match attention capacity to task demands (Lamb & Robertson, 1987). For example, impairment on simple (automatic skill) tasks can be overcome at low alcohol doses by investing greater effort (George, Raynor, & Nochajski, 1990). Conversely, the alcohol impairment evident for more complex (controlled skill) tasks may increase with the introduction of additional task demands (Maylor, Rabbitt, James, & Kerr, 1992).

There is evidence to suggest that alcohol may exacerbate the impairment effect of other distractions in non-fatal crashes (Wang, Knipling, & Goodman, 1996). Alcohol may reduce normal visual search patterns (Green, 2003), reduce processing speed (Jones, Chronister, & Kennedy, 1998), restrict attention focus (Bartholow et al., 2003), or impair attention-switching ability within multi-task environments which require controlled skills (Holloway, 1995; Moskowitz & Fiorentino, 2000). Thus, the impairment effect of alcohol on driver behavior may result from a reduced ability to divide attention between several task demands in the driving environment. This may also imply a reduced capacity to focus effort on processing information at each separate level of the information processing system when performing a single task. Obviously, these are not independent sources of impairment such that aspects of both may occur simultaneously. In general, the measurement of driver impairment involves the specification of behaviors that are relevant to crash risk and the formulation of metrics to quantify those behavioral measurements. Typically, these measures quantify behaviors of operating the vehicle controls and the resulting tactical behavior of the vehicle in relation to other traffic and the road environment with respect to lateral and longitudinal control. There are a number of measures that have some empirically demonstrated sensitivity to alcohol effects and theoretical relevance to crashes.

Table 44.1 outlines driving-related measures that reflect impairment along the sensory, perceptual, motor and cognitive skill components of behavior. For example, speed maintenance requires a visual assessment of the speedometer and the outside world (sensory-perceptual skills) and appropriate use of pedal inputs (motor skills). Responding to a hazard event requires

TABLE 44.1 Alcohol Impairment for Specific Human Performance Tests

BAC (g/dl)	Ranking by Lowest BAC Level at Which Significant Impairment was Reported for a Specific Test	Ranking by BAC Level above Which 50% of Studies Reported Impairment for a Specific Test
≥0.10	Critical Flicker Fusion	Simple Reaction Time, Critical Flicker Fusion
0.090 – 0.079		
0.060 – 0.069		Cognitive Tasks (e.g., Digital Symbol Substitution), Psychomotor Skills (e.g. balance, finger tapping, manual assembly), Choice Reaction Time
0.050 – 0.059		Adaptive, Pursuit, Compensatory Tracking
0.040 – 0.049	Simple Reaction Time	Perception (e.g., visual search), Visual Functions (e.g., visual acuity)
0.030 – 0.039	Vigilance, Perception	Vigilance
0.020 – 0.029	Choice Reaction Time, Visual Functions	
0.010 – 0.019	Drowsiness, Psychomotor Skills, Cognitive Tasks, Tracking	Drowsiness (e.g., Multiple Sleep Latency Test, Repeated Test of Sustained Wakefulness)
0.001 – 0.009	Driving, Flying, Divided Attention	Driving, Flying, Divided Attention (e.g., combined central and peripheral task)

Source: Adapted from Moskowitz, H., & Florentino, D. (2000). *A review of the literature on the effects of low doses of alcohol on driving-related skills* (DOT HS 809 028). Washington, DC: National Highway Traffic Safety Administration, US Department of Transportation.

perception of the event and an appropriate behavioral response to avoid the hazard (i.e., correct decision to handle the situation). (1995) reported that a high percentage of simulator studies of driving and flying show sensitivity to alcohol impairment, which may be expected because they comprise complex activities necessitating shared attention to multiple-task features with information processing at multiple levels. Therefore, it is likely that alcohol impairment of one or more skills while driving may result in an increased crash risk.

44.2 Driving Simulators and Alcohol Research

Driving simulators have been used for almost 40 years to test the effect of alcohol on driving performance. Simulators provide a safe environment in which to test a number of driving situations while impaired. Gawron and Ranney (1988) summarized the literature of alcohol impairment on simulated driving tasks in comparison to results obtained from closed-course test tracks. This included investigations on increases in the risk-taking associated with the psychomotor and perceptual effects of alcohol (e.g., Allen, Schwartz, Hogge, & Stein, 1978), steering and braking performance (e.g., Rimm, Siniger, Faherty, Whitley, & Perl, 1982), vehicle position, velocity and reaction time (e.g., Sugarman, Cozad, & Zavala, 1973), and the effects of wind gusts, isolated curves and obstacles during night-time driving (e.g., Stein & Allen, 1984). Overall, they discovered that driving simulation studies and closed-course test tracks were both sensitive for identifying the effects of alcohol in relation to speed variability, lateral-position errors, lateral-position variability, reaction time, and coarse steering-wheel reversals. Differences in results were noted for fine steering control and for increases or decreases in speed. More recent research shows increases in the standard deviation of lane position (SDLP), in brake reaction times (BRT) and changes in coherence (see Brookhuis & de Waard, 1994, for a definition of coherence) when following a lead vehicle on a test track when impaired at a BAC of 0.06% (Kuypers, Samyn, & Ramaekers, 2006). SDLP and BRT have both been demonstrated in simulator research of alcohol impairment (see Table 44.2), but one study that used coherence measures for car following did not find an effect of alcohol (average BAC 0.07%; Rakauskas, Ward, Boer, Bernet, & Cadwallander, 2008). This suggests that simulators are able to capture real-world driving behaviors related to alcohol impairment. However, the ability to capture alcohol-related impairment measures consistently in a simulator or on a test track is likely to be dependent on a number of factors related to study design and implementation.

44.2.1 Scenarios and Measures

The reviewed research covers a number of variables that have been shown to be sensitive to alcohol impairment and these variables can be employed in a number of scenarios. Table 44.2 summarizes the effects of increasing BAC levels on driving simulator variables from a number of studies. The results displayed in Table 44.2 were obtained from a variety of simulators with varying degrees of system complexity. System complexity refers to the *number and range* of components that provide mapping to real-world sensory experiences. This is preferred over the value-laden and somewhat subjective term "fidelity" that is concerned with the *accuracy* of mapping between the simulator and real-world system.

Low-complexity driving simulators, such as the York Driving Simulator (Arnedt, Wilde, Munt, & MacLean, 2001) or the STISIM (Harrison & Fillmore, 2005; Moskowitz, Burns, Fioretino, Smiley, & Zador, 2000), may use one or more video monitors to display the driving scene. For example, the STISIM used by Moskowitz and Fiorentino (2000) had three video monitors to achieve 110 degrees of viewing angle whereas the STISIM used by Harrison and Fillmore (2005) and the York Driving Simulator only had one video monitor each. Medium- to high-complexity simulator systems such as the Transport Research Laboratory simulator (Burns, Parkes, Burton, Smith, & Burch, 2002) and the HumanFIRST Driving Simulator (Rakauskas et al., 2008), provide a more immersive environment with up to 210 degrees forward-field of view, 60-degree rear views, and the use of side mirrors. Additionally, higher-complexity systems may include partial motion-bases to simulate pitch and roll, and audio systems that provide vehicle and road noise, as well as traffic sounds. Higher-complexity simulators can provide the ability to test more complex driving scenarios than can be typically completed on a closed-course, such as scenarios involving traffic or complex urban environments.

Overall, this group of studies shows that driving performance impairment occurs in simulated driving and similar measures can be observed from lower and higher-complexity simulator systems. However, Gawron and Ranney's (1988) research indicated that certain measures may not be sensitive to alcohol effects in lower-complexity simulators. This may be due to the lower task demand imposed by a low-complexity system that does not require attention management between multiple sensory components analogous to the real world. For vehicle control, Gawron and Ranney indicated that speed choice for curve negotiation was more impaired for the closed-course study than the simulator and suggested that the limited results in the simulator may be due to the lack of lateral acceleration cues available in the simulator. When these cues are available, such as on the closed-course, increases in speed variability and speed violations occur at higher BAC levels. Therefore, higher-complexity simulators with motion bases that help indicate lateral acceleration through curves may be more successful at identifying problems with curve negotiation than lower-complexity simulators.

In addition to direct measures of performance in the simulator, it is also possible to collect indirect measures of impairment through psychophysiological measures. For example Brookhuis, de Waard and Fairclough (2003) discuss a variety of measures, such as heart rate variability and EEG to identify impairment due to fatigue or alcohol. Rakauskas, Ward, Bernat, Cadwallader and de Waard (2005) also found that intoxication caused a significant reduction in amplification of the P300 evoked response potential (ERP) in intoxicated drivers. Therefore, psychophysiological

TABLE 44.2 Alcohol Effects in Simulator Research

Variable	BAC Effect[a]	Study
Speed Variability	Increased speed variability (SD speed)	Stein & Allen, 1984 Moskowitz et al., 2000 Arnedt et al., 2001 Weafer et al., 2008
	Speed deviation from the posted speed decreased as subjects drove faster	Arnedt et al., 2001
Speed Violations	Increased number of violations (exceeding the speed limit)	Gawron & Ranney, 1988 Moskowitz et al., 2000
Steering Movements	More steering instability (variability in steering wheel position)	Rakauskas et al., 2008 Weiler et al., 2000 Weafer et al., 2008 Rimm et al., 1982
	Increased steering error	Roehrs et al., 1994
	Increased steering deviations	Weafer et al., 2008
Headway	Smaller minimum following distances to lead vehicle	Weiler et al., 2000
Lane position deviations (lateral position deviations)	Increased lane position variability (SDLP)	Burns et al., 2002 Moskowitz et al., 2000 Rakauskas et al., 2008 Stein & Allen, 1984 Weafer et al., 2008
	Increased tracking variability (SD of vehicle from the center of the lane)	Arnedt et al., 2001
	Increased lateral position errors; increased heading errors at 0.12%	Gawron & Ranney, 1988
	Change of within-lane deviation as measured from the center of the lane (SD of within-lane position)	Harrison & Fillmore, 2005 Harrison, Marczinski, & Fillmore, 2007
Roadway/lane departures	More roadway departures	Stein & Allen, 1984 Arnedt et al., 2001
	More crossings of center line	Weiler et al., 2000
	More lane crossings than controls (center and side)	Fairclough & Graham, 1999
Reaction Time/ Response Time	Detection times of vehicles slower on curves	Leung & Starmer, 2005
	Slower reaction times to road signs	Burns et al., 2002
	Slower brake reaction time	Kerr & Hindmarch, 1998
Crashes	More crashes than in baseline (or placebo) driving condition	Gawron & Ranney, 1988 Moskowitz et al., 2000 Roehrs et al., 1994 Harrison & Fillmore, 2005 Kerr & Hindmarch, 1998

[a] Definitions of the variables are included where possible.

measures may provide an indirect measure of potential alcohol impairment regardless of the presence of overt behavioral changes.

A standard set of driving scenarios does not exist for alcohol-impaired driving, as the goal of previous research has been to identify all areas of driving skill that are impaired by alcohol. Overall, Table 44.2 indicates a number of measures for common driving scenarios used in all types of driving simulation research, such as car following, responding to emergency events (e.g., lead vehicle braking, vehicle incursions, roadway obstructions) and general driving on a variety of roads (e.g., rural, urban) where speed and lane keeping are measured. Measures that are observable in both test track and simulator studies may hold the most validity for alcohol research and these include speed variability, lateral position deviations, and responses to hazards in the driving environment. However, not all measures have been tested in

both settings. Therefore, it is important to select measures that reflect the research question or for which impaired performance may lead to a safety-critical situation.

44.2.2 Research Topics

Because alcohol has been well studied for basic driving skills, it is suggested that future research examine more complex issues associated with impairment. This includes examining alcohol in combination or comparison with factors such as fatigue or with a distracting task (e.g., talking on a cell phone), or examining the contribution of individual differences to driving choices and degrees of impairment.

For example, the rate of impairment among drivers is four times higher at night-time as compared to day-time driving

(36% vs. 9%) (NHTSA, 2008). Therefore, it is likely that fatigue may impact the overall effect of alcohol on behavior and performance while driving. Roehrs, Beare, Zorick and Roth (1994) found that the impairing effect of alcohol is enhanced in fatigued individuals, but the extent of this effect for different populations and levels of fatigue has not been fully examined. Drivers are also more likely to experience sedation-like effects after they have peaked at their maximum BAC and begin sobering up (Martin, Earleywine, Musty, Perrine, & Swift, 1993). This implies that impairment effects could be more noticeable at lower BACs than typically found in studies where drivers are tested either as they ascend towards their peak BAC or as soon as they reach their peak BAC. The questions associated with fatigue could address the crash risk of drivers who are below the legal limit but may be unable to perform optimally due to the combined effects of alcohol and fatigue.

Additionally, the contribution of individual differences to alcohol impairment effects is frequently discussed, but is not always extensively examined in driving research studies. Many studies use small samples that are frequently made up of healthy young men. Many study designs also control for any potential effects of individual differences through the use of covariates or within-subjects designs, but do not always address the true correlation or contribution of differences to observed impairment effects. Because alcohol may affect individuals differentially along performance and behavior continuums, it may be necessary to collect data on drivers' personality (e.g., risk-taking, aggression) and/or their attitudes or experiences with alcohol (e.g., how much they drink per week) in order to calibrate their performance on the driving task. The choice of these measures will be determined by the question being asked. For example, taking psychological measures of participants' reported levels of aggression or risk-taking while sober may help identify increases in aggressive or risky behaviors while driving when intoxicated. A number of individual differences known to affect alcohol impairment are discussed in Section 44.3.2.

Finally, because the basic effects of alcohol have been extensively examined in the past, researchers are beginning to use alcohol as a benchmark for other forms of impairment, such as fatigue and distraction. This is because demonstrable limits have been set for alcohol impairment (i.e., BAC 0.08) and it is already widely regulated (National Council on Alcoholism and Drug Dependence, 2000). The risk imposed by legislated BAC limits can be considered a permitted baseline for sanctioned risk of impairment to which other types of impairment can be compared. The key question for these types of comparisons is not if other impairment types impose a risk but rather if the amount of risk is unacceptable. An acceptable risk threshold can be assessed in relative terms by comparing performance at known alcohol impairment levels to performance under other forms of impairment. The result is a useful assessment that the novel impairment presents a lower, higher, or equivalent risk to performance as driving under the influence of alcohol. From the research studies cited above, two investigated the effects of alcohol in comparison to cell phone use (Burns, Parkes, Burton, Smith, & Burch, 2002; Rakauskas et al.,

2008), two compared alcohol to fatigue (Roehrs, Beare, Zorick & Roth, 1994; Fairclough & Graham, 1999), and one investigated alcohol impairment in comparison to over-the-counter (OTC) antihistamines known to cause sedation (Weiler, Bloomfield, & Woodworth, 2000).

44.3 Experimental Design Issues

Impaired driving research is subject to the same experimental considerations in driving simulation research as it is in other contexts. Because the results of alcohol research are susceptible to a number of factors, choices about study design, participant characteristics, dosing options and the testing situation should be considered carefully in relation to the research question.

44.3.1 Study Designs

44.3.1.1 Within-Subjects Designs

Typically, most alcohol studies use a within-subjects design where the participant acts as their own control and the presentation of alcohol conditions (including placebo) are counterbalanced (Moskowitz & Fiorentino, 2000). Within-subjects designs allow for increased power for statistical analyses using a smaller sample size and also result in a reduction in the error variance associated with individual differences. Practice effects and fatigue effects are both weaknesses of within-subjects designs and can be exacerbated in alcohol research if correct study design is not employed. One of the main criticisms made by Moskowitz and Fiorentino in their review of alcohol effects in driving skills was that true effects of alcohol could not necessarily be separated form potential practice effects in the studies that employed within-subjects designs.

Practice effects can occur because participants complete the experimental tasks more than once, meaning they can improve their performance on a task by their last session regardless of the condition. In general, practice effects can be mitigated by ensuring that participants in the study are well-practiced on the experimental tasks before beginning the test trials. Therefore, any differences in performance between conditions are likely due to the condition and not a practice effect because participants are skilled at the task. Fatigue effects can occur if trials are too close together or if it takes a long time to complete all sets of trials. Participants may not perform optimally by the last set of tasks because they become bored or tired. Both of these problems can affect the results of a study and some previous alcohol and driving research has employed dosing schedules that may be more susceptible to these problems.

For example, using the descending alcohol curve as a method for testing multiple alcohol levels on a single participant in one testing session (e.g., Moskowitz et al., 2000) is problematic. In this method, participants are dosed to the highest desired alcohol level first and complete the experimental tasks at this level. They then wait to come down to lower levels and complete the tasks again at each alcohol level. Although this method is efficient in that participants can experience multiple alcohol levels in one session, the potential for practice or fatigue effects is exacerbated by the order of alcohol level presentations. Participants will complete the experimental tasks at the highest BAC first and, more specifically, at the BAC level most likely to result in impaired performance. Because this is the first set of experimental tasks, it is possible that any observed impairment may be due to both the alcohol effect and less practice with that task. When participants arrive at the lowest alcohol level, or the level least likely to result in impaired performance, their performance may be better than expected because they have more practice. Therefore, performance impairment at the highest alcohol level may be larger than expected and performance at the lowest alcohol level may be less than expected due to these practice effects.

Fatigue effects may also adversely affect the descending curve method because participants are more likely to experience sedation-like effects on the descending alcohol curve (Martin & Sayette, 1993). This means participants may be more tired when performing the experimental tasks at the lower alcohol levels, which could result in poorer performance at the lower levels and smaller differences being observed from high- to low-alcohol conditions.

The descending curve method is not recommended for alcohol and driving research. Instead, it is recommended that confounds be addressed for within-subjects designs by having participants complete each alcohol session on a different day and then controlling the time of day and week for participation (e.g., participant comes in at noon on three consecutive Mondays). Additionally, the use of a placebo for sober trials, random assignment to a random or counterbalanced (e.g., Latin Square) presentation of alcohol dosing levels, and extensive practice on tasks prior to beginning experimental testing sessions can also be included in any alcohol-related study design. The choice of controls will depend on the research question.

44.3.1.2 Between-Subjects Design

A between-subjects design can reduce the learning effects associated with a within-subjects design. In this design each participant is assigned to only one alcohol level and performs the experimental tasks once. The use of a between-subjects design eliminates the practice effects inherent in the within-subjects design. However, because of the between-subjects variability in impairment, it is necessary to ensure that each alcohol level contains a large enough sample of participants to account for this variability. Power analyses can be conducted a priori to ensure that samples are large enough for all comparisons. Many of the protocols and controls employed for within-subjects designs are also necessary for between-subjects designs (e.g., random assignment to conditions, placebo, practice, etc.).

44.3.1.3 Mixed Designs

A mixed-design may be required if it is necessary to test impairment across different groups of participants (e.g., gender, age). In

this case, alcohol condition would be a within-subjects design and the group would be between-subjects.

44.3.2 Participant Characteristics

Gender (Mumenthaler, Taylor, O'Hara, & Yesavage, 1999), drinking experience (Holdstock, King, & de Wit, 2000), fatigue (Roehrs, Beare, Zorick, & Roth, 1994), driving experience (Harrison &, Fillmore, 2005), and the stage of drinking (i.e., ascending versus descending impairment curve; Martin & Sayette, 1993) may all affect the performance and behavior of participants in an alcohol study. Women frequently achieve higher BAC levels with the same alcohol dose (even when doses are controlled for body weight) than men and may be more susceptible to alcohol's effects on cognitive functions (e.g., Mumenthaler et al., 1999). Feeling stimulated is more often reported on the rising BAC curve and sedation more often reported on the descending curve (Martin & Sayette, 1993). Moderate/heavy drinkers have been shown to experience more stimulant-like effects and fewer sedative and aversive effects from alcohol than light drinkers (Holdstock, King, & de Wit, 2000). As already discussed, impairment may be more evident in conjunction with fatigue (Roehrs, Beare, Zorick, & Roth, 1994), meaning that testing sessions that take place later in the day may show more pronounced effects than those occurring earlier in the day. Finally, drivers with poorer driving skills may suffer larger alcohol effects at the same BACs as more experienced drivers (Harrison & Fillmore, 2005). Therefore, it is important to consider these factors when developing an alcohol study and to be aware of any limitations in the study design when analyzing the results of alcohol impairment on driving performance.

Controls for participant characteristics can be employed in the study design if necessary, such as choosing what gender and level of drinking experience will be required to participate in the study. Control can also occur during analysis by using baseline performance on tasks and other measures known to influence the effect of alcohol (e.g., drinking experience, risk-taking, baseline driving performance) as covariates in the analyses.

In some cases, ethical concerns limit the types of participants who can be tested. For example, most alcohol studies screen out individuals who exhibit signs of alcohol abuse using questionnaires such as the CAGE (Ewing, 1984) or the Michigan Alcohol Screening Test (MAST) (Selzer, 1971). Additionally, participants with health concerns or those who take medications that react with alcohol may also be screened out of alcohol studies. This means that issues which can potentially change or interact with the effects of alcohol to affect driving performance may not be testable among certain groups of individuals.

44.3.3 Testing Situation

Situational variables can have an effect on driver behavior. One of the main criticisms of simulator research is that it is not the real world and that the behaviors observed may not be similar to those which would be observed in the real world. This applies to alcohol research as well. The perceived level of risk experienced by drivers in a simulator is different from that experienced in the real world and may affect their motivations to avoid risk during a simulated drive (Brookhuis & de Waard, 2003). Drinking is frequently a social activity and this social aspect can influence behavior. For example, marijuana is also a social drug and researchers have argued that the sterile conditions of a clinical trial do not capture the social effects of the drug experience. In this regard, Smiley, Noy and Tostowaryk (1987) has conducted one of the few social drug studies (marijuana) of driver impairment in which the drug dosing took the form of non-regimented smoking in a social environment comprised of multiple study participants in a party-like atmosphere (large canopy tent set up aside the airport landing strip on which the driving impairment study was conducted). During a test track study of alcohol impairment on motorcycle riding skills, Creaser et al. (2007) found that overlapping participant schedules where two or more participants were present during dosing created a more relaxed social environment for the participants.

44.3.4 Dosing

Given that the primary treatment variable in alcohol research is the administration of alcohol, alcohol dosing is a critical procedure in alcohol research. Previous research has shown a deterioration in driving skills at BAC 0.05 (Council on Scientific Affairs, 1986) and even starting as low as BAC 0.013 (Moskowitz & Fiorentino, 2000). The volume of beverage administered to each individual to reach a particular level of intoxication may be calculated fairly accurately using the 8/10 version of the Widmark formula (National Highway Traffic Safety Administration, 1994). This frequently-used formula only requires the experimenter to measure the participant's body-weight and can be modified easily to target the peak BAC to occur at any (reasonable) time after consumption.

However, all dosing methodologies should take into account the large number of individual tolerance differences inherent in experimental dosing, as exemplified in Friel, Baer and Logan (1995). Although these differences may result in large variations of peak and time of peak BAC level, a window of peak intoxication can be approximated based on previous findings (Friel, Logan, O'Malley, & Baer, 1999). After consumption of alcohol, it is recommended that experimenters regularly monitor participants' BAC level and engage participants in the experimental methodology when they reach the desired intoxication level.

44.3.4.1 The Placebo (Sober) Condition

Most alcohol studies employ some form of blinding or placebo for the sober condition. However, some research has suggested that blinding or placebo is not necessary because subjects cannot be deceived about beverage content when doses are intended to raise the BAC above 0.05 (Martin & Sayette, 1993; Sayette, Breslin, Wilson, & Rosenblum, 1994). The argument is further continued that unblinding subjects to the nature of the alcohol condition is more ecologically valid because people typically know when they are drinking alcohol (Arnedt, Wilde, Munt, & MacLean, 2001). Ultimately, the choice to use a placebo/blinding technique or to openly indicate alcohol consumption to participants may depend on the research question. For example, when examining the social effects of drinking alcohol on behavior it is likely to be useful for the participants in the group to be aware that they are drinking alcohol and how much they have consumed. However, if the goal is to identify alcohol effects subjectively, the use of a placebo or blinding for the sober condition may avoid bias in self-reported indications of intoxication. Overall, blinding is useful for obscuring the objective alcohol dose rather than hiding the presence of alcohol altogether.

44.4 Conclusions

Driving simulation can provide the scenario control and ethical management of risk to support experimentation of impairment effects across levels of alcohol. Whereas there are some common scenario features and measures reported in the literature, not every measure is used by all studies, and sometimes the definitions for the same measure can differ between studies. This variance may be due in part to differences in research hypotheses, simulator capabilities and the lack of a consistent reporting methodology. Several measures (e.g., speed variability, lane position deviations, hazard response) have been validated in both closed-course and simulator settings. Overall, there are many advantages to using driving simulators to perform controlled experiments of alcohol impairment on driving behavior and performance. Additionally, the extensive research on the effect of alcohol on basic driving skills allows future research to address more complex questions related to alcohol impairment and driving performance, and to use alcohol impairment as a benchmark for other forms of impairment (e.g., fatigue, distraction).

Key Points

- Physiological changes caused by alcohol result in psychological and perceptual impairments that affect driving behavior.
- In general, the measurement of driver impairment involves the specification of behaviors that are relevant to crash risk and the formulation of metrics to quantify those behavioral measurements.
- A comparison of test track studies to simulator studies show that many of the same variables that are affected by alcohol in real-world driving settings are also affected in simulated driving settings.
- Because alcohol impairment effects are well established and documented, alcohol impairment can be used as a benchmark for other forms of driving impairment, such as fatigue, or in comparison to the effects of other drugs.
- Because the results of alcohol research are susceptible to a number of factors, choices about study design, participant characteristics, dosing options, and the testing situation should be considered carefully in relation to the research question.

Keywords: Alcohol, Alcohol Dosing, Impaired Driving, Impairment Measures, Study Design

Key Readings

Brookhuis, K., & de Waard, D. (2003). Criteria for driver impairment. *Ergonomics, 46*(5), 433–445.

Gawron, V. J., & Ranney, T. A. (1988). The effects of alcohol on driving performance on a closed-course and in a driving simulator. *Ergonomics, 31*(9), 1219–1244.

Holloway, F. A. (1995). Low-dose alcohol effects on human behavior and performance. *Alcohol, Drugs and Driving, 11*, 39–56.

Moskowitz, H., & Florentino, D. (2000). *A review of the literature on the effects of low doses of alcohol on driving-related skills* (DOT HS 809 028). Washington, DC: National Highway Traffic Safety Administration, US Department of Transportation.

Sayette, M. A., Breslin, F. C., Wilson, G. T., & Rosenblum, G. D. (1994). An evaluation of the balanced placebo design in alcohol administration research. *Addictive Behaviors, 19*, 333–342.

References

Allen, R. W., Schwartz, S. H., Hogge, J. R., & Stein, A. C. (1978). *The effects of alcohol on the driver's decision-making behavior*. Executive Summary and Technical Report (DOT HS 803 608)(Vol. 1). Washington, DC: US Department of Transportation.

Arnedt, J. T., Wilde, G. J. S., Munt, P. W., & MacLean, A. W. (2001). How do prolonged wakefulness and alcohol compare in the decrements they produce on a simulated driving task? *Accident Analysis & Prevention, 33*, 337–344.

Bartholow, B. D., Pearson, M., Sher, K. J., Wieman, L. C., Fabiani, M., & Gratton, G. (2003). Effects of alcohol consumption and alcohol susceptibility on cognition: A psychological examination. *Biological Psychology, 64*, 167–190.

Blincoe, L., Seay, A., Zaloshnja, E., Miller, T., Romano, E., Luchter, S., & Spicer, R. (2002). *The economic impact of motor vehicle crashes, 2000* (DOT HS 809 446). Washington, DC: National Highway Traffic Safety Administration.

Brookhuis, K. A., & de Waard, D. (1994). Measuring driving performance by car following in traffic. *Ergonomics, 37*(3), 427–434.

Brookhuis, K. A., & de Waard, D. (2003). *On the assessment of criteria for driver impairment: In search of the golden yardstick for driving performance*. Proceedings of the Second International Driving Symposium on Human Factors in Driver Assessment, Training and Vehicle Design. Iowa City, IA: University of Iowa.

Brookhuis, K. A., de Waard, D., & Fairclough, S. H. (2003). Criteria for driver impairment. *Ergonomics, 46*, 433–445.

Burns, P. C., Parkes, A., Burton, S., Smith, R. K., & Burch, D. (2002). *How dangerous is driving with a mobile phone? Benchmarking the impairment to alcohol*. Crowthorne, England: Transport Research Laboratory.

Council on Scientific Affairs. (1986). Alcohol and the driver. *JAMA, 255*, 522–527.

Creaser, J. I., Ward, N. J., Rakauskas, M. E., Boer, E., Shankwitz, C., & Nardi, F. (2007). *Effects of alcohol on motorcycle riding* (DOT-HS-810-877). Washington, DC: US Department of Transportation.

Ewing, J. A. (1984). Detecting alcoholism: The CAGE Questionnaire. *JAMA, 252*, 1905–1907.

Fairclough, S. H., & Graham, R. (1999). Impairment of driving performance caused by sleep deprivation or alcohol: A comparative study. *Human Factors, 41*(1), 118–128.

Friel, P. N., Logan, B. K., O'Malley, D., & Baer, J. S. (1999). Development of dosing guidelines for reaching selected target breath alcohol concentrations. *Journal of Studies on Alcohol, 60*, 555–565.

Friel, P. N., Baer, J. S., & Logan, B. K. (1995). Variability of ethanol absorption and breath concentrations during a large-scale alcohol administration study. *Alcoholism: Clinical and Experimental Research, 19*(4), 1055–1060.

Gawron, V. J., & Ranney, T. A. (1988). The effects of alcohol on driving performance on a closed-course and in a driving simulator. *Ergonomics, 31*(9), 1219–1244.

George, W. H., Raynor, J. O., & Nochajski, T. H. (1990). Resistance to alcohol of visual-motor performance II: Effects for attentional set and self-reported concentration. *Pharmacology, Biochemistry and Behavior, 36*, 261–266.

Green, P. (2003). Where do drivers look while driving (and for how long)? In R. E. Dewar & P. L. Olson (Eds.), *Human factors in traffic safety*. Tuscan, AZ: Lawyers & Judges Publishing Company, Inc.

Harrison, E. L. R., & Fillmore, M. T. (2005). Are bad drivers more impaired by alcohol? Sober driving precision predicts impairment from alcohol in a simulated driving task. *Accident Analysis & Prevention, 37*, 882–889.

Holdstock, L., King, A. C., & de Wit, H. (2000). Subjective and objective responses to ethanol in moderate/heavy and light social drinkers. *Alcoholism: Clinical and Experimental Research, 24*(6), 789–794.

Holloway, F. A. (1995). Low-dose alcohol effects on human behavior and performance. *Alcohol, Drugs and Driving, 11*, 39–56.

Jones, M. B., Chronister, J. L., & Kennedy, R. S. (1998). Effects of alcohol on perceptual speed. *Perceptual and Motor Skills, 87*(3, pt 2), 1247–1255.

Kuypers, K. P. C., Samyn, N., & Ramaekers, J. G. (2006). MDMA and alcohol effects, combined and alone, on objective and subjective measures of actual driving performance and psychomotor function. *Psychopharmacology, 187*, 467–475.

Lamb, M. R., & Robertson, L. C. (1987). Effect of acute alcohol on attention and the processing of hierarchical patterns. *Alcoholism: Clinical and Experimental Research, 11*, 243–248.

Leung, S., & Starmer, G. (2005). Gap acceptance and risk-taking by young and mature drivers, both sober and alcohol-intoxicated, in a simulated driving task. *Accident Analysis & Prevention, 37*, 1056–1065.

Martin, C. S., Earleywine, M., Musty, R. E., Perrine, M. W., & Swift, R. M. (1993). Development of the biphasic alcohol effects scale. *Alcoholism: Clinical and Experimental Research, 17*(1), 140–146.

Martin, C. S., & Sayette, M. A. (1993). Experimental design in alcohol administration research: Limitations and alternatives in the manipulation of dosage-set. *Journal of Studies on Alcohol, 54,* 750–761.

Maylor, E. A., Rabbitt, P. M., James, G. H., & Kerr, S. A. (1992). Effects of alcohol, practice, and task complexity on reaction time distributions. *Quarterly Journal of Experimental Psychology, 44A,* 119–139.

Moskowitz, H., Burns, M., Fiorentino, D., Smiley, A., & Zador, P. (2000). *Driver characteristics and impairment at various BACs* (DTNH-22-95-C-05000). Washington, DC: National Highway Traffic Safety Administration, US Department of Transportation.

Moskowitz, H., & Florentino, D. (2000). *A review of the literature on the effects of low doses of alcohol on driving-related skills* (DOT HS 809 028). Washington, DC: National Highway Traffic Safety Administration, US Department of Transportation.

Mumenthaler, M. S., Taylor, J. L., O'Hara, R., & Yesavage, J. A. (1999). Gender differences in moderate drinking effects. *Alcohol Research and Health, 23*(1), 55–61.

National Council on Alcoholism and Drug Dependence. (2000). *Facts and information: Alcoholism and alcohol related problems.* New York: Author. Retrieved June 3, 2004 from http://www.ncadd.org/facts/problems.html

National Highway Traffic Safety Administration. (1994). *Computing a BAC estimate.* National Highway Traffic Safety Administration, US Department of Transportation. Retrieved June 16, 2004 from http://www.nhtsa.dot.gov/people/injury/alcohol/bacreport.html

National Highway Traffic Safety Administration. (2001). *Alcohol and Highway Safety 2001: A Review of the State of Knowledge* (DOT HS 809 383). Washington, DC: National Highway Traffic Safety Administration, US Department of Transportation.

National Highway Transportation Safety Administration. (2007). *2006 Traffic safety annual assessment—Alcohol-related fatalities. Traffic safety facts* (DOT HS 810 821). Washington DC: National Center for Statistics and Analysis, US Department of Transportation.

National Highway Transportation Safety Administration. (2008). *Alcohol-impaired driving. Traffic safety facts* (DOT HS 810 801). Washington DC: National Center for Statistics and Analysis, US Department of Transportation.

Rakauskas, M., Ward, N., Bernat, E., Cadwallader, M., & de Waard, D. (2005). *Driving performance during cell phone conversation and common in-vehicle tasks while sober and drunk* (MnDOT and CTS contract 81655, Work Order 41). HumanFIRST Program, ITS Institute, University of Minnesota.

Rakauskas, M. E., Ward, N. J., Boer, E., Bernat, E., Cadwallader, M., & Patrick, C. (2008). Car following performance during conventional distractions and alcohol intoxication. *Accident Analysis & Prevention, 40,* 1742–49.

Rimm, D. C., Sininger, R. A., Faherty, J. D., Whitley, M. D., & Perl, M. B. (1982). A balanced placebo investigation of the effects of alcohol versus alcohol expectancy on simulated driving behavior. *Addictive Behaviors, 7,* 27–32.

Roehrs, T., Beare, D., Zorick, F., & Roth, T. (1994). Sleepiness and ethanol effects on simulated driving. *Alcoholism: Clinical and Experimental Research, 18*(1), 154–158.

Sayette, M. A., Breslin, F. C., Wilson, G. T., & Rosenblum, G. D. (1994). An evaluation of the balanced placebo design in alcohol administration research. *Addictive Behaviors, 19,* 333–342.

Selzer, M. L. (1971). The Michigan Alcoholism Screening Test (MAST): The quest for a new diagnostic instrument. *American Journal of Psychiatry, 127,* 1653–1658.

Skog, O.-J. (2001). Alcohol consumption and mortality rates from traffic accidents, accidental falls, and other accidents in 14 European countries. *Addiction, 96*(Suppl. 1), S49–S58.

Smiley, A. M., Noy, Y. I., & Tostowaryk, W. (1987). The effects of marijuana, alone and in conjunction with alcohol, on driving an instrumented car. In P. C. Noordzj & R. Roszback (Eds.), *Proceedings of the tenth international conference on alcohol, drugs and traffic safety* (pp. 203–206). Amsterdam, the Netherlands: Excerpta Medica.

Smith, G. S., Branas, C. C., & Miller, T. R. (1999). Fatal non-traffic injuries involving alcohol: A meta-analysis. *Annals of Emergency Medicine, 33*(6): 659–668.

Stein, A. C., & Allen, R. W. (1984). The combined effects of alcohol and marijuana on driving behavior. *Proceedings 28th conference of the American association for automotive medicine* (pp. 289–304). Lake Bluff, IL: American Association for Automotive Medicine. (Conference held in Denver, CO).

Sugarman, R. C., Cozad, C. P., & Zavala, A. (1973). *Alcohol-induced degradation of performance on simulated driving tasks* (Paper 730099). Warrendale, PA: Society of Automotive Engineers.

Weiler, J. M., Bloomfield, J. R., Woodworth, G. G., et al. (2000). Effects of fexofenadine, diphenhydramine, and alcohol on driving performance: A randomized, placebo-control trial in the Iowa Driving Simulator. *Annals of Internal Medicine, 132,* 354–363.

Wang, J., Knipling, R. R., & Goodman, M. J. (1996). *The role of driver inattention in crashes: New statistics from the 1995 crashworthiness data system.* Proceedings of the 40th Annual Association for the Advancement of Automotive Medicine. Vancouver, BC, Canada.

Validity of Three Experimental Performance Tests for Predicting Risk of Cannabis-Induced Road Crashes

Jan G. Ramaekers
Maastricht University

Manfred R. Moeller
Saarland University

Eef L. Theunissen
Maastricht University

Gerold F. Kauert
Goethe University of Frankfurt

Abstract

The Problem. The predictive validity of performance test in drugs and driving research is usually unknown, primarily due to a lack of real-life epidemiological (crash risk) data in general. The absence of such data has made attempts to correlate laboratory or driving data to real-life driving accidents extremely difficult. *Measures of Driving Performance.* This chapter will focus on the validity of tests measuring skills related to driving to predict drug-induced driving impairment and provide a practical example of how to calculate predictive validity of performance tests. The practical example will focus on laboratory tasks for measuring tetrahydrocannabinol (THC) effects on skills related to driving, but can easily be generalized to other drugs and driving parameters measured in a driving simulator or actual driving test. *Key Results.* Two recent publications on performance impairment and crash (culpability) risk as a function of THC in blood have provided a unique opportunity to calculate the validity of performance tests for predicting cannabis-induced crash risk. Curve estimation shows that data pairs of performance impairment and culpability risk within five successive THC concentration ranges can be perfectly fitted in exponential curves for performance tests measuring tracking performance, impulse control, and cognitive function. In addition, THC-induced performance impairments and THC-induced culpability risks were highly correlated. *Conclusion.* The implication is that performance tests of skills related to driving can be taken as valid measures to predict THC-induced crash risk in real life.

45.1 Introduction

A major problem in assessing the impact of drugs on driving is the fact that the driving variables being measured can vary significantly. In studies being reported, the basic parameters being assessed are often not comparable due to a lack of standardization in the field. Due to these methodological differences in the field researchers have tried to develop a consensus on research standards that will increase comparability between drug effects measured in laboratory tests, driving simulators, and actual driving tests. Such consensus reports have most notably been issued by the International Council on Alcohol, Drugs and Traffic Safety (ICADTS). A first consensus report was issued by the ICADTS working group on "the standardization of experimental studies of drugs and drivers" (1999). The second consensus report appeared in 2007 by an ICADTS working group on "illegal drugs and driving". Excerpts from these consensus reports related to the operationalization of driving tests will be given below. Full copies of these reports can be found at http://www.icadts.org.

45.2 ICADTS Guidelines on Experimental Studies to Determine Drug Effects on Driving (1999)

The following recommendations for operationalization of driver fitness are taken from the consensus report.

45.2.1 Operationalization of Driver Fitness

Studies for establishing the driving hazard potential of a particular medicinal drug should proceed from conventional laboratory testing to sophisticated driving simulators. Finally, actual driving tests (over-the-road tests in normal traffic conditions or tests on closed-circuits under controlled conditions) should be carried out as far as they can safely be applied and will be allowed by the ethic commission. The final evidence that the drug in question would be safe or hazardous to a specified degree should be based on the combined results of all tests in the program. Combining the results from programmatic research should follow scientific guidelines.

45.2.2 Laboratory Tests

A performance test battery as a sound operationalization of the construct "driver fitness" should possess content validity. That means, the test battery as a whole should be representative of the mental and/or behavioral functions relevant to driving and, simultaneously, should be representative of the pharmacological effects of the drug under study. Concerning the elements of the test battery, studies should include a test to measure divided attention or continuous perceptual-motor coordination and a test to measure sedation or drowsiness. Furthermore, the test battery should comprise performance areas like discrete perceptual

motor response, speed and accuracy of decision-making, sustained attention (vigilance), dynamic visual acuity, short-term spatial memory, risk avoidance and attentional resistance to distraction. This list is not closed. The duration of the test battery will depend upon, among other things, the kind and number of tests included, the medicinal drug, and the question of whether a time-course effect should be studied. There should be at least one test in the battery of longer duration (e.g., one hour).

45.2.3 Individual Tests

Many representative test procedures exist to test a single mental and/or behavioral function. The choice of an adequate test is left to the investigator who should choose a standard procedure or, if using a new one, the investigator should demonstrate that the procedure is able to discover the impairment addressed by the mental and/or behavioral function under study. In any case, the test selected must meet the requirements outlined in the following:

- Performance tests included in a test battery should possess construct validity. That means that the measures of the tests should be simultaneously relevant to driving performance and sensitive to the pharmacological drug effects.
- A validation of medicinal drug-induced changes in performance with changes in actual driving performance should be made before using laboratory tests in categorization procedures or other legal/regulatory affaires. Laboratory tests should possess either a test-retest reliability coefficient for raw scores measured in the absence of drug effects of $r >= 0.70$ or, preferably, a test-retest reliability coefficient for drug-placebo change scores of $r >= 0.50$.
- A standardization of the measurement procedure of the dependent variables of a test is very important due to reasons of comparability between studies. The training of the subjects on the test should be continued to individual stability of performance (determined by group means, standard deviation, and intertrial correlations).

45.2.4 Simulators and Real Driving

The term "simulator" contains a wide range of constructions from very simple to highly-sophisticated devices. Independent of the complexity, they should include tests of reactions on traffic signals, compliance with traffic control devices, passing maneuvers, and turning of intersections. A similar state of affairs is valid for real driving. Independent of the variety of designing the procedure, closed-course driving should include decision-making, responses to changes in traffic control devices, and interactions with other experimentally controlled road-users. Over-the-road driving tests have to include combined city and highway driving tests. The fundamental aspects of the driver-vehicle-road interaction (e.g., road tracking, speed maintenance, car following, etc.) have to be involved. The ethical implications must be considered with great care.

45.3 ICADTS Recommendations for Drugged Driving Research (2007)

The ICADTS consensus group postulated a large number of recommendations on behavioral, epidemiological and toxicological research on drugs and driving. In general, the working group indicated that there are three hierarchical levels of control that should be measured to assess the effects of drug on driving or crash risk: strategic level, maneuvering level, and automotive level (Michon, 1985). At the top level, the strategic level, executive decisions are made. These involve behaviors such as route-choice and planning, setting of trip goals (e.g., avoid peak hour), observation, judgment and understanding of traffic and risk assessment. At the intermediate level, the maneuvering level, negotiations of common driving situations occur. At this level behaviors are described as controlled and require conscious processing. They include reactions to the behavior of other drivers, distance keeping, speed adjustment, negotiation of curves and intersections, gap acceptance and obstacle avoidance. At the lowest level, the operational level, the basic vehicle-control processes occur. These include tracking, steering, braking and gear shifting and are described as automotive behaviors. Automotive behaviors require little conscious mental activity and develop following extended practice. A consensus was reached that behavioral studies should focus on a particular drug and a number of doses of that drug. Drugs of interest that are most common in DUI arrests or motor vehicle crash victims are: cannabis, benzodiazepines or other tranquilizing agents, opiates, stimulants (amphetamines, methamphetamine, cocaine, Methylenedioxymethamphetamine (MDMA)), antidepressants and antihistamines. The consensus group issued a large number of recommendations for behavioral drug and driving research. Many of them were in line with recommendations from the previous guidelines issued by ICADTS. However, the working group also issued two distinctive recommendations for behavioral drugs and driving research that deserve some special attention. These recommendations are:

- Researchers should use tests that have been validated to be sensitive to drug effects on driver performance, and to the extent possible, have demonstrated predictive validity of driving impairment.
- The behavioral effects of drugs should be related to dose and drug concentration in the brain. Currently, the best available indicator of drug concentration in the brain is the drug concentration in the blood.

These recommendations are a true challenge to current drugs and driving research. Most researchers do not design their studies to establish the relation between driving impairment and drug concentration in blood. At best they will include several doses of a drug in order to establish a dose-effect relation. Likewise, most researchers have not documented the validity of their tests for predicting drug-induced crash risk. The remainder of this chapter will focus on the validity of tests measuring skills related to driving to predict drug-induced driving impairment and provide a practical example of how to calculate predictive validity of performance tests. The practical example will focus on laboratory tasks for measuring THC effects on skills related to driving, but can easily be generalized to other drugs and driving parameters measured in a driving simulator or actual driving test.

45.4 Predictive Validity of Performance Tests

Predictive validity is the ability of a measure to predict something it should theoretically be able to predict. A high correlation between changes in the measure and changes in the construct that it is designed to predict would provide good evidence for its predictive validity. Thus in the field of experimental drugs and driving research the question that needs to be answered can be formulated as: Do actual driving tests or laboratory tests of skills related to driving predict crash risk in real life?

The predictive validity of performance test in drugs and driving research is usually unknown, primarily due to a lack of real-life epidemiological (crash risk) data in general. The absence of such data have made attempts to correlate laboratory data to real-life driving accidents extremely difficult. However, it should also be noted that in the past investigators have often neglected to calculate the predictive validity of their performance tests for alcohol-induced crash risk, even though a wealth of epidemiological data are available. This can be considered a major negligence in experimental drugs and driving research, for any performance task with demonstrated predictive validity for alcohol-induced crash risk is also likely to be sensitive to drug-induced crash risk.

45.4.1 Predictive Validity: A Practical Application in THC Research

The influence of Δ^9- tetrahydrocannabinol (THC) on driving performance has been the subject of a long-standing debate between researchers in fields of experimental psychopharmacology and forensic toxicology. The former have consistently demonstrated a dose-related impairment of driving ability in experimental studies measuring psychomotor skills, simulated driving and actual on-the-road driving after controlled cannabis administration. The latter, however, have argued that cannabis does not increase crash culpability or crash risk because epidemiological culpability surveys have repeatedly shown that cannabis use is not overrepresented in drivers who were responsible for their accident as compared to drivers who were not (Bates & Blakely, 1999; Fergusson, 2005; Kelly, Darke, & Ross, 2004; Ramaekers, Berghaus, van Laar, & Drummer, 2004).

A range of research issues can be raised that may have accounted for the apparent discrepancy between experimental performance data and epidemiological crash risk data. First, the validity of experimental performance studies for predicting a real-life event such as a car crash is not known (Berghaus, Ramaekers, & Drummer, 2006; De Raedt &

Ponjaert-Kristoffersen, 2001; Edwards, Vance, Wadley, Cissell, Roenker, & Ball, 2005). Second, both experimental and culpability studies have neglected to describe driver impairment or crash risk as a function of THC concentration in blood (Ramaekers et al., 2004). Third, misclassification of outcome measures may have introduced a systematic bias in culpability surveys. That is, most culpability studies have identified cannabis use among drivers by measuring THC-COOH—an inactive carboxy metabolite of THC—in blood or urine. Following the use of cannabis, THC-COOH may be present in blood or urine for days, which may lead to exposure misclassification of drivers at the time of the crash (Drummer et al., 2003).

Recently, these issues were overcome by researchers in the fields of psychopharmacology and forensic toxicology who undertook experimental (Ramaekers, Wingen, & Theunissen, 2006) and culpability studies (Laumon, Gadegbeku, Martin, & Biecheler, 2005) to assess THC-induced driver impairment in laboratory performance tests and THC-induced crash risk in real life as a function of THC concentration in blood. The experimental and culpability studies were mutually supportive and, for the first time, demonstrated a gradual increase in performance impairment and crash risk with rising THC concentrations in blood. Moreover, integration of the results from both studies now offers a unique and novel research focus: that is to calculate the validity of experimental performance tests for predicting cannabis-induced crash risk.

45.4.2 Data Sets From Single Culpability and Experimental Performance Studies

The culpability study by Laumon et al. (2005) and the experimental performance study by Ramaekers et al. (2006) served as input for calculating the association between data sets between both studies. The culpability study provided crash risk data as a function of THC concentration in blood. Likewise, the experimental study provided performance data in three laboratory tasks as a function of THC concentrations in blood. THC concentrations in both studies spanned an identical range but were reported in different units. That is, the culpability study reported THC concentrations in whole blood whereas the experimental study reported THC in serum. In blood, THC is almost completely bound to plasma proteins and is very poorly distributed into red blood cells. Consequently, the distribution of THC in serum or plasma is higher than in whole blood. The exact THC distribution ratio of whole blood to serum is not known but its median has been shown to fluctuate around two (Giroud et al, 2001). In order to increase comparability between THC concentrations in both studies, we multiplied whole blood concentrations from the culpability study by a factor 2 to obtain serum equivalents.

The culpability study by Laumon et al. (2005) arose from a unique political situation in France. The French government intended to install drug-driving legislation and funded a large-scale epidemiological study to determine the association between cannabis and crash risk. In addition, it temporarily

allowed compulsory urine and blood testing of all drivers (N = 10748) who were involved in fatal crashes in France from October 2001 to September 2003. The cases were 6766 drivers considered at fault in their crash and the controls were 3006 drivers selected from those who were not at fault. Their data demonstrated that cannabis use was significantly associated with a concentration-dependent increment in culpability risk. Adjusted odds ratios (OR) of culpability risk for serum THC concentrations ranging between 0–2 ng/ml, 2–5 ng/ml and 5–10 ng/ml were 1.89, 2.04 and 2.78 respectively. At serum THC concentrations > 10 ng/ml the odds ratio increased to 3.06. Based on these data, the authors concluded that driving under the influence of cannabis increases the risk of becoming involved in a crash.

The experimental performance study by Ramaekers et al. (2006) involved 20 recreational users of cannabis who participated in a double-blind, placebo controlled, three-way crossover study. Subjects were administered single doses of 0, 250 and 500 µg/kg THC by smoked route. Performance tests measuring skills related to driving were conducted at regular intervals between 15 minutes and six hours post-smoking, and included measures at the automotive level (critical tracking task), maneuvering level (stop signal task) and the strategical level (tower of London).

The critical tracking task (CTT) measures the subject's ability to control a displayed error signal in a first-order compensatory tracking task. Error is displayed as a horizontal deviation of a cursor from the midpoint on a horizontal, linear scale. Compensatory joystick movements null the error by returning the cursor to the midpoint. The frequency at which the subject loses the control is the critical frequency or Lambda-c. The critical tracking task measures the perceptual-motor delay lag (i.e., psychomotor control) during a closed-loop operation (Jex, McDonnell, & Phatak, 1966) and is the closest laboratory analogue to on-the-road tracking performance as measured in real-life driving (Ramaekers, 2003).

The stop signal task (SST) measures motor impulsivity, which is defined as the inability to inhibit an activated or pre-cued response, leading to errors of commission. The current test is adapted from an earlier version of Fillmore, Rush and Hays (2002) and has been validated for showing stimulant and sedative drug effects (Ramaekers & Kuypers, 2006). The task requires subjects to make quick key responses to visual go signals, such as the letters ABCD presented one at a time in the middle of the screen, and to inhibit any response when a visual stop signal, for example "*" in one of the four corners of the screen, is presented at predefined delays. The main dependent variable is the stop reaction time on stop signal trials that represents the estimated mean time required to inhibit a response. Stop reaction time was calculated by subtracting the stop signal delay from the reaction time on go-trials associated with n-th percentile of the reaction time (RT) distribution. The n-th percentile corresponds to the percentage of commission errors (Logan, 1994).

The tower of London (TOL) is a decision-making task that measures executive function and planning (Shallice, 1982). The task consists of computer-generated images of begin- and end-arrangements of three colored balls on three sticks. The subject's task is to determine, as quickly as possible, whether the end-arrangement can be accomplished by "moving" the balls in two, three, four, or five steps from the beginning arrangement by pushing the corresponding number-coded button. The total number of correct decisions is the main performance measure.

Blood was collected throughout testing. Individual THC concentrations in serum prior to performance assessments in each of the THC conditions were divided over six mutually-exclusive categories covering the full range of THC concentrations. Corresponding change scores—that is, performance scores during THC treatment minus performance scores during placebo treatment—of task performance were then classified either as showing "impairment" or "no impairment" for all individual cases within each of the THC concentration categories. Results showed that the proportion of observations showing impairment progressively increased as a function of serum THC in every performance task. The authors concluded that performance impairment selectively emerged at serum THC concentrations between 2–5 ng/ml and became truly prominent

across all performance domains at serum THC concentrations between 5–10 ng/ml.

The THC concentration-effect curves for culpability risk and performance impairment are given in Figure 45.1. It shows a remarkable correspondence between THC-induced impairments in three performance tasks and THC-induced culpability risk. Paired means of performance impairments in each of the experimental tasks and culpability risk within five serum THC concentration ranges (i.e., 0–1, 1–2, 3–5, 5–10 and >10 ng/ml) were taken as input for further statistical analyses.

45.4.3 Results: Curve Fitting of Behavioral and Epidemiological Data

Paired means of culpability risk and performance impairment in each of the given THC concentration ranges were analyzed by curve-fitting and correlation analyses for each of the three laboratory tasks using SPSS 12.0. Curve estimation shows that mean data pairs of performance impairment (PI) and culpability risk (CR) within the given THC ranges can be perfectly fitted in exponential curves for the critical tracking task ($CR = 0.67e^{0.0177PI}$; $p = 0.042$), the stop signal task ($CR = 0.88e^{0.0157PI}$; $p = 0.007$) and the tower of London task ($CR = 1.21e^{0.01PI}$; $p = 0.003$), This was

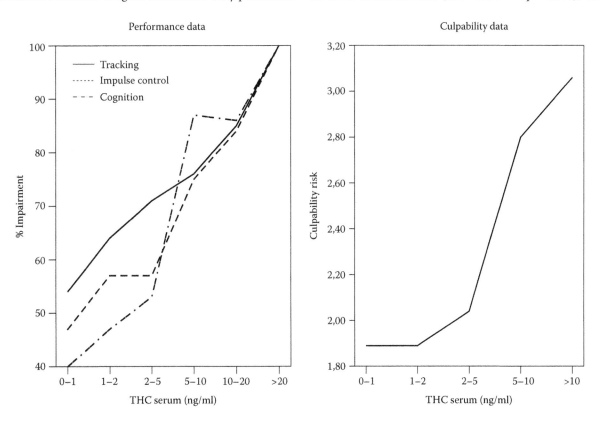

FIGURE 45.1 The left panel shows mean performance impairments (%) observed in the critical tracking task (tracking), stop signal task (impulse control) and the tower of London task (cognition) as function of serum THC (Adapted from Ramaekers, J. G., Moeller, M. R., van Ruitenbeek, P., Theunissen, E. L., Schneider, E., & Kauert, G. (2006). Cognition and motor control as a function of Delta9-THC concentration in serum and oral fluid: Limits of impairment. *Drug and Alcohol Dependence, 85*, 114–122.) The right panel shows mean crash risk as a function of serum THC (Adapted from Laumon, B., Gadegbeku, B., Martin, J. L., & Biecheler, M. B. (2005). Cannabis intoxication and fatal road crashes in France: population based case-control study. *British Medical Journal, 331*, 1371).

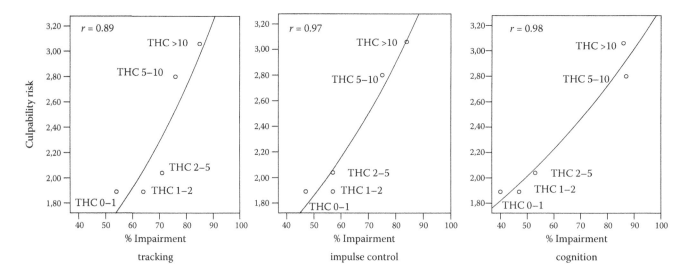

FIGURE 45.2 THC (ng/ml)-induced culpability risk as a function of THC (ng/ml)-induced performance impairment observed in the critical tracking task (tracking), stop signal task (impulse control) and the tower of London task (cognition).

also evinced by the fact that the correlation between mean pairs of performance impairment and culpability risk were highly correlated in the critical tracking task ($r = 0.89$, $p = 0.04$), the stop signal task ($r = 0.97$, $p = 0.006$) and the tower of London task ($r = 0.98$, $p = 0.005$).

Mean THC-induced culpability risk as a function of THC-induced performance impairment observed in three performance tasks is shown in Figure 45.2.

45.5 Discussion

The implication of the high correspondence between THC-induced performance impairments and THC-induced crash risk is that experimental performance tests can be taken as valid measures to predict drug-induced crash risk in real life. This notion offers the potential to further evaluate the predictive validities of a large range of performance and driving tests that are currently being employed in experimental drugs and driving research. Their predictive validities are usually not known, but should be established in order to define their relevance in the field.

Previous attempts to determine predictive validities have been hampered by a general lack of epidemiological data demonstrating a conclusive relation between drug use and traffic accidents. For a long time, alcohol has been the only drug for which an exponential relation between crash risk and blood alcohol concentration has been conclusively shown (Borkenstein, 1974; Kufera, Soderstrom, Dischinger, Ho, & Shepard, 2006; Movig et al., 2004; Ogden & Moskowitz, 2004). That situation changed after the culpability study by Laumon et al. (2005). Though previous studies had demonstrated that recent cannabis use increases culpability risk (Drummer et al., 2003), this was never shown over such a wide range of THC levels in blood. The latter information is indispensable for calculating predictive validities of performance test. Likewise, elaborated descriptions of THC concentration-effects relations were missing prior to the experimental performance study by Ramaekers et al. (2006). Both publications thus offered the unique opportunity to calculate the validity of three performance tests to predict crash risk. The finding that THC effects on performance measures and culpability risk are highly correlated indicates that performance on these three measures—the critical tracking task, the stop signal task and the tower of London task—can be taken as valid predictors of THC-induced crash risk in real life.

The construction of a well-founded task battery for evaluating drug effects on performance always has been, and still is, a major research priority. Yet, researchers have never been able to truly define the basic components of the driving task and their underlying psychological and pharmacological principles that should be reflected in a task battery. Instead, investigators have turned to the multi-faceted approach for measuring isolated skills in a wide range of laboratory task, driving simulators or on-the-road driving. The increasing availability of epidemiological crash risk data for specific drugs such as cannabis now offers the opportunity to further evaluate the predictive validities of all performance and driving tests that are currently being employed in experimental drugs and driving research. Such attempts will help to distinguish performance tests that are relevant to drug-induced crash risk and help determine the basic components of the driver task. For example, the present attempt has shown that tracking ability, impulse control and cognition/judgment all matter to driving safety. It indicates these psychological functions play an important role in driving and should be implemented in any model of the driving task. Obviously, there are more tasks and functions that could be included in such models and these should be identified by future alignments of epidemiological crash risk data and performance data for cannabis or any other drug.

The primary conclusion from the present comparison between performance and culpability data is that such data sets can be constructively used to reliably determine the validity of performance tests for predicting drug-induced crash risk.

Key Points

- Researchers should use tests that have been validated to be sensitive to drug effects on driver performance, and to the extent possible, have demonstrated predictive validity of driving impairment.
- The effects of drugs on driving should be related to dose and drug concentration in the brain.
- Performance and culpability data sets can be constructively used to determine the validity of driving performance tests for predicting drug-induced crash risk.

Keywords: Crash Risk, Performance Tests, Predictive Validity, THC

Key Readings

Grotenhermen, F., Leson, G., Berghaus, G., Drummer, O. H., Krüger, H. P., Longo, M., ... Tunbridge, R. (2007). Developing limits for driving under the influence of cannabis. *Addiction, 102,* 1910–1917.

Owens, K., & Ramaekers, J. G. (2009). Drugs, driving, and models to measure impairment. In J. C. Verster, S. R. Pandi-Perumal, J. G. Ramaekers, & J. J. De Gier (Eds.), *Drugs, driving and traffic safety* (pp. 43–58). Basel-Boston-Berlin: Birkhäuser Verlag AG.

Ramaekers, J. G., Wingen, M., & Theunissen, E. L. (2006). Performance and behavioral effects of medicinal drugs. In M. Burns (Ed.), *Medical legal aspects of drugs* (2nd ed., pp. 1–112). Tuscon, AZ: Lawyers & Judges Publishing.

Verster, J. C., & Ramaekers, J. G. (2009). The on-the-road driving test. *Drugs, driving and traffic safety* (pp. 83–92). Basel-Boston-Berlin: Birkhäuser Verlag AG.

References

Bates, M. N., & Blakely, T. A. (1999). Role of cannabis in motor vehicle crashes. *Epidemiology Review, 21,* 222–232.

Berghaus, G., Ramaekers, J. G., & Drummer, O. H. (2006). Demands on scientific studies in different fields of forensic medicine and forensic sciences traffic medicine-impaired driver: Alcohol, drugs, diseases. *Forensic Science International, 165*(2), 233–237.

Borkenstein, R. F. (1974). The role of the drinking driver in traffic accidents (the Grand Rapids study). *Blutalkohol, 11*(Suppl. 1), 1–131.

De Raedt, R., & Ponjaert-Kristoffersen, I. (2001). Predicting at-fault car accidents of older drivers. *Accident Analysis & Prevention, 33,* 809–819.

Drummer, O. H., Gerostamoulos, J., Batziris, H., et al. (2003). The incidence of drugs in drivers killed in Australian road traffic crashes. *Forensic Science International, 134,* 154–162.

Edwards, J. D., Vance, D. E., Wadley, V. G., Cissell, G. M., Roenker, D. L., & Ball, K. K. (2005). Reliability and validity of useful field of view test scores as administered by personal computer. *Journal of Clinical and Experimental Neuropsychology, 27,* 529–543.

Fergusson, D. M. (2005). Marijuana use and driver risks: The role of epidemiology and experimentation. *Addiction, 100,* 577–578.

Fillmore, M. T., Rush, C. R., & Hays, L. (2002). Acute effects of oral cocaine on inhibitory control of behavior in humans. *Drug and Alcohol Dependence, 67,* 157–167.

Giroud, C., Menetrey, A., Augsburger, M., Buclin, T., Sanchez-Mazas, P., & Mangin, P. (2001). Delta(9)-THC, 11-OH-Delta(9)-THC and Delta(9)-THCCOOH plasma or serum to whole blood concentrations distribution ratios in blood samples taken from living and dead people. *Forensic Science International, 123,*159–164.

International Council on Alcohol, Drugs and Traffic Safety (ICADTS). (1999). *Guidelines on experimental studies undertaken to determine a medicinal drug's effect on driving or skills related to driving.* Available at http://www.icadts.org

International Council on Alcohol, Drugs and Traffic Safety (ICADTS). (2007). *Guidelines and recommendations for drugged driving research.* Available at http://www.icadts.org

Jex, H. R., McDonnell, J. D., & Phatak, A. V. (1966). A "critical" tracking task for manual control research, *IEEE Transactions on Human Factors in Electronics, 7,* 138–145.

Kelly, E., Darke, S., & Ross, J. (2004). A review of drug use and driving: Epidemiology, impairment, risk factors and risk perceptions. *Drug and Alcohol Review, 23,* 319–344.

Kufera, J. A., Soderstrom, C. A., Dischinger, P. C., Ho, S. M., & Shepard, A. (2006). Crash culpability and the role of driver blood alcohol levels. *Annual Proceedings of the Association for the Advancement of Automotive Medicine, 50,* 91–106.

Laumon, B., Gadegbeku, B., Martin, J. L., & Biecheler, M. B. (2005). Cannabis intoxication and fatal road crashes in France: Population based case-control study. *British Medical Journal, 331,* 1371.

Logan, G. D. (1994). On the ability to inhibit though and action: A user's guide to the stop signal paradigm. In D. Dagenbach & T. H. Carr (Eds.), *Inhibitory processes in attention, memory and language* (pp. 189–239). San Diego, CA: Academic Press.

Michon, J. A. (1985). A critical view of driver behavior models. What do we know, what should we do? In L. Evans & R. Schwing (Eds.), *Human behavior and traffic safety* (pp. 485–525). New York: Plenum Press.

Movig, K. L., Mathijssen, M. P., & Nagel, P. H., et al. (2004). Psychoactive substance use and the risk of motor vehicle accidents. *Accident Analysis & Prevention, 36,* 631–636.

Ogden, E. J., & Moskowitz, H. (2004). Effects of alcohol and other drugs on driver performance. *Traffic Injury Prevention, 5,* 185–198.

Ramaekers, J. G. (2003). Antidepressants and driver impairment: Empirical evidence from a standard on-the-road test. *Journal of Clinical Psychiatry, 64,* 20–29.

Ramaekers, J. G., Berghaus, G., van Laar, M., & Drummer, O. H. (2004). Dose related risk of motor vehicle crashes after cannabis use. *Drug and Alcohol Dependence, 73,* 109–119.

Ramaekers, J. G., & Kuypers, K. P. (2006). Acute effects of 3, 4-methylenedioxy-methamphetamine (MDMA) on behavioral measures of impulsivity: Alone and in combination with alcohol. *Neuropsychopharmacology, 31,* 1048–1055.

Ramaekers, J. G., Moeller, M. R., van Ruitenbeek, P., Theunissen, E. L., Schneider, E., & Kauert, G. (2006). Cognition and motor control as a function of Delta9-THC concentration in serum and oral fluid: Limits of impairment. *Drug and Alcohol Dependence, 85,* 114–122.

Shallice, T. (1982) Specific impairments of planning. *Philosophical Transactions of the Royal Society of London Biological Sciences, 298,* 199–209.

46

Medical Disorders

Matthew Rizzo
The University of Iowa

Abstract

The Problem. Medical disorders impair driver perception, cognition, personality, and mood, which in turn can reduce driver performance and increase the risk of driver errors that lead to vehicle crashes. Minimizing crash risk and maximizing mobility depend on understanding how medical impairments affect particular aspects of driving behavior. *Role of Driving Simulators.* Studies of medically impaired operators in controlled circumstances in a driving simulator can reveal valuable information on the abilities required to safely drive a car. Simulation also provides a safe platform for driver training and for testing interventions designed to reduce driver errors and injuries, and increase mobility. This includes driver performance monitoring devices, and collision alerting and warning systems. *Key Results of Driving Simulator Studies.* Simulation has been successfully applied to assess performance profiles in drivers who are at increased risk for a crash due to a variety of conditions including Alzheimer's disease, Parkinson's disease, stroke, HIV, multiple sclerosis, traumatic brain injury, sleep apnea, hepatic encephalopathy, alcohol and other drug effects, visual disorders, psychiatric disorders, and age-related performance decline. *Scenarios and Dependent Variables.* The scenarios can focus on how specific medical deficits map onto driving tasks. Dependent measures can be determined using task analyses. Physiologic measures of health and disease, including brain activity, can be assessed in synchrony with driving task performance measures. *Platform Specificity and Equipment Limitations.* An "evergreen" issue is how much fidelity is needed and how the results compare across different simulators and real driving outcomes. Low-fidelity or "surrealistic" simulations can address focused questions on driver capacity. Austere versions of the driving task can be implemented in brain scanners to relate physiology and performance. It is unclear as yet if results in a simulator can be used for purposes of licensure or driving restriction. Simulator discomfort is a limiting factor. Predictions of driver safety and crash risk from driving simulator outcome measures depend on understanding the intervening patterns of real-world exposure, restriction, and other complex factors.

46.1 Introduction

This chapter provides an overview of a growing body of research on driving simulation in patients with medical disorders. Rather than covering detailed mechanisms of disease as a medical textbook would, we provide examples of translational research in driving simulation at the borders between medicine, neuropsychology, cognitive science, and human factors engineering. Other examples of driving simulation in medical disorders are covered in the *Handbook* chapters on neurodegenerative disorders (see, this book, chap. 48 by Uc & Rizzo, "Driving in Alzheimer's Disease, Parkinson's Disease, and Stroke"), psychiatric disorders (see, this book, chap. 47 by Moller, "Psychiatric Disorders and Driving Performance"), and drugs (see, this book, chap. 45 by Ramaekers, Moeller, Theunissen, & Kauert, "Validity of Three Experimental Performance Tests for Predicting Risk of Cannabis-Induced Road Crashes", and chap. 44 by Creaser, Ward, & Rakauskas, "Acute Alcohol Impairment Research in Driving Simulators").

46.2 Medical Disorders and Driving

A variety of systemic, neurological, psychiatric, and developmental disorders put drivers at particular risk for a car crash (Tables 46.1 and 46.2). However, medical diagnosis alone is often an unreliable criterion for driver certification. Thousands of different medical diseases affect the organ systems (circulatory, digestive, endocrine, integumentary, immune, lymphatic, musculoskeletal, nervous, reproductive, respiratory, urinary), the brain, and the mind (Table 46.3) in many different ways. Although many diseases impair performance and behavior, only a handful have been studied in any detail. Absence of evidence of unsafe driving in these understudied conditions is not evidence of absence of safety problems. Minimizing crash risk and maximizing mobility requires an understanding of how medical disorders affect driver performance and behavior. By

understanding the demographics, performance, and behavior patterns that lead to driver errors that cause crashes and injuries, it may be possible to design interventions to reduce these errors and injuries and preserve mobility. These interventions include driver performance monitoring devices, collision alerting and warning systems, road design, graded licensure strategies, and driver education, training, and rehabilitation.

Driving a motor vehicle safely requires a driver to perceive and attend to stimulus evidence, interpret the driving situation, generate a plan (based on previous driving experience and memory), execute an appropriate action (e.g., steer, accelerate, brake, disengage from a secondary task), and monitor the results of the action as feedback for corrective actions. Medical disorders can increase the risk of driver errors at one or more of these stages due to deficits of attention, perception (vision, audition, vestibular, and haptic), situation awareness and sensemaking, response selection (which depends on memory and decision-making), response implementation (a.k.a., executive functions), emotion, arousal (or sleepiness), psychomotor functions, general mobility, and self-awareness of ability (a.k.a., metacognition). Drivers with reduced awareness of impairment (a.k.a., anosognosia) may witlessly expose themselves to extra risk.

The impairments caused by medical disorders (Table 46.1) can be acute or chronic, inherited or acquired, reversible or irreversible. Central nervous system disorders such as stroke, epilepsy, and brain tumor, or degenerative disorders such as Alzheimer's disease and Parkinson's disease, are important neurological risk factors for driving impairment. Drugs, toxins, and systemic diseases (e.g., diabetes, kidney [renal] disease, liver [hepatic] disease, thyroid disease, cancer, and infections) impair driver performance through changes in brain function and body chemistry. These resulting disturbances are often included under the term encephalopathy, a broad term for a clinical syndrome caused by any brain disorder.

Systemic diseases can produce unrecognized encephalopathy (dysfunction of the brain) with serious consequences for driving

TABLE 46.1   Medical Factors Which May Increase Driving Risk

Conditions	Problem
Disorders of Childhood Learning and Development (ADD, Autism, Asperger's) and Chronic Diseases (e.g. Cancer, Epilepsy, Diabetes, Muscular Dystrophy)	These disorders may produce a range of subtle and relatively specific cognitive deficits in the context of normal intellect, which may affect driving behavior
Cardiac Disease	Syncope; distraction by cardiac sensations
Cervical Arthritis	Trouble checking the mirrors; distraction by pain
Deafness Hearing Loss	Lack of response to hazard cues in the auditory panorama; loss of binaural localization
Encephalopathy (see also Table 46.3)	A broad concept referring to a clinical syndrome caused by any disorder of the brain; can be acute or chronic, inherited or acquired, and reversible or irreversible, and may affect multiple aspects of driver performance and behavior
Medications, Alcohol, Illegal Drugs, Metabolic Disease	Sedation; psychosis on-the-road
Psychiatric Disease	Hostile or paranoid driving, response to hallucinations or delusions; suicidal intent
Syncope (e.g., Fainting)	Can cause loss of vehicular control
Vestibular Disease	Vertigo (as spinning sensation) on-the-road
Visual Dysfunction	Failure to resolve roadway objects or signs; failure to cope with glare or low ambient light conditions

TABLE 46.2 Neurological Syndromes Which May Increase Driving Risk

Syndrome	Problem
Agnosia	Inability to interpret road symbols
Amnesia	Failure to remember the rules of the road, vehicle operations, or the location of nearby vehicles
Aphasia	Trouble deciphering road maps or signs
Epilepsy	Loss of contact or consciousness
"Executive Dysfunction"	Strategic and tactical errors (e.g., driving in bad weather or in an unsafe vehicle), failure to formulate and implement plans for evasive maneuvers in emergency situations
Hemianopia	Failure to detect objects in one visual field
Hemineglect Syndrome	Failure to orient to objects on one side, often accompanied by other visuospatial disorders
Sleep Disorders	Falling asleep at the wheel
Topographic Disorientation	Getting lost
Transient Ischemic Attacks	Loss of control, contact, or consciousness during an attack
Visualspatial Disorders	Failure to perceive the location, speed, and direction of one's own vehicle, the structure of the roadway, the changing distance from other road objects, and the time-to-collision
Weakness, Incoordination, Ataxia, Loss of Limb Sensation	Inability to operate the steering, accelerator pedal, brake pedal, or other controls

(see the review of Rizzo & Kellison, 2009). For example, some diabetics may fail to recognize symptoms of insulin overdosage and drive while hypoglycemic (e.g., Cox, Kovatchev et al. 2004). Patients with liver disease may have undiagnosed "minimal hepatic encephalopathy" (a pre-clinical stage of hepatic encephalopathy that can be diagnosed by cognitive (neuropsychological) testing). Patients with kidney disease may have encephalopathy due to uremic poisoning or to fluid and acute electrolyte shifts with dialysis. Cardiac (heart) patients may be preoccupied with their cardiac sensations, or may have syncope (fainting) on the road. Patients with arthritis of the spine may have difficulty checking the mirrors or side windows and operating the controls because of reduced mobility or may be distracted by pain caused by vibrations and bumps in the car. Psychiatric patients with personality, thought, or mood disorders may exhibit hostile or paranoid driving, react to hallucinations or delusions, or drive with aggressive or suicidal intent. Fatigue, alcohol, and illicit drugs pose known road risks, as do many prescription medicines, (e.g., benzodiazepines, anticonvulsants, antihistamines, antidepressants, antipsychotics and narcotics; Tables 46.4 and 46.5), although the dose-response effects upon driving are mostly ill-defined. Some of these conditions are particularly amenable to testing in the safe and controlled environment of a driving simulator.

46.3 Advantages of Driving Simulation

Driving simulator studies can yield objective observations on fitness to drive under optimal stimulus and response control in tasks that are challenging and safe, and provide greater levels of context than do standard clinical or laboratory tests (e.g., visual, neuropsychological). The results can be compared with driver demographic and cognitive data, state records of citations and crashes, and field data in natural and naturalistic settings using instrumented vehicles, to inform public policy and guide development of in-vehicle countermeasures to avert real-world car crashes, injuries, and death.

Simulation can provide windows on the relationship between ubiquitous low-severity errors (e.g., failing to check the rear-view

TABLE 46.3 Metabolic Causes of Encephalopathy

Metabolic Cause	Specific Types
Circulatory Disturbances	Hypoxia-Ischemia, Hypertensive Encephalopathy
Pulmonary Disease	COPD, Asthma, Pulmonary Fibrosis
Disorders of Glucose Metabolism	Hypoglycemia (insulin induced), Hyperglycemia (diabetes)
Endocrinopathies	Hypothyroidism, Hyperthyroidism, Hypoparathyroidism, Hyperparathyroidism, Addison's Disease, Cushing's Disease
Fluid and Electrolyte Disturbances	Hyponatremia, Hypernatremia, Central Pontine Myelinolysis, Hypercalcemia
Hepatic Encephalopathy	Alcoholic, Infectious, Autoimmune
Renal Failure	Uremic Encephalopathy, Dialysis Dysequilibrium Syndrome, Dialysis Dementia, Transplantation
Vitamin Deficiency	Thiamine, Cobalamin, Niacin
Septic Encephalopathy	Delirium due to Blood Borne Systemic Infection
Organ Failure	
Post-Surgical Encephalopathy	Associated with Anesthesia, Pain Medicines and other Factors

TABLE 46.4 Therapeutic Agents Associated With Encephalopathy

Therapeutic Agent	Specific Types
Antibiotics	Penicillin, Cephalosporins, Quinolones, Amphotericin B
Anticholinergics	Tricyclic antidepressants, Antihistamines, Meclizine, Atropine, Scopolamine, Benztropine, Trihexyphenidyl
Antiepileptics	Phenobarbital, Clonazepam, Phenytoin, Carbamazepine
Antihypertensives	Methyldopa, Clonidine, Reserpine
Barbiturates	Butalbital, Secobarbital
Benzodiazepines	Diazepam, Chlordiazepoxide
Cancer chemotherapy drugs	Methotrexate, BCNU
Cardiac drugs	Digoxin, Quinidine, Disopyramide, Lidocaine
Cranial Irradiation	
Corticosteroids	Prednisone, Methylprednisolone, Dexamethasone
H2 receptor antagonists	Cimetidine, Ranitidine
Immunosuppressive drugs	Cyclosporine, Tacrolimus
Narcotics	Morphine, Hydrocodone, Oxycodone, Codeine, Meperidine, Other Schedule II Drugs
Neuroleptics	Phenothiazines, Butyrophenones, Others
Non-steroidal anti-inflammatory drugs	Ibuprofen, Naproxen, Diclofenac

mirror on a remote rural road) and rare high-severity errors that may lead to vehicle crashes. Simulation is important because real-world crashes are sporadic, uncontrolled events during which few objective observations can be made. Personal accounts and even state crash records may be incomplete and crashes are under-reported. Most state road tests are geared to test if novice drivers know and can apply the rules of the road, not to predict crash involvement in skilled drivers who may now be impaired.

Simulation can be used to assess performance profiles in drivers who are at risk for a crash due to a variety of different conditions such as sleep apnea, drowsiness, alcohol and other drug effects, old age, Alzheimer's disease, Parkinson's disease, HIV, or traumatic brain injury. The underlying assumption is that driver performance in the simulator, which reflects perceptual, cognitive and motor limits of what drivers can do under controlled conditions, is related to driver behavior—what drivers actually do on the road. In an ideal world, good performance predicts good behavior and bad performance predicts bad behavior. However, some drivers may not take driving in a driving simulator seriously because they recognize that the costs of a poor performance on a driving simulator differ from real driving where life, limb and license may be at stake (see, this book, chap. 9 by Ranney,

"Psychological Fidelity: Perception of Risk"). The best performing drivers are not always the best behaved and may act unsafely or illegally in the real world due to confidence, overconfidence, risk-acceptance or risk-seeking, such as a race car driver who shows winning skills at a racetrack, but chooses to speed or drive under the influence of drugs or alcohol on the highway. Conversely, some individuals with poor performance on a road test or in a simulator may self-restrict appropriately and manage to avoid traffic citations and crashes. Driving simulation, like standard road tests, tends to evaluate behavior over a period of an hour or less, and does not typically address strategic behaviors that tend to evolve over extended time frames. A more detailed discussion of these and similar issues is contained in the chapter by Mullen, Charlton, Devlin and Bédard (this book, chap. 13, "Simulator Validity: Behaviors Observed on the Simulator and on the Road").

Driving simulation can be a useful tool in addressing issues of prediction of driver safety and development of countermeasures. One useful application is for older drivers. Licensed drivers aged 60 years old and above have among the highest crash rates per mile of all age groups, approaching the rate of the risky and much less experienced cohort of drivers less than 25 years old (NHTSA, 2008). The aging Baby Boom is generating more older drivers, who

TABLE 46.5 Drugs of Abuse Associated With Encephalopathy

Drugs of Abuse	Specific Types
Ethanol	Beer, wine, hard liquor
Sedative-hypnotics	Benzodiazepines, Barbiturates
Opiates	Heroin, Prescription opiates
Stimulants	Dextroamphetamine, Methamphetamine
Cocaine	Blow, Coke, Crack, Crank, etc.
Ecstacy	aka MDMA
Hallucinogens	LSD, Psilocybin, Mescaline
Phencyclidine	aka PCP, Angel Dust
Marijuana	aka THC
Inhalants	Toluene, Other volatile agents

travel more miles per driver year than previous older cohorts. By 2004, there were over 28 million licensed drivers aged 65 years and older—up 17% from 1994, during which time the total number of licensed drivers in the U.S. increased by 13% (NHTSA, 2006). Older drivers are more likely to be killed or injured in a car crash than are younger adults. In the U.S., 3,355 occupants aged 65 and older died in motor vehicle crashes during 2004 (Centers for Disease Control and Prevention (CDC), 2006a). More than 177,000 adults aged 65 years and older suffered nonfatal injuries as occupants in vehicle crashes during 2005 (CDC, 2006). Drivers aged 80 years old and above have higher crash death rates per miles driven than all groups except teenage drivers (Insurance Institute for Highway Safety, 2006). During 2005, most traffic fatalities involving older drivers occurred during the daytime (79%) and on weekdays (73%), and 73% of the crashes involved another vehicle (NHTSA, 2006). Older adults in the U.S. favor driving as the main means of travel (Fonda, Wallace, & Herzog, 2001; Jette & Branch, 1992) and they are at increased risk of depression and decreased quality of life when they stop driving (DeCarlo, Scilley, Wells, & Owsley, 2003; Marottoli et al., 1997). To identify ways to enhance older driver safety and mobility, it is necessary to obtain accurate and objective measures of older driver behavior and safety errors and to develop the means of mitigating these errors.

As another example, drivers with diabetes have a significantly increased (19% greater) risk of a vehicle crash compared to drivers without diabetes (Tregear et al., 2007). Drivers who have insulin-induced hypoglycemia (typically Type 1 diabetics who are insulin-deficient but remain insulin-sensitive) may perform poorly in a driving simulator compared to drivers without hypoglycemia. There is evidence that drivers with insulin-dependent diabetes have a significantly greater crash risk than the general population of drivers (Skurtveit et al., 2009).

Some of these insulin-dependent drivers may be aware of hypoglycemic symptoms, enabling them to pull off the road or even abort a trip, and a driving simulator environment may not capture these self-regulatory behaviors, which evolve over days and weeks. These behaviors could potentially be addressed by combining continuous glucose monitoring and insulin administration records with extended samples of real-world activity (e.g., self-report using diaries and event logs; objective measures derived from actigraphy watches worn by the subject to capture diurnal patterns associated with waking activity and sleep) and instrumented vehicles, including event recorders (see below).

Real-world driver behavior varies with context, motivation, and goals—for example, driving fast and citation free to an appointment versus adhering to the speed limit versus driving slowly while "rubbernecking" near a distraction. The minutes-to-hours time frame of a simulation or even a standard road test may be insufficient to capture the epochs and contexts over which some behaviors evolve, particularly executive functions and decision-making. These include behaviors such as choosing to drive at night, during rush hour, or under poor road conditions, how to plan a trip, get gas or service a car, and whether to drive after a poor night of sleep. In a similar vein, most drivers with episodic disorders do not have the problem frequently enough to

capture the events in a driving simulator test, and if they do they may have already been identified as unfit to drive. For example, a simulation might capture effects of a disease that causes epileptic seizures, or adverse effects of drugs used to treat the seizures, but not the seizure itself. A patient with bipolar disorder may not be manic or depressed when tested. A driver with antisocial personality disorder may be on his best behavior, sober, and free of customary myriad irritants or conflicts of home and work.

46.4 Selecting a Simulator Scenario

There are several different types of simulators (film, non-interactive, interactive, fixed versus motion-based, desktop, full cab, and so forth) for implementing driving scenarios, and no one is necessarily better than another (see, this book, chap. 7 by Greenberg & Blommer, "Physical Fidelity of Driving Simulators"). In a practical vein, it is an open-ended and impractical exercise to attempt to list the simulator platforms, scenarios, dependent measures, and other details to use for medical disorders that may impair driving, of which there are far too many, or even all the details of simulation for just one disease. This depends on the evaluation goal and specific questions that need to be addressed. In a general sense, one needs to use the right tools for the right job. If you want to know what has been done lately, search the literature and grade the evidence according to standard protocols (e.g., using the Newcastle-Ottawa grading schema [Wells, 2006], among many others).

Similarly, the particular simulator scenario to use depends on the specific clinical question being asked. For instance, the evaluation of sleepy drivers such as those with obstructive sleep apnea (OSA) might use long monotonous drives together with equipment that captures relevant physiologic data such as electroencephalography (EEG) recordings (to capture microsleeps) or eye monitoring recordings to capture measures of percent lid closure (PERCLOS: Dingus, Hardy, & Wierwille, 1987). A study on diabetic drivers might monitor blood glucose during a drive following a dose of insulin to test effects of hypoglycemia (low blood sugar) and driver awareness thereof (Tregear et al., 2007). Driving performance in patients with pulmonary (lung) disorders such as pulmonary fibrosis or chronic obstructive pulmonary disease (often from chronic cigarette smoking) can be combined with oximetry (to non-invasively monitor blood oxygen levels). A study on the effects of a drug might focus on a scenario that can capture a continuous measure, such as car following (see Figure 46.1; Web Figure 46.1 on *Handbook* Web site for color version). A study of novice or developmentally delayed drivers might combine eye gaze recording with a scenario that focuses on expert perception of occult hazards, such as a pedestrian crossing hidden behind a parked truck (as in the work of Pradhan et al., 2005). A study of drivers with visual field loss might evaluate the response to a car pulling out from the side of the road. A study of drivers with executive dysfunction due to stroke, head injury, or a neurodegenerative impairment (such as Alzheimer's disease or frontotemporal dementia) might be tested on a driver's ability to make go/no-go decisions at traffic lights or

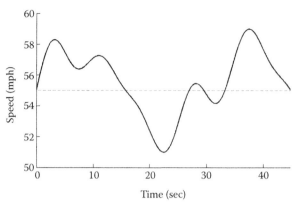

FIGURE 46.1 Car following. Left Panel: A car following scenario (depicted from the driver's view) is useful for evaluation of continuous attention and risk-taking in drivers who may be at risk for rear-end collisions due to disorders of alertness, vigilance, visuoperception, and judgment. Right Panel: To challenge the driver, the velocity of the lead vehicle can be made to change unpredictably.

intersections. Simulators can test in-vehicle devices before they are deployed on the road; they also can be used to train drivers with medical disorders in the aftermath of an illness in order to prepare them for return to driving. This is a rich area for ongoing research and clinical trials and knowing what to do depends on knowledge of disease pathophysiology and the bodily impairment it produces.

In a practical sense, the approach to driving simulator scenarios can adopt both fixed and flexible approaches, by analogy with the Iowa-Benton approach to neuropsychological testing (Tranel, 1996; Jones & Rizzo, 2004). An economical core battery of tasks can be administered, and the remaining examination is guided by referral questions, patient (driver) complaints, the underlying disease or condition, and results of the core battery and other tests. In the case of driving, there are core measures of vehicle control (speed, steering, lane position) on a few road types that one may wish to measure in all drivers. There are also situations (car following, merging, intersection behavior) that may be desirable to assess because these are key situations in which driver errors (and crashes, if one goes by the epidemiologic record) are more likely to occur. The core simulator assessment can proceed relatively rapidly, taking as little as 15 minutes. Specialized tests (e.g., serum glucose manipulations using intravenous glucose-insulin infusions in drivers with diabetes; electroencelphalographic recordings in drivers with microsleeps associated with a possible sleep disorder; use of prisms in clinical trials of optical devices in patients with visual field loss) aim to address special questions and are not necessary for most drivers. That is, they are not part of a general core simulation battery. A large set of tests is possible to assess complex and technical questions and such comprehensive assessments may take hours and require testing at inconvenient times (e.g., middle-of-the-night, one hour after ingesting alcohol or a drug, etc.). Note that most neuropsychological measures rely on empirical normative data from individuals without known neurological

condition or cognitive impairments (Spreen & Strauss, 1998), and are interpreted in the context of known statistical and psychometric properties such as reliability (test-retest, inter-rater, internal consistency) and validity (face, construct, predictive, concurrent). Performances on specific tests are adjusted for age, education, and other relevant demographic factors that bear on performance, and a given patient's scores are compared with the normative data. It would be desirable to apply these statistical and psychometric principles to emerging simulator tools for clinical applications.

It is not feasible to recommend a detailed list of scenarios to test every possible risky driving situation in every medical disorder because there are thousands of disorders and a myriad of driver error types. A hypothesis-based, deductive approach to behavioral diagnoses (such as unsafe driving) is needed for developing appropriate simulator scenarios, depending on the questions to be addressed. Investigations gather clinical observations, historical, and collateral data and results of tests (including simulation) that have varying sensitivity and specificity for behavioral conditions or outcomes (e.g., traffic citations, crashes, injuries, stopping driving). This combined evidence triggers differential hypotheses regarding diagnoses (e.g., unsafe driving status in certain situations) that are operationalized as working diagnoses. Working diagnoses are probed with further tests (including additional driving simulator scenarios) and follow-up to provide explanation, validation, verification, decision-making, interventions, and predictions. If the test data (from the simulator and other sources) do not support or validate a hypothesis, they can be combined with clinical data to generate new hypotheses.

The ability to distinguish a behavioral disorder from a normal state (e.g., unsafe versus safe driving) may employ receiver operator characteristic (ROC) curves that plot true positives (e.g., detecting unsafe driving when it is present) versus false positives (declaring a safe driver to be unsafe). Criteria for ruling in or ruling out a condition depend on the costs and benefits of making or missing a diagnosis. For example, a physician or department of transportation official may be sensitive to diagnosing unsafe driving in a commercial motor vehicle driver who presents with "spells" (e.g., possible epilepsy) if the person drives a school bus or hazardous materials because the risk of an epileptic seizure while driving may be catastrophic and the benefits of treatment are great. However, the physician or administrator may set more specific criteria for the diagnosis if the spells are non-specific and the driver appears to be a low-risk-taking driver with a private license who drives few miles and under limited circumstances.

46.5 Impairment of Sensation and Perception

Driving and Perceptual Judgments. Some medical illnesses impair perceptual judgments. For instance, hearing-impaired drivers may fail to appreciate in-vehicle alerts, sounds of approaching traffic or abnormal vehicle operations, or warning signals such as the horns of approaching vehicles. Patients with vestibular disease may develop acute vertigo (a spinning sensation), causing them to veer in traffic. Drivers with neuropathy (a disorder of the peripheral nerves) may have trouble finding, feeling and controlling the pedals.

Visual cues are especially important to driving because they convey long-range information about driver self-trajectory (egomotion), changes in the terrain, and the trajectories of objects. A range of static and dynamic visual cues provide information on the structure, distance, and time-to-contact with objects that might strike the driver. Our visual fields with both eyes open normally span about 180 degrees. The fovea has the highest acuity and spans about 3 degrees around fixation; the macula spans about 10 degrees and participates in detail-oriented tasks such as map reading and sign localization. The peripheral visual fields have low visual acuity, but good temporal resolution and movement detection. The chapter by Andersen (this book, chap. 8, "Sensory and Perceptual Factors in the Design of Driving Simulation Displays") discusses these issues in much more detail.

Medical Disorders and Evaluating Loss of Perceptual Judgments. Common patterns of visual loss correspond to different diseases and lesions in visual pathways and create a variety of risks for drivers. Common eye disorders cause visual sensitivity loss and visual field impairments that can impair driver safety (American Academy of Ophthalmology, 2006; Rizzo & Kellison, 2004). Strabismics (individuals with misaligned eyes) may have double vision and mistake which roadway image is real. Cataract, diabetic retinopathy, or senile macular degeneration, may impair drivers' resolution of roadway objects, especially in the presence of glare, low contrast, or low light. Cataracts are a risk factor for car crashes (Owsley, Stalvey, Wells, Sloane, & McGwin, 2001) and are treatable with surgery. They are common in aging eyes, and cause a reduction in acuity and the creation of distracting reflections (e.g., halos around lights) or glare. Glaucoma can affect driving by producing visual impairments and visual field loss. Macular degeneration affects areas of high-detail vision around fixation. Retinitis pigmentosa, an inherited condition that tends to affect younger drivers,

constricts the peripheral visual fields, causing inability to detect objects approaching from the side. Even prescription glasses may cause trouble while driving due to reflections, distortions, or discontinuities (e.g., caused by looking across bifocal lenses or glare). Effects of poor CS (contrast sensitivity) can be tested in low light (which is inherent in current simulator displays) and by adding fog (see, this book, chap. 28 by Wood & Chaparro, "Night Driving: How Low Illumination Affects Driving and the Challenges of Simulation"). Effects of visual field loss can be tested in a simulator by introducing unexpected challenges (e.g., an incurring vehicle) within the area of the impaired fields. This assessment would be aided by driving scenarios and simulators that allow the ability to introduce gaze contingent events (see, this book, chap. 18 by Fisher, Pollatsek, & Horrey, "Eye Behaviors: How Driving Simulators Can Expand Their Role in Science and Engineering").

Drivers with lesions of visual areas in the occipital lobe and adjacent temporal and parietal lobes have various visual field defects (such as homonymous hemianopia, the loss of vision in comparable fields in each eye due to damage in the cerebral pathways for vision) and may fail to perceive objects or events in the defective fields (Rizzo & Barton, 2005). Search strategies to compensate for the visual defect may create extra work that distracts from the driving task. Lesions in the occipital and parietal lobes (in the dorsal or "where" pathway) may have greater adverse effects on driving performance than do lesions in the occipital and temporal lobes (in the ventral or "what" pathway). Dorsal lesions cause visual loss in the ventral or lower visual fields, which may obscure the driver's view of the vehicle controls and of the road ahead. They can also impair processing of movement cues as in cerebral akinetopsia (motion blindness) and reduce visuospatial processing and attention abilities. Some of these patients have impairments of visual search and the useful field of view (i.e., the visual area from which information can be acquired without moving the eyes or head; Figure 46.2; Web Figure 46.2)

FIGURE 46.2 Driver with UFOV impairment. Visual field defects in patients with eye or brain lesions, or systemic effects (e.g., sleepiness, medications, encephalopathy) can impair visual search and processing speed, thereby reducing the useful or effective visual fields, and increasing the risk of vehicle crashes at intersections, as depicted in the figure.

and are clearly unfit to drive. One example is the hemineglect syndrome, the failure to orient to targets in the left visual hemifields in patients with right parietal lobe lesions (Rizzo & Barton, 2005). Another is Bálint's syndrome (simultanagnosia, spatial disorientation, optic ataxia and ocular apraxia), which generally involves bilateral dorsal, visual pathway lesions and affects visual attention, visual search, and visually-guided eye and hand movements. Ventrally located lesions produce upper visual field defects which may be less problematic for driving, but may also cause impairments of object recognition (visual agnosia), affecting interpretation of roadway targets; reading (pure alexia), affecting roadway sign and map comprehension; and color perception (cerebral achromatopsia), impeding the use of color cues in traffic signals and road signs and detection of roadway boundaries and objects defined by hue contrast (Rizzo & Barton, 2005). Sleepy drivers likewise have difficulty processing information in the visual panorama (Figure 46.3; Web Figure 46.3).

Driving and Collision Detection. The brain employs multiple cues on object structure and depth because this information is critical for interacting with moving objects and obstacles (Palmer, 1999). Binocular stereopsis (stereo vision) and motion parallax provide unambiguous cues to relative depth (Nawrot, 2001). Detecting and avoiding potential collisions require information on approaching objects and the driver's vehicle. Objects set to collide with the driver stay at a fixed location in the driver's field of view while "safe" objects move to the left or right (Figure 46.4; Web

Figure 46.4). Time-to-contact (TTC) is estimated from the expanding retinal image of the approaching object. Older drivers are less accurate than younger drivers at detecting an impending collision during braking and judging if an approaching object will crash into them (Andersen, Saidpour, & Enriquez, 2001). Performance is worse for longer TTC conditions, possibly due to a greater difficulty in detecting the motion of small objects in the road scene ahead of the driver (Andersen et al., 2001).

Displacement of images across the retina during travel produces optic flow patterns (Gibson, 1979) that can specify the trajectory of self-motion (egomotion) with accuracy (Warren & Hannon, 1988; Warren, Mestre, & Morris, 1991). On curved roads, drivers tend to fixate the information flowing from the inside edge of the road where the curve changes direction (Land & Lee, 1994). The findings are relevant to detection of collisions, design of roads, and positioning of traffic warnings within a driver's dynamic visual environment. Figure 46.5 depicts a display used to test heading perceptual deficits of drivers at risk for driving impairment due to drug use.

46.6 Executive Dysfunction and Driver Behavior

Executive functions provide control over information processing and are a key determinant of driver strategies, tactics, and safety. These functions include decision-making, impulse control, judgment, task switching, and planning. Executive functions strongly interact with working memory (i.e., the process of briefly storing information so it is available for use) and attention, which

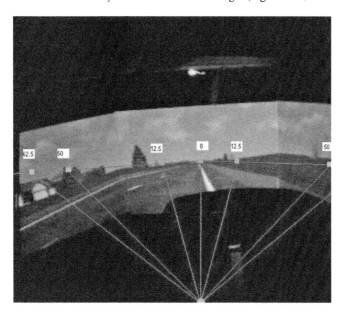

FIGURE 46.3 Vigilance task. Impairments of visual processing associated with sleepiness may also affect visual search, processing speed and vigilance. Drivers with obstructive sleep apnea (OSA) were asked to click the high beams in response to targets flashed at unpredictable intervals and locations along the horizon during a one-hour drive in the SIREN driving simulator. Drivers responded faster and more accurately to central targets than to peripheral targets. Sleepiness (SSS index) increased over the drive, correlating with increasing RT on the vigilance task. Vigilance declines as a function of sleepiness. Sleepiness decreases efficiency with which drivers extract information from a cluttered scene.

FIGURE 46.4 Collision detection task. The test displays simulate a roadway scene with objects translating with constant speed and linear trajectory toward the observer. Subjects are asked to judge whether approaching objects (the balls) will collide with them or pass by them. Subjects with neurodegenerative disease were less sensitive in detecting collisions than elderly controls, and sensitivity worsened with increasing number of objects in the display and increasing time to contact of those objects. (Adapted from Vaux, L. M., et al. (2010). _Accident Analysis and Prevention_, 42. 852–858.)

FIGURE 46.5 Heading perception. Perception of self-motion, or heading, from optical flow patterns is an important driver feedback cue and can be tested in an abstract environment using stimuli comprising random dot ground planes (see, this book, chap. 8 by Andersen). Subjects report if direction of virtual travel is to the left or the right relative to a vertical post whose position varied with respect to body midline. Abstinent users of psychoactive drugs (MDMA/THC) had difficulty on this task compared to non-drug users. (From Rizzo, M., Lamers, C. T., Sauer, C. G., Ramaekers, J. G., Bechara, A., & Andersen, G. J. (2005). Impaired perception of self-motion (heading) in abstinent ecstasy and marijuana users. Psychopharmacology, 179(3), 559–566.)

operates on the contents of working memory. The mapping between executive functions measured in laboratory settings and real-life driving performance can be addressed with a set of theoretically motivated tasks (Rizzo & Kellison, 2009).

Causes of impaired decision-making include acquired brain lesions affecting prefrontal areas (due to stroke, trauma, or neurodegenerative impairment), antisocial personality disorder, and the effects of drugs and alcohol. The brain's decision-making and other executive function capabilities may not yet be fully developed in teenage drivers (Rizzo & Eslinger, 2004). Driving outcomes in impaired decision-makers could include traffic violations (e.g., speeding), unsafe vehicle maneuvers, engaging in extraneous and unsafe behavior while driving, and crashes. Abstract virtual environments can be used to assess go/no-go decision-making behavior in a driving-like task. Using a personal computer equipped with a steering wheel and pedals, subjects drove through intersections that had gates that opened and closed (Rizzo, Severson, Cremer, & Price, 2003). A green "Go" or red "Stop" signal appeared at the bottom of the display as the subject approached the gate, and a gate-closing trigger point was computed. Most of the time the signal provided accurate advice, but a small proportion of the time the signal was misleading. Cognitively impaired drivers who had frontal lobe damage (see Figure 46.6; Web Figure 46.5) had more crashes into closed gates, more failures to go at open gates, and longer times to complete the task. This suggests a failure of response selection criteria based on prior experience, as previously reported in individuals

with decision-making impairments on a gambling-related task (e.g., Bechara, Damasio, Tranel, & Damasio, 1997). Drivers who had lesions in areas that did not produce executive dysfunction performed well on the go/no-go task, supporting the specificity of this task in localizing decision-making impairments in a driving-like task.

A car following task can also be used to evaluate driver behaviors linked to acceptance of risk, as in a recent study of abstinent users of MDMA ("Ecstasy") and marijuana (Dastrup, Lees, Bechara, Dawson, & Rizzo, 2010; see Figure 46.1) The authors in the latter study employed a simulated car following task that required continuous attention and vigilance. Drivers were asked to follow two car lengths behind a lead vehicle (LV). Three sinusoids generated unpredictable LV velocity changes. Drivers could mitigate risk by following further behind the erratic LV. From vehicle trajectory data the authors performed a Fourier analysis to derive measures of coherence, gain, and delay. These measures and headway distance were compared between the different groups. All MDMA drivers met coherence criteria indicating cooperation in the car following task. They matched periodic changes in LV velocity similar to controls (abstinent THC users, abstinent alcohol users, and non-drug users), militating against worse vigilance. While all participants traveled at approximately 55 mph (89 kph), the MDMA drivers followed 64 m closer to the LV and demonstrated 1.04 seconds shorter delays to LV velocity changes than other driver groups. The simulated

FIGURE 46.6 A non-conventional display design approach aimed at testing go/no-go decision-making by drivers with cognitive impairments. Drivers passed through a series of gates that closed completely or partway. Gate status was predicted by warning signals that were usually correct. Dependent measures were time-to-completion, number of crashes into closed gates, number of stops at open gates, and number of successes (i.e., stopping at closed gates and going at open gates). In one rendition we used unusual objects (such as the Venus de Milo) as distracters. This surrealistic task discriminated between drivers with executive dysfunction from frontal lobe brain injury (depicted in the dark areas of the 3D MR surface reconstruction of a brain injured subject).

car following task safely discriminated between driving behavior in abstinent MDMA users and controls. Abstinent MDMA users do not perform worse than controls, but may assume extra risk. The control theory framework used in this simulator study revealed behaviors that might not otherwise be evident.

46.7 Attention and Working Memory Impairment

Direction of attention to relevant features of the environment is an executive function of key importance to driving. We remember and act upon attended stimuli. Specific regions of the prefrontal cortex are essential in directing cognitive resources to accomplish tasks with a range of memory demands (Cabeza et al., 2003: Nyberg et al., 2003). Defects of attention clearly impair driver decisions and affect a variety of processes (e.g., Parasuraman & Davies, 1984). Automatic attention processes are fast and involuntary and contribute to subconscious corrections, including control of steering wheel or accelerator pedal position during routine driving. Controlled attention processes are slow and operate during capacity-demanding tasks and conscious decision-making. Examples include glancing between the road and rear-view mirror while maneuvering in and out of traffic, or making deliberate head and eye movements during go/no-go decisions at traffic signals, busy intersections, merging, or overtaking.

Owsley, Ball, Sloane, Roenker and Bruni (1991) and Ball, Owsley, Sloane, Roenker and Bruni (1993) linked driving impairment with reduction in the useful field of view (UFOV), the visual area from which information can be acquired without moving the eyes or head (Ball, Beard, Roenker, Miller, & Griggs, 1988). Performance on the UFOV task depends on speed of processing and divided and selective attention. The UFOV begins to deteriorate by age 20. This deterioration may reveal itself as a shrinking of the field of view or a decrease in efficiency with which drivers extract information from a cluttered scene (Sekuler, Bennett, & Mamelak, 2000; Ball et al., 2005). Diminished efficiency in advancing age is worse when attention is divided between central and peripheral visual tasks. Driver performance may also change when attention is divided between the road and an onboard task, and gaps of >2 seconds may interrupt scanning of the road (Wierwille, Hulse, Fischer, & Dingus, 1988). Underlying mechanisms and effects can be assessed in a simulator, under greater control than is possible in a roadway environment (e.g., see, this book, chap. 19 by Gugerty, "Situation Awareness in Driving", and chap. 27 by Strayer, Cooper, & Drews, "Profiles in Cell Phone-Induced Driver Distraction").

46.7.1 Executive Functions and Attention

Executive functions control the focus of attention (Vecera & Luck, 2003). Without focused attention, we can be unaware of marked changes in an object or a scene made during a saccade, flicker, blink, or movie cut; this is known as "change blindness"

(O'Regan, Rensink, & Clark, 1999; Rensink, O'Regan, & Clark, 1997, 2000; Simons & Levin, 1997; Simons & Rensink, 2005; Caird, Edwards, Creaser, & Horrey, 2005). Traces of retinal images in visual working memory fade without being consciously perceived or remembered ("inattentional amnesia"). The very act of perceiving one item in a rapid series of images briefly inhibits ability to perceive another image- the "attentional blink" (Rizzo, Akutsu, & Dawson, 2001). Focused attention is thought to permit consolidation of information temporarily stored in visual working memory. Perceptual errors are likely if working memory is still occupied by one item when another item arrives, due to interference or limitations in capacity, which at a bottleneck stage admits only one item at a time. Failure to detect roadway events increases when information load is high, as at complex traffic intersections with high traffic and visual clutter (Batchelder, Rizzo, Vanderleest, & Vecera, 2003; Caird et al., 2005). Driver errors occur when attention is focused away from a critical roadway event in which vehicles, traffic signals, and signs are seen but not acted upon, or are missed. Sometimes eye gaze is captured by irrelevant distracters such as "mudsplashes" on a windscreen that prevent a driver from seeing a critical event (O'Regan et al., 1999) like an incurring vehicle or a pedestrian entering a crosswalk (see Figure 46.7; Web Figure 46.6). Drivers with cerebral lesions are liable to be "looking but not seeing" despite low information load (Rizzo, Reinach, McGehee, & Dawson, 1997), like air-traffic controllers who fail to detect aircraft during prolonged monitoring of radar displays (Broadbent, 1958).

46.7.2 Switching Attention

Driving and Switching Attention. Safe driving requires executive control to switch the focus of attention between critical tasks such as tracking the road terrain, monitoring the changing locations of neighboring vehicles, reading signs, maps, traffic signals, and dashboard displays, and checking the mirrors (Owsley et al., 1991). This requires switching attention between disparate spatial locations, local and global object details, and different visual tasks. Drivers must also switch attention between modalities when they drive while conversing with other vehicle occupants, listening to the radio or tapes, using a cell phone, and interacting with in-vehicle devices (Kantowitz, 2001).

Medical Disorders and Evaluating Switching Attention. These attentional abilities can fail in drivers with visual processing impairments caused by cerebral lesions (Rizzo, 2004; Vecera & Rizzo, 2004). Executive control over attention is thought to depend on frontal lobe structures and their connections to other brain areas, particularly parietal, but also others. These abilities can be operationalized and tested in healthy and impaired drivers using cognitive science paradigms, which can be used in a range of settings from a computer monitor to a simulator to an MRI scanner. Studies using the UFOV task mentioned above have demonstrated deficits in attention function in at-risk drivers, who are less able to extract visual information from the roadway environment

(a)

(b)

FIGURE 46.7 Change detection task. Subjects were asked to respond to changes (via touch screen) in traffic related images that were created from photographs taken from inside a motor vehicle. The original image (a) progressively faded into the modified image, so that the participant was continuously viewing the same global scene, except for the changing feature (in this case pedestrians crossing the street) that was gradually blending in and out; (b) outcomes on this traffic image-based change detection task discriminated between participants with aging and cognitive impairment. (From Rizzo, M. et al. (2009). *J Clin Expl Neuropsychology*, 31, 245–256.)

efficiently. Because drivers must rapidly shift attention to multiple locations in order to monitor hazards or information of interest (e.g., pedestrians, traffic, signs, and navigational landmarks simultaneously surrounding them), such a deficit in attention can undermine driving performance and safety. While the UFOV task provides a useful measure for identifying at-risk drivers, it does not pinpoint which mechanisms of attention function are central to driving. This is because "attention" refers to a set of processes that allow an observer to shift the focus of processing in time and space, control the breadth of information processing, select relevant information, and suppress irrelevant information (see Knudsen, 2007; Posner & Cohen, 1984), each of which has been associated with different underlying neural systems (see Fan, McCandliss,

Sommer, Raz, & Posner, 2002). Selective deficits in some, but not others, of these processes may play out in different ways in operational settings (on the road).

Using UFOV deficits as a starting point, attempted to assess what specific mechanisms are linked with attention functions responsible for impaired performance on the UFOV task (Lees, Cosman, Vecera, Lee, & Rizzo, in press). Following classification, drivers performed both a visual search task and attentional cueing task. In the visual search task, the drivers must find a target item embedded in an array of distracters. This task measures the ability to voluntarily shift attention in space in the presence of distracting information. The attentional cuing task measures the efficiency of stimulus-driven attentional orienting and shifting in the absence of distracting information (Posner, Snyder, & Davidson, 1980). These scenarios can be presented in the same settings in which driving simulator scenarios are presented. With special arrangements, they can also be presented in a functional magnetic resonance imaging (fMRI) scanner. Driving scenarios which have evaluated attention switching amid greater contextual cues include studies on the effects of cell phones, use of on-board navigation and communication devices, use of entertainment and "infotainment" systems, and the ability to avoid hazards in complex traffic environments, among others. The neurocognitive mechanisms explored by Lees et al., are related to performance in these tasks, and defining these relationships is an aspect of translational brain and behavior research relevant to automobile driving.

Functional neuroimaging studies in driving-like tasks (in the restrictive setting of an fMRI scanner) show changes in frontal lobe activity with driver modulation of vehicle speed (Web Figure 46.7; see also Calhoun et al., 2002; Peres et al., 2000) and with alcohol-impaired driving (Calhoun, Pekar, & Pearlson, 2004). The relationships between driver performance measures and physiological outcomes measured using an austere, context-diminished driving task (to subjects lying flat and performing a tracking [steering] task while listening to someone speak) within an MR scanner (e.g., Just, Keller, & Cynkar, 2008) and real-world behavior and physiology associated with unfettered driving on a road remain to be determined.

Note that a subset of patients with neurological disorders have focal brain lesions or cognitive disturbances that can be used to probe the neural underpinnings of complex behaviors such as automobile driving or the key functions (attention, memory, executive functions) that are critical to automobile driving. The lesion method shows which brain structures are necessary for a certain function (Rorden & Karnath, 2004). This method successfully addresses the nature of mental representations and the organization of brain processes ("psychoanatomy") in animals and humans with brain lesions and is highly relevant to understanding the organization of cognition in normal individuals, including drivers. A useful and fruitful approach has been to develop and maintain a registry of subjects with lesions in virtually every cerebral area (Palca, 1990; Fellows, Stark, Berg, Chatterjee, 2008). Important sources of brain lesions include traumatic brain injury, stroke, infections, and tumors. Also, transient experimental lesions can be produced using transcranial magnetic stimulation. Some lesions subjects show "dissociation"—that is, a state in which performance in one experimental task changes much less than performance in a second task. A double dissociation is evident when damage to another area affects performance of the second task but not the first, suggesting that the two tasks are handled by separate systems or differing sets of neurons within the same system. The resulting patterns of cognitive deficits are sometimes described using information processing metaphors including serial processing, parallel processing, feedback, feedforward, and distributed processing. Complementary methods measure regional cerebral activity during cognitive tasks in neurologically normal and in brain-damaged individuals. The main techniques used to measure brain activity are functional neuroimaging using PET, fMRI, and related techniques such as electrical brain evoked potentials (EPs) and magnetoencephalography. A weakness of these techniques is their inability to resolve issues concerning the mechanisms or the necessary role of activated areas in the performance of the cognitive function being studied. In some studies the approach seems not so far from Gall's phrenology (Uttal, 2002; Vul, Harris, Winkielman, & Pashler, 2009). Knowing that a specific area is active during a specific task does not prove its criticality or that there is a point-to-point map of task-to-brain area; rather the activity may be an epiphenomenon. Lesion analysis in brain-injured populations complements what can be learned from functional neuroimaging; this combined approach is novel. Voxel-based approaches to mapping lesion-behavior correlations in brain-injured populations can leverage image analysis methods drawn from fMRI. Functional MRI provides a surrogate for neural activity- the BOLD (brain oxygen level dependent) response- and shows activation in areas that may be marginally involved or not necessary. Patterns of activity depend very much on thresholds drawn by investigators (Vul et al., 2009). Also, importantly in the context of automobile driving, the brain lesion method allows measurements of behavior in operational settings (in a driving simulator or on the road) that cannot be measured in the laboratory fMRI environment in which a subject is required to lie motionless on his or her back in a setting that is highly controlled and much reduced in context, content, and complexity.

46.8 Emotions and Personality

Driving a motor vehicle can be a major annoyance, especially to aggressive drivers, who may curse, shout, gesticulate, speed, ignore traffic signals, tailgate, drive under the influence of drugs and alcohol, or use their car to block, strike or intimidate another car or a pedestrian. "Road-rage" is a media term used to describe extremely aggressive and often criminal events (Brewer, 2000). Aggressive drivers are more likely to be young, male, and single, use alcohol or drugs, have a pre-morbid personality disorder or tendencies (usually antisocial), and have increased stress at home, at work, and in a car. Stresses can include high passenger compartment temperature, crowded roadways, vehicle breakdowns, getting lost, and slow drivers ahead. Personality factors associated with aggressive driver crash involvement are thrill

seeking, impulsiveness, hostility/aggression, emotional instability (Dahlen, Martin, Ragan, & Kuhlman, 2005; Schwebel, Severson, Ball, & Rizzo, 2006), antisocial personality disorder, depression, and psychosis (Noyes, 1985; Tsuang, Boor, & Fleming, 1985). A depressed driver may fail to focus attention adequately on the road. A schizophrenic driver may be distracted by pathological thoughts or hallucinations. Antipsychotic, antidepressant, and anxiolytic medications may slow driver reaction times and decrease driver arousal. The roles of personality and personality disorders, aggression, risk-taking, and psychiatric disorders and drugs on driver safety and crash risk all require further study (see, this book, chap. 47 by Moller, "Psychiatric Disorders and Driving Performance").

46.9 Arousal, Alertness and Fatigue

Attention, perception, memory, and executive functions that are crucial to the driving task are critically affected by drugs and fatigue (Barger et al., 2005; Mallis, Banks, & Dinges, 2007). Drivers with sleep disorders such as obstructive sleep apnea syndrome (OSA), are at particular risk for a crash (Figure 46.8; see also, in this book, chap. 49 by Tippin, "Driving Simulation in Epilepsy and Sleep Disorders"). Affected individuals may minimize the degree to which they are sleepy and fail to recognize they have trouble driving. Symptom minimization may be intentional or due to unawareness of sleepiness, possibly caused by an altered frame of reference for fatigue. Sleep disturbances can accompany the hallmark motor, cognitive, psychiatric, and autonomic disturbances in Parkinson's disease, due to varied involvement of noradrenergic, cholinergic, and serotoninergic systems (see, this book, chap. 48 by Uc & Rizzo, "Driving in Alzheimer's Disease, Parkinson's Disease, and Stroke"). Excessive daytime sleepiness or "sleep attacks" have been reported with multiple dopaminergic medications used to treat PD (Olanow, Schapira, & Roth, 2000).

Falling asleep can be conceived as a process characterized by decreasing arousal, lengthening response time, and intermittent response failure (Ogilvie, Wilkinson, & Allison, 1989). The EEG may show progression from wakefulness to stages I and II sleep, or be preceded by "microsleeps" in which the EEG shows five or more seconds of alpha drop-out and an increase in theta activity (Harrison & Horne, 1996). These periods of approaching sleep onset have been correlated with deteriorating driving simulator performance in healthy, sleep-deprived drivers (Horne & Reyner, 1996; Reyner & Horne, 1998) and OSA patients (Paul, Boyle, Rizzo, & Tippin, 2005; Risser, Ware, & Freeman, 2000).

Driving simulator research is a promising tool for addressing how cognitive errors in the driving task increase as a function of severity of sleep disturbances. Physiologic indices of impending sleep can be measured aboard a vehicle or simulator, using a variety of techniques. These include EEG (e.g., of drowsiness or microsleeps), decreased galvanic skin response (GSR), increased respiratory rate, increased heart rate variability, reduced EMG activity (e.g., cervical paraspinous muscles), and percent lid closure (PERCLOS). PERCLOS scores of 80% or greater are highly correlated with falling asleep (Dinges, 2000). The lid closure can be used to trigger auditory

Study name	Statistics for each study					Rate ratio and 95% CI
	Rate ratio	Lower limit	Upper limit	Z-value	p-value	
Barbe	2.570	1.304	5.065	2.727	0.006	
Shiomi	2.342	0.237	23.159	0.728	0.467	
Horstmann	8.719	6.179	12.303	12.326	0.000	
Lloberes	2.720	0.342	21.658	0.945	0.345	
Findley 2000	6.195	0.373	102.902	1.272	0.203	
George	1.306	0.791	2.158	1.043	0.297	
Stoohs	1.848	0.865	3.947	1.586	0.113	
Haraldsson	1.551	0.641	3.756	0.973	0.330	
Findley 1988	6.833	0.257	181.694	1.148	0.251	
	2.722	1.295	5.722	2.642	0.008	

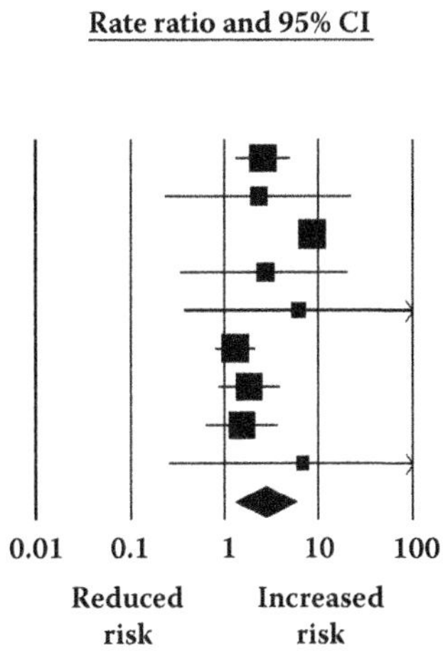

0.01 0.1 1 10 100

Reduced risk   Increased risk

FIGURE 46.8 Meta-analysis: Individuals with OSA are at increased risk for a MVC compared to drivers without OSA (Crash RR = 2.72, 95% CI: 1.30–5.72: *p* = 0.008). If driver crash risk is 0.08 crashes per person-year, OSA driver crash risk is ~0.21 (95% CI: 0.10 to 0.46) crashes per person-year. (From Tregear, S. J., Rizzo, M., Tiller, M., Schoelles, K., Hegmann, K. T., Greenberg, M. I., ... Anderson, G. (2007). *Diabetes and motor vehicle crashes: A systematic evidence-based review and meta-analysis.* In Proceedings of the Fourth International Driving Symposium on Human Factors in Driving Assessment, Training, and Vehicle Design, Stevenson, WA; Tregear, S. J., Reston, J., Schoeles, K., & Phillips, B. (2009). Obstructive sleep apnea and risk of motor vehicle crash: Systematic review and meta-analysis. *Journal of Clinical Sleep Medicine, 5,* 573–581.)

or haptic warnings (e.g., vibrating seats in a long haul truck) in attempts to reduce injuries caused by sleep- or drowsiness-related crashes. Positive airway pressure (PAP) therapy in drivers with OSA should lead to improvement in driving performance and awareness of impairment, and is another area for driving simulation research.

46.10 Drug Effects

Certain medications have been associated with greater relative risk of automobile crashes in the epidemiologic record (e.g., Ray, Thapa, & Shorr, 1993). Antidepressants, pain medications, antihistamines, anticonvulsants, antihypertensives, antilipemics, hypoglycemic agents, sedatives, and hypnotics have all been implicated (see Table 46.4). Specific mechanisms besides general drowsiness, whereby these medications impair driving performance, remain unclear. Drug use, including use of prescription medications, may cause as many fatal accidents as alcohol consumption (Centers for Disease Control and Prevention, 2006b). Alcohol and illicit drugs such as marijuana (e.g., Lamers & Ramaekers, 2001) and MDMA (e.g., Logan & Couper, 2001) also pose serious driving safety risks. One study examining the effects of alcohol alone, marijuana alone, and their combined effects reported significant impairment in driving ability following administration of alcohol or marijuana alone, whereas combining the two substances resulted in "dramatic" impairments in such driving-related phenomena as time driven out of lane and standard deviation of lateral position (Ramaekers, Robbe, & O'Hanlon, 2000). Driving performance is often impaired at legally defined cut-offs for sobriety (usually 0.8–1.0 mg/dl of ethanol in the U.S. and 0.5 mg/dl in Europe; see, this book, chap. 45 by Ramaekers et al., "Validity of Three Experimental Performance Tests for Predicting Risk of Cannabis-Induced Road Crashes").

The effects of pharmacologic agents on driving performance can be studied most safely in a driving simulator (Lamers, Bechara,

Rizzo, & Ramaekers, 2006). Deleterious effects of drugs on driving seem likely to depend in part on neurotransmitter systems involved in "executive functions" that are known to be critical for driving: decision-making and working memory. Elucidation of the chemical substrates (e.g., serotonin, dopamine, and acetylcholine) that modulate frontal lobe functions may help guide development of pharmacological interventions that improve driving task.

Regarding the scenarios and dependent variables that one should be using to study drug effects, these will vary greatly, and the selection can follow the general advice of core scenarios and special scenarios as dictated by experimental needs and hypotheses.

46.11 Linking Behavior and Physiology in Medical Simulation

Driving simulator set-ups provide a controlled environment for examining links between human physiology and behavior. Sensors can be fastened to a subject or positioned remotely to record physiological data from a driver to assess body temperature, heart rate, respiratory rate, galvanic skin response, EEG activity, electromyographic activity, eyelid closure (as an index of alertness/arousal), eye movements (as an index of cognitive processing), and even brain metabolic activity (e.g., using near-infrared spectroscopy [NIRS], or functional magnetic resonance imaging [fMRI]). These instruments differ on spatial and timing accuracy, ease of calibration, ability to accommodate paraphernalia (such as spectacles), susceptibility to vibration and lighting, measurement variability and reliability, ability to connect with auxiliary devices, synchronization with simulator performance measures, and ability to automate analysis of large data files and to visualize and analyze these data in software and statistical packages.

Physiological measurements are generally easier to make in a driving simulator than in a real car on a real road where movement, lighting, and other artifacts can be a nuisance and a source of confounding artifacts. Synchronous measures of operator physiology and performance in simulator scenarios can illuminate relationships between cognition and performance in drivers with a variety of medical disorders. Common technical difficulties facing researchers who record physiological measures include how specific measures are analyzed and reported (see, this book, chap. 17 by Brookhuis & de Waard, "Measuring Physiology in Simulators"). Common problems include the spatial and timing precision of the physiological recording hardware, post-processing of recorded measurements to find physiologically meaningful events (e.g., localization of fixation points from raw eye movement data or identification of microsleeps from EEG recordings), and synchronization of the stream of data on subject performance with the timing of simulation events and activities. Measurement devices, such as those that record galvanic skin response or eye movements, are temperamental and require frequent calibration and clearly defined dependent measures.

46.12 Simulation for Intervention in Medical Cohorts

One potential application of simulation is to retrain drivers who are recovering from an acute illness such as a stroke that has impaired their driving (see also, in this book, chap. 32 by Singh, Barbour, & Cox, "Driving Rehabilitation as Delivered by Driving Simulation"). A simulator can also be used to assess whether drivers are ready to be tested or trained on the road. Ongoing research in our laboratory aims to determine an optimal set of signals for alerting drivers to unsafe behavior and impending traffic conflicts using a driving simulator and then an instrumented vehicle, and to estimate the benefits of the proposed safety interventions across the U.S. in terms of crashes averted. A key aspect of this research is to develop collision warning algorithms and display parameters. Effective warning systems must promote a timely and appropriate driver response and minimize annoyance from nuisance warnings. System success depends on how well the algorithm and driver interface match driver capabilities and preferences (Brown, Lee, & McGehee, 2001; Lee, McGehee, Brown, & Reyes, 2002). The driver interface is also important because it influences how quickly the driver responds and whether the driver will accept the system. Another key interface characteristic is warning modality. Several studies have found that haptic displays (e.g., a vibrating seat, pedals, or steering wheel) improve driver reactions to collision situations (Raby, McGehee, Lee, & Norse, 2000). Lees et al., applied a cognitive science task (known as Posner's paradigm) in a driving simulator environment to determine whether visual cues, as opposed to haptic and auditory cues, provide effective alerts to drivers at increased risk for a vehicle crash due to UFOV impairments (Lees et al., submitted). Twenty-nine older drivers with UFOV impairments and 32 older drivers without impairments participated. Alerts were presented in a single modality or a combination of modalities (visual, auditory, haptic). Four cue types (valid, invalid, neutral, uncued) were employed. Following each cue, drivers discriminated the direction of a target (a Landolt

square with a gap facing up or down) in the visual panorama. Drivers with and without UFOV impairments showed comparable response times across the different cue types and alert modalities. Both groups benefited most from auditory and auditory/haptic alerts. Overall, drivers responded faster when cued. The results suggest that temporal aspects of cue presentation dominate spatial aspects of cues in their ability to speed responses to external targets. The combination of a cognitive science approach in a driving simulator environment provides a fruitful method for examining how different alerts influence speed of processing and attention in at-risk drivers in actual automobiles.

46.13 Practical Assessment

Demographic and health factors may impact driving ability and should be documented in simulation studies. Relevant factors are age, education, gender, general health, vision status, mobility, and driving frequency. Frequency of driving can be assessed using a Driving Habits Questionnaire (Ball, Beard, Roenker Miller & Griggs, 1988; Ball, Owley, Stalvey, Roenker, Sloane, & Graves, 1998). Health status information can be obtained using a checklist of medical conditions (e.g., heart disease, cancer) which includes the ability to enter information on when they occur. Certain medical factors (e.g., use of some medications, having a history of falls, back pain, kidney disease, heart disease, diabetes, stroke, bursitis, visual impairment, sleep apnea, etc.) are associated with increased risk of driving errors (see Hu, Trumble, Foley, Eberhard, & Wallace, 1998). Psychological state of the drivers can be collected using the General Health Questionnaire (GHQ) (Goldberg, 1972). Medication use can be assessed by asking drivers to bring all prescription and over-the-counter medication to the clinic. A driver's chronic sleep disturbance can be assessed from the driver's self-report on the Epworth Sleepiness Scale. A relevant visual assessment can include tests of letter acuity—for example, the ETDRS chart (Ferris, Kassoff, Bresnick, & Bailey, 1982), contrast sensitivity (Pelli, Robson, & Wilkins, 1988), and visual field sensitivity, which is often assessed using automated perimetry (Trick, 2003). UFOV reduction in patients who have normal visual fields can be demonstrated using visual tasks under differing attention loads (Ball et al., 1993). Overall visual health can be assessed with the National Eye Institute (NEI) Visual Functioning Questionnaire—25 items (Mangione et al., 2001). Several standardized tasks can be used to assess cognitive abilities that are essential to the driving task (see below). Impaired performance on some of these tasks (e.g. CFT, Trail Making Test Part B) may be especially predictive of driving safety risk (Dawson, Anderson, Uc, Dastrup, & Rizzo, 2009). No single test is sufficiently reliable to base judgments on fitness to drive, and a variety of sources and approaches are needed (Reger, et al., 2004) (see Figure 46.9; Web Figure 46.8).

46.14 Clinical Trials

Driving simulation provides a marvelous opportunity for conducting clinical trials of new drugs and devices. For example, do

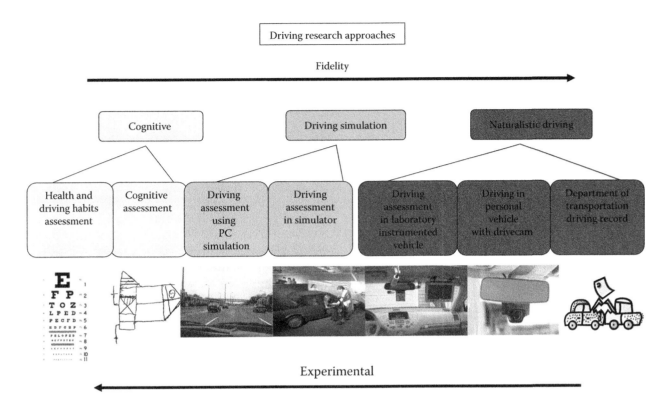

FIGURE 46.9 Sources of evidence on driving. Simulation is one of a range of tools that provides evidence on driver behavior and performance. These tools range (left to right) from simple tests of reaction time and sensory perception, to standardized neuropsychological tests which are capable of measuring levels of cognitive performance in health or disease, but which are not as good at predicting real-world driving performance, to computer-based tests that may use world images (e.g., change detection and hazard perception tasks), to various types of simulators (task focused, non-motion based, multicapable motion-based), to driving in the real world (on a test track, on a state road test, in an instrumented vehicle), and naturalistic driving (in the driver's own car, over extended time frames). Each tool has its advantages and drawbacks and the approach in medical disordered drivers depends on the question being asked.

intraocular lenses improve real-world driving performance? Do cognitive activators such as cholinesterase inhibitors improve driving performance in Alzheimer's disease or mild cognitive impairment (MCI)? Do medicines for attention deficit disorder (Cox, Merkel et al. 2004) or fatigue improve attention, alertness, and driving performance, or do they cause anxiety or psychosis that worsens driving behavior? Do in-vehicle devices which are intended to aid the hearing, vision, and attention impaired or paretic drivers help them or distract and hinder them? New treatments and devices must be proven to be safe and effective before they can be offered to patients. Clinical research trials in driving simulators can address such issues.

A clinical trial has "some formal structure of an experiment, particularly control over treatment assignment by the investigator" (Piantadosi, 1997, p. 581.). According to the National Institutes of Health, 2008), "A clinical trial . . . is a research study in human volunteers to answer specific health questions. [Clinical trials] are the fastest and safest way to find treatments that . . . improve health. Interventional trials determine whether experimental treatments . . . are safe and effective under controlled environments."

In this vein, driving simulator scenarios are becoming key elements in state and federal efforts to probe: (1) the efficacy of novice and older driver training programs; (2) fitness to drive in patients with performance declines due to aging and mild cognitive impairment, traumatic brain injury, neurodegenerative disorders such as Alzheimer's and Parkinson's diseases, and sleep disturbances; and (3) acute and chronic effects of many medications such as analgesics, antidepressants, psychostimulants, antidepressants, and cancer chemotherapy agents. To make meaningful comparisons between research studies conducted on differing platforms, it is essential that the scenarios used be specified in sufficient detail to implement them on simulators that depend on different hardware, visual database and scenario development tools, image rendering software, and data collection systems.

While simulators are an important means for the safe and objective assessment of performance capabilities in normal and medically-impaired automobile drivers, key issues limit clinical research using driving simulators. Key problems concern simulator graphics, audio, and movement (see also, in this book, chap. 8 by Andersen, "Sensory and Perceptual Factors in the Design of Driving Simulation Displays", and chap. 7 by Greenberg & Blommer, "Physical Fidelity of Driving Simulators"); scenario design and validation (see, this book, chap. 6 by Kearney & Grechkin, "Scenario Authoring"); subject adaptation and comfort (see, this book, chap. 9 by Ranney, "Psychological Fidelity: Perception of Risk"); selection of performance measures; and technical standards for reporting experimental set-up and results. For these reasons it is important to consider the benefits of standard scenarios in the assessments of medically-impaired drivers.

Standard scenarios enable comparisons across different simulator platforms, facilitate collection of large amounts of data at different institutions with greater power for addressing worldwide public health issues related to driving, and can help reveal missing descriptors and other weaknesses in mounting numbers of driving simulator research reports (Editors, this book, chap. 1, "*Handbook of Driving Simulation for Engineering, Medicine, and Psychology*: An Overview"). Consequently, the Simulator Users Group developed a driving wiki (Seppelt, Lees, Lee, & Rizzo, 2007) and proposed standard scenarios (http://www.simusers.com/tiki-index.php?page=Standard%20Scenarios, accessed 11/28/08).

Standard scenarios facilitate research across different populations in simulators of different levels of fidelity and may allow researchers to collectively and effectively appraise simulator and scenario designs. A scenario (a driving task or condition in a simulator drive) may comprise one or more events or stimuli intended to probe a driver response. Issues in the development of standard scenarios include: How well can key simulator scenarios be described within a common framework? Should researchers be expected to implement a key scenario just one way? Can levels of fidelity be adequately specified for cross-platform comparisons? Details to consider include selection of scenarios and dependent measures, standards of reporting in journals, subject selection criteria, reporting of independent measures, standards for physiologic recording, properties of visuals, audio and haptic cues, training criteria (for using the simulator), simulator adaptation, abstractness and contextual cues, and criteria for validation (i.e., cross-platform comparisons between the simulator and other tools and data).

Another key issue in clinical trials concerns the rationale for selecting standard scenarios, which include the ability to assess specific mechanisms of impairment as well as the importance of the simulated scenario in the epidemiologic record. These are not competing, but rather complementary and converging, approaches that address the problem of driver safety from both the bottom-up (neural or cognitive mechanisms) and top-down (epidemiology or systems) levels. A defensible and pragmatic strategy is to focus on key situations conveying potential high crash risk, for example: run off the road on curves, car following and rear-end collisions, intersection incursion avoidance, interaction with emergency vehicles/pedestrians, and merging with the potential for side impact collisions. Scenarios can also address behavioral effects of using in-vehicle devices, such as cell phones or information displays.

Clinical trials of driving simulation may also consider physiologic measures. The HASTE (Human Machine Interface and the Safety of Traffic in Europe) group proposed several physiological indices to assess user workload in VR (virtual reality) environments for driving. Mandatory workload measures (collected at all sites) included a rating scale of mental effort, eye glance measures (glance frequency, glance duration), measures of vehicle control while using vehicle information system performance, and a measure of situation awareness. Optional workload measures (which could be collected at sites with appropriate facilities) included heart rate and heart rate variability, respiration, and skin conductance.

46.15 Additional Challenges to Medical Applications

Another issue in simulator-based research is the need to test the validity of the simulation. This may involve detailed comparisons of driver performance in a simulator with performance in an instrumented vehicle and with state records of crashes and moving violations in each population of drivers being studied. The apparent face validity of the simulation—that is, that the driver appears to be driving a car and is immersed in the task—does not guarantee a life-like performance (see, this book, chap. 9 by Ranney, "Psychological Fidelity: Perception of Risk"). Drivers may behave differently in a simulator where no injury can occur, compared to real-life driving situations where it can. They may even perform differently when the same scenario is implemented on different simulator platforms, motivating calls for development of standard scenarios outlined above.

A relative drawback to simulation research is simulator adaptation syndrome (SAS), characterized by autonomic symptoms including nausea and sweating (Stanney, 2002). The discomfort may be due to a mismatch between visual cues of movement, which are plentiful, and inertial cues, which are lacking or imperfect, even in simulators with a motion base (Rizzo, Sheffield, & Stierman, 2003). In our experience, SAS is more likely with crowded displays (as in simulated urban traffic), advanced age, female gender, and history of migraine or motion sickness. SAS may also result from conflicts between visual cues that are represented with differing success in modern computer displays (see, this book, chap. 8 by Andersen, "Sensory and Perceptual Factors in the Design of Driving Simulation Displays", and chap. 14 by Stoner, Fisher, & Mollenhauer, "Simulator and Scenario Factors Influencing Simulator Sickness"). Choice of equipment and scenario design (e.g., avoiding sharp curves, left turns, frequent stops, and crowded scenes) can minimize SAS. SAS might conceivably pose a health threat to patients with cardiovascular disorders, although to our knowledge, no such complication has been reported.

As yet, the federal government and insurers have no clear policy on the reimbursement of clinical procedures that use driving simulation for patient assessment or interventions, and the default approach is often denial. The burden of proof remains on the researchers and clinicians to provide convincing evidence on the validity and utility of emerging simulator tools.

Many individuals now perform driving simulation work in academic settings and conduct behavioral research using techniques from experimental psychology, psychophysics and psychophysiology, motor control, neuroimaging, pharmacology, and neuroepidemiology (Batchelor & Cudkowicz, 2001). They would benefit from understanding differences between main elements of clinical design including Phase I, Phase II, Phase III, and Phase IV trials, elements of experimental design including sample size, clinico-metric methods including principles and development of cognitive instruments. It is important for behavioral clinicians to understand the fundamental issues that underlie accuracy of behavioral assessment and diagnosis. These include:

1. Reliability—the extent to which a test instrument yields the same result on repeated trials under identical circumstances, with different examiners, and with different forms;
2. Internal consistency—the correlation of items within a test;
3. Sensitivity of tests to change; and
4. Validity—the extent to which a test measures what it is theoretically supposed to measure (construct validity); correlates with other validated measures (concurrent validity); provides additional information beyond available measures (incremental validity); measures something that has functional implications for the person's life (ecological validity).

In summary, results using cognitive tests, instrumented vehicles, and driving simulators can help specify linkages between decline in key cognitive domains and crash risk. Results can help standardize assessment of fitness-to-drive in persons with a variety of medical impairments. By understanding patterns of driver safety errors that cause crashes, it may be possible to design interventions (e.g., driver performance monitoring devices and collision warning systems) that will reduce these errors.

Key Points

- Results using cognitive tests, instrumented vehicles, and driving simulators can help specify linkages between decline in key cognitive domains and crash risk.
- By understanding patterns of driver safety errors that cause crashes, it may be possible to design interventions (e.g., driver performance monitoring devices and collision warning systems) that will reduce these errors.
- Simulators can be deployed in clinical trials to evaluate in-vehicle devices before they are deployed on the road and to train drivers with medical disorders in the aftermath of an illness in order to prepare them for potential return to driving.
- Results of simulator studies can help standardize assessment of fitness-to-drive in persons with medical impairments.
- Predictions of driver safety and crash risk from driving simulator outcome measures depend on understanding the intervening patterns of real-world exposure, restriction, and other complex factors.

Keywords: Attention, Clinical Trials, Dementia, Emotion, Encephalopathy, Executive Functions, Memory, Mobility, Sleep Apnea, Systemic Disorders

Web Resources

The *Handbook* Web site contains supplemental materials for this chapter including color versions of a number of the printed figures, and a Web-only figure (Web Figure 46.7) showing fMRI activity during driving tasks.

Web Figure 46.1: Car following (Color version of Figure 46.1).

Web Figure 46.2: Driver with UFOV impairment (Color version of Figure 46.2).

Web Figure 46.3: Vigilance Task (Color version of Figure 46.3).

Web Figure 46.4: Collision Detection Task (Color version of Figure 46.4).

Web Figure 46.5: A non-conventional display design approach aimed at testing go/no-go decision-making by drivers with cognitive impairments (Color version of Figure 46.6).

Web Figure 46.6: Change detection task. (a) Original image; (b) modified image (Color version of Figure 46.7).

Web Figure 46.7: Functional magnetic resonance imaging (fMRI). Comparing brain activity during fixation, passive viewing (watching) and active participation in a driving-like task presented in the confines of an fMRI scanner.

Web Figure 46.8: Sources of Evidence on Driving (Color version of Figure 46.9).

Key Readings

Bacon, D., Fisher, R. S., Morris, J. D., Rizzo, M., & Spanaki, M. V. (2007). Position statement on physician reporting of medical conditions that may affect driving competence. *Neurology, 68,* 174–177. Accessible at: http://www.aan.com/advocacy/federal/pdfs/2006-59_PhysicianReporting_PositionStatement.pdf

Barrash, J., Stillman, A., Anderson, S. W., Uc, E. Y., Dawson, J. D., & Rizzo, M. (accepted). Prediction of driving ability with neuropsychological tests: Demographic adjustments diminish accuracy. *Journal of the International Neuropsychology Society.*

Boyle, L., Tippin, J., Paul, A., & Rizzo, M. (2008). Driver performance in the moments surrounding a microsleep. *Transportation Research Part F: Traffic Psychology and Behaviour, 11*(2), 126–136.

Dastrup, E., Lees, M. N., Bechara, A., Dawson, J. D., & Rizzo, M. (2010). Risky car following in abstinent users of MDMA. *Accident Analysis & Prevention* (in press).

Dawson, J. D., Anderson, S. W., Uc, E. Y., Dastrup, E., & Rizzo, M. (2009). Predictors of driving safety in early Alzheimer's disease. *Neurology, 72,* 521–527.

Iverson, D. J., Gronseth, G., Reger, R., Classen, S., Dubinsky, R. M., & Rizzo, M. (2010). Practice parameter update: Evaluation and management of driving risk in dementia (an evidence-based review). *Neurology, 74*(15), first published online on April 12, 2010.

Rizzo, M., Shi, Q., Dawson, J., Anderson, S. W., Kellison, I., & Pietras, T. A. (2005). Stops for cops: Impaired response implementation in older drivers with cognitive decline. *Transportation Research Record, 1922,* 1–8.

Rizzo, M., A 71-year-old man trying to decide if he should continue driving. (2010). *Journal of the American Medical Association.* Clinical Crossroads (accepted).

Rizzo, M., McGehee, D., Dawson, J., & Anderson, S. W. (2001). Simulated car crashes at intersections in drivers with Alzheimer disease. *Alzheimer Disease and Associated Disorders, 15,* 10–20.

Tippin, J., Sparks, J. D., & Rizzo, M. (2009). Visual vigilance in drivers with obstructive sleep apnea. *Journal of Psychosomatic Research, 67*(2), 143–151. doi:10.1016/j.jpsychores.2009.03.015

Tregear, S. J., Rizzo, M., Tiller, M., Schoelles, K., Hegmann, K., Greenberg, M., . . . Anderson, G. (2007). Diabetes and motor vehicle crashes: A systematic evidence-based review and meta-analysis. *The fourth international driving symposium on human factors in driving assessment, training and vehicle design* (pp. 343–350). Iowa City, IA: University of Iowa.

Turner, B. M., Rizzo, M., Block, R., Pearlson, G., & O'Leary, D. S. (2010). Sex differences in the effects of marijuana on simulated driving performance. *Journal of Psychoactive Drugs.* (in press).

References

American Academy of Ophthalmology. (2006). *Policy statement: Vision requirements for driving.* Retrieved November 28, 2008, from http://www.aao.org/aao/member/policy/driving.cfm

Andersen, G. J., Saidpour, A., & Enriquez, A. (2001). *Detection of collision events by older and younger drivers.* Paper presented at the First International Driving Symposium on Human Factors in Driver Assessment, Training and Vehicle Design, Aspen, CO.

Ball, K. K., Beard, B. L., Roenker, D. L., Miller, R. L., & Griggs, D. S. (1988). Age and visual search: Expanding the useful field of view. *Journal of the Optical Society of America A, 5,* 2210–2219.

Ball, K. K., Clay, O. J., Wadley, V. G., Roth, D. L., Edwards, J. D., & Roenker, D. L. (2005). Predicting driving performance in older adults with the useful field of view test: A meta-analysis. *Proceedings of the third international driving symposium on human factors in driver assessment, training and vehicle design* (pp. 51–57). Rockport, ME.

Ball, K., Owsley, C., Sloane, M. E., Roenker, D. L., & Bruni, J. R. (1993). Visual attention problems as a predictor of vehicle crashes in older drivers. *Investigative Ophthalmology & Visual Science, 34,* 3110–3123.

Ball, K., Owsley, C., Stalvey, B., Roenker, D. L., Sloane, M. E., & Graves, M. (1998). Driving avoidance and functional impairment in older drivers. *Accident Analysis & Prevention, 30,* 313–322.

Barger, L. K., Cade, B. E., Ayas, N. T., Cronin, J. W., Rosner, B., Speizer, F. E., et al., and Harvard Work Hours, Health, and Safety Group (2005). Extended work shifts and the risk of motor vehicle crashes among interns. *The New England Journal of Medicine, 352,* 125–134.

Batchelder, S., Rizzo, M., Vanderleest, R., & Vecera, S. P. (2003). Traffic scene related change blindness in older drivers. In M. Rizzo, J. D. Lee, & D. McGehee (Eds.), *Proceedings of the second international driving symposium on human factors in driver assessment, training and vehicle design,* Park City, UT (pp. 177–181). Iowa City, IA: The University of Iowa.

Batchelor, T., & Cudkowicz, M. (Eds.). (2001). *Principles of neuroepidemiology.* Boston, MA: Butterworth Heinemann.

Bechara, A., Damasio, H., Tranel, D., & Damasio, A. R. (1997). Deciding advantageously before knowing the advantageous strategy. *Science, 275*(5304), 1293–1295.

Brewer, A. M. (2000). Road rage: What, who, when, where and how. *Transport Reviews, 20,* 49–64.

Broadbent, D. (1958). *Perception and communication.* London: Pergamon Press.

Brown, T. L., Lee, J. D., & McGehee, D. V. (2001). Human performance models and rear-end collision avoidance algorithms. *Human Factors, 43*(3), 462–482.

Cabeza, R., Dolcos, F., Prince, S. E., Rice, H. J., Weissman, D. H., & Nyberg, L. (2003). Attention-related activity during episodic memory retrieval: A cross-function fMRI study. *Neuropsychologia, 41,* 390–399.

Caird, J. K., Edwards, C. J., Creaser, J. I., & Horrey, W. J. (2005). Older driver failures of attention at intersections: Using change blindness methods to assess turn decision accuracy. *Human Factors, 47*(2), 235–249.

Calhoun, V. D., Pekar, J. J., & Pearlson, G. D. (2004). Alcohol intoxication effects on simulated driving: Exploring alcohol-dose effects on brain activation using functional MRI. *Neuropsychopharmacology, 29*(11), 2097–2117.

Calhoun, V. D., Pekar, J. J., McGinty, V. B., Adali, T., Watson, T. D., & Pearlson, G. D. (2002). Different activation dynamics in multiple neural systems during simulated driving. *Human Brain Mapping, 16*(3), 158–167.

Centers for Disease Control and Prevention (CDC). (2006a). *Morbidity and Mortality Weekly Report, 55*(48), 1293–1296.

Centers for Disease Control and Prevention (CDC). (2006b). *Web-based injury statistics query and reporting system (WISQARS)* [Database]. Available from http://www.cdc.gov/injury/wisqars/index.html

Cox, D. J., Kovatchev, B., Koev, D., Koeva, L., Dachev, S., Tcharaktchiev, D., . . . Clarke, W. (2004). Hypoglycemia anticipation, awareness and treatment training (HAATT) reduces occurrence of severe hypoglycemia among adults with type 1 diabetes mellitus. *International Journal of Behavioral Medicine, 11*(4), 212–8.

Cox, D. J., Merkel, R. L., Penberthy, J. K., Kovatchev, B., & Hankins, C. (2004). Impact of methylphenidate delivery profiles on driving performance of adolescents with attention-deficit/hyperactivity disorder (ADHD). *Academy of the American Association of Child and Adolescent Psychiatry, 43*(3), 269–275.

Dahlen, E. R., Martin, R. C., Ragan, K., & Kuhlman, M. M. (2005). Driving anger, sensation seeking, impulsiveness, and boredom proneness in the prediction of unsafe driving. *Accident Analysis & Prevention, 37*(2), 341–348.

Dastrup, E., Lees, M., Bechara, A., Dawson, J. D., & Rizzo, M. (2010). Risky car following in abstinent users of MDMA. *Accident Analysis & Prevention* (in press).

Dawson, J. D., Anderson, S. W., Uc, E. Y., Dastrup, E., & Rizzo, M. (2009). Predictors of driving safety in early Alzheimer's disease. *Neurology, 72*, 521–527.

DeCarlo, D. K., Scilley, K., Wells, J., & Owsley, C. (2003). Driving habits and health-related quality of life in patients with age-related maculopathy. *Optometry and Vision Science, 80*(3), 207.

Dinges, D. (2000). Accidents and fatigue. *Proceedings of the international conference: The sleepy driver and pilot.* Stockholm, Sweden: National Institute for Psychosocial Factors and Health.

Dingus, T. A., Hardee, H. L., & Wierwille, W. W. (1987). Development of models for on-board detection of driver impairment. *Accident Analysis & Prevention, 19*(4), 271–283.

Fan, J., McCandliss, B. D., Sommer, T., Raz, M., & Posner, M. I. (2002). Testing the efficiency and independence of attentional networks. *Journal of Cognitive Neuroscience, 3*(14) 340–347.

Fellows, L. K., Stark, M., Berg, A., & Chatterjee, A. (2008). Patient registries in cognitive neuroscience research: Advantages, challenges, and practical advice. *Journal of Cognitive Neuroscience, 20*(6), 1107–1113.

Ferris, F. L., III, Kassoff, A., Bresnick, G. H., & Bailey, I. (1982). New visual acuity charts for clinical research. *American Journal of Ophthalmology, 94*(1), 91–96.

Fonda, S. J., Wallace, R. B., & Herzog, A. R. (2001). Changes in driving patterns and worsening depressive symptoms among older adults. *The Journals of Gerontology Series B: Psychological Sciences and Social Sciences, 56*, S343.

Gibson, J. J. (1979). *The ecological approach to visual perception.* Boston, MA: Houghton Mifflin.

Goldberg, D. (1972). *GHQ: The selection of psychiatric illness by questionnaire.* London: Oxford University Press.

Harrison, Y., & Horne, J. A. (1996). Occurrence of "microsleeps" during daytime sleep onset in normal subjects. *Electroencephalography and Clinical Neurophysiology, 98*(5), 411–416.

Horne, J. A., & Reyner, L. A. (1996). Counteracting driver sleepiness: Effects of napping, caffeine, and placebo. *Psychophysiology, 33*(3), 306–309.

Hu, P. S., Trumble, D. A., Foley, D. J., Eberhard, J. W., & Wallace, R. B. (1998). Crash risks of older drivers: A panel data analysis. *Accident Analysis & Prevention, 30*(5), 569–581.

Insurance Institute for Highway Safety. (2006). *Fatality facts, older people.* Available from http://www.iihs.org/research/fatality_facts_2006/olderpeople.html

Jette, A. M., & Branch, L. G. A ten-year follow-up of driving patterns among community-dwelling elderly. *Human Factors, 34*(1), 25–31.

Just, M. A., Keller, T. A., & Cynkar, J. (2008). A decrease in brain activation associated with driving when listening to someone speak. *Brain Research, 1205*, 70–80.

Kantowitz, B. H. (2001). Using microworlds to design intelligent interfaces that minimize driver distraction. *Proceedings of the first international driving symposium on human factors in driver assessment, training and vehicle design.* Aspen, CO (pp. 42–57). Iowa City, IA: University of Iowa.

Knudsen, E. I. (2007). Fundamental component of attention. *Annual Review of Neuroscience, 30*, 57–78. doi:10.1146/annurev.neuro.30.051606.094256

Lamers, C. T., & Ramaekers, J. G. (2001). Visual search and urban driving under the influence of marijuana and alcohol. *Human Psychopharmacology, 16*(5), 393–401.

Lamers, C. T., Bechara, A., Rizzo, M., & Ramaekers, J. G. (2006). Cognitive function and mood in MDMA/THC users, THC users and non-drug using controls. *Journal of Psychopharmacology, 20*(2), 302–311.

Land, M. F., & Lee, D. N. (1994). Where we look when we steer. *Nature, 369*(6483), 742–744.

Lee, J. D., McGehee, D. V., Brown, T. L., & Reyes, M. L. (2002). Collision warning timing, driver distraction, and driver response to imminent rear-end collisions in a high-fidelity driving simulator. *Human Factors, 44*(2), 314–334.

Lees, M. N., Cosman, J., Lee, J. D., Vecera, S., Jang, M., Dawson, J. D., & Rizzo, M. (submitted). Cross-modal alerts for orienting of attention in attention impaired.

Logan, B. K., & Couper, F. J. (2001). 3, 4-Methylenedioxymethamphetamine (MDMA, ecstasy) and driving impairment. *Journal of Forensic Sciences, 46*(6), 1426–1433.

Mallis, M., Banks, S., & Dinges, D. (2007). Sleep and circadian control of neurobehavioral function. In R. Parasuraman & M. Rizzo (Eds.), *Neuroergonomics: The brain at work* (pp. 207–220). New York: Oxford University Press.

Mangione, C. M., Lee, P. P., Gutierrez, P. R., Spritzer, K., Berry, S., & Hays, R. D. (2001). Development of the 25-item National Eye Institute visual function questionnaire. *Archives of Ophthalmology, 119*(7), 1050–1058.

Marottoli, R. A., Allore, H., Araujo, K. L. B., Iannone, L. P., Acampora, D., Gottschalk, M., . . . Peduzzi, P. (2007). A randomized trial of a physical conditioning program to enhance the driving performance of older persons. *Journal of Geriatric Internal Medicine, 22*, 590–597.

National Highway Traffic Safety Administration. (2006). *Traffic safety facts (2004 data): Older drivers* (DOT HS 810 622). Washington, DC: National Highway Traffic Safety Administration, National Center for Statistics and Analysis.

National Highway Traffic Safety Administration. (2008). *2007 traffic safety annual assessment—Highlights* (DOT HS 811 017). Washington, DC: National Highway Traffic Safety Administration, National Center for Statistics and Analysis.

National Institutes of Health. (2008, March). *Glossary of clinical trials terms.* Retrieved November 25, 2008 from http://clinicaltrials.gov/ct2/info/glossary

Nawrot, M. (2001). *Depth perception in driving: Alcohol intoxication, eye movement changes, and the disruption of motion parallax.* Paper presented at the First International Driving Symposium on Human Factors in Driver Assessment, Training and Vehicle Design, Aspen, CO.

Noyes, R., Jr. (1985). Motor vehicle accidents related to psychiatric impairment. *Psychosomatics, 26*(7), 569–580.

Nyberg, L., Marklund, P., Persson, J., Cabeza, R., Forkstam, C., Petersson, K. M., et al. (2003). Common prefrontal activations during working memory, episodic memory, and semantic memory. *Neuropsychologia, 41*, 317–377.

O'Regan, J. K., Rensink, R. A., & Clark, J. J. (1999). Change-Blindness as a result of "mudsplashes". *Nature, 398*, 34.

Ogilvie, R. D., Wilkinson, R. T., & Allison, S. (1989). The detection of sleep onset: Behavioral, physiological, and subjective convergence. *Sleep, 12*(5), 458–474.

Olanow, C. W., Schapira, A. H., & Roth, T. (2000). Waking up to sleep episodes in Parkinson's disease. *Movement Disorders, 15*(2), 212–215.

Owsley, C., Ball, K., Sloane, M. E., Roenker, D. L., & Bruni, J. R. (1991). Visual/cognitive correlates of vehicle accidents in older drivers. *Psychology and Aging, 6*, 403–415.

Owsley, C., Stalvey, B. T., Wells, J., Sloane, M. E., & McGwin, G., Jr. (2001). Visual risk factors for crash involvement in older drivers with cataract. *Archives of Ophthalmology, 119*, 881–887.

Palca J. (1990). Insights from broken brains [news]. *Science, 248*, 812–814.

Palmer, S. E. (1999). *Vision science: Photons to phenomenology.* Cambridge, MA: MIT Press.

Parasuraman, R., & Davies, D. R. (1984). *Varieties of attention.* Orlando: Academic Press.

Paul, A., Boyle, L., Rizzo, M., & Tippin, J. (2005). *Variability of driving performance during microsleeps.* Paper presented at the Third International Driving Symposium on Human Factors in Driving Assessment, Training, and Vehicle Design, Rockport, ME.

Pelli, D. G., Robson, J. G., & Wilkins, A. J. (1988). The design of a new letter chart for measuring contrast sensitivity. *Clinical Vision Sciences, 2*, 187–199.

Peres, M., van de Moortele, P. F., Pierard, C., Lehericy, S., LeBihan, D., & Guezennez, C. Y. (2000). fMRI of mental strategy in a simulated aviation performance task. *Aviation, Space, and Environmental Medicine, 71*, 1218–1231.

Piantadosi, S. (1997). *Clinical trials: A methodological perspective.* New York: Wiley.

Posner M. I., & Cohen, Y. (1984). Components of visual orienting. In H. Bouma & D. Bouwhuis (Eds.), *Attention and performance X* (pp. 531–556). Hillsdale, NJ: Lawrence Erlbaum Associates.

Posner, M. I., Snyder, C. R. R., & Davidson, B. J. (1980). Attention and the detection of signals. *Journal of Experimental Psychology: General 109*, 160–174.

Pradhan, A. K., Hammel, K. R., DeRamus, R., Pollatsek, A., Noyce, D. A., & Fisher, D. L (2005). The use of eye movements to

evaluate the effects of driver age on risk perception in an advanced driving simulator. *Human Factors, 47*, 840–852

Raby, M., McGehee, D. V., Lee, J. D., & Norse, G. E. (2000). Defining the interface for a snowplow lane tracking device using a systems-based approach. *Proceedings of the Human Factors and Ergonomics Society, 3*, 369–372.

Ramaekers, J. G., Robbe, H. W. J., & O'Hanlon, J. F. (2000). Marijuana, alcohol and actual driving performance. *Human Psychopharmacology: Clinical and Experimental, 15*(7), 551–558.

Ray, W. A., Thapa, P. B., & Shorr, R. I. (1993). Medications and the older driver. *Clinics in Geriatric Medicine, 9*(2), 413–438.

Reger, M. A., Welsh, R. K., Watson, G. S., Cholerton, B., Baker, L. D., & Craft, S. (2004). The relationship between neuropsychological functioning and driving ability in dementia: A meta-analysis. *Neuropsychology, 18*(1), 85–93.

Reyner, L. A., & Horne, J. A. (1998). Falling asleep whilst driving: Are drivers aware of prior sleepiness? *International Journal of Legal Medicine, 111*(3), 120–123.

Risser, M. R., Ware, J. C., & Freeman, F. G. (2000). Driving simulation with EEG monitoring in normal and obstructive sleep apnea patients. *Sleep, 23*(3), 393–398.

Rizzo, M. (2004). Safe and unsafe driving. In M. Rizzo & P. J. Eslinger (Eds.), *Principles and practice of behavioral neurology and neuropsychology* (pp. 197–222). Philadelphia, PA: WB Saunders.

Rizzo, M., & Barton, J. J. S. (2005). Central disorders of visual function. In N. R. Miller & N. J. Newman (Eds.), *Walsh and Hoyt's neuro-ophthalmology* (6th ed.). Baltimore, MD: Williams & Wilkins.

Rizzo, M., & Eslinger, P. J. (Eds.). (2004). *Principles and practice of behavioral neurology and neuropsychology.* Philadelphia, PA: WB Saunders.

Rizzo, M., & Kellison, I. L. (2004). Eyes, brains, and autos. *Archives of Ophthalmology, 122*(4), 641–647.

Rizzo, M., & Kellison, I. (2009). The brain on the road. In T. Marcotte & I. Grant (Eds.), *Everyday functioning: Translating laboratory performance to the real world.* New York: Guilford Press.

Rizzo, M., Akutsu, H., & Dawson, J. (2001). Increased attentional blink after focal cerebral lesions. *Neurology, 57*(5), 795–800.

Rizzo, M., Lamers, C. T., Sauer, C. G., Ramaekers, J. G., Bechara, A., & Andersen, G. J. (2005). Impaired perception of self-motion (heading) in abstinent ecstasy and marijuana users. *Psychopharmacology, 179*(3), 559–566.

Rizzo, M., Reinach, S., McGehee, D., & Dawson, J. (1997). Simulated car crashes and crash predictors in drivers with Alzheimer's disease. *Archives of Neurology, 54*, 545–553.

Rizzo, M., Severson, J., Cremer, J., & Price, K. (2003). An abstract virtual environment tool to assess decision-making in impaired drivers. *Proceedings of the second international driving symposium on human factors in driver assessment, training, and vehicle design* (pp. 40–47), Park City, UT.

Rizzo, M., Sheffield, R., & Stierman, L. (2003). Demographic and driving performance factors in simulator adaptation syndrome. In M. Rizzo, J. D. Lee., & D. McGehee (Eds.), *Proceedings of the second international driving symposium on human factors in driver assessment, training and vehicle design* (pp. 201–208). Iowa City, IA: University of Iowa.

Rizzo, M., Sparks, J. D., McEvoy, S., Viamonte, S., Kellison, I., & Vecera, S. (2009). Change blindness, aging and cognition. *Journal of Clinical and Experimental Neuropsychology, 31*, 245–256.

Rorden, C., & Karnath, H.-O. (2004). Using human brain lesions to infer function—A relic from a past era in the fMRI age? *Nature Reviews Neuroscience, 5*, 813–819.

Schwebel, D. C., Severson, J., Ball, K. K., & Rizzo, M. (2006). Individual difference factors in risky driving: The roles of anger/hostility, conscientiousness, and sensation-seeking. *Accident Analysis & Prevention, 38*(4), 801–810.

Sekuler, A. B., Bennett, P. J., & Mamelak, M. (2000). Effects of aging on the useful field of view. *Experimental Aging Research, 26*(2), 103–120.

Seppelt, B., Lees, M., Lee, J., & Rizzo, M. (2007). *Promoting repeatable research and collaboration—The benefits of a driving simulation wiki.* Driving Simulation North America Conference 2007, University of Iowa, Iowa City, IA.

Simons, D. L., & Levin, D. T. (1997). Change blindness. *Trends in Cognitive Science, 1*, 261–267.

Simons, D. J., & Rensink, R. A. (2005). Change blindness: Past, present, and future. *Trends in Cognitive Sciences, 9*(1), 16–20.

Skurtveit, S., Strøm, H., Skrivarhaug, T., Mørland, J., Bramness, J., & Engeland, A. (2009). Original Article: Complications Road traffic accident risk in patients with diabetes mellitus receiving blood glucose-lowering drugs. Prospective follow-up study. *Diabetic Medicine, 26*, 404–408.

Spreen, O., & Strauss, E. (1998). *A compendium of neuropsychological tests: Administration, norms, and commentary* (2nd ed.). New York: Oxford University Press, 1998.

Stalvey, B., Owsley, C., Sloan, M. E., & Ball, K. (1999). The Life Space Questionnaire: A measure of the extent of mobility of older adults. *Journal of Applied Gerontology, 18*, 479–498.

Stanney, K. M. (2002). *Handbook of virtual environments: Design, implementation, and applications.* Mahwah, NJ: Lawrence Erlbaum Associates.

Tranel, D. (1996). The Iowa-Benton School of neuropsychological assessment. In I. Grant & K. M. Adams (Eds.), *Neuropsychological assessment of neuropsychiatric disorders* (2nd ed., pp. 81–101). New York: Oxford University Press.

Tregear, S. J., Reston, J., Schoeles, K., & Phillips, B. (2009). Obstructive sleep apnea and risk of motor vehicle crash: Systematic review and meta-analysis. *Journal of Clinical Sleep Medicine, 5*, 573–581.

Tregear, S. J., Rizzo, M., Tiller, M., Schoelles, K., Hegmann, K. T., Greenberg, M. I., … Anderson, G. (2007). *Diabetes and motor vehicle crashes: A systematic evidence-based review and meta-analysis.* In Proceedings of the Fourth International Driving Symposium on Human Factors in Driving Assessment, Training, and Vehicle Design, Stevenson, WA.

Trick, G. (2003). Beyond visual acuity: New and complementary tests of visual function. In J. J. S. Barton & M. Rizzo (Eds.), *Neuro-ophthalmology: Vision and brain.* Neurologic Clinics of North American Series. Philadelphia: WB Saunders.

Tsuang, M. T., Boor, M., & Fleming, J. A. (1985). Psychiatric aspects of traffic accidents. *American Journal of Psychiatry, 142*(5), 538–546.

Uttal, W. R. (2002). Précis of the new phrenology: The limits of localizing cognitive processes in the brain. *Brain and Mind, 3*(2), 221–228.

Vaux, L. M., Ni, R., Rizzo, M., Uc, E. Y., & Anderson, G. J. (2010). Detection of imminent collisions by drivers with Alzheimer's disease, and Parkinson's disease: A preliminary study. *Accident Analysis and Prevention, 42*, 852–858.

Vecera, S. P., & Luck, S. J. (2003). Attention. In V. S. Ramachandran (Ed.), *Encyclopedia of the human brain* (Vol. 1, pp. 269–284). San Diego, CA: Academic Press.

Vecera, S. P., & Rizzo, M. (2004). What are you looking at? Impaired "social attention" following frontal-lobe damage. *Neuropsychologia, 42*(12), 1657–1665.

Vul, E., Harris, C., Winkielman, P., & Pashler, H. (2009). Puzzlingly high correlations in fMRI studies of emotion, personality, and social cognition. *Perspectives on Psychological Science, 4*(3), 274–290.

Warren, W. H., & Hannon, D. J. (1988). Direction of self-motion is perceived from optical flow. *Nature, 336*, 162–163.

Warren, W. H. J., Mestre, D. R., & Morris, M. W. (1991). Perception of circular heading from optical flow. *Journal of Experimental Psychology: Human Perception & Performance, 17*, 28–43.

Wells, G. S., O'Connell, D., Peterson, J., Welch, V., Losos, M., & Tugwell, P. (2006). *The Newcastle-Ottawa Scale (NOS) for assessing the quality of non-randomized studies in meta-analysis.* Retrieved January 5, 2010, from http://www.ohri.ca/programs/clinical_epidemiology/oxford.htm

Wierwille, W. W., Hulse, M. C., Fischer, T. J., & Dingus, T. A. (1988). *Effects of variation in driving task attentional demand on in-car navigation system usage.* Virginia Polytechnic Institute and State University. Contract Report. Warren, MI: General Motors Research Laboratories.

47

Psychiatric Disorders and Driving Performance

Henry Moller
University of Toronto

Abstract

The Problem. This chapter discusses the continuum of driving performance in healthy individuals under adverse conditions and specific psychopathologies ranging from anxiety disorders to mood disorders, psychotic disorders, attention deficit hyperactivity disorder (ADHD), and personality disorders. *Role of Driving Simulators.* Driving simulators are being used in a number of domains in psychiatric research and practice: (1) design of experimental protocols addressing epidemiological findings related to driving safety for specific conditions in a controlled setting; (2) diagnostic use of simulators as baseline performance-based "probes" of neurocognitive function in untreated patients; (3) related to this, tracking of performance over a treatment period, i.e., in a clinical trial context; (4) assessment of safety and therapeutic effect of medications used to treat psychiatric illness with respect to psychomotor function; (5) medicolegal assessments of fitness-to-drive; and (6) skills training and rehabilitation. *Key Results of Driving Simulator Studies.* Skills training and phobia desensitization models have been shown to generalize to improved clinical outcome and on-road performance in anxiety disorders. Targeted training of strategy application and spatial navigation may improve frontal/executive impairments found in ADHD, and mood and psychotic disorders. Pharmacologic studies have been useful in the study of acute psychomotor and next-day sedative effects of psychotropics; generally, studies have demonstrated superiority of novel antidepressant and antipsychotic medication with respect to psychomotor function. Studies on use of stimulants in ADHD suggest improved short-term performance, however data on long-term effects are needed and trends related to multitasking while driving need to be taken into account. *Scenarios,*

Dependent Variables, and Limitations. This chapter highlights a continuum-based model of "Type 1/Type 2" crash risk scenarios, related to underarousal versus overarousal of the human perceptual apparatus and low versus high presence states. In this model, "Type 1" crashes are provoked by excessively monotonous driving conditions resulting in perceptual disengagement and "Type 2" crashes occur under excessively complex driving conditions; the more variable, challenging, and stimulus-rich a scenario is, the greater the "Type 2" crash risk. Relevant variables include reaction time, braking accuracy, road position, which can be continuously monitored, and specific driving errors such as lane incursions and collisions. While scenario design may be modeled based on presumed functional impairments in specific psychiatric disorders, simulator models need to individualize platforms when testing complex scenarios assaying attentional and executive function or interpersonal aspects of driving such as road rage.

47.1 Introduction

47.1.1 General Comments

Driving is the quintessential twentieth century phenomenon and has become an instrumental activity of daily living for the majority adults in the developed world. In America, the driver's license has been the de facto identification card for years and from a self-psychology perspective, the automobile in contemporary society might be interpreted as the outermost shell of our fragile personality structures (Kohut, 1972). Yet psychiatrists have historically shown only a passing interest in driving ability as an example of human interaction with machines. Application of driving simulators in psychiatric research is a recent development, as the impact of mental illness on functional abilities such as driving or work performance is increasingly appreciated. The term "accident" has long fallen out of favor in the medical and human factor/error literature, as this term would see a collision as a random/unpredictable event, as opposed to a possible predictable and/or preventable entity, based on particular risk factors (O'Neill, 1998; O'Neill, 1993; Haddon, 1968). These risk factors might be driver/operator-specific, machine/vehicle-specific, or relate to the human-machine interface. The concept of "accident proneness" in an industrial setting was first described in 1919 (Greenwood & Woods, 1919), although this was not extrapolated to traffic safety research for several decades. An article entitled "The accident prone automobile driver" appeared in the *American Journal of Psychiatry* in 1949 which attempted to correlate individual psychopathology with driving records in taxi-drivers (Tilman & Hobbs, 1949). However, research into the epidemiology, psychology and neurophysiology of driving is slowly emerging, and this chapter will focus on potential relevance to driving simulation.

In this chapter, the differential effects of psychiatric disorders on neurocognitive functional impairments relevant to traffic safety will be addressed, to highlight potential application of driving simulator paradigms to assess and remediate these. The importance of attentional and strategic processes in relation to driving not only in attention-deficit disorder (ADHD), but for most disorders affecting neuropsychiatric function is emphasized. Separate sections in this chapter are provided for anxiety disorders, mood disorders, psychotic disorders, ADHD and personality disorders, with a format focusing on functional impairments of each, and psychological, neuroanatomic, ergonomic and/or pharmacological models that may be useful for investigators contemplating driving simulator application for diagnostic or therapeutic assessment in each diagnostic cluster of disorder.

47.1.2 Epidemiologic Estimates Informing Simulator Studies

The potential benefit of simulator studies should be considered in the context of emerging epidemiologic data linking psychiatric disorders and traffic safety risk. A 2003 survey of driving safety practices sponsored by the Transportation Research Board cited National Institute of Mental Health figures indicating that about 22% of adult Americans suffer from a diagnosable mental disorder (Knipling, Hickman, & Bergoffen, 2003), yet found that in the commercial trucking and bus industry, these mental health problems are relatively underappreciated with respect to their impact on road safety. A 2004 estimate provided by Australia's Monash University Accident Research Center gave a population prevalence of 25% for psychiatric disorders of clinically significant magnitude, if substance/alcohol-related disorders were included (Charlton et al., 2004). On a global level, recent World Health Organization projections pegging major depression and road traffic accidents as, respectively, the third and fourth leading causes of disability by 2030 similarly highlight the need to consider the interface of mental health and traffic medicine (Mathers & Loncar, 2006).

47.1.3 Driver Psychopathology in Adverse Situations and Mental Illness

As shown in Figure 47.1, in a cybernetic model of driving, driver psychopathology is only one of a multitude of factors that may affect driving output. Furthermore, many psychiatrically-based impairments of driving ability can be understood in terms of exaggerated limitations of psychological and neurophysiologic functions that could also affect driving performance in healthy individuals under adverse or extreme conditions. For example, a driver may not be able to fully focus on the road because they have recently suffered a personal tragedy, such as the death of a loved one, loss of a job, failure in an important examination, or monetary loss (Mohan, 2001). An individual may also be disturbed due to problems in personal relationships with a spouse, parent, sibling or close friend. Such situational impairments may both overlap and be differentiated from psychiatric illnesses:

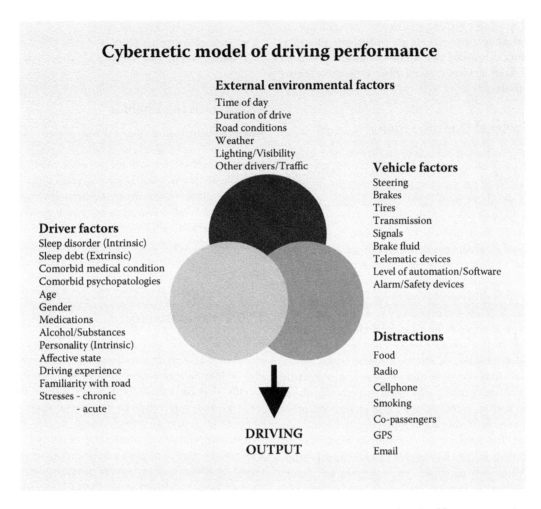

FIGURE 47.1   In a cybernetic model of driving output, driver psychopathology is only one of a multitude of factors to consider.

Clinically recognizable symptoms or behaviors characterized by abnormalities in cognition or emotion and often associated with distress and interference with personal functions (World Health Organization, 2001).

The heterogeneous, episodic, situational and/or interpersonally-specific nature of psychiatric disorders creates particular challenges for design of simulator-based assessment models (Niveau & Kelley-Puskas, 2001; Moller, 2008); performance on a simulator may therefore vary significantly, and can best be viewed as a cross-sectional psychometric profile at one point in time, rather than an enduring, longitudinal feature of a particular condition. Description of simulator applications in psychiatric research and practice is framed in the context of functional impairment inherent to each condition relevant to traffic safety.

47.1.4 Functional Impairments

Driving simulators have the potential for ecologically valid neuroergonomic assessment and, possibly, remediation of functional impairments in psychiatric illness. Commonly found areas of functional impairment correlating with psychiatric disorders include (Charlton et al., 2004):

- impaired information-processing ability, including attention, concentration, and memory components;
- reduced sustained attention—that is, alertness/vigilance;
- impaired visual-spatial functioning;
- increased psychomotor response latency;
- impaired impulse control, including increased degree of risk-taking;
- poor judgment and problem-solving, including the ability to predict/anticipate; and
- reduced ability to multi-task and respond to simultaneous stimuli in a changing perceptual-motor environment or crisis context.

As the nature of psychiatric illnesses typically entails fluctuations in functional impairment, their precise effect on driving ability may be unclear. This is an issue of particular significance in jurisdictions mandating duty-to-report by physicians and other health-care providers (Niveau et al., 2001; Canadian Medical Association, 2006). In a 2006 Canadian survey of 248 psychiatrists, the majority (64.1%) reported agreeing that addressing patients' fitness to drive is an important issue (Menard et al., 2006). However, only 18.0% of respondents were always aware of whether their patients were active drivers and only one-quarter of respondents felt that they were confident in their ability to evaluate fitness to drive. The authors concluded that a knowledge gap and need for education about evidence-based fitness to drive exists among psychiatrists.

47.1.5 Medicolegal Considerations

At this time, no published standards or guidelines exist regarding the use of driving simulators in medicolegal or forensic psychiatry contexts. Use of on-road testing as well as standardized simulator-based protocols potentially offers a more objective evaluation of skills, and the costs of such assessments must be weighed against the consequences of false-positive and false-negative functional diagnoses of impairment. In general, grossly impaired driving simulator performance is a useful "red flag" for further ergonomic/on-road evaluation (in the case of patients whose fitness-to-drive is in question but have a license at the time of testing), or can increase the confidence of a clinician who is reluctant to give in to pressure by a patient or family to have a license reinstated. In this sense, there is a role for use of standardized simulator scenarios in a forensic medicolegal context. However, normal simulator performance does not necessarily translate to normal "real-world" performance, where difficult-to-control variables such as traffic volume, interaction with other drivers and distractions cause fluctuating cognitive and task-specific demands. The development of simulator protocols that are individualized enough to be relevant to individual psychopathologies affecting driving ability, yet which are sufficiently broad-based to allow for standardized ergonomic testing and training of affected individuals is particularly challenging in mental disorders. Use of a variable "menu" of driving conditions, hazards, and other challenging stochastic events relevant to individual psychopathology would be useful in addition to standardized protocols.

47.2 Anxiety Disorders: Phobias, Panic Disorders and Post-Traumatic Stress Disorder

47.2.1 Functional Impairments

Anxiety disorders may represent a handicap to safe and effective driving skills, and can exist in both novice and experienced drivers (Taylor, Deane, & Podd, 2002). Patients who present with a fear of driving may be diagnosed with a specific phobia, panic disorder (with or without agoraphobia), or PTSD. While some individuals may be chiefly concerned about having a collision and about the consequences of a crash, others may worry about being suddenly incapacitated by anxiety while driving, similar to the concerns of patients with panic disorder. Due to features of anxiety such as hyperarousal, heightened alertness for threat and the tendency to worry excessively, individuals with anxiety disorders typically have neurocognitive symptoms affecting driving capacity including (Charlton et al., 2004):

- decreased working memory;
- increased distractibility; and
- lowered cognitive reserve and attentional capacity.

47.2.2 Driving Phobias

Anxiety regarding driving may also present as an impediment to driving in individuals who have delayed obtaining driver's licenses until well into adulthood due to avoidant behavior. In this model, driving fear can be thought of as a performance-related fear that is specific to the driving situation, thus overlapping in relevant concerns with social phobia. An analogy to social anxiety disorder has been hypothesized (Taylor, Deane, & Podd, 2007), with one focus including a fear of one's driving performance being criticized or negatively evaluated by others.

47.2.3 Panic Disorder

Panic attacks, typically involving respiratory and autonomic sensations as well as psychological sensations such as feelings of intense fearfulness or impending doom, can occur in specific driving situations evoking motion-induced mental loading, and may relate to a hypersensitive or imbalanced optokinetic and/or autonomic response (Shioiri et al., 2004). In a driving context, agoraphobia, which typically consists of a fear or avoidance of places or situations from which escape would be difficult, may involve avoidance of tunnels, causeways or bridges (Shioiri et al., 2004). In a more general context, it is common for drivers to be apprehensive of high-speed, high-volume routes (Roth, 2005), where the realistic risk of serious injury is greatest.

47.2.4 Post-Traumatic Stress Disorder

Drivers who have experienced serious collisions may develop complex fear responses following motor vehicle collisions, including PTSD. Driving phobias may be so severe that individuals are limited to drive very short distances or, in some cases, they may be unable to drive at all. In any given year, approximately 1% of the American population will be injured in a collision (Walshe, Lewis, Kim, O'Sullivan, & Wiederhold, 2003) and between 20–50% of crash survivors develop PTSD (Mayou, Bryant, & Duthie, 1993), with 11% continuing to meet criteria at three-year follow-up (Mayou, Ehlers, & Bryant, 2002). These individuals may have recurrent flashbacks to the accident, nightmares as well as other sleep disturbances, avoidant behaviors, and a generalized increase in anxiety.

47.2.5 Therapeutic Issues

Unless the key fear of re-engaging in an accident is addressed and desensitized with emotional processing, the crash survivor with PTSD-related driving phobias remains at high risk of developing a recurrence of relapse with reoccurrence of a collision (Walshe, Lewis, O'Sullivan, & Kim, 2005). If an individual avoids driving situations for prolonged periods, routine-driving skills previously learned begin to deteriorate, causing a secondary deconditioning of psychomotor skills. Simulation practice allows treatment to occur in the privacy of a clinic or laboratory setting, and involves graded exposure behavioral treatment models of anxiety and phobic responses, typically using a combination of standardized and individualized protocols (Walshe et al., 2003). Driving tasks of increasing difficulty and under a variety of conditions are assigned to the patient, while the patient's reactions are observed and measured.

Therapist-guided "virtual" driver training is gaining acceptance as a form of exposure therapy by mainstream psychiatry and psychology. In a simulator-training model, graded exposure therapy for driving-related anxiety overlaps with psychomotor skills training/retraining. Alongside simulator-tracked performance, monitoring of autonomic measures such as respiratory rate, galvanic skin response and heart rate variability, as well as subjective and clinician rating scales can be used to assess impairment and progress during treatment (Jang et al., 2002; Walshe et al., 2003). A course of simulator-based therapy can either serve as a segue to on-road assisted driving retraining or be used as standalone therapy leading to autonomous driving. Rigorously conducted outcome studies with large population samples are needed, although treatment of phobic and anxiety states is clearly one of the most promising clinical applications for simulators (Klingberg, 1998).

47.2.6 Simulation Design Issues

A key issue in using simulator-based assessment and treatment of driving anxiety is the degree of immersion or "presence" (Walshe et al., 2005; Simone, Schultheis, Rebimbas, & Millis, 2006; Schultheis, Rebimbas, Mourant, & Millis, 2007), allowing for the fear response to be evoked, which can subsequently be treated. Aside from the pictorial realism of graphics, it is likely that external factors such as: a) projecting images onto a large screen; b) using car seats for both driver and passenger; and c) increasing vibration sense through use of subwoofer speakers, can enhance presence and add to therapeutic efficacy (Jang et al., 2002).

47.3 Mood Disorders, Antidepressants and Related Psychotropic Therapeutics

47.3.1 Functional Impairments

As noted previously, clinical depression is a common illness and can be very debilitating; it is frequently comorbid with other psychiatric disorders including PTSD, panic disorder and alcohol or substance abuse; mood-related symptoms typically include diminished feelings of pleasure, hopelessness, as well as lowered motivation and interest in everyday tasks. Similarly, cognitive deficiencies including indecisiveness and a decreased ability to concentrate are correlated with impairment on psychomotor functioning (Bulmash et al., 2006; Edwards, 1995). Specifically, research has shown that individuals who suffer from major depression may demonstrate (Charlton et al., 2004):

- disturbances in attention;
- impaired information processing and judgment;
- psychomotor retardation or agitation;
- diminished concentration and memory ability;
- decreased reaction time;
- sleep disturbances and fatigue; and
- suicidal ideation.

Drivers with severe depression are at increased risk of being involved in a motor vehicle crash on several counts: Firstly, in a relatively rare but more dramatic context, suicidal thinking may cause them to crash their car in an attempt (whether planned or unplanned) to end their lives (Murray & de Leo, 2007); secondly, slowed reactivity to the environment and poor concentration put them at risk from a vehicle-handling point of view (Silverstone, 1988; Bulmash et al., 2006); and finally, potential contribution of psychotropics may play a role that is of relevance to simulator-based research.

One model of impaired psychomotor performance in depressed patients on any human-machine interaction task sees this as a condition of "absence" of psychomotor reactivity on the absence-presence immersion spectrum (Moller, 2008). Thus, low-fidelity graphics or failure to include salient cues within the simulation may be factors that decrease presence, but factors of mood, motivation and distracting internal states are always relevant to consider in a performance-based task such as driving (Figure 47.2).

47.3.2 Antidepressant Medications and Simulator Performance Variables

In 1985, Linnoila and Seppala completed a meta-analysis concluding that successful medication treatment of depressive symptoms often resulted in normalized performance on various indices of psychomotor performance (Linnoila & Seppala, 1985). In fact, driving simulator assessment in mood disorders has generally focused on the assessment of psychomotor/neurocognitive skills and the effects of antidepressant treatments. While a complete summary of the effects of psychotropics on driving performance

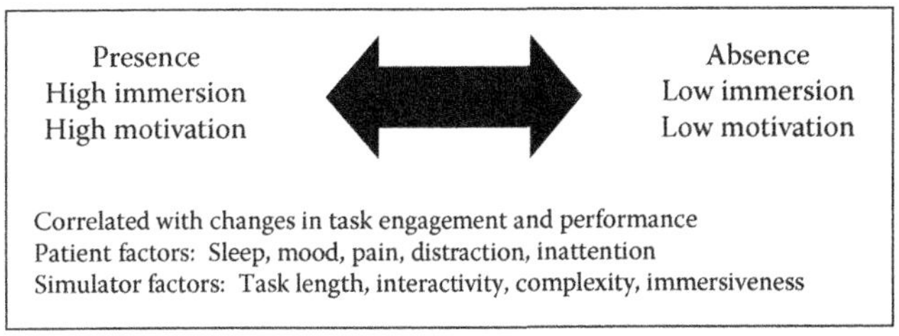

Presence		Absence
High immersion		Low immersion
High motivation		Low motivation

Correlated with changes in task engagement and performance
Patient factors: Sleep, mood, pain, distraction, inattention
Simulator factors: Task length, interactivity, complexity, immersiveness

FIGURE 47.2 The presence-absence-spectrum immersion spectrum affects sustained task performance in a human-machine interaction task.

is beyond the scope of this chapter, some generalizations can be made with respect to both simulator and on-road research. Rapoport and Banina (2007) recently reviewed the literature regarding the varieties of outcome measures used in studies employing driving simulators to assess effects of psychotropics, highlighting limitations of currently available data (lack of generalizability, standardization and small sample sizes).

Most frequently-used outcome measures have been variations in speed, steering, deviation from lateral position (tracking, lane drifting), reaction time or braking accuracy, driving errors (e.g., errors in turning, coordination, gap acceptance, signaling, following distance) and vehicle collisions. Most commonly found drug-related impairments tend to be those of tracking and reaction time, though there may be a bias towards sedation-specific effects as compared to agitation. As reviewed by Moller, Shapiro and Kayumov (2006) and Ramaekers (2003), tricyclic antidepressants (TCAs) are thought to impose significant potential risk compared to newer agents; amitriptyline, imipramine and doxepin can produce levels of driving impairment measurable as an increased "weaving tendency" with respect to tracking (i.e., standard deviation of lane position—SDLP) comparable to a blood-alcohol concentration of 0.8 mg/mL or more. Novel antidepressants such as SSRIs generally appear to be less impairing, although it is important to consider dosing time-of-day (Moller, Shapiro et al., 2006; Vermeeren, 2004). Furthermore, while benzodiazepines and TCAs have been most commonly associated with impairment in depression, the level of impairment is dependent on the specific population studied and dosing issues (Linnoila et al., 1985; Moller, Devins, Shen, & Shapiro, 2006; Rapoport & Banina, 2007). The role of residual "hangover" effects (i.e., residual daytime sleepiness and impairment of psychomotor and cognitive functioning the day after bedtime administration) on driving performance has been the subject of considerable scrutiny with hypnotics/sedatives, which are commonly prescribed in combination with antidepressants (Vermeeren, 2004; Moller, Devins, et al., 2006; O'Neill, 1998).

47.3.3 Potential Confounding Factors in Protocol Design

As shown in Table 47.1, collision risk can be understood in the context of the spectrum of perceptual engagement in a driving task, ranging from excessively low ("Type 1") to excessively high ("Type 2"). This engagement is both internally generated by mood, motivation, fatigue and alertness, and externally driven by the external environment of the simulation or actual driving task (Moller, Devins, et al., 2006). It is counterintuitive that depression alone produces impairment patterns of increased weaving tendency similar to alcohol, except when accompanied by hypersomnolence. Thus, SDLP patterns may be useful to assess the sedating effects of psychotropics including antidepressant medications, but less so for core psychomotor deficits related to the illness itself. An important confounder is the role of sleep disturbance on mood disorders. Hypersomnolence or insomnia may both be features of depressive illness, and antidepressant medications may have psychotropic effects ranging from sedating to activating. While increased daytime somnolence typically presents a higher Type 1 crash risk in a single-vehicle context, particularly if there is an element of sleep loss or if sedating psychotropics are involved, insomnia patients more typically demonstrate fatigue and reduced vigilance without frank sleepiness. Such patients may be more prone to demonstrate a risk in Type 2 traffic situations requiring intensive mental workload, rapid shifts in cognitive set, and decision-making—all of which rely on executive function and working memory (Moller, Devins et al., 2006). An appropriate global assessment battery for mood-related driving impairment should therefore include components of both Type 1 (underarousal) and Type 2 (overarousal) scenarios. This implies that the ability to maintain performance on monotonous drives as well as scenarios involving unpredictable hazards requiring rapid decision-making is desirable. Also, investigators should consider a protocol administered at multiple times of the day to address issues relating to circadian variation in mood (Bulmash et al., 2006).

47.3.4 Neurocognitive Abnormalities in Mood Disorders and Potential Simulator-Based Therapeutic Applications

The act of driving, particularly in urban settings, requires spatial memory and frequent shifts in executive function sets, making a well-designed driving simulation task a potential probe for cognitive function in mood disorders (Robbins, 2007). Spatial memory may be of particular interest because of reports of hippocampal

TABLE 47.1 "Type 1/Type 2" Stratification of Crash Risk Based on Underarousal or Overarousal of the Human Perceptual System

Type 1	Type 2
• Single vehicle	• Multi-vehicle
• Lack of external stimulation	• Overstimulation results when mental tasks excessively complex and sustained
• Circadian component	• Multi-tasking/overarousal
• Boredom/underarousal leads to lowering of brain activity	• Urban/suburban intersection
• Rural/long monotonous stretch of highway	• More often related to traffic volume

Source: Moller, H. J., Devins, G. M., Shen, J., & Shapiro, C. M. (2006). Sleepiness is not the inverse of alertness: Evidence from four sleep disorder patient groups. *Experimental Brain Research, 3(2)*, 258.

volume reduction in patients with mood disorders (Lucassen et al., 2006; Colla et al., 2007; Porter, Bourke, & Gallagher, 2007). Patients with unipolar and bipolar mood disorder have been shown to have deficits in spatial navigation (Gould et al., 2007); this may be related to underlying deficits in hippocampal and frontostriatal function (Lucassen et al., 2006; Colla et al., 2007). Recent research suggests that sleep deprivation may increase risk-taking behaviors (Killgore, Balkin, & Wesensten, 2006); this may in part account for crashes due to impulsive errors of commission associated with manic or mixed-mood disorder states.

Maguire et al. showed increased posterior hippocampal gray-matter volume in a group of healthy and experienced London taxi-drivers compared to non-professional drivers (Maguire et al., 2003; Maguire et al., 2000). The implication of these findings is that practicing spatial navigation and knowledge in the simulated practice of driving skill may in itself serve as a neurorehabilitative exercise through the process of neuroplasticity. Simulation-based practice exercises may be a safe and fairly specific neuroergonomic task to act as a retraining exercise of neurocognitive function which has been impaired by a wide array of brain diseases ranging from stroke to major depressive disorder (Akinwuntan et al., 2005). To date, the effects of intensive training in a driving simulation regimen on hippocampal and frontal-executive function has not been documented, and this may be a promising route of future research that might also be applied to residual cognitive deficits in some psychotic illnesses.

47.4 Schizophrenia and Psychotic Disorders

47.4.1 Functional Impairments

Untreated psychotic disorders are generally seen as an absolute contraindication to driving (Laux, 2002; Canadian Medical Association, 2006). The 2004 Monash report lists schizophrenia—an illness with a prevalence of 1% across cultures—as imposing a moderately high crash risk, and is inconclusive with respect to fitness-to-drive in medication-treated patients (Charlton et al., 2004). Research has shown that individuals with psychotic illnesses have widespread and multifaceted impairments in many domains of cognitive function. Individuals with schizophrenia may demonstrate a number of functional impairments that have obvious consequences for driving ability:

- reduced ability to selectively filter relevant from irrelevant information;
- diminished ability to sustain concentration or attention;
- reduced cognitive and perceptual processing speeds, including reaction time;
- a reduced ability to perform in more complex, distracting environments; and
- possible impaired judgment and decision-making if under the influence of delusions or hallucinations during an acute episode, causing a patient to misperceive the driving environment or misattribute intentions to other drivers.

(St Germain, Kurtz, Pearlson, & Astur, 2005).

Research on fitness-to-drive in psychotic patients has typically extrapolated findings from neuropsychological test batteries as opposed to driving simulator or on-road assessments. Szlyk reviewed the correlation of a variety of classic pencil-and-paper neuropsychological tests with driving simulator performance in a cohort of elderly drivers at risk of dementia, and found best correlations with the trail-making "A" and "B" tests (Szlyk, Myers, Zhang, Wetzel, & Shapiro, 2002). Unlike progressive cognitive impairment occurring in late life, functional impairments associated with schizophrenia are particularly difficult to determine because the degree of impairment fluctuates between the acute and residual phase of the illness. St Germain and colleagues reported on a small study comparing healthy controls and schizophrenic patients with a desktop simulator involving a rural drive with a single incursion hazard (St Germain et al., 2005); psychotic patients displayed driving impairments relative to controls as evidenced by lane errors, driving speed and collision probability.

Schizophrenia reduces performance even during tasks placing minimal cognitive load on working memory and perceptual processing (Birkett et al., 2007). In bipolar affective disorder with psychotic features, normal inter-episode functioning is typically described, while in schizophrenia, residual symptoms typically consist of more disabling and enduring neurocognitive impairments (Czobor, Jaeger, Berns, Gonzalez, & Loftus, 2007).

Birkett et al. (2007) postulate that increased reaction time in the disease and its predisposition are due to changes in working memory and response distribution rather than by a limitation of maximum response speed (Birkett et al., 2007), which suggests that a simulation incorporating elements of a continuous performance task (CPT) might be useful in any simulator performance task probing neurocognitive function. An interactive CPT-based simulator task that assays the ability to monitor and react to stochastic factors such as windgusts or obstacles occurring at regular but unpredictable intervals has been used a probe for reaction time for a variety of neuropsychiatric paradigms (Moller, 2008).

47.4.2 Antipsychotic Medications and Simulator Performance

An early Danish study from 1975 using simulator methodology suggested that the conventional antipsychotic haloperidol did not impair, but in fact improved driving performance (measured by brake time) in schizophrenic inpatients (Bech, 1975). More current studies have focused on comparisons between various neuroleptic medications (Kagerer, Winter, Moller, & Soyka, 2003), and suggest the superiority of newer agents. Soyka et al. recently studied psychomotor performance relevant to driving using the interactive computer-based Act and React Test (which assesses visual perception, attention, reaction time, and sensorimotor performance) among a group of schizophrenic inpatients approaching discharge; use of the novel atypical neuroleptics resulted in improved driving performance compared to the conventional antipsychotic haloperidol taken at comparable doses

(Soyka et al., 2005). Similarly, a recent German sample of 120 consecutively admitted psychotic inpatients treated with atypical neuroleptics or clozapine showed superior Act and React test performance on skills related to driving ability when compared with patients on typical neuroleptics (Brunnauer, Laux, Geiger, & Moller, 2004). Differences were most pronounced in neurocognitive measures of divided attention, stress tolerance, and attention. Similar to risperidone, treatment with clozapine had an overall positive impact on measures of reactivity and stress tolerance (Soyka et al., 2005).

47.4.3 Therapeutic Applications

As with mood disorders, neurocognitive deficits and the potential to retrain these has been a topic of some interest in researching rehabilitation strategies for schizophrenia and related psychotic disorders (McGurk, Mueser, Feldman, Wolfe, & Pascaris, 2007). Palmer et al. (2002) recently reported that out of a sample of American middle-aged and elderly outpatients with schizophrenia, 43% were current drivers. These findings contradict notions that functional impairment is inevitable in older schizophrenia patients, and indeed highlight the importance of assessment of functional skills by possibly targeting them as a treatment focus in some individuals, depending on vocational need.

47.5 Attention-Deficit Hyperactivity Disorder (ADHD)

47.5.1 Functional Impairments

The reported prevalence of ADHD in school age children varies from 3–10% depending on the criteria used, with males being over represented by a ratio, on average, of 3:1 (Barkley, 1997; Barkley & Cox, 2007). Individuals diagnosed with ADHD may display the following impairments:

- difficulty in planning, organizing and prioritizing tasks;
- difficulty estimating time;
- difficulty sustaining focus, shifting focus from one task to another, or filtering out
- distractions;
- inability to persist on a task in the face of temptation, frustration, or interruption;
- difficulty managing frustration and modulating emotions;
- impaired processing speed;
- difficulty utilizing working memory and accessing recall; and
- difficulty self-monitoring and regulating impulsivity.

While traditionally viewed as a childhood disorder, ADHD is now viewed as a disorder with direct implications on driving safety, as there has been a greater focus on the persistence of ADHD symptoms to adulthood in a significant subset of patients (Barkley & Cox, 2007). Furthermore, in the past decade, researchers have begun to conceptualize ADHD not simply as a problem with paying attention, but rather as a developmental

impairment of a complex range of executive functions related to goal-directed behavior (Barkley, 1997; Doyle, 2006).

47.5.2 Neurocognitive Abnormalities of Attentional Function Affecting Simulator Performance

Rasmussen proposed in his classic model of human control and behavior (Rasmussen, 1987) that during training of a task such as driving, control moves from knowledge or rule-based levels towards a skill-based level, resulting in the reduction in mental/cognitive workload required for the operations involved in the driving task and, thereby, inherently accommodating a larger amount of available attention that can be allocated to other tasks or operations. In applying this to ADHD, Barkley analogously conceptualizes driving as involving at least three levels, including basic cognitive abilities necessary for driving (operational), skills for maneuvering the vehicle in traffic (tactical), and the more executive, goal-directed aspects of driving (strategic) (Barkley, 2004). In this model, ADHD may interfere with the basic operational components of driving via impairment of sustained attention, resistance to distraction, response inhibition, slower and more variable reaction time, and the capacity to follow rules that may compete with ongoing sensory information (Schachar et al., 2004).

Functional brain differences in ADHD that may relate to driving are inherent to the phenomenon of multi-tasking itself, implying ADHD could be seen as a "strategy application disorder" as described by Burgess, Veitch, de Lacy and Shallice (2000). The prefrontal cortex (PFC) is important for sustaining attention over a delay, inhibiting distraction, and dividing attention, while more posterior cortical areas are essential for perception and allocation of attentional resources. The PFC in the right hemisphere is especially important for behavioral inhibition. Lesions to the PFC produce a profile of distractibility, forgetfulness, impulsivity, poor planning, and locomotor hyperactivity (Dickstein, Bannon, Xavier, & Milham, 2006). Executive functions such as sustained attention (Oken, Salinsky, & Elsas, 2006; Smid, de Witte, Homminga, & van den Bosch, 2006), set-shifting (Konishi et al., 2002; Braver, Reynolds, & Donaldson, 2003), working memory (Klingberg, 1998; Konishi et al., 1999) and response inhibition (Schachar et al., 2007; Rubia, Smith, Brammer, Toone, & Taylor, 2005) have also been attributed to frontal lobe function. Dysregulation of frontal or prefrontal lobe function has been described as an anatomical model for cognitive dysfunction in numerous other conditions including stroke (Lincoln, Majid, & Weyman, 2000), mood disorders (Holmes et al., 2005; Galynker et al., 1998), schizophrenia (Rubia et al., 2001; Holmes et al., 2005; Semkovska, Bédard, & Stip, 2001), obstructive sleep apnea (Alchanatis et al., 2005) and chronic insomnia (Nofzinger et al., 2004). This highlights the primacy of attentional processes across healthy and psychopathological states in assessing driving safety. Thus, while ADHD may be considered the "flagship disorder" of disturbed attention, a variety of other psychopathologies should be considered in the differential diagnosis of ADHD.

47.5.3 Other Ergonomic Factors to Consider in Impairment of Attention Processes

Public health officials and the insurance industry have long been aware of the elevated incidence of traffic incidents in young male drivers (Jonah, 1986), an epidemiologic cohort where adult ADHD is more likely to be diagnosed. This does not necessarily imply that ADHD itself is the chief cause of this elevated crash risk. Young drivers, whether "normal" or attention-disordered are also most likely to be users of distracting telematic devices. As Lee points out, "infotainment" technology stresses the same vulnerabilities that already lead young drivers to crash more frequently than other drivers, without the specific diagnosis of ADHD (Lee, 2007). Cell phones, text messaging, MP3 players, and other portable media devices all present a threat because young drivers may lack the spare attentional capacity for vehicle control and the ability to anticipate and manage hazards.

The issue of insight, judgment and self-awareness of driving abilities and impairments in ADHD also remains controversial. Individuals with ADHD may underappreciate deficits related to driving safety, due to deficits in self-monitoring and meta-cognitive abilities. A study by Fischer et al. of driving instructor assessments for 147 ADHD-diagnosed drivers found significantly more impulsive errors on behind-the-wheel road tests compared to controls (Fischer, Barkley, Smallish, & Fletcher, 2007). Specifically, on a simulator driving task performed adjunctively to on-road testing, ADHD-drivers showed slower and more variable reaction times, more errors of impulsiveness (false alarms, poor rule following), greater steering variability, and more crashes of the simulated vehicle.

On the spectrum of novelty-seeking/harm avoidance (Cloninger, Svrakic, & Przybeck, 1993), ADHD drivers may be understood as being on the opposite end of the spectrum from those with anxiety disorders (see Figure 47.3). Issues that have been identified as associated with ADHD include speeding, passing, weaving between traffic, reckless driving, driving without a license, hit-and-run crashes, and license suspensions (Biederman et al., 2007; Fischer et al., 2007; Reimer, D'Ambrosio, Coughlin, Kafrissen, & Biederman, 2006). This can make differentiation from personality pathologies such as antisocial

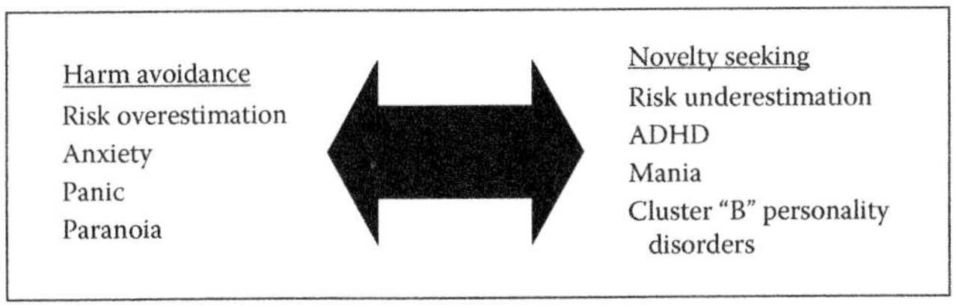

FIGURE 47.3 Dimensional spectrum of psychiatric disturbances related to driving. (Adapted from Cloninger, C. R., Svrakic, D. M., & Przybeck, T. R. (1993). A psychobiological model of temperament and character. *Archives of General Psychiatry, 50,* 975–990.)

personality disorder challenging at times (Malta, Blanchard, & Freidenberg, 2005).

47.5.4 Stimulants and ADHD in Simulator Models

One focus of driving simulator research in ADHD has been on providing a safe and reliable mechanism for assessment of stimulant medication (Cox, Merkel, Kovatchev, & Seward, 2000; Cox et al., 2006). It has been suggested that aside from control of core ADHD symptoms, stimulant medication treatment may reduce traffic risks in this population (Barkley & Cox, 2007). While stimulant efficacy in smaller short-term studies of driving simulator has been demonstrated, prospective studies with adequate long-term follow-up are needed to clarify this important clinical question. An important research avenue also relates to the added value of a simulator-based educational and skills training program in drivers with ADHD and in younger drivers in general.

47.6 Personality Disorders, Aggressive Driving and Road Rage

47.6.1 Functional Impairments

The incidence of personality disorders has been estimated from 1–10% of the general population in North America, depending on the criterion being used (Malta et al., 2005). Individuals with severe cluster "B" personality disorders such as antisocial personality disorder (more commonly diagnosed in men) or histrionic and borderline personality disorders (more commonly diagnosed in women) are at high risk of alcohol or drug abuse, and violent or self-destructive behaviors (Dumais et al., 2005). Furthermore, specific personality traits have been shown by several researchers to be risk factors for traffic accidents (Noyes, 1985; Tsuang, Boor, & Fleming, 1985), with obvious consequences for driving ability:

- aggression;
- egocentricity;
- impulsiveness;
- resentment of authority;
- intolerance of frustration; and
- irresponsibility.

Borderline and antisocial personality disorders, in which impulsive-aggressive behaviors play a central role, and substance use disorders appear to be risk factors for traffic fatalities in young adults (Dumais et al., 2005). One should note that the term "functional impairment" is largely used to describe personality characteristics over which individuals are assumed to have limited levels of control compared to non-affected individuals. Personality disorders have also been linked to descriptions of "road rage", although assigning a volitional attribute or excusing behaviors as caused by illness somewhat risks extrapolating simplistic and controversial "Mad/Bad" dichotomies of forensic psychiatry (Fong, Frost, & Stansfeld, 2001).

47.6.2 Relationship to Road Rage

Any driving simulator designed to study or treat road rage should in part consider interpersonal/cognitive attributions that drivers make of each other. However, road rage can also in part be understood on a neuropsychiatric basis. One plausible neurocognitive model for road rage links the stop-and-start traffic inherently related to sociodemographic trends in commuting length and congested traffic density (Moller, 2005; Parker, Lajunen, & Summala, 2002), causing impairments in set-shift of frontostriatal systems (Robbins, 2007). As discussed in the "Type 1/Type 2" model of crash risk, the recurrent shifts from cognitive understimulation to overarousal may challenge frontal lobe executive function pathways, with resultant symptoms of anger and irritability mediated by monoaminergic transmitters such as dopamine, serotonin and norepinephrine (Fava & Rosenbaum, 1999). No literature on the use of SSRI or other antidepressants to treat road rage is available, and skills training of the ability to effectively cope with stress-inducing traffic situations might have a longer-lasting cognitive effect. A recent review of road rage prevention focuses on sociolegal and vehicle-design as possible remedies (Asbridge, Smart, & Mann, 2006). The design of skills-based simulator protocols to address coping mechanisms related to typical frustrations of driving and commuting would be welcome and of significant potential public health benefit. Use in a forensic/medicolegal context would also appear to be a possible assessment application in assessing traffic offenders with or without psychopathologies. The obvious limitation in a simulation, however, remains the relative paucity of interpersonal contexts underlying hostile cognitive attributions that may underlie road rage (Britt & Garrity, 2006).

47.7 Conclusion

Neuroergonomic assessment of driving behavior in the context of psychopathology is useful if one includes the automobile as a physical extension of the human behavioral repertoire. It may be difficult at times to differentiate generic stresses in healthy drivers exposed to extreme or unusual conditions from symptomatic changes in cognitive and emotional processing described in psychiatric disorders. Simulator testing offers the promise of a reliable, ecologically valid and replicable environment for testing of psychomotor performance and skills training relevant to psychiatric illness.

This is relevant in the medicolegal context of fitness-to-drive assessments, repeated-measure skills training programs, and determination of the acute and chronic effects of medications used to treat psychiatric illness. Experimental use of simulator testing in cross-sectional or prospective studies is useful to test hypotheses generated by correlational/epidemiologic models associating psychiatric disorders with traffic risk. Specific conditions discussed in this chapter have included anxiety disorders, mood disorders, psychotic illnesses, ADHD, and

personality disorders, with an emphasis on linking neuropsychiatric models with expected and demonstrated impairments for each condition in a simulation context. Diagnostic as well as rehabilitative/therapeutic uses of simulators, including their limitations have been outlined, and a number of theoretical models have been proposed for simulators as an adjunct in modern psychiatry. Future development of low-cost, flexible scenarios that allow for a high sense of presence/immersion will allow for improved understanding of traffic safety by researchers and clinicians caring for individuals with psychiatric disease.

Key Points

- Psychiatric disorders frequently lead to functional impairments in activities of daily living, including driving.
- Due to the fluctuating severity of symptoms of many psychiatric illnesses, driving performance at one point in time is difficult to generalize.
- Relevant variables included reaction time (steering/braking), road position, speed, and specific errors (e.g., crashes/lane incursions).
- There is an emerging body of literature regarding the use of simulators to assess effects of medications used to treat psychiatric illness. Simulators may potentially be used to sequentially track psychomotor performance during a treatment period.
- There is evidence for the potential use of simulators in skills training and rehabilitation in neuropsychiatric conditions.
- Use of simulators in the context of psychiatric disorders should take into account driving conditions and mental states that are most likely related to driving errors (i.e., monotony versus high complexity/overarousal).

Keywords: ADHD, Anxiety Disorders, Driving Simulator, Mood Disorders, Psychiatric Disorders, Psychotic Disorders, PTSD, Road Rage, Traffic Safety

Glossary

Continuous performance task (CPT): A psychological test which measures a person's sustained and selective attention and impulsivity.

Executive functions: A term used by neuroscientists to describe brain processes whose role is to guide thought and behavior in accordance with internally-generated goals or plans.

Neurocognitive deficit: A reduction or impairment of cognitive function in neural pathways, or cortical networks in the brain, occurring as a consequence of neurological illness, mental illness, or drug use.

Phobia: An irrational, intense, persistent fear of certain people, situations, activities, or objects. Phobias are the most common forms of anxiety disorders.

Presence: A theoretical concept describing the effect that people experience when they interact with a computer-mediated or computer-generated environment. A fluctuating spectrum of "absence" to presence is theorized to underlie a sense of "being there" in interactions with virtual and simulated environments.

Psychotropic: A chemical substance that acts primarily upon the central nervous system where it alters brain function, resulting in changes in consciousness, perception, mood, and behavior. This may include psychiatric medications as well as drugs of abuse.

Key Readings

Biederman, J., Fried, R., Monuteaux, M. C., Reimer, B., Coughlin, J. F., Surman, C. B. et al. (2007). A laboratory driving simulation for assessment of driving behavior in adults with ADHD: A controlled study. *Annals of General Psychiatry, 6,* 4.

Moller, H. J., Shapiro, C. M., & Kayumov, L. (2006). Effects of psychotropics on driving performance. In M. Lader, D. P. Cardinali, & S. R. Pandi-Perumal (Eds.), *Sleep and sleep disorders: Neuropsychopharmacological approach* (pp. 121–125). Georgetown, TX: Landes Biosciences.

Moller, H. J. (2008). Neural correlates of "Absence" in interactive simulator protocols. *Cyberpsychology & Behavior, 11*(2), 181–187.

Ramaekers, J. G. (2003). Antidepressants and driver impairment: Empirical evidence from a standard on-the-road test. *Journal of Clinical Psychiatry, 64*(1), 20–29.

Silverstone, T. (1988). The influence of psychiatric disease and its treatment on driving performance. *International Clinical Psychopharmacology, 3*(Supplement 1), 59–66.

Walshe, D., Lewis, E., O'Sullivan, K., & Kim, S. I. (2005). Virtually driving: Are the driving environments "real enough" for exposure therapy with accident victims? *Cyberpsychology & Behavior, 8*(6), 532–537.

References

Akinwuntan, A. E., de Weert, W., Feys, H., Pauwels, J., Baten, G., Arno, P. et al. (2005). Effect of simulator training on driving after stroke: A randomized controlled trial. *Neurology, 65,* 843–850.

Alchanatis, M., Zias, N., Deligiorgis, N., Amfilochiou, A., Dionellis, G., & Orphanidou, D. (2005). Sleep apnea-related cognitive deficits and intelligence: An implication of cognitive reserve theory. *Journal of Sleep Research, 14,* 69–75.

Asbridge, M., Smart, R. G., & Mann, R. E. (2006). Can we prevent road rage? *Trauma Violence Abuse, 7,* 109–121.

Barkley, R. A. (1997). Behavioral inhibition, sustained attention, and executive functions: Constructing a unifying theory of ADHD. *Psychological Bulletin, 121,* 65–94.

Barkley, R. A. (2004). Driving impairments in teens and adults with attention-deficit/hyperactivity disorder. *Psychiatric Clinics of North America, 27,* 233–260.

Barkley, R. A., & Cox, D. (2007). A review of driving risks and impairments associated with attention-deficit/hyperactivity disorder and the effects of stimulant medication on driving performance. *Journal of Safety Research, 38,* 113–128.

Bech, P. (1975). Mental illness and simulated driving: Before and during treatment. *Pharmakopsychiatry Neuropsychopharmakology, 8,* 143–150.

Biederman, J., Fried, R., Monuteaux, M. C., Reimer, B., Coughlin, J. F., Surman, C. B. et al. (2007). A laboratory driving simulation for assessment of driving behavior in adults with ADHD: A controlled study. *Annals of General Psychiatry, 6,* 4.

Birkett, P., Sigmundsson, T., Sharma, T., Toulopoulou, T., Griffiths, T. D., Reveley, A. et al. (2007). Reaction time and sustained attention in schizophrenia and its genetic predisposition. *Schizophrenia Research, 95*(1), 75–85.

Braver, T. S., Reynolds, J. R., & Donaldson, D. I. (2003). Neural mechanisms of transient and sustained cognitive control during task switching. *Neuron, 39,* 713–726.

Britt, T. W., & Garrity, M. J. (2006). Attributions and personality as predictors of the road rage response. *British Journal of Social Psychology, 45,* 127–147.

Brunnauer, A., Laux, G., Geiger, E., & Moller, H. J. (2004). The impact of antipsychotics on psychomotor performance with regards to car driving skills. *Journal of Clinical Psychopharmacology, 24,* 155–160.

Bulmash, E. L., Moller, H. J., Kayumov, L., Shen, J., Wang, X., & Shapiro, C. M. (2006). Psychomotor disturbance in depression: Assessment using a driving simulator paradigm. *Journal of Affective Disorders, 93,* 213–218.

Burgess, P. W., Veitch, E., de Lacy, C. A., & Shallice, T. (2000). The cognitive and neuroanatomical correlates of multitasking. *Neuropsychologia, 38,* 848–863.

Canadian Medical Association. (2006). *Determining medical fitness to operate motor vehicles: CMA driver's guide,* 7th ed. Ottawa, Canada: CMA.

Charlton, J., Koppel, S., O'Hare, M., Andrea, D., Smith, G., Khodr, B. et al. (2004). *Influence of chronic illness on crash involvement of motor vehicle drivers* (Report 213). Victoria, Australia: Monash University Accident Research Center.

Cloninger, C. R., Svrakic, D. M., & Przybeck, T. R. (1993). A psychobiological model of temperament and character. *Archives of General Psychiatry, 50,* 975–990.

Colla, M., Kronenberg, G., Deuschle, M., Meichel, K., Hagen, T., Bohrer, M. et al. (2007). Hippocampal volume reduction and HPA-system activity in major depression. *Journal of Psychiatric Research, 41,* 553–560.

Cox, D. J., Merkel, R. L., Kovatchev, B., & Seward, R. (2000). Effect of stimulant medication on driving performance of young adults with attention-deficit hyperactivity disorder: A preliminary double-blind placebo controlled trial. *Journal of Mental and Nervous Disease, 188,* 230–234.

Cox, D. J., Merkel, R. L., Moore, M., Thorndike, F., Muller, C., & Kovatchev, B. (2006). Relative benefits of stimulant therapy with OROS methylphenidate versus mixed amphetamine salts extended release in improving the driving performance of adolescent drivers with attention-deficit/hyperactivity disorder. *Pediatrics, 118,* 704–710.

Czobor, P., Jaeger, J., Berns, S. M., Gonzalez, C., & Loftus, S. (2007). Neuropsychological symptom dimensions in bipolar disorder and schizophrenia. *Bipolar Disorders, 9,* 71–92.

Dickstein, S. G., Bannon, K., Xavier, C. F., & Milham, M. P. (2006). The neural correlates of attention-deficit hyperactivity disorder: An ALE meta-analysis. *Journal of Child Psychology and Psychiatry, 47,* 1051–1062.

Doyle, A. E. (2006). Executive functions in attention-deficit/hyperactivity disorder. *Journal of Clinical Psychiatry, 67 Suppl 8,* 21–26.

Dumais, A., Lesage, A. D., Boyer, R., Lalovic, A., Chawky, N., Menard-Buteau, C. et al. (2005). Psychiatric risk factors for motor vehicle fatalities in young men. *Canadian Journal of Psychiatry, 50,* 838–844.

Edwards, J. G. (1995). Depression, antidepressants, and accidents. *British Medical Journal, 311,* 887–888.

Fava, M., & Rosenbaum, J. F. (1999). Anger attacks in patients with depression. *Journal of Clinical Psychiatry, 60*(Suppl. 15), 21–24.

Fischer, M., Barkley, R. A., Smallish, L., & Fletcher, K. (2007). Hyperactive children as young adults: Driving abilities, safe driving behavior, and adverse driving outcomes. *Accident Analysis & Prevention, 39,* 94–105.

Fong, G., Frost, D., & Stansfeld, S. (2001). Road rage: A psychiatric phenomenon? *Social Psychiatry and Psychiatric Epidemiology, 36,* 277–286.

Galynker, I. I., Cai, J., Ongseng, F., Finestone, H., Dutta, E., & Serseni, D. (1998). Hypofrontality and negative symptoms in major depressive disorder. *Journal of Nuclear Medicine, 39,* 608–612.

Gould, N. F., Holmes, M. K., Fantie, B. D., Luckenbaugh, D. A., Pine, D. S., Gould, T. D. et al. (2007). Performance on a virtual reality spatial memory navigation task in depressed patients. *American Journal of Psychiatry, 164,* 516–519.

Greenwood, M., & Woods, H. (1919). The incidence of industrial accidents. *Reports Industrial Research Board, 10,* 4–20.

Haddon, W. (1968). The changing approach to the epidemiology, prevention and amelioration of trauma. *American Journal of Public Health, 58,* 1431–1438.

Holmes, A. J., MacDonald, A., III., Carter, C. S., Barch, D. M., Andrew, S. V., & Cohen, J. D. (2005). Prefrontal functioning during context processing in schizophrenia and major depression: An event-related fMRI study. *Schizophrenia Research, 76,* 199–206.

Jang, D. P., Kim, I. Y., Nam, S. W., Wiederhold, B. K., Wiederhold, M. D., & Kim, S. I. (2002). Analysis of physiological response to two virtual environments: Driving and flying simulation. *Cyberpsychology & Behavior, 5,* 11–18.

Jonah, B. A. (1986). Accident risk and risk-taking behavior among young drivers. *Accident Analysis & Prevention, 18,* 255–271.

Kagerer, S., Winter, C., Moller, H. J., & Soyka, M. (2003). Effects of haloperidol and atypical neuroleptics on psychomotor performance and driving ability in schizophrenic patients. Results from an experimental study. *Neuropsychobiology, 47,* 212–218.

Killgore, W. D., Balkin, T. J., & Wesensten, N. J. (2006). Impaired decision-making following 49 h of sleep deprivation. *Journal of Sleep Research, 15,* 7–13.

Klingberg, T. (1998). Concurrent performance of two working memory tasks: Potential mechanisms of interference. *Cerebral Cortex, 8,* 593–601.

Knipling, R., Hickman, J. S., & Bergoffen, G. (2003). *CTBSSP synthesis of safety practice 1: Effective commercial truck and bus safety management techniques.* Washington, DC: Transportation Research Board of the National Academies.

Kohut, H. (1972). Thoughts on narcissism and narcissistic rage. *Psychoanalytical Study of the Child, 27,* 360–400.

Konishi, S., Hayashi, T., Uchida, I., Kikyo, H., Takahashi, E., & Miyashita, Y. (2002). Hemispheric asymmetry in human lateral prefrontal cortex during cognitive set shifting. *Proceedings of the National Academy of Sciences U.S.A., 99,* 7803–7808.

Konishi, S., Kawazu, M., Uchida, I., Kikyo, H., Asakura, I., & Miyashita, Y. (1999). Contribution of working memory to transient activation in human inferior prefrontal cortex during performance of the Wisconsin Card Sorting Test. *Cerebral Cortex, 9,* 745–753.

Laux, G. (2002). Psychiatric disorders and fitness to drive an automobile: An overview. *Nervenarzt, 73,* 231–238.

Lee, J. D. (2007). Technology and teen drivers. *Journal of Safety Research, 38,* 203–213.

Lincoln, N. B., Majid, M. J., & Weyman, N. (2000). Cognitive rehabilitation for attention deficits following stroke. *Cochrane Database of Systematic Reviews, 4,* CD002842.

Linnoila, M., & Seppala, T. (1985). Antidepressants and driving. *Accident Analysis &.Prevention, 17,* 297–301.

Lucassen, P. J., Heine, V. M., Muller, M. B., van der Beek, E. M., Wiegant, V. M., De Kloet, E. R. et al. (2006). Stress, depression and hippocampal apoptosis. *CNS Neurological Disorders—Drug Targets, 5,* 531–546.

Maguire, E. A., Gadian, D. G., Johnsrude, I. S., Good, C. D., Ashburner, J., Frackowiak, R. S. et al. (2000). Navigation-related structural change in the hippocampi of taxi drivers. *Proceedings of the National Academy of Sciences, U.S.A., 97,* 4398–4403.

Maguire, E. A., Spiers, H. J., Good, C. D., Hartley, T., Frackowiak, R. S., & Burgess, N. (2003). Navigation expertise and the human hippocampus: A structural brain imaging analysis. *Hippocampus, 13,* 250–259.

Malta, L. S., Blanchard E. B., & Freidenberg, B. M. (2005). Psychiatric and behavioral problems in aggressive drivers. *Behavior Research and Therapy, 43,* 1467–1484.

Mathers, C. D., & Loncar, D. (2006). Projections of global mortality and burden of disease from 2002 to 2030. *PLoS Medicine, 3*, 442–446.

Mayou, R., Bryant, B., & Duthie, R. (1993). Psychiatric consequences of road traffic accidents. *British Medical Journal, 307*, 647–651.

Mayou, R. A., Ehlers, A., & Bryant, B. (2002). Post-traumatic stress disorder after motor vehicle accidents: 3-year follow-up of a prospective longitudinal study. *Behavior Research and Therapy, 40*, 665–675.

McGurk, S. R., Mueser, K. T., Feldman, K., Wolfe, R., & Pascaris, A. (2007). Cognitive training for supported employment: 2–3 year outcomes of a randomized controlled trial. *American Journal of Psychiatry, 164*, 437–441.

Menard, I., Korner-Bitensky, N., Dobbs, B., Casacalenda, N., Beck, P. R., Dippsych, C. M. et al. (2006). Canadian psychiatrists' current attitudes, practices, and knowledge regarding fitness to drive in individuals with mental illness: A cross-Canada survey. *Canadian Journal of Psychiatry, 51*, 836–846.

Mohan, D. (2001). Road traffic crashes, injuries and public health. *Proceedings of the World Health Organization meeting to develop a 5-year strategy for road traffic injury prevention* (pp. 2–13). Geneva, Switzerland: World Health Organization.

Moller, H. J. (2005). Task fatigue and driving performance: Implications on the 21st century commute. *Occupational Safety, 43*, 15.

Moller, H. J. (2008). Neural Correlates of "Absence" in Interactive Simulator Protocols. *Cyberpsychology & Behavior, 11*, 181–187.

Moller, H. J., Devins, G. M., Shen, J., & Shapiro, C. M. (2006). Sleepiness is not the inverse of alertness: Evidence from four sleep disorder patient groups. *Experimental Brain Research, 3*(2), 258.

Moller, H. J., Shapiro, C. M., & Kayumov, L. (2006). Effects of psychotropics on driving performance. In M. Lader, D. P. Cardinali, & S. R. Pandi-Perumal (Eds.), *Sleep and sleep disorders: Neuropsychopharmacological approach* (pp. 121–125). Georgetown, TX: Landes Biosciences.

Murray, D., & de Leo, D. (2007). Suicidal behavior by motor vehicle collision. *Traffic Injury Prevention, 8*, 244–247.

Niveau, G., & Kelley-Puskas, M. (2001). Psychiatric disorders and fitness to drive. *Journal of Medical Ethics, 27*, 36–39.

Nofzinger, E. A., Buysse, D. J., Germain, A., Price, J. C., Miewald, J. M., & Kupfer, D. J. (2004). Functional neuroimaging evidence for hyperarousal in insomnia. *American Journal of Psychiatry, 161*, 2126–2128.

Noyes, R., Jr. (1985). Motor vehicle accidents related to psychiatric impairment. *Psychosomatics, 26*, 569–6, 579.

O'Neill, D. (1993). Driving and psychiatric illness. *American Journal of Psychiatry, 150*, 351.

O'Neill, D. (1998). Benzodiazepines and driver safety. *Lancet, 352*, 1324–1325.

Oken, B. S., Salinsky, M. C., & Elsas, S. M. (2006). Vigilance, alertness, or sustained attention: Physiological basis and measurement. *Clinical Neurophysiology, 117*(9), 1885–1901.

Palmer, B. W., Heaton, R. K., Gladsjo, J. A., Evans, J. D., Patterson, T. L., Golshan, S. et al. (2002). Heterogeneity in functional status among older outpatients with schizophrenia: Employment history, living situation, and driving. *Schizophrenia Research, 55*, 205–215.

Parker, D., Lajunen, T., & Summala, H. (2002). Anger and aggression among drivers in three European countries. *Accident Analysis & Prevention, 34*, 229–235.

Porter, R. J., Bourke, C., & Gallagher, P. (2007). Neuropsychological impairment in major depression: Its nature, origin and clinical significance. *Australia and New Zealand Journal of Psychiatry, 41*, 115–128.

Ramaekers, J. G. (2003). Antidepressants and driver impairment: Empirical evidence from a standard on-the-road test. *Journal of Clinical Psychiatry, 64*, 20–29.

Rapoport, M. J., & Banina, M. C. (2007). Impact of psychotropic medications on simulated driving: A critical review. *Central Nervous System. Drugs, 21*, 503–519.

Rasmussen, J. (1987). Cognitive control and human error mechanisms. In J. Rasmussen, K. Duncan, & J. Leplat (Eds.), *New technology and human error* (pp. 53–62). Chichester: John Wiley & Sons.

Reimer, B., D'Ambrosio, L. A., Coughlin, J. F., Kafrissen, M. E., & Biederman, J. (2006). Using self-reported data to assess the validity of driving simulation data. *Behavior Research Methods, 38*, 314–324.

Robbins, T. W. (2007). Shifting and stopping: Fronto-striatal substrates, neurochemical modulation and clinical implications. *Philosophical Transactions of the Royal Society of London B Biological Sciences, 362*, 917–932.

Roth, W. T. (2005). Physiological markers for anxiety: Panic disorder and phobias. *International Journal of Psychophysiology, 58*, 190–198.

Rubia, K., Russell, T., Bullmore, E. T., Soni, W., Brammer, M. J., Simmons, A., et al. (2001). An fMRI study of reduced left prefrontal activation in schizophrenia during normal inhibitory function. *Schizophrenia Research, 52*, 47–55.

Rubia, K., Smith, A. B., Brammer, M. J., Toone, B., & Taylor, E. (2005). Abnormal brain activation during inhibition and error detection in medication-naive adolescents with ADHD. *American Journal of Psychiatry, 162*, 1067–1075.

Schachar, R. J., Chen, S., Logan, G. D., Ornstein, T. J., Crosbie, J., Ickowicz, A. et al. (2004). Evidence for an error monitoring deficit in attention-deficit hyperactivity disorder. *Journal of Abnormal Child Psychology, 32*, 285–293.

Schachar, R., Logan, G. D., Robaey, P., Chen, S., Ickowicz, A., & Barr, C. (2007). Restraint and cancellation: Multiple inhibition deficits in attention-deficit hyperactivity disorder. *Journal of Abnormal Child Psychology, 35*, 229–238.

Schultheis, M. T., Rebimbas, J., Mourant, R., & Millis, S. R. (2007). Examining the usability of a virtual reality driving simulator. *Assistive Technology, 19*, 1–8.

Semkovska, M., Bédard, M. A., & Stip, E. (2001). Hypofrontality and negative symptoms in schizophrenia: Synthesis of anatomic and neuropsychological knowledge and ecological perspectives. *Encephale, 27,* 405–415.

Shioiri, T., Kojima, M., Hosoki, T., Kitamura, H., Tanaka, A., Bando, T. et al. (2004). Momentary changes in the cardiovascular autonomic system during mental loading in patients with panic disorder: A new physiological index "rho(max)". *Journal of Affective Disorders, 82,* 395–401.

Silverstone, T. (1988). The influence of psychiatric disease and its treatment on driving performance. *International Clinical Psychopharmacology, 3*(Suppl. 1), 59–66.

Simone, L. K., Schultheis, M. T., Rebimbas, J., & Millis, S. R. (2006). Head-mounted displays for clinical virtual reality applications: Pitfalls in understanding user behavior while using technology. *Cyberpsychology & Behavior, 9,* 591–602.

Smid, H. G., de Witte, M. R., Homminga, I., & van den Bosch, R. J. (2006). Sustained and transient attention in the continuous performance task. *Journal of Clinical and Experimental Neuropsychology, 28,* 859–883.

Soyka, M., Winter, C., Kagerer, S., Brunnauer, M., Laux, G., & Moller, H. J. (2005). Effects of haloperidol and risperidone on psychomotor performance relevant to driving ability in schizophrenic patients compared to healthy controls. *Journal of Psychiatric Research, 39,* 101–108.

St Germain, S. A., Kurtz, M. M., Pearlson, G. D., & Astur, R. S. (2005). Driving simulator performance in schizophrenia. *Schizophrenia Research, 74,* 121–122.

Szlyk, J. P., Myers, L., Zhang, Y., Wetzel, L., & Shapiro, R. (2002). Development and assessment of a neuropsychological battery to aid in predicting driving performance. *Journal of Rehabilitation Research and Development, 39,* 483–496.

Taylor, J., Deane, F., & Podd, J. (2002). Driving-related fear: A review. *Clinical Psychology Review, 22,* 631–645.

Taylor, J. E., Deane, F. P., & Podd, J. V. (2007). Driving fear and driving skills: Comparison between fearful and control samples using standardized on-road assessment. *Behavior Research and Therapy, 45,* 805–818.

Tilman, W. A., & Hobbs G. E. (1949). The accident-prone automobile driver. *American Journal of Psychiatry, 106,* 321–331.

Tsuang, M. T., Boor, M., & Fleming, J. A. (1985). Psychiatric aspects of traffic accidents. *American Journal of Psychiatry, 142,* 538–546.

Vermeeren, A. (2004). Residual effects of hypnotics: Epidemiology and clinical implications. *Central Nervous System Drugs, 18,* 297–328.

Walshe, D., Lewis, E., O'Sullivan, K., & Kim, S. I. (2005). Virtually driving: Are the driving environments "real enough" for exposure therapy with accident victims? An explorative study. *Cyberpsychology & Behavior, 8,* 532–537.

Walshe, D. G., Lewis, E. J., Kim, S. I., O'Sullivan, K., & Wiederhold, B. K. (2003). Exploring the use of computer games and virtual reality in exposure therapy for fear of driving following a motor vehicle accident. *Cyberpsychology & Behavior, 6,* 329–334.

World Health Organization. (2001). *Mental health problems: The undefined and hidden burden.* World Health Organization.

48

Driving in Alzheimer's Disease, Parkinson's Disease and Stroke

Ergun Y. Uc
The University of Iowa

Matthew Rizzo
The University of Iowa

Abstract

The Problem. The proportion of elderly in the general population is projected to rise, with corresponding increases in age-related neurodegenerative conditions such as Alzheimer's disease (AD) or Parkinson's disease (PD), and stroke. These disorders impair perception, cognition, and motor function, leading to reduced driver fitness and increased crash risk. Medical diagnosis or age alone is not reliable enough to predict driver safety or crashes and there is no effective rehabilitation for these drivers. *Role of Driving Simulators.* Driving simulators can help characterize patterns of driver error in a roadway environment that are challenging but safe compared to driving on the open road. There is preliminary evidence that simulators can be used in driver rehabilitation in neurological disorders (e.g., stroke). *Key Results of Driving Simulator Studies.* Driving simulators have been used to discriminate between the performance of drivers with AD, PD, and stroke and comparison drivers of similar age without neurological disease. Impairments in cognition, visual perception, and motor function have been the main predictors of poor simulated driving outcomes. *Scenarios and Dependent Variables.* A number of scenarios and the dependent variables can be used to differentiate between normal older drivers and those with PD, AD, and stroke, depending on the specific questions to be answered. *Platform Specificity and Equipment Limitations.* Scenarios to discriminate between safe and unsafe drivers with neurological disease can be implemented on a variety of simulators. The relationships between performance on these different platforms and real-world driving performance is under study.

48.1 Introduction

With increased longevity, the number of older drivers at risk for Alzheimer's disease (AD), Parkinson's disease (PD), and stroke is increasing. These conditions impair perception, cognition and motor function, which can reduce driver performance and increase the risk of errors that lead to crashes. Table 48.1, 48.3, and 48.4 show our approach to testing cognitive, visual, and motor testing and the comparisons of drivers with AD or PD to elderly controls. A clinician may be able to identify potentially hazardous drivers, but clinical judgment may fail, particularly in drivers with mild cognitive decline (Ott et al., 2005). Type and

48-1

degree of cognitive, visual, and motor impairments increase the accuracy of predictions of driving skills in older drivers with medical disorders (Fitten et al., 1995) and can be combined with evidence from state driving records and road tests including those administered in instrumented vehicles. Driving simulators provide an ideal platform for assessing the driving ability of drivers with PD, AD, and stroke. There is preliminary evidence to suggest that drivers with stroke can benefit from a rehabilitation program using simulator training (Akinwuntan et al., 2006).

Medical disorders affect several abilities that are important for safely driving a motor vehicle (see, this book, chap. 46 by Rizzo on medical disorders). The risk of driver errors increases with deficits of attention, perception, memory, decision-making, response implementation and awareness. Impairments in alertness occur in sleep disorders or with medication side effects (e.g., dopaminergic agonists in PD [Meindorfner et al., 2005; Moller et al., 2002] or anxiolytic agents).

Useful scenarios to test the consequences of cognitive, visual, and motor deficits in AD, PD, and stroke can be developed based on epidemiologic evidence of at-risk situations or to probe specific mechanisms of driver error. For example, drivers with visual perception, attention, and executive function problems are particularly challenged by intersections (Rizzo, McGehee, Dawson, & Anderson, 2001), where they may need to react to other vehicles or changing traffic signals. Dependent measures may include a behavioral outcome (e.g., crash or not, obeying the signal or not), reaction times (time-to-braking, releasing gas) and distances, qualities of the reaction (e.g., abrupt versus gradual), or speed at collision if one occurs. Some other useful scenarios include driving on curves, merging and car following (e.g., Dastrup, Lees, Dawson, Lee, & Rizzo, 2009). The scenarios can be made more challenging by decreasing the visual contrast in the environment (Uc et al., 2009a) or adding secondary tasks such as arithmetic tasks, which create a processing load tantamount to engaging in a conversation (Rizzo et al., 2004; Uc et al., 2006b), and route finding tasks that resemble the real-world situation in which a driver must follow verbal directions to a destination and put demands on driver memory, attention, executive functions, and perception (Uc et al., 2004a, b, 2007, 2009b). In such studies comparison drivers of similar age with no neurological disease are needed to discern if the performances of diseased drivers are abnormal and what the characteristics of their driving impairments are. Correlational analysis and modeling of outcomes can be performed using demographic data (e.g., age, education, gender) and indices of disease (e.g., neuropsychological, visual, motor tests) as independent variables to determine the subset of patients at greatest risk (see, this book, chap. 22 by Dawson; chap. 15 by McGwin, and chap. 21 by Boyle). Our general approach to prediction of driver safety is shown in Figure 48.1.

In the following sections, we will present the general characteristics of AD, PD, and stroke, and discuss the results and implications of driving research (road and simulator testing) in these populations.

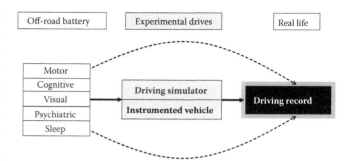

FIGURE 48.1 Flow diagram of the hypothetical relationships between the predictor measures in the *off-road battery*, performance on the *experimental drives*, and driving safety in *real life*, as represented in the state crash records. The simulator allows testing of driving behavior under high-risk conditions, whereas the instrumented vehicle provides insights into natural driving circumstances and tasks.

48.2 Alzheimer's Disease

48.2.1 General Information and Road Driving Performance

AD is the most common form of dementia with a much-increased prevalence with older age (Blennow, de Leon, & Zetterberg, 2006). AD increases crash rates compared to similar aged non-demented drivers, particularly in the few years surrounding diagnosis (Drachman & Swearer, 1993). Prospective longitudinal studies show that drivers with AD are more likely to stop driving compared to controls (Duchek, Hunt, Ball, Buckles, & Morris, 1998; Ott et al., 2008). Greater dementia severity, increased age, and lower education were associated with higher rates of failure and poorer performance on the road test (Ott et al., 2008).

Drivers with AD have also been shown to perform poorly on specific tasks. Standardized road tests in an instrumented vehicle have shown that drivers with AD had more difficulty multitasking (Rizzo et al., 2004), following a route (Uc, Rizzo, Anderson, Shi, & Dawson, 2004b), and identifying landmarks and traffic signs (Uc, Rizzo, Anderson, Shi, & Dawson, 2005c), and committed more at-fault safety errors than controls during the drive (Dawson, Anderson, Uc, Dastrup, & Rizzo, 2009), although their basic vehicular control abilities were comparable. The performance and errors of drivers with AD (also of PD and stroke) during these tasks were predicted by scores on standardized visual, cognitive, and motor tests conducted off-road (see Table 48.1). However, these predictors usually explained less than 50% of variability, consistent with meta-analysis of (Reger et al., 2004), which found significant relationships between cognitive measures and driving measures across dementia and control driver groups. However, within dementia there were moderate correlations between visuospatial skills and attention/concentration and driving outcomes (Reger et al., 2004).

48.2.2 Simulation

This section discusses use of simulators in AD. Driving simulators offer a more controlled and safer testing environment than testing on the road. Most studies cited on AD below (Table 48.2), as in PD

TABLE 48.1   Cognitive and Visual Test Battery

Abbreviation	Full Name of the Test	Measured Function
AVLT	Rey Auditory Verbal Learning Test (Rey, 1964; Taylor, 1959)	Anterograde Verbal Memory
BLOCKS	Block Design subtest from the WAIS-R (Wechsler Adult Intelligence Scale-Revised) (Wechsler, 1981)	Nonverbal Intellect
BVRT	Benton Visual Retention Test (Benton, 1974)	Visual Perception, Visual Memory, and Visuoconstructive Abilities
CFT-COPY	Rey-Osterreith Complex Figure Test (CFT) Copy version (Osterrieth, 1944)	Visuoconstructional Ability
CFT-RECALL	Rey-Osterreith Complex Figure Test Recall version (Osterrieth, 1944)	Nonverbal Anterograde Memory
COWA	Controlled Oral Word Association Test (Benton & Hamsher, 1978)	Executive Functions Using a Verbal Task
CS	Contrast sensitivity (Pelli et al., 1988)	Low to Medium Spatial Frequency Sensitivity
NVA	Near visual acuity (Ferris III, Kassoff, Bresnick, & Bailey, 1982)	
JLO	Judgment of Line Orientation (Benton, Hannay, & Varney, 1975)	Visual Perception & Spatial Abilities
SFM	Perception of 3-dimensional structure-from-motion and of motion direction (Rizzo & Nawrot, 1998)	
TMT-B	Trail-Making Test subtest B (Reitan & Davison, 1974)	Executive Functions Using a Nonverbal Task
UFOV	Useful Field of View (correlates with increased crash risk in a simulated driving scenarios and real life crashes) (Ball, Owsley, Sloane, Roenker, & Bruni, 1993; Owsley, Ball, Sloane, Roenker, & Bruni, 1991)	Speed of Visual Processing, Divided Attention, and Selective Attention
GUG	Get-Up-Go	Mobility
FR	Functional Reach	Mobility
PEG	Grooved Pegboard task	Motor Dexterity

Source: Uc, E. Y., Rizzo, M., Anderson, S. W., Shi, Q., & Dawson, J. D. (2005b). Driver identification of landmarks and traffic signs after a stroke. *Transportation Research Record*, 1922, 9–14. With permission.

Note: See Lezak 1995 and Spreen and Strauss 1991 for details.

and stroke, have used fixed-base simulators with varying degrees of fidelity. The authors reported either outcomes of a whole drive or selected scenarios from a drive and tried to find associations between predictor variables characterizing cognitive, visual, motor dysfunction in the respective disorders and driving outcomes. Associations between simulator outcomes and road tests or prospective crash/citation records were generally not covered.

In a group of 21 licensed drivers with AD and 18 controls, 29% of those with AD experienced crashes versus none of the control participants in the Iowa Driving Simulator (Rizzo, Reinach, McGehee, & Dawson, 1997). (Please see the *Handbook* web site

for a more detailed description of the Iowa Driving Simulator and the specific driving scenarios). Drivers with AD were more than twice as likely to experience close calls. Visuospatial impairment, reduction in the useful field of view, and reduced perception of three-dimensional structure-from-motion predicted crashes (Rizzo, Reinach, McGehee, & Dawson, 1997). In a follow-up study, during an illegal intersection incursion scenario (Figure 48.2; Web Figure 48.1), 6 out of 18 drivers with AD experienced crashes versus none of 12 non-demented drivers of similar age. Plots of control over steering wheel position, brake and accelerator pedals, vehicle speed, and vehicle position during the five seconds preceding a

TABLE 48.2   Simulator Studies in Alzheimer's Disease

Scenario	Dependent Measure	Subjects	Outcome	Reference
Intersection behavior	Crash/close call	21 AD, 18 controls	Increased crashes in AD	(Rizzo et al., 1997)
Illegal intersection incursion	Crash	18 AD, 12 controls	Increased crashes in AD	(Rizzo et al., 2001)
Common driving tasks	Task time, errors	29 AD, 21 controls	Poorer vehicle control in AD	(Cox et al., 1998)
Whole drive with several scenarios	Composite score	70 AD, 132 controls	Worse composite score in AD	(Anderson et al., 2005).
Rear-end collision avoidance	Crash, unsafe reaction	61 AD, 115 controls	Crash count similar, but more unsafe reactions in AD	(Uc et al., 2006a)
Approaching an emergency scene	Reaction time, behavior	48 AD, 101 controls	Slower RT and more improper behavior in AD	(Rizzo et al., 2005)

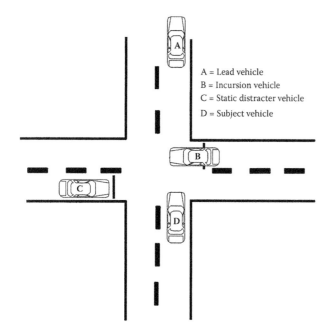

A = Lead vehicle
B = Incursion vehicle
C = Static distracter vehicle
D = Subject vehicle

FIGURE 48.2 Schematic depiction of a simulated intersection incursion. The incurring vehicle began on the subject's right side with its front bumper 48.5 ft from the yellow centerline, and accelerated at 13.8 ft/sec² for 2.31 seconds. It decelerated and stopped with its front bumper on the yellow centerline three seconds after beginning the incursion. This completely blocked the subject's lane. (From Rizzo, M., McGehee, D. V., Dawson, J. D., & Anderson, S. N. (2001). Simulated car crashes at intersections in drivers with Alzheimer disease. *Alzheimer Disease & Associated Disorders, 15,* 10–20. With permission.)

crash event showed inattention and inappropriate or slow control responses (Figure 48.3). Predictors of combined crashes in the illegal intersection incursion and rear-end collision (REC) avoidance scenarios included visuospatial impairment, disordered attention, reduced processing of visual motion cues, and overall cognitive decline (Rizzo, McGehee, Dawson, & Anderson, 2001).

A comparison of 29 outpatients with probable AD with 21 age-matched control participants on a driving simulator revealed that patients with AD had cognitive difficulty operating the simulator, drove off the road more often, spent more time driving slower than the posted speed limit, applied less brake pressure in stop zones, spent more time negotiating left turns, and drove more poorly overall. The AD patients who could not drive the simulator because of confusion and disorientation or had stopped driving already had lower Mini-Mental State Examination (MMSE) scores. The computed total driving score correlated significantly with MMSE scores (Cox, Quillian, Thorndike, Kovatchev, & Hanna, 1998).

A study of 70 drivers with mild AD and 132 elderly controls showed that the simulator composite score, reflecting overall driving ability, was significantly correlated with overall cognitive ability, as indexed by the neuropsychology composite score, as well as with individual cognitive tests of attention, memory, visuospatial and visuomotor abilities. Drivers who crashed during an intersection incursion scenario performed significantly

worse on the composite measure of cognitive function than those who successfully steered around the incurring vehicle. Memory test performances for both verbal information and visual material were associated with subsequent on-road crashes over a follow-up period of two years as confirmed by DOT records (Anderson, Rizzo, Shi, Uc, & Dawson, 2005).

Response to an emergency vehicle requires detection and recognition of an object in peripheral vision, situation recognition, and a rapid response to execute a safety maneuver to decrease the potential for crashing into the vehicle or striking people situated near it (Rizzo et al., 2005). These demands require an intact visual perception, attention, motor function, and executive functions, all potentially abnormal in AD. Forty-eight drivers with AD and 101 controls were tested using a high-fidelity interactive driving simulator, SIREN (Simulator for Interdisciplinary Research in Ergonomics and Neuroscience: see the *Handbook* web site for details on the simulator and scenario) in a scenario in which drivers encountered a police car on the shoulder of the road (Figure 48.4). Compared with controls, drivers with Alzheimer's reacted more slowly—with abrupt decelerations resulting—or failed to steer clear of the police car. Several older drivers stopped in the middle of the road. Examples of poor performance are seen in Figure 48.3. Poorer performance on tests of perception, attention, memory, and executive function predicted slower first reactions and increased the risk of inappropriate and potentially unsafe reactions. The findings suggest that there is decreased situation awareness or poor executive control over response implementation in older drivers with cognitive decline. Other simulator scenarios on the effect of decreased situation awareness and executive dysfunction in cognitively-impaired drivers can be designed to further explore the consequences of these deficits on driving safety and the predictors of the associated poor outcomes.

Rear-end collision (REC) avoidance was tested in 61 drivers with mild Alzheimer's disease (AD) and 115 elderly controls in the SIREN. After a segment of uneventful driving, each driver suddenly encountered a lead vehicle stopped at an intersection, creating the potential for a collision with the lead vehicle or with another vehicle following closely behind the driver (see the *Handbook* web site for further description). Eighty-nine percent of drivers with AD had unsafe outcomes, either an REC or risky avoidance behavior (defined as slowing down abruptly or prematurely, or swerving out of the traffic lane) compared to 65% of controls. Crash rates were similar in AD (5%) and controls (3%) (see Figure 48.5 for a crash depiction), yet a greater proportion of drivers with AD slowed abruptly (70% versus 37%) or prematurely (66% versus 45%). Abrupt slowing might increase the odds of REC by the following vehicle (Figure 48.4). Unsafe outcomes were predicted by tests of visual perception, attention, memory, visuospatial/constructional abilities, and executive functions, as well as vehicular control measures during an uneventful driving segment (Uc, Rizzo, Anderson, Shi, & Dawson, 2006a). The most important of unsafe outcomes were the Copy and Recall subtests of the Rey-Osterrieth Complex Figure Test, showing the importance of visuospatial abilities in avoiding a crash or other

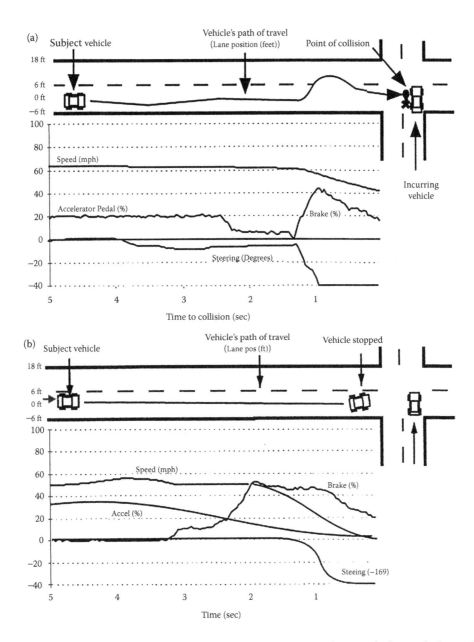

FIGURE 48.3 Plots of driver control in the simulated intersection incursion. Common ordinate scale shows vehicle speed, percentage of pedal application for accelerator and brake, and steering wheel rotations in degrees (upward deflections are CCW rotations). Path and lane positions of driver and other vehicles are depicted to scale at top. (a) S52, a patient with Alzheimer disease. (b) S105, a control subject. The control subject began braking immediately after letting off the accelerator. (From Rizzo, M., McGehee, D. V., Dawson, J. D., & Anderson, S. N. (2001). Simulated car crashes at intersections in drivers with *Alzheimer disease*. *Alzheimer Disease & Associated Disorders, 15*, 10–20. With permission.)

unsafe response in a complex driving situation approaching an intersection.

48.3 Parkinson's Disease

48.3.1 General Information and Road Driving Performance

This section reviews the performance and safety of drivers with PD, using an approach similar to the one used to evaluate drivers

with AD. In addition to hallmark motor impairments, patients with PD have cognitive and visual dysfunction (Table 48.3) (Uc et al., 2007), similar to, but milder than those in AD (Table 48.4) (Uc et al., 2006a).

PD has a prevalence of ~0.3% in the general population and of 3% in people over the age of 65 years (Lang & Lozano, 1998). PD has been recognized primarily as a motor disorder causing tremors, bradykinesia, rigidity, gait and balance problems. However, many patients with PD, including those in the early stages of the disease, suffer from cognitive (especially executive and visuospatial)

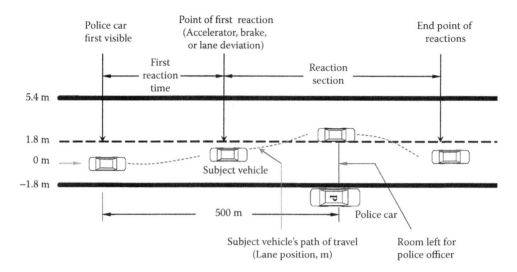

FIGURE 48.4   Response to emergency vehicle parked roadside. (From Rizzo, M., Shi, Q., Dawson, J. D., Anderson, S. W., Kellison, I., & Pietras, T. (2005). Stops for cops: Impaired response implementation for older drivers with cognitive decline. *Transportation Research Record*, 1992, 1–8. With permission.)

dysfunction (Uc, Rizzo, Anderson, Shi, & Dawson, 2005a). Even in the absence of dementia, cognitive dysfunction in PD can impair critical activities of daily living such as driving (Uc et al., 2007).

There are no well established epidemiologic data on crash risk in PD (Homann et al., 2003); however, PD appears to be associated with decreased driving and increased crashes, especially in those with worse motor and cognitive dysfunction (Dubinsky et al., 1991).

Excessive daytime sleepiness (EDS) can occur with dopaminergic medications and in PD itself, and affects a substantial proportion of PD patients (Comella, 2002). Isolated sleep attacks without any prior sleepiness are rare. EDS is associated with increased crashes in PD (Meindorfner et al., 2005) (see also, this book, chap. 49 by Tippin).

Studies testing road driving performance of patients with PD found increased safety errors (Heikkila, Turkka, Korpelainen, Kallanranta, & Summala, 1998; Wood, Worringham, Kerr, Mallon, & Silburn, 2005; Worringham, Wood, Kerr, & Silburn, 2006). Compared to controls, drivers with PD performed worse on tasks of route following, search for roadside landmarks and traffic signs, and driving during auditory-verbal distraction (Uc et al., 2007; Uc et al., 2006b, c). The decrease in road driving performance and safety in PD is related to poor performances on cognitive and visual tests (Heikkila, Turkka, Korpelainen, Kallanranta, & Summala, 1998; Wood, Worringham, Kerr, Mallon, & Silburn 2005; Worringham et al., 2006; Uc et al., 2007; Uc et al., 2006b, c; Devos et al., 2007) and by motor dysfunction (Radford, Lincoln, & Lennox, 2004; Devos et al., 2007).

48.3.2 Simulation

A summary of driving simulation studies can be found in Table 48.5 Earlier studies found decreased steering accuracy, slower driving reaction times, and traffic signal errors (e.g., missing red lights) in drivers with PD compared to controls (Madeley, Hulley,

Wildgust, & Mindham, 1990; Lings & Dupont, 1992). Moller et al. (2002) evaluated six PD patients with a history of sudden onset of sleep using a sleep-wake diary, the ESS, polysomnography, multiple sleep latency tests (MSLT), a vigilance test, and driving simulation. In all patients, ESS scores were increased and polysomnography showed sleep problems. Pathological results in the MSLT or the vigilance test were obtained in five cases. Three of these five patients had increased mean standard deviation of lane position (SDLP) values on the simulator (Moller et al., 2002).

Zesiewicz et al. (2002) compared the driving ability of 39 PD subjects (including seven non-drivers) with 25 control participants using a driving simulator, manufactured by FAAC Corp (Ann Arbor, MI). The PD group (including those who limited or stopped driving) had more collisions on the driving simulator than control drivers. PD subjects who were still driving (with and without limitations) had more simulator collisions than control drivers. PD drivers who reported no change in driving showed a trend toward having more simulator collisions than control drivers. Collisions correlated with higher motor severity and lower MMSE scores (Zesiewicz et al., 2002).

One of the key features of cognitive impairment in PD is the difficulty to adapt to a new situation without external guidance due to attentional and executive deficits, a defect in "self-directed task specific planning" or "internal cuing" (Brown & Marsden, 1988). Stolwyk, Triggs, Charlton, Iansek and Bradshaw (2005) examined the impact of impaired internal cuing (the ability to generate cues internally based on memory, rather than upon external cues such as warnings and signs) in PD on specific driving behaviors (see the *Handbook* web site for details on the simulation and study design). A number of curves and traffic signals were included in each simulation, either with a preceding warning (e.g., arrow road sign before curves-external cue condition) or without a warning (internal cue condition). Drivers with PD tended to approach traffic signals at lower speeds, decelerated

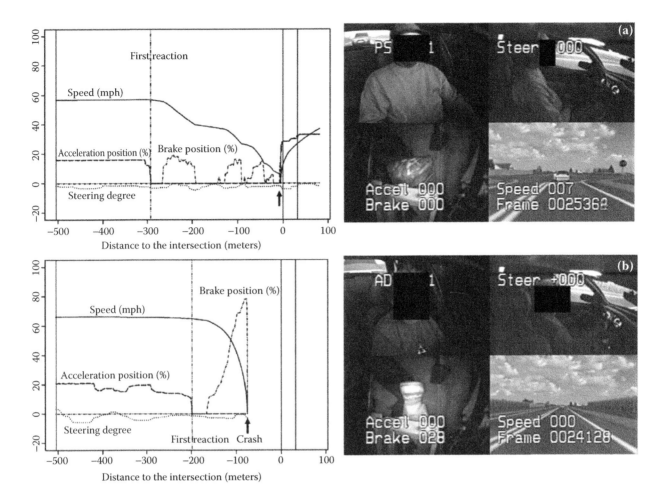

FIGURE 48.5 Plots of driver control over the vehicle and multiplex view in the moments preceding the end of rear-end collision avoidance scenario are shown in two different drivers. The arrow (↑) indicates the temporal position of the multiplex image in the rear-end collision avoidance scenario. In each case, the driver is approaching an intersection where a stopped lead vehicle is signaling to turn to the left. A sports utility vehicle (the following vehicle—FV) is following the driver. The common ordinate scale shows the driver's vehicle speed, percentage of pedal application for accelerator and brake, and steering wheel rotations in degrees (upward deflections are counter-clockwise rotations). The multiplex view shows four channels of video data with superimposed digital driving data. The upper left panel shows the driver's face and the view through the rear window (as seen in the rear-view mirror). The upper-right panel provides an over-the-shoulder view of the driver's actions with superimposed steering wheel position (degrees). The lower left panel provides a record of the subject's control of the foot pedals with superimposed % application of the brake and accelerator pedals. The lower right panel displays the view of the forward roadway that the driver should see, with superimposed speed (mph). In a, a control driver starts reacting about 300 m before the intersection by gradually releasing the accelerator and later pumping the brake resulting in a smooth deceleration and stop. The multiplex view shows he is gliding at 7 mph to a full stop while the lead vehicle is entering the intersection and the SUV is following at a safe distance. In b, a driver with AD releases the accelerator suddenly only 200 m before the intersection. She slams on her brakes, resulting in an abrupt and premature stop (80 m before the intersection) and is struck by the SUV which looms behind. (Reprinted from Uc, E. Y., Rizzo, M., Anderson, S. W., Shi, Q., & Dawson, J. D. Unsafe rear-end collision avoidance in Alzheimer's disease, *Journal of Science, 251*, 35–43, 2006, with permission from Elsevier.)

significantly later, and experienced significantly more difficulty stopping in time. Despite driving slower on curves, drivers with PD had higher variability of lane position. However, their performance improved when external cues were provided, showing the effect of attentional and executive dysfunction on the regulation of driving behavior (Stolwyk, Triggs, Charlton, Iansek, & Bradshaw, 2005).

The same group of subjects was compared using a concurrent task which was manipulated between conditions during simulated driving. Although groups were similarly affected by the concurrent task

on most driving measures, participants with PD started to decelerate later (i.e., closer to the traffic signal) compared to control participants (Stolwyk et al., 2006b). Within the PD group, executive dysfunction affecting working memory, planning, and set shifting were associated with reduced speed adaptation and curve navigation. Impaired information processing, visual attention, and visual perception in people with PD was associated with reacting to road obstacles and maintaining constant lane position. Few correlations were found between measures of physical mobility and psychomotor speed with driving measures (Stolwyk, Charlton, Triggs, Iansek, & Bradshaw, 2006a).

TABLE 48.3 Comparison of Patients With Parkinson's Disease (*n* = 76) and Controls (*n* = 152) Using Wilcoxon Rank Sum Test. Values Represent Mean ± SD (Median)

Off-Road Battery			Parkinson's Disease	Controls
	Demographics	Age	65.9 ± 8.6 (66.0)	65.3 ± 11.5 (68.0)
		Education (years)	14.8 ± 2.9 (13.0)*	15.6 ± 2.6 (16.0)
		Gender	(62 Male, 15 Female)***	(75 Male, 77 Female)
		Familiarity	25 Familiar***	110 Familiar
Category	Function	Measure		
Basic Visual Sensory Functions	Near Visual Acuity	Snellen Chart- logMAR	0.069 ± 0.070 (0.073)***	0.021 ± 0.041 (0.000)
	Far Visual Acuity	ETDRS Chart- logMAR	−0.029 ± 0155 (0.000)***	−0.085 ± 0.124 (-0.080)
	Contrast Sensitivity	Pelli-Robson Chart (↑)	1.70 ± 0.15 (1.65)***	1.84 ± 0.15 (1.95)
Visual Perception	Motion Perception	SFM % (↓)	12.6 ± 5.3 (11.9)**	10.4 ± 2.8 (10.1)
	Attention	UFOV (↓)	881 ± 371 (824)***	634 ± 247 (610)
	Spatial Perception	JLO (↑)	24.4 ± 4.3 (25.0)*	25.7 ± 3.9 (27.0)
	Construction	BLOCKS (↑)	33.1 ± 11.9 (30.0)***	39.9 ± 10.7 (40.0)
Visual Cognition		CFT-COPY (↑)	26.9 ± 5.1 (28.0)***	31.8 ± 3.8 (33.0)
	Memory	CFT-RECALL (↑)	13.2 ± 5.4 (13.3)**	15.8 ± 5.9 (15.0)
		BVRT-error (↓)	7.4 ± 4.2 (7.0)***	4.6 ± 2.6 (4.0)
Executive Functions	Set Shifting	TMT(B-A) (↓)	83.8 ± 80.3 (59.6)***	46.4 ± 34.6 (37.5)
	Verbal Fluency	COWA (↑)	34.5 ± 11.3 (33.0)**	39.1 ± 11.1 (40.0)
Verbal Memory		AVLT-RECALL (↑)	7.3 ± 3.6 (7.0)***	10.2 ± 3.2 (11.0)
Depression		GDS (↓)	6.0 ± 5.5 (5.0)***	3.3 ± 3.6 (2.0)
Balance		FR (↑)	11.5 ± 3.2 (11.0)***	13.4 ± 2.7 (13.5)

Source: Reprinted from Uc, E. Y., Rizzo, M., Anderson, S. W., Shi, Q., & Dawson, J. D. Unsafe rear-end collision avoidance in Alzheimer's disease, *Journal of Science, 251*, 35–43, 2006, with permission from Elsevier.

↑ = Higher Score Better, ↓ = Lower Score Better. *$p < 0.05$, **$p < 0.01$, ***$p < 0.001$

The addition of driving simulation assessment to a clinical screening battery (disease duration, contrast sensitivity, Clinical Dementia Rating, and motor part of the Unified Parkinson's Disease Rating Scale) increased the sensitivity and specificity of off-road testing in predicting pass/fail status of drivers with PD on an official road test (Devos et al., 2007). The details of this simulator and the drive, which was also used in a stroke study (Akinwuntan et al., 2005), are described below and in the supplemental materials on the *Handbook* web site.

To assess crash risk of drivers with Parkinson's disease (PD) during driving simulation using SIREN (Uc et al., 2006a) under low-visual contrast conditions, 67 drivers with mild-moderate PD and elderly controls (*n* = 51) drove under foggy conditions in the simulator, where the driver's approach to an intersection triggered an illegal incursion by another vehicle causing crash risk. Compared to 37.3% in controls, 76.1% of drivers with PD crashed. The time to first reaction (releasing accelerator, braking, or steering away) of drivers with PD was slower than that of controls (median 2.7 versus 2.1 seconds), which correlated with worse scores on tests of visual perception and memory, visuospatial abilities, executive functions, motor severity, and postural stability within the PD group. Prolonged reaction times and decreased performance on tests of contrast sensitivity, perception of structure from motion, visual speed of processing and attention (useful field of view), and visuospatial perception predicted crashes within the PD group. Motor dysfunction contributed to slowing of time to first reaction (Uc et al., 2009a).

In summary, patients with PD suffer from a varying combination of motor, cognitive, and visual dysfunction. Although PD is most recognized for its motor disorder with typical tremors, impairments of cognition and visual perception were the main determinants of performance and safety of drivers with PD.

48.4 Stroke

48.4.1 General Information and Road Driving Performance

Stroke is the third most frequent cause of death after heart disease and cancer in the U.S.; each year 700,000 people experience a stroke, and the prevalence of stroke is 2.5–3% in the general population. Eighty-seven percent of strokes are ischemic (i.e., infarction due to occlusion of an artery in the brain) with the remainder being brain hemorrhage (Rosamond et al., 2007). The length of time to recover from a stroke depends on its severity. From 50% to 70% of stroke survivors regain functional independence, but 15% to 30% are left permanently disabled, and 20% require institutional care at three months after onset. The proportion of individuals who are disabled after a stroke increases after the age of 65. The functional limitations depend on the location and size of the stroke and include hemiparesis, aphasia, visual field loss, cognitive impairment, gait/balance difficulties, and depression.

A community-based study found that 58% of drivers who suffered a stroke did not resume driving one year post-stroke

TABLE 48.4 Comparison of AD and Control Groups Using Wilcoxon Rank Sum Test. Values Expressed as Means (SD). M = Male, F = Female

	AD	Normal	P-values
Sample size	61	115	
	44 M, 17 F	58 M, 57 F	0.0064
Demographics	mean (SD)		
Age	73.5(8.5)	69.4(6.7)	0.0002
Education (years)	14.9(3.1)	15.5(2.6)	0.0966
Driving exposure (miles/week)	72(74)	159(189)	<0.0001
Cognition			
COGSTAT	285(59)	393(44)	<0.0001
AVLT-RECALL	3.0(2.7)	9.8(3.0)	<0.0001
BVRT	9.6(3.6)	5.0(2.7)	<0.0001
CFT-RECALL	8.5(4.5)	15.3(5.4)	<0.0001
JLO	22.3(4.9)	25.2(4.1)	<0.0001
BLOCKS	22.2(12.8)	37.9(10.2)	<0.0001
CFT-COPY	27.3(6.5)	31.5(4.1)	<0.0001
TMT-A	61(32)	37(13)	<0.0001
TMT-B	174(86)	86(43)	<0.0001
COWA	30.1(12.5)	38.0(11.5)	<0.0001
Vision			
NVA	0.058(.061)	0.024(.044)	<0.0001
FVA	0.030(.144)	−0.060(.124)	<0.0001
CS	1.64(.21)	1.82(.15)	<0.0001
SFM	12.7(7.3)	10.2(2.9)	0.1933
UFOVTOT	1279(312)	700(223)	<0.0001
Motor Function			
PEG	117(47)	86(18)	<0.0001
GUG	11.2(2.4)	8.8(2.6)	<0.0001
FR	11.5(2.7)	13.3(2.7)	<0.0001

Source: Uc, E. Y., Rizzo, M., Anderson, S. W., Sparks, J. D., Rodnitzky, R. L., & Dawson, J. D. Impaired navigation in drivers with Parkinson's disease, *Brain*, 2007 by permission of Oxford University Press.

due to mobility and cognitive issues. Stopping driving was associated with a loss of social activities and with a higher frequency of depression among ex-drivers, although 79% of ex-drivers had easy access to alternative car transport (Legh-Smith, Wade, & Hewer, 1986). Besides limb weakness, hemianopsia due to stroke impairs driving ability (Szlyk, Brigell, & Seiple, 1993).

Objective driving evaluations are needed in stroke survivors, as the patients themselves and their family may have a tendency to overestimate driving ability (Heikkila, Korpelainen, Turkka, Kallanranta, & Summala, 1999). The navigational abilities in a route-following task (Uc, Rizzo, Anderson, Shi, & Dawson, 2004a) and visual search abilities in a landmark/traffic sign identification task (Uc et al., 2005c) during an instrumented vehicle drive were impaired in chronic stroke patients. As with AD and PD groups, performance on these tasks and safety errors were predicted deficits in vision and cognition, rather than in motor function.

The Stroke Drivers Screening Assessment (SDSA) consists of subtests such as dot cancellation test, square matrix test, and the road sign recognition test, and was found to have an accuracy of 81% in predicting driver fitness in stroke patients

(Nouri & Lincoln 1993). A combination of visual neglect, complex figure of Rey, and on-road tests was found to be the model that best predicted fitness to drive after stroke during a later official assessment (Akinwuntan et al., 2006).

48.4.2 Simulation

Hemiparesis (weakness on one side) after stroke was associated with longer reaction times in a simulator, not merely for the paretic but also for the contralateral extremities. Neuropsychological deficits correlated with the number of driving errors, but to only a minor extent with reaction times (Lings & Jensen, 1991). Patients with anterior circulation strokes were found to have increased crash rates and poorer performance in a driving simulator compared to posterior circulation strokes and controls (Kotterba, Widdig, Brylak, & Orth, 2005).

Thirty-eight drivers with stroke were assessed in a driving simulator using the P-Drive (Performance Analysis of Driving Ability) tool (Patomella, Tham, & Kottorp, 2006) and with an Assessment of Awareness of Disability tool, to measure the discrepancy between observed driving actions and self-report of problems after a driving simulator evaluation. Lack of awareness of driving impairment and cognitive screening outcome explained 74% of the variance in driving ability, suggesting that stroke patients who fail a driving evaluation often have limited awareness of their impairment (Patomella, Tham, & Kottorp, 2008).

48.5 Challenges and Future Directions

48.5.1 Technical Issues

There are no reports of increased simulator adaptation syndrome in AD, PD, or stroke. Using a Simulator Adaptation Questionnaire tool (Rizzo et al., 2003) we found no significant differences in simulator adaptation in between drivers with AD or PD and controls (Uc et al., 2006a, 2009a). Avoiding cluttered scenes and sudden turns and stops and acclimatization before testing may help prepare drivers to complete simulator testing (Rizzo, 2004). Elderly patients usually have limited computer or virtual reality exposure and it helps to explain the circumstances of simulation. Subjects in later stages of PD may have diurnal fluctuations in their function. Laboratory personnel may inquire about these fluctuations to enable testing when the patient would feel ready to drive in natural settings (Uc et al., 2007).

48.5.2 Effect of Therapeutics

The development of central acetylcholine esterase inhibitors in the 1990s represented a modest first step in the treatment of cognitive impairments in AD and PD (Emre et al., 2004). Driving simulation could help assess if drugs used to treat dementia improve driver performance or if the side effects of psychoactive drugs degrade driving ability. For example, dopaminergic medications in PD may worsen sleepiness or impulsivity; systematic studies to evaluate these hypothetical effects could be conducted in a simulator.

TABLE 48.5 Simulator Studies in PD. RT = Reaction Time, SDLP = Standard Deviation of Lateral Position

Scenario	Dependent Measure	Subjects	Outcome	Reference
Common driving tasks	Reaction time, errors	10 PD, 110 controls	Poor steering, slower RT, signal violation in PD	(Madeley, Hulley, Wildgust, & Mindham, 1990)
General driving	Vigilance task, SDLP	6 PD	Decreased vigilance, increased SDLP	(Moller et al., 2002)
Collision hazards	Collision count	39 PD, 25 controls	Increased in PD	(Zesiewicz et al., 2002)
Driving on curves and approach to traffic signals with or without warning signs	Lane behavior in curves, stopping behavior at traffic signals	18 PD, 18 controls	Drivers with PD had increased SDLP on curves and difficulty with stopping at signals; more dependent on external cues	(Stolwyk et al., 2005)
Driving on curves and approach to traffic signals with or without concurrent task	Lane behavior in curves, stopping behavior at traffic signals	18 PD, 18 controls	Closer deceleration points in PD drivers before traffic signals with distracting concurrent task	(Stolwyk et al., 2006b)
Intersection behavior	Crash	67 PD, 51 controls	Increased crashes and RT in PD	(Uc et al., 2009a)

48.5.3 Use in Driving Rehabilitation

Unlike progressive neurodegenerative diseases such as AD and PD, stroke can be a monophasic illness with gradual functional recovery if underlying risk factors are controlled (e.g., high blood pressure or cholesterol, tobacco abuse) and is a target of driving rehabilitation efforts. A road-training program helped 85% of stroke patients to pass the second evaluation after initial failure (Soderstrom, Pettersson, & Leppert, 2006). A randomized study in 97 stroke survivors (UFOV training versus computerized visuoperception retraining) showed improved UFOV scores, but there were no group differences in on-road driving evaluation and tests of visual perception and attention (Mazer et al., 2003). However, there was a tendency for road test improvement after UFOV training in subjects with right-sided lesions.

Eighty-three first-ever subacute stroke patients were randomized to a five-week 15-hour training in a simulator or on cognitive tasks. The simulator training was administered in a fixed-base simulator (STI Drive System version 1.03; Systems Technology Inc.). Adaptive aids such as left-sided accelerator pedal, right-sided indicator stick, and steering spinner were used when required. The driving scenario started on a two-lane road with urban-like traffic and progressed to a four-lane highway with a 120 km/h speed restriction and possibilities to overtake other cars before terminating on a two-lane road in a rural setting. Simulator training subjects were administered different 5 km training scenarios containing common traffic demands such as lane tracking, speed control, overtaking, road sign recognition, and multitasking. Controls received standardized training by performing driving-related cognitive tasks such as route finding on a map or memory training with numbers. Both groups improved in visual and neuropsychological evaluations and in the on-road test after training. There were improvements in simulator measures after training such as decreased number of collisions, pedestrian strikes, and faults; 73% of simulator-trained and 42% of controls passed the follow-up official pre-driving assessment at six to nine months. Changes in some simulator outcome measures predicted improvements in performance on the

road. Higher level of education and lower disability from stroke predicted driving benefit from simulator training. However, the results should be interpreted cautiously due to ~30% dropout in each group and the potential of neurologic recovery unrelated to training (Akinwuntan et al., 2005).

Behaviors such as learned non-use hinder hemiplegic stroke survivors from the full use of both arms and add to disability. Active force-feedback cues to restrain the use of the less-affected arm and to increase the productive use of their impaired arm during driving simulation counteracted the tendency of hemiplegic subjects to produce counter-productive torques during steering tasks (Johnson, Van der Loos, Burgar, Shor, & Leifer, 2005).

To our knowledge, there are no published reports on driver rehabilitation in AD and PD using simulation. The main challenge is the progressive nature of these degenerative illnesses. In particular, a substantial proportion (~80%) of drivers with mild AD (CDR = 1) are likely to cease driving within two years after the baseline evaluation (Ott et al., 2008; Duchek et al., 2003). Drivers with very mild dementia (CDR = 0.5) or mild cognitive impairment may have better potential for driving rehabilitation, which remains to be tested.

PD patients usually survive for longer than AD patients (15–20 years versus 8–10 years). PD has a heterogeneous course and severe cognitive impairment occurs in a minority of patients, usually after many years of PD (Hely, Reid, Adena, Halliday, & Morris, 2008). There is potential for driving rehabilitation using simulators in PD, but there are no published studies yet. The challenge is to identify the drivers that would benefit from such intervention, to determine the remediable components of driving impairment, and to design training scenarios that are feasible and useful.

48.5.4 Medicolegal Aspects

Many people with early AD are capable of driving safely, with evidence suggesting that the risk of crashes in drivers with dementia is low for up to three years after disease onset (Breen, Breen, Moore, Breen, & O'Neill, 2007). Medical diagnosis or a clinician's assessment alone are not accurate enough to determine driving competence in those with dementia (Ott et al., 2005; Ott et al., 2008). Although neuropsychologic testing helps one to understand the associations of driver performance with cognitive impairment, a general lack of validated cutoff scores makes it impossible to employ these tests in a standardized fashion to advise patients (Molnar, Patel, Marshall, Man-Song-Hing, & Wilson, 2006a). Furthermore, there are no established guidelines on the timing of follow-up for drivers with mild dementia; recommendations range from six months or less to a full year (Molnar, Patel, Marshall, Man-Son-Hing, & Wilson, 2006b).

Although typically identified as a motor disorder, impairments in cognition and visual perception and increased daytime sleepiness are present in drivers with PD and contribute substantially to driving problems. Further research needs to be carried out in order to establish the natural course of driving abilities and to develop a reliable and efficient evaluation battery to predict driver safety in PD. For both AD and PD, there are no evidence-based useful methods for driver rehabilitation (Uc & Rizzo, 2008).

Reporting requirements for impaired drivers are not uniform across the U.S. The American Academy of Neurology (AAN) "supports optional reporting of individuals with medical conditions that may impact one's ability to drive safety, especially in cases where public safety has already been compromised, or it is clear that the person no longer has the skills needed to drive safely" (p. 1176) and advocates immunity for physicians "both for reporting and not reporting a patient's condition when such action is taken in good faith, when the patient is reasonably informed of his or her driving risks, and when such actions are documented by the physician in good faith" (p. 1177) (Bacon, Fisher, Morris, Rizzo, & Spanaki, 2007).

In summary, cognitive and visual testing, patient and care-giver attestation, driving history, and road testing help a clinician to decide on the driving fitness of a patient with AD, PD, and stroke on a case-by-case basis. Although there is potential and some preliminary data (e.g., improvement of road fitness prediction when simulator testing is added to the visual and cognitive battery in PD [Devos et al., 2007]), driving simulators alone do not have proven predictive validity in AD, PD, or stroke for real-life driving safety and outcomes. Neither, for that matter, do road tests.

Key Points

- AD, PD, and stroke affect driving performance by impairing cognition, visual perception, mood, and motor function. Medication side effects may complicate the driving performance by increasing sleepiness, and causing thought and impulse problems.
- Scenario design should aim at discerning the effect of cognitive, visual, and motor deficits on driving in these conditions and should take practical difficulties of implementation into consideration.
- Driving simulators have been able to differentiate between controls and drivers with AD, PD, or stroke, and have helped to characterize driving impairments and understand the mechanisms of driver error in these disorders.

- Driving simulators may help in assessing driver fitness and rehabilitating impaired drivers, however further research is needed to validate their predictive ability in real-life driving and driver rehabilitation potential.

Keywords: Alzheimer's Disease, Cerebrovascular Disease, Cognition, Dementia, Parkinson's Disease, Stroke, Visual Perception

Acknowledgment

Supported by NIH R01 grants NS044930, AG 17717, and 15071.

Glossary

Bradykinesia: Slowness of movement, as in Parkinson's disease or severe depression.

Cognitive impairment: Impairment in specific abilities such as memory, executive functions, attention, visual processing, and language. A driver can have cognitive impairment but not dementia. Common causes include neurological disorders, sleepiness and drug effects.

Dementia: A clinical state characterized by loss of function in multiple cognitive domains. The decline must represent a decline from a previously higher level of functioning and be severe enough to impair social and occupational functioning. The most commonly-used criteria for diagnoses of dementia is DSM-IV (Diagnostic and Statistical Manual for Mental Disorders, American Psychiatric Association).

Web Resources

The *Handbook*'s web site contains additional descriptions, and related figures, for the following:

- Iowa Driving Simulator (IDS)
- IDS illegal intersection incursion scenario description
- SIREN (Simulator for Interdisciplinary Research in Ergonomics and Neuroscience)
- Rear-end collision scenario
- Stolwyk et al. (2005): The simulator description and study design
- The CARA simulator used in articles on Parkinson's Disease (Devos et al., 2007) and stroke (Akinwuntan et al., 2005)

Key Readings

Akinwuntan, A. E., De Weerdt, W., Feys, H., Pauwels, J., Baten, G., Orno, P., et al. (2005). Effect of simulator training on driving after stroke: A randomized controlled trial. *Neurology, 65,* 843–850.

Rizzo, M., & Eslinger, P. J. (Eds.). (2004). *Principles and practice of behavioral neurology and neuropsychology.* Philadelphia, PA: WB Saunders.

Stolwyk, R. J., Triggs, T. J., Charlton, J. L., Iansek, R., Bradshaw, J. L. (2005). Impact of internal versus external cueing on driving performance in people with Parkinson's disease. *Movement Disorders, 20*, 846–857.

Uc, E. Y., Rizzo, M., Anderson, S. W., Shi, Q., & Dawson, J. D. (2006a). Unsafe rear-end collision avoidance in Alzheimer's disease. *Journal of Neurological Sciences, 251*, 35–43.

References

Akinwuntan, A. E., De Weerdt, W., Feys, H., Pauwels, J., Baten, G., Orno, P., et al. (2005). Effect of simulator training on driving after stroke: A randomized controlled trial. *Neurology, 65*, 843–850.

Akinwuntan, A. E., Feys, H., De Weerdt, W., Baten, G., Arno, P., & Kiekens, C. (2006). Prediction of driving after stroke: A prospective study. *Neurorehabilitation and Neural Repair, 20*, 417–423.

Anderson, S. W., Rizzo, M., Shi, Q., Uc, E. Y., & Dawson, J. D. (2005). Cognitive abilities related to driving performance in a simulator and crashing on the road. University of Iowa. *Proceedings of the third international driving symposium on human factors in driver assessment, training and vehicle design* (pp. 286–292). Iowa City, IA: University of Iowa.

Bacon, D., Fisher, R. S., Morris, J. C., Rizzo, M., & Spanaki, M. V. (2007). American Academy of Neurology position statement on physician reporting of medical conditions that may affect driving competence. *Neurology, 68*, 1174–1177.

Ball, K., Owsley, C., Sloane, M. E., Roenker, D. L., & Bruni, J. R. (1993). Visual attention problems as a predictor of vehicle crashes in older drivers. *Investigative Ophthalmology & Visual Science, 34*, 3110–3123.

Benton, A., Hannay, H. J., & Varney, N. R. (1975). Visual perception of line direction in patients with unilateral brain disease. *Neurology, 25*, 907–910.

Benton, A. L. (1974). *The revised visual retention test.* New York: Psychological Corporation.

Benton, A. L., & Hamsher, K. (1978). *Multilingual aphasia examination.* Iowa City, IA: University of Iowa Hospitals.

Blennow, K., de Leon, M. J., & Zetterberg, H. (2006) Alzheimer's disease. *Lancet, 368*, 387–403.

Breen, D. A., Breen, D. P., Moore, J. W., Breen, P. A., & O'Neill, D. (2007). Driving and dementia. *British Medical Journal, 334*, 1365–1369.

Brown, R. G., & Marsden, C. D. (1988). Internal versus external cues and the control of attention in Parkinson's disease. *Brain, 111*(Pt 2), 323–345.

Comella, C. L. (2002). Daytime sleepiness, agonist therapy, and driving in Parkinson disease. *Journal of the American Medical Association, 287*, 509–511.

Cox, D. J., Quillian, W. C., Thorndike, F. P., Kovatchev, B. P., & Hanna, G. (1998). Evaluating driving performance of outpatients with Alzheimer disease. *Journal of the American Board of Family Medicine, 11*, 264–271.

Dastrup, E., Lees, M. N., Dawson, J. D., Lee J. D., & Rizzo, M. (2009). Differences in simulated car following behavior of younger and older drivers. *Proceedings of the second international driving symposium on human factors in driver assessment, training and vehicle design* (76–82). Iowa City, IA: University of Iowa.

Dawson, J. D., Anderson, S. W., Uc, E. Y., Dastrup, E., & Rizzo, M. (2009). Predictors of driving safety in early Alzheimer disease. *Neurology, 72*, 521–527.

De Raedt, R., & Ponjaert-Kristoffersen, I. (2001). Predicting at-fault car accidents of older drivers. *Accident Analysis & Prevention, 33*, 809–819.

Devos, H., Vandenberghe, W., Nieuwboer, A., Tant, M., Baten, G., & De Weerdt, W. (2007). Predictors of fitness to drive in people with Parkinson disease. *Neurology, 69*, 1434–1441.

Drachman, D. A., & Swearer, J. M. (1993). Driving and Alzheimer's disease: The risk of crashes. *Neurology, 43*, 2448–2456.

Dubinsky, R. M., Gray, C., Husted, D., Busenbark, K., Vetere-Overfield, B., Wiltfong, D., et al. (1991). Driving in Parkinson's disease. *Neurology, 41*, 517–520.

Duchek, J. M., Carr, D. B., Hunt, L., Roe, C. M., Xiong, C., Shah, K., et al. (2003). Longitudinal driving performance in early-stage dementia of the Alzheimer type. *Journal of the American Geriatrics Society, 51*, 1342–1347.

Duchek, J. M., Hunt, L., Ball, K., Buckles, V., & Morris, J. C. (1998). Attention and driving performance in Alzheimer's disease. *Journals of Gerontology Series B: Psychological Sciences, 53*, 130–141.

Emre, M., Aarsland, D., Albanese, A., Bryne, E. J., Deuschl, G., De Deyn, P. P., et al. (2004). Rivastigmine for dementia associated with Parkinson's disease. *New England Journal of Medicine, 351*, 2509–2518.

Ferris F. L., III, Kassoff, A., Bresnick, G. H., & Bailey, I. (1982). New visual acuity charts for clinical research. *American Journal of Ophthalmology, 94*, 91–96.

Fitten, L. J., Perryman, K. M., Wilkinson, C. J., Little, R. J., Burns, M. M., Pachana, N., et al. (1995). Alzheimer and vascular dementias and driving. A prospective road and laboratory study. *JAMA, 273*, 1360–1365.

Heikkila, V. M., Korpelainen, J., Turkka, J., Kallanranta, T., & Summala, H. (1999). Clinical evaluation of the driving ability in stroke patients. *Acta Neurologica Scandinavica, 99*, 349–355.

Heikkila, V. M., Turkka, J., Korpelainen, J., Kallanranta, T., & Summala, H. (1998). Decreased driving ability in people with Parkinson's disease. *Journal of Neurology, Neurosurgery & Psychiatry, 64*, 325–330.

Hely, M. A., Reid, W. G., Adena, M. A., Halliday, G. M., & Morris, J. G. (2008). The Sydney multicenter study of Parkinson's disease: The inevitability of dementia at 20 years. *Movement Disorders, 23*, 837–844.

Homann, C. N., Suppan, K., Homann, B., Crevenna, R., Ivanic, G., & Ruzicka, E. (2003). Driving in Parkinson's disease—A health hazard? *Journal of Neurology, 250*, 1439–1446.

Johnson, M. J., Van der Loos, H. F., Burgar, C. G., Shor, P., & Leifer, L. J. (2005). Experimental results using force-feedback cueing in robot-assisted stroke therapy. *IEEE Transactions of Neural Systems and Rehabilitation Engineering, 13*, 335–348.

Kotterba, S., Widdig, W., Brylak, S., & Orth, M. (2005). Driving after cerebral ischemia—A driving simulator investigation. *Wien Med Wochenschr, 155*, 348–353.

Lang, A. E., & Lozano, A. M. (1998). Parkinson's disease. First of two parts. *New England Journal of Medicine, 339*, 1044–1053.

Legh-Smith, J., Wade, D. T., & Hewer, R. L. (1986). Driving after a stroke. *Journal of the Royal Society of Medicine, 79*, 200–203.

Lezak, M. D. (1995). *Neuropsychological assessment*. New York, NY: Oxford University Press.

Lings, S., & Dupont, E. (1992). Driving with Parkinson's disease. A controlled laboratory investigation. *Acta Neurologica Scandinavica, 86*, 33–39.

Lings, S., & Jensen, P. B. (1991). Driving after stroke: A controlled laboratory investigation. *International Disability Studies, 13*, 74–82.

Madeley, P., Hulley, J. L., Wildgust, H., & Mindham, R. H. (1990). Parkinson's disease and driving ability. *Journal of Neurology, Neurosurgery & Psychiatry, 53*, 580–582.

Mazer, B. L., Sofer, S., Korner-Bitensky, N., Gelinas, I., Hanley, J., & Wood-Dauphinee, S. (2003). Effectiveness of a visual attention retraining program on the driving performance of clients with stroke. *Archives of Physical Medicine and Rehabilitation, 84*, 541–550.

Meindorfner, C., Korner, Y., Moller, J. C., Stiasny-Kolster, K., Oertel, W. H., & Kruger, H. P. (2005). Driving in Parkinson's disease: Mobility, accidents, and sudden onset of sleep at the wheel. *Movement Disorders, 20*, 832–842.

Moller, J. C., Stiasny, K., Hargutt, V., Cassel, W., Tietze, H., Hermann, P. J., et al. (2002). Evaluation of sleep and driving performance in six patients with Parkinson's disease reporting sudden onset of sleep under dopaminergic medication: A pilot study. *Movement Disorders, 17*, 474–481.

Molnar, F. J., Patel, A., Marshall, S. C., Man-Son-Hing, M., & Wilson, K. G. (2006a). Clinical utility of office-based cognitive predictors of fitness to drive in persons with dementia: A systematic review. *Journal of the American Geriatrics Society, 54*, 1809–1824.

Molnar, F. J., Patel, A., Marshall, S. C., Man-Son-Hing, M., & Wilson, K. G. (2006b). Systematic review of the optimal frequency of follow-up in persons with mild dementia who continue to drive. *Alzheimer Disease & Associated Disorders, 20*, 295–297.

Nouri, F. M., & Lincoln, N. B. (1993). Predicting driving performance after stroke. *BMJ, 307*, 482–483.

Osterrieth, P. A. (1944). Le test de copie d'une figure complex: Contribution a l'etude de la perception et de la memoire. *Archives de Psychologie, 30*, 286–356.

Ott, B. R., Anthony, D., Papandonatos, G. D., D'Abreu, A., Burock, J., Curtin, A., et al. (2005). Clinician assessment of the driving competence of patients with dementia. *Journal of the American Geriatrics Society, 53*, 829–833.

Ott, B. R., Heindel, W. C., Papandonatos, G. D., Festa, K., Davis, J. D., Daiello, L. A., & Morris, J. C.(2008). A longitudinal study of drivers with Alzheimer disease. *Neurology, 70*, 1171–1178.

Owsley, C., Ball, K., Sloane, M. E., Roenker, D. L., & Bruni, J. R. (1991). Visual/cognitive correlates of vehicle accidents in older drivers. *Psychology and Aging, 6*, 403–415.

Patomella, A. H., Kottorp, A., & Tham, K. (2008). Awareness of driving disability in people with stroke tested in a simulator. *Scandinavian Journal of Occupational Therapy, 15*, 184–192.

Patomella, A. H., Tham, K., & Kottorp, A. (2006). P-drive: Assessment of driving performance after stroke. *Journal of Rehabilitation Medicine, 38*, 273–279.

Pelli, D. G., Robson, J. G., & Wilkins, A. J. (1988). The design of a new letter chart for measuring contrast sensitivity. *Clinical Vision Sciences, 2*, 187–199.

Radford, K., Lincoln, N., & Lennox, G. (2004). The effects of cognitive abilities on driving in people with Parkinson's disease. *Disability Rehabilitation, 26*, 65–70.

Reger, M. A., Welsh, R. K., Watson, G. S., Cholerton, B., Baker, L. D., & Craft, S. (2004). The relationship between neuropsychological functioning and driving ability in dementia: A meta-analysis. *Neuropsychology, 18*, 85–93.

Reitan, R. M. & Davison, L. A. (1974). *Clinical neuropsychology: Current status and applications*. New York: Hemisphere Publishing.

Rey, A. (1964). *L'examen clinique en psychologie*. Paris: Universitaires de France.

Rizzo, M. (2004). Safe and unsafe driving. In M. Rizzo & P. J. Eslinger (Eds.), *Principles and practice of behavioral neurology and neuropsychology* (pp. 197–222). Philadelphia, PA: WB Saunders.

Rizzo, M., McGehee, D. V., Dawson, J. D., & Anderson, S. N. (2001). Simulated car crashes at intersections in drivers with Alzheimer disease. *Alzheimer Disease & Associated Disorders, 15*, 10–20.

Rizzo, M., & Nawrot, M. (1998). Perception of movement and shape in Alzheimer's disease. *Brain, 121*, 2259–2270.

Rizzo, M., Reinach, S., McGehee, D., & Dawson, J. (1997). Simulated car crashes and crash predictors in drivers with Alzheimer disease. *Archives of Neurology, 54*, 545–551.

Rizzo, M., Sheffield, R. A., Stierman, L., et al. (2003). Demographic and driving performance factors in simulator adaptation syndrome. *Proceedings of the second international driving symposium on human factors in driver assessment, training and vehicle design* (pp. 201–208). Iowa City, IA: University of Iowa.

Rizzo, M., Shi, Q., Dawson, J. D., Anderson, S. W., Kellison, I., & Pietras, T. (2005). Stops for cops: Impaired response implementation for older drivers with cognitive decline. *Transportation Research Record, 1992*, 1–8.

Rizzo, M., Stierman, L., Skaar, N., Dawson, J. D., Anderson, S. W., & Vecera, S. P. (2004). Effects of a controlled auditory-verbal distraction task an older driver vehicle control. *Transportation Research Record, 1865*, 1–6.

Rosamond, W., Flegal, K., Friday, G., Furie, K., Go, A., & Greenlund, K., et al. (2007). Heart disease and stroke statistics—2007 update: A report from the American Heart Association Statistics Committee and Stroke Statistics Subcommittee. *Circulation, 115*, 69–171.

Soderstrom, S. T, Pettersson, R. P., & Leppert, J. (2006). Prediction of driving ability after stroke and the effect of behind-the-wheel training. *Scandinavian Journal of Psychology, 47*, 419–429.

Spreen, O., & Strauss, E. A. (1991). *Compendium of neuropsychological tests*. New York: Oxford University Press.

Stolwyk, R. J., Charlton, J. L., Triggs, T. J., Iansek, R., & Bradshaw, J. L. (2006a). Neuropsychological function and driving ability in people with Parkinson's disease. *Journal of Clinical and Experimental Neuropsychology, 28*, 898–913.

Stolwyk, R. J., Triggs, T. J., Charlton, J. L., Iansek, R., Bradshaw, J. L. (2005). Impact of internal versus external cueing on driving performance in people with Parkinson's disease. *Movement Disorders, 20*, 846–857.

Stolwyk, R. J., Triggs, T. J., Charlton, J. L., Moss, S., Iansek, R., & Bradshaw, J. L. (2006b). Effect of a concurrent task on driving performance in people with Parkinson's disease. *Movement Disorders, 21*, 2096–2100.

Szlyk, J. P., Brigell, M., & Seiple, W. (1993). Effects of age and hemianopic visual field loss on driving. *Optometry & Vision Science, 70*, 1031–1037.

Taylor, E. M. (1959). *Psychological appraisal of children with cerebral deficits*. Cambridge, MA: Harvard University Press.

Uc, E. Y., & Rizzo, M. (2008). Driving and neurodegenerative diseases. *Current Neurology and Neuroscience Reports, 8*, 377–383.

Uc, E. Y., Rizzo, M., Anderson, S. W., Shi, Q., & Dawson, J. D. (2004a). Route-following and safety errors by drivers with stroke. *Transportation Research Record, 1899*, 90–95.

Uc, E. Y., Rizzo, M., Anderson, S. W., Shi, Q., & Dawson, J. D. (2004b). Driver route-following and safety errors in early Alzheimer disease. *Neurology, 63*, 832–837.

Uc, E. Y., Rizzo, M., Anderson, S. W., Shi, Q., & Dawson, J. D. (2005a). Visual dysfunction in Parkinson disease without dementia. *Neurology, 65*, 1907–1913.

Uc, E. Y., Rizzo, M., Anderson, S. W., Shi, Q., & Dawson, J. D. (2005b). Driver identification of landmarks and traffic signs after a stroke. *Transportation Research Record, 1922*, 9–14.

Uc, E. Y., Rizzo, M., Anderson, S. W., Shi, Q., & Dawson, J. D. (2005c). Driver landmark and traffic sign identification in early Alzheimer's disease. *Journal of Neurology, Neurosurgery & Psychiatry, 76*, 764–768.

Uc, E. Y., Rizzo, M., Anderson, S. W., Shi, Q., & Dawson, J. D. (2006a). Unsafe rear-end collision avoidance in Alzheimer's disease. *Journal of the Neurological Sciences, 251*, 35–43.

Uc, E. Y., Rizzo, M., Anderson, S. W., Sparks, J., Rodnitzky, R. L., & Dawson, J. D. (2006b). Impaired visual search in drivers with Parkinson's disease. *Annals of Neurology, 60*, 407–413.

Uc, E. Y., Rizzo, M., Anderson, S. W., Sparks, J. D., Rodnitzky, R. L., & Dawson, J. D. (2006c). Driving with distraction in Parkinson disease. *Neurology, 67*, 1774–1780.

Uc, E. Y., Rizzo, M., Anderson, S. W., Sparks, J. D., Rodnitzky, R. L., & Dawson, J. D. (2007). Impaired navigation in drivers with Parkinson's disease. *Brain, 130*, 2433–2440.

Uc, E. Y., Rizzo, M., Anderson, S. W., Dastrup, E., Sparks, J., & Dawson, J. D. (2009a). Driving under low contrast visibility conditions in Parkinson's disease. *Neurology, 73*, 1103–1110.

Uc, E. Y., Rizzo, M., Johnson, A. M., Dastrup, E., Anderson, S. W., & Dawson, J. D. (2009b). Road safety in drivers with Parkinson disease. *Neurology, 73*, 2112–2119.

Wechsler D. (1981). *Wechsler adult intelligence scale-revised (WAIS-R®)*. The Psychological Corporation.

Wood, J. M., Worringham, C., Kerr, G., Mallon, K., & Silburn, P. (2005). Quantitative assessment of driving performance in Parkinson's disease. *Journal of Neurology, Neurosurgery & Psychiatry, 76*, 176–180.

Worringham, C. J., Wood, J. M., Kerr, G. K., & Silburn, P. A. (2006). Predictors of driving assessment outcome in Parkinson's disease. *Movement Disorders, 21*, 230–235.

Zesiewicz, T. A., Cimino, C. R., Malek, A. R., Gardner, N., Leaverton, P. L., Dunne, P. B., & Hauser, A. (2002). Driving safety in Parkinson's disease. *Neurology, 59*, 1787–1788.

49

Driving Simulation in Epilepsy and Sleep Disorders

Jon M. Tippin
The University of Iowa Hospitals and Clinics

Abstract

The Problem. Drivers with sleep disorders, such as obstructive sleep apnea (OSA), have an increased crash risk that can be difficult to predict in individual drivers. Epileptics may also have an increased risk, although the extent of that risk is not completely defined. Because seizures are intermittent events, and impairment from sleepiness varies based upon a variety of factors, determining crash risk is difficult in these patient populations. *Role of Driving Simulators.* Driving simulators offer an opportunity to study drivers with episodic impairment of function due to sleep disorders or seizures in a safe, controlled environment. The addition of divided attention tasks and physiological measures, such as the electroencephalogram (EEG), may enhance the ability to detect dangerous driving behaviors. *Key Results of Driving Simulator Studies.* Studies with driving simulators show that drivers with OSA and narcolepsy perform worse than those without neurological and sleep disorders, and their performance improves with treatment. Few data are available on drivers with epilepsy, although some preliminary studies have shown deteriorated performance due to medications and EEG-defined "subclinical" seizures. *Scenarios and Dependent Variables.* Studies of drivers with sleep disorders typically use scenarios that are monotonous, with few obstacles. Some have incorporated tests of divided attention. Standard deviation of lane position (SDLP) is the most frequently used dependent variable to differentiate drivers with sleep disorders from normal controls, although other measures, such as crashes or off-road events are also used. Little is known about appropriate scenarios and dependent variables in epileptic drivers. *Platform Specificity and Equipment Limitations.* All currently available simulators have the scenario development tools and data-recording capabilities required to perform the necessary assessments. However, standardization of these procedures does not exist at present.

49.1 Introduction

Patients with episodic neurological dysfunction form a group of particular interest to clinicians, as well as investigators studying driving performance. Of these, the two general categories that will be discussed in this chapter are sleep disorders and epilepsy. The term epilepsy comprises a number of discrete clinical syndromes that share in common the characteristic of recurrent seizures, which vary in type, severity and frequency of recurrence. Seizures, as defined here, may be thought of as discrete episodes of altered neurological function caused by paroxysmal cortical electrical discharges. Seizures may affect the entire cortex or just a part of it, and are therefore called either generalized or partial, respectively (International League Against Epilepsy [ILAE], 1981). Generalized seizures are always associated with loss of or alteration in consciousness, while partial seizures may or may not. Partial seizures unaccompanied by altered consciousness are termed simple (i.e., "simple partial seizures"), while those in which consciousness or memory are affected are called complex (i.e., "complex partial seizures"). Epilepsy is a relatively common condition that affects over two million Americans, with about 44 new cases per 100,000 reported annually (Hauser, Annengers,

& Kurland, 1993). Although epilepsy is not uncommon, chronic sleepiness is a much more pervasive problem, especially in industrialized societies. More than 20% of 532 non-shift daytime workers indicated on a self-assessment questionnaire that they were excessively sleepy, and these respondents had twice as many work-related accidents as their less sleepy co-workers (Melamed & Oksenberg, 2002). Most of these individuals do not have a sleep disorder, but are simply sleep-deprived. However, a substantial portion of sleepy individuals suffer from a sleep disorder, the most common of which that is associated with excessive daytime sleepiness (EDS) is obstructive sleep apnea (OSA). This disorder, which affects more than 2–4% of the middle-aged population (Young et al., 1993), is caused by intermittent obstruction of the upper airway during sleep leading to fragmented sleep and recurrent hypoxemia. A much less common cause of EDS is narcolepsy. Narcolepsy is a neurological disorder that—in addition to having sleepiness and sleep attacks as cardinal features—is often associated with cataplexy (episodes of muscle weakness provoked by sudden, strong emotional stimuli). Approximately 0.05% of Americans are afflicted with this disorder (Silber, Krahn, Olson, & Pankratz, 2002). As most studies investigating the relationship between sleep disorders and driving have focused on OSA and narcolepsy, the discussion in this chapter regarding sleep disorders will be limited to these two conditions. Understanding the ways in which driving performance deteriorates in patients with sleep disorders may permit generalization of these findings to the much larger population of healthy, but sleep-deprived drivers.

49.2 Sleep Disorders

49.2.1 Crash Risk in Sleepy Drivers

Driver sleepiness is a major cause of motor vehicle crashes (MVCs). Each year MVCs result in approximately 40,000 injuries and 1,500 deaths in the U.S. alone (Knipling & Wang, 1994). Royal (2003) estimated that 1.35 million drivers were involved in a drowsy driving related crash in the five years prior to a 2002 U.S. Gallup poll, and McCartt, Ribner, Pack and Hammer (1996) found that approximately 55% of 1000 surveyed drivers indicated that they had driven while drowsy, while 23% had fallen asleep at the wheel. Sleepiness may also play a role in crashes that are erroneously attributed to other causes (Connor, Whitlock, Norton, & Jackson, 2001). Although most of these crashes probably occur in otherwise healthy sleep-deprived drivers, a portion occur in those with sleep disorders, such as OSA. Drivers with OSA have a crash risk that is more than twice that of the general population (Sassani et al., 2004; Tregear et al., 2009). Fragmentation of sleep leads to chronic sleep-deprivation and EDS, which is a major cause of the cognitive dysfunction found in this population (Engleman, Kingshott, Martin, & Douglas, 2000). However, there are data suggesting that the cumulative effects of repeated episodes of nocturnal hypoxemia also cause cognitive deficits, some of which may be irreversible (Nowak, Kornhuber, & Meyrer, 2006). Regardless of the mechanism by which these deficits occur, improperly treated OSA may lead

to more than \$11 billion in MVC costs annually (Sassani et al., 2004). An additional problem is that some individuals with OSA may be unaware of their sleepiness and cognitive impairment (Engleman, Hirst, & Douglas, 1997), leading them to unwittingly engage in risky driving behavior.

49.2.2 Simulators in OSA

Driving simulators have been used to study performance in drivers impaired by OSA by a number of investigators. Early studies used simple computer game-like devices, such as "Steer Clear", in which the image of a car driving down a two-lane highway is displayed on a conventional computer monitor. During the 30-minute test, 780 obstacles appear intermittently in the vehicle's lane at intervals ranging from two seconds to two minutes. To avoid hitting an obstacle, the subject is instructed to move the car to the other lane by pressing the space bar on the keyboard. Using this device, Findley (Findley et al., 1995; Findley, Suratt, & Dinges, 1999) showed that OSA subjects made more than twice as many errors as controls, with the variance in errors in these drivers being 27 times greater than those without OSA. "Steer Clear" is primarily a vigilance task. However, safe driving requires the continuous coordination of several cognitive abilities in addition to vigilance, such as visual perception and executive functions (decision-making and implementation) (Rizzo, 2004). In an attempt to better assess these functions, devices that incorporate target tracking and divided attention tasks have been developed (George, Boudreau, & Smiley, 1996a; George, Boudreau, & Smiley, 1997). The divided attention driving test (DADT) is performed by having subjects track a randomly moving target (a small box) on a computer monitor, which randomly shifts from one side to the other. Subjects are instructed to center the cursor on the box by using the steering wheel to correct for the lateral movements. As a divided attention task, subjects are instructed to search for a target digit (e.g., "2") appearing in the corners of the monitor screen (Figure 49.1) and to push the wheel-mounted buttons corresponding to the target digit

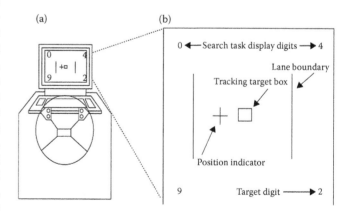

FIGURE 49.1 The divided attention driving task (DADT). See text for details. (From George, C. F. P., Boudreau, A. C., & Smiley, A. (1997). Effects of nasal CPAP on simulated driving performance in patients with obstructive sleep apnea. *Thorax, 52*, 648–653. With permission.)

when it appears. In addition to assessing tracking errors, divided attention is indexed by reaction time (RT) and divided attention index or DAI (the number of incorrect responses divided by the total). OSA drivers make significantly more tracking errors and have longer RTs than controls on the DADT, a few of whom perform even worse than some subjects intoxicated with alcohol (mean blood alcohol concentration 95 +/− 25 mg/dl) (George, Boudreau, & Smiley, 1996b).

Poor driving performance in OSA has also been demonstrated with more advanced simulators of varying sophistication, as shown in Table 49.1. Several different simulator systems and driving scenarios have been used but, as a rule, drive times typically range between 25 and 60 minutes and scenarios are usually monotonous with either few or no turns, obstacles, or

on-coming vehicles, while subjects attempt to maintain a fixed speed. Some systems utilize sound effects, such as engine noise and tire squeals, and most incorporate some type of divided attention task, like the previously described DADT. Dependent measures of performance typically include the standard deviation of lane position (SDLP), speed variability, steering rate variability, number of crashes or off-road events, RT and DAI. One of the most commonly used simulators is the Divided Attention Steering Stimulator (DASS), sometimes referred to as the Divided Attention Driving Simulator (DADS), which consists of a computer-generated image of the moving edges of a pseudo-randomly winding road portrayed on a display monitor. The image of a car hood is displayed on the bottom of the screen and the subject steers using a standard computer game steering

TABLE 49.1 Summary of Studies Showing Impaired Simulated Driving Performance in OSA. See Text for Details

Study	Subjects	Simulator	Results	Comments
Hack et al., 2000	OSA on CPAP = 59 (26 therapeutic, 33 subtherapeutic pressures)	DASS	Inc. SDLP Inc. RT	Studied 1 month after CPAP
Haraldsson et al., 1991	OSA = 15 Controls = 15	SRTRI	Inc. SDLP Inc. brake RT Inc. off-road events	Only brake RT improved after surgery
Hoeckma et al., 2007	OSA = 20 (10 CPAP, 10 OA) Controls = 16	See text	Inc. SDLP	Both treatment groups improved after 8 weeks
Juniper et al., 2000	OSA = 12 Controls = 12	DASS	Inc. SDLP Inc. RT Inc. off-road events	N/A
Orth et al., 2005	OSA = 31	CAR	Inc. "tracking errors" Inc. "concentration faults"	Improvement by day 2 of CPAP, some further improvement by day 42
Pizza et al., 2008	OSA = 30	STISIM	Inc. SDLP Inc. speed variability Inc. crashes Inc. RT Inc. DAI	Performance errors primarily in drivers with AHI >40 and elevated ESS
Pizza et al, 2009	OSA = 24	STSIM	As above	SDLP, crashes correlated with MWT > MSLT
Risser et al., 2000	OSA = 15 Controls = 15	STSIM	Inc. SDLP Inc. speed variability Inc. steering rate variability Inc. crashes	Performance inversely correlated with number of EEG-defined "attention lapses"
Sagaspe et al., 2007	OSA = 30	DASS	Inc. SDLP	Performance correlated with MSLT
Tippin et al., 2009	OSA = 25 Controls = 41	SIREN	+/− inc. SDLP No difference lane position, speed errors	OSA performed worse on vigilance task, vigilance correlated with driving in entire group, not OSA alone
Turkington et al., 2001	OSA = 150	DADS (DASS)	Inc. SDLP Inc. RT Inc. off-road events	Performance correlated with older age, female gender, alcohol use and history of "near misses"
Turkington et al., 2004	OSA = 18 (on/off CPAP) OSA = 18 (off CPAP)	DADS (DASS)	Same as above	Performance improved by day 7 of CPAP, deteriorated by day 7

wheel. Other investigators have used the Systems Technology, Incorporated Driving Simulator (STISIM®), which includes gas and brake pedals, a steering unit, and a 21-inch roadway monitor displaying a two-lane highway with occasional, long and wide curves. Orth et al. (2005) used the Computer-Aided Risk (CAR) stimulator wherein subjects sit in a standard car seat enclosed within a structure equipped with complete dashboard instrumentation, steering wheel, brake pedal and accelerator. Various driving scenarios can be displayed on the computer screen, including different lighting and weather conditions. Haraldsson, Carenfelt, Persson, Sachs and Tornos (1991) used the Swedish Road and Traffic Research Institute (SRTRI) simulator, a fully interactive system with the driver's cabin of a SAAB 900 set upon a moving base with a wide-angle visual display generated by three color projectors. Finally, in the system employed by Hoekema et al. (2007) subjects sit in an office chair behind a 21-inch computer screen on which the image of a straight road is projected. Lateral position is controlled by means of a computer-game steering wheel, and the only objective is to drive in the middle of the right lane, while compensating for the effects of unpredictable "side-wind" pulses.

Taking the notion of the DADT a step further, we have looked at divided attention with a visual vigilance task in a group of untreated OSA drivers in SIREN (Simulator for Interdisciplinary Research in Ergonomics and Neuroscience) (Tippin et al., 2009). SIREN is a high-fidelity, interactive driving simulator that creates an immersive, real-time virtual environment which permits the assessment of driver risk in a medical setting (Rizzo, McGehee, & Jermeland, 2000). We studied 25 drivers with OSA (mean age 45.0 +/− 9.7 years) and 41 comparison drivers (40.6 +/− 9.4 years) with no neurological

or sleep disorders. Each driver responded to light targets flashed at unpredictable temporal intervals at seven locations across the forward horizon by clicking the high beams (Figure 49.2). OSA drivers showed reduced vigilance based on lower HR (hit rate, or percent correct), especially for peripheral targets (80.7% +/− 14.8 versus 86.7% +/− 8.8, $p = 0.03$). Self-assessment of sleepiness with the Stanford Sleepiness Scale (SSS) was completed before and after the drive. Not only were the OSA drivers sleepier at the end of the drive, but increased sleepiness inversely correlated with decreased HR ($r = −0.49$, $p = 0.01$) only for this group. Poorer vigilance scores predicted greater numbers of lane crossing and speed maintenance errors in all drivers. However, there was only a trend for higher SDLP in the OSA drivers, while no differences were found in the number of lane deviations or speed errors. The majority of our drivers had mild-to-moderate OSA, which probably explains why we did not find impaired driving performance in these drivers, as has been reported by the other investigators listed in Table 49.1 who have studied more severely affected patients.

Performance in driving simulators shows, at most, a weak association with objective measures of disease severity, such as the apnea-hypopnea index (AHI) (Findley et al., 1999; George et al., 1996a, b; Orth et al., 2005; Tippin et al., 2009). However, some investigators have found that driving performance is strongly correlated with mean sleep latency (MSL) in this population, especially on the Maintenance of Wakefulness Test, which has been used in some situations to assess driver fitness (Pizza, Contardi, Ferlisi, Mondini, & Cirignotta, 2008; Pizza, Contardi, Mondini, Trentin, & Cirignotta, 2009; Sagaspe et al., 2007). There are data suggesting that nocturnal hypoxia (indexed by the minimum oxygen saturation or oxygen desaturation index on polysomnography) may predict some

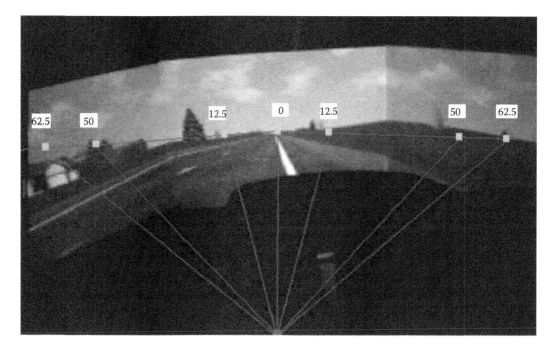

FIGURE 49.2 Vigilance task during simulated driving in SIREN. Squares represent locations of potential targets (values are in degrees from center).

performance decrements (Findley et al., 1995; Tippin et al., 2009). While those patients with severe OSA generally perform worse than less severely affected individuals, it is usually difficult to predict performance based solely upon objective measures of disease severity. George et al. (1996a,b) showed that AHI and MSL accounted for only 25-30% of the variance in DADT performance. Simulator performance is also difficult to predict based upon the driver's perception of sleepiness. Subjective measures of self-reported sleepiness, such as the SSS (which asks subjects to rate their perception of sleepiness at that moment) and the ESS (Epworth Sleepiness Scale, a metric of global sleepiness) show no consistent relationship with simulated driving performance (Findley et al., 1999; Risser, Ware, & Freeman, 2000; Tippin et al., 2009). Of interest, however, we found that the SSS assessed after an hour-long simulated drive correlated with degraded vigilance in OSA drivers, and with poorer driving performance in OSA and control drivers combined.

Although overall driving performance has been shown to be worse in OSA than controls, most investigators have not attempted to assess momentary changes in performance. Sleep onset is not a discrete event that can be precisely identified. Rather than going from full wakefulness to sleep at a single point in time, there is a "sleep onset period" during which one becomes increasingly unresponsive until a point is reached where there is complete failure to respond to external stimuli (Ogilvie, Wilkinson, & Allison, 1989). During this period, wakefulness is often interrupted by periods of decreased responsiveness termed "microsleeps" (MSs), which can be identified on the EEG by brief periods where the normal, quietly awake alpha background drops out and is replaced by slower activity (Harrison & Horne, 1996). Our laboratory has looked at driving performance in the moments surrounding EEG-defined MSs by comparing performance in a case crossover design with OSA drivers acting as their own controls (Boyle, Tippin, Paul, & Rizzo, 2008). We found that steering and lane position variability increased, and minimum time-to-lane crossing decreased during the MSs. These findings were more striking when the MSs occurred on a curved rather than straight road segment and the degree of deterioration in driving performance correlated with MS duration (Figure 49.3). Rather than studying typical MSs, Risser et al. (2000) found that lane position variability and crash frequency correlated with the frequency and duration of EEG-defined "attention lapses" (defined as "bursts" of increased alpha or theta activity lasting for at least three seconds). OSA drivers had more bursts, which were also of longer duration than those in controls. In addition, OSA drivers had greater speed variability, greater variability in lane position, steering rate and speed, and a greater number of "crashes" than controls had. Microsleeps have also been shown to correlate with simulated driving performance in normal drivers (Moller, Kayumov, Bulmash, Nhan, & Shappiro, 2006). It should be mentioned, however, that recording EEGs can be technically difficult in a driving simulator due to electrode artifacts, as well as artifacts caused by patient movement, chewing, eye movement/blinking, and the like. Little data is available concerning recording of other physiological data during simulated driving in patients with OSA.

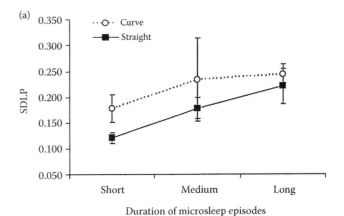

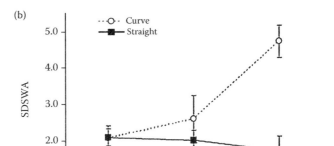

FIGURE 49.3 The effect of microsleep duration on lateral controls: (a) SDLP and (b) SDSWA. (From Boyle, L. N., Tippin, J., Paul, A., & Rizzo, M. (2008). Driver performance in the moments surrounding a microsleep. *Transportation Research, Part F: Traffic Psychology and Behaviour, 11*(2), 126–136. With permission.)

OSA drivers perform better after CPAP (continuous positive airway pressure) therapy (George et al., 1997; Hack et al., 2000; Orth et al., 2005), surgery (Haraldsson et al., 1991) and use of oral appliances (Hoekema et al., 2007), which has led some authors to suggest that simulators may be an effective tool for determining fitness to drive in treated patients (Orth et al., 2005). Simulators may be especially useful for detecting high-risk drivers if they also have a history of prior MVCs and "near misses", and subjective sleepiness (Turkington, Sircar, Allgar, & Elliott, 2001). Simulated driving performance improves within a few days of starting CPAP and the improvement may linger for up to a week after discontinuing treatment (Orth et al., 2005; Turkington, Sircar, Allgar, & Elliott, 2004). However, performance may not approach that of normal, matched control drivers, even after a month of CPAP (Hack et al., 2000).

Given the general lack of correlation between MVCs and objective measures of disease severity or self-reported sleepiness in this population, simulators may provide unique information about driver safety. The use of simulators to determine fitness is intuitively attractive; they can replicate real-world driving in a safe environment and are more comprehensive than simple vigilance tests. It is not yet clear, however, how well simulated driving performance predicts real-world driving and MVC risk in individual drivers with OSA. There are conflicting data on the association between simulated driving performance and crash history in OSA (George, 2004; Pizza et al., 2008; Turkington et al., 2001). If simulators are to become a means of determining driver fitness in this population, a number of problems will have to be resolved. For example, there is as yet no standardized protocol to use. How long should the drive be? What type of driving scenario should be used (rural versus city driving, night versus day, etc)? Also, how does one determine what constitutes a "failing" performance? Recently published guidelines for commercial drivers with OSA (Hartenbaum et al., 2006) stratify driving restrictions based upon adequacy of therapy, presence of sleepiness and disease severity with the greatest restriction reserved those with AHI > 30. According to the currently available data, assessment of driver fitness for most OSA drivers should be based upon similar information. The potential role for driving simulators in assisting this determination has yet to be identified.

49.2.3 Simulators in Narcolepsy

As in drivers with OSA, narcoleptic drivers perform more poorly on "Steer Clear" (Findley et al., 1995; Findley et al., 1999), the DADT (George et al., 1996) and high-fidelity simulators (Kotterba et al., 2004) than control drivers. What is more, narcoleptics have four times the variability in performance on "Steer Clear" as OSA drivers (Findley et al., 1999). While narcoleptics perform less well than normal controls as a group, as many as one-half of narcoleptic drivers do as well as controls (George et al., 1996), making it difficult to make broad generalizations about their driving behavior. Also, similar to OSA drivers, it is difficult to predict performance based upon objective measures of disease severity. George et al. (1996a,b) performed the DADT concomitantly with the multiple sleep latency test (MSLT), arguably the "gold standard" for confirming the diagnosis of narcolepsy. However, only weak relationships between the MSLT and tracking error and response time were found ($r = -0.32$, $p < 0.05$). Performance improves after treatment with stimulants (Kotterba et al., 2004), raising the possibility that simulators may be useful in determining fitness to drive. Additional support for this contention is given by the fact that narcoleptics who perform poorly on "Steer Clear" have been shown to have more MVCs than those who perform normally; 0.20 crashes/driver/five years versus 0.05 crashes/driver/five years (Findley et al., 1995). However, as with OSA drivers, the optimal testing procedure, how to define a failing performance and the like, have not been determined.

In addition to sleepiness, narcoleptics are at risk for crashing because of sudden generalized weakness caused by cataplexy. As noted earlier, cataplexy is provoked by strong emotions, such as laughter, fear, anger and surprise. Because cataplexy is unpredictable (it is not always precipitated by these stimuli), its effect on driving has been difficult to study. However, Augustine and colleagues at the Mayo Clinic (Augustine, Cameron, Camp, Krahn, & Robb, 2002) have developed a virtual reality/simulator technology that they call "CatNAP" (Cataplexy/Narcolepsy Activation Program). In CatNAP, subjects wear a stereoscopic head-mounted display that shows a "virtual" town through which they are instructed to drive. Control of the vehicle is provided through a steering wheel and pedal cluster. While subjects drive through the town, various stimuli are presented which are designed to provoke a startle response, frustration or laughter. In addition to recording driving performance, various physiological parameters, such as electromyography (EMG) are monitored. Although still in the development phase, this virtual reality technology may allow investigators to study the effect of cataplexy on driving in a controlled and safe environment, which has previously been very difficult to do.

49.3 Epilepsy

49.3.1 MVC Risk in Epilepsy

As epilepsy is a condition commonly associated with loss of or alteration in consciousness, it would seem logical to assume that drivers with epilepsy have an increased crash risk relative to the general population. This may be true, although some authors have found no increased risk, or if increased, the relative risk is actually less than in drivers with cardiovascular disease or healthy drivers less than 25 years of age (Hansotia & Broste, 1991; Sheth, Krauss, Krumholz, & Li, 2004). Seemingly counterintuitive, most of the MVCs in epileptic drivers are not directly related to having a seizure; only about 11% of crashes can be attributed as such (Hansotia & Broste, 1991). The cause of the remainder is often difficult to determine, although medication-induced cognitive impairments could play a key role in a substantial proportion (Ramaekers et al., 2002). Anti-epileptic drugs (AEDs) may affect multiple spheres of cognition, especially attention and memory, even in therapeutic doses (Meador et al., 1995). Given that safe driving is dependent upon the continuous implementation of these cognitive skills, one might anticipate that AEDs could negatively impact performance in some drivers. Unfortunately, there are few data concerning the direct effects of these medications on driving and crash risk.

Despite the relatively small increased crash risk, all states and the District of Columbia, as well as the European Union, impose some restrictions on epileptic drivers. Interestingly, it was not until after World War II that epileptics were permitted to drive at all in the United States (Wisconsin was the first state to grant driver's licenses to this group of patients). Typically, a specified seizure-free interval (SFI) ranging from 3 to 12 months is required before epileptics may drive, or in some situations, be granted a license. A specific SFI is mandated by 24 states, while in the remainder the ability to drive is determined by a Medical Advisory Board (MAB) or the patient's own treating physician. Six states have laws that enforce mandatory physician reporting to the state Department of Motor Vehicles. This confusing picture is only aggravated by the fact that there is little evidence that the length of the SFI in any way impacts the risk for further crashes. If one compares the crash mortality rates for states with SFIs of 3, 6 or 12 months with those with a more flexible system (e.g., determined by the MAB), there is actually a trend for the rate to be lowest in those states with the shortest requirement (Sheth, Krauss, Krumholz, & Li, 2004). An interesting "accident of legislature" occurred in Arizona when the state legislature changed the required SFI from 12 to 3 months in 1994. Despite the fact that a longer SFI is known to be a strong predictor for individual epileptic drivers not having a seizure-related crash (Krauss, Krumholz, Carter, Li, & Kaplan, 1999), no change in the overall rate of crash-related deaths in Arizona was detected (Drazkowski et al., 2003). Along similar lines, Gastaut and Zifkin (1987) found that states with more restrictive SFIs actually had higher incidences of crashes, and the severity of these crashes was often greater. The reason for this apparent contradiction is that epileptic drivers are notoriously non-compliant with following imposed driving restrictions, and the majority never informs their physicians about recent seizures (Polychronopoulos et al., 2006). Longer mandatory SFIs may actually make matters worse in the long run by encouraging non-compliance.

49.3.2 Simulators in Epilepsy

There are no published studies concerning simulator performance in either epileptics or normal subjects taking AEDs. However, our laboratory has completed a pilot study of five patients with complex partial epilepsy treated with a variety of commonly prescribed AEDs. Subjects drove on a simulated rural two-lane highway with interactive traffic for 60 minutes in SIREN. The simulation consisted of six "events" associated with potential crashes interspersed with uneventful highway segments. All subjects made safety errors (mean events = 9, range 6–16), and an assortment of additional errors occurred during incursion events. For example, two subjects failed to slow for a left-turning vehicle, and two subjects struck a dog that ran across the road in front of the vehicle. Also, a variety of dangerous driving strategies were seen in response to the incursion of a vehicle illegally crossing an intersection from the driver's right side. All drivers crashed. Two drivers showed no change in speed or steering, and the other three braked inadequately in response to the vehicle. Two of these steered the vehicle into on-coming traffic, resulting in a collision. It is interesting to note that this pattern of degraded performance is similar to what has been described in some patients with early Alzheimer's disease (Rizzo, McGehee, Dawson, & Anderson, 2001).

Expanding on the limited information about simulated driving in epilepsy, a few studies have investigated driving performance in instrumented vehicles (IVs) and off-road tasks linked to driving performance. Kasteleijn-Nolst Trenite, Riemersma, Binnie, Smit and Meinardi (1987) monitored six epileptic drivers with electroencephalography (EEG) in an IV. Driving was clearly impaired during EEG-defined "subclinical" epileptiform discharges in three drivers and was possibly impaired in another. Of interest, two of these drivers had been seizure-free

for up to four years, suggesting that even those without active epilepsy may have an increased crash risk. Regarding the effects of AEDs, Ramaekers et al. (2002) showed that healthy volunteers receiving the AED carbamazepine have mild but significantly impaired lateral position maintenance as indexed by SDLP, as well as speed adaptation in a vehicle-following task. An as yet unpublished multi-center study funded by one of the pharmaceutical companies is looking at the differential effects of two AEDs on performance in a desktop computer-based selective visual attention task (Hoyler, Blum, Hammer, & Caldwell, 2002). Performance on this type of task, which is particularly affected by sedating AEDs such as Phenobarbital, has been said to correlate with on-road driving by one group of investigators (Barcs, Vitrai, & Halasz, 1997), although they provided no details about which driving measures were assessed.

49.4 Challenges and Future Directions

Driving simulator studies demonstrate poorer performance in drivers with OSA and narcolepsy compared to those without neurological or sleep disorders. Simulators appear to provide novel information about driving performance that is not easily predictable based upon subjective sleepiness or objective measures of disease severity, suggesting a potential role in determining driver fitness. However, their usefulness in this regard is as yet unconfirmed. Additional studies are needed to define appropriate testing procedures and to delineate what constitutes adequate performance. Commonly-used scenarios are monotonous with few, if any, distractions or obstacles. However, real-world driving is replete with such events. Optimal drive duration, as well as timing and number of obstacles and other such events are currently unknown as procedures often vary considerably from one investigator to the next. Other important, unanswered questions include: Can simulators be useful in determining adequacy of treatment with either CPAP or stimulant medications? What is the minimum degree of compliance with CPAP that is necessary to normalize driving performance? Is performance enhanced by the addition of stimulants such as methylphenidate or modafinil? In light of these unresolved questions, simulators appear to have a limited role in determining driver fitness in sleepy drivers at the present time, and for this reason should not be used in situations where legal challenges exist. Although similar questions remain concerning the effects of epilepsy and AEDs on driving performance, the situation is made even worse by a tremendous paucity of data. Which medications are more likely to degrade performance? Is there a dose response? Are patients with specific epileptic syndromes more likely to crash? Can simulators help identify those patients who are at risk for crashing because of "subclinical" epileptiform EEG activity? As seizures are intermittent events, is it possible to design a scenario that is likely to "catch" such episodes? What does it mean if no seizures occur within a specified drive duration? Is that person safe to drive? Physiological data, such as EEG, may provide supplementary information about factors affecting performance, but recording such data can be technically difficult. The measures and techniques that are most reliable and predictive have not yet been determined. Only after these questions are answered will it be possible for driving simulators to become an integral component of the process for determining driver fitness in drivers with epilepsy, sleep disorders or simple sleep-deprivation/fatigue.

Key Points

- Drivers with sleep disorders, such as OSA, have increased crash risk as a group, but determining risk in individual drivers is difficult. Simulators may play a role in assessing crash risk for these drivers.
- Dependent measures commonly used in simulator studies of drivers with either OSA or narcolepsy include SDLP and crashes or off-road events. Drivers with sleep disorders perform worse than control drivers before treatment and improve after a variety of therapies.
- Few data are available about the use of simulators in assessing drivers with epilepsy. This is a potentially fruitful area for future research.

Keywords: Continuous Positive Airway Pressure, Driving Simulators, Epilepsy, Narcolepsy, Obstructive Sleep Apnea

Glossary

Apnea-hypopnea index (AHI): The number of apneas (absent airflow) and hypopneas (reduced airflow) divided by the total sleep time recorded during polysomnography. Both types of respiratory events have a minimum duration of 10 seconds. Hypopneas must have accompanying reduction in the blood oxygen level ("oxygen desaturation") or electroencephalographic evidence of arousal, while apneas do not require these associated physiological changes.

Electroencepaholgram (EEG): Brain electrical activity recorded from the scalp with surface electrodes. EEG activity is generated by the cerebral cortex, which is influenced by deeper "subcortical" brain structures. The EEG can be interpreted by visual review of the "raw" data, or subjected to quantitative analysis of voltage or frequency.

Epworth sleepiness scale (ESS): A subjective measure of global or "trait" sleepiness. Subjects are instructed to rate their likelihood of dozing in eight specific situations on a zero to three scale.

Hypoxemia: Abnormal reduction in blood oxygen level.

Mean sleep latency (MSL): The mean time from "lights out" to the first EEG appearance of sleep in four to five nap attempts during the multiple sleep latency test (MSLT). The MSLT is considered an objective measure of manifest sleepiness.

Stanford sleepiness scale (SSS): A subjective measure of momentary or "state" sleepiness. Subjects rate their

current level of arousal from full wakefulness to impending sleep on a one to seven scale.

Vigilance: A state in which attention is maintained over time; sustained attention.

Key Readings

George, C. F. P., Boudreau, A. C., & Smiley, A. (1996). Comparison of simulated driving performance in narcolepsy and sleep apnea patients. *Sleep, 19*(9), 711–717.

Orth, M., Duchna, M. W., Leidag, M., Widding, W., Rasche, K., Bauer, T. T., . . . Kotterba, S. (2005). Driving simulator and neuropsychological testing in OSA before and under CPAP therapy. *European Respiratory Journal*, 26, 989–903.

Risser, M. R., Ware, J. C., & Freeman, F. G. (2000). Driving simulation with EEG monitoring in normal and obstructive sleep apnea patients. *Sleep, 23*(3), 393–398.

Sassani, A., Findley, L., Kryger, M., Goldlust, E., George, C., & Davidson, T. (2004). Reducing motor-vehicle collisions, costs, and fatalities by treating obstructive sleep apnea syndrome. *Sleep, 27*(3), 453–458.

Sheth, S. G., Krauss, G., Krumholz, A., & Li, G. (2004). Mortality in epilepsy: Driving fatalities vs. other causes of death in patients with epilepsy. *Neurology, 63(6)*, 1002–1007.

Tregear, S. J., Reston, J., Schoelles, K., & Phillips, B. (2009). Obstructive sleep apnea and risk of motor vehicle crash: systemic review and meta-analysis. *Journal of Clinical Sleep Medicine, 5*, 573–581.

References

Augustine, K., Cameron, B., Camp, J., Krahn, L., & Robb, R. (2002). An immersive system for provoking and analyzing cataplexy. *Studies in Health Technology and Informatics, 85*, 31–37.

Barcs, G., Vitrai, J., & Halász, P. (1997). Investigation of vehicle driving ability in two diagnostic groups of epileptic patients with special neuropsychological approach. *Medicine and Law, 6*(2), 277–287.

Boyle, L. N., Tippin, J., Paul, A., & Rizzo, M. (2008). Driver performance in the moments surrounding a microsleep. *Transportation Research, Part F: Traffic Psychology and Behavior, 11*(2), 126–136.

Connor, J., Whitlock, G., Norton, R., & Jackson, R. (2001). The role of driver sleepiness in car crashes: A systematic review of epidemiological studies. *Accident Analysis & Prevention, 33*, 31–41.

Drazkowski, J. F., Fisher, R. S., Sirven, J. I., Demaerschalk, B. M., Uber-Zak, L., Hentz, J. G., & Labiner, D. (2003). Seizure-related motor vehicle crashes in Arizona before and after reducing the driving restriction from 12 to 3 months. *Mayo Clinic Proceedings, 78*(7), 819–825.

Engleman, H. M., Hirst, W., & Douglas, N. J. (1997). Under reporting of sleepiness and driving impairment in patients with sleep apnea/hypopnea syndrome. *Journal of Sleep Research, 6*, 272–275.

Engleman, H. M., Kingshott, R. N., Martin, S. E., & Douglas, N. J. (2000). Cognitive function in the sleep apnea/hypopnea syndrome. *Sleep 23*(Suppl. 4), S102–S108.

Findley, L., Unverzagt, M., Guchu, R., Fabrizio, M., Buckner, J., & Suratt, P. (1995). Vigilance and automobile accidents in patients with sleep apnea or narcolepsy. *Chest, 108*(3), 619–624.

Findley, L. J., Suratt, P. M., & Dinges, D. F. (1999). Time-on-task decrements in "Steer Clear" performance of patients with sleep apnea and narcolepsy. *Sleep, 22*(6), 804–809.

Gastaut, J., & Zifkin, B. G. (1987). The risk of automobile accidents with seizures occurring while driving: Relation to seizure type. *Neurology, 37*(10), 1613–1616.

George, C. F. P. (2004). Driving and automobile crashes in patients with obstructive sleep apnea/hypopnea syndrome. *Thorax, 59*, 804–807.

George, C. F. P., Boudreau, A. C., & Smiley, A. (1996a). Comparison of simulated driving performance in narcolepsy and sleep apnea patients. *Sleep, 19*(9), 711–717.

George, C. F., Boudreau, A. C., & Smiley, A. (1996b). Simulated driving performance in patients with obstructive sleep apnea. *American Journal of Respiratory and Critical Care Medicine, 154*(1), 175–181.

George, C. F. P., Boudreau, A. C., & Smiley, A. (1997). Effects of nasal CPAP on simulated driving performance in patients with obstructive sleep apnea. *Thorax, 52*, 648–653.

Hack, M., Davies, R. J. O., Mullins, R., Choi, S. J., Ramdassingh-Dow, S., Jenkinson, C., & Stradling, J. R. (2000). Randomized prospective parallel trial of therapeutic versus subtherapeutic nasal continuous positive airway pressure on simulated steering performance in patients with obstructive sleep apnea. *Thorax, 55*, 224–231.

Hansotia, P., & Broste, S. K. (1991). The effect of epilepsy or diabetes mellitus on the risk of automobile accidents. *New England Journal of Medicine, 324*(1), 22–6.

Haraldsson, P.-O., Carenfelt, C., Persson, H. E., Sachs, C., & Tornros, J. (1991). Simulated long-term driving performance before and after uvulopalatopharyngoplasty. *ORL: Journal for Oto-Rhino-Laryngology and its Related Specialties, 53*(2), 106–110.

Harrison, Y., & Horne, J. A. (1996). Occurrence of "microsleeps" during daytime sleep onset in normal subjects. *Electroencephalography and Clinical Neurophysiology, 98*(5), 411–416.

Hartenbaum, N., Collop, N., Rosen, I., Phillips, B., George, C. F. P., Rowley, J. A., . . . Moffitt, G. L. (2006). Sleep apnea and commercial motor vehicle operators: Statement from the Joint Task Force of the American College of Chest Physicians, the American College of Occupational and Environmental Medicine, and the National Sleep Foundation. *Chest, 130*, 902–905.

Hauser, W. A., Annengers, J. F., & Kurland, L. T. (1993). Incidence of epilepsy and unprovoked seizures in Rochester, Minnesota: 1935–1984. *Epilepsia, 34*(3), 453–468.

Hoekema, A., Stegenga, B., Bakker, M., Brouwer, W. H., de Bont, L. G. M., Wijkstra, P. J., & van der Hoeven, J. H. (2007). Simulated driving in obstructive sleep apnea-hypopnea; effects of oral appliances and continuous positive airway pressure. *Sleep and Breathing, 11*, 129–138.

Hoyler, S. L., Blum, D., Hammer, A., & Caldwell, P. (2002). *Double-blind randomized trial of cognitive effects of Lamictal (lamotrigine) versus topiramate in epilepsy.* (GlaxoSmithKline Protocol LAM40112). GlaxoSmithKline.

International League Against Epilepsy (ILAE). (1981). Commission on classification and terminology of the International League against Epilepsy: Proposal for revised clinical and electroencephalographic classification of epileptic seizures. *Epilepsia, 22*, 489–501.

Juniper, M., Hack, M. A., George, C. F., Davies, R. J. O., & Stradling, J. R. (2000). Steering simulation performance in patients with obstructive sleep apnea and matched control subjects. *European Respiratory Journal, 15*, 590–595.

Kasteleijn-Nolst Trenite, D. G., Riemersma, J. B., Binnie, C. D., Smit, A. M., & Meinardi, H. (1987). The influence of subclinical epileptiform EEG discharges on driving behavior. *Electroencephalography and Clinical Neurophysiology, 67*(2), 167–170.

Knipling, R. R., & Wang, J.-S. (1994). *Crashes and fatalities related to driver drowsiness/fatigue. Research Note from the Office of Crash Avoidance Research* (pp. 1–8). Washington, DC: US Department of Transportation, National Highway Traffic Safety Administration.

Kotterba, S., Mueller, N., Leidag, M., Widdig, W., Rasche, K., Malin, J. P., . . . Orth, M. (2004). Comparison of driving simulator performance and neuropsychological testing in narcolepsy. *Clinical Neurology and Neurosurgery, 106*(4), 275–279.

Krauss, G. L., Krumholz, A., Carter, R. C., Li, G., & Kaplan, P. (1999). Risk factors for seizure-related motor vehicle crashes in patients with epilepsy. *Neurology, 52*(7), 1324–1329.

McCartt, A. T., Ribner, S. A., Pack, A. I., & Hammer, M. C. (1996). The scope and nature of the drowsy driving problem in New York State. *Accident Analysis & Prevention, 28*(4), 511–517.

Meador, K. J., Loring, D. W., Moore, E. E., Thompson, W. O., Nichols, M. E., Oberzan, R. E., . . . King, D. W. (1995). Comparative cognitive effects of Phenobarbital, phenytoin, and valproate in healthy adults. *Neurology, 45*, 1494–1499.

Melamed, S., & Oksenberg, A. (2002). Excessive daytime sleepiness and risk of occupational injuries in non-shift daytime workers. *Sleep, 25*(3), 315–322.

Moller, H. J., Kayumov, L., Bulmash, E. L., Nhan, J., & Shapiro, C. M. (2006). Simulator performance, microsleep episodes, and subjective sleepiness: Normative data using convergent methodologies to assess driver drowsiness. *Journal of Psychosomatic Research, 61*, 335–342.

Nowak, M., Kornhuber, J., & Meyrer, R. (2006). Daytime impairment and neurodegeneration in OSAS. *Sleep, 29*(12), 1521–1530.

Ogilvie, R. D., Wilkinson, R. T., & Allison, S. (1989). The detection of sleep onset: Behavioral, physiological, and subjective convergence. *Sleep, 12*, 458–474.

Orth, M., Duchna, M. W., Leidag, M., Widding, W., Rasche, K., Bauer, T. T., ... Kotterba, S. (2005). Driving simulator and neuropsychological testing in OSA before and under CPAP therapy. *European Respiratory Journal, 26*, 989–903.

Pizza, F., Contardi, S., Ferlisi, M., Mondini, S., & Cirignotta, F. (2008). Daytime driving performance and sleepiness in obstructive sleep apnea patients. *Accident Analysis, & Prevention, 40*, 602–609.

Pizza, F., Contardi, S., Mondini, S., Trentin, L., & Cirignotta, F., (2009). Daytime sleepiness and driving performance in patients with obstructive sleep apnea: Comparison of the MSLT, the MWT, and a simulated driving task. *Sleep, 32*, 382–391.

Polychronopoulos, P., Argyriou, A. A., Huliara, V., Sirrou, V., Gourzis, P., & Chroni, E. (2006). Factors associated with poor compliance of patients with epilepsy driving restrictions. *Neurology, 67*(5), 869–871.

Ramaekers, J. G., Lamers, C. T. J., Verhey, F., Muntjewerff, N. D., Mobbs, E., Sanders, N., ... Lockton, J. A. (2002). A comparative study of the effects of carbamazepine and the NMDA receptor agonist remacemide on road tracking and car-following performance in actual traffic. *Psychopharmacology, 159*(2), 203–210.

Risser, M. R., Ware, J. C., & Freeman, F. G. (2000). Driving simulation with EEG monitoring in normal and obstructive sleep apnea patients. *Sleep, 23*(3), 393–398.

Rizzo, M., McGehee, D., & Jermeland, J. (2000). Design and installation of a driving simulator in a hospital environment. In K. Brookhuis, D. DeWaard, & C. Weikert (Eds.), *Human system interaction: Education, research and application in the 21st century* (pp. 69–77). Maastricht, The Netherlands: Shaker Publishing.

Rizzo, M., McGehee, D. V., Dawson, J. D., & Anderson, S. N. (2001). Simulated car crashes at intersections in drivers with Alzheimer's disease. *Alzheimer's Disease and Associated Disorders, 15*(1), 10–20.

Rizzo, M. (2004). Safe and unsafe driving. In: M. Rizzo & P. J. Eslinger (Eds.), *Principles and practice of behavioral neurology and neuropsychology* (pp. 197–222). Philadelphia, PA: W. B. Saunders.

Royal, D. (2003). *Volume I: Findings for national survey of distracted and drowsy driving attitudes and behavior: 2002.* US Department of Transportation, National Highway Traffic Safety Administration. Retrieved from http://www.nhtsa.dot.gov/People/injury/drowsy_driving1/survey-distractive03/index.htm

Sagaspe, P., Taillard, J., Chaumet, G., Guilleminault, C., Coste, O., Moore, N., ... Philip, P. (2007). Maintenance of wakefulness test as a predictor or driving performance in patients with untreated obstructive sleep apnea. *Sleep, 30*, 327–330.

Sassani, A., Findley, L., Kryger, M., Goldlust, E., George, C., & Davidson, T. (2004). Reducing motor-vehicle collisions, costs, and fatalities by treating obstructive sleep apnea syndrome. *Sleep, 27*(3), 453–458.

Sheth, S. G., Krauss, G., Krumholz, A., & Li, G. (2004). Mortality in epilepsy: Driving fatalities vs. other causes of death in patients with epilepsy. *Neurology, 63*(6), 1002–1007.

Silber, M. H., Krahn, L. E., Olson, E. J., & Pankratz, V. S. (2002). The epidemiology of narcolepsy in Olmstead County, Minnesota: A population-based study. *Sleep, 25*, 197–202.

Tippin, J., Sparks, J., & Rizzo, M. (2009). Visual vigilance in drivers with obstructive sleep apnea. *Journal of Psychosomatic Research, 67*, 143–151.

Turkington, P. M., Sircar, M., Allgar, V., & Elliott, M. W. (2001). Relationship between obstructive sleep apnea, driving simulator performance, and risk of road traffic accidents. *Thorax, 56*, 800–805.

Turkington, P. M., Sircar, M., Allgar, V., & Elliott, M. W. (2004). Time course of changes in driving performance with and without treatment in patients with sleep apnea hypopnea syndrome. *Thorax, 59*, 56–59.

Young, T., Palta, M., Dempsey, J., Skatrud, J., Weber, S., & Badr, S. (1993). The occurrence of sleep-disordered breathing among middle-aged adults. *New England Journal of Medicine, 328*, 1230–1235.

50

Traumatic Brain Injury: Tests in a Driving Simulator as Part of the Neuropsychological Assessment of Fitness to Drive

Wiebo Brouwer
*University Medical
Center Groningen*

Rens B. Busscher
*University Medical
Center Groningen*

Ragnhild J. Davidse
*SWOV Institute for Road
Safety Research*

Harie Pot
*University Medical
Center Groningen*

Peter C. van Wolffelaar
University of Groningen

Abstract

The Problem. Very severe traumatic brain injury (TBI) can cause persistent functional limitations in visual-motor and cognitive functions which may have consequences for fitness to drive. On-road driving performance in this population is significantly related to performance on neuropsychological tests of attention and information processing. This particularly concerns basic "operational" driving tasks like brake reaction time, visual search time, and lateral position control (Van Wolffelaar, Brouwer & van Zomeren, 1990). Many drivers in this population, however, succeed in compensating effectively for these functional limitations, presumably because higher-order "tactical" aspects of driving allow them to anticipate and avoid driving situations which would be too demanding. *Role of Driving Simulators.* To be able to assess compensatory potential, two test drives in an advanced driving simulator are included in the neuropsychological assessment of fitness to drive. The tests and the context in which they are used are described. By means of three case studies with very severe TBI the additional value of the driving simulator tests is illustrated and discussed. *Key Results of Driving Simulator Studies.* Test drives in a driving simulator for the assessment of brain-damaged drivers have not yet been properly validated and standardized. As an addition to a neuropsychological examination, a driving test in a realistic driving simulator can be a useful tool to assess (lacking) compensatory driving skill in brain-damaged drivers. *Scenarios and Dependent Variables.* In the Swing Drive we assess the operational skill of lateral position control and the tactical skill of controlling speed in response to changing motivation and operator capacities. The test requires steering at various speeds on a winding country road with alternating left-right curves and a continuous stream of traffic from the opposing direction. In various conditions the average lateral position, the standard deviation of lateral position, lane crossings left and right, and collisions with other traffic are computed. In the City Drive we use a city and route developed in research on intersection complexity and ADAS in older drivers (Davidse, 2007). We assess practical fitness to drive based on observation of speed choice, lateral and longitudinal position control, viewing behavior, gap acceptance, application of priority rules, etc. in a variety of road and traffic situations as they develop during the drive. ***Platform Specificity and Equipment Limitations.*** The simulator system of ST Software B.V. is used (http://www. stsoftware.nl/). It contains the design tools to create simulated road environments and the runtime modules required for real-time simulation. Simulation of traffic is based on 'autonomous agents' technology which models the simulated participants (i.e., all traffic surrounding the simulator driver) as self-governing intelligent objects that show natural and normative driver behavior.

50.1 Introduction

50.1.1 Traumatic Brain Injury

Traumatic brain injury (TBI) is the most frequently acquired neurological disorder in people under the age of fifty. The vast majority are closed head injuries (CHI) caused by a blunt impact not penetrating the skull. Typically the injury is caused by a traffic crash or a fall that very suddenly decelerates the head. The impact causes the brain to move violently within the skull, resulting in contusion and laceration of brain tissue, particularly diffuse axonal injuries in the white matter and focal contusions in the frontal and temporal areas. Further damage to the brain can result from complications during recovery (e.g., Long & Ross, 1992; Richardson, 2001).

The immediate effect of the impact is a loss of consciousness and a state of coma. This is followed by a transition period of post-traumatic amnesia (PTA), a state of disturbed consciousness between coma and normal wakefulness. Depending on the severity of the injuries, coma is "deeper" and consciousness remains disturbed for longer. Clinically, the severity of TBI is often graded on the basis of the duration of the period of disturbed consciousness after trauma retrospectively indicated by the patient when questioned about the return of his/her continuous everyday memory.

After the return of continuous everyday memory, significant cognitive limitations persist, in more severe cases for the rest of TBI patients' lives. Characteristic impairments concern memory, attention, and speed of information processing (e.g., Brooks, 1984; Richardson, 2001; Granacher, 2003). Also, dysexecutive problems are frequently reported, both in neuropsychological test performance and in everyday functioning. This latter area borders on personality changes, such as decreased social judgment and loss of empathy (e.g., Granacher, 2003). All of the above limitations have negative effects on fitness to drive, particularly if severe and chronic. Another common complication in severe TBI is post-traumatic epilepsy, which is especially prevalent in penetrating injuries (Agrawal, Timothy, Pandit, &

Manju, 2006). In cases where patients use anti-epileptic medication, it may further slow their information processing.

Fortunately most cases of CHI are slight, mild and moderate, and not compromising cognitive and social function to a degree relevant for driving safety (Brouwer & Withaar, 1997). The latter authors argue that, because of psychometric objections, the official assessment of medical fitness to drive should not be applied in these milder cases. A further complicating factor in research about driving behavior after TBI is that pre-morbid characteristics of this group, in terms of unsafe traffic behavior and personality, may differ from the general population so that one should be cautious in relating abnormal driving performance and behavior to brain injury (Pietrapiana, et al., 2005).

In the category of very severe CHI (see Table 50.1), however, most persons have objectively demonstrable brain injury and significant remaining impairments as a sequel of their injury. This is a population with a relatively high proportion of problematic fitness to drive, often unable to meet the requirements for safe and smooth driving or requiring car adaptations and special training (Brouwer & Withaar, 1997; Pietrapiana et al., 2005; Leon-Carrion, Dominguez-Morales, Barroso, & Martin, 2005).

Many severe TBI patients suffer from depression (Kreutzer, Seel, & Gourley, 2001). Driver rehabilitation can play a significant role in preventing social isolation. Apart from that, driving is often very important for self esteem, especially in the young male population where CHI is over-represented. The assessment of fitness to drive should therefore not only distinguish between fit and unfit drivers but should also deliver recommendations for interventions (Brouwer & Withaar, 1997; Brouwer, Withaar, Tant, & Van Zomeren, 2002).

50.1.1.1 Traumatic Brain Injury and Fitness to Drive

On-road driving performance in this population is significantly related to performance on neuropsychological tests of attention and information processing. This particularly shows up in basic "operational" driving tasks such as brake reaction time,

TABLE 50.1 Severity of Chi Based on the Duration of Post-Traumatic Amnesia (PTA)

PTA duration	Severity of Injury	
	After Russell and Smith (1961)	After Long and Ross (1991)
<5 minutes	Slight	Minimal
5–60 minutes		Very mild
1–24 hours	Moderate	Mild
1–7 days	Severe	Moderate
1–4 weeks	Very severe	Severe
>4 weeks		Very severe

visual search time, and lateral position control (Van Wolffelaar et al., 1990; Brouwer & Withaar, 1997). However, some drivers in this category succeed in compensating very successfully for these impairments, presumably by using higher order "tactical" aspects of driving that allow them to anticipate and avoid driving situations which would be too demanding for their abilities. Also, the effects of driving experience (mileage) are considerable.

50.1.1.1.1 An Early Driving Simulator Study

Van Wolffelaar et al. (1990) studied 18 CHI patients with PTA durations between 11 and 124 days (very severe) 5 to 10 years after injury with a driving simulator test of flexible adaptation to changing external circumstances. Performance was also assessed in on-road driving tests and on neuropsychological tests of information processing. All patients had returned to driving and all volunteered to participate in the study, so it was not a random sample of very severe CHI patients.

A road test considered representative for normal everyday driving conditions was administered by an expert observer of the Dutch automobile club (ANWB). It consisted of a one-hour drive during daytime in various road and traffic situations in the subject's own car. The experts give an overall score of between one (optimal) and six (insufficient skill) which is statistically quite reliable (Brouwer, Rothengatter, & Van Wolffelaar, 1988). They also give separate scores for specific skills like handling the car mechanics, traffic maneuvers, and traffic perception and insight. The latter score has a strong positive correlation with the overall score.

Performance on this classical on-road test could be predicted very well by driving experience (total mileage, estimated in the interview with the driver), and by a driving simulator score, namely the degree to which subjects had learned to tune speed and steering to delayed feedback information on the simulator screen. Compared to real-life road and traffic situations, the simulator that we used in 1990 was simple and abstract. Nevertheless, it yielded important information outperforming classical tests of reaction time and visual-motor skills in the prediction of on-road driving quality.

Two structured on-road tests which allowed less compensatory behavior were also administered, namely a test of lateral position control (LPC) when driving a straight highway section at constant speed, and a test of merging decisions. Scores on these tests showed quite a different pattern of correlation, lower with driving experience but moderately high with tests of information processing, LPC particularly with visual-motor speed (and with lateral position control in the driving simulator), and merging decisions particularly with cognitive speed.

From this and comparable studies in the domain of aging (e.g., Brouwer et al., 1988) we have learned that slowness of information processing is predictive of the quality of driving performance when the demand is excessive and/or when a driving test allows limited opportunity for compensation. If slowness is not excessive and the task allows compensatory behaviors, then other characteristics of the driver play a role, particularly driving experience (total mileage and recent mileage, again estimated

on the basis of the patient interview) and, perhaps, higher order cognitive functions relevant for learning to cope with changing circumstances.

50.1.2 Recent Developments

We concluded from the early studies that in order to improve the validity of the neuropsychological assessment in the sense of its potential to predict safe driving performance and behavior, we should include test conditions where patients could also demonstrate their compensatory potential. Also, we believed that driving simulators could play an important role there. The developments in driving simulation have allowed us to elaborate these plans.

An important development since our early attempts is that modern advanced driving simulators in principle allow the creation of naturalistic road and traffic situations, including other traffic behaving and interacting as if controlled by human operators. This new step was facilitated by the E. C.-funded research about (generic) intelligent driver support (GIDS) initiated by Michon (1993). GIDS required a real-time driver information processing and decision-making model to be able to determine at which moments and in which situations drivers should be supported with information. Once this model was developed in the form of a production system, Van Winsum applied it as a control mechanism for the autonomous agents (Van Wolffelaar & Van Winsum, 1995).

Because of the increased realism—also of that in the reactions of other traffic—driving simulators have the potential to trigger the real-world visual and visual-motor schemata underlying driving expertise (procedural knowledge). Whether the promise has become reality should be argued separately for each specific driving test in the simulator. For the two tests that will be described below, we have theoretical and anecdotal arguments (see below) that they are valid and yield additional information which is not available in regular neuropsychological tests. Quantitative validation against on-road tests and specific traffic-psychological tests awaits further investigation.

For a different set of driving tests in a simulator Lew (2005) showed that in a small sample of moderate to severe TBI patients, an automated Simulator Performance Index (SPI) predicted family observations of poor and dangerous driving better than did on-road driving test performance, qualitatively assessed by an expert. Unfortunately they did not compare the predictive quality of the SPI with neuropsychological tests or computerized tests of divided attention. The driving simulator system they used according to their description does not appear to be very realistic (one small screen; no simulator sickness, high amounts of violations and crashes; high reported difficulty; additional artificial divided attention task), and might be viewed as a complex test of divided attention. Based on a literature review on driving and attention in various clinical populations, Brouwer (2002) observed that in a number of studies time-pressed divided attention tasks had moderately strong relationships with measures of driving quality, so from that perspective it is not surprising that the SPI did a good predictive job.

Our present approach to driving simulator assessment is to focus on driving tasks which subjects find easy, easier than neuropsychological tests, and almost as easy as they (claim to) find regular driving. Because these tasks are in reality very complex, subjective easiness can only emerge if over-learned schemata for driving skill are triggered. The main reason for our focus on easy tasks is that in samples of moderately-impaired neurological patients, a well-chosen neuropsychological test battery is already quite sensitive with regard to predicting on-road limitations in time-pressured divided attention situations (Nouri & Lincoln, 1992; Withaar, 2000; De Raedt, 2000; Tant, Brouwer, Cornelissen, & Kooijman, 2002; Brouwer, 2002; McKenna, Jefferies, Dobson, & Frude, 2004; Clay et al., 2005; Coleman-Bryer, Rapport, & Hanks, 2006; McKenna, 2007) so we may not need a driving simulator for that. However, there are also studies (e.g., Larsson & Falkmer, 2007) finding poorer prediction of on-road driving from neuropsychological test performance.*

Because they have been designed to measure impairments and not activity limitations, neuropsychological tests are not sensitive with regard to the effects of driving experience and domain-specific compensatory skills. It is particularly the latter that we hope to assess with driving simulator tests. In terms of the International Classification of Functioning (ICF) proposed by the World Health Organization (http://www3.who.int/icf), we use the driving simulator for measurement on the level of activities, not the level of impairments.

Because of this philosophy, and also because of the presently lacking evidence base, we always combine driving simulator assessment with a conventional neuropsychological test battery consisting of tests and test-domains which, according to the literature and our own results, have good validity with regard to predicting on-road driving performance and crashes. Also, we often have or seek opportunities to obtain information about on-road driving performance—for example, if someone is referred to our department because of crash circumstances suggesting a problem with lateral position control, we will particularly observe steering performance in various road conditions. Also, many patients come with a family member from whom we will always seek views and observations (heteroanamnesis). This may provide hypotheses with regard to specific problems expected in the simulator.

---

* One must also keep in mind that on-road test scores only have limited reliability. Inter-rater reliability based on the same test drive is acceptable (see, e.g., Akinwuntan et al., 2003, 2005) but the limited information that we have found about test-retest correlations between different drives with the same drivers, even with the same observer, is not very encouraging and values do not appear to greatly exceed the correlations between aggregated neuropsychological test scores and test drive scores (Klebelsberg, 1982; Brouwer et al., 1988; Goldenbeld, Baughan, & Hatacka, 1999; Coeckelbergh, Brouwer, Cornelissen, & Kooijman, 2004).

50.1.2.1.1 Individual Differences and Co-morbidity

We have also learned from our work with TBI patients and stroke patients that, besides general information processing characteristics, as assessed with tests of attention and visual-motor speed, qualitative syndrome characteristics of brain injury must be taken into account. For example, visual field defects and dysexecutive syndromes are not at all uncommon in very severe and extremely severe TBI patients (Brouwer et al., 2002). The latter authors argue that the combination of hemianopia and limitations of attention functions (perceptual speed, visual working memory) is problematic, because the compensation requires time and visual imagery (see also Tant et al., 2002). In stroke patients, hemineglect, agnosia and apraxia are other important syndrome categories to consider.

Obvious as this lesson may seem, these syndromes do occur in sub-clinical forms in the sense of not having been noticed by anyone before we test them. In retrospect they sometimes appear to be the explanation for previously not understood accident-proneness or repeated failures of a driving test.

Finally, our work with very old drivers with cognitive limitations has taught us that there is frequent co-morbidity with visual limitations, such as acuity problems from cataracts and macular degeneration, contrast sensitivity problems in Parkinson's disease, and peripheral field loss in diabetes. As in the case of the neurological syndromes, it is important to know the visual limitations in order to be able to understand test scores and their consequences.

50.1.2.1.2 The 12DRIVE Protocol

We have argued that for the assessment of fitness to drive, a comprehensive protocol is required. It involves screening for neurological syndromes and visual functional limitations, administering neuropsychological tests with cut-off scores validated against road tests and/or crash involvement, and finally it involves the administration of realistic driving simulator tests allowing subjects to demonstrate their driving skills, including how they compensate for functional limitations.

At the department of Neuropsychology of the University Medical Center Groningen (UMCG) the comprehensive protocol used is called 12DRIVE. This stands for the three elements of the protocol. Part 1 comprises the anamnesis and heteroanamnesis (interview with a relative), a short and sensitive neuropsychological test battery, a screening for neurological syndromes and a screening for visual functional limitations. Part 2 is an individually tailored neuropsychological assessment depending on the results obtained in Part 1. *DRIVE* is the simulator assessment. Parts 1 and 2 are described in more detail below.

Part 1 of the 12DRIVE Protocol. Part 1 takes 1–1.5 hours and comprises the anamnesis (and heteroanamnesis), screening for neurological syndromes (dementia, apraxia, agnosia, dysexecutive syndrome), screening for visual functional limitations (acuity, contrast sensitivity and binocular visual field) and a short and sensitive neurospsychological battery. Elements of the battery are Constructional Drawings (Withaar, 2000), the Trail Making Test of visual search speed and divided attention

FIGURE 50.1 View from the mock-up in the Swing Drive. The simulator car is approximately in the middle of the right lane here.

(Reitan, 1958), and the Mannheim Tachistoscopic Traffic Test (Schufried, 2003a) to assess perceptual speed and the functional field of view. In older drivers with (suspected) cognitive impairment, these tests and test domains have been found to be quite predictive of performance in the test drive of practical fitness to drive which currently is the official Dutch standard for fitness to drive in brain-impaired drivers (Withaar, 2000). Also in the international literature, these tests and test domains have been found to be moderately good predictors of on-road performance (e.g., Coleman-Bryer et al., 2006). Cut-off scores indicating problems to be assessed further in Part 2 and Part DRIVE, were derived from Withaar (2000).

Part 2 of the 12DRIVE Protocol. Part 2, also taking approximately 1–1.5 hours, is individually tailored depending on the (hetero) anamnesis and test results obtained in Part 1. In a right hemisphere stroke it might involve administering a series of neglect tests (Tant et al., 2002). In a slow and paretic TBI patient it would include reaction time tests and multimodal tests of divided attention, such as the Wiener Determinations Test from the Vienna test system (Schufried, 2003b). In a socially-disinhibited TBI patient awareness might be assessed by comparing DEX questionnaires of patient and partner. The DEX questionnaires are part of the test for the Behavioral Assesment of the Desexecutive Syndrome (BADS) (Wilson, Evans, Alderman, Burgess, & Emslie, 1997) and the partner might be systematically interviewed about personality change with the Neuropsychiatric Inventory (Cummings et al., 1994; Dutch version by De Jonghe, Borkent, & Kat, 1997).

50.2 DRIVE: The Driving Tests in the Simulator

50.2.1 Hardware

The configuration consists of two computers, a projection screen stage and an open cabin "mock-up" containing a force-feedback steering wheel, accelerator, clutch and brake pedals standing within this stage. The projection screen stage holds three video projectors and a large flat-screen of 6 m length, bent 60 degrees inwards at 1/3 and 2/3 of its length, which presents the driver with a wide horizontal view. The simulator is situated in an air-conditioned room of approximately 25 square meters at the University Medical Center Groningen (Figure 50.1).

For projection, the fast high-quality graphic computers are connected to three high-contrast LCD projectors together

rendering a 180-degree* and 45-degree vertical projection at 60 Hz frame rate. The computers draw to the graphical screen (rendering), sample and regulate the mock-up, compute the simulated traffic, perform scenario control, process and store performance data, and run a graphical user-interface for the simulator operator. A graphical dashboard and cabin overlay projection is used to cover part of the out-window information on the screen so that from the driver's seat a realistic view emerges, as if looking from the inside of an actual car (see Figure 50.1). The actual form and structure of the cabin overlay are designed to reduce the probability of simulator sickness (Busscher, Van Wolffelaar, & Brouwer, 2009). Whether simulator sickness actually occurs is also highly dependent on the nature of the drive: Sharp angles, high-contrast housing and fences, and sudden speed changes increase the probability of simulator sickness.

50.2.2 Some General Aspects of the Simulator Software

The simulator system of ST software B. V. (http://www.stsoftware.nl/) has been developed in cooperation with Dutch universities and research institutes (University of Groningen, University of Delft, Stichting Wetenschappelijk Onderzoek Verkeersveiligheid [SWOV], TNO). ST software has designed a general research simulator platform that standardizes various aspects of simulator-based research. It contains the design tools to create simulated road environments and the runtime modules required for real-time simulation. Simulation of traffic is based on "autonomous agents" technology, which models the simulated participants (i.e., all traffic surrounding the simulator driver) as self-governing intelligent objects that show natural and normative driver behavior. Normative behavior implies basically safe and correct traffic behavior as a default condition. The agents possess functions to perceive the world around them, process this information according to their behavioral rules and act adequately on the perceived

---

* We have deliberately chosen a large screen approach instead of a head-mounted VR display because, in our view, a continuously present 180-degree visual field is typical for real-life locomotion tasks like driving. Some authors, however, have been experimenting with driving simulation in head injury rehabilitation using a head-mounted display (e.g., Schultheis, Rebimbas, Mourant, & Millis, 2007) which is more cost-effective in terms of equipment and space but which gives subjects a tunnel vision. To our knowledge no comparison has been made between both approaches in a rehabilitation setting.

circumstances (Van Wolffelaar & Van Winsum, 1995; Kappé, Van Winsum, & Van Wolffelaar, 2002). This means, for example, that the simulated cars that surround the simulator driver recognize roads on which they have the right of way and that, as a result of this, they will not yield for traffic from the right on these roads unless the respective traffic does not yield and a crash would occur if they themselves would not react.

Script commands may overrule the basically normative behavior. Scenario scripts, defined in ASCII readable script language (ST Scenario script), are lists of formal instructions in which all aspects of a simulation task are defined. In fact ST Scenario constitutes a programming language specifically tailored to access and control all internal functionality of the simulator. Each simulator application is governed by its own ST Scenario script file. Traffic participants, once created in script, behave in a natural and safe way but can at the same time be controlled by other scenarios to perform the specific maneuvers required for the test drive. The script language also enables definition of functions for real-time analyses of driver performance and corresponding specific reactions of the environmental traffic or feedback information to the driver as a consequence of behavior. One can create identical experimental initial conditions and situations for all simulator drivers and at the same time let the system respond individually, depending on the reaction of the driver. The autonomous nature of the traffic participants will always result in a natural reaction unless, of course, one intentionally writes scripts that force another response. An example is a car that follows the simulator driver's car at too close a distance.

50.2.3 The Swing Drive

This test is called swing drive because one drives along a steadily-winding road.

50.2.3.1 Construct

We measure the operational driving skill of lateral position control and the tactical driving skill of safely controlling speed, taking into account the driver's own operational driving skill and changing motivation. Motivation is operationally defined by instructions to the driver, either driving as usual for a road like this or as if in a hurry. The test requires driving on a winding two-lane country road with alternating left-right curves of 40 degrees and 500 m radius with a continuous stream of traffic from the opposing direction but no traffic on the lane of the simulator driver. The first part (*paced-up*), where speed is controlled by the computer, consists of six 700 m sections. The first section is driven at 50 km/h and increasing to 100 km/h in steps of 10 km/h. In the second part (*free speed*) the driver controls the speed. For the first 2100 m, the task is to drive at a comfortable speed, and in the last 2100 m the task is to drive as fast as possible while still driving safely (as if in a hurry).

50.2.3.2 Dependent Measures and Results

For all sections the average lateral position, the standard deviation of lateral position, lane crossings left and right, and

collisions are computed. After some data reduction they are compared with norms derived in healthy control subjects. With few exceptions, people tend to drive somewhat right of the middle of the lane (negative AVLP) with a SDLP of approximately 20–25 cm dependent on speed. These values appear comparable to what would be found in actual traffic (Van Zomeren, Brouwer, Rothengatter, & Snoek, 1988). One or more crossings of the right road marking occur in about half of the subjects, but crossing the mid-line only seldom occurs. In the Swing Drive simulator sickness is very infrequent, even in persons without driving simulator experience (1–5%). Subjectively, the test is easy to understand and to perform. Subjects with severe memory and language impairment may have difficulty with understanding or remembering the instruction to adapt speed in the free speed condition. In the case of memory impairment, the instruction to drive as if in a hurry (though still safely) is repeated halfway through the test drive. Safe tactical adjustment of speed requires that a speed is chosen which a driver can afford in view of his/her personal steering abilities, and which is not too fast or too slow for the road and traffic situation. What is too fast or too slow is empirically based on norms in healthy drivers in various age groups.

50.2.4 The City Drive

This drive is called the city drive because it uses the city and route developed by Davidse in her research on the effects of intersection complexity and ADAS in older drivers (Davidse, 2007; Davidse, Van Wolffelaar, Hagenzieker, & Brouwer, 2009).

50.2.4.1 Construct

The drive is a measure of practical fitness to drive (see glossary). The driver is instructed to drive at his or her own preferred speed, following the route instructions spoken by the computer, taking into account traffic signs and signals and obeying the traffic rules. It is also explained to the driver that other traffic (only consisting of passenger cars) will obey the rules. The route leads participants across several intersections, which differ in terms of layout (three-way intersection, four-way intersection, dual carriageway, roundabout), priority regulation, view of the intersection, traffic density on the road or lane to cross or join, and speed of traffic driving on the road or lane to cross or join. Subjects can state a preference for automatic or manual (five-speed) transmission.

50.2.4.2 Dependent Variables

We make a qualitative assessment of practical fitness to drive from a multi-screen DVD recording by observing speed choice, lateral and longitudinal position control, viewing behavior, application of priority rules, gap acceptance, and so forth, in a variety of road and traffic situations as they develop during the drive. For a number of quantifiable elements, results are compared to a norm group consisting of healthy middle-aged and older drivers.

50.3 Case Studies

50.3.1 Slowness Only?

Mr. A is a 55-year old truck driver who fell from his roof while repairing the chimney. He was admitted in a coma to the university hospital with a left-sided skull fracture and a focal contusion in the right temporal lobe. No further information is available about the location and severity of his injuries. He remained in a coma for three days. He was referred eight months after the fall by his neurologist to objectify his cognitive complaints and to advise about his fitness to drive, both privately and as a professional driver.

Retrospectively he estimated his PTA duration as approximately eight days (see Table 50.1). At the time of testing his most important mental complaints concerned memory and mental slowness. He also complained about irritability and anxiety. On the Hospital Anxiety and Depression Scale his anxiety score was high. One source of his anxiety concerned the time pressure he felt in his work as a truck driver after his head injury. He started driving small routes some three months after injury. He noticed that the paperwork and finding destinations cost him much more time than before for which he tried to compensate by leaving home hours too early. Also, he sometimes needed to park the truck when traffic was too busy for him. Because of these problems he stopped working as a truck driver. He did not complain or worry about time pressure and anxiety with regard to driving his private passenger car. At the time of testing, Mr. A did not use medication.

On Part 1 of 12DRIVE, Mr. A. did well except on the Mannheim Tachistoscopic Traffic Test (Schufried, 2003a) where he scored in the low to very low range for his age (correct responses: 2nd percentile; overview: 11th percentile). In the Mannheim test, static traffic scenes are shown on a computer screen for one second only; after the disappearance of the scene, subjects must indicate the presence of various traffic elements (traffic signs, pedestrians, etc.) in a multiple-choice display. In Part 2 of 12DRIVE, reaction time and divided attention were assessed with the Wiener Determination Test (Schufried, 2003b) and he also performed very low on this test (number of correct reactions: 2nd percentile). In both tests he made only very few errors (accuracy 86th and 93rd percentile). As the Wiener Determination Test is adaptive, the poor performance on the number of correct reactions in combination with the high accuracy means that he reacted very slowly but with few errors. Also, in the rest of the Part 2 battery, mental and motor slowness and memory impairment (particularly verbal memory) were obvious. In tests where he had time to think and plan, performance was average (e.g., WAIS subtests Similarity and Matrix Reasoning) and unimpaired (BADS subtests 6-Elements and ZOO-Map). Based on tests of social cognition and questionnaires not described here, we further found out that his skills in coping with social stress were not very effective.

Mr. A drove the Swing Drive and the City Drive. In the paced-up condition of the Swing Drive the Average Lateral Position was average while his SDLP was better than average at all speeds.

In the free speed condition, he adapted well to the instruction, always keeping a safe road position and stable course without any line crossing. In the City Drive he initially had some trouble in handling the (slightly awkward) manual transmission, but as illustrated by the smile on his face, he experienced this as a positive challenge and quite soon he managed to drive smoothly and safely. His looking strategy and interaction with other traffic were flawless, he accurately followed the traffic signs and signals, and made no violations.

Based on the information obtained we reported that his post-traumatic complaints about memory and mental slowness can be objectified consistently. His memory and speed of information processing are clearly impaired as a result of his head injury. Based on the anamnesis, the unimpaired neuropsychological test performance without time pressure, and his performance in the driving simulator, we think him sufficiently fit to drive a car privately. On the other hand, we advised against him working as professional driver because of his mental and motor slowness and because he cannot handle the time pressure.

50.3.2 Dysexecutive Syndrome

Mr. B is 50 years old and sustained an extremely severe TBI in a motor cycle crash when he was 35. After treatment in a rehabilitation hospital he eventually returned to his profession and family but his marriage did not last and he lost his former profession. He lived alone, owned and drove a car, but could not manage independently and got into serious personal and social problems. He was admitted voluntarily to a psychiatric hospital (about 18 months before the neuropsychological assessment of fitness to drive). At the psychiatric hospital he was diagnosed with post-traumatic frontal syndrome with impaired impulse control. In the course of his treatment they convinced him that he should turn in his driving license, which he did. After treatment he has lived on his own again for some months but now receives support in house-keeping and other daily activities. He lives some 10 km from the town where he wants to go regularly for shopping and social activities. However, he does not often succeed in getting there because, according to him, public transport is poor and it is too far to go by bicycle. Therefore he thinks he needs a car and wants his driving license back. Thus, he indicated to the driver licensing authority that he is fit to drive again. To check his claim they referred him to our department where we assessed him with the 12DRIVE protocol.

During Part 1 and Part 2 of the assessment his behavior was conspicuous. He was impulsive and verbally disinhibited towards the test assistant. For example, he often started a test without listening to the instruction and made various verbal threats. Furthermore, after returning from lunch he seated himself behind the wrong computer at the wrong department where he started an animated discussion with the nurses. It took quite some time before they understood where he should go.

Test performance was not always very poor but it was also conspicuous in a qualitative sense. For example, in the Mannheim Tachistoscopic Traffic Test he correctly detected 49 items (56th percentile) but also made 11 errors (3rd percentile) which strongly suggests that the correct detection score is an overestimation of his actual capacity. Also, in the Wiener Determination Test he made many omissions and errors (22nd and 10th percentile respectively); this time he also had a very poor correct reaction score (1st percentile), showing a serious problem of divided attention consistent with his (dysexecutive) frontal syndrome.

In the driving simulator Mr. B drove the Swing Drive, which he described as a fairground attraction and which he greatly enjoyed. In the paced-up condition the AVLP was average but he was very slow and inaccurate in controlling steering yielding a SDLP outside the normal range with six line crossings left and two right. In the free speed condition, he did not adapt sufficiently to the instruction and drove too fast for his ability, showing four right line crossings and nine left, the last one so extreme that he crashed with an approaching car. Although he was generally not concerned about his own performance, he worried about the crash.

Based on the severe dysexecutive syndrome and the slow reactions observed during testing we advised the licensing authority to declare Mr. B unfit to drive. It might be asked whether other negative factors such as the anti-epileptic medication he uses and/or the lack of recent driving experience influenced his performance in the simulator. These factors are not crucial in this case because they would have influenced driving performance ("can do") and not so much driving behavior ("do"), which is obviously disturbed in this case.

50.3.3 Homonymous Hemianopia

Mr. C is a 35-year old gardener who sustained a severe TBI when he was 10 years old. He was referred to our department by his rehabilitation doctor who treated him for ankle and leg injuries caused by a recent car crash. During the consultation Mr. C mentioned that he had been involved in two recent car crashes and the doctor and he wondered whether his brain injury might be a contributing factor. This is the reason why he was referred.

After having been admitted (as a child) to the university hospital with a skull fracture he remained in a coma for seven days. One month after injury he returned home and, although he lost a school year, he was able to normally finish primary and secondary school. After finishing his theoretically-oriented education he went to a practical agricultural training center but was not able to graduate there because, according to him, he was too slow and could not do two things simultaneously. Mr. C learned to drive when 21 and failed the driving test two or three times before passing which is normal. No official assessment of his medical fitness to drive was made then; he was treated like any healthy person and passed the regular examination.

Mr. C described two recent crashes. When turning right at an intersection, the left front of his car hit the right rear side of a vehicle waiting to cross the intersection. In the second crash, in which he seriously injured his legs, again the left front side of his car was damaged. In this case, his car drifted into the lane of opposing traffic where it crashed with an approaching car. He believes that he had fallen asleep in the latter crash because he remembers the crash but not what went on before it.

In this case, Part 1 of the 12DRIVE procedure revealed unexpected visual problems. The two primarily visual subtests, namely Trail Making A and the Mannheim Tachistoscopic Traffic Test, were performed very poorly (respectively 1st and 3rd percentile) while the cognitively more demanding Trail Making B version hardly took him more time than the simple A version (also, other tests of executive function were performed well). The test assistant further observed that he did not immediately oversee what was on the table—for example, the elements of the six-elements test—while in the Wiener Determination Test he frequently missed the visual stimuli for the pedal reaction on the lower left side of the screen. Before going into the driving simulator, the horizontal visual fields are assessed by confrontation, which in his case revealed a complete loss of the left visual field. Visual acuity was optimal and so was contrast sensitivity.

In Mr. C's case, the drives in the simulator seemed unproblematic. In the Swing Drive his road position on average was slightly right from the center (AVLP 50th-75th percentile) and quite stable (SDLP 50th-75th percentile). In the free speed condition he followed the instruction well but drove relatively slowly in both conditions (10th-25th percentile). In the City Drive (with automatic transmission) he drove carefully but not too slowly, and looked to the left and right well at intersections, sometimes very far to the left, which is required in a case of left hemianopia. During regular straight-ahead driving, however, he did not take his hemianopia into account as he did not make the required compensatory eye and head movements into the area normally covered by the left visual field. In retrospect, the poorly-compensated visual field defect may well have contributed to the first crash reported above. In the second crash, the relationship is less obvious. We advised further assessment in a visual rehabilitation center and a report to the driver licensing authority asking for an on-road assessment.

50.4 Discussion

The original reason for including driving simulation as part of the neuropsychological assessment was the observation that very severely head-injured individuals can be much safer and smoother drivers than expected on the basis of their information processing capacities as assessed with clinical and laboratory test methods. The hypothetical explanation was that on-road driving allows tactical adaptation. Based on learned regularities in the driving task and preserved central executive functions, drivers can adjust their driving performance and behavior to their reduced capacities. Based on this analysis the simulator tests have been explicitly designed to allow this compensation.

The simulator drives were thus included to allow cognitively-impaired subjects to show their learned driving skills and compensation strategies in a naturalistic driving task and served as an addition to the neuropsychological tests, which typically assess performance in unfamiliar tasks with little opportunity

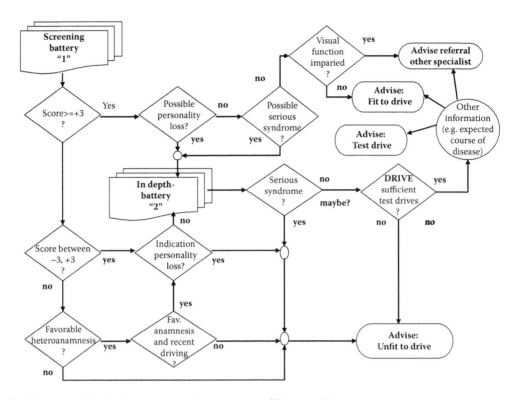

FIGURE 50.2 Block diagram of the decision process in the assessment of fitness to drive.

for compensation. With this aim in mind, the simulator tests in our department serve their purpose well as we regularly observe mildly demented and/or aphasic experienced drivers, who perform poorly on most neuropsychological tests, bloom up behind the wheel of the simulator car where they can show something they are good at. If their performances are indeed average or better, we use this information clinically as one argument in the advice to allow an on-road test even if the neuropsychological test scores argue against it. We also take the heteroanamnesis and behavioral observations into account in the advice. In Figure 50.2 the flow diagram indicating the diagnostic process is presented.

Interestingly, the simulator tests—also the seemingly simple Swing Drive—can sometimes elicit a very abnormal performance which does not show up, or does not show up as clearly, in any other test. An example of this is the occurrence of crashes in the Swing Drive. Until now, in only seven drivers did a crash with opposing traffic occur. All of these were brain injured patients who were suspected to be unfit to drive based on serious problems in everyday activities and the presence of a specific neurological syndrome. The cause and nature of the crashes appears to be different in different syndromes. In the right hemisphere (RH) stroke cases, the AVLP was too much to the left. Four patients with subcortical disorders were excessively slow in correcting the vehicular course at higher speeds and one patient with a dysexecutive syndrome was very slow and got distracted (see case study). Diagnostically, an interesting point is that in the RH strokes and in two of the subcortical cases, performances on the neuropsychological tests in the Part 1 of the 12DRIVE were sufficient, but in all seven cases there was

clear evidence (from the referral, heteroanamnesis and/or behavior observation) of severe driving-related problems in everyday life.

Key Points

- Severe traumatic brain injury (TBI) impairs information processing functions required for driving, thus posing a potential limit to *fitness to drive*. It is argued that this particularly concerns the operational level.
- Some severely impaired drivers with TBI effectively compensate by adapting the driving task on the tactical level. They realize *practical fitness to drive* by adapting driving skill and driving behavior in such a way that time pressure and the need for divided attention are reduced.
- Conventional neuropsychological assessment puts most emphasis on demonstrating impairment instead of on potential for rehabilitation and recovery. It is illustrated with case studies that driving simulator tests can redress this bias, at least if they can realistically tap into the tactical level of the driving task.

Keywords: Driving, Driving Simulator, Fitness to Drive, Rehabilitation, Traumatic Brain Injury

Acknowledgments

We gratefully acknowledge the financial support of the Dutch road licensing association CBR (Centraal Bureau Rijvaardigheidsbewijzen) which has allowed us to set up a clinically applicable realistic driving simulator within the University

Medical Center Groningen. We cordially thank Dr Marie Vanier at the Université de Montréal for allowing us to demonstrate our test drives on the Driving Simulators in Rehabilitation web site (http://www.driving-simulator-rehabilitation.com).

Glossary

Driving behavior: The actual driving maneuvers undertaken in various road and traffic situations (actual performance: "do" instead of "can do"). This definition is on the level of activities according to the ICF.

Driving skill: To have the learned procedural and declarative knowledge to be able to steer a car through various road and traffic situations according to the traffic regulations and in interaction with other traffic participants, resulting in smooth and safe traffic participation (maximum performance). This definition is on the level of activities according to the International Classification of Functioning (ICF) of the World Health Organization.

Fitness to drive: To have the physical and mental functions and personality characteristics required to learn and apply driving skill as determined by body structure and pathology (maximum performance). This definition is on the level of (impaired) body functions.

Practical fitness to drive: To be able to compensate impaired fitness to drive by consistent adaptations in driving skill and/or by using technical aids, rendering safe and smooth traffic participation in spite of significant impairments (maximum performance). This definition is on the level of activities according to the ICF.

Web Resources

The test drives can be viewed on the Driving Simulators in Rehabilitation web site: http://www.driving-simulator-rehabilitation.com/.

The *Handbook* web site also contains supplemental material for the chapter, including a color version of Figure 50.1.

Web Figure 50.1: View from the mock-up in the Swing Drive.

Key Readings

Brouwer, W. H., & Withaar, F. K. (1997). Fitness to drive after traumatic brain injury. *Neuropsychological Rehabilitation, 7*(3), 177–193.

Coleman-Bryer, R., Rapport, P., & Hanks, R. A. (2006). Determining fitness to drive: Neuro-psychological and psychological considerations. In J. M. Pellerito, Jr. (Ed.), *Driver rehabilitation and community mobility* (pp. 165–184). St Louis, MO: Elsevier Mosby.

McKenna, P., Jefferies, L., Dobson, A., & Frude, N. (2004). The use of a cognitive battery to predict who will fail an on-road driving test. *British Journal of Clinical Psychology, 43*, 325–336.

References

Agrawal, A., Timothy, J., Pandit, L., & Manju, M. (2006). Post-traumatic epilepsy: An overview. *Clinical Neurology and Neurosurgery, 108*(5), 433–439.

Akinwuntan, A., DeWeerdt, W., Feys, P. T., Baten, G., Arno, P., & Kiekens, C. (2003). Reliability of a road test after stroke. *Archives of Physical Medicine and Rehabilitation, 84*, 1792–1796.

Akinwuntan, A., DeWeerdt, W., Feys, P. T., Baten, G., Arno, P., & Kiekens, C. (2005). Validity of a road test after stroke. *Archives of Physical Medicine and Rehabilitation, 86*, 421–426.

Brooks, N., (Ed.). (1984). *Closed head injury: Psychological, social and family consequences.* Oxford, United Kingdom: Oxford University Press.

Brouwer, W. H. (2002). Attention and driving: A cognitive neuropsychological approach. In P. Zimmermann & M. Leclercq (Eds.), *Applied neuropsychology of attention* (pp. 223–248). Hove, United Kingdom: Psychology Press.

Brouwer, W. H., Rothengatter, J. A., & Van Wolffelaar, P. C. (1988). Compensatory potential in elderly drivers. In J. A. Rothengatter & R. A. de Bruin (Eds.), *Road user behavior: Theory and research* (pp. 296–301). Assen, The Netherlands: Van Gorcum.

Brouwer, W. H., & Withaar, F. K. (1997). Fitness to drive after traumatic brain injury. *Neuropsychological Rehabilitation, 7*(3), 177–193.

Brouwer, W. H., Withaar, F. K., Tant, M. L. M., & Van Zomeren, A. H. (2002). Attention and driving in traumatic brain injury: A question of coping with time-pressure. *Journal of Head Trauma Rehabilitation, 17*, 1–15.

Busscher, R. B., Van Wolffelaar, P. C., & Brouwer, W. H. (2009). *Reducing simulator sickness.* Proceedings of the 21th International Congress of the International Traffic Medicine Association. The Hague. International Traffic Medicine Association (ITMA).

Clay, O. J., Wadley, V. G., Edwards, J. D., Roth, D. L., Roenker, D. L., & Ball, K. K. (2005). Cumulative meta-analysis of the relationship between useful field of view and driving performance in older adults: Current and future implications. *Optometry and Vision Science, 82*(8), 724–731.

Coeckelbergh, T. R. M., Brouwer, W. H., Cornelissen, F. W., & Kooijman, A. C. (2004). Predicting practical fitness to drive in drivers with visual field defects caused by ocular pathology. *Human Factors, 46*, 748–760.

Coleman-Bryer, R., Rapport, P., & Hanks, R. A. (2006). Determining fitness to drive: Neuropsychological and psychological considerations. In J. M. Pellerito, Jr. (Ed.), *Driver rehabilitation and community mobility* (pp. 165–184). St Louis, MO: Elsevier Mosby.

Cummings, J. L., Mega, M., Gray, K., Rosenberg-Thompson, S., Carusi, D. A., & Gornbein, J. (1994). The Neuropsychiatric Inventory: Comprehensive assessment of psychopathology in dementia. *Neurology, 44*, 2308–2314.

Davidse, R. J. (2007). *Assisting the older driver: Intersection design and in-car devices to improve the safety of the older driver.* Doctoral dissertation. University of Groningen, The Netherlands.

Davidse, R. J., van Wolffelaar, P. C., Hagenzieker, M. P., & Brouwer, W. H. (2009). Effects of in-car support on mental workload and driving performance of older drivers. *Human Factors, 51,* 463–476.

De Jonghe, J. F. M., Borkent, L. M., & Kat, M. G. (1997). *Neuropsychiatrische Vragenlijst* (Dutch translation of *Neuropsychiatric Inventory*). Alkmaar, The Netherlands: Medical Center Alkmaar, Department of Clinical Psychology.

De Raedt, R. (2000). *Cognitive/neuropsychological functioning and compensation related to car driving performance in older adults.* Doctoral dissertation, Vrije Universiteit, Brussels, Belgium.

Goldenbeld, C., Baughan, C. J., & Hatakka, M. (1999). Driver testing. In S. Siegrist (Ed.), *Driver training, testing and licensing—Towards theory-based management of young driver's injury risk in road traffic* (pp. 257). Berne, Switzerland: Schoch & Co.

Granacher, R. P., Jr. (2003). *Traumatic brain injury: Methods for clinical and forensic neuropsychiatric assessment.* Boca Raton, FL: CRC Press.

Kappé, B., Van Winsum, W., & Van Wolffelaar, P. C. (2002). A cost-effective driving-simulator. *Proceedings of the driving simulation conference 2002* (pp. 123–133). Paris, France.

Klebelsberg, D. (1982). *Verkehrspsychologie.* Berlin, Germany: Springer Verlag.

Kreutzer, J. S., Seel, R. T., & Gourley, E. (2001) The prevalence and symptom rates of depression after traumatic brain injury: A comprehensive examination. *Brain Injury, 15,* 563–576.

Larsson, H., & Falkmer, T. (2007). Off-road and on-road assessment methods. What do they say? A clinical sample. In L. Boyle et al. (Eds.), *Proceedings of the fourth international driving symposium on human factors in driving assessment, training, and vehicle design* (pp. 335–342).

Leon-Carrion, J., Dominguez-Morales, M. R., Barroso, J. M., & Martin, Y. (2005) Driving with cognitive deficits: Neurorehabilitation and legal measures are needed for driving again after severe traumatic brain injury. *Brain Injury, 19,* 213–219.

Lew, H. L., Poole, J. H., Lee, E.-H., Jaffe, D. L., Huang, H.-C., & Brodd, E. (2005). Predictive validity of driving-simulator assessments following traumatic brain injury: A preliminary study. *Brain Injury, 19,* 177–188.

Long, C. J., & Ross, L. K., (Eds.). (1992). *Handbook of head trauma: Acute care to recovery.* New York: Springer.

McKenna, P. (2007). At home with the Rookwood driving battery—Theory and practice in assessing fitness to drive in neurologically impaired people. In R. Risser & W. Rüdiger Nickel (Eds.), *Fit to drive: Proceedings of the second international traffic expert congress Vienna 2007* (pp. 106–110). Bonn, Germany: Kirschbaum Verlag.

McKenna, P., Jefferies, L., Dobson, A., & Frude, N. (2004). The use of a cognitive battery to predict who will fail an on-road driving test. *British Journal of Clinical Psychology, 43,* 325–336.

Michon, J. A., (Ed.). (1993). *Generic intelligent driver support system.* London: Taylor & Francis.

Nouri, F. M., & Lincoln, N. B. (1992). Validation of a cognitive assessment: Predicting driving performance after stroke. *Clinical Rehabilitation, 6,* 275–281.

Pietrapiana, P., Tamietto, M., Torrini, G., Mezzanato, T., Rago, R., & Perino, C. (2005). Role of premorbid factors in predicting safe return to driving after severe TBI. *Brain Injury, 19*(3), 197–211.

Reitan, R. M. (1958). Validity of the Trail Making Tests as an indication of organic brain damage. *Perceptual and Motor Skills, 8,* 271–276.

Richardson, J. T. E. (2001) *Clinical and neuropsychological aspects of closed head injury.* Hove, United Kingdom: Psychology Press.

Russell, W. R., & Smith, A. (1961). Post-traumatic amnesia after closed head injury. *Archives of Neurology 5,* 16–29.

Schufried, G. (2003a). *Test manual of the Tachistoscopic Traffic Test Mannheim for screen.* Mödling, Austria: Dr Gernot Schufried Gesellschaft m.b.H.

Schufried, G. (2003b). *Test manual of the Wiener Determination Tests.* Mödling, Austria: Dr Gernot Schufried Gesellschaft m.b.H.

Schultheis, M. T., Rebimbas, M. T., Mourant, J., & Millis, S. R. (2007). Examining the usability of a virtual reality driving simulator. *Assistive Technology, 19*(1), 1–10.

Tant, M. L. M., Brouwer, W. H., Cornelissen, F. W., & Kooijman, A. C. (2002). Driving and visuospatial performance in people with hemianopia. *Neuropsychological Rehabilitation, 12*(5), 419–437.

Van Wolffelaar, P. C., Brouwer, W. H., & Van Zomeren, A. H. (1990). Driving ability 5–10 years after severe head injury. In T. Benjamin (Ed.), *Driving behavior in a social context* (pp. 564–574). Caen, France: Paradigme.

Van Wolffelaar, P. C., & Van Winsum, W. (1995). Traffic modeling and driving simulation—An integrated approach. *Proceedings of the driving simulation conference 1995* (pp. 236–244). Toulouse, France: Teknea.

Van Zomeren, A. H., Brouwer, W. H., Rothengatter, J. A., & Snoek, J. W. (1988). Fitness to drive a car after recovery from severe head injury. *Archives of Physical Medicine and Rehabilitation, 69,* 90–96.

Wilson, B. A., Evans, J. J., Alderman, N., Burgess, P. W., & Emslie, H. (1997). Behavioral assessment of the dysexecutive syndrome. In P. Rabbitt (Ed.), *Methodology of frontal and executive function* (pp. 239–250). Hove, United Kingdom: Psychology Press.

Withaar, F. K. (2000). *Divided attention and driving: The effects of aging and brain injury.* Doctoral dissertation, University of Groningen, Groningen, the Netherlands.

Index

Printed and bound by CPI Group (UK) Ltd, Croydon, CR0 4YY

18/10/2024

01776254-0013